## CONVERSION FACTORS

**Length**

1 m = 39.37 in. = 3.281 ft

1 in. = 2.54 cm (exact)

1 km = 0.621 mi

1 mi = 5 280 ft = 1.609 km

1 lightyear (ly) = $9.461 \times 10^{15}$ m

1 angstrom (Å) = $10^{-10}$ m

**Mass**

1 kg = $10^3$ g = $6.85 \times 10^{-2}$ slug

1 slug = 14.59 kg

1 u = $1.66 \times 10^{-27}$ kg = 931.5 MeV/$c^2$

**Time**

1 min = 60 s

1 h = 3 600 s

1 day = 24 h = $1.44 \times 10^3$ min = $8.64 \times 10^4$ s

1 yr = 365.242 days = $3.156 \times 10^7$ s

**Volume**

1 L = 1 000 $cm^3$ = 0.035 3 $ft^3$

1 $ft^3$ = $2.832 \times 10^{-2}$ $m^3$

1 gal = 3.786 L = 231 $in.^3$

**Angle**

180° = $\pi$ rad

1 rad = 57.30°

1° = 60 min = $1.745 \times 10^{-2}$ rad

**Speed**

1 km/h = 0.278 m/s = 0.621 mi/h

1 m/s = 2.237 mi/h = 3.281 ft/s

1 mi/h = 1.61 km/h = 0.447 m/s = 1.47 ft/s

**Force**

1 N = 0.224 8 lb = $10^5$ dynes

1 lb = 4.448 N

1 dyne = $10^{-5}$ N = $2.248 \times 10^{-6}$ lb

**Work and energy**

1 J = $10^7$ erg = 0.738 ft·lb = 0.239 cal

1 cal = 4.186 J

1 ft·lb = 1.356 J

1 Btu = $1.054 \times 10^3$ J = 252 cal

1 J = $6.24 \times 10^{18}$ eV

1 eV = $1.602 \times 10^{-19}$ J

1 kWh = $3.60 \times 10^6$ J

**Pressure**

1 atm = $1.013 \times 10^5$ N/$m^2$ (or Pa) = 14.70 lb/$in.^2$

1 Pa = 1 N/$m^2$ = $1.45 \times 10^{-4}$ lb/$in.^2$

1 lb/$in.^2$ = $6.895 \times 10^3$ N/$m^2$

**Power**

1 hp = 550 ft·lb/s = 0.746 kW

1 W = 1 J/s = 0.738 ft·lb/s

1 Btu/h = 0.293 W

© Mike Chew/CORBIS

Tenth Edition

# College Physics

## Volume 2

**Raymond A. Serway** | *Emeritus, James Madison University*

**Chris Vuille** | *Embry-Riddle Aeronautical University*

**With contributions from John Hughes** | *Embry-Riddle Aeronautical University*

Australia • Brazil • Mexico • Singapore • United Kingdom • United States

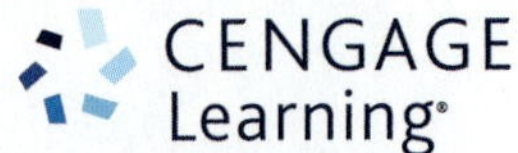

***College Physics*, Tenth Edition**
**Volume 2**
**Raymond A. Serway and Chris Vuille**

Product Director: Mary Finch
Senior Product Manager: Charlie Hartford
Development Editor: Ed Dodd
Product Assistant: Chris Robinson
Managing Developer: Peter McGahey
Senior Market Development Manager: Rebecca Berardy Schwartz
Media Developer: Andrew Coppola
Senior Marketing Development Manager: Janet del Mundo
Brand Manager: Nicole Hamm
Content Project Manager: Alison Eigel Zade
Senior Art Director: Cate Barr
Manufacturing Planner: Sandee Milewski
Rights Acquisition Specialist: Shalice Shah-Caldwell
Production Service and Compositor: Graphic World Inc.
Text & Cover Designer: Dare Porter
Cover Image: Street luge
Cover Image Credit: © Mike Chew/CORBIS



Library of Congress Control Number: 2013948341

ISBN-13: 978-1-285-73704-1

ISBN-10: 1-285-73704-0

**Cengage Learning**
200 First Stamford Place, 4th Floor
Stamford, CT 06902
USA

Cengage Learning is a leading provider of customized learning solutions with office locations around the globe, including Singapore, the United Kingdom, Australia, Mexico, Brazil, and Japan. Locate your local office at **www.cengage.com/global.**

Cengage Learning products are represented in Canada by Nelson Education, Ltd.

To learn more about Cengage Learning Solutions, visit **www.cengage.com.**

Purchase any of our products at your local college store or at our preferred online store **www.cengagebrain.com.**
**Instructors:** Please visit **login.cengage.com** and log in to access instructor-specific resources.

Printed in the United States of America
2 3 4 5 6 7 17 16 15 14

We dedicate this book to our wives, children, grandchildren, relatives, and friends who have provided so much love, support, and understanding through the years, and to the students for whom this book was written.

# Contents Overview

## VOLUME 2

### PART 4 | Electricity and Magnetism



### PART 5 | Light and Optics



### PART 6 | Modern Physics

# Contents

# About the Authors

**Raymond A. Serway** received his doctorate at Illinois Institute of Technology and is Professor Emeritus at James Madison University. In 2011, he was awarded an honorary doctorate degree from his alma mater, Utica College. He received the 1990 Madison Scholar Award at James Madison University, where he taught for 17 years. Dr. Serway began his teaching career at Clarkson University, where he conducted research and taught from 1967 to 1980. He was the recipient of the Distinguished Teaching Award at Clarkson University in 1977 and the Alumni Achievement Award from Utica College in 1985. As Guest Scientist at the IBM Research Laboratory in Zurich, Switzerland, he worked with K. Alex Müller, 1987 Nobel Prize recipient. Dr. Serway was also a visiting scientist at Argonne National Laboratory, where he collaborated with his mentor and friend, the late Sam Marshall. Early in his career, he was employed as a research scientist at the Rome Air Development Center from 1961 to 1963 and at the IIT Research Institute from 1963 to 1967. Dr. Serway is also the coauthor of *Physics for Scientists and Engineers*, ninth edition; *Principles of Physics: A Calculus-Based Text*, fifth edition; *Essentials of College Physics*, *Modern Physics*, third edition; and the high school textbook *Physics*, published by Holt, Rinehart and Winston. In addition, Dr. Serway has published more than 40 research papers in the field of condensed matter physics and has given more than 60 presentations at professional meetings. Dr. Serway and his wife Elizabeth enjoy traveling, playing golf, fishing, gardening, singing in the church choir, and especially spending quality time with their four children, nine grandchildren, and a recent great grandson.

**Chris Vuille** is an associate professor of physics at Embry-Riddle Aeronautical University (ERAU), Daytona Beach, Florida, the world's premier institution for aviation higher education. He received his doctorate in physics from the University of Florida in 1989 and moved to Daytona after a year at ERAU's Prescott, Arizona, campus. Although he has taught courses at all levels, including postgraduate, his primary interest has been instruction at the level of introductory physics. He has received several awards for teaching excellence, including the Senior Class Appreciation Award (three times). He conducts research in general relativity and quantum theory and was a participant in the JOVE program, a special three-year NASA grant program during which he studied neutron stars. His work has appeared in a number of scientific journals, and he has been a featured science writer in Analog Science Fiction/Science Fact magazine. In addition to this textbook, he is coauthor of *Essentials of College Physics.* Dr. Vuille enjoys tennis, swimming, and playing classical piano, and he is a former chess champion of St. Petersburg and Atlanta. In his spare time he writes fiction and goes to the beach. His wife, Dianne Kowing, is Chief of Optometry for a local Veterans' Administration clinic. They have a daughter, Kira, and two sons, Christopher and James.

# Preface

*College Physics* is written for a one-year course in introductory physics usually taken by students majoring in biology, the health professions, or other disciplines, including environmental, earth, and social sciences, and technical fields such as architecture. The mathematical techniques used in this book include algebra, geometry, and trigonometry, but not calculus. Drawing on positive feedback from users of the ninth edition, analytics gathered from both professors and students who use Enhanced WebAssign, as well as reviewers' suggestions, we have refined the text to better meet the needs of students and teachers.

This textbook, which covers the standard topics in classical physics and twentieth-century physics, is divided into six parts. Part 1 (Chapters 1–9) deals with Newtonian mechanics and the physics of fluids; Part 2 (Chapters 10–12) is concerned with heat and thermodynamics; Part 3 (Chapters 13 and 14) covers wave motion and sound; Part 4 (Chapters 15–21) develops the concepts of electricity and magnetism; Part 5 (Chapters 22–25) treats the properties of light and the field of geometric and wave optics; and Part 6 (Chapters 26–30) provides an introduction to special relativity, quantum physics, atomic physics, and nuclear physics.

## Objectives

The main objectives of this introductory textbook are twofold: to provide the student with a clear and logical presentation of the basic concepts and principles of physics and to strengthen an understanding of those concepts and principles through a broad range of interesting, real-world applications. To meet those objectives, we have emphasized sound physical arguments and problem-solving methodology. At the same time we have attempted to motivate the student through practical examples that demonstrate the role of physics in other disciplines.

## Changes to the Tenth Edition

Several changes and improvements have been made in preparing the tenth edition of this text. Some of the new features are based on our experiences and on current trends in science education. Other changes have been incorporated in response to comments and suggestions offered by users of the ninth edition. The features listed here represent the major changes made for the tenth edition.

### New Learning Objectives Added for Every Section

In response to a growing trend across the discipline (and the request of many users), we have added learning objectives for every section of the tenth edition. The learning objectives identify the major concepts in a given section and also identify the specific skills/outcomes students should be able to demonstrate once they have a solid understanding of those concepts. It is hoped that these learning objectives will assist those professors who are transitioning their course to a more outcomes-based approach.

### New Online Tutorials

The new online tutorials (available via Enhanced WebAssign) offer students another training tool to assist them in understanding how to apply certain key concepts presented in a given chapter. The tutorials first present a brief review of the necessary concepts from the text, together with advice on how to solve problems involving them. The student can then attempt to solve one or two such problems, guided by questions presented in the tutorial. The tutorial automatically scores student responses and presents correct solutions together with discussion. Students can then practice on several additional problems of a similar level and, in some cases go to higher level or related problems, depending on the concepts covered in the tutorial.

## New Warm-Up Exercises in Every Chapter

Warm-up exercises (over 320 are included in the full book) appear at the beginning of each chapter's problems set, and were inspired by one of the author's (Vuille) classroom experiences. The idea behind warm-up exercises is to review mathematical and physical concepts that are prerequisites for a given chapter's problems set, and also to provide students with a general preview of the new physics concepts covered in a given chapter. By doing the warm-up exercises first, students will have an easier time getting comfortable with the new concepts of a chapter before tackling harder problems.

## New Algorithmic Solutions in Enhanced WebAssign

All quantitative end-of-chapter problems in Enhanced WebAssign now feature *algorithmic solutions.* Fully worked-out solutions are available to students with quantitative parameters exactly matching the version of the problem assigned to individual students. As always for all "Hints" features, Enhanced WebAssign offers great flexibility to instructors regarding when to enable algorithmic solutions.

## Chapter-by-Chapter Changes

The text has been carefully edited to improve clarity of presentation and precision of language. We hope that the result is a book both accurate and enjoyable to read. Although the overall content and organization of the textbook are similar to the ninth edition, a few changes were implemented. The list below highlights some of the major changes for the tenth edition.

**Chapter 15 Electric Forces and Electric Fields**
- Fourteen new warm-up exercises have been added.
- Two new tutorials (*Coulomb's law and the electric field* and *Applying Gauss's law to distributions of charge*) have been added in Enhanced WebAssign.

**Chapter 16 Electrical Energy and Capacitance**
- Twelve new warm-up exercises have been added.
- Two new tutorials (*Applying the work-energy theorem to systems of charges* and *Evaluating the equivalent capacitance of systems of capacitors*) have been added in Enhanced WebAssign.

**Chapter 17 Current and Resistance**
- Ten new warm-up exercises have been added.
- One new tutorial (*Exploring electrical current, energy, and power*) has been added in Enhanced WebAssign.

**Chapter 18 Direct-Current Circuits**
- Thirteen new warm-up exercises have been added.
- Two new tutorials (*Simplifying circuits with both series and parallel resistors* and *Applying Kirchhoff's rules to complex DC circuits*) have been added in Enhanced WebAssign.

**Chapter 19 Magnetism**
- Nine new warm-up exercises have been added.
- Two new tutorials (*The motion of charged particles in a uniform magnetic field* and *Applying Ampere's law to current-carrying wires and cylinders*) have been added in Enhanced WebAssign.

**Chapter 20 Induced Voltages and Inductance**
- Ten new warm-up exercises have been added.
- Three new tutorials (*Using magnetic flux and Faraday's* law, *Calculating motional EMF,* and *RL circuits*) have been added in Enhanced WebAssign.

**Chapter 21 Alternating-Current Circuits and Electromagnetic Waves**
- Eleven new warm-up exercises have been added.
- Two new tutorials (*Purely resistive, capacitive, and inductive AC circuits* and *Analyzing series RLC AC circuits*) have been added in Enhanced WebAssign.

**Chapter 22 Reflection and Refraction of Light**

- Five new warm-up exercises have been added.
- One new tutorial (*Reflection, refraction, and Snell's law*) has been added in Enhanced WebAssign.

**Chapter 23 Mirrors and Lenses**

- Seven new warm-up exercises have been added.
- Two new tutorials (*Analyze the optical properties of concave and convex mirrors* and *Analyze the optical properties of convergent and divergent thin lenses*) have been added in Enhanced WebAssign.

**Chapter 24 Wave Optics**

- Ten new warm-up exercises have been added.
- Three new tutorials (*Solving problems involving Young's experiment, Interference in thin films*, and *Diffraction and diffraction gratings*) have been added in Enhanced WebAssign.

**Chapter 25 Optical Instruments**

- Twelve new warm-up exercises have been added.
- Two new tutorials (*Prescribing corrective lenses* and *Optical instruments*) have been added in Enhanced WebAssign.

**Chapter 26 Relativity**

- Ten new warm-up exercises have been added.
- One new tutorial (*Comparing space and time measurements by different observers in relativity*) has been added in Enhanced WebAssign.

**Chapter 27 Quantum Physics**

- Eleven new warm-up exercises have been added.
- One new tutorial (*Quantum physics*) has been added in Enhanced WebAssign.

**Chapter 28 Atomic Physics**

- Ten new warm-up exercises have been added.
- One new tutorial (*Hydrogen and hydrogen-like atoms*) has been added in Enhanced WebAssign.

**Chapter 29 Nuclear Physics**

- Ten new warm-up exercises have been added.
- One new tutorial (*Analyzing radioactivity*) has been added in Enhanced WebAssign.

**Chapter 30 Nuclear Energy and Elementary Particles**

- Ten new warm-up exercises have been added.
- One new tutorial (*Calculating the energy released in nuclear reactions*) has been added in Enhanced WebAssign.

## Textbook Features

Most instructors would agree that the textbook assigned in a course should be the student's primary guide for understanding and learning the subject matter. Further, the textbook should be easily accessible and written in a style that facilitates instruction and learning. With that in mind, we have included many pedagogical features that are intended to enhance the textbook's usefulness to both students and instructors. The following features are included.

**Examples** For this tenth edition we have reviewed all the worked examples and made numerous improvements. Every effort has been made to ensure the collection of examples, as a whole, is comprehensive in covering all the physical concepts, physics problem types, and required mathematical techniques. The Questions usually require a conceptual response or determination, but they also include estimates requiring knowledge of the relationships between concepts. The answers for the Questions can be found at the back of the book. The examples are in a two-column format for a pedagogic purpose: students can study the example, then cover up the right column and attempt to solve the problem using the cues in the left column. Once successful in that exercise, the student can cover up both solution columns and attempt to solve the problem using only the strategy statement, and finally just the problem statement. Here is a sample of an in-text worked example, with an explanation of each of the example's main parts:

The **Goal** describes the physical concepts being explored within the worked example.

The **Problem** statement presents the problem itself.

The **Strategy** section helps students analyze the problem and create a framework for working out the solution.

The **Solution** section uses a two-column format that gives the explanation for each step of the solution in the left-hand column, while giving each accompanying mathematical step in the right-hand column. This layout facilitates matching the idea with its execution and helps students learn how to organize their work. Another benefit: students can easily use this format as a training tool, covering up the solution on the right and solving the problem using the comments on the left as a guide.

**EXAMPLE 13.7 Measuring the Value of $g$**

**GOAL** Determine $g$ from pendulum motion.

**PROBLEM** Using a small pendulum of length 0.171 m, a geophysicist counts 72.0 complete swings in a time of 60.0 s. What is the value of $g$ in this location?

**STRATEGY** First calculate the period of the pendulum by dividing the total time by the number of complete swings. Solve Equation 13.15 for $g$ and substitute values.

**SOLUTION**

Calculate the period by dividing the total elapsed time by the number of complete oscillations:

$$T = \frac{\text{time}}{\#\text{ of oscillations}} = \frac{60.0\text{ s}}{72.0} = 0.833\text{ s}$$

Solve Equation 13.15 for $g$ and substitute values:

$$T = 2\pi\sqrt{\frac{L}{g}} \quad \rightarrow \quad T^2 = 4\pi^2\frac{L}{g}$$

$$g = \frac{4\pi^2 L}{T^2} = \frac{(39.5)(0.171\text{ m})}{(0.833\text{ s})^2} = 9.73\text{ m/s}^2$$

**REMARKS** Measuring such a vibration is a good way of determining the local value of the acceleration of gravity.

**QUESTION 13.7** True or False: A simple pendulum of length 0.50 m has a larger frequency of vibration than a simple pendulum of length 1.0 m.

**EXERCISE 13.7** What would be the period of the 0.171-m pendulum on the Moon, where the acceleration of gravity is $1.62\text{ m/s}^2$?

**ANSWER** 2.04 s

**Remarks** follow each Solution and highlight some of the underlying concepts and methodology used in arriving at a correct solution. In addition, the remarks are often used to put the problem into a larger, real-world context.

**Question** Each worked example features a conceptual question that promotes student understanding of the underlying concepts contained in the example.

**Exercise/Answer** Every Question is followed immediately by an exercise with an answer. These exercises allow students to reinforce their understanding by working a similar or related problem, with the answers giving them instant feedback. At the option of the instructor, the exercises can also be assigned as homework. Students who work through these exercises on a regular basis will find the end-of-chapter problems less intimidating.

Many Worked Examples are also available to be assigned in the Enhanced WebAssign homework management system (visit **www.cengage.com/physics/serway** for more details).

**Integration with Enhanced WebAssign** The textbook's tight integration with Enhanced WebAssign content facilitates an online learning environment that helps students improve their problem-solving skills and gives them a variety of tools to meet their individual learning styles. Extensive user data gathered by WebAssign were used to ensure that the problems most often assigned were retained for this new edition. In each chapter's problems set, the top quartile of problems that were assigned in WebAssign have cyan-shaded problem numbers for easy identification, allowing professors to quickly and easily find the most popular problems that were assigned in Enhanced WebAssign. Master It tutorials help students solve problems by having them work through a stepped-out solution. Problems with Master It tutorials are indicated in each chapter's problem set with a M icon. In addition, Watch It solution videos (indicated by a W icon) explain fundamental problem-solving strategies to help students step through selected problems. The problems most often assigned in Enhanced WebAssign (shaded in blue) have feedback to address student misconceptions, helping students avoid common pitfalls.

**Artwork** Every piece of artwork in the tenth edition is in a modern style that helps express the physics principles at work in a clearer and more precise fashion. Every piece of art is also drawn to make certain that the physical situations presented correspond exactly to the text discussion at hand.

*Guidance labels* are included with many figures in the text; these point out important features of the figure and guide students through figures without having to go back and forth from the figure legend to the figure itself. This format also helps those students who are visual learners. An example of this kind of figure appears below.

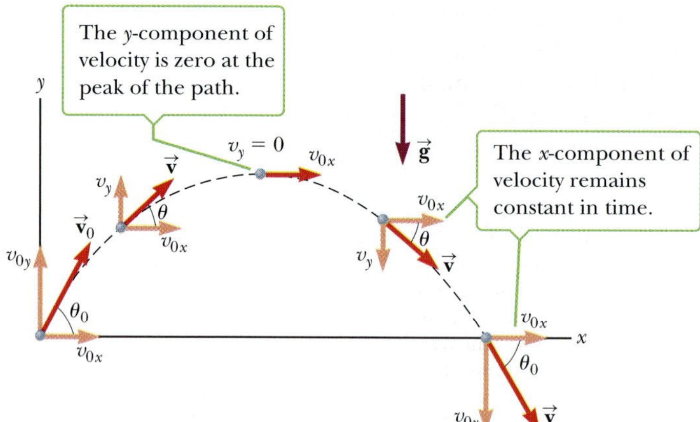


**Figure 3.14**
The parabolic trajectory of a particle that leaves the origin with a velocity of $\vec{v}_0$. Note that $\vec{v}$ changes with time. However, the $x$-component of the velocity, $v_x$, remains constant in time, equal to its initial velocity, $v_{0x}$. Also, $v_y = 0$ at the peak of the trajectory, but the acceleration is always equal to the free-fall acceleration and acts vertically downward.

**Warm-Up Exercises** As discussed earlier, these new exercises (over 320 are included in the full book) were inspired by one of the author's (Vuille) classroom experiences. Warm-up exercises review mathematical and physical concepts that are prerequisites for a given chapter's problems set and also provide students with a general preview of the new physics concepts covered in a given chapter. By doing the warm-up exercises first, students will have an easier time getting comfortable with the new concepts of a chapter before tackling harder problems. Answers to odd-numbered warm-up exercises are included in the Answers section at the end of the book. Answers to all warm-up exercises are in the *Instructor's Solutions Manual.*

**Conceptual Questions** At the end of each chapter are approximately a dozen conceptual questions. The Applying Physics examples presented in the text serve as models for students when conceptual questions are assigned and show how the concepts can be applied to understanding the physical world. The conceptual questions provide the student with a means of self-testing the concepts presented in the chapter. Some conceptual questions are appropriate for initiating classroom discussions. Answers to odd-numbered conceptual questions are included in the Answers section at the end of the book. Answers to all conceptual questions are in the *Instructor's Solutions Manual.*

**Problems** All questions and problems for this revision were carefully reviewed to improve their variety, interest, and pedagogical value while maintaining their clarity and quality. An extensive set of problems is included at the end of each chapter (in all, more than 2 000 problems are provided in the tenth edition). Answers to odd-numbered problems are given at the end of the book. For the convenience of both the student and instructor, about two-thirds of the problems are keyed to specific sections of the chapter. The remaining problems, labeled "Additional Problems," are not keyed to specific sections. The three levels of problems are graded according to their difficulty. Straightforward problems are numbered in **black**, intermediate level problems are numbered in **blue**, and the most challenging problems are numbered in **red**. The BIO icon identifies problems dealing with applications to the life sciences and medicine. Solutions to approximately 12 problems in each chapter are in the *Student Solutions Manual and Study Guide.*

There are three other types of problems we think instructors and students will find interesting as they work through the text:

- **S** **Symbolic problems** require the student to obtain an answer in terms of symbols. In general, some guidance is built into the problem statement. The goal is to better train the student to deal with mathematics at a level appropriate to this course. Most students at this level are uncomfortable with symbolic equations, which is unfortunate because symbolic equations are the most efficient vehicle for presenting relationships between physics concepts. Once students understand the physical concepts, their ability to solve problems is greatly enhanced. As soon as the numbers are substituted into an equation, however, all the concepts and their relationships to one another are lost, melded together in the student's calculator. Symbolic problems train the student to postpone substitution of values, facilitating their ability to think conceptually using the equations. An example of a symbolic problem is provided here:

> 14. **S** An object of mass $m$ is dropped from the roof of a building of height $h$. While the object is falling, a wind blowing parallel to the face of the building exerts a constant horizontal force $F$ on the object. (a) How long does it take the object to strike the ground? Express the time $t$ in terms of $g$ and $h$. (b) Find an expression in terms of $m$ and $F$ for the acceleration $a_x$ of the object in the horizontal direction (taken as the positive $x$-direction). (c) How far is the object displaced horizontally before hitting the ground? Answer in terms of $m$, $g$, $F$, and $h$. (d) Find the magnitude of the object's acceleration while it is falling, using the variables $F$, $m$, and $g$.

- **Q|C** **Quantitative/conceptual problems** encourage the student to think conceptually about a given physics problem rather than rely solely on computational skills. Research in physics education suggests that standard physics problems requiring calculations may not be entirely adequate in training students to think conceptually. Students learn to substitute numbers for symbols in the equations without fully understanding what they are doing or what the symbols mean. Quantitative/conceptual problems combat this tendency by asking for answers that require something other than a number or a calculation. An example of a quantitative/conceptual problem is provided here:

> 5. **Q|C** Starting from rest, a 5.00-kg block slides 2.50 m down a rough 30.0° incline. The coefficient of kinetic friction between the block and the incline is $\mu_k$ = 0.436. Determine (a) the work done by the force of gravity, (b) the work done by the friction force between block and incline, and (c) the work done by the normal force. (d) Qualitatively, how would the answers change if a shorter ramp at a steeper angle were used to span the same vertical height?

- **GP** **Guided problems** help students break problems into steps. A physics problem typically asks for one physical quantity in a given context. Often, however, several concepts must be used and a number of calculations are required to get that final answer. Many students are not accustomed to this level of complexity and often don't know where to start. A guided problem breaks a problem into smaller steps, enabling students to grasp all the concepts and strategies required to arrive at a correct solution. Unlike standard physics problems, guidance is often built into the problem statement. For example, the problem might say "Find the speed using conservation of energy" rather than asking only for the speed. In any given chapter there are usually two or three problem types that are particularly suited to this problem form. The problem must have a certain level of complexity, with a similar problem-solving strategy involved each time it appears. Guided problems are reminiscent of how a student might interact with a professor in an office visit. These problems help train students to break down complex problems into a series of simpler problems, an essential problem-solving skill. An example of a guided problem is provided here:

32. **GP** Two blocks of masses $m_1$ and $m_2$ ($m_1 > m_2$) are placed on a frictionless table in contact with each other. A horizontal force of magnitude $F$ is applied to the block of mass $m_1$ in Figure P4.32. (a) If $P$ is the magnitude of the contact force between the blocks, draw the free-body diagrams for each block. (b) What is the net force on the system consisting of both blocks? (c) What is the net force acting on $m_1$? (d) What is the net force acting on $m_2$? (e) Write the $x$-component of Newton's second law for each block. (f) Solve the resulting system of two equations and two unknowns, expressing the acceleration $a$ and contact force $P$ in terms of the masses and force. (g) How would the answers change if the force had been applied to $m_2$ instead? (*Hint:* Use symmetry; don't calculate!) Is the contact force larger, smaller, or the same in this case? Why?

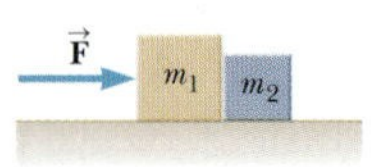

**Figure P4.32**

**Quick Quizzes** All the Quick Quizzes (see example below) are cast in an objective format, including multiple-choice, true–false, matching, and ranking questions. Quick Quizzes provide students with opportunities to test their understanding of the physical concepts presented. The questions require students to make decisions on the basis of sound reasoning, and some have been written to help students overcome common misconceptions. Answers to all Quick Quiz questions are found at the end of the textbook, and answers with detailed explanations are provided in the *Instructor's Solutions Manual*. Many instructors choose to use Quick Quiz questions in a "peer instruction" teaching style.

■ *Quick Quiz*

**4.4** A small sports car collides head-on with a massive truck. The greater impact force (in magnitude) acts on (a) the car, (b) the truck, (c) neither, the force is the same on both. Which vehicle undergoes the greater magnitude acceleration? (d) the car, (e) the truck, (f) the accelerations are the same.

**Problem-Solving Strategies** A general problem-solving strategy to be followed by the student is outlined at the end of Chapter 1. This strategy provides students with a structured process for solving problems. In most chapters, more specific strategies and suggestions (see example below) are included for solving the types of problems featured in both the worked examples and the end-of-chapter problems. This feature helps students identify the essential steps in solving problems and increases their skills as problem solvers.

■ **PROBLEM-SOLVING STRATEGY**

**Newton's Second Law**

*Problems involving Newton's second law can be very complex. The following protocol breaks the solution process down into smaller, intermediate goals:*

1. **Read** the problem carefully at least once.
2. **Draw** a picture of the system, identify the object of primary interest, and indicate forces with arrows.
3. **Label** each force in the picture in a way that will bring to mind what physical quantity the label stands for (e.g., $T$ for tension).
4. **Draw** a free-body diagram of the object of interest, based on the labeled picture. If additional objects are involved, draw separate free-body diagrams for them. Choose convenient coordinates for each object.
5. **Apply Newton's second law.** The $x$- and $y$-components of Newton's second law should be taken from the vector equation and written individually. This usually results in two equations and two unknowns.
6. **Solve** for the desired unknown quantity, and substitute the numbers.

**Biomedical Applications** For biology and pre-med students, BIO icons point the way to various practical and interesting applications of physical principles to biology and medicine.

**MCAT Skill Builder Study Guide** The tenth edition of *College Physics* has a special skill-building Appendix (Appendix E) available via CengageCompose to help pre-med students prepare for the MCAT exam. The appendix contains examples written by the text authors that help students build conceptual and quantitative skills. These skill-building examples are followed by MCAT-style questions written by test prep experts to make sure students are ready to ace the exam.

**MCAT Test Preparation Guide** Located at the front of the book, this guide outlines the six content categories related to physics on the new MCAT exam that will be administered starting in 2015. Students can use the guide to prepare for the MCAT exam, class tests, or homework assignments.

**Applying Physics** The Applying Physics features provide students with an additional means of reviewing concepts presented in that section. Some Applying Physics examples demonstrate the connection between the concepts presented in that chapter and other scientific disciplines. These examples also serve as models for students when assigned the task of responding to the Conceptual Questions presented at the end of each chapter. For examples of Applying Physics boxes, see Applying Physics 9.5 (Home Plumbing) on page 313 and Applying Physics 13.1 (Bungee Jumping) on page 456.

**Tips** Placed in the margins of the text, Tips address common student misconceptions and situations in which students often follow unproductive paths (see example at the right). More than 95 Tips are provided in this edition to help students avoid common mistakes and misunderstandings.

**Tip 4.3 Newton's Second Law Is a *Vector* Equation**

In applying Newton's second law, add all of the forces on the object as vectors and then find the resultant vector acceleration by dividing by $m$. Don't find the individual magnitudes of the forces and add them like scalars.

**Marginal Notes** Comments and notes appearing in the margin (see example at the right) can be used to locate important statements, equations, and concepts in the text.

◀ Newton's third law

**Applications** Although physics is relevant to so much in our modern lives, it may not be obvious to students in an introductory course. Application margin notes (see example to the right) make the relevance of physics to everyday life more obvious by pointing out specific applications in the text. Some of these applications pertain to the life sciences and are marked with a BIO icon. A list of the Applications in Volume 2 appears after this Preface.

BIO **APPLICATION**

Diet Versus Exercise in Weight-loss Programs

**Style** To facilitate rapid comprehension, we have attempted to write the book in a style that is clear, logical, relaxed, and engaging. The somewhat informal and relaxed writing style is designed to connect better with students and enhance their reading enjoyment. New terms are carefully defined, and we have tried to avoid the use of jargon.

**Introductions** All chapters begin with a brief preview that includes a discussion of the chapter's objectives and content.

**Units** The international system of units (SI) is used throughout the text. The U.S. customary system of units is used only to a limited extent in the chapters on mechanics and thermodynamics.

**Pedagogical Use of Color** Readers should consult the pedagogical color chart (inside the front cover) for a listing of the color-coded symbols used in the text diagrams. This system is followed consistently throughout the text.

**Important Statements and Equations** Most important statements and definitions are set in **boldface** type or are highlighted with a background screen for added emphasis and ease of review. Similarly, important equations are highlighted with a tan background to facilitate location.

**Illustrations and Tables** The readability and effectiveness of the text material, worked examples, and end-of-chapter conceptual questions and problems are enhanced by the large number of figures, diagrams, photographs, and tables. Full color adds clarity to the artwork and makes illustrations as realistic as possible. Three-dimensional effects are rendered with the use of shaded and lightened areas where appropriate. Vectors are color coded, and curves in graphs are drawn in color. Color photographs have been carefully selected, and their accompanying captions have been written to serve as an added instructional tool. A complete description of the pedagogical use of color appears on the inside front cover.

**Summary** The end-of-chapter Summary is organized by individual section heading for ease of reference. Most chapter summaries also feature key figures from the chapter.

**Significant Figures** Significant figures in both worked examples and end-of-chapter problems have been handled with care. Most numerical examples and problems are worked out to either two or three significant figures, depending on the accuracy of the data provided. Intermediate results presented in the examples are rounded to the proper number of significant figures, and only those digits are carried forward.

**Appendices and Endpapers** Several appendices are provided at the end of the textbook. Most of the appendix material (Appendix A) represents a review of mathematical concepts and techniques used in the text, including scientific notation, algebra, geometry, and trigonometry. Reference to these appendices is made as needed throughout the text. Most of the mathematical review sections include worked examples and exercises with answers. In addition to the mathematical review, some appendices contain useful tables that supplement textual information. For easy reference, the front endpapers contain a chart explaining the use of color throughout the book and a list of frequently used conversion factors.

## Teaching Options

This book contains more than enough material for a one-year course in introductory physics, which serves two purposes. First, it gives the instructor more flexibility in choosing topics for a specific course. Second, the book becomes more useful as a resource for students. On average, it should be possible to cover about one chapter each week for a class that meets three hours per week. Those sections, examples, and end-of-chapter problems dealing with applications of physics to life sciences are identified with the BIO icon. We offer the following suggestions for shorter courses for those instructors who choose to move at a slower pace through the year.

***Option A:*** If you choose to place more emphasis on contemporary topics in physics, you could omit all or parts of Chapter 8 (Rotational Equilibrium and Rotational Dynamics), Chapter 21 (Alternating-Current Circuits and Electromagnetic Waves), and Chapter 25 (Optical Instruments).

***Option B:*** If you choose to place more emphasis on classical physics, you could omit all or parts of Part 6 of the textbook, which deals with special relativity and other topics in twentieth-century physics.

The *Instructor's Solutions Manual* offers additional suggestions for specific sections and topics that may be omitted without loss of continuity if time presses.

### CengageCompose Options for *College Physics*

Would you like to easily create your own personalized text, selecting the elements that meet your specific learning objectives?

**CengageCompose** puts the power of the vast Cengage Learning library of learning content at your fingertips to create exactly the text you need. The all-new, Web-based CengageCompose site lets you quickly scan content and review materials to pick what you need for your text. Site tools let you easily assemble the modular learning units into the order you want and immediately provide you with an online copy for review. Add enrichment content like case studies, exercises, and lab materials to further build your ideal learning materials. Even choose from hundreds of vivid, art-rich, customizable, full-color covers.

Cengage Learning offers the fastest and easiest way to create unique customized learning materials delivered the way you want. For more information about custom publishing options, visit **www.cengage.com/custom** or contact your local Cengage Learning representative.

## Course Solutions That Fit Your Teaching Goals and Your Students' Learning Needs

Recent advances in educational technology have made homework management systems and audience response systems powerful and affordable tools to enhance the way you teach your course. Whether you offer a more traditional text-based course, are interested in using or are currently using an online homework management system such as Enhanced WebAssign, or are ready to turn your lecture into an interactive learning environment with JoinIn™, you can be confident that the text's proven content provides the foundation for each and every component of our technology and ancillary package.

### Homework Management Systems

**Enhanced WebAssign for *College Physics*, Tenth Edition.** Exclusively from Cengage Learning, Enhanced WebAssign offers an extensive online program for physics to encourage the practice that's so critical for concept mastery. The meticulously crafted pedagogy and exercises in our proven texts become even more effective in Enhanced WebAssign. Enhanced WebAssign includes the Cengage YouBook, a highly customizable, interactive eBook. WebAssign includes:

- **All of the quantitative end-of-chapter problems, now including worked out solutions, matching the algorithmic version of the question assigned to each student.**
- **Selected problems enhanced with targeted feedback.** An example of targeted feedback appears below:

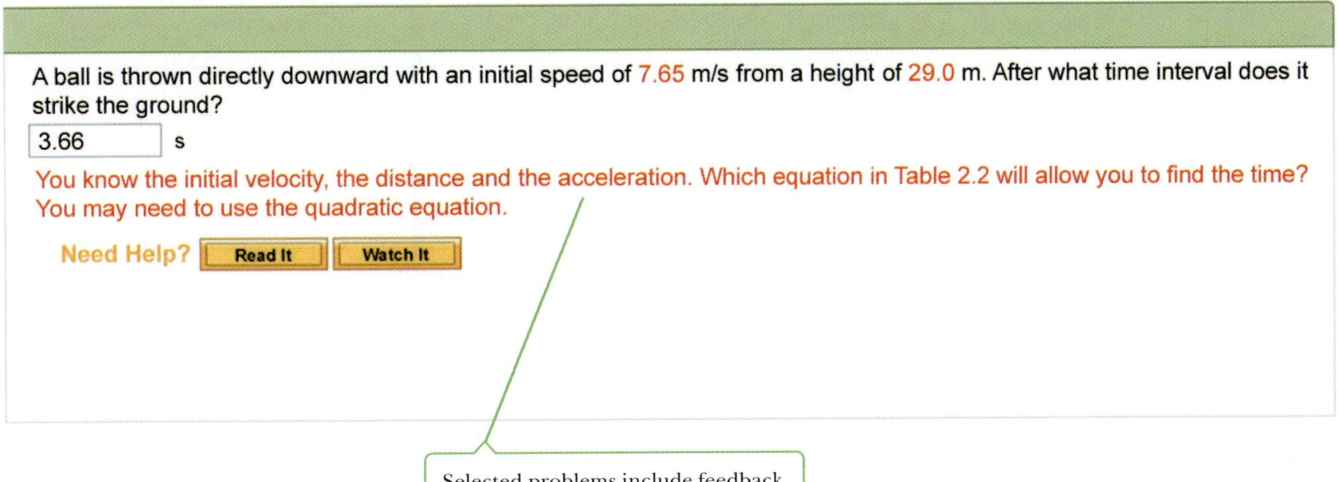

A ball is thrown directly downward with an initial speed of 7.65 m/s from a height of 29.0 m. After what time interval does it strike the ground?
3.66 s
You know the initial velocity, the distance and the acceleration. Which equation in Table 2.2 will allow you to find the time? You may need to use the quadratic equation.
Need Help? Read It Watch It


Selected problems include feedback to address common mistakes that students make. This feedback was developed by professors with years of classroom experience.

- **Master It tutorials** (indicated in the text by an **M** icon), to help students work through the problem one step at a time. An example of a Master It tutorial appears below:

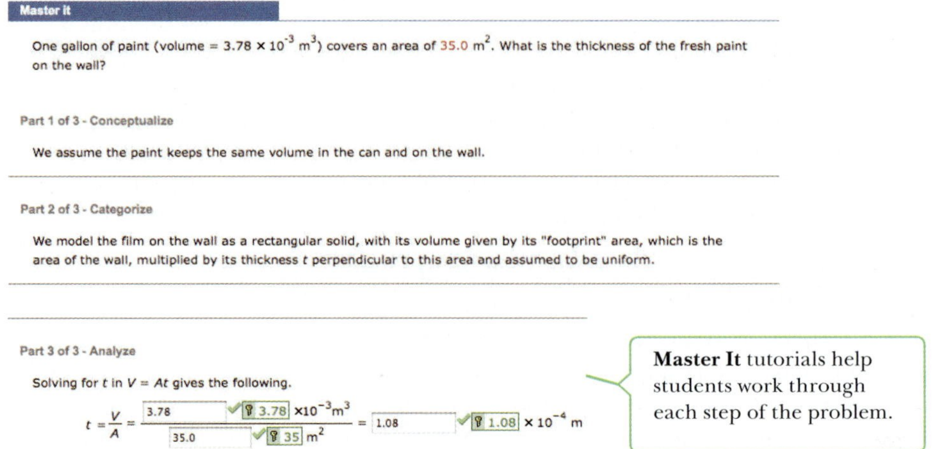

Master It
One gallon of paint (volume = 3.78 × 10^-3 m^3) covers an area of 35.0 m^2. What is the thickness of the fresh paint on the wall?
Part 1 of 3 - Conceptualize
We assume the paint keeps the same volume in the can and on the wall.
Part 2 of 3 - Categorize
We model the film on the wall as a rectangular solid, with its volume given by its "footprint" area, which is the area of the wall, multiplied by its thickness t perpendicular to this area and assumed to be uniform.
Part 3 of 3 - Analyze
Solving for t in V = At gives the following.
Master It tutorials help students work through each step of the problem.


- **Watch It solution videos** (indicated in the text by a **W** icon) that explain fundamental problem-solving strategies, to help students step through the problem. In addition, instructors can choose to include video hints of problem-solving strategies. A screen shot from a Watch It solution video appears below:

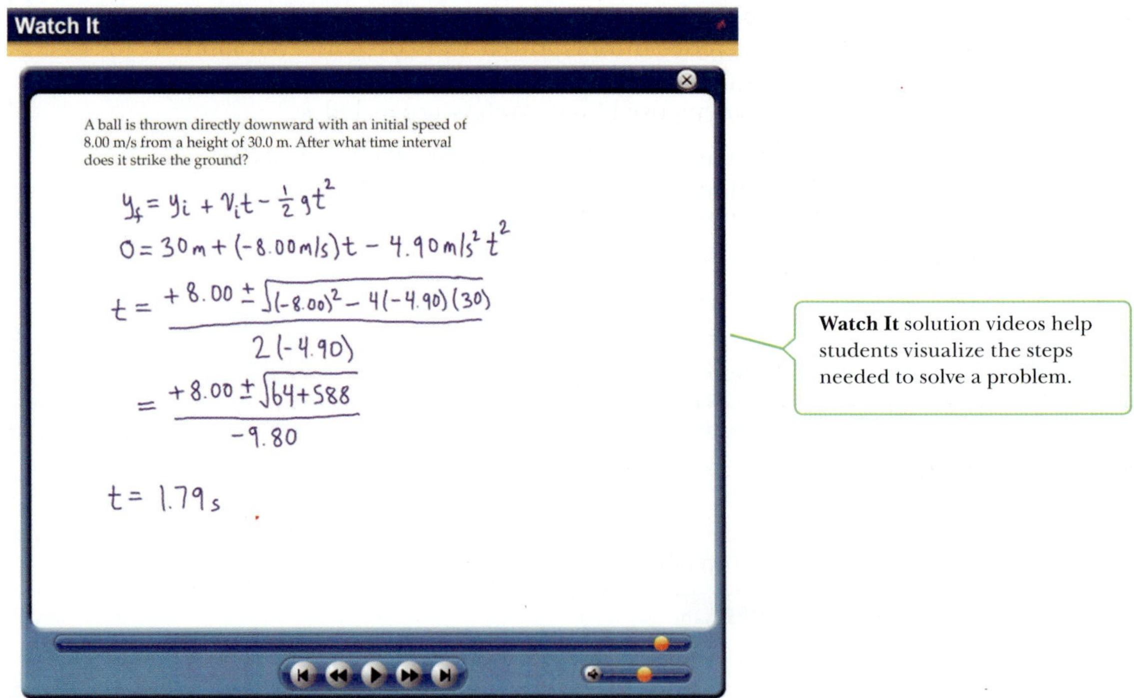

Watch It
A ball is thrown directly downward with an initial speed of 8.00 m/s from a height of 30.0 m. After what time interval does it strike the ground?
t = 1.79 s
Watch It solution videos help students visualize the steps needed to solve a problem.


- **Concept Checks**
- **PhET simulations**
- **Most worked examples**, enhanced with hints and feedback, to help strengthen students' problem-solving skills
- **Every Quick Quiz**, giving your students ample opportunity to test their conceptual understanding

- **Personalized Study Plan.** The Personal Study Plan in Enhanced WebAssign provides chapter and section assessments that show students what material they know and what areas require more work. For items that they answer incorrectly, students can click on links to related study resources such as videos, tutorials, or reading materials. Color-coded progress indicators let them see how well they are doing on different topics. You decide what chapters and sections to include—and whether to include the plan as part of the final grade or as a study guide with no scoring involved.
- **The Cengage YouBook.** WebAssign has a customizable and interactive eBook, the Cengage YouBook, that lets you tailor the textbook to fit your course and connect with your students. You can remove and rearrange chapters in the table of contents and tailor assigned readings that match your syllabus exactly. Powerful editing tools let you change as much as you'd like—or leave it just like it is. You can highlight key passages or add sticky notes to pages to comment on a concept in the reading, and then share any of these individual notes and highlights with your students, or keep them personal. You can also edit narrative content in the textbook by adding a text box or striking out text. With a handy link tool, you can drop in an icon at any point in the eBook that lets you link to your own lecture notes, audio summaries, video lectures, or other files on a personal Web site or anywhere on the Web. A simple YouTube widget lets you easily find and embed videos from YouTube directly into eBook pages. The Cengage YouBook helps students go beyond just reading the textbook. Students can also highlight the text and add their own notes or bookmarks. Animations play right on the page at the point of learning so that they're not speed bumps to reading but true enhancements. Please visit **www.webassign.net/brookscole** to view an interactive demonstration of Enhanced WebAssign.
- Offered exclusively in WebAssign, **Quick Prep** for physics is algebra and trigonometry math remediation within the context of physics applications and principles. Quick Prep helps students succeed by using narratives illustrated throughout with video examples. The Master It tutorial problems allow students to assess and retune their understanding of the material. The Practice Problems that go along with each tutorial allow both the student and the instructor to test the student's understanding of the material.

Quick Prep includes the following features:

- 67 interactive tutorials
- 67 additional practice problems
- A thorough overview of each topic, including video examples
- Can be taken before the semester begins or during the first few weeks of the course
- Can also be assigned alongside each chapter for "just in time" remediation

Topics include units, scientific notation, and significant figures; the motion of objects along a line; functions; approximation and graphing; probability and error; vectors, displacement, and velocity; spheres; and force and vector projections.

### MindTap™: The Personal Learning Experience

MindTap for Serway and Vuille *College Physics* is a personalized, fully online digital learning platform of authoritative textbook content, WebAssign assignments, and services that engages your students with interactivity while also offering choices in the configuration of coursework and enhancement of the curriculum via complimentary Web apps known as MindApps. MindApps range from WebAssign, ReadSpeaker (which reads the text out loud to students), to Kaltura (allowing you to insert inline video and audio into your curriculum), to ConnectYard (allowing you to create digital "yards" through social media—all without "friending" your students). MindTap is well beyond an eBook, a homework solution or digital supplement, a resource center Web site, a course delivery platform, or a Learning Management System. It is the first in a new category—the Personal Learning Experience.

### CengageBrain.com

On **CengageBrain.com** students will be able to save up to 60% on their course materials through our full spectrum of options. Students will have the option to rent their textbooks or purchase print textbooks, e-textbooks, or individual e-chapters and audio books all for substantial savings over average retail prices. **CengageBrain.com** also includes access to Cengage Learning's broad range of homework and study tools and features a selection of free content.

### Lecture Presentation Resources

**Instructor's Companion Site for *College Physics*, Tenth Edition.** Bringing physics principles and concepts to life in your lectures has never been easier! The full-featured Instructor's Companion Site provides everything you need for *College Physics*, tenth edition. Key content includes the *Instructor's Solutions Manual*, art and images from the text, premade chapter-specific PowerPoint lectures, Cengage Learning Testing Powered by Cognero with pre-loaded test questions, JoinIn response-system "clickers," Active Figures animations, a physics movie library, and more.

**Cengage Learning Testing Powered by Cognero** is a flexible, online system that allows you to author, edit, and manage test bank content, create multiple test versions in an instant, and deliver tests from your LMS, your classroom, or wherever you want. No special installs or downloads needed, you can create tests from anywhere with internet access. Cognero brings simplicity at every step, with a desktop-inspired interface, a full-featured test generator, and cross-platform compatibility.

**JoinIn.** *Assessing to Learn in the Classroom* questions developed at the University of Massachusetts Amherst. This collection of 250 advanced conceptual questions has been tested in the classroom for more than ten years and takes peer learning to a new level. JoinIn helps you turn your lectures into an interactive learning environment that promotes conceptual understanding. Available exclusively for higher education from our partnership with Turning Technologies, JoinIn is the easiest way to turn your lecture hall into a personal, fully interactive experience for your students!

### Assessment and Course Preparation Resources

A number of resources listed below will assist with your assessment and preparation processes.

***Instructor's Solutions Manual*** This manual contains complete worked solutions to all end-of-chapter warm-up exercises, conceptual questions, and problems in the text, and full answers with explanations to the Quick Quizzes. Volume 1 contains Chapters 1 through 14, and Volume 2 contains Chapters 15 through 30. Electronic files of the *Instructor's Solutions Manual* are available on the Instructor's Companion Site.

***Test Bank*** by Ed Oberhofer (University of North Carolina at Charlotte and Lake-Sumter Community College). The test bank is available on the Instructor's Companion Site. This two-volume test bank contains approximately 1 750 multiple-choice questions. Instructors may print and duplicate pages for distribution to students. The test bank is available in the Cognero test-generator, or in PDF, Word, WebCT, or Blackboard versions on the instructor's companion site at **www.CengageBrain.com**.

## Supporting Materials for the Instructor

Supporting instructor materials are available to qualified adopters. Please consult your local Cengage Learning representative for details. Visit **www.CengageBrain.com** to

- request a desk copy
- locate your local representative
- download electronic files of select support materials

## Student Resources

Visit the *College Physics* website at **www.CengageBrain.com** to see samples of select student supplements. Go to **CengageBrain.com** to purchase and access this product at Cengage Learning's preferred online store.

***Student Solutions Manual and Study Guide*** Now offered in two volumes, the *Student Solutions Manual and Study Guide* features detailed solutions to approximately 12 problems per chapter. Boxed numbers identify those problems in the textbook for which complete solutions are found in the manual. The manual also features a skills section, important notes from key sections of the text, and a list of important equations and concepts. Volume 1 contains Chapters 1 through 14, and Volume 2 contains Chapters 15 through 30.

***Physics Laboratory Manual,* Third Edition** by David Loyd (Angelo State University) supplements the learning of basic physical principles while introducing laboratory procedures and equipment. Each chapter includes a prelaboratory assignment, objectives, an equipment list, the theory behind the experiment, experimental procedures, graphing exercises, and questions. A laboratory report form is included with each experiment so that the student can record data, calculations, and experimental results. Students are encouraged to apply statistical analysis to their data. A complete *Instructor's Manual* is also available to facilitate use of this lab manual.

***Physics Laboratory Experiments,* Seventh Edition** by Jerry D. Wilson (Lander College) and Cecilia A. Hernández (American River College). This market-leading manual for the first-year physics laboratory course offers a wide range of class-tested experiments designed specifically for use in small to midsize lab programs. A series of integrated experiments emphasizes the use of computerized instrumentation and includes a set of "computer-assisted experiments" to allow students and instructors to gain experience with modern equipment. This option also enables instructors to determine the appropriate balance between traditional and computer-based experiments for their courses. By analyzing data through two different methods, students gain a greater understanding of the concepts behind the experiments. The seventh edition is updated with the latest information and techniques involving state-of-the-art equipment and a new Guided Learning feature addresses the growing interest in guided-inquiry pedagogy. Fourteen additional experiments are also available through custom printing.

# Acknowledgments

In preparing the tenth edition of this textbook, we have been guided by the expertise of many people who have reviewed manuscript or provided suggestions. Prior to our work on this revision, we conducted a survey of over 250 professors who teach the course; their collective feedback helped shape this revision, and we thank them. We also wish to acknowledge the following reviewers of recent editions, and express our sincere appreciation for their helpful suggestions, criticism, and encouragement.

Gary B. Adams, *Arizona State University*; Ricardo Alarcon, *Arizona State University*; Natalie Batalha, *San Jose State University*; Gary Blanpied, *University of South Carolina*; Thomas K. Bolland, *The Ohio State University*; Kevin R. Carter, *School of Science and Engineering Magnet*; Kapila Calara Castoldi, *Oakland University*; David Cinabro, *Wayne State University*; Andrew Cornelius, *University of Nevada–Las Vegas*; Yesim Darici, *Florida International University*; N. John DiNardo, *Drexel University*; Steve Ellis, *University of Kentucky*; Hasan Fakhruddin, *Ball State University/The Indiana Academy*; Emily Flynn; Lewis Ford, *Texas A & M University;* Gardner Friedlander, *University School of Milwaukee*; Dolores Gende, *Parish Episcopal School*; Mark Giroux, *East Tennessee State University*; James R. Goff, *Pima Community College*; Yadin Y. Goldschmidt, *University of Pittsburgh*; Torgny Gustafsson, *Rutgers* University; Steve Hagen, *University of Florida*; Raymond Hall, *California State University–Fresno*; Patrick Hamill, *San Jose State University*; Joel Handley; Grant W. Hart, *Brigham Young University*; James E. Heath, *Austin Community College*; Grady Hendricks, *Blinn College*; Rhett Herman, *Radford University*; Aleksey Holloway, *University of Nebraska at Omaha*; Joey Huston, *Michigan State University*; Mark James, *Northern Arizona University*; Randall Jones, *Loyola College Maryland*; Teruki Kamon, *Texas A & M University*; Joseph Keane, *St. Thomas Aquinas College*; Dorina Kosztin, *University of Missouri–Columbia*; Martha Lietz, *Niles West High School*; Edwin Lo; Rafael Lopez-Mobilia, *University of Texas at San Antonio*; Mark Lucas, *Ohio University*; Mark E. Mattson, *James Madison University*; Sylvio May, *North Dakota State University*; John A. Milsom, *University of Arizona*; Monty Mola, *Humboldt State University*; Charles W. Myles, *Texas Tech University*; Ed Oberhofer, *Lake Sumter Community College*; Chris Pearson, *University of Michigan–Flint*; Alexey A. Petrov, *Wayne State University*; J. Patrick Polley, *Beloit College*; Scott Pratt, *Michigan State University;* M. Anthony Reynolds, *Embry-Riddle Aeronautical University*; Dubravka Rupnik, *Louisiana State University*; Scott Saltman, *Phillips Exeter Academy*; Surajit Sen, *State University of New York at Buffalo*; Bartlett M. Sheinberg, *Houston Community College*; Marllin L. Simon, *Auburn University*; Matthew Sirocky; Gay Stewart, *University of Arkansas*; George Strobel, *University of Georgia*; Eugene Surdutovich, *Oakland University*; Marshall Thomsen, *Eastern Michigan University*; James Wanliss, *Presbyterian College*; Michael Willis, *Glen Burnie High School*; David P. Young, *Louisiana State University*

*College Physics*, tenth edition, was carefully checked for accuracy by Mark L. Giroux, *East Tennessee State University*; Grant W. Hart, *Brigham Young University*; Mark James, *Northern Arizona University*; Randall Jones, *Loyola University Maryland*; Ed Oberhofer, *Lake Sumter Community College*; M. Anthony Reynolds, *Embry-Riddle Aeronautical University*; Phillip Sprunger, *Louisiana State University*; and Eugene Surdutovich, *Oakland University*. Although responsibility for any remaining errors rests with us, we thank them for their dedication and vigilance.

Gerd Kortemeyer and Randall Jones contributed several end-of-chapter problems, especially those of interest to the life sciences. Edward F. Redish of the University of Maryland graciously allowed us to list some of his problems from the Activity Based Physics Project.

Special thanks and recognition go to the professional staff at Cengage Learning—in particular, Mary Finch, Charlie Hartford, Ed Dodd, Andrew Coppola, Alison Eigel Zade, Janet del Mundo, Nicole Molica, Cate Barr, Chris Robinson, and Karolina Kiwak—for their fine work during the development, production, and promotion of this textbook. We recognize the skilled production service provided by the staff at Graphic World Inc., and the dedicated photo research efforts of Vignesh Sadhasivam and Abbey Stebing at PreMediaGlobal.

Finally, we are deeply indebted to our wives and children for their love, support, and long-term sacrifices.

**Raymond A. Serway**
St. Petersburg, Florida

**Chris Vuille**
Daytona Beach, Florida

# Engaging Applications

Although physics is relevant to so much in our lives, it may not be obvious to students in an introductory course. In this tenth edition of *College Physics,* we continue a design feature begun in the seventh edition. This feature makes the relevance of physics to everyday life more obvious by pointing out specific applications in the form of a marginal note. Some of these applications pertain to the life sciences and are marked with the BIO icon. The list below is not intended to be a complete listing of all the applications of the principles of physics found in this textbook. Many other applications are to be found within the text and especially in the worked examples, conceptual questions, and end-of-chapter problems.

**Chapter 30**

# To the Student

As a student, it's important that you understand how to use this book most effectively and how best to go about learning physics. Scanning through the Preface will acquaint you with the various features available, both in the book and online. Awareness of your educational resources and how to use them is essential. Although physics is challenging, it can be mastered with the correct approach.

## How to Study

Students often ask how best to study physics and prepare for examinations. There is no simple answer to this question, but we'd like to offer some suggestions based on our own experiences in learning and teaching over the years.

First and foremost, maintain a positive attitude toward the subject matter. Like learning a language, physics takes time. Those who keep applying themselves on a *daily basis* can expect to reach understanding and succeed in the course. Keep in mind that physics is the most fundamental of all natural sciences. Other science courses that follow will use the same physical principles, so it is important that you understand and are able to apply the various concepts and theories discussed in the text. They're relevant!

## Concepts and Principles

Students often try to do their homework without first studying the basic concepts. It is essential that you understand the basic concepts and principles *before* attempting to solve assigned problems. You can best accomplish this goal by carefully reading the textbook *before* you attend your lecture on the covered material. When reading the text, you should jot down those points that are not clear to you. Also be sure to make a diligent attempt at answering the questions in the Quick Quizzes as you come to them in your reading. We have worked hard to prepare questions that help you judge for yourself how well you understand the material. Pay careful attention to the many Tips throughout the text. They will help you avoid misconceptions, mistakes, and misunderstandings as well as maximize the efficiency of your time by minimizing adventures along fruitless paths. During class, take careful notes and ask questions about those ideas that are unclear to you. Keep in mind that few people are able to absorb the full meaning of scientific material after only one reading. Your lectures and laboratory work supplement your textbook and should clarify some of the more difficult material. You should minimize rote memorization of material. Successful memorization of passages from the text, equations, and derivations does not necessarily indicate that you understand the fundamental principles.

Your understanding will be enhanced through a combination of efficient study habits, discussions with other students and with instructors, and your ability to solve the problems presented in the textbook. Ask questions whenever you think clarification of a concept is necessary.

## Study Schedule

It is important for you to set up a regular study schedule, preferably a daily one. Make sure you read the syllabus for the course and adhere to the schedule set by your instructor. As a general rule, you should devote about two hours of study time for every one hour you are in class. If you are having trouble with the course, seek the advice of the instructor or other students who have taken the course. You may find it necessary to seek further instruction from experienced students. Very often, instructors offer review sessions in addition to regular class periods. It is important that you avoid the practice of delaying study until a day or two before an exam. One hour of study a day for 14 days is far more effective than 14 hours the day before the exam. "Cramming" usually produces disastrous results, especially in science. Rather than attempting an all-night study session immediately before an exam, briefly review the basic concepts and equations and get a good night's rest. If you think you need additional help in understanding the concepts, in preparing for exams, or in problem solving, we suggest you acquire a copy of the *Student Solutions Manual and Study Guide* that accompanies this textbook; this manual should be available at your college bookstore.

Visit the *College Physics* website at **www.CengageBrain.com** to see samples of select student supplements. Go to **CengageBrain.com** to purchase and access this product at Cengage Learning's preferred online store.

## Use the Features

You should make full use of the various features of the text discussed in the preface. For example, marginal notes are useful for locating and describing important equations and concepts, and **boldfaced** type indicates important statements and definitions. Many useful tables are contained in the appendices, but most tables are incorporated in the text where they are most often referenced. Appendix A is a convenient review of mathematical techniques.

Answers to all Quick Quizzes and Example Questions, as well as odd-numbered multiple-choice questions, conceptual questions, and problems, are given at the end of the textbook. Answers to selected end-of-chapter problems are provided in the *Student Solutions Manual and Study Guide.* Problem-Solving Strategies included in selected chapters throughout the text give you additional information about how you should solve problems. The contents provide an overview of the entire text, and the index enables you to locate specific material quickly. Footnotes sometimes are used to supplement the text or to cite other references on the subject discussed.

After reading a chapter, you should be able to define any new quantities introduced in that chapter and to discuss the principles and assumptions used to arrive at certain key relations. The chapter summaries and the review sections of the *Student Solutions Manual and Study Guide* should help you in this regard. In some cases, it may be necessary for you to refer to the index of the text to locate certain topics. You should be able to correctly associate each physical quantity with the symbol used to represent that quantity and the unit in which the quantity is specified. Further, you should be able to express each important relation in a concise and accurate prose statement.

## Problem Solving

R. P. Feynman, Nobel laureate in physics, once said, "You do not know anything until you have practiced." In keeping with this statement, we strongly advise that you develop the skills necessary to solve a wide range of problems. Your ability to solve problems will be one of the main tests of your knowledge of physics, so you should try to solve as many problems as possible. It is essential that you understand basic concepts and principles before attempting to solve problems. It is good practice to try to find alternate solutions to the same problem. For example, you can solve problems in mechanics using Newton's laws, but very often an alternate method that draws on energy considerations is more direct. You should not deceive yourself into thinking you understand a problem merely because you have seen it solved in class. You must be able to solve the problem and similar problems on your own. We have cast the examples in this book in a special, two-column format to help you in this regard. After studying an example, see if you can cover up the right-hand side and do it yourself, using only the written descriptions on the left as hints. Once you succeed at that, try solving the example using only the strategy statement as a guide. Finally, try to solve the problem completely on your own. At this point you are ready to answer the associated question and solve the exercise. Once you have accomplished all these steps, you will have a good mastery of the problem, its concepts, and mathematical technique. After studying all the Example Problems in this way, you are ready to tackle the problems at the end of the chapter. Of those, the guided problems provide another aid to learning how to solve some of the more complex problems.

The approach to solving problems should be carefully planned. A systematic plan is especially important when a problem involves several concepts. First, read the problem several times until you are confident you understand what is being asked. Look for any key words that will help you interpret the problem and perhaps allow you to make certain assumptions. Your ability to interpret a question properly is an integral part of problem solving. Second, you should acquire the habit of writing down the information given in a problem and those quantities that need to be found; for example, you might construct a table listing both the quantities given and the quantities to be found. This procedure is sometimes used in the worked examples of the textbook. After you have decided on the method you think is appropriate for a given problem, proceed with your solution. Finally, check your results to see if they are reasonable and consistent with your initial understanding of the problem. General problem-solving strategies of this type are included in the text and are highlighted with a surrounding box. If you follow the steps of this procedure, you will find it easier to come up with a solution and will also gain more from your efforts.

Often, students fail to recognize the limitations of certain equations or physical laws in a particular situation. It is very important that you understand and remember the assumptions underlying a particular theory or formalism. For example, certain equations in kinematics apply only to a particle moving with constant acceleration. These equations are not valid for describing motion whose acceleration is not constant, such as the motion of an object connected to a spring or the motion of an object through a fluid.

## Experiments

Because physics is a science based on experimental observations, we recommend that you supplement the text by performing various types of "hands-on" experiments, either at home or in the laboratory. For example, the common Slinky™ toy is excellent for studying traveling waves, a ball swinging on the end of a long string can be used to investigate pendulum motion, various masses attached to the end of a vertical spring or rubber band can be used to determine their elastic nature, an old pair of Polaroid sunglasses and some discarded lenses and a magnifying glass are the components of various experiments in optics, and the approximate measure of the free-fall acceleration can be determined simply by measuring with a stopwatch the time it takes for a ball to drop from a known height. The list of such experiments is endless. When physical models are not available, be imaginative and try to develop models of your own.

## New Media

If available, we strongly encourage you to use the **Enhanced WebAssign** product that is available with this textbook. It is far easier to understand physics if you see it in action, and the materials available in Enhanced WebAssign will enable you to become a part of that action. Enhanced WebAssign is described in the Preface.

## An Invitation to Physics

It is our hope that you too will find physics an exciting and enjoyable experience and that you will profit from this experience, regardless of your chosen profession. Welcome to the exciting world of physics!

*To see the World in a Grain of Sand*
*And a Heaven in a Wild Flower,*
*Hold infinity in the palm of your hand*
*And Eternity in an hour.*

**William Blake, "Auguries of Innocence"**

# Welcome to Your MCAT Test Preparation Guide

The MCAT Test Preparation Guide makes your copy of *College Physics*, tenth edition, the most comprehensive MCAT study tool and classroom resource in introductory physics. Starting with the Spring 2015 test, the MCAT will be thoroughly revised (see **www.aamc.org/students/applying/mcat/mcat2015** for more details). The new test section that will include problems related to physics is *Chemical and Physical Foundations of Biological Systems.* Of the ~65 test questions in this section, approximately 25% will relate to introductory physics topics from the six content categories shown below:

**Content Category 4A:** Translational motion, forces, work, energy, and equilibrium in living systems

**Review Plan**

**Motion**

- **Chapter 1, Sections 1.1, 1.3, and 1.5**
  Examples 1.1–1.2 and 1.4–1.5
  Chapter problems 1–6 and 15–27
- **Chapter 2, Sections 2.2 and 2.3**
  Quick Quizzes 2.1–2.3
  Examples 2.1–2.3
  Chapter problems 1–25
- **Chapter 3, Sections 3.1 and 3.2**
  Quick Quizzes 3.1–3.3
  Examples 3.1–3.3
  Chapter problems 1–21

**Equilibrium**

- **Chapter 4, Sections 4.1–4.5**
  Quick Quizzes 4.1–4.6
  Examples 4.1–4.11
  Chapter problems 1–38
- **Chapter 8, Sections 8.1–8.5**
  Quick Quizzes 8.1–8.3
  Examples 8.1–8.11
  Chapter problems 1–41

**Work**

- **Chapter 5, Sections 5.1 and 5.2**
  Quick Quiz 5.1
  Examples 5.1–5.3
  Chapter problems 1–18
- **Chapter 12, Section 12.1**
  Quick Quiz 12.1
  Examples 12.1–12.2
  Chapter problems 1–10

**Energy**

- **Chapter 5, Sections 5.2–5.6**
  Quick Quizzes 5.2–5.4
  Examples 5.3–5.14
  Chapter problems 9–58

**Content Category 4B:** Importance of fluids for the circulation of blood, gas movement, and gas exchange

**Review Plan**

**Fluids**

- **Chapter 9, Sections 9.2, 9.4–9.7, and 9.9**
  Quick Quizzes 9.1–9.7
  Examples 9.1, 9.5–9.14, and 9.16–9.19
  Chapter problems 1–7 and 20–72

**Gas phase**

- **Chapter 9, Section 9.5**
  Quick Quizzes 9.3–9.4
  Chapter problems 20–28
- **Chapter 10, Sections 10.2, 10.4, and 10.5**
  Quick Quiz 10.6
  Examples 10.1–10.2 and 10.6–10.10
  Chapter problems 1–10 and 29–46

**Content Category 4C:** Electrochemistry and electrical circuits and their elements.

### Review Plan

#### Electrostatics

- **Chapter 15, Sections 15.1–15.2 and 15.4**
  Quick Quizzes 15.1 and 15.3–15.5
  Examples 15.4 and 15.5
  Chapter problems 17–29
- **Chapter 16, Sections 16.1–16.3**
  Quick Quizzes 16.1–16.7
  Examples 16.1–16.5
  Chapter problems 1–24

#### Circuit elements

- **Chapter 15, Sections 15.1 and 15.6**
  Chapter problems 30–35
- **Chapter 16, Sections 16.7–16.10**
  Quick Quizzes 16.8–16.11
  Examples 16.6–16.12
  Chapter problems 25–53
- **Chapter 17, Sections 17.1 and 17.3–17.5**
  Quick Quizzes 17.1 and 17.3–17.6
  Examples 17.1 and 17.3–17.4
  Chapter problems 1–32
- **Chapter 18, Sections 18.1–18.3**
  Quick Quizzes 18.1–18.8
  Examples 18.1–18.3
  Chapter problems 1–15

**Content Category 4D:** How light and sound interact with matter

### Review Plan

#### Sound

- **Chapter 13, Sections 13.6 and 13.8**
  Examples 13.8–13.9
  Chapter problems 41–49
- **Chapter 14, Sections 14.1–14.4, 14.6, 14.9–14.10, and 14.12**
  Quick Quizzes 14.1–14.3 and 14.5–14.6
  Examples 14.1–14.2, 14.4–14.5, and 14.9–14.10
  Chapter problems 1–32, 48–54

#### Light, electromagnetic radiation

- **Chapter 21, Sections 21.11–21.12**
  Quick Quizzes 21.7 and 21.8
  Examples 21.8 and 21.9
  Chapter problems 49–63
- **Chapter 22, Sections 22.1 and 22.4**
  Example 22.5
  Chapter problems 1–7 and 28–33
- **Chapter 24, Sections 24.1–24.2, 24.4, 24.6–24.9**
  Quick Quizzes 24.1–24.6
  Examples 24.1–24.4 and 24.6–24.8
  Chapter problems 1–61
- **Chapter 27, Section 27.3**
  Chapter problems 15–17

#### Geometrical optics

- **Chapter 22, Sections 22.2–22.4 and 22.7**
  Quick Quizzes 22.2–22.4
  Examples 22.1–22.6
  Chapter problems 8–44
- **Chapter 23, Sections 23.1–23.4 and 23.6–23.7**
  Quick Quizzes 23.1–23.6
  Examples 23.1–23.10
  Chapter problems 1–46
- **Chapter 25, Sections 25.1–25.6**
  Quick Quizzes 25.1–25.2
  Examples 25.1–25.8
  Chapter problems 1–46

**Content Category 4E:** Atoms, nuclear decay, electronic structure, and atomic chemical behavior

### Review Plan

#### Atomic nucleus

- **Chapter 19, Section 19.6**
  Quick Quiz 19.4
  Examples 19.5 and 19.6
  Chapter problems 33–42
- **Chapter 29, Sections 29.1–29.4**
  Quick Quizzes 29.1–29.3
  Examples 29.1–29.5
  Chapter problems 1–31

#### Electronic structure

- **Chapter 19, Section 19.10**
- **Chapter 27, Sections 27.2 and 27.8**
  Examples 27.1 and 27.5
  Chapter problems 9–14 and 33–38
- **Chapter 28, Sections 28.2–28.3, 28.5, and 28.7**
  Quick Quizzes 28.1 and 28.3
  Examples 28.1 and 28.2
  Chapter problems 1–26 and 30–33

**Content Category 5E:** Principles of chemical thermodynamics and kinetics

### Review Plan

#### Energy changes in chemical reactions

- **Chapter 10, Sections 10.1 and 10.3**
  Quick Quizzes 10.1–10.5
  Examples 10.3–10.5
  Chapter problems 11–28
- **Chapter 11, Sections 11.1–11.5**
  Quick Quizzes 11.1–11.5
  Examples 11.1–11.11
  Chapter problems 1–50
- **Chapter 12, Sections 12.2 and 12.4–12.5**
  Quick Quizzes 12.3–12.5
  Examples 12.3, 12.10–12.12, and 12.14–12.16
  Chapter problems 11–54

Shvaygert Ekaterina/Shutterstock.com

View of lightning over a city at night. During a thunderstorm, a high concentration of electrical charge in a thundercloud creates a higher-than-normal electric field between the thundercloud and the negatively charged Earth's surface. This strong electric field creates an electric discharge—an enormous spark—between the charged cloud and the ground. Other discharges observed in the sky include cloud-to-cloud discharges and the more frequent intracloud discharges.

# Electric Forces and Electric Fields 15



Electricity is the lifeblood of technological civilization and modern society. Without it, we revert to the mid-nineteenth century: no telephones, no television, none of the household appliances that we take for granted. Modern medicine would be a fantasy, and due to the lack of sophisticated experimental equipment and fast computers—and especially the slow dissemination of information—science and technology would grow at a glacial pace.

Instead, with the discovery and harnessing of electric forces and fields, we can view arrangements of atoms, probe the inner workings of the cell, and send spacecraft beyond the limits of the solar system. All this has become possible in just the last few generations of human life, a blink of the eye compared to the million years our kind spent foraging the savannahs of Africa.

Around 700 B.C. the ancient Greeks conducted the earliest known study of electricity. It all began when someone noticed that a fossil material called amber would attract small objects after being rubbed with wool. Since then we have learned that this phenomenon is not restricted to amber and wool, but occurs (to some degree) when almost any two nonconducting substances are rubbed together.

In this chapter we use the effect of charging by friction to begin an investigation of electric forces. We then discuss Coulomb's law, which is the fundamental law of force between any two stationary charged particles. The concept of an electric field associated with charges is introduced and its effects on other charged particles described. We end with discussions of the Van de Graaff generator and Gauss's law.

**Benjamin Franklin (1706–1790)**
Franklin was a printer, author, physical scientist, inventor, diplomat, and a founding father of the United States. His work on electricity in the late 1740s changed a jumbled, unrelated set of observations into a coherent science.

# 15.1 Properties of Electric Charges

**LEARNING OBJECTIVES**

1. Define the SI unit of electric charge and identify the basic carriers of positive and negative charge.
2. Discuss the concept of charge conservation and the role of experiments in early studies of electricity.

After running a plastic comb through your hair, you will find that the comb attracts bits of paper. The attractive force is often strong enough to suspend the paper from the comb, defying the gravitational pull of the entire Earth. The same effect occurs with other rubbed materials, such as glass and hard rubber.

Another simple experiment is to rub an inflated balloon against wool (or across your hair). On a dry day, the rubbed balloon will then stick to the wall of a room, often for hours. These materials have become electrically charged. You can give your body an electric charge by vigorously rubbing your shoes on a wool rug or by sliding across a car seat. You can then surprise and annoy a friend or coworker with a light touch on the arm, delivering a slight shock to both yourself and your victim. (If the coworker is your boss, don't expect a promotion!) These experiments work best on a dry day because excessive moisture can facilitate a leaking away of the charge.

Experiments also demonstrate that there are two kinds of electric charge, which Benjamin Franklin (1706–1790) named **positive** and **negative.** Figure 15.1 illustrates the interaction of the two charges. A hard rubber (or plastic) rod that has been rubbed with fur is suspended by a piece of string. When a glass rod that has been rubbed with silk is brought near the rubber rod, the rubber rod is attracted toward the glass rod (Fig. 15.1a). If two charged rubber rods (or two charged glass rods) are brought near each other, as in Figure 15.1b, the force between them is repulsive. These observations may be explained by assuming the rubber and glass rods have acquired different kinds of excess charge. We use the convention suggested by Franklin, where the excess electric charge on the glass rod is called positive and that on the rubber rod is called negative. On the basis of such observations, we conclude that **like charges repel one another and unlike charges attract one another.** Objects usually contain equal amounts of positive and negative charge; electrical forces between objects arise when those objects have net negative or positive charges.

Like charges repel; unlike charges attract. ▶

Nature's basic carriers of positive charge are protons, which, along with neutrons, are located in the nuclei of atoms. The nucleus, about $10^{-15}$ m in radius, is surrounded by a cloud of negatively charged electrons with a radius about ten

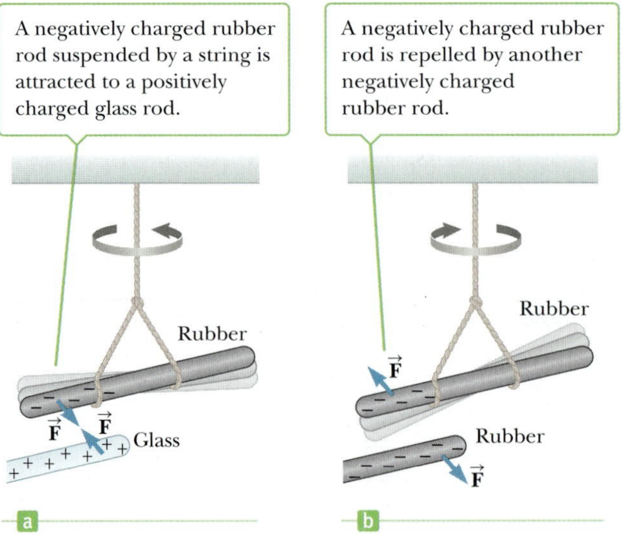
A negatively charged rubber rod suspended by a string is attracted to a positively charged glass rod.
A negatively charged rubber rod is repelled by another negatively charged rubber rod.
Rubber
Glass
Rubber
Rubber
a
b

**Figure 15.1** An experimental setup for observing the electrical force between two charged objects.

thousand times larger. An electron has the same magnitude charge as a proton, but the opposite sign. In a gram of matter there are approximately $10^{23}$ positively charged protons and just as many negatively charged electrons, so the net charge is zero. Because the nucleus of an atom is held firmly in place inside a solid, protons never move from one material to another. Electrons are far lighter than protons and hence more easily accelerated by forces. Further, they occupy the outer regions of the atom. Consequently, objects become charged by gaining or losing electrons.

Charge transfers readily from one type of material to another. Rubbing the two materials together serves to increase the area of contact, facilitating the transfer process.

◀ Charge is conserved

An important characteristic of charge is that **electric charge is always conserved.** Charge isn't *created* when two neutral objects are rubbed together; rather, the objects become charged because **negative charge is transferred from one object to the other.** One object gains a negative charge while the other loses an equal amount of negative charge and hence is left with a net positive charge. When a glass rod is rubbed with silk, as in Figure 15.2, electrons are transferred from the rod to the silk. As a result, the glass rod carries a net positive charge, the silk a net negative charge. Likewise, when rubber is rubbed with fur, electrons are transferred from the fur to the rubber.

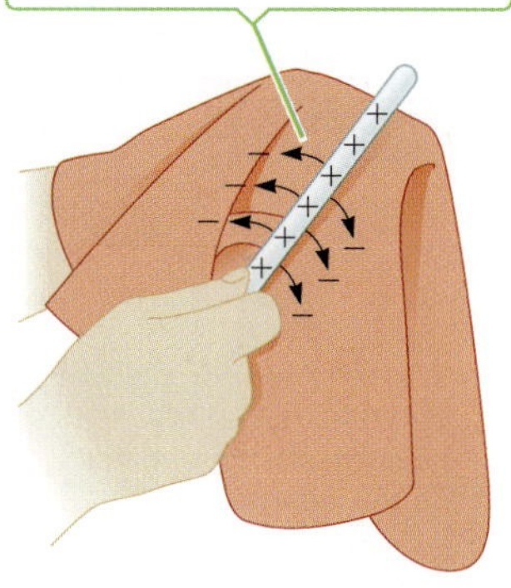

Each (negatively-charged) electron transferred from the rod to the silk leaves an equal positive charge on the rod.


**Figure 15.2** When a glass rod is rubbed with silk, electrons are transferred from the glass to the silk. Because the charges are transferred in discrete bundles, the charges on the two objects are $\pm e$, $\pm 2e$, $\pm 3e$, and so on.

In 1909 Robert Millikan (1886–1953) discovered that if an object is charged, its charge is always a multiple of a fundamental unit of charge, designated by the symbol $e$. In modern terms, the charge is said to be **quantized,** meaning that charge occurs in discrete chunks that can't be further subdivided. An object may have a charge of $\pm e$, $\pm 2e$, $\pm 3e$, and so on, but never[1] a fractional charge of $\pm 0.5e$ or $\pm 0.22e$. Other experiments in Millikan's time showed that the electron has a charge of $-e$ and the proton has an equal and opposite charge of $+e$. Some particles, such as a neutron, have no net charge. A neutral atom (an atom with no net charge) contains as many protons as electrons. The value of $e$ is now known to be $1.602\ 19 \times 10^{-19}$ C. (The SI unit of electric charge is the **coulomb,** or C.)

# 15.2 Insulators and Conductors

**LEARNING OBJECTIVES**

1. Describe conductors, insulators, and semi-conductors on the basis of their relative abilities to conduct electric charge.
2. Describe the physical processes of polarization and charging by conduction or induction.

Substances can be classified in terms of their ability to conduct electric charge.

In **conductors,** electric charges move freely in response to an electric force. All other materials are called **insulators.**

Glass and rubber are insulators. When such materials are charged by rubbing, only the rubbed area becomes charged, and there is no tendency for the charge to move into other regions of the material. In contrast, materials such as copper, aluminum, and silver are good conductors. When such materials are charged in some small region, the charge readily distributes itself over the entire surface of the material. If you hold a copper rod in your bare hand and rub the rod with wool or fur, it will not attract a piece of paper. This might suggest that a metal can't be charged. However, if you hold the copper rod with an insulator and then rub it with wool or fur, the rod remains charged and attracts the paper. In the first case,

[1]There is strong evidence for the existence of fundamental particles called **quarks** that have charges of $\pm e/3$ or $\pm 2e/3$. The charge is *still* quantized, but in units of $\pm e/3$ rather than $\pm e$. A more complete discussion of quarks and their properties is presented in Chapter 30.

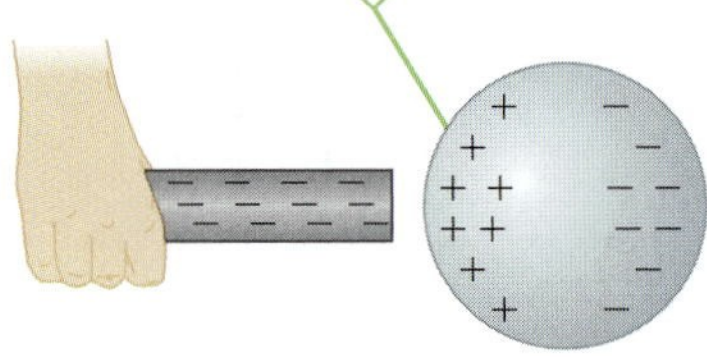

Before contact, the negative rod repels the sphere's electrons, inducing a local positive charge.
a


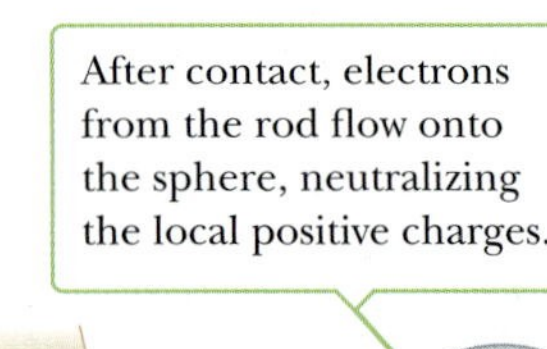

After contact, electrons from the rod flow onto the sphere, neutralizing the local positive charges.


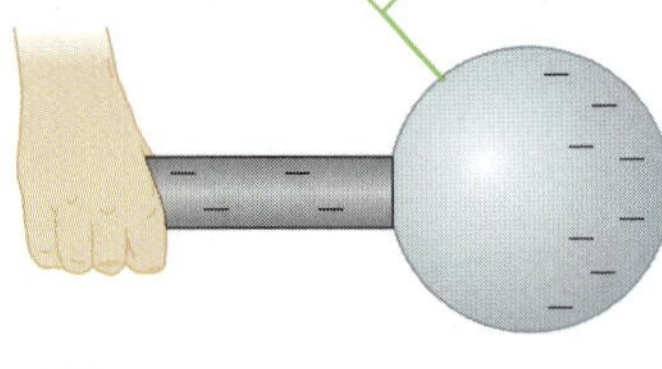


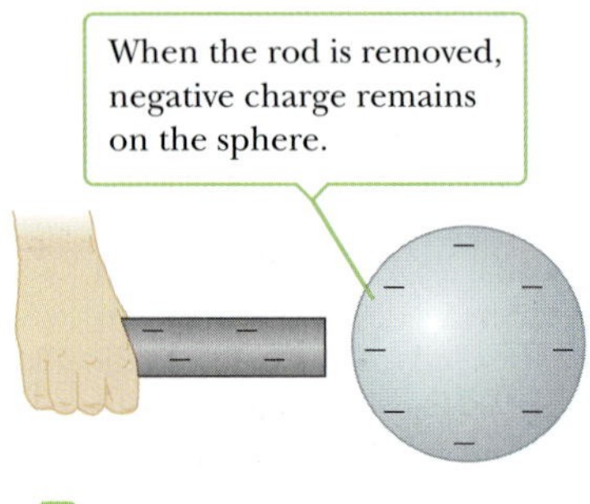

When the rod is removed, negative charge remains on the sphere.
c


**Figure 15.3** Charging a metallic object by conduction.

the electric charges produced by rubbing readily move from the copper through your body and finally to ground. In the second case, the insulating handle prevents the flow of charge to ground.

*Semiconductors* are a third class of materials, and their electrical properties are somewhere between those of insulators and those of conductors. Silicon and germanium are well-known semiconductors that are widely used in the fabrication of a variety of electronic devices.

## Charging by Conduction

Consider a negatively charged rubber rod brought into contact with an insulated neutral conducting sphere. The excess electrons on the rod repel electrons on the sphere, creating local positive charges on the neutral sphere. On contact, some electrons on the rod are now able to move onto the sphere, as in Figure 15.3, neutralizing the positive charges. When the rod is removed, the sphere is left with a net negative charge. This process is referred to as charging by **conduction.** The object being charged in such a process (the sphere) is always left with a charge having the same sign as the object doing the charging (the rubber rod).

## Charging by Induction

An object connected to a conducting wire or copper pipe buried in the Earth is said to be **grounded.** The Earth can be considered an infinite reservoir for electrons; in effect, it can accept or supply an unlimited number of electrons. With this idea in mind, we can understand the charging of a conductor by a process known as **induction.**

Consider a negatively charged rubber rod brought near a neutral (uncharged) conducting sphere that is insulated, so there is no conducting path to ground (Fig. 15.4). Initially the sphere is electrically neutral (Fig. 15.4a). When the negatively charged rod is brought close to the sphere, the repulsive force between the electrons in the rod and those in the sphere causes some electrons to move to the side of the sphere farthest away from the rod (Fig. 15.4b). The region of the sphere nearest the negatively charged rod has an excess of positive charge because of the migration of electrons away from that location. If a grounded conducting wire is then connected to the sphere, as in Figure 15.4c, some of the electrons leave the sphere and travel to ground. If the wire to ground is then removed (Fig. 15.4d), the conducting sphere is left with an excess of induced positive charge. Finally, when the rubber rod is removed from the vicinity of the sphere (Fig. 15.4e), the induced positive charge remains on the ungrounded sphere. Even though the positively charged atomic nuclei remain fixed, this excess positive charge becomes uniformly distributed over the surface of the ungrounded sphere because of the repulsive forces among the like charges and the high mobility of electrons in a metal.

In the process of inducing a charge on the sphere, the charged rubber rod doesn't lose any of its negative charge because it never comes in contact with the sphere. Further, the sphere is left with a charge opposite that of the rubber rod. **Charging an object by induction requires no contact with the object inducing the charge.**

A process similar to charging by induction in conductors also takes place in insulators. In most neutral atoms or molecules, the center of positive charge coincides with the center of negative charge. In the presence of a charged object, however, these centers may separate slightly, resulting in more positive charge on one side of the molecule than on the other. This effect is known as **polarization.** The realignment of charge within individual molecules produces an induced charge on the surface of the insulator, as shown in Figure 15.5a. This property explains why a balloon charged through rubbing will stick to an electrically neutral wall or why the comb you just used on your hair attracts tiny bits of neutral paper.

**Quick Quiz**

**15.1** A suspended object $A$ is attracted to a neutral wall. It's also attracted to a positively charged object $B$. Which of the following is true about object $A$? (a) It is uncharged. (b) It has a negative charge. (c) It has a positive charge. (d) It may be either charged or uncharged.

# 15.3 Coulomb's Law

**LEARNING OBJECTIVES**

1. State Coulomb's law and summarize the properties of the electric force between two charged particles.
2. Apply Coulomb's law and the superposition principle to systems of charged particles.

In 1785 Charles Coulomb (1736–1806) experimentally established the fundamental law of electric force between two stationary charged particles.

An **electric force** has the following properties:

1. It is directed along a line joining the two particles and is inversely proportional to the square of the separation distance $r$, between them.
2. It is proportional to the product of the magnitudes of the charges, $|q_1|$ and $|q_2|$, of the two particles.
3. It is attractive if the charges are of opposite sign and repulsive if the charges have the same sign.

From these observations, Coulomb proposed the following mathematical form for the electric force between two charges:

The magnitude of the electric force $F$ between charges $q_1$ and $q_2$ separated by a distance $r$ is given by

$$F = k_e \frac{|q_1||q_2|}{r^2} \qquad [15.1]$$

where $k_e$ is a constant called the *Coulomb constant.*

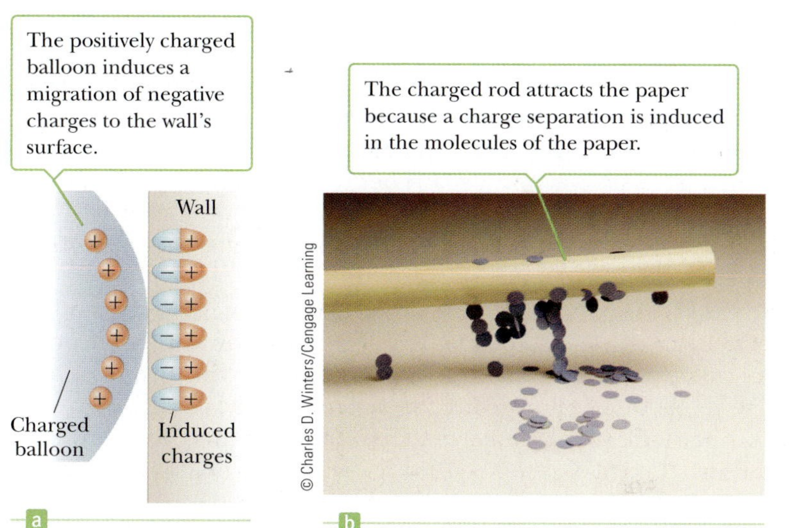


**Figure 15.5** (a) A charged balloon is brought near an insulating wall. (b) A charged rod is brought close to bits of paper.

The neutral sphere has equal numbers of positive and negative charges.

a

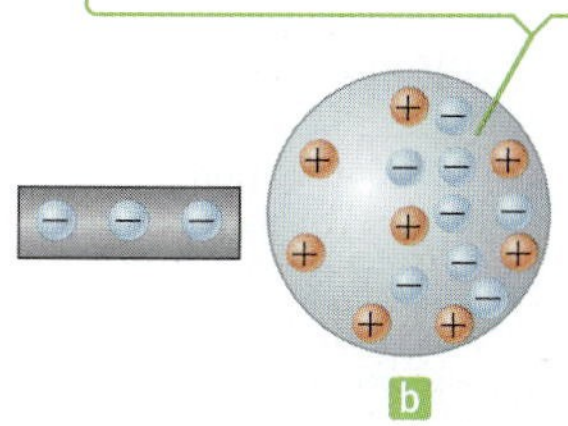

Electrons redistribute when a charged rod is brought close.
b


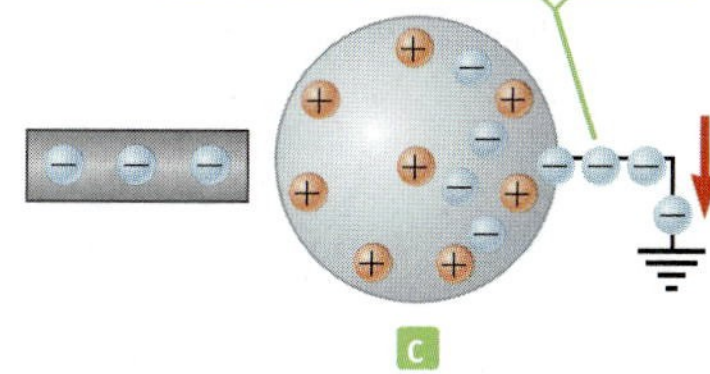

Some electrons leave the grounded sphere through the ground wire.
c


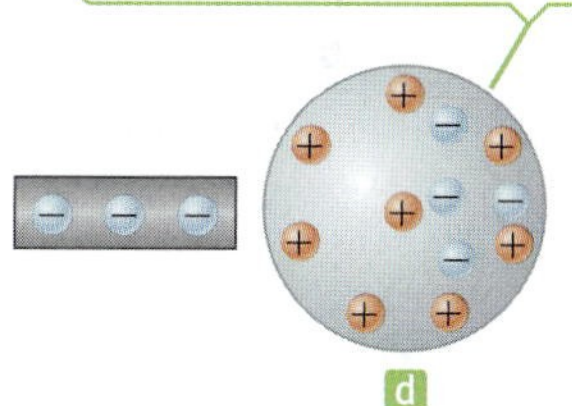

The excess positive charge is nonuniformly distributed.
d


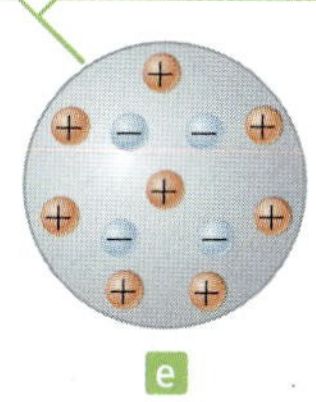

The remaining electrons redistribute uniformly, and there is a net uniform distribution of positive charge on the sphere's surface.
e


**Figure 15.4** Charging a metallic object by induction. (a) A neutral metallic sphere. (b) A charged rubber rod is placed near the sphere. (c) The sphere is grounded. (d) The ground connection is removed. (e) The rod is removed.

**Charles Coulomb**
**(1736–1806)**
Coulomb's major contribution to science was in the field of electrostatics and magnetism. During his lifetime, he also investigated the strengths of materials and identified the forces that affect objects on beams, thereby contributing to the field of structural mechanics.

**Table 15.1** Charge and Mass of the Electron, Proton, and Neutron

| Particle | Charge (C) | Mass (kg) |
|---|---|---|
| Electron | $-1.60 \times 10^{-19}$ | $9.11 \times 10^{-31}$ |
| Proton | $+1.60 \times 10^{-19}$ | $1.67 \times 10^{-27}$ |
| Neutron | 0 | $1.67 \times 10^{-27}$ |

Equation 15.1, known as **Coulomb's law,** applies exactly only to point charges and to spherical distributions of charges, in which case $r$ is the distance between the two centers of charge. Electric forces between unmoving charges are called *electrostatic* forces. Moving charges, in addition, create magnetic forces, studied in Chapter 19.

The value of the Coulomb constant in Equation 15.1 depends on the choice of units. The SI unit of charge is the **coulomb** (C). From experiment, we know that the **Coulomb constant** in SI units has the value

$$k_e = 8.987\,5 \times 10^9 \text{ N} \cdot \text{m}^2/\text{C}^2 \quad [15.2]$$

This number can be rounded, depending on the accuracy of other quantities in a given problem. We'll use either two or three significant digits, as usual.

The charge on the proton has a magnitude of $e = 1.6 \times 10^{-19}$ C. Therefore, it would take $1/e = 6.3 \times 10^{18}$ protons to create a total charge of +1.0 C. Likewise, $6.3 \times 10^{18}$ electrons would have a total charge of −1.0 C. Compare this charge with the number of free electrons in 1 cm$^3$ of copper, which is on the order of $10^{23}$. Even so, 1.0 C is a very large amount of charge. In typical electrostatic experiments in which a rubber or glass rod is charged by friction, there is a net charge on the order of $10^{-6}$ C (= 1 $\mu$C). Only a very small fraction of the total available charge is transferred between the rod and the rubbing material. Table 15.1 lists the charges and masses of the electron, proton, and neutron.

When using Coulomb's force law, remember that force is a vector quantity and must be treated accordingly. Figure 15.6a shows the electric force of repulsion between two positively charged particles. Like other forces, electric forces obey Newton's third law; hence, the forces $\vec{\mathbf{F}}_{12}$ and $\vec{\mathbf{F}}_{21}$ are equal in magnitude but opposite in direction. (The notation $\vec{\mathbf{F}}_{12}$ denotes the force exerted by particle 1 on particle 2; likewise, $\vec{\mathbf{F}}_{21}$ is the force exerted by particle 2 on particle 1.) From Newton's third law, $F_{12}$ and $F_{21}$ are always equal regardless of whether $q_1$ and $q_2$ have the same magnitude.

**Quick Quiz**

**15.2** Object $A$ has a charge of +2 $\mu$C, and object $B$ has a charge of +6 $\mu$C. Which statement is true?
(a) $\vec{\mathbf{F}}_{AB} = -3\vec{\mathbf{F}}_{BA}$ (b) $\vec{\mathbf{F}}_{AB} = -\vec{\mathbf{F}}_{BA}$ (c) $3\vec{\mathbf{F}}_{AB} = -\vec{\mathbf{F}}_{BA}$

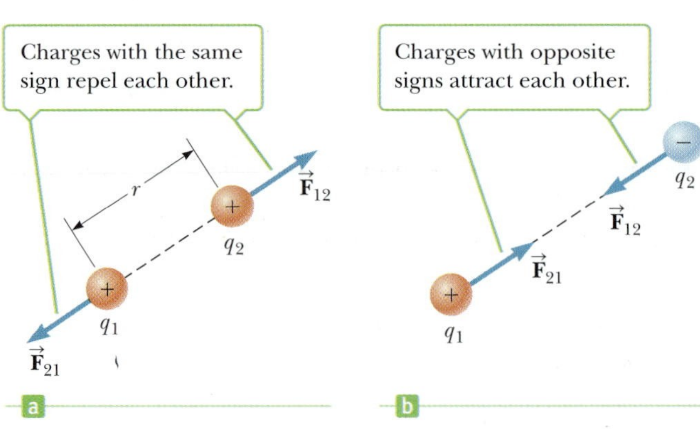


**Figure 15.6** Two point charges separated by a distance $r$ exert a force on each other given by Coulomb's law. The force on $q_1$ is equal in magnitude and opposite in direction to the force on $q_2$.

The Coulomb force is similar to the gravitational force. Both act at a distance without direct contact. Both are inversely proportional to the distance squared, with the force directed along a line connecting the two bodies. The mathematical form is the same, with the masses $m_1$ and $m_2$ in Newton's law replaced by $q_1$ and $q_2$ in Coulomb's law and with Newton's constant $G$ replaced by Coulomb's constant $k_e$. There are two important differences: (1) electric forces can be either attractive or repulsive, but gravitational forces are always attractive, and (2) the electric force between charged elementary particles is far stronger than the gravitational force between the same particles, as the next example shows.

## ■ EXAMPLE 15.1 Forces in a Hydrogen Atom

**GOAL** Contrast the magnitudes of an electric force and a gravitational force.

**PROBLEM** The electron and proton of a hydrogen atom are separated (on the average) by a distance of about $5.3 \times 10^{-11}$ m. **(a)** Find the magnitudes of the electric force and the gravitational force that each particle exerts on the other, and the ratio of the electric force $F_e$ to the gravitational force $F_g$. **(b)** Compute the acceleration caused by the electric force of the proton on the electron. Repeat for the gravitational acceleration.

**STRATEGY** Solving this problem is just a matter of substituting known quantities into the two force laws and then finding the ratio.

**SOLUTION**

**(a)** Compute the magnitudes of the electric and gravitational forces, and find the ration $F_e/F_g$.

Substitute $|q_1| = |q_2| = e$ and the distance into Coulomb's law to find the electric force:

$$F_e = k_e\frac{|e|^2}{r^2} = \left(8.99 \times 10^9 \,\frac{\text{N}\cdot\text{m}^2}{\text{C}^2}\right)\frac{(1.6 \times 10^{-19}\,\text{C})^2}{(5.3 \times 10^{-11}\,\text{m})^2}$$

$$= 8.2 \times 10^{-8}\,\text{N}$$

Substitute the masses and distance into Newton's law of gravity to find the gravitational force:

$$F_g = G\frac{m_e m_p}{r^2}$$

$$= \left(6.67 \times 10^{-11}\,\frac{\text{N}\cdot\text{m}^2}{\text{kg}^2}\right)\frac{(9.11 \times 10^{-31}\,\text{kg})(1.67 \times 10^{-27}\,\text{kg})}{(5.3 \times 10^{-11}\,\text{m})^2}$$

$$= 3.6 \times 10^{-47}\,\text{N}$$

Find the ratio of the two forces:

$$\frac{F_e}{F_g} = 2.3 \times 10^{39}$$

**(b)** Compute the acceleration of the electron caused by the electric force. Repeat for the gravitational acceleration.

Use Newton's second law and the electric force found in part (a):

$$m_e a_e = F_e \quad\rightarrow\quad a_e = \frac{F_e}{m_e} = \frac{8.2 \times 10^{-8}\,\text{N}}{9.11 \times 10^{-31}\,\text{kg}} = 9.0 \times 10^{22}\,\text{m/s}^2$$

Use Newton's second law and the gravitational force found in part (a):

$$m_e a_g = F_g \quad\rightarrow\quad a_g = \frac{F_g}{m_e} = \frac{3.6 \times 10^{-47}\,\text{N}}{9.11 \times 10^{-31}\,\text{kg}} = 4.0 \times 10^{-17}\,\text{m/s}^2$$

**REMARKS** The gravitational force between the charged constituents of the atom is negligible compared with the electric force between them. The electric force is so strong, however, that any net charge on an object quickly attracts nearby opposite charges, neutralizing the object. As a result, gravity plays a greater role in the mechanics of moving objects in everyday life.

**QUESTION 15.1** If the distance between two charges is doubled, by what factor is the magnitude of the electric force changed?

*(Continued)*

**EXERCISE 15.1** (a) Find the magnitude of the electric force between two protons separated by 1 femtometer ($10^{-15}$ m), approximately the distance between two protons in the nucleus of a helium atom. (b) If the protons were not held together by the strong nuclear force, what would be their initial acceleration due to the electric force between them?

**ANSWERS** (a) $2 \times 10^2$ N (b) $1 \times 10^{29}$ m/s$^2$

## The Superposition Principle

When a number of separate charges act on the charge of interest, each exerts an electric force. These electric forces can all be computed separately, one at a time, then added as vectors. This is another example of the **superposition principle.** The following example illustrates this procedure in one dimension.

### EXAMPLE 15.2 Finding Electrostatic Equilibrium

**GOAL** Apply Coulomb's law in one dimension.

**PROBLEM** Three charges lie along the $x$-axis as in Figure 15.7. The positive charge $q_1 = 15\ \mu$C is at $x = 2.0$ m, and the positive charge $q_2 = 6.0\ \mu$C is at the origin. Where must a *negative* charge $q_3$ be placed on the $x$-axis so that the resultant electric force on it is zero?

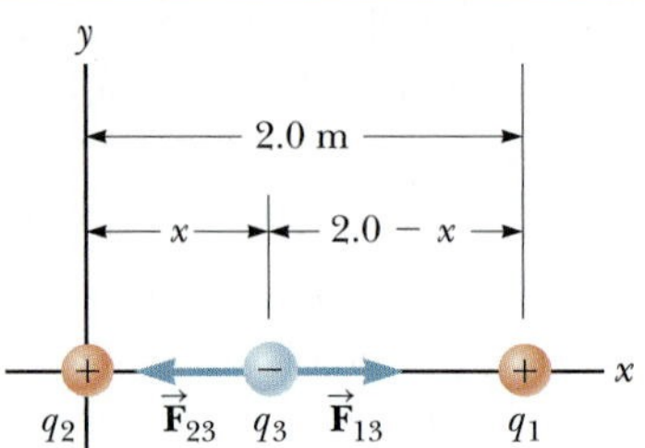


**Figure 15.7** (Example 15.2) Three point charges are placed along the $x$-axis. The charge $q_3$ is negative, whereas $q_1$ and $q_2$ are positive. If the resultant force on $q_3$ is zero, the force $\vec{F}_{13}$ exerted by $q_1$ on $q_3$ must be equal in magnitude and opposite the force $\vec{F}_{23}$ exerted by $q_2$ on $q_3$.

**STRATEGY** If $q_3$ is to the right or left of the other two charges, the net force on $q_3$ can't be zero because then $\vec{F}_{13}$ and $\vec{F}_{23}$ act in the same direction. Consequently, $q_3$ must lie between the two other charges. Write $\vec{F}_{13}$ and $\vec{F}_{23}$ in terms of the unknown coordinate position $x$, then sum them and set them equal to zero, solving for the unknown. The solution can be obtained with the quadratic formula.

**SOLUTION**

Write the $x$-component of $\vec{F}_{13}$:

$$F_{13x} = +k_e \frac{(15 \times 10^{-6}\ \text{C})|q_3|}{(2.0\ \text{m} - x)^2}$$

Write the $x$-component of $\vec{F}_{23}$:

$$F_{23x} = -k_e \frac{(6.0 \times 10^{-6}\ \text{C})|q_3|}{x^2}$$

Set the sum equal to zero:

$$k_e \frac{(15 \times 10^{-6}\ \text{C})|q_3|}{(2.0\ \text{m} - x)^2} - k_e \frac{(6.0 \times 10^{-6}\ \text{C})|q_3|}{x^2} = 0$$

Cancel $k_e$, $10^{-6}$, and $q_3$ from the equation and rearrange terms (explicit significant figures and units are temporarily suspended for clarity):

$$\textbf{(1)}\quad 6(2 - x)^2 = 15x^2$$

Put this equation into standard quadratic form, $ax^2 + bx + c = 0$:

$$6(4 - 4x + x^2) = 15x^2 \quad \rightarrow \quad 2(4 - 4x + x^2) = 5x^2$$

$$3x^2 + 8x - 8 = 0$$

Apply the quadratic formula:

$$x = \frac{-8 \pm \sqrt{64 - (4)(3)(-8)}}{2 \cdot 3} = \frac{-4 \pm 2\sqrt{10}}{3}$$

Only the positive root makes sense:

$$x = \boxed{0.77\ \text{m}}$$

**REMARKS** Notice that physical reasoning was required to choose between the two possible answers for $x$, which is nearly always the case when quadratic equations are involved. Use of the quadratic formula could have been avoided by taking the square root of both sides of Equation (1); however, this shortcut is often unavailable.

**QUESTION 15.2** If $q_1$ has the same magnitude as before but is negative, in what region along the $x$-axis would it be possible for the net electric force on $q_3$ to be zero? (a) $x < 0$ (b) $0 < x < 2$ m (c) 2 m $< x$

**EXERCISE 15.2** Three charges lie along the $x$-axis. A positive charge $q_1 = 10.0\ \mu$C is at $x = 1.00$ m, and a negative charge $q_2 = -2.00\ \mu$C is at the origin. Where must a positive charge $q_3$ be placed on the $x$-axis so that the resultant force on it is zero?

**ANSWER** $x = -0.809$ m

---

## EXAMPLE 15.3 A Charge Triangle

**GOAL** Apply Coulomb's law in two dimensions.

**PROBLEM** Consider three point charges at the corners of a triangle, as shown in Figure 15.8, where $q_1 = 6.00 \times 10^{-9}$ C, $q_2 = -2.00 \times 10^{-9}$ C, and $q_3 = 5.00 \times 10^{-9}$ C. **(a)** Find the components of the force $\vec{\mathbf{F}}_{23}$ exerted by $q_2$ on $q_3$. **(b)** Find the components of the force $\vec{\mathbf{F}}_{13}$ exerted by $q_1$ on $q_3$. **(c)** Find the resultant force on $q_3$, in terms of components and also in terms of magnitude and direction.

**STRATEGY** Coulomb's law gives the magnitude of each force, which can be split with right-triangle trigonometry into $x$- and $y$-components. Sum the vectors componentwise and then find the magnitude and direction of the resultant vector.

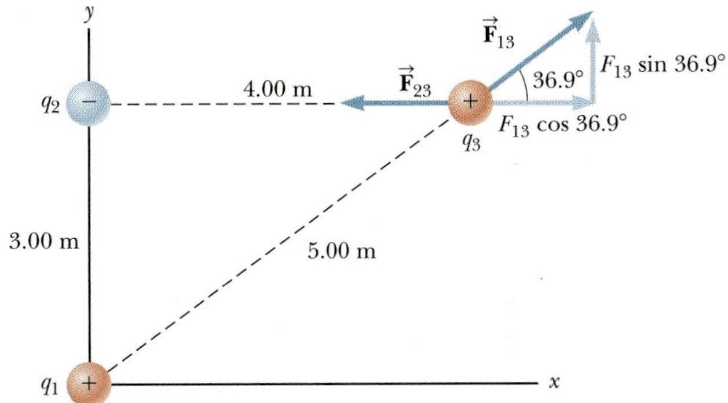


**Figure 15.8** (Example 15.3) The force exerted by $q_1$ on $q_3$ is $\vec{\mathbf{F}}_{13}$. The force exerted by $q_2$ on $q_3$ is $\vec{\mathbf{F}}_{23}$. The *resultant force* $\vec{\mathbf{F}}_3$ exerted on $q_3$ is the *vector* sum $\vec{\mathbf{F}}_{13} + \vec{\mathbf{F}}_{23}$.

**SOLUTION**

**(a)** Find the components of the force exerted by $q_2$ on $q_3$.

Find the magnitude of $\vec{\mathbf{F}}_{23}$ with Coulomb's law:

$$F_{23} = k_e \frac{|q_2||q_3|}{r^2}$$

$$= (8.99 \times 10^9\ \mathrm{N \cdot m^2/C^2}) \frac{(2.00 \times 10^{-9}\ \mathrm{C})(5.00 \times 10^{-9}\ \mathrm{C})}{(4.00\ \mathrm{m})^2}$$

$$F_{23} = 5.62 \times 10^{-9}\ \mathrm{N}$$

Because $\vec{\mathbf{F}}_{23}$ is horizontal and points in the negative $x$-direction, the negative of the magnitude gives the $x$-component, and the $y$-component is zero:

$$F_{23x} = -5.62 \times 10^{-9}\ \mathrm{N}$$

$$F_{23y} = 0$$

**(b)** Find the components of the force exerted by $q_1$ on $q_3$.

Find the magnitude of $\vec{\mathbf{F}}_{13}$:

$$F_{13} = k_e \frac{|q_1||q_3|}{r^2}$$

$$= (8.99 \times 10^9\ \mathrm{N \cdot m^2/C^2}) \frac{(6.00 \times 10^{-9}\ \mathrm{C})(5.00 \times 10^{-9}\ \mathrm{C})}{(5.00\ \mathrm{m})^2}$$

$$F_{13} = 1.08 \times 10^{-8}\ \mathrm{N}$$

Use the given triangle to find the components of $\vec{\mathbf{F}}_{13}$:

$$F_{13x} = F_{13} \cos\theta = (1.08 \times 10^{-8}\ \mathrm{N}) \cos(36.9°)$$

$$= 8.64 \times 10^{-9}\ \mathrm{N}$$

$$F_{13y} = F_{13} \sin\theta = (1.08 \times 10^{-8}\ \mathrm{N}) \sin(36.9°)$$

$$= 6.48 \times 10^{-9}\ \mathrm{N}$$

**(c)** Find the components of the resultant vector.

Sum the $x$-components to find the resultant $F_x$:

$$F_x = -5.62 \times 10^{-9}\ \mathrm{N} + 8.64 \times 10^{-9}\ \mathrm{N}$$

$$= 3.02 \times 10^{-9}\ \mathrm{N}$$

*(Continued)*

Sum the $y$-components to find the resultant $F_y$:

$$F_y = 0 + 6.48 \times 10^{-9}\text{ N} = 6.48 \times 10^{-9}\text{ N}$$

Find the magnitude of the resultant force on the charge $q_3$, using the Pythagorean theorem:

$$|\vec{\mathbf{F}}| = \sqrt{F_x^{\,2} + F_y^{\,2}}$$
$$= \sqrt{(3.02 \times 10^{-9}\text{ N})^2 + (6.48 \times 10^{-9}\text{ N})^2}$$
$$= 7.15 \times 10^{-9}\text{ N}$$

Find the angle the resultant force makes with respect to the positive $x$-axis:

$$\theta = \tan^{-1}\left(\frac{F_y}{F_x}\right) = \tan^{-1}\left(\frac{6.48 \times 10^{-9}\text{ N}}{3.02 \times 10^{-9}\text{ N}}\right) = 65.0°$$

**REMARKS** The methods used here are just like those used with Newton's law of gravity in two dimensions.

**QUESTION 15.3** Without actually calculating the electric force on $q_2$, determine the quadrant into which the electric force vector points.

**EXERCISE 15.3** Using the same triangle, find the vector components of the electric force on $q_1$ and the vector's magnitude and direction.

**ANSWERS** $F_x = -8.64 \times 10^{-9}$ N, $F_y = 5.52 \times 10^{-9}$ N, $F = 1.03 \times 10^{-8}$ N, $\theta = 147°$

## 15.4 The Electric Field

**LEARNING OBJECTIVE**

1. Define the electric field and apply it to systems of charged particles.

The gravitational force and the electrostatic force are both capable of acting through space, producing an effect even when there isn't any physical contact between the objects involved. Field forces can be discussed in a variety of ways, but an approach developed by Michael Faraday (1791–1867) is the most practical. In this approach an **electric field** is said to exist in the region of space around a charged object. The electric field exerts an electric force on any other charged object within the field. This differs from the Coulomb's law concept of a force exerted at a distance in that the force is now exerted by something—the field—that is in the same location as the charged object.

Figure 15.9 shows an object with a small positive charge $q_0$ placed near a second object with a much larger positive charge $Q$.

The electric field $\vec{\mathbf{E}}$ produced by a charge $Q$ at the location of a small "test" charge $q_0$ is defined as the electric force $\vec{\mathbf{F}}$ exerted by $Q$ on $q_0$ divided by the test charge $q_0$:

$$\vec{\mathbf{E}} \equiv \frac{\vec{\mathbf{F}}}{q_0} \qquad [15.3]$$

**SI unit: newton per coulomb (N/C)**

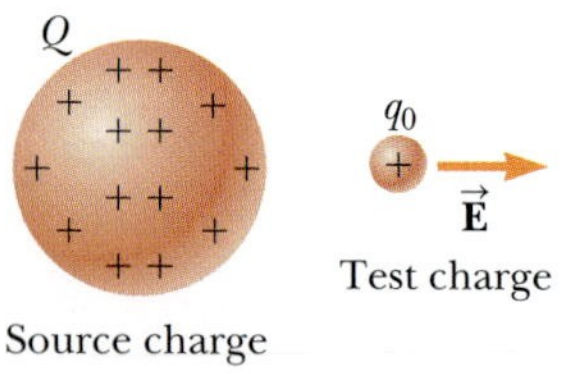

Q

$q_0$

$\vec{\mathbf{E}}$

Test charge

Source charge

**Figure 15.9** A small object with a positive charge $q_0$ placed near an object with a larger positive charge $Q$ is subject to an electric field $\vec{\mathbf{E}}$ directed as shown. The magnitude of the electric field at the location of $q_0$ is defined as the electric force on $q_0$ divided by the charge $q_0$.

Conceptually and experimentally, the test charge $q_0$ is required to be very small (arbitrarily small, in fact), so it doesn't cause any significant rearrangement of the charge creating the electric field $\vec{\mathbf{E}}$. Mathematically, however, the size of the test charge makes no difference: the calculation comes out the same, regardless. In view of this, using $q_0 = 1$ C in Equation 15.3 can be convenient if not rigorous.

When a positive test charge is used, the electric field always has the same direction as the electric force on the test charge, which follows from Equation 15.3. Hence, in Figure 15.9, the direction of the electric field is horizontal and to the right. The electric field at point $A$ in Figure 15.10a is vertical and downward because at that point a positive test charge would be attracted toward the negatively charged sphere.

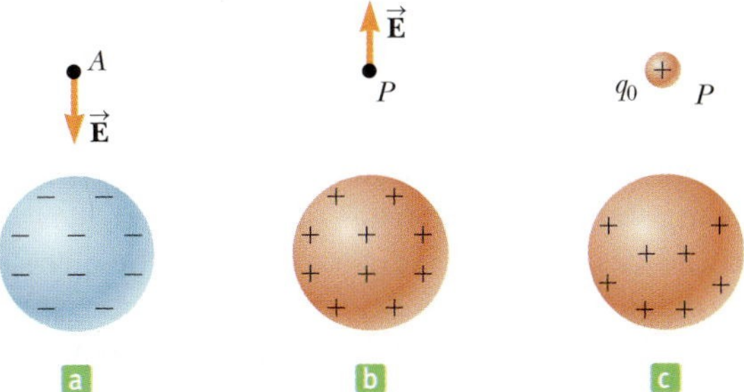
A
E
P
$q_0$
P
a
b
c

**Figure 15.10** (a) The electric field at $A$ due to the negatively charged sphere is downward, toward the negative charge. (b) The electric field at $P$ due to the positively charged conducting sphere is upward, away from the positive charge. (c) A test charge $q_0$ placed at $P$ will cause a rearrangement of charge on the sphere unless $q_0$ is negligibly small compared with the charge on the sphere.

Once the electric field due to a given arrangement of charges is known at some point, the force on *any* particle with charge $q$ placed at that point can be calculated from a rearrangement of Equation 15.3:

$$\vec{\mathbf{F}} = q\vec{\mathbf{E}} \qquad [15.4]$$

Here $q_0$ has been replaced by $q$, which need not be a mere test charge.

As shown in Figure 15.11, the direction of $\vec{\mathbf{E}}$ is the direction of the force that acts on a positive test charge $q_0$ placed in the field. We say that **an electric field exists at a point if a test charge at that point is subject to an electric force.**

Consider a point charge $q$ located a distance $r$ from a test charge $q_0$. According to Coulomb's law, the *magnitude* of the electric force of the charge $q$ on the test charge is

$$F = k_e \frac{|q||q_0|}{r^2} \qquad [15.5]$$

Because the magnitude of the electric field at the position of the test charge is defined as $E = F/q_0$, we see that the *magnitude* of the electric field due to the charge $q$ at the position of $q_0$ is

$$E = k_e \frac{|q|}{r^2} \qquad [15.6]$$

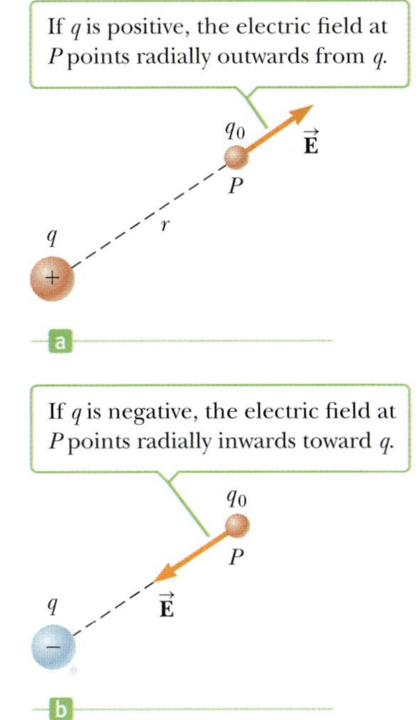
If $q$ is positive, the electric field at $P$ points radially outwards from $q$.
$q_0$
E
P
r
q
a
If $q$ is negative, the electric field at $P$ points radially inwards toward $q$.
$q_0$
P
q
E
b

**Figure 15.11** A test charge $q_0$ at $P$ is a distance $r$ from a point charge $q$.

Equation 15.6 points out an important property of electric fields that makes them useful quantities for describing electrical phenomena. As the equation indicates, an electric field at a given point depends only on the charge $q$ on the object setting up the field and the distance $r$ from that object to a specific point in space. As a result, we can say that an electric field exists at point $P$ in Figure 15.11 whether or not there is a test charge at $P$.

The principle of superposition holds when the electric field due to a group of point charges is calculated. We first use Equation 15.6 to calculate the electric field produced by each charge individually at a point and then add the electric fields together as vectors.

It's also important to exploit any symmetry of the charge distribution. For example, if equal charges are placed at $x = a$ and at $x = -a$, the electric field is zero at the origin, by symmetry. Similarly, if the $x$-axis has a uniform distribution of positive charge, it can be guessed by symmetry that the electric field points away from the $x$ axis and is zero parallel to that axis.

■ *Quick Quiz*

**15.3** A test charge of $+3\ \mu C$ is at a point $P$ where the electric field due to other charges is directed to the right and has a magnitude of $4 \times 10^6$ N/C. If the test charge is replaced with a charge of $-3\ \mu C$, the electric field at $P$ (a) has the same magnitude as before, but changes direction, (b) increases in magnitude and changes direction, (c) remains the same, or (d) decreases in magnitude and changes direction.

**15.4** A circular ring of charge of radius $b$ has a total charge $q$ uniformly distributed around it. Find the magnitude of the electric field in the center of the ring.

(a) 0 (b) $k_e q/b^2$ (c) $k_e q^2/b^2$ (d) $k_e q^2/b$ (e) None of these answers is correct.

**15.5** A "free" electron and a "free" proton are placed in an identical electric field. Which of the following statements are true? (a) Each particle is acted upon by the same electric force and has the same acceleration. (b) The electric force on the proton is greater in magnitude than the electric force on the electron, but in the opposite direction. (c) The electric force on the proton is equal in magnitude to the electric force on the electron, but in the opposite direction. (d) The magnitude of the acceleration of the electron is greater than that of the proton. (e) Both particles have the same acceleration.

## EXAMPLE 15.4 Electrified Oil

**GOAL** Use electric forces and fields together with Newton's second law in a one-dimensional problem.

**PROBLEM** Tiny droplets of oil acquire a small negative charge while dropping through a vacuum (pressure = 0) in an experiment. An electric field of magnitude $5.92 \times 10^4$ N/C points straight down. **(a)** One particular droplet is observed to remain suspended against gravity. If the mass of the droplet is $2.93 \times 10^{-15}$ kg, find the charge carried by the droplet. **(b)** Another droplet of the same mass falls 10.3 cm from rest in 0.250 s, again moving through a vacuum. Find the charge carried by the droplet.

**STRATEGY** We use Newton's second law with both gravitational and electric forces. In both parts the electric field $\vec{\mathbf{E}}$ is pointing down, taken as the negative direction, as usual. In part (a) the acceleration is equal to zero. In part (b) the acceleration is uniform, so the kinematic equations yield the acceleration. Newton's law can then be solved for $q$.

**SOLUTION**

**(a)** Find the charge on the suspended droplet.

Apply Newton's second law to the droplet in the vertical direction:

$$\textbf{(1)}\quad ma = \sum F = -mg + Eq$$

$E$ points downward, hence is negative. Set $a = 0$ in Equation (1) and solve for $q$:

$$q = \frac{mg}{E} = \frac{(2.93 \times 10^{-15}\ \text{kg})(9.80\ \text{m/s}^2)}{-5.92 \times 10^4\ \text{N/C}}$$

$$= \boxed{-4.85 \times 10^{-19}\ \text{C}}$$

**(b)** Find the charge on the falling droplet.

Use the kinematic displacement equation to find the acceleration:

$$\Delta y = \tfrac{1}{2}at^2 + v_0 t$$

Substitute $\Delta y = -0.103$ m, $t = 0.250$ s, and $v_0 = 0$:

$$-0.103\ \text{m} = \tfrac{1}{2}\,a(0.250\ \text{s})^2 \quad\rightarrow\quad a = -3.30\ \text{m/s}^2$$

Solve Equation (1) for $q$ and substitute:

$$q = \frac{m(a+g)}{E}$$

$$= \frac{(2.93 \times 10^{-15}\ \text{kg})(-3.30\ \text{m/s}^2 + 9.80\ \text{m/s}^2)}{-5.92 \times 10^4\ \text{N/C}}$$

$$= \boxed{-3.22 \times 10^{-19}\ \text{C}}$$

**REMARKS** This example exhibits features similar to the Millikan Oil-Drop experiment discussed in Section 15.7, which determined the value of the fundamental electric charge $e$. Notice that in both parts of the example, the charge is very nearly a multiple of $e$.

**QUESTION 15.4** What would be the acceleration of the oil droplet in part (a) if the electric field suddenly reversed direction without changing in magnitude?

**EXERCISE 15.4** Suppose a droplet of unknown mass remains suspended against gravity when $E = -2.70 \times 10^5$ N/C. What is the minimum mass of the droplet?

**ANSWER** $4.41 \times 10^{-15}$ kg

## PROBLEM-SOLVING STRATEGY

### Calculating Electric Forces and Fields

*The following procedure is used to calculate electric forces. The same procedure can be used to calculate an electric field, a simple matter of replacing the charge of interest, q, with a convenient test charge and dividing by the test charge at the end:*

1. **Draw** a diagram of the charges in the problem.
2. **Identify** the charge of interest, $q$, and circle it.
3. **Convert all units** to SI, with charges in coulombs and distances in meters, so as to be consistent with the SI value of the Coulomb constant $k_e$.
4. **Apply Coulomb's law.** For each charge $Q$, find the electric force on the charge of interest, $q$. The magnitude of the force can be found using Coulomb's law. The vector direction of the electric force is along the line of the two charges, directed away from $Q$ if the charges have the same sign, toward $Q$ if the charges have the opposite sign. Find the angle $\theta$ this vector makes with the positive $x$-axis. The $x$-component of the electric force exerted by $Q$ on $q$ will be $F\cos\theta$, and the $y$-component will be $F\sin\theta$.
5. **Sum all the $x$-components,** getting the $x$-component of the resultant electric force.
6. **Sum all the $y$-components,** getting the $y$-component of the resultant electric force.
7. **Use the Pythagorean theorem and trigonometry** to find the magnitude and direction of the resultant force if desired.

## EXAMPLE 15.5 Electric Field Due to Two Point Charges

**GOAL** Use the superposition principle to calculate the electric field due to two point charges.

**PROBLEM** Charge $q_1 = 7.00\ \mu\text{C}$ is at the origin, and charge $q_2 = -5.00\ \mu\text{C}$ is on the $x$-axis, 0.300 m from the origin (Fig. 15.12). **(a)** Find the magnitude and direction of the electric field at point $P$, which has coordinates (0, 0.400) m. **(b)** Find the force on a charge of $2.00 \times 10^{-8}$ C placed at $P$.

**STRATEGY** Follow the problem-solving strategy, finding the electric field at point $P$ due to each individual charge in terms of $x$- and $y$-components, then adding the components of each type to get the $x$- and $y$-components of the resultant electric field at $P$. The magnitude of the force in part (b) can be found by simply multiplying the magnitude of the electric field by the charge.

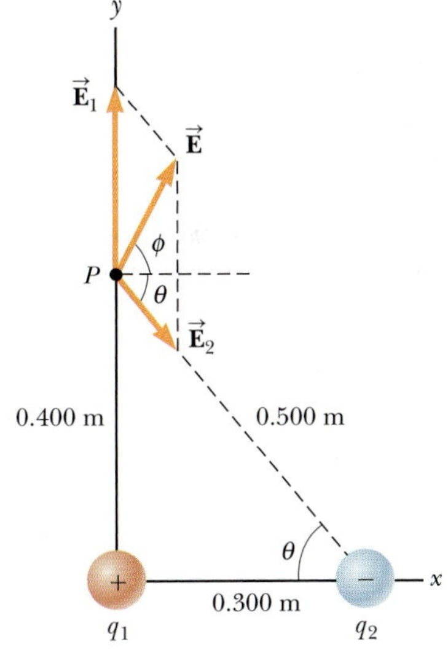


**Figure 15.12** (Example 15.5) The resultant electric field $\vec{E}$ at $P$ equals the vector sum $\vec{E}_1 + \vec{E}_2$, where $\vec{E}_1$ is the field due to the positive charge $q_1$ and $\vec{E}_2$ is the field due to the negative charge $q_2$.

**SOLUTION**

**(a)** Calculate the electric field at $P$.

Find the magnitude of $\vec{E}_1$ with Equation 15.6:

$$E_1 = k_e\frac{|q_1|}{r_1^{\,2}} = (8.99 \times 10^9\ \text{N}\cdot\text{m}^2/\text{C}^2)\frac{(7.00 \times 10^{-6}\ \text{C})}{(0.400\ \text{m})^2}$$
$$= 3.93 \times 10^5\ \text{N/C}$$

The vector $\vec{E}_1$ is vertical, making an angle of 90° with respect to the positive $x$-axis. Use this fact to find its components:

$$E_{1x} = E_1 \cos(90°) = 0$$
$$E_{1y} = E_1 \sin(90°) = 3.93 \times 10^5\ \text{N/C}$$

Next, find the magnitude of $\vec{E}_2$, again with Equation 15.6:

$$E_2 = k_e\frac{|q_2|}{r_2^{\,2}} = (8.99 \times 10^9\ \text{N}\cdot\text{m}^2/\text{C}^2)\frac{(5.00 \times 10^{-6}\ \text{C})}{(0.500\ \text{m})^2}$$
$$= 1.80 \times 10^5\ \text{N/C}$$

*(Continued)*

Obtain the $x$-component of $\vec{\mathbf{E}}_2$, using the triangle in Figure 15.12 to find $\cos\theta$:

$$\cos\theta = \frac{\text{adj}}{\text{hyp}} = \frac{0.300}{0.500} = 0.600$$

$$E_{2x} = E_2\cos\theta = (1.80 \times 10^5 \text{ N/C})(0.600)$$
$$= 1.08 \times 10^5 \text{ N/C}$$

Obtain the $y$-component in the same way, but a minus sign has to be provided for $\sin\theta$ because this component is directed downwards:

$$\sin\theta = \frac{\text{opp}}{\text{hyp}} = \frac{0.400}{0.500} = 0.800$$

$$E_{2y} = E_2\sin\theta = (1.80 \times 10^5 \text{ N/C})(-0.800)$$
$$= -1.44 \times 10^5 \text{ N/C}$$

Sum the $x$-components to get the $x$-component of the resultant vector:

$$E_x = E_{1x} + E_{2x} = 0 + 1.08 \times 10^5 \text{ N/C} = 1.08 \times 10^5 \text{ N/C}$$

Sum the $y$-components to get the $y$-component of the resultant vector:

$$E_y = E_{1y} + E_{2y} = 3.93 \times 10^5 \text{ N/C} - 1.44 \times 10^5 \text{ N/C}$$
$$E_y = 2.49 \times 10^5 \text{ N/C}$$

Use the Pythagorean theorem to find the magnitude of the resultant vector:

$$E = \sqrt{E_x^{\,2} + E_y^{\,2}} = \boxed{2.71 \times 10^5 \text{ N/C}}$$

The inverse tangent function yields the direction of the resultant vector:

$$\phi = \tan^{-1}\left(\frac{E_y}{E_x}\right) = \tan^{-1}\left(\frac{2.49 \times 10^5 \text{ N/C}}{1.08 \times 10^5 \text{ N/C}}\right) = \boxed{66.6^\circ}$$

**(b)** Find the force on a charge of $2.00 \times 10^{-8}$ C placed at $P$.

Calculate the magnitude of the force (the direction is the same as that of $\vec{\mathbf{E}}$ because the charge is positive):

$$F = Eq = (2.71 \times 10^5 \text{ N/C})(2.00 \times 10^{-8} \text{ C})$$
$$= \boxed{5.42 \times 10^{-3} \text{ N}}$$

**REMARKS** There were numerous steps to this problem, but each was very short. When attacking such problems, it's important to focus on one small step at a time. The solution comes not from a leap of genius, but from the assembly of a number of relatively easy parts.

**QUESTION 15.5** Suppose $q_2$ were moved slowly to the right. What would happen to the angle $\phi$?

**EXERCISE 15.5** (a) Place a charge of $-7.00\ \mu$C at point $P$ and find the magnitude and direction of the electric field at the location of $q_2$ due to $q_1$ and the charge at $P$. (b) Find the magnitude and direction of the force on $q_2$.

**ANSWERS** (a) $5.84 \times 10^5$ N/C, $\phi = 20.2^\circ$ (b) $F = 2.92$ N, $\phi = 200.^\circ$

## 15.5 Electric Field Lines

**LEARNING OBJECTIVES**

1. Relate the electric field to electric field lines.
2. State the rules for drawing electric field lines and sketch field lines for simple charge configurations.

**Tip 15.1 Electric Field Lines Aren't Paths of Particles**

Electric field lines are *not* material objects. They are used only as a pictorial representation of the electric field at various locations. Except in special cases, they *do not* represent the path of a charged particle released in an electric field.

A convenient aid for visualizing electric field patterns is to draw lines pointing in the direction of the electric field vector at any point. These lines, introduced by Michael Faraday and called **electric field lines,** are related to the electric field in any region of space in the following way:

1. The electric field vector $\vec{\mathbf{E}}$ is tangent to the electric field lines at each point.
2. The number of lines per unit area through a surface perpendicular to the lines is proportional to the strength of the electric field in a given region.

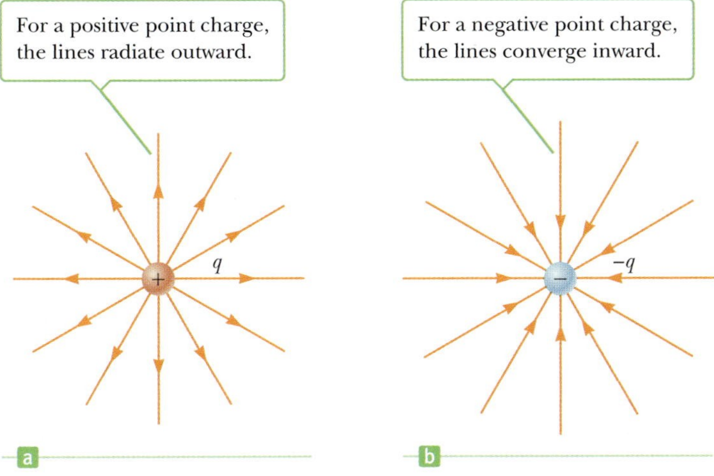


**Figure 15.13** (a), (b) The electric field lines for a point charge. Notice that the figures show only those field lines that lie in the plate of the page.

Note that $\vec{\mathbf{E}}$ is large when the field lines are close together and small when the lines are far apart.

Figure 15.13a shows some representative electric field lines for a single positive point charge. This two-dimensional drawing contains only the field lines that lie in the plane containing the point charge. The lines are actually directed radially outward from the charge in *all* directions, somewhat like the quills of an angry porcupine. Because a positive test charge placed in this field would be repelled by the charge $q$, the lines are directed radially away from the positive charge. The electric field lines for a single negative point charge are directed toward the charge (Fig. 15.13b) because a positive test charge is attracted by a negative charge. In either case the lines are radial and extend all the way to infinity. Note that the lines are closer together as they get near the charge, indicating that the strength of the field is increasing. Equation 15.6 verifies that this is indeed the case.

The rules for drawing electric field lines for any charge distribution follow directly from the relationship between electric field lines and electric field vectors:

1. The lines for a group of point charges must begin on positive charges and end on negative charges. In the case of an excess of charge, some lines will begin or end infinitely far away.
2. The number of lines drawn leaving a positive charge or ending on a negative charge is proportional to the magnitude of the charge.
3. No two field lines can cross each other.

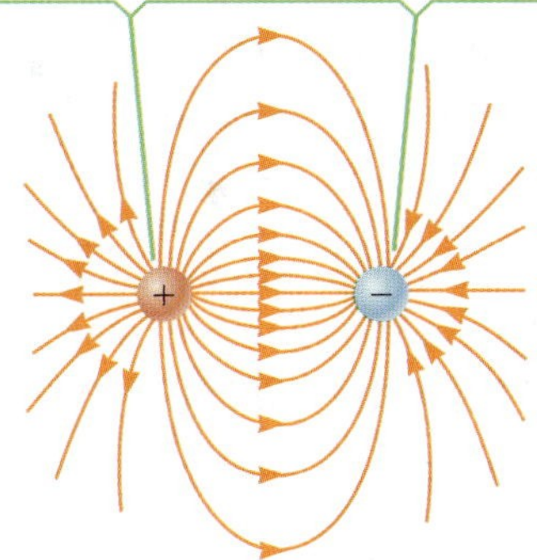


**Figure 15.14** The electric field lines for two equal and opposite point charges (an electric dipole).

Figure 15.14 shows the beautifully symmetric electric field lines for two point charges of equal magnitude but opposite sign. This charge configuration is called an **electric dipole.** Note that the number of lines that begin at the positive charge must equal the number that terminate at the negative charge. At points very near either charge, the lines are nearly radial. The high density of lines between the charges indicates a strong electric field in this region.

Figure 15.15 shows the electric field lines in the vicinity of two equal positive point charges. Again, close to either charge the lines are nearly radial. The same number of lines emerges from each charge because the charges are equal in magnitude. At great distances from the charges, the field is approximately equal to that of a single point charge of magnitude $2q$. The bulging out of the electric field lines between the charges reflects the repulsive nature of the electric force between like charges. Also, the low density of field lines between the charges indicates a weak field in this region, unlike the dipole.

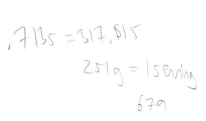

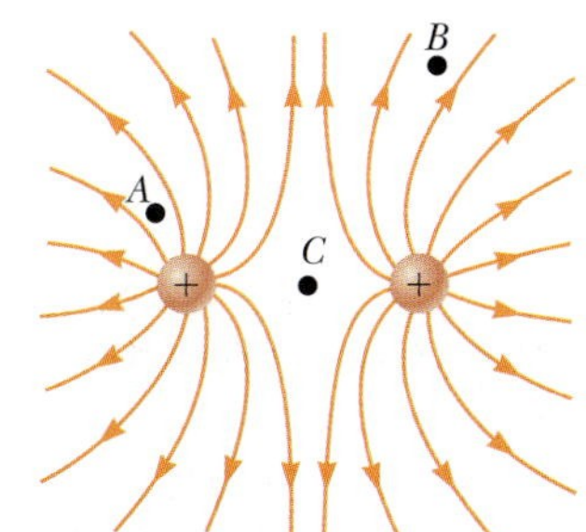


**Figure 15.15** The electric field lines for two positive point charges. The points $A$, $B$, and $C$ are discussed in Quick Quiz 15.6.

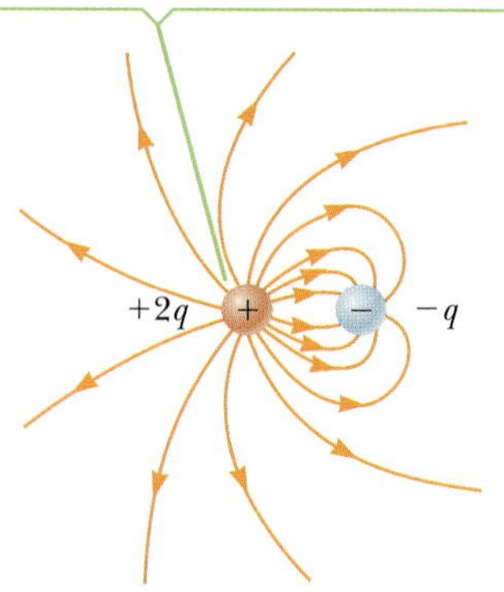
Two field lines leave $+2q$ for every one that terminates on $-q$.
$+2q$
$-q$

**Figure 15.16** The electric field lines for a point charge of $+2q$ and a second point charge of $-q$.

Finally, Figure 15.16 is a sketch of the electric field lines associated with the positive charge $+2q$ and the negative charge $-q$. In this case the number of lines leaving charge $+2q$ is twice the number terminating on charge $-q$. Hence, only half of the lines that leave the positive charge end at the negative charge. The remaining half terminate on negative charges that we assume to be located at infinity. At great distances from the charges (great compared with the charge separation), the electric field lines are equivalent to those of a single charge $+q$.

### Quick Quiz

**15.6** Rank the magnitudes of the electric field at points $A$, $B$, and $C$ in Figure 15.15, with the largest magnitude first.

(a) $A$, $B$, $C$ (b) $A$, $C$, $B$ (c) $C$, $A$, $B$ (d) The answer can't be determined by visual inspection.

## APPLYING PHYSICS 15.1 Measuring Atmospheric Electric Fields

The electric field near the surface of the Earth in fair weather is about 100 N/C downward. Under a thundercloud, the electric field can be very large, on the order of 20 000 N/C. How are these electric fields measured?

**EXPLANATION** A device for measuring these fields is called the *field mill.* Figure 15.17 shows the fundamental components of a field mill: two metal plates parallel to the ground. Each plate is connected to ground with a wire, with an ammeter (a low-resistance device for measuring the flow of charge, to be discussed in Section 17.3) in one path. Consider first just the lower plate. Because it's connected to ground and the ground carries a negative charge, the plate is negatively charged. The electric field lines are therefore directed downward, ending on the plate as in Figure 15.17a. Now imagine that the upper plate is suddenly moved over the lower plate, as in Figure 15.17b. This plate is also connected to ground and is also negatively charged, so the field lines now end on the upper plate. The negative charges in the lower plate are repelled by those on the upper plate and must pass through the ammeter, registering a flow of charge. The amount of charge that was on the lower plate is related to the strength of the electric field. In this way, the flow of charge through the ammeter can be calibrated to measure the electric field. The plates are normally designed like the blades of a fan, with the upper plate rotating so that the lower plate is alternately covered and uncovered. As a result, charges flow back and forth continually through the ammeter, and the reading can be related to the electric field strength. ■

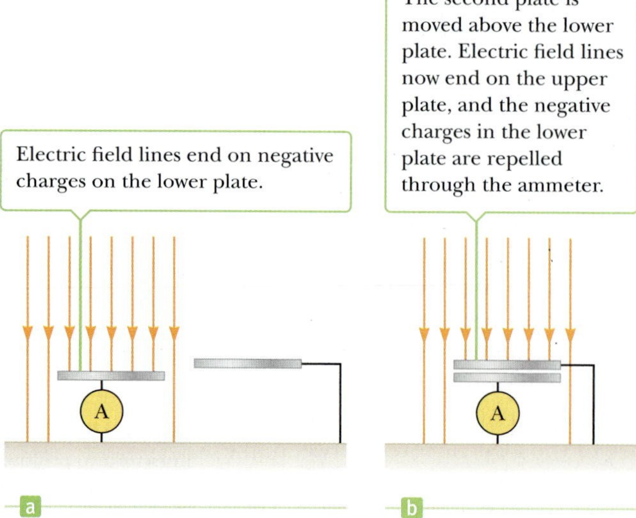
Electric field lines end on negative charges on the lower plate.
The second plate is moved above the lower plate. Electric field lines now end on the upper plate, and the negative charges in the lower plate are repelled through the ammeter.
A
A
a
b

**Figure 15.17** Experimental setup for Applying Physics 15.1.

# 15.6 Conductors in Electrostatic Equilibrium

LEARNING OBJECTIVE

1. Discuss the properties of an isolated conductor in electrostatic equilibrium.

A good electric conductor like copper, although electrically neutral, contains charges (electrons) that aren't bound to any atom and are free to move about within the material. When no net motion of charge occurs within a conductor, the conductor is said to be in **electrostatic equilibrium.** An isolated conductor (one that is insulated from ground) has the following properties:

Properties of an isolated conductor

1. The electric field is zero everywhere inside the conducting material.
2. Any excess charge on an isolated conductor resides entirely on its surface.
3. The electric field just outside a charged conductor is perpendicular to the conductor's surface.
4. On an irregularly shaped conductor, the charge accumulates at sharp points, where the radius of curvature of the surface is smallest.

The first property can be understood by examining what would happen if it were *not* true. If there were an electric field inside a conductor, the free charge there would move and a flow of charge, or current, would be created. If there were a net movement of charge, however, the conductor would no longer be in electrostatic equilibrium.

Property 2 is a direct result of the $1/r^2$ repulsion between like charges described by Coulomb's law. If by some means an excess of charge is placed inside a conductor, the repulsive forces between the like charges push them as far apart as possible, causing them to quickly migrate to the surface. (We won't prove it here, but the excess charge resides on the surface because Coulomb's law is an inverse-square law. With any other power law, an excess of charge would exist on the surface, but there would be a distribution of charge, of either the same or opposite sign, inside the conductor.)

Property 3 can be understood by again considering what would happen if it were not true. If the electric field in Figure 15.18 were not perpendicular to the surface, it would have a component along the surface, which would cause the free charges of the conductor to move (to the left in the figure). If the charges moved, however, a current would be created and the conductor would no longer be in electrostatic equilibrium. Therefore, $\vec{\mathbf{E}}$ must be perpendicular to the surface.

To see why property 4 must be true, consider Figure 15.19a (page 540), which shows a conductor that is fairly flat at one end and relatively pointed at the other. Any excess charge placed on the object moves to its surface. Figure 15.19b shows the forces between two such charges at the flatter end of the object. These forces

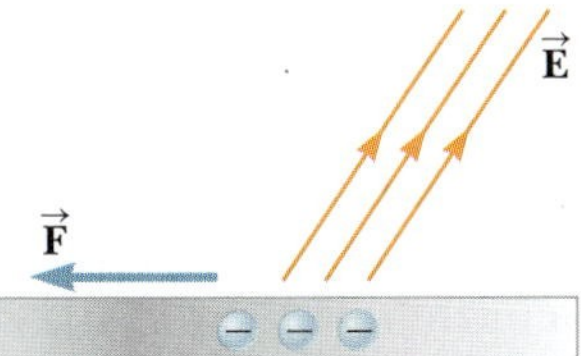


**Figure 15.18** This situation is *impossible* if the conductor is in electrostatic equilibrium. If the electric field $\vec{\mathbf{E}}$ had a component parallel to the surface, an electric force would be exerted on the charges along the surface and they would move to the left.

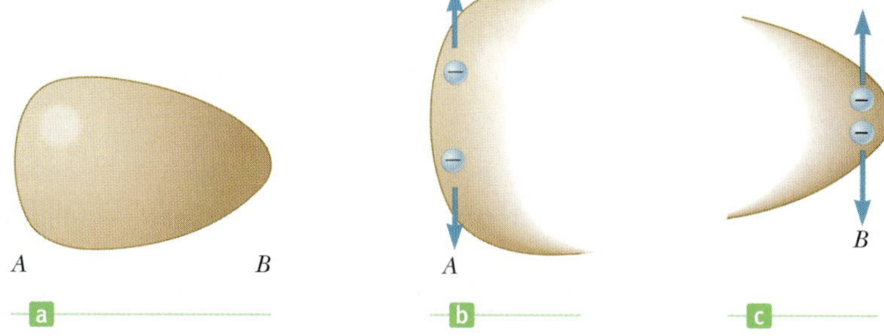


**Figure 15.19** (a) A conductor with a flatter end *A* and a relatively sharp end *B*. Excess charge placed on this conductor resides entirely at its surface and is distributed so that (b) there is less charge per unit area on the flatter end and (c) there is a large charge per unit area on the sharper end.

are predominantly directed parallel to the surface, so the charges move apart until repulsive forces from other nearby charges establish an equilibrium. At the sharp end, however, the forces of repulsion between two charges are directed predominantly away from the surface, as in Figure 15.19c. As a result, there is less tendency for the charges to move apart along the surface here, and the amount of charge per unit area is greater than at the flat end. The cumulative effect of many such outward forces from nearby charges at the sharp end produces a large resultant force directed away from the surface that can be great enough to cause charges to leap from the surface into the surrounding air.

Many experiments have shown that the net charge on a conductor resides on its surface. One such experiment was first performed by Michael Faraday and is referred to as *Faraday's ice pail experiment.* Faraday lowered a negatively-charged metal ball at the end of a silk thread (an insulator) into an uncharged hollow conductor insulated from ground, a metal ice pail as in Figure 15.20a. As the ball entered the pail, the needle on an electrometer attached to the outer surface of the pail was observed to deflect. (An electrometer is a device used to measure charge.) The needle deflected because the charged ball induced a positive charge on the inner wall of the pail, which left an equal negative charge on the outer wall (Fig. 15.20b).

Faraday next touched the inner surface of the pail with the ball and noted that the deflection of the needle did not change, either when the ball touched the inner surface of the pail (Fig. 15.20c) or when it was removed (Fig. 15.20d). Further, he found that the ball was now uncharged because when it touched the inside of the pail, the excess negative charge on the ball had been drawn off, neutralizing the induced positive charge on the inner surface of the pail. In this way Faraday discovered the useful result that *all* the excess charge on an object can be transferred to an already charged metal shell if the object is touched to the *inside* of the shell. As we will see, this result is the principle of operation of the Van de Graaff generator.

Faraday concluded that because the deflection of the needle in the electrometer didn't change when the charged ball touched the inside of the pail, the positive charge induced on the inside surface of the pail was just enough to neutralize the negative charge on the ball. As a result of his investigations, he concluded that a charged object suspended inside a metal container rearranged the charge on the container so that the sign of the charge on its inside surface was *opposite* the sign of the charge on the suspended object. This produced a charge on the outside surface of the container of the same sign as that on the suspended object.

Faraday also found that if the electrometer was connected to the inside surface of the pail after the experiment had been run, the needle showed no deflection. Thus, the *excess* charge acquired by the pail when contact was made between ball and pail appeared on the outer surface of the pail.

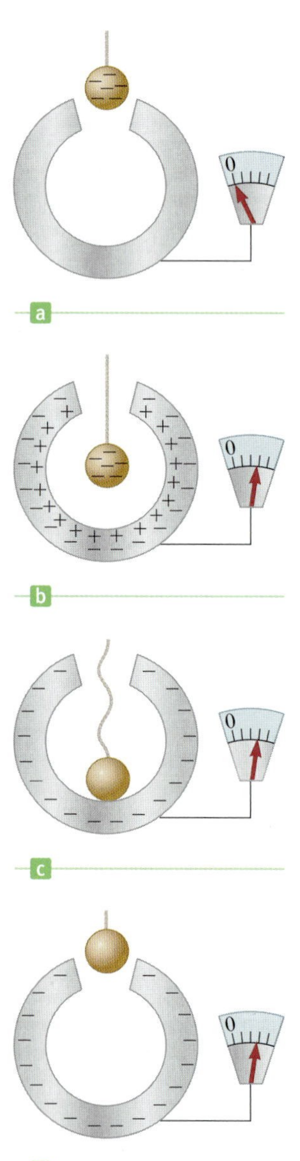


**Figure 15.20** An experiment showing that any charge transferred to a conductor resides on its surface in electrostatic equilibrium. The hollow conductor is insulated from ground, and the small metal ball is supported by an insulating thread.

If a metal rod having sharp points is attached to a house, most of the charge on the house passes through these points, eliminating the induced charge on the house produced by storm clouds. In addition, a lightning discharge striking the house passes through the metal rod and is safely carried to the ground through wires leading from the rod to the Earth. Lightning rods using this principle were first developed by Benjamin Franklin. Some European countries couldn't accept the fact that such a worthwhile idea could have originated in the New World, so they "improved" the design by eliminating the sharp points!

**APPLICATION**
Lightning Rods

### APPLYING PHYSICS 15.2 Conductors and Field Lines

Suppose a point charge $+Q$ is in empty space. Wearing rubber gloves, you proceed to surround the charge with a concentric spherical conducting shell. What effect does that have on the field lines from the charge?

**EXPLANATION** When the spherical shell is placed around the charge, the charges in the shell rearrange to satisfy the rules for a conductor in equilibrium. A net charge of $-Q$ moves to the interior surface of the conductor, so the electric field inside the conductor becomes zero. This means the field lines originating on the $+Q$ charge now terminate on the negative charges. The movement of the negative charges to the inner surface of the sphere leaves a net charge of $+Q$ on the outer surface of the sphere. Then the field lines outside the sphere look just as before: the only change, overall, is the absence of field lines within the conductor. ■

### APPLYING PHYSICS 15.3 Driver Safety During Electrical Storms

Why is it safe to stay inside an automobile during a lightning storm?

**EXPLANATION** Many people believe that staying inside the car is safe because of the insulating characteristics of the rubber tires, but in fact that isn't true. Lightning can travel through several kilometers of air, so it can certainly penetrate a centimeter of rubber. The safety of remaining in the car is due to the fact that charges on the metal shell of the car will reside on the outer surface of the car, as noted in property 2 discussed earlier. As a result, an occupant in the automobile touching the inner surfaces is not in danger. ■

## 15.7 The Millikan Oil-Drop Experiment

**LEARNING OBJECTIVE**

1. Describe Robert Millikan's oil-drop experiment and how it measured the electron's charge.

From 1909 to 1913, Robert Andrews Millikan (1868–1953) performed a brilliant set of experiments at the University of Chicago in which he measured the elementary charge $e$ of the electron and demonstrated the quantized nature of the electronic charge. The apparatus he used, diagrammed in Figure 15.21, contains two parallel metal plates. Oil droplets that have been charged by friction in an atomizer are allowed to pass through a small hole in the upper plate. A horizontal light beam is used to illuminate the droplets, which are viewed by a telescope with axis at right angles to the beam. The droplets then appear as shining stars against a dark background, and the rate of fall of individual drops can be determined.

We assume a single drop having a mass of $m$ and carrying a charge of $q$ is being viewed and its charge is negative. If no electric field is present between the plates,

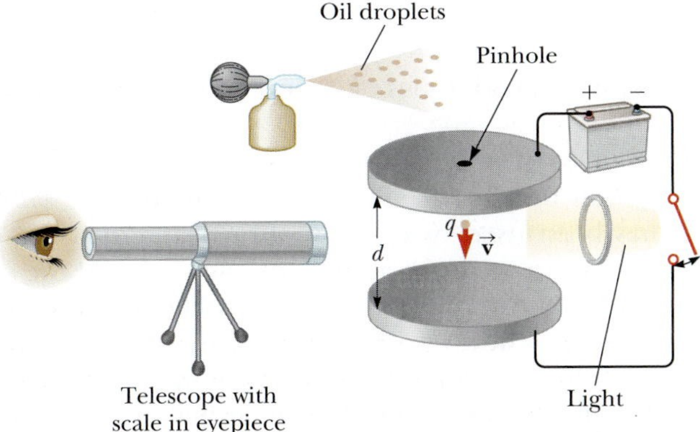


**Figure 15.21** A schematic view of Millikan's oil-drop apparatus.

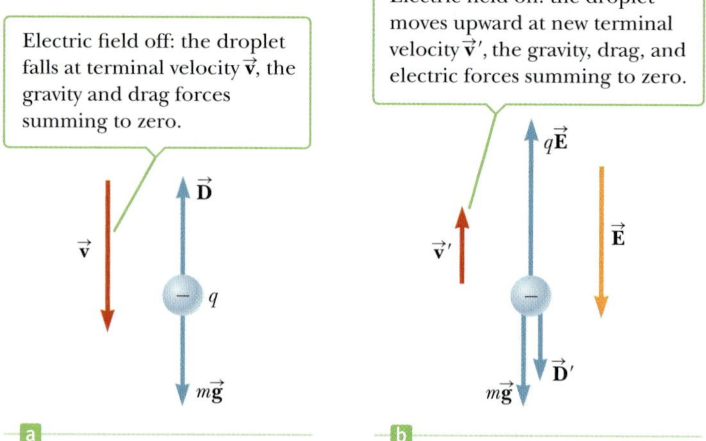


**Figure 15.22** The forces on a negatively charged oil droplet in Millikan's experiment.

the two forces acting on the charge are the force of gravity, $m\vec{g}$, acting downward, and an upward viscous drag force $\vec{D}$ (Fig. 15.22a). The drag force is proportional to the speed of the drop. When the drop reaches its terminal speed, $v$, the two forces balance each other ($mg = D$).

Now suppose an electric field is set up between the plates by a battery connected so that the upper plate is positively charged. In this case a third force, $q\vec{E}$, acts on the charged drop. Because $q$ is negative and $\vec{E}$ is downward, the electric force is *upward* as in Figure 15.22b. If this force is great enough, the drop moves upward and the drag force $\vec{D}'$ acts downward. When the upward electric force, $q\vec{E}$, balances the sum of the force of gravity and the drag force, both acting downward, the drop reaches a new terminal speed $v'$.

With the field turned on, a drop moves slowly upward, typically at a rate of *hundredths* of a centimeter per second. The rate of fall in the absence of a field is comparable. Hence, a single droplet with constant mass and radius can be followed for hours as it alternately rises and falls, simply by turning the electric field on and off.

After making measurements on thousands of droplets, Millikan and his coworkers found that, to within about 1% precision, every drop had a charge equal to some positive or negative integer multiple of the elementary charge $e$,

$$q = ne \quad n = 0, \pm 1, \pm 2, \pm 3, \ldots \qquad [15.7]$$

where $e = 1.60 \times 10^{-19}$ C. It was later established that positive integer multiples of $e$ would arise when an oil droplet had lost one or more electrons. Likewise, negative integer multiples of $e$ would arise when a drop had gained one or more electrons. Gains or losses in integral numbers provide conclusive evidence that charge is quantized. In 1923 Millikan was awarded the Nobel Prize in Physics for this work.

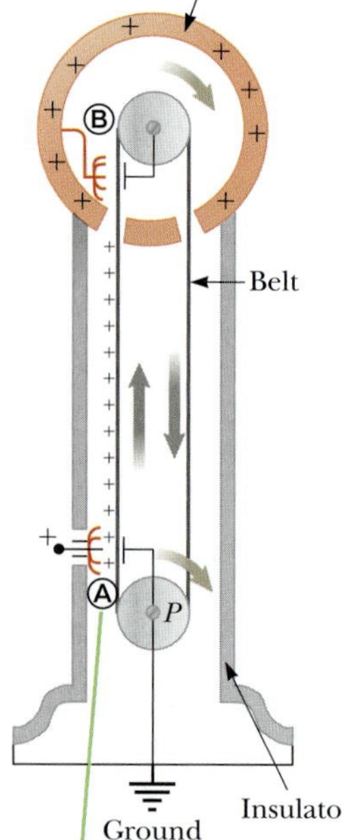


**Figure 15.23** A schematic diagram of a Van de Graaff generator. Charge is transferred to the dome by means of a rotating belt.

## 15.8 The Van de Graaff Generator

**LEARNING OBJECTIVE**

1. Describe the operating principles of Robert Van de Graaff's electrostatic generator.

In 1929 Robert J. Van de Graaff (1901–1967) designed and built an electrostatic generator that has been used extensively in nuclear physics research. The principles of its operation can be understood with knowledge of the properties of electric fields and charges already presented in this chapter. Figure 15.23 shows the basic construction of this device. A motor-driven pulley $P$ moves a belt past positively charged comb-like metallic needles positioned at $A$. Negative charges are attracted to these needles from the belt, leaving the left side of the belt with a net positive charge. The positive charges attract electrons onto the belt as it moves past a second comb of needles at $B$, increasing the excess positive charge on the dome. Because the electric field inside the metal dome is negligible, the positive charge on it can easily be increased regardless of how much charge is already present. The result is that the dome is left with a large amount of positive charge.

This accumulation of charge on the dome can't continue indefinitely. As more and more charge appears on the surface of the dome, the magnitude of the electric field at that surface also increases. Finally, the strength of the field becomes great enough to partially ionize the air near the surface, increasing the conductivity of the air. Charges on the dome now have a pathway to leak off into the air, producing some spectacular "lightning bolts" as the discharge occurs. As noted earlier, charges find it easier to leap off a surface at points where the curvature is great. As a result, one way to inhibit the electric discharge, and to increase the amount of charge that can be stored on the dome, is to increase its radius. Another method for inhibiting discharge is to place the entire system in a container filled with a high-pressure gas, which is significantly more difficult to ionize than air at atmospheric pressure.

If protons (or other charged particles) are introduced into a tube attached to the dome, the large electric field of the dome exerts a repulsive force on the protons, causing them to accelerate to energies high enough to initiate nuclear reactions between the protons and various target nuclei.

## 15.9 Electric Flux and Gauss's Law

**LEARNING OBJECTIVES**

1. Define electric flux and calculate it in elementary contexts.
2. State Gauss's law relating electric flux through a closed surface to the charge inside the surface.
3. Apply Gauss's law to distributions of charge.

Gauss's law is essentially a technique for calculating the average electric field on a closed surface, developed by Karl Friedrich Gauss (1777–1855). When the electric field, because of its symmetry, is constant everywhere on that surface and perpendicular to it, the exact electric field can be found. In such special cases, Gauss's law is far easier to apply than Coulomb's law.

Gauss's law relates the electric flux through a closed surface and the total charge inside that surface. A *closed surface* has an inside and an outside: an example is a sphere. *Electric flux* is a measure of how much the electric field vectors penetrate through a given surface. If the electric field vectors are tangent to the surface at all points, for example, they don't penetrate the surface and the electric flux through the surface is zero. These concepts will be discussed more fully in the next two subsections. As we'll see, Gauss's law states that the electric flux through a closed surface is proportional to the charge contained *inside* the surface.

### Electric Flux

Consider an electric field that is uniform in both magnitude and direction, as in Figure 15.24. The electric field lines penetrate a surface of area $A$, which is perpendicular to the field. The technique used for drawing a figure such as Figure 15.24 is that the number of lines per unit area, $N/A$, is proportional to the magnitude of the electric field, or $E \propto N/A$. We can rewrite this proportion as $N \propto EA$, which means that the number of field lines is proportional to the *product* of $E$ and $A$, called the **electric flux** and represented by the symbol $\Phi_E$:

$$\Phi_E = EA \qquad [15.8]$$

Note that $\Phi_E$ has SI units of $\mathrm{N \cdot m^2/C}$ and is proportional to the number of field lines that pass through some area $A$ oriented perpendicular to the field. (It's called

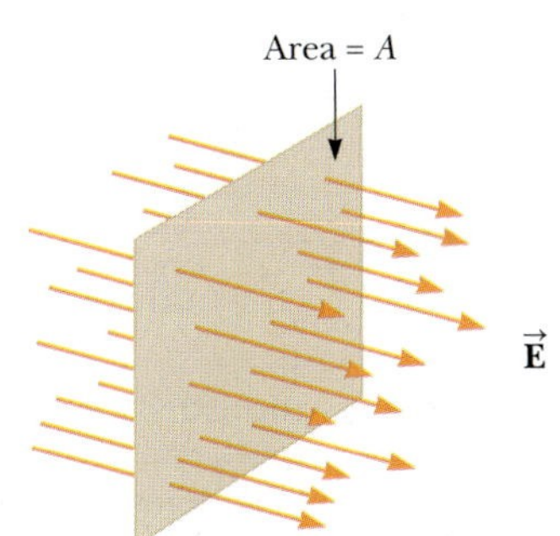


**Figure 15.24** Field lines of a uniform electric field penetrating a plane of area $A$ perpendicular to the field. The electric flux $\Phi_E$ through this area is equal to $EA$.

flux by analogy with the term *flux* in fluid flow, which is the volume of liquid flowing through a perpendicular area per second.) If the surface under consideration is not perpendicular to the field, as in Figure 15.25, the expression for the electric flux is

Electric flux ▶

$$\Phi_E = EA \cos \theta \qquad [15.9]$$

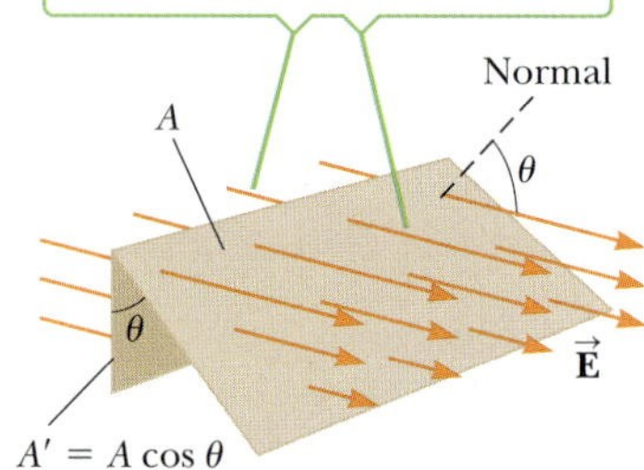
The number of field lines that go through the area $A'$ is the same as the number that go through area $A$.
Normal
$A$
$\theta$
$\theta$
$\vec{E}$
$A' = A \cos \theta$

**Figure 15.25** Field lines for a uniform electric field through an area $A$ that is at an angle of $(90° - \theta)$ to the field.

where a vector perpendicular to the area $A$ is at an angle $\theta$ with respect to the field. This vector is often said to be *normal* to the surface, and we will refer to it as "the normal vector to the surface." The number of lines that cross this area is equal to the number that cross the projected area $A'$, which is perpendicular to the field. We see that the two areas are related by $A' = A \cos \theta$. From Equation 15.9, we see that the flux through a surface of fixed area has the maximum value $EA$ when the surface is perpendicular to the field (when $\theta = 0°$) and that the flux is zero when the surface is parallel to the field (when $\theta = 90°$). **By convention, for a closed surface, the flux lines passing into the interior of the volume are negative and those passing out of the interior of the volume are positive.** This convention is equivalent to requiring the normal vector of the surface to point outward when computing the flux through a closed surface.

**Quick Quiz**

**15.7** Calculate the magnitude of the flux of a constant electric field of 5.00 N/C in the $z$-direction through a rectangle with area 4.00 m$^2$ in the $xy$-plane. (a) 0 (b) 10.0 N · m$^2$/C (c) 20.0 N · m$^2$/C (d) More information is needed

**15.8** Suppose the electric field of Quick Quiz 15.7 is tilted 60° away from the positive $z$-direction. Calculate the magnitude of the flux through the same area. (a) 0 (b) 10.0 N · m$^2$/C (c) 20.0 N · m$^2$/C (d) More information is needed

## EXAMPLE 15.6 Flux Through a Cube

**GOAL** Calculate the electric flux through a closed surface.

**PROBLEM** Consider a uniform electric field oriented in the $x$-direction. Find the electric flux through each surface of a cube with edges $L$ oriented as shown in Figure 15.26, and the net flux.

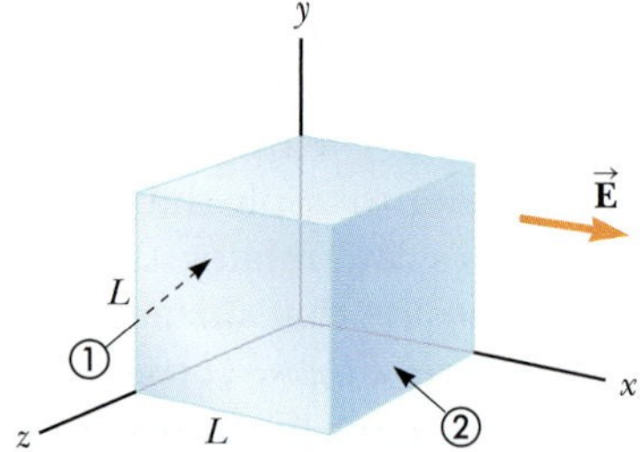
$y$
$\vec{E}$
$L$
①
$x$
$z$
$L$
②

**Figure 15.26** (Example 15.6) A hypothetical surface in the shape of a cube in a uniform electric field parallel to the $x$-axis. The net flux through the surface is zero when the net charge inside the cube is zero.

**STRATEGY** This problem involves substituting into the definition of electric flux given by Equation 15.9. In each case $E$ and $A = L^2$ are the same; the only difference is the angle $\theta$ that the electric field makes with respect to a vector perpendicular to a given surface and pointing outward (the normal vector to the surface). The angles can be determined by inspection. The flux through a surface parallel to the $xy$-plane will be labeled $\Phi_{xy}$ and further designated by position (front, back); others will be labeled similarly: $\Phi_{xz}$ top or bottom, and $\Phi_{yz}$ left or right.

**SOLUTION**

The normal vector to the $xy$-plane points in the negative $z$-direction. This, in turn, is perpendicular to $\vec{E}$, so $\theta = 90°$. (The opposite side works similarly.)

$$\Phi_{xy} = EA \cos(90°) = 0 \text{ (back and front surfaces)}$$

The normal vector to the $xz$-plane points in the negative $y$-direction. This, in turn, is perpendicular to $\vec{E}$, so again $\theta = 90°$. (The opposite side works similarly.)

$$\Phi_{xz} = EA \cos(90°) = 0 \text{ (top and bottom surfaces)}$$

The normal vector to surface ① (the $yz$-plane) points in the negative $x$-direction. This is antiparallel to $\vec{E}$, so $\theta = 180°$.

$$\Phi_{yz} = EA \cos(180°) = -EL^2 \text{ (surface ①)}$$

Surface ② has normal vector pointing in the positive $x$-direction, so $\theta = 0°$.

$$\Phi_{yz} = EA\cos(0°) = EL^2 \text{ (surface ②)}$$

We calculate the net flux by summing:

$$\Phi_{\text{net}} = 0 + 0 + 0 + 0 - EL^2 + EL^2 = 0$$

**REMARKS** In doing this calculation, it is necessary to remember that the angle in the definition of flux is measured from the normal vector to the surface and that this vector must point outwards for a closed surface. As a result, the normal vector for the $yz$-plane on the left points in the negative $x$-direction, and the normal vector to the plane parallel to the $yz$-plane on the right points in the positive $x$-direction. Notice that there aren't any charges in the box. The net electric flux is always zero for closed surfaces that contain a net charge of zero.

**QUESTION 15.6** If the surface in Figure 15.26 were spherical, would the answer be (a) greater than, (b) less than, or (c) the same as the net electric flux found for the cubical surface?

**EXERCISE 15.6** Suppose the constant electric field in Example 15.6 points in the positive $y$-direction instead. Calculate the flux through the $xz$-plane and the surface parallel to it. What's the net electric flux through the surface of the cube?

**ANSWERS** $\Phi_{xz} = -EL^2$ (bottom surface), $\Phi_{xz} = +EL^2$ (top surface). The net flux is still zero.

## Gauss's Law

Consider a point charge $q$ surrounded by a spherical surface of radius $r$ centered on the charge, as in Figure 15.27a. The magnitude of the electric field everywhere on the surface of the sphere is

$$E = k_e \frac{q}{r^2}$$

Note that the electric field is perpendicular to the spherical surface at all points on the surface. The electric flux through the surface is therefore $EA$, where $A = 4\pi r^2$ is the surface area of the sphere:

$$\Phi_E = EA = k_e \frac{q}{r^2}(4\pi r^2) = 4\pi k_e q$$

It's sometimes convenient to express $k_e$ in terms of another constant, $\epsilon_0$, as $k_e = 1/(4\pi\epsilon_0)$. The constant $\epsilon_0$ is called the **permittivity of free space** and has the value

$$\epsilon_0 = \frac{1}{4\pi k_e} = 8.85 \times 10^{-12}\ \text{C}^2/\text{N}\cdot\text{m}^2 \qquad [15.10]$$

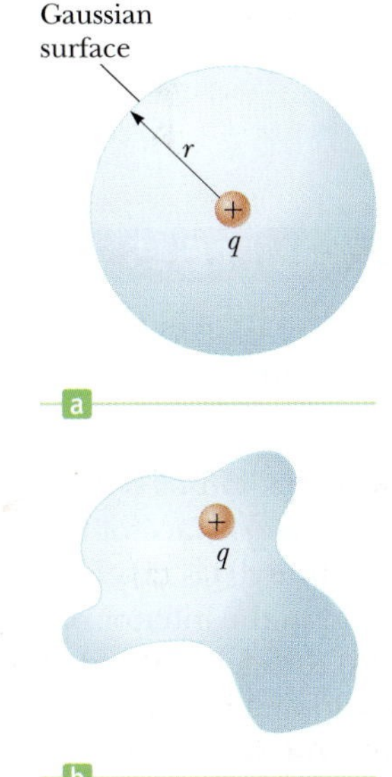


**Figure 15.27** (a) The flux through a spherical surface of radius $r$ surrounding a point charge $q$ is $\Phi_E = q/\epsilon_0$. (b) The flux through any arbitrary surface surrounding the charge is also equal to $q/\epsilon_0$.

The use of $k_e$ or $\epsilon_0$ is strictly a matter of taste. The electric flux through the closed spherical surface that surrounds the charge $q$ can now be expressed as

$$\Phi_E = 4\pi k_e q = \frac{q}{\epsilon_0}$$

This result says that the electric flux through a sphere that surrounds a charge $q$ is equal to the charge divided by the constant $\epsilon_0$. Using calculus, this result can be proven for *any* closed surface that surrounds the charge $q$. For example, if the surface surrounding $q$ is irregular, as in Figure 15.27b, the flux through that surface is also $q/\epsilon_0$. This leads to the following general result, known as Gauss's law:

◀ Gauss's Law

The electric flux $\Phi_E$ through any closed surface is equal to the net charge inside the surface, $Q_{\text{inside}}$, divided by $\epsilon_0$:

$$\Phi_E = \frac{Q_{\text{inside}}}{\epsilon_0} \qquad [15.11]$$

Although it's not obvious, Gauss's law describes how charges create electric fields. In principle it can always be used to calculate the electric field of a system of charges or a continuous distribution of charge. In practice, the technique is

**Tip 15.2 Gaussian Surfaces Aren't Real**

A Gaussian surface is an imaginary surface, created solely to facilitate a mathematical calculation. It doesn't necessarily coincide with the surface of a physical object.

useful only in a limited number of cases in which there is a high degree of symmetry, such as spheres, cylinders, or planes. With the symmetry of these special shapes, the charges can be surrounded by an imaginary surface, called a Gaussian surface. This imaginary surface is used strictly for mathematical calculation, and need not be an actual, physical surface. If the imaginary surface is chosen so that the electric field is constant everywhere on it, the electric field can be computed with

$$EA = \Phi_E = \frac{Q_{\text{inside}}}{\epsilon_0} \quad [15.12]$$

as will be seen in the examples. Although Gauss's law in this form can be used to obtain the electric field only for problems with a lot of symmetry, it can *always* be used to obtain the *average* electric field on *any* surface.

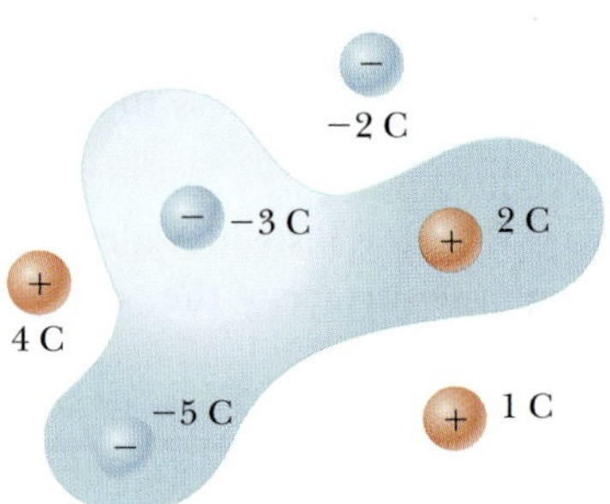


**Figure 15.28** (Quick Quiz 15.9)

**Quick Quiz**

**15.9** Find the electric flux through the surface in Figure 15.28. Assume all charges in the shaded area are inside the surface. (a) $-(3\text{ C})/\epsilon_0$ (b) $(3\text{ C})/\epsilon_0$ (c) 0 (d) $-(6\text{ C})/\epsilon_0$

**15.10** For a closed surface through which the net flux is zero, each of the following four statements *could* be true. Which of the statements *must* be true? (There may be more than one.) (a) There are no charges inside the surface. (b) The net charge inside the surface is zero. (c) The electric field is zero everywhere on the surface. (d) The number of electric field lines entering the surface equals the number leaving the surface.

## EXAMPLE 15.7 The Electric Field of a Charged Spherical Shell

**GOAL** Use Gauss's law to determine electric fields when the symmetry is spherical.

**PROBLEM** A spherical conducting shell of inner radius $a$ and outer radius $b$ carries a total charge $+Q$ distributed on the surface of a conducting shell (Fig. 15.29a). The quantity $Q$ is taken to be positive. **(a)** Find the electric field in the interior of the conducting shell, for $r < a$, and **(b)** the electric field outside the shell, for $r > b$. **(c)** If an additional charge of $-2Q$ is placed at the center, find the electric field for $r > b$. **(d)** What is the distribution of charge on the sphere in part (c)?

**STRATEGY** For each part, draw a spherical Gaussian surface in the region of interest. Add up the charge inside the Gaussian surface, substitute it and the area into Gauss's law, and solve for the electric field. To find the distribution of charge in part (c), use Gauss's law in reverse: the charge distribution must be such that the electrostatic field is zero inside a conductor.

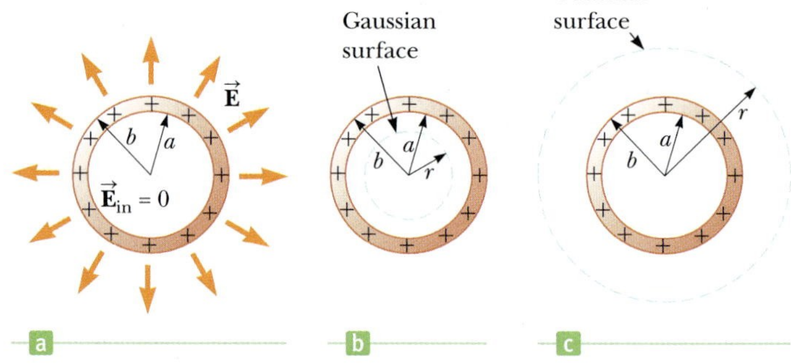


**Figure 15.29** (Example 15.7) (a) The electric field inside a uniformly charged spherical shell is *zero.* It is also zero for the conducting material in the region $a < r < b$. The field outside is the same as that of a point charge having a total charge $Q$ located at the center of the shell. (b) The construction of a Gaussian surface for calculating the electric field *inside* a spherical shell. (c) The construction of a Gaussian surface for calculating the electric field *outside* a spherical shell.

**SOLUTION**

**(a)** Find the electric field for $r < a$.

Apply Gauss's law, Equation 15.12, to the Gaussian surface illustrated in Figure 15.29b (note that there isn't any charge inside this surface):

$$EA = E(4\pi r^2) = \frac{Q_{\text{inside}}}{\epsilon_0} = 0 \quad \rightarrow \quad E = 0$$

**(b)** Find the electric field for $r > b$.

Apply Gauss's law, Equation 15.12, to the Gaussian surface illustrated in Figure 15.29c:

$$EA = E(4\pi r^2) = \frac{Q_{\text{inside}}}{\epsilon_0} = \frac{Q}{\epsilon_0}$$

Divide by the area: $E = \dfrac{Q}{4\pi\epsilon_0 r^2}$

**(c)** Now an additional charge of $-2Q$ is placed at the center of the sphere. Compute the new electric field outside the sphere, for $r > b$.

Apply Gauss's law as in part (b), including the new charge in $Q_{\text{inside}}$: $EA = E(4\pi r^2) = \dfrac{Q_{\text{inside}}}{\epsilon_0} = \dfrac{+Q - 2Q}{\epsilon_0}$

Solve for the electric field: $E = -\dfrac{Q}{4\pi\epsilon_0 r^2}$

**(d)** Find the charge distribution on the sphere for part (c).

Write Gauss's law for the interior of the shell: $EA = \dfrac{Q_{\text{inside}}}{\epsilon_0} = \dfrac{Q_{\text{center}} + Q_{\text{inner surface}}}{\epsilon_0}$

Find the charge on the inner surface of the shell, noting that the electric field in the conductor is zero:

$$Q_{\text{center}} + Q_{\text{inner surface}} = 0$$

$$Q_{\text{inner surface}} = -Q_{\text{center}} = +2Q$$

Find the charge on the outer surface, noting that the inner and outer surface charges must sum to $+Q$:

$$Q_{\text{outer surface}} + Q_{\text{inner surface}} = Q$$

$$Q_{\text{outer surface}} = -Q_{\text{inner surface}} + Q = -Q$$

**REMARKS** The important thing to notice is that in each case, the charge is spread out over a region with spherical symmetry or is located at the exact center. That is what allows the computation of a value for the electric field.

**QUESTION 15.7** If the charge at the center of the sphere is made positive, how is the charge on the inner surface of the sphere affected?

**EXERCISE 15.7** Suppose the charge at the center is now increased to $+2Q$, while the surface of the conductor still retains a charge of $+Q$. (a) Find the electric field exterior to the sphere, for $r > b$. (b) What's the electric field inside the conductor, for $a < r < b$? (c) Find the charge distribution on the conductor.

**ANSWERS** (a) $E = 3Q/4\pi\epsilon_0 r^2$ (b) $E = 0$, which is always the case when charges are not moving in a conductor. (c) Inner surface: $-2Q$; outer surface: $+3Q$

Problems like Example 15.7 sometimes involve "thin, nonconducting shells" carrying a uniformly distributed charge. In these cases no distinction need be made between the outer surface and inner surface of the shell. The next example makes that implicit assumption.

## EXAMPLE 15.8 A Nonconducting Plane Sheet of Charge

**GOAL** Apply Gauss's law to a problem with plane symmetry.

**PROBLEM** Find the electric field above and below a nonconducting infinite plane sheet of charge with uniform positive charge per unit area $\sigma$ (Fig. 15.30a, page 548).

**STRATEGY** By symmetry, the electric field must be perpendicular to the plane and directed away from it on either side, as shown in Figure 15.30b. For the Gaussian surface, choose a small cylinder with axis perpendicular to the plane, each end having area $A_0$. No electric field lines pass through the curved surface of the cylinder, only through the two ends, which have total area $2A_0$. Apply Gauss's law, using Figure 15.30b.

*(Continued)*

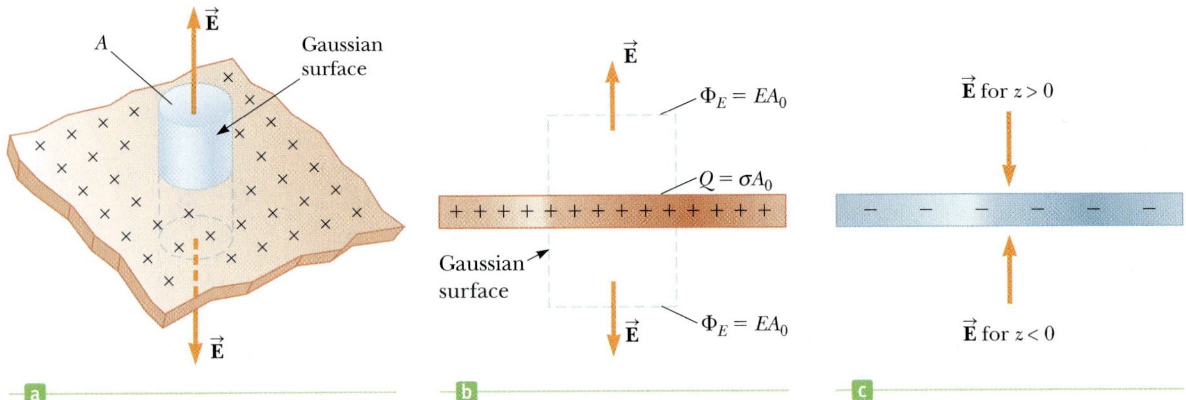


**Figure 15.30** (Example 15.8) (a) A cylindrical Gaussian surface penetrating an infinite sheet of charge. (b) A cross section of the same Gaussian cylinder. The flux through each end of the Gaussian surface is $EA_0$. There is no flux through the cylindrical surface. (c) (Exercise 15.8).

**SOLUTION**

Find the electric field above and below a plane of uniform charge.

Apply Gauss's law, Equation 15.12:

$$EA = \frac{Q_{\text{inside}}}{\epsilon_0}$$

The total charge inside the Gaussian cylinder is the charge density times the cross-sectional area:

$$Q_{\text{inside}} = \sigma A_0$$

The electric flux comes entirely from the two ends, each having area $A_0$. Substitute $A = 2A_0$ and $Q_{\text{inside}}$ and solve for $E$.

$$E = \frac{\sigma A_0}{(2A_0)\epsilon_0} = \frac{\sigma}{2\epsilon_0}$$

This is the *magnitude* of the electric field. Find the $z$-component of the field above and below the plane. The electric field points away from the plane, so it's positive above the plane and negative below the plane.

$$E_z = \frac{\sigma}{2\epsilon_0} \quad z > 0$$

$$E_z = -\frac{\sigma}{2\epsilon_0} \quad z < 0$$

**REMARKS** Notice here that the plate was taken to be a thin, nonconducting shell. If it's made of metal, of course, the electric field inside it is zero, with half the charge on the upper surface and half on the lower surface.

**QUESTION 15.8** In reality, the sheet carrying charge would likely be metallic and have a small but nonzero thickness. If it carries the same charge per unit area, what is the electric field inside the sheet between the two surfaces?

**EXERCISE 15.8** Suppose an infinite nonconducting plane of charge as in Example 15.8 has a uniform negative charge density of $-\sigma$. Find the electric field above and below the plate. Sketch the field.

**ANSWERS** $E_z = \frac{-\sigma}{2\epsilon_0}$, $z > 0$; $E_z = \frac{\sigma}{2\epsilon_0}$, $z < 0$. See Figure 15.30c for the sketch.

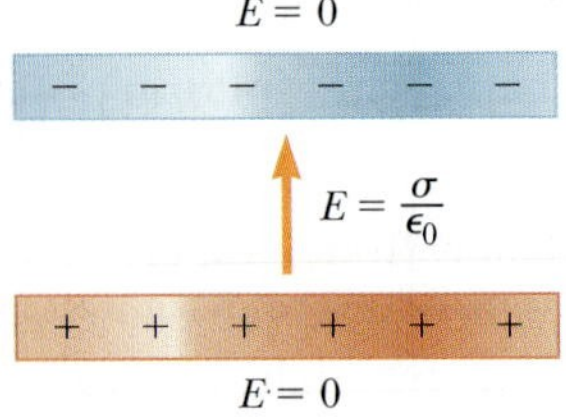


**Figure 15.31** Cross section of an idealized parallel-plate capacitor. Electric field vector contributions sum together in between the plates, but cancel outside.

An important circuit element that will be studied extensively in the next chapter is the parallel-plate capacitor. The device consists of a plate of positive charge, as in Example 15.8, with the negative plate of Exercise 15.8 placed above it. The sum of these two fields is illustrated in Figure 15.31. The result is an electric field with double the magnitude in between the two plates:

$$E = \frac{\sigma}{\epsilon_0} \qquad [15.13]$$

Outside the plates, the electric fields cancel.

# SUMMARY

## 15.1 Properties of Electric Charges

**Electric charges** have the following properties:

1. Unlike charges attract one another and like charges repel one another.
2. Electric charge is always conserved.
3. Charge comes in discrete packets that are integral multiples of the basic electric charge $e = 1.6 \times 10^{-19}$ C.
4. The force between two charged particles is proportional to the inverse square of the distance between them.

## 15.2 Insulators and Conductors

**Conductors** are materials in which charges move freely in response to an electric field. All other materials are called **insulators.**

## 15.3 Coulomb's Law

**Coulomb's law** states that the electric force between two stationary charged particles separated by a distance $r$ has the magnitude

$$F = k_e \frac{|q_1||q_2|}{r^2} \quad [15.1]$$

where $|q_1|$ and $|q_2|$ are the magnitudes of the charges on the particles in coulombs and

$$k_e \approx 8.99 \times 10^9 \text{ N} \cdot \text{m}^2/\text{C}^2 \quad [15.2]$$

is the **Coulomb constant.**

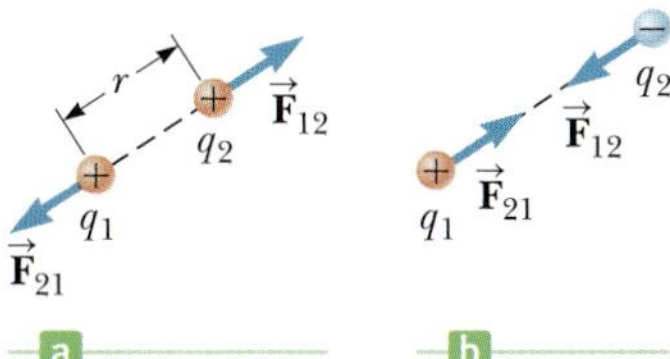

(a) The electric force between two charges with the same sign is repulsive, and (b) it is attractive when the charges have opposite signs.

## 15.4 The Electric Field

An electric field $\vec{\mathbf{E}}$ exists at some point in space if a small test charge $q_0$ placed at that point is acted upon by an electric force $\vec{\mathbf{F}}$. The electric field is defined as

$$\vec{\mathbf{E}} \equiv \frac{\vec{\mathbf{F}}}{q_0} \quad [15.3]$$

The **direction** of the electric field at a point in space is defined to be the direction of the electric force that would be exerted on a small positive charge placed at that point.

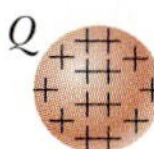
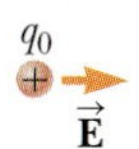

Source charge Test charge

The electric force of charge $Q$ on a test charge $q_0$ divided by $q_0$ gives the electric field $\vec{\mathbf{E}}$ of $Q$ at that point.

The magnitude of the electric field due to a *point charge* $q$ at a distance $r$ from the point charge is

$$E = k_e \frac{|q|}{r^2} \quad [15.6]$$

## 15.5 Electric Field Lines

**Electric field lines** are useful for visualizing the electric field in any region of space. The electric field vector $\vec{\mathbf{E}}$ is tangent to the electric field lines at every point. Further, the number of electric field lines per unit area through a surface perpendicular to the lines is proportional to the strength of the electric field at that surface.

## 15.6 Conductors in Electrostatic Equilibrium

A **conductor in electrostatic equilibrium** has the following properties:

1. The electric field is zero everywhere inside the conducting material.
2. Any excess charge on an isolated conductor must reside entirely on its surface.
3. The electric field just outside a charged conductor is perpendicular to the conductor's surface.
4. On an irregularly shaped conductor, charge accumulates where the radius of curvature of the surface is smallest, at sharp points.

## 15.9 Electric Flux and Gauss's Law

**Gauss's law** states that the electric flux through any closed surface is equal to the net charge $Q$ inside the surface divided by the permittivity of free space, $\epsilon_0$:

$$EA = \Phi_E = \frac{Q_{\text{inside}}}{\epsilon_0} \quad [15.12]$$

For highly symmetric distributions of charge, Gauss's law can be used to calculate electric fields.

The electric flux $\Phi$ through any arbitrary surface surrounding a charge $q$ is $q/\epsilon_0$.

# WARM-UP EXERCISES

ENHANCED WebAssign The warm-up exercises in this chapter may be assigned online in Enhanced WebAssign.

1. **Math Review** A force vector has components given by $F_x = -7.00$ N and $F_y = 4.50$ N. Find (a) the magnitude and (b) the direction of the force, measured counterclockwise from the positive $x$-axis.

2. **Math Review** The force acting on a particle has a magnitude of 125 N and is directed 30.0° above the positive $x$-axis. Determine (a) the $x$-component and (b) $y$-component of the force.

3. **Math Review** Two electric force vectors act on a particle. Their $x$-components are 15.0 N and −7.50 N and their $y$-components are −11.5 N and −4.50 N, respectively. For the resultant electric force, find (a) the $x$-component, (b) the $y$-component, and the (c) magnitude and (d) direction of the resultant electric force, measured counterclockwise from the positive $x$-axis. (See also Section 15.4.)

4. **Physics Review** A large firecracker explodes under a soda can, sending it straight upward with an initial speed of 7.27 m/s. Neglecting air friction, find (a) the can's maximum height, (b) the first time at which it reaches a height of 1.00 m, and (c) the velocity at that time. (See Section 2.5.)

5. **Physics Review** A simple pendulum with mass 0.250 kg hangs from string 1.40 m long. (See Fig. WU15.5.) A leaf blower blows air at it, exerting a constant horizontal drag force $\vec{\mathbf{F}}_{drag}$ of magnitude 1.67 N. Find (a) the tension in the string and (b) the angle the pendulum makes with respect to the vertical. (See Section 4.5.)

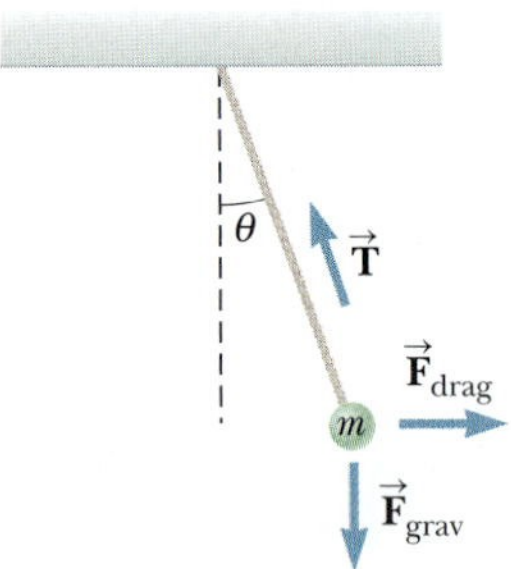

**Figure WU15.5**

6. **Physics Review** The Falcon 9 rocket, manufactured by the commercial SpaceX Corporation, has a takeoff mass of $4.80 \times 10^5$ kg and an initial thrust of $5.88 \times 10^6$ N. At takeoff, determine (a) the net upward force acting on the rocket, and (b) the upward acceleration of the rocket. (See Section 4.5.)

7. **Physics Review** A horizontal spring with force constant $k = 57.0$ N/m is used to accelerate a 0.500-kg mass on a frictionless surface. If the spring is compressed by $5.00 \times 10^{-2}$ m and released, find (a) the initial magnitude of the spring force when the mass is released and (b) the initial acceleration of the mass. (See Section 5.4.)

8. Two protons are separated by a distance of 0.100 m. Given the proton charge of $1.60 \times 10^{-19}$ C, determine the magnitude of the electric force that one proton exerts on the other. (See Section 15.3.)

9. A charge of 16.0 nC is located at the origin. Find (a) the magnitude and (b) the direction of the displacement vector pointing from the origin to (3.00, 4.00) m. Calculate (c) the magnitude, and (d) the $x$- and (e) $y$-components of the electric field at that point. (See Section 15.4.)

10. Consider a system of two charges where the first charge is on the positive $x$-axis and the second charge is on the positive $y$-axis. If these two charges produce electric fields at the origin given by (2.50, 0) N/C and (0, −2.50) N/C, respectively, determine (a) the magnitude and (b) the direction of the resultant electric field at the origin, measured counterclockwise from the positive $x$-axis. (c) What would be the magnitude of the force on a $1.41 \times 10^{-9}$ C charge placed at the origin? (See Sections 15.3 and 15.4.)

11. A tiny droplet of oil of mass $1.50 \times 10^{-15}$ kg having a charge of $-2.56 \times 10^{-18}$ C accelerates straight down inside a vacuum chamber. If the electric field is −5 625 N/C, find (a) the force of gravity on the droplet, (b) the electric force, and (c) the droplet's acceleration. (Give the correct signs, with down the negative direction.)

12. A constant electric field of magnitude 5.00 N/C and pointing in the positive $z$-direction passes through a square with sides of length 0.500 m, located in the $xy$-plane. (a) Calculate the magnitude of the electric flux through the square. (b) Repeat the calculation if the electric field is oriented at an angle of 30.0° with respect to the positive $z$-direction. (See Section 15.9.)

13. Three charges are placed inside a basketball. If the charges are 2.50 nC, −1.25 nC and 0.500 nC, respectively, what is the electric flux through the basketball? (See Section 15.9.)

14. A thin, hollow sphere has a charge $q = -2.00$ nC evenly distributed over its surface. The radius of the charged sphere is 1.00 m and the radii of two spherical Gaussian surfaces are $a = 0.500$ m and $b = 1.50$ m, respectively. (See Figure WU15.14.) A charge of $Q = 5.00$ nC is placed at the center of the charged sphere. (a) How much charge is inside the inner Gaussian surface? (b) Find the electric flux through the inner Gaussian surface, and (c) the electric field there, by dividing by the sphere's area. (d) How much charge is inside the outer dashed sphere? Find (e) the electric flux through the outer Gaussian surface and (f) the electric field strength on that surface. (See Section 15.9.)

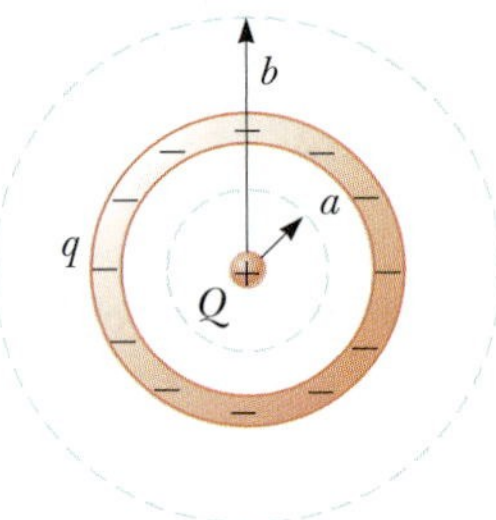

**Figure WU15.14**

## CONCEPTUAL QUESTIONS

ENHANCED WebAssign The conceptual questions in this chapter may be assigned online in Enhanced WebAssign.

1. A glass object receives a positive charge of +3 nC by rubbing it with a silk cloth. In the rubbing process, have protons been added to the object or have electrons been removed from it?

2. Explain from an atomic viewpoint why charge is usually transferred by electrons.

3. A person is placed in a large, hollow metallic sphere that is insulated from ground. If a large charge is placed on the sphere, will the person be harmed upon touching the inside of the sphere?

4. Why must hospital personnel wear special conducting shoes while working around oxygen in an operating

room? What might happen if the personnel wore shoes with rubber soles?

5. (a) Would life be different if the electron were positively charged and the proton were negatively charged? (b) Does the choice of signs have any bearing on physical and chemical interactions? Explain your answers.

6. If a suspended object *A* is attracted to a charged object *B*, can we conclude that *A* is charged? Explain.

7. Explain how a positively charged object can be used to leave another metallic object with a net negative charge. Discuss the motion of charges during the process.

8. Consider point *A* in Figure CQ15.8 located an arbitrary distance from two point charges in otherwise empty space. (a) Is it possible for an electric field to exist at point *A* in empty space? (b) Does charge exist at this point? (c) Does a force exist at this point?

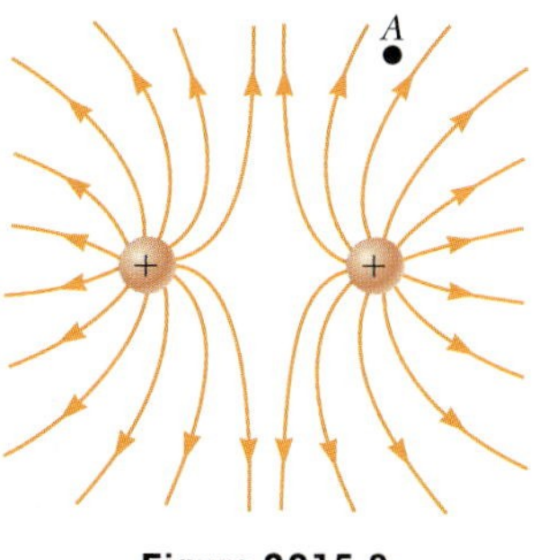


**Figure CQ15.8**

9. A student stands on a thick piece of insulating material, places her hand on top of a Van de Graaff generator, and then turns on the generator. Does she receive a shock?

10. In fair weather, there is an electric field at the surface of the Earth, pointing down into the ground. What is the sign of the electric charge on the ground in this situation?

11. A charged comb often attracts small bits of dry paper that then fly away when they touch the comb. Explain why that occurs.

12. Why should a ground wire be connected to the metal support rod for a television antenna?

13. There are great similarities between electric and gravitational fields. A room can be electrically shielded so that there are no electric fields in the room by surrounding it with a conductor. Can a room be gravitationally shielded? Explain.

14. A spherical surface surrounds a point charge $q$. Describe what happens to the total flux through the surface if (a) the charge is tripled, (b) the volume of the sphere is doubled, (c) the surface is changed to a cube, (d) the charge is moved to another location inside the surface, and (e) the charge is moved outside the surface.

15. If more electric field lines leave a Gaussian surface than enter it, what can you conclude about the net charge enclosed by that surface?

16. A student who grew up in a tropical country and is studying in the United States may have no experience with static electricity sparks and shocks until his or her first American winter. Explain.

17. What happens when a charged insulator is placed near an uncharged metallic object? (a) They repel each other. (b) They attract each other. (c) They may attract or repel each other, depending on whether the charge on the insulator is positive or negative. (d) They exert no electrostatic force on each other. (e) The charged insulator always spontaneously discharges.

## PROBLEMS

ENHANCED WebAssign The problems in this chapter may be assigned online in Enhanced WebAssign.

**1.** denotes straightforward problem; **2.** denotes intermediate problem;

**3.** denotes challenging problem

**1.** denotes full solution available in *Student Solutions Manual/ Study Guide*

**1.** denotes problems most often assigned in Enhanced WebAssign

BIO denotes biomedical problems

GP denotes guided problems

M denotes Master It tutorial available in Enhanced WebAssign

Q|C denotes asking for quantitative and conceptual reasoning

S denotes symbolic reasoning problem

W denotes Watch It video solution available in Enhanced WebAssign

### 15.3 Coulomb's Law

1. Q|C A 7.50-nC charge is located 1.80 m from a 4.20-nC charge. (a) Find the magnitude of the electrostatic force that one particle exerts on the other. (b) Is the force attractive or repulsive?

2. A charged particle *A* exerts a force of 2.62 N to the right on charged particle *B* when the particles are 13.7 mm apart. Particle *B* moves straight away from *A* to make the distance between them 17.7 mm. What vector force does particle *B* then exert on *A*?

3. Q|C Two metal balls *A* and *B* of negligible radius are floating at rest on Space Station Freedom between two metal bulkheads, connected by a taut nonconducting thread of length 2.00 m. Ball *A* carries charge $q$, and ball *B* carries charge $2q$. Each ball is 1.00 m away from a bulkhead. (a) If the tension in the string is 2.50 N, what is the magnitude of $q$? (b) What happens to the system as time passes? Explain.

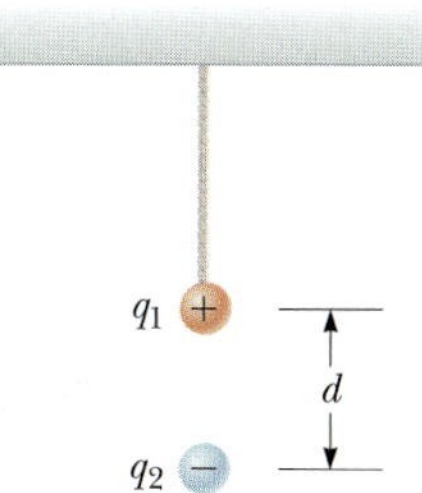


**Figure P15.4**

4. A small sphere of mass $m$ = 7.50 g and charge $q_1$ = 32.0 nC is attached to the end of a string and hangs vertically as in Figure P15.4. A second charge of equal mass and charge $q_2$ = −58.0 nC is located below the first charge a

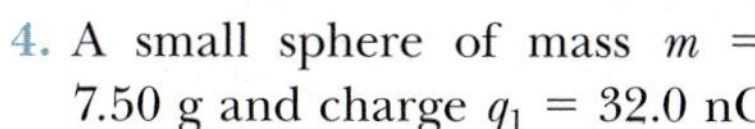

distance $d = 2.00$ cm below the first charge as in Figure P15.4. (a) Find the tension in the string. (b) If the string can withstand a maximum tension of 0.180 N, what is the smallest value $d$ can have before the string breaks?

**5.** The nucleus of $^8$Be, which consists of 4 protons and 4 neutrons, is very unstable and spontaneously breaks into two alpha particles (helium nuclei, each consisting of 2 protons and 2 neutrons). (a) What is the force between the two alpha particles when they are $5.00 \times 10^{-15}$ m apart, and (b) what is the initial magnitude of the acceleration of the alpha particles due to this force? Note that the mass of an alpha particle is 4.002 6 u.

**6.** BIO A molecule of DNA (deoxyribonucleic acid) is 2.17 $\mu$m long. The ends of the molecule become singly ionized: negative on one end, positive on the other. The helical molecule acts like a spring and compresses 1.00% upon becoming charged. Determine the effective spring constant of the molecule.

**7.** A small sphere of charge 0.800 $\mu$C hangs from the end of a spring as in Figure P15.7a. When another small sphere of charge $-0.600$ $\mu$C is held beneath the first sphere as in Figure P15.7b, the spring stretches by $d = 3.50$ cm from its original length and reaches a new equilibrium position with a separation between the charges of $r = 5.00$ cm. What is the force constant of the spring?

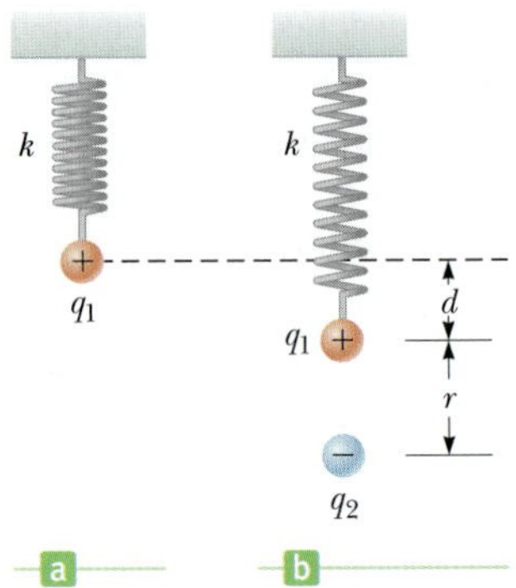


**Figure P15.7**

**8.** S Four point charges are at the corners of a square of side $a$ as shown in Figure P15.8. Determine the magnitude and direction of the resultant electric force on $q$, with $k_e$, $q$, and $a$ left in symbolic form.

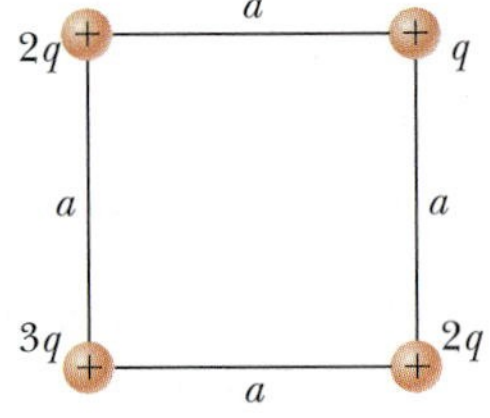


**Figure P15.8**

**9.** Two small identical conducting spheres are placed with their centers 0.30 m apart. One is given a charge of $12 \times 10^{-9}$ C, the other a charge of $-18 \times 10^{-9}$ C. (a) Find the electrostatic force exerted on one sphere by the other. (b) The spheres are connected by a conducting wire. Find the electrostatic force between the two after equilibrium is reached, where both spheres have the same charge.

**10.** Calculate the magnitude and direction of the Coulomb force on each of the three charges shown in Figure P15.10.

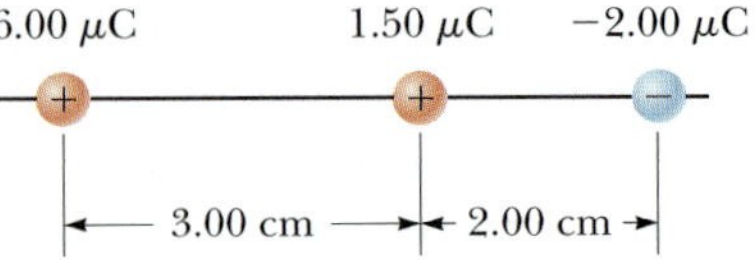


**Figure P15.10** Problems 10 and 18.

**11.** M Three charges are arranged as shown in Figure P15.11. Find the magnitude and direction of the electrostatic force on the charge at the origin.

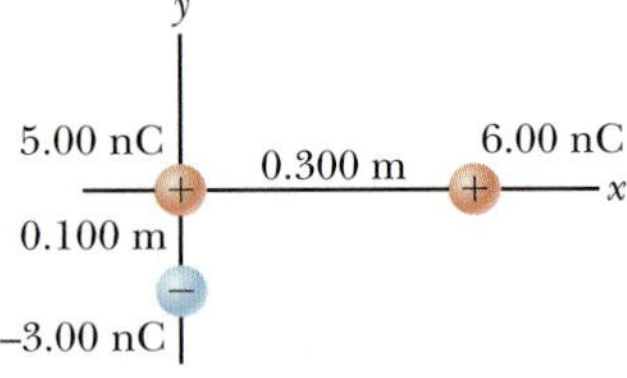


**Figure P15.11**

**12.** A positive charge $q_1 = 2.70$ $\mu$C on a frictionless horizontal surface is attached to a spring of force constant $k$ as in Figure P15.12. When a charge of $q_2 = -8.60$ $\mu$C is placed 9.50 cm away from the positive charge, the spring stretches by 5.00 mm, reducing the distance between charges to $d = 9.00$ cm. Find the value of $k$.

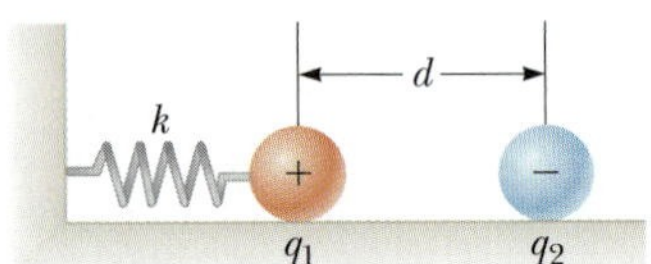


**Figure P15.12**

**13.** Three point charges are located at the corners of an equilateral triangle as in Figure P15.13. Find the magnitude and direction of the net electric force on the 2.00 $\mu$C charge.

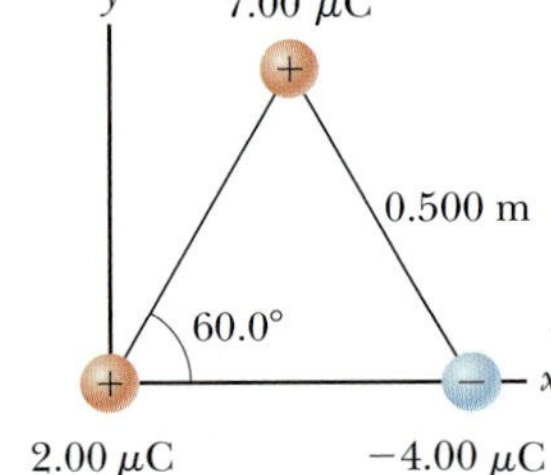


**Figure P15.13** Problems 13 and 24.

**14.** Two identical metal blocks resting on a frictionless horizontal surface are connected by a light metal spring having constant $k = 100$ N/m and unstretched length $L_i = 0.400$ m as in Figure P15.14a. A charge $Q$ is slowly placed on each block causing the spring to stretch to an equilibrium length $L = 0.500$ m as in Figure P15.14b. Determine the value of $Q$, modeling the blocks as charged particles.

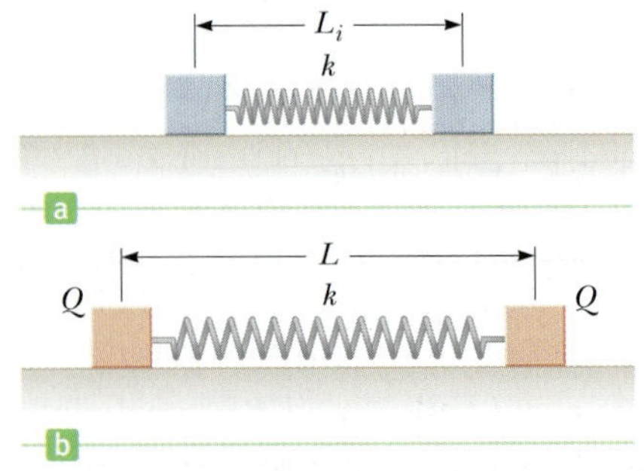


**Figure P15.14**

15. W Two small metallic spheres, each of mass $m = 0.20$ g, are suspended as pendulums by light strings from a common point as shown in Figure P15.15. The spheres are given the same electric charge, and it is found that they come to equilibrium when each string is at an angle of $\theta = 5.0°$ with the vertical. If each string has length $L = 30.0$ cm, what is the magnitude of the charge on each sphere?

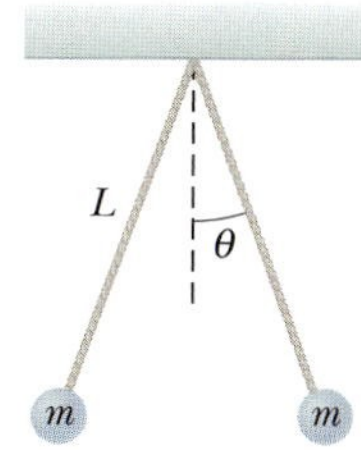

**Figure P15.15**

16. GP Particle $A$ of charge $3.00 \times 10^{-4}$ C is at the origin, particle $B$ of charge $-6.00 \times 10^{-4}$ C is at (4.00 m, 0) and particle $C$ of charge $1.00 \times 10^{-4}$ C is at (0, 3.00 m). (a) What is the $x$-component of the electric force exerted by $A$ on $C$? (b) What is the $y$-component of the force exerted by $A$ on $C$? (c) Find the magnitude of the force exerted by $B$ on $C$. (d) Calculate the $x$-component of the force exerted by $B$ on $C$. (e) Calculate the $y$-component of the force exerted by $B$ on $C$. (f) Sum the two $x$-components to obtain the resultant $x$-component of the electric force acting on $C$. (g) Repeat part (f) for the $y$-component. (h) Find the magnitude and direction of the resultant electric force acting on $C$.

## 15.4 The Electric Field

17. A small object of mass 3.80 g and charge $-18\ \mu$C is suspended motionless above the ground when immersed in a uniform electric field perpendicular to the ground. What is the magnitude and direction of the electric field?

18. (a) Determine the electric field strength at a point 1.00 cm to the left of the middle charge shown in Figure P15.10. (b) If a charge of $-2.00\ \mu$C is placed at this point, what are the magnitude and direction of the force on it?

19. An electric field of magnitude $5.25 \times 10^5$ N/C points due south at a certain location. Find the magnitude and direction of the force on a $-6.00\ \mu$C charge at this location.

20. W An electron is accelerated by a constant electric field of magnitude 300 N/C. (a) Find the acceleration of the electron. (b) Use the equations of motion with constant acceleration to find the electron's speed after $1.00 \times 10^{-8}$ s, assuming it starts from rest.

21. A small block of mass $m$ and charge $Q$ is placed on an insulated, frictionless, inclined plane of angle $\theta$ as in Figure P15.21. An electric field is applied parallel to the incline. (a) Find an expression for the magnitude of the electric field that enables the block to remain at rest. (b) If $m = 5.40$ g, $Q = -7.00\ \mu$C, and $\theta = 25.0°$, determine the magnitude and direction of the electric field that enables the block to remain at rest on the incline.

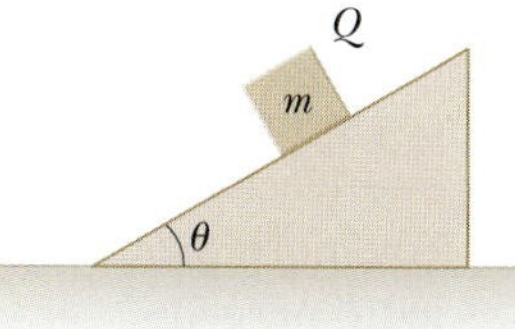

**Figure P15.21**

22. A small sphere of charge $q = +68\ \mu$C and mass $m = 5.8$ g is attached to a light string and placed in a uniform electric field $\vec{\mathbf{E}}$ that makes an angle $\theta = 37°$ with the horizontal. The opposite end of the string is attached to a wall and the sphere is in static equilibrium when the string is horizontal as in Figure P15.22. (a) Construct a free body diagram for the sphere. Find (b) the magnitude of the electric field and (c) the tension in the string.

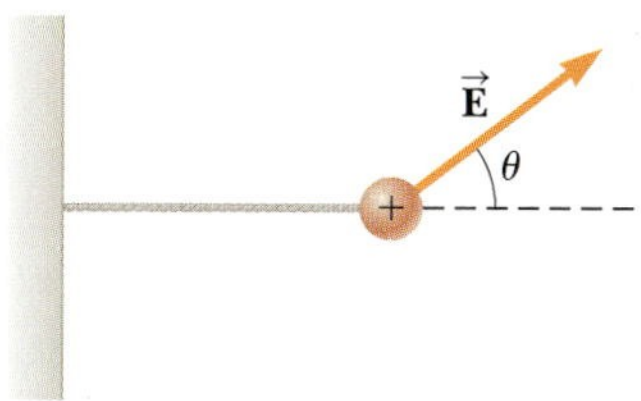

**Figure P15.22**

23. M A proton accelerates from rest in a uniform electric field of 640 N/C. At some later time, its speed is $1.20 \times 10^6$ m/s. (a) Find the magnitude of the acceleration of the proton. (b) How long does it take the proton to reach this speed? (c) How far has it moved in that interval? (d) What is its kinetic energy at the later time?

24. Q|C (a) Find the magnitude and direction of the electric field at the position of the 2.00 $\mu$C charge in Figure P15.13. (b) How would the electric field at that point be affected if the charge there were doubled? Would the magnitude of the electric force be affected?

25. S Two equal positive charges are at opposite corners of a trapezoid as in Figure P15.25. Find symbolic expressions for the components of the electric field at the point $P$.

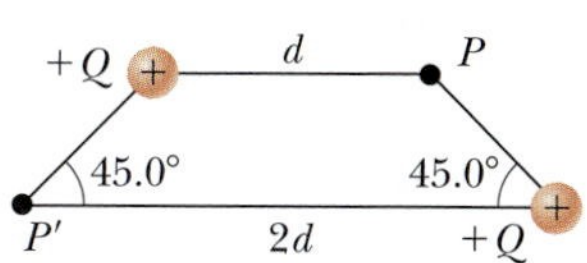

**Figure P15.25**

26. Three point charges are located on a circular arc as shown in Figure P15.26. (a) What is the total electric field at $P$, the center of the arc? (b) Find the electric force that would be exerted on a $-5.00$-nC charge placed at $P$.

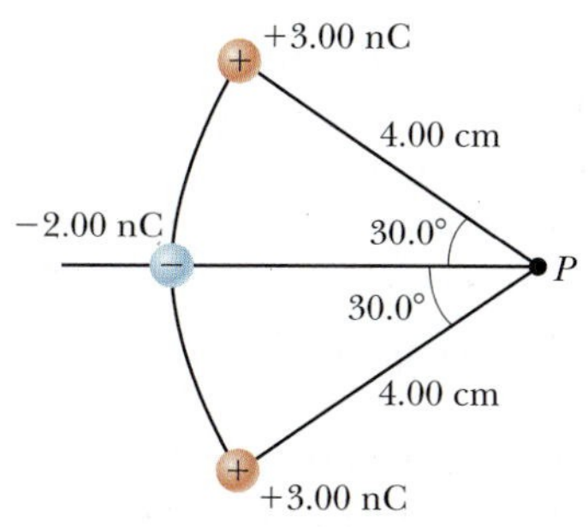

**Figure P15.26**

27. In Figure P15.27 determine the point (other than infinity) at which the total electric field is zero.

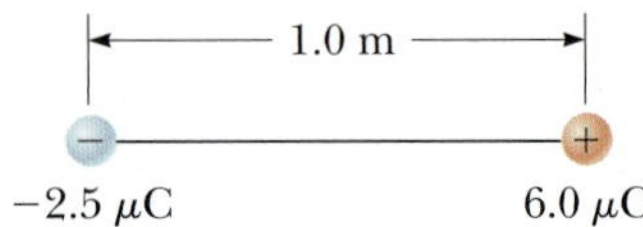


**Figure P15.27**

28. Three charges are at the corners of an equilateral triangle, as shown in Figure P15.28. Calculate the electric field at a point midway between the two charges on the $x$-axis.

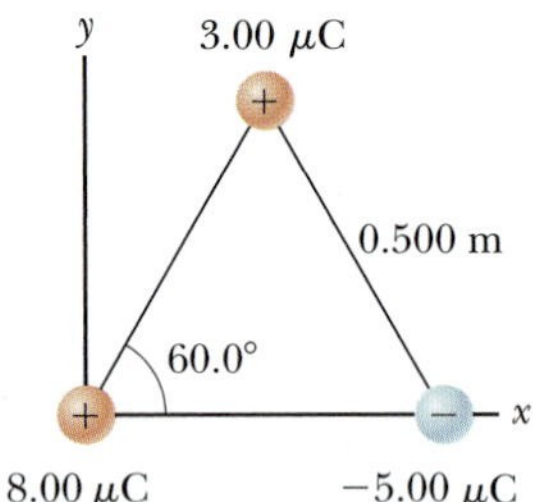


**Figure P15.28**

29. Three identical charges ($q = -5.0\ \mu$C) lie along a circle of radius 2.0 m at angles of 30°, 150°, and 270°, as shown in Figure P15.29. What is the resultant electric field at the center of the circle?

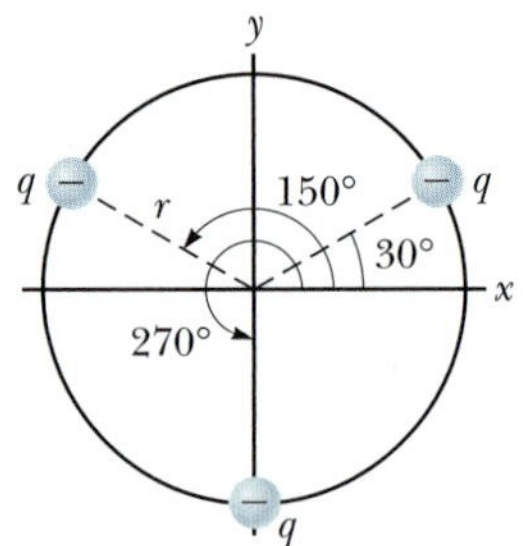


**Figure P15.29**

## 15.5 Electric Field Lines

## 15.6 Conductors in Electrostatic Equilibrium

30. Figure P15.30 shows the electric field lines for two point charges separated by a small distance. (a) Determine the ratio $q_1/q_2$. (b) What are the signs of $q_1$ and $q_2$?

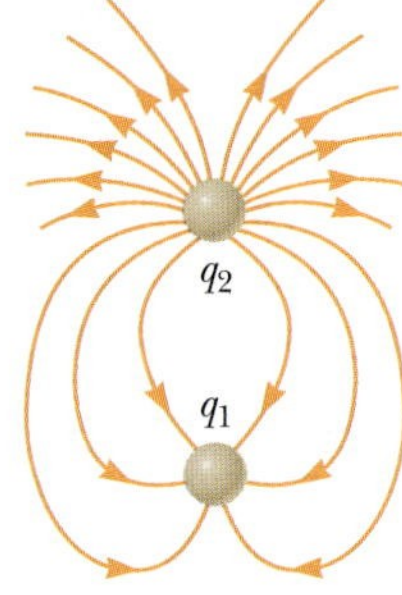


**Figure P15.30**

31. (a) Sketch the electric field lines around an isolated point charge $q > 0$. (b) Sketch the electric field pattern around an isolated negative point charge of magnitude $-2q$.

32. (a) Sketch the electric field pattern around two positive point charges of magnitude 1 $\mu$C placed close together. (b) Sketch the electric field pattern around two negative point charges of $-2\ \mu$C, placed close together. (c) Sketch the pattern around two point charges of $+1\ \mu$C and $-2\ \mu$C, placed close together.

33. Two point charges are a small distance apart. (a) Sketch the electric field lines for the two if one has a charge four times that of the other and both charges are positive. (b) Repeat for the case in which both charges are negative.

34. S Three equal positive charges are at the corners of an equilateral triangle of side $a$ as in Figure P15.34. Assume the three charges together create an electric field. (a) Sketch the electric field lines in the plane of the charges. (b) Find the location of one point (other than ∞) where the electric field is zero. What are (c) the magnitude and (d) the direction of the electric field at $P$ due to the two charges at the base?

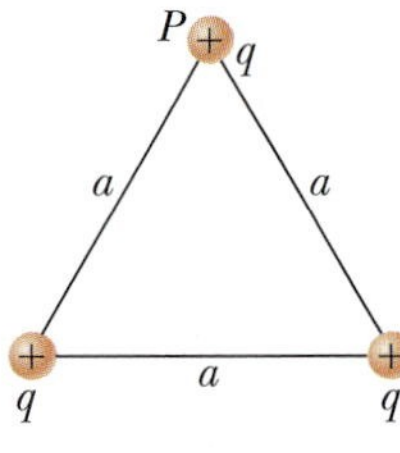


**Figure P15.34**

35. Refer to Figure 15.20. The charge lowered into the center of the hollow conductor has a magnitude of 5 $\mu$C. Find the magnitude and sign of the charge on the inside and outside of the hollow conductor when the charge is as shown in (a) Figure 15.20a, (b) Figure 15.20b, (c) Figure 15.20c, and (d) Figure 15.20d.

## 15.8 The Van de Graaff Generator

36. The dome of a Van de Graaff generator receives a charge of $2.0 \times 10^{-4}$ C. Find the strength of the electric field (a) inside the dome, (b) at the surface of the dome, assuming it has a radius of 1.0 m, and (c) 4.0 m from the center of the dome. *Hint:* See Section 15.6 to review properties of conductors in electrostatic equilibrium. Also, note that the points on the surface are outside a spherically symmetric charge distribution; the total charge may be considered to be located at the center of the sphere.

37. If the electric field strength in air exceeds $3.0 \times 10^6$ N/C, the air becomes a conductor. Using this fact, determine the maximum amount of charge that can be carried by a metal sphere 2.0 m in radius. (See the hint in Problem 36.)

38. In the Millikan oil-drop experiment illustrated in Figure 15.21, an atomizer (a sprayer with a fine nozzle) is used to introduce many tiny droplets of oil between two oppositely charged parallel metal plates. Some of the droplets pick up one or more excess electrons. The charge on the plates is adjusted so that the electric force on the excess electrons exactly balances the weight of the droplet. The idea is to look for a droplet that has the smallest electric force and assume it has only one excess electron. This strategy lets the observer measure the charge on the electron. Suppose we are using an electric field of $3 \times 10^4$ N/C. The charge on one electron is about $1.6 \times 10^{-19}$ C. Estimate the radius of an oil drop of density 858 kg/m$^3$ for which its weight could be balanced by the electric force of this field on one electron. (Problem 38 is courtesy of E. F. Redish. For more problems of this type, visit www.physics.umd.edu/perg/.)

39. W A Van de Graaff generator is charged so that a proton at its surface accelerates radially outward at $1.52 \times 10^{12}$ m/s$^2$. Find (a) the magnitude of the electric force on the proton at that instant and (b) the

magnitude and direction of the electric field at the surface of the generator.

## 15.9 Electric Flux and Gauss's Law

**40.** A uniform electric field of magnitude $E = 435$ N/C makes an angle of $\theta = 65.0°$ with a plane surface of area $A = 3.50$ m$^2$ as in Figure P15.40. Find the electric flux through this surface.

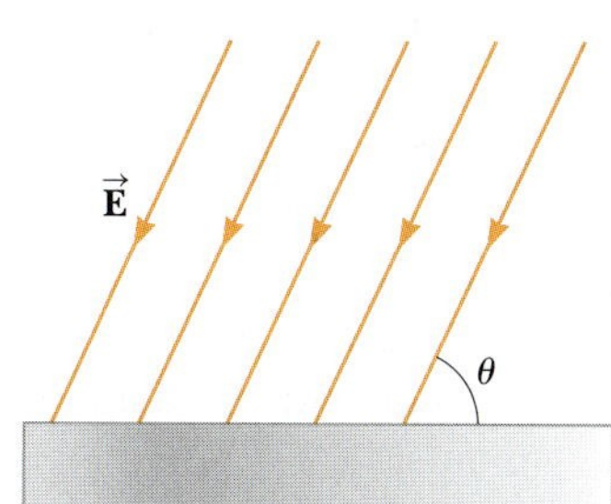


**Figure P15.40**

**41.** M An electric field of intensity 3.50 kN/C is applied along the $x$-axis. Calculate the electric flux through a rectangular plane 0.350 m wide and 0.700 m long if (a) the plane is parallel to the $yz$-plane, (b) the plane is parallel to the $xy$-plane, and (c) the plane contains the $y$-axis and its normal makes an angle of 40.0° with the $x$-axis.

**42.** Q|C The electric field everywhere on the surface of a charged sphere of radius 0.230 m has a magnitude of 575 N/C and points radially outward from the center of the sphere. (a) What is the net charge on the sphere? (b) What can you conclude about the nature and distribution of charge inside the sphere?

**43.** S Four closed surfaces, $S_1$ through $S_4$, together with the charges $-2Q$, $Q$, and $-Q$, are sketched in Figure P15.43. (The colored lines are the intersections of the surfaces with the page.) Find the electric flux through each surface.

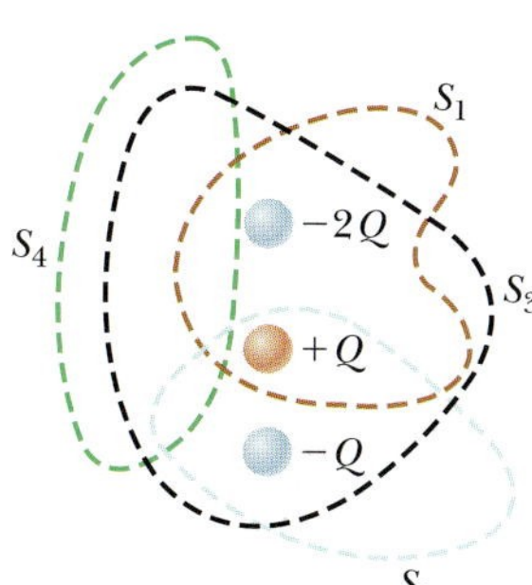


**Figure P15.43**

**44.** A charge $q = +5.80$ $\mu$C is located at the center of a regular tetrahedron (a four-sided surface) as in Figure P15.44. Find (a) the total electric flux through the tetrahedron and (b) the electric flux through one face of the tetrahedron.

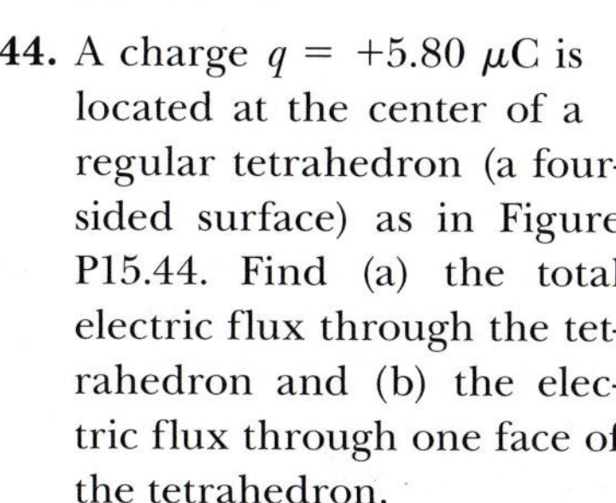


**Figure P15.44**

**45.** A point charge $q$ is located at the center of a spherical shell of radius $a$ that has a charge $-q$ uniformly distributed on its surface. Find the electric field (a) for all points outside the spherical shell and (b) for a point inside the shell a distance $r$ from the center.

**46.** Q|C A charge of $1.70 \times 10^2$ $\mu$C is at the center of a cube of edge 80.0 cm. No other charges are nearby. (a) Find the flux through the whole surface of the cube. (b) Find the flux through each face of the cube. (c) Would your answers to parts (a) or (b) change if the charge were not at the center? Explain.

**47.** Suppose the conducting spherical shell of Figure 15.29 carries a charge of 3.00 nC and that a charge of $-2.00$ nC is at the center of the sphere. If $a = 2.00$ m and $b = 2.40$ m, find the electric field at (a) $r = 1.50$ m, (b) $r = 2.20$ m, and (c) $r = 2.50$ m. (d) What is the charge distribution on the sphere?

**48.** S A very large nonconducting plate lying in the $xy$-plane carries a charge per unit area of $\sigma$. A second such plate located at $z = 2.00$ cm and oriented parallel to the $xy$-plane carries a charge per unit area of $-2\sigma$. Find the electric field (a) for $z < 0$, (b) $0 < z < 2.00$ cm, and (c) $z > 2.00$ cm.

## Additional Problems

**49.** In deep space two spheres each of radius 5.00 m are connected by a $3.00 \times 10^2$ m nonconducting cord. If a uniformly distributed charge of 35.0 mC resides on the surface of each sphere, calculate the tension in the cord.

**50.** Q|C A nonconducting, thin plane sheet of charge carries a uniform charge per unit area of 5.20 $\mu$C/m$^2$ as in Figure 15.30. (a) Find the electric field at a distance of 8.70 cm from the plate. (b) Explain whether your result changes as the distance from the sheet is varied.

**51.** W Three point charges are aligned along the $x$-axis as shown in Figure P15.51. Find the electric field at the position $x = +2.0$ m, $y = 0$.

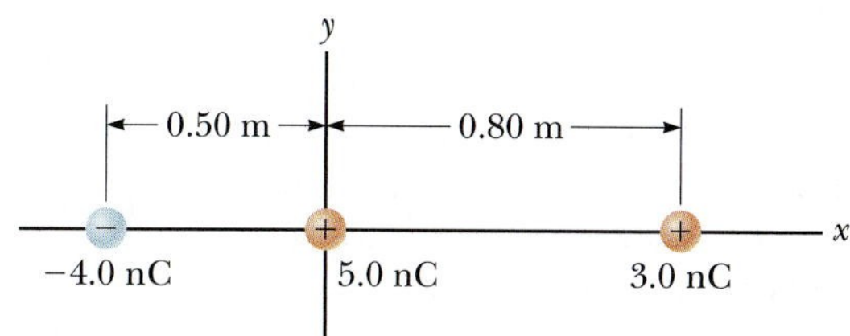


**Figure P15.51**

**52.** M A small plastic ball of mass $m = 2.00$ g is suspended by a string of length $L = 20.0$ cm in a uniform electric field, as shown in Figure P15.52. If the ball is in

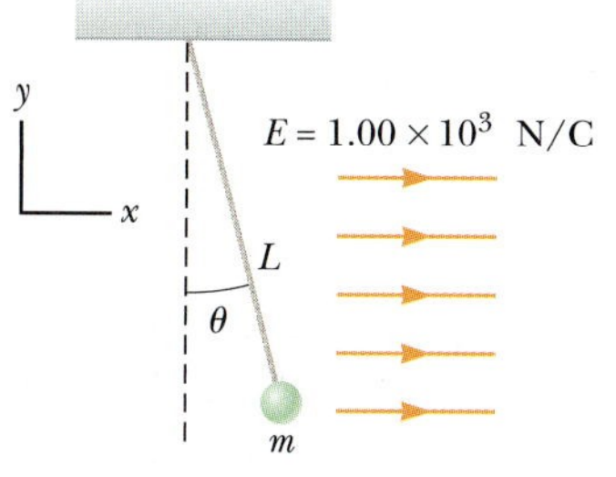


**Figure P15.52**

equilibrium when the string makes a $\theta = 15.0°$ angle with the vertical as indicated, what is the net charge on the ball?

53. (a) Two identical point charges $+q$ are located on the $y$-axis at $y = +a$ and $y = -a$. What is the electric field along the $x$-axis at $x = b$? (b) A circular ring of charge of radius $a$ has a total positive charge $Q$ distributed uniformly around it. The ring is in the $x = 0$ plane with its center at the origin. What is the electric field along the $x$-axis at $x = b$ due to the ring of charge? *Hint:* Consider the charge $Q$ to consist of many pairs of identical point charges positioned at the ends of diameters of the ring.

54. **S** The electrons in a particle beam each have a kinetic energy $K$. Find the magnitude of the electric field that will stop these electrons in a distance $d$, expressing the answer symbolically in terms of $K$, $e$, and $d$. Should the electric field point in the direction of the motion of the electron, or should it point in the opposite direction?

55. **S** A point charge $+2Q$ is at the origin and a point charge $-Q$ is located along the $x$-axis at $x = d$ as in Figure P15.55. Find symbolic expressions for the components of the net force on a third point charge $+Q$ located along the $y$-axis at $y = d$.

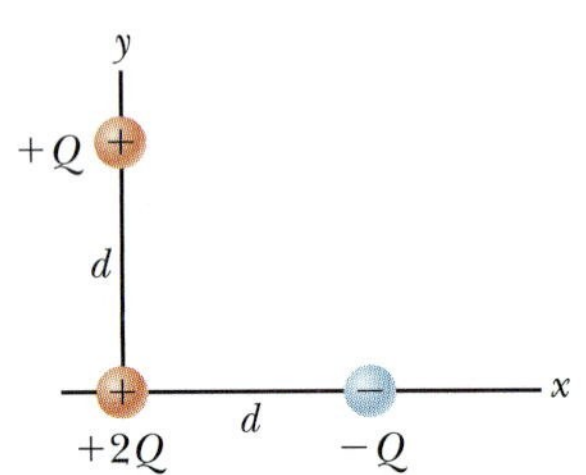

Figure P15.55

56. **Q|C** A 1.00-g cork ball having a positive charge of 2.00 mC is suspended vertically on a 0.500-m-long light string in the presence of a uniform downward-directed electric field of magnitude $E = 1.00 \times 10^5$ N/C as in Figure P15.56. If the ball is displaced slightly from the vertical, it oscillates like a simple pendulum. (a) Determine the period of the ball's oscillation. (b) Should gravity be included in the calculation for part (a)? Explain.

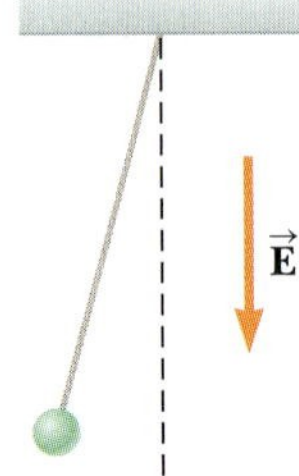

Figure P15.56

57. Two 2.0-g spheres are suspended by 10.0-cm-long light strings (Fig. P15.57). A uniform electric field is applied in the $x$-direction. If the spheres have charges of $-5.0 \times 10^{-8}$ C and $+5.0 \times 10^{-8}$ C, determine the electric field intensity that enables the spheres to be in equilibrium at $\theta = 10°$.

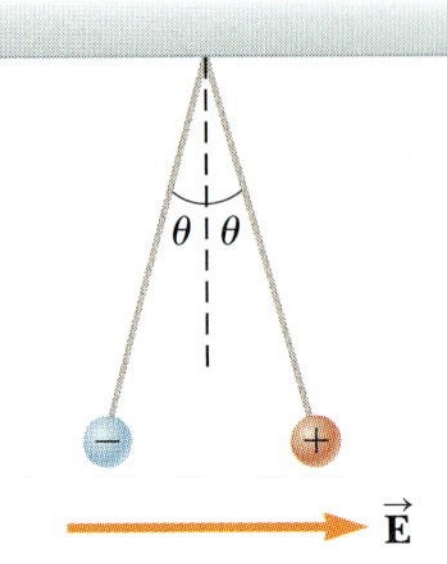

Figure P15.57

58. **Q|C** A point charge of magnitude 5.00 $\mu$C is at the origin of a coordinate system, and a charge of $-4.00$ $\mu$C is at the point $x = 1.00$ m. There is a point on the $x$-axis, at $x$ less than infinity, where the electric field goes to zero. (a) Show by conceptual arguments that this point cannot be located between the charges. (b) Show by conceptual arguments that the point cannot be at any location between $x = 0$ and negative infinity. (c) Show by conceptual arguments that the point must be between $x = 1.00$ m and $x =$ positive infinity. (d) Use the values given to find the point and show that it is consistent with your conceptual argument.

59. Two hard rubber spheres, each of mass $m = 15.0$ g, are rubbed with fur on a dry day and are then suspended with two insulating strings of length $L = 5.00$ cm whose support points are a distance $d = 3.00$ cm from each other as shown in Figure P15.59. During the rubbing process, one sphere receives exactly twice the charge of the other. They are observed to hang at equilibrium, each at an angle of $\theta = 10.0°$ with the vertical. Find the amount of charge on each sphere.

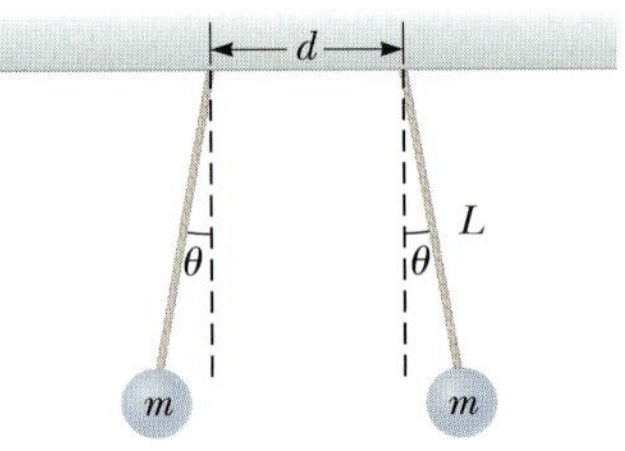

Figure P15.59

60. Two small beads having positive charges $q_1 = 3q$ and $q_2 = q$ are fixed at the opposite ends of a horizontal insulating rod of length $d = 1.50$ m. The bead with charge $q_1$ is at the origin. As shown in Figure P15.60, a third small charged bead is free to slide on the rod. At what position $x$ is the third bead in equilibrium?

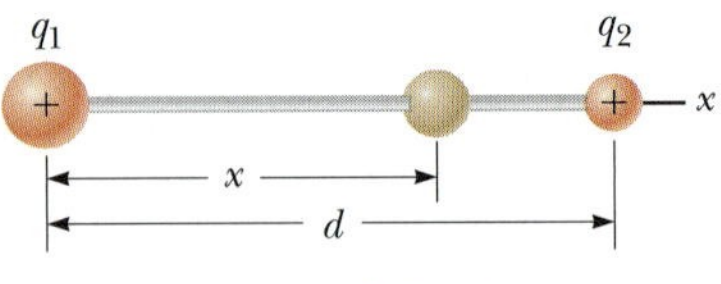

Figure P15.60

61. **M** A solid conducting sphere of radius 2.00 cm has a charge of 8.00 $\mu$C. A conducting spherical shell of inner radius 4.00 cm and outer radius 5.00 cm is concentric with the solid sphere and has a charge of $-4.00$ $\mu$C. Find the electric field at (a) $r = 1.00$ cm, (b) $r = 3.00$ cm, (c) $r = 4.50$ cm, and (d) $r = 7.00$ cm from the center of this charge configuration.

62. Three identical point charges, each of mass $m = 0.100$ kg, hang from three strings, as shown in Figure P15.62. If the lengths of the left and right strings are

each $L = 30.0$ cm and if the angle $\theta$ is 45.0°, determine the value of $q$.

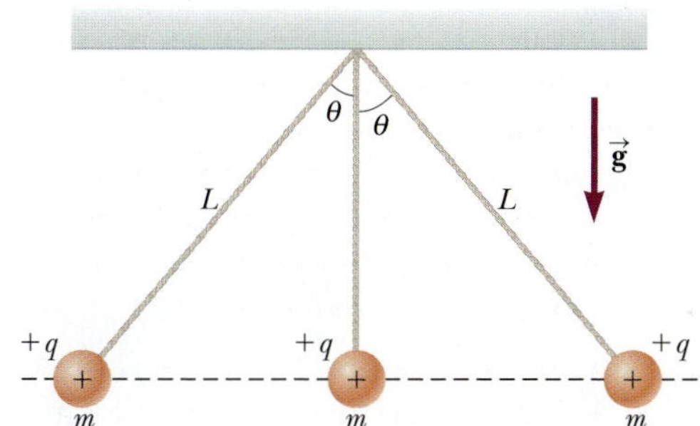


Figure P15.62

63. Each of the electrons in a particle beam has a kinetic energy of $1.60 \times 10^{-17}$ J. (a) What is the magnitude of the uniform electric field (pointing in the direction of the electrons' movement) that will stop these electrons in a distance of 10.0 cm? (b) How long will it take to stop the electrons? (c) After the electrons stop, what will they do? Explain.

64. Protons are projected with an initial speed $v_0 = 9\,550$ m/s into a region where a uniform electric field of magnitude $E = 720$ N/C is present (Fig. P15.64). The protons are to hit a target that lies a horizontal distance of 1.27 mm from the point where the protons are launched. Find (a) the two projection angles $\theta$ that will result in a hit and (b) the total duration of flight for each of the two trajectories.

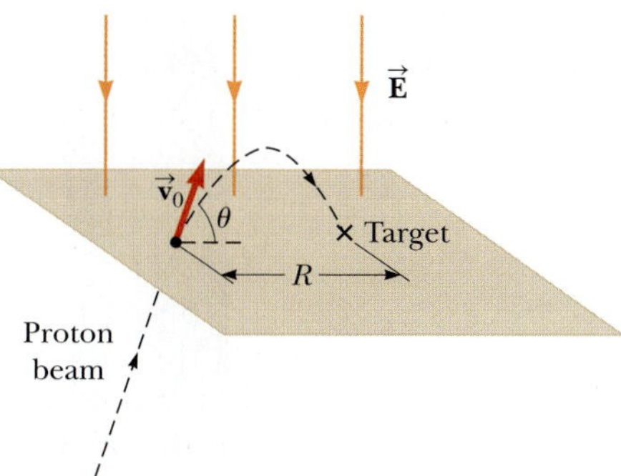


Figure P15.64

The only effective treatment for a patient with a heart in ventricular fibrillation, a spastic quivering of the heart muscle that is fatal in minutes, is an electrical shock delivered by a defibrillator. A capacitor in the defibrillator stores a large charge at high voltage and delivers it rapidly, shocking the heart and restoring a normal heart beat.

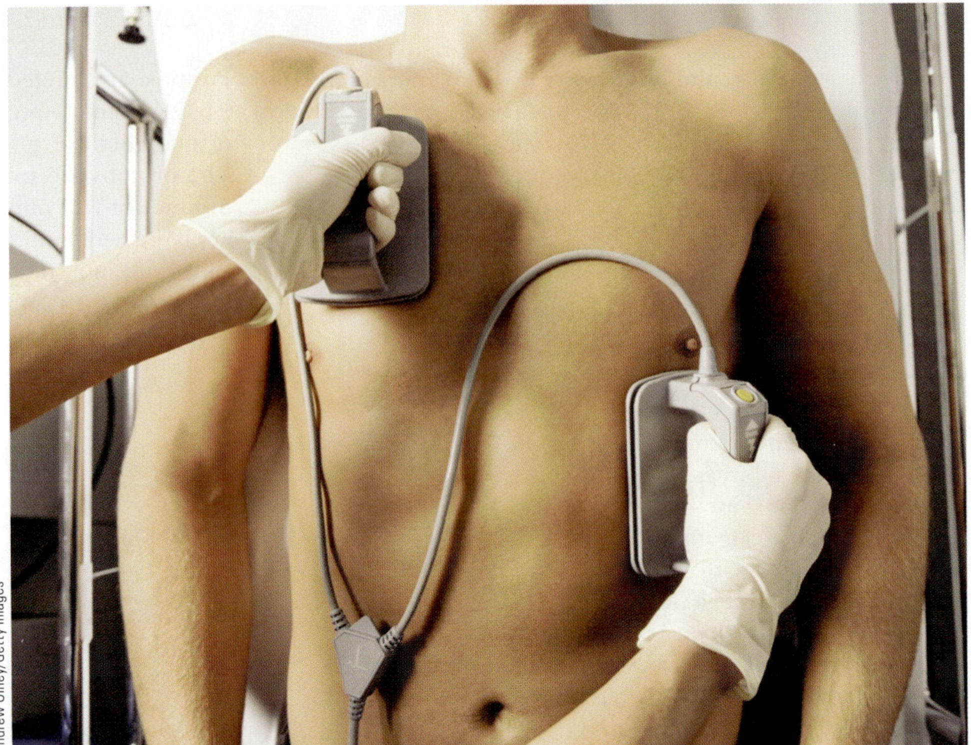

Andrew Olney/Getty Images

# 16 Electrical Energy and Capacitance



The concept of potential energy was first introduced in Chapter 5 in connection with the conservative forces of gravity and springs. By using the principle of conservation of energy, we were often able to avoid working directly with forces when solving problems. Here we learn that the potential energy concept is also useful in the study of electricity. Because the Coulomb force is conservative, we can define an electric potential energy corresponding to that force. In addition, we define an electric potential—the potential energy per unit charge—corresponding to the electric field.

With the concept of electric potential in hand, we can begin to understand electric circuits, starting with an investigation of common circuit elements called capacitors. These simple devices store electrical energy and have found uses virtually everywhere, from etched circuits on a microchip to the creation of enormous bursts of power in fusion experiments.

## 16.1 Electric Potential Energy and Electric Potential

**LEARNING OBJECTIVES**

1. Define electric potential energy difference for a constant electric field in terms of the work done by the field.
2. Contrast the concept of electric potential energy with that of electric potential.
3. Apply the work-energy theorem to systems involving electric potential energy and electric potential.

Electric potential energy and electric potential are closely related concepts. The electric potential turns out to be just the electric potential energy per unit charge. This relationship is similar to that between electric force and the electric field, which is the electric force per unit charge.

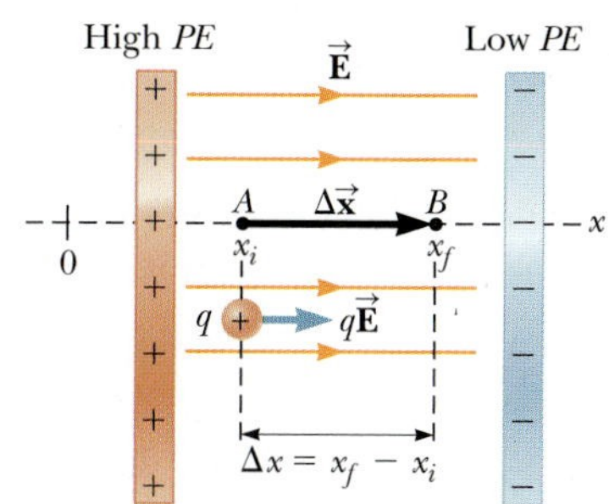


**Figure 16.1** When a charge $q$ moves in a uniform electric field $\vec{\mathbf{E}}$ from point $A$ to point $B$, the work done on the charge by the electric force is $qE_x\,\Delta x$.

## Work and Electric Potential Energy

Recall from Chapter 5 that the work done by a conservative force $\vec{\mathbf{F}}$ on an object depends only on the initial and final positions of the object and not on the path taken between those two points. This, in turn, means that a potential energy function $PE$ exists. As we have seen, potential energy is a scalar quantity with the change in potential energy equal by definition to the negative of the work done by the conservative force: $\Delta PE = PE_f - PE_i = -W_F$.

Both the Coulomb force law and the universal law of gravity are proportional to $1/r^2$. Because they have the same mathematical form and because the gravity force is conservative, it follows that **the Coulomb force is also conservative.** As with gravity, an electrical potential energy function can be associated with this force.

To make these ideas more quantitative, imagine a small positive charge placed at point $A$ in a *uniform* electric field $\vec{\mathbf{E}}$, as in Figure 16.1. For simplicity, we first consider only constant electric fields and charges that move parallel to that field in one dimension (taken to be the $x$-axis). The electric field between equally and oppositely charged parallel plates is an example of a field that is approximately constant. (See Chapter 15.) As the charge moves from point $A$ to point $B$ under the influence of the electric field $\vec{\mathbf{E}}$, the work done on the charge by the electric field is equal to the part of the electric force $q\vec{\mathbf{E}}$ acting parallel to the displacement times the displacement $\Delta x = x_f - x_i$:

$$W_{AB} = F_x\,\Delta x = qE_x(x_f - x_i)$$

In this expression $q$ is the charge and $E_x$ is the vector component of $\vec{\mathbf{E}}$ in the $x$-direction (*not* the magnitude of $\vec{\mathbf{E}}$). Unlike the magnitude of $\vec{\mathbf{E}}$, the component $E_x$ can be positive or negative, depending on the direction of $\vec{\mathbf{E}}$, although in Figure 16.1 $E_x$ is positive. Finally, note that the displacement, like $q$ and $E_x$, can also be either positive or negative, depending on the direction of the displacement.

The preceding expression for the work done by an electric field on a charge moving in one dimension is valid for both positive and negative charges and for constant electric fields pointing in *any* direction. When numbers are substituted with correct signs, the overall correct sign automatically results. In some books the expression $W = qEd$ is used, instead, where $E$ is the magnitude of the electric field and $d$ is the distance the particle travels. The weakness of this formulation is that it doesn't allow, mathematically, for negative electric work on positive charges, nor for positive electric work on negative charges! Nonetheless, the expression is easy to remember and useful for finding magnitudes: the magnitude of the work done by a constant electric field on a charge moving parallel to the field is always given by $|W| = |q|Ed$.

We can substitute our definition of electric work into the work–energy theorem (assume other forces are absent):

$$W = qE_x\,\Delta x = \Delta KE$$

The electric force is conservative, so the electric work depends only on the endpoints of the path, $A$ and $B$, not on the path taken. Therefore, as the charge accelerates to the right in Figure 16.1, it gains kinetic energy and loses an equal amount of potential energy. Recall from Chapter 5 that **the work done by a conservative force can be reinterpreted as the negative of the change in a potential energy**

**associated with that force**. This interpretation motivates the definition of the change in electric potential energy:

Change in electric potential energy ▶

The change in the electric potential energy, $\Delta PE$, of a system consisting of an object of charge $q$ moving through a displacement $\Delta x$ in a constant electric field $\vec{\mathbf{E}}$ is given by

$$\Delta PE = -W_{AB} = -qE_x \Delta x \quad [16.1]$$

where $E_x$ is the $x$-component of the electric field and $\Delta x = x_f - x_i$ is the displacement of the charge along the $x$-axis.

**SI unit: joule (J)**

Although potential energy can be defined for any electric field, **Equation 16.1 is valid only for the case of a uniform (i.e., constant) electric field, for a particle that undergoes a displacement along a given axis (here called the *x*-axis)**. Because the electric field is conservative, the change in potential energy doesn't depend on the path. Consequently, it's unimportant whether or not the charge remains on the axis at all times during the displacement: the change in potential energy will be the same. In subsequent sections we will examine situations in which the electric field is not uniform.

Electric and gravitational potential energy can be compared in Figure 16.2. In this figure the electric and gravitational fields are both directed downwards. We see that positive charge in an electric field acts very much like mass in a gravity field: a positive charge at point $A$ falls in the direction of the electric field, just as a positive mass falls in the direction of the gravity field. Let point $B$ be the zero point for potential energy in both Figures 16.2a and 16.2b. From conservation of energy, in falling from point $A$ to point $B$ the positive charge gains kinetic energy equal in magnitude to the loss of electric potential energy:

$$\Delta KE + \Delta PE_{\text{el}} = \Delta KE + (0 - |q|Ed) = 0 \quad \rightarrow \quad \Delta KE = |q|Ed$$

The absolute-value signs on $q$ are there only to make explicit that the charge is positive in this case. Similarly, the object in Figure 16.2b gains kinetic energy equal in magnitude to the loss of gravitational potential energy:

$$\Delta KE + \Delta PE_g = \Delta KE + (0 - mgd) = 0 \quad \rightarrow \quad \Delta KE = mgd$$

So for positive charges, electric potential energy works very much like gravitational potential energy. In both cases moving an object opposite the direction of the field results in a gain of potential energy, and upon release, the potential energy is converted to the object's kinetic energy.

Electric potential energy differs significantly from gravitational potential energy, however, in that there are two kinds of electrical charge—positive and negative—whereas gravity has only positive "gravitational charge" (i.e., mass). A negatively charged particle at rest at point $A$ in Figure 16.2a would have to be *pushed* down to point $B$. To see why, apply the work–energy theorem to a negative charge at rest at point $A$ and assumed to have some speed $v$ on arriving at point $B$:

$$W = \Delta KE + \Delta PE_{\text{el}} = \left(\tfrac{1}{2}mv^2 - 0\right) + \left[0 - (-|q|Ed)\right]$$

$$W = \tfrac{1}{2}mv^2 + |q|Ed$$

Notice that the negative charge, $-|q|$, unlike the positive charge, had a positive change in electric potential energy in moving from point $A$ to point $B$. If the negative charge has any speed at point $B$, the kinetic energy corresponding to that speed is also positive. Because both terms on the right-hand side of the work–energy equation are positive, there is no way of getting the negative charge from point $A$ to point $B$ without doing positive work $W$ on it. In fact, if the negative charge is simply released at point $A$, it will "fall" upwards against the direction of the field!

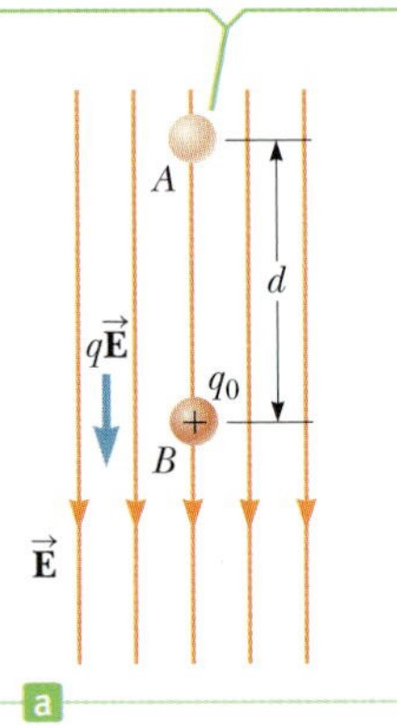
When a positive test charge moves from $A$ to $B$, the electric potential energy decreases.
$A$
$d$
$q\vec{\mathbf{E}}$
$q_0$
$B$
$\vec{\mathbf{E}}$
a

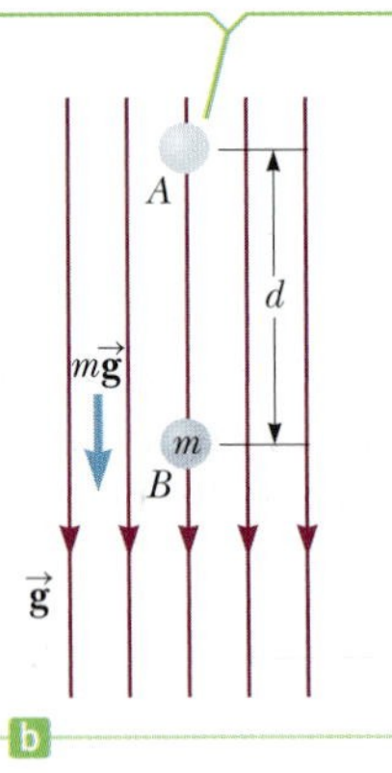
When an object with mass moves from $A$ to $B$, the gravitational potential energy decreases.
$A$
$d$
$m\vec{\mathbf{g}}$
$m$
$B$
$\vec{\mathbf{g}}$
b

**Figure 16.2** (a) When the electric field $\vec{\mathbf{E}}$ is directed downward, point $B$ is at a lower electric potential than point $A$. (b) An object of mass $m$ moves in the direction of the gravitational field $\vec{\mathbf{g}}$.

**Quick Quiz**

**16.1** If an electron is released from rest in a uniform electric field, does the electric potential energy of the charge–field system (a) increase, (b) decrease, or (c) remain the same?

## EXAMPLE 16.1 Potential Energy Differences in an Electric Field

**GOAL** Illustrate the concept of electric potential energy.

**PROBLEM** A proton is released from rest at $x = -2.00$ cm in a constant electric field with magnitude $1.50 \times 10^3$ N/C, pointing in the positive $x$-direction. **(a)** Calculate the change in the electric potential energy associated with the proton when it reaches $x = 5.00$ cm. **(b)** An electron is now fired in the same direction from the same position. What is the change in electric potential energy associated with the electron if it reaches $x = 12.0$ cm? **(c)** If the direction of the electric field is reversed and an electron is released from rest at $x = 3.00$ cm, by how much has the electric potential energy changed when the electron reaches $x = 7.00$ cm?

**STRATEGY** This problem requires a straightforward substitution of given values into the definition of electric potential energy, Equation 16.1.

**SOLUTION**

**(a)** Calculate the change in the electric potential energy associated with the proton.

Apply Equation 16.1:

$$\Delta PE = -qE_x\,\Delta x = -qE_x(x_f - x_i)$$
$$= -(1.60 \times 10^{-19}\text{ C})(1.50 \times 10^3\text{ N/C}) \times [0.050\,0\text{ m} - (-0.020\,0\text{ m})]$$
$$= -1.68 \times 10^{-17}\text{ J}$$

**(b)** Find the change in electric potential energy associated with an electron fired from $x = -0.020\,0$ m and reaching $x = 0.120$ m.

Apply Equation 16.1, but in this case note that the electric charge $q$ is negative:

$$\Delta PE = -qE_x\,\Delta x = -qE_x(x_f - x_i)$$
$$= -(-1.60 \times 10^{-19}\text{ C})(1.50 \times 10^3\text{ N/C}) \times [(0.120\text{ m} - (-0.020\,0\text{ m})]$$
$$= +3.36 \times 10^{-17}\text{ J}$$

**(c)** Find the change in potential energy associated with an electron traveling from $x = 3.00$ cm to $x = 7.00$ cm if the direction of the electric field is reversed.

Substitute, but now the electric field points in the negative $x$-direction, hence carries a minus sign:

$$\Delta PE = -qE_x\,\Delta x = -qE_x\,(x_f - x_i)$$
$$= -(-1.60 \times 10^{-19}\text{ C})(-1.50 \times 10^3\text{ N/C}) \times (0.070\text{ m} - 0.030\text{ m})$$
$$= -9.60 \times 10^{-18}\text{ J}$$

**REMARKS** Notice that the proton (actually the proton–field system) lost potential energy when it moved in the positive $x$-direction, whereas the electron gained potential energy when it moved in the same direction. Finding changes in potential energy with the field reversed was only a matter of supplying a minus sign, bringing the total number in this case to three! It's important not to drop any of the signs.

**QUESTION 16.1** True or False: When an electron is released from rest in a constant electric field, the change in the electric potential energy associated with the electron becomes more negative with time.

**EXERCISE 16.1** Find the change in electric potential energy associated with the electron in part (b) as it goes on from $x = 0.120$ m to $x = -0.180$ m. (Note that the electron must turn around and go back at some point. The location of the turning point is unimportant because changes in potential energy depend only on the endpoints of the path.)

**ANSWER** $-7.20 \times 10^{-17}$ J

## EXAMPLE 16.2 Dynamics of Charged Particles

**GOAL** Use electric potential energy in conservation of energy problems.

**PROBLEM** **(a)** Find the speed of the proton at $x = 0.050\,0$ m in part (a) of Example 16.1. **(b)** Find the initial speed of the electron (at $x = -2.00$ cm) in part (b) of Example 16.1 given that its speed has fallen by half when it reaches $x = 0.120$ m.

**STRATEGY** Apply conservation of energy, solving for the unknown speeds. Part (b) involves two equations: the conservation of energy equation and the condition $v_f = \frac{1}{2}v_i$ for the unknown initial and final speeds. The changes in electric potential energy have already been calculated in Example 16.1.

**SOLUTION**

**(a)** Calculate the proton's speed at $x = 0.050$ m.

Use conservation of energy, with an initial speed of zero:

$$\Delta KE + \Delta PE = 0 \quad \rightarrow \quad (\tfrac{1}{2}m_p v^2 - 0) + \Delta PE = 0$$

Solve for $v$ and substitute the change in potential energy found in Example 16.1a:

$$v^2 = -\frac{2}{m_p}\Delta PE$$

$$v = \sqrt{-\frac{2}{m_p}\Delta PE}$$

$$= \sqrt{-\frac{2}{(1.67 \times 10^{-27}\ \text{kg})}(-1.68 \times 10^{-17}\ \text{J})}$$

$$= \boxed{1.42 \times 10^5\ \text{m/s}}$$

**(b)** Find the electron's initial speed (at $x = -2.00$ cm) given that its speed has fallen by half at $x = 0.120$ m.

Apply conservation of energy once again, substituting expressions for the initial and final kinetic energies:

$$\Delta KE + \Delta PE = 0$$

$$(\tfrac{1}{2}m_e v_f^{\,2} - \tfrac{1}{2}m_e v_i^{\,2}) + \Delta PE = 0$$

Substitute the condition $v_f = \frac{1}{2}v_i$ and subtract the change in potential energy from both sides:

$$\tfrac{1}{2}m_e(\tfrac{1}{2}v_i)^2 - \tfrac{1}{2}m_e v_i^{\,2} = -\Delta PE$$

Combine terms and solve for $v_i$, the initial speed, and substitute the change in potential energy found in Example 16.1b:

$$-\tfrac{3}{8}m_e v_i^{\,2} = -\Delta PE$$

$$v_i = \sqrt{\frac{8\Delta PE}{3m_e}} = \sqrt{\frac{8(3.36 \times 10^{-17}\ \text{J})}{3(9.11 \times 10^{-31}\ \text{kg})}}$$

$$= \boxed{9.92 \times 10^6\ \text{m/s}}$$

**REMARKS** Although the changes in potential energy associated with the proton and electron were similar in magnitude, the effect on their speeds differed dramatically. The change in potential energy had a far greater effect on the much lighter electron than on the proton.

**QUESTION 16.2** True or False: If a proton and electron both move through the same displacement in an electric field, the change in potential energy associated with the proton must be equal in magnitude and opposite in sign to the change in potential energy associated with the electron.

**EXERCISE 16.2** Refer to Exercise 16.1. Find the electron's speed at $x = -0.180$ m. *Note*: Use the initial velocity from part (b) of Example 16.2.

**ANSWER** $1.35 \times 10^7$ m/s The answer is 4.5% of the speed of light.

## Electric Potential

In Chapter 15 it was convenient to define an electric field $\vec{\mathbf{E}}$ related to the electric force $\vec{\mathbf{F}} = q\vec{\mathbf{E}}$. In this way the properties of fixed collections of charges could be easily studied, and the force on any particle in the electric field could be obtained simply by multiplying by the particle's charge $q$. For the same reasons, it's useful to define an *electric potential difference* $\Delta V$ related to the potential energy by $\Delta PE = q\Delta V$:

The electric potential difference $\Delta V$ between points $A$ and $B$ is the change in electric potential energy as a charge $q$ moves from $A$ to $B$ divided by the charge $q$:

$$\Delta V = V_B - V_A = \frac{\Delta PE}{q} \qquad [16.2]$$

**SI unit: joule per coulomb, or volt (J/C, or V)**

◀ Potential difference between two points

This definition is completely general, although in many cases calculus would be required to compute the change in potential energy of the system. Because electric potential energy is a scalar quantity, **electric potential is also a scalar quantity**. From Equation 16.2, we see that electric potential difference is a measure of the change in electric potential energy per unit charge. Alternately, the electric potential difference is the work per unit charge that would have to be done by some force to move a charge from point $A$ to point $B$ in the electric field. The SI unit of electric potential is the joule per coulomb, called the volt (V). From the definition of that unit, 1 J of work must be done to move a 1-C charge between two points that are at a potential difference of 1 V. In the process of moving through a potential difference of 1 V, the 1-C charge gains 1 J of energy.

For the special case of a uniform electric field such as that between charged parallel plates, dividing Equation 16.1 by $q$ gives

$$\frac{\Delta PE}{q} = -E_x \Delta x$$

Comparing this equation with Equation 16.2, we find that

$$\Delta V = -E_x \Delta x \qquad [16.3]$$

Equation 16.3 shows that potential difference also has units of electric field times distance. It then follows that the SI unit of the electric field, the newton per coulomb, can also be expressed as volts per meter:

$$1 \text{ N/C} = 1 \text{ V/m}$$

Because Equation 16.3 is directly related to Equation 16.1, remember that it's valid only for the system consisting of a uniform electric field and a charge moving in one dimension.

Released from rest, positive charges accelerate spontaneously from regions of high potential to low potential. If a positive charge is given some initial velocity in the direction of high potential, it can move in that direction, but will slow and finally turn around, just like a ball tossed upwards in a gravity field. Negative charges do exactly the opposite: released from rest, they accelerate from regions of low potential toward regions of high potential. Work must be done on negative charges to make them go in the direction of lower electric potential.

**Tip 16.1 Potential and Potential Energy**

Electric potential is characteristic of the field only, independent of a test charge that may be placed in that field. On the other hand, potential energy is a characteristic of the charge-field system due to an interaction between the field and a charge placed in the field.

## Quick Quiz

**16.2** If a negatively charged particle is placed at rest in an electric potential field that increases in the positive $x$-direction, will the particle (a) accelerate in the positive $x$-direction, (b) accelerate in the negative $x$-direction, or (c) remain at rest?

**16.3** Figure 16.3 is a graph of an electric potential as a function of position. If a positively charged particle is placed at point $A$, what will its subsequent motion be? Will it (a) go to the right, (b) go to the left, (c) remain at point $A$, or (d) oscillate around point $B$?

(*Continued*)

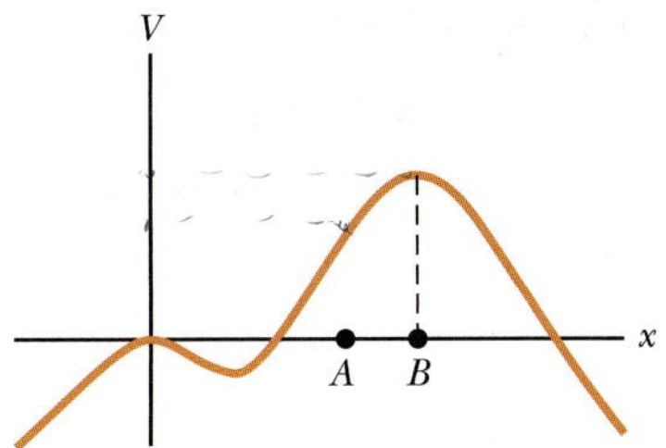


**Figure 16.3** (Quick Quizzes 16.3 and 16.4)

**16.4** If a negatively charged particle is placed at point $B$ in Figure 16.3 and given a very small kick to the right, what will its subsequent motion be? Will it (a) go to the right and not return, (b) go to the left, (c) remain at point $B$, or (d) oscillate around point $B$?

**APPLICATION**
Automobile Batteries

An application of potential difference is the 12-V battery found in an automobile. Such a battery maintains a potential difference across its terminals, with the positive terminal 12 V higher in potential than the negative terminal. In practice the negative terminal is usually connected to the metal body of the car, which can be considered to be at a potential of zero volts. The battery provides the electrical current necessary to operate headlights, a radio, power windows, motors, and so forth. Now consider a charge of +1 C, to be moved around a circuit that contains the battery and some of these external devices. As the charge is moved inside the battery from the negative terminal (at 0 V) to the positive terminal (at 12 V), the work done on the charge by the battery is 12 J. Every coulomb of positive charge that leaves the positive terminal of the battery carries an energy of 12 J. As the charge moves through the external circuit toward the negative terminal, it gives up its 12 J of electrical energy to the external devices. When the charge reaches the negative terminal, its electrical energy is zero again. At this point, the battery takes over and restores 12 J of energy to the charge as it is moved from the negative to the positive terminal, enabling it to make another transit of the circuit. The actual amount of charge that leaves the battery each second and traverses the circuit depends on the properties of the external devices, as seen in the next chapter.

## EXAMPLE 16.3 TV Tubes and Atom Smashers

**GOAL** Relate electric potential to an electric field and conservation of energy.

**PROBLEM** In atom smashers (also known as cyclotrons and linear accelerators) charged particles are accelerated in much the same way they are accelerated in TV tubes: through potential differences. Suppose a proton is injected at a speed of $1.00 \times 10^6$ m/s between two plates 5.00 cm apart, as shown in Figure 16.4. The proton subsequently accelerates across the gap and exits through the opening. **(a)** What must the electric potential difference be if the exit speed is to be $3.00 \times 10^6$ m/s? **(b)** What is the electric field between the plates, assuming it's constant? The positive $x$-direction is to the right.

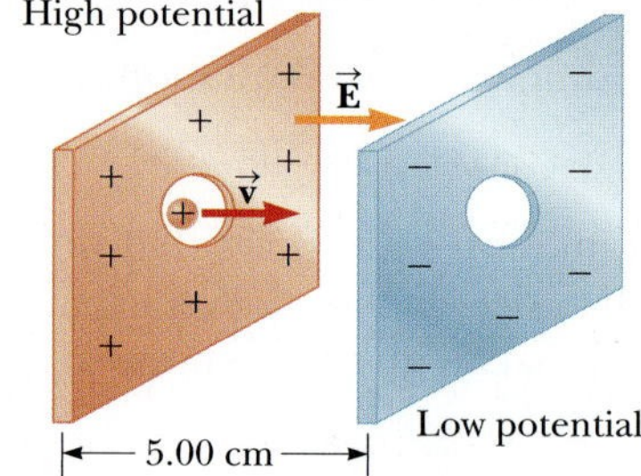


**Figure 16.4** (Example 16.3) A proton enters a cavity and accelerates from one charged plate toward the other in an electric field $\vec{E}$.

**STRATEGY** Use conservation of energy, writing the change in potential energy in terms of the change in electric potential, $\Delta V$, and solve for $\Delta V$. For part (b), solve Equation 16.3 for the electric field.

**SOLUTION**

**(a)** Find the electric potential yielding the desired exit speed of the proton.

Apply conservation of energy, writing the potential energy in terms of the electric potential:

$$\Delta KE + \Delta PE = \Delta KE + q\,\Delta V = 0$$

Solve the energy equation for the change in potential:

$$\Delta V = -\frac{\Delta KE}{q} = -\frac{\frac{1}{2}m_p{v_f}^2 - \frac{1}{2}m_p{v_i}^2}{q} = -\frac{m_p}{2q}({v_f}^2 - {v_i}^2)$$

Substitute the given values, obtaining the necessary potential difference:

$$\Delta V = -\frac{(1.67 \times 10^{-27}\text{ kg})}{2(1.60 \times 10^{-19}\text{ C})}[(3.00 \times 10^6\text{ m/s})^2 - (1.00 \times 10^6\text{ m/s})^2]$$

$$\Delta V = \boxed{-4.18 \times 10^4\text{ V}}$$

**(b)** What electric field must exist between the plates?

Solve Equation 16.3 for the electric field and substitute:

$$E = -\frac{\Delta V}{\Delta x} = \frac{4.18 \times 10^4\text{ V}}{0.050\,0\text{ m}} = \boxed{8.36 \times 10^5\text{ N/C}}$$

**REMARKS** Systems of such cavities, consisting of alternating positive and negative plates, are used to accelerate charged particles to high speed before smashing them into targets. To prevent a slowing of, say, a positively charged particle after it passes through the negative plate of one cavity and enters the next, the charges on the plates are reversed. Otherwise, the particle would be traveling from the negative plate to a positive plate in the second cavity, and the kinetic energy gained in the previous cavity would be lost in the second.

**QUESTION 16.3** True or False: A more massive particle gains less energy in traversing a given potential difference than does a lighter particle having the same charge.

**EXERCISE 16.3** Suppose electrons in a TV tube are accelerated through a potential difference of $2.00 \times 10^4$ V from the heated cathode (negative electrode), where they are produced, toward the screen, which also serves as the anode (positive electrode), 25.0 cm away. (a) At what speed would the electrons impact the phosphors on the screen? Assume they accelerate from rest and ignore relativistic effects (Chapter 26). (b) What's the magnitude of the electric field, if it is assumed constant?

**ANSWERS** (a) $8.38 \times 10^7$ m/s (b) $8.00 \times 10^4$ V/m

## 16.2 Electric Potential and Potential Energy Due to Point Charges

**LEARNING OBJECTIVES**

1. Define the electric potential of a point charge and the potential energy of a pair of point charges.
2. Apply electric potential and electric potential energy to systems of charged particles.

In electric circuits a point of zero electric potential is often defined by grounding (connecting to the Earth) some point in the circuit. For example, if the negative terminal of a 12-V battery were connected to ground, it would be considered to have a potential of zero, whereas the positive terminal would have a potential of +12 V. The potential difference created by the battery, however, is only locally defined. In this section we describe the electric potential of a point charge, which is defined throughout space.

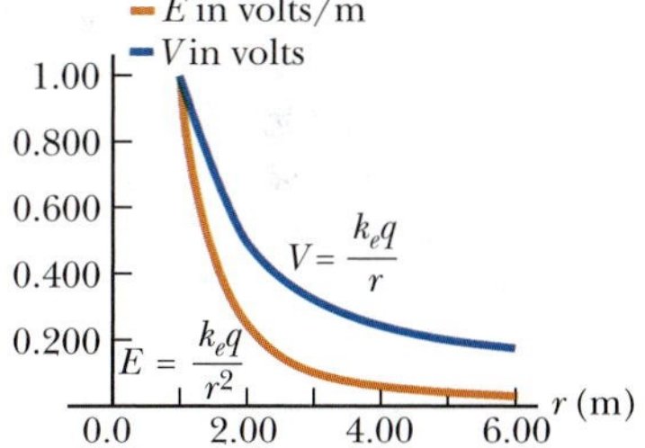


**Figure 16.5** Electric field and electric potential versus distance from a point charge of $1.11 \times 10^{-10}$ C. Note that $V$ is proportional to $1/r$, whereas $E$ is proportional to $1/r^2$.

The electric field of a point charge extends throughout space, so its electric potential does, also. The zero point of electric potential could be taken anywhere, but is usually taken to be an infinite distance from the charge, far from its influence and the influence of any other charges. With this choice, the methods of calculus can be used to show that the electric potential created by a point charge $q$ at any distance $r$ from the charge is given by

◀ Electric potential created by a point charge

$$V = k_e \frac{q}{r} \quad [16.4]$$

Equation 16.4 shows that the electric potential, or work per unit charge, required to move a positive test charge in from infinity to a distance $r$ from a positive point charge $q$ increases as the test charge moves closer to $q$. A plot of Equation 16.4 in Figure 16.5 shows that the potential associated with a point charge decreases as $1/r$ with increasing $r$, in contrast to the magnitude of the charge's electric field, which decreases as $1/r^2$.

◀ Superposition principle

The electric potential of two or more charges is obtained by applying the **superposition principle: the total electric potential at some point $P$ due to several point charges is the algebraic sum of the electric potentials due to the individual charges.** This method is similar to the one used in Chapter 15 to find the resultant electric field at a point in space. Unlike electric field superposition, which involves a sum of vectors, the superposition of electric potentials requires evaluating a sum

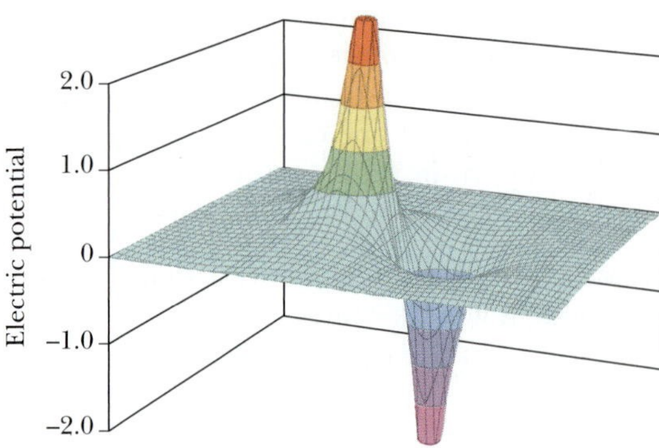


**Figure 16.6** The electric potential (in arbitrary units) in the plane containing an electric dipole. Potential is plotted in the vertical dimension.

of scalars. As a result, it's much easier to evaluate the electric potential at some point due to several charges than to evaluate the electric field, which is a vector quantity.

Figure 16.6 is a computer-generated plot of the electric potential associated with an electric dipole, which consists of two charges of equal magnitude but opposite in sign. The charges lie in a horizontal plane at the center of the potential spikes. The value of the potential is plotted in the vertical dimension. The computer program has added the potential of each charge to arrive at total values of the potential.

Just as in the case of constant electric fields, there is a relationship between electric potential and electric potential energy. If $V_1$ is the electric potential due to charge $q_1$ at a point $P$ (Fig. 16.7a) the work required to bring charge $q_2$ from infinity to $P$ without acceleration is $q_2V_1$. By definition, this work equals the potential energy $PE$ of the two-particle system when the particles are separated by a distance $r$ (Fig. 16.7b).

We can therefore express the electrical potential energy of the *pair* of charges as

Potential energy of a pair of charges ▶

$$PE = q_2V_1 = k_e\frac{q_1q_2}{r} \qquad [16.5]$$

If the charges are of the *same* sign, $PE$ is positive. Because like charges repel, positive work must be done on the system by an external agent to force the two charges near each other. Conversely, if the charges are of *opposite* sign, the force is attractive and $PE$ is negative. This means that negative work must be done to prevent unlike charges from accelerating toward each other as they are brought close together.

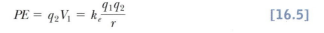

P $V_1 = \frac{k_eq_1}{r}$

r

+

$q_1$

a

+

$q_2$

r

$PE = \frac{k_eq_1q_2}{r}$

+

$q_1$

b

**Figure 16.7** (a) The electric potential $V_1$ at $P$ due to the point charge $q_1$ is $V_1 = k_eq_1/r$. (b) If a second charge, $q_2$, is brought from infinity to $P$, the potential energy of the pair is $PE = k_eq_1q_2/r$.

■ *Quick Quiz*

**16.5** Consider a collection of charges in a given region and suppose all other charges are distant and have a negligible effect. Further, the electric potential is taken to be zero at infinity. If the electric potential at a given point in the region is zero, which of the following statements must be true? (a) The electric field is zero at that point. (b) The electric potential energy is a minimum at that point. (c) There is no net charge in the region. (d) Some charges in the region are positive, and some are negative. (e) The charges have the same sign and are symmetrically arranged around the given point.

**16.6** A spherical balloon contains a positively charged particle at its center. As the balloon is inflated to a larger volume while the charged particle remains at the center, which of the following are true? (a) The electric potential at the surface of the balloon increases. (b) The magnitude of the electric field at the surface of the balloon increases. (c) The electric flux through the balloon remains the same. (d) None of these.

## PROBLEM-SOLVING STRATEGY

### Electric Potential

1. Draw a diagram of all charges and circle the point of interest.
2. Calculate the distance from each charge to the point of interest, labeling it on the diagram.
3. For each charge $q$, calculate the scalar quantity $V = \frac{k_e q}{r}$. *The sign of each charge must be included in your calculations!*
4. Sum all the numbers found in the previous step, obtaining the electric potential at the point of interest.

## EXAMPLE 16.4 Finding the Electric Potential

**GOAL** Calculate the electric potential due to a collection of point charges.

**PROBLEM** A 5.00-$\mu$C point charge is at the origin, and a point charge $q_2 = -2.00\ \mu$C is on the $x$-axis at (3.00, 0) m, as in Figure 16.8. **(a)** If the electric potential is taken to be zero at infinity, find the electric potential due to these charges at point $P$ with coordinates (0, 4.00) m. **(b)** How much work is required to bring a third point charge of 4.00 $\mu$C from infinity to $P$?

**STRATEGY** For part (a), the electric potential at $P$ due to each charge can be calculated from $V = k_e q/r$. The electric potential at $P$ is the sum of these two quantities. For part (b), use the work–energy theorem, together with Equation 16.5, recalling that the potential at infinity is taken to be zero.

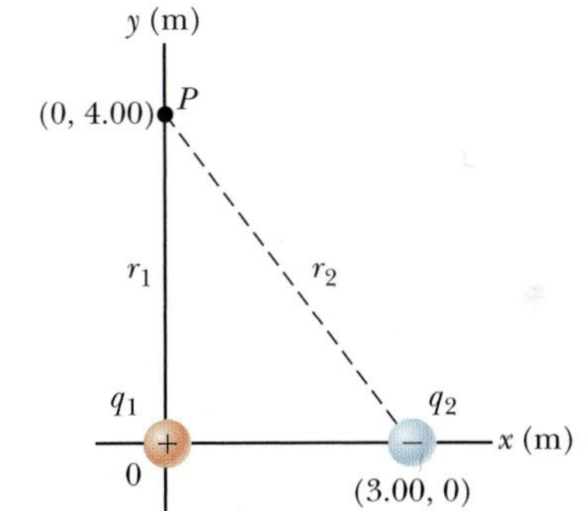


**Figure 16.8** (Example 16.4) The electric potential at point $P$ due to the point charges $q_1$ and $q_2$ is the algebraic sum of the potentials due to the individual charges.

**SOLUTION**

**(a)** Find the electric potential at point $P$.

Calculate the electric potential at $P$ due to the 5.00-$\mu$C charge:

$$V_1 = k_e \frac{q_1}{r_1} = \left(8.99 \times 10^9 \frac{\text{N}\cdot\text{m}^2}{\text{C}^2}\right)\left(\frac{5.00 \times 10^{-6}\ \text{C}}{4.00\ \text{m}}\right)$$
$$= 1.12 \times 10^4\ \text{V}$$

Find the electric potential at $P$ due to the $-2.00$-$\mu$C charge:

$$V_2 = k_e \frac{q_2}{r_2} = \left(8.99 \times 10^9 \frac{\text{N}\cdot\text{m}^2}{\text{C}^2}\right)\left(\frac{-2.00 \times 10^{-6}\ \text{C}}{5.00\ \text{m}}\right)$$
$$= -0.360 \times 10^4\ \text{V}$$

Sum the two quantities to find the total electric potential at $P$:

$$V_P = V_1 + V_2 = 1.12 \times 10^4\ \text{V} + (-0.360 \times 10^4\ \text{V})$$
$$= 7.6 \times 10^3\ \text{V}$$

**(b)** Find the work needed to bring the 4.00-$\mu$C charge from infinity to $P$.

Apply the work–energy theorem, with Equation 16.5:

$$W = \Delta PE = q_3\,\Delta V = q_3(V_P - V_\infty)$$
$$= (4.00 \times 10^{-6}\ \text{C})(7.6 \times 10^3\ \text{V} - 0)$$
$$W = 3.0 \times 10^{-2}\ \text{J}$$

**REMARKS** Unlike the electric field, where vector addition is required, the electric potential due to more than one charge can be found with ordinary addition of scalars. Further, notice that the work required to move the charge is equal to the change in electric potential energy. The sum of the work done moving the particle plus the work done by the electric field is zero ($W_{\text{other}} + W_{\text{electric}} = 0$) because the particle starts and ends at rest. Therefore, $W_{\text{other}} = -W_{\text{electric}} = \Delta U_{\text{electric}} = q\,\Delta V$.

*(Continued)*

**QUESTION 16.4** If $q_2$ were moved to the right, what would happen to the electric potential $V_p$ at point $P$? (a) It would increase. (b) It would decrease. (c) It would remain the same.

**EXERCISE 16.4** Suppose a charge of $-2.00\ \mu C$ is at the origin and a charge of $3.00\ \mu C$ is at the point (0, 3.00) m. (a) Find the electric potential at (4.00, 0) m, assuming the electric potential is zero at infinity, and (b) find the work necessary to bring a $4.00\ \mu C$ charge from infinity to the point (4.00, 0) m.

**ANSWERS** (a) $8.99 \times 10^2$ V (b) $3.60 \times 10^{-3}$ J

## EXAMPLE 16.5 Electric Potential Energy and Dynamics

**GOAL** Apply conservation of energy and electrical potential energy to a configuration of charges.

**PROBLEM** Suppose three protons lie on the *x*-axis, at rest relative to one another at a given instant of time, as in Figure 16.9. If proton $q_3$ on the right is released while the others are held fixed in place, find a symbolic expression for the proton's speed at infinity and evaluate this speed when $r_0 = 2.00$ fm. (*Note:* 1 fm = $10^{-15}$ m.)

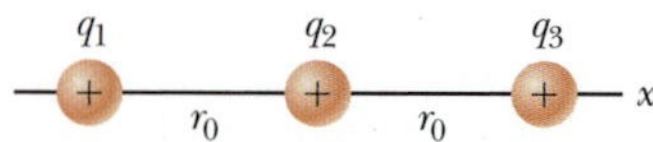


**Figure 16.9** (Example 16.5)

**STRATEGY** First calculate the initial electric potential energy associated with the system of three particles. There will be three terms, one for each interacting pair. Then calculate the final electric potential energy associated with the system when the proton on the right is arbitrarily far away. Because the electric potential energy falls off as $1/r$, two of the terms will vanish. Using conservation of energy then yields the speed of the particle in question.

**SOLUTION**

Calculate the electric potential energy associated with the initial configuration of charges:

$$PE_i = \frac{k_e q_1 q_2}{r_{12}} + \frac{k_e q_1 q_3}{r_{13}} + \frac{k_e q_2 q_3}{r_{23}} = \frac{k_e e^2}{r_0} + \frac{k_e e^2}{2r_0} + \frac{k_e e^2}{r_0}$$

Calculate the electric potential energy associated with the final configuration of charges:

$$PE_f = \frac{k_e q_1 q_2}{r_{12}} = \frac{k_e e^2}{r_0}$$

Write the conservation of energy equation:

$$\Delta KE + \Delta PE = KE_f - KE_i + PE_f - PE_i = 0$$

Substitute appropriate terms:

$$\tfrac{1}{2}m_3 v_3^{\ 2} - 0 + \frac{k_e e^2}{r_0} - \left(\frac{k_e e^2}{r_0} + \frac{k_e e^2}{2r_0} + \frac{k_e e^2}{r_0}\right) = 0$$

$$\tfrac{1}{2}m_3 v_3^{\ 2} - \left(\frac{k_e e^2}{2r_0} + \frac{k_e e^2}{r_0}\right) = 0$$

Solve for $v_3$ after combining the two remaining potential energy terms:

$$v_3 = \boxed{\sqrt{\frac{3k_e e^2}{m_3 r_0}}}$$

Evaluate taking $r_0 = 2.00$ fm:

$$v_3 = \sqrt{\frac{3(8.99 \times 10^9\ \text{N} \cdot \text{m}^2/\text{C}^2)(1.60 \times 10^{-19}\ \text{C})^2}{(1.67 \times 10^{-27}\ \text{kg})(2.00 \times 10^{-15}\ \text{m})}} = 1.44 \times 10^7\ \text{m/s}$$

**REMARKS** The difference in the initial and final potential energies yields the energy available for motion. This calculation is somewhat contrived because it would be difficult, although not impossible, to arrange such a configuration of protons; it could conceivably occur by chance inside a star.

**QUESTION 16.5** If a fourth proton were placed to the right of $q_3$, how many additional potential energy terms would have to be calculated in the initial configuration?

**EXERCISE 16.5** Starting from the initial configuration of three protons, suppose the end two particles are released simultaneously and the middle particle is fixed. Obtain a numerical answer for the speed of the two particles at infinity. (Note that their speeds, by symmetry, must be the same.)

**ANSWER** $1.31 \times 10^7$ m/s

# 16.3 Potentials and Charged Conductors

**LEARNING OBJECTIVES**

1. Discuss the electric potential of a perfect conductor.
2. Define the electron volt unit of energy.

The electric potential at all points on a charged conductor can be determined by combining Equations 16.1 and 16.2. From Equation 16.1, we see that the work done on a charge by electric forces is related to the change in electrical potential energy of the charge by

$$W = -\Delta PE$$

From Equation 16.2, we see that the change in electric potential energy between two points $A$ and $B$ is related to the potential difference between those points by

$$\Delta PE = q(V_B - V_A)$$

Combining these two equations, we find that

$$W = -q(V_B - V_A) \qquad [16.6]$$

Using this equation, we obtain the following general result: **No net work is required to move a charge between two points that are at the same electric potential.** In mathematical terms this result says that $W = 0$ whenever $V_B = V_A$.

In Chapter 15 we found that when a conductor is in electrostatic equilibrium, a net charge placed on it resides entirely on its surface. Further, we showed that the electric field just outside the surface of a charged conductor in electrostatic equilibrium is perpendicular to the surface and that the field inside the conductor is zero. We now show that **all points on the surface of a charged conductor in electrostatic equilibrium are at the same potential.**

Consider a surface path connecting any points $A$ and $B$ on a charged conductor, as in Figure 16.10. The charges on the conductor are assumed to be in equilibrium with each other, so none are moving. In this case the electric field $\vec{\mathbf{E}}$ is always perpendicular to the displacement along this path. This must be so, for otherwise the part of the electric field tangent to the surface would move the charges. Because $\vec{\mathbf{E}}$ is perpendicular to the path, no work is done by the electric field if a charge is moved between the given two points. From Equation 16.6 we see that if the work done is zero, the difference in electric potential, $V_B - V_A$, is also zero. It follows that **the electric potential is a constant everywhere on the surface of a charged conductor in equilibrium**. Further, because the electric field inside a conductor is zero, no work is required to move a charge between two points inside the conductor. Again, Equation 16.6 shows that if the work done is zero, the difference in electric potential between any two points inside a conductor must also be zero. We conclude that the electric potential is constant everywhere inside a conductor.

Finally, because one of the points inside the conductor could be arbitrarily close to the surface of the conductor, we conclude that **the electric potential is constant everywhere inside a conductor and equal to that same value at the surface**. As a consequence, no work is required to move a charge from the interior of a charged conductor to its surface. (It's important to realize that the potential inside a conductor is not necessarily zero, even though the interior electric field is zero.)

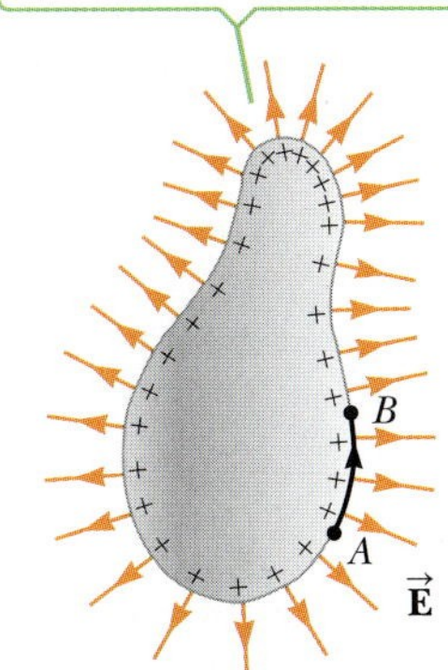
Notice from the spacing of the positive signs that the surface charge density is nonuniform.
B
A
$\vec{\mathbf{E}}$

**Figure 16.10** An arbitrarily shaped conductor with an excess positive charge. When the conductor is in electrostatic equilibrium, all the charge resides at the surface, $\vec{\mathbf{E}} = 0$, inside the conductor, and the electric field just outside the conductor is perpendicular to the surface. The potential is constant inside the conductor and is equal to the potential at the surface.

## The Electron Volt

An appropriately sized unit of energy commonly used in atomic and nuclear physics is the electron volt (eV). For example, electrons in normal atoms typically have energies of tens of eV's, excited electrons in atoms emitting $x$-rays have energies of thousands of eV's, and high-energy gamma rays (electromagnetic waves) emitted by the nucleus have energies of millions of eV's.

Definition of the electron volt ▶

The **electron volt** is defined as the kinetic energy that an electron gains when accelerated through a potential difference of 1 V.

Because 1 V = 1 J/C and because the magnitude of the charge on the electron is $1.60 \times 10^{-19}$ C, we see that the electron volt is related to the joule by

$$1 \text{ eV} = 1.60 \times 10^{-19} \text{ C} \cdot \text{V} = 1.60 \times 10^{-19} \text{ J} \qquad [16.7]$$

■ *Quick Quiz*

**16.7** An electron initially at rest accelerates through a potential difference of 1 V, gaining kinetic energy $KE_e$, whereas a proton, also initially at rest, accelerates through a potential difference of −1 V, gaining kinetic energy $KE_p$. Which of the following relationships holds? (a) $KE_e = KE_p$ (b) $KE_e < KE_p$ (c) $KE_e > KE_p$ (d) The answer can't be determined from the given information.

# 16.4 Equipotential Surfaces

**LEARNING OBJECTIVE**

1. Define an equipotential surface and discuss its electrical properties.

A surface on which all points are at the same potential is called an **equipotential surface**. The potential difference between any two points on an equipotential surface is zero. Hence, **no work is required to move a charge at constant speed on an equipotential surface**.

Equipotential surfaces have a simple relationship to the electric field: **The electric field at every point of an equipotential surface is perpendicular to the surface**. If the electric field $\vec{\mathbf{E}}$ had a component parallel to the surface, that component would produce an electric force on a charge placed on the surface. This force would do work on the charge as it moved from one point to another, in contradiction to the definition of an equipotential surface.

Equipotential surfaces can be represented on a diagram by drawing equipotential contours, which are two-dimensional views of the intersections of the equipotential surfaces with the plane of the drawing. These equipotential contours are generally referred to simply as **equipotentials**. Figure 16.11a shows the equipotentials (in blue) associated with a positive point charge. Note that the equipotentials are perpendicular to the electric field lines (in orange) at all points. Recall that the electric potential created by a point charge $q$ is given by $V = k_e q/r$. This relation shows that, for a single point charge, the potential is constant on any surface on which $r$ is constant. It follows that the equipotentials of a point charge are a family

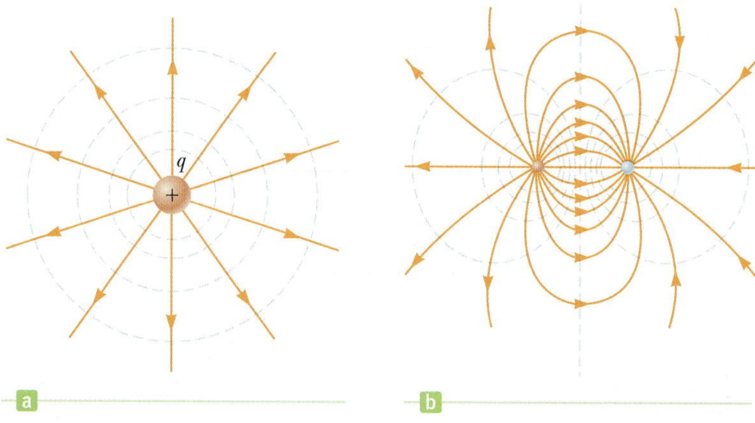


**Figure 16.11** Equipotentials (dashed blue lines) and electric field lines (orange lines) for (a) a positive point charge and (b) two point charges of equal magnitude and opposite sign. In all cases the equipotentials are *perpendicular* to the electric field lines at every point.

of spheres centered on the point charge. Figure 16.11b shows the equipotentials associated with two charges of equal magnitude but opposite sign.

## 16.5 Applications

**LEARNING OBJECTIVE**

1. Describe several important applications of electrically charged systems.

### The Electrostatic Precipitator

**APPLICATION**
The Electrostatic Precipitator

One important application of electric discharge in gases is a device called an *electrostatic precipitator*. This device removes particulate matter from combustion gases, thereby reducing air pollution. It's especially useful in coal-burning power plants and in industrial operations that generate large quantities of smoke. Systems currently in use can eliminate approximately 90% by mass of the ash and dust from smoke. Unfortunately, a very high percentage of the lighter particles still escape, and they contribute significantly to smog and haze.

Figure 16.12 illustrates the basic idea of the electrostatic precipitator. A high voltage (typically 40 kV to 100 kV) is maintained between a wire running down the center of a duct and the outer wall, which is grounded. The wire is maintained at a negative electric potential with respect to the wall, so the electric field is directed toward the wire. The electric field near the wire reaches a high enough value to cause a discharge around the wire and the formation of positive ions, electrons, and negative ions, such as $O_2^-$. As the electrons and negative ions are accelerated toward the outer wall by the nonuniform electric field, the dirt particles in the streaming gas become charged by collisions and ion capture. Because most of the charged dirt particles are negative, they are also drawn to the outer wall by the electric field. When the duct is shaken, the particles fall loose and are collected at the bottom.

In addition to reducing the amounts of harmful gases and particulate matter in the atmosphere, the electrostatic precipitator recovers valuable metal oxides from the stack.

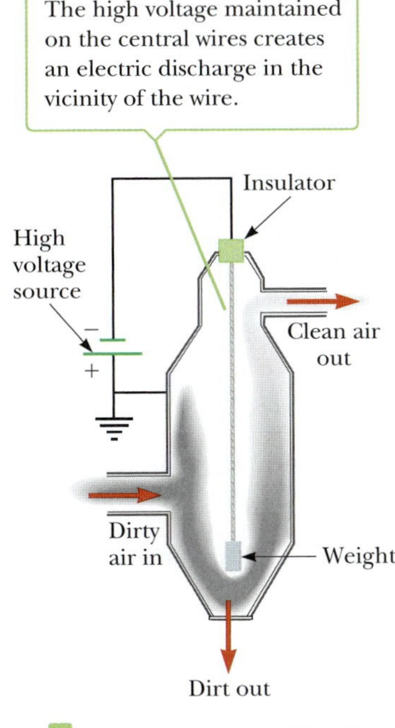

**Figure 16.12** (a) A schematic diagram of an electrostatic precipitator. Compare the air pollution when the precipitator is (b) operating and (c) turned off.

**APPLICATION**
The Electrostatic Air Cleaner

A similar device called an *electrostatic air cleaner* is used in homes to relieve the discomfort of allergy sufferers. Air laden with dust and pollen is drawn into the device across a positively charged mesh screen. The airborne particles become positively charged when they make contact with the screen, and then they pass through a second, negatively charged mesh screen. The electrostatic force of attraction between the positively charged particles in the air and the negatively charged screen causes the particles to precipitate out on the surface of the screen, removing a very high percentage of contaminants from the air stream.

**APPLICATION**
Xerographic Copiers

## Xerography and Laser Printers

Xerography is widely used to make photocopies of printed materials. The basic idea behind the process was developed by Chester Carlson, who was granted a patent for his invention in 1940. In 1947 the Xerox Corporation launched a full-scale program to develop automated duplicating machines using Carlson's process. The huge success of that development is evident: today, practically all offices and libraries have one or more duplicating machines, and the capabilities of these machines continue to evolve.

Some features of the xerographic process involve simple concepts from electrostatics and optics. The one idea that makes the process unique, however, is the use of photoconductive material to form an image. A photoconductor is a material that is a poor conductor of electricity in the dark, but a reasonably good conductor when exposed to light.

Figure 16.13 illustrates the steps in the xerographic process. First, the surface of a plate or drum is coated with a thin film of the photoconductive material (usually selenium or some compound of selenium), and the photoconductive surface is given a positive electrostatic charge in the dark (Fig. 16.13a). The page to be copied is then projected onto the charged surface (Fig. 16.13b). The photoconducting surface becomes conducting only in areas where light strikes; there the light produces charge carriers in the photoconductor that neutralize the positively charged surface. The charges remain on those areas of the photoconductor not exposed to light, however, leaving a hidden image of the object in the form of a positive distribution of surface charge.

**APPLICATION**
Laser Printers

Next, a negatively charged powder called a *toner* is dusted onto the photoconducting surface (Fig. 16.13c). The charged powder adheres only to the areas that contain the positively charged image. At this point, the image becomes visible. It is then transferred to the surface of a sheet of positively charged paper. Finally,

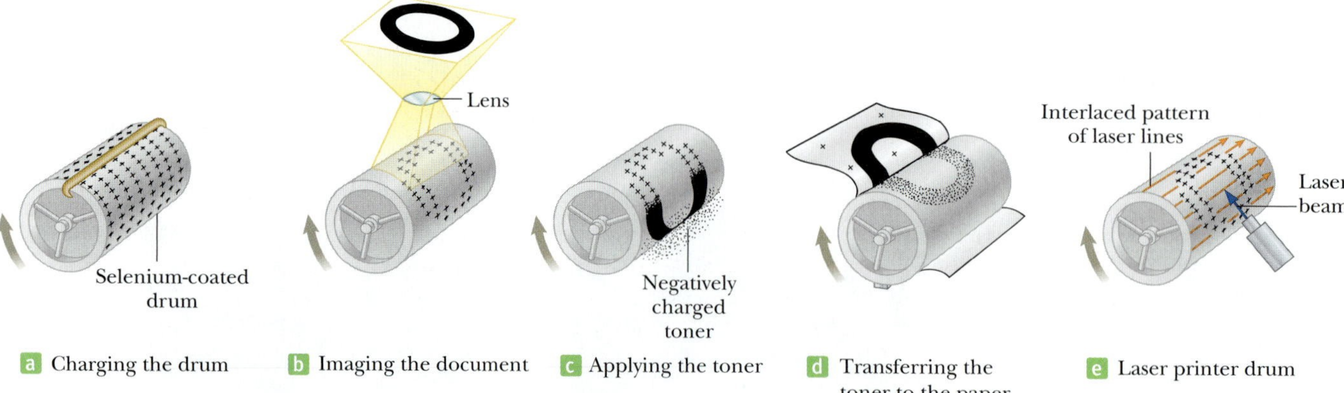


**Figure 16.13** The xerographic process. (a) The photoconductive surface is positively charged. (b) Through the use of a light source and a lens, a hidden image is formed on the charged surface in the form of positive charges. (c) The surface containing the image is covered with a negatively charged powder, which adheres only to the image area. (d) A piece of paper is placed over the surface and given a charge. This transfers the image to the paper, which is then heated to "fix" the powder to the paper. (e) The image on the drum of a laser printer is produced by turning a laser beam on and off as it sweeps across the selenium-coated drum.

the toner is "fixed" to the surface of the paper by heat (Fig. 16.13d), resulting in a permanent copy of the original.

The steps for producing a document on a laser printer are similar to those used in a photocopy machine in that parts (a), (c), and (d) of Figure 16.13 remain essentially the same. The difference between the two techniques lies in the way the image is formed on the selenium-coated drum. In a laser printer the command to print the letter *O*, for instance, is sent to a laser from the memory of a computer. A rotating mirror inside the printer causes the beam of the laser to sweep across the selenium-coated drum in an interlaced pattern (Fig. 16.13e). Electrical signals generated by the printer turn the laser beam on and off in a pattern that traces out the letter *O* in the form of positive charges on the selenium. Toner is then applied to the drum, and the transfer to paper is accomplished as in a photocopy machine.

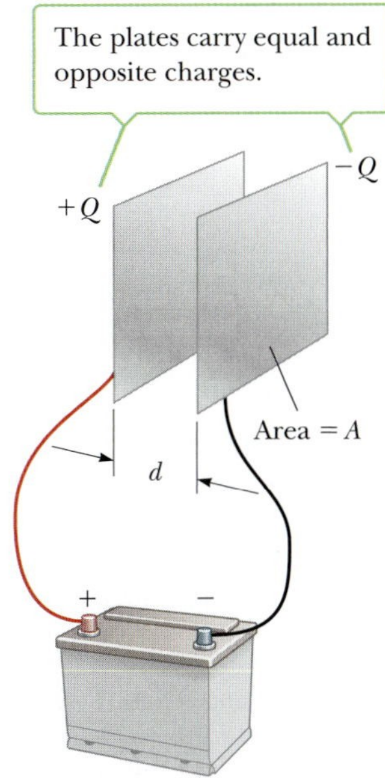


**Figure 16.14** A parallel-plate capacitor consists of two parallel plates, each of area *A*, separated by a distance *d*.

# 16.6 Capacitance

**LEARNING OBJECTIVE**

1. Describe a capacitor and define capacitance.

A **capacitor** is a device used in a variety of electric circuits, such as to tune the frequency of radio receivers, eliminate sparking in automobile ignition systems, or store short-term energy for rapid release in electronic flash units. Figure 16.14 shows a typical design for a capacitor. It consists of two parallel metal plates separated by a distance *d*. Used in an electric circuit, the plates are connected to the positive and negative terminals of a battery or some other voltage source. When this connection is made, electrons are pulled off one of the plates, leaving it with a charge of $+Q$, and are transferred through the battery to the other plate, leaving it with a charge of $-Q$, as shown in the figure. The transfer of charge stops when the potential difference across the plates equals the potential difference of the battery. A charged capacitor is a device that stores energy that can be reclaimed when needed for a specific application.

The capacitance *C* of a capacitor is the ratio of the magnitude of the charge on either conductor (plate) to the magnitude of the potential difference between the conductors (plates):

$$C \equiv \frac{Q}{\Delta V} \qquad [16.8]$$

◀ Capacitance of a pair of conductors

**SI unit: farad (F) = coulomb per volt (C/V)**

The quantities *Q* and $\Delta V$ are always taken to be positive when used in Equation 16.8. For example, if a 3.0-$\mu$F capacitor is connected to a 12-V battery, the magnitude of the charge on each plate of the capacitor is

$$Q = C\Delta V = (3.0 \times 10^{-6}\ \text{F})(12\ \text{V}) = 36\ \mu\text{C}$$

From Equation 16.8, we see that a large capacitance is needed to store a large amount of charge for a given applied voltage. The farad is a very large unit of capacitance. In practice, most typical capacitors have capacitances ranging from microfarads ($1\ \mu\text{F} = 1 \times 10^{-6}$ F) to picofarads ($1\ \text{pF} = 1 \times 10^{-12}$ F).

# 16.7 The Parallel-Plate Capacitor

**LEARNING OBJECTIVES**

1. Derive an expression for the capacitance of a parallel-plate capacitor.
2. Calculate the fundamental physical properties of a parallel-plate capacitor.

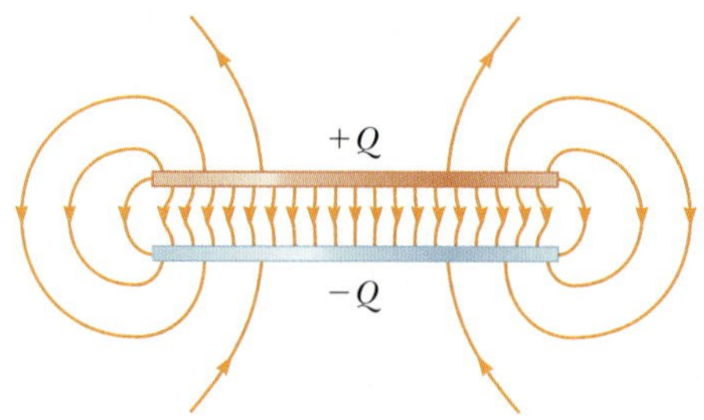


**Figure 16.15** The electric field between the plates of a parallel-plate capacitor is uniform near the center, but nonuniform near the edges.

**Tip 16.2 Potential Difference Is $\Delta V$, Not $V$**

Use the symbol $\Delta V$ for the potential difference across a circuit element or a device (many other books use simply $V$ for potential difference). The dual use of $V$ to represent potential in one place and a potential difference in another can lead to unnecessary confusion.

The capacitance of a device depends on the geometric arrangement of the conductors. The capacitance of a parallel-plate capacitor with plates separated by air (see Fig. 16.14) can be easily calculated from three facts. First, recall from Chapter 15 that the magnitude of the electric field between two plates is given by $E = \sigma/\epsilon_0$, where $\sigma$ is the magnitude of the charge per unit area on each plate. Second, we found earlier in this chapter that the potential difference between two plates is $\Delta V = Ed$, where $d$ is the distance between the plates. Third, the charge on one plate is given by $Q = \sigma A$, where $A$ is the area of the plate. Substituting these three facts into the definition of capacitance gives the desired result:

$$C = \frac{Q}{\Delta V} = \frac{\sigma A}{Ed} = \frac{\sigma A}{(\sigma/\epsilon_0)d}$$

Canceling the charge per unit area, $\sigma$, yields

Capacitance of a parallel-plate capacitor ▶

$$C = \epsilon_0 \frac{A}{d} \qquad [16.9]$$

where $A$ is the area of one of the plates, $d$ is the distance between the plates, and $\epsilon_0$ is the permittivity of free space.

From Equation 16.9, we see that plates with larger area can store more charge. The same is true for a small plate separation $d$ because then the positive charges on one plate exert a stronger force on the negative charges on the other plate, allowing more charge to be held on the plates.

Figure 16.15 shows the electric field lines of a more realistic parallel-plate capacitor. The electric field is very nearly constant in the center between the plates, but becomes less so when approaching the edges. For most purposes, however, the field may be taken as constant throughout the region between the plates.

**APPLICATION**
Camera Flash Attachments

One practical device that uses a capacitor is the flash attachment on a camera. A battery is used to charge the capacitor, and the stored charge is then released when the shutter-release button is pressed to take a picture. The stored charge is delivered to a flash tube very quickly, illuminating the subject at the instant more light is needed.

**APPLICATION**
Computer Keyboards

Computers make use of capacitors in many ways. For example, one type of computer keyboard has capacitors at the bases of its keys, as in Figure 16.16. Each key is connected to a movable plate, which represents one side of the capacitor; the fixed plate on the bottom of the keyboard represents the other side of the capacitor. When a key is pressed, the capacitor spacing decreases, causing an increase in capacitance. External electronic circuits recognize each key by the *change* in its capacitance when it is pressed.

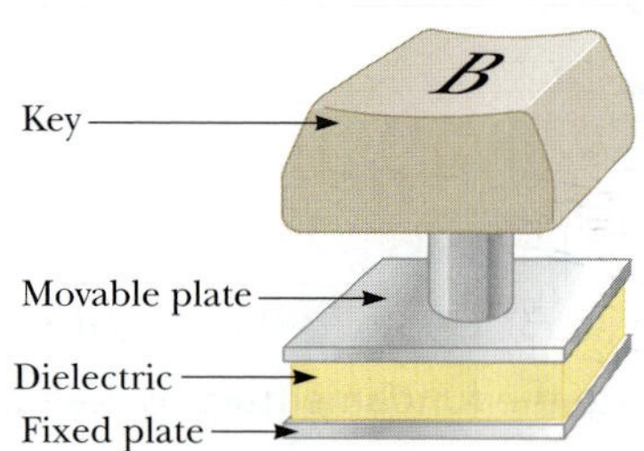


**Figure 16.16** When the key of one type of keyboard is pressed, the capacitance of a parallel-plate capacitor increases as the plate spacing decreases. The substance labeled "dielectric" is an insulating material, as described in Section 16.10.

Capacitors are useful for storing a large amount of charge that needs to be delivered quickly. A good example on the forefront of fusion research is electrostatic confinement. In this role capacitors discharge their electrons through a grid. The negatively charged electrons in the grid draw positively charged particles to them and therefore to each other, causing some particles to fuse and release energy in the process.

**APPLICATION**
Electrostatic Confinement

### EXAMPLE 16.6 A Parallel-Plate Capacitor

**GOAL** Calculate fundamental physical properties of a parallel-plate capacitor.

**PROBLEM** A parallel-plate capacitor has an area $A = 2.00 \times 10^{-4}\ \text{m}^2$ and a plate separation $d = 1.00 \times 10^{-3}$ m. **(a)** Find its capacitance. **(b)** How much charge is on the positive plate if the capacitor is connected to a 3.00-V battery? Calculate **(c)** the charge density on the positive plate, assuming the density is uniform, and **(d)** the magnitude of the electric field between the plates.

**STRATEGY** Parts (a) and (b) can be solved by substituting into the basic equations for capacitance. In part (c) use the definition of charge density, and in part (d) use the fact that the voltage difference equals the electric field times the distance.

**SOLUTION**

**(a)** Find the capacitance.

Substitute into Equation 16.9:

$$C = \epsilon_0 \frac{A}{d} = (8.85 \times 10^{-12}\ \text{C}^2/\text{N}\cdot\text{m}^2)\left(\frac{2.00 \times 10^{-4}\ \text{m}^2}{1.00 \times 10^{-3}\ \text{m}}\right)$$

$$C = 1.77 \times 10^{-12}\ \text{F} = 1.77\ \text{pF}$$

**(b)** Find the charge on the positive plate after the capacitor is connected to a 3.00-V battery.

Substitute into Equation 16.8:

$$C = \frac{Q}{\Delta V} \quad \rightarrow \quad Q = C\,\Delta V = (1.77 \times 10^{-12}\ \text{F})(3.00\ \text{V})$$

$$= 5.31 \times 10^{-12}\ \text{C}$$

**(c)** Calculate the charge density on the positive plate.

Charge density is charge divided by area:

$$\sigma = \frac{Q}{A} = \frac{5.31 \times 10^{-12}\ \text{C}}{2.00 \times 10^{-4}\ \text{m}^2} = 2.66 \times 10^{-8}\ \text{C/m}^2$$

**(d)** Calculate the magnitude of the electric field between the plates.

Apply $\Delta V = Ed$:

$$E = \frac{\Delta V}{d} = \frac{3.00\ \text{V}}{1.00 \times 10^{-3}\ \text{m}} = 3.00 \times 10^3\ \text{V/m}$$

**REMARKS** The answer to part (d) could also have been obtained from the electric field derived for a parallel plate capacitor, Equation 15.13, $E = \sigma/\epsilon_0$.

**QUESTION 16.6** How do the answers change if the distance between the plates is doubled?

**EXERCISE 16.6** Two plates, each of area $3.00 \times 10^{-4}\ \text{m}^2$, are used to construct a parallel-plate capacitor with capacitance 1.00 pF. (a) Find the necessary separation distance. (b) If the positive plate is to hold a charge of $5.00 \times 10^{-12}$ C, find the charge density. (c) Find the electric field between the plates. (d) What voltage battery should be attached to the plate to obtain the preceding results?

**ANSWERS** (a) $2.66 \times 10^{-3}$ m (b) $1.67 \times 10^{-8}\ \text{C/m}^2$ (c) $1.89 \times 10^3$ N/C (d) 5.00 V

## Symbols for Circuit Elements and Circuits

The symbol that is commonly used to represent a capacitor in a circuit is —||— or sometimes —|(—. Don't confuse either of these symbols with the circuit symbol, —|‖— (+ −) which is used to designate a battery (or any other source of

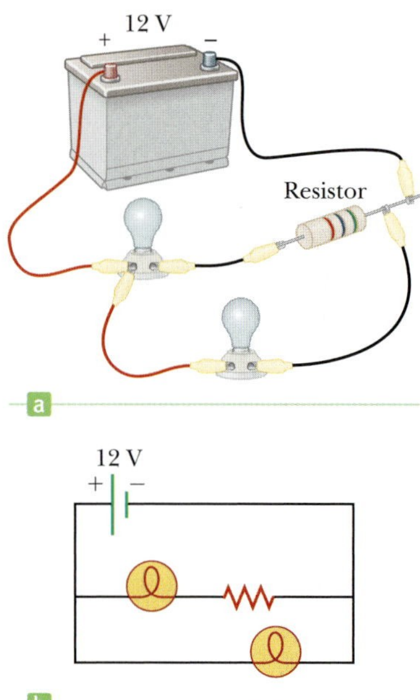


**Figure 16.17** (a) A real circuit and (b) its equivalent circuit diagram.

direct current). The positive terminal of the battery is at the higher potential and is represented by the longer vertical line in the battery symbol. In the next chapter we discuss another circuit element, called a resistor, represented by the symbol —WW—. When wires in a circuit don't have appreciable resistance compared with the resistance of other elements in the circuit, the wires are represented by straight lines.

It's important to realize that a circuit is a collection of real objects, usually containing a source of electrical energy (such as a battery) connected to elements that convert electrical energy to other forms (light, heat, sound) or store the energy in electric or magnetic fields for later retrieval. A real circuit and its schematic diagram are sketched in Figure 16.17. The circuit symbol for a lightbulb shown in Figure 16.17b is —(ℓ)—.

If you are not familiar with circuit diagrams, trace the path of the real circuit with your finger to see that it is equivalent to the geometrically regular schematic diagram.

## 16.8 Combinations of Capacitors

### LEARNING OBJECTIVES

1. Derive the equivalent capacitance of capacitors in parallel.
2. Analyze a circuit with capacitors in parallel.
3. Derive the equivalent capacitance of capacitors in series.
4. Analyze a circuit with a series of capacitors.
5. Analyze a circuit with combinations of both parallel and series capacitors.

Two or more capacitors can be combined in circuits in several ways, but most reduce to two simple configurations, called *parallel* and *series*. The idea, then, is to find the single equivalent capacitance due to a combination of several different capacitors that are in parallel or in series with each other. Capacitors are manufactured with a number of different standard capacitances, and by combining them in different ways, any desired value of capacitance can be obtained.

### Capacitors in Parallel

Two capacitors connected as shown in Figure 16.18a are said to be in *parallel*. The left plate of each capacitor is connected to the positive terminal of the battery by a conducting wire, so the left plates are at the same potential. In the same way, the right plates, both connected to the negative terminal of the battery, are also at the same potential. This means that **capacitors in parallel both have the**

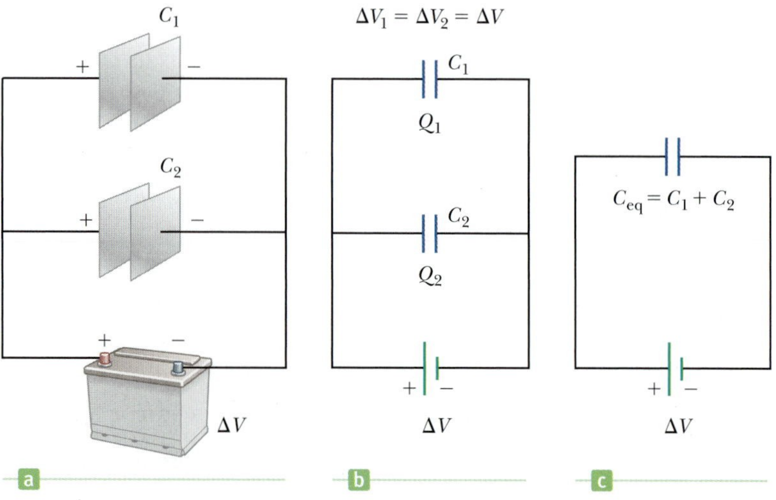


**Figure 16.18** (a) A parallel connection of two capacitors. (b) The circuit diagram for the parallel combination. (c) The potential differences across the capacitors are the same, and the equivalent capacitance is $C_{eq} = C_1 + C_2$.

**same potential difference $\Delta V$ across them.** Capacitors in parallel are illustrated in Figure 16.18b.

When the capacitors are first connected in the circuit, electrons are transferred from the left plates through the battery to the right plates, leaving the left plates positively charged and the right plates negatively charged. The energy source for this transfer of charge is the internal chemical energy stored in the battery, which is converted to electrical energy. The flow of charge stops when the voltage across the capacitors equals the voltage of the battery, at which time the capacitors have their maximum charges. If the maximum charges on the two capacitors are $Q_1$ and $Q_2$, respectively, the *total charge*, $Q$, stored by the two capacitors is

$$Q = Q_1 + Q_2 \qquad [16.10]$$

We can replace these two capacitors with one equivalent capacitor having a capacitance of $C_{eq}$. This equivalent capacitor must have exactly the same external effect on the circuit as the original two, so it must store $Q$ units of charge and have the same potential difference across it. The respective charges on each capacitor are

$$Q_1 = C_1\,\Delta V \quad \text{and} \quad Q_2 = C_2\,\Delta V$$

The charge on the equivalent capacitor is

$$Q = C_{eq}\,\Delta V$$

Substituting these relationships into Equation 16.10 gives

$$C_{eq}\,\Delta V = C_1\,\Delta V + C_2\,\Delta V$$

or

$$C_{eq} = C_1 + C_2 \quad \begin{pmatrix}\text{parallel}\\ \text{combination}\end{pmatrix} \qquad [16.11]$$

If we extend this treatment to three or more capacitors connected in parallel, the equivalent capacitance is found to be

$$C_{eq} = C_1 + C_2 + C_3 + \cdots \quad \begin{pmatrix}\text{parallel}\\ \text{combination}\end{pmatrix} \qquad [16.12]$$

We see that **the equivalent capacitance of a parallel combination of capacitors is larger than any of the individual capacitances.**

**Tip 16.3 Voltage Is the Same as Potential Difference**

A voltage *across* a device, such as a capacitor, has the same meaning as the potential difference across the device. For example, if we say that the voltage across a capacitor is 12 V, we mean that the potential difference between its plates is 12 V.

## ■ EXAMPLE 16.7 Four Capacitors Connected in Parallel

**GOAL** Analyze a circuit with several capacitors in parallel.

**PROBLEM** **(a)** Determine the capacitance of the single capacitor that is equivalent to the parallel combination of capacitors shown in Figure 16.19. Find **(b)** the charge on the 12.0-$\mu$F capacitor and **(c)** the total charge contained in the configuration. **(d)** Derive a symbolic expression for the fraction of the total charge contained on one of the capacitors.

**STRATEGY** For part (a), add the individual capacitances. For part (b), apply the formula $C = Q/\Delta V$ to the 12.0-$\mu$F capacitor. The voltage difference is the same as the difference across the battery. To find the total charge contained in all four capacitors, use the equivalent capacitance in the same formula.

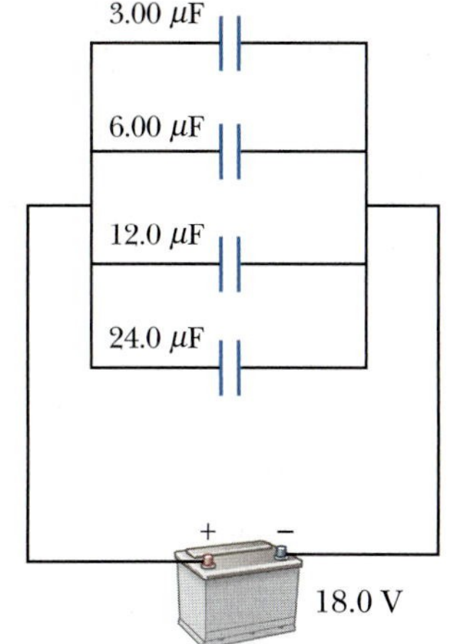


**Figure 16.19** (Example 16.7) Four capacitors connected in parallel.

*(Continued)*

**SOLUTION**

**(a)** Find the equivalent capacitance.

Apply Equation 16.12:

$$C_{eq} = C_1 + C_2 + C_3 + C_4$$
$$= 3.00\ \mu F + 6.00\ \mu F + 12.0\ \mu F + 24.0\ \mu F$$
$$= \boxed{45.0\ \mu F}$$

**(b)** Find the charge on the 12-$\mu$F capacitor (designated $C_3$).

Solve the capacitance equation for $Q$ and substitute:

$$Q = C_3\,\Delta V = (12.0 \times 10^{-6}\ F)(18.0\ V) = 216 \times 10^{-6}\ C$$
$$= \boxed{216\ \mu C}$$

**(c)** Find the total charge contained in the configuration.

Use the equivalent capacitance:

$$C_{eq} = \frac{Q}{\Delta V} \quad \rightarrow \quad Q = C_{eq}\Delta V = (45.0\ \mu F)(18.0\ V) = \boxed{8.10 \times 10^2\ \mu C}$$

**(d)** Derive a symbolic expression for the fraction of the total charge contained in one of the capacitors.

Write a symbolic expression for the fractional charge in the *i*th capacitor and use the capacitor definition:

$$\boxed{\frac{Q_i}{Q_{tot}} = \frac{C_i \Delta V}{C_{eq} \Delta V} = \frac{C_i}{C_{eq}}}$$

**REMARKS** The charge on any one of the parallel capacitors can be found as in part (b) because the potential difference is the same. Notice that finding the total charge does not require finding the charge on each individual capacitor and adding. It's easier to use the equivalent capacitance in the capacitance definition.

**QUESTION 16.7** If all four capacitors had the same capacitance, what fraction of the total charge would be held by each?

**EXERCISE 16.7** Find the charge on the 24.0-$\mu$F capacitor.

**ANSWER** 432 $\mu$C

## Capacitors in Series

$Q$ is the same for all capacitors connected in series ▶

Now consider two capacitors connected in *series*, as illustrated in Figure 16.20a. **For a series combination of capacitors, the magnitude of the charge must be the same on all the plates.** To understand this principle, consider the charge transfer process in some detail. When a battery is connected to the circuit, electrons with total charge $-Q$ are transferred from the left plate of $C_1$ to the right plate of $C_2$ through the battery, leaving the left plate of $C_1$ with a charge of $+Q$. As a consequence, the magnitudes of the charges on the left plate of $C_1$ and the right plate of $C_2$ must be the same. Now consider the right plate of $C_1$ and the left plate of $C_2$, in the middle. These plates are not connected to the battery (because of the gap across the plates) and, taken together, are electrically neutral. The charge of $+Q$ on the left plate of $C_1$, however, attracts negative charges to the right plate of $C_1$. These charges will continue to accumulate until the left and right plates of $C_1$, taken together, become electrically neutral, which means that the charge on the right plate of $C_1$ is $-Q$. This negative charge could only have come from the left plate of $C_2$, so $C_2$ has a charge of $+Q$.

Therefore, regardless of how many capacitors are in series or what their capacitances are, **all the right plates gain charges of $-Q$ and all the left plates have charges of $+Q$** (a consequence of the conservation of charge).

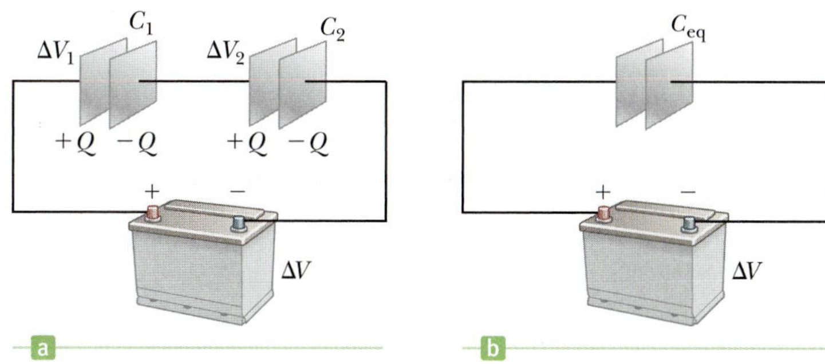


**Figure 16.20** A series combination of two capacitors. The charges on the capacitors are the same, and the equivalent capacitance can be calculated from the reciprocal relationship $1/C_{eq} = (1/C_1) + (1/C_2)$.

After an equivalent capacitor for a series of capacitors is fully charged, **the equivalent capacitor must end up with a charge of $-Q$ on its right plate and a charge of $+Q$ on its left plate.** Applying the definition of capacitance to the circuit in Figure 16.20b, we have

$$\Delta V = \frac{Q}{C_{eq}}$$

where $\Delta V$ is the potential difference between the terminals of the battery and $C_{eq}$ is the equivalent capacitance. Because $Q = C\Delta V$ can be applied to each capacitor, the potential differences across them are given by

$$\Delta V_1 = \frac{Q}{C_1} \qquad \Delta V_2 = \frac{Q}{C_2}$$

From Figure 16.20a, we see that

$$\Delta V = \Delta V_1 + \Delta V_2 \qquad [16.13]$$

where $\Delta V_1$ and $\Delta V_2$ are the potential differences across capacitors $C_1$ and $C_2$ (a consequence of the conservation of energy).

The potential difference across any number of capacitors (or other circuit elements) in series equals the sum of the potential differences across the individual capacitors. Substituting these expressions into Equation 16.13 and noting that $\Delta V = Q/C_{eq}$, we have

$$\frac{Q}{C_{eq}} = \frac{Q}{C_1} + \frac{Q}{C_2}$$

Canceling $Q$, we arrive at the following relationship:

$$\frac{1}{C_{eq}} = \frac{1}{C_1} + \frac{1}{C_2} \quad \left(\begin{array}{l}\text{series}\\ \text{combination}\end{array}\right) \qquad [16.14]$$

If this analysis is applied to three or more capacitors connected in series, the equivalent capacitance is found to be

$$\frac{1}{C_{eq}} = \frac{1}{C_1} + \frac{1}{C_2} + \frac{1}{C_3} + \cdots \quad \left(\begin{array}{l}\text{series}\\ \text{combination}\end{array}\right) \qquad [16.15]$$

As we will show in Example 16.8, Equation 16.15 implies that **the equivalent capacitance of a series combination is always smaller than any individual capacitance in the combination.**

**Quick Quiz**

**16.8** A capacitor is designed so that one plate is large and the other is small. If the plates are connected to a battery, (a) the large plate has a greater charge than the small plate, (b) the large plate has less charge than the small plate, or (c) the plates have equal, but opposite, charge.

## EXAMPLE 16.8 Four Capacitors Connected in Series

**GOAL** Find an equivalent capacitance of capacitors in series, and the charge and voltage on each capacitor.

**PROBLEM** Four capacitors are connected in series with a battery, as in Figure 16.21. **(a)** Calculate the capacitance of the equivalent capacitor. **(b)** Compute the charge on the 12-$\mu$F capacitor. **(c)** Find the voltage drop across the 12-$\mu$F capacitor.

**STRATEGY** Combine all the capacitors into a single, equivalent capacitor using Equation 16.15. Find the charge on this equivalent capacitor using $C = Q/\Delta V$. This charge is the same as on the individual capacitors. Use this same equation again to find the voltage drop across the 12-$\mu$F capacitor.

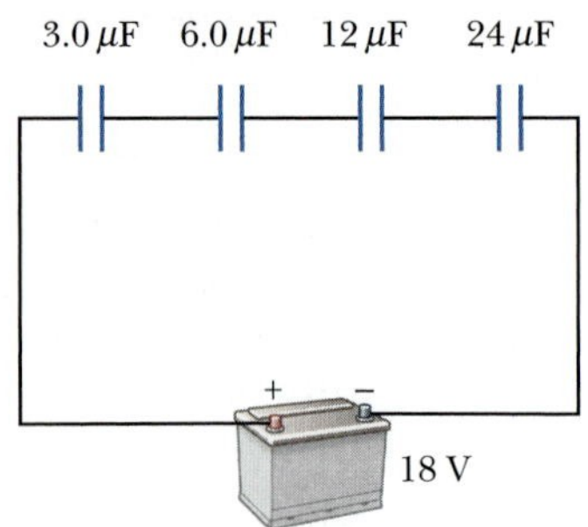


**Figure 16.21** (Example 16.8) Four capacitors connected in series.

**SOLUTION**

**(a)** Calculate the equivalent capacitance of the series.

Apply Equation 16.15:

$$\frac{1}{C_{\text{eq}}} = \frac{1}{3.0\ \mu\text{F}} + \frac{1}{6.0\ \mu\text{F}} + \frac{1}{12\ \mu\text{F}} + \frac{1}{24\ \mu\text{F}}$$

$$C_{\text{eq}} = 1.6\ \mu\text{F}$$

**(b)** Compute the charge on the 12-$\mu$F capacitor.

The desired charge equals the charge on the equivalent capacitor:

$$Q = C_{\text{eq}}\,\Delta V = (1.6 \times 10^{-6}\ \text{F})(18\ \text{V}) = 29\ \mu\text{C}$$

**(c)** Find the voltage drop across the 12-$\mu$F capacitor.

Apply the basic capacitance equation:

$$C = \frac{Q}{\Delta V} \quad \rightarrow \quad \Delta V = \frac{Q}{C} = \frac{29\ \mu\text{C}}{12\ \mu\text{F}} = 2.4\ \text{V}$$

**REMARKS** Notice that the equivalent capacitance is less than that of any of the individual capacitors. The relationship $C = Q/\Delta V$ can be used to find the voltage drops on the other capacitors, just as in part (c).

**QUESTION 16.8** Over which capacitor is the voltage drop the smallest? The largest?

**EXERCISE 16.8** The 24-$\mu$F capacitor is removed from the circuit, leaving only three capacitors in series. Find (a) the equivalent capacitance, (b) the charge on the 6-$\mu$F capacitor, and (c) the voltage drop across the 6-$\mu$F capacitor.

**ANSWERS** (a) 1.7 $\mu$F (b) 31 $\mu$C (c) 5.2 V

## PROBLEM-SOLVING STRATEGY

### Complex Capacitor Combinations

1. **Combine** capacitors that are in series or in parallel, following the derived formulas.
2. **Redraw** the circuit after every combination.
3. **Repeat** the first two steps until there is only a single equivalent capacitor.
4. **Find the charge** on the single equivalent capacitor, using $C = Q/\Delta V$.
5. **Work backwards** through the diagrams to the original one, finding the charge and voltage drop across each capacitor along the way. To do this, use the following collection of facts:
   A. The capacitor equation: $C = Q/\Delta V$
   B. Capacitors in parallel: $C_{\text{eq}} = C_1 + C_2$
   C. Capacitors in parallel all have the same voltage difference, $\Delta V$, as does their equivalent capacitor.
   D. Capacitors in series: $\dfrac{1}{C_{\text{eq}}} = \dfrac{1}{C_1} + \dfrac{1}{C_2}$
   E. Capacitors in series all have the same charge, $Q$, as does their equivalent capacitor.

## EXAMPLE 16.9 Equivalent Capacitance

**GOAL** Solve a complex combination of series and parallel capacitors.

**PROBLEM (a)** Calculate the equivalent capacitance between $a$ and $b$ for the combination of capacitors shown in Figure 16.22a. All capacitances are in microfarads. **(b)** If a 12-V battery is connected across the system between points $a$ and $b$, find the charge on the 4.0-$\mu$F capacitor in the first diagram and the voltage drop across it.

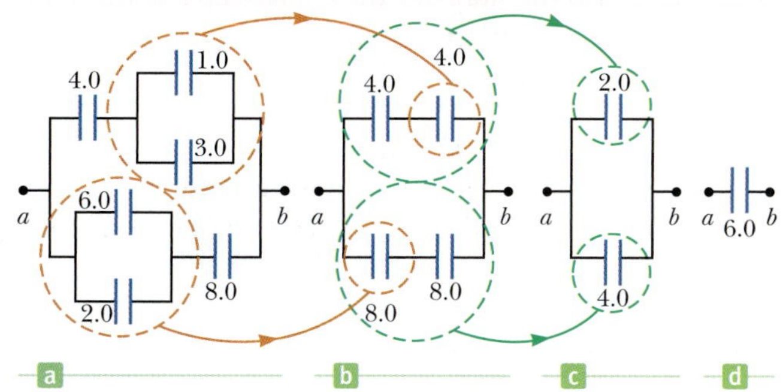


**Figure 16.22** (Example 16.9) To find the equivalent capacitance of the circuit in (a), use the series and parallel rules described in the text to successively reduce the circuit as indicated in (b), (c), and (d). All capacitances are in microfarads.

**STRATEGY** For part (a), use Equations 16.12 and 16.15 to reduce the combination step by step, as indicated in the figure. For part (b), to find the charge on the 4.0-$\mu$F capacitor, start with Figure 16.22c, finding the charge on the 2.0-$\mu$F capacitor. This same charge is on each of the 4.0-$\mu$F capacitors in the second diagram, by fact 5E of the Problem-Solving Strategy. One of these 4.0-$\mu$F capacitors in the second diagram is simply the original 4.0-$\mu$F capacitor in the first diagram.

**SOLUTION**

**(a)** Calculate the equivalent capacitance.

Find the equivalent capacitance of the parallel 1.0-$\mu$F and 3.0-$\mu$F capacitors in Figure 16.22a:

$$C_{\text{eq}} = C_1 + C_2 = 1.0\ \mu\text{F} + 3.0\ \mu\text{F} = 4.0\ \mu\text{F}$$

Find the equivalent capacitance of the parallel 2.0-$\mu$F and 6.0-$\mu$F capacitors in Figure 16.22a:

$$C_{\text{eq}} = C_1 + C_2 = 2.0\ \mu\text{F} + 6.0\ \mu\text{F} = 8.0\ \mu\text{F}$$

Combine the two series 4.0-$\mu$F capacitors in Figure 16.22b:

$$\frac{1}{C_{\text{eq}}} = \frac{1}{C_1} + \frac{1}{C_2} = \frac{1}{4.0\ \mu\text{F}} + \frac{1}{4.0\ \mu\text{F}}$$
$$= \frac{1}{2.0\ \mu\text{F}} \quad \rightarrow \quad C_{\text{eq}} = 2.0\ \mu\text{F}$$

Combine the two series 8.0-$\mu$F capacitors in Figure 16.22b:

$$\frac{1}{C_{\text{eq}}} = \frac{1}{C_1} + \frac{1}{C_2} = \frac{1}{8.0\ \mu\text{F}} + \frac{1}{8.0\ \mu\text{F}}$$
$$= \frac{1}{4.0\ \mu\text{F}} \quad \rightarrow \quad C_{\text{eq}} = 4.0\ \mu\text{F}$$

Finally, combine the two parallel capacitors in Figure 16.22c to find the equivalent capacitance between $a$ and $b$:

$$C_{\text{eq}} = C_1 + C_2 = 2.0\ \mu\text{F} + 4.0\ \mu\text{F} = \boxed{6.0\ \mu\text{F}}$$

**(b)** Find the charge on the 4.0-$\mu$F capacitor and the voltage drop across it.

Compute the charge on the 2.0-$\mu$F capacitor in Figure 16.22c, which is the same as the charge on the 4.0-$\mu$F capacitor in Figure 16.22a:

$$C = \frac{Q}{\Delta V} \quad \rightarrow \quad Q = C\Delta V = (2.0\ \mu\text{F})(12\ \text{V}) = \boxed{24\ \mu\text{C}}$$

Use the basic capacitance equation to find the voltage drop across the 4.0-$\mu$F capacitor in Figure 16.22a:

$$C = \frac{Q}{\Delta V} \quad \rightarrow \quad \Delta V = \frac{Q}{C} = \frac{24\ \mu\text{C}}{4.0\ \mu\text{F}} = \boxed{6.0\ \text{V}}$$

**REMARKS** To find the rest of the charges and voltage drops, it's just a matter of using $C = Q/\Delta V$ repeatedly, together with facts 5C and 5E in the Problem-Solving Strategy. The voltage drop across the 4.0-$\mu$F capacitor could also have been found by noticing, in Figure 16.22b, that both capacitors had the same value and so by symmetry would split the total drop of 12 volts between them.

**QUESTION 16.9** Which capacitor holds more charge, the 1.0-$\mu$F capacitor or the 3.0-$\mu$F capacitor?

*(Continued)*

**EXERCISE 16.9** (a) In Example 16.9 find the charge on the 8.0-$\mu$F capacitor in Figure 16.22a and the voltage drop across it. (b) Do the same for the 6.0-$\mu$F capacitor in Figure 16.22a.

**ANSWERS** (a) 48 $\mu$C, 6.0 V (b) 36 $\mu$C, 6.0 V

## 16.9 Energy Stored in a Charged Capacitor

**LEARNING OBJECTIVES**

1. Obtain an expression for the energy stored by a capacitor.
2. Apply energy and power concepts to a capacitor.

Almost everyone who works with electronic equipment has at some time verified that a capacitor can store energy. If the plates of a charged capacitor are connected by a conductor such as a wire, charge transfers from one plate to the other until the two are uncharged. The discharge can often be observed as a visible spark. If you accidentally touched the opposite plates of a charged capacitor, your fingers would act as a pathway by which the capacitor could discharge, inflicting an electric shock. The degree of shock would depend on the capacitance and voltage applied to the capacitor. *Where high voltages and large quantities of charge are present, as in the power supply of a television set, such a shock can be fatal.*

Capacitors store electrical energy, and that energy is the same as the work required to move charge onto the plates. If a capacitor is initially uncharged (both plates are neutral) so that the plates are at the same potential, very little work is required to transfer a small amount of charge $\Delta Q$ from one plate to the other. Once this charge has been transferred, however, a small potential difference $\Delta V = \Delta Q/C$ appears between the plates, so work must be done to transfer additional charge against this potential difference. From Equation 16.6, if the potential difference at any instant during the charging process is $\Delta V$, the work $\Delta W$ required to move more charge $\Delta Q$ through this potential difference is given by

$$\Delta W = \Delta V \Delta Q$$

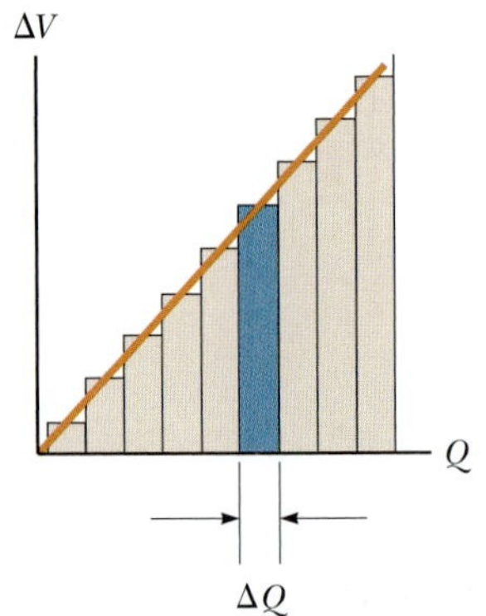


**Figure 16.23** A plot of voltage vs. charge for a capacitor is a straight line with slope $1/C$. The work required to move a charge of $\Delta Q$ through a potential difference of $\Delta V$ across the capacitor plates is $\Delta W = \Delta V \Delta Q$, which equals the area of the blue rectangle. The *total work* required to charge the capacitor to a final charge of $Q$ is the area under the straight line, which equals $Q\Delta V/2$.

We know that $\Delta V = Q/C$ for a capacitor that has a total charge of $Q$. Therefore, a plot of voltage versus total charge gives a straight line with a slope of $1/C$, as shown in Figure 16.23. The work $\Delta W$, for a particular $\Delta V$, is the area of the blue rectangle. Adding up all the rectangles gives an approximation of the total work needed to fill the capacitor. In the limit as $\Delta Q$ is taken to be infinitesimally small, the total work needed to charge the capacitor to a final charge $Q$ and voltage $\Delta V$ is the area under the line. This is just the area of a triangle, one-half the base times the height, so it follows that

$$W = \tfrac{1}{2} Q \Delta V \qquad [16.16]$$

As previously stated, $W$ is also the energy stored in the capacitor. From the definition of capacitance, we have $Q = C\,\Delta V$; hence, we can express the energy stored three different ways:

$$\text{Energy stored} = \tfrac{1}{2}Q\Delta V = \tfrac{1}{2}C\,(\Delta V)^2 = \frac{Q^2}{2C} \qquad [16.17]$$

For example, the amount of energy stored in a 5.0-$\mu$F capacitor when it is connected across a 120-V battery is

$$\text{Energy stored} = \tfrac{1}{2}C\,(\Delta V)^2 = \tfrac{1}{2}(5.0 \times 10^{-6}\,\text{F})(120\,\text{V})^2 = 3.6 \times 10^{-2}\,\text{J}$$

In practice, there is a limit to the maximum energy (or charge) that can be stored in a capacitor. At some point, the Coulomb forces between the charges on the plates become so strong that electrons jump across the gap, discharging the capacitor. For this reason, capacitors are usually labeled with a maximum operating voltage. (This physical fact can actually be exploited to yield a circuit with a regularly blinking light.)

Large capacitors can store enough electrical energy to cause severe burns or even death if they are discharged so that the flow of charge can pass through the heart. Under the proper conditions, however, they can be used to *sustain* life by stopping cardiac fibrillation in heart attack victims. When fibrillation occurs, the heart produces a rapid, irregular pattern of beats. A fast discharge of electrical energy through the heart can return the organ to its normal beat pattern. Emergency medical teams use portable defibrillators that contain batteries capable of charging a capacitor to a high voltage. (The circuitry actually permits the capacitor to be charged to a much higher voltage than the battery.) In this case and others (camera flash units and lasers used for fusion experiments), capacitors serve as energy reservoirs that can be slowly charged and then quickly discharged to provide large amounts of energy in a short pulse. The stored electrical energy is released through the heart by conducting electrodes, called paddles, placed on both sides of the victim's chest. The paramedics must wait between applications of electrical energy because of the time it takes the capacitors to become fully charged. The high voltage on the capacitor can be obtained from a low-voltage battery in a portable machine through the phenomenon of *electromagnetic induction*, to be studied in Chapter 20.

**APPLICATION BIO**
Defibrillators

## EXAMPLE 16.10 Typical Voltage, Energy, and Discharge Time for a Defibrillator

**GOAL** Apply energy and power concepts to a capacitor.

**PROBLEM** A fully charged defibrillator contains 1.20 kJ of energy stored in a $1.10 \times 10^{-4}$ F capacitor. In a discharge through a patient, $6.00 \times 10^2$ J of electrical energy is delivered in 2.50 ms. **(a)** Find the voltage needed to store 1.20 kJ in the unit. **(b)** What average power is delivered to the patient?

**STRATEGY** Because we know the energy stored and the capacitance, we can use Equation 16.17 to find the required voltage in part (a). For part (b), dividing the energy delivered by the time gives the average power.

**SOLUTION**

**(a)** Find the voltage needed to store 1.20 kJ in the unit.

Solve Equation 16.17 for $\Delta V$:

$$\text{Energy stored} = \tfrac{1}{2}C\Delta V^2$$

$$\Delta V = \sqrt{\frac{2 \times (\text{energy stored})}{C}} = \sqrt{\frac{2(1.20 \times 10^3\ \text{J})}{1.10 \times 10^{-4}\ \text{F}}} = 4.67 \times 10^3\ \text{V}$$

**(b)** What average power is delivered to the patient?

Divide the energy delivered by the time:

$$P_{\text{av}} = \frac{\text{energy delivered}}{\Delta t} = \frac{6.00 \times 10^2\ \text{J}}{2.50 \times 10^{-3}\ \text{s}} = 2.40 \times 10^5\ \text{W}$$

**REMARKS** The power delivered by a draining capacitor isn't constant, as we'll find in the study of *RC* circuits in Chapter 18. For that reason, we were able to find only an average power. Capacitors are necessary in defibrillators because they can deliver energy far more quickly than batteries. Batteries provide current through relatively slow chemical reactions, whereas capacitors can quickly release charge that has already been produced and stored.

**QUESTION 16.10** If the voltage across the capacitor were doubled, would the energy stored be (a) halved, (b) doubled, or (c) quadrupled?

*(Continued)*

**EXERCISE 16.10** (a) Find the energy contained in a $2.50 \times 10^{-5}$ F parallel-plate capacitor if it holds $1.75 \times 10^{-3}$ C of charge. (b) What's the voltage between the plates? (c) What new voltage will result in a doubling of the stored energy?

**ANSWERS** (a) $6.13 \times 10^{-2}$ J (b) 70.0 V (c) 99.0 V

### APPLYING PHYSICS 16.1 Maximum Energy Design

How should three capacitors and two batteries be connected so that the capacitors will store the maximum possible energy?

**EXPLANATION** The energy stored in the capacitor is proportional to the capacitance and the square of the potential difference, so we would like to maximize each of these quantities. If the three capacitors are connected in parallel, their capacitances add, and if the batteries are in series, their potential differences, similarly, also add together. ■

**Quick Quiz**

**16.9** A parallel-plate capacitor is disconnected from a battery, and the plates are pulled a small distance farther apart. Do the following quantities increase, decrease, or stay the same?
(a) $C$ (b) $Q$ (c) $E$ between the plates (d) $\Delta V$ (e) energy stored in the capacitor

## 16.10 Capacitors with Dielectrics

**LEARNING OBJECTIVES**

1. Discuss the physical origins and practical applications of using dielectric materials in capacitors.
2. Evaluate the capacitance for capacitors with dielectrics.
3. Describe the physical origins of the polarization of molecules.

A **dielectric** is an insulating material, such as rubber, plastic, or waxed paper. When a dielectric is inserted between the plates of a capacitor, the capacitance increases. If the dielectric completely fills the space between the plates, the capacitance is multiplied by the factor $\kappa$, called the **dielectric constant**.

The following experiment illustrates the effect of a dielectric in a capacitor. Consider a parallel-plate capacitor of charge $Q_0$ and capacitance $C_0$ in the absence of a dielectric. The potential difference across the capacitor plates can be measured, and is given by $\Delta V_0 = Q_0/C_0$ (Fig. 16.24a). Because the capacitor is not connected to an external circuit, there is no pathway for charge to leave or be added to the plates. If a dielectric is now inserted between the plates as in Figure 16.24b, the voltage across the plates is *reduced* by the factor $\kappa$ to the value

$$\Delta V = \frac{\Delta V_0}{\kappa}$$

Because $\kappa > 1$, $\Delta V$ is less than $\Delta V_0$. Because the charge $Q_0$ on the capacitor doesn't change, we conclude that the capacitance in the presence of the dielectric must change to the value

$$C = \frac{Q_0}{\Delta V} = \frac{Q_0}{\Delta V_0/\kappa} = \frac{\kappa Q_0}{\Delta V_0}$$

or

$$C = \kappa C_0 \qquad [16.18]$$

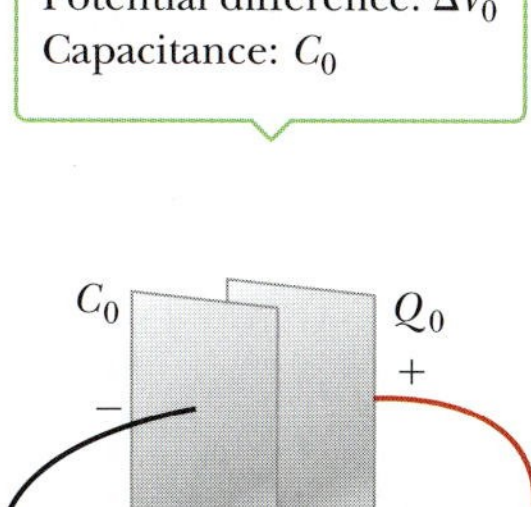


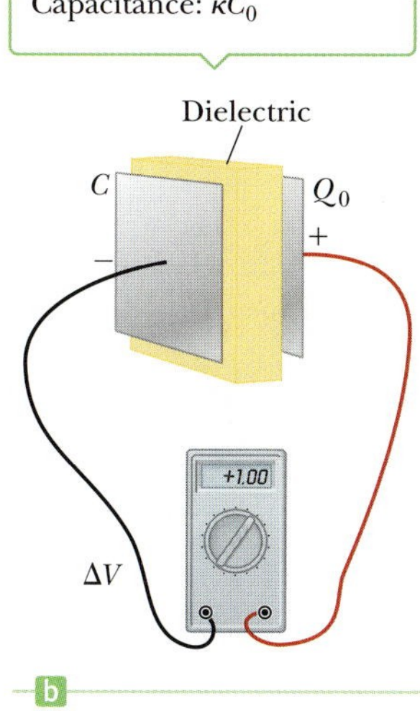

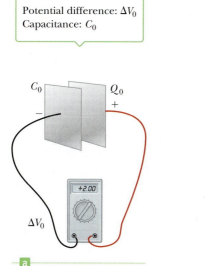

**Figure 16.24** When a dielectric with dielectric constant $\kappa$ is inserted in a charged capacitor that is *not* connected to a battery, the potential difference is reduced to $\Delta V = \Delta V_0/\kappa$ and the capacitance increases to $C = \kappa C_0$.

According to this result, the capacitance is *multiplied* by the factor $\kappa$ when the dielectric fills the region between the plates. For a parallel-plate capacitor, where the capacitance in the absence of a dielectric is $C_0 = \epsilon_0 A/d$, we can express the capacitance in the presence of a dielectric as

$$C = \kappa\epsilon_0 \frac{A}{d} \qquad [16.19]$$

From this result, it appears that the capacitance could be made very large by decreasing $d$, the separation between the plates. In practice the lowest value of $d$ is limited by the electric discharge that can occur through the dielectric material separating the plates. For any given plate separation, there is a maximum electric field that can be produced in the dielectric before it breaks down and begins to conduct. This maximum electric field is called the **dielectric strength**, and for air its value is about $3 \times 10^6$ V/m. Most insulating materials have dielectric strengths greater than that of air, as indicated by the values listed in Table 16.1. Figure 16.25 shows an instance of dielectric breakdown in air.

Visuals Unlimited/Corbis

**Figure 16.25** Dielectric breakdown in air. Sparks are produced when a large alternating voltage is applied across the wires by a high-voltage induction coil power supply.

Commercial capacitors are often made by using metal foil interlaced with thin sheets of paraffin-impregnated paper or Mylar®, which serves as the dielectric material. These alternate layers of metal foil and dielectric are rolled into a small cylinder (Fig. 16.26a on page 586). One type of a high-voltage capacitor consists of

**Table 16.1** Dielectric Constants and Dielectric Strengths of Various Materials at Room Temperature

| Material | Dielectric Constant $\kappa$ | Dielectric Strength (V/m) |
|---|---|---|
| Air | 1.000 59 | $3 \times 10^6$ |
| Bakelite® | 4.9 | $24 \times 10^6$ |
| Fused quartz | 3.78 | $8 \times 10^6$ |
| Neoprene rubber | 6.7 | $12 \times 10^6$ |
| Nylon | 3.4 | $14 \times 10^6$ |
| Paper | 3.7 | $16 \times 10^6$ |
| Polystyrene | 2.56 | $24 \times 10^6$ |
| Pyrex® glass | 5.6 | $14 \times 10^6$ |
| Silicone oil | 2.5 | $15 \times 10^6$ |
| Strontium titanate | 233 | $8 \times 10^6$ |
| Teflon® | 2.1 | $60 \times 10^6$ |
| Vacuum | 1.000 00 | — |
| Water | 80 | — |

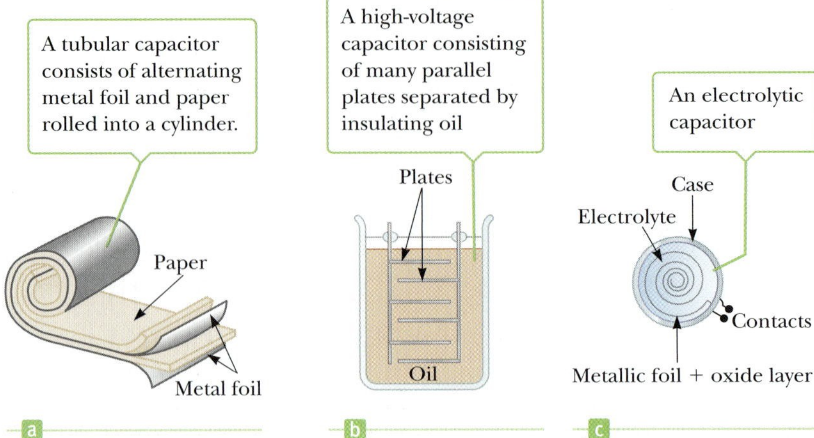
A tubular capacitor consists of alternating metal foil and paper rolled into a cylinder.
Paper
Metal foil
a
A high-voltage capacitor consisting of many parallel plates separated by insulating oil
Plates
Oil
b
An electrolytic capacitor
Case
Electrolyte
Contacts
Metallic foil + oxide layer
c

**Figure 16.26** Three commercial capacitor designs.

Chris Vuille

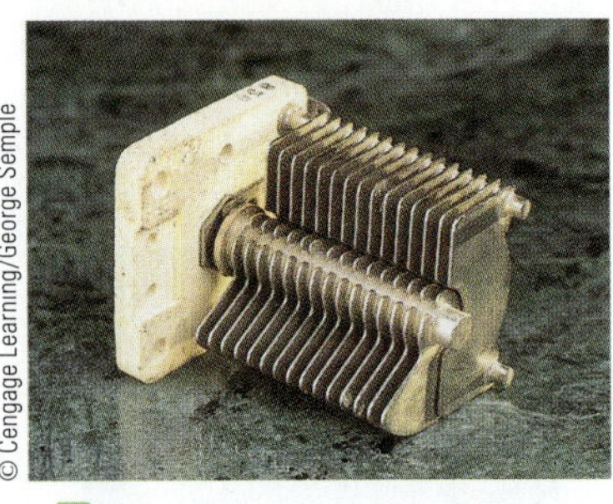
© Cengage Learning/George Semple

**Figure 16.27** (a) A collection of capacitors used in a variety of applications. (b) A variable capacitor. When one set of metal plates is rotated so as to lie between a fixed set of plates, the capacitance of the device changes.

a number of interwoven metal plates immersed in silicone oil (Fig. 16.26b). Small capacitors are often constructed from ceramic materials. Variable capacitors (typically 10 pF to 500 pF) usually consist of two interwoven sets of metal plates, one fixed and the other movable, with air as the dielectric.

An electrolytic capacitor (Fig. 16.26c) is often used to store large amounts of charge at relatively low voltages. It consists of a metal foil in contact with an electrolyte—a solution that conducts charge by virtue of the motion of the ions contained in it. When a voltage is applied between the foil and the electrolyte, a thin layer of metal oxide (an insulator) is formed on the foil, and this layer serves as the dielectric. Enormous capacitances can be attained because the dielectric layer is very thin.

Figure 16.27 shows a variety of commercially available capacitors. Variable capacitors are used in radios to adjust the frequency.

When electrolytic capacitors are used in circuits, the polarity (the plus and minus signs on the device) must be observed. If the polarity of the applied voltage is opposite that intended, the oxide layer will be removed and the capacitor will conduct rather than store charge. Further, reversing the polarity can result in such a large current that the capacitor may either burn or produce steam and explode.

## APPLYING PHYSICS 16.2 Stud Finders

If you have ever tried to hang a picture on a wall securely, you know that it can be difficult to locate a wooden stud in which to anchor your nail or screw. The principles discussed in this section can be used to detect a stud electronically. The primary element of an electronic stud finder is a capacitor with its plates arranged side by side instead of facing one another, as in Figure 16.28. How does this device work?

**EXPLANATION** As the detector is moved along a wall, its capacitance changes when it passes across a stud because the dielectric constant of the material "between" the plates changes. The change in capacitance can be used to cause a light to come on, signaling the presence of the stud. ■

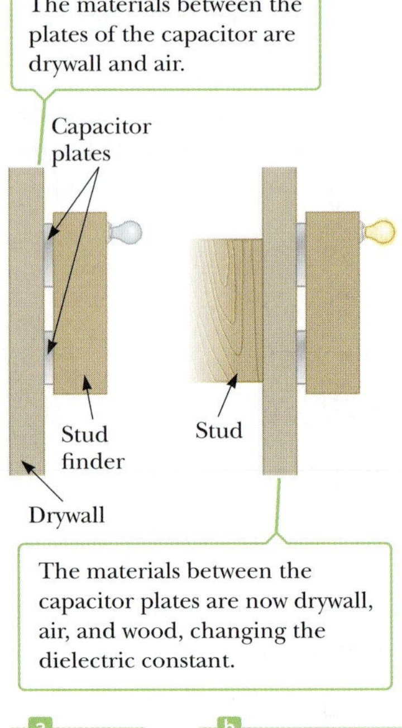
The materials between the plates of the capacitor are drywall and air.
Capacitor plates
Stud finder
Stud
Drywall
The materials between the capacitor plates are now drywall, air, and wood, changing the dielectric constant.
a
b

**Figure 16.28** (Applying Physics 16.2) A stud finder produces an electric field that is affected by the dielectric constant of the materials placed in its field. When the device moves across a stud, the change in dielectric constant activates a signal light.

**Quick Quiz**

**16.10** A fully charged parallel-plate capacitor remains connected to a battery while a dielectric is slid between the plates. Do the following quantities increase, decrease, or stay the same? (a) $C$ (b) $Q$ (c) $E$ between the plates (d) $\Delta V$ (e) energy stored in the capacitor

## EXAMPLE 16.11 A Paper-Filled Capacitor

**GOAL** Calculate fundamental physical properties of a parallel-plate capacitor with a dielectric.

**PROBLEM** A parallel-plate capacitor has plates 2.0 cm by 3.0 cm. The plates are separated by a 1.0-mm thickness of paper. Find **(a)** the capacitance of this device and **(b)** the maximum charge that can be placed on the capacitor. **(c)** After the fully charged capacitor is disconnected from the battery, the dielectric is subsequently removed. Find the new electric field across the capacitor. Does the capacitor discharge?

**STRATEGY** For part (a), obtain the dielectric constant for paper from Table 16.1 and substitute, with other given quantities, into Equation 16.19. For part (b), note that Table 16.1 also gives the dielectric strength of paper, which is the maximum electric field that can be applied before electrical breakdown occurs. Use Equation 16.3, $\Delta V = Ed$, to obtain the maximum voltage and substitute into the basic capacitance equation. For part (c), remember that disconnecting the battery traps the extra charge on the plates, which must remain even after the dielectric is removed. Find the charge density on the plates and use Gauss's law to find the new electric field between the plates.

**SOLUTION**

**(a)** Find the capacitance of this device.

Substitute into Equation 16.19:

$$C = \kappa\epsilon_0 \frac{A}{d}$$

$$= 3.7\left(8.85 \times 10^{-12}\,\frac{\text{C}^2}{\text{N}\cdot\text{m}^2}\right)\left(\frac{6.0 \times 10^{-4}\,\text{m}^2}{1.0 \times 10^{-3}\,\text{m}}\right)$$

$$= 2.0 \times 10^{-11}\,\text{F}$$

**(b)** Find the maximum charge that can be placed on the capacitor.

Calculate the maximum applied voltage, using the dielectric strength of paper, $E_{max}$:

$$\Delta V_{max} = E_{max}d = (16 \times 10^6\,\text{V/m})(1.0 \times 10^{-3}\,\text{m})$$

$$= 1.6 \times 10^4\,\text{V}$$

Solve the basic capacitance equation for $Q_{max}$ and substitute $\Delta V_{max}$ and $C$:

$$Q_{max} = C\,\Delta V_{max} = (2.0 \times 10^{-11}\,\text{F})(1.6 \times 10^4\,\text{V})$$

$$= 0.32\,\mu\text{C}$$

**(c)** Suppose the fully charged capacitor is disconnected from the battery and the dielectric is subsequently removed. Find the new electric field between the plates of the capacitor. Does the capacitor discharge?

Compute the charge density on the plates:

$$\sigma = \frac{Q_{max}}{A} = \frac{3.2 \times 10^{-7}\,\text{C}}{6.0 \times 10^{-4}\,\text{m}^2} = 5.3 \times 10^{-4}\,\text{C/m}^2$$

Calculate the electric field from the charge density:

$$E = \frac{\sigma}{\epsilon_0} = \frac{5.3 \times 10^{-4}\,\text{C/m}^2}{8.85 \times 10^{-12}\,\text{C}^2/\text{m}^2\cdot\text{N}} = 6.0 \times 10^7\,\text{N/C}$$

Because the electric field without the dielectric exceeds the value of the dielectric strength of air, the capacitor discharges across the gap.

**REMARKS** Dielectrics allow $\kappa$ times as much charge to be stored on a capacitor for a given voltage. They also allow an increase in the applied voltage by increasing the threshold of electrical breakdown.

**QUESTION 16.11** Without the paper dielectric, is the maximum charge that can be stored on this capacitor (a) larger than, (b) smaller than, or (c) the same as found in part (b)?

*(Continued)*

**EXERCISE 16.11** A parallel-plate capacitor has plate area of $2.50 \times 10^{-3}$ m$^2$ and distance between the plates of 2.00 mm. (a) Find the maximum charge that can be placed on the capacitor if air is between the plates. (b) Find the maximum charge if the air is replaced by polystyrene.

**ANSWERS** (a) $7 \times 10^{-8}$ C (b) $1.4 \times 10^{-6}$ C

## EXAMPLE 16.12 Capacitors with Two Dielectrics

**GOAL** Derive a symbolic expression for a parallel-plate capacitor with two dielectrics.

**PROBLEM** A parallel-plate capacitor has dielectrics with constants $\kappa_1$ and $\kappa_2$ between the two plates, as shown in Figure 16.29. Each dielectric fills exactly half the volume between the plates. Derive expressions for **(a)** the potential difference between the two plates and **(b)** the resulting capacitance of the system.

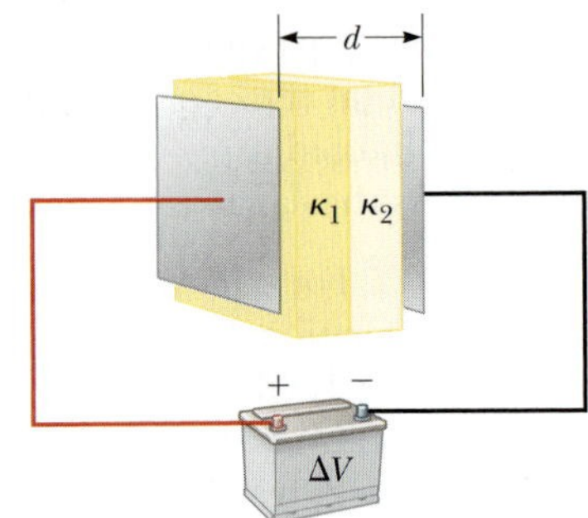


**Figure 16.29** (Exercise 16.12)

**STRATEGY** The magnitude of the potential difference between the two plates of a capacitor is equal to the electric field multiplied by the plate separation. The electric field in a region is reduced by a factor of $1/\kappa$ when a dielectric is introduced, so $E = \sigma/\epsilon = \sigma/\kappa\epsilon_0$. Add the potential difference across each dielectric to find the total potential difference $\Delta V$ between the plates. The voltage difference across each dielectric is given by $\Delta V = Ed$, where $E$ is the electric field and $d$ the displacement. Obtain the capacitance from the relationship $C = Q/\Delta V$.

**SOLUTION**

**(a)** Derive an expression for the potential difference between the two plates.

Write a general expression for the potential difference across both slabs:

$$\Delta V = \Delta V_1 + \Delta V_2 = E_1 d_1 + E_2 d_2$$

Substitute expressions for the electric fields and dielectric thicknesses, $d_1 = d_2 = d/2$:

$$\Delta V = \frac{\sigma}{\kappa_1\epsilon_0}\frac{d}{2} + \frac{\sigma}{\kappa_2\epsilon_0}\frac{d}{2} = \frac{\sigma d}{2\epsilon_0}\left(\frac{1}{\kappa_1} + \frac{1}{\kappa_2}\right)$$

**(b)** Derive an expression for the resulting capacitance of the system.

Write the general expression for capacitance:

$$C = \frac{Q}{\Delta V}$$

Substitute $Q = \sigma A$ and the expression for the potential difference from part (a):

$$C = \frac{\sigma A}{\dfrac{\sigma d}{2\epsilon_0}\left(\dfrac{1}{\kappa_1} + \dfrac{1}{\kappa_2}\right)} = \frac{2\epsilon_0 A}{d}\frac{\kappa_1\kappa_2}{\kappa_1 + \kappa_2}$$

**REMARKS** The answer is the same as if there had been two capacitors in series with the respective dielectrics. When a capacitor consists of two dielectrics as shown in Figure 16.30, however, it's equivalent to two different capacitors in parallel.

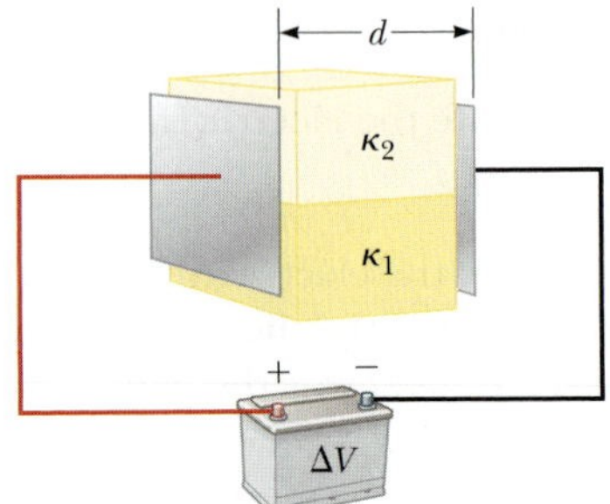


**Figure 16.30** (Exercise 16.12)

**QUESTION 16.12** What answer is obtained when the two dielectrics are removed so there is vacuum between the plates?

**EXERCISE 16.12** Suppose a capacitor has two dielectrics arranged as shown in Figure 16.30, each dielectric filling exactly half of the volume between the two plates. Derive an expression for the capacitance if each dielectric fills exactly half the volume between the plates.

**ANSWER** $C = \dfrac{\kappa_1 + \kappa_2}{2}\dfrac{\epsilon_0 A}{d}$

## An Atomic Description of Dielectrics

The explanation of why a dielectric increases the capacitance of a capacitor is based on an atomic description of the material, which in turn involves a property of some molecules called **polarization.** A molecule is said to be polarized when there is a separation between the average positions of its negative charge and its positive charge. In some molecules, such as water, this condition is always present. To see why, consider the geometry of a water molecule (Fig. 16.31).

The molecule is arranged so that the negative oxygen atom is bonded to the positively charged hydrogen atoms with a 105° angle between the two bonds. The center of negative charge is at the oxygen atom, and the center of positive charge lies at a point midway along the line joining the hydrogen atoms (point $x$ in the diagram). Materials composed of molecules that are permanently polarized in this way have large dielectric constants, and indeed, Table 16.1 shows that the dielectric constant of water is large ($\kappa = 80$) compared with other common substances.

A symmetric molecule (Fig. 16.32a) can have no permanent polarization, but a polarization can be induced in it by an external electric field. A field directed to the left, as in Figure 16.32b, would cause the center of positive charge to shift to the left from its initial position and the center of negative charge to shift to the right. This *induced polarization* is the effect that predominates in most materials used as dielectrics in capacitors.

To understand why the polarization of a dielectric can affect capacitance, consider the slab of dielectric shown in Figure 16.33. Before placing the slab between the plates of the capacitor, the polar molecules are randomly oriented (Fig. 16.33a). The polar molecules are dipoles, and each creates a dipole electric field, but because of their random orientation, this field averages to zero.

After insertion of the dielectric slab into the electric field $\vec{\mathbf{E}}_0$ between the plates (Fig. 16.33b), the positive plate attracts the negative ends of the dipoles and the negative plate attracts the positive ends of the dipoles. These forces exert a torque on the molecules making up the dielectric, reorienting them so that on average the negative pole is more inclined toward the positive plate and the positive pole is more aligned toward the negative plate. The positive and negative charges in the middle still cancel each other, but there is a net accumulation of negative charge in the dielectric next to the positive plate and a net accumulation of positive charge next to the negative plate. This configuration can be modeled as an additional pair of charged plates, as in Figure 16.33c, creating an induced electric field $\vec{\mathbf{E}}_{\text{ind}}$ that partly cancels the original electric field $\vec{\mathbf{E}}_0$. If the battery is not connected when the dielectric is inserted, the potential difference $\Delta V_0$ across the plates is reduced to $\Delta V_0/\kappa$.

If the capacitor is still connected to the battery, however, the negative poles push more electrons off the positive plate, making it more positive. Meanwhile, the

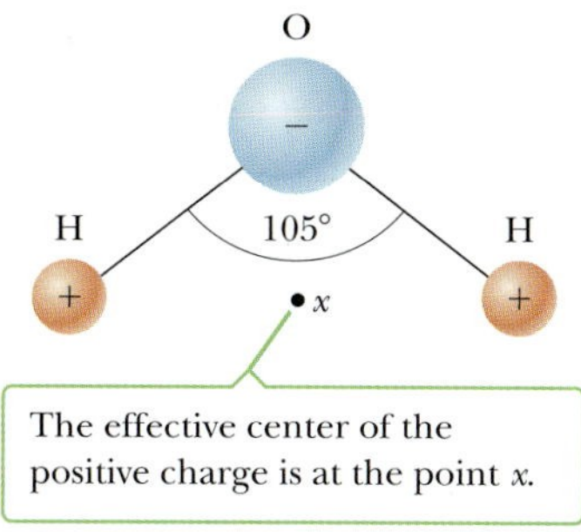


**Figure 16.31** The water molecule, $H_2O$, has a permanent polarization resulting from its bent geometry.

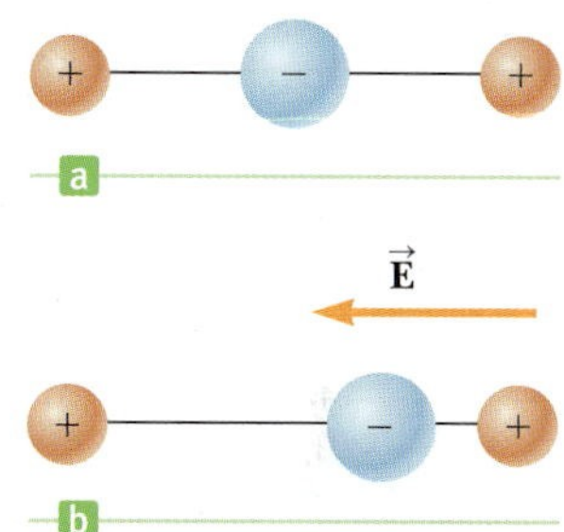


**Figure 16.32** (a) A symmetric molecule has no permanent polarization. (b) An external electric field induces a polarization in the molecule.

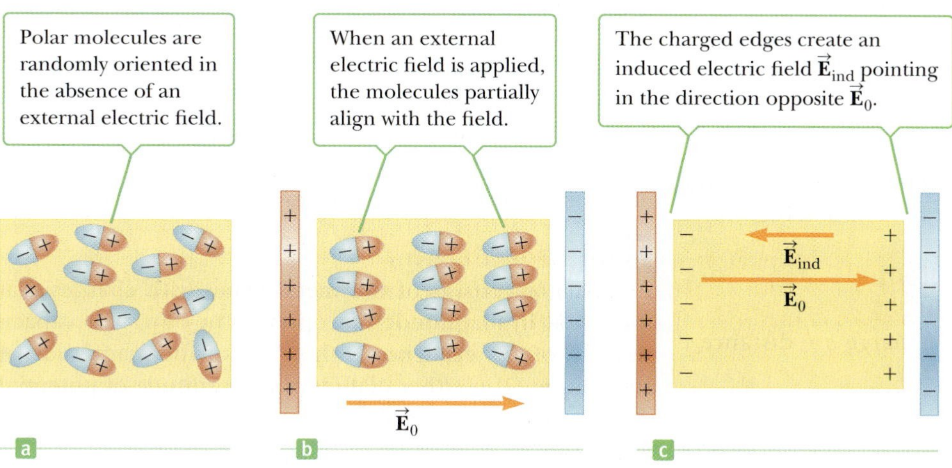

Polar molecules are randomly oriented in the absence of an external electric field.
When an external electric field is applied, the molecules partially align with the field.
The charged edges create an induced electric field E_ind pointing in the direction opposite E_0.
E_ind
E_0
E_0
a
b
c


**Figure 16.33** (a) Polar molecules are randomly oriented in a dielectric. (b) An electric field is applied to the dielectric. (c) The charged edges of the dielectric act like an additional pair of parallel plates, reducing the overall field between the actual plates. The interior of the dielectric is still neutral.

positive poles attract more electrons onto the negative plate. This situation continues until the potential difference across the battery reaches its original magnitude, equal to the potential gain across the battery. The net effect is an increase in the amount of charge stored on the capacitor. Because the plates can store more charge for a given voltage, it follows from $C = Q\Delta V$ that the capacitance must increase.

**Quick Quiz**

**16.11** Consider a parallel-plate capacitor with a dielectric material between the plates. If the temperature of the dielectric increases, does the capacitance (a) decrease, (b) increase, or (c) remain the same?

## SUMMARY

### 16.1 Potential Difference and Electric Potential

The change in the electric potential energy of a system consisting of an object of charge $q$ moving through a displacement $\Delta x$ in a constant electric field $\vec{\mathbf{E}}$ is given by

$$\Delta PE = -W_{AB} = -qE_x\,\Delta x \qquad [16.1]$$

When a charge $q$ moves in a uniform electric field $\vec{\mathbf{E}}$ from point $A$ to point $B$, the work done on the charge by the electric force is $qE_x\,\Delta x$.

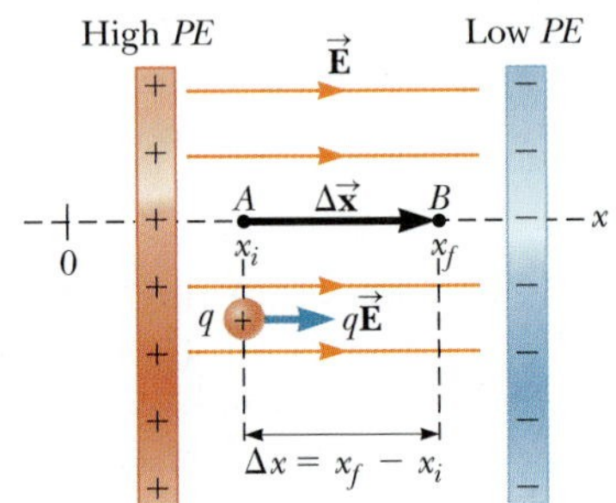


where $E_x$ is the component of the electric field in the $x$-direction and $\Delta x = x_f - x_i$. The **difference in electric potential** between two points $A$ and $B$ is

$$\Delta V = V_B - V_A = \frac{\Delta PE}{q} \qquad [16.2]$$

where $\Delta PE$ is the *change* in electrical potential energy as a charge $q$ moves between $A$ and $B$. The units of potential difference are joules per coulomb, or **volts;** 1 J/C = 1 V.

The **electric potential difference** between two points $A$ and $B$ in a *uniform* electric field $\vec{\mathbf{E}}$ is

$$\Delta V = -E_x\,\Delta x \qquad [16.3]$$

where $\Delta x = x_f - x_i$ is the displacement between $A$ and $B$ and $E_x$ is the $x$-component of the electric field in that region.

### 16.2 Electric Potential and Potential Energy Due to Point Charges

The **electric potential** due to a point charge $q$ at distance $r$ from the point charge is

$$V = k_e\frac{q}{r} \qquad [16.4]$$

Electric field and electric potential versus distance from a point charge of $1.11 \times 10^{-10}$ C. Note that $V$ is proportional to $1/r$, whereas $E$ is proportional to $1/r^2$.

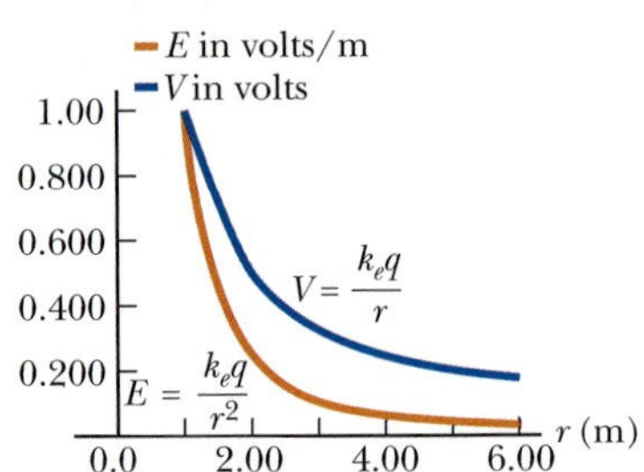


The **electric potential energy** of a pair of point charges separated by distance $r$ is

$$PE = k_e\frac{q_1q_2}{r} \qquad [16.5]$$

These equations can be used in the solution of conservation of energy problems and in the work–energy theorem.

### 16.3 Potentials and Charged Conductors

### 16.4 Equipotential Surfaces

Every point on the surface of a charged conductor in electrostatic equilibrium is at the same potential. Further, the potential is constant everywhere inside the conductor and equals its value on the surface.

The **electron volt** is defined as the energy that an electron (or proton) gains when accelerated through a potential difference of 1 V. The conversion between electron volts and joules is

$$1\text{ eV} = 1.60 \times 10^{-19}\text{ C}\cdot\text{V} = 1.60 \times 10^{-19}\text{ J} \qquad [16.7]$$

Any surface on which the potential is the same at every point is called an equipotential surface. The electric field is always oriented perpendicular to an equipotential surface.

### 16.6 Capacitance

A capacitor consists of two metal plates with charges that are equal in magnitude but opposite in sign. The capacitance $C$ of any capacitor is the ratio of the magnitude of the charge $Q$ on either plate to the magnitude of potential difference $\Delta V$ between them:

$$C \equiv \frac{Q}{\Delta V} \qquad [16.8]$$

Capacitance has the units coulombs per volt, or farads; 1 C/V = 1 F.

## 16.7 The Parallel-Plate Capacitor

The capacitance of two parallel metal plates of area $A$ separated by distance $d$ is

$$C = \epsilon_0 \frac{A}{d} \qquad [16.9]$$

where $\epsilon_0 = 8.85 \times 10^{-12}\ C^2/N \cdot m^2$ is a constant called the **permittivity of free space.**

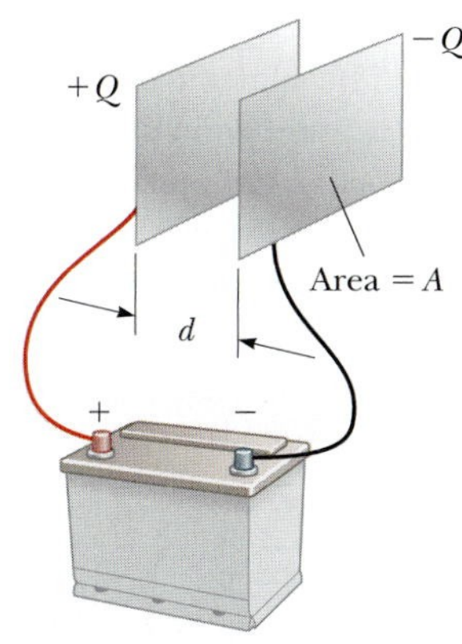


A parallel-plate capacitor consists of two parallel plates, each of area $A$, separated by a distance $d$.

## 16.8 Combinations of Capacitors

The **equivalent capacitance of a parallel combination** of capacitors is

$$C_{eq} = C_1 + C_2 + C_3 + \cdots \qquad [16.12]$$

If two or more capacitors are connected in series, the **equivalent capacitance of the series combination** is

$$\frac{1}{C_{eq}} = \frac{1}{C_1} + \frac{1}{C_2} + \frac{1}{C_3} + \cdots \qquad [16.15]$$

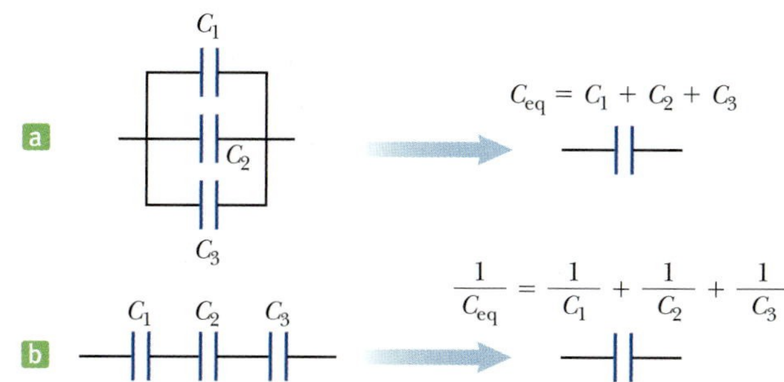


Capacitors in (a) parallel or in (b) series can be written as a single equivalent capacitor.

Problems involving a combination of capacitors can be solved by applying Equations 16.12 and 16.15 repeatedly to a circuit diagram, simplifying it as much as possible. This step is followed by working backwards to the original diagram, applying $C = Q/\Delta V$, that parallel capacitors have the same voltage drop, and that series capacitors have the same charge.

## 16.9 Energy Stored in a Charged Capacitor

Three equivalent expressions for calculating the **energy stored** in a charged capacitor are

$$\text{Energy stored} = \tfrac{1}{2}Q\,\Delta V = \tfrac{1}{2}C(\Delta V)^2 = \frac{Q^2}{2C} \qquad [16.17]$$

## 16.10 Capacitors with Dielectrics

When a nonconducting material, called a **dielectric,** is placed between the plates of a capacitor, the capacitance is multiplied by the factor $\kappa$, which is called the **dielectric constant**, a property of the dielectric material. The capacitance of a parallel-plate capacitor filled with a dielectric is

$$C = \kappa\epsilon_0 \frac{A}{d} \qquad [16.19]$$

## WARM-UP EXERCISES

ENHANCED WebAssign The warm-up exercises in this chapter may be assigned online in Enhanced WebAssign.

1. **Math Review** Determine the value of $C_{eq}$ in the following equation if $C_1 = 5.00\ \Omega$, $C_2 = 6.00\ \Omega$, and $C_3 = 7.00\ \Omega$: $\frac{1}{C_{eq}} = \frac{1}{C_1} + \frac{1}{C_2} + \frac{1}{C_3}$. (See also Section 16.8.)

2. **Physics Review** A 2.50-kg object initially at rest has a gravitational potential energy of 72.0 J. Determine the speed of the object when it has moved under the influence of gravity to a location where its gravitational potential energy is 45.0 J. (See Section 5.3.)

3. A uniform electric field of magnitude 3.00 N/C is directed along the $+x$-axis. If a 2.00 $\mu$C charge moves from (1.00, 0) m to (2.50, 0) m in this field, determine (a) the work done by the electric force, (b) the change in the electric potential energy of the particle, and (c) the electric potential difference between the particle's initial and final points. (See Section 16.1.)

4. A +4.00 $\mu$C charge is located at the origin. Determine (a) the electric potential at a distance of 2.00 m from the charge, and (b) the electric potential energy of a −2.03 $\mu$C charge located at (0, 2.00) m. (See Section 16.2.)

5. Two protons are located at (1.00, 0) m and (0,1.50) m, respectively. Determine (a) the electric potential at the origin, and (b) the electric potential energy of a third proton located at the origin. (See Section 16.2.)

6. (a) A hydrogen atom can be ionized by a photon having an energy of 13.6 eV. Convert that quantity to joules. (b) It requires 4.186 J of thermal energy to warm a gram of water by 1.00 K. Convert that energy to electron volts. (See Section 16.3.)

7. A capacitor with capacitance 3.00 $\mu$F is connected to a 9.00-V battery. (a) Find the charge on the capacitor in coulombs. (b) What voltage battery would be required to store $7.20 \times 10^{-5}$ C on the capacitor? (See Section 16.6.)

8. A parallel-plate capacitor with no dielectric is made from two plates separated by $1.15 \times 10^{-3}$ m. (a) Determine the device's capacitance if each plate has an area of 0.250 m$^2$. (See Section 16.6.) (b) What magnitude of charge is stored on each plate when the capacitor is connected to a 12.0-V battery? (See Section 16.6.) (c) If a material of dielectric constant 12.5 is inserted between the plates, what are the new values of the capacitance and (d) the magnitude of charge stored on each plate? (See Section 16.10.)

9. Two capacitors have capacitance 2.00 $\mu$F and 3.00 $\mu$F, respectively. Calculate the equivalent capacitance in microfarads if the capacitors are put (a) in parallel with each other, and (b) in series with each other. (See Section 16.8.)

10. Two capacitors are connected in parallel across a 12.0-V battery. If their capacitances are 12.0 $\mu$F and 25.0 $\mu$F, determine (a) the voltage across each capacitor, (b) the magnitude of charge stored on each plate of the 12.0 $\mu$F capacitor, (c) the magnitude of charge stored on each plate of the 25.0 $\mu$F capacitor, and (d) the equivalent capacitance of the system. (See Section 16.8.)

11. Two capacitors are connected in series between the terminals of a 12.0-V battery. If their capacitances are 12.0 $\mu$F and 25.0 $\mu$F, determine (a) the equivalent capacitance of the system, (b) the magnitude of charge stored on each plate of either capacitor, (c) the voltage across the 12.0 $\mu$F capacitor, and (d) the voltage across the 25.0 $\mu$F capacitor. (See Section 16.8.)

12. The plates of a parallel-plate capacitor are charged to a potential difference of 12.0 V. If the capacitance is 15.0 $\mu$F, calculate (a) the energy stored in the capacitor and (b) the magnitude of charge stored on each plate of the capacitor. (See Section 16.9.)

## CONCEPTUAL QUESTIONS

ENHANCED WebAssign The conceptual questions in this chapter may be assigned online in Enhanced WebAssign.

1. (a) Describe the motion of a proton after it is released from rest in a uniform electric field. (b) Describe the changes (if any) in its kinetic energy and the electric potential energy associated with the proton.

2. Rank the potential energies of the four systems of particles shown in Figure CQ16.2 from largest to smallest. Include equalities if appropriate.

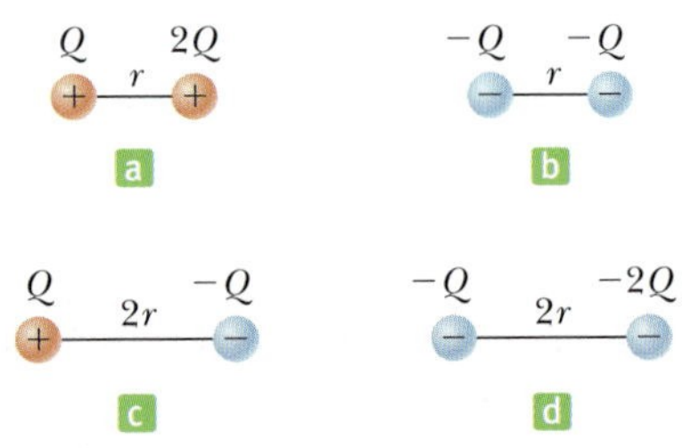

**Figure CQ16.2**

3. A parallel-plate capacitor is charged by a battery, and the battery is then disconnected from the capacitor. Because the charges on the capacitor plates are opposite in sign, they attract each other. Hence, it takes positive work to increase the plate separation. Show that the external work done when the plate separation is increased leads to an increase in the energy stored in the capacitor.

4. When charged particles are separated by an infinite distance, the electric potential energy of the pair is zero. When the particles are brought close, the electric potential energy of a pair with the same sign is positive, whereas the electric potential energy of a pair with opposite signs is negative. Explain.

5. Suppose you are sitting in a car and a 20-kV power line drops across the car. Should you stay in the car or get out? The power line potential is 20 kV compared to the potential of the ground.

6. Why is it important to avoid sharp edges or points on conductors used in high-voltage equipment?

7. Explain why, under static conditions, all points in a conductor must be at the same electric potential.

8. If you are given three different capacitors $C_1$, $C_2$, and $C_3$, how many different combinations of capacitance can you produce, using all capacitors in your circuits?

9. (a) Why is it dangerous to touch the terminals of a high-voltage capacitor even after the voltage source that charged the battery is disconnected from the capacitor? (b) What can be done to make the capacitor safe to handle after the voltage source has been removed?

10. The plates of a capacitor are connected to a battery. (a) What happens to the charge on the plates if the connecting wires are removed from the battery? (b) What happens to the charge if the wires are removed from the battery and connected to each other?

11. Rank the electric potentials at the four points shown in Figure CQ16.11 from largest to smallest.

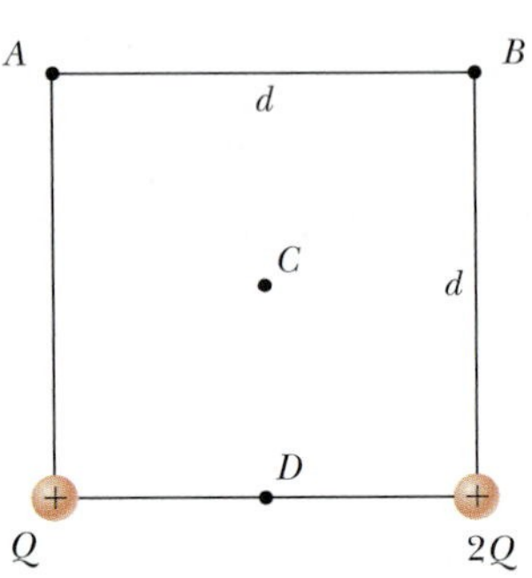

**Figure CQ16.11**

12. If you were asked to design a capacitor in which small size and large capacitance were required, what would be the two most important factors in your design?

13. Is it always possible to reduce a combination of capacitors to one equivalent capacitor with the rules developed in this chapter? Explain.

14. Explain why a dielectric increases the maximum operating voltage of a capacitor even though the physical size of the capacitor doesn't change.

## PROBLEMS

ENHANCED WebAssign The problems in this chapter may be assigned online in Enhanced WebAssign.

1. denotes straightforward problem; 2. denotes intermediate problem;
3. denotes challenging problem
1. denotes full solution available in *Student Solutions Manual/Study Guide*
1. denotes problems most often assigned in Enhanced WebAssign

BIO denotes biomedical problems
GP denotes guided problems
M denotes Master It tutorial available in Enhanced WebAssign
Q|C denotes asking for quantitative and conceptual reasoning
S denotes symbolic reasoning problem
W denotes Watch It video solution available in Enhanced WebAssign

### 16.1 Potential Difference and Electric Potential

1. A uniform electric field of magnitude 375 N/C pointing in the positive $x$-direction acts on an electron, which is initially at rest. After the electron has moved 3.20 cm, what is (a) the work done by the field on the electron, (b) the change in potential energy associated with the electron, and (c) the velocity of the electron?

2. A proton is released from rest in a uniform electric field of magnitude 385 N/C. Find (a) the electric force on the proton, (b) the acceleration of the proton, and (c) the distance it travels in 2.00 $\mu$s.

3. BIO W A potential difference of 90 mV exists between the inner and outer surfaces of a cell membrane. The inner surface is negative relative to the outer surface. How much work is required to eject a positive sodium ion ($Na^+$) from the interior of the cell?

4. A metal sphere of radius 5.00 cm is initially uncharged. How many electrons would have to be placed on the sphere to produce an electric field of magnitude $1.50 \times 10^5$ N/C at a point 8.00 cm from the center of the sphere?

5. The potential difference between the accelerating plates of a TV set is about 25 kV. If the distance between the plates is 1.5 cm, find the magnitude of the uniform electric field in the region between the plates.

6. A point charge $q = +40.0$ $\mu$C moves from $A$ to $B$ separated by a distance $d = 0.180$ m in the presence of an external electric field $\vec{\mathbf{E}}$ of magnitude 275 N/C directed toward the right as in Figure P16.6. Find (a) the electric force exerted on the charge, (b) the work done by the electric force, (c) the change in the electric potential energy of the charge, and (d) the potential difference between $A$ and $B$.

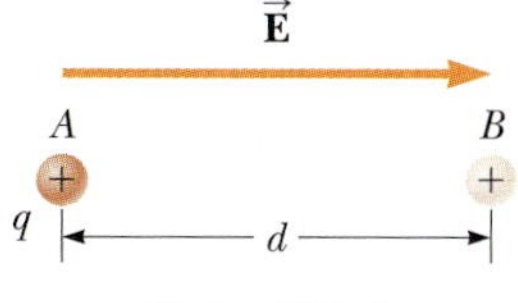

**Figure P16.6**

7. M Oppositely charged parallel plates are separated by 5.33 mm. A potential difference of 600 V exists between the plates. (a) What is the magnitude of the electric field between the plates? (b) What is the magnitude of the force on an electron between the plates? (c) How much work must be done on the electron to move it to the negative plate if it is initially positioned 2.90 mm from the positive plate?

8. Q|C S (a) Find the potential difference $\Delta V_e$ required to stop an electron (called a "stopping potential") moving with an initial speed of $2.85 \times 10^7$ m/s. (b) Would a proton traveling at the same speed require a greater or lesser magnitude potential difference? Explain. (c) Find a symbolic expression for the ratio of the proton stopping potential and the electron stopping potential, $\Delta V_p/\Delta V_e$. The answer should be in terms of the proton mass $m_p$ and electron mass $m_e$.

9. Q|C A 74.0-g block carrying a charge $Q = 35.0$ $\mu$C is connected to a spring for which $k = 78.0$ N/m. The block lies on a frictionless, horizontal surface and is immersed in a uniform electric field of magnitude $E = 4.86 \times 10^4$ N/C directed as shown in Figure P16.9. If the block is released from rest when the spring is unstretched ($x = 0$), (a) by what maximum distance does the block move from its initial position? (b) Find the subsequent equilibrium position of the block and the amplitude of its motion. (c) Using conservation of energy, find a symbolic relationship giving the potential difference between its initial position and the point of maximum extension in terms of the spring constant $k$, the amplitude $A$, and the charge $Q$.

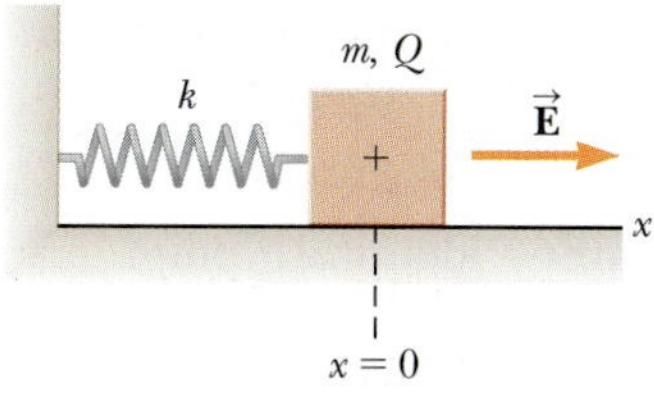

**Figure P16.9**

10. On planet Tehar, the free-fall acceleration is the same as that on the Earth, but there is also a strong downward electric field that is uniform close to the planet's surface. A 2.00-kg ball having a charge of 5.00 $\mu$C is thrown upward at a speed of 20.1 m/s. It hits the ground after an interval of 4.10 s. What is the potential difference between the starting point and the top point of the trajectory?

### 16.2 Electric Potential and Potential Energy Due to Point Charges

### 16.3 Potentials and Charged Conductors

### 16.4 Equipotential Surfaces

11. **QIC** An electron is at the origin. (a) Calculate the electric potential $V_A$ at point $A$, $x = 0.250$ cm. (b) Calculate the electric potential $V_B$ at point $B$, $x = 0.750$ cm. What is the potential difference $V_B - V_A$? (c) Would a negatively charged particle placed at point $A$ necessarily go through this same potential difference upon reaching point $B$? Explain.

12. The two charges in Figure P16.12 are separated by $d = 2.00$ cm. Find the electric potential at (a) point $A$ and (b) point $B$, which is halfway between the charges.

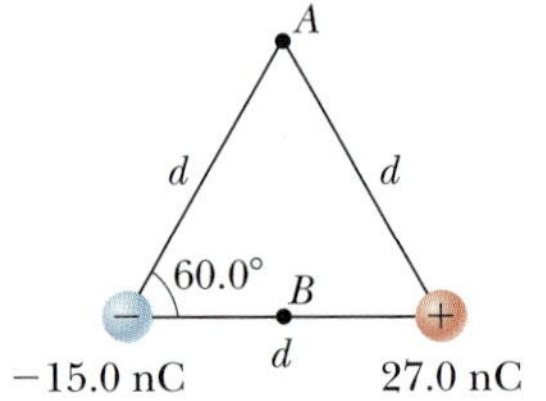

**Figure P16.12**

13. (a) Find the electric potential, taking zero at infinity, at the upper right corner (the corner without a charge) of the rectangle in Figure P16.13. (b) Repeat if the 2.00-$\mu$C charge is replaced with a charge of $-2.00$ $\mu$C.

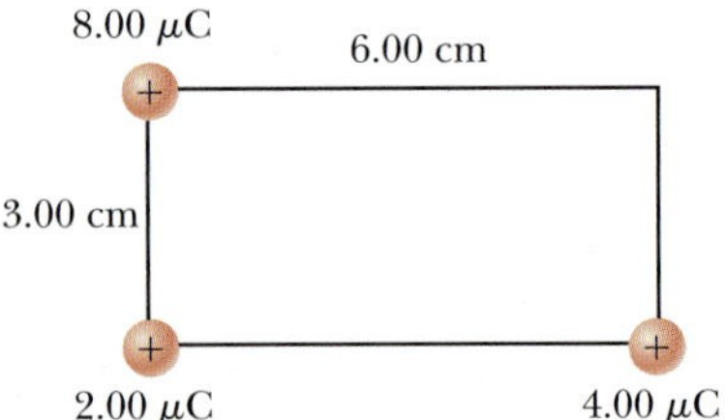

**Figure P16.13** Problems 13 and 14.

14. Three charges are situated at corners of a rectangle as in Figure P16.13. How much work must an external agent do to move the 8.00-$\mu$C charge to infinity?

15. **M** **QIC** Two point charges $Q_1 = +5.00$ nC and $Q_2 = -3.00$ nC are separated by 35.0 cm. (a) What is the electric potential at a point midway between the charges? (b) What is the potential energy of the pair of charges? What is the significance of the algebraic sign of your answer?

16. **QIC** **S** Three identical point charges each of charge $q$ are located at the vertices of an equilateral triangle as in Figure P16.16. The distance from the center of the triangle to each vertex is $a$. (a) Show that the electric field at the center of the triangle is zero. (b) Find a symbolic expression for the electric potential at the center of the triangle. (c) Give a physical explanation of the fact that the electric potential is not zero, yet the electric field is zero at the center.

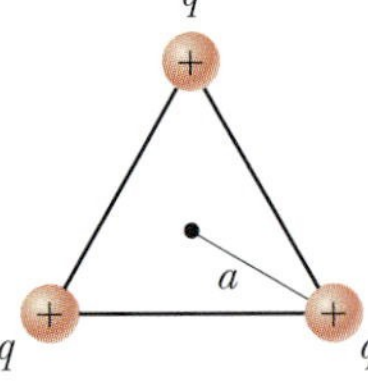

**Figure P16.16**

17. The three charges in Figure P16.17 are at the vertices of an isosceles triangle. Let $q = 7.00$ nC and calculate the electric potential at the midpoint of the base.

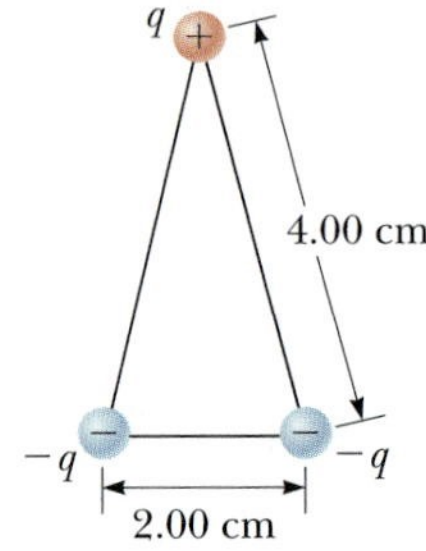

**Figure P16.17**

18. **S** A positive point charge $q = +2.50$ nC is located at $x = 1.20$ m and a negative charge of $-2q = -5.00$ nC is located at the origin as in Figure P16.18. (a) Sketch the electric potential versus $x$ for points along the $x$-axis in the range $-1.50$ m $< x <$ 1.50 m. (b) Find a symbolic expression for the potential on the $x$-axis at an arbitrary point $P$ between the two charges. (c) Find the electric potential at $x = 0.600$ m. (d) Find the point along the $x$-axis between the two charges where the electric potential is zero.

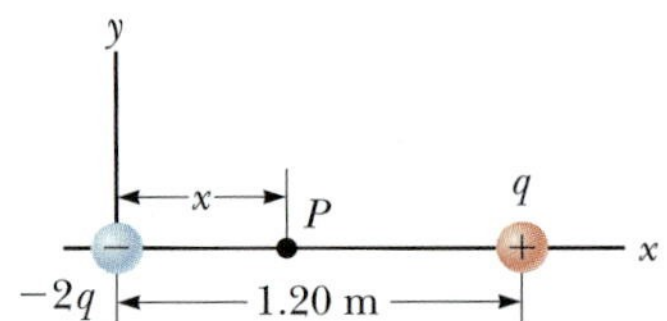

**Figure P16.18**

19. **GP** A proton is located at the origin, and a second proton is located on the $x$-axis at $x = 6.00$ fm (1 fm $= 10^{-15}$ m). (a) Calculate the electric potential energy associated with this configuration. (b) An alpha particle (charge $= 2e$, mass $= 6.64 \times 10^{-27}$ kg) is now placed at $(x, y) = (3.00, 3.00)$ fm. Calculate the electric potential energy associated with this configuration. (c) Starting with the three-particle system, find the change in electric potential energy if the alpha particle is allowed to escape to infinity while the two protons remain fixed in place. (Throughout, neglect any radiation effects.) (d) Use conservation of energy to calculate the speed of the alpha particle at infinity. (e) If the two protons are released from rest and the alpha particle remains fixed, calculate the speed of the protons at infinity.

20. **QIC** A proton and an alpha particle (charge $= 2e$, mass $= 6.64 \times 10^{-27}$ kg) are initially at rest, separated by $4.00 \times 10^{-15}$ m. (a) If they are both released simultaneously, explain why you can't find their velocities at infinity using only conservation of energy. (b) What other conservation law can be applied in this case? (c) Find the speeds of the proton and alpha particle, respectively, at infinity.

21. A tiny sphere of mass 8.00 $\mu$g and charge $-2.80$ nC is initially at a distance of 1.60 $\mu$m from a fixed charge of +8.50 nC. If the 8.00-mg sphere is released from rest, find (a) its kinetic energy when it is 0.500 $\mu$m from the fixed charge and (b) its speed when it is 0.500 $\mu$m from the fixed charge.

22. The metal sphere of a small Van de Graaff generator illustrated in Figure 15.23 has a radius of 18 cm. When the electric field at the surface of the sphere reaches $3.0 \times 10^6$ V/m, the air breaks down, and the generator discharges. What is the maximum potential the sphere can have before breakdown occurs?

23. W In Rutherford's famous scattering experiments that led to the planetary model of the atom, alpha particles (having charges of $+2e$ and masses of $6.64 \times 10^{-27}$ kg) were fired toward a gold nucleus with charge $+79e$. An alpha particle, initially very far from the gold nucleus, is fired at $2.00 \times 10^7$ m/s directly toward the nucleus, as in Figure P16.23. How close does the alpha particle get to the gold nucleus before turning around? Assume the gold nucleus remains stationary.

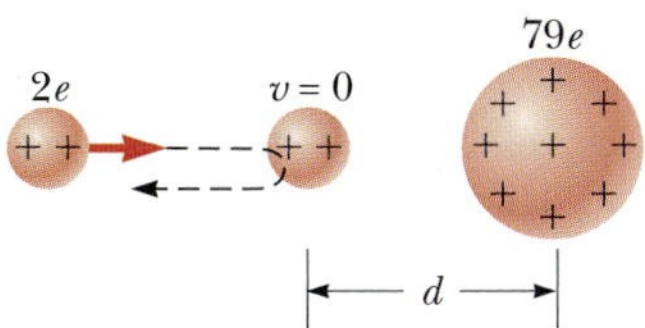


Figure P16.23

24. S Four point charges each having charge $Q$ are located at the corners of a square having sides of length $a$. Find symbolic expressions for (a) the total electric potential at the center of the square due to the four charges and (b) the work required to bring a fifth charge $q$ from infinity to the center of the square.

## 16.6 Capacitance

## 16.7 The Parallel-Plate Capacitor

25. W Consider the Earth and a cloud layer 800 m above the planet to be the plates of a parallel-plate capacitor. (a) If the cloud layer has an area of 1.0 km$^2$ = $1.0 \times 10^6$ m$^2$, what is the capacitance? (b) If an electric field strength greater than $3.0 \times 10^6$ N/C causes the air to break down and conduct charge (lightning), what is the maximum charge the cloud can hold?

26. (a) When a 9.00-V battery is connected to the plates of a capacitor, it stores a charge of 27.0 $\mu$C. What is the value of the capacitance? (b) If the same capacitor is connected to a 12.0-V battery, what charge is stored?

27. An air-filled parallel-plate capacitor has plates of area 2.30 cm$^2$ separated by 1.50 mm. The capacitor is connected to a 12.0-V battery. (a) Find the value of its capacitance. (b) What is the charge on the capacitor? (c) What is the magnitude of the uniform electric field between the plates?

28. Two conductors having net charges of +10.0 $\mu$C and $-10.0$ $\mu$C have a potential difference of 10.0 V between them. (a) Determine the capacitance of the system. (b) What is the potential difference between the two conductors if the charges on each are increased to +100 $\mu$C and $-100$ $\mu$C?

29. M An air-filled capacitor consists of two parallel plates, each with an area of 7.60 cm$^2$ and separated by a distance of 1.80 mm. If a 20.0-V potential difference is applied to these plates, calculate (a) the electric field between the plates, (b) the capacitance, and (c) the charge on each plate.

30. A 1-megabit computer memory chip contains many $60.0 \times 10^{-15}$-F capacitors. Each capacitor has a plate area of $21.0 \times 10^{-12}$ m$^2$. Determine the plate separation of such a capacitor. (Assume a parallel-plate configuration.) The diameter of an atom is on the order of $10^{-10}$ m = 1 Å. Express the plate separation in angstroms.

31. Q|C A parallel-plate capacitor with area 0.200 m$^2$ and plate separation of 3.00 mm is connected to a 6.00-V battery. (a) What is the capacitance? (b) How much charge is stored on the plates? (c) What is the electric field between the plates? (d) Find the magnitude of the charge density on each plate. (e) Without disconnecting the battery, the plates are moved farther apart. Qualitatively, what happens to each of the previous answers?

32. A small object with a mass of 350 $\mu$g carries a charge of 30.0 nC and is suspended by a thread between the vertical plates of a parallel-plate capacitor. The plates are separated by 4.00 cm. If the thread makes an angle of 15.0° with the vertical, what is the potential difference between the plates?

## 16.8 Combinations of Capacitors

33. Given a 2.50-$\mu$F capacitor, a 6.25-$\mu$F capacitor, and a 6.00-V battery, find the charge on each capacitor if you connect them (a) in series across the battery and (b) in parallel across the battery.

34. Two capacitors, $C_1$ = 5.00 $\mu$F and $C_2$ = 12.0 $\mu$F, are connected in parallel, and the resulting combination is connected to a 9.00-V battery. Find (a) the equivalent capacitance of the combination, (b) the potential difference across each capacitor, and (c) the charge stored on each capacitor.

35. Find (a) the equivalent capacitance of the capacitors in Figure P16.35, (b) the charge on each capacitor, and (c) the potential difference across each capacitor.

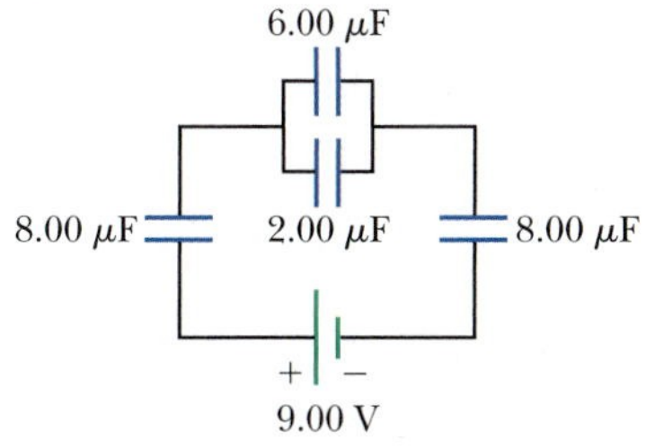


Figure P16.35

36. **W** Two capacitors give an equivalent capacitance of 9.00 pF when connected in parallel and an equivalent capacitance of 2.00 pF when connected in series. What is the capacitance of each capacitor?

37. For the system of capacitors shown in Figure P16.37, find (a) the equivalent capacitance of the system, (b) the charge on each capacitor, and (c) the potential difference across each capacitor.

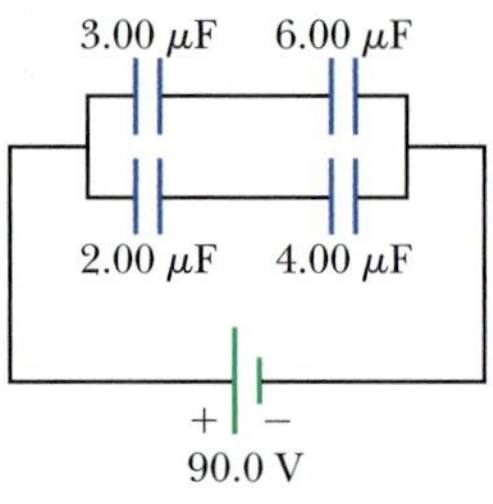

3.00 $\mu$F 6.00 $\mu$F
2.00 $\mu$F 4.00 $\mu$F
+ −
90.0 V

**Figure P16.37** Problems 37 and 56.

38. **GP** Consider the combination of capacitors in Figure P16.38. (a) Find the equivalent single capacitance of the two capacitors in series and redraw the diagram (called diagram 1) with this equivalent capacitance. (b) In diagram 1 find the equivalent capacitance of the three capacitors in parallel and redraw the diagram as a single battery and single capacitor in a loop. (c) Compute the charge on the single equivalent capacitor. (d) Returning to diagram 1, compute the charge on each individual capacitor. Does the sum agree with the value found in part (c)? (e) What is the charge on the 24.0-$\mu$F capacitor and on the 8.00-$\mu$F capacitor? Compute the voltage drop across (f) the 24.0-$\mu$F capacitor and (g) the 8.00-$\mu$F capacitor.

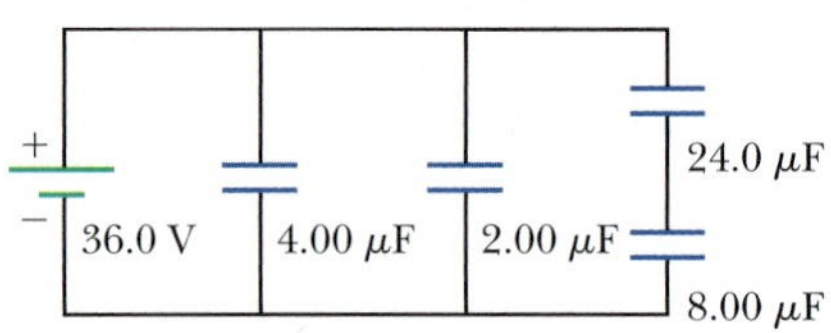


**Figure P16.38**

39. Find the charge on each of the capacitors in Figure P16.39.

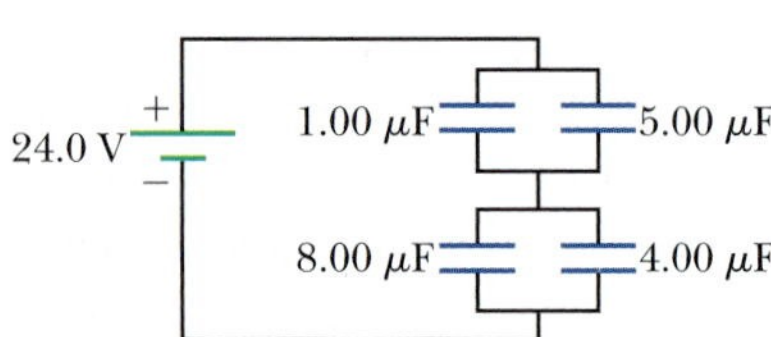


**Figure P16.39**

40. **QIC** **S** Three capacitors are connected to a battery as shown in Figure P16.40. Their capacitances are $C_1 = 3C$, $C_2 = C$, and $C_3 = 5C$. (a) What is the equivalent capacitance of this set of capacitors? (b) State the ranking of the capacitors according to the charge they store from largest to smallest. (c) Rank the capacitors according to the potential differences across them from largest to smallest. (d) Assume $C_3$ is increased. Explain what happens to the charge stored by each capacitor.

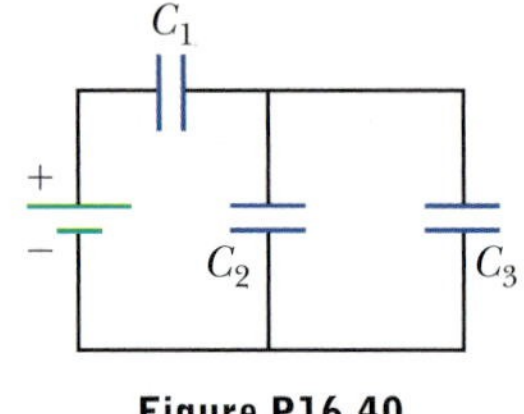


**Figure P16.40**

41. A 25.0-$\mu$F capacitor and a 40.0-$\mu$F capacitor are charged by being connected across separate 50.0-V batteries. (a) Determine the resulting charge on each capacitor. (b) The capacitors are then disconnected from their batteries and connected to each other, with each negative plate connected to the other positive plate. What is the final charge of each capacitor? (c) What is the final potential difference across the 40.0-$\mu$F capacitor?

42. (a) Find the equivalent capacitance between points $a$ and $b$ for the group of capacitors connected as shown in Figure P16.42 if $C_1 = 5.00\ \mu$F, $C_2 = 10.00\ \mu$F, and $C_3 = 2.00\ \mu$F. (b) If the potential between points $a$ and $b$ is 60.0 V, what charge is stored on $C_3$?

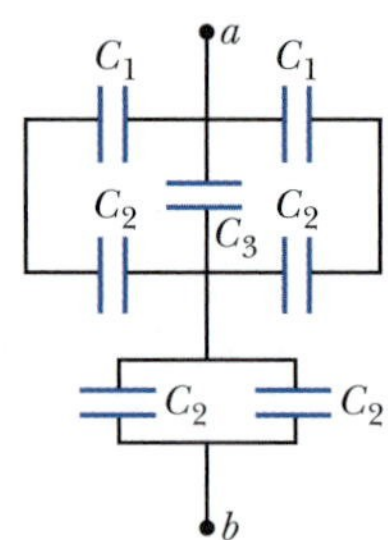


**Figure P16.42**

43. A 1.00-$\mu$F capacitor is charged by being connected across a 10.0-V battery. It is then disconnected from the battery and connected across an uncharged 2.00-$\mu$F capacitor. Determine the resulting charge on each capacitor.

44. **M** Four capacitors are connected as shown in Figure P16.44. (a) Find the equivalent capacitance between points $a$ and $b$. (b) Calculate the charge on each capacitor, taking $\Delta V_{ab} = 15.0$ V.

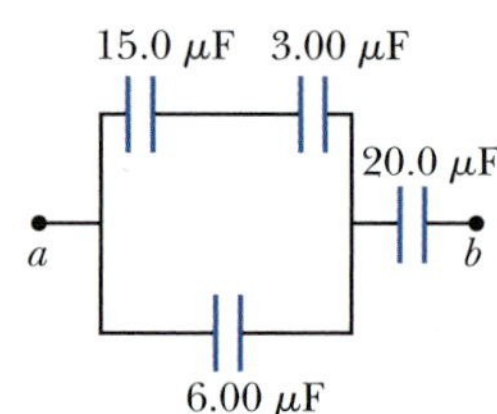


**Figure P16.44**

## 16.9 Energy Stored in a Charged Capacitor

45. A 12.0-V battery is connected to a 4.50-$\mu$F capacitor. How much energy is stored in the capacitor?

46. **QIC** Two capacitors, $C_1 = 18.0\ \mu$F and $C_2 = 36.0\ \mu$F, are connected in series, and a 12.0-V battery is connected across them. (a) Find the equivalent capacitance, and the energy contained in this equivalent capacitor. (b) Find the energy stored in each individual capacitor. Show that the sum of these two energies is the same as the energy found in part (a). Will this equality always be true, or does it depend on the number of capacitors and their capacitances? (c) If the same capacitors were connected in parallel, what potential difference would be required across them so that the combination stores the same energy as in part (a)? Which capacitor stores more energy in this situation, $C_1$ or $C_2$?

47. A parallel-plate capacitor has capacitance 3.00 $\mu$F. (a) How much energy is stored in the capacitor if it is connected to a 6.00-V battery? (b) If the battery is disconnected and the distance between the charged

plates doubled, what is the energy stored? (c) The battery is subsequently reattached to the capacitor, but the plate separation remains as in part (b). How much energy is stored? (Answer each part in microjoules.)

48. A certain storm cloud has a potential difference of $1.00 \times 10^8$ V relative to a tree. If, during a lightning storm, 50.0 C of charge is transferred through this potential difference and 1.00% of the energy is absorbed by the tree, how much water (sap in the tree) initially at 30.0°C can be boiled away? Water has a specific heat of 4 186 J/kg·°C, a boiling point of 100°C, and a heat of vaporization of $2.26 \times 10^6$ J/kg.

## 16.10 Capacitors with Dielectrics

49. Q|C The voltage across an air-filled parallel-plate capacitor is measured to be 85.0 V. When a dielectric is inserted and completely fills the space between the plates as in Figure P16.49, the voltage drops to 25.0 V. (a) What is the dielectric constant of the inserted material? Can you identify the dielectric? (b) If the dielectric doesn't completely fill the space between the plates, what could you conclude about the voltage across the plates?

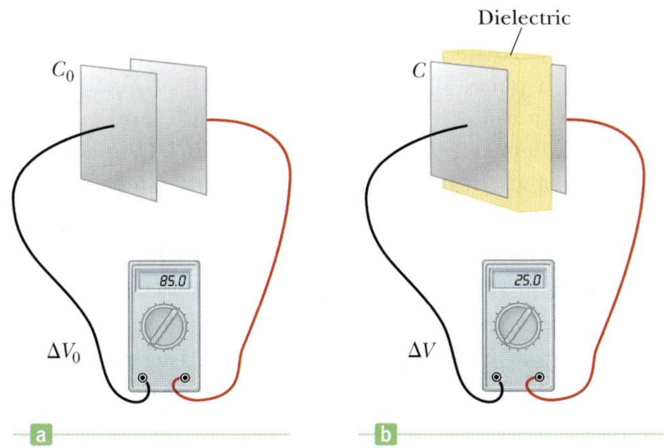


**Figure P16.49**

50. (a) How much charge can be placed on a capacitor with air between the plates before it breaks down if the area of each plate is 5.00 cm²? (b) Find the maximum charge if polystyrene is used between the plates instead of air. Assume the dielectric strength of air is $3.00 \times 10^6$ V/m and that of polystyrene is $24.0 \times 10^6$ V/m.

51. Determine (a) the capacitance and (b) the maximum voltage that can be applied to a Teflon-filled parallel-plate capacitor having a plate area of 175 cm² and an insulation thickness of 0.040 0 mm.

52. Consider a plane parallel-plate capacitor made of two strips of aluminum foil separated by a layer of paraffin-coated paper. Each strip of foil and paper is 7.00 cm wide. The foil is 0.004 00 mm thick, and the paper is 0.025 0 mm thick and has a dielectric constant of 3.70. What length should the strips be if a capacitance of $9.50 \times 10^{-8}$ F is desired? (If, after this plane capacitor is formed, a second paper strip can be added below the foil-paper-foil stack and the resulting assembly rolled into a cylindrical form—similar to that shown in Figure 16.26—the capacitance can be doubled because both surfaces of each foil strip would then store charge. Without the second strip of paper, however, rolling the layers would result in a short circuit.)

53. BIO A model of a red blood cell portrays the cell as a spherical capacitor, a positively charged liquid sphere of surface area $A$ separated from the surrounding negatively charged fluid by a membrane of thickness $t$. Tiny electrodes introduced into the interior of the cell show a potential difference of 100 mV across the membrane. The membrane's thickness is estimated to be 100 nm and has a dielectric constant of 5.00. (a) If an average red blood cell has a mass of $1.00 \times 10^{-12}$ kg, estimate the volume of the cell and thus find its surface area. The density of blood is 1 100 kg/m³. (b) Estimate the capacitance of the cell by assuming the membrane surfaces act as parallel plates. (c) Calculate the charge on the surface of the membrane. How many electronic charges does the surface charge represent?

## Additional Problems

54. When a potential difference of 150 V is applied to the plates of an air-filled parallel-plate capacitor, the plates carry a surface charge density of $3.00 \times 10^{-10}$ C/cm². What is the spacing between the plates?

55. S Three parallel-plate capacitors are constructed, each having the same plate area $A$ and with $C_1$ having plate spacing $d_1$, $C_2$ having plate spacing $d_2$, and $C_3$ having plate spacing $d_3$. Show that the total capacitance $C$ of the three capacitors connected in series is the same as a capacitor of plate area $A$ and with plate spacing $d = d_1 + d_2 + d_3$.

56. Q|C For the system of four capacitors shown in Figure P16.37, find (a) the total energy stored in the system and (b) the energy stored by each capacitor. (c) Compare the sum of the answers in part (b) with your result to part (a) and explain your observation.

57. S A parallel-plate capacitor with a plate separation $d$ has a capacitance $C_0$ in the absence of a dielectric. A slab of dielectric material of dielectric constant $\kappa$ and thickness $d/3$ is then inserted between the plates as in Figure P16.57a. Show that the capacitance of this partially filled capacitor is given by

$$C = \left(\frac{3\kappa}{2\kappa + 1}\right)C_0$$

*Hint:* Treat the system as two capacitors connected in series as in Figure P16.57b, one with dielectric in it and the other one empty.

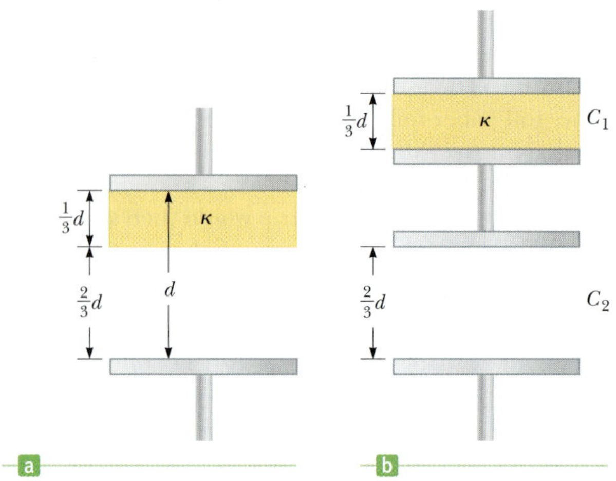

Figure P16.57

58. **S** Two capacitors give an equivalent capacitance of $C_p$ when connected in parallel and an equivalent capacitance of $C_s$ when connected in series. What is the capacitance of each capacitor?

59. **M** A parallel-plate capacitor is constructed using a dielectric material whose dielectric constant is 3.00 and whose dielectric strength is $2.00 \times 10^8$ V/m. The desired capacitance is 0.250 $\mu$F, and the capacitor must withstand a maximum potential difference of 4.00 kV. Find the minimum area of the capacitor plates.

60. Two charges of 1.0 $\mu$C and $-2.0$ $\mu$C are 0.50 m apart at two vertices of an equilateral triangle as in Figure P16.60. (a) What is the electric potential due to the 1.0-$\mu$C charge at the third vertex, point $P$? (b) What is the electric potential due to the $-2.0$-$\mu$C charge at $P$? (c) Find the total electric potential at $P$. (d) What is the work required to move a 3.0-$\mu$C charge from infinity to $P$?

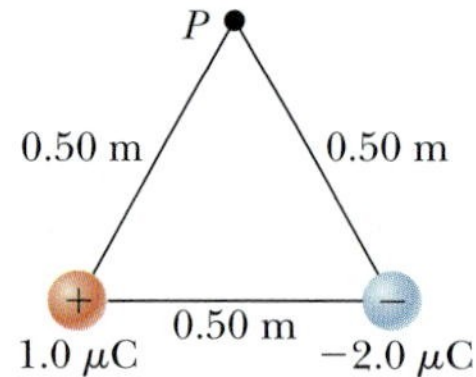

Figure P16.60

61. Find the equivalent capacitance of the group of capacitors shown in Figure P16.61.

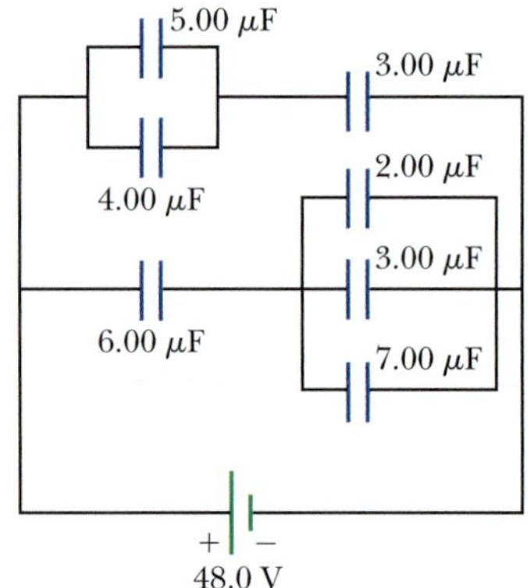

Figure P16.61

62. A spherical capacitor consists of a spherical conducting shell of radius $b$ and charge $-Q$ concentric with a smaller conducting sphere of radius $a$ and charge $Q$. (a) Find the capacitance of this device. (b) Show that as the radius $b$ of the outer sphere approaches infinity, the capacitance approaches the value $a/k_e = 4\pi\epsilon_0 a$.

63. **BIO** The immediate cause of many deaths is ventricular fibrillation, an uncoordinated quivering of the heart, as opposed to proper beating. An electric shock to the chest can cause momentary paralysis of the heart muscle, after which the heart will sometimes start organized beating again. A *defibrillator* is a device that applies a strong electric shock to the chest over a time of a few milliseconds. The device contains a capacitor of a few microfarads, charged to several thousand volts. Electrodes called paddles, about 8 cm across and coated with conducting paste, are held against the chest on both sides of the heart. Their handles are insulated to prevent injury to the operator, who calls "Clear!" and pushes a button on one paddle to discharge the capacitor through the patient's chest. Assume an energy of 300 W · s is to be delivered from a 30.0-$\mu$F capacitor. To what potential difference must it be charged?

64. When a certain air-filled parallel-plate capacitor is connected across a battery, it acquires a charge of 150 $\mu$C on each plate. While the battery connection is maintained, a dielectric slab is inserted into, and fills, the region between the plates. This results in the accumulation of an additional charge of 200 $\mu$C on each plate. What is the dielectric constant of the slab?

65. Capacitors $C_1 = 6.0$ $\mu$F and $C_2 = 2.0$ $\mu$F are charged as a parallel combination across a 250-V battery. The capacitors are disconnected from the battery and from each other. They are then connected positive plate to negative plate and negative plate to positive plate. Calculate the resulting charge on each capacitor.

66. **S** Two positive charges each of charge $q$ are fixed on the $y$-axis, one at $y = d$ and the other at $y = -d$ as in Figure P16.66. A third positive charge $2q$ located on the $x$-axis at $x = 2d$ is released from rest. Find symbolic expressions for (a) the total electric potential due to the first two charges at the location of the charge $2q$, (b) the electric potential energy of the charge $2q$, (c) the kinetic energy of the charge $2q$ after it has moved infinitely far from the other charges, and (d) the speed of the charge $2q$ after it has moved infinitely far from the other charges if its mass is $m$.

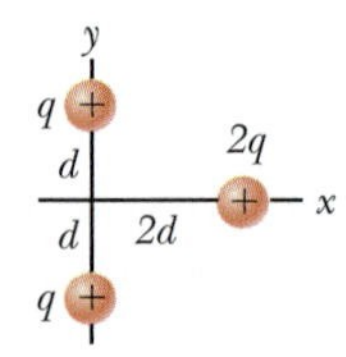

Figure P16.66

67. Metal sphere A of radius 12.0 cm carries 6.00 $\mu$C of charge, and metal sphere B of radius 18.0 cm carries $-4.00$ $\mu$C of charge. If the two spheres are attached by

a very long conducting thread, what is the final distribution of charge on the two spheres?

**68.** An electron is fired at a speed $v_0 = 5.6 \times 10^6$ m/s and at an angle $\theta_0 = -45°$ between two parallel conducting plates that are $D = 2.0$ mm apart, as in Figure P16.68. If the voltage difference between the plates is $\Delta V = 100$ V, determine (a) how close, $d$, the electron will get to the bottom plate and (b) where the electron will strike the top plate.

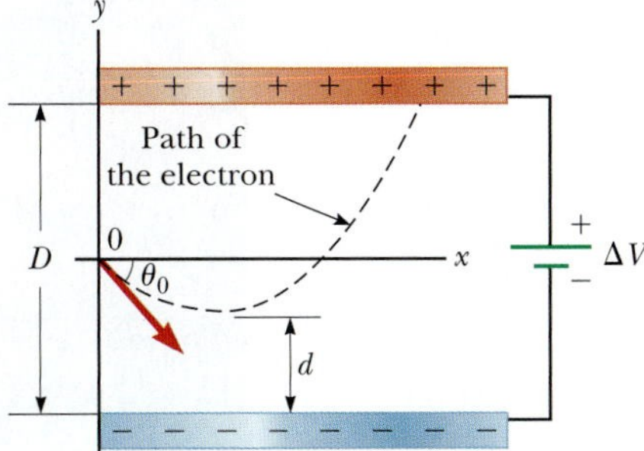


Figure P16.68

The blue glow comes from positively-charged xenon atoms that are electrostatically accelerated, then expelled from an ion engine prototype. The current of ions produces ninety millinewtons of thrust continuously for months at a time. Electrons must be fed back into the exhaust to prevent a buildup of negative charge. Such engines are highly efficient and suitable for extended deep space missions.

# 17 Current and Resistance



Many practical applications and devices are based on the principles of static electricity, but electricity was destined to become an inseparable part of our daily lives when scientists learned how to produce a continuous flow of charge for relatively long periods of time using batteries. The battery or voltaic cell was invented in 1800 by Italian physicist Alessandro Volta. Batteries supplied a continuous flow of charge at low potential, in contrast to earlier electrostatic devices that produced a tiny flow of charge at high potential for brief periods. This steady source of electric current allowed scientists to perform experiments to learn how to control the flow of electric charges in circuits. Today, electric currents power our lights, radios, television sets, air conditioners, computers, and refrigerators. They ignite the gasoline in automobile engines, travel through miniature components making up the chips of microcomputers, and provide the power for countless other invaluable tasks.

In this chapter we define current and discuss some of the factors that contribute to the resistance to the flow of charge in conductors. We also discuss energy transformations in electric circuits. These topics will be the foundation for additional work with circuits in later chapters.

# 17.1 Electric Current

**LEARNING OBJECTIVES**

1. Define both average and instantaneous electrical current and discuss their physical meaning.
2. Apply the concept of current to simple electrical systems.

**Tip 17.1 *Current Flow* Is Redundant**

The phrases *flow of current* and *current flow* are commonly used, but here the word *flow* is redundant because current is already defined as a flow (of charge). Avoid this construction!

In Figure 17.1 charges move in a direction perpendicular to a surface of area $A$. (That area could be the cross-sectional area of a wire, for example.) **The current is the rate at which charge flows through this surface.**

Suppose $\Delta Q$ is the amount of charge that flows through an area $A$ in a time interval $\Delta t$ and that the direction of flow is perpendicular to the area. Then the **average current $I_{av}$** is equal to the amount of charge divided by the time interval:

$$I_{av} \equiv \frac{\Delta Q}{\Delta t} \quad [17.1a]$$

**SI unit: coulomb/second (C/s), or the ampere (A)**

Current is composed of individual moving charges, so for an extremely low current, it is conceivable that a single charge could pass through area $A$ in one instant and no charge in the next instant. All currents, then, are essentially averages over time. Given the very large number of charges usually involved, however, it makes sense to define an instantaneous current.

The **instantaneous current $I$** is the limit of the average current as the time interval goes to zero:

$$I = \lim_{\Delta t \to 0} I_{av} = \lim_{\Delta t \to 0} \frac{\Delta Q}{\Delta t} \quad [17.1b]$$

**SI unit: coulomb/second (C/s), or the ampere (A)**

When the current is steady, the average and instantaneous currents are the same. Note that one ampere of current is equivalent to one coulomb of charge passing through a given area $A$ in a time interval of 1 s.

When charges flow through a surface as in Figure 17.1, they can be positive, negative, or both. **The direction of conventional current used in this book is the direction positive charges flow.** (This historical convention originated about 200 years ago, when the ideas of positive and negative charges were introduced.) In a common conductor such as copper, the current is due to the motion of negatively charged electrons, so the direction of the current is opposite the direction of motion of the electrons. On the other hand, for a beam of positively charged protons in an accelerator, the current is in the same direction as the motion of the protons. In some cases—gases and electrolytes, for example—the current is the result of the flows of both positive and negative charges. Moving charges, whether positive or negative, are referred to as *charge carriers*. In a metal, for example, the charge carriers are electrons.

In electrostatics, where charges are stationary, the electric potential is the same everywhere in a conductor. That is no longer true for conductors carrying current: as charges move along a wire, the electric potential is continually decreasing (except in the special case of superconductors). The decreasing electric potential means that the moving charges lose energy according to the relationship $\Delta U_{charges} = q\Delta V$, while

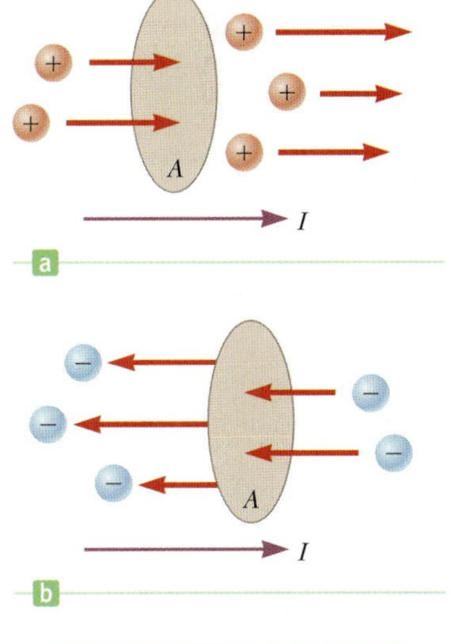


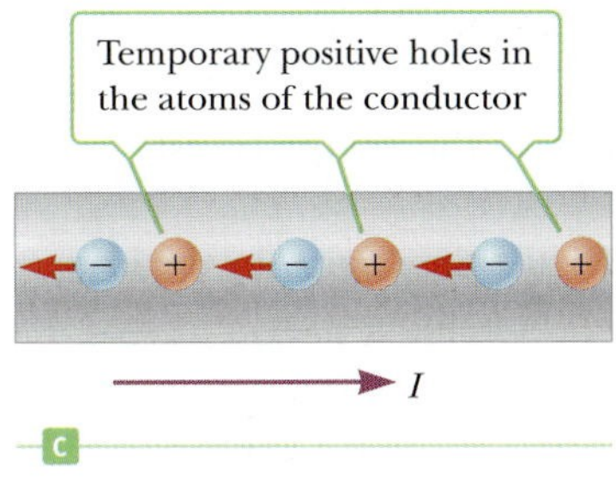


**Figure 17.1** The time rate of flow of charge through area $A$ is the current $I$. (a) The direction of current is the same as the flow of positive charge. (b) Negative charge flowing to the left is equivalent to an equal amount of positive charge flowing to the right. (c) In a conductor, positive holes open in the lattice of the conductor's atoms as electrons move in response to a potential. Negative electrons moving actively to the left are equivalent to positive holes migrating to the right.

an energy $\Delta U_{\text{wire}} = -q\Delta V$ is deposited in the current-carrying wire. (Those expressions derive from Equation 16.2.) If $q$ is taken to be positive, corresponding to the convention of positive current, then $\Delta V = V_f - V_i$ is negative because, in a circuit, positive charges move from regions of high potential to regions of low potential. That in turn means $\Delta U_{\text{charges}} = q\Delta V$ is negative, as it should be, because the moving charges lose energy. Often only the magnitude is desired, however, in which case absolute values are substituted into $q$ and $\Delta V$. If the current is constant, then dividing the energy by the elapsed time yields the power delivered to the circuit element, such as a lightbulb filament.

## EXAMPLE 17.1 Turn On the Light

**GOAL** Apply the concept of current.

**PROBLEM** The amount of charge that passes through the filament of a certain lightbulb in 2.00 s is 1.67 C. Find **(a)** the average current in the lightbulb and **(b)** the number of electrons that pass through the filament in 5.00 s. **(c)** If the current is supplied by a 12.0-V battery, what total energy is delivered to the lightbulb filament during 2.00 s? What is the average power?

**STRATEGY** Substitute into Equation 17.1a for part (a), then multiply the answer by the time given in part (b) to get the total charge that passes in that time. The total charge equals the number $N$ of electrons going through the circuit times the charge per electron. To obtain the energy delivered to the filament, multiply the potential difference, $\Delta V$, by the total charge. Dividing the energy by time yields the average power.

**SOLUTION**

**(a)** Compute the average current in the lightbulb.

Substitute the charge and time into Equation 17.1a:

$$I_{\text{av}} = \frac{\Delta Q}{\Delta t} = \frac{1.67\text{ C}}{2.00\text{ s}} = \boxed{0.835\text{ A}}$$

**(b)** Find the number of electrons passing through the filament in 5.00 s.

The total number $N$ of electrons times the charge per electron equals the total charge, $I_{\text{av}}\,\Delta t$:

$$\textbf{(1)}\quad Nq = I_{\text{av}}\,\Delta t$$

Substitute and solve for $N$:

$$N(1.60 \times 10^{-19}\text{ C/electron}) = (0.835\text{ A})(5.00\text{ s})$$

$$N = \boxed{2.61 \times 10^{19}\text{ electrons}}$$

**(c)** What total energy is delivered to the lightbulb filament? What is the average power?

Multiply the potential difference by the total charge to obtain the energy transferred to the filament:

$$\textbf{(2)}\quad \Delta U = q\Delta V = (1.67\text{ C})(12.0\text{ V}) = \boxed{20.0\text{ J}}$$

Divide the energy by the elapsed time to calculate the average power:

$$P_{\text{av}} = \frac{\Delta U}{\Delta t} = \frac{20.0\text{ J}}{2.00\text{ s}} = \boxed{10.0\text{ W}}$$

**REMARKS** It's important to use units to ensure the correctness of equations such as Equation (1). Notice the enormous number of electrons that pass through a given point in a typical circuit. Magnitudes were used in calculating the energies in Equation (2). Technically, the charge carriers are electrons with negative charge moving from a lower potential to a higher potential, so the change in their energy is $\Delta U_{\text{charge}} = q\Delta V = (-1.67\text{ C})(+12.0\text{ V}) = -20.0\text{ J}$, a loss of energy that is delivered to the filament, $\Delta U_{\text{fil}} = -\Delta U_{\text{charge}} = +20.0\text{ J}$. The energy and power, calculated here using the definitions of Chapter 16, will be further addressed in Section 17.6.

**QUESTION 17.1** Is it possible to have an instantaneous current of $e/2$ per second? Explain. Can the average current take this value?

**EXERCISE 17.1** A 9.00-V battery delivers a current of 1.34 A to the lightbulb filament of a pocket flashlight. (a) How much charge passes through the filament in 2.00 min? (b) How many electrons pass through the filament? Calculate (c) the energy delivered to the filament during that time and (d) the power delivered by the battery.

**ANSWERS** (a) 161 C (b) $1.01 \times 10^{21}$ electrons (c) $1.45 \times 10^{3}$ J (d) 12.1 W

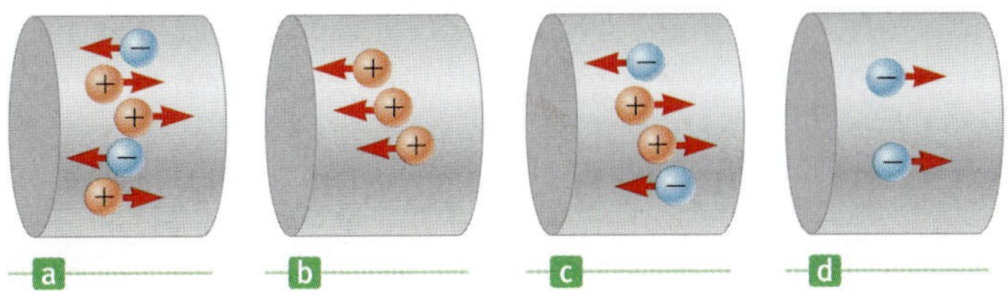

**Figure 17.2** (Quick Quiz 17.1)

**Quick Quiz**

**17.1** Consider positive and negative charges all moving horizontally with the same speed through the four regions in Figure 17.2. Rank the magnitudes of the currents in these four regions from lowest to highest. ($I_a$ is the current in Figure 17.2a, $I_b$ the current in Figure 17.2b, etc.) (a) $I_d$, $I_a$, $I_c$, $I_b$ (b) $I_a$, $I_c$, $I_b$, $I_d$ (c) $I_c$, $I_a$, $I_d$, $I_b$ (d) $I_d$, $I_b$, $I_c$, $I_a$ (e) $I_a$, $I_b$, $I_c$, $I_d$ (f) None of these

# 17.2 A Microscopic View: Current and Drift Speed

**LEARNING OBJECTIVES**

1. Relate electrical current to the drift speed of charge carriers.
2. Evaluate the drift speed in typical electrical conductors.

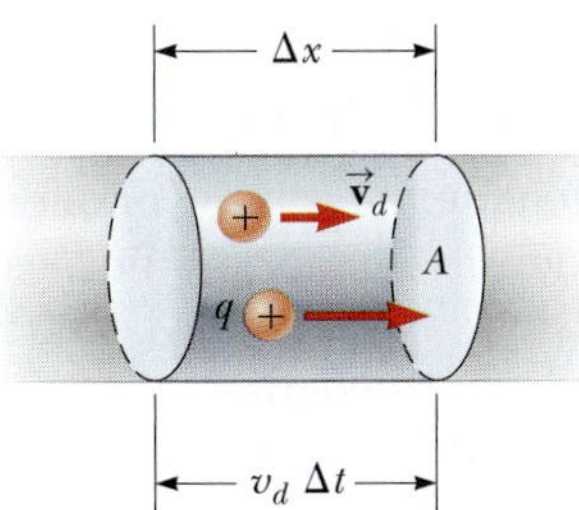

Δx
v⃗d
A
q
vd Δt

**Figure 17.3** A section of a uniform conductor of cross-sectional area $A$. The charge carriers move with a speed $v_d$, and the distance they travel in time $\Delta t$ is given by $\Delta x = v_d \Delta t$. The number of mobile charge carriers in the section of length $\Delta x$ is given by $nAv_d\Delta t$, where $n$ is the number of mobile carriers per unit volume.

Macroscopic currents can be related to the motion of the microscopic charge carriers making up the current. It turns out that current depends on the average speed of the charge carriers in the direction of the current, the number of charge carriers per unit volume, and the of the charge carried by each charge carrier.

Consider identically charged particles moving in a conductor of cross-sectional area $A$ (Fig. 17.3). The volume of an element of length $\Delta x$ of the conductor is $A\Delta x$. If $n$ represents the number of mobile charge carriers per unit volume, the number of carriers in the volume element is $nA\Delta x$. The mobile charge $\Delta Q$ in this element is therefore

$$\Delta Q = \text{number of carriers} \times \text{charge per carrier} = (nA\Delta x)q$$

where $q$ is the charge on each carrier. If the carriers move with a constant average speed called the **drift speed** $v_d$, the distance they move in the time interval $\Delta t$ is $\Delta x = v_d \Delta t$. We can therefore write

$$\Delta Q = (nAv_d \Delta t)q$$

If we divide both sides of this equation by $\Delta t$ and take the limit as $\Delta t$ goes to zero, we see that the current in the conductor is

$$I = \lim_{\Delta t \to 0} \frac{\Delta Q}{\Delta t} = nqv_d A \qquad [17.2]$$

To understand the meaning of drift speed, consider a conductor in which the charge carriers are free electrons. If the conductor is isolated, these electrons undergo random motion similar to the motion of the molecules in a gas. The drift speed is normally much smaller than the free electrons' average speed between collisions with the fixed atoms of the conductor. When a potential difference is applied between the ends of the conductor (say, with a battery), an electric field is set up in the conductor, creating an electric force on the electrons and hence a current. In reality, the electrons don't simply move in straight lines along the conductor. Instead, they undergo repeated collisions with the atoms of the metal, and the result is a complicated zigzag motion with only a small average drift speed along the wire (Fig. 17.4). The energy transferred from the electrons to the metal atoms during a collision increases the vibrational energy of the atoms and causes

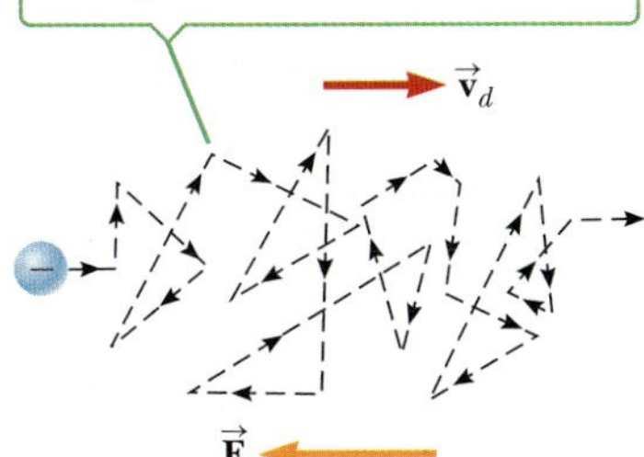

Although electrons move with average velocity v⃗d, collisions with atoms cause sharp, momentary changes of direction.
v⃗d
E⃗

**Figure 17.4** A schematic representation of the zigzag motion of a charge carrier in a conductor. Notice that the drift velocity $\vec{\mathbf{v}}_d$ is opposite the direction of the electric field.

a corresponding increase in the temperature of the conductor. Despite the collisions, however, the electrons move slowly along the conductor in a direction opposite $\vec{\mathbf{E}}$ with the drift velocity $\vec{\mathbf{v}}_d$.

## EXAMPLE 17.2 Drift Speed of Electrons

**GOAL** Calculate a drift speed and compare it with the rms speed of an electron gas.

**PROBLEM** A copper wire of cross-sectional area $3.00 \times 10^{-6}\ \text{m}^2$ carries a current of 10.0 A. **(a)** Assuming each copper atom contributes one free electron to the metal, find the drift speed of the electrons in this wire. **(b)** Use the ideal gas model to compare the drift speed with the random rms speed an electron would have at 20.0°C. The density of copper is $8.92\ \text{g/cm}^3$, and its atomic mass is 63.5 u.

**STRATEGY** All the variables in Equation 17.2 are known except for $n$, the number of free charge carriers per unit volume. We can find $n$ by recalling that one mole of copper contains an Avogadro's number ($6.02 \times 10^{23}$) of atoms and each atom contributes one charge carrier to the metal. The volume of one mole can be found from copper's known density and atomic mass. The atomic mass is the same, numerically, as the number of grams in a mole of the substance.

**SOLUTION**

**(a)** Find the drift speed of the electrons.

Calculate the volume of one mole of copper from its density and its atomic mass:

$$V = \frac{m}{\rho} = \frac{63.5\ \text{g/mol}}{8.92\ \text{g/cm}^3} = 7.12\ \text{cm}^3/\text{mol}$$

Convert the volume from $\text{cm}^3$ to $\text{m}^3$:

$$7.12\ \text{cm}^3/\text{mol}\left(\frac{1\ \text{m}}{10^2\ \text{cm}}\right)^3 = 7.12 \times 10^{-6}\ \text{m}$$

Divide Avogadro's number (the number of electrons in one mole) by the volume per mole to obtain the number density:

$$n = \frac{6.02 \times 10^{23}\ \text{electrons/mole}}{7.12 \times 10^{-6}\ \text{m}^3/\text{mole}}$$

$$= 8.46 \times 10^{28}\ \text{electrons/m}^3$$

Solve Equation 17.2 for the drift speed and substitute:

$$v_d = \frac{I}{nqA}$$

$$= \frac{10.0\ \text{C/s}}{(8.46 \times 10^{28}\ \text{electrons/m}^3)(1.60 \times 10^{-19}\ \text{C})(3.00 \times 10^{-6}\ \text{m}^2)}$$

$$v_d = 2.46 \times 10^{-4}\ \text{m/s}$$

**(b)** Find the rms speed of a gas of electrons at 20.0°C.

Apply Equation 10.18:

$$v_{\text{rms}} = \sqrt{\frac{3k_BT}{m_e}}$$

Convert the temperature to the Kelvin scale and substitute values:

$$v_{\text{rms}} = \sqrt{\frac{3(1.38 \times 10^{-23}\ \text{J/K})(293\ \text{K})}{9.11 \times 10^{-31}\ \text{kg}}}$$

$$= 1.15 \times 10^5\ \text{m/s}$$

**REMARKS** The drift speed of an electron in a wire is very small, only about one-billionth of its random thermal speed.

**QUESTION 17.2** True or False: The drift velocity in a wire of a given composition is inversely proportional to the number density of charge carriers.

**EXERCISE 17.2** What current in a copper wire with a cross-sectional area of $7.50 \times 10^{-7}\ \text{m}^2$ would result in a drift speed equal to $5.00 \times 10^{-4}$ m/s?

**ANSWER** 5.08 A

Example 17.2 shows that drift speeds are typically very small. In fact, the drift speed is much smaller than the average speed between collisions. Electrons traveling at $2.46 \times 10^{-4}$ m/s, as in the example, would take about 68 min to travel 1 m! In view of this low speed, why does a lightbulb turn on almost instantaneously when a switch is thrown? Think of the flow of water through a pipe. If a drop of water is forced into one end of a pipe that is already filled with water, a drop must be pushed out the other end of the pipe. Although it may take an individual drop a long time to make it through the pipe, a flow initiated at one end produces a similar flow at the other end very quickly. Another familiar analogy is the motion of a bicycle chain. When the sprocket moves one link, the other links all move more or less immediately, even though it takes a given link some time to make a complete rotation. In a conductor, the change in the electric field that drives the free electrons travels at a speed close to that of light, so when you flip a light switch, the message for the electrons to start moving through the wire (the electric field) reaches them at a speed on the order of $10^8$ m/s!

**Tip 17.2 Electrons Are Everywhere in the Circuit**

Electrons don't have to travel from the light switch to the lightbulb for the lightbulb to operate. Electrons already in the filament of the lightbulb move in response to the electric field set up by the battery. Also, the battery does *not* provide electrons to the circuit; it provides *energy* to the existing electrons.

**Quick Quiz**

**17.2** Suppose a current-carrying wire has a cross-sectional area that gradually becomes smaller along the wire so that the wire has the shape of a very long, truncated cone. How does the drift speed vary along the wire? (a) It slows down as the cross section becomes smaller. (b) It speeds up as the cross section becomes smaller. (c) It doesn't change. (d) More information is needed.

## 17.3 Current and Voltage Measurements In Circuits

**LEARNING OBJECTIVES**

1. Discuss the concept of an electrical circuit.
2. Discuss ammeters and voltmeters, instruments used to measure currents and potential differences in circuits.

To study electric current in circuits, we need to understand how to measure currents and voltages.

The circuit shown in Figure 17.5a is a drawing of the actual circuit necessary for measuring the current in Example 17.1. Figure 17.5b shows a stylized figure called a circuit diagram that represents the actual circuit of Figure 17.5a. This circuit consists of only a battery and a lightbulb. The word *circuit* means "a closed loop of some sort around which current circulates." The battery pumps charge through the bulb and around the loop. No charge would flow without a complete conducting path from the positive terminal of the battery into one side of the bulb, out the other side, and

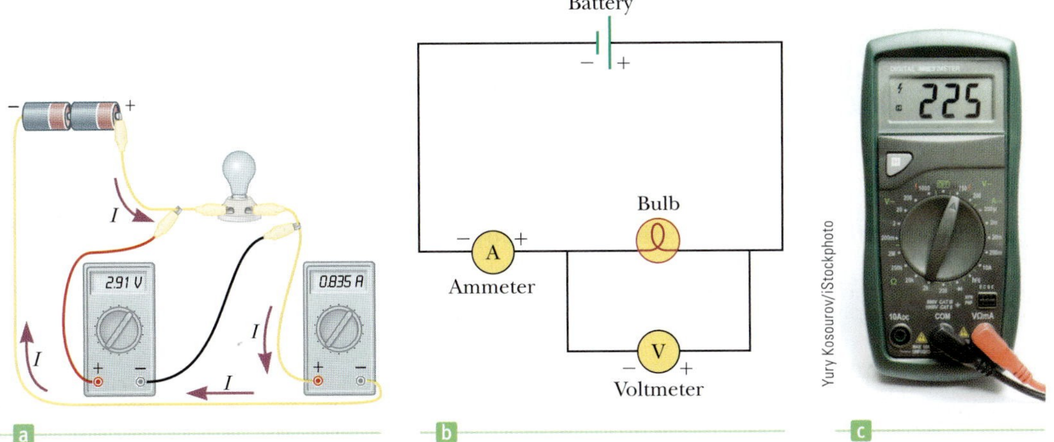


**Figure 17.5** (a) A sketch of an actual circuit used to measure the current in a flashlight bulb and the potential difference across it. (b) A schematic diagram of the circuit shown in (a). (c) A digital multimeter can be used to measure both current and potential difference.

through the copper conducting wires back to the negative terminal of the battery. The most important quantities that characterize how the bulb works in different situations are the current $I$ in the bulb and the potential difference $\Delta V$ across the bulb. To measure the current in the bulb, we place an ammeter, the device for measuring current, in line with the bulb so there is no path for the current to bypass the meter; all the charge passing through the bulb must also pass through the ammeter. The voltmeter measures the potential difference, or voltage, between the two ends of the bulb's filament. If we use two meters simultaneously as in Figure 17.5a, we can remove the voltmeter and see if its presence affects the current reading. Figure 17.5c shows a digital multimeter, a convenient device, with a digital readout, that can be used to measure voltage, current, or resistance. An advantage of using a digital multimeter as a voltmeter is that it will usually not affect the current because a digital meter has enormous resistance to the flow of charge in the voltmeter mode.

At this point, you can measure the current as a function of voltage (an $I$–$\Delta V$ curve) of various devices in the lab. All you need is a variable voltage supply (an adjustable battery) capable of supplying potential differences from about $-5$ V to $+5$ V, a bulb, a resistor, some wires and alligator clips, and a couple of multimeters. Be sure to always start your measurements using the highest multimeter scales (say, 10 A and 1 000 V), and increase the sensitivity one scale at a time to obtain the highest accuracy without overloading the meters. (Increasing the sensitivity means lowering the maximum current or voltage that the scale reads.) Note that the meters must be connected with the proper polarity with respect to the voltage supply, as shown in Figure 17.5b. Finally, follow your instructor's directions carefully to avoid damaging the meters and incurring a soaring lab fee.

**Quick Quiz**

**17.3** Look at the four "circuits" shown in Figure 17.6 and select those that will light the bulb.

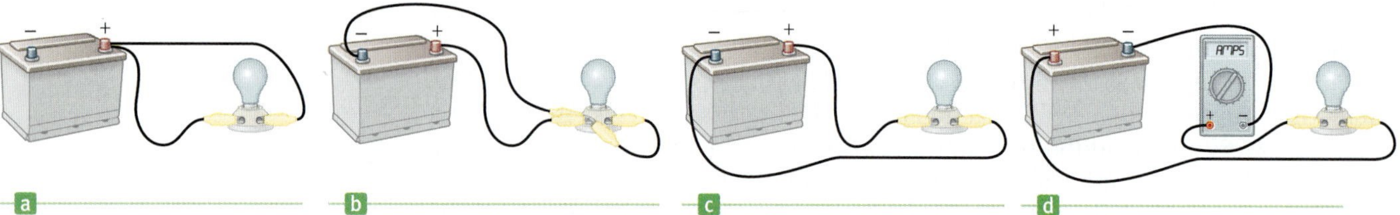


**Figure 17.6** (Quick Quiz 17.3)

# 17.4 Resistance, Resistivity, and Ohm's Law

**LEARNING OBJECTIVES**

1. Define electrical resistance and discuss its physical origins.
2. Relate resistance to resistivity.
3. Apply the concepts of resistivity and resistance to electrical systems.

## Resistance and Ohm's Law

When a voltage (potential difference) $\Delta V$ is applied across the ends of a metallic conductor as in Figure 17.7, the current in the conductor is found to be proportional to the applied voltage; $I \propto \Delta V$. If the proportionality holds, we can write $\Delta V = IR$, where the proportionality constant $R$ is called the *resistance* of the conductor. In fact, we define the **resistance** as the ratio of the voltage across the conductor to the current it carries:

$$R \equiv \frac{\Delta V}{I} \qquad [17.3]$$

◀ Resistance

Resistance has SI units of volts per ampere, called **ohms** ($\Omega$). If a potential difference of 1 V across a conductor produces a current of 1 A, the resistance of the conductor is 1 $\Omega$. For example, if an electrical appliance connected to a 120-V source carries a current of 6 A, its resistance is 20 $\Omega$.

The concepts of electric current, voltage, and resistance can be compared to the flow of water in a river. As water flows downhill in a river of constant width and depth, the flow rate (water current) depends on the steepness of descent of the river and the effects of rocks, the riverbank, and other obstructions. The voltage difference is analogous to the steepness, and the resistance to the obstructions. Based on this analogy, it seems reasonable that increasing the voltage applied to a circuit should increase the current in the circuit, just as increasing the steepness of descent increases the water current. Also, increasing the obstructions in the river's path will reduce the water current, just as increasing the resistance in a circuit will lower the electric current. Resistance in a circuit arises due to collisions between the electrons carrying the current with fixed atoms inside the conductor. These collisions inhibit the movement of charges in much the same way as would a force of friction. For many materials, including most metals, experiments show that **the resistance remains constant over a wide range of applied voltages or currents**. This statement is known as **Ohm's law**, after Georg Simon Ohm (1789–1854), who was the first to conduct a systematic study of electrical resistance.

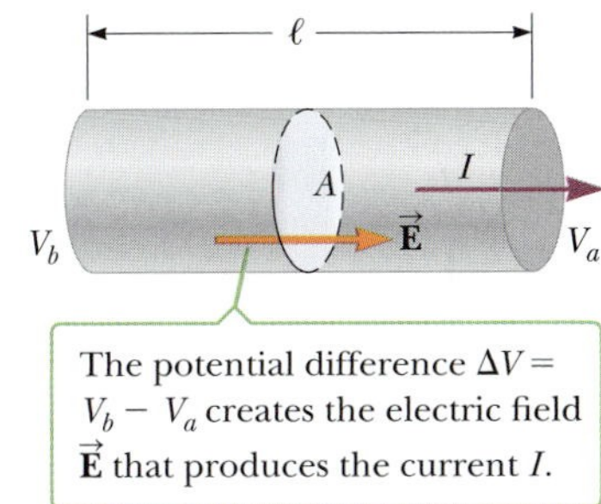


**Figure 17.7** A uniform conductor of length $\ell$ and cross-sectional area $A$. The current $I$ is proportional to the potential difference or, equivalently, to the electric field and length.

Ohm's law is given by

$$\Delta V = IR \qquad [17.4]$$

where $R$ is understood to be independent of $\Delta V$, the potential drop across the resistor, and $I$, the current in the resistor. We will continue to use this traditional form of Ohm's law when discussing electrical circuits. A **resistor** is a conductor that provides a specified resistance in an electric circuit. The symbol for a resistor in circuit diagrams is a zigzag line: —⩘⩘—.

© Bettmann/CORBIS

**Georg Simon Ohm**
**(1787–1854)**
A high school teacher in Cologne and later a professor at Munich, Ohm formulated the concept of resistance and discovered the proportionalities expressed in Equation 17.5.

Ohm's law is an empirical relationship valid only for certain materials. Materials that obey Ohm's law, and hence have a constant resistance over a wide range of voltages, are said to be **ohmic**. Materials having resistance that changes with voltage or current are **nonohmic**. Ohmic materials have a linear current–voltage relationship over a large range of applied voltages (Fig. 17.8a page 608). Nonohmic materials have a nonlinear current–voltage relationship (Fig. 17.8b). One common semiconducting device that is nonohmic is the *diode*, a circuit element that acts like a one-way valve for current. Its resistance is small for currents in one direction (positive $\Delta V$) and large for currents in the reverse direction (negative $\Delta V$). Most modern electronic devices, such as transistors, have nonlinear current–voltage relationships; their operation depends on the particular ways in which they violate Ohm's law.

■ *Quick Quiz*

**17.4** In Figure 17.8b does the resistance of the diode (a) increase or (b) decrease as the positive voltage $\Delta V$ increases?

**17.5** All electric devices are required to have identifying plates that specify their electrical characteristics. The plate on a certain steam iron states that the iron carries a current of 6.00 A when connected to a source of $1.20 \times 10^2$ V. What is the resistance of the steam iron? (a) 0.050 0 $\Omega$ (b) 20.0 $\Omega$ (c) 36.0 $\Omega$

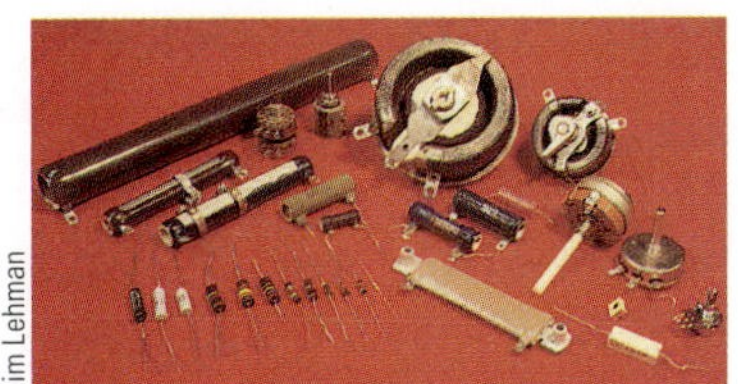
Courtesy of Henry Leap and Jim Lehman

An assortment of resistors used for a variety of applications in electronic circuits.

## Resistivity

Electrons don't move in straight-line paths through a conductor. Instead, they undergo repeated collisions with the metal atoms. Consider a conductor with a voltage applied across its ends. An electron gains speed as the electric force associated with the internal electric field accelerates it, giving it a velocity in the direction

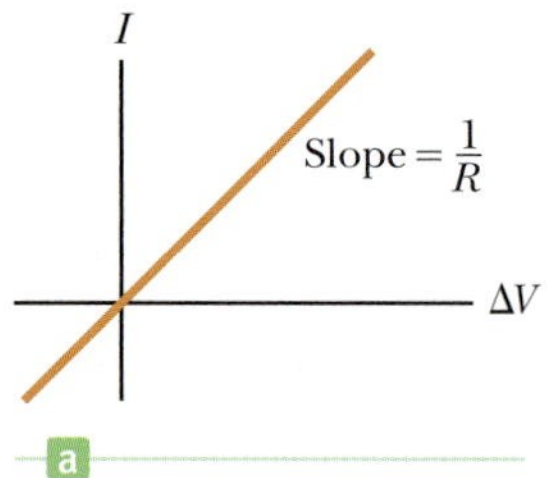


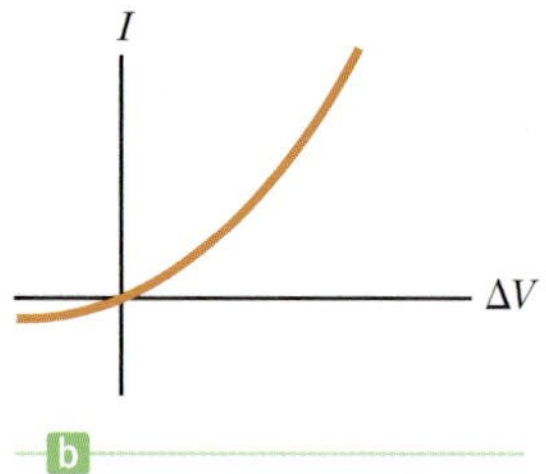


**Figure 17.8** (a) The current–voltage curve for an ohmic material. The curve is linear, and the slope gives the resistance of the conductor. (b) A nonlinear current–voltage curve for a semiconducting diode. This device doesn't obey Ohm's law.

opposite that of the electric field. A collision with an atom randomizes the electron's velocity, reducing it in the direction opposite the field. The process then repeats itself. Together, these collisions affect the electron somewhat as a force of internal friction would. This step is the origin of a material's resistance.

The resistance of an ohmic conductor increases with length, which makes sense because the electrons going through it must undergo more collisions in a longer conductor. A smaller cross-sectional area also increases the resistance of a conductor, just as a smaller pipe slows the fluid moving through it. The resistance, then, is proportional to the conductor's length $\ell$ and inversely proportional to its cross-sectional area $A$,

$$R = \rho \frac{\ell}{A} \qquad [17.5]$$

where the constant of proportionality, $\rho$, is called the **resistivity** of the material. Every material has a characteristic resistivity that depends on its electronic structure and on temperature. Good electric conductors have very low resistivities, and good insulators have very high resistivities. Table 17.1 lists the resistivities of various materials at 20°C. Because resistance values are in ohms, resistivity values must be in ohm-meters ($\Omega \cdot$ m).

**Table 17.1** Resistivities and Temperature Coefficients of Resistivity for Various Materials (at 20°C)

| Material | Resistivity ($\Omega \cdot$ m) | Temperature Coefficient of Resistivity $[(°C)^{-1}]$ |
|---|---|---|
| Silver | $1.59 \times 10^{-8}$ | $3.8 \times 10^{-3}$ |
| Copper | $1.7 \times 10^{-8}$ | $3.9 \times 10^{-3}$ |
| Gold | $2.44 \times 10^{-8}$ | $3.4 \times 10^{-3}$ |
| Aluminum | $2.82 \times 10^{-8}$ | $3.9 \times 10^{-3}$ |
| Tungsten | $5.6 \times 10^{-8}$ | $4.5 \times 10^{-3}$ |
| Iron | $10.0 \times 10^{-8}$ | $5.0 \times 10^{-3}$ |
| Platinum | $11 \times 10^{-8}$ | $3.92 \times 10^{-3}$ |
| Lead | $22 \times 10^{-8}$ | $3.9 \times 10^{-3}$ |
| Nichrome[a] | $150 \times 10^{-8}$ | $0.4 \times 10^{-3}$ |
| Carbon | $3.5 \times 10^{-5}$ | $-0.5 \times 10^{-3}$ |
| Germanium | 0.46 | $-48 \times 10^{-3}$ |
| Silicon | 640 | $-75 \times 10^{-3}$ |
| Glass | $10^{10}$–$10^{14}$ | |
| Hard rubber | $\approx 10^{13}$ | |
| Sulfur | $10^{15}$ | |
| Quartz (fused) | $75 \times 10^{16}$ | |

[a]A nickel-chromium alloy commonly used in heating elements.

## APPLYING PHYSICS 17.1 Dimming of Aging Lightbulbs

As a lightbulb ages, why does it gives off less light than when new?

**EXPLANATION** There are two reasons for the lightbulb's behavior, one electrical and one optical, but both are related to the same phenomenon occurring within the bulb. The filament of an old lightbulb is made of a tungsten wire that has been kept at a high temperature for many hours. High temperatures evaporate tungsten from the filament, decreasing its radius. From $R = \rho\ell/A$, we see that a decreased cross-sectional area leads to an increase in the resistance of the filament. This increasing resistance with age means that the filament will carry less current for the same applied voltage. With less current in the filament, there is less light output, and the filament glows more dimly.

At the high operating temperature of the filament, tungsten atoms leave its surface, much as water molecules evaporate from a puddle of water. The atoms are carried away by convection currents in the gas in the bulb and are deposited on the inner surface of the glass. In time, the glass becomes less transparent because of the tungsten coating, which decreases the amount of light that passes through the glass. ■

## EXAMPLE 17.3 The Resistance of Nichrome Wire

**GOAL** Combine the concept of resistivity with Ohm's law.

**PROBLEM** **(a)** Calculate the resistance per unit length of a 22-gauge Nichrome wire of radius 0.321 mm. **(b)** If a potential difference of 10.0 V is maintained across a 1.00-m length of the Nichrome wire, what is the current in the wire? **(c)** The wire is melted down and recast with twice its original length. Find the new resistance $R_N$ as a multiple of the old resistance $R_O$.

**STRATEGY** Part (a) requires substitution into Equation 17.5, after calculating the cross-sectional area, whereas part (b) is a matter of substitution into Ohm's law. Part (c) requires some algebra. The idea is to take the expression for the new resistance and substitute expressions for $\ell_N$ and $A_N$, the new length and cross-sectional area, in terms of the old length and cross section. For the area substitution, remember that the volumes of the old and new wires are the same.

**SOLUTION**

**(a)** Calculate the resistance per unit length.

Find the cross-sectional area of the wire:

$$A = \pi r^2 = \pi(0.321 \times 10^{-3}\ \text{m})^2 = 3.24 \times 10^{-7}\ \text{m}^2$$

Obtain the resistivity of Nichrome from Table 17.1, solve Equation 17.5 for $R/\ell$, and substitute:

$$\frac{R}{\ell} = \frac{\rho}{A} = \frac{1.5 \times 10^{-6}\ \Omega \cdot \text{m}}{3.24 \times 10^{-7}\ \text{m}^2} = \boxed{4.6\ \Omega/\text{m}}$$

**(b)** Find the current in a 1.00-m segment of the wire if the potential difference across it is 10.0 V.

Substitute given values into Ohm's law:

$$I = \frac{\Delta V}{R} = \frac{10.0\ \text{V}}{4.6\ \Omega} = \boxed{2.2\ \text{A}}$$

**(c)** If the wire is melted down and recast with twice its original length, find the new resistance as a multiple of the old.

Find the new area $A_N$ in terms of the old area $A_O$, using the fact the volume doesn't change and $\ell_N = 2\ell_O$:

$$V_N = V_O \rightarrow A_N\ell_N = A_O\ell_O \rightarrow A_N = A_O(\ell_O/\ell_N)$$
$$A_N = A_O(\ell_O/2\ell_O) = A_O/2$$

Substitute into Equation 17.5:

$$R_N = \frac{\rho\ell_N}{A_N} = \frac{\rho(2\ell_O)}{(A_O/2)} = 4\frac{\rho\ell_O}{A_O} = \boxed{4R_O}$$

**REMARKS** From Table 17.1, the resistivity of Nichrome is about 100 times that of copper, a typical good conductor. Therefore, a copper wire of the same radius would have a resistance per unit length of only 0.052 Ω/m, and a 1.00-m length of copper wire of the same radius would carry the same current (2.2 A) with an applied voltage of only 0.115 V.

Because of its resistance to oxidation, Nichrome is often used for heating elements in toasters, irons, and electric heaters.

**QUESTION 17.3** Would replacing the Nichrome with copper result in a higher current or lower current?

**EXERCISE 17.3** What is the resistance of a 6.0-m length of Nichrome wire that has a radius 0.321 mm? How much current does it carry when connected to a 120-V source?

**ANSWERS** 28 Ω; 4.3 A

### Quick Quiz

**17.6** Suppose an electrical wire is replaced with one having every linear dimension doubled (i.e., the length and radius have twice their original values). Does the wire now have (a) more resistance than before, (b) less resistance, or (c) the same resistance?

In an old-fashioned carbon filament incandescent lamp, the electrical resistance is typically 10 Ω, but changes with temperature.

# 17.5 Temperature Variation of Resistance

**LEARNING OBJECTIVES**

1. Explain the physical reasons for the increase in resistivity and resistance with increasing temperature.
2. State the dependence of resistivity on temperature for limited temperature ranges, and extend it to the temperature dependence of resistance.
3. Apply the temperature dependence of resistance to electrical systems.

The resistivity $\rho$, and hence the resistance, of a conductor depends on a number of factors. One of the most important is the temperature of the metal. For most metals, resistivity increases with increasing temperature. This correlation can be understood as follows: as the temperature of the material increases, its constituent atoms vibrate with greater amplitudes. As a result, the electrons find it more difficult to get by those atoms, just as it is more difficult to weave through a crowded room when the people are in motion than when they are standing still. The increased electron scattering with increasing temperature results in increased resistivity. Technically, thermal expansion also affects resistance; however, this is a very small effect.

Over a limited temperature range, the resistivity of most metals increases linearly with increasing temperature according to the expression

$$\rho = \rho_0[1 + \alpha(T - T_0)] \quad [17.6]$$

where $\rho$ is the resistivity at some temperature $T$ (in Celsius degrees), $\rho_0$ is the resistivity at some reference temperature $T_0$ (usually taken to be 20°C), and $\alpha$ is a parameter called the **temperature coefficient of resistivity**. Temperature coefficients for various materials are provided in Table 17.1. The interesting negative values of $\alpha$ for semiconductors arise because these materials possess weakly bound charge carriers that become free to move and contribute to the current as the temperature rises.

Because the resistance of a conductor with a uniform cross section is proportional to the resistivity according to Equation 17.5 ($R = \rho l/A$), the temperature variation of resistance can be written

$$R = R_0[1 + \alpha(T - T_0)] \quad [17.7]$$

Precise temperature measurements are often made using this property, as shown by the following example.

## EXAMPLE 17.4 A Platinum Resistance Thermometer

**GOAL** Apply the temperature dependence of resistance.

**PROBLEM** A resistance thermometer, which measures temperature by measuring the change in resistance of a conductor, is made of platinum and has a resistance of 50.0 Ω at 20.0°C. **(a)** When the device is immersed in a vessel containing indium at its melting point, its resistance increases to 76.8 Ω. From this information, find the melting point of indium. **(b)** The indium is heated further until it reaches a temperature of 235°C. Estimate the ratio of the new current in the platinum to the current $I_{mp}$ at the melting point, assuming the coefficient of resistivity for platinum doesn't change significantly with temperature.

**STRATEGY** For part (a), solve Equation 17.7 for $T - T_0$ and get $\alpha$ for platinum from Table 17.1, substituting known quantities. For part (b), use Ohm's law in Equation 17.7.

**SOLUTION**

**(a)** Find the melting point of indium.

Solve Equation 17.7 for $T - T_0$:

$$T - T_0 = \frac{R - R_0}{\alpha R_0} = \frac{76.8\ \Omega - 50.0\ \Omega}{[3.92 \times 10^{-3}\ (°\text{C})^{-1}][50.0\ \Omega]}$$

$$= 137°\text{C}$$

Substitute $T_0 = 20.0°\text{C}$ and obtain the melting point of indium:

$$T = 157°\text{C}$$

**(b)** Estimate the ratio of the new current to the old when the temperature rises from 157°C to 235°C.

Write Equation 17.7, with $R_0$ and $T_0$ replaced by $R_{mp}$ and $T_{mp}$, the resistance and temperature at the melting point.

$$R = R_{mp}[1 + \alpha(T - T_{mp})]$$

According to Ohm's law, $R = \Delta V/I$ and $R_{mp} = \Delta V/I_{mp}$. Substitute these expressions into Equation 17.7:

$$\frac{\Delta V}{I} = \frac{\Delta V}{I_{mp}}[1 + \alpha(T - T_{mp})]$$

Cancel the voltage differences, invert the two expressions, and then divide both sides by $I_{mp}$:

$$\frac{I}{I_{mp}} = \frac{1}{1 + \alpha(T - T_{mp})}$$

Substitute $T = 235°\text{C}$, $T_{mp} = 157°\text{C}$, and the value for $\alpha$, obtaining the desired ratio:

$$\frac{I}{I_{mp}} = 0.766$$

**REMARKS** The answer to part (b) is only an estimate because the temperature coefficient is, in fact, temperature dependent. As the temperature rises, both the rms speed of the electrons in the metal and the resistance increase.

**QUESTION 17.4** What happens to the drift speed of the electrons as the temperature rises? (a) It becomes larger. (b) It becomes smaller. (c) It remains unchanged.

**EXERCISE 17.4** Suppose a wire made of an unknown alloy and having a temperature of 20.0°C carries a current of 0.450 A. At 52.0°C the current is 0.370 A for the same potential difference. Find the temperature coefficient of resistivity of the alloy.

**ANSWER** $6.76 \times 10^{-3}\ (°\text{C})^{-1}$

## 17.6 Electrical Energy and Power

**LEARNING OBJECTIVES**

1. Derive an expression for the electrical power delivered to a resistor and discuss the transfer of energy in a simple circuit.
2. Apply the concept of electrical power to practical systems.

If a battery is used to establish an electric current in a conductor, chemical energy stored in the battery is continuously transformed into kinetic energy of the charge carriers. This kinetic energy is quickly lost as a result of collisions between the charge carriers and fixed atoms in the conductor, causing an increase in the temperature of the conductor. In this way the chemical energy stored in the battery is continuously transformed into thermal energy.

To understand the process of energy transfer in a simple circuit, consider a battery with terminals connected to a resistor (Fig. 17.9; remember that the positive terminal of the battery is always at the higher potential). Now imagine following a quantity of positive charge $\Delta Q$ around the circuit from point $A$, through the battery and resistor, and back to $A$. Point $A$ is a reference point that

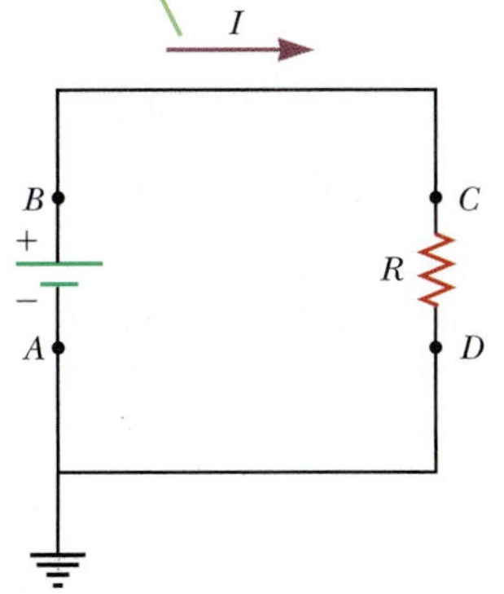
Positive current travels clockwise from the positive to the negative terminal of the battery.
I
B
C
+
−
R
A
D

**Figure 17.9** A circuit consisting of a battery and a resistance $R$. Point $A$ is grounded.

is grounded (the ground symbol is ⏚), and its potential is taken to be zero. As the charge $\Delta Q$ moves from $A$ to $B$ through the battery, the electrical potential energy of the system increases by the amount $\Delta Q\,\Delta V$ and the chemical potential energy in the battery decreases by the same amount. (Recall from Chapter 16 that $\Delta PE = q\,\Delta V$.) As the charge moves from $C$ to $D$ through the resistor, however, it loses this electrical potential energy during collisions with atoms in the resistor. In the process the energy is transformed to internal energy corresponding to increased vibrational motion of those atoms. Because we can ignore the very small resistance of the interconnecting wires, no energy transformation occurs for paths $BC$ and $DA$. When the charge returns to point $A$, the net result is that some of the chemical energy in the battery has been delivered to the resistor and has caused its temperature to rise.

The charge $\Delta Q$ loses energy $\Delta Q\,\Delta V$ as it passes through the resistor. If $\Delta t$ is the time it takes the charge to pass through the resistor, the instantaneous rate at which it loses electric potential energy is

$$\lim_{\Delta t \to 0} \frac{\Delta Q}{\Delta t}\Delta V = I\Delta V$$

where $I$ is the current in the resistor and $\Delta V$ is the potential difference across it. Of course, the charge regains this energy when it passes through the battery, at the expense of chemical energy in the battery. The rate at which the system loses potential energy as the charge passes through the resistor is equal to the rate at which the system gains internal energy in the resistor. Therefore, the power $P$, representing the rate at which energy is delivered to the resistor, is

Power ▶

$$P = I\Delta V \qquad [17.8]$$

Although this result was developed by considering a battery delivering energy to a resistor, Equation 17.8 can be used to determine the power transferred from a voltage source to *any* device carrying a current $I$ and having a potential difference $\Delta V$ between its terminals.

Using Equation 17.8 and the fact that $\Delta V = IR$ for a resistor, we can express the power delivered to the resistor in the alternate forms

Power delivered to ▶ a resistor

$$P = I^2R = \frac{\Delta V^2}{R} \qquad [17.9]$$

**Tip 17.3 Misconception About Current**

Current is *not* "used up" in a resistor. Rather, some of the energy the charges have received from the voltage source is delivered to the resistor, making it hot and causing it to radiate. Also, the current doesn't slow down when going through the resistor: it's the same throughout the circuit.

When $I$ is in amperes, $\Delta V$ in volts, and $R$ in ohms, the SI unit of power is the watt (introduced in Chapter 5). The power delivered to a conductor of resistance $R$ is often referred to as an *$I^2R$ loss.* Note that Equation 17.9 applies only to resistors and not to nonohmic devices such as lightbulbs and diodes.

Regardless of the ways in which you use electrical energy in your home, you ultimately must pay for it or risk having your power turned off. The unit of energy used by electric companies to calculate consumption, the **kilowatt-hour**, is defined in terms of the unit of power and the amount of time it's supplied. One kilowatt-hour (kWh) is the energy converted or consumed in 1 h at the constant rate of 1 kW. It has the numerical value

$$1 \text{ kWh} = (10^3 \text{ W})(3\,600 \text{ s}) = 3.60 \times 10^6 \text{ J} \qquad [17.10]$$

On an electric bill, the amount of electricity used in a given period is usually stated in multiples of kilowatt-hours.

## APPLYING PHYSICS 17.2 Lightbulb Failures

Why do lightbulbs fail so often immediately after they're turned on?

**EXPLANATION** Once the switch is closed, the line voltage is applied across the bulb. As the voltage is applied across the cold filament when the bulb is first turned on, the resistance of the filament is low, the current is high, and a relatively large amount of power is delivered to the bulb. This current spike at the beginning of operation is the reason lightbulbs often fail immediately after they are turned on. As the filament warms, its resistance rises and the current decreases. As a result, the power delivered to the bulb decreases and the bulb is less likely to burn out. ■

### Quick Quiz

**17.7** A voltage $\Delta V$ is applied across the ends of a Nichrome heater wire having a cross-sectional area $A$ and length $L$. The same voltage is applied across the ends of a second Nichrome heater wire having a cross-sectional area $A$ and length $2L$. Which wire gets hotter? (a) The shorter wire does. (b) The longer wire does. (c) More information is needed.

**17.8** For the two resistors shown in Figure 17.10, rank the currents at points $a$ through $f$ from largest to smallest. (a) $I_a = I_b > I_e = I_f > I_c = I_d$ (b) $I_a = I_b > I_c = I_d > I_e = I_f$ (c) $I_e = I_f > I_c = I_d > I_a = I_b$

**17.9** Two resistors, A and B, are connected in a series circuit with a battery. The resistance of A is twice that of B. Which resistor dissipates more power? (a) Resistor A does. (b) Resistor B does. (c) More information is needed.

**17.10** The diameter of wire A is greater than the diameter of wire B, but their lengths and resistivities are identical. For a given voltage difference across the ends, what is the relationship between $P_A$ and $P_B$, the dissipated power for wires A and B, respectively? (a) $P_A = P_B$ (b) $P_A < P_B$ (c) $P_A > P_B$

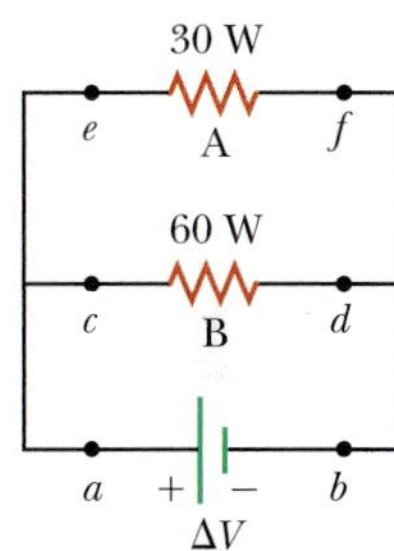


**Figure 17.10** (Quick Quiz 17.8)

## EXAMPLE 17.5 The Cost of Lighting Up Your Life

**GOAL** Apply the electric power concept and calculate the cost of power usage using kilowatt-hours.

**PROBLEM** A circuit provides a maximum current of 20.0 A at an operating voltage of $1.20 \times 10^2$ V. **(a)** How many 75 W bulbs can operate with this voltage source? **(b)** At \$0.120 per kilowatt-hour, how much does it cost to operate these bulbs for 8.00 h?

**STRATEGY** Find the necessary power with $P = I\Delta V$ then divide by 75.0 W per bulb to get the total number of bulbs. To find the cost, convert power to kilowatts and multiply by the number of hours, then multiply by the cost per-kilowatt-hour.

**SOLUTION**

**(a)** Find the number of bulbs that can be lighted.

Substitute into Equation 17.8 to get the total power:

$$P_{\text{total}} = I\Delta V = (20.0\text{ A})(1.20 \times 10^2\text{ V}) = 2.40 \times 10^3\text{ W}$$

Divide the total power by the power per bulb to get the number of bulbs:

$$\text{Number of bulbs} = \frac{P_{\text{total}}}{P_{\text{bulb}}} = \frac{2.40 \times 10^3\text{ W}}{75.0\text{ W}} = \boxed{32.0}$$

**(b)** Calculate the cost of this electricity for an 8.00-h day.

Find the energy in kilowatt-hours:

$$\text{Energy} = Pt = (2.40 \times 10^3\text{ W})\left(\frac{1.00\text{ kW}}{1.00 \times 10^3\text{ W}}\right)(8.00\text{ h})$$

$$= 19.2\text{ kWh}$$

Multiply the energy by the cost per kilowatt-hour:

$$\text{Cost} = (19.2\text{ kWh})(\$0.12/\text{kWh}) = \boxed{\$2.30}$$

*(Continued)*

**REMARKS** This amount of energy might correspond to what a small office uses in a working day, taking into account all power requirements (not just lighting). In general, resistive devices can have variable power output, depending on how the circuit is wired. Here, power outputs were specified, so such considerations were unnecessary.

**QUESTION 17.5** Considering how hot the parts of an incandescent light bulb get during operation, guess what fraction of the energy emitted by an incandescent lightbulb is in the form of visible light. (a) 10% (b) 50% (c) 80%

**EXERCISE 17.5** (a) How many Christmas tree lights drawing 5.00 W of power each could be run on a circuit operating at $1.20 \times 10^2$ V and providing 15.0 A of current? (b) Find the cost to operate one such string 24.0 h per day for the Christmas season (two weeks), using the rate \$0.12/kWh.

**ANSWERS** (a) $3.60 \times 10^2$ bulbs (b) \$72.60

## EXAMPLE 17.6 The Power Converted by an Electric Heater

**GOAL** Calculate an electrical power output and link to its effect on the environment through the first law of thermodynamics.

**PROBLEM** An electric heater is operated by applying a potential difference of 50.0 V to a Nichrome wire of total resistance 8.00 Ω. **(a)** Find the current carried by the wire and the power rating of the heater. **(b)** Using this heater, how long would it take to heat $2.50 \times 10^3$ moles of diatomic gas (e.g., a mixture of oxygen and nitrogen, or air) from a chilly 10.0°C to 25.0°C? Take the molar specific heat at constant volume of air to be $\frac{5}{2}R$. **(c)** How many kilowatt-hours of electricity are used during the time calculated in part (b) and at what cost, at \$0.12 per kilowatt-hour?

**STRATEGY** For part (a), find the current with Ohm's law and substitute into the expression for power. Part (b) is an isovolumetric process, so the thermal energy provided by the heater all goes into the change in internal energy, $\Delta U$. Calculate this quantity using the first law of thermodynamics and divide by the power to get the time. Finding the number of kilowatt-hours used requires a simple unit conversion technique. Multiplying by the cost per kilowatt-hour yields the total cost of operating the heater for the given time.

**SOLUTION**

**(a)** Compute the current and power output.

Apply Ohm's law to get the current:

$$I = \frac{\Delta V}{R} = \frac{50.0\text{ V}}{8.00\ \Omega} = \boxed{6.25\text{ A}}$$

Substitute into Equation 17.9 to find the power:

$$P = I^2R = (6.25\text{ A})^2(8.00\ \Omega) = \boxed{313\text{ W}}$$

**(b)** How long does it take to heat the gas?

Calculate the thermal energy transfer from the first law. Note that $W = 0$ because the volume doesn't change.

$$Q = \Delta U = nC_v\Delta T$$
$$= (2.50 \times 10^3\text{ mol})(\tfrac{5}{2} \cdot 8.31\text{ J/mol} \cdot \text{K})(298\text{ K} - 283\text{ K})$$
$$= 7.79 \times 10^5\text{ J}$$

Divide the thermal energy by the power to get the time:

$$t = \frac{Q}{P} = \frac{7.79 \times 10^5\text{ J}}{313\text{ W}} = \boxed{2.49 \times 10^3\text{ s}}$$

**(c)** Calculate the kilowatt-hours of electricity used and the cost.

Convert the energy to kilowatt-hours, noting that $1\text{ J} = 1\text{ W} \cdot \text{s}$:

$$U = (7.79 \times 10^5\text{ W} \cdot \text{s})\left(\frac{1.00\text{ kW}}{1.00 \times 10^3\text{ W}}\right)\left(\frac{1.00\text{ h}}{3.60 \times 10^3\text{ s}}\right) = \boxed{0.216\text{ kWh}}$$

Multiply by \$0.12/kWh to obtain the total cost of operation:

$$\text{Cost} = (0.216\text{ kWh})(\$0.12/\text{kWh}) = \boxed{\$0.026}$$

**REMARKS** The number of moles of gas given here is approximately what would be found in a bedroom. Warming the air with this space heater requires only about 40 minutes. The calculation, however, doesn't take into account conduction losses. Recall that a 20-cm-thick concrete wall, as calculated in Chapter 11, permitted the loss of more than 2 megajoules an hour by conduction!

**QUESTION 17.6** If the heater wire is replaced by a wire with lower resistance, is the time required to heat the gas (a) unchanged, (b) increased, or (c) decreased?

**EXERCISE 17.6** A hot-water heater is rated at $4.50 \times 10^3$ W and operates at $2.40 \times 10^2$ V. (a) Find the resistance in the heating element and the current. (b) How long does it take to heat 125 L of water from 20.0°C to 50.0°C, neglecting conduction and other losses? (c) How much does it cost at \$0.12/kWh?

**ANSWERS** (a) 12.8 Ω, 18.8 A (b) $3.49 \times 10^3$ s (c) \$0.52

## 17.7 Superconductors

### LEARNING OBJECTIVE

1. Discuss superconductivity and describe important applications.

There is a class of metals and compounds with resistances that fall virtually to *zero* below a certain temperature $T_c$ called the *critical temperature*. These materials are known as **superconductors**. The resistance vs. temperature graph for a superconductor follows that of a normal metal at temperatures above $T_c$ (Fig. 17.11). When the temperature is at or below $T_c$, however, the resistance suddenly drops to zero. This phenomenon was discovered in 1911 by Dutch physicist H. Kamerlingh Onnes as he and a graduate student worked with mercury, which is a superconductor below 4.1 K. Recent measurements have shown that the resistivities of superconductors below $T_c$ are less than $4 \times 10^{-25}\ \Omega \cdot \text{m}$, around $10^{17}$ times smaller than the resistivity of copper and in practice considered to be zero.

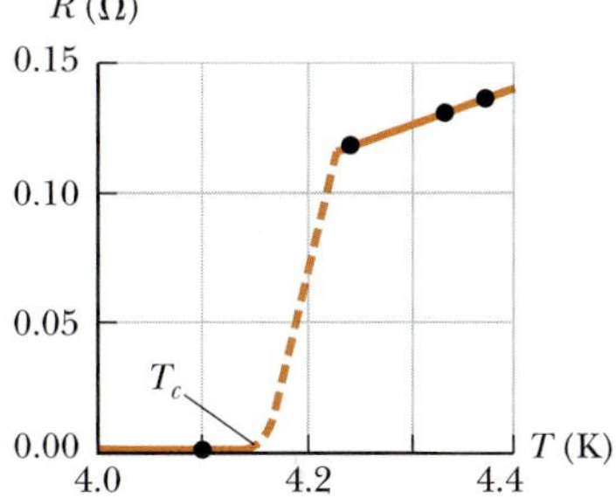


**Figure 17.11** Resistance versus temperature for a sample of mercury (Hg). The graph follows that of a normal metal above the critical temperature $T_c$. The resistance drops to zero at the critical temperature, which is 4.2 K for mercury, and remains at zero for lower temperatures.

Today, thousands of superconductors are known, including such common metals as aluminum, tin, lead, zinc, and indium. Table 17.2 lists the critical temperatures of several superconductors. The value of $T_c$ is sensitive to chemical composition, pressure, and crystalline structure. Interestingly, copper, silver, and gold, which are excellent conductors, don't exhibit superconductivity.

**Table 17.2** Critical Temperatures for Various Superconductors

| Material | $T_c$(K) |
|---|---|
| Zn | 0.88 |
| Al | 1.19 |
| Sn | 3.72 |
| Hg | 4.15 |
| Pb | 7.18 |
| Nb | 9.46 |
| $Nb_3Sn$ | 18.05 |
| $Nb_3Ge$ | 23.2 |
| $YBa_2Cu_3O_7$ | 92 |
| Bi–Sr–Ca–Cu–O | 105 |
| Tl–Ba–Ca–Cu–O | 125 |
| $HgBa_2Ca_2Cu_3O_8$ | 134 |

A truly remarkable feature of superconductors is that once a current is set up in them, it persists *without any applied voltage* (because $R = 0$). In fact, steady currents in superconducting loops have been observed to persist for years with no apparent decay!

An important development in physics that created much excitement in the scientific community was the discovery of high-temperature copper-oxide-based superconductors. The excitement began with a 1986 publication by J. Georg Bednorz and K. Alex Müller, scientists at the IBM Zurich Research Laboratory in Switzerland, in which they reported evidence for superconductivity at a temperature near 30 K in an oxide of barium, lanthanum, and copper. Bednorz and Müller were awarded the Nobel Prize in Physics in 1987 for their important discovery. The discovery was remarkable because the critical temperature was significantly higher than that of any previously known superconductor. Shortly thereafter a new family of compounds was investigated, and research activity in the field of superconductivity proceeded vigorously. In early 1987 groups at the University of Alabama at Huntsville and the University of Houston announced the discovery of superconductivity at about 92 K in an oxide of yttrium, barium, and copper ($YBa_2Cu_3O_7$). Late in 1987, teams of scientists from Japan and the United States reported superconductivity at 105 K in an oxide of bismuth, strontium, calcium, and copper. More recently, scientists have reported superconductivity at temperatures as high as 150 K in an oxide containing mercury. The search for novel superconducting materials continues, with the hope of someday obtaining a room-temperature superconducting material. This research is

**Figure 17.12** A small permanent magnet floats freely above a nitrogen-cooled ceramic superconductor. The superconductor has zero resistance and expels any magnetic field from its interior by creating a mirror image of the magnetic poles of the permanent magnet. This "Meissner effect" results in magnetic levitation of the permanent magnet.

important both for scientific reasons and for practical applications. A superconducting ceramic levitates a permanent magnet in Figure 17.12.

An important and useful application is the construction of superconducting magnets in which the magnetic field intensities are about ten times greater than those of the best normal electromagnets. Such magnets are being considered as a means of storing energy. The idea of using superconducting power lines to transmit power efficiently is also receiving serious consideration. Modern superconducting electronic devices consisting of two thin-film superconductors separated by a thin insulator have been constructed. Among these devices are magnetometers (magnetic-field measuring devices) and various microwave devices.

## 17.8 Electrical Activity in the Heart BIO

**LEARNING OBJECTIVES**

1. Discuss the critical role of electrical activity in the human body.
2. Discuss medical instruments used to monitor and regulate the body's electrical activity.

### Electrocardiograms

**APPLICATION**
Electrocardiograms

Every action involving the body's muscles is initiated by electrical activity. The voltages produced by muscular action in the heart are particularly important to physicians. Voltage pulses cause the heart to beat, and the waves of electrical excitation that sweep across the heart associated with the heartbeat are conducted through the body via the body fluids. These voltage pulses are large enough to be detected by suitable monitoring equipment attached to the skin. A sensitive voltmeter making good electrical contact with the skin by means of contacts attached with conducting paste can be used to measure heart pulses, which are typically of the order of 1 mV at the surface of the body. The voltage pulses can be recorded on an instrument called an **electrocardiograph**, and the pattern recorded by this instrument is called an **electrocardiogram** (EKG). To understand the information contained in an EKG pattern, it is necessary first to describe the underlying principles concerning electrical activity in the heart.

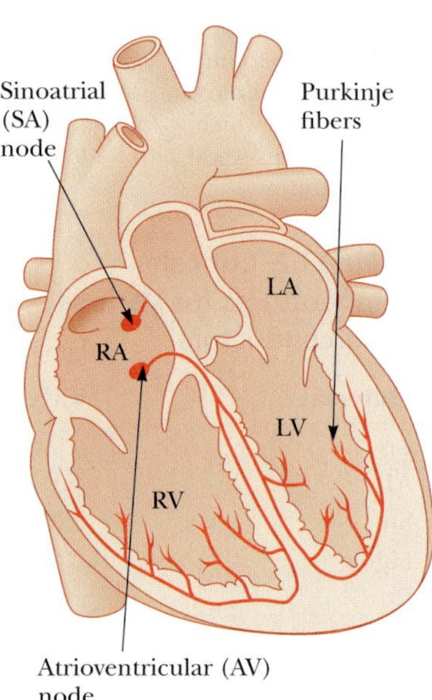


**Figure 17.13** The electrical conduction system of the human heart. (RA: right atrium; LA: left atrium; RV: right ventricle; LV: left ventricle.)

The right atrium of the heart contains a specialized set of muscle fibers called the SA (sinoatrial) node that initiates the heartbeat (Fig. 17.13). Electric impulses that originate in these fibers gradually spread from cell to cell throughout the right and left atrial muscles, causing them to contract. The pulse that passes through the muscle cells is often called a *depolarization wave* because of its effect on individual cells. If an individual muscle cell were examined in its resting state, a double-layer electric charge distribution would be found on its surface, as shown in Figure 17.14a. The impulse generated by the SA node momentarily and locally allows positive charge on the outside of the

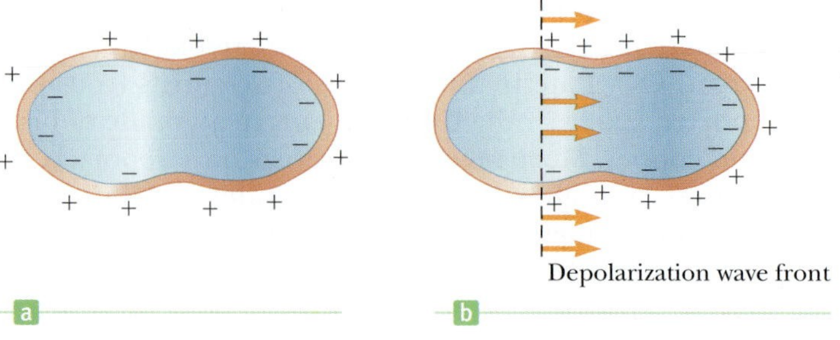


**Figure 17.14** (a) Charge distribution of a muscle cell in the atrium before a depolarization wave has passed through the cell. (b) Charge distribution as the wave passes.

cell to flow in and neutralize the negative charge on the inside layer. This effect changes the cell's charge distribution to that shown in Figure 17.14b. Once the depolarization wave has passed through an individual heart muscle cell, the cell recovers the resting-state charge distribution (positive out, negative in) shown in Figure 17.14a in about 250 ms. When the impulse reaches the atrioventricular (AV) node (Fig. 17.13), the muscles of the atria begin to relax, and the pulse is directed to the ventricular muscles by the AV node. The muscles of the ventricles contract as the depolarization wave spreads through the ventricles along a group of fibers called the *Purkinje fibers.* The ventricles then relax after the pulse has passed through. At this point, the SA node is triggered again and the cycle is repeated.

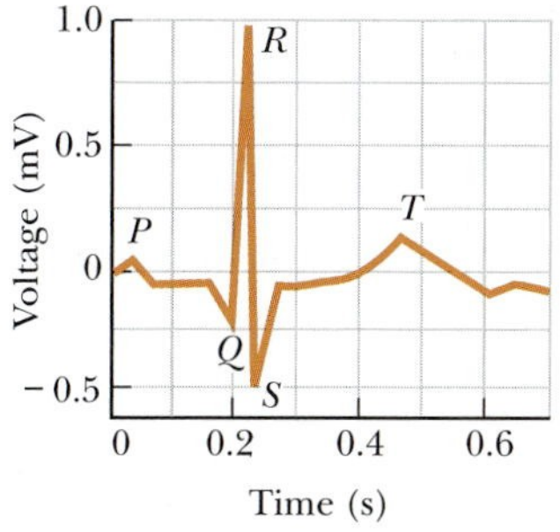


**Figure 17.15** An EKG response for a normal heart.

A sketch of the electrical activity registered on an EKG for one beat of a normal heart is shown in Figure 17.15. The pulse indicated by *P* occurs just before the atria begin to contract. The *QRS* pulse occurs in the ventricles just before they contract, and the *T* pulse occurs when the cells in the ventricles begin to recover. EKGs for an abnormal heart are shown in Figure 17.16. The *QRS* portion of the pattern shown in Figure 17.16a is wider than normal, indicating that the patient may have an enlarged heart. (Why?) Figure 17.16b indicates that there is no constant relationship between the *P* pulse and the *QRS* pulse. This suggests a blockage in the electrical conduction path between the SA and AV nodes that results in the atria and ventricles beating independently and inefficient heart pumping. Finally, Figure 17.16c shows a situation in which there is no *P* pulse and an irregular spacing between the *QRS* pulses. This is symptomatic of irregular atrial contraction, which is called *fibrillation.* In this condition, the atrial and ventricular contractions are irregular.

**BIO APPLICATION**
Cardiac Pacemakers

As noted previously, the sinoatrial node directs the heart to beat at the appropriate rate, usually about 72 beats per minute. Disease or the aging process, however, can damage the heart and slow its beating, and a medical assist may be necessary in the form of a *cardiac pacemaker* attached to the heart. This matchbox-sized electrical device implanted under the skin has a lead that is connected to the wall of the right ventricle. Pulses from this lead stimulate the heart to maintain its proper rhythm. In general, a pacemaker is designed to produce pulses at a rate of about 60 per minute, slightly slower than the normal number of beats per minute, but sufficient to maintain life. The circuitry

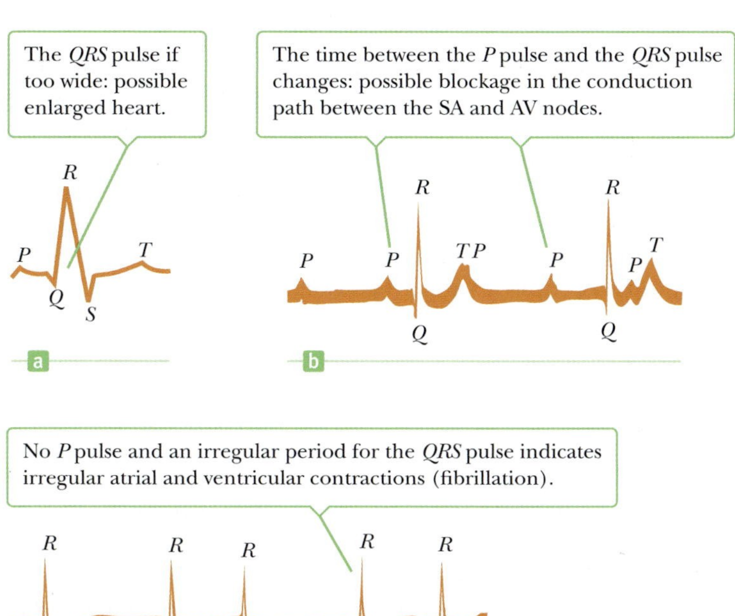


**Figure 17.16** Abnormal EKGs.

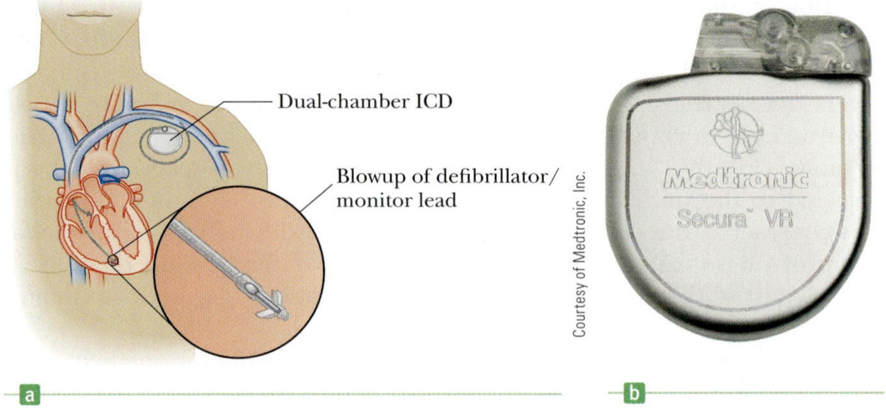


**Figure 17.17** (a) A dual-chamber ICD with leads in the heart. One lead monitors and stimulates the right atrium, and the other monitors and stimulates the right ventricle. (b) Medtronic Dual Chamber ICD.

consists of a capacitor charging up to a certain voltage from a lithium battery and then discharging. The design of the circuit is such that if the heart is beating normally, the capacitor is not allowed to charge completely and send pulses to the heart.

## An Emergency Room in Your Chest

**BIO APPLICATION**
Implanted Cardioverter Defibrillators

In June 2001 an operation on then–Vice President Dick Cheney focused attention on the progress in treating heart problems with tiny implanted electrical devices. Aptly called "an emergency room in your chest" by Cheney's attending physician, devices called *I*mplanted *C*ardioverter *D*efibrillators (**ICDs**) can monitor, record, and logically process heart signals and then supply different corrective signals to hearts beating too slowly, too rapidly, or irregularly. ICDs can even monitor and send signals to the atria and ventricles independently! Figure 17.17a shows a sketch of an ICD with conducting leads that are implanted in the heart. Figure 17.17b shows an actual titanium-encapsulated dual-chamber ICD.

The latest ICDs are sophisticated devices capable of a number of functions:

1. monitoring both atrial and ventricular chambers to differentiate between atrial and potentially fatal ventricular arrhythmias, which require prompt regulation
2. storing about a half hour of heart signals that can easily be read out by a physician
3. being easily reprogrammed with an external magnetic wand
4. performing complicated signal analysis and comparison
5. supplying 0.25- to 10-V repetitive pacing signals to speed up or slow down a malfunctioning heart, or a high-voltage pulse of about 800 V to halt the potentially fatal condition of ventricular fibrillation, in which the heart quivers rapidly rather than beats (people who have experienced such a high-voltage jolt say that it feels like a kick or a bomb going off in the chest)
6. automatically adjusting the number of pacing pulses per minute to match the patient's activity

ICDs are powered by lithium batteries and have implanted lifetimes of 4 to 6 years. Some basic properties of these adjustable ICDs are given in Table 17.3. In the table *tachycardia* means "rapid heartbeat" and *bradycardia* means "slow heartbeat." A key factor in developing tiny electrical implants that serve as defibrillators is the development of capacitors with relatively large capacitance (125 $\mu$F) and small physical size.

**Table 17.3** Properties of Implanted Cardioverter Defibrillators

| **Physical Specifications** | |
|---|---|
| Mass (g) | 85 |
| Size (cm) | 7.3 × 6.2 × 1.3 (about five stacked silver dollars) |
| **Antitachycardia Pacing** | ICD delivers a burst of critically timed low-energy pulses |
| Number of bursts | 1–15 |
| Burst cycle length (ms) | 200–552 |
| Number of pulses per burst | 2–20 |
| Pulse amplitude (V) | 7.5 or 10 |
| Pulse width (ms) | 1.0 or 1.9 |
| **High-Voltage Defibrillation** | |
| Pulse energy (J) | 37 stored/33 delivered |
| Pulse amplitude (V) | 801 |
| **Bradycardia Pacing** | A dual-chamber ICD can steadily deliver repetitive pulses to both the atrium and the ventricle |
| Base frequency (beats/minute) | 40–100 |
| Pulse amplitude (V) | 0.25–7.5 |
| Pulse width (ms) | 0.05, 0.1–1.5, 1.9 |

*Note:* For more information, go to **www.photonicd.com/specs.html.**

## SUMMARY

### 17.1 Electric Current

The **average electric current** $I$ in a conductor is defined as

$$I_{\text{av}} \equiv \frac{\Delta Q}{\Delta t} \quad \textbf{[17.1a]}$$

Current is the time rate of flow of charge through a surface.

where $\Delta Q$ is the charge that passes through a cross section of the conductor in time $\Delta t$. The SI unit of current is the **ampere** (A); 1 A = 1 C/s. By convention, the direction of current is the direction of flow of positive charge.

The **instantaneous current** I is the limit of the average current as the time interval goes to zero:

$$I = \lim_{\Delta t \to 0} I_{\text{av}} = \lim_{\Delta t \to 0} \frac{\Delta Q}{\Delta t} \quad \textbf{[17.1b]}$$

### 17.2 A Microscopic View: Current and Drift Speed

The current in a conductor is related to the motion of the charge carriers by

$$I = nqv_dA \quad \textbf{[17.2]}$$

where $n$ is the number of mobile charge carriers per unit volume, $q$ is the charge on each carrier, $v_d$ is the drift speed of the charges, and $A$ is the cross-sectional area of the conductor.

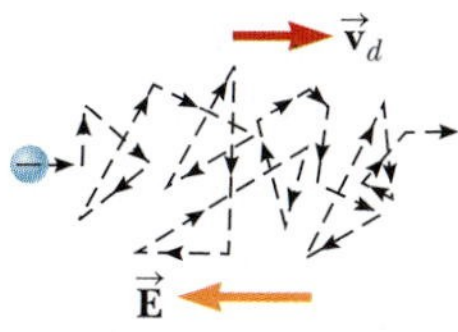


The current $I$ in a conductor is related to the number density $n$ of charge carriers, the charge $q$ per carrier, the drift speed $\vec{\mathbf{v}}_d$, and the cross-sectional area of the conductor, $A$.

### 17.4 Resistance, Resistivity, and Ohm's Law

The **resistance** $R$ of a conductor is defined as the ratio of the potential difference across the conductor to the current in it:

$$R \equiv \frac{\Delta V}{I} \quad \textbf{[17.3]}$$

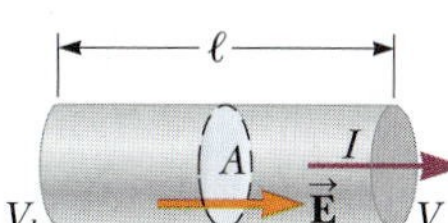


The potential difference $\Delta V$ is proportional to the current, $I$.

The SI units of resistance are volts per ampere, or **ohms** (Ω); 1 Ω = 1 V/A.

**Ohm's law** describes many conductors for which the applied voltage is directly proportional to the current it causes. The proportionality constant is the resistance:

$$\Delta V = IR \quad \textbf{[17.4]}$$

If a conductor has length $\ell$ and cross-sectional area $A$, its **resistance** is

$$R = \rho \frac{\ell}{A} \quad \textbf{[17.5]}$$

where $\rho$ is an intrinsic property of the conductor called the **electrical resistivity**. The SI unit of resistivity is the **ohm-meter** ($\Omega \cdot$ m).

## 17.5 Temperature Variation of Resistance

Over a limited temperature range, the resistivity of a conductor varies with temperature according to the expression

$$\rho = \rho_0[1 + \alpha(T - T_0)] \qquad [17.6]$$

where $\alpha$ is the **temperature coefficient of resistivity** and $\rho_0$ is the resistivity at some reference temperature $T_0$ (usually taken to be 20°C).

The resistance of a conductor varies with temperature according to the expression

$$R = R_0[1 + \alpha(T - T_0)] \qquad [17.7]$$

## 17.6 Electrical Energy and Power

If a potential difference $\Delta V$ is maintained across an electrical device, the **power,** or rate at which energy is supplied to the device, is

$$P = I\Delta V \qquad [17.8]$$

Because the potential difference across a resistor is $\Delta V = IR$, the **power delivered to a resistor** can be expressed as

$$P = I^2R = \frac{\Delta V^2}{R} \qquad [17.9]$$

A **kilowatt-hour** is the amount of energy converted or consumed in one hour by a device supplied with power at the rate of 1 kW. It is equivalent to

$$1 \text{ kWh} = 3.60 \times 10^6 \text{ J} \qquad [17.10]$$

## WARM-UP EXERCISES

ENHANCED WebAssign The warm-up exercises in this chapter may be assigned online in Enhanced WebAssign.

1. **Physics Review** A 5.00-kg box slides across a rough horizontal floor, initially at 2.50 m/s. If friction brings the box to rest after 1.50 s, determine the magnitude of the average rate in watts at which friction dissipates the block's mechanical energy. (See Sections 5.5 and 5.6.)
2. **Physics Review** Suppose an isolated conductor is given −5.00 nC of electrical charge. Determine the number of excess electrons on the conductor. (See Section 15.1.)
3. **Physics Review** A proton, initially at a location where the electric potential is 2.00 V, moves past a point where the potential is −4.00 V. (a) Calculate the change in the proton's speed at that point if it is initially at rest. (The mass of a proton is $1.673 \times 10^{-27}$ kg.) (See Section 16.1.) (b) Repeat the calculation if the proton is initially traveling at $1.20 \times 10^4$ m/s.
4. A wire carries a current of 1.60 A. How many electrons per second pass a given point in the wire? (See Section 17.1.)
5. Copper contains approximately $8.46 \times 10^{28}$ electrons/m$^3$ available to carry current. Suppose a copper wire has a diameter of 2.05 mm. (a) Calculate the cross-sectional area of the wire. (b) If the wire carries a current of 10.0 A, determine the drift speed of electrons through the wire. (See Section 17.2.)
6. A simple circuit consists of a light bulb connected to the terminals of a battery. A voltmeter shows a potential difference of 12.0 V across the light bulb, whereas an ammeter registers a current of 0.500 A in the circuit. Determine the resistance of the bulb. (See Section 17.4.)
7. Determine the length of copper wire having a resistance of 2.00 Ω at 20°C if the wire's cross-sectional area is $3.31 \times 10^{-6}$ m$^2$. (See Section 17.4.)
8. Suppose a spool of copper wire is measured to have a resistance of 0.750 Ω at room temperature (21.0°C). Find its resistance on a cold winter day when the temperature is −25.0°C. (See Section 17.5.)
9. A simple circuit is constructed so that a 12.0-V source supplies current to a $1.00 \times 10^3$ Ω resistor. (a) Find the current in the resistor. (b) Calculate the power delivered to the resistor. (c) Determine the energy delivered to the resistor in 1.00 h. (See Section 17.6.)
10. A typical hair dryer consumes $1.20 \times 10^3$ W of electrical power. If the hair dryer runs for 10.0 minutes, determine (a) the amount of energy consumed, (b) the amount of energy consumed in kWh, (c) the cost of this energy, assuming the electrical power company charges $0.120 per kWh. (See Section 17.6.)

## CONCEPTUAL QUESTIONS

ENHANCED WebAssign The conceptual questions in this chapter may be assigned online in Enhanced WebAssign.

1. We have seen that an electric field must exist inside a conductor that carries a current. How is that possible in view of the fact that in electrostatics we concluded that the electric field must be zero inside a conductor?
2. Use the atomic theory of matter to explain why the resistance of a material should increase as its temperature increases.

3. If charges flow very slowly through a metal, why does it not require several hours for a light to come on when you throw a switch?

4. In an analogy between traffic flow and electrical current, (a) what would correspond to the charge $Q$? (b) What would correspond to the current $I$?

5. When the voltage across a certain conductor is doubled, the current is observed to triple. What can you conclude about the conductor?

6. Two lightbulbs are each connected to a voltage of 120 V. One has a power of 25 W, the other 100 W. (a) Which lightbulb has the higher resistance? (b) Which lightbulb carries more current?

7. Newspaper articles often have statements such as "10 000 volts of electricity surged *through* the victim's body." What is wrong with this statement?

8. There is an old admonition given to experimenters to "keep one hand in the pocket" when working around high voltages. Why is this warning a good idea?

9. What could happen to the drift velocity of the electrons in a wire and to the current in the wire if the electrons could move through it freely without resistance?

10. Some homes have light dimmers that are operated by rotating a knob. What is being changed in the electric circuit when the knob is rotated?

11. When is more power delivered to a lightbulb, immediately after it is turned on and the glow of the filament is increasing or after it has been on for a few seconds and the glow is steady?

## PROBLEMS

ENHANCED WebAssign The problems in this chapter may be assigned online in Enhanced WebAssign.

1. denotes straightforward problem; 2. denotes intermediate problem;
3. denotes challenging problem
[1.] denotes full solution available in *Student Solutions Manual/ Study Guide*
1. denotes problems most often assigned in Enhanced WebAssign

BIO denotes biomedical problems
GP denotes guided problems
M denotes Master It tutorial available in Enhanced WebAssign
Q|C denotes asking for quantitative and conceptual reasoning
S denotes symbolic reasoning problem
W denotes Watch It video solution available in Enhanced WebAssign

### 17.1 Electric Current

### 17.2 A Microscopic View: Current and Drift Speed

1. W If a current of 80.0 mA exists in a metal wire, (a) how many electrons flow past a given cross section of the wire in 10.0 min? (b) In what direction do the electrons travel with respect to the current?

2. Q|C A copper wire has a circular cross section with a radius of 1.25 mm. (a) If the wire carries a current of 3.70 A, find the drift speed of the electrons in the wire. (See Example 17.2 for relevant data on copper.) (b) All other things being equal, what happens to the drift speed in wires made of metal having a larger number of conduction electrons per atom than copper? Explain.

[3.] In the Bohr model of the hydrogen atom, an electron in the lowest energy state moves at a speed equal to $2.19 \times 10^6$ m/s in a circular path having a radius of $5.29 \times 10^{-11}$ m. What is the effective current associated with this orbiting electron?

4. The current supplied by a battery in a portable device is typically 0.15 A. Find the number of electrons passing through the device in one hour.

5. A certain laboratory experiment requires an aluminum wire of length of 32.0 m and a resistance of 2.50 Ω at 20.0°C. What diameter wire must be used?

6. If $3.25 \times 10^{-3}$ kg of gold is deposited on the negative electrode of an electrolytic cell in a period of 2.78 h, what is the current in the cell during that period? Assume the gold ions carry one elementary unit of positive charge.

7. M A 200-km-long high-voltage transmission line 2.0 cm in diameter carries a steady current of 1 000 A. If the conductor is copper with a free charge density of $8.5 \times 10^{28}$ electrons per cubic meter, how many years does it take one electron to travel the full length of the cable?

[8.] An aluminum wire having a cross-sectional area of $4.00 \times 10^{-6}$ m$^2$ carries a current of 5.00 A. The density of aluminum is 2.70 g/cm$^3$. Assume each aluminum atom supplies one conduction electron per atom. Find the drift speed of the electrons in the wire.

9. GP An iron wire has a cross-sectional area of $5.00 \times 10^{-6}$ m$^2$. Carry out steps (a) through (e) to compute the drift speed of the conduction electrons in the wire. (a) How many kilograms are there in 1 mole of iron? (b) Starting with the density of iron and the result of part (a), compute the molar density of iron (the number of moles of iron per cubic meter). (c) Calculate the number density of iron atoms using Avogadro's number. (d) Obtain the number density of conduction electrons given that there are two conduction electrons per

iron atom. (e) If the wire carries a current of 30.0 A, calculate the drift speed of conduction electrons.

## 17.4 Resistance, Resistivity, and Ohm's Law

**10.** An electric heater carries a current of 13.5 A when operating at a voltage of $1.20 \times 10^2$ V. What is the resistance of the heater?

**11.** BIO A person notices a mild shock if the current along a path through the thumb and index finger exceeds 80 $\mu$A. Compare the maximum possible voltage without shock across the thumb and index finger with a dry-skin resistance of $4.0 \times 10^5\ \Omega$ and a wet-skin resistance of 2 000 $\Omega$.

**12.** W Suppose you wish to fabricate a uniform wire out of 1.00 g of copper. If the wire is to have a resistance $R = 0.500\ \Omega$, and if all the copper is to be used, what will be (a) the length and (b) the diameter of the wire?

**13.** Nichrome wire of cross-sectional radius 0.791 mm is to be used in winding a heating coil. If the coil must carry a current of 9.25 A when a voltage of $1.20 \times 10^2$ V is applied across its ends, find (a) the required resistance of the coil and (b) the length of wire you must use to wind the coil.

**14.** A wire of diameter 0.800 mm and length 25.0 m has a measured resistance of 1.60 $\Omega$. What is the resistivity of the wire?

**15.** A potential difference of 12 V is found to produce a current of 0.40 A in a 3.2-m length of wire with a uniform radius of 0.40 cm. What is (a) the resistance of the wire? (b) The resistivity of the wire?

**16.** S Aluminum and copper wires of equal length are found to have the same resistance. What is the ratio of their radii?

**17.** M A wire 50.0 m long and 2.00 mm in diameter is connected to a source with a potential difference of 9.11 V, and the current is found to be 36.0 A. Assume a temperature of 20°C and, using Table 17.1, identify the metal out of which the wire is made.

**18.** A rectangular block of copper has sides of length 10 cm, 20 cm, and 40 cm. If the block is connected to a 6.0-V source across two of its opposite faces, what are (a) the maximum current and (b) the minimum current the block can carry?

**19.** A wire of initial length $L_0$ and radius $r_0$ has a measured resistance of 1.0 $\Omega$. The wire is drawn under tensile stress to a new uniform radius of $r = 0.25r_0$. What is the new resistance of the wire?

**20.** BIO Q|C The human body can exhibit a wide range of resistances to current depending on the path of the current, contact area, and sweatiness of the skin. Suppose the resistance across the chest from the left hand to the right hand is $1.0 \times 10^6\ \Omega$. (a) How much voltage is required to cause possible heart fibrillation in a man, which corresponds to 500 mA of direct current? (b) Why should rubber-soled shoes and rubber gloves be worn when working around electricity?

**21.** S Starting from Ohm's law, show that $E = J\rho$, where $E$ is the magnitude of the electric field (assumed constant) and $J = I/A$ is called the current density. The result is in fact true in general.

## 17.5 Temperature Variation of Resistance

**22.** If a certain silver wire has a resistance of 6.00 $\Omega$ at 20.0°C, what resistance will it have at 34.0°C?

**23.** While taking photographs in Death Valley on a day when the temperature is 58.0°C, Bill Hiker finds that a certain voltage applied to a copper wire produces a current of 1.000 A. Bill then travels to Antarctica and applies the same voltage to the same wire. What current does he register there if the temperature is −88.0°C? Assume no change occurs in the wire's shape and size.

**24.** Q|C A length of aluminum wire has a resistance of 30.0 $\Omega$ at 20.0°C. When the wire is warmed in an oven and reaches thermal equilibrium, the resistance of the wire increases to 46.2 $\Omega$. (a) Neglecting thermal expansion, find the temperature of the oven. (b) Qualitatively, how would thermal expansion be expected to affect the answer?

**25.** Gold is the most ductile of all metals. For example, one gram of gold can be drawn into a wire 2.40 km long. The density of gold is $19.3 \times 10^3\ \text{kg/m}^3$, and its resistivity is $2.44 \times 10^{-8}\ \Omega \cdot \text{m}$. What is the resistance of such a wire at 20.0°C?

**26.** At what temperature will aluminum have a resistivity that is three times the resistivity of copper at room temperature?

**27.** At 20°C, the carbon resistor in an electric circuit connected to a 5.0-V battery has a resistance of $2.0 \times 10^2\ \Omega$. What is the current in the circuit when the temperature of the carbon rises to 80°C?

**28.** A wire 3.00 m long and 0.450 mm$^2$ in cross-sectional area has a resistance of 41.0 $\Omega$ at 20°C. If its resistance increases to 41.4 $\Omega$ at 29.0°C, what is the temperature coefficient of resistivity?

**29.** M (a) A 34.5-m length of copper wire at 20.0°C has a radius of 0.25 mm. If a potential difference of 9.0 V is applied across the length of the wire, determine the current in the wire. (b) If the wire is heated to 30.0°C while the 9.0-V potential difference is maintained, what is the resulting current in the wire?

**30.** An engineer needs a resistor with a zero overall temperature coefficient of resistance at 20.0°C. She designs a pair of circular cylinders, one of carbon and one of Nichrome as shown in Figure P17.30. The device must have an overall resistance of $R_1 + R_2 = 10.0\ \Omega$ independent of temperature and a uniform radius of $r = 1.50$ mm. Ignore thermal expansion of the cylinders and assume both are always at the same temperature. (a) Can she meet the design goal with this method? (b) If so, state what you can determine about the lengths $L_1$ and $L_2$ of each segment. If not, explain.

$L_1$ $L_2$

**Figure P17.30**

31. BIO In one form of plethysmograph (a device for measuring volume), a rubber capillary tube with an inside diameter of 1.00 mm is filled with mercury at 20°C. The resistance of the mercury is measured with the aid of electrodes sealed into the ends of the tube. If 100.00 cm of the tube is wound in a spiral around a patient's upper arm, the blood flow during a heartbeat causes the arm to expand, stretching the tube to a length of 100.04 cm. From this observation, and assuming cylindrical symmetry, you can find the change in volume of the arm, which gives an indication of blood flow. (a) Calculate the resistance of the mercury. (b) Calculate the fractional change in resistance during the heartbeat. Take $\rho_{Hg} = 9.4 \times 10^{-7}\ \Omega \cdot \text{m}$. *Hint:* Because the cylindrical volume is constant, $V = A_iL_i = A_fL_f$ and $A_f = A_i(L_i/L_f)$.

32. A platinum resistance thermometer has resistances of 200.0 Ω when placed in a 0°C ice bath and 253.8 Ω when immersed in a crucible containing melting potassium. What is the melting point of potassium? *Hint:* First determine the resistance of the platinum resistance thermometer at room temperature, 20°C.

## 17.6 Electrical Energy and Power

33. Suppose your waffle iron is rated at 1.00 kW when connected to a $1.20 \times 10^2$ V source. (a) What current does the waffle iron carry? (b) What is its resistance?

34. If electrical energy costs $0.12 per kilowatt-hour, how much does it cost to (a) burn a 100-W lightbulb for 24 h? (b) Operate an electric oven for 5.0 h if it carries a current of 20.0 A at 220 V?

35. Q|C Residential building codes typically require the use of 12-gauge copper wire (diameter 0.205 cm) for wiring receptacles. Such circuits carry currents as large as 20.0 A. If a wire of smaller diameter (with a higher gauge number) carried that much current, the wire could rise to a high temperature and cause a fire. (a) Calculate the rate at which internal energy is produced in 1.00 m of 12-gauge copper wire carrying 20.0 A. (b) Repeat the calculation for a 12-gauge aluminum wire. (c) Explain whether a 12-gauge aluminum wire would be as safe as a copper wire.

36. A high-voltage transmission line with a resistance of 0.31 Ω/km carries a current of 1 000 A. The line is at a potential of 700 kV at the power station and carries the current to a city located 160 km from the station. (a) What is the power loss due to resistance in the line? (b) What fraction of the transmitted power does this loss represent?

37. M The heating element of a coffeemaker operates at 120 V and carries a current of 2.00 A. Assuming the water absorbs all the energy converted by the resistor, calculate how long it takes to heat 0.500 kg of water from room temperature (23.0°C) to the boiling point.

38. W The power supplied to a typical black-and-white television set is 90 W when the set is connected to 120 V. (a) How much electrical energy does this set consume in 1 hour? (b) A color television set draws about 2.5 A when connected to 120 V. How much time is required for it to consume the same energy as the black-and-white model consumes in 1 hour?

39. Q|C Lightbulb A is marked "25.0 W 120 V," and lightbulb B is marked "100 W 120 V." These labels mean that each lightbulb has its respective power delivered to it when it is connected to a constant 120-V source. (a) Find the resistance of each lightbulb. (b) During what time interval does 1.00 C pass into lightbulb A? (c) Is this charge different upon its exit versus its entry into the lightbulb? Explain. (d) In what time interval does 1.00 J pass into lightbulb A? (e) By what mechanisms does this energy enter and exit the lightbulb? Explain. (f) Find the cost of running lightbulb A continuously for 30.0 days, assuming the electric company sells its product at $0.110 per kWh.

40. A certain toaster has a heating element made of Nichrome resistance wire. When the toaster is first connected to a 120-V source of potential difference (and the wire is at a temperature of 20.0°C), the initial current is 1.80 A but the current begins to decrease as the resistive element warms up. When the toaster reaches its final operating temperature, the current has dropped to 1.53 A. (a) Find the power the toaster converts when it is at its operating temperature. (b) What is the final temperature of the heating element?

41. A copper cable is designed to carry a current of 300 A with a power loss of 2.00 W/m. What is the required radius of this cable?

42. Batteries are rated in terms of ampere-hours (A · h). For example, a battery that can deliver a current of 3.0 A for 5.0 h is rated at 15 A · h. (a) What is the total energy, in kilowatt-hours, stored in a 12-V battery rated at 55 A · h? (b) At $0.12 per kilowatt-hour, what is the value of the electricity that can be produced by this battery?

43. BIO The potential difference across a resting neuron in the human body is about 75.0 mV and carries a current of about 0.200 mA. How much power does the neuron release?

44. The cost of electricity varies widely throughout the United States; $0.120/kWh is a typical value. At this unit price, calculate the cost of (a) leaving a 40.0-W porch light on for 2 weeks while you are on vacation, (b) making a piece of dark toast in 3.00 min with a 970-W toaster, and (c) drying a load of clothes in 40.0 min in a 5 200-W dryer.

45. An electric utility company supplies a customer's house from the main power lines (120 V) with two copper wires, each of which is 50.0 m long and has a resistance of 0.108 Ω per 300 m. (a) Find the potential difference at the customer's house for a load current of 110 A. For this load current, find (b) the power delivered to the customer.

46. **Q|C** An office worker uses an immersion heater to warm 250 g of water in a light, covered, insulated cup from 20°C to 100°C in 4.00 minutes. The heater is a Nichrome resistance wire connected to a 120-V power supply. Assume the wire is at 100°C throughout the 4.00-min time interval. (a) Calculate the average power required to warm the water to 100°C in 4.00 min. (b) Calculate the required resistance in the heating element at 100°C. (c) Calculate the resistance of the heating element at 20.0°C. (d) Derive a relationship between the diameter of the wire, the resistivity at 20.0°C, $\rho_0$, the resistance at 20.0°C, $R_0$, and the length $L$. (e) If $L = 3.00$ m, what is the diameter of the wire?

47. **S** Two wires $A$ and $B$ made of the same material and having the same lengths are connected across the same voltage source. If the power supplied to wire $A$ is three times the power supplied to wire $B$, what is the ratio of their diameters?

48. **Q|C** A tungsten wire in a vacuum has length 15.0 cm and radius 1.00 mm. A potential difference is applied across it. (a) What is the resistance of the wire at 293 K? (b) Suppose the wire reaches an equilibrium temperature such that it emits 75.0 W in the form of radiation. Neglecting absorption of any radiation from its environment, what is the temperature of the wire? (*Note:* $e = 0.320$ for tungsten.) (c) What is the resistance of the wire at the temperature found in part (b)? Assume the temperature changes linearly over this temperature range. (d) What voltage drop is required across the wire? (e) Why are tungsten lightbulbs energetically inefficient as light sources?

## Additional Problems

49. If a battery is rated at 60.0 A · h, how much total charge can it deliver before it goes "dead"?

50. A car owner forgets to turn off the headlights of his car while it is parked in his garage. If the 12.0-V battery in his car is rated at 90.0 A · h and each headlight requires 36.0 W of power, how long will it take the battery to completely discharge?

51. Consider an aluminum wire of diameter 0.600 mm and length 15.0 m. The resistivity of aluminum at 20.0°C is $2.82 \times 10^{-8}\ \Omega \cdot$ m. (a) Find the resistance of this wire at 20.0°C. (b) If a 9.00 V battery is connected across the ends of the wire, find the current in the wire.

52. A given copper wire has a resistance of 5.00 Ω at 20.0°C while a tungsten wire of the same diameter has a resistance of 4.75 Ω at 20.0°C. At what temperature will the two wires have the same resistance?

53. **M** A particular wire has a resistivity of $3.0 \times 10^{-8}\ \Omega \cdot$ m and a cross-sectional area of $4.0 \times 10^{-6}$ m$^2$. A length of this wire is to be used as a resistor that will develop 48 W of power when connected across a 20-V battery. What length of wire is required?

54. Birds resting on high-voltage power lines are a common sight. The copper wire on which a bird stands is 2.2 cm in diameter and carries a current of 50 A. If the bird's feet are 4.0 cm apart, calculate the potential difference between its feet.

55. An experiment is conducted to measure the electrical resistivity of Nichrome in the form of wires with different lengths and cross-sectional areas. For one set of measurements, a student uses 30-gauge wire, which has a cross-sectional area of $7.30 \times 10^{-8}$ m$^2$. The student measures the potential difference across the wire and the current in the wire with a voltmeter and an ammeter, respectively. For each of the measurements given in the following table taken on wires of three different lengths, calculate the resistance of the wires and the corresponding value of the resistivity.

| $L$ (m) | $\Delta V$(V) | $I$(A) | $R$ (Ω) | $\rho$ (Ω · m) |
|---|---|---|---|---|
| 0.540 | 5.22 | 0.500 | | |
| 1.028 | 5.82 | 0.276 | | |
| 1.543 | 5.94 | 0.187 | | |

What is the average value of the resistivity, and how does this value compare with the value given in Table 17.1?

56. A 50.0-g sample of a conducting material is all that is available. The resistivity of the material is measured to be $11 \times 10^{-8}\ \Omega \cdot$ m, and the density is 7.86 g/cm$^3$. The material is to be shaped into a solid cylindrical wire that has a total resistance of 1.5 Ω. (a) What length of wire is required? (b) What must be the diameter of the wire?

57. You are cooking breakfast for yourself and a friend using a 1 200-W waffle iron and a 500-W coffeepot. Usually, you operate these appliances from a 110-V outlet for 0.500 h each day. (a) At 12 cents per kWh, how much do you spend to cook breakfast during a 30.0-day period? (b) You find yourself addicted to waffles and would like to upgrade to a 2 400-W waffle iron that will enable you to cook twice as many waffles during a half-hour period, but you know that the circuit breaker in your kitchen is a 20-A breaker. Can you do the upgrade?

58. The current in a conductor varies in time as shown in Figure P17.58. (a) How many coulombs of charge pass through a cross section of the conductor in the interval from $t = 0$ to $t = 5.0$ s? (b) What constant current would transport the same total charge during the 5.0-s interval as does the actual current?

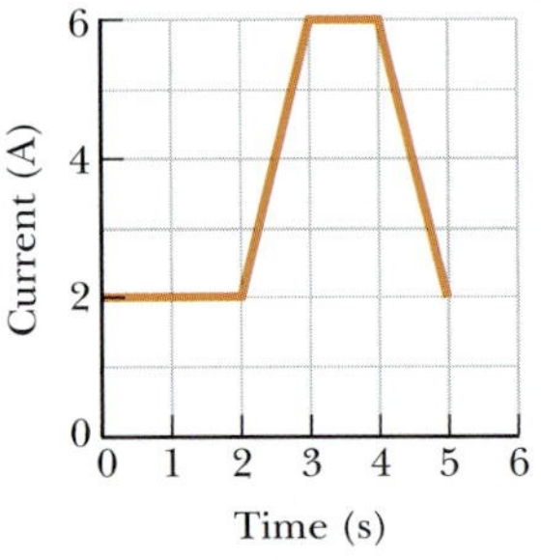

**Figure P17.58**

59. A 120-V motor has mechanical power output of 2.50 hp. It is 90.0% efficient in converting power that it takes in by electrical transmission into mechanical power. (a) Find the current in the motor. (b) Find the energy delivered to the motor by electrical transmission in 3.00 h of operation. (c) If the electric company charges \$0.110/kWh, what does it cost to run the motor for 3.00 h?

60. **Q|C** A Nichrome heating element in an oven has a resistance of 8.0 Ω at 20°C. (a) What is its resistance at 350°C? (b) What assumption did you have to make to obtain your answer to part (a)?

61. A length of metal wire has a radius of $5.00 \times 10^{-3}$ m and a resistance of 0.100 Ω. When the potential difference across the wire is 15.0 V, the electron drift speed is found to be $3.17 \times 10^{-4}$ m/s. On the basis of these data, calculate the density of free electrons in the wire.

62. In a certain stereo system, each speaker has a resistance of 4.00 Ω. The system is rated at 60.0 W in each channel. Each speaker circuit includes a fuse rated at a maximum current of 4.00 A. Is this system adequately protected against overload?

63. A resistor is constructed by forming a material of resistivity $3.5 \times 10^5\ \Omega \cdot \text{m}$ into the shape of a hollow cylinder of length 4.0 cm and inner and outer radii 0.50 cm and 1.2 cm, respectively. In use, a potential difference is applied between the ends of the cylinder, producing a current parallel to the length of the cylinder. Find the resistance of the cylinder.

64. **S** When a straight wire is heated, its resistance changes according to the equation

$$R = R_0[1 + \alpha(T - T_0)]$$

(Eq. 17.7), where $\alpha$ is the temperature coefficient of resistivity. (a) Show that a more precise result, which includes the length and area of a wire change when it is heated, is

$$R = \frac{R_0[1 + \alpha(T - T_0)][1 + \alpha'(T - T_0)]}{[1 + 2\alpha'(T - T_{0)})]}$$

where $\alpha'$ is the coefficient of linear expansion. (See Chapter 10.) (b) Compare the two results for a 2.00-m-long copper wire of radius 0.100 mm, starting at 20.0°C and heated to 100.0°C.

65. **BIO** An x-ray tube used for cancer therapy operates at 4.0 MV, with a beam current of 25 mA striking the metal target. Nearly all the power in the beam is transferred to a stream of water flowing through holes drilled in the target. What rate of flow, in kilograms per second, is needed if the rise in temperature ($\Delta T$) of the water is not to exceed 50°C?

66. A man wishes to vacuum his car with a canister vacuum cleaner marked 535 W at 120 V. The car is parked far from the building, so he uses an extension cord 15.0 m long to plug the cleaner into a 120-V source. Assume the cleaner has constant resistance. (a) If the resistance of each of the two conductors of the extension cord is 0.900 Ω, what is the actual power delivered to the cleaner? (b) If, instead, the power is to be at least 525 W, what must be the diameter of each of two identical copper conductors in the cord the young man buys? (c) Repeat part (b) if the power is to be at least 532 W. *Suggestion:* A symbolic solution can simplify the calculations.

This vision-impaired patient is wearing an artificial vision device. A camera on the lens sends video to a portable computer. The computer then converts the video to signals that stimulate implants in the patient's visual cortex, giving him enough sight to move around on his own.

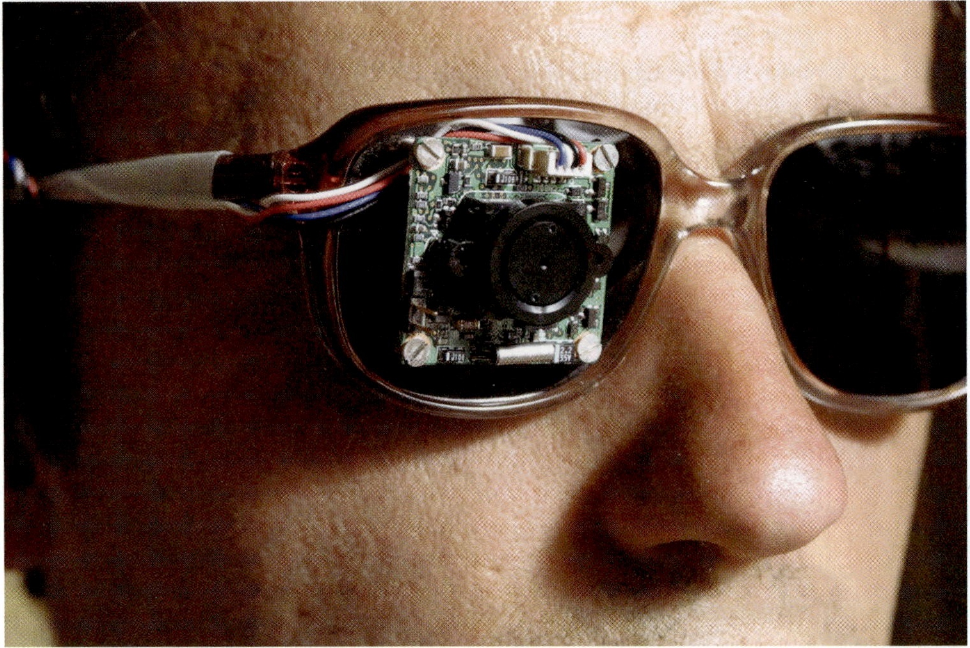



# 18 Direct-Current Circuits



Batteries, resistors, and capacitors can be used in various combinations to construct electric circuits, which direct and control the flow of electricity and the energy it conveys. Such circuits make possible all the modern conveniences in a home: electric lights, electric stove tops and ovens, washing machines, and a host of other appliances and tools. Electric circuits are also found in our cars, in tractors that increase farming productivity, and in all types of medical equipment that saves so many lives every day.

In this chapter we study and analyze a number of simple direct-current circuits. The analysis is simplified by the use of two rules known as Kirchhoff's rules, which follow from the principle of conservation of energy and the law of conservation of charge. Most of the circuits are assumed to be in *steady state,* which means that the currents are constant in magnitude and direction. We close the chapter with a discussion of circuits containing resistors and capacitors, in which current varies with time.

## 18.1 Sources of emf

**LEARNING OBJECTIVES**

1. Describe in physical terms the concepts of emf, terminal voltage, internal resistance, and load resistance.
2. Evaluate the current and total power output of an emf source.

A current is maintained in a closed circuit by a source of emf.[1] Among such sources are any devices (for example, batteries and generators) that increase the potential

[1]The term was originally an abbreviation for *electromotive force,* but emf is not really a force, so the long form is discouraged.

energy of the circulating charges. A source of emf can be thought of as a "charge pump" that forces electrons to move in a direction opposite the electrostatic field inside the source. The emf $\mathcal{E}$ of a source is the work done per unit charge; hence the SI unit of emf is the volt.

Consider the circuit in Figure 18.1a consisting of a battery connected to a resistor. We assume the connecting wires have no resistance. If we neglect the internal resistance of the battery, the potential drop across the battery (the terminal voltage) equals the emf of the battery. Because a real battery always has some internal resistance $r$, however, the terminal voltage is not equal to the emf. The circuit of Figure 18.1a can be described schematically by the diagram in Figure 18.1b. The battery, represented by the dashed rectangle, consists of a source of emf $\mathcal{E}$ in series with an internal resistance $r$. Now imagine a positive charge moving through the battery from $a$ to $b$ in the figure. As the charge passes from the negative to the positive terminal of the battery, the potential of the charge increases by $\mathcal{E}$. As the charge moves through the resistance $r$, however, its potential decreases by the amount $Ir$, where $I$ is the current in the circuit. The terminal voltage of the battery, $\Delta V = V_b - V_a$, is therefore given by

$$\Delta V = \mathcal{E} - Ir \qquad [18.1]$$

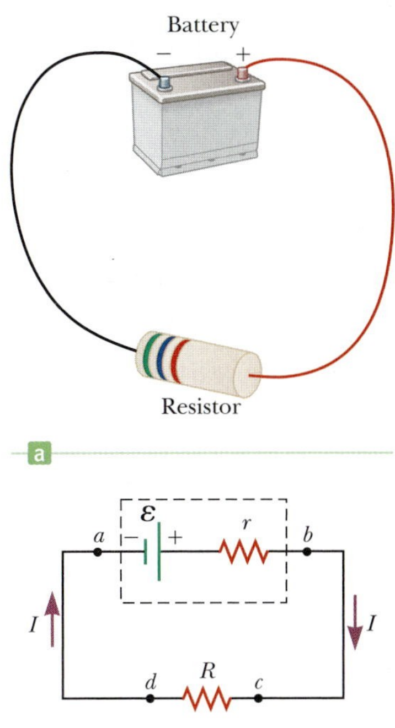


**Figure 18.1** (a) A circuit consisting of a resistor connected to the terminals of a battery. (b) A circuit diagram of a source of emf $\mathcal{E}$ having internal resistance $r$ connected to an external resistor $R$.

From this expression, we see that **$\mathcal{E}$ is equal to the terminal voltage when the current is zero**, called the **open-circuit voltage**. By inspecting Figure 18.1b, we find that the terminal voltage $\Delta V$ must also equal the potential difference across the external resistance $R$, often called the **load resistance**; that is, $\Delta V = IR$. Combining this relationship with Equation 18.1, we arrive at

$$\mathcal{E} = IR + Ir \qquad [18.2]$$

Solving for the current gives

$$I = \frac{\mathcal{E}}{R + r}$$

The preceding equation shows that the current in this simple circuit depends on both the resistance external to the battery and the internal resistance of the battery. If $R$ is much greater than $r$, we can neglect $r$ in our analysis (an option we usually select).

If we multiply Equation 18.2 by the current $I$, we get

$$I\mathcal{E} = I^2R + I^2r$$

© Cengage Learning/George Semple

An assortment of batteries.

This equation tells us that the total power output $I\mathcal{E}$ of the source of emf is converted at the rate $I^2R$ at which energy is delivered to the load resistance, *plus* the rate $I^2r$ at which energy is delivered to the internal resistance. Again, if $r << R$, most of the power delivered by the battery is transferred to the load resistance.

Unless otherwise stated, in our examples and end-of-chapter problems we will assume the internal resistance of a battery in a circuit is negligible.

## Quick Quiz

**18.1** True or False: While discharging, the terminal voltage of a battery can never be greater than the emf of the battery.

**18.2** Why does a battery get warm while in use?

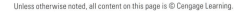

**Tip 18.1 What's Constant in a Battery?**

Equation 18.2 shows that the current in a circuit depends on the resistance of the battery, so a battery can't be considered a source of constant current. Even the terminal voltage of a battery given by Equation 18.1 can't be considered constant because the internal resistance can change (due to warming, for example, during the operation of the battery). A battery is, however, a source of constant emf.

## 18.2 Resistors in Series

**LEARNING OBJECTIVES**

1. Derive an expression for the equivalent resistance of a series of resistors.
2. Analyze circuits having resistors in series.

When two or more resistors are connected end to end as in Figure 18.2, they are said to be in *series*. The resistors could be simple devices, such as lightbulbs or heating elements. When two resistors $R_1$ and $R_2$ are connected to a battery as in Figure 18.2, **the current is the same in the two resistors because any charge that flows through $R_1$ must also flow through $R_2$.** This is analogous to water flowing through a pipe with two constrictions, corresponding to $R_1$ and $R_2$. Whatever volume of water flows in one end in a given time interval must exit the opposite end.

Because the potential difference between $a$ and $b$ in Figure 18.2b equals $IR_1$ and the potential difference between $b$ and $c$ equals $IR_2$, the potential difference between $a$ and $c$ is

$$\Delta V = IR_1 + IR_2 = I(R_1 + R_2)$$

Regardless of how many resistors we have in series, the sum of the potential differences across the resistors is equal to the total potential difference across the combination. As we will show later, this result is a consequence of the conservation of energy. Figure 18.2c shows an equivalent resistor $R_{eq}$ that can replace the two resistors of the original circuit. The equivalent resistor has the same effect on the circuit because it results in the same current in the circuit as the two resistors. Applying Ohm's law to this equivalent resistor, we have

$$\Delta V = IR_{eq}$$

Equating the preceding two expressions, we have

$$IR_{eq} = I(R_1 + R_2)$$

or

$$R_{eq} = R_1 + R_2 \text{ (series combination)} \quad [18.3]$$

An extension of the preceding analysis shows that the equivalent resistance of three or more resistors connected in series is

Equivalent resistance of a series combination of resistors ▶

$$R_{eq} = R_1 + R_2 + R_3 + \cdots \quad [18.4]$$

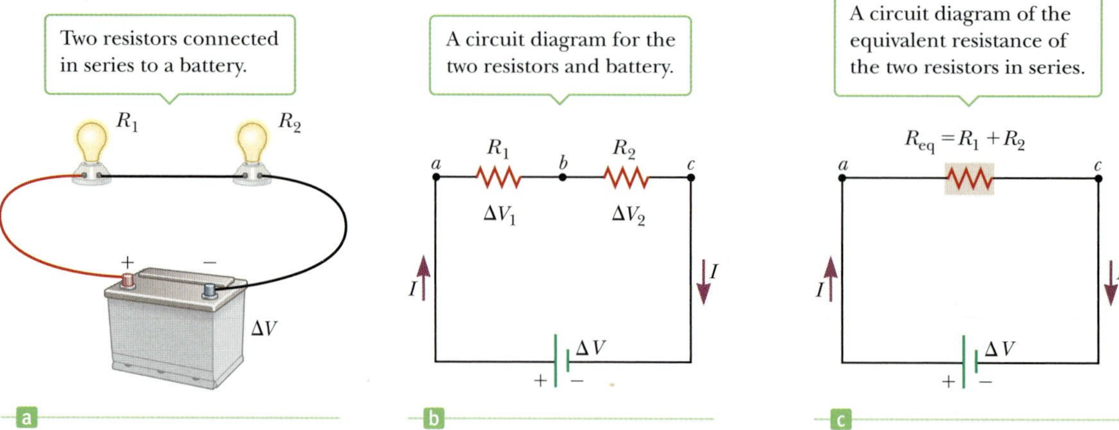


**Figure 18.2** Two resistors, $R_1$ and $R_2$, in the form of incandescent light bulbs, in series with a battery. The currents in the resistors are the same, and the equivalent resistance of the combination is $R_{eq} = R_1 + R_2$.

Therefore, **the equivalent resistance of a series combination of resistors is the algebraic sum of the individual resistances and is always greater than any individual resistance**.

Note that if the filament of one lightbulb in Figure 18.2 were to fail, the circuit would no longer be complete (an open-circuit condition would exist) and the second bulb would also go out.

## APPLYING PHYSICS 18.1 Christmas Lights in Series

A new design for Christmas lights allows them to be connected in series. A failed bulb in such a string would result in an open circuit, and all the bulbs would go out. How can the bulbs be redesigned to prevent that from happening?

**EXPLANATION** If the string of lights contained the usual kind of bulbs, a failed bulb would be hard to locate. Each bulb would have to be replaced with a good bulb, one by one, until the failed bulb was found. If there happened to be two or more failed bulbs in the string of lights, finding them would be a lengthy and annoying task.

Christmas lights use special bulbs that have an insulated loop of wire (a jumper) across the conducting supports to the bulb filaments (Fig. 18.3). If the filament breaks and the bulb fails, the bulb's resistance increases dramatically. As a result, most of the applied voltage appears across the loop of wire. This voltage causes the insulation around the loop of wire to burn, causing the metal wire to make electrical contact with the supports. This produces a conducting path through the bulb, so the other bulbs remain lit. ■

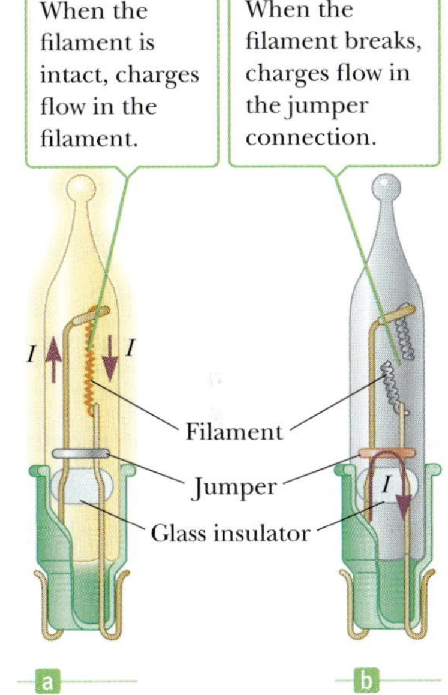


**Figure 18.3** (Applying Physics 18.1) (a) Schematic diagram of a modern "miniature" holiday lightbulb, with a jumper connection to provide a current path if the filament breaks. (b) A holiday lightbulb with a broken filament.

### Quick Quiz

**18.3** In Figure 18.4 the current is measured with the ammeter at the bottom of the circuit. When the switch is opened, does the reading on the ammeter (a) increase, (b) decrease, or (c) not change?

**18.4** The circuit in Figure 18.4 consists of two resistors, a switch, an ammeter, and a battery. When the switch is closed, power $P_c$ is delivered to resistor $R_1$. When the switch is opened, which of the following statements is true about the power $P_o$ delivered to $R_1$? (a) $P_o < P_c$ (b) $P_o = P_c$ (c) $P_o > P_c$

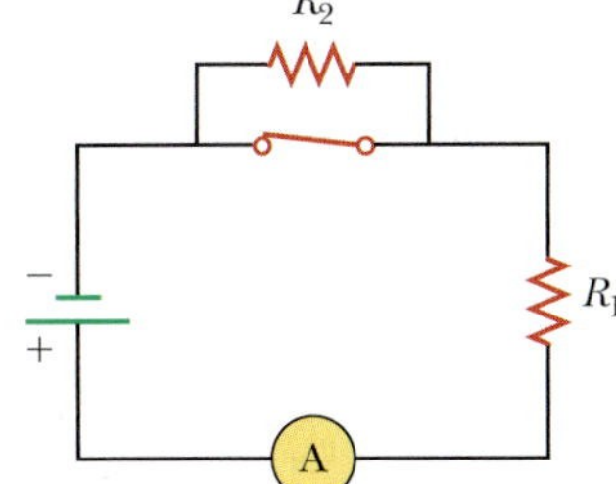


**Figure 18.4** (Quick Quizzes 18.3 and 18.4)

## EXAMPLE 18.1 Four Resistors in Series

**GOAL** Analyze several resistors connected in series.

**PROBLEM** Four resistors are arranged as shown in Figure 18.5a. Find **(a)** the equivalent resistance of the circuit and **(b)** the current in the circuit if the closed-circuit terminal voltage of the battery is 6.0 V. **(c)** Calculate the electric potential at point $A$ if the potential at the positive terminal is 6.0 V. **(d)** Suppose the open circuit voltage, or emf $\mathcal{E}$, is 6.2 V. Calculate the battery's internal resistance. **(e)** What fraction $f$ of the battery's power is delivered to the load resistors?

*(Continued)*

**STRATEGY** Because the resistors are connected in series, summing their resistances gives the equivalent resistance. Ohm's law can then be used to find the current. To find the electric potential at point $A$, calculate the voltage drop $\Delta V$ across the 2.0-Ω resistor and subtract the result from 6.0 V. In part (d) use Equation 18.1 to find the internal resistance of the battery. The fraction of the power delivered to the load resistance is just the power delivered to load, $I\Delta V$, divided by the total power, $I\mathcal{E}$.

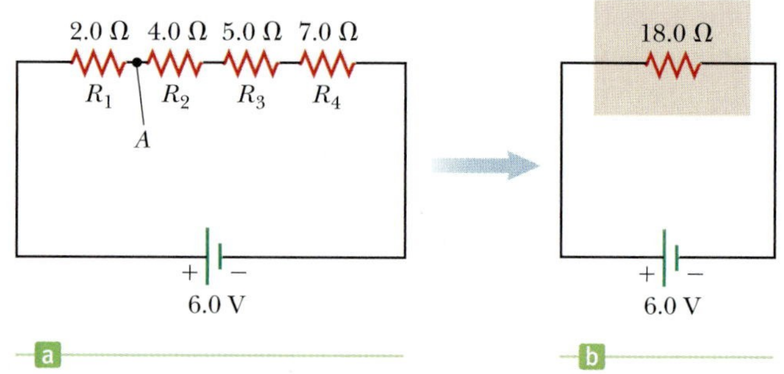


**Figure 18.5** (Example 18.1) (a) Four resistors connected in series. (b) The equivalent resistance of the circuit in (a).

**SOLUTION**

**(a)** Find the equivalent resistance of the circuit.

Apply Equation 18.4, summing the resistances:

$$R_{eq} = R_1 + R_2 + R_3 + R_4 = 2.0\ \Omega + 4.0\ \Omega + 5.0\ \Omega + 7.0\ \Omega = \boxed{18.0\ \Omega}$$

**(b)** Find the current in the circuit.

Apply Ohm's law to the equivalent resistor in Figure 18.5b, solving for the current:

$$I = \frac{\Delta V}{R_{eq}} = \frac{6.0\ \text{V}}{18.0\ \Omega} = \boxed{0.33\ \text{A}}$$

**(c)** Calculate the electric potential at point $A$.

Apply Ohm's law to the 2.0-Ω resistor to find the voltage drop across it:

$$\Delta V = IR = (0.33\ \text{A})(2.0\ \Omega) = 0.66\ \text{V}$$

To find the potential at $A$, subtract the voltage drop from the potential at the positive terminal:

$$V_A = 6.0\ \text{V} - 0.66\ \text{V} = \boxed{5.3\ \text{V}}$$

**(d)** Calculate the battery's internal resistance if the battery's emf is 6.2 V.

Write Equation 18.1:

$$\Delta V = \mathcal{E} - Ir$$

Solve for the internal resistance $r$ and substitute values:

$$r = \frac{\mathcal{E} - \Delta V}{I} = \frac{6.2\ \text{V} - 6.0\ \text{V}}{0.33\ \text{A}} = \boxed{0.6\ \Omega}$$

**(e)** What fraction $f$ of the battery's power is delivered to the load resistors?

Divide the power delivered to the load by the total power output:

$$f = \frac{I\Delta V}{I\mathcal{E}} = \frac{\Delta V}{\mathcal{E}} = \frac{6.0\ \text{V}}{6.2\ \text{V}} = \boxed{0.97}$$

**REMARKS** A common misconception is that the current is "used up" and steadily declines as it progresses through a series of resistors. That would be a violation of the conservation of charge. What is actually used up is the electric potential energy of the charge carriers, some of which is delivered to each resistor.

**QUESTION 18.1** Explain why the current in a real circuit very slowly decreases with time compared to its initial value.

**EXERCISE 18.1** A closed circuit consists of a battery with a terminal voltage of 12.0 V, and 3.0-Ω, 6.0-Ω, 8.0-Ω, and 9.0-Ω resistors connected in series, oriented as in Figure 18.5a, with the battery in the bottom of the loop, positive terminal on the left, and resistors in increasing order, left to right, in the top of the loop. Calculate (a) the equivalent resistance of the circuit, (b) the current, and (c) the total power dissipated by the load resistors. (d) What is the electric potential at a point between the 6.0-Ω and 8.0-Ω resistors, if the electric potential at the positive terminal is 12.0 V? (e) If the battery has an emf of 12.1 V, find the battery's internal resistance.

**ANSWERS** (a) 26.0 Ω (b) 0.462 A (c) 5.54 W (d) 7.84 V (e) 0.2 Ω

# 18.3 Resistors in Parallel

**LEARNING OBJECTIVES**

1. Derive an expression for the equivalent resistance of resistors in parallel.
2. Analyze circuits having resistors in parallel.
3. Simplify circuits having combinations of parallel and series resistors.

Now consider two resistors connected in parallel, as in Figure 18.6. In this case **the potential differences across the resistors are the same because each is connected directly across the battery terminals.** The currents are generally not the same. When charges reach point $a$ (called a junction) in Figure 18.6b, the current splits into two parts: $I_1$, flowing through $R_1$; and $I_2$, flowing through $R_2$. If $R_1$ is greater than $R_2$, then $I_1$ is less than $I_2$. In general, more charge travels through the path with less resistance. **Because charge is conserved, the current $I$ that enters point $a$ must equal the total current $I_1 + I_2$ leaving that point.** Mathematically, this is written

$$I = I_1 + I_2$$

The potential drop must be the same for the two resistors and must also equal the potential drop across the battery. Ohm's law applied to each resistor yields

$$I_1 = \frac{\Delta V}{R_1} \qquad I_2 = \frac{\Delta V}{R_2}$$

Ohm's law applied to the equivalent resistor in Figure 18.6c gives

$$I = \frac{\Delta V}{R_{eq}}$$

When these expressions for the currents are substituted into the equation $I = I_1 + I_2$ and the $\Delta V$'s are canceled, we obtain

$$\frac{1}{R_{eq}} = \frac{1}{R_1} + \frac{1}{R_2} \quad \text{(parallel combination)} \qquad [18.5]$$

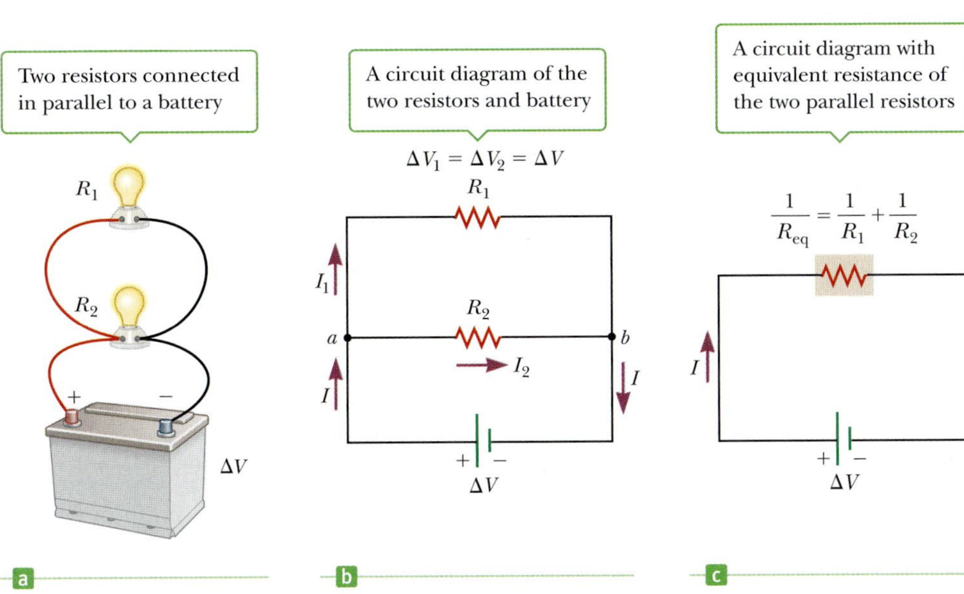


**Figure 18.6** Two resistors, $R_1$ and $R_2$, in the form of incandescent light bulbs, in parallel with a battery. The potential differences across $R_1$ and $R_2$ are the same. Currents in the resistors are the same, and the equivalent resistance of the combination is $1/R_{eq} = 1/R_1 + 1/R_2$.

An extension of this analysis to three or more resistors in parallel produces the following general expression for the equivalent resistance:

Equivalent resistance of a parallel combination of resistors

$$\frac{1}{R_{eq}} = \frac{1}{R_1} + \frac{1}{R_2} + \frac{1}{R_3} + \cdots \qquad [18.6]$$

From this expression, we see that **the inverse of the equivalent resistance of two or more resistors connected in parallel is the sum of the inverses of the individual resistances and is always less than the smallest resistance in the group.**

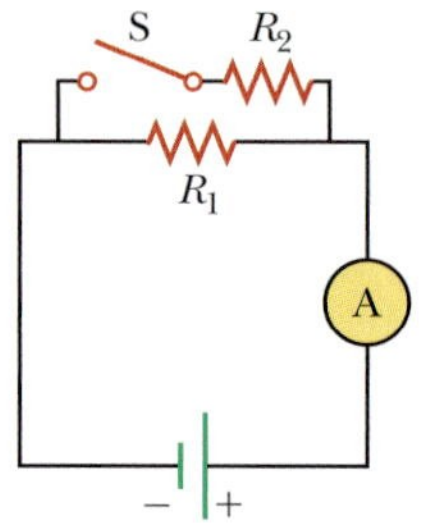


**Figure 18.7** (Quick Quizzes 18.5 and 18.6)

## Quick Quiz

**18.5** In Figure 18.7 the current is measured with the ammeter on the right side of the circuit diagram. When the switch is closed, does the reading on the ammeter (a) increase, (b) decrease, or (c) remain the same?

**18.6** When the switch is open in Figure 18.7, power $P_o$ is delivered to the resistor $R_1$. When the switch is closed, which of the following is true about the power $P_c$ delivered to $R_1$? (Neglect the internal resistance of the battery.) (a) $P_c < P_o$ (b) $P_c = P_o$ (c) $P_c > P_o$

## EXAMPLE 18.2 Three Resistors in Parallel

**GOAL** Analyze a circuit that contains resistors connected in parallel.

**PROBLEM** Three resistors are connected in parallel as in Figure 18.8. A potential difference of 18 V is maintained between points *a* and *b*. **(a)** Find the current in each resistor. **(b)** Calculate the power delivered to each resistor and the total power. **(c)** Find the equivalent resistance of the circuit. **(d)** Find the total power delivered to the equivalent resistance.

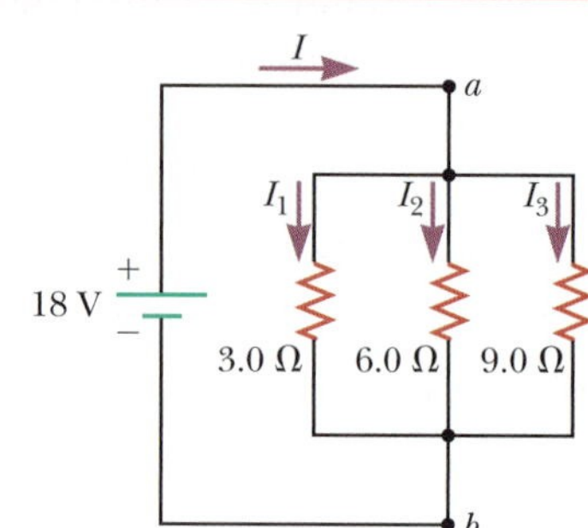


**Figure 18.8** (Example 18.2) Three resistors connected in parallel. The voltage across each resistor is 18 V.

**STRATEGY** To get the current in each resistor we can use Ohm's law and the fact that the voltage drops across parallel resistors are all the same. The rest of the problem just requires substitution into the equation for power delivered to a resistor, $P = I^2R$, and the reciprocal-sum law for parallel resistors.

**SOLUTION**

**(a)** Find the current in each resistor.

Apply Ohm's law, solved for the current $I$ delivered by the battery to find the current in each resistor:

$$I_1 = \frac{\Delta V}{R_1} = \frac{18\text{ V}}{3.0\ \Omega} = 6.0\text{ A}$$

$$I_2 = \frac{\Delta V}{R_2} = \frac{18\text{ V}}{6.0\ \Omega} = 3.0\text{ A}$$

$$I_3 = \frac{\Delta V}{R_3} = \frac{18\text{ V}}{9.0\ \Omega} = 2.0\text{ A}$$

**(b)** Calculate the power delivered to each resistor and the total power.

Apply $P = I^2R$ to each resistor, substituting the results from part (a):

$$3\ \Omega\text{: } P_1 = I_1^2R_1 = (6.0\text{ A})^2(3.0\ \Omega) = 110\text{ W}$$

$$6\ \Omega\text{: } P_2 = I_2^2R_2 = (3.0\text{ A})^2(6.0\ \Omega) = 54\text{ W}$$

$$9\ \Omega\text{: } P_3 = I_3^2R_3 = (2.0\text{ A})^2(9.0\ \Omega) = 36\text{ W}$$

Sum to get the total power:

$$P_{tot} = 110\text{ W} + 54\text{ W} + 36\text{ W} = 2.0 \times 10^2\text{ W}$$

**(c)** Find the equivalent resistance of the circuit.

Apply the reciprocal-sum rule, Equation 18.6:

$$\frac{1}{R_{eq}} = \frac{1}{R_1} + \frac{1}{R_2} + \frac{1}{R_3}$$

$$\frac{1}{R_{eq}} = \frac{1}{3.0\ \Omega} + \frac{1}{6.0\ \Omega} + \frac{1}{9.0\ \Omega} = \frac{11}{18\ \Omega}$$

$$R_{eq} = \frac{18}{11}\ \Omega = \boxed{1.6\ \Omega}$$

**(d)** Compute the power dissipated by the equivalent resistance.

Use the alternate power equation:

$$P = \frac{(\Delta V)^2}{R_{eq}} = \frac{(18\ \text{V})^2}{1.6\ \Omega} = \boxed{2.0 \times 10^2\ \text{W}}$$

**REMARKS** There's something important to notice in part (a): the smallest 3.0 Ω resistor carries the largest current, whereas the other, larger resistors of 6.0 Ω and 9.0 Ω carry smaller currents. The largest current is always found in the path of least resistance. In part (b) the power could also be found with $P = (\Delta V)^2/R$. Note that $P_1 = 108$ W, but is rounded to 110 W because there are only two significant figures. Finally, notice that the total power dissipated in the equivalent resistor is the same as the sum of the power dissipated in the individual resistors, as it should be.

**Tip 18.2 Don't Forget to Flip It!**

The most common mistake in calculating the equivalent resistance for resistors in parallel is to forget to invert the answer after summing the reciprocals. Don't forget to flip it!

**QUESTION 18.2** If a fourth resistor were added in parallel to the other three, how would the equivalent resistance change? (a) It would be larger. (b) It would be smaller. (c) More information is required to determine the effect.

**EXERCISE 18.2** Suppose the resistances in the example are 1.0 Ω, 2.0 Ω, and 3.0 Ω, respectively, and a new voltage source is provided. If the current measured in the 3.0-Ω resistor is 2.0 A, find (a) the potential difference provided by the new battery and the currents in each of the remaining resistors, (b) the power delivered to each resistor and the total power, (c) the equivalent resistance, and (d) the total current and the power dissipated by the equivalent resistor.

**ANSWERS** (a) $\mathcal{E} = 6.0$ V, $I_1 = 6.0$ A, $I_2 = 3.0$ A (b) $P_1 = 36$ W, $P_2 = 18$ W, $P_3 = 12$ W, $P_{tot} = 66$ W (c) $\frac{6}{11}$ Ω (d) $I = 11$ A, $P_{eq} = 66$ W

**Quick Quiz**

**18.7** Suppose you have three identical lightbulbs, some wire, and a battery. You connect one lightbulb to the battery and take note of its brightness. You add a second lightbulb, connecting it in parallel with the previous lightbulbs, and again take note of the brightness. Repeat the process with the third lightbulb, connecting it in parallel with the other two. As the lightbulbs are added, what happens to (a) the brightness of the lightbulbs? (b) The individual currents in the lightbulbs? (c) The power delivered by the battery? (d) The lifetime of the battery? (Neglect the battery's internal resistance.)

**18.8** If the lightbulbs in Quick Quiz 18.7 are connected one by one in series instead of in parallel, what happens to (a) the brightness of the lightbulbs? (b) The individual currents in the lightbulbs? (c) The power delivered by the battery? (d) The lifetime of the battery? (Again, neglect the battery's internal resistance.)

Household circuits are always wired so that the electrical devices are connected in parallel, as in Figure 18.6a. In this way each device operates independently of the others so that if one is switched off, the others remain on. For example, if one of the lightbulbs in Figure 18.6 were removed from its socket, the other would continue to operate. Equally important is that each device operates at the same voltage. If the devices were connected in series, the voltage across each one would depend on how many there were in the combination and on their individual resistances.

**APPLICATION**
Circuit Breakers

In many household circuits, circuit breakers are used in series with other circuit elements for safety purposes. A circuit breaker is designed to switch off and open the circuit at some maximum value of current (typically 15 A or 20 A) that depends on the nature of the circuit. If a circuit breaker were not used, excessive currents caused by operating several devices simultaneously could result in excessive wire temperatures, perhaps causing a fire. In older homes, fuses were used in place of circuit breakers. When the current in a circuit exceeded some value, the conductor in a fuse melted and opened the circuit. The disadvantage of fuses is that they are destroyed in the process of opening the circuit, whereas circuit breakers can be reset.

## APPLYING PHYSICS 18.2 Lightbulb Combinations

Compare the brightness of the four identical lightbulbs shown in Figure 18.9. What happens if bulb A fails and so cannot conduct current? What if C fails? What if D fails?

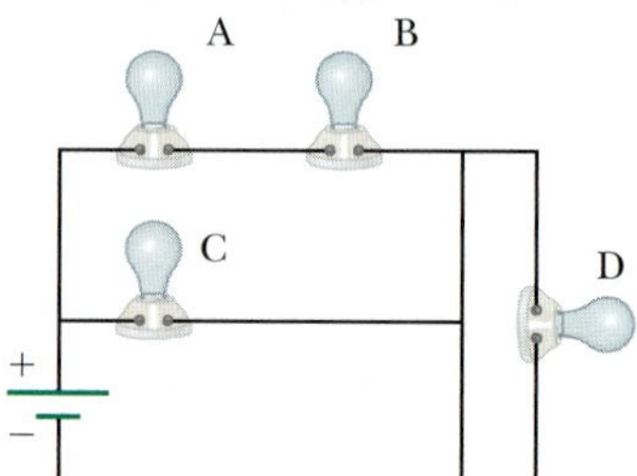

**Figure 18.9** (Applying Physics 18.2)

**EXPLANATION** Bulbs A and B are connected in series across the emf of the battery, whereas bulb C is connected by itself across the battery. This means the voltage drop across C has the same magnitude as the battery emf, whereas this same emf is split between bulbs A and B. As a result, bulb C will glow more brightly than either of bulbs A and B, which will glow equally brightly. Bulb D has a wire connected across it—a short circuit—so the potential difference across bulb D is zero and it doesn't glow. If bulb A fails, B goes out, but C stays lit. If C fails, there is no effect on the other bulbs. If D fails, the event is undetectable because D was not glowing initially. ■

## APPLYING PHYSICS 18.3 Three-Way Lightbulbs

Figure 18.10 illustrates how a three-way lightbulb is constructed to provide three levels of light intensity. The socket of the lamp is equipped with a three-way switch for selecting different light intensities. The bulb contains two filaments. Why are the filaments connected in parallel? Explain how the two filaments are used to provide the three different light intensities.

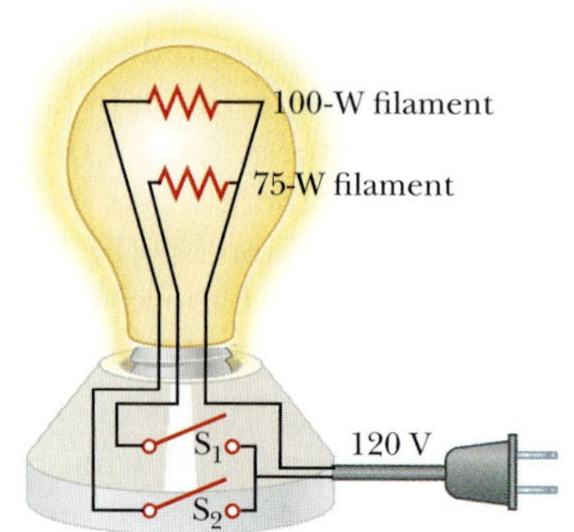

**Figure 18.10** (Applying Physics 18.3)

**EXPLANATION** If the filaments were connected in series and one of them were to fail, there would be no current in the bulb and the bulb would not glow, regardless of the position of the switch. When the filaments are connected in parallel and one of them (say, the 75-W filament) fails, however, the bulb will still operate in one of the switch positions because there is current in the other (100-W) filament. The three light intensities are made possible by selecting one of three values of filament resistance, using a single value of 120 V for the applied voltage. The 75-W filament offers one value of resistance, the 100-W filament offers a second value, and the third resistance is obtained by combining the two filaments in parallel. When switch $S_1$ is closed and switch $S_2$ is opened, only the 75-W filament carries current. When switch $S_1$ is opened and switch $S_2$ is closed, only the 100-W filament carries current. When both switches are closed, both filaments carry current and a total illumination corresponding to 175 W is obtained. ■

## PROBLEM-SOLVING STRATEGY

### Simplifying Circuits with Resistors

1. **Combine all resistors in series** by summing the individual resistances and draw the new, simplified circuit diagram.
   **Useful facts:** $R_{eq} = R_1 + R_2 + R_3 + \cdots$
   The current in each resistor is the same.
2. **Combine all resistors in parallel** by summing the reciprocals of the resistances and then taking the reciprocal of the result. Draw the new, simplified circuit diagram.
   **Useful facts:** $\frac{1}{R_{eq}} = \frac{1}{R_1} + \frac{1}{R_2} + \frac{1}{R_3} + \cdots$
   The potential difference across each resistor is the same.
3. **Repeat** the first two steps as necessary, until no further combinations can be made. If there is only a single battery in the circuit, the result will usually be a single equivalent resistor in series with the battery.
4. **Use Ohm's law,** $\Delta V = IR$, to determine the current in the equivalent resistor. Then work backwards through the diagrams, applying the useful facts listed in step 1 or step 2 to find the currents in the other resistors. (In more complex circuits, Kirchhoff's rules will be needed, as described in the next section.)

## EXAMPLE 18.3 Equivalent Resistance

**GOAL** Solve a problem involving both series and parallel resistors.

**PROBLEM** Four resistors are connected as shown in Figure 18.11a. **(a)** Find the equivalent resistance between points *a* and *c*. **(b)** What is the current in each resistor if a 42-V battery is connected between *a* and *c*?

**STRATEGY** Reduce the circuit in steps, as shown in Figures 18.11b and 18.11c, using the sum rule for resistors in series and the reciprocal-sum rule for resistors in parallel. Finding the currents is a matter of applying Ohm's law while working backwards through the diagrams.

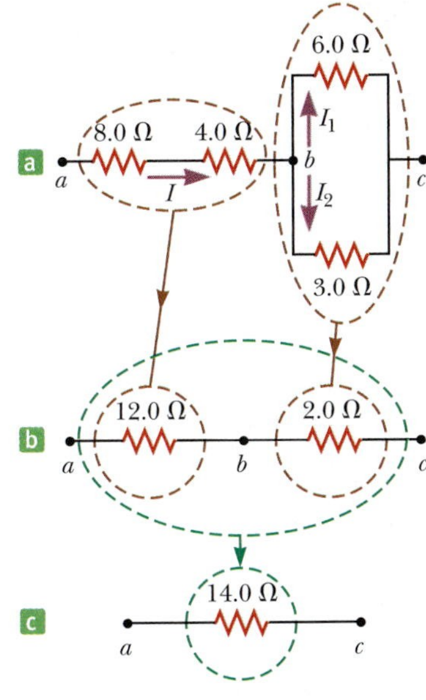


**Figure 18.11** (Example 18.3) The four resistors shown in (a) can be reduced in steps to an equivalent 14-Ω resistor.

**SOLUTION**

**(a)** Find the equivalent resistance of the circuit.

The 8.0-Ω and 4.0-Ω resistors are in series, so use the sum rule to find the equivalent resistance between *a* and *b*:

$$R_{eq} = R_1 + R_2 = 8.0\ \Omega + 4.0\ \Omega = 12.0\ \Omega$$

The 6.0-Ω and 3.0-Ω resistors are in parallel, so use the reciprocal-sum rule to find the equivalent resistance between *b* and *c* (don't forget to invert!):

$$\frac{1}{R_{eq}} = \frac{1}{R_1} + \frac{1}{R_2} = \frac{1}{6.0\ \Omega} + \frac{1}{3.0\ \Omega} = \frac{1}{2.0\ \Omega}$$

$$R_{eq} = 2.0\ \Omega$$

*(Continued)*

In the new diagram, 18.11b, there are now two resistors in series. Combine them with the sum rule to find the equivalent resistance of the circuit:

$$R_{eq} = R_1 + R_2 = 12.0\ \Omega + 2.0\ \Omega = \boxed{14.0\ \Omega}$$

**(b)** Find the current in each resistor if a 42-V battery is connected between points *a* and *c*.

Find the current in the equivalent resistor in Figure 18.11c, which is the total current. Resistors in series all carry the same current, so this value is the current in the 12-Ω resistor in Figure 18.11b and also in the 8.0-Ω and 4.0-Ω resistors in Figure 18.11a.

$$I = \frac{\Delta V_{ac}}{R_{eq}} = \frac{42\ \text{V}}{14\ \Omega} = \boxed{3.0\ \text{A}}$$

Calculate the voltage drop $\Delta V_{para}$ across the parallel circuit, which has an equivalent resistance of 2.0 Ω:

$$\Delta V_{para} = IR = (3.0\ \text{A})(2.0\ \Omega) = 6.0\ \text{V}$$

Apply Ohm's law again to find the currents in each resistor of the parallel circuit:

$$I_1 = \frac{\Delta V_{para}}{R_{6.0\ \Omega}} = \frac{6.0\ \text{V}}{6.0\ \Omega} = \boxed{1.0\ \text{A}}$$

$$I_2 = \frac{\Delta V_{para}}{R_{3.0\ \Omega}} = \frac{6.0\ \text{V}}{3.0\ \Omega} = \boxed{2.0\ \text{A}}$$

**REMARKS** As a final check, note that $\Delta V_{bc} = (6.0\ \Omega)I_1 = (3.0\ \Omega)I_2 = 6.0$ V and $\Delta V_{ab} = (12\ \Omega)I_1 = 36$ V; therefore, $\Delta V_{ac} = \Delta V_{ab} + \Delta V_{bc} = 42$ V, as expected.

**QUESTION 18.3** Which of the original resistors dissipates energy at the greatest rate?

**EXERCISE 18.3** Suppose the series resistors in Example 18.3 are now 6.00 Ω and 3.00 Ω while the parallel resistors are 8.00 Ω (top) and 4.00 Ω (bottom), and the battery provides an emf of 27.0 V. Find (a) the equivalent resistance and (b) the currents $I$, $I_1$, and $I_2$.

**ANSWERS** (a) 11.7 Ω (b) $I = 2.31$ A, $I_1 = 0.770$ A, $I_2 = 1.54$ A

---

The current $I_1$ entering the junction must equal the sum of the currents $I_2$ and $I_3$ leaving the junction.

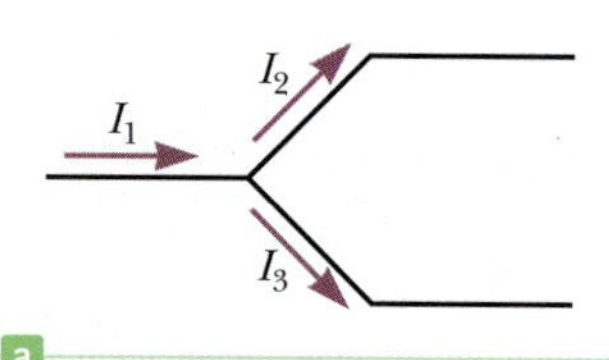


The net volume flow rate in must equal the net volume flow rate out.

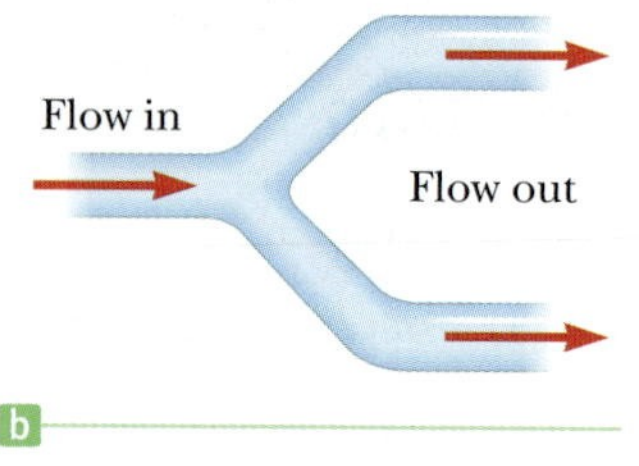


**Figure 18.12** (a) Kirchoff's junction rule. (b) A mechanical analog of the junction rule.

# 18.4 Kirchhoff's Rules and Complex DC Circuits

**LEARNING OBJECTIVES**

1. State Kirchoff's rules and discuss their physical origins.
2. Apply Kirchoff's rules to DC circuits.

As demonstrated in the preceding section, we can analyze simple circuits using Ohm's law and the rules for series and parallel combinations of resistors. There are, however, many ways in which resistors can be connected so that the circuits formed can't be reduced to a single equivalent resistor. The procedure for analyzing more complex circuits can be facilitated by the use of two simple rules called **Kirchhoff's rules**:

1. The sum of the currents entering any junction must equal the sum of the currents leaving that junction. (This rule is often referred to as the **junction rule.**)
2. The sum of the potential differences across all the elements around any closed circuit loop must be zero. (This rule is usually called the **loop rule.**)

The junction rule is a statement of *conservation of charge.* Whatever current enters a given point in a circuit must leave that point because charge can't build up or disappear at a point. If we apply this rule to the junction in Figure 18.12a, we get

$$I_1 = I_2 + I_3$$

Figure 18.12b represents a mechanical analog of the circuit shown in Figure 18.12a. In this analog, water flows through a branched pipe with no leaks. The flow rate into the pipe equals the total flow rate out of the two branches.

The loop rule is equivalent to the principle of *conservation of energy*. Any charge that moves around any closed loop in a circuit (starting and ending at the same point) must gain as much energy as it loses. It gains energy as it is pumped through a source of emf. Its energy may decrease in the form of a potential drop $-IR$ across a resistor or as a result of flowing backward through a source of emf, from the positive to the negative terminal inside the battery. In the latter case, electrical energy is converted to chemical energy as the battery is charged.

When applying Kirchhoff's rules, you must make two decisions at the beginning of the problem:

1. Assign symbols and directions to the currents in all branches of the circuit. Don't worry about guessing the direction of a current incorrectly; the resulting answer will be negative, but *its magnitude will be correct.* (Because the equations are *linear* in the currents, all currents are to the first power.)
2. When applying the loop rule, you must choose a direction for traversing the loop and be consistent in going either clockwise or counterclockwise. As you traverse the loop, record voltage drops and rises according to the following rules (summarized in Fig. 18.13, where it is assumed that movement is from point $a$ toward point $b$):
   **(a)** If a resistor is traversed in the direction of the current, the change in electric potential across the resistor is $-IR$ (Fig. 18.13a).
   **(b)** If a resistor is traversed in the direction opposite the current, the change in electric potential across the resistor is $+IR$ (Fig. 18.13b).
   **(c)** If a source of emf is traversed in the direction of the emf (from $-$ to $+$ on the terminals), the change in electric potential is $+\mathcal{E}$ (Fig. 18.13c).
   **(d)** If a source of emf is traversed in the direction opposite the emf (from $+$ to $-$ on the terminals), the change in electric potential is $-\mathcal{E}$ (Fig. 18.13d).

In each diagram, $\Delta V = V_b - V_a$ and the circuit element is traversed from $a$ to $b$, left to right.

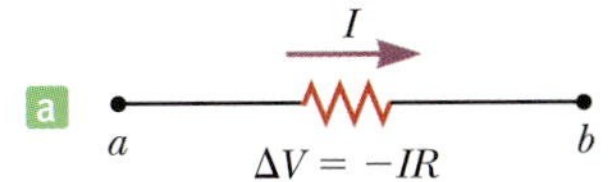


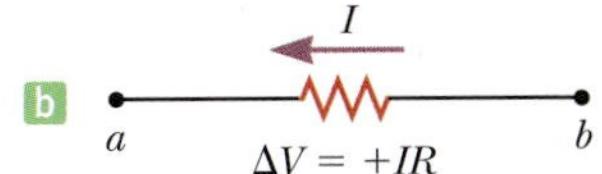


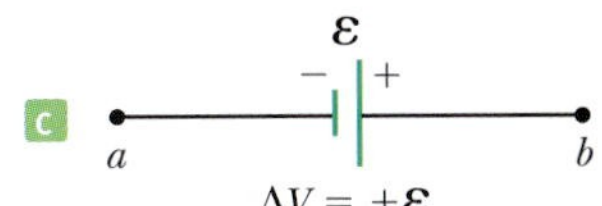


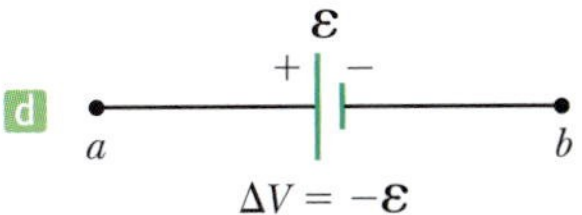


**Figure 18.13** Rules for determining the potential differences across a resistor and a battery, assuming the battery has no internal resistance.

There are limits to the number of times the junction rule and the loop rule can be used. You can use the junction rule as often as needed as long as you include a current in each new junction equation that has not been used in a previous junction-rule equation. (If this procedure isn't followed, the new equation will just be a combination of two other equations that you already have.) In general, the number of times the junction rule can be used is one fewer than the number of junction points in the circuit. The loop rule can also be used as often as needed, so long as a new circuit element (resistor or battery) or a new current appears in each new equation. **To solve a particular circuit problem, you need as many independent equations as you have unknowns.**

**Gustav Kirchhoff**
**German Physicist (1824–1887)**
Together with German chemist Robert Bunsen, Kirchhoff, a professor at Heidelberg, invented the spectroscopy that we study in Chapter 28. He also formulated another rule that states, "A cool substance will absorb light of the same wavelengths that it emits when hot."

## PROBLEM-SOLVING STRATEGY

### Applying Kirchhoff's Rules to a Circuit

1. **Assign labels and symbols** to all the known and unknown quantities.
2. **Assign *directions* to the currents** in each part of the circuit. Although the assignment of current directions is arbitrary, you must stick with your original choices throughout the problem as you apply Kirchhoff's rules.

*(Continued)*

3. **Apply the junction rule** to any junction in the circuit. The rule may be applied as many times as a new current (one not used in a previously found equation) appears in the resulting equation.
4. **Apply Kirchhoff's loop rule** to as many loops in the circuit as are needed to solve for the unknowns. To apply this rule, you must correctly identify the change in electric potential as you cross each element in traversing the closed loop. Watch out for signs!
5. **Solve the equations** simultaneously for the unknown quantities, using substitution or any other method familiar to the student.
6. **Check your answers** by substituting them into the original equations.

## EXAMPLE 18.4 Applying Kirchhoff's Rules

**GOAL** Use Kirchhoff's rules to find currents in a circuit with three currents and one battery.

**PROBLEM** Find the currents in the circuit shown in Figure 18.14 by using Kirchhoff's rules.

**STRATEGY** There are three unknown currents in this circuit, so we must obtain three independent equations, which then can be solved by substitution. We can find the equations with one application of the junction rule and two applications of the loop rule. We choose junction *c*. (Junction *d* gives the same equation.) For the loops, we choose the bottom loop and the top loop, both shown by blue arrows, which indicate the direction we are going to traverse the circuit mathematically (not necessarily the direction of the current). The third loop gives an equation that can be obtained by a linear combination of the other two, so it provides no additional information and isn't used.

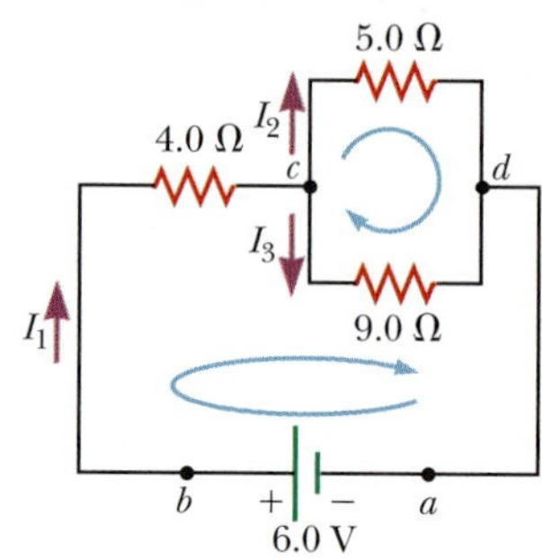


**Figure 18.14** (Example 18.4)

**SOLUTION**

Apply the junction rule to point *c*. $I_1$ is directed into the junction, $I_2$ and $I_3$ are directed out of the junction.

$$I_1 = I_2 + I_3$$

Select the bottom loop and traverse it clockwise starting at point *a*, generating an equation with the loop rule:

$$\sum \Delta V = \Delta V_{\text{bat}} + \Delta V_{4.0\,\Omega} + \Delta V_{9.0\,\Omega} = 0$$

$$6.0\text{ V} - (4.0\,\Omega)I_1 - (9.0\,\Omega)I_3 = 0$$

Select the top loop and traverse it clockwise from point *c*. Notice the gain across the 9.0-Ω resistor because it is traversed *against* the direction of the current!

$$\sum \Delta V = \Delta V_{5.0\,\Omega} + \Delta V_{9.0\,\Omega} = 0$$

$$-(5.0\,\Omega)I_2 + (9.0\,\Omega)I_3 = 0$$

Rewrite the three equations, rearranging terms and dropping units for the moment, for convenience:

$$\textbf{(1)}\quad I_1 = I_2 + I_3$$

$$\textbf{(2)}\quad 4.0I_1 + 9.0I_3 = 6.0$$

$$\textbf{(3)}\quad -5.0I_2 + 9.0I_3 = 0$$

Solve Equation (3) for $I_2$ and substitute into Equation (1):

$$I_2 = 1.8I_3$$

$$I_1 = I_2 + I_3 = 1.8I_3 + I_3 = 2.8I_3$$

Substitute the latter expression into Equation (2) and solve for $I_3$:

$$4.0(2.8I_3) + 9.0I_3 = 6.0 \quad \rightarrow \quad I_3 = 0.30\text{ A}$$

Substitute $I_3$ back into Equation (3) to get $I_2$:

$$-5.0I_2 + 9.0(0.30\text{ A}) = 0 \quad \rightarrow \quad I_2 = 0.54\text{ A}$$

Substitute $I_3$ into Equation (2) to get $I_1$:

$$4.0I_1 + 9.0(0.30\text{ A}) = 6.0 \quad \rightarrow \quad I_1 = 0.83\text{ A}$$

**REMARKS** Substituting these values back into the original equations verifies that they are correct, with any small discrepancies due to rounding. The problem can also be solved by first combining resistors.

**QUESTION 18.4** How would the answers change if the indicated directions of the currents in Figure 18.14 were all reversed?

**EXERCISE 18.4** Suppose the 6.0-V battery is replaced by a battery of unknown emf and an ammeter measures $I_1 = 1.5$ A. Find the other two currents and the emf of the battery.

**ANSWERS** $I_2 = 0.96$ A, $I_3 = 0.54$ A, $\mathcal{E} = 11$ V

**Tip 18.3 More Current Goes in the Path of Less Resistance**

You may have heard the statement "Current takes the path of least resistance." For a parallel combination of resistors, this statement is inaccurate because current actually follows all paths. The most current, however, travels in the path of least resistance.

## EXAMPLE 18.5 Another Application of Kirchhoff's Rules

**GOAL** Find the currents in a circuit with three currents and two batteries when some current directions are chosen inaccurately.

**PROBLEM** Find $I_1$, $I_2$, and $I_3$ in Figure 18.15a.

**STRATEGY** Use Kirchhoff's two rules, the junction rule once and the loop rule twice, to develop three equations for the three unknown currents. Solve the equations simultaneously.

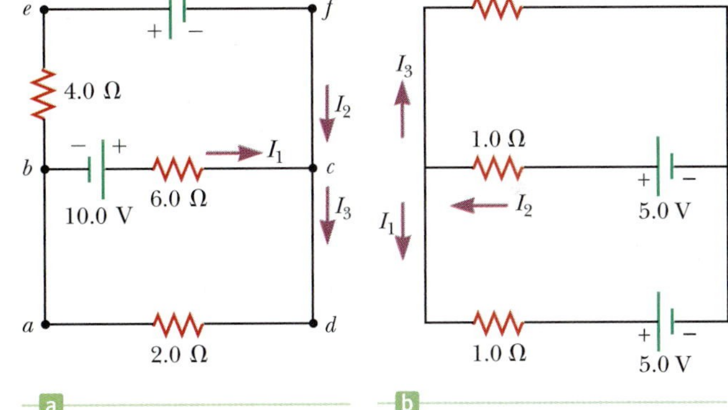


**Figure 18.15** (a) (Example 18.5) (b) (Exercise 18.5)

**SOLUTION**

Apply Kirchhoff's junction rule to junction *c*. Because of the chosen current directions, $I_1$ and $I_2$ are directed into the junction and $I_3$ is directed out of the junction.

$$\text{(1)} \quad I_3 = I_1 + I_2$$

Apply Kirchhoff's loop rule to the loops *abcda* and *befcb*. (Loop *aefda* gives no new information.) In loop *befcb*, a positive sign is obtained when the 6.0-Ω resistor is traversed because the direction of the path is opposite the direction of the current $I_1$.

$$\text{(2)} \quad \text{Loop } abcda\text{: } 10 \text{ V} - (6.0\ \Omega)I_1 - (2.0\ \Omega)I_3 = 0$$

$$\text{(3)} \quad \text{Loop } befcb\text{: } -(4.0\ \Omega)I_2 - 14 \text{ V} + (6.0\ \Omega)I_1 - 10 \text{ V} = 0$$

Using Equation (1), eliminate $I_3$ from Equation (2) (ignore units for the moment):

$$10 - 6.0I_1 - 2.0(I_1 + I_2) = 0$$

$$\text{(4)} \quad 10 = 8.0I_1 + 2.0I_2$$

Divide each term in Equation (3) by 2 and rearrange the equation so that the currents are on the right side:

$$\text{(5)} \quad -12 = -3.0I_1 + 2.0I_2$$

Subtracting Equation (5) from Equation (4) eliminates $I_2$ and gives $I_1$:

$$22 = 11I_1 \quad \rightarrow \quad I_1 = \boxed{2.0 \text{ A}}$$

Substituting this value of $I_1$ into Equation (5) gives $I_2$:

$$2.0I_2 = 3.0I_1 - 12 = 3.0(2.0) - 12 = -6.0 \text{ A}$$

$$I_2 = \boxed{-3.0 \text{ A}}$$

Finally, substitute the values found for $I_1$ and $I_2$ into Equation (1) to obtain $I_3$:

$$I_3 = I_1 + I_2 = 2.0 \text{ A} - 3.0 \text{ A} = \boxed{-1.0 \text{ A}}$$

*(Continued)*

**REMARKS** The fact that $I_2$ and $I_3$ are both negative indicates that the wrong directions were chosen for these currents. Nonetheless, the magnitudes are correct. Choosing the right directions of the currents at the outset is unimportant because the equations are linear, and wrong choices result only in a minus sign in the answer.

**QUESTION 18.5** Is it possible for the current in a battery to be directed from the positive terminal toward the negative terminal?

**EXERCISE 18.5** Find the three currents in Figure 18.15b. (Note that the direction of one current was deliberately chosen wrongly!)

**ANSWERS** $I_1 = -1.0\text{ A}$, $I_2 = 1.0\text{ A}$, $I_3 = 2.0\text{ A}$

## 18.5 RC Circuits

**LEARNING OBJECTIVES**

1. Describe the time dependence of charge on a capacitor in an *RC* circuit.
2. Define the time constant and discuss its physical significance.
3. Evaluate elementary properties of *RC* circuits.

So far, we have been concerned with circuits with constant currents. We now consider direct-current circuits containing capacitors, in which the currents vary with time. Consider the series circuit in Figure 18.16. We assume the capacitor is initially uncharged with the switch opened. After the switch is closed, the battery begins to charge the plates of the capacitor and the charge passes through the resistor. As the capacitor is being charged, the circuit carries a changing current. The charging process continues until the capacitor is charged to its maximum equilibrium value, $Q = C\mathcal{E}$, where $\mathcal{E}$ is the maximum voltage across the capacitor. Once the capacitor is fully charged, the current in the circuit is zero. If we assume the capacitor is uncharged before the switch is closed, and if the switch is closed at $t = 0$, we find that the charge on the capacitor varies with time according to the equation

$$q = Q(1 - e^{-t/RC}) \qquad [18.7]$$

where $e = 2.718\ldots$ is Euler's constant, the base of the natural logarithm. Figure 18.16b is a graph of this equation. The charge is zero at $t = 0$ and

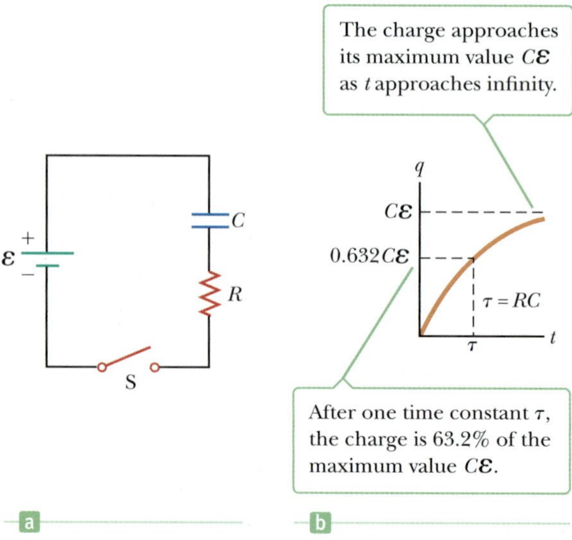
The charge approaches its maximum value $C\mathcal{E}$ as $t$ approaches infinity.
q
$C\mathcal{E}$
$0.632C\mathcal{E}$
$\tau = RC$
$\tau$
t
C
$\mathcal{E}$
+
−
R
S
After one time constant $\tau$, the charge is 63.2% of the maximum value $C\mathcal{E}$.
a
b

**Figure 18.16** (a) A capacitor in series with a resistor, a battery, and a switch. (b) A plot of the charge on the capacitor versus time after the switch on the circuit is closed.

approaches its maximum value, $Q$, as $t$ approaches infinity. The voltage $\Delta V$ across the capacitor at any time is obtained by dividing the charge by the capacitance: $\Delta V = q/C$.

As you can see from Equation 18.7, it would take an infinite amount of time, in this model, for the capacitor to become fully charged. The reason is mathematical: in obtaining that equation, charges are assumed to be infinitely small, whereas in reality the smallest charge is that of an electron, with a magnitude equal to $1.60 \times 10^{-19}$ C. For all practical purposes, the capacitor is fully charged after a finite amount of time. The term $RC$ that appears in Equation 18.7 is called the **time constant** $\tau$ (Greek letter tau), so

$$\tau = RC \qquad [18.8]$$

The time constant represents the time required for the charge to increase from zero to 63.2% of its maximum equilibrium value. This means that in a period of time equal to one time constant, the charge on the capacitor increases from zero to $0.632Q$. This can be seen by substituting $t = \tau = RC$ into Equation 18.7 and solving for $q$. (Note that $1 - e^{-1} = 0.632$.) It's important to note that a capacitor charges very slowly in a circuit with a long time constant, whereas it charges very rapidly in a circuit with a short time constant. After a time equal to ten time constants, the capacitor is more than 99.99% charged.

Now consider the circuit in Figure 18.17a, consisting of a capacitor with an initial charge $Q$, a resistor, and a switch. Before the switch is closed, the potential difference across the charged capacitor is $Q/C$. Once the switch is closed, the charge begins to flow through the resistor from one capacitor plate to the other until the capacitor is fully discharged. If the switch is closed at $t = 0$, it can be shown that the charge $q$ on the capacitor varies with time according to the equation

$$q = Qe^{-t/RC} \qquad [18.9]$$

The charge decreases exponentially with time, as shown in Figure 18.17b. In the interval $t = \tau = RC$, the charge decreases from its initial value $Q$ to $0.368Q$. In other words, in a time equal to one time constant, the capacitor loses 63.2% of its initial charge. Because $\Delta V = q/C$, the voltage across the capacitor also decreases exponentially with time according to the equation $\Delta V = \mathcal{E}e^{-t/RC}$, where $\mathcal{E}$ (which equals $Q/C$) is the initial voltage across the fully charged capacitor.

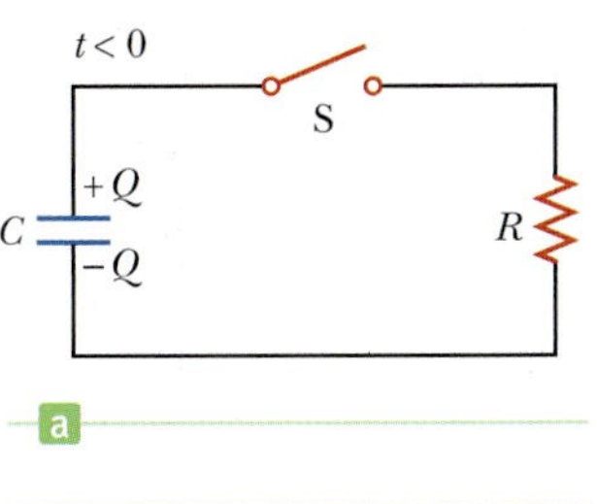


The charge has its maximum value $Q$ at $t = 0$ and decays to zero exponentially as $t$ approaches infinity.

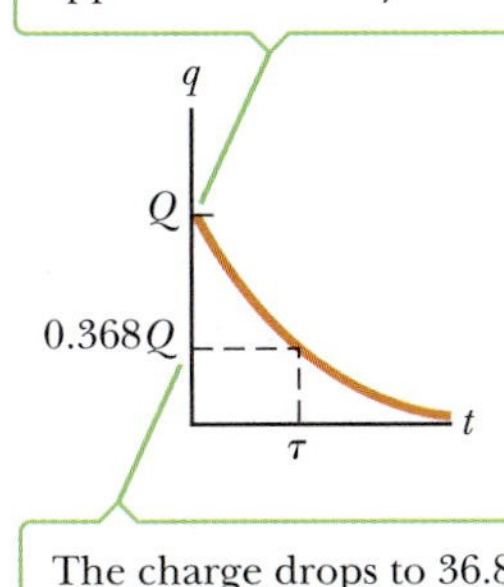


The charge drops to 36.8% of its initial value when one time constant has elapsed.

**Figure 18.17** (a) A charged capacitor connected to a resistor and a switch. (b) A graph of the charge on the capacitor versus time after the switch is closed.

## APPLYING PHYSICS 18.4 Timed Windshield Wipers

Many automobiles are equipped with windshield wipers that can be used intermittently during a light rainfall. How does the operation of this feature depend on the charging and discharging of a capacitor?

**EXPLANATION** The wipers are part of an $RC$ circuit with time constant that can be varied by selecting different values of $R$ through a multiposition switch. The brief time that the wipers remain on and the time they are off are determined by the value of the time constant of the circuit. ■

## APPLYING PHYSICS 18.5 Bacterial Growth

In biological research concerning population growth, an equation is used that is similar to the exponential equations encountered in the analysis of $RC$ circuits. Applied to a number of bacteria, this equation is

$$N_f = N_i 2^n$$

where $N_f$ is the number of bacteria present after $n$ doubling times, $N_i$ is the number present initially, and $n$ is the number of growth cycles or doubling times. Doubling times vary according to the organism. The doubling time is about 30 days for the bacteria responsible for leprosy, and about 20 minutes for the salmonella bacteria responsible for food poisoning. Suppose only 10 salmonella bacteria find their way onto a turkey leg after your Thanksgiving meal. Four hours later you come back for a midnight snack. How many bacteria are present now?

**EXPLANATION** The number of doubling times is 240 min/20 min = 12. Thus,

$$N_f = N_i 2^n = (10 \text{ bacteria})(2^{12}) = 40\,960 \text{ bacteria}$$

So your system will have to deal with an invading host of about 41 000 bacteria, which are going to continue to double in a very promising environment. ■

## APPLYING PHYSICS 18.6 Roadway Flashers

Many roadway construction sites have flashing yellow lights to warn motorists of possible dangers. What causes the lights to flash?

**EXPLANATION** A typical circuit for such a flasher is shown in Figure 18.18. The lamp L is a gas-filled lamp that acts as an open circuit until a large potential difference causes a discharge, which gives off a bright light. During this discharge, charge flows through the gas between the electrodes of the lamp. When the switch is closed, the battery charges the capacitor. At the beginning, the current is high and the charge on the capacitor is low, so most of the potential difference appears across the resistance $R$. As the capacitor charges, more potential difference appears across it, reflecting the lower current and lower potential difference across the resistor. Eventually, the potential difference across the capacitor reaches a value at which the lamp will conduct, causing a flash. This flash discharges the capacitor through the lamp, and the process of charging begins again. The period between flashes can be adjusted by changing the time constant of the $RC$ circuit. ■

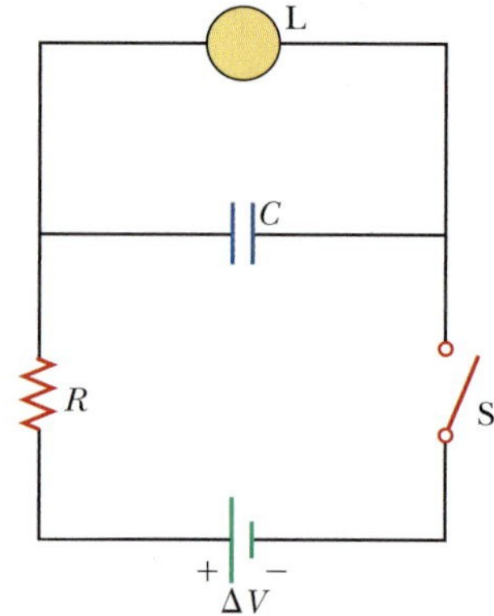


**Figure 18.18** (Applying Physics 18.6)

### Quick Quiz

**18.9** The switch is closed in Figure 18.19. After a long time compared with the time constant of the circuit, what will the current be in the 2-Ω resistor? (a) 4 A (b) 3 A (c) 2 A (d) 1 A (e) More information is needed.

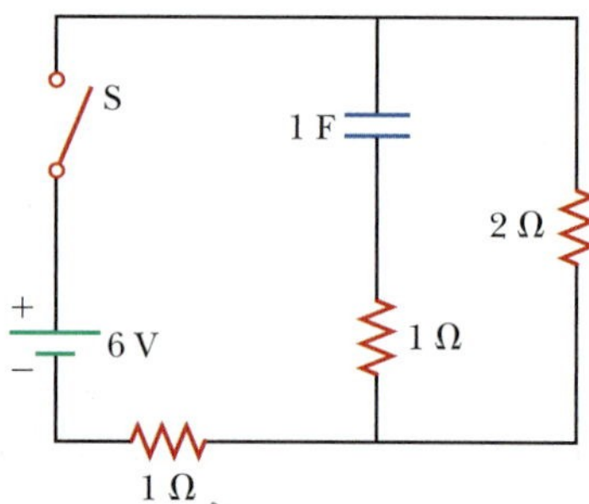


**Figure 18.19** (Quick Quiz 18.9)

## EXAMPLE 18.6 Charging a Capacitor in an *RC* Circuit

**GOAL** Calculate elementary properties of a simple $RC$ circuit.

**PROBLEM** An uncharged capacitor and a resistor are connected in series to a battery, as in Figure 18.16a. If $\mathcal{E} = 12.0$ V, $C = 5.00\ \mu$F, and $R = 8.00 \times 10^5\ \Omega$, find **(a)** the time constant of the circuit, **(b)** the maximum charge on the capacitor, **(c)** the charge on the capacitor after 6.00 s, **(d)** the potential difference across the resistor after 6.00 s, and **(e)** the current in the resistor at that time.

**STRATEGY** Finding the time constant in part (a) requires substitution into Equation 18.8. For part (b), the maximum charge occurs after a long time, when the current has dropped to zero. By Ohm's law, $\Delta V = IR$, the potential difference across the resistor is also zero at that time, and Kirchhoff's loop rule then gives the maximum charge. Finding the charge at some particular time, as in part (c), is a matter of substituting into Equation 18.7. Kirchhoff's loop rule and the capacitance equation can be used to indirectly find the potential drop across the resistor in part (d), and then Ohm's law yields the current.

**SOLUTION**

**(a)** Find the time constant of the circuit.

Use the definition of the time constant, Equation 18.8:

$$\tau = RC = (8.00 \times 10^5\ \Omega)(5.00 \times 10^{-6}\ \text{F}) = 4.00\ \text{s}$$

**(b)** Calculate the maximum charge on the capacitor.

Apply Kirchhoff's loop rule to the $RC$ circuit, going clockwise, which means that the voltage difference across the battery is positive and the differences across the capacitor and resistor are negative:

$$\text{(1)}\quad \Delta V_{\text{bat}} + \Delta V_C + \Delta V_R = 0$$

From the definition of capacitance (Eq. 16.8) and Ohm's law, we have $\Delta V_C = -q/C$ and $\Delta V_R = -IR$. These are voltage drops, so they're negative. Also, $\Delta V_{\text{bat}} = +\mathcal{E}$.

$$\text{(2)}\quad \mathcal{E} - \frac{q}{C} - IR = 0$$

When the maximum charge $q = Q$ is reached, $I = 0$. Solve Equation (2) for the maximum charge:

$$\mathcal{E} - \frac{Q}{C} = 0 \quad \rightarrow \quad Q = C\mathcal{E}$$

Substitute to find the maximum charge:

$$Q = (5.00 \times 10^{-6}\text{ F})(12.0\text{ V}) = \boxed{60.0\ \mu\text{C}}$$

**(c)** Find the charge on the capacitor after 6.00 s.

Substitute into Equation 18.7:

$$q = Q(1 - e^{-t/\tau}) = (60.0\ \mu\text{C})(1 - e^{-6.00\text{ s}/4.00\text{ s}})$$
$$= \boxed{46.6\ \mu\text{C}}$$

**(d)** Compute the potential difference across the resistor after 6.00 s.

Compute the voltage drop $\Delta V_C$ across the capacitor at that time:

$$\Delta V_C = -\frac{q}{C} = -\frac{46.6\ \mu\text{C}}{5.00\ \mu\text{F}} = -9.32\text{ V}$$

Solve Equation (1) for $\Delta V_R$ and substitute:

$$\Delta V_R = -\Delta V_{\text{bat}} - \Delta V_C = -12.0\text{ V} - (-9.32\text{ V})$$
$$= \boxed{-2.68\text{ V}}$$

**(e)** Find the current in the resistor after 6.00 s.

Apply Ohm's law, using the results of part (d) (remember that $\Delta V_R = -IR$ here):

$$I = \frac{-\Delta V_R}{R} = \frac{-(-2.68\text{ V})}{(8.00 \times 10^5\ \Omega)}$$
$$= \boxed{3.35 \times 10^{-6}\text{ A}}$$

**REMARKS** In solving this problem, we paid scrupulous attention to signs. These signs must always be chosen when applying Kirchhoff's loop rule and must remain consistent throughout the problem. Alternately, magnitudes can be used and the signs chosen by physical intuition. For example, the magnitude of the potential difference across the resistor must equal the magnitude of the potential difference across the battery minus the magnitude of the potential difference across the capacitor.

**QUESTION 18.6** In an *RC* circuit as depicted in Figure 18.16a, what happens to the time required for the capacitor to be charged to half its maximum value if either the resistance or capacitance is increased? (a) It increases. (b) It decreases. (c) It remains the same.

**EXERCISE 18.6** Find (a) the charge on the capacitor after 2.00 s have elapsed, (b) the magnitude of the potential difference across the capacitor after 2.00 s, and (c) the magnitude of the potential difference across the resistor at that same time.

**ANSWERS** (a) 23.6 $\mu$C (b) 4.72 V (c) 7.28 V

## EXAMPLE 18.7 Discharging a Capacitor in an *RC* Circuit

**GOAL** Calculate some elementary properties of a discharging capacitor in an *RC* circuit.

**PROBLEM** Consider a capacitor $C$ being discharged through a resistor $R$ as in Figure 18.17a (page 641). **(a)** How long does it take the charge on the capacitor to drop to one-fourth its initial value? Answer as a multiple of $\tau$. **(b)** Compute the initial charge and time constant, and **(c)** the time it takes to discharge all but the last quantum of charge, $1.60 \times 10^{-19}$ C, if the initial potential difference across the capacitor is 12.0 V, the capacitance is equal to $3.50 \times 10^{-6}$ F, and the resistance is 2.00 Ω. (Assume an exponential decrease during the entire discharge process.)

**STRATEGY** This problem requires substituting given values into various equations, as well as a few algebraic manipulations involving the natural logarithm. In part (a) set $q = \frac{1}{4}Q$ in Equation 18.9 for a discharging capacitor, where $Q$ is the initial charge, and solve for time $t$. In part (b) substitute into Equations 16.8 and 18.8 to find the initial capacitor charge and time constant, respectively. In part (c) substitute the results of part (b) and the final charge $q = 1.60 \times 10^{-19}$ C into the discharging-capacitor equation, again solving for time.

**SOLUTION**

**(a)** How long does it take the charge on the capacitor to reduce to one-fourth its initial value?

Apply Equation 18.9:

$$q(t) = Qe^{-t/RC}$$

(*Continued*)

Substitute $q(t) = Q/4$ into the preceding equation and cancel $Q$:

$$\tfrac{1}{4}Q = Qe^{-t/RC} \quad \rightarrow \quad \tfrac{1}{4} = e^{-t/RC}$$

Take natural logarithms of both sides and solve for the time $t$:

$$\ln\left(\tfrac{1}{4}\right) = -t/RC$$

$$t = -RC\ln\left(\tfrac{1}{4}\right) = 1.39RC = \boxed{1.39\tau}$$

**(b)** Compute the initial charge and time constant from the given data.

Use the capacitance equation to find the initial charge:

$$C = \frac{Q}{\Delta V} \quad \rightarrow \quad Q = C\,\Delta V = (3.50 \times 10^{-6}\text{ F})(12.0\text{ V})$$

$$Q = \boxed{4.20 \times 10^{-5}\text{ C}}$$

Now calculate the time constant:

$$\tau = RC = (2.00\ \Omega)(3.50 \times 10^{-6}\text{ F}) = \boxed{7.00 \times 10^{-6}\text{ s}}$$

**(c)** How long does it take to drain all but the last quantum of charge?

Apply Equation 18.9, divide by $Q$, and take natural logarithms of both sides:

$$q(t) = Qe^{-t/\tau} \quad \rightarrow \quad e^{-t/\tau} = \frac{q}{Q}$$

Take the natural logs of both sides:

$$-t/\tau = \ln\left(\frac{q}{Q}\right) \quad \rightarrow \quad t = -\tau\ln\left(\frac{q}{Q}\right)$$

Substitute $q = 1.60 \times 10^{-19}$ C and the values for $Q$ and $\tau$ found in part (b):

$$t = -(7.00 \times 10^{-6}\text{ s})\ln\left(\frac{1.60 \times 10^{-19}\text{ C}}{4.20 \times 10^{-5}\text{ C}}\right)$$

$$= \boxed{2.32 \times 10^{-4}\text{ s}}$$

**REMARKS** Part (a) shows how useful information can often be obtained even when no details concerning capacitances, resistances, or voltages are known. Part (c) demonstrates that capacitors can be rapidly discharged (or conversely, charged), despite the mathematical form of Equations 18.7 and 18.9, which indicate an infinite time would be required.

**QUESTION 18.7** Suppose the initial voltage used to charge the capacitor were doubled. Would the time required for discharging all but the last quantum of charge (a) increase, (b) decrease, (c) remain the same?

**EXERCISE 18.7** Suppose the same type of series circuit has $R = 8.00 \times 10^{4}\ \Omega$, $C = 5.00\ \mu$F, and an initial voltage across the capacitor of 6.00 V. (a) How long does it take the capacitor to lose half its initial charge? (b) How long does it take to lose all but the last 10 electrons on the negative plate?

**ANSWERS** (a) 0.277 s (b) 12.2 s

## 18.6 Household Circuits

**LEARNING OBJECTIVE**

1. Describe fundamental properties of household circuits.

Household circuits are a practical application of some of the ideas presented in this chapter. In a typical installation the utility company distributes electric power to individual houses with a pair of wires, or power lines. Electrical devices in a house are then connected in parallel to these lines, as shown in Figure 18.20. The potential difference between the two wires is about 120 V. (These currents and voltages are actually alternating currents and voltages, but for the present discussion we will assume they are direct currents and voltages.) One of the wires is connected to ground, and the other wire, sometimes called the "hot" wire, is at a potential of 120 V. A meter and a circuit breaker (or a fuse) are connected in series with the wire entering the house, as indicated in the figure.

In modern homes, circuit breakers are used in place of fuses. When the current in a circuit exceeds some value (typically 15 A or 20 A), the circuit breaker acts as a switch and opens the circuit. Figure 18.21 shows one design for a circuit breaker. Current passes through a bimetallic strip, the top of which bends to the left when excessive current heats it. If the strip bends far enough to the left, it settles into a groove in the spring-loaded metal bar. When this settling occurs, the bar drops enough to open the circuit at the contact point. The bar also flips a switch that indicates that the circuit breaker is not operational. (After the overload is removed, the switch can be flipped back on.) Circuit breakers based on this design have the disadvantage that some time is required for the heating of the strip, so the circuit may not be opened rapidly enough when it is overloaded. Therefore, many circuit breakers are now designed to use electromagnets (discussed in Chapter 19).

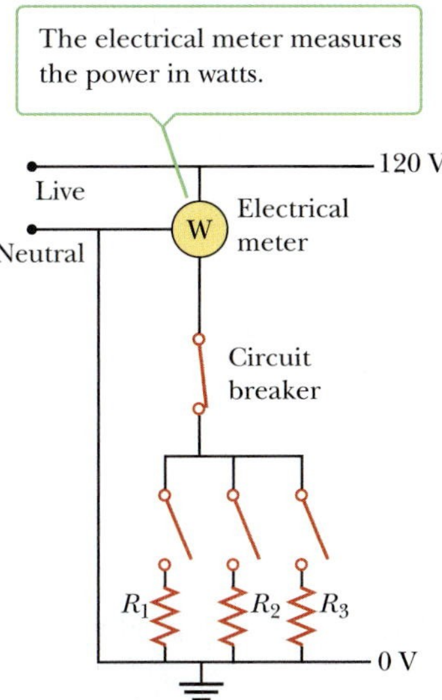


**Figure 18.20** A wiring diagram for a household circuit. The resistances $R_1$, $R_2$, and $R_3$ represent appliances or other electrical devices that operate at an applied voltage of 120 V.

The wire and circuit breaker are carefully selected to meet the current demands of a circuit. If the circuit is to carry currents as large as 30 A, a heavy-duty wire and an appropriate circuit breaker must be used. Household circuits that are normally used to power lamps and small appliances often require only 20 A. Each circuit has its own circuit breaker to accommodate its maximum safe load.

As an example, consider a circuit that powers a toaster, a microwave oven, and a heater (represented by $R_1$, $R_2$, and $R_3$ in Fig. 18.20). Using the equation $P = I\Delta V$, we can calculate the current carried by each appliance. The toaster, rated at 1 000 W, draws a current of 1 000/120 = 8.33 A. The microwave oven, rated at 800 W, draws a current of 6.67 A, and the heater, rated at 1 300 W, draws a current of 10.8 A. If the three appliances are operated simultaneously, they draw a total current of 25.8 A. Therefore, the breaker should be able to handle at least this much current, or else it will be tripped. As an alternative, the toaster and microwave oven could operate on one 20-A circuit with the heater on a separate 20-A circuit.

**APPLICATION**

Fuses and Circuit Breakers

Many heavy-duty appliances, such as electric ranges and clothes dryers, require 240 V to operate. The power company supplies this voltage by providing, in addition to a live wire that is 120 V above ground potential, another wire, also considered live, that is 120 V below ground potential (Fig. 18.22, page 646). Therefore, the potential drop across the two live wires is 240 V. An appliance operating from a 240-V line requires half the current of one operating from a 120-V line; consequently, smaller wires can be used in the higher-voltage circuit without becoming overheated.

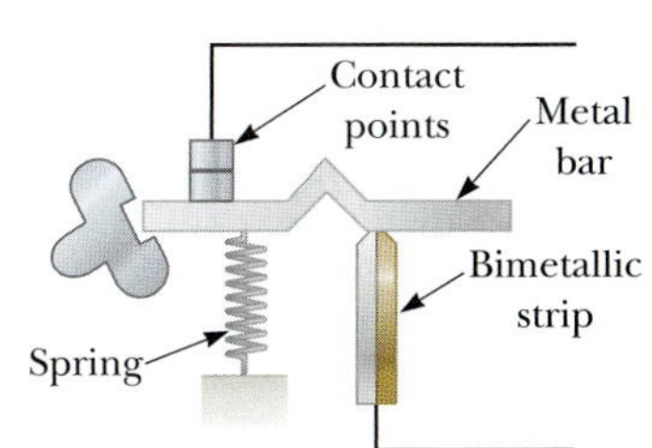


**Figure 18.21** A circuit breaker that uses a bimetallic strip for its operation.

# 18.7 Electrical Safety

**LEARNING OBJECTIVE**

1. Describe several important aspects of electrical safety.

A person can be electrocuted by touching a live wire while in contact with ground. Such a hazard is often due to frayed insulation that exposes the conducting wire. The ground contact might be made by touching a water pipe (which is normally at ground potential) or by standing on the ground with wet feet because impure water is a good conductor. Obviously, such situations should be avoided at all costs.

Electric shock can result in fatal burns or cause the muscles of vital organs, such as the heart, to malfunction. The degree of damage to the body depends on the magnitude of the current, the length of time it acts, and the part of the body through which it passes. Currents of 5 mA or less can cause a sensation of shock, but ordinarily do little or no damage. If the current is larger than about 10 mA, the hand muscles contract and the person may be unable to let go of the live wire. If a current of about 100 mA passes through the body for just a few seconds, it

**APPLICATION**
Third Wire on Consumer Appliances

can be fatal. Such large currents paralyze the respiratory muscles. In some cases, currents of about 1 A through the body produce serious (and sometimes fatal) burns.

As an additional safety feature for consumers, electrical equipment manufacturers now use electrical cords that have a third wire, called a *case ground.* To understand how this works, consider the drill being used in Figure 18.23. A two-wire device has one wire, called the "hot" wire, connected to the high-potential (120-V) side of the input power line, while the second wire is connected to ground (0 V). If the high-voltage wire comes in contact with the case of the drill (Fig. 18.23a), a short circuit occurs. In this undesirable circumstance, the pathway for the current is from the high-voltage wire through the person holding the drill and to Earth, a pathway that can be fatal. Protection is provided by a third wire, connected to the case of the drill (Fig. 18.23b). In this case, if a short occurs, the path of least resistance for the current is from the high-voltage wire through the case and back to ground through the third wire. The resulting high current produced will blow a fuse or trip a circuit breaker before the consumer is injured.

Special power outlets called ground-fault interrupters (GFIs) are now being used in kitchens, bathrooms, basements, and other hazardous areas of new homes. They are designed to protect people from electrical shock by sensing small currents—approximately 5 mA and greater—leaking to ground. When current above this level is detected, the device shuts off (interrupts) the current in less than a millisecond. (Ground-fault interrupters will be discussed in Chapter 20.)

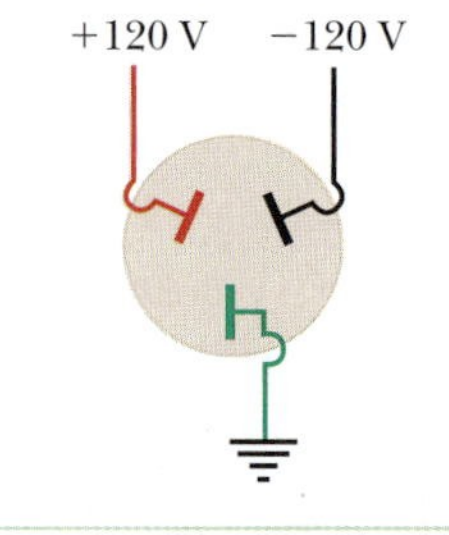

**Figure 18.22** (a) The connections for each of the openings in a 240-V outlet. (b) An outlet for connection to a 240-V supply.

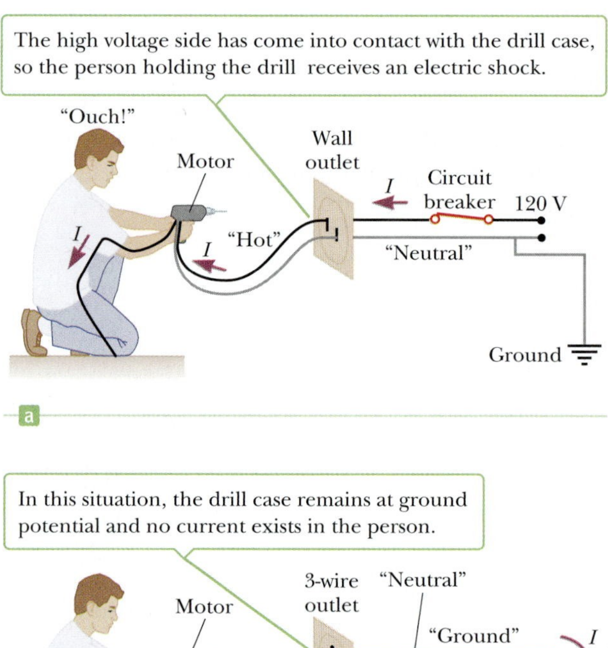


**Figure 18.23** The "hot" (or "live") wire, at 120 V, always includes a circuit breaker for safety. (a) When the drill is operated with two wires, the normal current path is from the "hot" wire, through the motor connections, and back to ground through the "neutral" wire. (b) Shock can be prevented by a third wire running from the drill case to the ground. The wire colors represent electrical standards in the United States: the "hot" wire is black, the ground wire is green, and the neutral wire is white (shown as gray in the figure).

# 18.8 Conduction of Electrical Signals by Neurons[2] BIO

**LEARNING OBJECTIVES**

1. Describe the conduction of electrical signals by neurons.
2. Describe the structure and operational principles of neurons.

The most remarkable use of electrical phenomena in living organisms is found in the nervous system of animals. Specialized cells in the body called **neurons** form a complex network that receives, processes, and transmits information from one part of the body to another. The center of this network is located in the brain, which has the ability to store and analyze information. On the basis of this information, the nervous system controls parts of the body.

The nervous system is highly complex and consists of about $10^{10}$ interconnected neurons. Some aspects of the nervous system are well known. Over the past several decades the method of signal propagation through the nervous system has been established. The messages transmitted by neurons are voltage pulses called *action potentials.* When a neuron receives a strong enough stimulus, it produces identical voltage pulses that are actively propagated along its structure. The strength of the stimulus is conveyed by the number of pulses produced. When the pulses reach the end of the neuron, they activate either muscle cells or other neurons. There is a "firing threshold" for neurons: action potentials propagate along a neuron only if the stimulus is sufficiently strong.

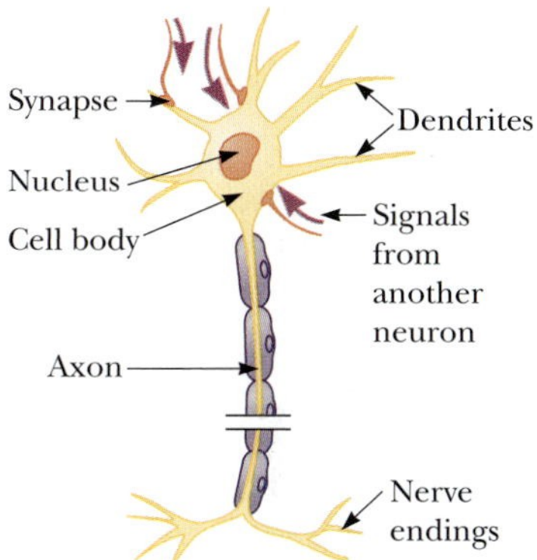


**Figure 18.24** Diagram of a neuron.

Neurons can be divided into three classes: sensory neurons, motor neurons, and interneurons. The sensory neurons receive stimuli from sensory organs that monitor the external and internal environment of the body. Depending on their specialized functions, the sensory neurons convey messages about factors such as light, temperature, pressure, muscle tension, and odor to higher centers in the nervous system. The motor neurons carry messages that control the muscle cells. Those messages are based on the information provided by the sensory neurons and by the brain. The interneurons transmit information from one neuron to another.

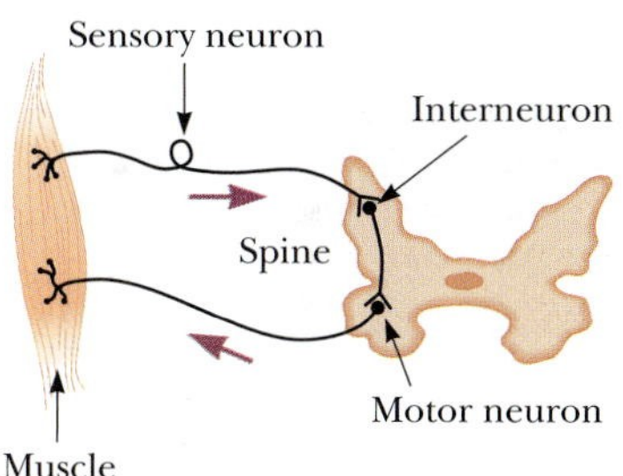


**Figure 18.25** A simple neural circuit.

Each neuron consists of a cell body to which are attached input ends called **dendrites** and a long tail called the **axon**, which transmits signals away from the cell (Fig. 18.24). The far end of the axon branches into nerve endings that transmit signals across small gaps to other neurons or to muscle cells. A simple sensorimotor neuron circuit is shown in Figure 18.25. A stimulus from a muscle produces nerve impulses that travel to the spine. Here the signal is transmitted to a motor neuron, which in turn sends impulses to control the muscle. Figure 18.26 shows an electron microscope image of neurons in the brain.

The axon, which is an extension of the neuron cell, conducts electric impulses away from the cell body. Some axons are extremely long. In humans, for example, the axons connecting the spine with the fingers and toes are more than 1 m long. The neuron can transmit messages because of the special active electrical characteristics of the axon. (The axon acts as an **active** source of energy like a battery, rather than like a **passive** stretch of resistive wire.) Much of the information about the electrical and chemical properties of the axon is obtained by inserting small needlelike probes into it. Figure 18.27 (page 648) shows an experimental setup.

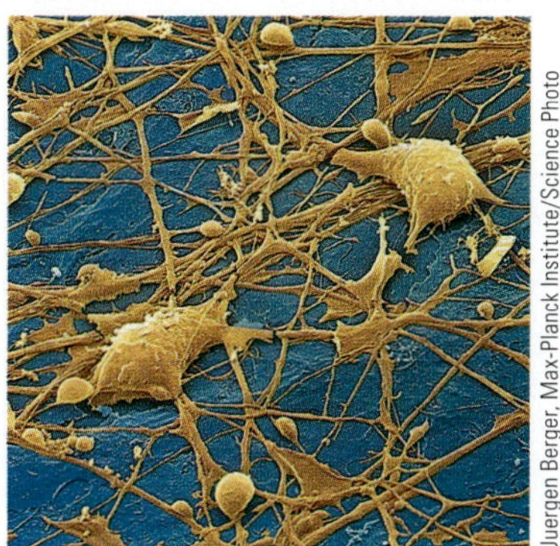


**Figure 18.26** Stellate neuron from human cortex.

Note that the outside of the axon is grounded, so all measured voltages are with respect to a zero potential on the outside. With these probes it is possible to inject current into the axon, measure the resulting action potential as a function of time at a fixed point, and sample the cell's chemical composition. Such experiments are usually difficult to run because the diameter of most axons is very small. Even the largest axons in the human nervous system have a diameter of only about $20 \times 10^{-4}$ cm. The

[2]This section is based on an essay by Paul Davidovits of Boston College.

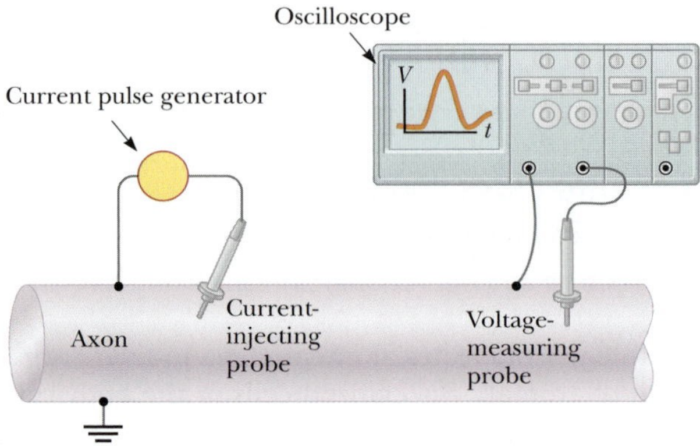


**Figure 18.27** An axon stimulated electrically. The left probe injects a short pulse of current, and the right probe measures the resulting action potential as a function of time.

giant squid, however, has an axon with a diameter of about 0.5 mm, which is large enough for the convenient insertion of probes. Much of the information about signal transmission in the nervous system has come from experiments with the squid axon.

In the aqueous environment of the body, salts and other molecules dissociate into positive and negative ions. As a result, body fluids are relatively good conductors of electricity. The inside of the axon is filled with an ionic fluid that is separated from the surrounding body fluid by a thin membrane that is only about 5 nm to 10 nm thick. The resistivities of the internal and external fluids are about the same, but their chemical compositions are substantially different. The external fluid is similar to seawater: its ionic solutes are mostly positive sodium ions and negative chloride ions. Inside the axon, the positive ions are mostly potassium ions and the negative ions are mostly large organic ions.

Ordinarily, the concentrations of sodium and potassium ions inside and outside the axon would be equalized by diffusion. The axon, however, is a living cell with an energy supply and can change the permeability of its membranes on a time scale of milliseconds.

When the axon is not conducting an electric pulse, the axon membrane is highly permeable to potassium ions, slightly permeable to sodium ions, and impermeable to large organic ions. Consequently, although sodium ions cannot easily enter the axon, potassium ions can leave it. As the potassium ions leave the axon, however, they leave behind large negative organic ions, which cannot follow them through the membrane. As a result, a negative potential builds up inside the axon with respect to the outside. The final negative potential reached, which has been measured at about $-70$ mV, holds back the outflow of potassium ions so that at equilibrium the concentration of ions remains as stated above.

The mechanism for the production of an electric signal by the neuron is conceptually simple, but was experimentally difficult to sort out. When a neuron changes its resting potential because of an appropriate stimulus, the properties of its membrane change locally. As a result, there is a sudden flow of sodium ions into the cell that lasts for about 2 ms. This flow produces the $+30$ mV peak in the action potential shown in Figure 18.28a. Immediately afterward, there is an increase in potassium ion flow out of the cell that restores the resting action potential of $-70$ mV in an additional 3 ms. Both the $Na^+$ and $K^+$ ion flows have been measured by using radioactive Na and K tracers. The nerve signal has been measured to travel along the axon at speeds from 50 m/s to about 150 m/s. This flow of charged particles (or signal transmission) in a nerve axon is *unlike* signal transmission in a metal wire. In an axon, charges move perpendicular to the direction of travel of the nerve signal, and the signal moves much more slowly than a voltage pulse traveling along a metallic wire.

Although the axon is a highly complex structure and much of how $Na^+$ and $K^+$ ion channels open and close is not understood, standard electric circuit concepts of current and capacitance can be used to analyze axons. It is left as a problem (Problem 41) to show that the axon, having equal and opposite charges separated by a thin dielectric membrane, acts like a capacitor.

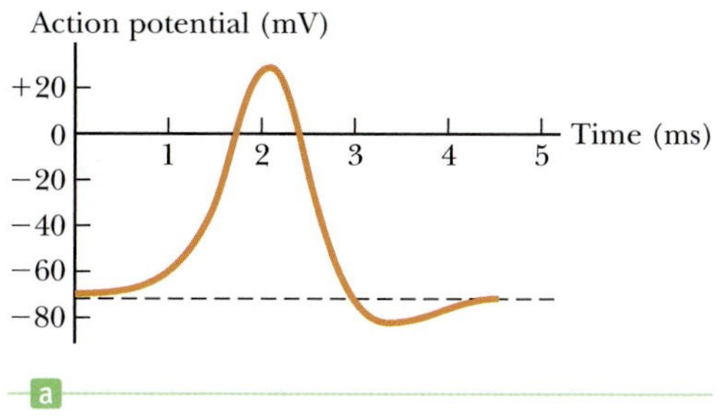


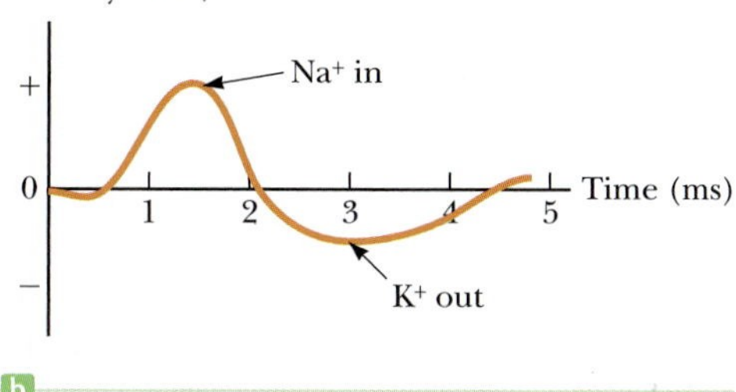


**Figure 18.28** (a) Typical action potential as a function of time. (b) Current in the axon membrane wall as a function of time.

## SUMMARY

### 18.1 Sources of emf

Any device, such as a battery or generator, that increases the electric potential energy of charges in an electric circuit is called a **source of emf**. Batteries convert chemical energy into electrical potential energy, and generators convert mechanical energy into electrical potential energy.

The terminal voltage $\Delta V$ of a battery is given by

$$\Delta V = \mathcal{E} - Ir \qquad [18.1]$$

where $\varepsilon$ is the emf of the battery, $I$ is the current, and $r$ is the internal resistance of the battery. Generally, the internal resistance is small enough to be neglected.

### 18.2 Resistors in Series

The **equivalent resistance** of a set of resistors connected in **series** is

$$R_{\text{eq}} = R_1 + R_2 + R_3 + \cdots \qquad [18.4]$$

$R_1$ $R_2$ $R_3$ → $R_{\text{eq}} = R_1 + R_2 + R_3$

Several resistors in series can be replaced with a single equivalent resistor.

The current remains at a constant value as it passes through a series of resistors. The potential difference across any two resistors in series is different, unless the resistors have the same resistance.

### 18.3 Resistors in Parallel

The **equivalent resistance** of a set of resistors connected in **parallel** is

$$\frac{1}{R_{\text{eq}}} = \frac{1}{R_1} + \frac{1}{R_2} + \frac{1}{R_3} + \cdots \qquad [18.6]$$

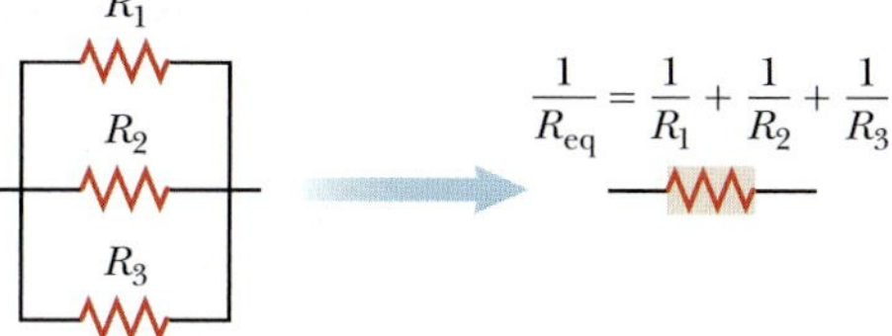


Several resistors in parallel can be replaced with a single equivalent resistor.

The potential difference across any two parallel resistors is the same; the current in each resistor, however, will be different unless the two resistances are equal.

### 18.4 Kirchhoff's Rules and Complex DC Circuits

Complex circuits can be analyzed using **Kirchhoff's rules**:

1. The sum of the currents entering any junction must equal the sum of the currents leaving that junction.
2. The sum of the potential differences across all the elements around any closed circuit loop must be zero.

The first rule, called the junction rule, is a statement of **conservation of charge**. The second rule, called the loop rule, is a statement of **conservation of energy**. Solving problems involves using these rules to generate as many equations as there are unknown currents. The equations can then be solved simultaneously.

## 18.5 *RC* Circuits

In a simple *RC* circuit with a battery, a resistor, and a capacitor in series, the charge on the capacitor increases according to the equation

$$q = Q(1 - e^{-t/RC}) \qquad [18.7]$$

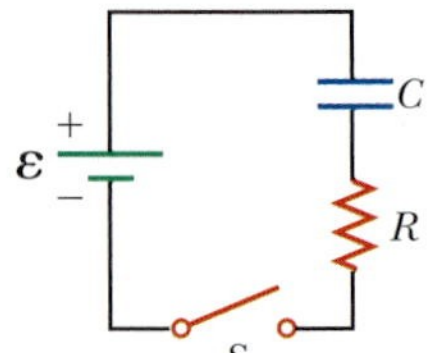


An *RC* circuit with a battery and a resistor charges a capacitor when the switch is closed.

The term *RC* in Equation 18.7 is called the **time constant** $\tau$ (Greek letter tau), so

$$\tau = RC \qquad [18.8]$$

The time constant represents the time required for the charge to increase from zero to 63.2% of its maximum equilibrium value.

A simple *RC* circuit consisting of a charged capacitor in series with a resistor discharges according to the expression

$$q = Qe^{-t/RC} \qquad [18.9]$$

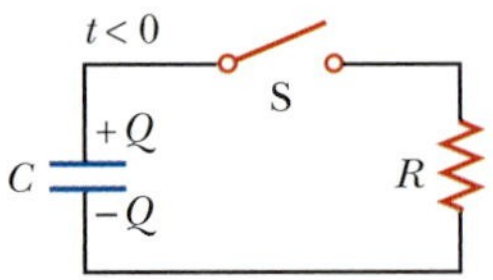


An *RC* circuit with charged capacitor discharges across a resistor when the switch is closed.

Problems can be solved by substituting values for $q$, $Q$, $t$ or $\tau$ into these equations. The voltage $\Delta V$ across the capacitor at any time is obtained by dividing the charge by the capacitance: $\Delta V = q/C$. Using Kirchhoff's loop rule yields the potential difference across the resistor. Ohm's law applied to the resistor then gives the current.

## WARM-UP EXERCISES

ENHANCED WebAssign The warm-up exercises in this chapter may be assigned online in Enhanced WebAssign.

1. **Math Review** Determine the values of (a) $I_1$ and (b) $I_2$ that satisfy the following two equations: $9.00 + 3.00\ I_1 - 2.00\ I_2 = 0$ and $-2.00\ I_1 + I_2 = 0$. (See also Section 18.4.)

2. **Math Review** Determine the values of (a) $I_1$, (b) $I_2$, and (c) $I_3$ that satisfy the following three equations: $-I_1 + I_2 + I_3 = 5.00$, $-3.00\ I_1 + I_2 = 0$, and $I_1 + I_3 = 0$.

3. **Math Review** Suppose the charge $q$ on a capacitor is given by $q = (3.00\ \text{C})\ e^{-t/(0.200\ \text{s})}$. Find (a) the value of $q$ when $t = 0.040\ 0$ s, and (b) the value of $t$ when $q = 1.20$ C. (See also Section 18.5.)

4. **Physics Review** Find the charge on a 25.0 $\mu$F capacitor when connected to a 12.0-V battery. (See Section 16.6.)

5. **Physics Review** Two capacitors have capacitances of 2.00 $\mu$F and 0.500 $\mu$F. Determine the equivalent capacitance if they are connected (a) in parallel, and (b) in series. (See Section 16.8.)

6. **Physics Review** A resistor with resistance 30.0 Ω is connected to the terminals of a voltage source. Calculate (a) the current through the resistor, and (b) the power dissipated by the resistor when the voltage is set at 6.00 V. (c) If the voltage is doubled, recalculate the power. (See Sections 17.4 and 17.6.)

7. A battery has an open-circuit voltage (emf) of 9.20 V and an internal resistance of 0.400 Ω. Determine (a) the current when the terminals are connected across a 3.00-Ω resistor, (b) the terminal voltage $\Delta V$, and (c) the total power supplied by the battery. (See Section 18.1.)

8. Two resistors have resistances of 2.00 Ω and 0.500 Ω. Determine the equivalent resistance if they are connected (a) in series, (b) in parallel. (See Sections 18.2 and 18.3.)

9. Three resistors have resistances of 2.00 Ω, 10.0 Ω and 25.0 Ω. Determine the equivalent resistance if they are connected (a) in parallel, and (b) in series. (See Sections 18.2 and 18.3.)

10. A simple circuit is constructed from a 12.0-V battery, a 3.00-Ω resistor and a 5.00-Ω resistor. Ignoring the internal resistance of the battery, determine the current supplied by the battery if the resistors are connected across the battery terminals (a) in series, and (b) in parallel. (See Sections 18.2 and 18.3.)

11. Consider the circuit shown in Figure WU18.11. Determine (a) the equivalent resistance of the three resistors in parallel, (b) the equivalent resistance of all four resistors in the circuit, (c) the current supplied by the battery, and (d) the magnitude of the potential difference across the 2.00-Ω resistor, (e) the magnitude of the potential difference across the 10.0-Ω resistor, and (f) the current through the 10.0-Ω resistor. (See Sections 18.2, 18.3 and 18.4)

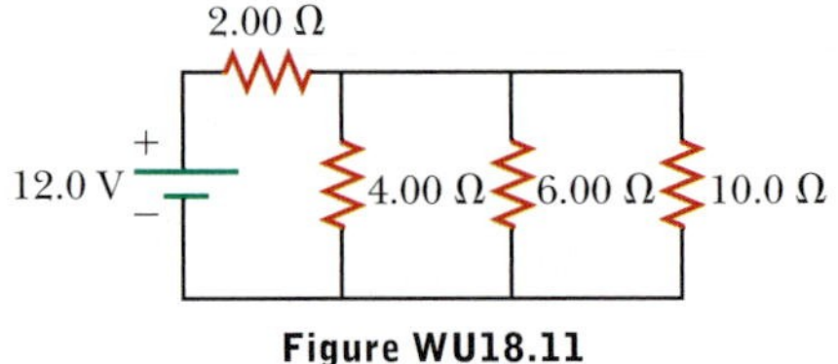


**Figure WU18.11**

12. A flashing light blub is driven by a simple series *RC* circuit with a time constant of 1.30 s. If the circuit's equivalent resistance is $5.00 \times 10^3\ \Omega$, determine its equivalent capacitance. (See Section 18.5.)

13. A 275-$\Omega$ resistor is in series with a 36.0 $\mu$F capacitor and a 12.0-V voltage source with the circuit switch open and the capacitor uncharged. (a) What is the time constant of this *RC* circuit? Calculate (b) the maximum charge the capacitor can accumulate, and (c) the charge on the capacitor 5.00 ms after the switch is closed. (See Section 18.5.)

## CONCEPTUAL QUESTIONS

ENHANCED WebAssign The conceptual questions in this chapter may be assigned online in Enhanced WebAssign.

1. Is the direction of current in a battery always from the negative terminal to the positive one? Explain.
2. Given three lightbulbs and a battery, sketch as many different circuits as you can.
3. Suppose the energy transferred to a dead battery during charging is *W*. The recharged battery is then used until fully discharged again. Is the total energy transferred out of the battery during use also *W*?
4. A short circuit is a circuit containing a path of very low resistance in parallel with some other part of the circuit. Discuss the effect of a short circuit on the portion of the circuit it parallels. Use a lamp with a frayed line cord as an example.
5. Connecting batteries in series increases the emf applied to a circuit. What advantage might there be to connecting them in parallel?
6. If electrical power is transmitted over long distances, the resistance of the wires becomes significant. Why? Which mode of transmission would result in less energy loss, high current and low voltage or low current and high voltage? Discuss.
7. If you have your headlights on while you start your car, why do they dim while the car is starting?
8. Two sets of Christmas lights are available. For set A, when one bulb is removed, the remaining bulbs remain illuminated. For set B, when one bulb is removed, the remaining bulbs do not operate. Explain the difference in wiring for the two sets.
9. Why is it possible for a bird to sit on a high-voltage wire without being electrocuted? (See Fig. CQ18.9.)

Superstock

**Figure CQ18.9**

10. (a) Two resistors are connected in series across a battery. Is the power delivered to each resistor (i) the same or (ii) not necessarily the same? (b) Two resistors are connected in parallel across a battery. Is the power delivered to each resistor (i) the same or (ii) not necessarily the same?
11. Suppose a parachutist lands on a high-voltage wire and grabs the wire as she prepares to be rescued. Will she be electrocuted? If the wire then breaks, should she continue to hold onto the wire as she falls to the ground?
12. A ski resort consists of a few chairlifts and several interconnected downhill runs on the side of a mountain, with a lodge at the bottom. The lifts are analogous to batteries, and the runs are analogous to resistors. Describe how two runs can be in series. Describe how three runs can be in parallel. Sketch a junction of one lift and two runs. One of the skiers is carrying an altimeter. State Kirchhoff's junction rule and Kirchhoff's loop rule for ski resorts.
13. Embodied in Kirchhoff's rules are two conservation laws. What are they?
14. Why is it dangerous to turn on a light when you are in a bathtub?

## PROBLEMS

ENHANCED WebAssign The problems in this chapter may be assigned online in Enhanced WebAssign.

**1.** denotes straightforward problem; **2.** denotes intermediate problem;
**3.** denotes challenging problem
**[1.]** denotes full solution available in *Student Solutions Manual/ Study Guide*
**1.** denotes problems most often assigned in Enhanced WebAssign

BIO denotes biomedical problems
GP denotes guided problems
M denotes Master It tutorial available in Enhanced WebAssign
Q|C denotes asking for quantitative and conceptual reasoning
S denotes symbolic reasoning problem
W denotes Watch It video solution available in Enhanced WebAssign

## 18.1 Sources of emf

## 18.2 Resistors in Series

## 18.3 Resistors in Parallel

1. A battery having an emf of 9.00 V delivers 117 mA when connected to a 72.0-Ω load. Determine the internal resistance of the battery.

2. Three 9.0-Ω resistors are connected in series with a 12-V battery. Find (a) the equivalent resistance of the circuit and (b) the current in each resistor. (c) Repeat for the case in which all three resistors are connected in parallel across the battery.

3. A lightbulb marked "75 W [at] 120 V" is screwed into a socket at one end of a long extension cord in which each of the two conductors has a resistance of 0.800 Ω. The other end of the extension cord is plugged into a 120-V outlet. (a) Draw a circuit diagram, and (b) find the actual power of the bulb in the circuit described.

4. (a) Find the current in an 8.00-Ω resistor connected to a battery that has an internal resistance of 0.15 Ω if the voltage across the battery (the terminal voltage) is 9.00 V. (b) What is the emf of the battery?

5. **M** (a) Find the equivalent resistance between points $a$ and $b$ in Figure P18.5. (b) Calculate the current in each resistor if a potential difference of 34.0 V is applied between points $a$ and $b$.

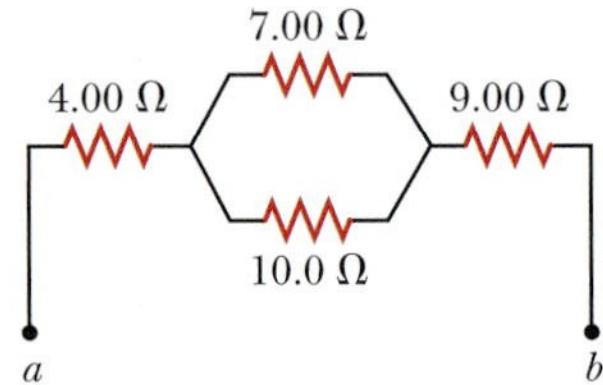


Figure P18.5

6. Consider the combination of resistors shown in Figure P18.6. (a) Find the equivalent resistance between point $a$ and $b$. (b) If a voltage of 35.0 V is applied between points $a$ and $b$, find the current in each resistor.

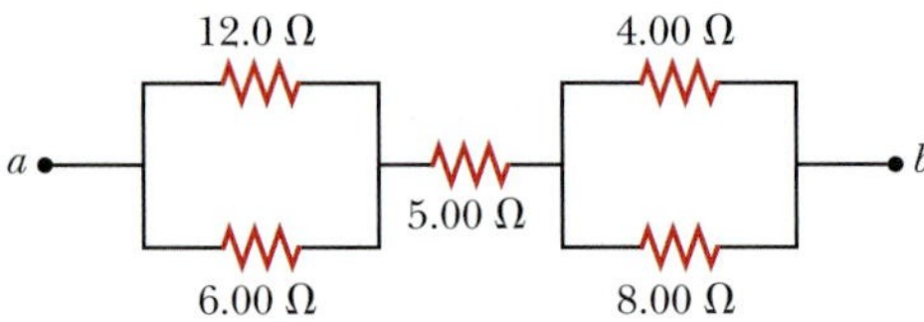


Figure P18.6

7. Two resistors connected in series have an equivalent resistance of 690 Ω. When they are connected in parallel, their equivalent resistance is 150 Ω. Find the resistance of each resistor.

8. **GP** Consider the circuit shown in Figure P18.8. (a) Calculate the equivalent resistance of the 10.0-Ω and 5.00-Ω resistors connected in parallel. (b) Using the result of part (a), calculate the combined resistance of the 10.0-Ω, 5.00-Ω, and 4.00-Ω resistors. (c) Calculate the equivalent resistance of the combined resistance found in part (b) and the parallel 3.00-Ω resistor. (d) Combine the equivalent resistance found in part (c) with the 2.00-Ω resistor. (e) Calculate the total current in the circuit. (f) What is the voltage drop across the 2.00-Ω resistor? (g) Subtracting the result of part (f) from the battery voltage, find the voltage across the 3.00-Ω resistor. (h) Calculate the current in the 3.00-Ω resistor.

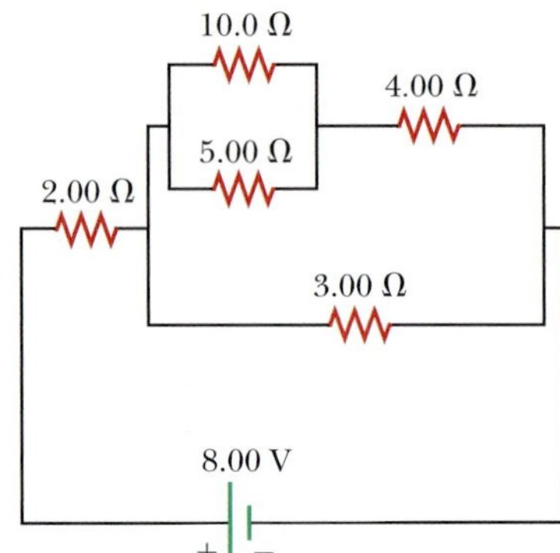


Figure P18.8

9. Consider the circuit shown in Figure P18.9. Find (a) the potential difference between points $a$ and $b$ and (b) the current in the 20.0-Ω resistor.

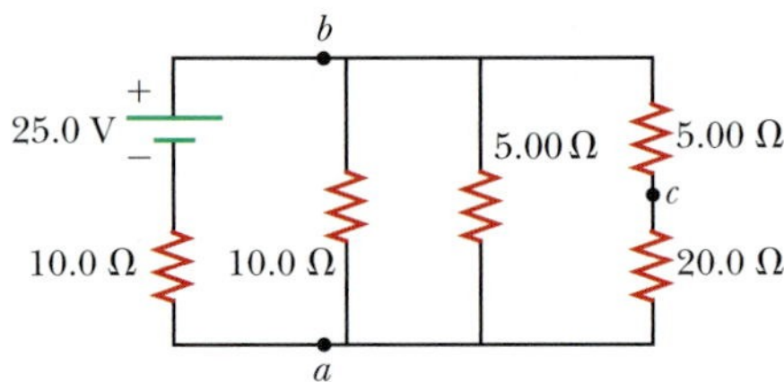


Figure P18.9

10. **S** Four resistors are connected to a battery as shown in Figure P18.10. (a) Determine the potential difference across each resistor in terms of $\mathcal{E}$. (b) Determine the current in each resistor in terms of $I$.

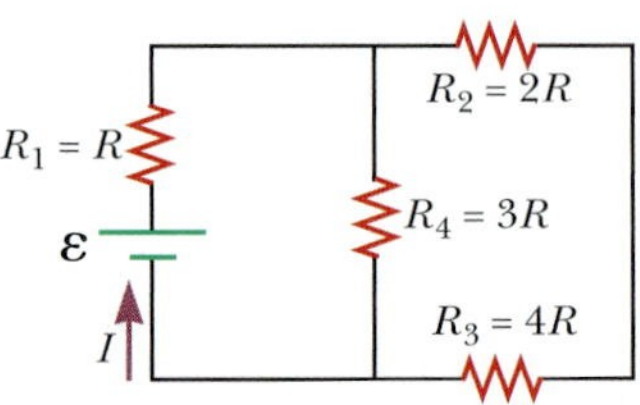


Figure P18.10

11. The resistance between terminals $a$ and $b$ in Figure P18.11 is 75 Ω. If the resistors labeled $R$ have the same value, determine $R$.

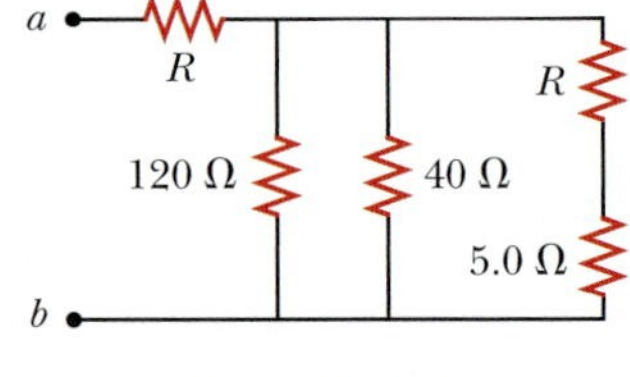


Figure P18.11

12. A battery with $\mathcal{E}$ = 6.00 V and no internal resistance supplies current to the circuit shown in Figure P18.12. When the double-throw switch S is open as shown in the figure, the current in the battery is 1.00 mA. When the switch is closed in position $a$, the current in the battery is 1.20 mA. When the switch is closed in position $b$, the current in the battery is 2.00 mA. Find the resistances (a) $R_1$, (b) $R_2$, and (c) $R_3$.

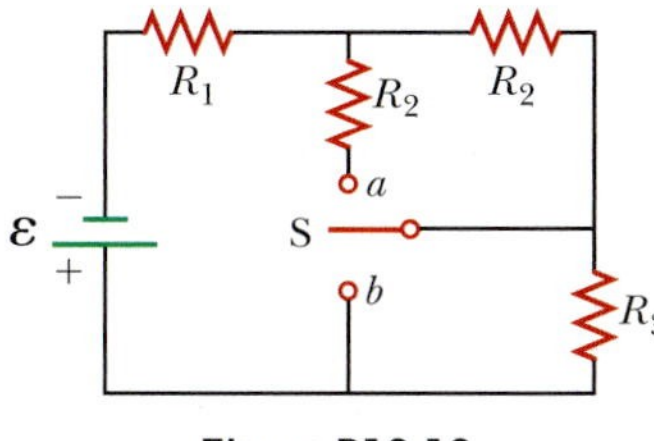


**Figure P18.12**

13. Find the current in the 12-Ω resistor in Figure P18.13.

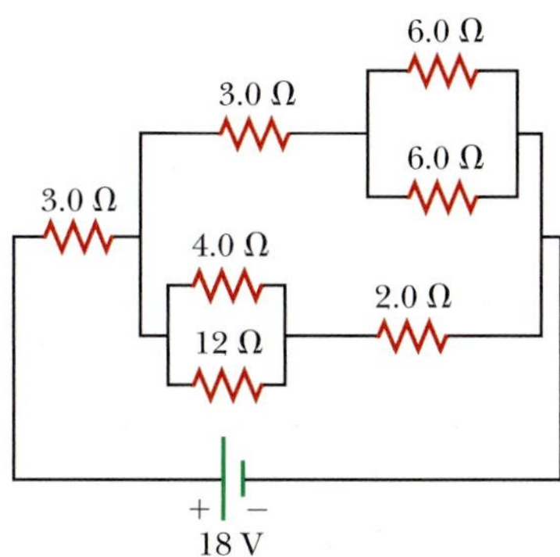


**Figure P18.13**

14. **Q|C** (a) Is it possible to reduce the circuit shown in Figure P18.14 to a single equivalent resistor connected across the battery? Explain. (b) Find the current in the 2.00-Ω resistor. (c) Calculate the power delivered by the battery to the circuit.

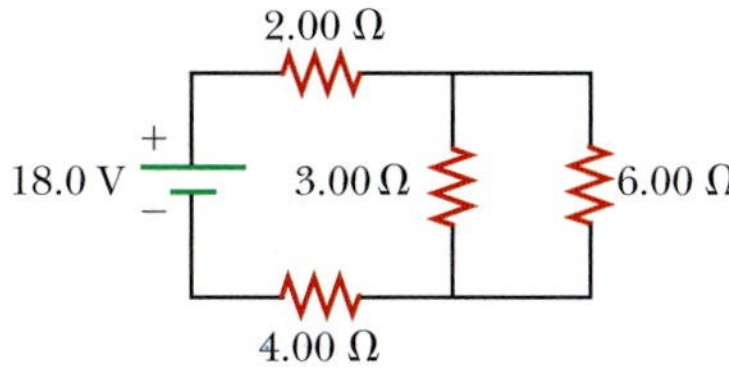


**Figure P18.14**

15. **W** (a) You need a 45-Ω resistor, but the stockroom has only 20-Ω and 50-Ω resistors. How can the desired resistance be achieved under these circumstances? (b) What can you do if you need a 35-Ω resistor?

## 18.4 Kirchhoff's Rules and Complex DC Circuits

> *Note:* For some circuits, the currents are not necessarily in the direction shown.

16. **S** (a) Find the current in each resistor of Figure P18.16 by using the rules for resistors in series and parallel. (b) Write three independent equations for the three currents using Kirchhoff's laws: one with the node rule; a second using the loop rule through the battery, the 6.0-Ω resistor, and the 24.0-Ω resistor; and the third using the loop rule through the 12.0-Ω and 24.0-Ω resistors. Solve to check the answers found in part (a).

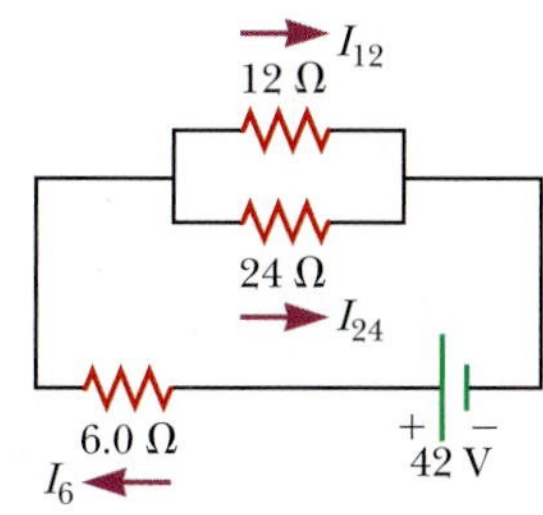


**Figure P18.16**

17. The ammeter shown in Figure P18.17 reads 2.00 A. Find $I_1$, $I_2$, and $\mathcal{E}$.

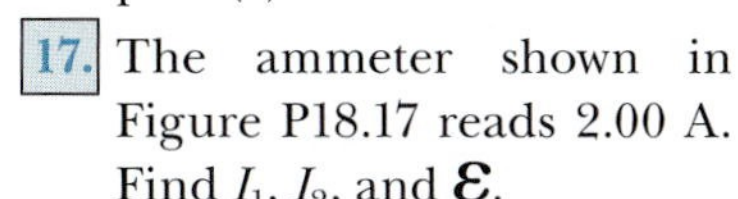

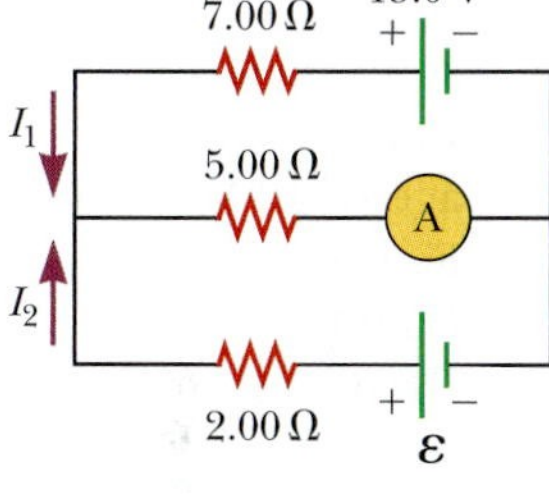


**Figure P18.17**

18. **W** For the circuit shown in Figure P18.18, calculate (a) the current in the 2.00-Ω resistor and (b) the potential difference between points $a$ and $b$, $\Delta V = V_b - V_a$.

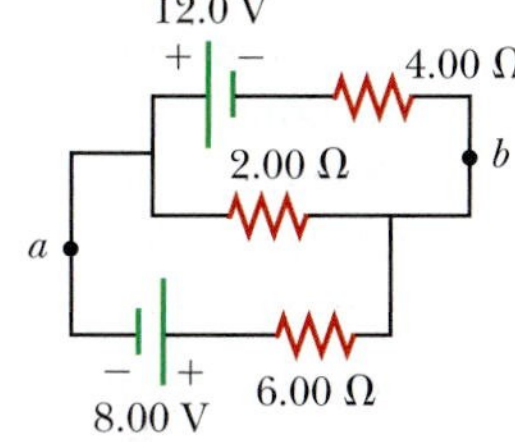


**Figure P18.18**

19. Taking $R$ = 1.00 kΩ and $\mathcal{E}$ = 250 V in Figure P18.19, determine the direction and magnitude of the current in the horizontal wire between $a$ and $e$.

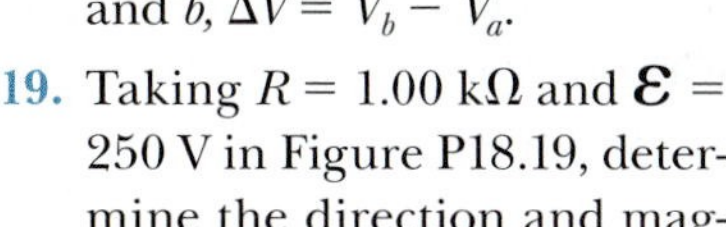

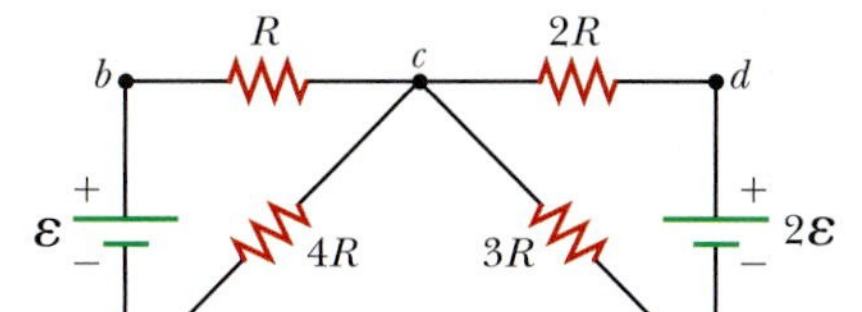


**Figure P18.19**

20. **W** In the circuit of Figure P18.20, the current $I_1$ is 3.0 A and the values of $\mathcal{E}$ and $R$ are unknown. What are the currents $I_2$ and $I_3$?

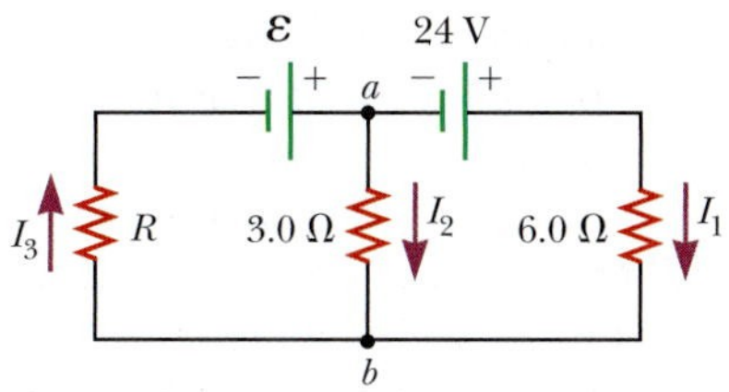


**Figure P18.20**

21. In the circuit of Figure P18.21, determine (a) the current in each resistor and (b) the potential difference across the 200-Ω resistor.

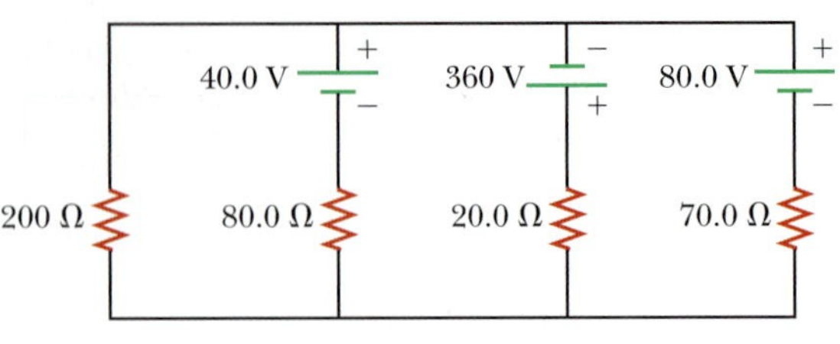


Figure P18.21

22. Q|C Four resistors are connected to a battery with a terminal voltage of 12 V, as shown in Figure P18.22. (a) How would you reduce the circuit to an equivalent single resistor connected to the battery? Use this procedure to find the equivalent resistance of the circuit. (b) Find the current delivered by the battery to this equivalent resistance. (c) Determine the power delivered by the battery. (d) Determine the power delivered to the 50.0-Ω resistor.

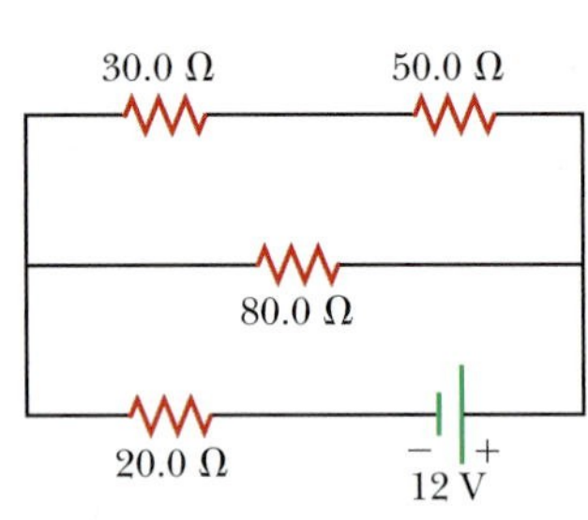


Figure P18.22

23. M Using Kirchhoff's rules, (a) find the current in each resistor shown in Figure P18.23 and (b) find the potential difference between points $c$ and $f$.

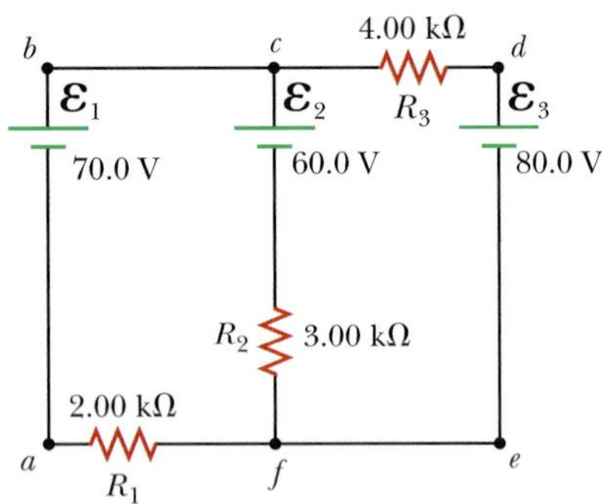


Figure P18.23

24. Two 1.50-V batteries—with their positive terminals in the same direction—are inserted in series into the barrel of a flashlight. One battery has an internal resistance of 0.255 Ω, the other an internal resistance of 0.153 Ω. When the switch is closed, a current of 0.600 A passes through the lamp. (a) What is the lamp's resistance? (b) What fraction of the power dissipated is dissipated in the batteries?

25. Q|C (a) Can the circuit shown in Figure P18.25 be reduced to a single resistor connected to the batteries? Explain. (b) Calculate each of the unknown currents $I_1$, $I_2$, and $I_3$ for the circuit.

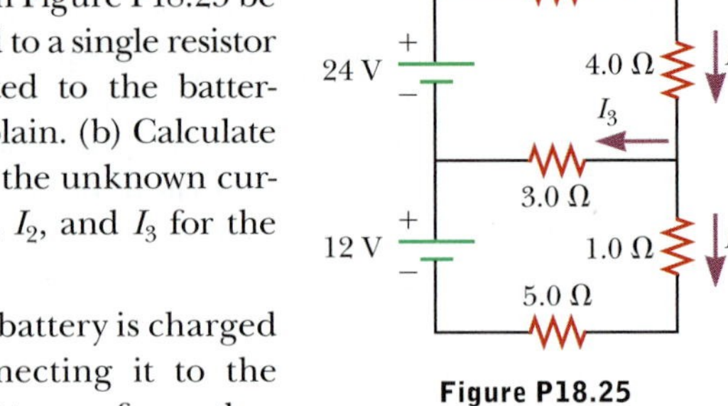


Figure P18.25

26. A dead battery is charged by connecting it to the live battery of another car with jumper cables (Fig. P18.26). Determine the current in (a) the starter and in (b) the dead battery.

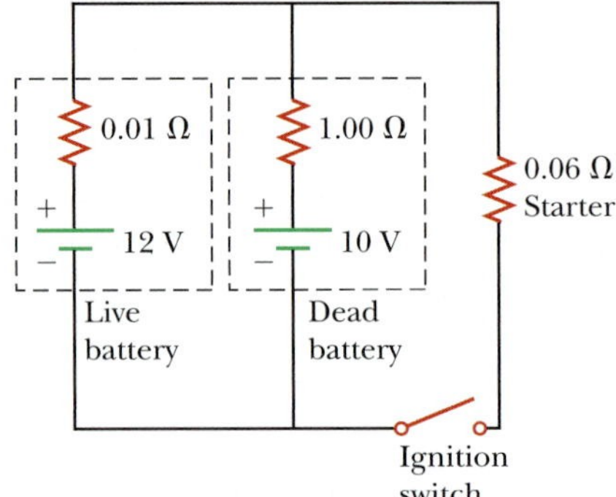


Figure P18.26

27. Q|C (a) Can the circuit shown in Figure P18.27 be reduced to a single resistor connected to the batteries? Explain. (b) Find the magnitude of the current and its direction in each resistor.

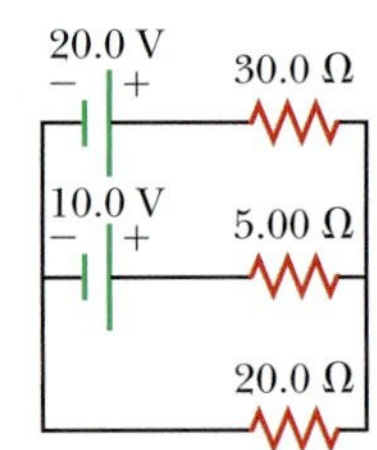


Figure P18.27

28. GP S For the circuit shown in Figure P18.28, use Kirchhoff's rules to obtain equations for (a) the upper loop, (b) the lower loop, and (c) the node on the left side. In each case suppress units for clarity and simplify, combining like terms. (d) Solve the node equation for $I_{36}$. (e) Using the equation found in (d), eliminate $I_{36}$ from the equation found in part (b). (f) Solve the equations found in part (a) and part (e) simultaneously for the two

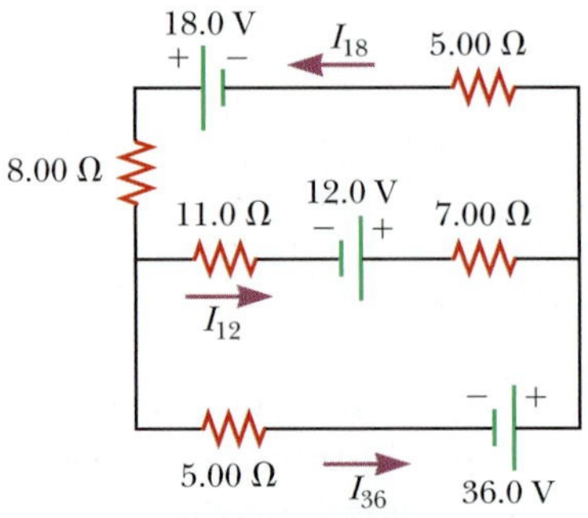


Figure P18.28

unknowns for $I_{18}$ and $I_{12}$, respectively. (g) Substitute the answers found in part (f) into the node equation found in part (d), solving for $I_{36}$. (h) What is the significance of the negative answer for $I_{12}$?

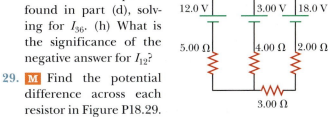

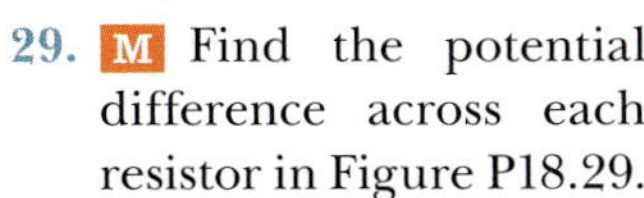

**29.** M Find the potential difference across each resistor in Figure P18.29.

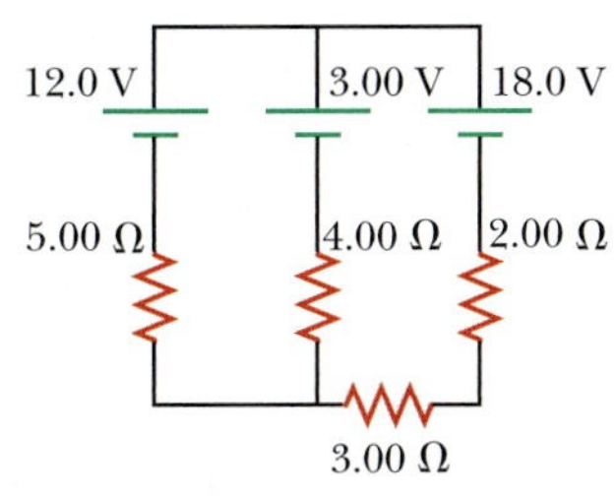


**Figure P18.29**

## 18.5 *RC* Circuits

**30.** Show that $\tau = RC$ has units of time.

**31.** Consider the series *RC*-circuit shown in Figure 18.16 for which $R = 75.0$ kΩ, $C = 25.0$ μF, and $\mathcal{E} = 12.0$ V. Find (a) the time constant of the circuit and (b) the charge on the capacitor one time constant after the switch is closed.

**32.** An uncharged capacitor and a resistor are connected in series to a source of emf. If $\mathcal{E} = 9.00$ V, $C = 20.0$ μF, and $R = 100$ Ω, find (a) the time constant of the circuit, (b) the maximum charge on the capacitor, and (c) the charge on the capacitor after one time constant.

**33.** Consider a series *RC* circuit as in Figure P18.33 for which $R = 1.00$ MΩ, $C = 5.00$ μF, and $\mathcal{E} = 30.0$ V. Find (a) the time constant of the circuit and (b) the maximum charge on the capacitor after the switch is thrown closed. (c) Find the current in the resistor 10.0 s after the switch is closed.

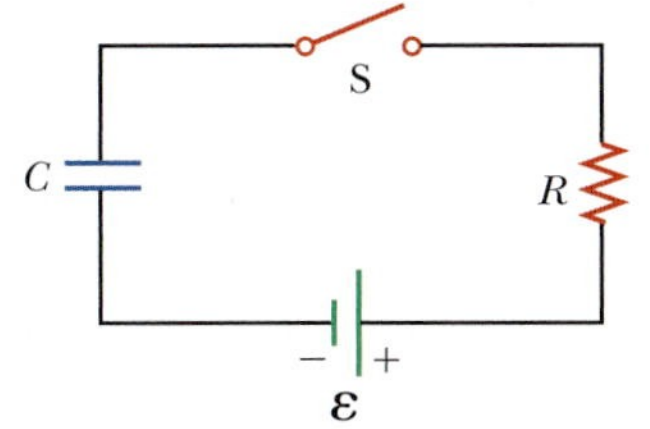


**Figure P18.33** Problems 33 and 34.

**34.** W Consider a series *RC* circuit for which $R = 1.0$ MΩ, $C = 5.0$ μF, and $\mathcal{E} = 30$ V as in Figure P18.33. Find the charge on the capacitor 10 s after the switch is closed.

**35.** A charged capacitor is connected to a resistor and a switch as in Figure P18.35. The circuit has a time constant of 1.50 s. Soon after the switch is closed, the charge on the capacitor is 75.0% of its initial charge. (a) Find the time interval required for the capacitor to reach this charge. (b) If $R = 250$ kΩ, what is the value of $C$?

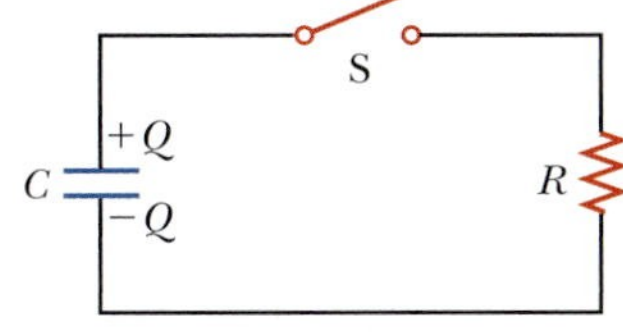


**Figure P18.35**

**36.** A 10.0-μF capacitor is charged by a 10.0-V battery through a resistance $R$. The capacitor reaches a potential difference of 4.00 V in a time interval of 3.00 s after charging begins. Find $R$.

## 18.6 Household Circuits

**37.** What minimum number of 75-W lightbulbs must be connected in parallel to a single 120-V household circuit to trip a 30.0-A circuit breaker?

**38.** A lamp ($R = 150$ Ω), an electric heater ($R = 25$ Ω), and a fan ($R = 50$ Ω) are connected in parallel across a 120-V line. (a) What total current is supplied to the circuit? (b) What is the voltage across the fan? (c) What is the current in the lamp? (d) What power is expended in the heater?

**39.** M A heating element in a stove is designed to dissipate 3 000 W when connected to 240 V. (a) Assuming the resistance is constant, calculate the current in the heating element if it is connected to 120 V. (b) Calculate the power it dissipates at that voltage.

**40.** Q|C A coffee maker is rated at 1 200 W, a toaster at 1 100 W, and a waffle maker at 1 400 W. The three appliances are connected in parallel to a common 120-V household circuit. (a) What is the current in each appliance when operating independently? (b) What total current is delivered to the appliances when all are operating simultaneously? (c) Is a 15-A circuit breaker sufficient in this situation? Explain.

## 18.8 Conduction of Electrical Signals by Neurons

**41.** BIO Assume a length of axon membrane of about 0.10 m is excited by an action potential (length excited = nerve speed × pulse duration = 50.0 m/s × $2.0 \times 10^{-3}$ s = 0.10 m). In the resting state, the outer surface of the axon wall is charged positively with $K^+$ ions and the inner wall has an equal and opposite charge of negative organic ions, as shown in Figure P18.41. Model the axon as a parallel-plate capacitor and take $C = \kappa\epsilon_0 A/d$ and $Q = C\Delta V$ to investigate the charge as follows. Use typical values for a cylindrical axon of cell wall thickness $d = 1.0 \times 10^{-8}$ m, axon radius $r = 1.0 \times 10^1$ μm, and cell-wall dielectric constant

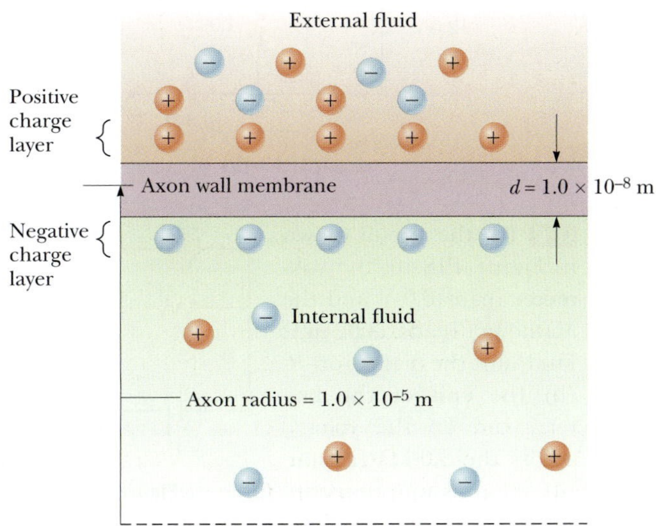


**Figure P18.41** Problems 41 and 42.

$\kappa = 3.0$. (a) Calculate the positive charge on the outside of a 0.10-m piece of axon when it is not conducting an electric pulse. How many $K^+$ ions are on the outside of the axon assuming an initial potential difference of $7.0 \times 10^{-2}$ V? Is this a large charge per unit area? *Hint:* Calculate the charge per unit area in terms of electronic charge $e$ per squared (Å$^2$). An atom has a cross section of about 1 Å$^2$ (1 Å $= 10^{-10}$ m). (b) How much positive charge must flow through the cell membrane to reach the excited state of $+3.0 \times 10^{-2}$ V from the resting state of $-7.0 \times 10^{-2}$ V? How many sodium ions ($Na^+$) is this? (c) If it takes 2.0 ms for the $Na^+$ ions to enter the axon, what is the average current in the axon wall in this process? (d) How much energy does it take to raise the potential of the inner axon wall to $+3.0 \times 10^{-2}$ V, starting from the resting potential of $-7.0 \times 10^{-2}$ V?

42. BIO Consider the model of the axon as a capacitor from Problem 41 and Figure P18.41. (a) How much energy does it take to restore the inner wall of the axon to $-7.0 \times 10^{-2}$ V, starting from $+3.0 \times 10^{-2}$ V? (b) Find the average current in the axon wall during this process.

43. BIO Using Figure 18.28b and the results of Problems 18.41d and 18.42a, find the power supplied by the axon per action potential.

## Additional Problems

44. How many different resistance values can be constructed from a 2.0-Ω, a 4.0-Ω, and a 6.0-Ω resistor? Show how you would get each resistance value either individually or by combining them.

45. (a) Calculate the potential difference between points $a$ and $b$ in Figure P18.45 and (b) identify which point is at the higher potential.

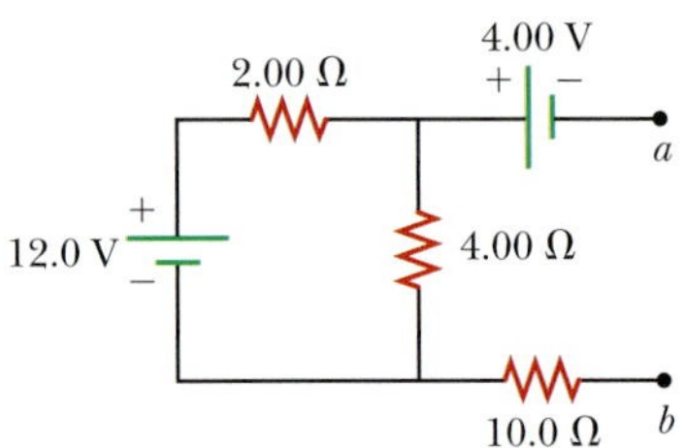


Figure P18.45

46. QIC For the circuit shown in Figure P18.46, the voltmeter reads 6.0 V and the ammeter reads 3.0 mA. Find (a) the value of $R$, (b) the emf of the battery, and (c) the voltage across the 3.0-kΩ resistor. (d) What assumptions did you have to make to solve this problem?

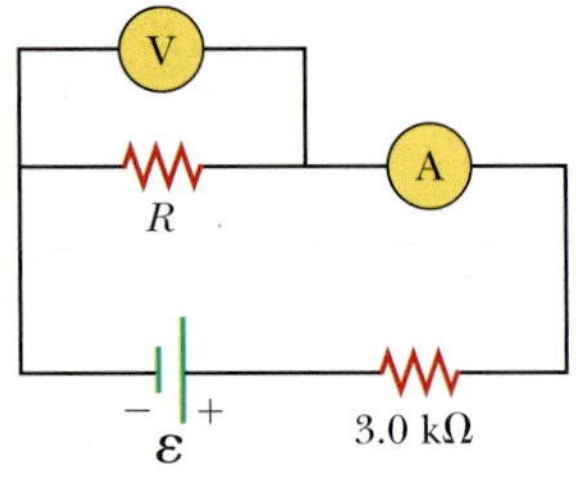


Figure P18.46

47. Find (a) the equivalent resistance of the circuit in Figure P18.47, (b) each current in the circuit, (c) the potential difference across each resistor, and (d) the power dissipated by each resistor.

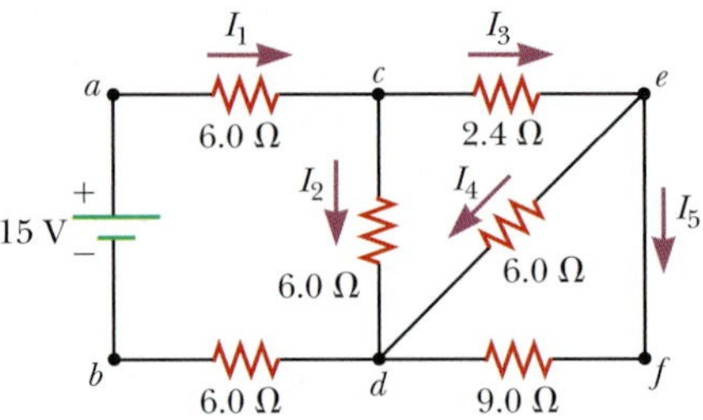


Figure P18.47

48. Three 60.0-W, 120-V lightbulbs are connected across a 120-V power source, as shown in Figure P18.48. Find (a) the total power delivered to the three bulbs and (b) the potential difference across each. Assume the resistance of each bulb is constant (even though, in reality, the resistance increases markedly with current).

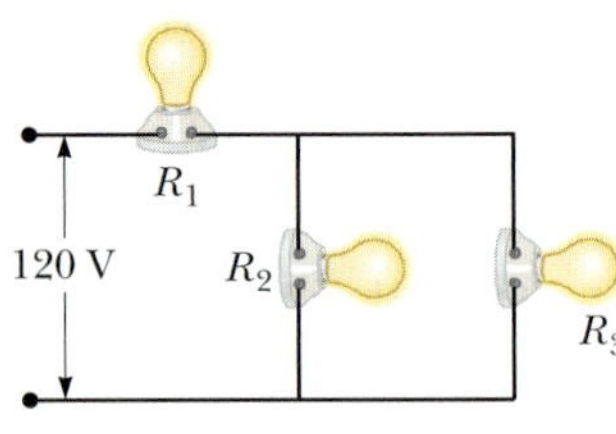


Figure P18.48

49. When two unknown resistors are connected in series with a battery, the battery delivers 225 W and carries a total current of 5.00 A. For the same total current, 50.0 W is delivered when the resistors are connected in parallel. Determine the value of each resistor.

50. QIC The circuit in Figure P18.50a consists of three resistors and one battery with no internal resistance. (a) Find the current in the 5.00-Ω resistor. (b) Find the power delivered to the 5.00-Ω resistor. (c) In each of the circuits in Figures P18.50b, P18.50c, and P18.50d, an additional 15.0-V battery has been inserted into the circuit. Which diagram or diagrams represent a circuit that requires the use of Kirchhoff's rules to find the

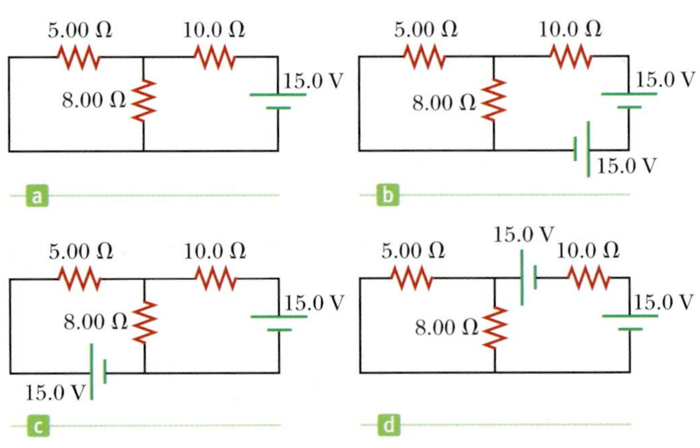


Figure P18.50

currents? Explain why. (d) In which of these three new circuits is the smallest amount of power delivered to the 10.0-Ω resistor? (You need not calculate the power in each circuit if you explain your answer.)

**51.** Q|C S A circuit consists of three identical lamps, each of resistance $R$, connected to a battery as in Figure P18.51. (a) Calculate an expression for the equivalent resistance of the circuit when the switch is open. Repeat the calculation when the switch is closed. (b) Write an expression for the power supplied by the battery when the switch is open. Repeat the calculation when the switch is closed. (c) Using the results already obtained, explain what happens to the brightness of the lamps when the switch is closed.

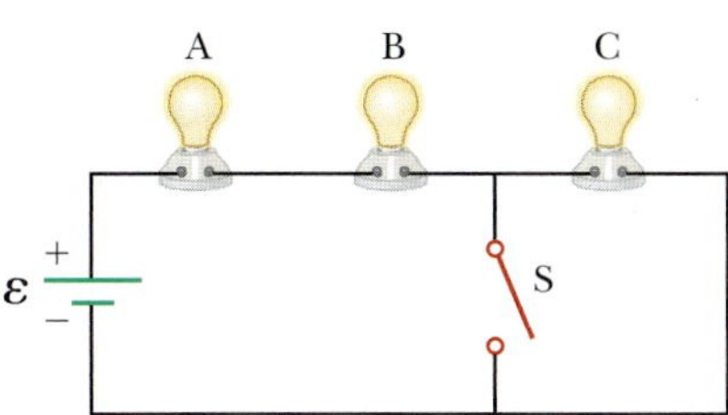


**Figure P18.51**

**52.** The resistance between points $a$ and $b$ in Figure P18.52 drops to one-half its original value when switch S is closed. Determine the value of $R$.

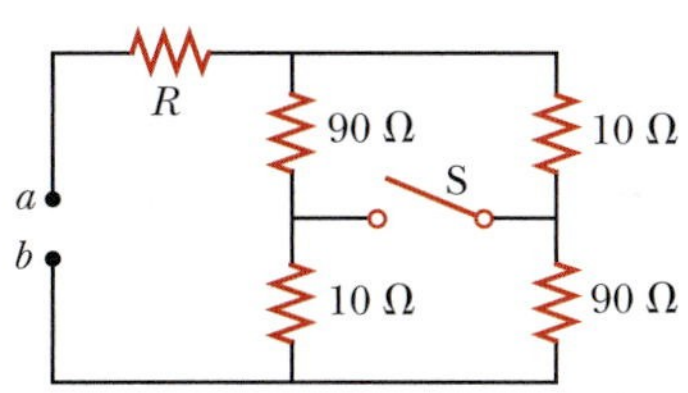


**Figure P18.52**

**53.** The circuit in Figure P18.53 has been connected for several seconds. Find the current (a) in the 4.00-V battery, (b) in the 3.00-Ω resistor, (c) in the 8.00-V battery, and (d) in the 3.00-V battery. (e) Find the charge on the capacitor.

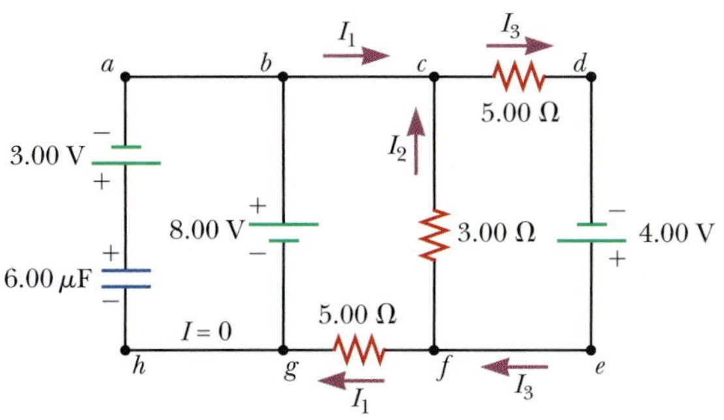


**Figure P18.53**

**54.** An emf of 10 V is connected to a series $RC$ circuit consisting of a resistor of $2.0 \times 10^6\ \Omega$ and an initially uncharged capacitor of 3.0 $\mu$F. Find the time required for the charge on the capacitor to reach 90% of its final value.

**55.** The student engineer of a campus radio station wishes to verify the effectiveness of the lightning rod on the antenna mast (Fig. P18.55). The unknown resistance $R_x$ is between points $C$ and $E$. Point $E$ is a "true ground," but is inaccessible for direct measurement because the stratum in which it is located is several meters below Earth's surface. Two identical rods are driven into the ground at $A$ and $B$, introducing an unknown resistance $R_y$. The procedure for finding the unknown resistance $R_x$ is as follows. Measure resistance $R_1$ between points $A$ and $B$. Then connect $A$ and $B$ with a heavy conducting wire and measure resistance $R_2$ between points $A$ and $C$. (a) Derive a formula for $R_x$ in terms of the observable resistances $R_1$ and $R_2$. (b) A satisfactory ground resistance would be $R_x < 2.0\ \Omega$. Is the grounding of the station adequate if measurements give $R_1 = 13\ \Omega$ and $R_2 = 6.0\ \Omega$?

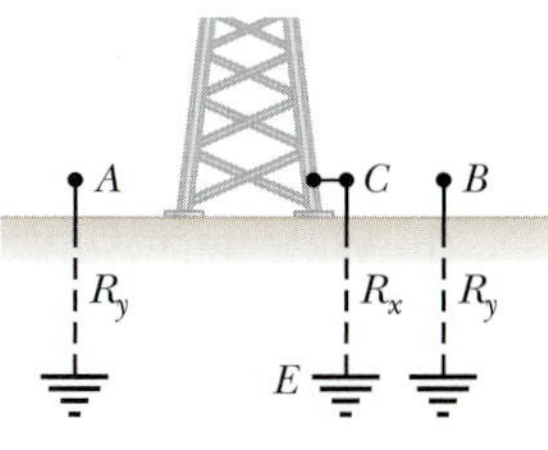


**Figure P18.55**

**56.** The resistor $R$ in Figure P18.56 dissipates 20 W of power. Determine the value of $R$.

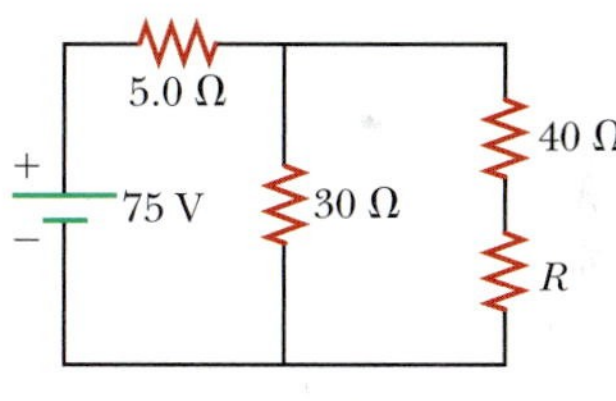


**Figure P18.56**

**57.** A voltage $\Delta V$ is applied to a series configuration of $n$ resistors, each of resistance $R$. The circuit components are reconnected in a parallel configuration, and voltage $\Delta V$ is again applied. Show that the power consumed by the series configuration is $1/n^2$ times the power consumed by the parallel configuration.

**58.** For the network in Figure P18.58, show that the resistance between points $a$ and $b$ is $R_{ab} = \frac{27}{17}\ \Omega$. (*Hint:* Connect a battery with emf $\mathcal{E}$ across points $a$ and $b$ and determine $\mathcal{E}/I$, where $I$ is the current in the battery.)

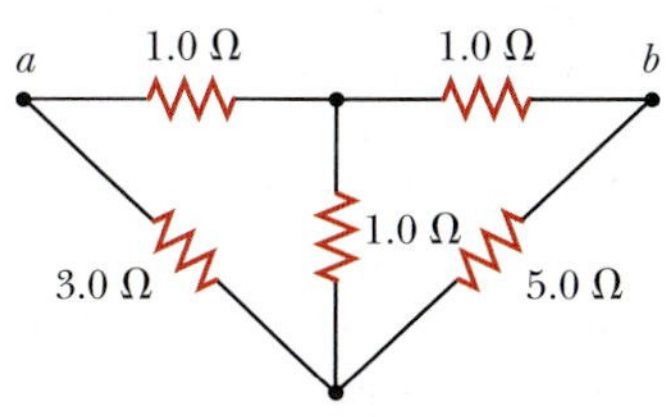


**Figure P18.58**

59. A battery with an internal resistance of 10.0 Ω produces an open circuit voltage of 12.0 V. A variable load resistance with a range from 0 to 30.0 Ω is connected across the battery. (*Note:* A battery has a resistance that depends on the condition of its chemicals and that increases as the battery ages. This internal resistance can be represented in a simple circuit diagram as a resistor in series with the battery.) (a) Graph the power dissipated in the load resistor as a function of the load resistance. (b) With your graph, demonstrate the following important theorem: *The power delivered to a load is a maximum if the load resistance equals the internal resistance of the source.*

60. The circuit in Figure P18.60 contains two resistors, $R_1 = 2.0\ \text{k}\Omega$ and $R_2 = 3.0\ \text{k}\Omega$, and two capacitors, $C_1 = 2.0\ \mu\text{F}$ and $C_2 = 3.0\ \mu\text{F}$, connected to a battery with emf $\mathcal{E} = 120$ V. If there are no charges on the capacitors before switch S is closed, determine the charges $q_1$ and $q_2$ on capacitors $C_1$ and $C_2$, respectively, as functions of time, after the switch is closed. *Hint:* First reconstruct the circuit so that it becomes a simple *RC* circuit containing a single resistor and single capacitor in series, connected to the battery, and then determine the total charge $q$ stored in the circuit.

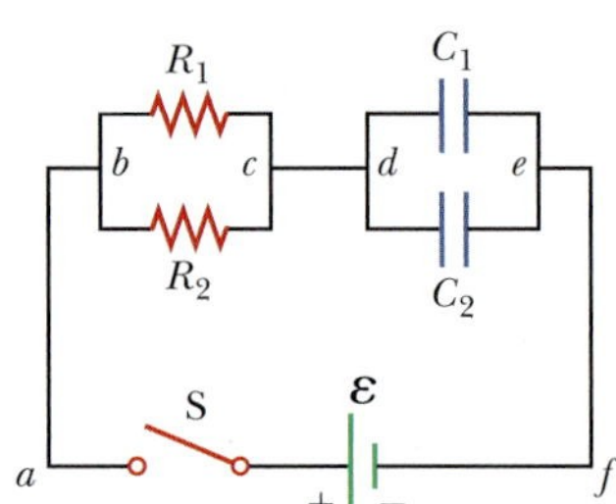

Figure P18.60

61. BIO An electric eel generates electric currents through its highly specialized Hunter's organ, in which thousands of disk-shaped cells called electrocytes are lined up in series, very much in the same way batteries are lined up inside a flashlight. When activated, each electrocyte can maintain a potential difference of about 150 mV at a current of 1 A for about 2.0 ms. Suppose a grown electric eel has $4.0 \times 10^3$ electrocytes and can deliver up to 300 shocks in rapid series over about 1 s. (a) What maximum electrical power can an electric eel generate? (b) Approximately how much energy does it release in one shock? (c) How high would a mass of 1 kg have to be lifted so that its gravitational potential energy equals the energy released in 300 such shocks?

62. In Figure P18.62, $R_1 = 0.100\ \Omega$, $R_2 = 1.00\ \Omega$, and $R_3 = 10.0\ \Omega$. Find the equivalent resistance of the circuit and the current in each resistor when a 5.00-V power supply is connected between (a) points $A$ and $B$, (b) points $A$ and $C$, and (c) points $A$ and $D$.

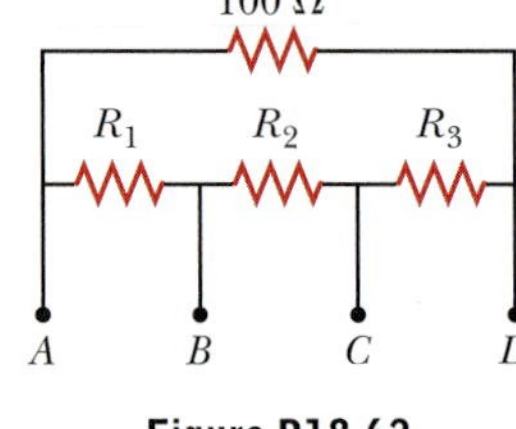

Figure P18.62

63. M What are the expected readings of the ammeter and voltmeter for the circuit in Figure P18.63?

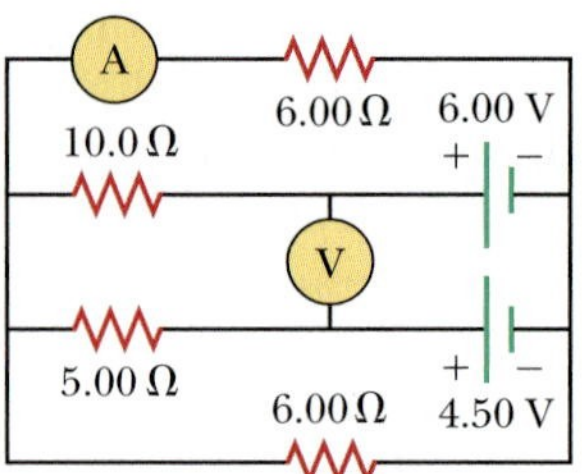

Figure P18.63

64. Consider the two arrangements of batteries and bulbs shown in Figure P18.64. The two bulbs are identical and have resistance $R$, and the two batteries are identical with output voltage $\Delta V$. (a) In case 1, with the two bulbs in series, compare the brightness of each bulb, the current in each bulb, and the power delivered to each bulb. (b) In case 2, with the two bulbs in parallel, compare the brightness of each bulb, the current in each bulb, and the power supplied to each bulb. (c) Which bulbs are brighter, those in case 1 or those in case 2? (d) In each case, if one bulb fails, will the other go out as well? If the other bulb doesn't fail, will it get brighter or stay the same? (Problem 64 is courtesy of E. F. Redish. For other problems of this type, visit http://www.physics.umd.edu/perg/.)

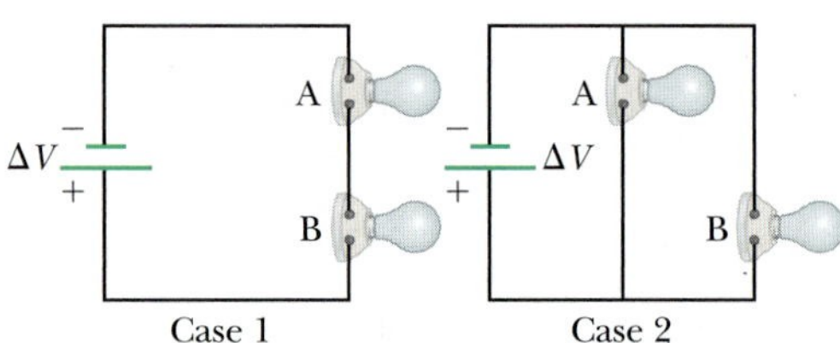

Figure P18.64

65. The given pair of capacitors in Figure P18.65 is fully charged by a 12.0-V battery. The battery is disconnected and the circuit closed. After 1.00 ms, how much charge remains on (a) the 3.00-$\mu$F capacitor? (b) The 2.00-$\mu$F capacitor? (c) What is the current in the resistor?

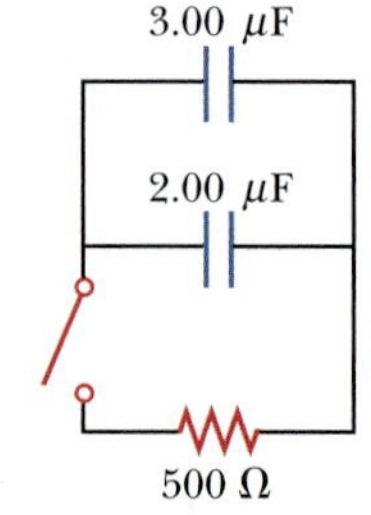

Figure P18.65

66. A 2.00-nF capacitor with an initial charge of 5.10 $\mu$C is discharged through a 1.30-k$\Omega$ resistor. (a) Calculate the magnitude of the current in the resistor 9.00 $\mu$s after the resistor is connected across the terminals of the capacitor. (b) What charge remains on the capacitor after 8.00 $\mu$s? (c) What is the maximum current in the resistor?

Dhanachote Vongprasert/Shutterstock.com

Aurora borealis, the northern lights. Displays such as this one are caused by cosmic ray particles trapped in Earth's magnetic field. When the particles collide with atoms in the atmosphere, they cause the atoms to emit visible light.

# Magnetism 19

In terms of applications, magnetism is one of the most important fields in physics. Large electromagnets are used to pick up heavy loads. Magnets are used in such devices as meters, motors, and loudspeakers. Magnetic tapes and disks are used routinely in sound- and video-recording equipment and to store computer data. Intense magnetic fields are used in magnetic resonance imaging (MRI) devices to explore the human body with better resolution and greater safety than x-rays can provide. Giant superconducting magnets are used in the cyclotrons that guide particles into targets at nearly the speed of light. Rail guns (Fig. 19.1) use magnetic forces to fire high-speed projectiles, and magnetic bottles hold antimatter, a possible key to future space propulsion systems.

Magnetism is closely linked with electricity. Magnetic fields affect moving charges, and moving charges produce magnetic fields. Changing magnetic fields can even create electric fields. These phenomena signify an underlying unity of electricity and magnetism, which James Clerk Maxwell first described in the 19th century. The ultimate source of any magnetic field is electric current.



## 19.1 Magnets

**LEARNING OBJECTIVES**

1. Discuss the basic properties of magnets and magnetic fields, contrasting them with electric charges and electric fields.
2. Characterize magnetic materials as hard or soft by the extent to which they retain their magnetism. Give examples of each kind.
3. Identify the origin of magnetic fields in moving electric charges. State some applications of magnets and their fields.

Defense Threat Reduction Agency [DTRA]

**Figure 19.1** Rail guns launch projectiles at high speed using magnetic force. Larger versions could send payloads into space or be used as thrusters to move metal-rich asteroids from deep space to Earth orbit. In this photo a rail gun at Sandia National Laboratories in Albuquerque, New Mexico, fires a projectile at over three kilometers per second. (For more information on this, see Applying Physics 20.2 on pages 709–710.)

Most people have had experience with some form of magnet. You are most likely familiar with the common iron horseshoe magnet that can pick up iron-containing objects such as paper clips and nails. In the discussion that follows, we assume the magnet has the shape of a bar. Iron objects are most strongly attracted to either end of such a bar magnet, called its **poles**. One end is called the **north pole** and the other the **south pole.** The names come from the behavior of a magnet in the presence of Earth's magnetic field. If a bar magnet is suspended from its midpoint by a piece of string so that it can swing freely in a horizontal plane, it will rotate until its north pole points to the north and its south pole points to the south. The same idea is used to construct a simple compass. Magnetic poles also exert attractive or repulsive forces on each other similar to the electrical forces between charged objects. In fact, simple experiments with two bar magnets show that **like poles repel each other and unlike poles attract each other**.

Although the force between opposite magnetic poles is similar to the force between positive and negative electric charges, there is an important difference: positive and negative electric charges can exist in isolation of each other; north and south poles don't. No matter how many times a permanent magnet is cut, each piece always has a north pole and a south pole. There is some theoretical basis, however, for the speculation that magnetic monopoles (isolated north or south poles) exist in nature, and the attempt to detect them is currently an active experimental field of investigation.

An unmagnetized piece of iron can be magnetized by stroking it with a magnet. Magnetism can also be induced in iron (and other materials) by other means. For example, if a piece of unmagnetized iron is placed near a strong permanent magnet, the piece of iron eventually becomes magnetized. The process can be accelerated by heating and then cooling the iron.

Naturally occurring magnetic materials such as magnetite are magnetized in this way because they have been subjected to Earth's magnetic field for long periods of time. The extent to which a piece of material retains its magnetism depends on whether it is classified as magnetically hard or soft. **Soft** magnetic materials, such as iron, are easily magnetized but tend to lose their magnetization easily. These materials are used in the cores of transformers, generators, and motors. Iron is the most common choice because it's inexpensive. Other magnetically soft materials include nickel, nickel-iron alloys, and ferrites. Ferrites are combinations of a divalent metal oxide of nickel or magnesium with ferric oxide. Ferrites are used in high-frequency applications, such as radar.

**Hard** magnetic materials are used in permanent magnets. Such magnets provide magnetic fields without the use of electricity. Permanent magnets are used in many devices, including loudspeakers, permanent-magnet motors, and the read/write heads of computer hard drives. There are a large number of different materials used in permanent magnets. Alnico is a generic name for various alloys of iron, cobalt, and nickel, together with smaller amounts of aluminum, copper, or other elements. Rare earths such as samarium and neodymium are also used in conjunction with other elements to make strong permanent magnets.

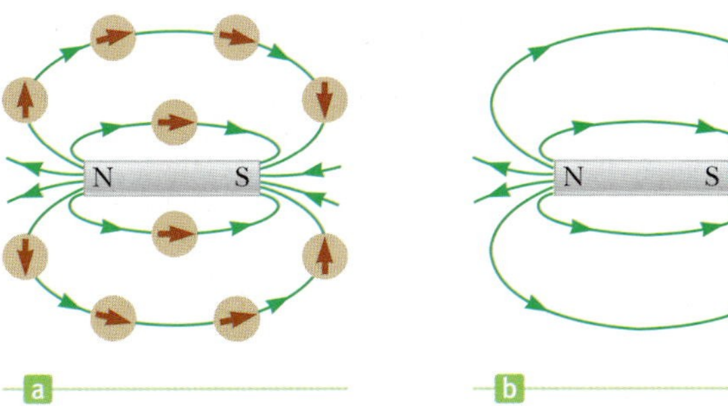


**Figure 19.2** (a) Tracing the magnetic field of a bar magnet with compasses. (b) Several magnetic field lines of a bar magnet.

In earlier chapters we described the interaction between charged objects in terms of electric fields. Recall that an electric field surrounds any stationary electric charge. The region of space surrounding a *moving* charge includes a magnetic field as well. A magnetic field also surrounds a properly magnetized magnetic material.

To describe any type of vector field, we must define its magnitude, or strength, and its direction. The direction of a magnetic field $\vec{\mathbf{B}}$ at any location is the direction in which the north pole of a compass needle points at that location. Figure 19.2a shows how the magnetic field of a bar magnet can be traced with the aid of a compass, defining a **magnetic field line**. Several magnetic field lines of a bar magnet traced out in this way appear in the two-dimensional representation in Figure 19.2b. Magnetic field patterns can be displayed by placing small iron filings in the vicinity of a magnet, as in Figure 19.3.

**APPLICATION**

Dusting for Fingerprints

Forensic scientists use a technique similar to that shown in Figure 19.3 to find fingerprints at a crime scene. One way to find latent, or invisible, prints is by sprinkling a powder of iron dust on a surface. The iron adheres to any perspiration or body oils that are present and can be spread around on the surface with a magnetic brush that never comes into contact with the powder or the surface.

## 19.2 Earth's Magnetic Field

**LEARNING OBJECTIVE**

1. Describe the Earth's magnetic field geographically, and discuss its possible origins and the ability of some life to sense and react to the field.

A small bar magnet is said to have north and south poles, but it's more accurate to say it has a "north-seeking" pole and a "south-seeking" pole. By these expressions, we mean that if such a magnet is used as a compass, one end will "seek," or point to, the geographic North Pole of Earth and the other end will "seek," or point to, the geographic South Pole of Earth. We conclude that **the geographic North Pole of Earth corresponds to a magnetic south pole, and the geographic**

**Tip 19.1 The Geographic North Pole Is the Magnetic South Pole**

The north pole of a magnet in a compass points north because it's attracted to Earth's *magnetic* south pole, located near Earth's *geographic* north pole.

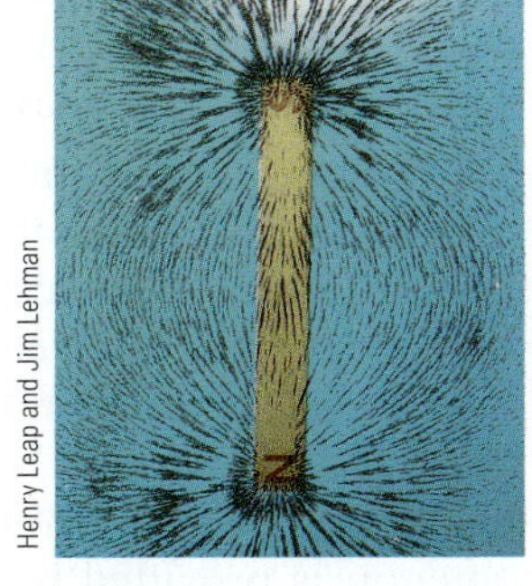

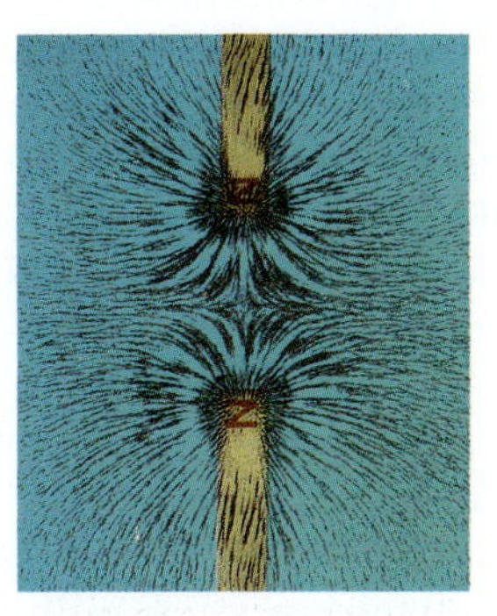


Henry Leap and Jim Lehman

**Figure 19.3** (a) The magnetic field pattern of a bar magnet, as displayed with iron filings on a sheet of paper. (b) The magnetic field pattern between *unlike* poles of two bar magnets, as displayed with iron filings. (c) The magnetic field pattern between two *like* poles.

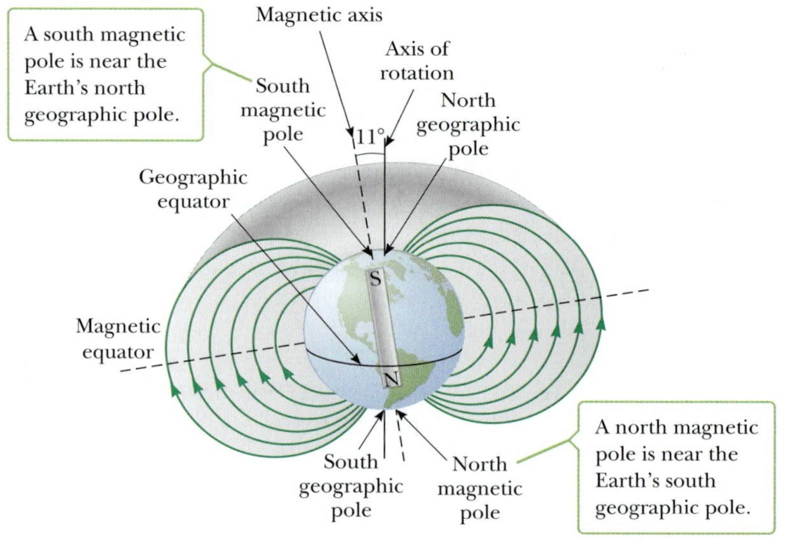


**Figure 19.4** Earth's magnetic field lines. The lines leading away from the immediate vicinity of the north magnetic pole and entering the vicinity of the south magnetic pole have been left out for clarity.

**South Pole of Earth corresponds to a magnetic north pole.** In fact, the configuration of Earth's magnetic field, pictured in Figure 19.4, very much resembles what would be observed if a huge bar magnet were buried deep in the Earth's interior.

If a compass needle is suspended in bearings that allow it to rotate in the vertical plane as well as in the horizontal plane, the needle is horizontal with respect to Earth's surface only near the equator. As the device is moved northward, the needle rotates so that it points more and more toward the surface of Earth. The angle between the direction of the magnetic field and the horizontal is called the **dip angle.** Finally, at a point just north of Hudson Bay in Canada, the north pole of the needle points directly downward, with a dip angle of 90°. That site, first found in 1832, is considered to be the location of the south magnetic pole of Earth. It is approximately 1 300 miles from Earth's geographic North Pole and varies with time. Similarly, Earth's magnetic north pole is about 1 200 miles from its geographic South Pole. This means that compass needles point only approximately north. The difference between true north, defined as the geographic North Pole, and north indicated by a compass varies from point to point on Earth, a difference referred to as *magnetic declination.* For example, along a line through South Carolina and the Great Lakes a compass indicates true north, whereas in Washington state it aligns 25° east of true north (Fig. 19.5).

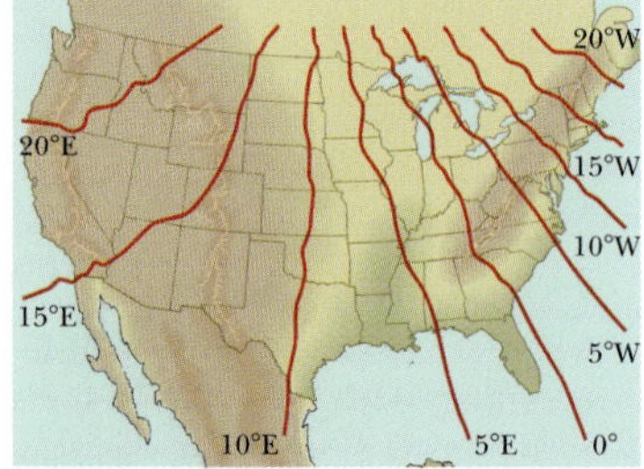


**Figure 19.5** A map of the continental United States showing the declination of a compass from true north.

Although the magnetic field pattern of Earth is similar to the pattern that would be set up by a bar magnet placed at its center, the source of Earth's field can't consist of large masses of permanently magnetized material. Earth does have large deposits of iron ore deep beneath its surface, but the high temperatures in the core prevent the iron from retaining any permanent magnetization. It's considered more likely that the true source of Earth's magnetic field is electric current in the liquid part of its core. This current, which is not well understood, may be driven by an interaction between the planet's rotation and convection in the hot liquid core. There is some evidence that the strength of a planet's magnetic field is related to the planet's rate of rotation. For example, Jupiter rotates faster than Earth, and recent space probes indicate that Jupiter's magnetic field is stronger than Earth's, even though Jupiter lacks an iron core. Venus, on the other hand, rotates more slowly than Earth, and its magnetic field is weaker. Investigation into the cause of Earth's magnetism continues.

An interesting fact concerning Earth's magnetic field is that its direction reverses every few million years. Evidence for this phenomenon is provided by basalt (an iron-containing rock) that is sometimes spewed forth by volcanic activity on the ocean floor. As the lava cools, it solidifies and retains a picture of the direction of Earth's magnetic field. When the basalt deposits are dated, they provide evidence for periodic reversals of the magnetic field. The cause of these field reversals is still not understood.

BIO APPLICATION
Magnetic Bacteria

It has long been speculated that some animals, such as birds, use the magnetic field of Earth to guide their migrations. Studies have shown that a type of anaerobic bacterium that lives in swamps has a magnetized chain of magnetite as part of its internal structure. (The term *anaerobic* means that these bacteria live and grow without oxygen; in fact, oxygen is toxic to them.) The magnetized chain acts as a compass needle that enables the bacteria to align with Earth's magnetic field. When they find themselves out of the mud on the bottom of the swamp, they return to their oxygen-free environment by following the magnetic field lines of Earth. Further evidence for their magnetic sensing ability is that bacteria found in the northern hemisphere have internal magnetite chains that are opposite in polarity to those of similar bacteria in the southern hemisphere. Similarly, in the northern hemisphere, Earth's field has a downward component, whereas in the southern hemisphere it has an upward component. Recently, a meteorite originating on Mars has been found to contain a chain of magnetite. NASA scientists believe it may be a fossil of ancient Martian bacterial life.

APPLICATION
Labeling Airport Runways

The magnetic field of Earth is used to label runways at airports according to their direction. A large number is painted on the end of the runway so that it can be read by the pilot of an incoming airplane. This number describes the direction in which the airplane is traveling, expressed as the magnetic heading, in degrees measured clockwise from magnetic north divided by 10. A runway marked 9 would be directed toward the east (90° divided by 10), whereas a runway marked 18 would be directed toward magnetic south.

### APPLYING PHYSICS 19.1 | Compasses Down Under

On a business trip to Australia, you take along your American-made compass that you may have used on a camping trip. Does this compass work correctly in Australia?

**EXPLANATION** There's no problem with using the compass in Australia. The north pole of the magnet in the compass will be attracted to the south magnetic pole near the geographic North Pole, just as it was in the United States. The only difference in the magnetic field lines is that they have an upward component in Australia, whereas they have a downward component in the United States. Held in a horizontal plane, your compass can't detect this difference; it only displays the direction of the horizontal component of the magnetic field. ■

## 19.3 Magnetic Fields

**LEARNING OBJECTIVES**

1. Define the magnetic force on a charged particle in terms of experimental observations.
2. Master the right hand rule that gives the direction of the magnetic force on a charged particle.
3. Apply the magnetic force in physical contexts.

Experiments show that a stationary charged particle doesn't interact with a static magnetic field. **When a charged particle is *moving* through a magnetic field, however, a magnetic force acts on it**. This force has its maximum value when the charge moves in a direction perpendicular to the magnetic field lines, decreases in value at other angles, and becomes zero when the particle moves along the field lines. This is quite different from the electric force, which exerts a force on a charged particle whether it's moving or at rest. Further, the electric force is directed parallel to the electric field whereas the magnetic force on a moving charge is directed perpendicular to the magnetic field.

In our discussion of electricity, the electric field at some point in space was defined as the electric force per unit charge acting on some test charge placed at that point. In a similar way, we can describe the properties of the magnetic field $\vec{\mathbf{B}}$ at some point in terms of the magnetic force exerted on a test charge at that point. Our test object is a charge $q$ moving with velocity $\vec{\mathbf{v}}$. It is found experimentally that the strength of the magnetic force on the particle is proportional to the magnitude of the charge $q$, the magnitude of the velocity $\vec{\mathbf{v}}$, the strength of the external magnetic field $\vec{\mathbf{B}}$, and the sine of the angle $\theta$ between the direction of $\vec{\mathbf{v}}$ and the direction of $\vec{\mathbf{B}}$. These observations can be summarized by writing the magnitude of the magnetic force as

$$F = qvB \sin \theta \qquad [19.1]$$

This expression is used to define the magnitude of the magnetic field as

$$B \equiv \frac{F}{qv \sin \theta} \qquad [19.2]$$

If $F$ is in newtons, $q$ in coulombs, and $v$ in meters per second, the SI unit of magnetic field is the **tesla** (T), also called the **weber** (Wb) **per square meter** (1 T = 1 Wb/m$^2$). If a 1-C charge moves in a direction perpendicular to a magnetic field of magnitude 1 T with a speed of 1 m/s, the magnetic force exerted on the charge is 1 N. We can express the units of $\vec{\mathbf{B}}$ as

$$[\mathrm{B}] = \mathrm{T} = \frac{\mathrm{Wb}}{\mathrm{m}^2} = \frac{\mathrm{N}}{\mathrm{C}\cdot\mathrm{m/s}} = \frac{\mathrm{N}}{\mathrm{A}\cdot\mathrm{m}} \qquad [19.3]$$

In practice, the cgs unit for magnetic field, the **gauss** (G), is often used. The gauss is related to the tesla through the conversion

$$1\ \mathrm{T} = 10^4\ \mathrm{G}$$

Conventional laboratory magnets can produce magnetic fields as large as about 25 000 G, or 2.5 T. Superconducting magnets that can generate magnetic fields as great as $3 \times 10^5$ G, or 30 T, have been constructed. These values can be compared with the small value of Earth's magnetic field near its surface, which is only about 0.5 G, or $0.5 \times 10^{-4}$ T.

From Equation 19.1 we see that the force on a charged particle moving in a magnetic field has its maximum value when the particle's motion is *perpendicular* to the magnetic field, corresponding to $\theta = 90°$, so that $\sin \theta = 1$. The magnitude of this maximum force has the value

$$F_{\mathrm{max}} = qvB \qquad [19.4]$$

Also from Equation 19.1, $F$ is zero when $\vec{\mathbf{v}}$ is parallel to $\vec{\mathbf{B}}$ (corresponding to $\theta = 0°$ or $180°$), so no magnetic force is exerted on a charged particle when it moves in the direction of the magnetic field or opposite the field.

Experiments show that the direction of the magnetic force is always perpendicular to both $\vec{\mathbf{v}}$ and $\vec{\mathbf{B}}$, as shown in Figure 19.6 for a positively charged particle. To determine the direction of the force, we employ **right-hand rule number 1**:

1. Point the fingers of your right hand in the direction of the velocity $\vec{\mathbf{v}}$.
2. Curl the fingers in the direction of the magnetic field $\vec{\mathbf{B}}$, moving through the smallest angle (as in Fig. 19.7).
3. Your thumb is now pointing in the direction of the magnetic force $\vec{\mathbf{F}}$ exerted on a positive charge.

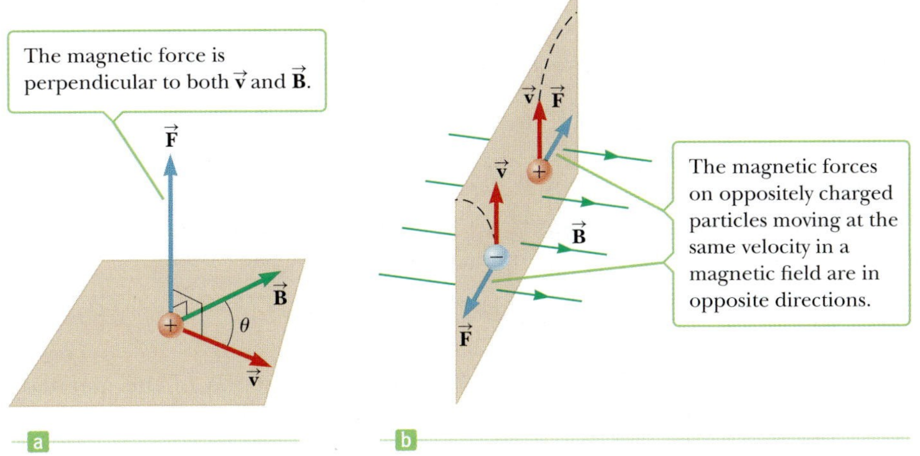


**Figure 19.6** (a) The direction of the magnetic force $\vec{F}$ acting on a charged particle moving with a velocity $\vec{v}$ in the presence of a magnetic field $\vec{B}$. (b) Magnetic forces on positive and negative charges. The dashed lines show the paths of the particles, which are investigated in Section 19.6.

If the charge is negative rather than positive, the force $\vec{F}$ is directed *opposite* to what's shown in Figures 19.6a and 19.7. So if $q$ is negative, simply use the right-hand rule to find the direction for positive $q$ and then reverse that direction for the negative charge. Figure 19.6b illustrates the effect of a magnetic field on charged particles with opposite signs.

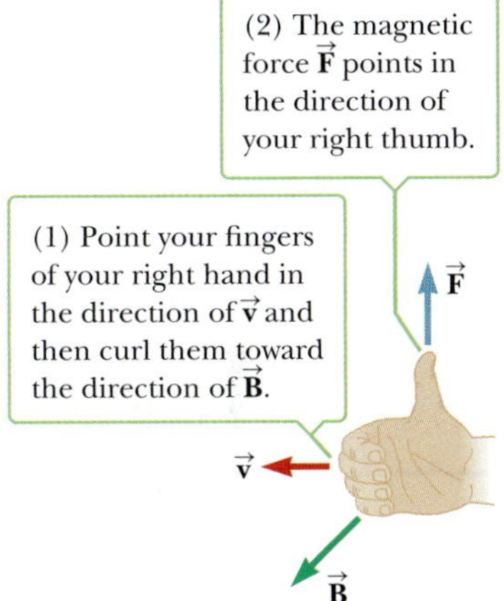

(2) The magnetic force $\vec{F}$ points in the direction of your right thumb.
(1) Point your fingers of your right hand in the direction of $\vec{v}$ and then curl them toward the direction of $\vec{B}$.
$\vec{F}$
$\vec{v}$
$\vec{B}$


**Figure 19.7** Right-hand rule number 1 for determining the direction of the magnetic force on a positive charge moving with a velocity $\vec{v}$ in a magnetic field $\vec{B}$.

### Quick Quiz

**19.1** A charged particle moves in a straight line through a region of space. Which of the following answers *must* be true? (Assume any other fields are negligible.) The magnetic field (a) has a magnitude of zero (b) has a zero component perpendicular to the particle's velocity (c) has a zero component parallel to the particle's velocity in that region.

**19.2** The north-pole end of a bar magnet is held near a stationary positively charged piece of plastic. Is the plastic (a) attracted, (b) repelled, or (c) unaffected by the magnet?

## EXAMPLE 19.1 A Proton Traveling in Earth's Magnetic Field

**GOAL** Calculate the magnitude and direction of a magnetic force.

**PROBLEM** A proton moves with a speed of $1.00 \times 10^5$ m/s through Earth's magnetic field, which has a value of 55.0 $\mu$T at a particular location. When the proton moves eastward, the magnetic force acting on it is directed straight upward, and when it moves northward, no magnetic force acts on it. **(a)** What is the direction of the magnetic field, and **(b)** what is the strength of the magnetic force when the proton moves eastward? **(c)** Calculate the gravitational force on the proton and compare it with the magnetic force. Compare it also with the electric force if there were an electric field with a magnitude equal to $E = 1.50 \times 10^2$ N/C at that location, a common value at Earth's surface. Note that the mass of the proton is $1.67 \times 10^{-27}$ kg.

**STRATEGY** The direction of the magnetic field can be found from an application of the right-hand rule, together with the fact that no force is exerted on the proton when it's traveling north. Substituting into Equation 19.1 yields the magnitude of the magnetic field.

**SOLUTION**

**(a)** Find the direction of the magnetic field.

No magnetic force acts on the proton when it's going north, so the angle such a proton makes with the magnetic field direction must be either 0° or 180°. Therefore, the magnetic field $\vec{B}$ must point either north or south. Now apply the right-hand rule. When the particle travels east, the magnetic force is directed upward. Point your thumb in the direction of the force and your fingers in the direction of the velocity eastward. When you curl your fingers, they point north, which must therefore be the direction of the magnetic field.

*(Continued)*

**(b)** Find the magnitude of the magnetic force.

Substitute the given values and the charge of a proton into Equation 19.1. From part (a), the angle between the velocity $\vec{\mathbf{v}}$ of the proton and the magnetic field $\vec{\mathbf{B}}$ is 90.0°.

$$F = qvB\sin\theta$$
$$= (1.60 \times 10^{-19}\text{ C})(1.00 \times 10^{5}\text{ m/s}) \times (55.0 \times 10^{-6}\text{ T})\sin(90.0°)$$
$$= 8.80 \times 10^{-19}\text{ N}$$

**(c)** Calculate the gravitational force on the proton and compare it with the magnetic force and also with the electric force if $E = 1.50 \times 10^{2}$ N/C.

$$F_{\text{grav}} = mg = (1.67 \times 10^{-27}\text{ kg})(9.80\text{ m/s}^2)$$
$$= 1.64 \times 10^{-26}\text{ N}$$
$$F_{\text{elec}} = qE = (1.60 \times 10^{-19}\text{ C})(1.50 \times 10^{2}\text{ N/C})$$
$$= 2.40 \times 10^{-17}\text{ N}$$

**REMARKS** The information regarding a proton moving north was necessary to fix the direction of the magnetic field. Otherwise, an upward magnetic force on an eastward-moving proton could be caused by a magnetic field pointing anywhere northeast or northwest. Notice in part (c) the relative strengths of the forces, with the electric force larger than the magnetic force and both much larger than the gravitational force, all for typical field values found in nature.

**QUESTION 19.1** An electron and proton moving with the same velocity enter a uniform magnetic field. In what two ways does the magnetic field affect the electron differently from the proton?

**EXERCISE 19.1** Suppose an electron is moving due west in the same magnetic field as in Example 19.1 at a speed of $2.50 \times 10^{5}$ m/s. Find the magnitude and direction of the magnetic force on the electron.

**ANSWER** $2.20 \times 10^{-18}$ N, straight up. (Don't forget, the electron is negatively charged!)

## ■ EXAMPLE 19.2 A Proton Moving in a Magnetic Field

**GOAL** Calculate the magnetic force and acceleration when a particle moves at an angle other than 90° to the field.

**PROBLEM** A proton moves at $8.00 \times 10^{6}$ m/s along the $x$-axis. It enters a region in which there is a magnetic field of magnitude 2.50 T, directed at an angle of 60.0° with the $x$-axis and lying in the $xy$-plane (Fig. 19.8). **(a)** Find the initial magnitude and direction of the magnetic force on the proton. **(b)** Calculate the proton's initial acceleration.

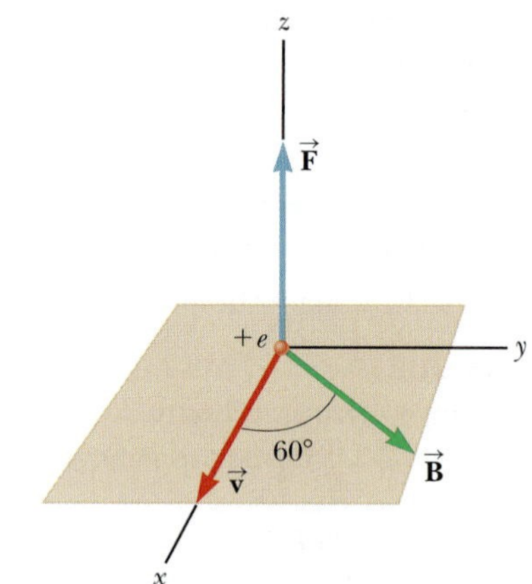


**Figure 19.8** (Example 19.2) The magnetic force $\vec{\mathbf{F}}$ on a proton is in the positive $z$-direction when $\vec{\mathbf{v}}$ and $\vec{\mathbf{B}}$ lie in the $xy$-plane.

**STRATEGY** Finding the magnitude and direction of the magnetic force requires substituting values into the equation for magnetic force, Equation 19.1, and using the right-hand rule. Applying Newton's second law solves part (b).

**SOLUTION**

**(a)** Find the magnitude and direction of the magnetic force on the proton.

Substitute $v = 8.00 \times 10^{6}$ m/s, the magnetic field strength $B = 2.50$ T, the angle, and the charge of a proton into Equation 19.1:

$$F = qvB\sin\theta$$
$$= (1.60 \times 10^{-19}\text{ C})(8.00 \times 10^{6}\text{ m/s})(2.50\text{ T})(\sin 60°)$$
$$F = 2.77 \times 10^{-12}\text{ N}$$

Apply right-hand rule number 1 to find the initial direction of the magnetic force:

Point the fingers of the right hand in the $x$-direction (the direction of $\vec{\mathbf{v}}$) and then curl them toward $\vec{\mathbf{B}}$. The thumb points upward, in the positive $z$-direction.

**(b)** Calculate the proton's initial acceleration.

Substitute the force and the mass of a proton into Newton's second law:

$$ma = F \rightarrow (1.67 \times 10^{-27}\text{ kg})a = 2.77 \times 10^{-12}\text{ N}$$

$$a = 1.66 \times 10^{15}\text{ m/s}^2$$

**REMARKS** The initial acceleration is also in the positive $z$-direction. Because the direction of $\vec{\mathbf{v}}$ changes, however, the subsequent direction of the magnetic force also changes. In applying right-hand rule number 1 to find the direction, it was important to take into consideration the charge. A negatively charged particle accelerates in the opposite direction.

**QUESTION 19.2** Can a constant magnetic field change the speed of a charged particle? Explain.

**EXERCISE 19.2** Calculate the acceleration of an electron that moves through the same magnetic field as in Example 19.2, at the same velocity as the proton. The mass of an electron is $9.11 \times 10^{-31}$ kg.

**ANSWER** $3.04 \times 10^{18}$ m/s² in the negative $z$-direction

## 19.4 Magnetic Force on a Current-Carrying Conductor

**LEARNING OBJECTIVES**

1. Derive the magnetic force exerted by a magnetic field on a long, current-carrying wire.
2. Calculate magnetic forces on current-carrying wires.

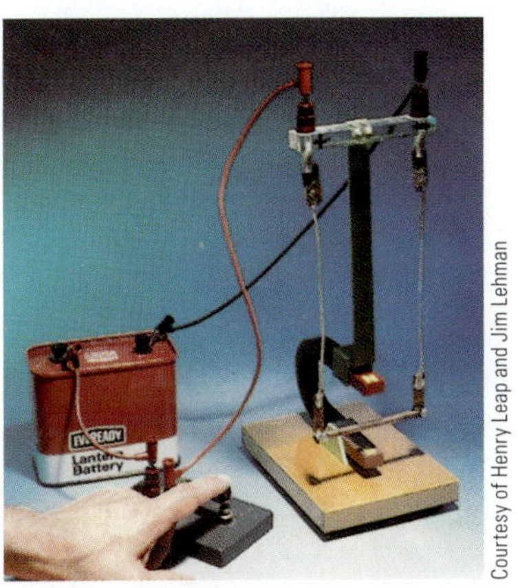


**Figure 19.9** This apparatus demonstrates the force on a current-carrying conductor in an external magnetic field. Why does the bar swing *toward* the magnet after the switch is closed?

If a magnetic field exerts a force on a single charged particle when it moves through a magnetic field, it should be no surprise that magnetic forces are exerted on a current-carrying wire as well (see Fig. 19.9). Because the current is a collection of many charged particles in motion, the resultant force on the wire is due to the sum of the individual forces on the charged particles. The force on the particles is transmitted to the "bulk" of the wire through collisions with the atoms making up the wire.

Some explanation is in order concerning notation in many of the figures. To indicate the direction of $\vec{\mathbf{B}}$, we use the following conventions:

> If $\vec{\mathbf{B}}$ is directed into the page, as in Figure 19.10, we use a series of green crosses, representing the tails of arrows. If $\vec{\mathbf{B}}$ is directed out of the page, we use a series of green dots, representing the tips of arrows. If $\vec{\mathbf{B}}$ lies in the plane of the page, we use a series of green field lines with arrowheads.

The force on a current-carrying conductor can be demonstrated by hanging a wire between the poles of a magnet, as in Figure 19.10 (page 668). In this figure, the magnetic field is directed into the page and covers the region within the shaded area. The wire deflects to the right or left when it carries a current.

We can quantify this discussion by considering a straight segment of wire of length $\ell$ and cross-sectional area $A$ carrying current $I$ in a uniform external magnetic field $\vec{\mathbf{B}}$, as in Figure 19.11 (page 668). We assume that the magnetic field is perpendicular to the wire and is directed into the page. A force of magnitude $F_{\max} = qv_dB$ is exerted on each charge carrier in the wire, where $v_d$ is the drift velocity of the charge. To find the total force on the wire, we multiply the force on one charge carrier by the number of carriers in the segment. Because the volume of the segment is $A\ell$, the number of carriers is $nA\ell$, where $n$ is the number of carriers per unit volume. Hence, the magnitude of the total magnetic force on the wire of length $\ell$ is as follows:

Total force = force on each charge carrier × total number of carriers

$$F_{\max} = (qv_dB)(nA\ell)$$

**Tip 19.2 The Origin of the Magnetic Force on a Wire**

When a magnetic field is applied at some angle to a wire carrying a current, a magnetic force is exerted on each moving charge in the wire. The total magnetic force on the wire is the sum of all the magnetic forces on the individual charges producing the current.

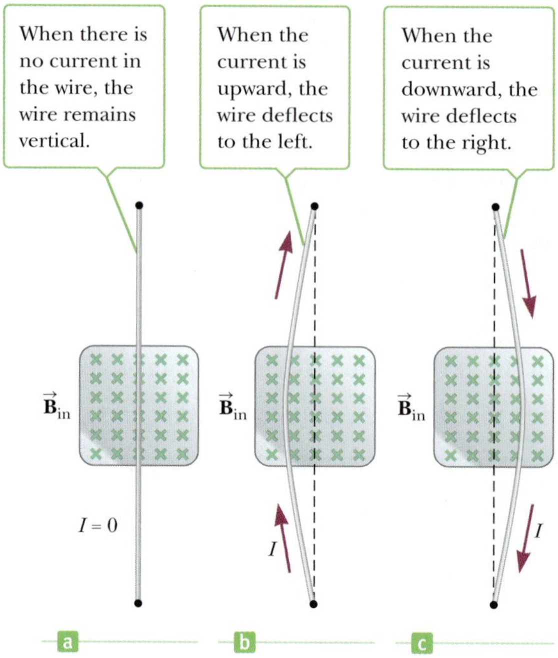

When there is no current in the wire, the wire remains vertical.

When the current is upward, the wire deflects to the left.

When the current is downward, the wire deflects to the right.

$\vec{\mathbf{B}}_{in}$ $I = 0$ $I$ a b c

**Figure 19.10** A segment of a flexible vertical wire partially stretched between the poles of a magnet, with the field (green crosses) directed into the page.

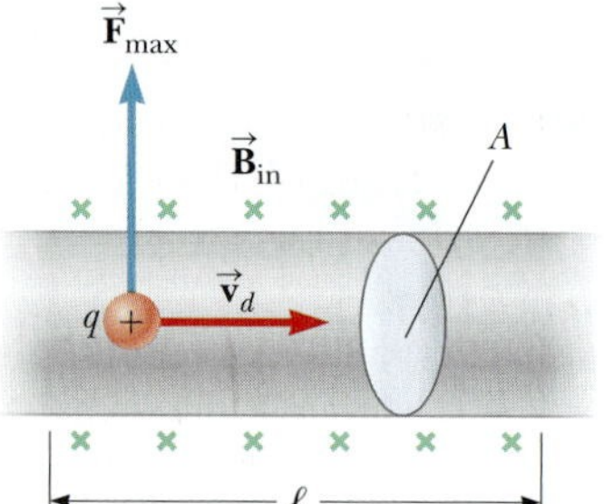


**Figure 19.11** A section of a wire containing moving charges in an external magnetic field $\vec{\mathbf{B}}$.

From Chapter 17, however, we know that the current in the wire is given by the expression $I = nqv_dA$, so

$$F_{max} = BI\ell \qquad [19.5]$$

**This equation can be used only when the current and the magnetic field are at right angles to each other.**

If the wire is not perpendicular to the field but is at some arbitrary angle, as in Figure 19.12, the magnitude of the magnetic force on the wire is

$$F = BI\ell \sin \theta \qquad [19.6]$$

where $\theta$ is the angle between $\vec{\mathbf{B}}$ and the direction of the current. The direction of this force can be obtained by the use of right-hand rule number 1. In this case, however, you must place your fingers in the direction of the positive current $I$, rather than in the direction of $\vec{\mathbf{v}}$, before curling them in the direction of $\vec{\mathbf{B}}$. The thumb then points in the direction of the force, as before. The current, naturally, is made up of charges moving at some velocity, so this really isn't a separate rule. In Figure 19.12 the direction of the magnetic force on the wire is out of the page.

Finally, when the current is either in the direction of the field or opposite the direction of the field, the magnetic force on the wire is zero.

**APPLICATION**

Loudspeaker Operation

The way a magnetic force acts on a current-carrying wire in a magnetic field is the operating principle of most speakers in sound systems. One speaker design, shown in Figure 19.13, consists of a coil of wire called the voice coil, a flexible paper cone, and a permanent magnet. The coil of wire surrounding the north

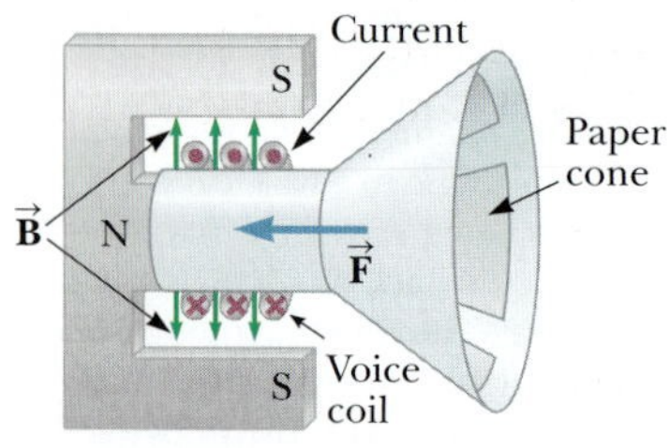


**Figure 19.13** A diagram of a loudspeaker.

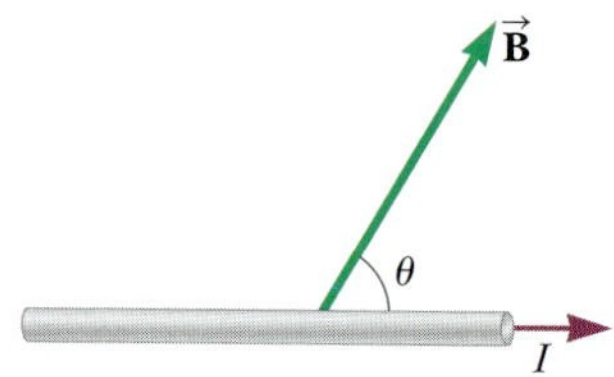


**Figure 19.12** A wire carrying a current $I$ in the presence of an external magnetic field $\vec{\mathbf{B}}$ that makes an angle $\theta$ with the wire. The magnetic force vector comes out of the page.

pole of the magnet is shaped so that the magnetic field lines are directed radially outward from the coil's axis. When an electrical signal is sent to the coil, producing a current in the coil as in Figure 19.13, a magnetic force to the left acts on the coil. (This can be seen by applying right-hand rule number 1 to each turn of wire.) When the current reverses direction, as it would for a current that varied sinusoidally, the magnetic force on the coil also reverses direction, and the cone, which is attached to the coil, accelerates to the right. An alternating current through the coil causes an alternating force on the coil, which results in vibrations of the cone. The vibrating cone creates sound waves as it pushes and pulls on the air in front of it. In this way, a 1-kHz electrical signal is converted to a 1-kHz sound wave.

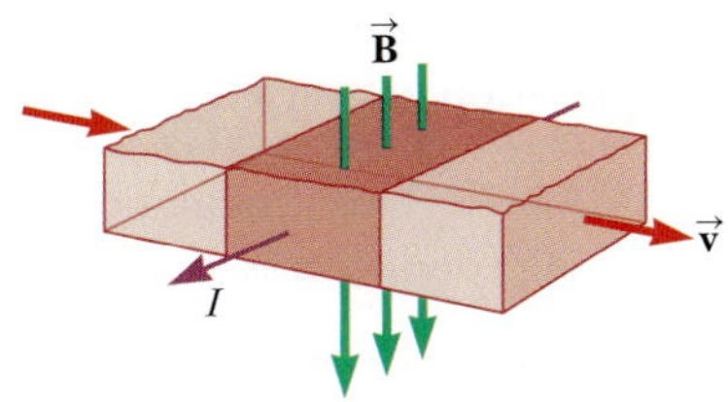


**Figure 19.14** A simple electromagnetic pump has no moving parts that might damage a conducting fluid, such as blood passing through it. Application of right-hand rule number 1 (right fingers in the direction of the current $I$, curl them in the direction of $\vec{\mathbf{B}}$, thumb points in the direction of the force) shows that the force on the current-carrying segment of the fluid is in the direction of the velocity.

An application of the force on a current-carrying conductor is illustrated by the electromagnetic pump shown in Figure 19.14. Artificial hearts require a pump to keep the blood flowing, and kidney dialysis machines also require a pump to assist the heart in pumping blood that is to be cleansed. Ordinary mechanical pumps create problems because they damage the blood cells as they move through the pump. The mechanism shown in the figure has demonstrated some promise in such applications. A magnetic field is established across a segment of the tube containing the blood, flowing in the direction of the velocity $\vec{\mathbf{v}}$. An electric current passing through the fluid in the direction shown has a magnetic force acting on it in the direction of $\vec{\mathbf{v}}$, as applying the right-hand rule shows. This force helps to keep the blood in motion.

**BIO APPLICATION**

Electromagnetic Pumps for Artificial Hearts and Kidneys

## APPLYING PHYSICS 19.2 Lightning Strikes

In a lightning strike there is a rapid movement of negative charge from a cloud to the ground. In what direction is a lightning strike deflected by Earth's magnetic field?

**EXPLANATION** The downward flow of negative charge in a lightning strike is equivalent to a current moving upward. Consequently, we have an upward-moving current in a northward-directed magnetic field. According to right-hand rule number 1, the lightning strike would be deflected toward the west. ■

## EXAMPLE 19.3 A Current-Carrying Wire in Earth's Magnetic Field

**GOAL** Compare the magnetic force on a current-carrying wire with the gravitational force exerted on the wire.

**PROBLEM** A wire carries a current of 22.0 A from west to east. Assume the magnetic field of Earth at this location is horizontal and directed from south to north and it has a magnitude of $0.500 \times 10^{-4}$ T. **(a)** Find the magnitude and direction of the magnetic force on a 36.0-m length of wire. **(b)** Calculate the gravitational force on the same length of wire if it's made of copper and has a cross-sectional area of $2.50 \times 10^{-6}$ m$^2$.

**SOLUTION**

**(a)** Calculate the magnetic force on the wire.

Substitute into Equation 19.6, using the fact that the magnetic field and the current are at right angles to each other:

$$F = BI\ell \sin\theta = (0.500 \times 10^{-4}\ \text{T})(22.0\ \text{A})(36.0\ \text{m}) \sin 90.0°$$
$$= 3.96 \times 10^{-2}\ \text{N}$$

Apply right-hand rule number 1 to find the direction of the magnetic force:

With the fingers of your right hand pointing west to east in the direction of the current, curl them north in the direction of the magnetic field. Your thumb points upward.

**(b)** Calculate the gravitational force on the wire segment.

First, obtain the mass of the wire from the density of copper, the length, and cross-sectional area of the wire:

$$m = \rho V = \rho(A\ell)$$
$$= (8.92 \times 10^3\ \text{kg/m}^3)(2.50 \times 10^{-6}\ \text{m}^2 \cdot 36.0\ \text{m})$$
$$= 0.803\ \text{kg}$$

*(Continued)*

To get the gravitational force, multiply the mass by the acceleration of gravity:

$$F_{grav} = mg = \boxed{7.87 \text{ N}}$$

**REMARKS** This calculation demonstrates that under normal circumstances, the gravitational force on a current-carrying conductor is much greater than the magnetic force due to Earth's magnetic field.

**QUESTION 19.3** What magnetic force is exerted on a wire carrying current parallel to the direction of the magnetic field?

**EXERCISE 19.3** What current would make the magnetic force in the example equal in magnitude to the gravitational force?

**ANSWER** $4.37 \times 10^3$ A, a large current that would very rapidly heat and melt the wire.

# 19.5 Torque on a Current Loop and Electric Motors

**LEARNING OBJECTIVES**

1. Derive the magnetic torque on a rectangular loop of current-carrying wire.
2. Define the magnetic moment of a current loop in a plane.
3. Calculate the magnetic torque on simple current loops.
4. Discuss the application of magnetic torque to electric motors.

In the preceding section we showed how a magnetic force is exerted on a current-carrying conductor when the conductor is placed in an external magnetic field. With this starting point, we now show that a torque is exerted on a current loop placed in a magnetic field. The results of this analysis will be of great practical value when we discuss generators and motors in Chapter 20.

Consider a rectangular loop carrying current $I$ in the presence of an external uniform magnetic field in the plane of the loop, as shown in Figure 19.15a.

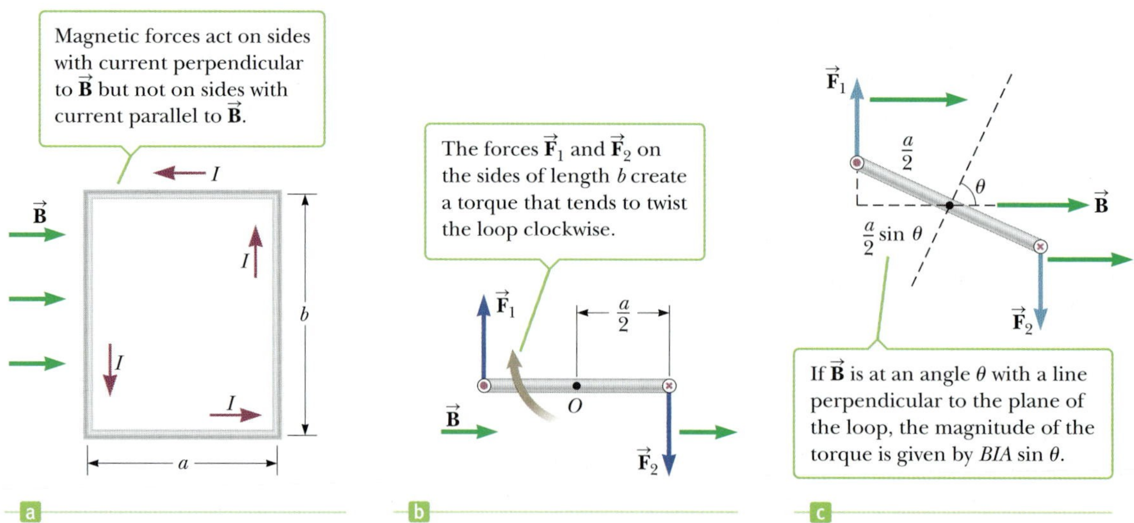


**Figure 19.15** (a) Top view of a rectangular loop in a uniform magnetic field. (b) An edge view of the rectangular loop in part (a). (c) An edge view of the loop in part (a) with the normal to the loop at angle $\theta$ with respect to the magnetic field.

The forces on the sides of length $a$ are zero because these wires are parallel to the field. The magnitudes of the magnetic forces on the sides of length $b$, however, are

$$F_1 = F_2 = BIb$$

The direction of $\vec{\mathbf{F}}_1$, the force on the left side of the loop, is out of the page and that of $\vec{\mathbf{F}}_2$, the force on the right side of the loop, is into the page. If we view the loop from the side, as in Figure 19.15b, the forces are directed as shown. If we assume the loop is pivoted so that it can rotate about point $O$, we see that these two forces produce a torque about $O$ that rotates the loop clockwise. The magnitude of this torque, $\tau_{\max}$, is

$$\tau_{\max} = F_1 \frac{a}{2} + F_2 \frac{a}{2} = (BIb)\frac{a}{2} + (BIb)\frac{a}{2} = BIab$$

where the moment arm about $O$ is $a/2$ for both forces. Because the area of the loop is $A = ab$, the maximum torque can be expressed as

$$\tau_{\max} = BIA \qquad [19.7]$$

This result is valid only when the magnetic field is *parallel* to the plane of the loop, as in Figure 19.15b. If the field makes an angle $\theta$ with a line perpendicular to the plane of the loop, as in Figure 19.15c, the moment arm for each force is given by $(a/2)\sin\theta$. An analysis similar to the previous one gives, for the magnitude of the torque,

$$\tau = BIA\sin\theta \qquad [19.8]$$

This result shows that the torque has the *maximum* value $BIA$ when the field is parallel to the plane of the loop ($\theta = 90°$) and is *zero* when the field is perpendicular to the plane of the loop ($\theta = 0$). As seen in Figure 19.15c, the loop tends to rotate to smaller values of $\theta$ (so that the normal to the plane of the loop rotates toward the direction of the magnetic field).

Although the foregoing analysis was for a rectangular loop, a more general derivation shows that Equation 19.8 applies regardless of the shape of the loop. Further, the torque on a coil with $N$ turns is

$$\tau = BIAN\sin\theta \qquad [19.9a]$$

The quantity $\mu = IAN$ is defined as the magnitude of a vector $\vec{\boldsymbol{\mu}}$ called the *magnetic moment* of the coil. The vector $\vec{\boldsymbol{\mu}}$ always points perpendicular to the plane of the loop(s) and is such that if the thumb of the right hand points in the direction of $\vec{\boldsymbol{\mu}}$, the fingers of the right hand point in the direction of the current. The angle $\theta$ in Equations 19.8 and 19.9 lies between the directions of the magnetic moment $\vec{\boldsymbol{\mu}}$ and the magnetic field $\vec{\mathbf{B}}$. The magnetic torque can then be written

$$\tau = \mu B\sin\theta \qquad [19.9b]$$

Note that the torque $\vec{\boldsymbol{\tau}}$ is always perpendicular to both the magnetic moment $\vec{\boldsymbol{\mu}}$ and the magnetic field $\vec{\mathbf{B}}$.

**Quick Quiz**

**19.3** A square and a circular loop with the same area lie in the $xy$-plane, where there is a uniform magnetic field $\vec{\mathbf{B}}$ pointing at some angle $\theta$ with respect to the positive $z$-direction. Each loop carries the same current, in the same direction. Which magnetic torque is larger? (a) the torque on the square loop (b) the torque on the circular loop (c) the torques are the same (d) more information is needed

## ■ EXAMPLE 19.4 The Torque on a Circular Loop in a Magetic Field

**GOAL** Calculate a magnetic torque on a loop of current.

**PROBLEM** A circular wire loop of radius 1.00 m is placed in a magnetic field of magnitude 0.500 T. The normal to the plane of the loop makes an angle of 30.0° with the magnetic field (Fig. 19.16a). The current in the loop is 2.00 A in the direction shown. **(a)** Find the magnetic moment of the loop and the magnitude of the torque at this instant. **(b)** The same current is carried by the rectangular 2.00-m by 3.00-m coil with three loops shown in Figure 19.16b. Find the magnetic moment of the coil and the magnitude of the torque acting on the coil at that instant.

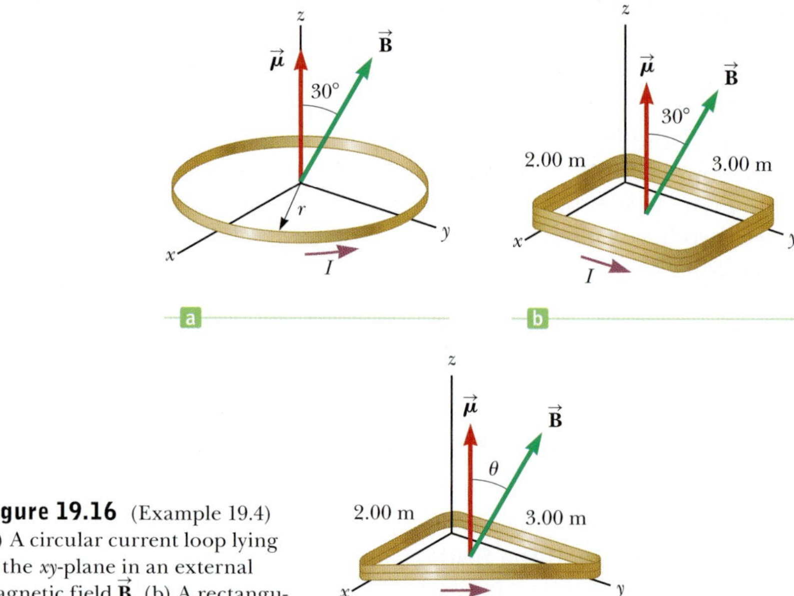


**Figure 19.16** (Example 19.4) (a) A circular current loop lying in the *xy*-plane in an external magnetic field $\vec{\mathbf{B}}$. (b) A rectangular coil lying in the *xy*-plane in the same field. (c) (Exercise 19.4)

**STRATEGY** For each part, we just have to calculate the area, use it in the calculation of the magnetic moment, and multiply the result by $B \sin \theta$. Altogether, this process amounts to substituting values into Equation 19.9b.

**SOLUTION**

**(a)** Find the magnetic moment of the circular loop and the magnetic torque exerted on it.

First, calculate the enclosed area of the circular loop:

$$A = \pi r^2 = \pi(1.00 \text{ m})^2 = 3.14 \text{ m}^2$$

Calculate the magnetic moment of the loop:

$$\mu = IAN = (2.00 \text{ A})(3.14 \text{ m}^2)(1) = \boxed{6.28 \text{ A} \cdot \text{m}^2}$$

Now substitute values for the magnetic moment, magnetic field, and $\theta$ into Equation 19.9b:

$$\tau = \mu B \sin \theta = (6.28 \text{ A} \cdot \text{m}^2)(0.500 \text{ T})(\sin 30.0°)$$
$$= \boxed{1.57 \text{ N} \cdot \text{m}}$$

**(b)** Find the magnetic moment of the rectangular coil and the magnetic torque exerted on it.

Calculate the area of the coil:

$$A = L \times H = (2.00 \text{ m})(3.00 \text{ m}) = 6.00 \text{ m}^2$$

Calculate the magnetic moment of the coil:

$$\mu = IAN = (2.00 \text{ A})(6.00 \text{ m}^2)(3) = \boxed{36.0 \text{ A} \cdot \text{m}^{2)}}$$

Substitute values into Equation 19.9b:

$$\tau = \mu B \sin \theta = (0.500 \text{ T})(36.0 \text{ A} \cdot \text{m}^2)(\sin 30.0°)$$
$$= \boxed{9.00 \text{ N} \cdot \text{m}}$$

**REMARKS** In calculating a magnetic torque, it's not strictly necessary to calculate the magnetic moment. Instead, Equation 19.9a can be used directly.

**QUESTION 19.4** What happens to the magnitude of the torque if the angle increases toward 90°? Goes beyond 90°?

**EXERCISE 19.4** Suppose a right triangular coil with base of 2.00 m and height 3.00 m having two loops carries a current of 2.00 A as shown in Figure 19.16c. Find the magnetic moment and the torque on the coil. The magnetic field is again 0.500 T and makes an angle of 30.0° with respect to the normal direction.

**ANSWERS** $\mu = 12.0 \text{ A} \cdot \text{m}^2$, $\tau = 3.00 \text{ N} \cdot \text{m}$

## Electric Motors

It's hard to imagine life in the 21st century without electric motors. Some appliances that contain motors include computer disk drives, CD players, DVD players, food processors and blenders, car starters, furnaces, and air conditioners. The motors convert electrical energy to kinetic energy of rotation and consist of a rigid current-carrying loop that rotates when placed in a magnetic field.

**APPLICATION**
Electric Motors

As we have just seen (Fig. 19.15), the torque on such a loop rotates the loop to smaller values of $\theta$ until the torque becomes zero, where the magnetic field is perpendicular to the plane of the loop and $\theta = 0$. If the loop turns past this angle and the current remains in the direction shown in the figure, the torque reverses direction and turns the loop in the opposite direction, that is, counterclockwise. To overcome this difficulty and provide continuous rotation in one direction, the current in the loop must periodically reverse direction. In alternating-current (AC) motors, such a reversal occurs naturally 120 times each second. In direct-current (DC) motors, the reversal is accomplished mechanically with split-ring contacts (commutators) and brushes, as shown in Figure 19.17.

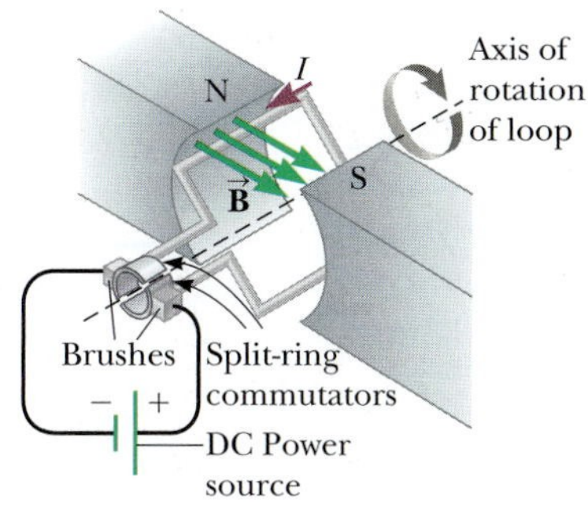


**Figure 19.17** Simplified sketch of a DC electric motor.

Although actual motors contain many current loops and commutators, for simplicity Figure 19.17 shows only a single loop and a single set of split-ring contacts rigidly attached to and rotating with the loop. Electrical stationary contacts called *brushes* are maintained in electrical contact with the rotating split ring. These brushes are usually made of graphite because it is a good electrical conductor as well as a good lubricant. Just as the loop becomes perpendicular to the magnetic field and the torque becomes zero, inertia carries the loop forward in the clockwise direction and the brushes cross the gaps in the ring, causing the loop current to reverse its direction. This reversal provides another pulse of torque in the clockwise direction for another 180°, the current reverses, and the process repeats itself. Figure 19.18 shows a modern motor used to power a hybrid gas–electric car.

# 19.6 Motion of a Charged Particle in a Magnetic Field

**LEARNING OBJECTIVES**

1. Describe the motion of a charged particle in a uniform magnetic field.
2. Apply the second law in polar coordinates to particles moving in a uniform magnetic field.

Consider the case of a positively charged particle moving in a uniform magnetic field so that the direction of the particle's velocity is *perpendicular to the field*, as in Figure 19.19 (page 674). The label $\vec{\mathbf{B}}_{\text{in}}$ and the crosses indicate that $\vec{\mathbf{B}}$ is directed into the page. Application of the right-hand rule to the particle at the bottom of the circle shows that the direction of the magnetic force $\vec{\mathbf{F}}$ at that location is upward. The force causes the particle to alter its direction of travel and to follow a curved path. Application of the right-hand rule to the particle at other points on the circle shows that **the magnetic force is always directed toward the center of the circular path**; therefore, the magnetic force causes a centripetal acceleration, which changes only the direction of $\vec{\mathbf{v}}$ and not its magnitude. Because $\vec{\mathbf{F}}$ produces the centripetal acceleration, we can equate its magnitude, $qvB$ in this case, to the mass of the particle multiplied by the centripetal acceleration $v^2/r$. From Newton's second law, we find that

$$F = qvB = \frac{mv^2}{r}$$

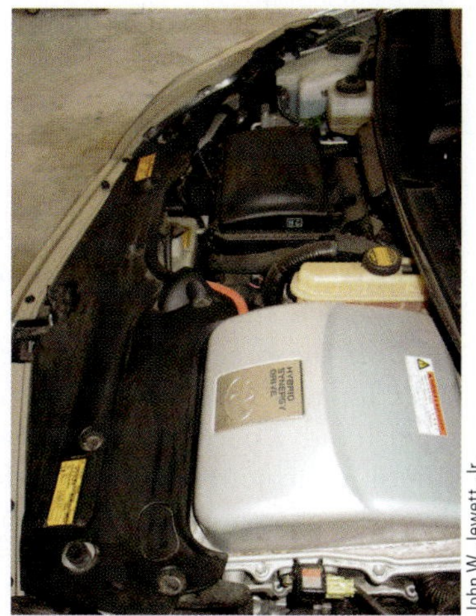


**Figure 19.18** The engine compartment of the Toyota Prius, a hybrid vehicle.

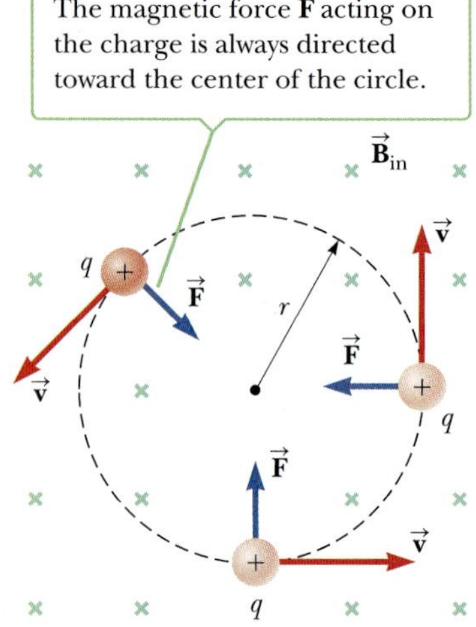


**Figure 19.19** When the velocity of a charged particle is perpendicular to a uniform magnetic field, the particle moves in a circle in a plane perpendicular to $\vec{B}$, which is directed into the page. (The crosses represent the tails of the magnetic field vectors.)

**Figure 19.20** A charged particle having a velocity directed at an angle with a uniform magnetic field moves in a helical path.

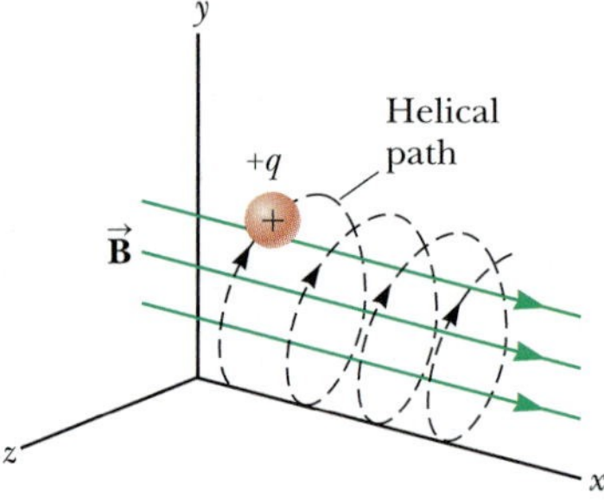


which gives

$$r = \frac{mv}{qB} \quad [19.10]$$

This equation says that the radius of the path is proportional to the momentum $mv$ of the particle and is inversely proportional to the charge and the magnetic field. Equation 19.10 is often called the *cyclotron equation* because it's used in the design of these instruments (popularly known as atom smashers).

If the initial direction of the velocity of the charged particle is not perpendicular to the magnetic field, as shown in Figure 19.20, the path followed by the particle is a spiral (called a helix) along the magnetic field lines.

## APPLYING PHYSICS 19.3 Trapping Charges

Storing charged particles is important for a variety of applications. Suppose a uniform magnetic field exists in a finite region of space. Can a charged particle be injected into this region from the outside and remain trapped in the region by magnetic force alone?

**EXPLANATION** It's best to consider separately the components of the particle velocity parallel and perpendicular to the field lines in the region. There is no magnetic force on the particle associated with the velocity component parallel to the field lines, so that velocity component remains unchanged.

Now consider the component of velocity perpendicular to the field lines. This component will result in a magnetic force that is perpendicular to both the field lines and the velocity component itself. The path of a particle for which the force is always perpendicular to the velocity is a circle. The particle therefore follows a circular arc and exits the field on the other side of the circle, as shown in Figure 19.21 for a particle with constant kinetic energy. On the other hand, a particle can become trapped if it loses some kinetic energy in a collision after entering the field, so that it turns in a smaller circle and stays within the field.

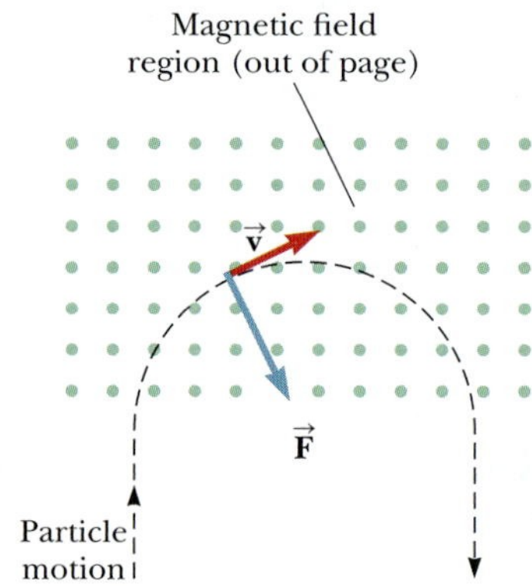


**Figure 19.21** (Applying Physics 19.3)

Particles *can* be injected and contained if, in addition to the magnetic field, electrostatic fields are involved. These fields are used in the *Penning trap*. With these devices, it's possible to store charged particles for extended periods. Such traps are useful, for example, in the storage of antimatter, which disintegrates completely on contact with ordinary matter. ■

■ Quick Quiz

**19.4** As a charged particle moves freely in a circular path in the presence of a constant magnetic field applied perpendicular to the particle's velocity, the particle's kinetic energy (a) remains constant, (b) increases, or (c) decreases.

## EXAMPLE 19.5 The Mass Spectrometer: Identifying Particles

**GOAL** Use the cyclotron equation to identify a particle.

**PROBLEM** A charged particle enters the magnetic field of a mass spectrometer at a speed of $1.79 \times 10^6$ m/s. It subsequently moves in a circular orbit with a radius of 16.0 cm in a uniform magnetic field of magnitude 0.350 T having a direction perpendicular to the particle's velocity. Find the particle's mass-to-charge ratio and identify it based on the table below.

**STRATEGY** After finding the mass-to-charge ratio with Equation 19.10, compare it with the values in the table, identifying the particle.

**SOLUTION**

Write the cyclotron equation:

$$r = \frac{mv}{qB}$$

Solve this equation for the mass divided by the charge, $m/q$, and substitute values:

$$\frac{m}{q} = \frac{rB}{v} = \frac{(0.160\text{ m})(0.350\text{ T})}{1.79 \times 10^6\text{ m/s}} = 3.13 \times 10^{-8}\ \frac{\text{kg}}{\text{C}}$$

Identify the particle from the table. All particles are completely ionized, bare nuclei.

| Nucleus | $m$ (kg) | $q$ (C) | $m/q$ (kg/C) |
|---|---|---|---|
| Hydrogen | $1.67 \times 10^{-27}$ | $1.60 \times 10^{-19}$ | $1.04 \times 10^{-8}$ |
| Deuterium | $3.35 \times 10^{-27}$ | $1.60 \times 10^{-19}$ | $2.09 \times 10^{-8}$ |
| Tritium | $5.01 \times 10^{-27}$ | $1.60 \times 10^{-19}$ | $3.13 \times 10^{-8}$ |
| Helium-3 | $5.01 \times 10^{-27}$ | $3.20 \times 10^{-19}$ | $1.57 \times 10^{-8}$ |

The particle is a tritium nucleus.

**REMARKS** The mass spectrometer is an important tool in both chemistry and physics. Unknown chemicals can be heated and ionized, and the resulting particles passed through the mass spectrometer and subsequently identified.

**QUESTION 19.5** What happens to the momentum of a charged particle in a uniform magnetic field?

**EXERCISE 19.5** Suppose a second charged particle enters the mass spectrometer at the same speed as the particle in Example 19.5. If it travels in a circle with radius 10.7 cm, find the mass-to-charge ratio and identify the particle from the table above.

**ANSWERS** $2.09 \times 10^{-8}$ kg/C; deuterium nucleus

## EXAMPLE 19.6 The Mass Spectrometer: Separating Isotopes

**GOAL** Apply the cyclotron equation to the process of separating isotopes.

**APPLICATION**

Mass Spectrometers

**PROBLEM** Two singly ionized atoms move out of a slit at point *S* in Figure 19.22 and into a magnetic field of magnitude 0.100 T pointing into the page. Each has a speed of $1.00 \times 10^6$ m/s. The nucleus of the first atom contains one proton and has a mass of $1.67 \times 10^{-27}$ kg, whereas the nucleus of the second atom contains a proton and a neutron and has a mass of $3.34 \times 10^{-27}$ kg. Atoms with the same number of protons in the nucleus but different masses are called isotopes. The two isotopes here are hydrogen and deuterium. Find their distance of separation when they strike a photographic plate at *P*.

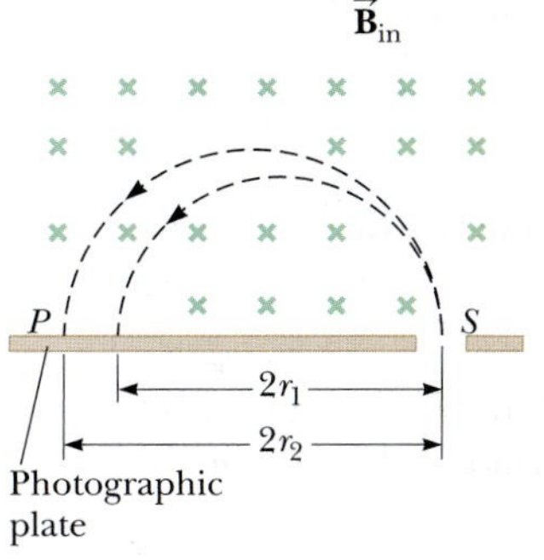


**Figure 19.22** (Example 19.6) Two isotopes leave the slit at point *S* and travel in different circular paths before striking a photographic plate at *P*.

**STRATEGY** Apply the cyclotron equation to each atom, finding the radius of the path of each. Double the radii to find the path diameters and then find their difference.

*(Continued)*

**SOLUTION**

Use Equation 19.10 to find the radius of the circular path followed by the lighter isotope, hydrogen:

$$r_1 = \frac{m_1 v}{qB} = \frac{(1.67 \times 10^{-27}\text{ kg})(1.00 \times 10^6\text{ m/s})}{(1.60 \times 10^{-19}\text{ C})(0.100\text{ T})}$$
$$= 0.104\text{ m}$$

Use the same equation to calculate the radius of the path of deuterium, the heavier isotope:

$$r_2 = \frac{m_2 v}{qB} = \frac{(3.34 \times 10^{-27}\text{ kg})(1.00 \times 10^6\text{ m/s})}{(1.60 \times 10^{-19}\text{ C})(0.100\text{ T})}$$
$$= 0.209\text{ m}$$

Multiply the radii by 2 to find the diameters and take the difference, getting the separation $x$ between the isotopes:

$$x = 2r_2 - 2r_1 = 0.210\text{ m}$$

**REMARKS** During World War II, mass spectrometers were used to separate the radioactive uranium isotope U-235 from its far more common isotope, U-238.

**QUESTION 19.6** Estimate the radius of the circle traced out by a singly ionized lead atom moving at the same speed.

**EXERCISE 19.6** Use the same mass spectrometer as in Example 19.6 to find the separation between two isotopes of helium: normal helium-4, which has a nucleus consisting of two protons and two neutrons, and helium-3, which has two protons and a single neutron. Assume both nuclei, doubly ionized (having a charge of $2e = 3.20 \times 10^{-19}$ C), enter the field at $1.00 \times 10^6$ m/s. The helium-4 nucleus has a mass of $6.64 \times 10^{-27}$ kg, and the helium-3 nucleus has a mass equal to $5.01 \times 10^{-27}$ kg.

**ANSWER** 0.102 m

## 19.7 Magnetic Field of a Long, Straight Wire and Ampère's Law

**LEARNING OBJECTIVES**

1. State the equation giving the magnitude of the magnetic field of a long straight wire and generalize it to Ampère's Law.
2. Apply Ampère's Law to determine the magnetic fields of wires and cylinders carrying current.

During a lecture demonstration in 1819, Danish scientist Hans Oersted (1777–1851) found that an electric current in a wire deflected a nearby compass needle. This momentous discovery, linking a magnetic field with an electric current for the first time, was the beginning of our understanding of the origin of magnetism.

A simple experiment first carried out by Oersted in 1820 clearly demonstrates that a current-carrying conductor produces a magnetic field. In this experiment, several compass needles are placed in a horizontal plane near a long vertical wire, as in Figure 19.23a. When there is no current in the wire, all needles point in the

When no current is present in the vertical wire, all compass needles point in the same direction (that of Earth's field).
$I = 0$
a
When the wire carries a strong current, the compass needles deflect in directions tangent to the circle, pointing in the direction of $\vec{\mathbf{B}}$, due to the current.
$I$
$\vec{\mathbf{B}}$
b



**Figure 19.23** (a), (b) Compasses shown the effects of the current in a nearby wire.

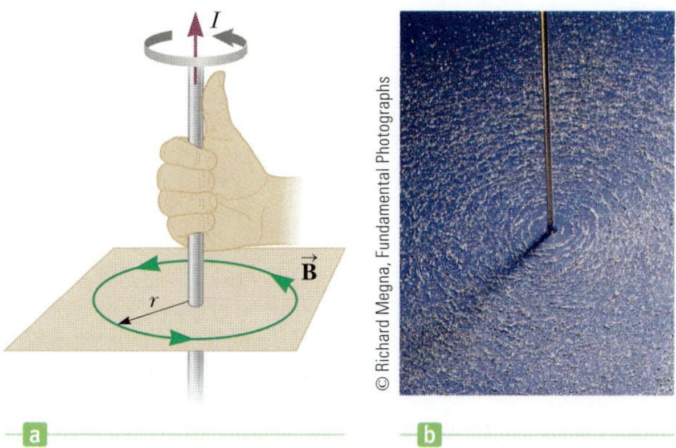


**Figure 19.24** (a) Right-hand rule number 2 for determining the direction of the magnetic field due to a long, straight wire carrying a current. Note that the magnetic field lines form circles around the wire. (b) Circular magnetic field lines surrounding a current-carrying wire, displayed by iron filings.

same direction (that of Earth's field), as one would expect. When the wire carries a strong, steady current, however, the needles all deflect in directions tangent to the circle, as in Figure 19.23b. These observations show that the direction of $\vec{\mathbf{B}}$ is consistent with the following convenient rule, **right-hand rule number 2**:

> Point the thumb of your right hand along a wire in the direction of positive current, as in Figure 19.24a. Your fingers then naturally curl in the direction of the magnetic field $\vec{\mathbf{B}}$.

When the current is reversed, the filings in Figure 19.24b also reverse.

Because the filings point in the direction of $\vec{\mathbf{B}}$, we conclude that the lines of $\vec{\mathbf{B}}$ form circles about the wire. By symmetry, the magnitude of $\vec{\mathbf{B}}$ is the same everywhere on a circular path centered on the wire and lying in a plane perpendicular to the wire. By varying the current and distance from the wire, it can be experimentally determined that $\vec{\mathbf{B}}$ is proportional to the current and inversely proportional to the distance from the wire. These observations lead to a mathematical expression for the strength of the magnetic field due to the current $I$ in a long, straight wire:

**Tip 19.3 Raise Your Right Hand!**

We have introduced two right-hand rules in this chapter. Be sure to use *only* your right hand when applying these rules.

$$B = \frac{\mu_0 I}{2\pi r} \qquad [19.11]$$

◀ Magnetic field due to a long, straight wire

The proportionality constant $\mu_0$, called the **permeability of free space,** has the value

$$\mu_0 \equiv 4\pi \times 10^{-7}\,\mathrm{T\cdot m/A} \qquad [19.12]$$

## Ampère's Law and a Long, Straight Wire

Equation 19.11 enables us to calculate the magnetic field due to a long, straight wire carrying a current. A general procedure for deriving such equations was proposed by French scientist André-Marie Ampère (1775–1836); it provides a relation between the current in an arbitrarily shaped wire and the magnetic field produced by the wire.

Consider an arbitrary closed path surrounding a current as in Figure 19.25 (page 678). The path consists of many short segments, each of length $\Delta\ell$. Multiply one of these lengths by the component of the magnetic field parallel to that segment, where the product is labeled $B_{\parallel}\,\Delta\ell$. According to Ampère, the sum of all such products over the closed path is equal to $\mu_0$ times the net current $I$ that passes through the surface bounded by the closed path. This statement, known as **Ampère's circuital law**, can be written

$$\sum B_{\parallel}\,\Delta\ell = \mu_0 I \qquad [19.13]$$

**Hans Christian Oersted**
**Danish Physicist and Chemist (1777–1851)**
Oersted is best known for observing that a compass needle deflects when placed near a wire carrying a current. This important discovery was the first evidence of the connection between electric and magnetic phenomena. Oersted was also the first to prepare pure aluminum.

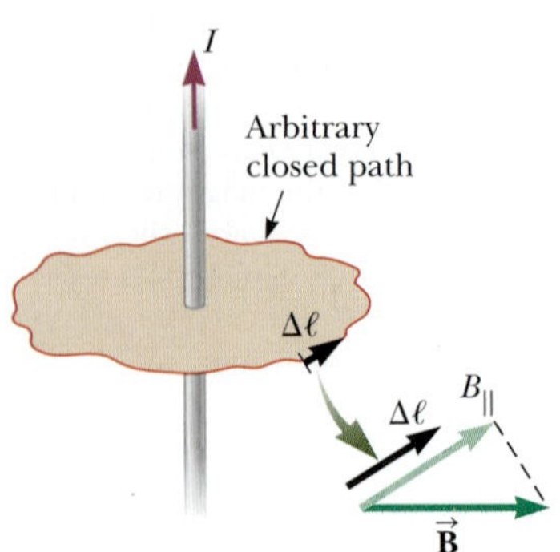


**Figure 19.25** An arbitrary closed path around a current is used to calculate the magnetic field of the current by the use of Ampère's rule.

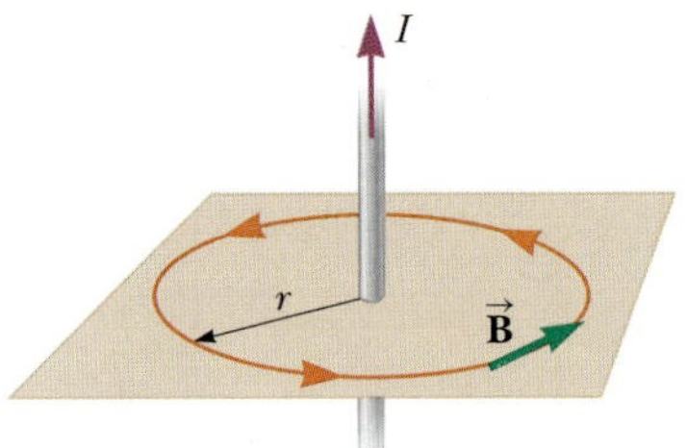


**Figure 19.26** A closed, circular path of radius $r$ around a long, straight, current-carrying wire is used to calculate the magnetic field set up by the wire.

where $B_{\|}$ is the component of $\vec{\mathbf{B}}$ parallel to the segment of length $\Delta\ell$ and $\Sigma B_{\|}\,\Delta\ell$ means that we take the sum over all the products $B_{\|}\,\Delta\ell$ around the closed path. Ampère's law is the fundamental law describing how electric currents create magnetic fields in the surrounding empty space.

We can use Ampère's circuital law to derive the magnetic field due to a long, straight wire carrying a current $I$. As discussed earlier, each of the magnetic field lines of this configuration forms a circle with the wire at its center. The magnetic field is tangent to this circle at every point, and its magnitude has the same value $B$ over the entire circumference of a circle of radius $r$, so $B_{\|} = B$, as shown in Figure 19.26. In calculating the sum $\sum B_{\|}\,\Delta\ell$ over the circular path, notice that $B_{\|}$ can be removed from the sum (because it has the same value $B$ for each element on the circle). Equation 19.13 then gives

$$\sum B_{\|}\,\Delta\ell = B_{\|} \sum \Delta\ell = B(2\pi r) = \mu_0 I$$

Dividing both sides by the circumference $2\pi r$, we obtain

$$B = \frac{\mu_0 I}{2\pi r}$$

This result is identical to Equation 19.11, which is the magnetic field due to the current $I$ in a long, straight wire.

Ampère's circuital law provides an elegant and simple method for calculating the magnetic fields of highly symmetric current configurations, but it can't easily be used to calculate magnetic fields for complex current configurations that lack symmetry. In addition, Ampère's circuital law in this form is valid only when the currents and fields don't change with time.

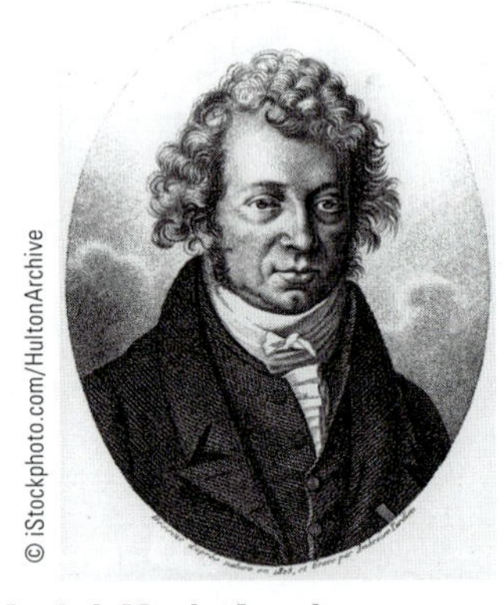


**André-Marie Ampère (1775–1836)**
Ampère, a Frenchman, is credited with the discovery of electromagnetism, the relationship between electric currents and magnetic fields.

## ■ EXAMPLE 19.7 The Magnetic Field of a Coaxial Cable

**GOAL** Use Ampère's law to calculate the magnetic field produced by current-carrying wires and cylinders.

**PROBLEM** A coaxial cable consists of an insulated wire carrying current $I_1 = 3.00$ A surrounded by a cylindrical conductor carrying current $I_2 = 1.00$ A in the opposite direction, as in Figure 19.27. **(a)** Calculate the magnetic field inside the cylindrical conductor at $r_{\text{int}} = 0.500$ cm. **(b)** Calculate the magnetic field outside the cylindrical conductor at $r_{\text{ext}} = 1.50$ cm.

**STRATEGY** Construct a circular path around the interior wire, as in Figure 19.27. Only the current inside that circle contributes to the magnetic field $B_{\text{int}}$ at points on the circle. To compute the magnetic field $B_{\text{ext}}$ exterior to the cylinder, construct a circular path outside the cylinder. Now both currents must be included in the calculation, but the current going down the page must be subtracted from the current in the wire.

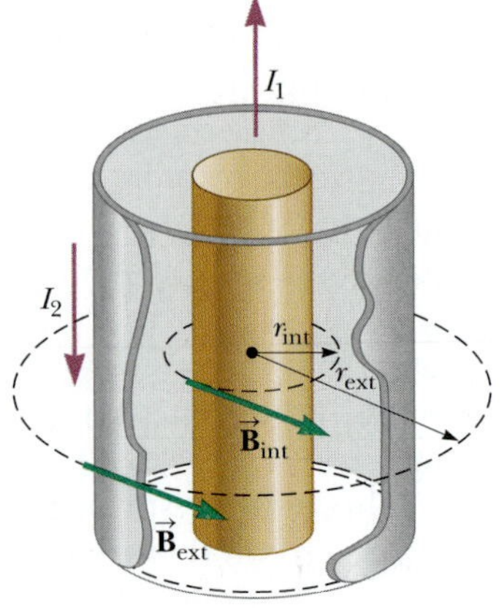


**Figure 19.27** (Example 19.7)

**SOLUTION**

**(a)** Calculate the magnetic field $B_{int}$ inside the cylindrical conductor at $r_{int} = 0.500$ cm.

Write Ampère's law:

$$\sum B_{\parallel}\,\Delta\ell = \mu_0 I$$

The magnetic field is constant on the given path and the total path length is $2\pi r_{int}$:

$$B_{int}\,(2\pi r_{int}) = \mu_0 I_1$$

Solve for $B_{int}$ and substitute values:

$$B_{int} = \frac{\mu_0 I_1}{2\pi r_{int}} = \frac{(4\pi \times 10^{-7}\,\mathrm{T\cdot m/A})(3.00\,\mathrm{A})}{2\pi(0.005\,\mathrm{m})}$$

$$= \boxed{1.20 \times 10^{-4}\,\mathrm{T}}$$

**(b)** Calculate the magnetic field $B_{ext}$ outside the cylindrical conductor at $r_{ext} = 1.50$ cm.

Write Ampère's law:

$$\sum B_{\parallel}\,\Delta\ell = \mu_0 I$$

The magnetic field is again constant on the given path and the total path length is $2\pi r_{ext}$:

$$B_{ext}(2\pi r_{ext}) = \mu_0(I_1 - I_2)$$

Solve for $B_{ext}$ and substitute values:

$$B_{ext} = \frac{\mu_0(I_1 - I_2)}{2\pi r_{ext}} = \frac{(4\pi \times 10^{-7}\,\mathrm{T\cdot m/A})(3.00\,\mathrm{A} - 1.00\,\mathrm{A})}{2\pi(0.015\,\mathrm{m})}$$

$$= \boxed{2.67 \times 10^{-5}\,\mathrm{T}}$$

**REMARKS** The direction of the field both inside and outside the cylinder is given by the right-hand rule, or counterclockwise in the perspective of Figure 19.27. Coaxial cables can be used to minimize the magnetic effects of current, provided that the currents inside the wire and the cylinder are equal in magnitude and opposite in direction.

**QUESTION 19.7** What direction is the magnetic force on a proton traveling up the page (a) to the right of the cable? (b) On the left side?

**EXERCISE 19.7** Suppose the current in the wire is 4.00 A downward and the current in the cylindrical conductor is 5.00 A upward. Find the magnitudes of the magnetic field (a) inside the cable at $r_{int} = 0.25$ cm and (b) outside the cable at $r_{ext} = 1.25$ cm.

**ANSWERS** (a) $3.20 \times 10^{-4}$ T (b) $1.60 \times 10^{-5}$ T

## 19.8 Magnetic Force Between Two Parallel Conductors

**LEARNING OBJECTIVE**

1. Calculate the magnetic force between two parallel current-carrying conductors.

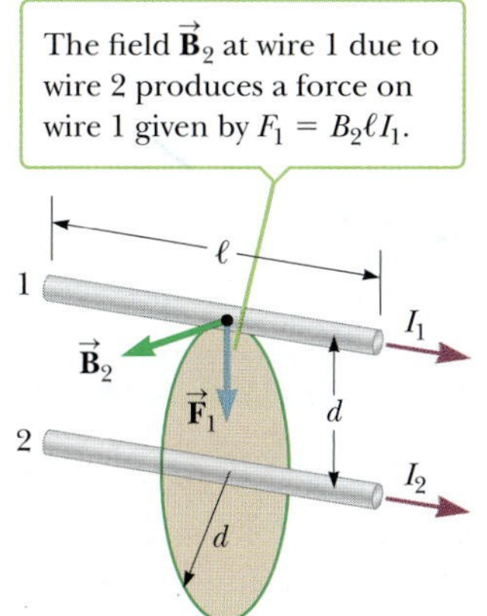

The field $\vec{B}_2$ at wire 1 due to wire 2 produces a force on wire 1 given by $F_1 = B_2\ell I_1$.
$\ell$
1
$I_1$
$\vec{B}_2$
$\vec{F}_1$
$d$
2
$I_2$
$d$

**Figure 19.28** Two parallel wires, oriented vertically, carry steady currents and exert forces on each other. The force is attractive if the currents have the same direction, as shown, and repulsive if the two currents have opposite directions.

As we have seen, a magnetic force acts on a current-carrying conductor when the conductor is placed in an external magnetic field. Because a conductor carrying a current creates a magnetic field around itself, it is easy to understand that two current-carrying wires placed close together exert magnetic forces on each other. Consider two long, straight, parallel wires separated by the distance $d$ and carrying currents $I_1$ and $I_2$ in the same direction, as shown in Figure 19.28. Wire 1 is directly above wire 2. What's the magnetic force on one wire due to a magnetic field set up by the other wire?

In this calculation we are finding the force on wire 1 due to the magnetic field of wire 2. The current $I_2$ sets up magnetic field $\vec{\mathbf{B}}_2$ at wire 1. The direction of $\vec{\mathbf{B}}_2$ is perpendicular to the wire, as shown in the figure. Using Equation 19.11, we find that the magnitude of this magnetic field is

$$B_2 = \frac{\mu_0 I_2}{2\pi d}$$

According to Equation 19.5, the magnitude of the magnetic force on wire 1 in the presence of field $\vec{\mathbf{B}}_2$ due to $I_2$ is

$$F_1 = B_2 I_1 \ell = \left(\frac{\mu_0 I_2}{2\pi d}\right) I_1 \ell = \frac{\mu_0 I_1 I_2 \ell}{2\pi d}$$

We can rewrite this relationship in terms of the force per unit length:

$$\frac{F_1}{\ell} = \frac{\mu_0 I_1 I_2}{2\pi d} \qquad [19.14]$$

The direction of $\vec{\mathbf{F}}_1$ is downward, toward wire 2, as indicated by right-hand rule number 1. This calculation is completely symmetric, which means that the force $\vec{\mathbf{F}}_2$ on wire 2 is equal to and opposite $\vec{\mathbf{F}}_1$, as expected from Newton's third law of action–reaction.

We have shown that parallel conductors carrying currents in the same direction *attract* each other. You should use the approach indicated by Figure 19.28 and the steps leading to Equation 19.14 to show that parallel conductors carrying currents in opposite directions *repel* each other.

The force between two parallel wires carrying a current is used to define the SI unit of current, the **ampere** (A), as follows:

Definition of the ampere ▶

If two long, parallel wires 1 m apart carry the same current and the magnetic force per unit length on each wire is $2 \times 10^{-7}$ N/m, the current is defined to be 1 A.

The SI unit of charge, the **coulomb** (C), can now be defined in terms of the ampere as follows:

Definition of the coulomb ▶

If a conductor carries a steady current of 1 A, the quantity of charge that flows through any cross section in 1 s is 1 C.

■ *Quick Quiz*

**19.5** Which of the following actions would double the magnitude of the magnetic force per unit length between two parallel current-carrying wires? Choose all correct answers. (a) Double one of the currents. (b) Double the distance between them. (c) Reduce the distance between them by half. (d) Double both currents.

**19.6** If, in Figure 19.28, $I_1 = 2$ A and $I_2 = 6$ A, which of the following is true? (Note that $F_2$ represents the magnitude of the force on wire 2.) (a) $F_1 = 3F_2$ (b) $F_1 = F_2$ (c) $F_1 = F_2/3$

## ■ EXAMPLE 19.8 Levitating a Wire

**GOAL** Calculate the magnetic force of one current-carrying wire on a parallel current-carrying wire.

**PROBLEM** Two wires, each having a weight per unit length of $1.00 \times 10^{-4}$ N/m, are parallel with one directly above the other. Assume the wires carry currents that are equal in magnitude and opposite in direction. The wires are 0.10 m apart, and the sum of the magnetic force and gravitational force on the upper wire is zero. Find the current in the wires. (Neglect Earth's magnetic field.)

**STRATEGY** The upper wire must be in equilibrium under the forces of magnetic repulsion and gravity. Set the sum of the forces equal to zero and solve for the unknown current, *I*.

**SOLUTION**

Set the sum of the forces equal to zero and substitute the appropriate expressions. Notice that the magnetic force between the wires is repulsive.

$$\vec{\mathbf{F}}_{\text{grav}} + \vec{\mathbf{F}}_{\text{mag}} = 0$$

$$-mg + \frac{\mu_0 I_1 I_2}{2\pi d}\ell = 0$$

The currents are equal, so $I_1 = I_2 = I$. Make these substitutions and solve for $I^2$:

$$\frac{\mu_0 I^2}{2\pi d}\ell = mg \quad \rightarrow \quad I^2 = \frac{(2\pi d)(mg/\ell)}{\mu_0}$$

Substitute given values, finding $I^2$, then take the square root. Notice that the weight per unit length, $mg/\ell$, is given.

$$I^2 = \frac{(2\pi \cdot 0.100\text{ m})(1.00 \times 10^{-4}\text{ N/m})}{(4\pi \times 10^{-7}\text{ T}\cdot\text{m})} = 50.0\text{ A}^2$$

$$I = \boxed{7.07\text{ A}}$$

**REMARKS** Exercise 19.3 showed that using Earth's magnetic field to levitate a wire required extremely large currents. Currents in wires can create much stronger magnetic fields than Earth's magnetic field in regions near the wire.

**QUESTION 19.8** Why can't cars be constructed that can magnetically levitate in Earth's magnetic field?

**EXERCISE 19.8** If the current in each wire is doubled, how far apart should the wires be placed if the magnitudes of the gravitational and magnetic forces on the upper wire are to be equal?

**ANSWER** 0.400 m

## 19.9 Magnetic Fields of Current Loops and Solenoids

**LEARNING OBJECTIVES**

1. Generalize the form of a magnetic field inside a loop to that inside a solenoid.
2. Calculate the magnetic field inside a solenoid.
3. Use Ampere's Law to derive the magnetic field inside a solenoid.

The strength of the magnetic field set up by a piece of wire carrying a current can be enhanced at a specific location if the wire is formed into a loop. You can understand this by considering the effect of several small segments of the current loop, as in Figure 19.29. The small segment at the bottom of the loop, labeled $\Delta x_1$, produces a magnetic field of magnitude $B_1$ at the loop's center, directed upward. The direction of $\vec{\mathbf{B}}$ can be verified using right-hand rule number 2 for a long, straight wire. Imagine holding the wire with your right hand, with your thumb pointing in the direction of the current. Your fingers then curl around in the direction of $\vec{\mathbf{B}}$.

A segment of length $\Delta x_2$ at the top of the loop also contributes to the field at the center, increasing its strength. The field produced at the center by the segment $\Delta x_2$ has the same magnitude as $B_1$ and is also directed upward. Similarly, all other such segments of the current loop contribute to the field. The net effect is a magnetic field for the current loop as pictured in Figure 19.30a.

Notice in Figure 19.30a that the magnetic field lines enter at the bottom of the current loop and exit at the top. Compare this figure with Figure 19.30b, illustrating the field of a bar magnet. The two fields are similar. One side of the loop acts as though it were the north pole of a magnet, and the other acts as a south pole. The similarity of these two fields will be used to discuss magnetism in matter in an upcoming section.

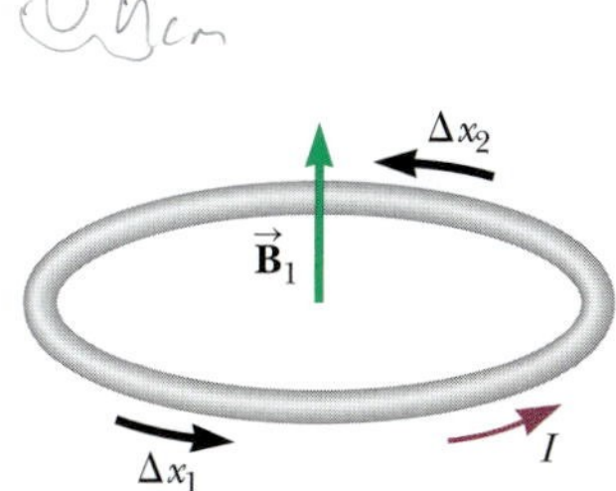


**Figure 19.29** All segments of the current loop produce a magnetic field at the center of the loop, directed *upward.*

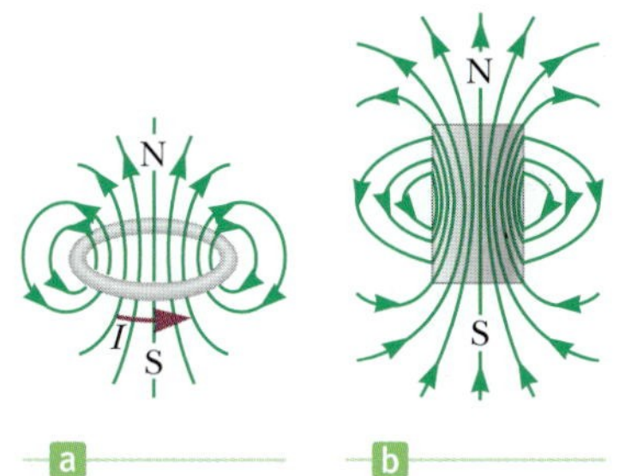


**Figure 19.30** (a) Magnetic field lines for a current loop. Note that the lines resemble those of a bar magnet. (b) The magnetic field of a bar magnet is similar to that of a current loop.

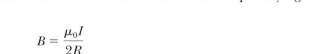

## APPLYING PHYSICS 19.4 Twisted Wires

In electrical circuits it is often the case that insulated wires carrying currents in opposite directions are twisted together. What is the advantage of doing this?

**EXPLANATION** If the wires are not twisted together, the combination of the two wires forms a current loop, which produces a relatively strong magnetic field. This magnetic field generated by the loop could be strong enough to affect adjacent circuits or components. When the wires are twisted together, their magnetic fields tend to cancel. ■

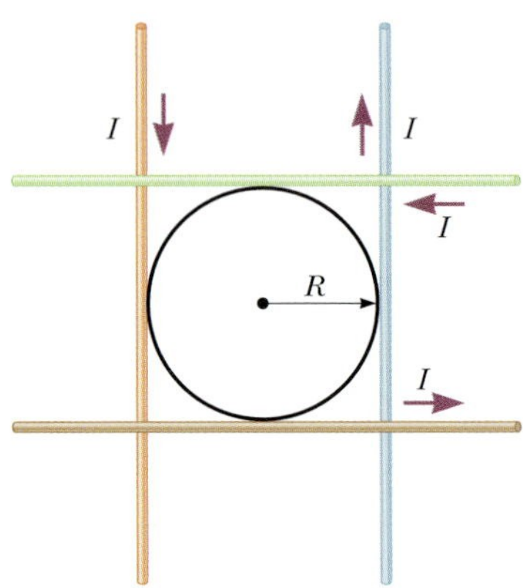
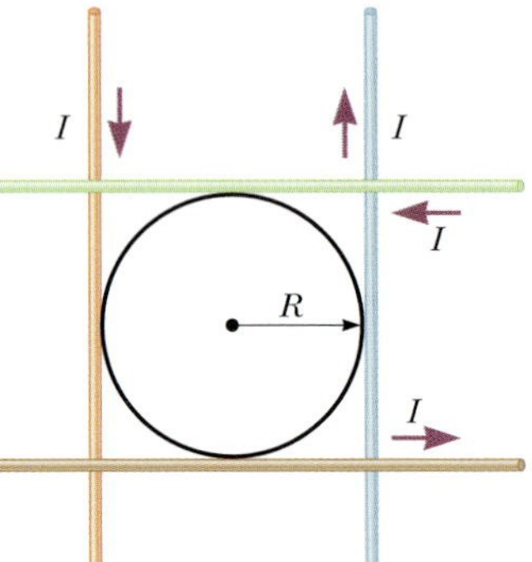


**Figure 19.31** The field of a circular loop carrying current $I$ can be approximated by the field due to four straight wires, each carrying current $I$.

The magnitude of the magnetic field at the center of a circular loop carrying current $I$ is given by

$$B = \frac{\mu_0 I}{2R}$$

This equation must be derived with calculus. It can be shown, however, to be reasonable by calculating the field at the center of four long wires, each carrying current $I$ and forming a square, as in Figure 19.31, with a circle of radius $R$ inscribed within it. Intuitively, this arrangement should give a magnetic field at the center that is similar in magnitude to the field produced by the circular loop. The current in the circular wire is closer to the center, so that wire would have a magnetic field somewhat stronger than just the four legs of the rectangle, but the lengths of the straight wires beyond the rectangle compensate for it. Each wire contributes the same magnetic field at the exact center, so the total field is given by

$$B = 4 \times \frac{\mu_0 I}{2\pi R} = \frac{4}{\pi}\left(\frac{\mu_0 I}{2R}\right) = (1.27)\left(\frac{\mu_0 I}{2R}\right)$$

This result is *approximately* the same as the field produced by the circular loop of current.

When the coil has $N$ loops, each carrying current $I$, the magnetic field at the center is given by

$$B = N\frac{\mu_0 I}{2R} \qquad [19.15]$$

## Magnetic Field of a Solenoid

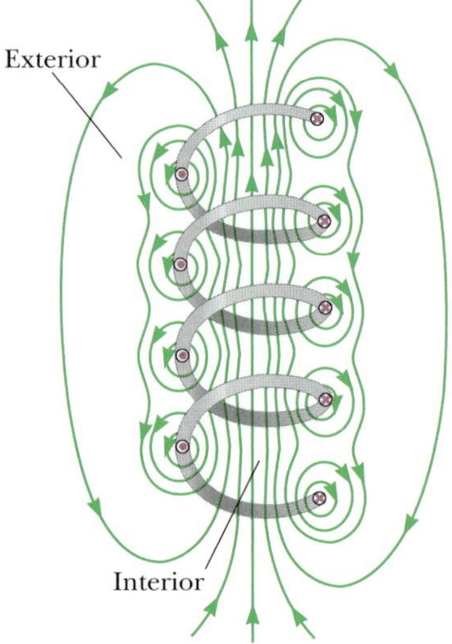


**Figure 19.32** The magnetic field lines for a loosely wound solenoid.

If a long, straight wire is bent into a coil of several closely spaced loops, the resulting device is a **solenoid**, often called an **electromagnet**. This device is important in many applications because it acts as a magnet only when it carries a current. The magnetic field inside a solenoid increases with the current and is proportional to the number of coils per unit length.

Figure 19.32 shows the magnetic field lines of a loosely wound solenoid of length $\ell$ and total number of turns $N$. Notice that the field lines inside the solenoid are nearly parallel, uniformly spaced, and close together. As a result, the field inside the solenoid is strong and approximately uniform. The exterior field at the sides of the solenoid is nonuniform, much weaker than the interior field, and *opposite in direction* to the field inside the solenoid.

If the turns are closely spaced, the field lines are as shown in Figure 19.33a, entering at one end of the solenoid and emerging at the other. One end of the solenoid acts as a north pole and the other end acts as a south pole. If the length of the solenoid is much greater than its radius, the lines that leave the north end of the solenoid spread out over a wide region before returning to enter the south end. The more widely separated the field lines are, the weaker the field. This is in contrast to a much stronger field *inside* the solenoid, where the lines are close together. Also, the field inside the solenoid has a constant magnitude at all points

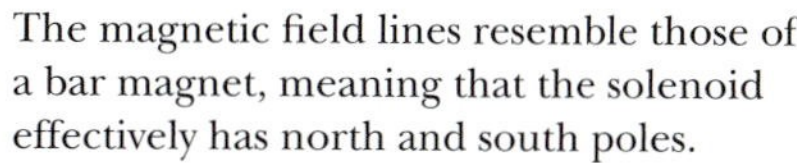

The magnetic field lines resemble those of a bar magnet, meaning that the solenoid effectively has north and south poles.


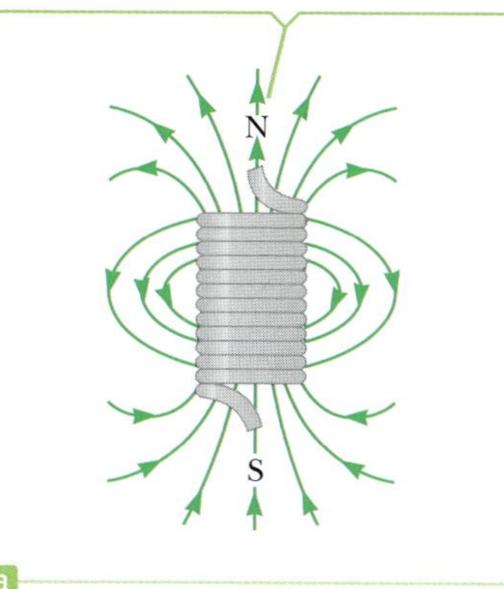


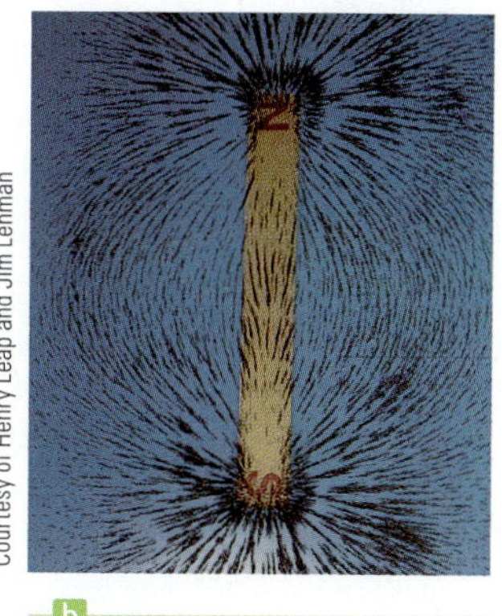


**Figure 19.33** (a) Magnetic field lines for a tightly wound solenoid of finite length carrying a steady current. The field inside the solenoid is nearly uniform and strong. (b) The magnetic field pattern of a bar magnet, displayed by small iron filings on a sheet of paper.

far from its ends. As will be shown subsequently, these considerations allow the application of Ampère's law to the solenoid, giving a result of

$$B = \mu_0 n I \qquad [19.16]$$

◀ The magnetic field inside a solenoid

for the field inside the solenoid, where $n = N/\ell$ is the number of turns per unit length of the solenoid.

Numerous devices create beams of charged particles for various purposes, and those particles are usually controlled and directed by electromagnetic fields. Old-style cathode ray TV sets use steering magnets that rapidly and accurately direct an electron beam across a screen of phosphors in a scanning motion, creating an illusion of a moving picture out of a series of bright dots. (See Fig. 19.34.) Electron microscopes (see Fig. 27.17b, page 937) use a similar gun and both electrostatic and electromagnetic lenses to focus the beam. Particle accelerators require very large electromagnets to turn particles moving at nearly the speed of light. Tokamaks, experimental devices used in fusion power research, use magnetic fields to contain hot plasmas. Figure 19.35 is a photograph of one such device.

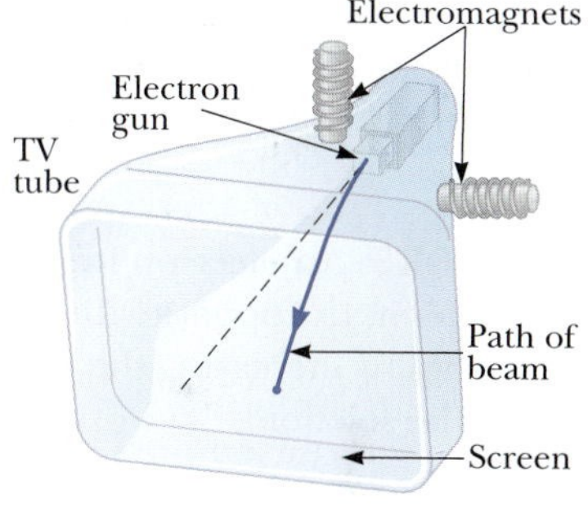


**Figure 19.34** Electromagnets are used to deflect electrons to desired positions on the screen of a television tube.

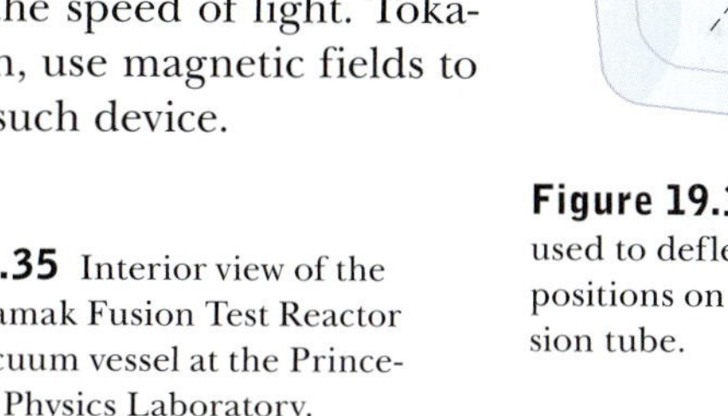

**Figure 19.35** Interior view of the closed Tokamak Fusion Test Reactor (TFTR) vacuum vessel at the Princeton Plasma Physics Laboratory.

## EXAMPLE 19.9 The Magnetic Field Inside a Solenoid

**GOAL** Calculate the magnetic field of a solenoid from given data and the momentum of a charged particle in this field.

**PROBLEM** A certain solenoid consists of 100 turns of wire and has a length of 10.0 cm. **(a)** Find the magnitude of the magnetic field inside the solenoid when it carries a current of 0.500 A. **(b)** What is the momentum of a proton orbiting inside the solenoid in a circle with a radius of 0.020 m? The axis of the solenoid is perpendicular to the plane of the orbit. **(c)** Approximately how much wire would be needed to build this solenoid? Assume the solenoid's radius is 5.00 cm.

**STRATEGY** In part (a) calculate the number of turns per meter and substitute that and given information into Equation 19.16, getting the magnitude of the magnetic field. Part (b) is an application of Newton's second law.

*(Continued)*

**SOLUTION**

**(a)** Find the magnitude of the magnetic field inside the solenoid when it carries a current of 0.500 A.

Calculate the number of turns per unit length:

$$n = \frac{N}{\ell} = \frac{100 \text{ turns}}{0.100 \text{ m}} = 1.00 \times 10^3 \text{ turns/m}$$

Substitute $n$ and $I$ into Equation 19.16 to find the magnitude of the magnetic field:

$$B = \mu_0 n I$$
$$= (4\pi \times 10^{-7} \text{ T} \cdot \text{m/A})(1.00 \times 10^3 \text{ turns/m})(0.500 \text{ A})$$
$$= 6.28 \times 10^{-4} \text{ T}$$

**(b)** Find the momentum of a proton orbiting in a circle of radius 0.020 m near the center of the solenoid.

Write Newton's second law for the proton:

$$ma = F = qvB$$

Substitute the centripetal acceleration $a = v^2/r$:

$$m\frac{v^2}{r} = qvB$$

Cancel one factor of $v$ on both sides and multiply by $r$, getting the momentum $mv$:

$$mv = rqB = (0.020 \text{ m})(1.60 \times 10^{-19} \text{ C})(6.28 \times 10^{-4} \text{ T})$$
$$p = mv = 2.01 \times 10^{-24} \text{ kg} \cdot \text{m/s}$$

**(c)** Approximately how much wire would be needed to build this solenoid?

Multiply the number of turns by the circumference of one loop:

$$\text{Length of wire} \approx (\text{number of turns})(2\pi r)$$
$$= (1.00 \times 10^2 \text{ turns})(2\pi \cdot 0.050\,0 \text{ m})$$
$$= 31.4 \text{ m}$$

**REMARKS** An electron in part (b) would have the same momentum as the proton, but a much higher speed. It would also orbit in the opposite direction. The length of wire in part (c) is only an estimate because the wire has a certain thickness, slightly increasing the size of each loop. In addition, the wire loops aren't perfect circles because they wind slowly up along the solenoid.

**QUESTION 19.9** What would happen to the orbiting proton if the solenoid were oriented vertically?

**EXERCISE 19.9** Suppose you have a 32.0-m length of copper wire. If the wire is wrapped into a solenoid 0.240 m long and having a radius of 0.040 0 m, how strong is the resulting magnetic field in its center when the current is 12.0 A?

**ANSWER** $8.00 \times 10^{-3}$ T

## Ampère's Law Applied to a Solenoid

We can use Ampère's law to obtain the expression for the magnetic field inside a solenoid carrying a current $I$. A cross section taken along the length of part of our solenoid is shown in Figure 19.36. $\vec{\mathbf{B}}$ inside the solenoid is uniform and parallel to the axis, and $\vec{\mathbf{B}}$ outside is approximately zero. Consider a rectangular path of length $L$ and width $w$, as shown in the figure. We can apply Ampère's law to this path by evaluating the sum of $B_{\parallel}\,\Delta\ell$ over each side of the rectangle. The contribution along side 3 is clearly zero because $\vec{\mathbf{B}} = 0$ in this region. The contributions from sides 2 and 4 are both zero because $\vec{\mathbf{B}}$ is perpendicular to $\Delta\ell$ along these paths. Side 1 of length $L$ gives a contribution $BL$ to the sum because $\vec{\mathbf{B}}$ is uniform along this path and parallel to $\Delta\ell$. Therefore, the sum over the closed rectangular path has the value

$$\sum B_{\parallel}\,\Delta\ell = BL$$

The right side of Ampère's law involves the total current that passes through the area bounded by the path chosen. In this case, the total current through the rectangular path equals the current through each turn of the solenoid, multiplied by the number of turns. If $N$ is the number of turns in the length $L$, then the total current through the rectangular path equals $NI$. Ampère's law applied to this path therefore gives

$$\sum B_{\parallel} \Delta \ell = BL = \mu_0 NI$$

or

$$B = \mu_0 \frac{N}{L} I = \mu_0 nI$$

where $n = N/L$ is the number of turns per unit length.

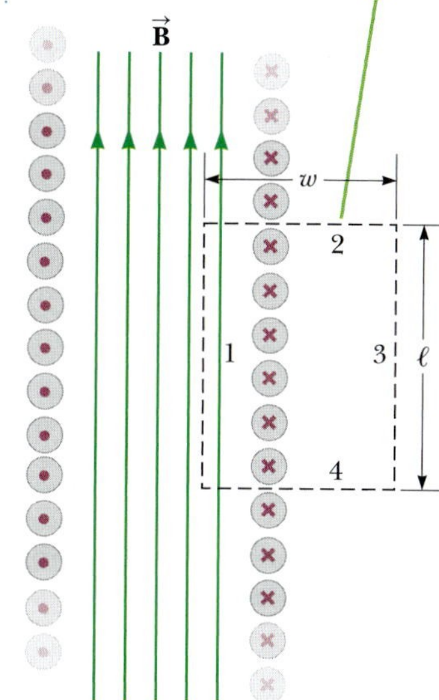
Ampère's law applied to the rectangular dashed path can be used to calculate the field inside the solenoid.
$\vec{B}$
$w$
1
2
3
4
$\ell$

**Figure 19.36** A cross-sectional view of a tightly wound solenoid. If the solenoid is long relative to its radius, we can assume the magnetic field inside is uniform and the field outside is zero.

# 19.10 Magnetic Domains

**LEARNING OBJECTIVES**

1. Discuss the creation of magnet fields at the atomic level by orbiting electrons and electron spin.
2. Define ferromagnetic materials and magnetic domains.
3. Contrast the magnetic properties of ferromagnetic, paramagnetic, and diamagnetic materials.

The magnetic field produced by a current in a coil of wire gives us a hint as to what might cause certain materials to exhibit strong magnetic properties. A single coil like that in Figure 19.30a has a north pole and a south pole, but if that is true for a coil of wire, it should also be true for any current confined to a circular path. In particular, *an individual atom should act as a magnet because of the motion of the electrons about the nucleus.* Each electron, with its charge of $1.6 \times 10^{-19}$ C, circles the atom once in about $10^{-16}$ s. If we divide the electric charge by this time interval, we see that the orbiting electron is equivalent to a current of $1.6 \times 10^{-3}$ A. Such a current produces a magnetic field on the order of 20 T at the center of the circular path. From this we see that a very strong magnetic field would be produced if several of these atomic magnets could be aligned inside a material. This doesn't occur, however, because the simple model we have described is not the complete story. A thorough analysis of atomic structure shows that the magnetic field produced by one electron in an atom is often canceled by an oppositely revolving electron in the same atom. The net result is that **the magnetic effect produced by the electrons orbiting the nucleus is either zero or very small for most materials**.

The magnetic properties of many materials can be explained by the fact that an electron not only circles in an orbit, but also spins on its axis like a top, with spin magnetic moment as shown (Fig. 19.37). (This classical description should not be taken too literally. The property of electron *spin* can be understood only in the context of quantum mechanics, which we will not discuss here.) The spinning electron represents a charge in motion that produces a magnetic field. The field due to the spinning is generally stronger than the field due to the orbital motion. In atoms containing many electrons, the electrons usually pair up with their spins opposite each other so that their fields cancel. That is why most substances are not magnets. In certain strongly magnetic materials, such as iron, cobalt, and nickel, however, the magnetic fields produced by the electron spins don't cancel completely. Such materials are said to be **ferromagnetic**. In ferromagnetic materials strong coupling occurs between neighboring atoms,

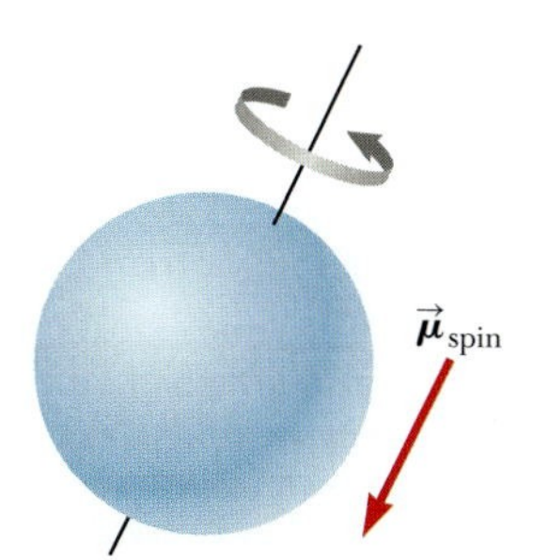
$\vec{\mu}_{\text{spin}}$

**Figure 19.37** Classical model of a spinning electron.

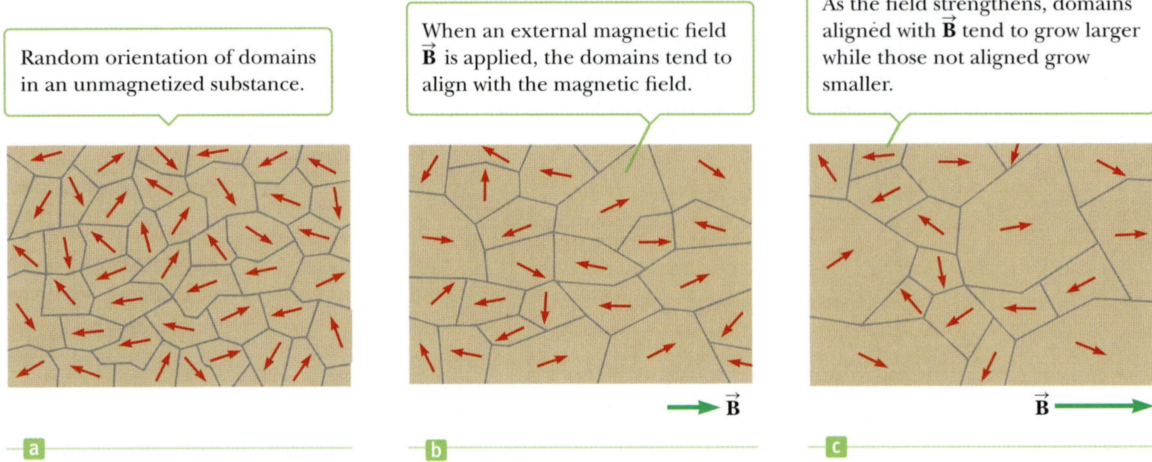

Random orientation of domains in an unmagnetized substance.
When an external magnetic field $\vec{B}$ is applied, the domains tend to align with the magnetic field.
As the field strengthens, domains alignėd with $\vec{B}$ tend to grow larger while those not aligned grow smaller.
$\vec{B}$
$\vec{B}$
a
b
c


**Figure 19.38** Orientation of magnetic dipoles before and after a magnetic field is applied to a ferromagnetic substance.

forming large groups of atoms with spins that are aligned. Called **domains**, the sizes of these groups typically range from about $10^{-4}$ cm to 0.1 cm. In an unmagnetized substance the domains are randomly oriented, as shown in Figure 19.38a. When an external field is applied, as in Figures 19.38b and 19.38c, the magnetic field of each domain tends to come nearer to alignment with the external field, resulting in magnetization.

In what are called hard magnetic materials, domains remain aligned even after the external field is removed; the result is a **permanent magnet**. In soft magnetic materials, such as iron, once the external field is removed, thermal agitation produces motion of the domains and the material quickly returns to an unmagnetized state.

The alignment of domains explains why the strength of an electromagnet is increased dramatically by the insertion of an iron core into the magnet's center. The magnetic field produced by the current in the loops causes the domains to align, thus producing a large net external field. The use of iron as a core is also advantageous because it is a soft magnetic material that loses its magnetism almost instantaneously after the current in the coils is turned off.

The formation of domains in ferromagnetic substances also explains why such substances are attracted to permanent magnets. The magnetic field of a permanent magnet realigns domains in a ferromagnetic object so that the object becomes temporarily magnetized. The object's poles are then attracted to the corresponding opposite poles of the permanent magnet. The object can similarly attract other ferromagnetic objects, as illustrated in Figure 19.39.

**Figure 19.39** The permanent magnet (red) temporarily magnetizes some paper clips, which then cling to each other through magnetic forces.

## Types of Magnetic Materials

Magnetic materials can be classified according to how they react to the application of a magnetic field. In **ferromagnetic** materials the atoms have permanent magnetic moments that align readily with an externally applied magnetic field. Examples of ferromagnetic materials are iron, cobalt, and nickel. Such substances can retain some of their magnetization even after the applied magnetic field is removed.

**Paramagnetic** materials also have magnetic moments that tend to align with an externally applied magnetic field, but the response is extremely weak compared with that of ferromagnetic materials. Examples of paramagnetic substances are aluminum, calcium, and platinum. A ferromagnetic material can

**Tip 19.4 The Electron Spins, but Doesn't!**

Even though we use the word *spin*, the electron, unlike a child's top, isn't physically spinning in this sense. The electron has an intrinsic angular momentum that causes it to act *as if it were spinning*, but the concept of spin angular momentum is actually a relativistic quantum effect.

**Figure 19.40** Diamagnetism. A frog is levitated in a 16-T magnetic field at the Nijmegen High Field Magnet Laboratory in the Netherlands. The levitation force is exerted on the diamagnetic water molecules in the frog's body. The frog suffered no ill effects from the levitation experience.

become paramagnetic when warmed to a certain critical temperature, the Curie temperature, that depends on the material.

In **diamagnetic** materials, an externally applied magnetic field induces a very weak magnetization that is opposite the applied field. Ordinarily diamagnetism isn't observed because paramagnetic and ferromagnetic effects are far stronger. In Figure 19.40, however, a very high magnetic field exerts a levitating force on the diamagnetic water molecules in a frog.

## SUMMARY

### 19.3 Magnetic Fields

The **magnetic force** that acts on a charge $q$ moving with velocity $\vec{\mathbf{v}}$ in a magnetic field $\vec{\mathbf{B}}$ has magnitude

$$F = qvB\sin\theta \qquad [19.1]$$

where $\theta$ is the angle between $\vec{\mathbf{v}}$ and $\vec{\mathbf{B}}$.

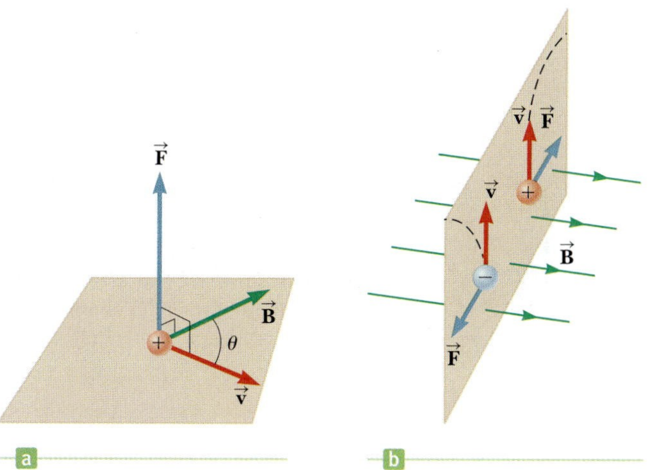


(a) The direction of the magnetic force $\vec{\mathbf{F}}$ acting on a charged particle moving with a velocity $\vec{\mathbf{v}}$ in the presence of a magnetic field $\vec{\mathbf{B}}$. The magnetic force is perpendicular to both $\vec{\mathbf{v}}$ and $\vec{\mathbf{B}}$. (b) Magnetic forces on positive and negative charges. The dashed lines show the paths of the particles, which are investigated in Section 19.6. The magnetic forces on oppositely charged particles moving at the same velocity in a magnetic field are in opposite directions.

To find the direction of this force, use **right-hand rule number 1**: point the fingers of your open right hand in the direction of $\vec{\mathbf{v}}$ and then curl them in the direction of $\vec{\mathbf{B}}$. Your thumb then points in the direction of the magnetic force $\vec{\mathbf{F}}$.

If the charge is *negative* rather than positive, the force is directed opposite the force given by the right-hand rule.

The SI unit of the magnetic field is the **tesla** (T), or weber per square meter (Wb/m$^2$). An additional commonly used unit for the magnetic field is the **gauss** (G); 1 T = $10^4$ G.

### 19.4 Magnetic Force on a Current-Carrying Conductor

If a straight conductor of length $\ell$ carries current $I$, the magnetic force on that conductor when it is placed in a uniform external magnetic field $\vec{\mathbf{B}}$ is

$$F = BI\ell\sin\theta \qquad [19.6]$$

where $\theta$ is the angle between the direction of the current and the direction of the magnetic field.

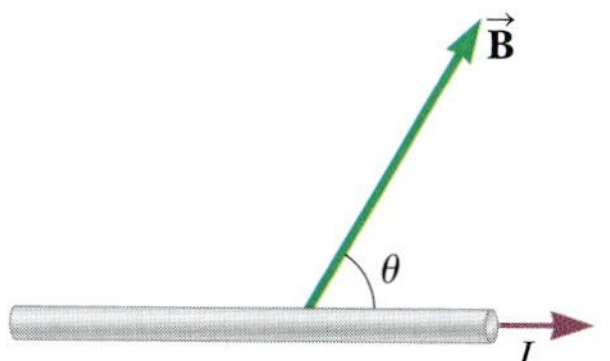


The magnetic force on this current-carrying conductor is directed straight up out of the page.

Right-hand rule number 1 also gives the direction of the magnetic force on the conductor. In this case, however, you must point your fingers in the direction of the current rather than in the direction of $\vec{\mathbf{v}}$.

## 19.5 Torque on a Current Loop and Electric Motors

The torque $\tau$ on a current-carrying loop of wire in a magnetic field $\vec{\mathbf{B}}$ has magnitude

$$\tau = BIA \sin \theta \qquad [19.8]$$

where $I$ is the current in the loop and $A$ is its cross-sectional area. The magnitude of the magnetic moment of a current-carrying coil is defined by $\mu = IAN$, where $N$ is the number of loops. The magnetic moment is considered a vector, $\vec{\boldsymbol{\mu}}$, that is perpendicular to the plane of the loop. The angle between $\vec{\mathbf{B}}$ and $\vec{\boldsymbol{\mu}}$ is $\theta$.

## 19.6 Motion of a Charged Particle in a Magnetic Field

If a charged particle moves in a uniform magnetic field so that its initial velocity is perpendicular to the field, it will move in a circular path in a plane perpendicular to the magnetic field. The radius $r$ of the circular path can be found from Newton's second law and centripetal acceleration, and is given by

$$r = \frac{mv}{qB} \qquad [19.10]$$

where $m$ is the mass of the particle and $q$ is its charge.

## 19.7 Magnetic Field of a Long, Straight Wire and Ampère's Law

The magnetic field at distance $r$ from a **long, straight wire** carrying current $I$ has the magnitude

$$B = \frac{\mu_0 I}{2\pi r} \qquad [19.11]$$

where $\mu_0 = 4\pi \times 10^{-7}\ \mathrm{T \cdot m/A}$ is the **permeability of free space**. The magnetic field lines around a long, straight wire are circles concentric with the wire.

**Ampère's law** can be used to find the magnetic field around certain simple current-carrying conductors. It can be written

$$\sum B_{\parallel}\, \Delta\ell = \mu_0 I \qquad [19.13]$$

where $B_{\parallel}$ is the component of $\vec{\mathbf{B}}$ tangent to a small current element of length $\Delta\ell$ that is part of a closed path and $I$ is the total current that penetrates the closed path.

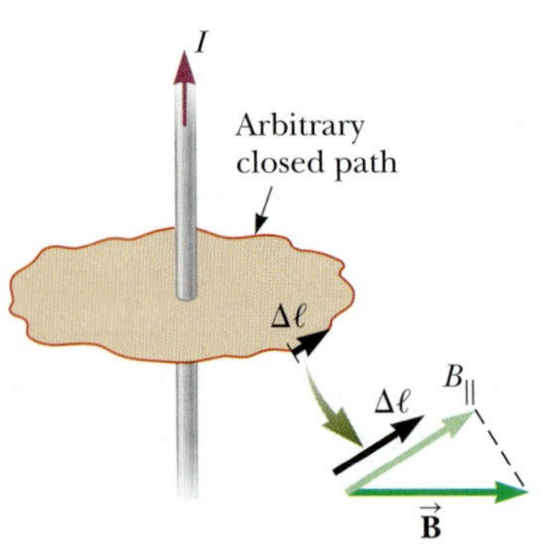


An arbitrary closed path around a current is used to calculate the magnetic field of the current by the use of Ampère's rule.

## 19.8 Magnetic Force Between Two Parallel Conductors

The force per unit length on each of two parallel wires separated by the distance $d$ and carrying currents $I_1$ and $I_2$ has the magnitude

$$\frac{F}{\ell} = \frac{\mu_0 I_1 I_2}{2\pi d} \qquad [19.14]$$

The forces are attractive if the currents are in the same direction and repulsive if they are in opposite directions.

## 19.9 Magnetic Field of Current Loops and Solenoids

The magnetic field at the center of a coil of $N$ circular loops of radius $R$, each carrying current $I$, is given by

$$B = N\frac{\mu_0 I}{2R} \qquad [19.15]$$

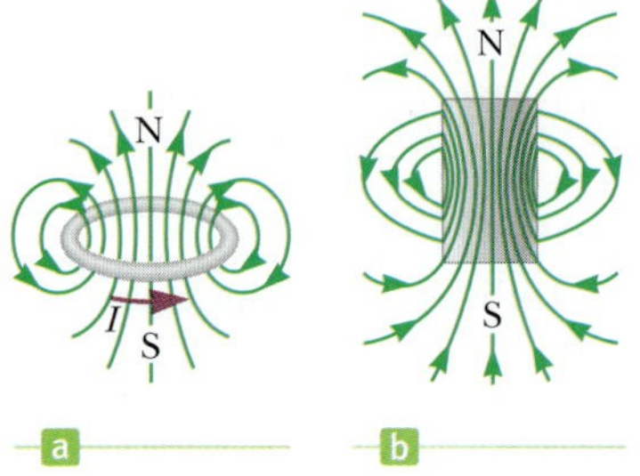


(a) Magnetic field lines for a current loop. Note that the lines resemble those of a bar magnet. (b) The magnetic field of a bar magnet is similar to that of a current loop.

The magnetic field inside a solenoid has the magnitude

$$B = \mu_0 n I \qquad [19.16]$$

where $n = N/\ell$ is the number of turns of wire per unit length.

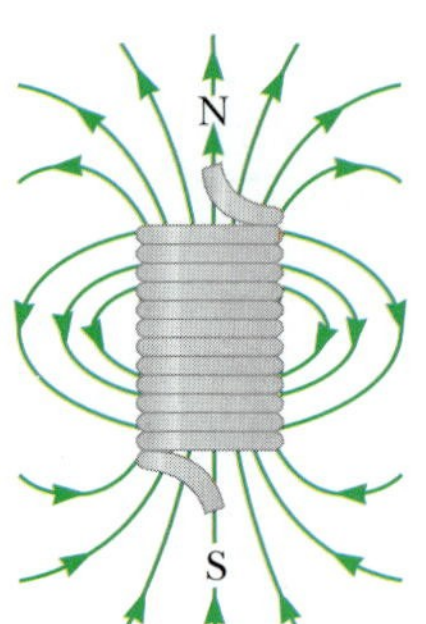


Magnetic field lines for a tightly wound solenoid of finite length carrying a steady current. The field inside the solenoid is nearly uniform and strong. Note that the field lines resemble those of a bar magnet, so the solenoid effectively has north and south poles.

## WARM-UP EXERCISES

WebAssign The warm-up exercises in this chapter may be assigned online in Enhanced WebAssign.

1. **Physics Review** A man of mass 70.0 kg sits on the end of a seesaw of length 4.00 m, pivoted in the middle. His end of the seesaw is on the ground. A standing woman grasps and exerts a force straight down on the raised end, which is initially 2.00 m off the ground. (a) What torque must she exceed if she is to begin lifting the man off the ground? (b) What minimum torque must she exert just as the seesaw becomes level? (See Section 8.1.)

2. **Physics Review** Suppose a futuristic tractor beam on a large space station exerts a constant force on a satellite of mass 2 420 kg, keeping it in a circular orbit of radius 785 m. If the satellite is traveling at 1 250 m/s, what magnitude force must the tractor beam exert on it? (See Section 7.4.)

3. A wire of length 0.500 m carries a current of 0.100 A in the positive $x$-direction, parallel to the ground. If the wire has a weight of $1.00 \times 10^{-2}$ N, what is the minimum magnitude magnetic field that exerts a magnetic force on the wire equal to its weight? (See Section 19.4.)

4. An electron moves across Earth's equator at a speed of $2.53 \times 10^{6}$ m/s and in a direction 35.0° N of E. At this point, Earth's magnetic field has a direction due north, is parallel to the surface, and has a magnitude of $0.100 \times 10^{-4}$ T. (a) What is the magnitude of the force acting on the electron due to its interaction with Earth's magnetic field? (b) Is the force toward, away, or parallel to the Earth's surface? (See Section 19.3.)

5. A rectangular coil of wire consisting of 10.0 loops, each with length 0.200 m and width 0.300 m, lies in the $xy$-plane. If the coil carries a current of 2.00 A, what is the magnitude of the torque exerted by a magnetic field of magnitude 0.010 0 T directed at an angle of 30.0° with respect to the positive $z$-axis? (See Section 19.5.)

6. A long wire carries a current of 0.500 A. Find the magnitude of the magnetic field 0.700 m away from the wire. (See Section 19.7.)

7. A proton enters a constant magnetic field of magnitude 0.050 0 T and traverses a semicircle of radius 1.00 mm before leaving the field. What is the proton's speed? (See Section 19.6.)

8. Calculate the magnitude of the magnetic force per unit length between a pair of parallel wires separated by 2.00 m if they each carry a current of 3.00 A. (See Section 19.8.)

9. What is the magnitude of the magnetic field at the core of a 120-turn solenoid of length 0.50 m carrying a current of 2.0 A? (See Section 19.9.)

## CONCEPTUAL QUESTIONS

WebAssign The conceptual questions in this chapter may be assigned online in Enhanced WebAssign.

1. In older television sets, a beam of electrons moves from the back of the picture tube to the screen, where it strikes a fluorescent dot that glows with a particular color when hit. Earth's magnetic field at the location of the television set is horizontal and toward the north. In which direction(s) should the set be oriented so that the beam undergoes the largest deflection?

2. Which way would a compass point if you were at Earth's north magnetic pole?

3. How can the motion of a charged particle be used to distinguish between a magnetic field and an electric field in a certain region? Give a specific example to justify your answer.

4. Can a constant magnetic field set a proton at rest into motion? Explain your answer.

5. Explain why two parallel wires carrying currents in opposite directions repel each other.

6. Will a nail be attracted to either pole of a magnet? Explain what is happening inside the nail when it is placed near the magnet.

7. A Hindu ruler once suggested that he be entombed in a magnetic coffin with the polarity arranged so that he could be forever suspended between heaven and Earth. Is such magnetic levitation possible? Discuss.

8. A magnet attracts a piece of iron. The iron can then attract another piece of iron. On the basis of domain alignment, explain what happens in each piece of iron.

9. Can you use a compass to detect the currents in wires in the walls near light switches in your home?

10. Is the magnetic field created by a current loop uniform? Explain.

11. Suppose you move along a wire at the same speed as the drift speed of the electrons in the wire. Do you now measure a magnetic field of zero?

12. Why do charged particles from outer space, called cosmic rays, strike Earth more frequently at the poles than at the equator?

13. A hanging Slinky® toy is attached to a powerful battery and a switch. When the switch is closed so that the toy now carries current, does the Slinky compress or expand?

14. How can a current loop be used to determine the presence of a magnetic field in a given region of space?

15. Parallel wires exert magnetic forces on each other. What about perpendicular wires? Imagine two wires oriented perpendicular to each other and almost touching. Each wire carries a current. Is there a force between the wires?

16. Figure CQ19.16 shows four permanent magnets, each having a hole through its center. Notice that the blue and yellow magnets are levitated above the red ones. (a) How does this levitation occur? (b) What purpose do the rods serve? (c) What can you say about the poles of the magnets from this observation? (d) If the upper magnet were inverted, what do you suppose would happen?

**Figure CQ19.16**

17. Two charged particles are projected in the same direction into a magnetic field perpendicular to their velocities. If the two particles are deflected in opposite directions, what can you say about them?

18. Two long, straight wires cross each other at right angles, and each carries the same current as in Figure CQ19.18. Which of the following statements are true regarding the total magnetic field at the various points due to the two wires? (There may be more than one correct statement.) (a) The field is strongest at points *B* and *D*. (b) The field is strongest at points *A* and *C*. (c) The field is out of the page at point *B* and into the page at point *D*. (d) The field is out of the page at point *C* and out of the page at point *D*. (e) The field has the same magnitude at all four points.

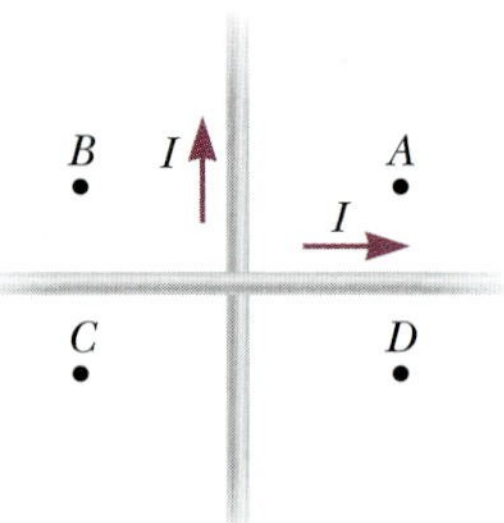


**Figure CQ19.18**

19. A magnetic field exerts a torque on each of the current-carrying single loops of wire shown in Figure CQ19.19. The loops lie in the *xy*-plane, each carrying the same magnitude current, and the uniform magnetic field points in the positive *x*-direction. Rank the coils by the magnitude of the torque exerted on them by the field, from largest to smallest. (a) A, B, C (b) A, C, B (c) B, A, C (d) B, C, A (e) C, A, B

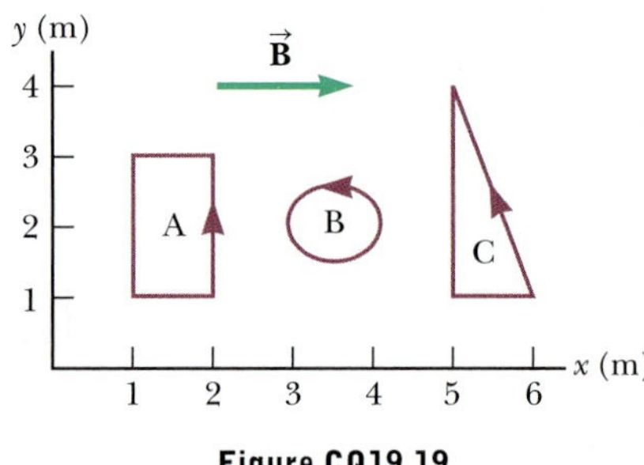


**Figure CQ19.19**

## PROBLEMS

WebAssign The problems in this chapter may be assigned online in Enhanced WebAssign.

1. denotes straightforward problem; 2. denotes intermediate problem;
3. denotes challenging problem
1. denotes full solution available in *Student Solutions Manual/Study Guide*
1. denotes problems most often assigned in Enhanced WebAssign

BIO denotes biomedical problems
GP denotes guided problems
M denotes Master It tutorial available in Enhanced WebAssign
Q|C denotes asking for quantitative and conceptual reasoning
S denotes symbolic reasoning problem
W denotes Watch It video solution available in Enhanced WebAssign

### 19.3 Magnetic Fields

1. Consider an electron near the Earth's equator. In which direction does it tend to deflect if its velocity is (a) directed downward? (b) Directed northward? (c) Directed westward? (d) Directed southeastward?

2. (a) Find the direction of the force on a proton (a positively charged particle) moving through the magnetic fields in Figure P19.2, as shown. (b) Repeat part (a), assuming the moving particle is an electron.

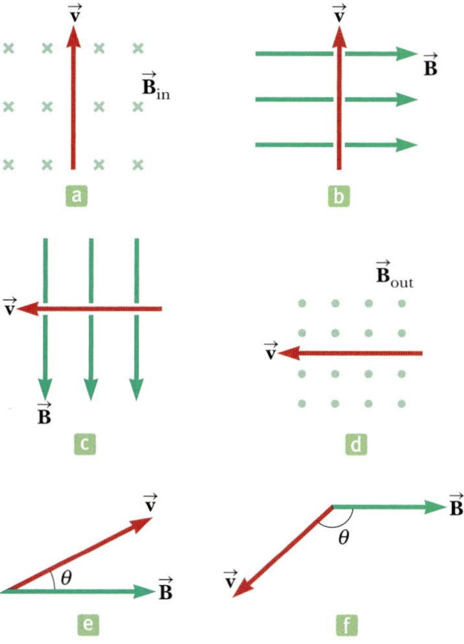


**Figure P19.2** Problems 2 and 12.

**3.** Find the direction of the magnetic field acting on the positively charged particle moving in the various situations shown in Figure P19.3 if the direction of the magnetic force acting on it is as indicated.

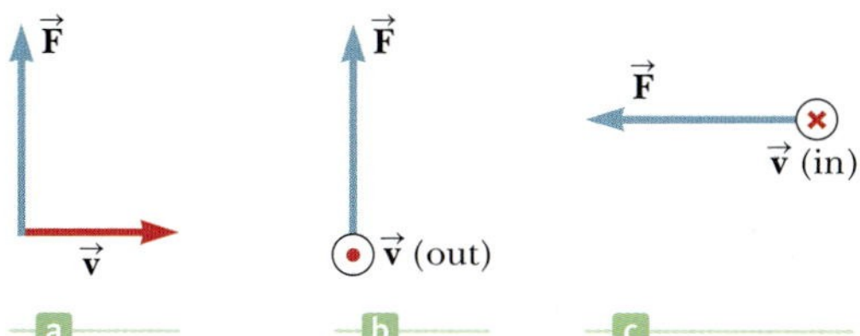


**Figure P19.3** (Problems 3 and 15) For Problem 15, replace the velocity vector with a current in that direction.

**4.** Determine the initial direction of the deflection of charged particles as they enter the magnetic fields, as shown in Figure P19.4.

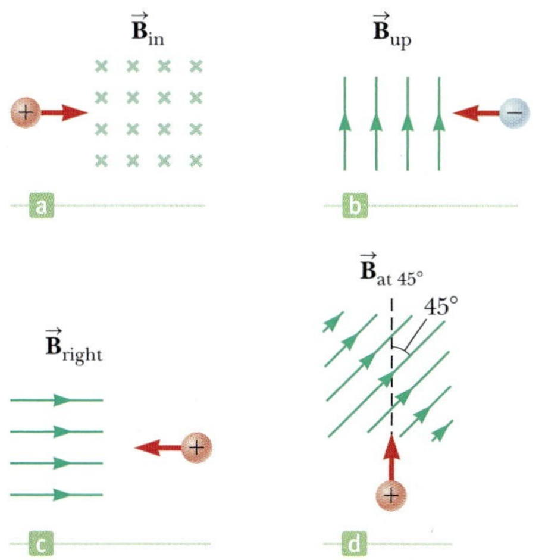


**Figure P19.4**

**5.** **QIC** A laboratory electromagnet produces a magnetic field of magnitude 1.50 T. A proton moves through this field with a speed of $6.00 \times 10^6$ m/s. (a) Find the magnitude of the maximum magnetic force that could be exerted on the proton. (b) What is the magnitude of the maximum acceleration of the proton? (c) Would the field exert the same magnetic force on an electron moving through the field with the same speed? (d) Would the electron undergo the same acceleration? Explain.

**6.** **M** A proton moves perpendicular to a uniform magnetic field $\vec{\mathbf{B}}$ at a speed of $1.00 \times 10^7$ m/s and experiences an acceleration of $2.00 \times 10^{13}$ m/s$^2$ in the positive $x$-direction when its velocity is in the positive $z$-direction. Determine the magnitude and direction of the field.

**7.** What velocity would a proton need to circle Earth 1 000 km above the magnetic equator, where Earth's magnetic field is directed horizontally north and has a magnitude of $4.00 \times 10^{-8}$ T?

**8.** **W** An electron is accelerated through 2 400 V from rest and then enters a region where there is a uniform 1.70-T magnetic field. What are (a) the maximum and (b) the minimum magnitudes of the magnetic force acting on this electron?

**9.** **M** A proton moving at $4.00 \times 10^6$ m/s through a magnetic field of magnitude 1.70 T experiences a magnetic force of magnitude $8.20 \times 10^{-13}$ N. What is the angle between the proton's velocity and the field?

**10.** **BIO** Sodium ions ($Na^+$) move at 0.851 m/s through a bloodstream in the arm of a person standing near a large magnet. The magnetic field has a strength of 0.254 T and makes an angle of 51.0° with the motion of the sodium ions. The arm contains 100 cm$^3$ of blood with a concentration of $3.00 \times 10^{20}$ $Na^+$ ions per cubic centimeter. If no other ions were present in the arm, what would be the magnetic force on the arm?

**11.** At the equator, near the surface of Earth, the magnetic field is approximately 50.0 $\mu$T northward, and the electric field is about 100 N/C downward in fair weather. Find the gravitational, electric, and magnetic forces on an electron with an instantaneous velocity of $6.00 \times 10^6$ m/s directed to the east in this environment.

## 19.4 Magnetic Force on a Current-Carrying Conductor

**12.** In Figure P19.2 assume in each case the velocity vector shown is replaced with a wire carrying a current in the direction of the velocity vector. For each case, find the direction of the magnetic force acting on the wire.

**13.** **W** A current $I = 15$ A is directed along the positive $x$-axis and perpendicular to a magnetic field. A magnetic force per unit length of 0.12 N/m acts on the conductor in the negative $y$-direction. Calculate the magnitude and direction of the magnetic field in the region through which the current passes.

14. **Q|C** A straight wire carrying a 3.0-A current is placed in a uniform magnetic field of magnitude 0.28 T directed perpendicular to the wire. (a) Find the magnitude of the magnetic force on a section of the wire having a length of 14 cm. (b) Explain why you can't determine the direction of the magnetic force from the information given in the problem.

15. In Figure P19.3 assume in each case the velocity vector shown is replaced with a wire carrying a current in the direction of the velocity vector. For each case, find the direction of the magnetic field that will produce the magnetic force shown.

16. A wire having a mass per unit length of 0.500 g/cm carries a 2.00-A current horizontally to the south. What are the direction and magnitude of the minimum magnetic field needed to lift this wire vertically upward?

17. A wire carries a current of 10.0 A in a direction that makes an angle of 30.0° with the direction of a magnetic field of strength 0.300 T. Find the magnetic force on a 5.00-m length of the wire.

18. At a certain location, Earth has a magnetic field of $0.60 \times 10^{-4}$ T, pointing 75° below the horizontal in a north–south plane. A 10.0-m-long straight wire carries a 15-A current. (a) If the current is directed horizontally toward the east, what are the magnitude and direction of the magnetic force on the wire? (b) What are the magnitude and direction of the force if the current is directed vertically upward?

19. A wire with a mass of 1.00 g/cm is placed on a horizontal surface with a coefficient of friction of 0.200. The wire carries a current of 1.50 A eastward and moves horizontally to the north. What are the magnitude and the direction of the *smallest* vertical magnetic field that enables the wire to move in this fashion?

20. A conductor suspended by two flexible wires as shown in Figure P19.20 has a mass per unit length of 0.040 0 kg/m. (a) What current must exist in the conductor for the tension in the supporting wires to be zero when the magnetic field is 3.60 T into the page? (b) What is the required direction for the current?

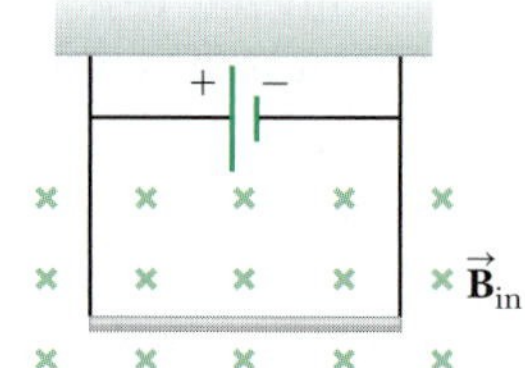

**Figure P19.20**

21. **Q|C** Consider the system pictured in Figure P19.21. A 15-cm length of conductor of mass 15 g, free to move vertically, is placed between two thin, vertical conductors, and a uniform magnetic field acts perpendicular to the page. When a 5.0-A current is directed as shown in the figure, the horizontal wire moves upward at constant velocity in the presence of gravity. (a) What forces act on the horizontal wire, and under what condition is the wire able to move upward at constant velocity? (b) Find the magnitude and direction of the minimum magnetic field required to move the wire at constant speed. (c) What happens if the magnetic field exceeds this minimum value? (The wire slides without friction on the two vertical conductors.)

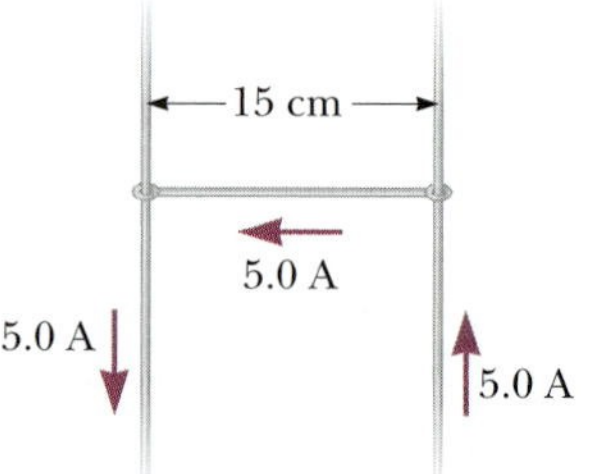

**Figure P19.21**

22. **S** A metal rod of mass $m$ carrying a current $I$ glides on two horizontal rails a distance $d$ apart. If the coefficient of kinetic friction between the rod and rails is $\mu_k$, what vertical magnetic field is required to keep the rod moving at a constant speed?

23. In Figure P19.23 the cube is 40.0 cm on each edge. Four straight segments of wire—$ab$, $bc$, $cd$, and $da$—form a closed loop that carries a current $I$ = 5.00 A in the direction shown. A uniform magnetic field of magnitude $B$ = 0.020 0 T is in the positive $y$-direction. Determine the magnitude and direction of the magnetic force on each segment.

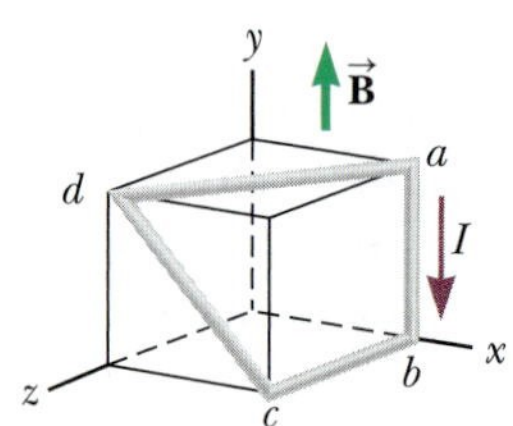

**Figure P19.23**

24. A horizontal power line of length 58 m carries a current of 2.2 kA as shown in Figure P19.24. Earth's magnetic field at this location has a magnitude equal to $5.0 \times 10^{-5}$ T and makes an angle of 65° with the power line. Find the magnitude and direction of the magnetic force on the power line.

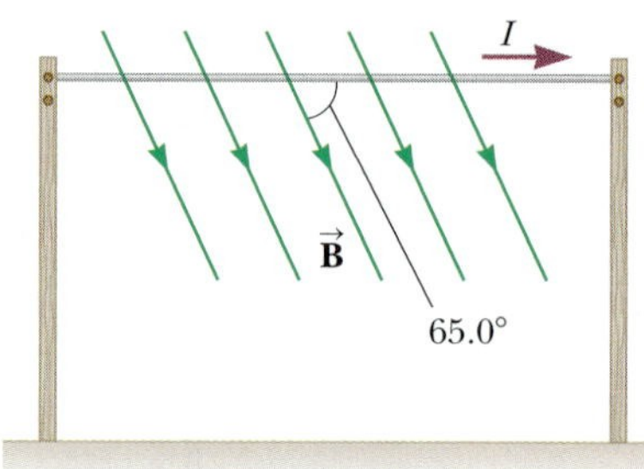

**Figure P19.24**

## 19.5 Torque on a Current Loop and Electric Motors

25. A wire is formed into a circle having a diameter of 10.0 cm and is placed in a uniform magnetic field of 3.00 mT. The wire carries a current of 5.00 A. Find the maximum torque on the wire.

26. A current of 17.0 mA is maintained in a single circular loop with a circumference of 2.00 m. A magnetic field of 0.800 T is directed parallel to the plane of the loop. What is the magnitude of the torque exerted by the magnetic field on the loop?

27. M An eight-turn coil encloses an elliptical area having a major axis of 40.0 cm and a minor axis of 30.0 cm (Fig. P19.27). The coil lies in the plane of the page and carries a clockwise current of 6.00 A. If the coil is in a uniform magnetic field of $2.00 \times 10^{-4}$ T directed toward the left of the page, what is the magnitude of the torque on the coil? *Hint:* The area of an ellipse is $A = \pi ab$, where $a$ and $b$ are, respectively, the semimajor and semiminor axes of the ellipse.

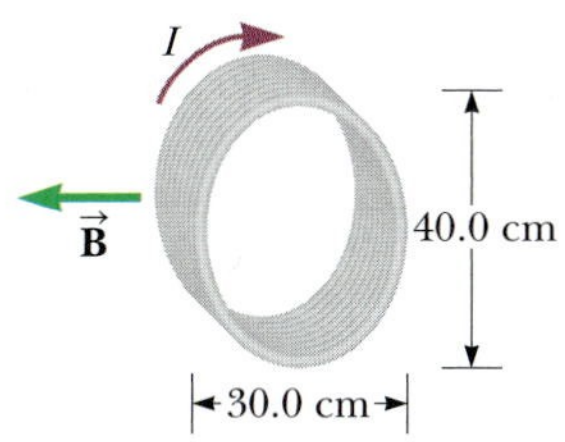

**Figure P19.27**

28. W A rectangular loop consists of 100 closely wrapped turns and has dimensions 0.400 m by 0.300 m. The loop is hinged along the $y$-axis, and the plane of the coil makes an angle of 30.0° with the $x$-axis (Fig. P19.28). What is the magnitude of the torque exerted on the loop by a uniform magnetic field of 0.800 T directed along the $x$-axis when the current in the windings has a value of 1.20 A in the direction shown? What is the expected direction of rotation of the loop?

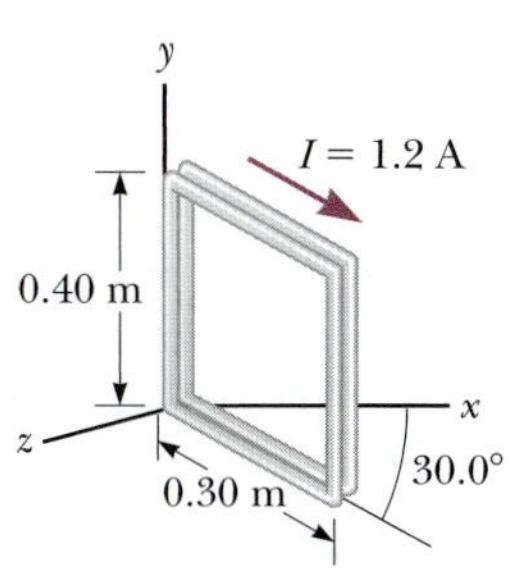

**Figure P19.28**

29. A 200-turn rectangular coil having dimensions of 3.0 cm by 5.0 cm is placed in a uniform magnetic field of magnitude 0.90 T. (a) Find the current in the coil if the maximum torque exerted on it by the magnetic field is 0.15 N · m. (b) Find the magnitude of the torque on the coil when the magnetic field makes an angle of 25° with the normal to the plane of the coil.

30. A copper wire is 8.00 m long and has a cross-sectional area of $1.00 \times 10^{-4}$ m². The wire forms a one-turn loop in the shape of a square and is then connected to a battery that applies a potential difference of 0.100 V. If the loop is placed in a uniform magnetic field of magnitude 0.400 T, what is the maximum torque that can act on it? The resistivity of copper is $1.70 \times 10^{-8}$ Ω · m.

31. A long piece of wire with a mass of 0.100 kg and a total length of 4.00 m is used to make a square coil with a side of 0.100 m. The coil is hinged along a horizontal side, carries a 3.40-A current, and is placed in a vertical magnetic field with a magnitude of 0.010 0 T. (a) Determine the angle that the plane of the coil makes with the vertical when the coil is in equilibrium. (b) Find the torque acting on the coil due to the magnetic force at equilibrium.

32. GP A rectangular loop has dimensions 0.500 m by 0.300 m. The loop is hinged along the $x$-axis and lies in the $xy$-plane (Fig. P19.32). A uniform magnetic field of 1.50 T is directed at an angle of 40.0° with respect to the positive $y$-axis and lies parallel everywhere to the $yz$-plane. The loop carries a current of 0.900 A in the direction shown. (Ignore gravitation.) (a) In what direction is magnetic force exerted on wire segment $ab$? What is the direction of the magnetic torque associated with this force, as computed with respect to the $x$-axis? (b) What is the direction of the magnetic force exerted on segment $cd$? What is the direction of the magnetic torque associated with this force, again computed with respect to the $x$-axis? (c) Can the forces examined in parts (a) and (b) combine to cause the loop to rotate around the $x$-axis? Can they affect the motion of the loop in any way? Explain. (d) What is the direction (in the $yz$-plane) of the magnetic force exerted on segment $bc$? Measuring torques with respect to the $x$-axis, what is the direction of the torque exerted by the force on segment $bc$? (e) Looking toward the origin along the positive $x$-axis, will the loop rotate clockwise or counterclockwise? (f) Compute the magnitude of the magnetic moment of the loop. (g) What is the angle between the magnetic moment vector and the magnetic field? (h) Compute the torque on the loop using the values found for the magnetic moment and magnetic field.

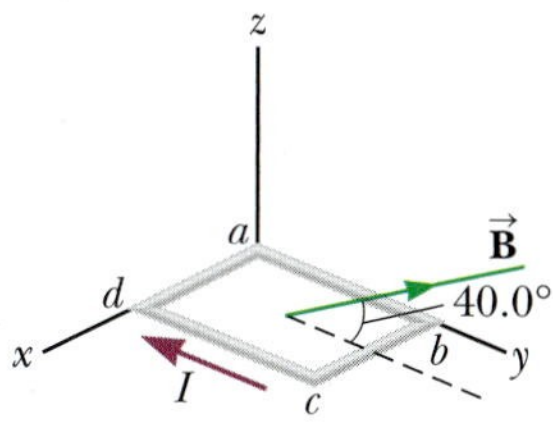

**Figure P19.32**

## 19.6 Motion of a Charged Particle in a Magnetic Field

33. An electron moves in a circular path perpendicular to a magnetic field of magnitude 0.235 T. If the kinetic energy of the electron is $3.30 \times 10^{-19}$ J, find (a) the speed of the electron and (b) the radius of the circular path.

34. A proton travels with a speed of $5.02 \times 10^{6}$ m/s at an angle of 60° with the direction of a magnetic field of magnitude 0.180 T in the positive $x$-direction. What are (a) the magnitude of the magnetic force on the proton and (b) the proton's acceleration?

35. Figure P19.35a is a diagram of a device called a velocity selector, in which particles of a specific velocity pass through undeflected while those with greater or lesser velocities are deflected either upwards or downwards. An electric field is directed perpendicular to a magnetic field, producing an electric force and a magnetic force on the charged particle that can be equal in magnitude and opposite in direction (Fig. P19.35b) and hence cancel. Show that particles with a speed of $v = E/B$ will pass through the velocity selector undeflected.

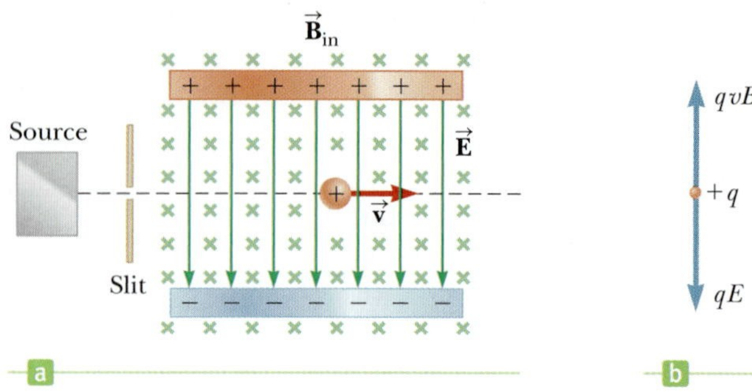


**Figure P19.35**

36. W Consider the mass spectrometer shown schematically in Figure P19.36. The electric field between the plates of the velocity selector is 950 V/m, and the magnetic fields in both the velocity selector and the deflection chamber have magnitudes of 0.930 T. Calculate the radius of the path in the system for a singly charged ion with mass $m = 2.18 \times 10^{-26}$ kg. *Hint:* See Problem 35.

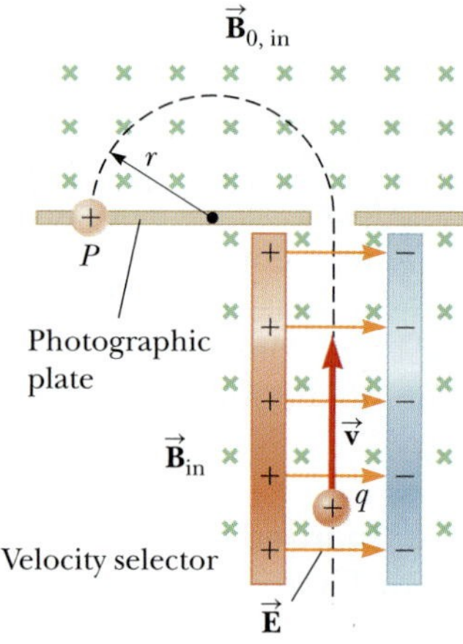


**Figure P19.36** (Problems 36 and 41) A mass spectrometer. Charged particles are first sent through a velocity selector. They then enter a region where a magnetic field $\vec{B}_0$ (directed inward) causes positive ions to move in a semicircular path and strike a photographic film at $P$.

37. A singly charged positive ion has a mass of $2.50 \times 10^{-26}$ kg. After being accelerated through a potential difference of 250 V, the ion enters a magnetic field of 0.500 T, in a direction perpendicular to the field. Calculate the radius of the path of the ion in the field.

38. A mass spectrometer is used to examine the isotopes of uranium. Ions in the beam emerge from the velocity selector at a speed of $3.00 \times 10^5$ m/s and enter a uniform magnetic field of 0.600 T directed perpendicularly to the velocity of the ions. What is the distance between the impact points formed on the photographic plate by singly charged ions of $^{235}$U and $^{238}$U?

39. A proton is at rest at the plane vertical boundary of a region containing a uniform vertical magnetic field $B$ (Fig. P19.39). An alpha particle moving horizontally makes a head-on elastic collision with the proton. Immediately after the collision, both particles enter the magnetic field, moving perpendicular to the direction of the field. The radius of the proton's trajectory is $R$. The mass of the alpha particle is four times that of the proton, and its charge is twice that of the proton. Find the radius of the alpha particle's trajectory.

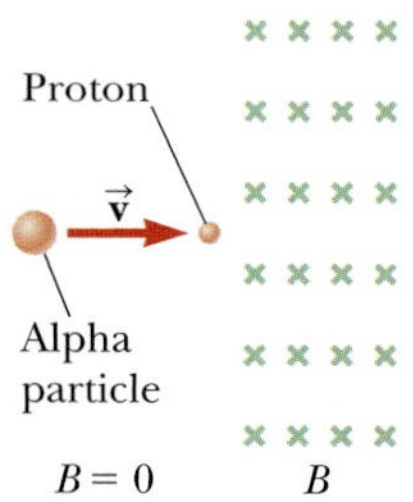


**Figure P19.39**

40. S A particle with charge $q$ and kinetic energy $KE$ travels in a uniform magnetic field of magnitude $B$. If the particle moves in a circular path of radius $R$, find expressions for (a) its speed and (b) its mass.

41. A particle passes through a mass spectrometer as illustrated in Figure P19.36. The electric field between the plates of the velocity selector has a magnitude of 8 250 V/m, and the magnetic fields in both the velocity selector and the deflection chamber have magnitudes of 0.093 1 T. In the deflection chamber the particle strikes a photographic plate 39.6 cm removed from its exit point after traveling in a semicircle. (a) What is the mass-to-charge ratio of the particle? (b) What is the mass of the particle if it is doubly ionized? (c) What is its identity, assuming it's an element?

42. S A proton (charge $+e$, mass $m_p$), a deuteron (charge $+e$, mass $2m_p$), and an alpha particle (charge $+2e$, mass $4m_p$) are accelerated from rest through a common potential difference $\Delta V$. Each of the particles enters a uniform magnetic field $\vec{B}$, with its velocity in a direction perpendicular to $\vec{B}$. The proton moves in a circular path of radius $r_p$. In terms of $r_p$, determine (a) the radius $r_d$ of the circular orbit for the deuteron and (b) the radius $r_\alpha$ for the alpha particle.

## 19.7 Magnetic Field of a Long, Straight Wire and Ampère's Law

43. A lightning bolt may carry a current of $1.00 \times 10^4$ A for a short time. What is the resulting magnetic field 100 m from the bolt? Suppose the bolt extends far above and below the point of observation.

44. In each of parts (a), (b), and (c) of Figure P19.44, find the direction of the current in the wire that would produce a magnetic field directed as shown.

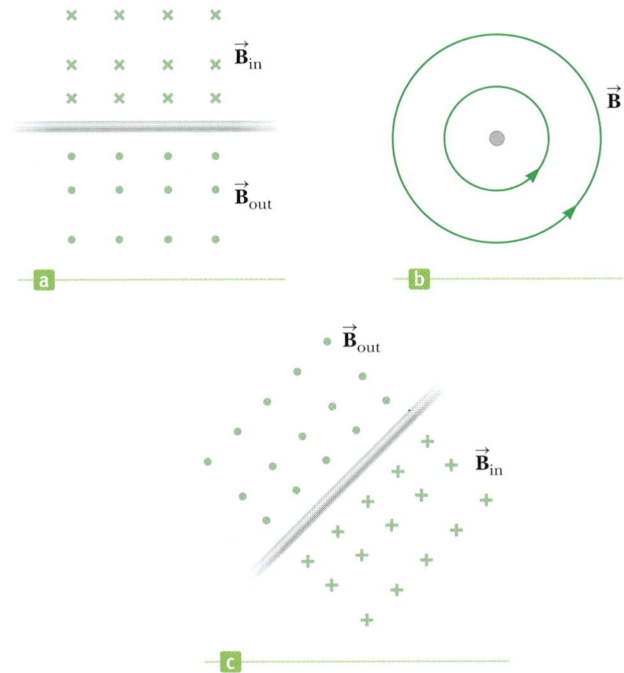

Figure P19.44

45. BIO Neurons in our bodies carry weak currents that produce detectable magnetic fields. A technique called *magnetoencephalography*, or MEG, is used to study electrical activity in the brain using this concept. This technique is capable of detecting magnetic fields as weak as $1.0 \times 10^{-15}$ T. Model the neuron as a long wire carrying a current and find the current it must carry to produce a field of this magnitude at a distance of 4.0 cm from the neuron.

46. In 1962 measurements of the magnetic field of a large tornado were made at the Geophysical Observatory in Tulsa, Oklahoma. If the magnitude of the tornado's field was $B = 1.50 \times 10^{-8}$ T pointing north when the tornado was 9.00 km east of the observatory, what current was carried up or down the funnel of the tornado? Model the vortex as a long, straight wire carrying a current.

47. BIO A cardiac pacemaker can be affected by a static magnetic field as small as 1.7 mT. How close can a pacemaker wearer come to a long, straight wire carrying 20 A?

48. The two wires shown in Figure P19.48 are separated by $d = 10.0$ cm and carry currents of $I = 5.00$ A in opposite directions. Find the magnitude and direction of the net magnetic field (a) at a point midway between the wires; (b) at point $P_1$, 10.0 cm to the right of the wire on the right; and (c) at point $P_2$, $2d = 20.0$ cm to the left of the wire on the left.

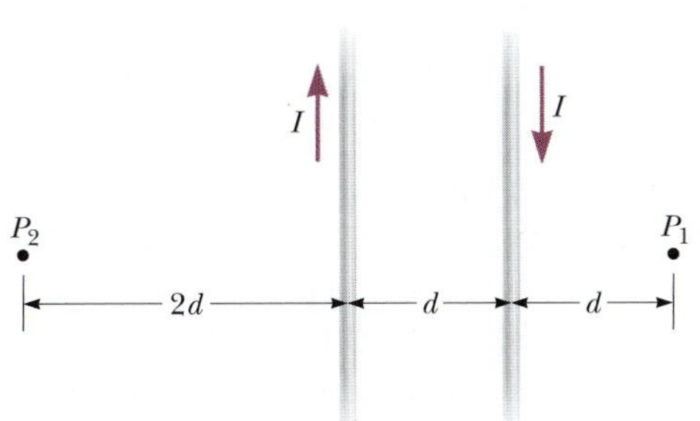

Figure P19.48

49. Four long, parallel conductors carry equal currents of $I = 5.00$ A. Figure P19.49 is an end view of the conductors. The direction of the current is into the page at points $A$ and $B$ (indicated by the crosses) and out of the page at $C$ and $D$ (indicated by the dots). Calculate the magnitude and direction of the magnetic field at point $P$, located at the center of the square with edge of length 0.200 m.

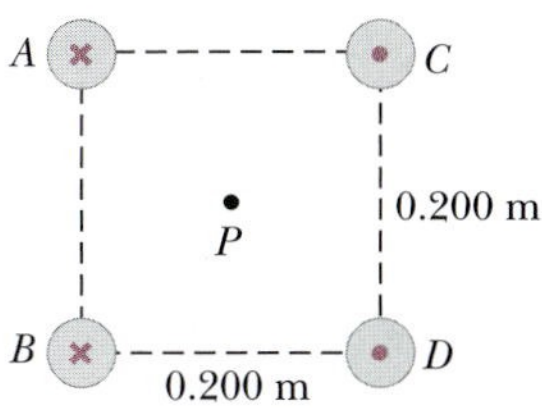

Figure P19.49

50. Two long, parallel wires carry currents of $I_1 = 3.00$ A and $I_2 = 5.00$ A in the direction indicated in Figure P19.50. (a) Find the magnitude and direction of the magnetic field at a point midway between the wires ($d = 20.0$ cm). (b) Find the magnitude and direction of the magnetic field at point $P$, located $d = 20.0$ cm above the wire carrying the 5.00-A current.

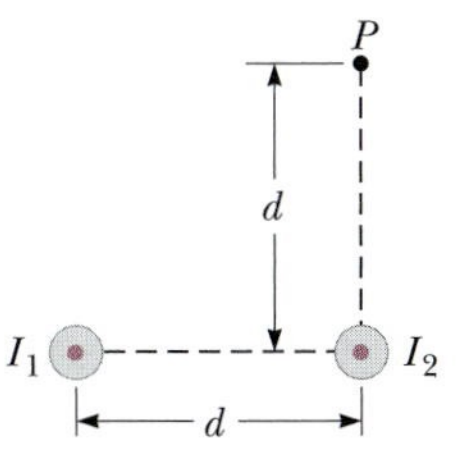

Figure P19.50

51. M A wire carries a 7.00-A current along the $x$-axis, and another wire carries a 6.00-A current along the $y$-axis, as shown in Figure P19.51. What is the magnetic field at point $P$, located at $x = 4.00$ m, $y = 3.00$ m?

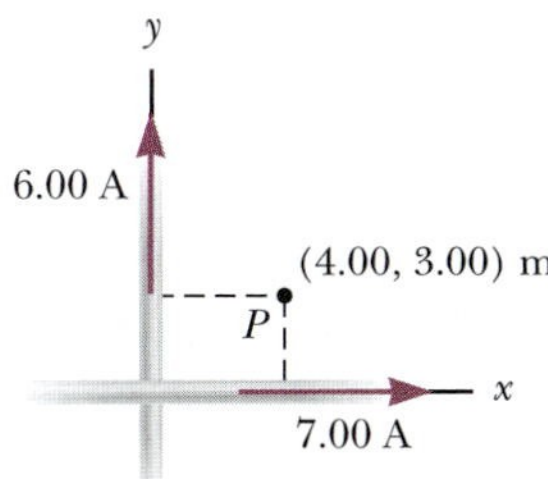

Figure P19.51

52. GP A long, straight wire lies on a horizontal table in the $xy$-plane and carries a current of 1.20 $\mu$A in the positive $x$-direction along the $x$-axis. A proton is traveling in the negative $x$-direction at speed $2.30 \times 10^4$ m/s a distance $d$ above the wire (i.e., $z = d$). (a) What is the direction of the magnetic field of the wire at the position of the proton? (b) What is the direction of the magnetic force acting on the proton? (c) Explain why the direction of the proton's motion doesn't change. (d) Using Newton's second law, find a symbolic expression for $d$ in terms of the acceleration of gravity $g$, the proton mass $m$, its speed $v$, charge $q$, and the current $I$.

(e) Find the numeric answer for the distance $d$ using the results of part (d).

53. The magnetic field 40.0 cm away from a long, straight wire carrying current 2.00 A is 1.00 $\mu$T. (a) At what distance is it 0.100 $\mu$T? (b) At one instant, the two conductors in a long household extension cord carry equal 2.00-A currents in opposite directions. The two wires are 3.00 mm apart. Find the magnetic field 40.0 cm away from the middle of the straight cord, in the plane of the two wires. (c) At what distance is it one-tenth as large? (d) The center wire in a coaxial cable carries current 2.00 A in one direction, and the sheath around it carries current 2.00 A in the opposite direction. What magnetic field does the cable create at points outside?

54. Q|C S Two long, parallel wires separated by a distance $2d$ carry equal currents in the same direction. An end view of the two wires is shown in Figure P19.54, where the currents are out of the page. (a) What is the direction of the magnetic field at $P$ on the $x$-axis set up by the two wires? (b) Find an expression for the magnitude of the field at $P$. (c) From your result to part (b), determine the field at a point midway between the two wires. Does your result meet with your expectation? Explain.

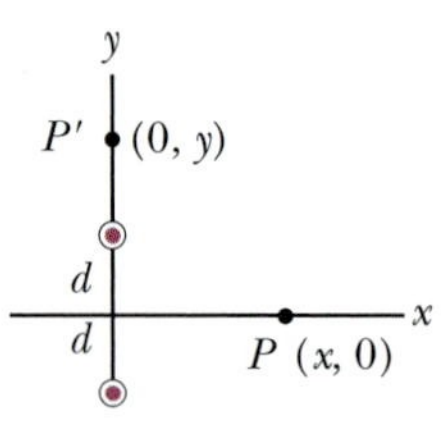

Figure P19.54

## 19.8 Magnetic Force Between Two Parallel Conductors

55. Two long, parallel wires separated by 2.50 cm carry currents in opposite directions. The current in one wire is 1.25 A, and the current in the other is 3.50 A. (a) Find the magnitude of the force per unit length that one wire exerts on the other. (b) Is the force attractive or repulsive?

56. Q|C Two parallel wires separated by 4.0 cm repel each other with a force per unit length of $2.0 \times 10^{-4}$ N/m. The current in one wire is 5.0 A. (a) Find the current in the other wire. (b) Are the currents in the same direction or in opposite directions? (c) What would happen if the direction of one current were reversed and doubled?

57. A wire with a weight per unit length of 0.080 N/m is suspended directly above a second wire. The top wire carries a current of 30.0 A, and the bottom wire carries a current of 60.0 A. Find the distance of separation between the wires so that the top wire will be held in place by magnetic repulsion.

58. In Figure P19.58 the current in the long, straight wire is $I_1 = 5.00$ A, and the wire lies in the plane of the rectangular loop, which carries 10.0 A. The dimensions shown are $c = 0.100$ m, $a = 0.150$ m, and $\ell = 0.450$ m. Find the magnitude and direction of the net force exerted by the magnetic field due to the straight wire on the loop.

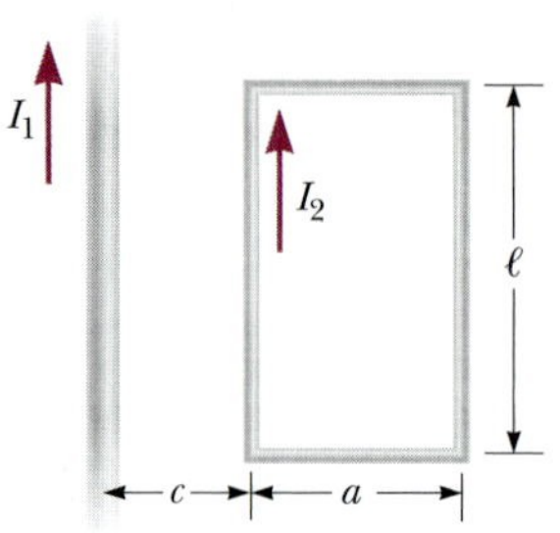

Figure P19.58

## 19.9 Magnetic Fields of Current Loops and Solenoids

59. M A long solenoid that has 1 000 turns uniformly distributed over a length of 0.400 m produces a magnetic field of magnitude $1.00 \times 10^{-4}$ T at its center. What current is required in the windings for that to occur?

60. A certain superconducting magnet in the form of a solenoid of length 0.50 m can generate a magnetic field of 9.0 T in its core when its coils carry a current of 75 A. The windings, made of a niobium–titanium alloy, must be cooled to 4.2 K. Find the number of turns in the solenoid.

61. It is desired to construct a solenoid that will have a resistance of 5.00 Ω (at 20°C) and produce a magnetic field of $4.00 \times 10^{-2}$ T at its center when it carries a current of 4.00 A. The solenoid is to be constructed from copper wire having a diameter of 0.500 mm. If the radius of the solenoid is to be 1.00 cm, determine (a) the number of turns of wire needed and (b) the length the solenoid should have.

62. A solenoid 10.0 cm in diameter and 75.0 cm long is made from copper wire of diameter 0.100 cm, with very thin insulation. The wire is wound onto a cardboard tube in a single layer, with adjacent turns touching each other. What power must be delivered to the solenoid if it is to produce a field of 8.00 mT at its center?

63. An electron is moving at a speed of $1.0 \times 10^4$ m/s in a circular path of radius 2.0 cm inside a solenoid. The magnetic field of the solenoid is perpendicular to the plane of the electron's path. Find (a) the strength of the magnetic field inside the solenoid and (b) the current in the solenoid if it has 25 turns per centimeter.

## Additional Problems

64. Q|C Figure P19.64 is a setup that can be used to measure magnetic fields. A rectangular coil of wire contains $N$ turns has a width $w$. The coil is attached to one arm of a balance and is suspended between the poles of a magnet. The field is uniform and perpendicular to the plane of the coil. The system is first balanced when the current in the coil is zero. When the switch is closed and the coil carries a current $I$, a mass $m$ must be added to the right side to balance the system. (a) Find an expression for the magnitude of the magnetic field

and determine its direction. (b) Why is the result independent of the vertical dimension of the coil? (c) Suppose the coil has 50 turns and width of 5.0 cm. When the switch is closed, the coil carries a current of 0.30 A, and a mass of 20.0 g must be added to the right side to balance the system. What is the magnitude of the magnetic field?

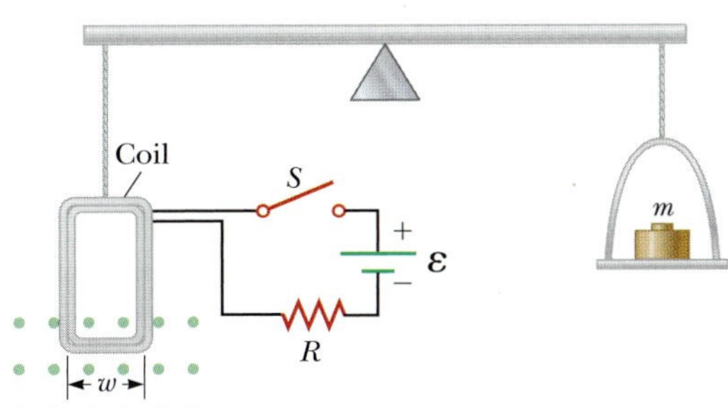


Figure P19.64

65. Two coplanar and concentric circular loops of wire carry currents of $I_1 = 5.00$ A and $I_2 = 3.00$ A in opposite directions as in Figure P19.65. (a) If $r_1 = 12.0$ cm and $r_2 = 9.00$ cm, what are (a) the magnitude and (b) the direction of the net magnetic field at the center of the two loops? (c) Let $r_1$ remain fixed at 12.0 cm and let $r_2$ be a variable. Determine the value of $r_2$ such that the net field at the center of the loop is zero.

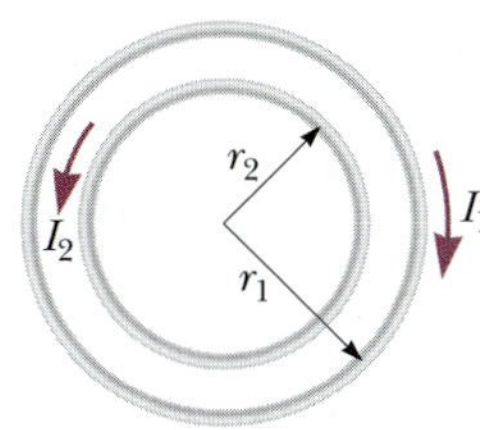


Figure P19.65

66. An electron moves in a circular path perpendicular to a constant magnetic field of magnitude 1.00 mT. The angular momentum of the electron about the center of the circle is $4.00 \times 10^{-25}$ kg · m²/s. Determine (a) the radius of the circular path and (b) the speed of the electron.

67. Two long, straight wires cross each other at right angles, as shown in Figure P19.67. (a) Find the direction and magnitude of the magnetic field at point *P*, which is in the same plane as the two wires. (b) Find the magnetic field at a point 30.0 cm above the point of intersection (30.0 cm out of the page, toward you).

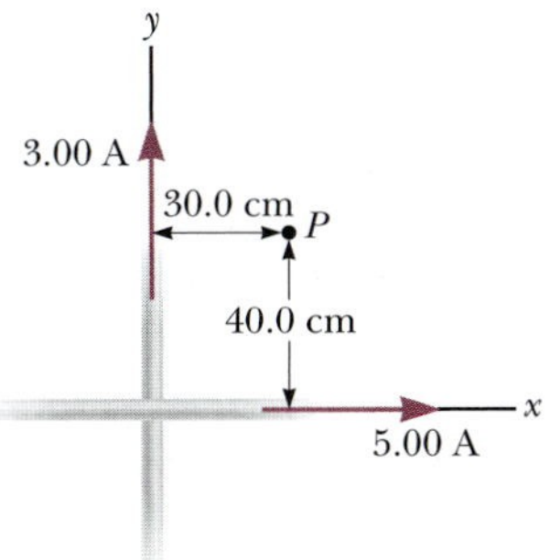


Figure P19.67

68. A 0.200-kg metal rod carrying a current of 10.0 A glides on two horizontal rails 0.500 m apart. What vertical magnetic field is required to keep the rod moving at a constant speed if the coefficient of kinetic friction between the rod and rails is 0.100?

69. BIO Using an electromagnetic flowmeter (Fig. P19.69), a heart surgeon monitors the flow rate of blood through an artery. Electrodes *A* and *B* make contact with the outer surface of the blood vessel, which has interior diameter 3.00 mm. (a) For a magnetic field magnitude of 0.040 0 T, a potential difference of 160 μV appears between the electrodes. Calculate the speed of the blood. (b) Verify that electrode *A* is positive, as shown. Does the sign of the emf depend on whether the mobile ions in the blood are predominantly positively or negatively charged? Explain.

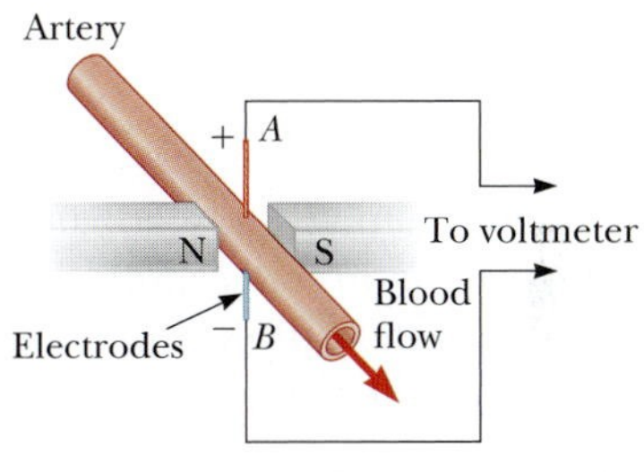


Figure P19.69

70. A uniform horizontal wire with a linear mass density of 0.50 g/m carries a 2.0-A current. It is placed in a constant magnetic field with a strength of $4.0 \times 10^{-3}$ T. The field is horizontal and perpendicular to the wire. As the wire moves upward starting from rest, (a) what is its acceleration and (b) how long does it take to rise 50 cm? Neglect the magnetic field of Earth.

71. Three long, parallel conductors carry currents of $I =$ 2.0 A. Figure P19.71 is an end view of the conductors, with each current coming out of the page. Given that $a = 1.0$ cm, determine the magnitude and direction of the magnetic field at points *A*, *B*, and *C*.

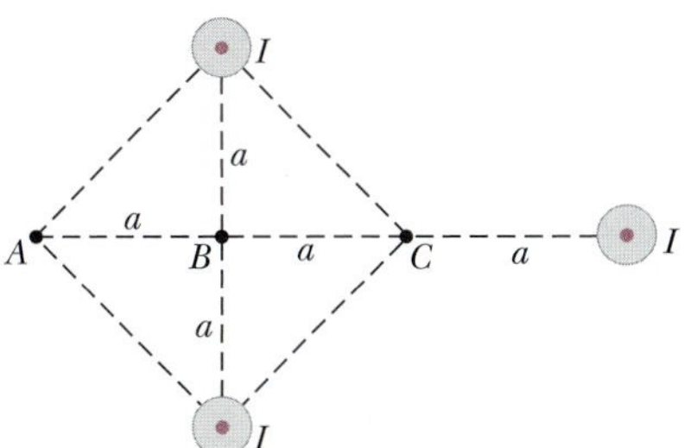


Figure P19.71

72. Two long, parallel wires, each with a mass per unit length of 40 g/m, are supported in a horizontal plane by 6.0-cm-long strings, as shown in Figure P19.72. Each wire carries the same current *I*, causing the wires to

repel each other so that the angle $\theta$ between the supporting strings is 16°. (a) Are the currents in the same or opposite directions? (b) Determine the magnitude of each current.

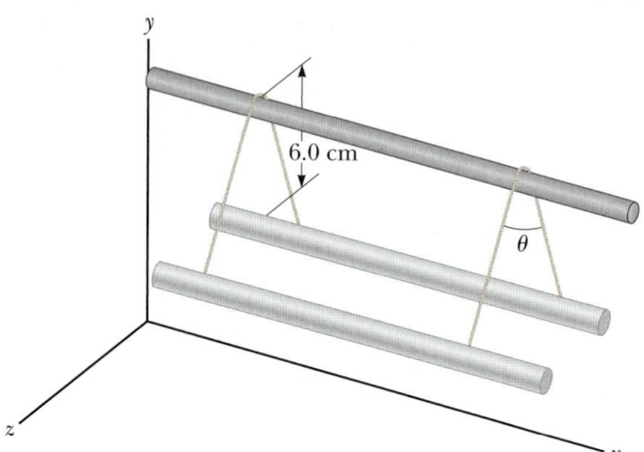

Figure P19.72

73. Protons having a kinetic energy of 5.00 MeV are moving in the positive $x$-direction and enter a magnetic field of 0.050 0 T in the $z$-direction, out of the plane of the page, and extending from $x = 0$ to $x = 1.00$ m as in Figure P19.73. (a) Calculate the $y$-component of the protons' momentum as they leave the magnetic field. (b) Find the angle $\alpha$ between the initial velocity vector of the proton beam and the velocity vector after the beam emerges from the field. *Hint:* Neglect relativistic effects and note that 1 eV $= 1.60 \times 10^{-19}$ J.

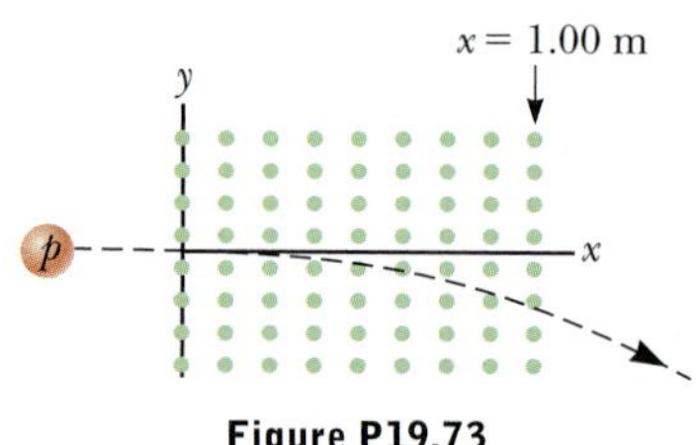

Figure P19.73

74. A straight wire of mass 10.0 g and length 5.0 cm is suspended from two identical springs that, in turn, form a closed circuit (Fig. P19.74). The springs stretch a distance of 0.50 cm under the weight of the wire. The circuit has a total resistance of 12 Ω. When a magnetic field directed out of the page (indicated by the dots in the figure) is turned on, the springs are observed to stretch an additional 0.30 cm. What is the strength of the magnetic field? (The upper portion of the circuit is fixed.)

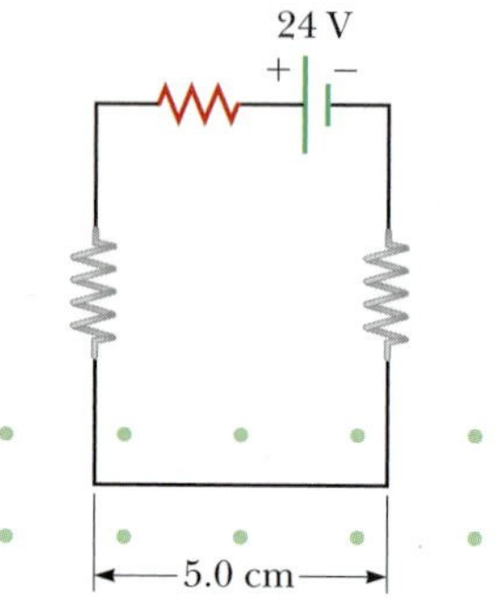

Figure P19.74

75. A 1.00-kg ball having net charge $Q = 5.00\ \mu$C is thrown out of a window horizontally at a speed $v = 20.0$ m/s. The window is at a height $h = 20.0$ m above the ground. A uniform horizontal magnetic field of magnitude $B = 0.010\ 0$ T is perpendicular to the plane of the ball's trajectory. Find the magnitude of the magnetic force acting on the ball just before it hits the ground. *Hint:* Ignore magnetic forces in finding the ball's final velocity.

76. Two long, parallel conductors separated by 10.0 cm carry currents in the same direction. The first wire carries a current $I_1 = 5.00$ A, and the second carries $I_2 = 8.00$ A. (a) What is the magnitude of the magnetic field created by $I_1$ at the location of $I_2$? (b) What is the force per unit length exerted by $I_1$ on $I_2$? (c) What is the magnitude of the magnetic field created by $I_2$ at the location of $I_1$? (d) What is the force per length exerted by $I_2$ on $I_1$?

South West News Service

A forest of fluorescent lights, not wired to any power source, are lighted by electromagnetic induction. Changing currents in the power wires overhead create time-dependent magnetic flux in the vicinity of the tubes, inducing a voltage across them.

# Induced Voltages and Inductance 20

In 1819 Hans Christian Oersted discovered that an electric current exerted a force on a magnetic compass. Although there had long been speculation that such a relationship existed, Oersted's finding was the first evidence of a link between electricity and magnetism. Because nature is often symmetric, the discovery that electric currents produce magnetic fields led scientists to suspect that magnetic fields could produce electric currents. Indeed, experiments conducted by Michael Faraday in England and independently by Joseph Henry in the United States in 1831 showed that a changing magnetic field could induce an electric current in a circuit. The results of these experiments led to a basic and important law known as Faraday's law. In this chapter we discuss Faraday's law and several practical applications, one of which is the production of electrical energy in power plants throughout the world.



## 20.1 Induced emf and Magnetic Flux

**LEARNING OBJECTIVES**

1. Define magnetic flux and discuss its role in producing an induced emf.
2. Evaluate the magnitude and change in the magnetic flux through a given area.

An experiment first conducted by Faraday demonstrated that a current can be produced by a changing magnetic field. The apparatus shown in Figure 20.1 consists of a coil connected to a switch and a battery. We call this coil the *primary coil* and the corresponding circuit the primary circuit. The coil is wrapped around an iron ring to intensify the magnetic field produced by the current in the coil. A

**Figure 20.1** Faraday's experiment.

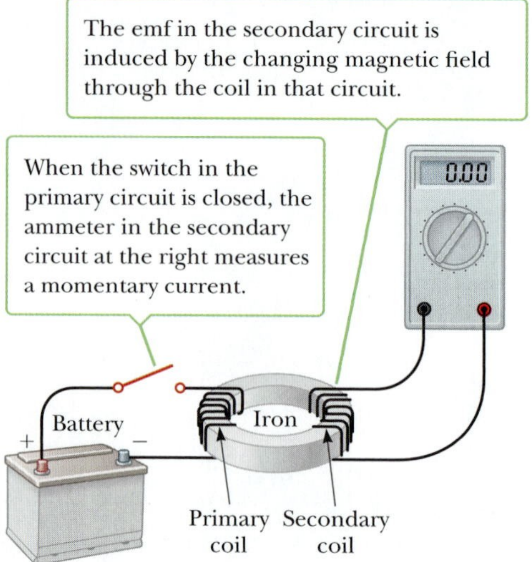

The emf in the secondary circuit is induced by the changing magnetic field through the coil in that circuit.
When the switch in the primary circuit is closed, the ammeter in the secondary circuit at the right measures a momentary current.
0.00
Battery
+
−
Iron
Primary coil
Secondary coil

**Michael Faraday**
**British physicist and chemist (1791–1867)**
Faraday is often regarded as the greatest experimental scientist of the 1800s. His many contributions to the study of electricity include the inventions of the electric motor, electric generator, and transformer, as well as the discovery of electromagnetic induction and the laws of electrolysis. Greatly influenced by religion, he refused to work on military poison gas for the British government.

*secondary coil*, at the right, is wrapped around the iron ring and is connected to an ammeter. The corresponding circuit is called the secondary circuit. It's important to notice that **there is no battery in the secondary circuit.**

At first glance, you might guess that no current would ever be detected in the secondary circuit. When the switch in the primary circuit in Figure 20.1 is suddenly closed, however, something amazing happens: the ammeter measures a current in the secondary circuit and then returns to zero! When the switch is opened again, the ammeter reads a current in the opposite direction and again returns to zero. Finally, whenever there is a steady current in the primary circuit, the ammeter reads zero.

From such observations, Faraday concluded that an electric current could be produced by a *changing* magnetic field. (A steady magnetic field doesn't produce a current unless the coil is moving, as explained below.) The current produced in the secondary circuit occurs for only an instant while the magnetic field through the secondary coil is changing. In effect, the secondary circuit behaves as though a source of emf were connected to it for a short time. It's customary to say that **an induced emf is produced in the secondary circuit by the changing magnetic field.**

## Magnetic Flux

To evaluate induced emfs quantitatively, we need to understand what factors affect the phenomenon. Although changing magnetic fields always induce electric fields, in other situations the magnetic field remains constant, yet an induced electric field is still produced. The best example of this is an electric generator: A loop of conductor rotating in a constant magnetic field creates an electric current.

The physical quantity associated with magnetism that creates an electric field is a **changing magnetic flux.** Magnetic flux is defined in the same way as electric flux (Section 15.9) and is proportional to both the strength of the magnetic field passing through the plane of a loop of wire and the area of the loop.

Magnetic flux ▶

The **magnetic flux $\mathbf{\Phi}_B$** through a loop of wire with area $A$ is defined by

$$\Phi_B \equiv B_{\perp}A = BA\cos\theta \qquad [20.1]$$

where $B_{\perp}$ is the component of a uniform magnetic field $\vec{\mathbf{B}}$ perpendicular to the plane of the loop, as in Figure 20.2a, and $\theta$ is the angle between $\vec{\mathbf{B}}$ and the normal (perpendicular) to the plane of the loop.

**SI unit: weber (Wb)**

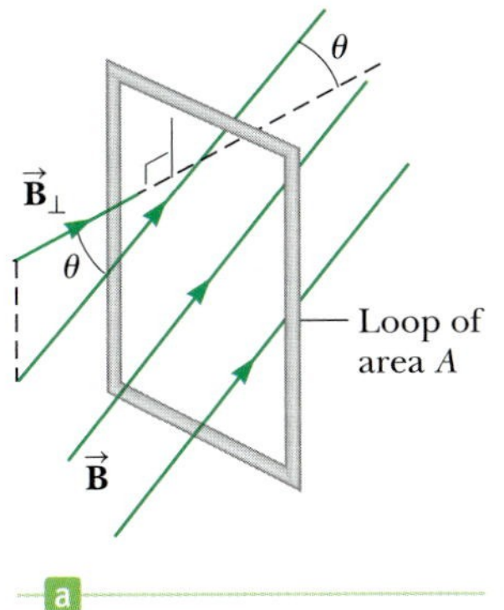


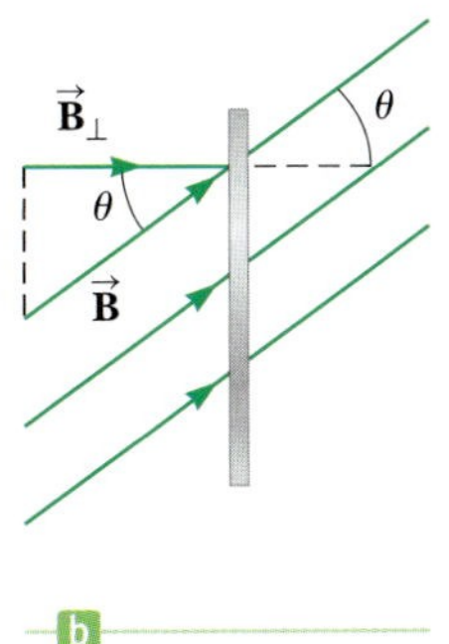


**Figure 20.2** (a) A uniform magnetic field $\vec{B}$ making an angle $\theta$ with a direction normal to the plane of a wire loop of area $A$. (b) An edge view of the loop.

Note that there are always two directions normal to a given plane surface. In Figure 20.2, for example, that direction could be chosen to be to the right, resulting in positive flux. The normal direction could also be chosen to point to the left, which would result in a negative flux of the same magnitude. The choice of normal direction is called the orientation of the surface. Once chosen in a given problem, the normal direction remains fixed. A good default is to choose the normal direction so that the initial angle between the magnetic field and the normal direction is less than 90°.

From Equation 20.1, it follows that $B_\perp = B \cos \theta$. The magnetic flux, in other words, is the magnitude of the part of $\vec{B}$ that is perpendicular to the plane of the loop times the area of the loop. Figure 20.2b is an edge view of the loop and the penetrating magnetic field lines. When the field is perpendicular to the plane of the loop as in Figure 20.3a, $\theta = 0$ and $\Phi_B$ has a maximum value, $\Phi_{B,\max} = BA$. When the plane of the loop is parallel to $\vec{B}$ as in Figure 20.3b, $\theta = 90°$ and $\Phi_B = 0$. The flux can also be negative. For example, when $\theta = 180°$, the flux is equal to $-BA$. Because the SI unit of $B$ is the tesla, or weber per square meter, the unit of flux is $\text{T} \cdot \text{m}^2$, or weber (Wb).

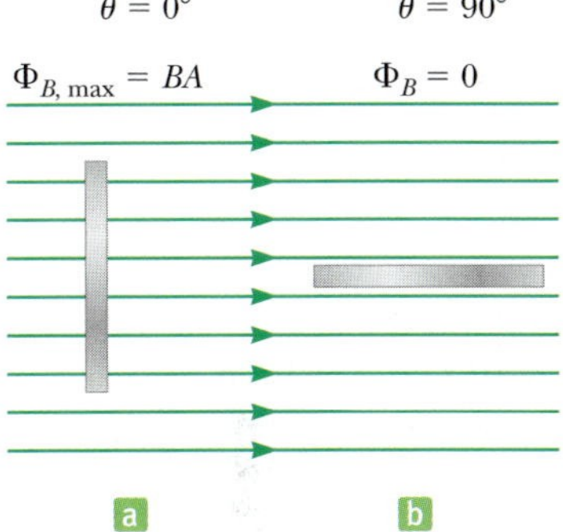


**Figure 20.3** An edge view of a loop in a uniform magnetic field. (a) When the field lines are perpendicular to the plane of the loop, the magnetic flux through the loop is a maximum and equal to $\Phi_B = BA$. (b) When the field lines are parallel to the plane of the loop, the magnetic flux through the loop is zero.

We can emphasize the qualitative meaning of Equation 20.1 by first drawing magnetic field lines, as in Figure 20.3. The number of lines per unit area increases as the field strength increases. **The value of the magnetic flux is proportional to the total number of lines passing through the loop.** We see that the most lines pass through the loop when its plane is perpendicular to the field, as in Figure 20.3a, so the flux has its maximum value at that time. As Figure 20.3b shows, no lines pass through the loop when its plane is parallel to the field, so in that case $\Phi_B = 0$.

## APPLYING PHYSICS 20.1 | Flux Compared

Argentina has more land area ($2.8 \times 10^6$ km²) than Greenland ($2.2 \times 10^6$ km²). Why is the magnetic flux of Earth's magnetic field larger through Greenland than through Argentina?

**EXPLANATION** Greenland (latitude 60° north to 80° north) is closer to a magnetic pole than Argentina (latitude 20° south to 50° south), so the magnetic field is stronger there. That in itself isn't sufficient to conclude that the magnetic flux is greater, but Greenland's proximity to a pole also means that the angle the magnetic field lines make with the vertical is smaller than in Argentina. As a result, more field lines penetrate the surface in Greenland, despite Argentina's slightly larger area. ■

## EXAMPLE 20.1 | Magnetic Flux

**GOAL** Calculate magnetic flux and a change in flux.

**PROBLEM** A conducting circular loop of radius 0.250 m is placed in the *xy*-plane in a uniform magnetic field of 0.360 T that points in the positive *z*-direction, the same direction as the normal to the plane. **(a)** Calculate the magnetic flux through the loop. **(b)** Suppose the loop is rotated clockwise around the *x*-axis, so the normal direction now points at a

*(Continued)*

45.0° angle with respect to the $z$-axis. Recalculate the magnetic flux through the loop. **(c)** What is the change in flux due to the rotation of the loop?

**STRATEGY** After finding the area, substitute values into the equation for magnetic flux for each part. Because the normal direction was chosen to be the same direction as the magnetic field, the angle between the magnetic field and the normal is initially 0°. After the rotation, that angle becomes 45°.

**SOLUTION**

**(a)** Calculate the initial magnetic flux through the loop.

First, calculate the area of the loop:

$$A = \pi r^2 = \pi(0.250 \text{ m})^2 = 0.196 \text{ m}^2$$

Substitute $A$, $B$, and $\theta = 0°$ into Equation 20.1 to find the initial magnetic flux:

$$\Phi_B = AB\cos\theta = (0.196 \text{ m}^2)(0.360 \text{ T})\cos(0°)$$
$$= 0.070\,6 \text{ T}\cdot\text{m}^2 = \boxed{0.070\,6 \text{ Wb}}$$

**(b)** Calculate the magnetic flux through the loop after it has rotated 45.0° around the $x$-axis.

Make the same substitutions as in part (a), except the angle between $\vec{\mathbf{B}}$ and the normal is now $\theta = 45.0°$:

$$\Phi_B = AB\cos\theta = (0.196 \text{ m}^2)(0.360 \text{ T})\cos(45.0°)$$
$$= 0.049\,9 \text{ T}\cdot\text{m}^2 = \boxed{0.049\,9 \text{ Wb}}$$

**(c)** Find the change in the magnetic flux due to the rotation of the loop.

Subtract the result of part (a) from the result of part (b):

$$\Delta\Phi_B = 0.049\,9 \text{ Wb} - 0.070\,6 \text{ Wb} = \boxed{-0.020\,7 \text{ Wb}}$$

**REMARKS** Notice that the rotation of the loop, not any change in the magnetic field, is responsible for the change in flux. This changing magnetic flux is essential in the functioning of electric motors and generators.

**QUESTION 20.1** True or False: If the loop is rotated in the opposite direction by the same amount, the change in magnetic flux has the same magnitude but opposite sign.

**EXERCISE 20.1** The loop, having rotated by 45°, rotates clockwise another 30°, so the normal to the plane points at an angle of 75° with respect to the direction of the magnetic field. Find (a) the magnetic flux through the loop when $\theta = 75°$ and (b) the change in magnetic flux during the rotation from 45° to 75°.

**ANSWERS** (a) 0.018 3 Wb (b) −0.031 6 Wb

## 20.2 Faraday's Law of Induction and Lenz's Law

**LEARNING OBJECTIVES**

1. State Faraday's and Lenz's laws and describe applications based on them.
2. Apply Faraday's and Lenz's laws to systems with changing magnetic flux.

The usefulness of the concept of magnetic flux can be made obvious by another simple experiment that demonstrates the basic idea of electromagnetic induction. Consider a wire loop connected to an ammeter as in Figure 20.4. If a magnet is moved toward the loop, the ammeter reads a current in one direction, as in Figure 20.4a. When the magnet is held stationary, as in Figure 20.4b, the ammeter reads zero current. If the magnet is moved away from the loop, the ammeter reads a current in the opposite direction, as in Figure 20.4c. If the magnet is held stationary and the loop is moved either toward or away from the magnet, the ammeter also reads a current. From these observations, it can be concluded that **a current is established in the circuit as long as there is relative motion between the magnet and the loop.** The same experimental results are found whether the loop moves or the magnet moves. We call such a current an **induced current** because it is produced by an **induced emf.**

**Tip 20.1 Induced Current Requires a Change in Magnetic Flux**

The existence of magnetic flux through an area is not sufficient to create an induced emf. A *change* in the magnetic flux over some time interval $\Delta t$ must occur for an emf to be induced.

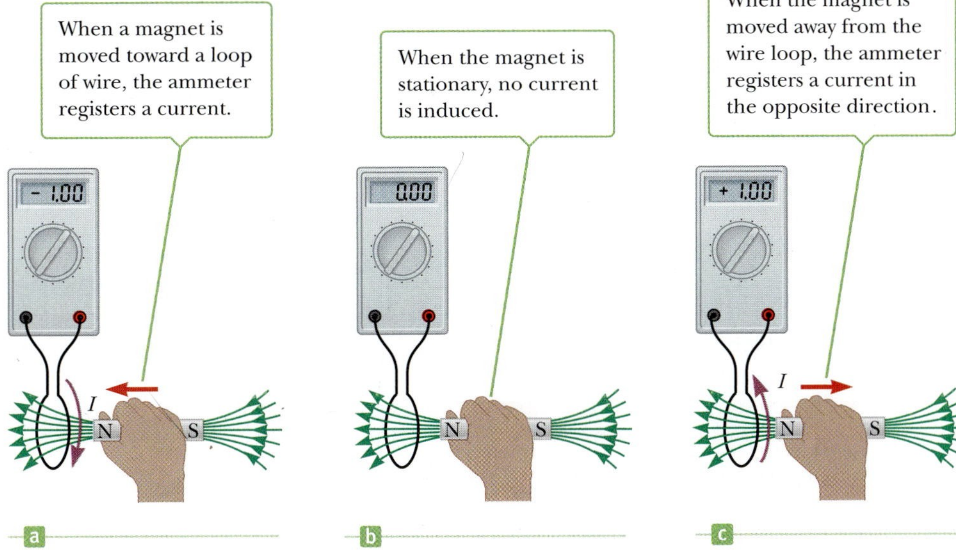
When a magnet is moved toward a loop of wire, the ammeter registers a current.
When the magnet is stationary, no current is induced.
When the magnet is moved away from the wire loop, the ammeter registers a current in the opposite direction.
- 1.00
0.00
+ 1.00
I
N
S
a
b
c

**Figure 20.4** A simple experiment showing that a current is induced in a loop when a magnet is moved toward or away from the loop.

### Tip 20.2 There Are Two Magnetic Fields to Consider

When applying Lenz's law, there are *two* magnetic fields to consider. The first is the external changing magnetic field that induces the current in a conducting loop. The second is the magnetic field produced by the induced current in the loop.

This experiment is similar to the Faraday experiment discussed in Section 20.1. In each case, an emf is induced in a circuit when the magnetic flux through the circuit changes with time. It turns out that the instantaneous emf induced in a circuit equals the negative of the rate of change of magnetic flux with respect to time through the circuit. This is **Faraday's law of magnetic induction**.

◀ Faraday's law

If a circuit contains $N$ tightly wound loops and the magnetic flux through each loop changes by the amount $\Delta\Phi_B$ during the interval $\Delta t$, the average emf induced in the circuit during time $\Delta t$ is

$$\mathcal{E} = -N\frac{\Delta\Phi_B}{\Delta t} \quad [20.2]$$

Because $\Phi_B = BA\cos\theta$, a change of any of the factors $B$, $A$, or $\theta$ with time produces an emf. We explore the effect of a change in each of these factors in the following sections. The minus sign in Equation 20.2 is included to indicate the polarity of the induced emf. This polarity determines the direction of the current in the loop, and is given by **Lenz's law**:

The current caused by the induced emf travels in the direction that creates a magnetic field with flux opposing the change in the original flux through the circuit.

Lenz's law says that if the magnetic flux through a loop is becoming more positive, say, then the induced emf creates a current and associated magnetic field that produces negative magnetic flux. Some mistakenly think this "counter magnetic field" created by the induced current, called $\vec{\mathbf{B}}_{\text{ind}}$ ("ind" for induced), will always point in a direction opposite the applied magnetic field $\vec{\mathbf{B}}$, but that is only true half the time! Figure 20.5a shows a field penetrating a loop. The graph in Figure 20.5b shows that the magnitude of the magnetic field $\vec{\mathbf{B}}$ shrinks with time, which means that the flux of $\vec{\mathbf{B}}$ is shrinking with time, so the induced field $\vec{\mathbf{B}}_{\text{ind}}$ will actually be in the same direction as $\vec{\mathbf{B}}$. In effect, $\vec{\mathbf{B}}_{\text{ind}}$ "shores up" the field $\vec{\mathbf{B}}$, slowing the loss of flux through the loop.

The direction of the current in Figure 20.5a can be determined by right-hand rule number 2: Point your right thumb in the direction that will cause the fingers

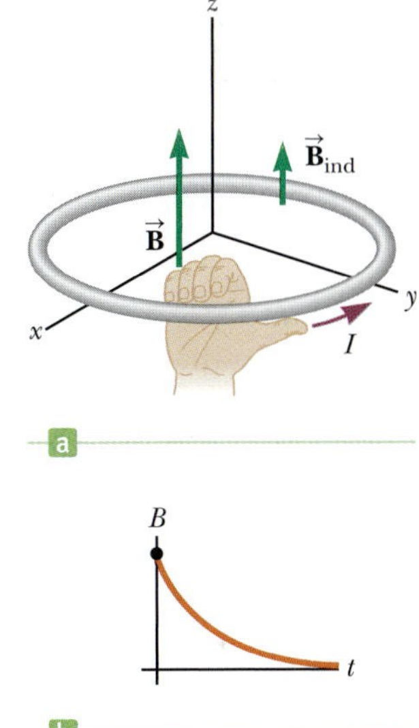
z
$\vec{\mathbf{B}}_{\text{ind}}$
$\vec{\mathbf{B}}$
x
y
I
a
B
t
b

**Figure 20.5** (a) The magnetic field $\vec{\mathbf{B}}$ becomes smaller with time, reducing the flux, so current is induced in a direction that creates an induced magnetic field $\vec{\mathbf{B}}_{\text{ind}}$ opposing the change in magnetic flux. (b) Graph of the magnitude of the magnetic field as a function of time.

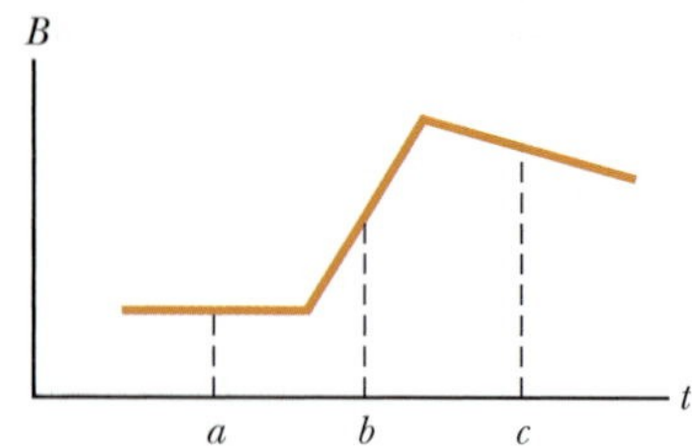


**Figure 20.6** (Quick Quiz 20.1)

on your right hand to curl in the direction of the induced field $\vec{\mathbf{B}}_{\text{ind}}$. In this case, that direction is counterclockwise: with the right thumb pointed in the direction of the current, your fingers curl down outside the loop and around and ***up through the inside of the loop***. Remember, inside the loop is where it's important for the induced magnetic field to be pointing up.

**Quick Quiz**

**20.1** Figure 20.6 is a graph of the magnitude $B$ versus time for a magnetic field that passes through a fixed loop and is oriented perpendicular to the plane of the loop. Rank the magnitudes of the emf generated in the loop from largest to smallest at the three instants indicated.

## EXAMPLE 20.2 Faraday and Lenz to the Rescue

**GOAL** Calculate an induced emf and current with Faraday's law and apply Lenz's law when the magnetic field changes with time.

**PROBLEM** A coil with 25 turns of wire is wrapped on a frame with a square cross section 1.80 cm on a side. Each turn has the same area, equal to that of the frame, and the total resistance of the coil is 0.350 Ω. An applied uniform magnetic field is perpendicular to the plane of the coil, as in Figure 20.7. **(a)** If the field changes uniformly from 0.00 T to 0.500 T in 0.800 s, what is the induced emf in the coil while the field is changing? Find **(b)** the magnitude and **(c)** the direction of the induced current in the coil while the field is changing.

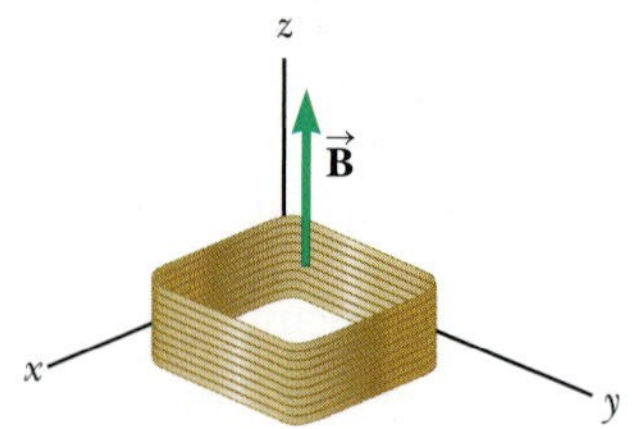


**Figure 20.7** (Example 20.2)

**STRATEGY** Part (a) requires substituting into Faraday's law, Equation 20.2. The necessary information is given, except for $\Delta\Phi_B$, the change in the magnetic flux during the elapsed time. Using the normal direction to coincide with the positive $z$-axis, compute the initial and final magnetic fluxes with Equation 20.1, find the difference, and assemble all terms in Faraday's law. The current can then be found with Ohm's law, and its direction with Lenz's law.

**SOLUTION**

**(a)** Find the induced emf in the coil.

To compute the flux, the area of the coil is needed:

$$A = L^2 = (0.018\,0 \text{ m})^2 = 3.24 \times 10^{-4} \text{ m}^2$$

The magnetic flux $\Phi_{B,i}$ through the coil at $t = 0$ is zero because $B = 0$. Calculate the flux at $t = 0.800$ s:

$$\Phi_{B,f} = BA\cos\theta = (0.500 \text{ T})(3.24 \times 10^{-4} \text{ m}^2)\cos(0°)$$
$$= 1.62 \times 10^{-4} \text{ Wb}$$

Compute the change in the magnetic flux through the cross section of the coil over the 0.800-s interval:

$$\Delta\Phi_B = \Phi_{B,f} - \Phi_{B,i} = 1.62 \times 10^{-4} \text{ Wb}$$

Substitute into Faraday's law of induction to find the induced emf in the coil:

$$\mathcal{E} = -N\frac{\Delta\Phi_B}{\Delta t} = -(25 \text{ turns})\left(\frac{1.62 \times 10^{-4} \text{ Wb}}{0.800 \text{ s}}\right)$$
$$= \boxed{-5.06 \times 10^{-3} \text{ V}}$$

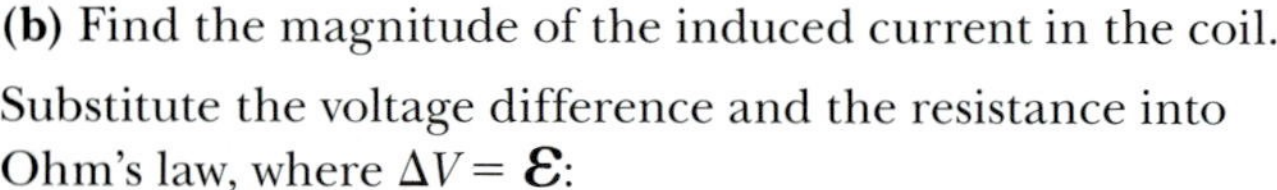

**(b)** Find the magnitude of the induced current in the coil.

Substitute the voltage difference and the resistance into Ohm's law, where $\Delta V = \mathcal{E}$:

$$I = \frac{\Delta V}{R} = \frac{5.06 \times 10^{-3} \text{ V}}{0.350\ \Omega} = \boxed{1.45 \times 10^{-2} \text{ A}}$$

**(c)** Find the direction of the induced current in the coil.

The magnetic field is increasing up through the loop, in the same direction as the normal to the plane; hence, the flux is positive and is also increasing. A downward-pointing induced magnetic field will create negative flux, opposing the change. If you point your right thumb in the clockwise direction along the loop as viewed from above, your fingers curl down through the loop, which is the correct direction for the counter magnetic field. Hence the current must proceed in a clockwise direction as viewed from above the coil.

**REMARKS** Lenz's law can best be handled by first sketching a diagram.

**QUESTION 20.2** What average emf is induced in the loop if, instead, the magnetic field changes uniformly from 0.500 T to 0 in 0.800 s? How would that affect the induced current?

**EXERCISE 20.2** Suppose the magnetic field changes uniformly from 0.500 T to 0.200 T in the next 0.600 s. Compute (a) the induced emf in the coil and (b) the magnitude and direction of the induced current.

**ANSWERS** (a) $4.05 \times 10^{-3}$ V (b) $1.16 \times 10^{-2}$ A (counterclockwise as viewed from above the coil)

## Finding the Direction of the Induced Current

Finding the direction of the induced current can be tricky. The following three examples illustrate how the direction is found using Lenz's law.

**Lenz's Law Example 1** The current in the wire of Figure 20.8 is steadily increasing in the direction indicated. Let's choose the normal direction to be out of the page so that magnetic field vectors coming out of the page will produce positive magnetic flux. The magnetic field created by the current $I$ circulates around the wire, going into the page on the right side of the long wire in the region of the rectangular coil and coming out of the page on the left side of the long straight wire. Therefore, the magnetic flux through the rectangular coil due to the current $I$ is negative. Because the current is increasing up the page, the magnetic field is becoming stronger, increasing the magnitude of the negative flux through the rectangular coil. By Lenz's law, the induced current in the coil must produce positive flux, countering the increasing negative flux. That requires an induced magnetic field pointing out of the page through the coil. Mentally curl the fingers of the right hand around the right branch of the rectangular coil, and note that the fingers come straight up out of the page through the coil, as required. The right thumb, meanwhile, points up the page, indicating the current direction in that part of the coil. Therefore the induced current in the coil is counterclockwise.

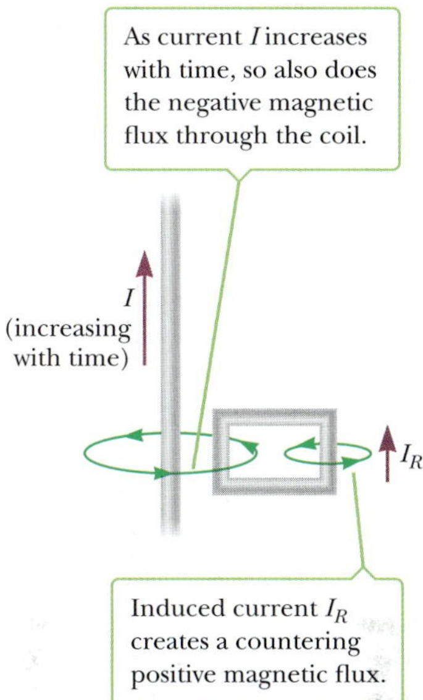
As current $I$ increases with time, so also does the negative magnetic flux through the coil.
$I$
(increasing with time)
$I_R$
Induced current $I_R$ creates a countering positive magnetic flux.

**Figure 20.8** (Lenz's Law Example 1) Current $I$ increases in magnitude with time, strengthening the magnetic field that circulates around the wire.

**Lenz's Law Example 2** In Figure 20.9a, the north pole of the magnet moves toward the coil. If the normal direction is chosen to the right, then the magnetic flux through the coil due to the magnet is positive and increases with time. A negative flux to the left must therefore be created by the induced current in the coil, so the induced magnetic field must also point to the left as indicated in Figure 20.9b. Imagine curling the fingers of the right hand around the coil so they point through the coil to the left. The right thumb then points upward, indicating the induced current is counterclockwise as viewed from the left side of the loop.

**Lenz's Law Example 3** Consider a coil of wire placed near a solenoid in Figure 20.10a (page 706). The wire is wrapped in such a way as to create a south magnetic pole at the right end when the switch is closed in Figure 20.10b. Choose left as the normal direction. When the switch is closed, the current in the solenoid begins to increase, and the magnetic flux through the coil is positive and increasing with time. Therefore the induced current in the coil must create negative magnetic flux to counteract the increasing positive flux created by the current in the

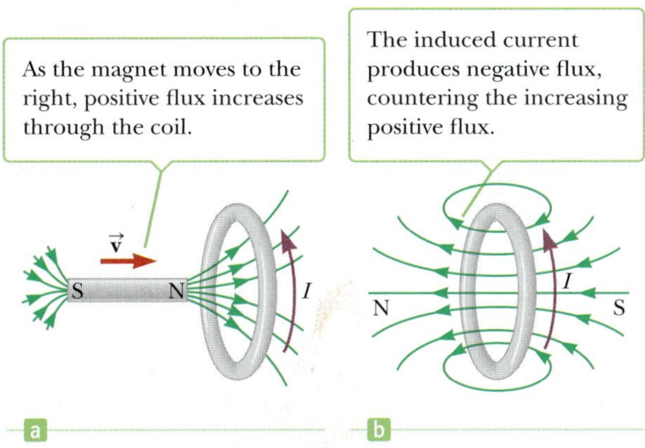
As the magnet moves to the right, positive flux increases through the coil.
The induced current produces negative flux, countering the increasing positive flux.
$\vec{v}$
S
N
$I$
N
$I$
S
a
b

**Figure 20.9** (Lenz's Law Example 2) (a) The north pole of the magnet approaches the coil from the left, with the normal direction taken to the right. (b) A current is induced in the coil.

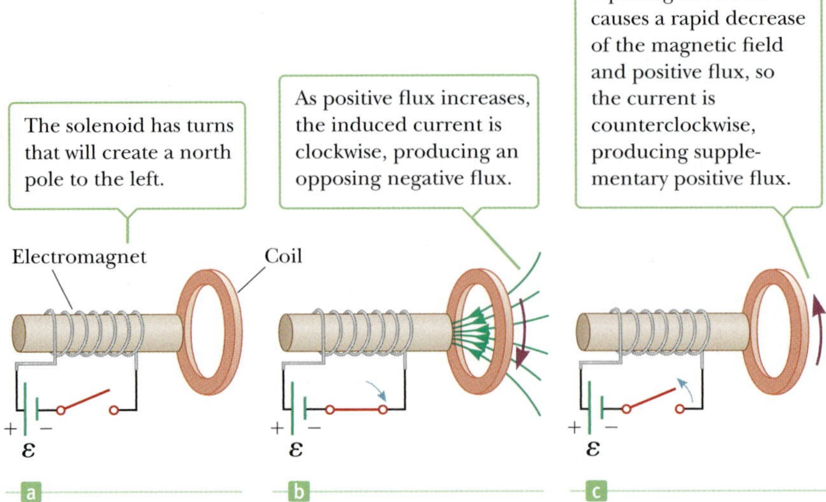

The solenoid has turns that will create a north pole to the left.
As positive flux increases, the induced current is clockwise, producing an opposing negative flux.
Opening the switch causes a rapid decrease of the magnetic field and positive flux, so the current is counterclockwise, producing supplementary positive flux.
Electromagnet
Coil
+ −
ε
a
b
c

**Figure 20.10** (Lenz's Law Example 3) (a) The turns of the solenoid create a magnetic field with north pole pointing left, which is also taken as the normal direction. (b) When the switch is closed, positive flux begins increasing through the coil as field lines converge on the solenoid's south pole. (c) Opening the switch causes the solenoid's field to rapidly decrease.

solenoid. That requires an induced magnetic field directed to the right through the coil. Turning the right hand so the thumb is pointed downward, the right fingers can curl through the coil and to the right. The induced current in the coil follows the direction of the right thumb, which is clockwise as viewed from the left end of the coil. When the switch is opened again in Figure 20.10c the current in the solenoid changes direction because the magnetic field and positive flux begin to decrease. Moving counterclockwise, the induced current creates positive flux through the coil, opposing the decrease in positive flux.

In all three of these examples, the critical idea is that a changing flux causes an induced current, and the associated induced magnetic field produces flux opposing the change in flux in accordance with Lenz's law. When the flux stops changing, the induced current stops. Although in each case the magnetic flux changed because of a changing magnetic field, induced currents can result even when the magnetic field is constant provided the flux through the loop changes. That fact will become clear when discussing motional emf in Section 20.3 and generators in Section 20.4.

■ *Quick Quiz*

**20.2** A bar magnet is falling toward the center of a loop of wire, with the north pole oriented downward. Viewed from the same side of the loop as the magnet, as the north pole approaches the loop, what is the direction of the induced current? (a) clockwise (b) zero (c) counterclockwise (d) along the length of the magnet

**20.3** Two circular loops are side by side and lie in the *xy*-plane. A switch is closed, starting a counterclockwise current in the left-hand loop, as viewed from a point on the positive *z*-axis passing through the center of the loop. Which of the following statements is true of the right-hand loop? (a) The current remains zero. (b) An induced current moves counterclockwise. (c) An induced current moves clockwise.

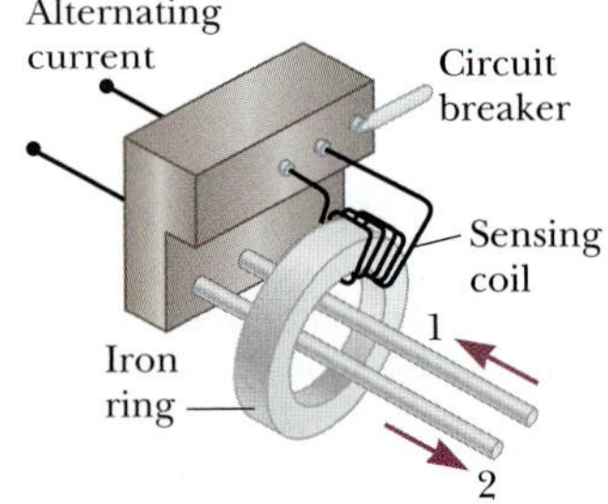

Alternating current
Circuit breaker
Sensing coil
Iron ring
1
2

**Figure 20.11** Essential components of a ground fault interrupter (contents of the gray box in Fig. 20.12a). In newer homes such devices are built directly into wall outlets. The purpose of the sensing coil and circuit breaker is to cut off the current before damage is done.

The ground fault interrupter (GFI) is an interesting safety device that protects people against electric shock when they touch appliances and power tools. Its operation makes use of Faraday's law. Figure 20.11 shows the essential parts of a ground fault interrupter. Wire 1 leads from the wall outlet to the appliance to be protected, and wire 2 leads from the appliance back to the wall outlet. An iron ring surrounds the two wires to confine the magnetic field set up by each wire. A sensing coil, which can activate a circuit breaker when changes in magnetic flux occur, is wrapped around part of the iron ring. Because the currents in the wires are in opposite directions, the net magnetic field through the sensing coil due to

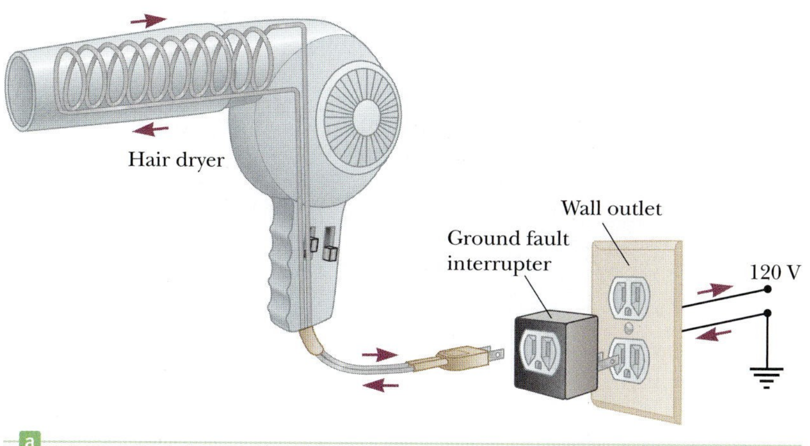

**Figure 20.12** (a) This hair dryer has been plugged into a ground fault interrupter that is in turn plugged into an unprotected wall outlet. (b) You likely have seen this kind of ground fault interrupter in a hotel bathroom, where hair dryers and electric shavers are often used by people just out of the shower or who might touch a water pipe, providing a ready path to ground in the event of a short circuit.

the currents is zero. If a short circuit occurs in the appliance so that there is no returning current, however, the net magnetic field through the sensing coil is no longer zero. A short circuit can happen if, for example, one of the wires loses its insulation, providing a path through you to ground if you happen to be touching the appliance and are grounded as in Figure 18.23a. Because the current is alternating, the magnetic flux through the sensing coil changes with time, producing an induced voltage in the coil. This induced voltage is used to trigger a circuit breaker, stopping the current quickly (in about 1 ms) before it reaches a level that might be harmful to the person using the appliance. A ground fault interrupter provides faster and more complete protection than even the case-ground-and-circuit-breaker combination shown in Figure 18.23b. For this reason, ground fault interrupters are commonly found in bathrooms, where electricity poses a hazard to people. (See Fig. 20.12.)

**APPLICATION**
Ground fault interrupters

Another interesting application of Faraday's law is the production of sound in an electric guitar. A vibrating string induces an emf in a coil (Fig. 20.13). The pickup coil is placed near the vibrating guitar string, which is made of a metal that can be magnetized. The permanent magnet inside the coil magnetizes the portion of the string nearest the coil. When the guitar string vibrates at some frequency, its magnetized segment produces a changing magnetic flux through the pickup coil. The changing flux induces a voltage in the coil, which is fed to an amplifier. The output of the amplifier is sent to the loudspeakers, producing the sound waves that we hear.

**APPLICATION**
Electric guitar pickups

Sudden infant death syndrome, or SIDS, is a devastating affliction in which a baby suddenly stops breathing during sleep without an apparent cause. One type of monitoring device, called an apnea monitor, is sometimes used to alert caregivers

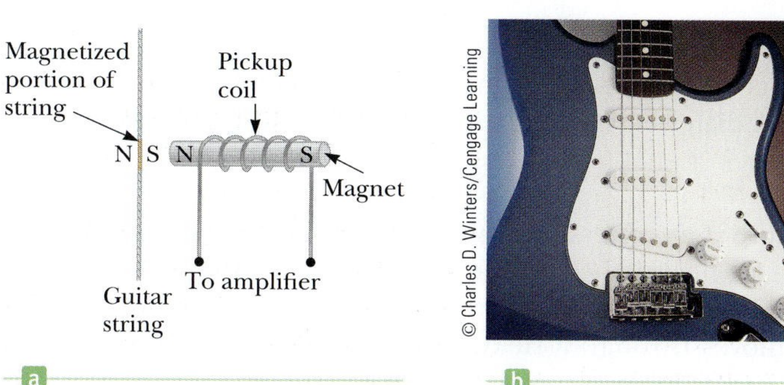


**Figure 20.13** (a) In an electric guitar a vibrating string induces a voltage in the pickup coil. (b) Several pickups allow the vibration to be detected from different portions of the string.

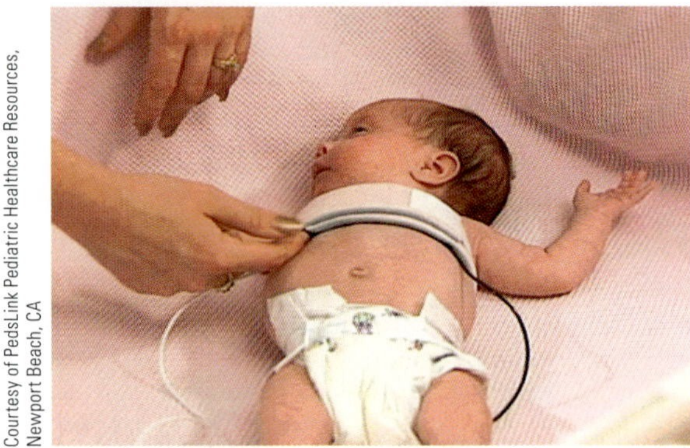

Courtesy of PedsLink Pediatric Healthcare Resources, Newport Beach, CA


**Figure 20.14** This infant is wearing a monitor designed to alert care-givers if breathing stops. Notice the two wires attached to opposite sides of the chest.

**BIO APPLICATION**
Apnea monitors

of the cessation of breathing. The device uses induced currents, as shown in Figure 20.14. A coil of wire attached to one side of the chest carries an alternating current. The varying magnetic flux produced by this current passes through a pickup coil attached to the opposite side of the chest. Expansion and contraction of the chest caused by breathing or movement change the strength of the voltage induced in the pickup coil. If breathing stops, however, the pattern of the induced voltage stabilizes, and external circuits monitoring the voltage sound an alarm to the caregivers after a momentary pause to ensure that a problem actually does exist.

## 20.3 Motional emf

**LEARNING OBJECTIVES**

1. Define motional emf and use Faraday's law to discuss the potential difference across a moving conductor.
2. Apply Faraday's law to systems involving a conductor moving through a magnetic field.

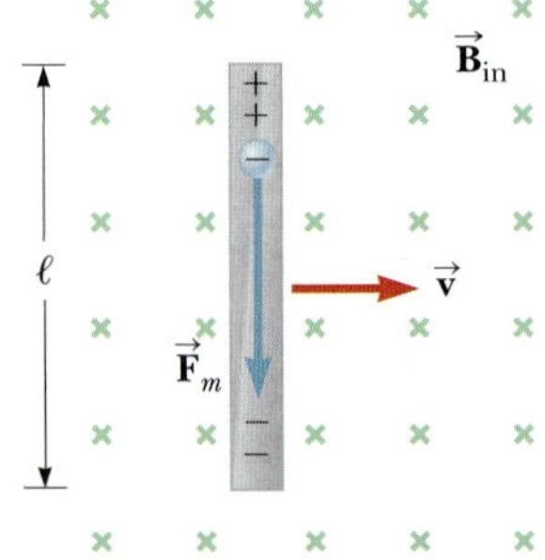


**Figure 20.15** A straight conductor of length $\ell$ moving with velocity $\vec{v}$ through a uniform magnetic field $\vec{B}$ directed perpendicular to $\vec{v}$. The vector $\vec{F}_m$ is the magnetic force on an electron in the conductor. An emf of $B\ell v$ is induced between the ends of the bar.

In Section 20.2 we considered emfs induced in a circuit when the magnetic field changes with time. In this section we describe a particular application of Faraday's law in which a so-called **motional emf** is produced. It is the emf induced in a conductor moving through a magnetic field.

First consider a straight conductor of length $\ell$ moving with constant velocity through a uniform magnetic field directed into the paper, as in Figure 20.15. For simplicity, we assume the conductor moves in a direction perpendicular to the field. A magnetic force of magnitude $F_m = qvB$, directed downward, acts on the electrons in the conductor. Because of this magnetic force, the free electrons move to the lower end of the conductor and accumulate there, leaving a net positive charge at the upper end. As a result of this charge separation, an electric field is produced in the conductor. The charge at the ends builds up until the downward magnetic force $qvB$ is balanced by the upward electric force $qE$. At this point, charge stops flowing and the condition for equilibrium requires that

$$qE = qvB \qquad \text{or} \qquad E = vB$$

Because the electric field is uniform, the field produced in the conductor is related to the potential difference across the ends by $\Delta V = E\ell$, giving

$$\Delta V = E\ell = B\ell v \qquad [20.3]$$

Because there is an excess of positive charge at the upper end of the conductor and an excess of negative charge at the lower end, the upper end is at a higher potential than the lower end. There is a potential difference across a conductor as long as it moves through a field. If the motion is reversed, the polarity of the potential difference is also reversed.

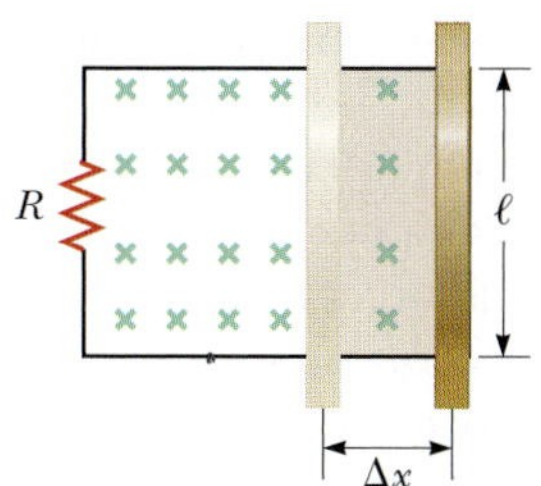


**Figure 20.17** As the bar moves to the right, the area of the loop increases by the amount $\ell \Delta x$ and the magnetic flux through the loop increases by $B\ell \Delta x$.

A more interesting situation occurs if the moving conductor is part of a closed conducting path. This situation is particularly useful for illustrating how a changing loop area induces a current in a closed circuit described by Faraday's law. Consider a circuit consisting of a conducting bar of length $\ell$, sliding along two fixed, parallel conducting rails, as in Figure 20.16a. For simplicity, assume the moving bar has zero resistance and the stationary part of the circuit has constant resistance $R$. Take the normal direction to coincide with the $z$-axis, out of the page. A uniform and constant magnetic field $\vec{\mathbf{B}}$ is applied perpendicular to the plane of the circuit. As the bar is pulled to the right in the positive $x$-direction with velocity $\vec{\mathbf{v}}$ under the influence of an applied force $\vec{\mathbf{F}}_{\text{app}}$, a magnetic force along the length of the bar acts on the free charges in the bar. This force in turn sets up an induced current because the charges are free to move in a closed conducting path. In this case, the changing magnetic flux through the loop and the corresponding induced emf across the moving bar arise from the *change in area of the loop* as the bar moves through the magnetic field. Because the flux into the page increases, by Lenz's law the induced current circulates counterclockwise, producing flux out of the page that opposes the change.

Assume the bar moves a distance $\Delta x$ in time $\Delta t$, as shown in Figure 20.17. The increase in flux $\Delta\Phi_B$ through the loop in that time is the amount of flux that now passes through the portion of the circuit that has area $\ell\ \Delta x$:

$$\Delta\Phi_B = BA = B\ell\ \Delta x$$

Using Faraday's law and noting that there is one loop ($N = 1$), we find that the magnitude of the induced emf is

$$|\mathcal{E}| = \frac{\Delta\Phi_B}{\Delta t} = B\ell \frac{\Delta x}{\Delta t} = B\ell v \qquad [20.4]$$

This induced emf is often called a **motional emf** because it arises from the motion of a conductor through a magnetic field.

Further, if the resistance of the circuit is $R$, the magnitude of the induced current in the circuit is

$$I = \frac{|\mathcal{E}|}{R} = \frac{B\ell v}{R} \qquad [20.5]$$

Figure 20.16b shows the equivalent circuit diagram for this example.

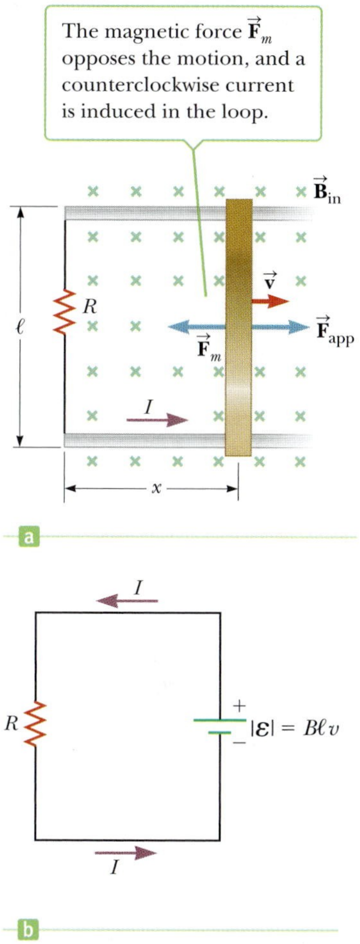

The magnetic force $\vec{\mathbf{F}}_m$ opposes the motion, and a counterclockwise current is induced in the loop.

**Figure 20.16** (a) A conducting bar sliding with velocity $\vec{\mathbf{v}}$ along two conducting rails under the action of an applied force $\vec{\mathbf{F}}_{\text{app}}$. (b) The equivalent circuit of that in (a).

## APPLYING PHYSICS 20.2 Space Catapult

Applying a force on the bar will result in an induced emf in the circuit shown in Figure 20.16. Suppose we remove the external magnetic field in the diagram and replace the resistor with a high-voltage source and a switch, as in Figure 20.18. What will happen when the switch is closed? Will the bar move, and does it matter which way we connect the high-voltage source?

**EXPLANATION** Suppose the source is capable of establishing high current. Then the two horizontal conducting rods will create a strong magnetic field in the area between

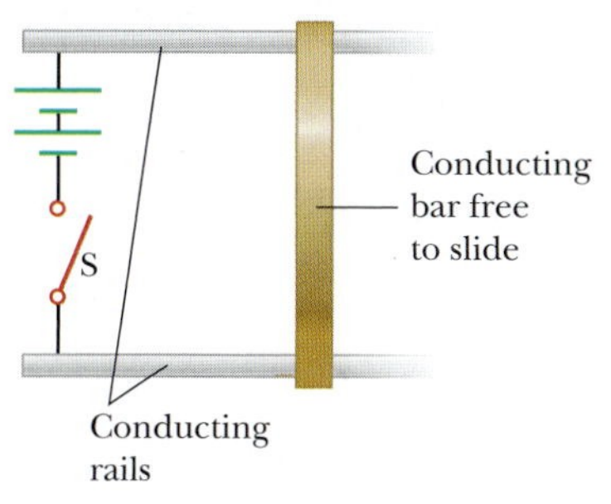


**Figure 20.18** (Applying Physics 20.2)

*(Continued)*

them, directed into the page. (The movable bar also creates a magnetic field, but this field can't exert force on the bar itself.) Because the moving bar carries a downward current, a magnetic force is exerted on the bar, directed to the right. Hence, the bar accelerates along the rails away from the power supply. If the polarity of the power were reversed, the magnetic field would be out of the page, the current in the bar would be upward, and the force on the bar would still be directed to the right. The $BI\ell$ force exerted by a magnetic field according to Equation 19.6 causes the bar to accelerate away from the voltage source. Studies have shown that it's possible to launch payloads into space with this technology. (This is the working principle of a rail gun.) Very large accelerations can be obtained with currently available technology, with payloads being accelerated to a speed of several kilometers per second in a fraction of a second. This acceleration is larger than humans can tolerate.

Rail guns have been proposed as propulsion systems for moving asteroids into more useful orbits. The material of an asteroid could be mined and launched off the surface by a rail gun, which would act like a rocket engine, modifying the velocity and hence the orbit of the asteroid. Some asteroids contain trillions of dollars worth of valuable metals. ■

### *Quick Quiz*

**20.4** A horizontal metal bar oriented east-west drops straight down in a location where Earth's magnetic field is due north. As a result, an emf develops between the ends. Which end is positively charged? (a) the east end (b) the west end (c) neither end carries a charge

**20.5** You intend to move a rectangular loop of wire into a region of uniform magnetic field at a given speed so as to induce an emf in the loop. The plane of the loop must remain perpendicular to the magnetic field lines. In which orientation should you hold the loop while you move it into the region with the magnetic field to generate the largest emf? (a) with the long dimension of the loop parallel to the velocity vector (b) with the short dimension of the loop parallel to the velocity vector (c) either way because the emf is the same regardless of orientation

## EXAMPLE 20.3 A Potential Difference Induced Across Airplane Wings

**GOAL** Find the emf induced by motion through a magnetic field.

**PROBLEM** An airplane with a wingspan of 30.0 m flies due north at a location where the downward component of Earth's magnetic field is $0.600 \times 10^{-4}$ T. There is also a component pointing due north that has a magnitude of $0.470 \times 10^{-4}$ T. **(a)** Find the difference in potential between the wingtips when the speed of the plane is $2.50 \times 10^{2}$ m/s. **(b)** Which wingtip is positive?

**STRATEGY** Because the plane is flying north, the northern component of magnetic field won't have any effect on the induced emf. The induced emf across the wing is caused solely by the downward component of the Earth's magnetic field. Substitute the given quantities into Equation 20.4. Use right-hand rule number 1 to find the direction positive charges would be propelled by the magnetic force.

**SOLUTION**

**(a)** Calculate the difference in potential across the wingtips.

Write the motional emf equation and substitute the given quantities:

$$\mathcal{E} = B\ell v = (0.600 \times 10^{-4}\ \text{T})(30.0\ \text{m})(2.50 \times 10^{2}\ \text{m/s})$$

$$= \boxed{0.450\ \text{V}}$$

**(b)** Which wingtip is positive?

Apply right-hand rule number 1:

Point the fingers of your right hand north, in the direction of the velocity, and curl them down, in the direction of the magnetic field. Your thumb points west. Therefore the west wingtip is positive.

**REMARKS** An induced emf such as this one can cause problems on an aircraft.

**QUESTION 20.3** In what directions are magnetic forces exerted on electrons in the metal aircraft if it is flying due west? (a) north (b) south (c) east (d) west (e) up (f) down

**EXERCISE 20.3** Suppose a space station is in orbit where the magnetic field is parallel to Earth's surface, points north, and has magnitude $1.80 \times 10^{-4}$ T. A metal cable attached to the space station stretches radially outwards 2.50 km. (a) Estimate the potential difference that develops between the ends of the cable if it's traveling eastward around Earth at a speed of $7.70 \times 10^3$ m/s. (b) Which end of the cable is positive, the lower end or the upper end?

**ANSWERS** (a) $3.47 \times 10^3$ V (b) The upper end is positive.

## EXAMPLE 20.4 Where Is the Energy Source?

**GOAL** Use motional emf to find an induced emf and a current.

**PROBLEM** **(a)** The sliding bar in Figure 20.16a has a length of 0.500 m and moves at 2.00 m/s in a magnetic field of magnitude 0.250 T. Using the concept of motional emf, find the induced voltage in the moving rod. **(b)** If the resistance in the circuit is 0.500 Ω, find the current in the circuit and the power delivered to the resistor. (*Note:* The current in this case goes counterclockwise around the loop.) **(c)** Calculate the magnetic force on the bar. **(d)** Use the concepts of work and power to calculate the applied force.

**STRATEGY** For part (a), substitute into Equation 20.4 for the motional emf. Once the emf is found, substitution into Ohm's law gives the current. In part (c), use Equation 19.6 for the magnetic force on a current-carrying conductor. In part (d), use the fact that the power dissipated by the resistor multiplied by the elapsed time must equal the work done by the applied force.

**SOLUTION**

**(a)** Find the induced emf with the concept of motional emf.

Substitute into Equation 20.4 to find the induced emf:

$$\mathcal{E} = B\ell v = (0.250\ \text{T})(0.500\ \text{m})(2.00\ \text{m/s}) = \boxed{0.250\ \text{V}}$$

**(b)** Find the induced current in the circuit and the power dissipated by the resistor.

Substitute the emf and the resistance into Ohm's law to find the induced current:

$$I = \frac{\mathcal{E}}{R} = \frac{0.250\ \text{V}}{0.500\ \Omega} = \boxed{0.500\ \text{A}}$$

Substitute $I = 0.500$ A and $\mathcal{E} = 0.250$ V into Equation 17.8 to find the power dissipated by the 0.500-Ω resistor:

$$P = I\Delta V = (0.500\ \text{A})(0.250\ \text{V}) = \boxed{0.125\ \text{W}}$$

**(c)** Calculate the magnitude and direction of the magnetic force on the bar.

Substitute values for $I$, $B$, and $\ell$ into Equation 19.6, with $\sin\theta = \sin(90°) = 1$, to find the magnitude of the force:

$$F_m = IB\ell = (0.500\ \text{A})(0.250\ \text{T})(0.500\ \text{m}) = \boxed{6.25 \times 10^{-2}\ \text{N}}$$

Apply right-hand rule number 1 to find the direction of the force:

Point the fingers of your right hand in the direction of the positive current, then curl them in the direction of the magnetic field. Your thumb points in the **negative $x$-direction.**

**(d)** Find the value of $F_{app}$, the applied force.

Set the work done by the applied force equal to the dissipated power times the elapsed time:

$$W_{app} = F_{app}d = P\Delta t$$

Solve for $F_{app}$ and substitute $d = v\,\Delta t$:

$$F_{app} = \frac{P\Delta t}{d} = \frac{P\Delta t}{v\Delta t} = \frac{P}{v} = \frac{0.125\ \text{W}}{2.00\ \text{m/s}} = \boxed{6.25 \times 10^{-2}\ \text{N}}$$

**REMARKS** Part (d) could be solved by using Newton's second law for an object in equilibrium: Two forces act horizontally on the bar and the acceleration of the bar is zero, so the forces must be equal in magnitude and opposite in direction. Notice the agreement between the answers for $F_m$ and $F_{app}$, despite the very different concepts used.

**QUESTION 20.4** Suppose the applied force and magnetic field in Figure 20.16a are removed, but a battery creates a current in the same direction as indicated. What happens to the bar?

*(Continued)*

**EXERCISE 20.4** Suppose the current suddenly increases to 1.25 A in the same direction as before due to an increase in speed of the bar. Find (a) the emf induced in the rod and (b) the new speed of the rod.

**ANSWERS** (a) 0.625 V (b) 5.00 m/s

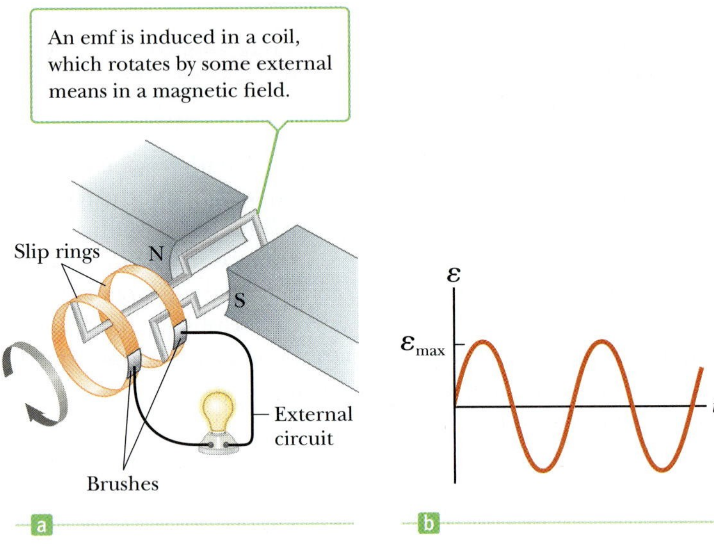

An emf is induced in a coil, which rotates by some external means in a magnetic field.
Slip rings
N
S
External circuit
Brushes
$\mathcal{E}$
$\mathcal{E}_{max}$
t
a
b

**Figure 20.19** (a) A schematic diagram of an AC generator. An emf is induced in a coil, which rotates by some external means in a magnetic field. (b) A plot of the alternating emf induced in the loop versus time.

## 20.4 Generators

**LEARNING OBJECTIVES**

1. Contrast the operating principles of AC and DC generators.
2. Describe the operating principles of motors and the phenomenon of back emf.
3. Apply Faraday's law to generators and motors.

**APPLICATION**
Alternating-current generators

Generators and motors are important practical devices that operate on the principle of electromagnetic induction. First, consider the **alternating-current** (AC) **generator**, a device that converts mechanical energy to electrical energy. In its simplest form, the AC generator consists of a wire loop rotated in a magnetic field by some external means (Fig. 20.19a). In commercial power plants, the energy required to rotate the loop can be derived from a variety of sources. In a hydroelectric plant, for example, falling water directed against the blades of a turbine produces the rotary motion; in a coal-fired plant, heat produced by burning coal is used to convert water to steam, and this steam is directed against the turbine blades. As the loop rotates, the magnetic flux through it changes with time, inducing an emf and a current in an external circuit. The ends of the loop are connected to slip rings that rotate with the loop. Connections to the external circuit are made by stationary brushes in contact with the slip rings.

We can derive an expression for the emf generated in the rotating loop by making use of the equation for motional emf, $\mathcal{E} = B\ell v$. Figure 20.20a shows a loop of wire rotating clockwise in a uniform magnetic field directed to the right. The magnetic force ($qvB$) on the charges in wires $AB$ and $CD$ is not along the lengths of the wires. (The force on the electrons in these wires is perpendicular to the wires.) Hence, an emf is generated only in wires $BC$ and $AD$. At any instant, wire $BC$ has velocity $\vec{\mathbf{v}}$ at an angle $\theta$ with the magnetic field, as shown in Figure 20.20b. (Note that the component of velocity parallel to the field has no effect on the charges in the wire, whereas the component of velocity perpendicular to the field produces a magnetic force on the charges that moves electrons from $C$ to $B$.) The emf generated in wire $BC$ equals $B\ell v_{\perp}$,

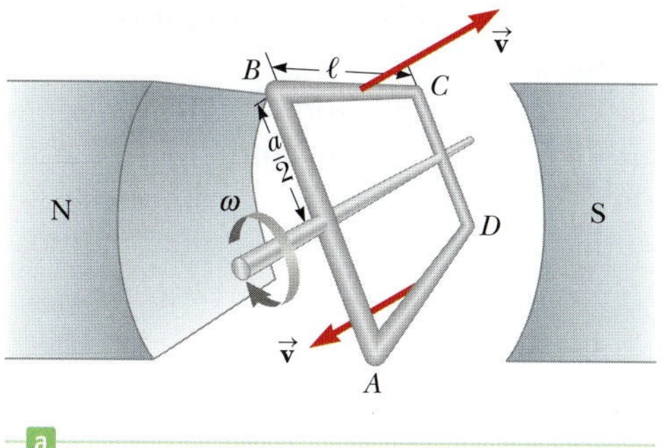


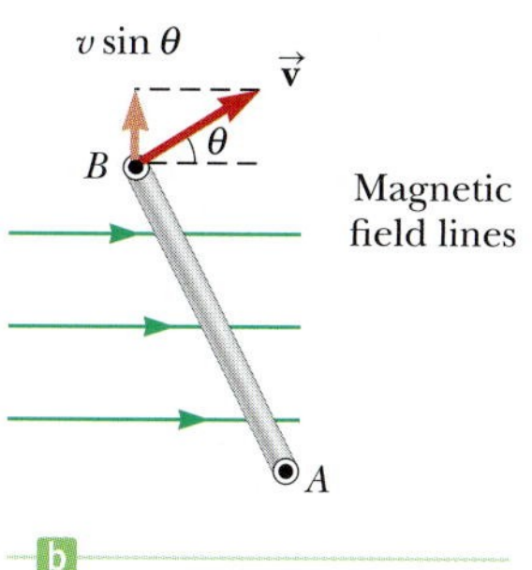


**Figure 20.20** (a) A loop rotating at a constant angular velocity in an external magnetic field. The emf induced in the loop varies sinusoidally with time. (b) An edge view of the rotating loop.

where $\ell$ is the length of the wire and $v_\perp$ is the component of velocity perpendicular to the field. An emf of $B\ell v_\perp$ is also generated in wire $DA$, and the sense of this emf is the same as that in wire $BC$. Because $v_\perp = v \sin\theta$, the total induced emf is

$$\mathcal{E} = 2B\ell v_\perp = 2B\ell v \sin\theta \qquad [20.6]$$

If the loop rotates with a constant angular speed $\omega$, we can use the relation $\theta = \omega t$ in Equation 20.6. Furthermore, because every point on the wires $BC$ and $DA$ rotates in a circle about the axis of rotation with the same angular speed $\omega$, we have $v = r\omega = (a/2)\omega$, where $a$ is the length of sides $AB$ and $CD$. Equation 20.6 therefore reduces to

$$\mathcal{E} = 2B\ell\left(\frac{a}{2}\right)\omega \sin \omega t = B\ell a\omega \sin \omega t$$

If a coil has $N$ turns, the emf is $N$ times as large because each loop has the same emf induced in it. Further, because the area of the loop is $A = \ell a$, the total emf is

$$\mathcal{E} = NBA\omega \sin \omega t \qquad [20.7]$$

©iStockphoto.com/KingWu

Turbines turn electric generators at a hydroelectric power plant.

This result shows that the emf varies sinusoidally with time, as plotted in Figure 20.19b. Note that the maximum emf has the value

$$\mathcal{E}_{\text{max}} = NBA\omega \qquad [20.8]$$

which occurs when $\omega t = 90°$ or $270°$. In other words, $\mathcal{E} = \mathcal{E}_{\text{max}}$ when the plane of the loop is parallel to the magnetic field. Further, the emf is zero when $\omega t = 0$ or $180°$, which happens whenever the magnetic field is perpendicular to the plane of the loop. In the United States and Canada the frequency of rotation for commercial generators is 60 Hz, whereas in some European countries 50 Hz is used. (Recall that $\omega = 2\pi f$, where $f$ is the frequency in hertz.)

**APPLICATION**
Direct-current generators

The **direct-current** (DC) **generator** is illustrated in Figure 20.21a. The components are essentially the same as those of the AC generator except that the contacts to the rotating loop are made by a split ring, or commutator. In this design the output voltage always has the same polarity and the current is a pulsating direct current, as in Figure 20.21b. Note that the contacts to the split ring reverse their

**Figure 20.21** (a) A schematic diagram of a DC generator. (b) The emf fluctuates in magnitude, but always has the same polarity.

roles every half cycle. At the same time, the polarity of the induced emf reverses. Hence, the polarity of the split ring remains the same.

A pulsating DC current is not suitable for most applications. To produce a steady DC current, commercial DC generators use many loops and commutators distributed around the axis of rotation so that the sinusoidal pulses from the loops overlap in phase. When these pulses are superimposed, the DC output is almost free of fluctuations.

## EXAMPLE 20.5 emf Induced in an AC Generator

**GOAL** Understand physical aspects of an AC generator.

**PROBLEM** An AC generator consists of eight turns of wire, each having area $A = 0.090\ 0\ \text{m}^2$, with a total resistance of 12.0 Ω. The coil rotates in a magnetic field of 0.500 T at a constant frequency of 60.0 Hz, with axis of rotation perpendicular to the direction of the magnetic field. **(a)** Find the maximum induced emf. **(b)** What is the maximum induced current? **(c)** Determine the induced emf and current as functions of time. **(d)** What maximum torque must be applied to keep the coil turning?

**STRATEGY** From the given frequency, calculate the angular frequency $\omega$ and substitute it, together with given quantities, into Equation 20.8. As functions of time, the emf and current have the form $A \sin \omega t$, where $A$ is the maximum emf or current, respectively. For part (d), calculate the magnetic torque on the coil when the current is at a maximum. (See Chapter 19.) The applied torque must do work against this magnetic torque to keep the coil turning.

**SOLUTION**

**(a)** Find the maximum induced emf.

First, calculate the angular frequency of the rotational motion:

$$\omega = 2\pi f = 2\pi(60.0\ \text{Hz}) = 377\ \text{rad/s}$$

Substitute the values for $N$, $A$, $B$, and $\omega$ into Equation 20.8, obtaining the maximum induced emf:

$$\mathcal{E}_{\text{max}} = NAB\omega = 8(0.090\ 0\ \text{m}^2)(0.500\ \text{T})(377\ \text{rad/s})$$
$$= \boxed{136\ \text{V}}$$

**(b)** What is the maximum induced current?

Substitute the maximum induced emf $\mathcal{E}_{\text{max}}$ and the resistance $R$ into Ohm's law to find the maximum induced current:

$$I_{\text{max}} = \frac{\mathcal{E}_{\text{max}}}{R} = \frac{136\ \text{V}}{12.0\ \Omega} = \boxed{11.3\ \text{A}}$$

**(c)** Determine the induced emf and the current as functions of time.

Substitute $\mathcal{E}_{\text{max}}$ and $\omega$ into Equation 20.7 to obtain the variation of $\mathcal{E}$ with time $t$ in seconds:

$$\mathcal{E} = \mathcal{E}_{\text{max}} \sin \omega t = \boxed{(136\ \text{V}) \sin 377t}$$

The time variation of the current looks just like this expression, except with the maximum current out in front:

$$I = \boxed{(11.3\ \text{A}) \sin 377t}$$

**(d)** Calculate the maximum applied torque necessary to keep the coil turning.

Write the equation for magnetic torque:

$$\tau = \mu B \sin \theta$$

Calculate the maximum magnetic moment of the coil, $\mu$:

$$\mu = I_{\text{max}}AN = (11.3\ \text{A})(0.090\ \text{m}^2)(8) = 8.14\ \text{A} \cdot \text{m}^2$$

Substitute into the magnetic torque equation, with $\theta = 90°$ to find the maximum applied torque:

$$\tau_{\text{max}} = (8.14\ \text{A} \cdot \text{m}^2)(0.500\ \text{T}) \sin 90° = \boxed{4.07\ \text{N} \cdot \text{m}}$$

**REMARKS** The number of loops, $N$, can't be arbitrary because there must be a force strong enough to turn the coil.

**QUESTION 20.5** What effect does doubling the frequency have on the maximum induced emf?

**EXERCISE 20.5** An AC generator is to have a maximum output of 301 V. Each circular turn of wire has an area of $0.100 \text{ m}^2$ and a resistance of 0.80 Ω. The coil rotates in a magnetic field of 0.600 T with a frequency of 40.0 Hz, with the axis of rotation perpendicular to the direction of the magnetic field. (a) How many turns of wire should the coil have to produce the desired emf? (b) Find the maximum current induced in the coil. (c) Determine the induced emf as a function of time.

**ANSWERS** (a) 20 turns (b) 18.8 A (c) $\mathcal{E} = (301 \text{ V}) \sin 251t$

## Motors and Back emf

**APPLICATION**
Motors

Motors are devices that convert electrical energy to mechanical energy. Essentially, **a motor is a generator run in reverse**: instead of a current being generated by a rotating loop, a current is supplied to the loop by a source of emf, and the magnetic torque on the current-carrying loop causes it to rotate.

A motor can perform useful mechanical work when a shaft connected to its rotating coil is attached to some external device. As the coil in the motor rotates, however, the changing magnetic flux through it induces an emf that acts to reduce the current in the coil. If it *increased* the current, Lenz's law would be violated. The phrase **back emf** is used for an emf that tends to reduce the applied current. The back emf increases in magnitude as the rotational speed of the coil increases. We can picture this state of affairs as the equivalent circuit in Figure 20.22. For illustrative purposes, assume the external power source supplying current in the coil of the motor has a voltage of 120 V, the coil has a resistance of 10 Ω, and the back emf induced in the coil at this instant is 70 V. The voltage available to supply current equals the difference between the applied voltage and the back emf, or 50 V in this case. The current is always reduced by the back emf.

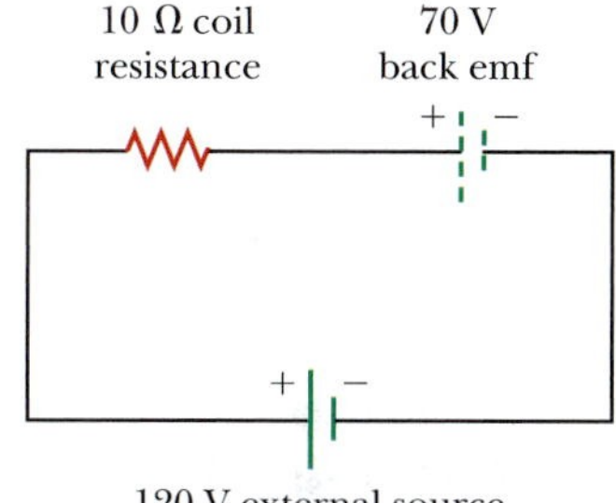


**Figure 20.22** A motor can be represented as a resistance plus a back emf.

When a motor is turned on, there is no back emf initially and the current is very large because it's limited only by the resistance of the coil. As the coil begins to rotate, the induced back emf opposes the applied voltage and the current in the coil is reduced. If the mechanical load increases, the motor slows down, which decreases the back emf. This reduction in the back emf increases the current in the coil and therefore also increases the power needed from the external voltage source. As a result, the power requirements for starting a motor and for running it under heavy loads are greater than those for running the motor under average loads. If the motor is allowed to run under no mechanical load, the back emf reduces the current to a value just large enough to balance energy losses by heat and friction.

### EXAMPLE 20.6 Induced Current in a Motor

**GOAL** Apply the concept of a back emf in calculating the induced current in a motor.

**PROBLEM** A motor has coils with a resistance of 10.0 Ω and is supplied by a voltage of $\Delta V = 1.20 \times 10^2$ V. When the motor is running at its maximum speed, the back emf is 70.0 V. Find the current in the coils **(a)** when the motor is first turned on and **(b)** when the motor has reached its maximum rotation rate.

**STRATEGY** For each part, find the net voltage, which is the applied voltage minus the induced emf. Divide the net voltage by the resistance to get the current.

**SOLUTION**

**(a)** Find the initial current, when the motor is first turned on.

If the coil isn't rotating, the back emf is zero and the current has its maximum value. Calculate the difference between the emf and the initial back emf and divide by the resistance $R$, obtaining the initial current:

$$I = \frac{\mathcal{E} - \mathcal{E}_{\text{back}}}{R} = \frac{1.20 \times 10^2 \text{ V} - 0}{10.0 \text{ }\Omega} = \boxed{12.0 \text{ A}}$$

*(Continued)*

**(b)** Find the current when the motor is rotating at its maximum rate.

Repeat the calculation, using the maximum value of the back emf:

$$I = \frac{\mathcal{E} - \mathcal{E}_{\text{back}}}{R} = \frac{1.20 \times 10^2 \text{ V} - 70.0 \text{ V}}{10.0\ \Omega} = \frac{50.0 \text{ V}}{10.0\ \Omega}$$

$$= 5.00 \text{ A}$$

**REMARKS** The phenomenon of back emf is one way in which the rotation rate of electric motors is limited.

**QUESTION 20.6** As a motor speeds up, what happens to the magnitude of the magnetic torque? (a) It increases. (b) It decreases. (c) It remains constant.

**EXERCISE 20.6** If the current in the motor is 8.00 A at some instant, what is the back emf at that time?

**ANSWER** 40.0 V

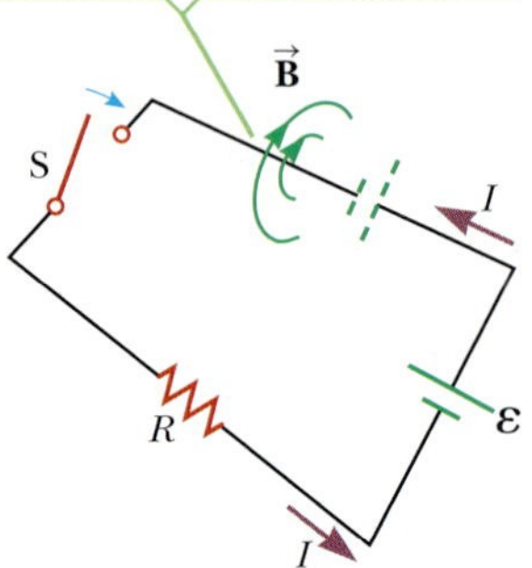


**Figure 20.23** After the switch in the circuit is closed, the current produces its own magnetic flux through the loop. An opposing emf is therefore induced, meaning the current relatively slowly increases toward its maximum value rather than jumping to that value right away. The battery with the dashed lines is a symbol for the self-induced emf.

## 20.5 Self-Inductance

**LEARNING OBJECTIVES**

1. Describe the concepts of self-induction and self-induced emf.
2. Evaluate the self-inductance and self-induced emf in simple electrical systems.

Consider a circuit consisting of a switch, a resistor, and a source of emf, as in Figure 20.23. When the switch is closed, the current doesn't immediately change from zero to its maximum value, $\mathcal{E}/R$. The law of electromagnetic induction, Faraday's law, prevents this change. What happens instead is the following: as the current increases with time, the magnetic flux through the loop due to this current also increases. The increasing flux induces an emf in the circuit that opposes the change in magnetic flux. By Lenz's law, the induced emf is in the direction indicated by the dashed battery in the figure. The net potential difference across the resistor is the emf of the battery minus the opposing induced emf. As the magnitude of the current increases, the *rate* of increase lessens and hence the induced emf decreases. This opposing emf results in a gradual increase in the current. For the same reason, when the switch is opened, the current doesn't immediately fall to zero. This effect is called **self-induction** because the changing flux through the circuit arises from the circuit itself. The emf that is set up in the circuit is called a **self-induced emf**.

As a second example of self-inductance, consider Figure 20.24, which shows a coil wound on a cylindrical iron core. (A practical device would have several hundred turns.) Assume the current changes with time. When the current is in the direction shown, a magnetic field is set up inside the coil, directed from right to left. As a result, some lines of magnetic flux pass through the cross-sectional area of the coil. As the current changes with time, the flux through the coil changes and induces an emf in the coil. Lenz's law shows that this induced emf has a direction that opposes the change in the current. If the current is increasing, the induced emf is as pictured in Figure 20.24b, and if the current is decreasing, the induced emf is as shown in Figure 20.24c.

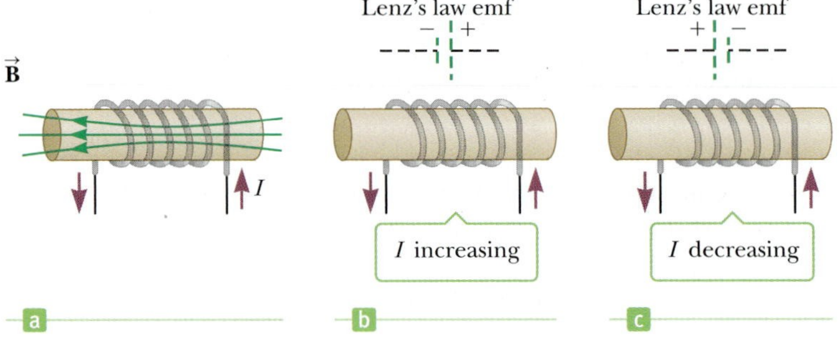


**Figure 20.24** (a) A current in the coil produces a magnetic field directed to the left. (b) If the current increases, the coil acts as a source of emf directed as shown by the dashed battery. (c) The induced emf in the coil changes its polarity if the current decreases. The battery symbols drawn with dashed lines represent the included emf in the coil.

To evaluate self-inductance quantitatively, first note that, according to Faraday's law, the induced emf is given by Equation 20.2:

$$\mathcal{E} = -N\frac{\Delta\Phi_B}{\Delta t}$$

The magnetic flux is proportional to the magnetic field, which is proportional to the current in the coil. Therefore, **the self-induced emf must be proportional to the rate of change of the current with time,** or

$$\mathcal{E} \equiv -L\frac{\Delta I}{\Delta t} \qquad [20.9]$$

where $L$ is a proportionality constant called the **inductance** of the device. The negative sign indicates that a changing current induces an emf in opposition to the change. In other words, if the current is increasing ($\Delta I$ positive), the induced emf is negative, indicating opposition to the increase in current. Likewise, if the current is decreasing ($\Delta I$ negative), the sign of the induced emf is positive, indicating that the emf is acting to oppose the decrease.

The inductance of a coil depends on the cross-sectional area of the coil and other quantities, which can all be grouped under the general heading of geometric factors. The SI unit of inductance is the **henry** (H), which, from Equation 20.9, is equal to 1 volt-second per ampere:

$$1\text{ H} = 1\text{ V}\cdot\text{s/A}$$

In the process of calculating self-inductance, it is often convenient to equate Equations 20.2 and 20.9 to find an expression for $L$:

$$N\frac{\Delta\Phi_B}{\Delta t} = L\frac{\Delta I}{\Delta t}$$

$$L = N\frac{\Delta\Phi_B}{\Delta I} = \frac{N\Phi_B}{I} \qquad [20.10]$$

◀ Inductance

Brady-Handy Collection, Library of Congress Prints and Photographs Division [LC-BH83-997]

**Joseph Henry**
**American physicist (1797–1878)**
Henry became the first director of the Smithsonian Institution and first president of the Academy of Natural Science. He was the first to produce an electric current with a magnetic field, but he failed to publish his results as early as Faraday because of his heavy teaching duties at the Albany Academy in New York State. He improved the design of the electromagnet and constructed one of the first motors. He also discovered the phenomenon of self-induction. The unit of inductance, the henry, is named in his honor.

## APPLYING PHYSICS 20.3 Making Sparks Fly

In some circuits a spark occurs between the poles of a switch when the switch is opened. Why isn't there a spark when the switch for this circuit is closed?

**EXPLANATION** According to Lenz's law, the direction of induced emfs is such that the induced magnetic field opposes change in the original magnetic flux. When the switch is opened, the sudden drop in the magnetic field in the circuit induces an emf in a direction that opposes change in the original current. This induced emf can cause a spark as the current bridges the air gap between the poles of the switch. The spark doesn't occur when the switch is closed, because the original current is zero and the induced emf opposes any change in that current. ■

In general, determining the inductance of a given current element can be challenging. Finding an expression for the inductance of a common solenoid, however, is straightforward. Let the solenoid have $N$ turns and length $\ell$. Assume $\ell$ is large compared with the radius and the core of the solenoid is air. We take the interior magnetic field to be uniform and given by Equation 19.16,

$$B = \mu_0 nI = \mu_0\frac{N}{\ell}I$$

where $n = N/\ell$ is the number of turns per unit length. The magnetic flux through each turn is therefore

$$\Phi_B = BA = \mu_0\frac{N}{\ell}AI$$

where $A$ is the cross-sectional area of the solenoid. From this expression and Equation 20.10, we find that

$$L = \frac{N\Phi_B}{I} = \frac{\mu_0 N^2 A}{\ell} \qquad [20.11a]$$

This equation shows that $L$ depends on the geometric factors $\ell$ and $A$ and on $\mu_0$ and is proportional to the square of the number of turns. Because $N = n\ell$, we can also express the result in the form

$$L = \mu_0 \frac{(n\ell)^2}{\ell} A = \mu_0 n^2 A\ell = \mu_0 n^2 V \qquad [20.11b]$$

where $V = A\ell$ is the volume of the solenoid.

## EXAMPLE 20.7 Inductance, Self-Induced emf, and Solenoids

**GOAL** Calculate the inductance and self-induced emf of a solenoid.

**PROBLEM** **(a)** Calculate the inductance of a solenoid containing 300 turns if the length of the solenoid is 25.0 cm and its cross-sectional area is $4.00 \times 10^{-4}$ m$^2$. **(b)** Calculate the self-induced emf in the solenoid described in part (a) if the current in the solenoid decreases at the rate of 50.0 A/s.

**STRATEGY** Substituting given quantities into Equation 20.11a gives the inductance $L$. For part (b), substitute the result of part (a) and $\Delta I/\Delta t = -50.0$ A/s into Equation 20.9 to get the self-induced emf.

**SOLUTION**

**(a)** Calculate the inductance of the solenoid.

Substitute the number $N$ of turns, the area $A$, and the length $\ell$ into Equation 20.11a to find the inductance:

$$L = \frac{\mu_0 N^2 A}{\ell}$$

$$= (4\pi \times 10^{-7}\,\mathrm{T\cdot m/A})\frac{(300)^2(4.00 \times 10^{-4}\,\mathrm{m^2})}{25.0 \times 10^{-2}\,\mathrm{m}}$$

$$= 1.81 \times 10^{-4}\,\mathrm{T\cdot m^2/A} = \boxed{0.181\,\mathrm{mH}}$$

**(b)** Calculate the self-induced emf in the solenoid.

Substitute $L$ and $\Delta I/\Delta t = -50.0$ A/s into Equation 20.9, finding the self-induced emf:

$$\mathcal{E} = -L\frac{\Delta I}{\Delta t} = -(1.81 \times 10^{-4}\,\mathrm{H})(-50.0\,\mathrm{A/s})$$

$$= \boxed{9.05\,\mathrm{mV}}$$

**REMARKS** Notice that $\Delta I/\Delta t$ is negative because the current is decreasing with time. The expression for the inductance in part (a) relies on the assumption that the radius of the solenoid is small compared to its length.

**QUESTION 20.7** If the solenoid were wrapped into a circle so as to become a toroidal solenoid, what would be true of its self-inductance? (a) It would be the same. (b) It would be larger. (c) It would be smaller.

**EXERCISE 20.7** A solenoid is to have an inductance of 0.285 mH, a cross-sectional area of $6.00 \times 10^{-4}$ m$^2$, and a length of 36.0 cm. (a) How many turns per unit length should it have? (b) If the self-induced emf is −12.5 mV at a given time, at what rate is the current changing at that instant?

**ANSWERS** (a) $1.02 \times 10^3$ turns/m (b) 43.9 A/s

# 20.6 *RL* Circuits

**LEARNING OBJECTIVES**

1. Describe the physical effect of introducing inductance into a circuit with changing current.
2. State the time constant for an *RL* circuit.
3. Apply the *RL* circuit time constant to circuits containing resistors and inductors.

A circuit element that has a large inductance, such as a closely wrapped coil of many turns, is called an **inductor**. The circuit symbol for an inductor is —ℓℓℓ—. We will always assume the self-inductance of the remainder of the circuit is negligible compared with that of the inductor in the circuit.

To gain some insight into the effect of an inductor in a circuit, consider the two circuits in Figure 20.25. Figure 20.25a shows a resistor connected to the terminals of a battery. For this circuit, Kirchhoff's loop rule is $\mathcal{E} - IR = 0$. The voltage drop across the resistor is

$$\Delta V_R = -IR \tag{20.12}$$

In this case, **we interpret resistance as a measure of opposition to the current**. Now consider the circuit in Figure 20.25b, consisting of an inductor connected to the terminals of a battery. At the instant the switch in this circuit is closed, because $IR = 0$, the emf of the battery equals the back emf generated in the coil. Hence, we have

$$\mathcal{E}_L = -L\frac{\Delta I}{\Delta t} \tag{20.13}$$

From this expression, **we can interpret $L$ as a measure of opposition to the rate of change of current**.

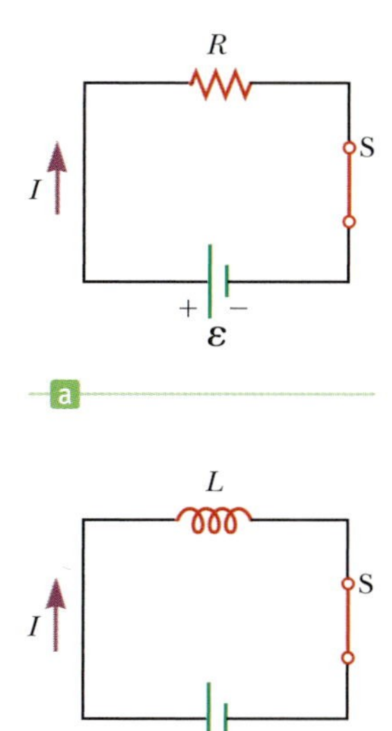


**Figure 20.25** A comparison of the effect of a resistor with that of an inductor in a simple circuit.

Figure 20.26 shows a circuit consisting of a resistor, an inductor, and a battery. Suppose the switch is closed at $t = 0$. The current begins to increase, but the inductor produces an emf that opposes the increasing current. As a result, the current can't change from zero to its maximum value of $\mathcal{E}/R$ instantaneously. Equation 20.13 shows that the induced emf is a maximum when the current is changing most rapidly, which occurs when the switch is first closed. As the current approaches its steady-state value, the back emf of the coil falls off because the current is changing more slowly. Finally, when the current reaches its steady-state value, the rate of change is zero and the back emf is also zero. Figure 20.27 plots current in the circuit as a function of time. This plot is similar to that of the charge on a capacitor as a function of time, discussed in Chapter 18, section 5, in connection with RC circuits. In that case, we found it convenient to introduce a quantity called the *time constant of the circuit*, which told us something about the time required for the capacitor to approach its steady-state charge. In the same way, time constants are defined for circuits containing resistors and inductors. The **time constant** $\tau$ for an *RL* circuit is the time required for the current in the circuit to reach 63.2% of its final value $\mathcal{E}/R$; the time constant of an *RL* circuit is given by

$$\tau = \frac{L}{R} \tag{20.14}$$

◀ Time constant for an *RL* circuit

Using methods of calculus, it can be shown that the current in such a circuit is given by

$$I = \frac{\mathcal{E}}{R}(1 - e^{-t/\tau}) \tag{20.15}$$

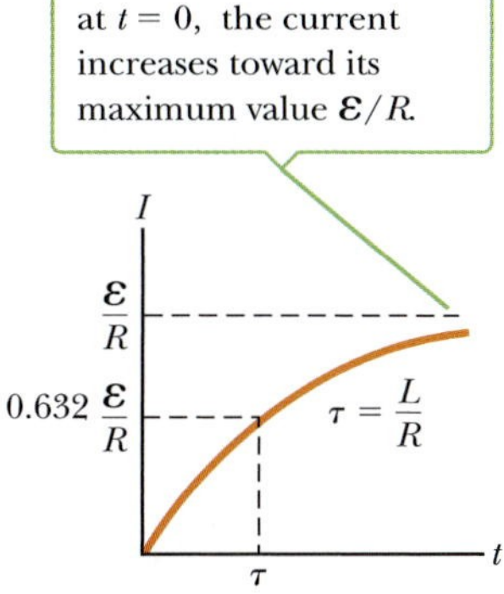


**Figure 20.27** A plot of current versus time for the *RL* circuit shown in Figure 20.26. The switch is closed at $t = 0$, and the current increases toward its maximum value $\mathcal{E}/R$. The time constant $\tau$ is the time it takes the current to reach 63.2% of its maximum value.

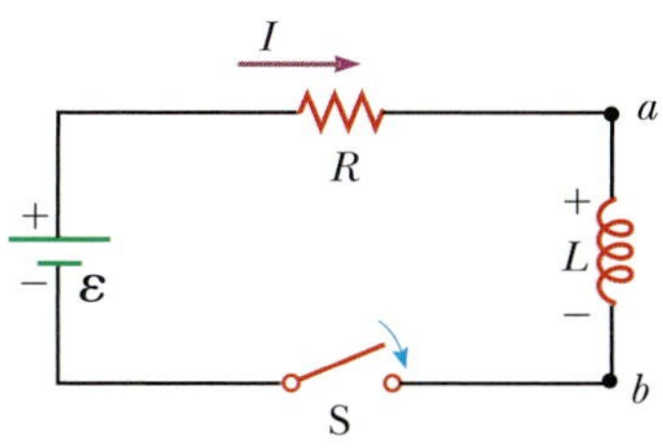


**Figure 20.26** A series *RL* circuit. As the current increases toward its maximum value, the inductor produces an emf that opposes the increasing current.

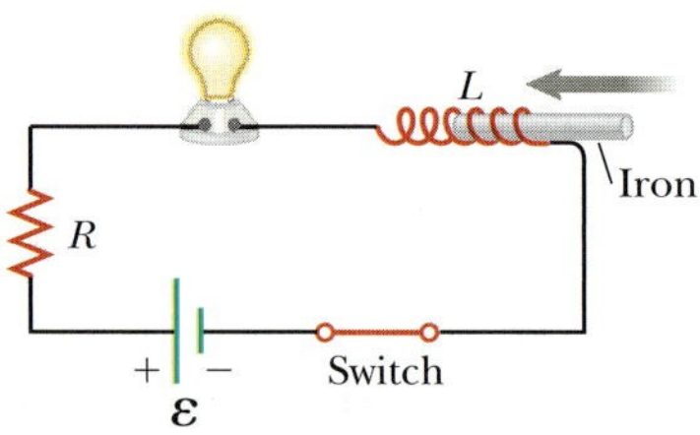


**Figure 20.28** (Quick Quiz 20.6)

This equation is consistent with our intuition: when the switch is closed at $t = 0$, the current is initially zero, rising with time to some maximum value. Notice the mathematical similarity between Equation 20.15 and Equation 18.7, which features a capacitor instead of an inductor. As in the case of a capacitor, the equation's form suggests an infinite amount of time is required for the current in the inductor to reach its maximum value. This is an artifact of assuming current is composed of moving charges that are infinitesimal, as will be demonstrated in Example 20.9.

**Quick Quiz**

**20.6** The switch in the circuit shown in Figure 20.28 is closed, and the lightbulb glows steadily. The inductor is a simple air-core solenoid. An iron rod is inserted into the interior of the solenoid, increasing the magnitude of the magnetic field in the solenoid. As the rod is inserted, the brightness of the lightbulb (a) increases, (b) decreases, or (c) remains the same.

## EXAMPLE 20.8 An *RL* Circuit

**GOAL** Calculate a time constant and relate it to current in an *RL* circuit.

**PROBLEM** A 12.6-V battery is in a circuit with a 30.0-mH inductor and a 0.150-Ω resistor, as in Figure 20.26. The switch is closed at $t = 0$. **(a)** Find the time constant of the circuit. **(b)** Find the current after one time constant has elapsed. **(c)** Find the voltage drops across the resistor when $t = 0$ and $t =$ one time constant, $\tau$. **(d)** What's the rate of change of the current after one time constant?

**STRATEGY** Part (a) requires only substitution into the definition of time constant. With this value and Ohm's law, the current after one time constant can be found, and multiplying this current by the resistance yields the voltage drop across the resistor after one time constant. With the voltage drop and Kirchhoff's loop law, the voltage across the inductor can be found. This value can be substituted into Equation 20.13 to obtain the rate of change of the current.

**SOLUTION**

**(a)** What's the time constant of the circuit?

Substitute the inductance $L$ and resistance $R$ into Equation 20.14, finding the time constant:

$$\tau = \frac{L}{R} = \frac{30.0 \times 10^{-3}\text{ H}}{0.150\ \Omega} = \boxed{0.200\text{ s}}$$

**(b)** Find the current after one time constant has elapsed.

First, use Ohm's law to compute the final value of the current after many time constants have elapsed:

$$I_{\text{max}} = \frac{\mathcal{E}}{R} = \frac{12.6\text{ V}}{0.150\ \Omega} = 84.0\text{ A}$$

After one time constant, the current rises to 63.2% of its final value:

$$I_{1\tau} = (0.632)I_{\text{max}} = (0.632)(84.0\text{ A}) = \boxed{53.1\text{ A}}$$

**(c)** Find the voltage drops across the resistance when $t = 0$ and $t =$ one time constant.

Initially, the current in the circuit is zero, so, from Ohm's law, the voltage across the resistor is zero:

$$\Delta V_R = IR$$

$$\Delta V_R\,(t = 0\text{ s}) = (0\text{ A})(0.150\ \Omega) = \boxed{0}$$

Next, using Ohm's law, find the magnitude of the voltage drop across the resistor after one time constant:

$$\Delta V_R\,(t = 0.200\text{ s}) = (53.1\text{ A})(0.150\ \Omega) = \boxed{7.97\text{ V}}$$

**(d)** What's the rate of change of the current after one time constant?

Using Kirchhoff's voltage rule, calculate the voltage drop across the inductor at that time:

$$\mathcal{E} + \Delta V_R + \Delta V_L = 0$$

Solve for $\Delta V_L$:

$$\Delta V_L = -\mathcal{E} - \Delta V_R = -12.6\text{ V} - (-7.97\text{ V}) = -4.6\text{ V}$$

Now solve Equation 20.13 for $\Delta I/\Delta t$ and substitute:

$$\Delta V_L = -L\frac{\Delta I}{\Delta t}$$

$$\frac{\Delta I}{\Delta t} = -\frac{\Delta V_L}{L} = -\frac{-4.6\text{ V}}{30.0 \times 10^{-3}\text{ H}} = \boxed{150\text{ A/s}}$$

**REMARKS** The values used in this problem were taken from actual components salvaged from the starter system of a car. Because the current in such an *RL* circuit is initially zero, inductors are sometimes referred to as "chokes" because they temporarily choke off the current. In solving part (d), we traversed the circuit in the direction of positive current, so the voltage difference across the battery was positive and the differences across the resistor and inductor were negative.

**QUESTION 20.8** Find the current in the circuit after two time constants.

**EXERCISE 20.8** A 12.6-V battery is in series with a resistance of 0.350 Ω and an inductor. (a) After a long time, what is the current in the circuit? (b) What is the current after one time constant? (c) What's the voltage drop across the inductor at this time? (d) Find the inductance if the time constant is 0.130 s.

**ANSWERS** (a) 36.0 A (b) 22.8 A (c) 4.62 V (d) $4.55 \times 10^{-2}$ H

## EXAMPLE 20.9 Formation of a Magnetic Field

**GOAL** Understand the role of time in setting up an inductor's magnetic field.

**PROBLEM** Given the *RL* circuit of Example 20.8, find the time required for the current to reach 99.9% of its maximum value after the switch is closed.

**STRATEGY** The solution requires solving Equation 20.15 for time followed by substitution. Notice that the maximum current is $I_{max} = \mathcal{E}/R$.

**SOLUTION**

Write Equation 20.15, with $I_f$ substituted for the current:

$$I_f = \frac{\mathcal{E}}{R}(1 - e^{-t/\tau}) = I_{max}(1 - e^{-t/\tau})$$

Divide both sides by $I_{max}$:

$$\frac{I_f}{I_{max}} = 1 - e^{-t/\tau}$$

Subtract 1 from both sides and then multiply both sides by −1:

$$1 - \frac{I_f}{I_{max}} = e^{-t/\tau}$$

Take the natural log of both sides:

$$\ln\left(1 - \frac{I_f}{I_{max}}\right) = \ln\left(e^{-t/\tau}\right) = -t/\tau$$

Solve for $t$ and substitute the expression for $\tau$ from Equation 20.14:

$$t = -\frac{L}{R}\ln\left(1 - \frac{I_f}{I_{max}}\right)$$

Substitute values, obtaining the desired time:

$$t = -\frac{30.0 \times 10^{-3}\text{ H}}{0.150\ \Omega}\ln(1 - 0.999) = \boxed{1.38\text{ s}}$$

**REMARKS** From this calculation, it's found that forming the magnetic field in an inductor and approaching the maximum current occurs relatively rapidly. Contrary to what might be expected from the mathematical form of Equation 20.15, an infinite amount of time is not actually required.

**QUESTION 20.9** If the inductance were doubled, by what factor would the length of time found be changed? (a) 1 (i.e., no change) (b) 2 (c) $\frac{1}{2}$

**EXERCISE 20.9** Suppose a series *RL* circuit is composed of a 2.00-Ω resistor, a 15.0 H inductor, and a 6.00 V battery. (a) What is the time constant for this circuit? (b) Once the switch is closed, how long does it take the current to reach half its maximum value?

**ANSWERS** (a) 7.50 s (b) 5.20 s

# 20.7 Energy Stored in a Magnetic Field

**LEARNING OBJECTIVES**

1. Define the energy stored in an inductor and contrast it with the energy stored in a capacitor.
2. Apply the energy stored by an inductor to *RL* circuits.

The emf induced by an inductor prevents a battery from establishing an instantaneous current in a circuit. The battery has to do work to produce a current. We can think of this needed work as energy stored in the inductor in its magnetic field. In a manner similar to that used in Section 16.9 to find the energy stored in a capacitor, we find that the energy stored by an inductor is

Energy stored in an inductor ▶

$$PE_L = \tfrac{1}{2}LI^2 \qquad [20.16]$$

Notice that the result is similar in form to the expression for the energy stored in a charged capacitor (Eq. 16.18):

Energy stored in a capacitor ▶

$$PE_C = \tfrac{1}{2}C(\Delta V)^2$$

## ■ EXAMPLE 20.10 Magnetic Energy

**GOAL** Relate the storage of magnetic energy to currents in an *RL* circuit.

**PROBLEM** A 12.0-V battery is connected in series to a 25.0-Ω resistor and a 5.00-H inductor. **(a)** Find the maximum current in the circuit. **(b)** Find the energy stored in the inductor at this time. **(c)** How much energy is stored in the inductor when the current is changing at a rate of 1.50 A/s?

**STRATEGY** In part (a) Ohm's law and Kirchhoff's voltage rule yield the maximum current because the voltage across the inductor is zero when the current is maximal. Substituting the current into Equation 20.16 gives the energy stored in the inductor. In part (c) the given rate of change of the current can be used to calculate the voltage drop across the inductor at the specified time. Kirchhoff's voltage rule and Ohm's law then give the current $I$ at that time, which can be used to find the energy stored in the inductor.

**SOLUTION**

**(a)** Find the maximum current in the circuit.

Apply Kirchhoff's voltage rule to the circuit:

$$\Delta V_{\text{batt}} + \Delta V_R + \Delta V_L = 0$$

$$\mathcal{E} - IR - L\frac{\Delta I}{\Delta t} = 0$$

When the maximum current is reached, $\Delta I/\Delta t$ is zero, so the voltage drop across the inductor is zero. Solve for the maximum current $I_{\text{max}}$:

$$I_{\text{max}} = \frac{\mathcal{E}}{R} = \frac{12.0\text{ V}}{25.0\ \Omega} = \boxed{0.480\text{ A}}$$

**(b)** Find the energy stored in the inductor at this time.

Substitute known values into Equation 20.16:

$$PE_L = \tfrac{1}{2}LI_{\text{max}}^2 = \tfrac{1}{2}(5.00\text{ H})(0.480\text{ A})^2 = \boxed{0.576\text{ J}}$$

**(c)** Find the energy in the inductor when the current changes at a rate of 1.50 A/s.

Apply Kirchhoff's voltage rule to the circuit once again:

$$\mathcal{E} - IR - L\frac{\Delta I}{\Delta t} = 0$$

Solve this equation for the current $I$ and substitute:

$$I = \frac{1}{R}\left(\mathcal{E} - L\frac{\Delta I}{\Delta t}\right)$$

$$= \frac{1}{25.0\ \Omega}[12.0\text{ V} - (5.00\text{ H})(1.50\text{ A/s})] = 0.180\text{ A}$$

Finally, substitute the value for the current into Equation 20.15, finding the energy stored in the inductor:

$$PE_L = \tfrac{1}{2}LI^2 = \tfrac{1}{2}(5.00\text{ H})(0.180\text{ A})^2 = \boxed{0.081\,0\text{ J}}$$

**REMARKS** Notice how important it is to combine concepts from previous chapters. Here, Ohm's law and Kirchhoff's loop rule were essential to the solution of the problem.

**QUESTION 20.10** True or False: The larger the value of the inductance in such an *RL* circuit, the larger the maximum current.

**EXERCISE 20.10** For the same circuit, find the energy stored in the inductor when the rate of change of the current is 1.00 A/s.

**ANSWER** 0.196 J

## SUMMARY

### 20.1 Induced emf and Magnetic Flux

The magnetic flux $\Phi_B$ through a closed loop is defined as

$$\Phi_B \equiv BA\cos\theta \qquad [20.1]$$

where $B$ is the strength of the uniform magnetic field, $A$ is the cross-sectional area of the loop, and $\theta$ is the angle between $\vec{\mathbf{B}}$ and a direction perpendicular to the plane of the loop.

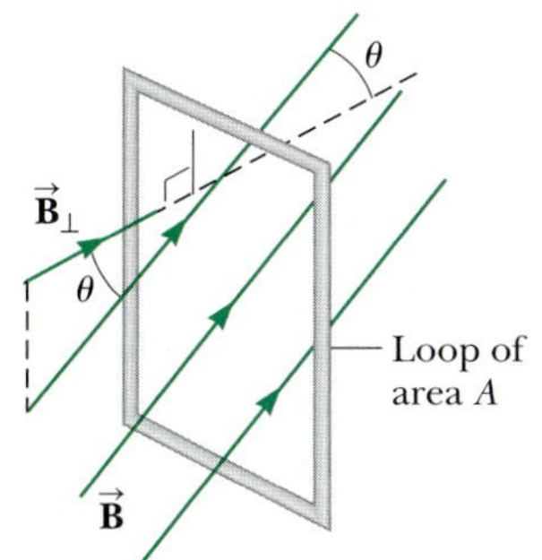


In this view of a loop of area $A$, the component of the magnetic field $\vec{\mathbf{B}}$ perpendicular to surface multiplied by the area gives the magnetic flux through the surface.

### 20.2 Faraday's Law of Induction and Lenz's Law

**Faraday's law of induction** states that the instantaneous emf induced in a circuit equals the negative of the rate of change of magnetic flux through the circuit,

$$\mathcal{E} = -N\frac{\Delta\Phi_B}{\Delta t} \qquad [20.2]$$

where $N$ is the number of loops in the circuit. The magnetic flux $\Phi_B$ can change with time whenever the magnetic field $\vec{\mathbf{B}}$, the area $A$, or the angle $\theta$ changes with time.

**Lenz's law** states that the current from the induced emf creates a magnetic field with flux opposing the *change* in magnetic flux through a circuit.

### 20.3 Motional emf

If a conducting bar of length $\ell$ moves through a magnetic field with a speed $v$ so that $\vec{\mathbf{B}}$ is perpendicular to the bar, the emf induced in the bar, often called a **motional emf**, is

$$|\mathcal{E}| = B\ell v \qquad [20.4]$$

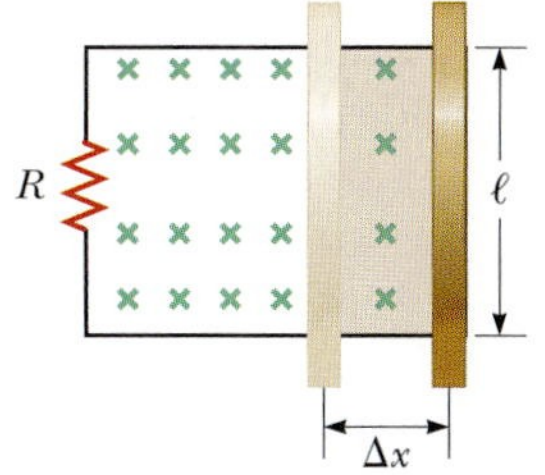


As the bar moves to the right, the area of the loop increases by the amount $\ell\Delta x$ and the magnetic flux through the loop increases by $B\ell\Delta x$.

### 20.4 Generators

When a coil of wire with $N$ turns, each of area $A$, rotates with constant angular speed $\omega$ in a uniform magnetic field $\vec{\mathbf{B}}$, the emf induced in the coil is

$$\mathcal{E} = NBA\omega\sin\omega t \qquad [20.7]$$

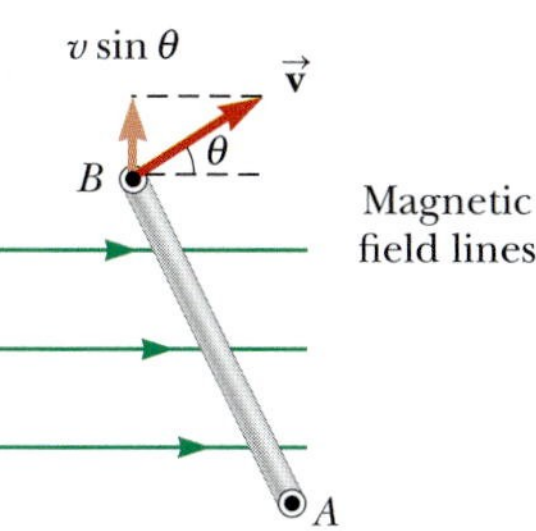


In this edge view of a rotating loop, the magnetic flux changes continuously, generating an alternating current in the loop.

Such generators naturally produce alternating current (AC), which changes direction with frequency $\omega/2\pi$. The AC current can be transformed to direct current.

### 20.5 Self-Inductance

### 20.6 *RL* Circuits

When the current in a coil changes with time, an emf is induced in the coil according to Faraday's law. This **self-induced emf** is defined by the expression

$$\mathcal{E} \equiv -L\frac{\Delta I}{\Delta t} \qquad [20.9]$$

where $L$ is the inductance of the coil. The SI unit for inductance is the henry (H); 1 H = 1 V · s/A.

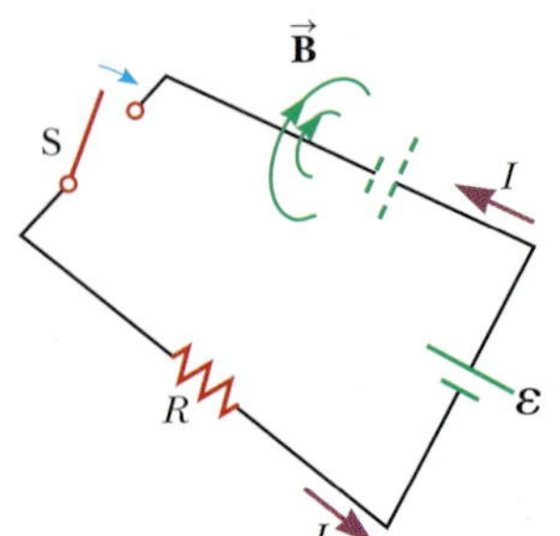


When the switch is closed, a magnetic field begins to develop as shown. The changing magnetic flux creates a self-induced emf in the opposite direction, represented by the dashed lines.

The **inductance** of a coil can be found from the expression

$$L = \frac{N\Phi_B}{I} \quad [20.10]$$

where $N$ is the number of turns on the coil, $I$ is the current in the coil, and $\Phi_B$ is the magnetic flux through the coil produced by that current. For a solenoid, the inductance is given by

$$L = \frac{\mu_0 N^2 A}{\ell} \quad [20.11a]$$

If a resistor and inductor are connected in series to a battery and a switch is closed at $t = 0$, the current in the circuit doesn't rise instantly to its maximum value. After one **time constant** $\tau = L/R$, the current in the circuit is 63.2% of its final value $\mathcal{E}/R$. As the current approaches its final, maximum value, the voltage drop across the inductor approaches zero. The current $I$ in such a circuit any time $t$ after the circuit is completed is

$$I = \frac{\mathcal{E}}{R}(1 - e^{-t/\tau}) \quad [20.15]$$

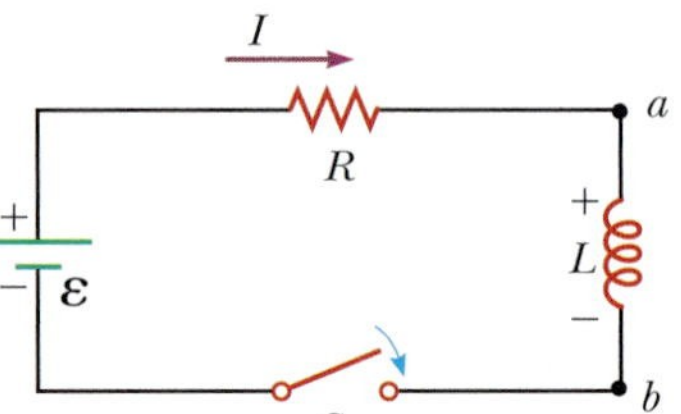


A series *RL* circuit. As the current increases toward its maximum value, the inductor produces an emf that opposes the increasing current.

## 20.7 Energy Stored in a Magnetic Field

The **energy stored** in the magnetic field of an inductor carrying current $I$ is

$$PE_L = \tfrac{1}{2}LI^2 \quad [20.16]$$

As the current in an *RL* circuit approaches its maximum value, the stored energy also approaches a maximum value.

## WARM-UP EXERCISES

ENHANCED WebAssign The warm-up exercises in this chapter may be assigned online in Enhanced WebAssign.

1. **Physics Review** A simple pendulum undergoes one complete oscillation in 4.30 s. Find (a) the frequency and (b) the angular frequency of the oscillation. (See Section 13.5.)

2. **Physics Review** An electric field of magnitude 2.75 V/m is oriented at an angle of 30.0° with respect to the positive *z*-direction. Determine the magnitude of the electric flux through a rectangular area of 2.00 m$^2$ in the *xy*-plane. (See Section 15.9.)

3. **Physics Review** Determine the time constant for an *RC* circuit having an effective resistance of $1.25 \times 10^4$ Ω and an effective capacitance of $2.22 \times 10^{-6}$ F. (See Section 18.5.)

4. A magnetic field of magnitude $5.00 \times 10^{-3}$ T is oriented at an angle of 30.0° with respect to the positive *z*-direction. Determine the magnitude of the magnetic flux through a rectangular surface in the *xy*-plane with an area of 3.00 m$^2$. (See Section 20.1.)

5. A coil of wire has 6.00 turns, each turn enclosing an area of $2.00 \times 10^{-3}$ m$^2$, while the total resistance of the coil is $1.20 \times 10^{-3}$ Ω. If a magnet is brought closer to the coil from a distance so that the magnetic field through each loop increases steadily from zero to $4.00 \times 10^{-2}$ T over the course of 0.750 s, determine the magnitude of (a) the induced emf in the coil and (b) the current in the coil while the flux is increasing. (See Section 20.2.)

6. In one of NASA's space tether experiments, a 20.0-km long conducting wire was deployed by the space shuttle as it orbited at $7.86 \times 10^3$ m/s around earth and across earth's magnetic field lines. The resulting motional emf was used as a power source. If the component of earth's magnetic field perpendicular to the tether was $1.50 \times 10^{-5}$ T, determine the maximum possible potential difference between the two ends of the tether. (See Section 20.3.)

7. Wind turbines use a generator to convert the wind's kinetic energy into electrical energy. Determine (a) the maximum emf of a generator with a 100-turn, 0.500 m$^2$ coil that is rotated by the wind at an angular frequency of 6.00 rad/s through a 0.450-T magnetic field and (b) the maximum induced current if the coil has a total resistance of 10.0 Ω. (See Section 20.4.)

8. Suppose insulated wire of length 5.28 m is coiled into a 0.015 0-m radius solenoid of length 0.060 0 m. Determine (a) the number of loops in the coil, (b) the inductance of the solenoid, and (c) the magnitude of the induced emf if the current is changing at a rate of $6.00 \times 10^2$ A/s. (See Section 20.5.)

9. Determine the time constant for an *RL* circuit having an effective resistance of $1.25 \times 10^4$ Ω and an inductance of $2.00 \times 10^{-2}$ H. (See Section 20.6.)

10. Determine the energy stored in a 2.50-mH inductor carrying a current of 10.0 A. (See Section 20.7.)

## CONCEPTUAL QUESTIONS

WebAssign The conceptual questions in this chapter may be assigned online in Enhanced WebAssign.

1. A spacecraft orbiting Earth has a coil of wire in it. An astronaut measures a small current in the coil, although there is no battery connected to it and there are no magnets in the spacecraft. What is causing the current?

2. Does dropping a magnet down a copper tube produce a current in the tube? Explain.

3. A circular loop is located in a uniform and constant magnetic field. Describe how an emf can be induced in the loop in this situation.

4. A loop of wire is placed in a uniform magnetic field. (a) For what orientation of the loop is the magnetic flux a maximum? (b) For what orientation is the flux zero?

5. As the conducting bar in Figure CQ20.5 moves to the right, an electric field directed downward is set up. If the bar were moving to the left, explain why the electric field would be upward.

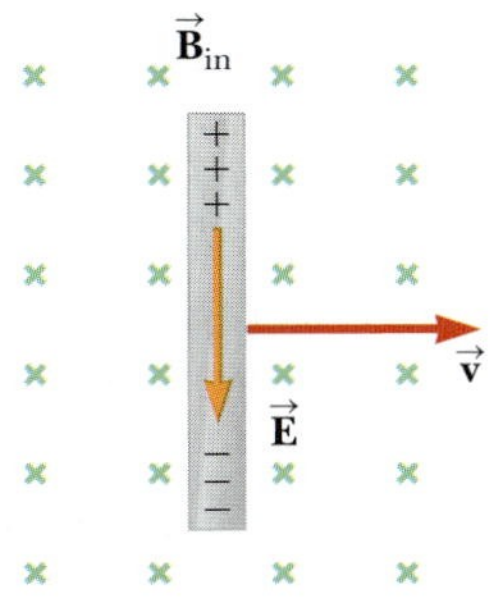


**Figure CQ20.5** Conceptual Questions 5 and 8.

6. How is electrical energy produced in dams? (That is, how is the energy of motion of the water converted to AC electricity?)

7. Wearing a metal bracelet in a region of strong magnetic field could be hazardous. Discuss this statement.

8. As the bar in Figure CQ20.5 moves perpendicular to the field, is an external force required to keep it moving with constant speed?

9. Eddy currents are induced currents set up in a piece of metal when it moves through a nonuniform magnetic field. For example, consider the flat metal plate swinging at the end of a bar as a pendulum, as shown in Figure CQ20.9. (a) At position 1, the pendulum is moving from a region where there is no magnetic field into a region where the field $\vec{\mathbf{B}}_{in}$ is directed into the paper. Show that at position 1 the direction of the eddy current is counterclockwise. (b) At position 2, the pendulum is moving out of the field into a region of zero field. Show that the direction of the eddy current is clockwise in this case. (c) Use right-hand rule number 2 to show that these eddy currents lead to a magnetic force on the plate directed as shown in the figure. Because the induced eddy current always produces a retarding force when the plate enters or leaves the field, the swinging plate quickly comes to rest.

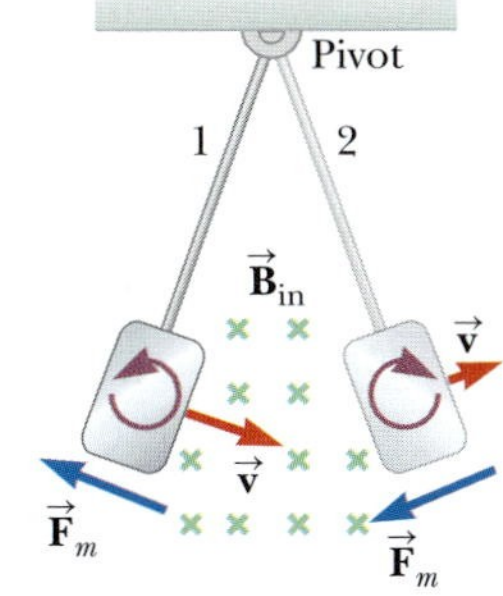


**Figure CQ20.9**

10. A bar magnet is dropped toward a conducting ring lying on the floor. As the magnet falls toward the ring, does it move as a freely falling object? Explain.

11. A piece of aluminum is dropped vertically downward between the poles of an electromagnet. Does the magnetic field affect the velocity of the aluminum? *Hint:* See Conceptual Question 9.

12. When the switch in Figure CQ20.12a is closed, a current is set up in the coil and the metal ring springs upward (Fig. CQ20.12b). Explain this behavior.

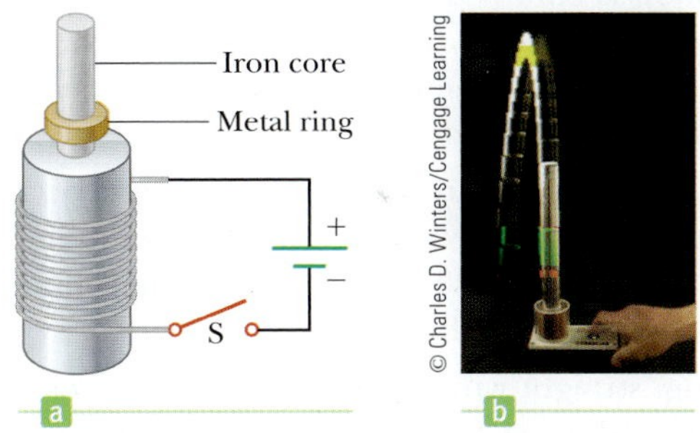


**Figure CQ20.12** Conceptual Questions 12 and 13.

13. Assume the battery in Figure CQ20.12a is replaced by an AC source and the switch is held closed. If held down, the metal ring on top of the solenoid becomes hot. Why?

14. A magneto is used to cause the spark in a spark plug in many lawn mowers today. A magneto consists of a permanent magnet mounted on a flywheel so that it spins past a fixed coil. Explain how this arrangement generates a large enough potential difference to cause the spark.

## PROBLEMS

ENHANCED WebAssign The problems in this chapter may be assigned online in Enhanced WebAssign.

1. denotes straightforward problem; 2. denotes intermediate problem;
3. denotes challenging problem

1. denotes full solution available in *Student Solutions Manual/Study Guide*

1. denotes problems most often assigned in Enhanced WebAssign

BIO denotes biomedical problems
GP denotes guided problems
M denotes Master It tutorial available in Enhanced WebAssign
Q|C denotes asking for quantitative and conceptual reasoning
S denotes symbolic reasoning problem
W denotes Watch It video solution available in Enhanced WebAssign

### 20.1 Induced EMF and Magnetic Flux

1. A uniform magnetic field of magnitude 0.50 T is directed perpendicular to the plane of a rectangular loop having dimensions 8.0 cm by 12 cm. Find the magnetic flux through the loop.

2. Find the flux of Earth's magnetic field of magnitude $5.00 \times 10^{-5}$ T through a square loop of area 20.0 cm$^2$ (a) when the field is perpendicular to the plane of the loop, (b) when the field makes a 30.0° angle with the normal to the plane of the loop, and (c) when the field makes a 90.0° angle with the normal to the plane.

3. A circular loop of radius 12.0 cm is placed in a uniform magnetic field. (a) If the field is directed perpendicular to the plane of the loop and the magnetic flux through the loop is $8.00 \times 10^{-3}$ T · m$^2$, what is the strength of the magnetic field? (b) If the magnetic field is directed parallel to the plane of the loop, what is the magnetic flux through the loop?

4. A long, straight wire carrying a current of 2.00 A is placed along the axis of a cylinder of radius 0.500 m and a length of 3.00 m. Determine the total magnetic flux through the cylinder.

5. A long, straight wire lies in the plane of a circular coil with a radius of 0.010 m. The wire carries a current of 2.0 A and is placed along a diameter of the coil. (a) What is the net flux through the coil? (b) If the wire passes through the center of the coil and is perpendicular to the plane of the coil, what is the net flux through the coil?

6. A 400-turn solenoid of length 36.0 cm and radius 3.00 cm carries a current of 5.00 A. Find (a) the magnetic field strength inside the coil at its midpoint and (b) the magnetic flux through a circular cross-sectional area of the solenoid at its midpoint.

7. M A cube of edge length $\ell$ = 2.5 cm is positioned as shown in Figure P20.7. There is a uniform magnetic field throughout the region with components $B_x$ = +5.0 T, $B_y$ = +4.0 T, and $B_z$ = +3.0 T. (a) Calculate the flux through the shaded face of the cube. (b) What is the total flux emerging from the volume enclosed by the cube (i.e., the total flux through all six faces)?

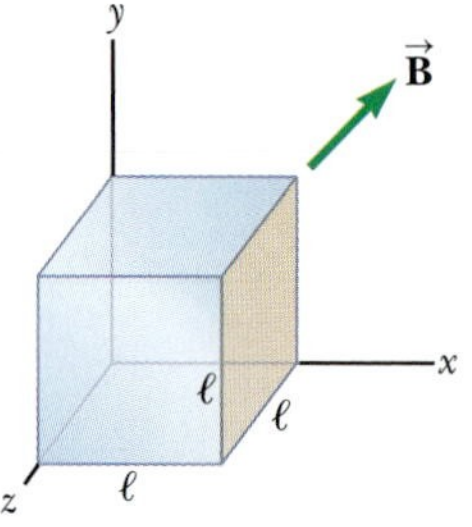

Figure P20.7

### 20.2 Faraday's Law of Induction and Lenz's Law

8. BIO Transcranial magnetic stimulation (TMS) is a noninvasive technique used to stimulate regions of the human brain. A small coil is placed on the scalp, and a brief burst of current in the coil produces a rapidly changing magnetic field inside the brain. The induced emf can be sufficient to stimulate neuronal activity. One such device generates a magnetic field within the brain that rises from zero to 1.5 T in 120 ms. Determine the induced emf within a circle of tissue of radius 1.6 mm and that is perpendicular to the direction of the field.

9. Three loops of wire move near a long straight wire carrying a current as in Figure P20.9. What is the direction of the induced current, if any, in (a) loop A, (b) loop B, and (c) loop C.

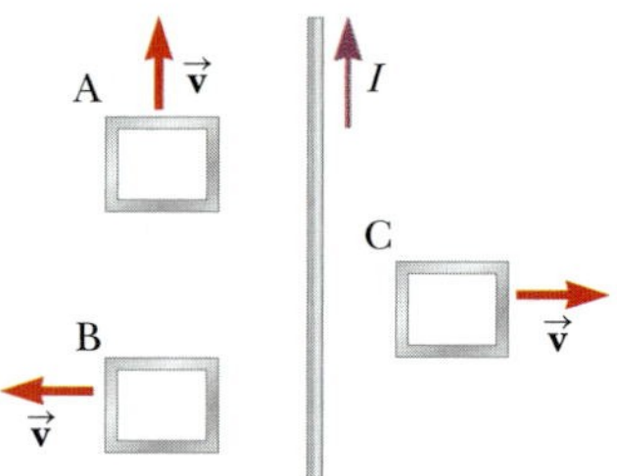

Figure P20.9

10. W The flexible loop in Figure P20.10 has a radius of 12 cm and is in a magnetic field of strength 0.15 T. The loop is grasped at points *A* and *B* and stretched until its area is nearly zero. If it takes 0.20 s to close the loop, what is the magnitude of the average induced emf in it during this time?

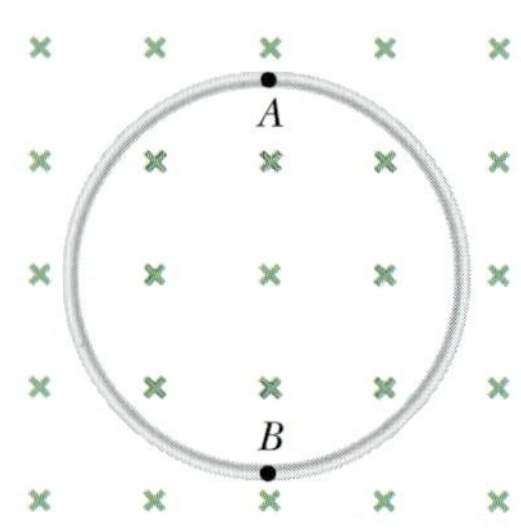

Figure P20.10
Problems 10, 12, and 22.

11. A wire loop of radius 0.30 m lies so that an external magnetic field of magnitude 0.30 T is perpendicular to the loop. The field reverses its direction, and its magnitude changes to 0.20 T in 1.5 s. Find the magnitude of the average induced emf in the loop during this time.

12. A circular loop of wire of radius 12.0 cm is placed in a magnetic field directed perpendicular to the plane of the loop, as shown in Figure P20.10. If the field decreases at the rate of 0.050 0 T/s in some time interval, what is the magnitude of the emf induced in the loop during this interval?

13. A technician wearing a circular metal band on his wrist moves his hand into a uniform magnetic field of magnitude 2.5 T in a time of 0.18 s. If the diameter of the band is 6.5 cm and the field is at an angle of 45° with the plane of the metal band while the hand is in the field, find the magnitude of the average emf induced in the band.

14. In Figure P20.14 what is the direction of the current induced in the resistor at the instant the switch is closed?

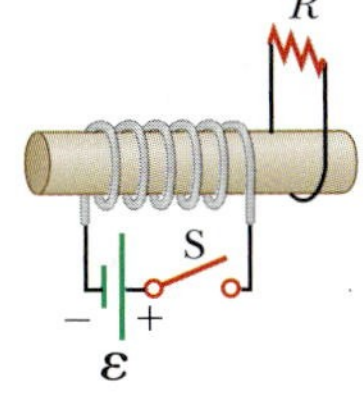


Figure P20.14

15. A bar magnet is positioned near a coil of wire, as shown in Figure P20.15. What is the direction of the current in the resistor when the magnet is moved (a) to the left and (b) to the right?

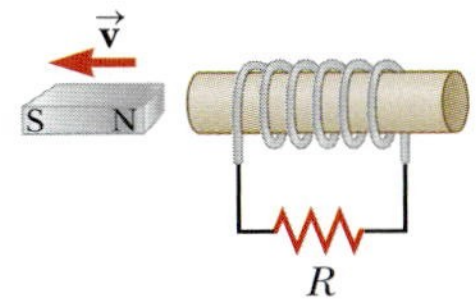


Figure P20.15

16. Find the direction of the current in the resistor shown in Figure P20.16 (a) at the instant the switch is closed, (b) after the switch has been closed for several minutes, and (c) at the instant the switch is opened.

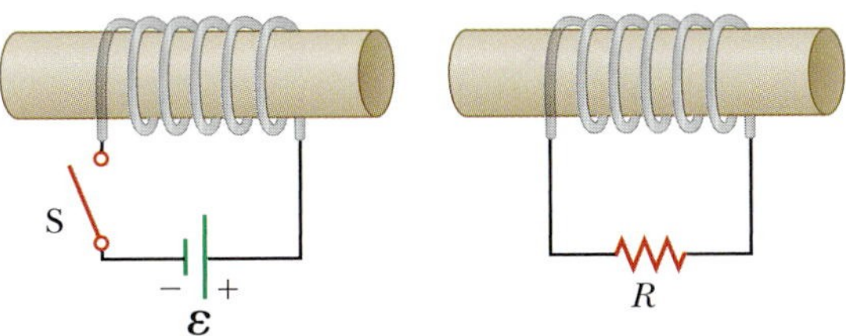


Figure P20.16

17. A circular loop of wire lies below a long wire carrying a current that is increasing as in Figure P20.17a. (a) What is the direction of the induced current in the loop, if any? (b) Now suppose the loop is next to the same wire as in P20.17b. What is the direction of the induced current in the loop, if any? Explain your answers.

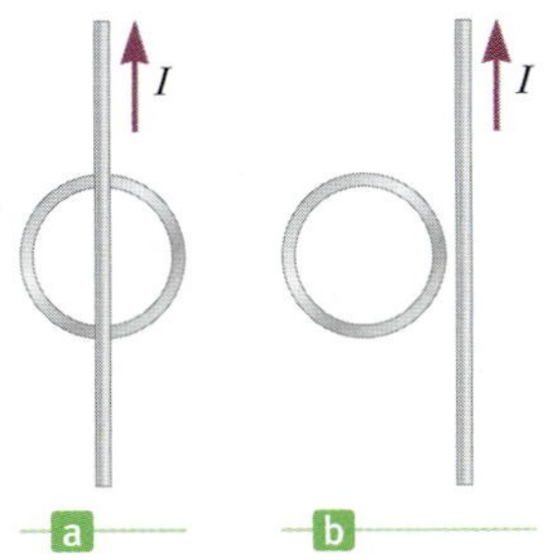


Figure P20.17

18. A square, single-turn wire loop $\ell = 1.00$ cm on a side is placed inside a solenoid that has a circular cross section of radius $r = 3.00$ cm, as shown in the end view of Figure P20.18. The solenoid is 20.0 cm long and wound with 100 turns of wire. (a) If the current in the solenoid is 3.00 A, what is the flux through the square loop? (b) If the current in the solenoid is reduced to zero in 3.00 s, what is the magnitude of the average induced emf in the square loop?

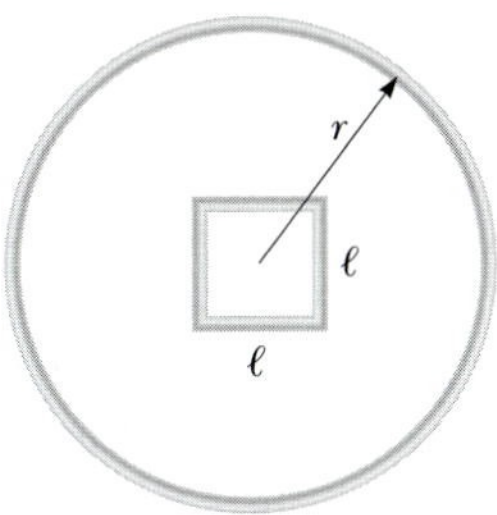


Figure P20.18

19. GP A 300-turn solenoid with a length of 20.0 cm and a radius of 1.50 cm carries a current of 2.00 A. A second coil of four turns is wrapped tightly around this solenoid, so it can be considered to have the same radius as the solenoid. The current in the 300-turn solenoid increases steadily to 5.00 A in 0.900 s. (a) Use Ampere's law to calculate the initial magnetic field in the middle of the 300-turn solenoid. (b) Calculate the magnetic field of the 300-turn solenoid after 0.900 s. (c) Calculate the area of the 4-turn coil. (d) Calculate the change in the magnetic flux through the 4-turn coil during the same period. (e) Calculate the average induced emf in the 4-turn coil. Is it equal to the instantaneous induced emf? Explain. (f) Why could contributions to the magnetic field by the current in the 4-turn coil be neglected in this calculation?

20. M A circular coil enclosing an area of 100 cm$^2$ is made of 200 turns of copper wire. The wire making up the coil has resistance of 5.0 Ω, and the ends of the wire are connected to form a closed circuit. Initially, a 1.1-T uniform magnetic field points perpendicularly upward through the plane of the coil. The direction of the field then reverses so that the final magnetic field has a magnitude of 1.1 T and points downward through the coil. If the time required for the field to reverse directions is 0.10 s, what is the average current in the coil during that time?

21. BIO To monitor the breathing of a hospital patient, a thin belt is girded around the patient's chest as in Figure P20.21. The belt is a 200-turn coil. When the patient inhales, the area encircled by the coil increases

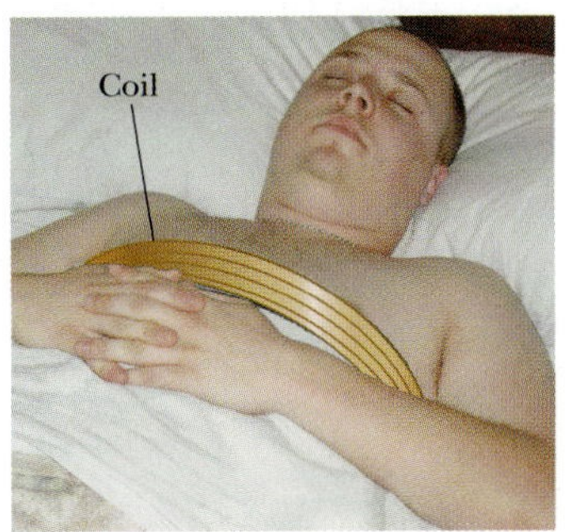


Figure P20.21

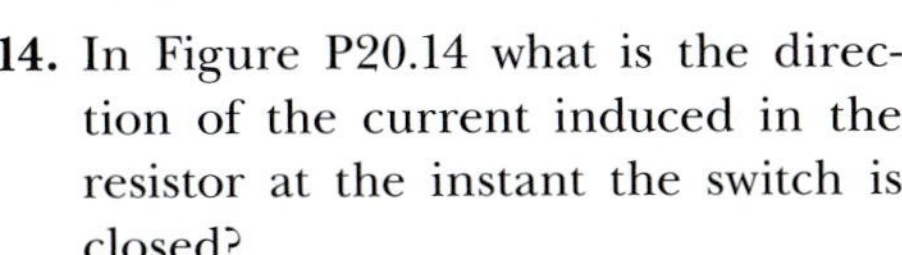

by 39.0 $cm^2$. The magnitude of Earth's magnetic field is 50.0 $\mu$T and makes an angle of 28.0° with the plane of the coil. Assuming a patient takes 1.80 s to inhale, find the magnitude of the average induced emf in the coil during that time.

22. Q|C S An $N$-turn circular wire coil of radius $r$ lies in the $xy$-plane (the plane of the page), as in Figure P20.10. A uniform magnetic field is turned on, increasing steadily from 0 to $B_0$ in the positive $z$-direction in $t$ seconds. (a) Find a symbolic expression for the emf, $\mathcal{E}$, induced in the coil in terms of the variables given. (b) Looking down on at the $xy$-plane from the positive $z$-axis, is the direction of the induced current clockwise or counterclockwise? (c) If each loop has resistance $R$, find an expression for the magnitude of the induced current, $I$.

## 20.3 Motional emf

23. A truck is carrying a steel beam of length 15.0 m on a freeway. An accident causes the beam to be dumped off the truck and slide horizontally along the ground at a speed of 25.0 m/s. The velocity of the center of mass of the beam is northward while the length of the beam maintains an east–west orientation. The vertical component of the Earth's magnetic field at this location has a magnitude of 35.0 $\mu$T. What is the magnitude of the induced emf between the ends of the beam?

24. A 2.00-m length of wire is held in an east–west direction and moves horizontally to the north with a speed of 15.0 m/s. The vertical component of Earth's magnetic field in this region is 40.0 $\mu$T directed downward. Calculate the induced emf between the ends of the wire and determine which end is positive.

25. A pickup truck has a width of 79.8 in. If it is traveling north at 37 m/s through a magnetic field with vertical component of 35 $\mu$T, what magnitude emf is induced between the driver and passenger sides of the truck?

26. Q|C A bar magnet is held stationary while a circular loop of wire is moved toward the magnet at constant velocity at position $A$ as in Figure P20.26. The loop passes over the magnet at position $B$ and moves away from the magnet at position $C$. Find the direction of the induced current in the loop using Lenz's law (a) at position $A$ and (b) at position $C$. (c) What is the induced current in the loop at position $B$? Explain.

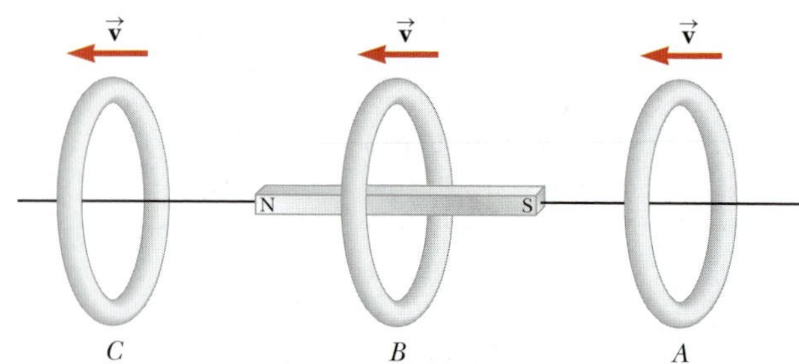

**Figure P20.26** Problems 26 and 43.

27. M An automobile has a vertical radio antenna 1.20 m long. The automobile travels at 65.0 km/h on a horizontal road where Earth's magnetic field is 50.0 $\mu$T, directed toward the north and downward at an angle of 65.0° below the horizontal. (a) Specify the direction the automobile should move so as to generate the maximum motional emf in the antenna, with the top of the antenna positive relative to the bottom. (b) Calculate the magnitude of this induced emf.

28. Q|C An astronaut is connected to her spacecraft by a 25-m-long tether cord as she and the spacecraft orbit Earth in a circular path at a speed of $3.0 \times 10^3$ m/s. At one instant, the voltage measured between the ends of a wire embedded in the cord is measured to be 0.45 V. Assume the long dimension of the cord is perpendicular to the vertical component of Earth's magnetic field at that instant. (a) What is the magnitude of the vertical component of Earth's field at this location? (b) Does the measured voltage change as the system moves from one location to another? Explain.

29. Figure P20.29 shows a bar of mass $m$ = 0.200 kg that can slide without friction on a pair of rails separated by a distance $\ell$ = 1.20 m and located on an inclined plane that makes an angle $\theta$ = 25.0° with respect to the ground. The resistance of the resistor is $R$ = 1.00 Ω, and a uniform magnetic field of magnitude $B$ = 0.500 T is directed downward, perpendicular to the ground, over the entire region through which the bar moves. With what constant speed $v$ does the bar slide along the rails?

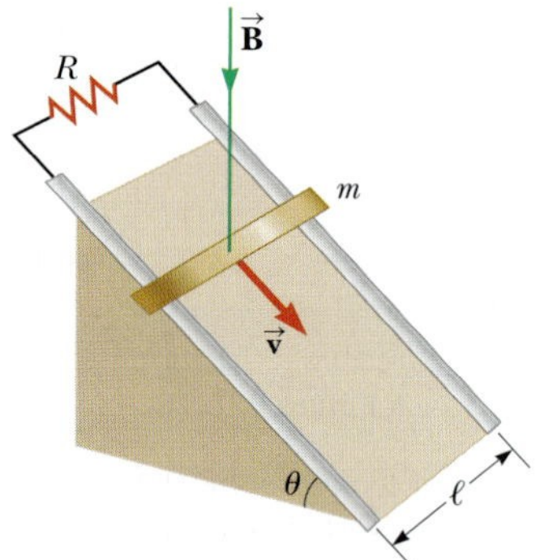

**Figure P20.29**

30. W Consider the arrangement shown in Figure P20.30. Assume $R$ = 6.00 Ω, $\ell$ = 1.20 m, and a uniform 2.50-T magnetic field is directed *into* the page. At what speed should the bar be moved to produce a current of 0.500 A in the resistor?

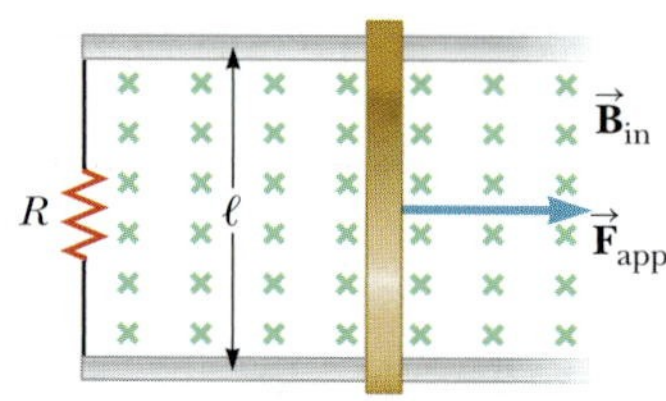

**Figure P20.30** Problems 30, 56, and 59.

## 20.4 Generators

**31.** A square coil of wire of side 2.80 cm is placed in a uniform magnetic field of magnitude 1.25 T directed into the page as in Figure P20.31. The coil has 28.0 turns and a resistance of 0.780 Ω. If the coil is rotated through an angle of 90.0° about the horizontal axis shown in 0.335 s, find (a) the magnitude of the average emf induced in the coil during this rotation and (b) the average current induced in the coil during this rotation.

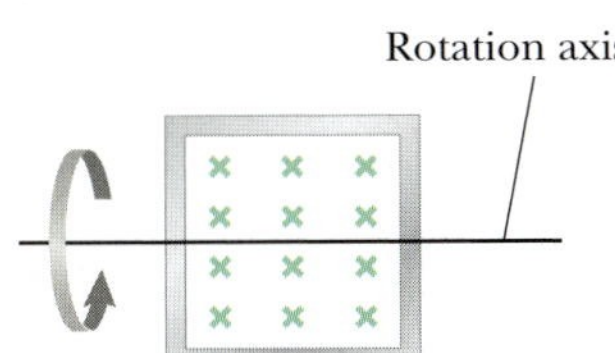


**Figure P20.31**

**32.** W A 100-turn square wire coil of area 0.040 m$^2$ rotates about a vertical axis at 1 500 rev/min, as indicated in Figure P20.32. The horizontal component of Earth's magnetic field at the location of the loop is $2.0 \times 10^{-5}$ T. Calculate the maximum emf induced in the coil by Earth's field.

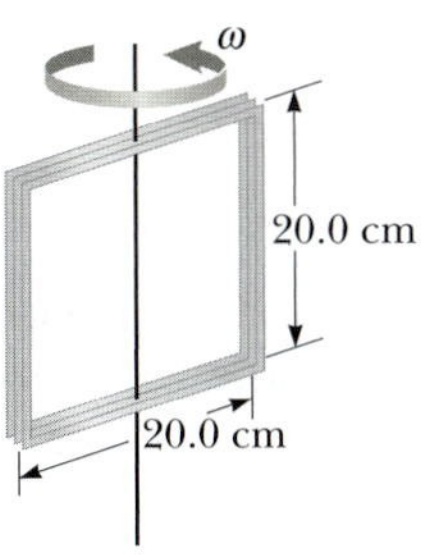


**Figure P20.32**

**33.** BIO Considerable scientific work is currently under way to determine whether weak oscillating magnetic fields such as those found near outdoor electric power lines can affect human health. One study indicated that a magnetic field of magnitude $1.0 \times 10^{-3}$ T, oscillating at 60 Hz, might stimulate red blood cells to become cancerous. If the diameter of a red blood cell is 8.0 μm, determine the maximum emf that can be generated around the perimeter of the cell.

**34.** A flat coil enclosing an area of 0.10 m$^2$ is rotating at 60 rev/s, with its axis of rotation perpendicular to a 0.20-T magnetic field. (a) If there are 1 000 turns on the coil, what is the maximum voltage induced in the coil? (b) When the maximum induced voltage occurs, what is the orientation of the coil with respect to the magnetic field?

**35.** In a model AC generator, a 500-turn rectangular coil 8.0 cm by 20 cm rotates at 120 rev/min in a uniform magnetic field of 0.60 T. (a) What is the maximum emf induced in the coil? (b) What is the instantaneous value of the emf in the coil at $t = (\pi/32)$ s? Assume the emf is zero at $t = 0$. (c) What is the smallest value of $t$ for which the emf will have its maximum value?

**36.** A motor has coils with a resistance of 30 Ω and operates from a voltage of 240 V. When the motor is operating at its maximum speed, the back emf is 145 V. Find the current in the coils (a) when the motor is first turned on and (b) when the motor has reached maximum speed. (c) If the current in the motor is 6.0 A at some instant, what is the back emf at that time?

**37.** A coil of 10.0 turns is in the shape of an ellipse having a major axis of 10.0 cm and a minor axis of 4.00 cm. The coil rotates at 100 rpm in a region in which the magnitude of Earth's magnetic field is 55 μT. What is the maximum voltage induced in the coil if the axis of rotation of the coil is along its major axis and is aligned (a) perpendicular to Earth's magnetic field and (b) parallel to Earth's magnetic field? *Note:* The area of an ellipse is given by $A = \pi ab$, where $a$ is the length of the *semi*major axis and $b$ is the length of the *semi*minor axis.

## 20.5 Self-Inductance

**38.** A technician wraps wire around a tube of length 36 cm having a diameter of 8.0 cm. When the windings are evenly spread over the full length of the tube, the result is a solenoid containing 580 turns of wire. (a) Find the self-inductance of this solenoid. (b) If the current in this solenoid increases at the rate of 4.0 A/s, what is the self-induced emf in the solenoid?

**39.** The current in a coil drops from 3.5 A to 2.0 A in 0.50 s. If the average emf induced in the coil is 12 mV, what is the self-inductance of the coil?

**40.** S Show that the two expressions for inductance given by

$$L = \frac{N\Phi_B}{I} \quad \text{and} \quad L = \frac{-\mathcal{E}}{\Delta I/\Delta t}$$

have the same units.

**41.** M A solenoid of radius 2.5 cm has 400 turns and a length of 20 cm. Find (a) its inductance and (b) the rate at which current must change through it to produce an emf of 75 mV.

**42.** An emf of 24.0 mV is induced in a 500-turn coil when the current is changing at a rate of 10.0 A/s. What is the magnetic flux through each turn of the coil at an instant when the current is 4.00 A?

## 20.6 *RL* Circuits

**43.** An electromagnet can be modeled as an inductor in series with a resistor. Consider a large electromagnet of inductance $L = 12.0$ H and resistance $R = 4.50$ Ω connected to a 24.0 V battery and switch as in Figure P20.43. After the switch is closed, find (a) the maximum current carried by the electromagnet, (b) the time constant of the circuit, and (c) the time it takes the current to reach 95.0% of its maximum value.

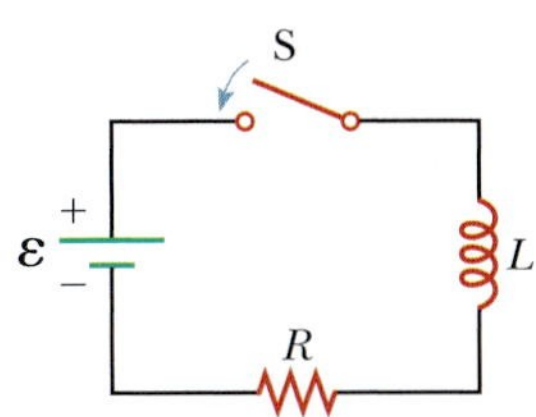


**Figure P20.43**
Problems 43, 45, 46, and 48

**44.** An *RL* circuit with $L = 3.00$ H and an *RC* circuit with $C = 3.00\ \mu$F have the same time constant. If the two circuits have the same resistance *R*, (a) what is the value of *R* and (b) what is this common time constant?

**45.** A battery is connected in series with a 0.30-Ω resistor and an inductor, as in Figure P20.43. The switch is closed at $t = 0$. The time constant of the circuit is 0.25 s, and the maximum current in the circuit is 8.0 A. Find (a) the emf of the battery, (b) the inductance of the circuit, (c) the current in the circuit after one time constant has elapsed, (d) the voltage across the resistor after one time constant has elapsed, and (e) the voltage across the inductor after one time constant has elapsed.

**46.** A 25-mH inductor, an 8.0-Ω resistor, and a 6.0-V battery are connected in series as in Figure P20.43. The switch is closed at $t = 0$. Find the voltage drop across the resistor (a) at $t = 0$ and (b) after one time constant has passed. Also, find the voltage drop across the inductor (c) at $t = 0$ and (d) after one time constant has elapsed.

**47.** Calculate the resistance in an *RL* circuit in which $L = 2.50$ H and the current increases to 90.0% of its final value in 3.00 s.

**48.** W Consider the circuit shown in Figure P20.43. Take $\mathcal{E} = 6.00$ V, $L = 8.00$ mH, and $R = 4.00\ \Omega$. (a) What is the inductive time constant of the circuit? (b) Calculate the current in the circuit 250 $\mu$s after the switch is closed. (c) What is the value of the final steady-state current? (d) How long does it take the current to reach 80.0% of its maximum value?

## 20.7 Energy Stored in a Magnetic Field

**49.** (a) If an inductor carrying a 1.70-A current stores an energy of 0.300 mJ, what is its inductance? (b) How much energy does the same inductor store if it carries a 3.0-A current?

**50.** A 300-turn solenoid has a radius of 5.00 cm and a length of 20.0 cm. Find (a) the inductance of the solenoid and (b) the energy stored in the solenoid when the current in its windings is 0.500 A.

**51.** M A 24-V battery is connected in series with a resistor and an inductor, with $R = 8.0\ \Omega$ and $L = 4.0$ H, respectively. Find the energy stored in the inductor (a) when the current reaches its maximum value and (b) one time constant after the switch is closed.

**52.** GP A 60.0-m length of insulated copper wire is wound to form a solenoid of radius 2.0 cm. The copper wire has a radius of 0.50 mm. (a) What is the resistance of the wire? (b) Treating each turn of the solenoid as a circle, how many turns can be made with the wire? (c) How long is the resulting solenoid? (d) What is the self-inductance of the solenoid? (e) If the solenoid is attached to a battery with an emf of 6.0 V and internal resistance of 350 mΩ, compute the time constant of the circuit. (f) What is the maximum current attained? (g) How long would it take to reach 99.9% of its maximum current? (h) What maximum energy is stored in the inductor?

## Additional Problems

**53.** Two circular loops of wire surround an insulating rod as in Figure P20.53. Loop 1 carries a current *I* in the clockwise direction when viewed from the left end. If loop 1 moves toward loop 2, which remains stationary, what is the direction of the induced current in loop 2 when viewed from the left end?

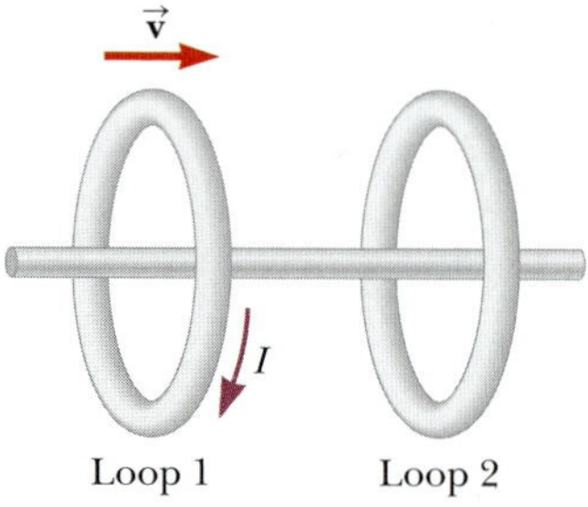


**Figure P20.53**

**54.** A circular loop of wire of resistance $R = 0.500\ \Omega$ and radius $r = 8.00$ cm is in a uniform magnetic field directed out of the page as in Figure P20.54. If a clockwise current of $I = 2.50$ mA is induced in the loop, (a) is the magnetic field increasing or decreasing in time? (b) Find the rate at which the field is changing with time.

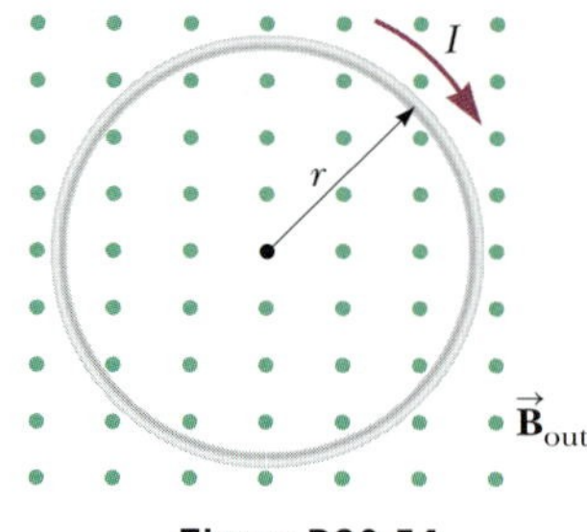


**Figure P20.54**

**55.** A rectangular coil with resistance *R* has *N* turns, each of length $\ell$ and width $w$, as shown in Figure P20.55. The coil moves into a uniform magnetic field $\vec{B}_{in}$ with constant velocity $\vec{v}$. What are the magnitude and direction of the total magnetic force on the coil (a) as it enters the magnetic field, (b) as it moves within the field, and (c) as it leaves the field?

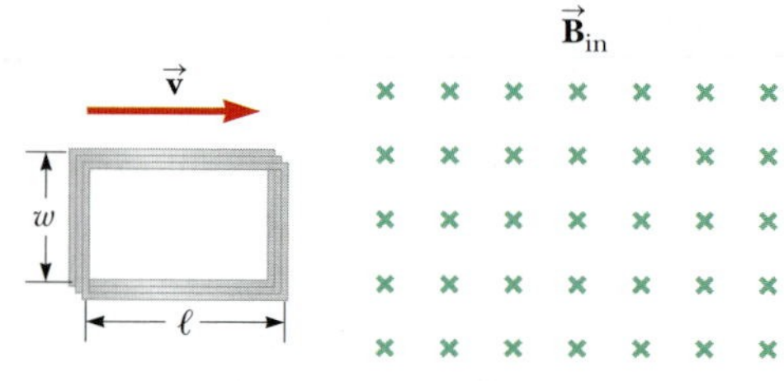


**Figure P20.55**

56. **Q|C** A conducting bar of length $\ell$ moves to the right on two frictionless rails, as shown in Figure P20.30. A uniform magnetic field directed into the page has a magnitude of 0.30 T. Assume $\ell = 35$ cm and $R = 9.0\ \Omega$. (a) At what constant speed should the bar move to produce an 8.5-mA current in the resistor? What is the direction of this induced current? (b) At what rate is energy delivered to the resistor? (c) Explain the origin of the energy being delivered to the resistor.

57. An 820-turn wire coil of resistance 24.0 Ω is placed on top of a 12 500-turn, 7.00-cm-long solenoid, as in Figure P20.57. Both coil and solenoid have cross-sectional areas of $1.00 \times 10^{-4}\ m^2$. (a) How long does it take the solenoid current to reach 0.632 times its maximum value? (b) Determine the average back emf caused by the self-inductance of the solenoid during this interval. The magnetic field produced by the solenoid at the location of the coil is one-half as strong as the field at the center of the solenoid. (c) Determine the average rate of change in magnetic flux through each turn of the coil during the stated interval. (d) Find the magnitude of the average induced current in the coil.

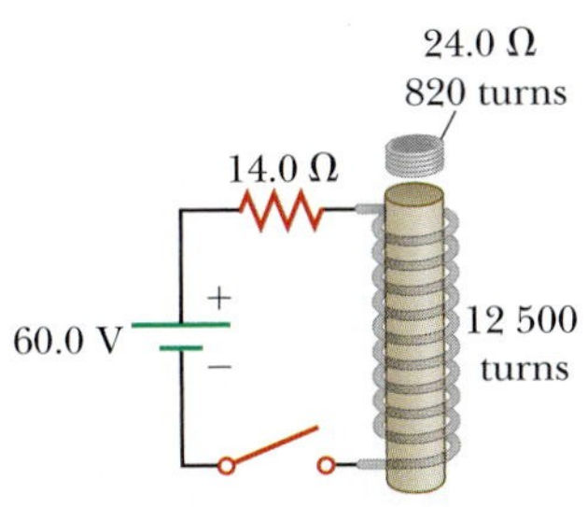


**Figure P20.57**

58. **Q|C** A spacecraft is in a circular orbit of radius equal to $3.0 \times 10^4$ km around a $2.0 \times 10^{30}$ kg pulsar. The magnetic field of the pulsar at that radial distance is $1.0 \times 10^2$ T directed perpendicular to the velocity of the spacecraft. The spacecraft is 0.20 km long with a radius of 0.040 km and moves counterclockwise in the *xy*-plane around the pulsar. (a) What is the speed of the spacecraft? (b) If the magnetic field points in the positive *z*-direction, is the emf induced from the back to the front of the spacecraft or from side to side? (c) Compute the induced emf. (d) Describe the hazards for astronauts inside any spacecraft moving in the vicinity of a pulsar.

59. **Q|C** A conducting rod of length $\ell$ moves on two horizontal frictionless rails, as in Figure P20.30. A constant force of magnitude 1.00 N moves the bar at a uniform speed of 2.00 m/s through a magnetic field $\vec{\mathbf{B}}$ that is directed into the page. (a) What is the current in an 8.00-Ω resistor *R*? (b) What is the rate of energy dissipation in the resistor? (c) What is the mechanical power delivered by the constant force?

60. A long solenoid of radius $r = 2.00$ cm is wound with $3.50 \times 10^3$ turns/m and carries a current that changes at the rate of 28.5 A/s as in Figure P20.60. What is the magnitude of the emf induced in the square conducting loop surrounding the center of the solenoid?

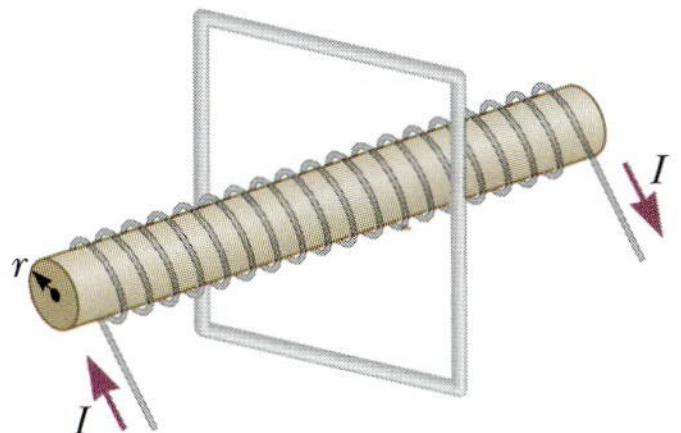


**Figure P20.60**

61. The bolt of lightning depicted in Figure P20.61 passes 200 m from a 100-turn coil oriented as shown. If the current in the lightning bolt falls from $6.02 \times 10^6$ A to zero in 10.5 μs, what is the average voltage induced in the coil? Assume the distance to the center of the coil determines the average magnetic field at the coil's position. Treat the lightning bolt as a long, vertical wire.

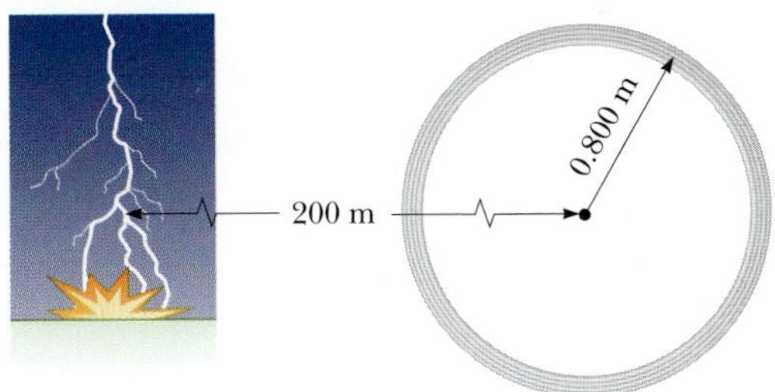


**Figure P20.61**

62. The square loop in Figure P20.62 is made of wires with a total series resistance of 10.0 Ω. It is placed in a uniform 0.100-T magnetic field directed perpendicular into the plane of the paper. The loop, which is hinged at each corner, is pulled as shown until the separation between points *A* and *B* is 3.00 m. If this process takes 0.100 s, what is the average current generated in the loop? What is the direction of the current?

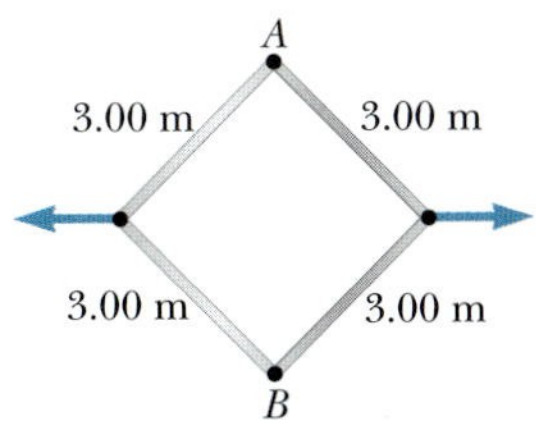


**Figure P20.62**

63. The magnetic field shown in Figure P20.63 (page 732) has a uniform magnitude of 25.0 mT directed into the paper. The initial diameter of the kink is 2.00 cm. (a) The wire is quickly pulled taut, and the kink shrinks to a diameter of zero in 50.0 ms. Determine the average voltage induced between endpoints *A* and *B*. Include the polarity. (b) Suppose the kink is undisturbed, but the magnetic field increases to 100 mT in $4.00 \times 10^{-3}$ s. Determine the average voltage across terminals *A* and *B*, including polarity, during this period.

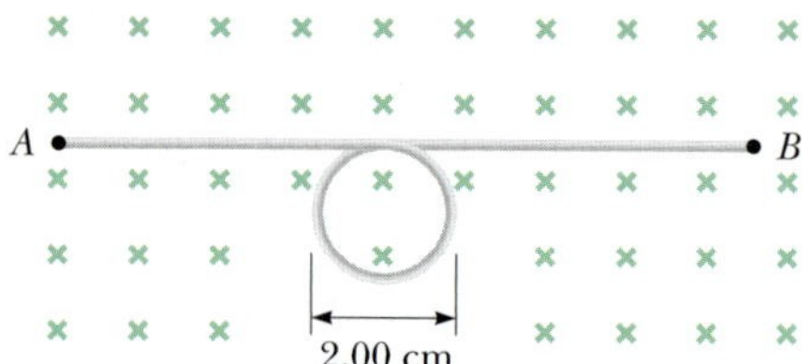


Figure P20.63

64. An aluminum ring of radius 5.00 cm and resistance $3.00 \times 10^{-4}\ \Omega$ is placed around the top of a long air-core solenoid with 1 000 turns per meter and a smaller radius of 3.00 cm, as in Figure P20.64. If the current in the solenoid is increasing at a constant rate of 270 A/s, what is the induced current in the ring? Assume the magnetic field produced by the solenoid over the area at the end of the solenoid is one-half as strong as the field at the center of the solenoid. Assume also the solenoid produces a negligible field outside its cross-sectional area.

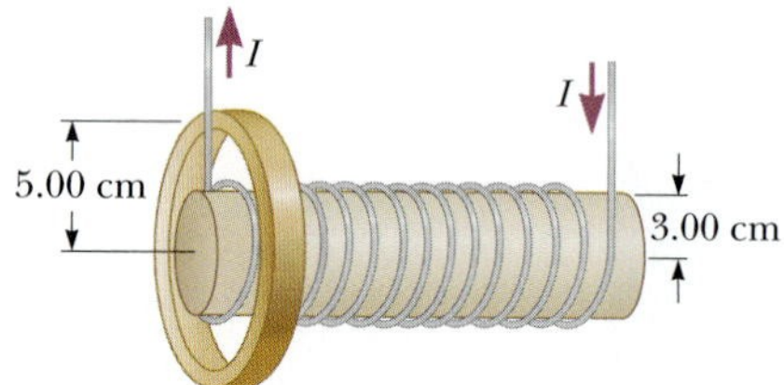


Figure P20.64

65. In Figure P20.65 the rolling axle of length 1.50 m is pushed along horizontal rails at a constant speed $v = 3.00$ m/s. A resistor $R = 0.400\ \Omega$ is connected to the rails at points $a$ and $b$, directly opposite each other. (The wheels make good electrical contact with the rails, so the axle, rails, and $R$ form a closed-loop circuit. The only significant resistance in the circuit is $R$.) A uniform magnetic field $B = 0.800$ T is directed vertically downward. (a) Find the induced current $I$ in the resistor. (b) What horizontal force $\vec{\mathbf{F}}$ is required to keep the axle rolling at constant speed? (c) Which end of the resistor, $a$ or $b$, is at the higher electric potential? (d) After the axle rolls past the resistor, does the current in $R$ reverse direction? Explain your answer.

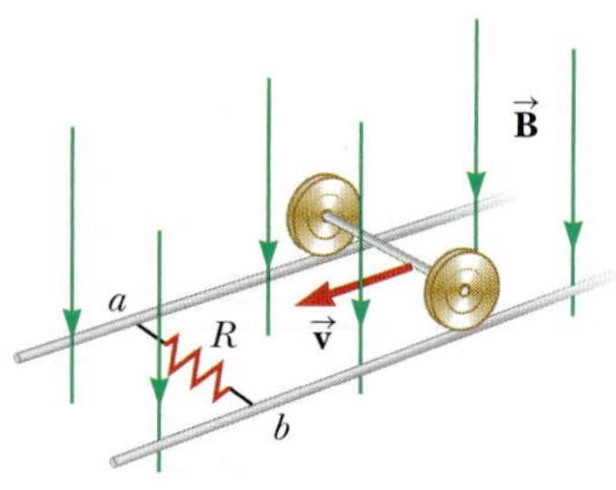


Figure P20.65

66. **QIC** **S** An $N$-turn square coil with side $\ell$ and resistance $R$ is pulled to the right at constant speed $v$ in the positive $x$-direction in the presence of a uniform magnetic field $B$ acting perpendicular to the coil, as shown in Figure P20.66. At $t = 0$, the right side of the coil is at the edge of the field. After a time $t$ has elapsed, the entire coil is in the region where $B = 0$. In terms of the quantities $N$, $B$, $\ell$, $v$, and $R$, find symbolic expressions for (a) the magnitude of the induced emf in the loop during the time interval $t$, (b) the magnitude of the induced current in the coil, (c) the power delivered to the coil, and (d) the force required to remove the coil from the field. (e) What is the direction of the induced current in the loop? (f) What is the direction of the magnetic force on the loop while it is being pulled out of the field?

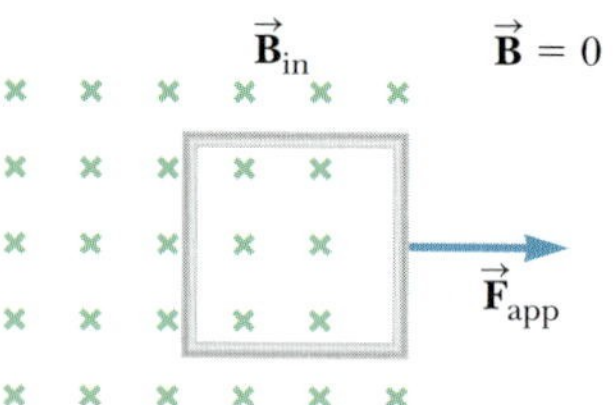


Figure P20.66

67. **S** A conducting rectangular loop of mass $M$, resistance $R$, and dimensions $w$ by $\ell$ falls from rest into a magnetic field $\vec{\mathbf{B}}$, as shown in Figure P20.67. During the time interval before the top edge of the loop reaches the field, the loop approaches a terminal speed $v_T$. (a) Show that

$$v_T = \frac{MgR}{B^2 w^2}$$

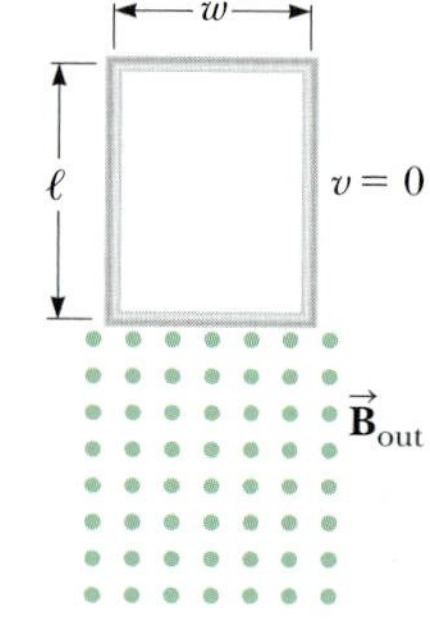


Figure P20.67

(b) Why is $v_T$ proportional to $R$? (c) Why is it inversely proportional to $B^2$?

Bruce Dale/National Geographic Creative

Arecibo, a large radio telescope in Puerto Rico, gathers electromagnetic radiation in the form of radio waves. These long wavelengths pass through obscuring dust clouds, allowing astronomers to create images of the core region of the Milky Way galaxy, which can't be observed in the visible spectrum.

# Alternating-Current Circuits and Electromagnetic Waves

# 21



Every time we turn on a television set, a stereo system, or any other electric appliances, we call on alternating currents (AC) to provide the power to operate them. We begin our study of AC circuits by examining the characteristics of a circuit containing a source of emf and one other circuit element: a resistor, a capacitor, or an inductor. Then we examine what happens when these elements are connected in combination with each other. Our discussion is limited to simple series configurations of the three kinds of elements.

We conclude this chapter with a discussion of **electromagnetic waves,** which are composed of fluctuating electric and magnetic fields. Electromagnetic waves in the form of visible light enable us to view the world around us; infrared waves warm our environment; radio-frequency waves carry our television and radio programs, as well as information about processes in the core of our galaxy; and X-rays allow us to perceive structures hidden inside our bodies and study properties of distant, collapsed stars. Light is key to our understanding of the universe.

## 21.1 Resistors in an AC Circuit

**LEARNING OBJECTIVES**

1. Define an AC circuit.
2. Define rms current, rms voltage differences, and power in circuits with alternating current.
3. Discuss the phase relationship between voltage difference and current in a resistive AC circuit.
4. Apply the concepts of rms voltage and current to resistive AC circuits.

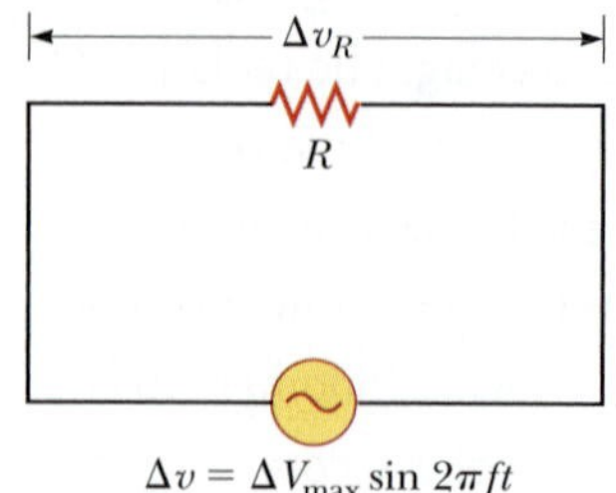
$\Delta v_R$
$R$
$\Delta v = \Delta V_{max} \sin 2\pi ft$

**Figure 21.1** A series circuit consisting of a resistor $R$ connected to an AC generator, designated by the symbol —(~)—

An AC circuit consists of combinations of circuit elements and an AC generator or an AC source, which provides the alternating current. We have seen that the output of an AC generator is sinusoidal and varies with time according to

$$\Delta v = \Delta V_{max} \sin 2\pi ft \qquad [21.1]$$

where $\Delta v$ is the instantaneous voltage, $\Delta V_{max}$ is the maximum voltage of the AC generator, and $f$ is the frequency at which the voltage changes, measured in hertz (Hz). (Compare Equations 20.7 and 20.8 with Equation 21.1, and recall that $\omega = 2\pi f$.) We first consider a simple circuit consisting of a resistor and an AC source (designated by the symbol —(~)—), as in Figure 21.1. The current and the voltage versus time across the resistor are shown in Figure 21.2.

To explain the concept of alternating current, we begin by discussing the current versus time curve in Figure 21.2. At point $a$ on the curve, the current has a maximum value in one direction, arbitrarily called the positive direction. Between points $a$ and $b$, the current is decreasing in magnitude but is still in the positive direction. At point $b$, the current is momentarily zero; it then begins to increase in the opposite (negative) direction between points $b$ and $c$. At point $c$, the current has reached its maximum value in the negative direction.

The current and voltage are in step with each other because they vary identically with time. **Because the current and the voltage reach their maximum values at the same time, they are said to be in phase.** Notice that **the average value of the current over one cycle is zero** because the current is maintained in one direction (the positive direction) for the same amount of time and at the same magnitude as it is in the opposite direction (the negative direction). The direction of the current, however, has no effect on the behavior of the resistor in the circuit: the collisions between electrons and the fixed atoms of the resistor result in an increase in the resistor's temperature regardless of the direction of the current.

We can quantify this discussion by recalling that the rate at which electrical energy is dissipated in a resistor, the power $P$, is

$$P = i^2R$$

where $i$ is the *instantaneous* current in the resistor. Because the heating effect of a current is proportional to the *square* of the current, it makes no difference whether the sign associated with the current is positive or negative. The heating effect produced by an alternating current with a maximum value of $I_{max}$ *is not the same* as that produced by a direct current of the same value, however. The reason is that the alternating current has this maximum value for only an instant of time during a cycle. The important quantity in an AC circuit is a special kind of average value of current, called the **rms current**: the direct current that dissipates the

**Figure 21.2** A plot of current and voltage difference in a resistor versus time.

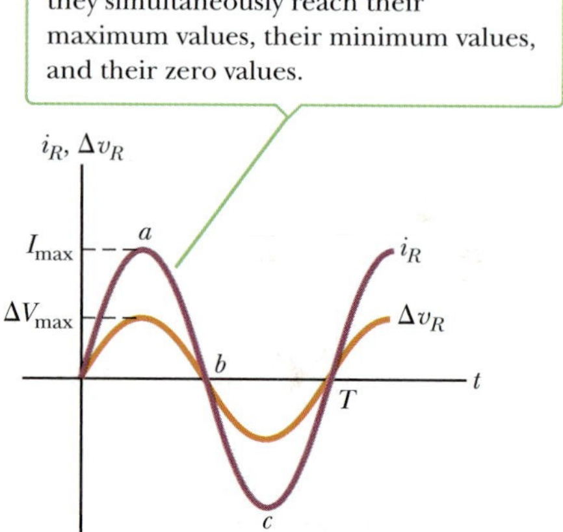
The current and the voltage are in phase: they simultaneously reach their maximum values, their minimum values, and their zero values.
$i_R, \Delta v_R$
$I_{max}$
$\Delta V_{max}$
a
b
c
T
t
$i_R$
$\Delta v_R$

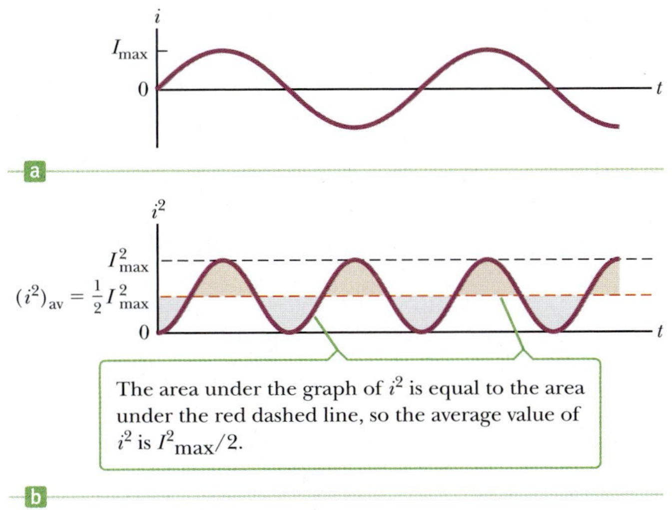


**Figure 21.3** (a) Plot of the current in a resistor as a function of time. (b) Plot of the square of the current in a resistor as a function of time. The area under the red-dashed line omits the tan-shaded regions, but that is made up by the inclusion of the gray-shaded regions with the same area. That shows the area under the curve of $i^2$ is the same as the area under the red dashed line, so $(i^2)_{av} = I^2_{max}/2$.

same amount of energy in a resistor as the actual alternating current. Finding the average of the alternating current $i$, depicted in Figure 21.3a, would not be useful because that average is zero, whereas the rms current is always positive. To find the rms current, we first square the current, then find its average value, and finally take the square root of this average value. Hence, the rms current is the square *root* of the average (*mean*) of the *square* of the current. Because $i^2$ varies as $\sin^2 2\pi ft$, the average value of $i^2$ is $\frac{1}{2}I^2_{max}$ (Fig. 21.3b).[1] Therefore, the rms current $I_{rms}$ is related to the maximum value of the alternating current $I_{max}$ by

$$I_{rms} = \frac{I_{max}}{\sqrt{2}} = 0.707 I_{max} \qquad [21.2]$$

This equation says that an alternating current with a maximum value of 3 A produces the same heating effect in a resistor as a direct current of $(3/\sqrt{2})$ A. We can therefore say that the average power dissipated in a resistor that carries alternating current $I$ is

$$P_{av} = I^2_{rms} R$$

Alternating voltages are also best discussed in terms of rms voltages, with a relationship identical to the preceding one,

$$\Delta V_{rms} = \frac{\Delta V_{max}}{\sqrt{2}} = 0.707\, \Delta V_{max} \qquad [21.3]$$

◀ rms voltage

where $\Delta V_{rms}$ is the rms voltage and $\Delta V_{max}$ is the maximum value of the alternating voltage.

[1]We can show that $(i^2)_{av} = I^2_{max}/2$ as follows: The current in the circuit varies with time according to the expression $i = I_{max} \sin 2\pi ft$, so $i^2 = I^2_{max} \sin^2 2\pi ft$. Therefore, we can find the average value of $i^2$ by calculating the average value of $\sin^2 2\pi ft$. Note that a graph of $\cos^2 2\pi ft$ versus time is identical to a graph of $\sin^2 2\pi ft$ versus time, except that the points are shifted on the time axis. Thus, the time average of $\sin^2 2\pi ft$ is equal to the time average of $\cos^2 2\pi ft$, taken over one or more cycles. That is,

$$(\sin^2 2\pi ft)_{av} = (\cos^2 2\pi ft)_{av}$$

With this fact and the trigonometric identity $\sin^2\theta + \cos^2\theta = 1$, we get

$$(\sin^2 2\pi ft)_{av} + (\cos^2 2\pi ft)_{av} = 2(\sin^2 2\pi ft)_{av} = 1$$

$$(\sin^2 2\pi ft)_{av} = \tfrac{1}{2}$$

When this result is substituted into the expression $i^2 = I^2_{max} \sin^2 2\pi ft$, we get $(i^2)_{av} = I^2_{rms} = I^2_{max}/2$, or $I_{rms} = I_{max}/\sqrt{2}$, where $I_{rms}$ is the rms current.

**Table 21.1** Notation Used in This Chapter

| | Voltage | Current |
|---|---|---|
| Instantaneous value | $\Delta v$ | $i$ |
| Maximum value | $\Delta V_{max}$ | $I_{max}$ |
| rms value | $\Delta V_{rms}$ | $I_{rms}$ |

When we speak of measuring an AC voltage of 120 V from an electric outlet, we actually mean an rms voltage of 120 V. A quick calculation using Equation 21.3 shows that such an AC voltage actually has a peak value of about 170 V. In this chapter we use rms values when discussing alternating currents and voltages. One reason is that AC ammeters and voltmeters are designed to read rms values. Further, if we use rms values, many of the equations for alternating current will have the same form as those used in the study of direct-current (DC) circuits. Table 21.1 summarizes the notations used throughout this chapter.

Consider the series circuit in Figure 21.1 (page 734), consisting of a resistor connected to an AC generator. A resistor impedes the current in an AC circuit, just as it does in a DC circuit. Ohm's law is therefore valid for an AC circuit, and we have

$$\Delta V_{R,rms} = I_{rms}R \qquad [21.4a]$$

**The rms voltage across a resistor is equal to the rms current in the circuit times the resistance.** This equation is also true if maximum values of current and voltage are used:

$$\Delta V_{R,max} = I_{max}R \qquad [21.4b]$$

**Quick Quiz**

**21.1** Which of the following statements can be true for a resistor connected in a simple series circuit to an operating AC generator? (a) $P_{av} = 0$ and $i_{av} = 0$ (b) $P_{av} = 0$ and $i_{av} > 0$ (c) $P_{av} > 0$ and $i_{av} = 0$ (d) $P_{av} > 0$ and $i_{av} > 0$

## EXAMPLE 21.1 What Is the rms Current?

**GOAL** Perform basic AC circuit calculations for a purely resistive circuit.

**PROBLEM** An AC voltage source has an output of $\Delta v = (2.00 \times 10^2\ \text{V}) \sin 2\pi ft$. This source is connected to a $1.00 \times 10^2\ \Omega$ resistor as in Figure 21.1. Find the rms voltage and rms current in the resistor.

**STRATEGY** Compare the expression for the voltage output just given with the general form, $\Delta v = \Delta V_{max} \sin 2\pi ft$, finding the maximum voltage. Substitute this result into the expression for the rms voltage.

**SOLUTION**

Obtain the maximum voltage by comparison of the given expression for the output with the general expression:

$$\Delta v = (2.00 \times 10^2\ \text{V}) \sin 2\pi ft \qquad \Delta v = \Delta V_{max} \sin 2\pi ft$$

$$\rightarrow \quad \Delta V_{max} = 2.00 \times 10^2\ \text{V}$$

Next, substitute into Equation 21.3 to find the rms voltage of the source:

$$\Delta V_{rms} = \frac{\Delta V_{max}}{\sqrt{2}} = \frac{2.00 \times 10^2\ \text{V}}{\sqrt{2}} = \boxed{141\ \text{V}}$$

Substitute this result into Ohm's law to find the rms current:

$$I_{rms} = \frac{\Delta V_{rms}}{R} = \frac{141\ \text{V}}{1.00 \times 10^2\ \Omega} = \boxed{1.41\ \text{A}}$$

**REMARKS** Notice how the concept of rms values allows the handling of an AC circuit quantitatively in much the same way as a DC circuit.

**QUESTION 21.1** True or False: The rms current in an AC circuit oscillates sinusoidally with time.

**EXERCISE 21.1** Find the maximum current in the circuit and the average power delivered to the circuit.

**ANSWER** 2.00 A; $2.00 \times 10^2$ W

**APPLYING PHYSICS 21.1 | Electric Fields and Cancer Treatment** BIO

Cancer cells multiply far more frequently than most normal cells, spreading throughout the body, using its resources and interfering with normal functioning. Most therapies damage both cancerous and healthy cells, so finding methods that target cancer cells is important in developing better treatments for the disease.

Because cancer cells multiply so rapidly, it's natural to consider treatments that prevent or disrupt cell division. Treatments such as chemotherapy interfere with the cell division cycle, but can also damage healthy cells. It has recently been found that alternating electric fields produced by alternating currents in the range of 100 kHz can disrupt the cell division cycle, either by slowing the division or by causing a dividing cell to disintegrate. Healthy cells that divide at only a very slow rate are less vulnerable than the rapidly dividing cancer cells, so such therapy holds out promise for certain types of cancer.

The alternating electric fields are thought to affect the process of mitosis, which is the dividing of the cell nucleus into two sets of identical chromosomes. Near the end of the first phase of mitosis, called the prophase, the mitotic spindle forms, a structure of fine filaments that guides the two replicated sets of chromosomes into separate daughter cells. The mitotic spindle is made up of a polymerization of dimers of tubulin, a protein with a large electric dipole moment. The alternating electric field exerts forces on these dipoles, disrupting their proper functioning.

Electric field therapy is especially promising for the treatment of brain tumors because healthy brain cells don't divide and therefore would be unharmed by the alternating electric fields. Research on such therapies is ongoing. ■

# 21.2 Capacitors in an AC Circuit

**LEARNING OBJECTIVES**

1. Define capacitive reactance.
2. Discuss the phase relationship between current and voltage difference in an AC circuit capacitor.
3. Calculate the capacitive reactance and rms current of capacitive AC circuits.

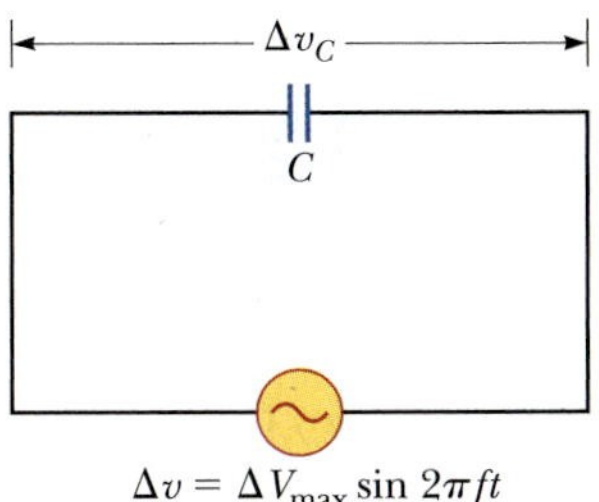
$\Delta v_C$
$C$
$\Delta v = \Delta V_{max} \sin 2\pi ft$

**Figure 21.4** A series circuit consisting of a capacitor $C$ connected to an AC generator.

To understand the effect of a capacitor on the behavior of a circuit containing an AC voltage source, we first review what happens when a capacitor is placed in a circuit containing a DC source, such as a battery. When the switch is closed in a series circuit containing a battery, a resistor, and a capacitor, the initial charge on the plates of the capacitor is zero. The motion of charge through the circuit is therefore relatively free, and there is a large current in the circuit. As more charge accumulates on the capacitor, the voltage across it increases, opposing the current. After some time interval, which depends on the time constant $RC$, the current approaches zero. Consequently, a capacitor in a DC circuit limits or impedes the current so that it approaches zero after a brief time.

Now consider the simple series circuit in Figure 21.4, consisting of a capacitor connected to an AC generator. We sketch curves of current versus time and voltage versus time, and then attempt to make the graphs seem reasonable. The curves are shown in Figure 21.5. First, notice that the segment of the current curve from $a$ to $b$ indicates that the current starts out at a rather large value. This large value can be understood by recognizing that there is no charge on the capacitor at $t = 0$; as a consequence, there is nothing in the circuit except the resistance of the wires to hinder the flow of charge at this instant. The current decreases, however, as the voltage across the capacitor increases from $c$ to $d$ on the voltage curve. When the voltage is at point $d$, the current reverses and begins to increase in the opposite direction (from $b$ to $e$ on the current curve). During this time, the voltage across the capacitor decreases from $d$ to $f$ because the plates are now losing the charge they accumulated earlier. The remainder of the cycle for both voltage and current is a repeat of what happened during the first half of the cycle. The current reaches a maximum value in the opposite direction at point $e$ on the current curve and then decreases as the voltage across the capacitor builds up.

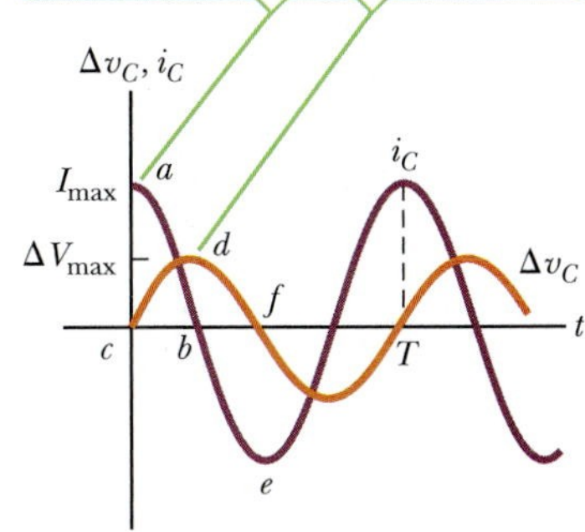
The voltage reaches its maximum value 90° after the current reaches its maximum value, so the voltage "lags" the current.
$\Delta v_C, i_C$
$I_{max}$
$\Delta V_{max}$
$a$
$d$
$i_C$
$\Delta v_C$
$f$
$t$
$c$
$b$
$T$
$e$

**Figure 21.5** Plots of current and voltage across a capacitor versus time in an AC circuit.

In a purely resistive circuit, the current and voltage are always in step with each other. That isn't the case when a capacitor is in the circuit. In Figure 21.5, when an alternating voltage is applied across a capacitor, the voltage reaches its maximum value one-quarter of a cycle after the current reaches its maximum value. We say that **the voltage across a capacitor always lags the current by 90°.**

The voltage across a capacitor lags the current by 90°

The impeding effect of a capacitor on the current in an AC circuit is expressed in terms of a factor called the **capacitive reactance** $X_C$, defined as

Capacitive reactance

$$X_C \equiv \frac{1}{2\pi f C} \quad [21.5]$$

When $C$ is in farads and $f$ is in hertz, the unit of $X_C$ is the ohm. Notice that $2\pi f = \omega$, the angular frequency.

From Equation 21.5, as the frequency $f$ of the voltage source increases, the capacitive reactance $X_C$ (the impeding effect of the capacitor) decreases, so the current increases. At high frequency, there is less time available to charge the capacitor, so less charge and voltage accumulate on the capacitor, which translates into less opposition to the flow of charge and, consequently, a higher current. The analogy between capacitive reactance and resistance means that we can write an equation of the same form as Ohm's law to describe AC circuits containing capacitors. This equation relates the rms voltage and rms current in the circuit to the capacitive reactance:

$$\Delta V_{C,\text{rms}} = I_{\text{rms}} X_C \quad [21.6]$$

## EXAMPLE 21.2 A Purely Capacitive AC Circuit

**GOAL** Perform basic AC circuit calculations for a capacitive circuit.

**PROBLEM** An 8.00-$\mu$F capacitor is connected to the terminals of an AC generator with an rms voltage of $1.50 \times 10^2$ V and a frequency of 60.0 Hz. Find the capacitive reactance and the rms current in the circuit.

**STRATEGY** Substitute values into Equations 21.5 and 21.6.

**SOLUTION**

Substitute the values of $f$ and $C$ into Equation 21.5:

$$X_C = \frac{1}{2\pi f C} = \frac{1}{2\pi(60.0\text{ Hz})(8.00 \times 10^{-6}\text{ F})} = \boxed{332\ \Omega}$$

Solve Equation 21.6 for the current and substitute the values for $X_C$ and the rms voltage to find the rms current:

$$I_{\text{rms}} = \frac{\Delta V_{C,\text{rms}}}{X_C} = \frac{1.50 \times 10^2\text{ V}}{332\ \Omega} = \boxed{0.452\text{ A}}$$

**REMARKS** Again, notice how similar the technique is to that of analyzing a DC circuit with a resistor.

**QUESTION 21.2** True or False: The larger the capacitance of a capacitor, the larger the capacitive reactance.

**EXERCISE 21.2** If the frequency is doubled, what happens to the capacitive reactance and the rms current?

**ANSWER** $X_C$ is halved, and $I_{\text{rms}}$ is doubled.

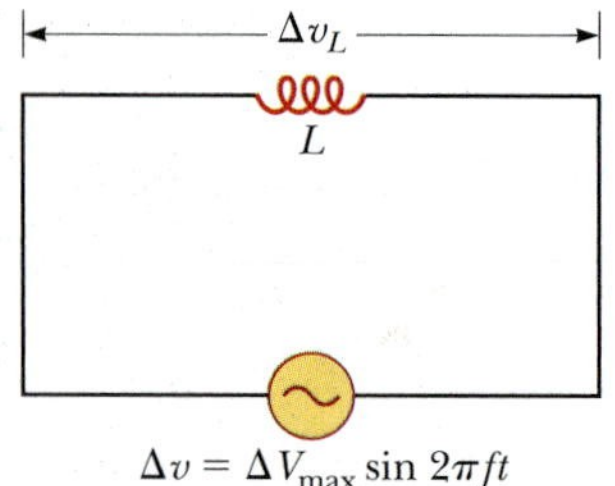


**Figure 21.6** A series circuit consisting of an inductor $L$ connected to an AC generator.

## 21.3 Inductors in an AC Circuit

**LEARNING OBJECTIVES**

1. Define inductive reactance.
2. Discuss the phase relationship between current and voltage difference in an AC circuit inductor.
3. Calculate the inductive reactance and rms current in inductive AC circuits.

Now consider an AC circuit consisting only of an inductor connected to the terminals of an AC source, as in Figure 21.6. (In any real circuit there is some resistance in the wire forming the inductive coil, but we ignore this consideration for now.) The changing current output of the generator produces a back emf that impedes the current in the circuit. The magnitude of this back emf is

$$\Delta v_L = L\frac{\Delta I}{\Delta t} \qquad [21.7]$$

The effective resistance of the coil in an AC circuit is measured by a quantity called the **inductive reactance, $X_L$:**

$$X_L \equiv 2\pi f L \qquad [21.8]$$

When $f$ is in hertz and $L$ is in henries, the unit of $X_L$ is the ohm. The inductive reactance *increases* with increasing frequency and increasing inductance. Contrast this with capacitors, where increasing frequency or capacitance *decreases* the capacitive reactance.

To understand the meaning of inductive reactance, compare Equation 21.8 with Equation 21.7. First, note from Equation 21.8 that the inductive reactance depends on the inductance $L$, which is reasonable because the back emf (Eq. 21.7) is large for large values of $L$. Second, note that the inductive reactance depends on the frequency $f$. This dependence, too, is reasonable because the back emf depends on $\Delta I/\Delta t$, a quantity that is large when the current changes rapidly, as it would for high frequencies.

With inductive reactance defined in this way, we can write an equation of the same form as Ohm's law for the voltage across the coil or inductor:

$$\Delta V_{L,\text{rms}} = I_{\text{rms}} X_L \qquad [21.9]$$

where $\Delta V_{L,\text{rms}}$ is the rms voltage across the coil and $I_{\text{rms}}$ is the rms current in the coil.

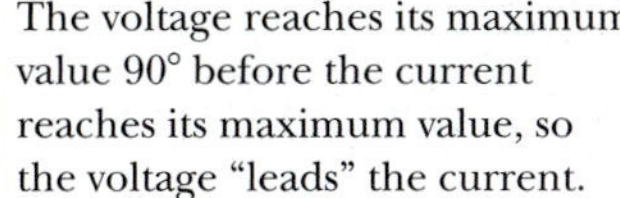
The voltage reaches its maximum value 90° before the current reaches its maximum value, so the voltage "leads" the current.

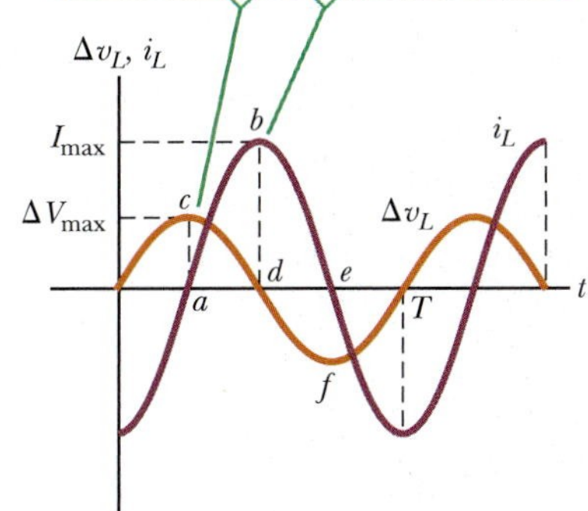
$\Delta v_L, i_L$
$I_{max}$
$\Delta V_{max}$
$i_L$
$\Delta v_L$
a
b
c
d
e
f
T
t

**Figure 21.7** Plots of current and voltage across an inductor versus time in an AC circuit.

Figure 21.7 shows the instantaneous voltage and instantaneous current across the coil as functions of time. When a sinusoidal voltage is applied across an inductor, the voltage reaches its maximum value one-quarter of an oscillation period before the current reaches its maximum value. In this situation we say that **the voltage across an inductor always leads the current by 90°.**

To see why there is a phase relationship between voltage and current, we examine a few points on the curves of Figure 21.7. At point $a$ on the current curve, the current is beginning to increase in the positive direction. At this instant the rate of change of current, $\Delta I/\Delta t$ (the slope of the current curve), is at a maximum, and we see from Equation 21.7 that the voltage across the inductor is consequently also at a maximum. As the current rises between points $a$ and $b$ on the curve, $\Delta I/\Delta t$ gradually decreases until it reaches zero at point $b$. As a result, the voltage across the inductor is decreasing during this same time interval, as the segment between $c$ and $d$ on the voltage curve indicates. Immediately after point $b$, the current begins to decrease, although it still has the same direction it had during the previous quarter cycle. As the current decreases to zero (from $b$ to $e$ on the curve), a voltage is again induced in the coil (from $d$ to $f$), but the polarity of this voltage is opposite the polarity of the voltage induced between $c$ and $d$. This occurs because back emfs always oppose the change in the current.

We could continue to examine other segments of the curves, but no new information would be gained because the current and voltage variations are repetitive.

## ■ EXAMPLE 21.3 | A Purely Inductive AC Circuit

**GOAL** Perform basic AC circuit calculations for an inductive circuit.

**PROBLEM** In a purely inductive AC circuit (see Fig. 21.6), $L = 25.0$ mH and the rms voltage is $1.50 \times 10^2$ V. Find the inductive reactance and rms current in the circuit if the frequency is 60.0 Hz.

*(Continued)*

**SOLUTION**

Substitute $L$ and $f$ into Equation 21.8 to get the inductive reactance:

$$X_L = 2\pi fL = 2\pi(60.0\ \text{s}^{-1})(25.0 \times 10^{-3}\ \text{H}) = \boxed{9.42\ \Omega}$$

Solve Equation 21.9 for the rms current and substitute:

$$I_{\text{rms}} = \frac{\Delta V_{L,\text{rms}}}{X_L} = \frac{1.50 \times 10^2\ \text{V}}{9.42\ \Omega} = \boxed{15.9\ \text{A}}$$

**REMARKS** The analogy with DC circuits is even closer than in the capacitive case because in the inductive equivalent of Ohm's law, the voltage across an inductor is *proportional* to the inductance $L$, just as the voltage across a resistor is proportional to $R$ in Ohm's law.

**QUESTION 21.3** True or False: A larger inductance or frequency results in a larger inductive reactance.

**EXERCISE 21.3** Calculate the inductive reactance and rms current in a similar circuit if the frequency is again 60.0 Hz, but the rms voltage is 85.0 V and the inductance is 47.0 mH.

**ANSWER** $X_L = 17.7\ \Omega$, $I_{\text{rms}} = 4.80$ A

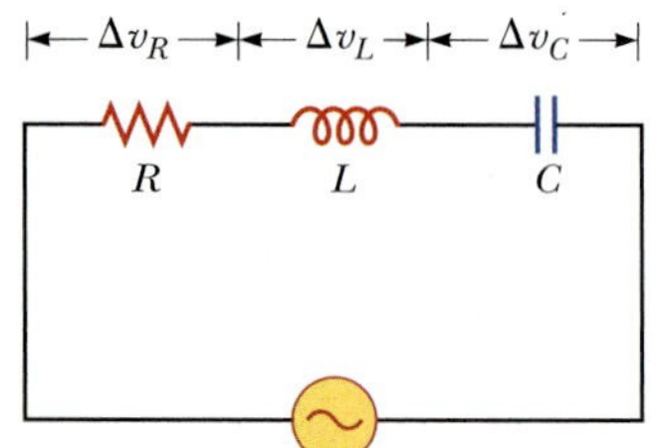


**Figure 21.8** A series circuit consisting of a resistor, an inductor, and a capacitor connected to an AC generator.

## 21.4 The *RLC* Series Circuit

**LEARNING OBJECTIVES**

1. Define impedance in an AC circuit.
2. Discuss the AC relationship between current and voltage differences for elements in a series *RLC* circuit.
3. Apply the concepts of impedance, phasors and phase diagrams to series *RLC* circuits.

In the foregoing sections we examined the effects of an inductor, a capacitor, and a resistor when they are connected separately across an AC voltage source. We now consider what happens when these elements are combined.

Figure 21.8 shows a circuit containing a resistor, an inductor, and a capacitor connected in series across an AC source that supplies a total voltage $\Delta v$ at some instant. The current in the circuit is the same at all points in the circuit at any instant and varies sinusoidally with time, as indicated in Figure 21.9a. This fact can be expressed mathematically as

$$i = I_{\text{max}} \sin 2\pi ft$$

Earlier, we learned that the voltage across each element may or may not be in phase with the current. The instantaneous voltages across the three elements, shown in Figure 21.9, have the following phase relations to the instantaneous current:

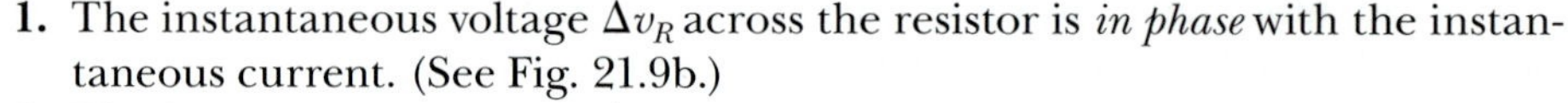

1. The instantaneous voltage $\Delta v_R$ across the resistor is *in phase* with the instantaneous current. (See Fig. 21.9b.)
2. The instantaneous voltage $\Delta v_L$ across the inductor *leads* the current by 90°. (See Fig. 21.9c.)
3. The instantaneous voltage $\Delta v_C$ across the capacitor *lags* the current by 90°. (See Fig. 21.9d.)

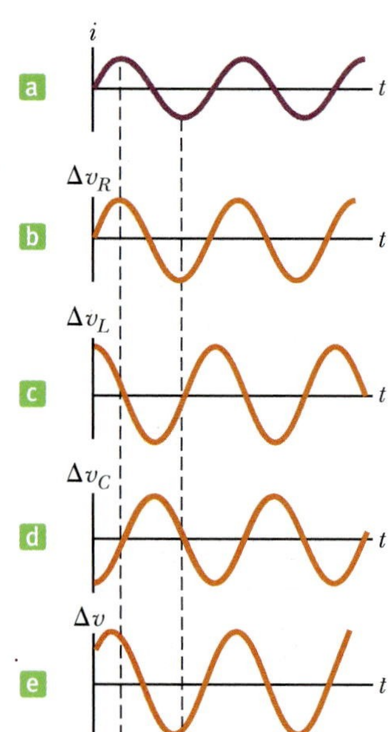


**Figure 21.9** Phase relations in the series *RLC* circuit shown in Figure 21.8.

The net instantaneous voltage $\Delta v$ supplied by the AC source equals the sum of the instantaneous voltages across the separate elements: $\Delta v = \Delta v_R + \Delta v_C + \Delta v_L$. This doesn't mean, however, that the voltages measured with an AC voltmeter across $R$, $C$, and $L$ sum to the measured source voltage! In fact, the measured voltages *don't* sum to the measured source voltage because the voltages across $R$, $C$, and $L$ all have different phases.

To account for the different phases of the voltage drops, we use a technique involving vectors. We represent the voltage across each element with a rotating vector, as in

Figure 21.10. The rotating vectors are referred to as **phasors**, and the diagram is called a **phasor diagram**. This particular diagram represents the circuit voltage given by the expression $\Delta v = \Delta V_{max} \sin (2\pi f t + \phi)$, where $\Delta V_{max}$ is the maximum voltage (the magnitude or length of the rotating vector or phasor) and $\phi$ is the angle between the phasor and the positive $x$-axis when $t = 0$. The phasor can be viewed as a vector of magnitude $\Delta V_{max}$ rotating at a constant frequency $f$ so that its projection along the $y$-axis is the instantaneous voltage in the circuit. Because $\phi$ is the phase angle between the voltage and current in the circuit, the phasor for the current (not shown in Fig. 21.10) lies along the positive $x$-axis when $t = 0$ and is expressed by the relation $i = I_{max} \sin (2\pi f t)$.

**Figure 21.10** A phasor diagram for the voltage in an AC circuit, where $\phi$ is the phase angle between the voltage and the current and $\Delta v$ is the instantaneous voltage.

The phasor diagrams in Figure 21.11 are useful for analyzing the *series RLC* circuit. Voltages in phase with the current are represented by vectors along the positive $x$-axis, and voltages out of phase with the current lie along other directions. $\Delta V_R$ is horizontal and to the right because it's in phase with the current. Likewise, $\Delta V_L$ is represented by a phasor along the positive $y$-axis because it leads the current by 90°. Finally, $\Delta V_C$ is along the negative $y$-axis because it lags the current[2] by 90°. If the phasors are added as vector quantities so as to account for the different phases of the voltages across $R$, $L$, and $C$, Figure 21.11a shows that the only $x$-component for the voltages is $\Delta V_R$ and the net $y$-component is $\Delta V_L - \Delta V_C$. We now add the phasors vectorially to find the phasor $\Delta V_{max}$ (Fig. 21.11b), which represents the maximum voltage. The right triangle in Figure 21.11b gives the following equations for the maximum voltage and the phase angle $\phi$ between the maximum voltage and the current:

$$\Delta V_{max} = \sqrt{\Delta V_R{}^2 + (\Delta V_L - \Delta V_C)^2} \quad [21.10]$$

$$\tan \phi = \frac{\Delta V_L - \Delta V_C}{\Delta V_R} \quad [21.11]$$

In these equations, all voltages are maximum values. Although we choose to use maximum voltages in our analysis, the preceding equations apply equally well to rms voltages because the two quantities are related to each other by the same factor for all circuit elements. The result for the maximum voltage $\Delta V_{max}$ given by Equation 21.10 reinforces the fact that **the voltages across the resistor, capacitor, and inductor are not in phase, so one cannot simply add them to get the voltage across the combination of element or to get the source voltage.**

## Quick Quiz

**21.2** For the circuit in Figure 21.8, is the instantaneous voltage of the source equal to (a) the sum of the maximum voltages across the elements, (b) the sum of the instantaneous voltages across the elements, or (c) the sum of the rms voltages across the elements?

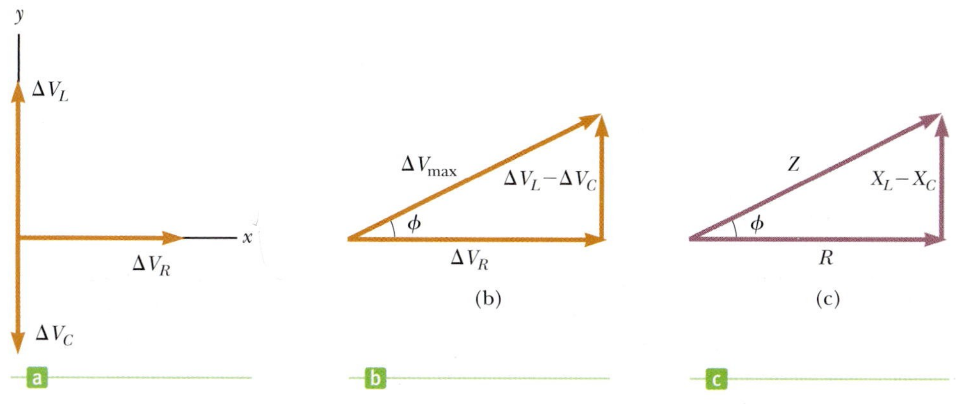


**Figure 21.11** (a) A phasor diagram for the *RLC* circuit. (b) Addition of the phasors as vectors gives $\Delta V_{max} = \sqrt{\Delta V_R{}^2 + (\Delta V_L - \Delta V_C)^2}$. (c) The reactance triangle that gives the impedance relation $Z = \sqrt{R^2 + (X_L - X_C)^2}$.

[2]A mnemonic to help you remember the phase relationships in *RLC* circuits is "*ELI* the *ICE* man." *E* represents the voltage $\mathcal{E}$, *I* the current, *L* the inductance, and *C* the capacitance. Thus, the name *ELI* means that in an inductive circuit, the voltage $\mathcal{E}$ leads the current *I*. In a capacitive circuit *ICE* means that the current leads the voltage.

We can write Equation 21.10 in the form of Ohm's law, using the relations $\Delta V_R = I_{max}R$, $\Delta V_L = I_{max}X_L$, and $\Delta V_C = I_{max}X_C$, where $I_{max}$ is the maximum current in the circuit:

$$\Delta V_{max} = I_{max}\sqrt{R^2 + (X_L - X_C)^2} \qquad [21.12]$$

It's convenient to define a parameter called the **impedance** $Z$ of the circuit as

Impedance ▶

$$Z \equiv \sqrt{R^2 + (X_L - X_C)^2} \qquad [21.13]$$

so that Equation 21.12 becomes

$$\Delta V_{max} = I_{max}Z \qquad [21.14]$$

Equation 21.14 is in the form of Ohm's law, $\Delta V = IR$, with $R$ replaced by the impedance in ohms. Indeed, Equation 21.14 can be regarded as a generalized form of Ohm's law applied to a series AC circuit. Both the impedance and therefore the current in an AC circuit depend on the resistance, the inductance, the capacitance, *and* the frequency (because the reactances are frequency dependent).

It's useful to represent the impedance $Z$ with a vector diagram such as the one depicted in Figure 21.11c. A right triangle is constructed with right side $X_L - X_C$, base $R$, and hypotenuse $Z$. Applying the Pythagorean theorem to this triangle, we see that

$$Z = \sqrt{R^2 + (X_L - X_C)^2}$$

which is Equation 21.13. Furthermore, we see from the vector diagram in Figure 21.11c that the phase angle $\phi$ between the current and the voltage obeys the relationship

Phase angle $\phi$ ▶

$$\tan\phi = \frac{X_L - X_C}{R} \qquad [21.15]$$

The physical significance of the phase angle will become apparent in Section 21.5.

Table 21.2 provides impedance values and phase angles for some series circuits containing different combinations of circuit elements.

**Nikola Tesla**
**(1856–1943)**
Tesla was born in Croatia, but spent most of his professional life as an inventor in the United States. He was a key figure in the development of alternating-current electricity, high-voltage transformers, and the transport of electrical power via AC transmission lines. Tesla's viewpoint was at odds with the ideas of Edison, who committed himself to the use of direct current in power transmission. Tesla's AC approach won out.

**Table 21.2** Impedance Values and Phase Angles for Various Combinations of Circuit Elements

| Circuit Elements | Impedance Z | Phase Angle $\phi$ |
|---|---|---|
| $R$ | $R$ | 0° |
| $C$ | $X_C$ | −90° |
| $L$ | $X_L$ | +90° |
| $R$, $C$ | $\sqrt{R^2 + X_C^2}$ | Negative, between −90° and 0° |
| $R$, $L$ | $\sqrt{R^2 + X_L^2}$ | Positive, between 0° and 90° |
| $R$, $L$, $C$ | $\sqrt{R^2 + (X_L - X_C)^2}$ | Negative if $X_C > X_L$<br>Positive if $X_C < X_L$ |

*Note:* In each case an AC voltage (not shown) is applied across the combination of elements (that is, across the dots).

Parallel alternating current circuits are also useful in everyday applications. We won't discuss them here, however, because their analysis is beyond the scope of this book.

**Quick Quiz**

**21.3** If switch A is closed in Figure 21.12, what happens to the impedance of the circuit? (a) It increases. (b) It decreases. (c) It doesn't change.

**21.4** Suppose $X_L > X_C$ in Figure 21.12. If switch A is closed, what happens to the phase angle? (a) It increases. (b) It decreases. (c) It doesn't change.

**21.5** Suppose $X_L > X_C$ in Figure 21.12. If switch A is left open and switch B is closed, what happens to the phase angle? (a) It increases. (b) It decreases. (c) It doesn't change.

**21.6** Suppose $X_L > X_C$ in Figure 21.12 and, with both switches open, a piece of iron is slipped into the inductor. During this process, what happens to the brightness of the bulb? (a) It increases. (b) It decreases. (c) It doesn't change.

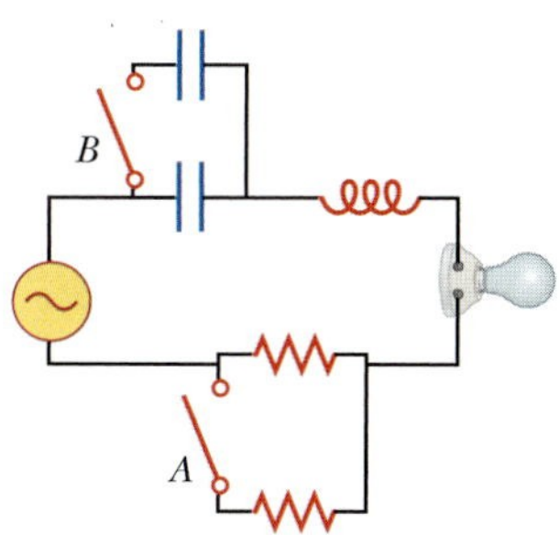


**Figure 21.12** (Quick Quizzes 21.3–21.6)

**PROBLEM-SOLVING STRATEGY**

***RLC* Circuits**

*The following procedure is recommended for solving series RLC circuit problems:*

1. Calculate the inductive and capacitive reactances, $X_L$ and $X_C$.
2. Use $X_L$ and $X_C$ together with the resistance $R$ to calculate the impedance $Z$ of the circuit.
3. Find the maximum current or maximum voltage drop with the equivalent of Ohm's law, $\Delta V_{\text{max}} = I_{\text{max}} Z$.
4. Calculate the voltage drops across the individual elements with the appropriate variations of Ohm's law: $\Delta V_{R,\text{max}} = I_{\text{max}} R$, $\Delta V_{L,\text{max}} = I_{\text{max}} X_L$, and $\Delta V_{C,\text{max}} = I_{\text{max}} X_C$.
5. Obtain the phase angle using $\tan \phi = (X_L - X_C)/R$.

**EXAMPLE 21.4 An *RLC* Circuit**

**GOAL** Analyze a series *RLC* AC circuit and find the phase angle.

**PROBLEM** A series *RLC* AC circuit has resistance $R = 2.50 \times 10^2\ \Omega$, inductance $L = 0.600$ H, capacitance $C = 3.50\ \mu\text{F}$, frequency $f = 60.0$ Hz, and maximum voltage $\Delta V_{\text{max}} = 1.50 \times 10^2$ V. Find **(a)** the impedance of the circuit, **(b)** the maximum current in the circuit, **(c)** the phase angle, and **(d)** the maximum voltages across the elements.

**STRATEGY** Calculate the inductive and capacitive reactances, which can be used with the resistance to calculate the impedance and phase angle. The impedance and Ohm's law yield the maximum current.

**SOLUTION**

**(a)** Find the impedance of the circuit.

First, calculate the inductive and capacitive reactances:

$$X_L = 2\pi f L = 226\ \Omega \qquad X_C = 1/2\pi f C = 758\ \Omega$$

Substitute these results and the resistance $R$ into Equation 21.13 to obtain the impedance of the circuit:

$$Z = \sqrt{R^2 + (X_L - X_C)^2}$$

$$= \sqrt{(2.50 \times 10^2\ \Omega)^2 + (226\ \Omega - 758\ \Omega)^2} = \boxed{588\ \Omega}$$

**(b)** Find the maximum current in the circuit.

Use Equation 21.12, the equivalent of Ohm's law, to find the maximum current:

$$I_{\text{max}} = \frac{\Delta V_{\text{max}}}{Z} = \frac{1.50 \times 10^2\ \text{V}}{588\ \Omega} = \boxed{0.255\ \text{A}}$$

*(Continued)*

**(c)** Find the phase angle.

Calculate the phase angle between the current and the voltage with Equation 21.15:

$$\phi = \tan^{-1}\frac{X_L - X_C}{R} = \tan^{-1}\left(\frac{226\ \Omega - 758\ \Omega}{2.50 \times 10^2\ \Omega}\right) = \boxed{-64.8^\circ}$$

**(d)** Find the maximum voltages across the elements.

Use the "Ohm's law" expressions for each individual type of current element:

$$\Delta V_{R,\text{max}} = I_{\text{max}}R = (0.255\text{ A})(2.50 \times 10^2\ \Omega) = \boxed{63.8\text{ V}}$$

$$\Delta V_{L,\text{max}} = I_{\text{max}}X_L = (0.255\text{ A})(2.26 \times 10^2\ \Omega) = \boxed{57.6\text{ V}}$$

$$\Delta V_{C,\text{max}} = I_{\text{max}}X_C = (0.255\text{ A})(7.58 \times 10^2\ \Omega) = \boxed{193\text{ V}}$$

**REMARKS** Because the circuit is more capacitive than inductive ($X_C > X_L$), $\phi$ is negative. A negative phase angle means that the current leads the applied voltage. Notice also that the sum of the maximum voltages across the elements is $\Delta V_R + \Delta V_L + \Delta V_C = 314$ V, which is much greater than the maximum voltage of the generator, 150 V. As we saw in Quick Quiz 21.2, the sum of the maximum voltages is a meaningless quantity because when alternating voltages are added, *both their amplitudes and their phases* must be taken into account. We know that the maximum voltages across the various elements occur at different times, so it doesn't make sense to add all the maximum values. The correct way to "add" the voltages is through Equation 21.10.

**QUESTION 21.4** True or False: In an *RLC* circuit, the impedance must always be greater than or equal to the resistance.

**EXERCISE 21.4** Analyze a series *RLC* AC circuit for which $R = 175\ \Omega$, $L = 0.500$ H, $C = 22.5\ \mu$F, $f = 60.0$ Hz, and $\Delta V_{\text{max}} = 325$ V. Find (a) the impedance, (b) the maximum current, (c) the phase angle, and (d) the maximum voltages across the elements.

**ANSWERS** (a) 189 Ω (b) 1.72 A (c) 22.0° (d) $\Delta V_{R,\text{max}} = 301$ V, $\Delta V_{L,\text{max}} = 324$ V, $\Delta V_{C,\text{max}} = 203$ V

## 21.5 Power in an AC Circuit

**LEARNING OBJECTIVES**

1. Discuss the conservation of energy in purely capacitive and purely inductive AC circuits.
2. Discuss the energy dissipated by resistive *RLC* circuits.
3. Evaluate the average power dissipated by an *RLC* circuit.

No power losses are associated with pure capacitors and pure inductors in an AC circuit. A pure capacitor, by definition, has no resistance or inductance, whereas a pure inductor has no resistance or capacitance. (These definitions are idealizations: in a real capacitor, for example, inductive effects could become important at high frequencies.) We begin by analyzing the power dissipated in an AC circuit that contains only a generator and a capacitor.

When the current increases in one direction in an AC circuit, charge accumulates on the capacitor and a voltage drop appears across it. When the voltage reaches its maximum value, the energy stored in the capacitor is

$$PE_C = \tfrac{1}{2}C(\Delta V_{\text{max}})^2$$

This energy storage is only momentary, however: When the current reverses direction, the charge leaves the capacitor plates and returns to the voltage source. During one-half of each cycle the capacitor is being charged, and during the other half the charge is being returned to the voltage source. Therefore, the average power supplied by the source is zero. In other words, **no power losses occur in a capacitor in an AC circuit.**

Similarly, the source must do work against the back emf of an inductor that is carrying a current. When the current reaches its maximum value, the energy stored in the inductor is a maximum and is given by

$$PE_L = \tfrac{1}{2}LI_{\text{max}}^2$$

When the current begins to decrease in the circuit, this stored energy is returned to the source as the inductor attempts to maintain the current in the circuit. The average power delivered to a resistor in an *RLC* circuit is

$$P_{av} = I_{rms}^2 R \quad [21.16]$$

**The average power delivered by the generator is converted to internal energy in the resistor. No power loss occurs in an ideal capacitor or inductor.**

An alternate equation for the average power loss in an AC circuit can be found by substituting (from Ohm's law) $R = \Delta V_{R,rms}/I_{rms}$ into Equation 21.16:

$$P_{av} = I_{rms}\,\Delta V_{R,rms}$$

It's convenient to refer to a voltage triangle that shows the relationship among $\Delta V_{rms}$, $\Delta V_{R,rms}$, and $\Delta V_{L,rms} - \Delta V_{C,rms}$, such as Figure 21.11b (page 741). (Remember that Fig. 21.11 applies to *both* maximum and rms voltages.) From this figure, we see that the voltage drop across a resistor can be written in terms of the voltage of the source, $\Delta V_{rms}$:

$$\Delta V_{R,rms} = \Delta V_{rms} \cos \phi$$

Hence, the average power delivered by a generator in an AC circuit is

$$P_{av} = I_{rms}\,\Delta V_{rms} \cos \phi \quad [21.17]$$

◀ Average power

where the quantity $\cos \phi$ is called the **power factor**.

Equation 21.17 shows that the power delivered by an AC source to any circuit depends on the phase difference between the source voltage and the resulting current. This fact has many interesting applications. For example, factories often use devices such as large motors in machines, generators, and transformers that have a large inductive load due to all the windings. To deliver greater power to such devices without using excessively high voltages, factory technicians introduce capacitance in the circuits to shift the phase.

**APPLICATION**

Shifting phase to deliver more power

## ■ EXAMPLE 21.5 Average Power in an *RLC* Series Circuit

**GOAL** Understand power in *RLC* series circuits.

**PROBLEM** Calculate the average power delivered to the series *RLC* circuit described in Example 21.4.

**STRATEGY** After finding the rms current and rms voltage with Equations 21.2 and 21.3, substitute into Equation 21.17, using the phase angle found in Example 21.4.

**SOLUTION**

First, use Equations 21.2 and 21.3 to calculate the rms current and rms voltage:

$$I_{rms} = \frac{I_{max}}{\sqrt{2}} = \frac{0.255\text{ A}}{\sqrt{2}} = 0.180\text{ A}$$

$$\Delta V_{rms} = \frac{\Delta V_{max}}{\sqrt{2}} = \frac{1.50 \times 10^2\text{ V}}{\sqrt{2}} = 106\text{ V}$$

Substitute these results and the phase angle $\phi = -64.8°$ into Equation 21.17 to find the average power:

$$P_{av} = I_{rms}\,\Delta V_{rms} \cos \phi = (0.180\text{ A})(106\text{ V}) \cos(-64.8°)$$

$$= 8.12\text{ W}$$

**REMARKS** The same result can be obtained from Equation 21.16, $P_{av} = I_{rms}^2 R$.

**QUESTION 21.5** Under what circumstance can the average power of an *RLC* circuit be zero?

**EXERCISE 21.5** Repeat this problem, using the system described in Exercise 21.4.

**ANSWER** 259 W

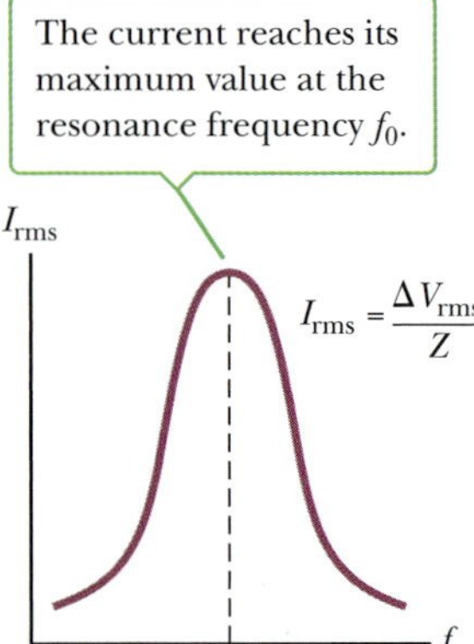
The current reaches its maximum value at the resonance frequency $f_0$.
$I_{rms}$
$I_{rms} = \frac{\Delta V_{rms}}{Z}$
$f_0$
$f$

**Figure 21.13** A plot of current amplitude in a series *RLC* circuit versus frequency of the generator voltage.

# 21.6 Resonance in a Series *RLC* Circuit

**LEARNING OBJECTIVES**

1. Discuss the concept of resonance in a series *RLC* circuit.
2. Evaluate and apply the resonance frequency of a series RLC circuit.

In general, the rms current in a series *RLC* circuit can be written

$$I_{rms} = \frac{\Delta V_{rms}}{Z} = \frac{\Delta V_{rms}}{\sqrt{R^2 + (X_L - X_C)^2}} \qquad [21.18]$$

From this equation, we see that if the frequency is varied, the current has its *maximum* value when the impedance has its *minimum* value, which occurs when $X_L = X_C$. In such a circumstance, the impedance of the circuit reduces to $Z = R$. The frequency $f_0$ at which this happens is called the **resonance frequency** of the circuit. To find $f_0$, we set $X_L = X_C$, which gives, from Equations 21.5 and 21.8,

$$2\pi f_0 L = \frac{1}{2\pi f_0 C}$$

$$f_0 = \frac{1}{2\pi\sqrt{LC}} \qquad [21.19]$$

Figure 21.13 is a plot of current as a function of frequency for a circuit containing a fixed value for both the capacitance and the inductance. From Equation 21.18, it must be concluded that the current would become infinite at resonance when $R = 0$. Although Equation 21.18 predicts this result, real circuits always have some resistance, which limits the value of the current.

**APPLICATION**
Tuning Your Radio

The tuning circuit of a radio is an important application of a series resonance circuit. The radio is tuned to a particular station (which transmits a specific radio-frequency signal) by varying a capacitor, which changes the resonance frequency of the tuning circuit. When this resonance frequency matches that of the incoming radio wave, the current in the tuning circuit increases.

## APPLYING PHYSICS 21.2 Metal Detectors at the Courthouse

When you walk through the doorway of a courthouse metal detector, as the person in Figure 21.14 is doing, you are really walking through a coil of many turns. How might the metal detector work?

**EXPLANATION** The metal detector is essentially a resonant circuit. The portal you step through is an inductor (a large loop of conducting wire) that is part of the circuit. The frequency of the circuit is tuned to the circuit's resonant frequency of the circuit when there is no metal in the inductor. When you walk through with metal in your pocket, you change the effective inductance of the resonance circuit, resulting in a change in the circuit's current. This change in current is detected, and an electronic circuit causes a sound to be emitted as an alarm. ■

Kira Vuille-Kowing

**Figure 21.14** (Applying Physics 21.2) A courthouse metal detector.

## EXAMPLE 21.6 A Circuit in Resonance

**GOAL** Understand resonance frequency and its relation to inductance, capacitance, and the rms current.

**PROBLEM** Consider a series *RLC* circuit for which $R = 1.50 \times 10^2\ \Omega$, $L = 20.0$ mH, $\Delta V_{rms} = 20.0$ V, and $f = 796\ \text{s}^{-1}$. **(a)** Determine the value of the capacitance for which the rms current is a maximum. **(b)** Find the maximum rms current in the circuit.

**STRATEGY** The current is a maximum at the resonance frequency $f_0$, which should be set equal to the driving frequency, 796 $s^{-1}$. The resulting equation can be solved for $C$. For part (b), substitute into Equation 21.18 to get the maximum rms current.

**SOLUTION**

**(a)** Find the capacitance giving the maximum current in the circuit (the resonance condition).

Solve the resonance frequency for the capacitance:

$$f_0 = \frac{1}{2\pi\sqrt{LC}} \rightarrow \sqrt{LC} = \frac{1}{2\pi f_0} \rightarrow LC = \frac{1}{4\pi^2 f_0^{\,2}}$$

$$C = \frac{1}{4\pi^2 f_0^{\,2} L}$$

Insert the given values, substituting the source frequency for the resonance frequency, $f_0$:

$$C = \frac{1}{4\pi^2(796\text{ Hz})^2(20.0\times 10^{-3}\text{ H})} = \boxed{2.00\times 10^{-6}\text{ F}}$$

**(b)** Find the maximum rms current in the circuit.

The capacitive and inductive reactances are equal, so $Z = R = 1.50 \times 10^2\ \Omega$. Substitute into Equation 21.18 to find the rms current:

$$I_{\text{rms}} = \frac{\Delta V_{\text{rms}}}{Z} = \frac{20.0\text{ V}}{1.50\times 10^2\ \Omega} = \boxed{0.133\text{ A}}$$

**REMARKS** Because the impedance $Z$ is in the denominator of Equation 21.18, the maximum current will always occur when $X_L = X_C$ because that yields the minimum value of $Z$.

**QUESTION 21.6** True or False: The magnitude of the current in an *RLC* circuit is never larger than the rms current.

**EXERCISE 21.6** Consider a series *RLC* circuit for which $R = 1.20 \times 10^2\ \Omega$, $C = 3.10 \times 10^{-5}$ F, $\Delta V_{\text{rms}} = 35.0$ V, and $f = 60.0\ s^{-1}$. (a) Determine the value of the inductance for which the rms current is a maximum. (b) Find the maximum rms current in the circuit.

**ANSWERS** (a) 0.227 H (b) 0.292 A

# 21.7 The Transformer

**LEARNING OBJECTIVES**

1. Discuss the concept and operation of an AC transformer.
2. Apply the current and voltage values on an AC transformer's primary coil to those on the secondary coil.

It's often necessary to change a small AC voltage to a larger one or vice versa. Such changes are effected with a device called a transformer.

In its simplest form the **AC transformer** consists of two coils of wire wound around a core of soft iron, as shown in Figure 21.15. The coil on the left, which is connected to the input AC voltage source and has $N_1$ turns, is called the primary winding, or the *primary*. The coil on the right, which is connected to a resistor $R$ and consists of $N_2$ turns, is the *secondary*. The common iron core is used to increase the magnetic flux and to provide a medium in which nearly all the flux through one coil passes through the other.

When an input AC voltage $\Delta V_1$ is applied to the primary, the induced voltage across it is given by

$$\Delta V_1 = -N_1 \frac{\Delta \Phi_B}{\Delta t} \qquad [21.20]$$

where $\Phi_B$ is the magnetic flux through each turn. If we assume that no flux leaks from the iron core, then the flux through each turn of the primary equals the

An AC voltage $\Delta V_1$ is applied to the primary coil, and the output voltage $\Delta V_2$ is observed across the load resistance $R$.

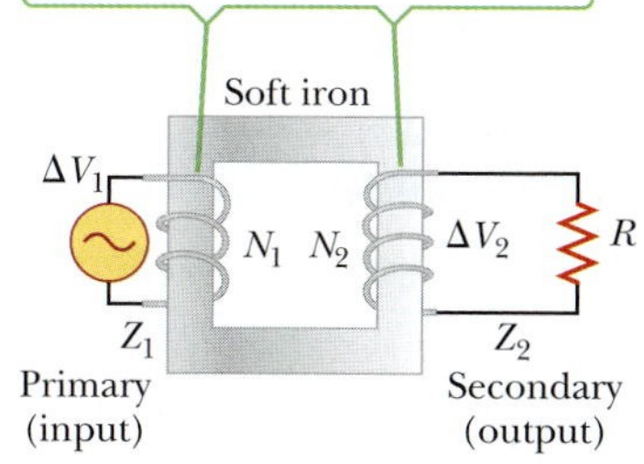


**Figure 21.15** An ideal transformer consists of two coils wound on the same soft iron core. An AC voltage $\Delta V_1$ is applied to the primary coil, and the output voltage $\Delta V_2$ is observed across the load resistance $R$ after the switch is closed.

flux through each turn of the secondary. Hence, the voltage across the secondary coil is

$$\Delta V_2 = -N_2 \frac{\Delta \Phi_B}{\Delta t} \quad [21.21]$$

The term $\Delta\Phi_B/\Delta t$ is common to Equations 21.20 and 21.21 and can be algebraically eliminated, giving

$$\Delta V_2 = \frac{N_2}{N_1}\Delta V_1 \quad [21.22]$$

When $N_2$ is greater than $N_1$, $\Delta V_2$ exceeds $\Delta V_1$ and the transformer is referred to as a *step-up transformer.* When $N_2$ is less than $N_1$, making $\Delta V_2$ less than $\Delta V_1$, we have a *step-down transformer.*

By Faraday's law, a voltage is generated across the secondary only when there is a *change* in the number of flux lines passing through the secondary. The input current in the primary must therefore change with time, which is what happens when an alternating current is used. When the input at the primary is a direct current, however, a voltage output occurs at the secondary only at the instant a switch in the primary circuit is opened or closed. Once the current in the primary reaches a steady value, the output voltage at the secondary is zero.

It may seem that a transformer is a device in which it is possible to get something for nothing. For example, a step-up transformer can change an input voltage from, say, 10 V to 100 V. This means that each coulomb of charge leaving the secondary has 100 J of energy, whereas each coulomb of charge entering the primary has only 10 J of energy. That is not the case, however, because **the power input to the primary equals the power output at the secondary**:

In an ideal transformer, the input power equals the output power ▶

$$I_1\,\Delta V_1 = I_2\,\Delta V_2 \quad [21.23]$$

Although the *voltage* at the secondary may be, say, ten times greater than the voltage at the primary, the *current* in the secondary will be smaller than the primary's current by a factor of ten. Equation 21.23 assumes an **ideal transformer** in which there are no power losses between the primary and the secondary. Real transformers typically have power efficiencies ranging from 90% to 99%. Power losses occur because of such factors as eddy currents induced in the iron core of the transformer, which dissipate energy in the form of $I^2R$ losses.

**APPLICATION**
Long-Distance Electric Power Transmission

When electric power is transmitted over large distances, it's economical to use a high voltage and a low current because the power lost via resistive heating in the transmission lines varies as $I^2R$. If a utility company can reduce the current by a factor of ten, for example, the power loss is reduced by a factor of one hundred. In practice, the voltage is stepped up to around 230 000 V at the generating station, then stepped down to around 20 000 V at a distribution station, and finally stepped down to 120 V at the customer's utility pole.

## ■ EXAMPLE 21.7 Distributing Power to a City

**GOAL** Understand transformers and their role in reducing power loss.

**PROBLEM** A generator at a utility company produces $1.00 \times 10^2$ A of current at $4.00 \times 10^3$ V. The voltage is stepped up to $2.40 \times 10^5$ V by a transformer before being sent on a high-voltage transmission line across a rural area to a city. Assume the effective resistance of the power line is 30.0 Ω and that the transformers are ideal. **(a)** Determine the percentage of power lost in the transmission line. **(b)** What percentage of the original power would be lost in the transmission line if the voltage were not stepped up?

**STRATEGY** Solving this problem is just a matter of substitution into the equation for transformers and the equation for power loss. To obtain the fraction of power lost, it's also necessary to compute the power output of the generator: the current times the potential difference created by the generator.

**SOLUTION**

**(a)** Determine the percentage of power lost in the line.

Substitute into Equation 21.23 to find the current in the transmission line:

$$I_2 = \frac{I_1 \Delta V_1}{\Delta V_2} = \frac{(1.00 \times 10^2 \text{ A})(4.00 \times 10^3 \text{ V})}{2.40 \times 10^5 \text{ V}} = 1.67 \text{ A}$$

Now use Equation 21.16 to find the power lost in the transmission line:

$$\textbf{(1)}\ P_{\text{lost}} = I_2{}^2 R = (1.67 \text{ A})^2(30.0\ \Omega) = 83.7 \text{ W}$$

Calculate the power output of the generator:

$$P = I_1\, \Delta V_1 = (1.00 \times 10^2 \text{ A})(4.00 \times 10^3 \text{ V}) = 4.00 \times 10^5 \text{ W}$$

Finally, divide $P_{\text{lost}}$ by the power output and multiply by 100 to find the percentage of power lost:

$$\%\text{ power lost} = \left(\frac{83.7 \text{ W}}{4.00 \times 10^5 \text{ W}}\right) \times 100 = \boxed{0.020\,9\%}$$

**(b)** What percentage of the original power would be lost in the transmission line if the voltage were not stepped up?

Replace the stepped-up current in Equation (1) by the original current of $1.00 \times 10^2$ A:

$$P_{\text{lost}} = I^2 R = (1.00 \times 10^2 \text{ A})^2(30.0\ \Omega) = 3.00 \times 10^5 \text{ W}$$

Calculate the percentage loss, as before:

$$\%\text{ power lost} = \left(\frac{3.00 \times 10^5 \text{ W}}{4.00 \times 10^5 \text{ W}}\right) \times 100 = \boxed{75\%}$$

**REMARKS** This example illustrates the advantage of high-voltage transmission lines. In the city, a transformer at a substation steps the voltage back down to about 4000 V, and this voltage is maintained across utility lines throughout the city. When the power is to be used at a home or business, a transformer on a utility pole near the establishment reduces the voltage to 240 V or 120 V.

This cylindrical step-down transformer drops the voltage from 4 000 V to 220 V for delivery to a group of residences.

**QUESTION 21.7** If the voltage is stepped up to double the amount in this problem, by what factor is the power loss changed? (a) 2 (b) no change (c) $\frac{1}{2}$ (d) $\frac{1}{4}$

**EXERCISE 21.7** Suppose the same generator has the voltage stepped up to only $7.50 \times 10^4$ V and the resistance of the line is 85.0 Ω. Find the percentage of power lost in this case.

**ANSWER** 0.604%

## 21.8 Maxwell's Predictions

**LEARNING OBJECTIVES**

1. State and discuss the four fundamental ideas underlying Maxwell's unifying theory of electricity and magnetism.
2. Discuss the concept of electromagnetic waves.

During the early stages of their study and development, electric and magnetic phenomena were thought to be unrelated. In 1865, however, James Clerk Maxwell (1831–1879) provided a mathematical theory that showed a close relationship between all electric and magnetic phenomena. In addition to unifying the formerly separate fields of electricity and magnetism, his brilliant theory predicted

**James Clerk Maxwell**
**Scottish Theoretical Physicist (1831–1879)**
Maxwell developed the electromagnetic theory of light and the kinetic theory of gases, and he explained the nature of Saturn's rings and color vision. Maxwell's successful interpretation of the electromagnetic field resulted in the equations that bear his name. Formidable mathematical ability combined with great insight enabled him to lead the way in the study of electromagnetism and kinetic theory.

that electric and magnetic fields can move through space as waves. The theory he developed is based on the following four pieces of information:

1. Electric field lines originate on positive charges and terminate on negative charges.
2. Magnetic field lines always form closed loops; they don't begin or end anywhere.
3. A varying magnetic field induces an emf and hence an electric field. This fact is a statement of Faraday's law (Chapter 20).
4. Magnetic fields are generated by moving charges (or currents), as summarized in Ampère's law (Chapter 19).

The first statement is a consequence of the nature of the electrostatic force between charged particles, given by Coulomb's law. It embodies the fact that **free charges (electric monopoles) exist in nature**.

The second statement—that magnetic fields form continuous loops—is exemplified by the magnetic field lines around a long, straight wire, which are closed circles, and the magnetic field lines of a bar magnet, which form closed loops. It says, in contrast to the first statement, that **free magnetic charges (magnetic monopoles) don't exist in nature**.

The third statement is equivalent to Faraday's law of induction, and the fourth is equivalent to Ampère's law.

In one of the greatest theoretical developments of the 19th century, Maxwell used these four statements within a corresponding mathematical framework to prove that electric and magnetic fields play symmetric roles in nature. It was already known from experiments that a changing magnetic field produced an electric field according to Faraday's law. Maxwell believed that nature was symmetric, and he therefore hypothesized that a changing electric field should produce a magnetic field. This hypothesis could not be proven experimentally at the time it was developed because the magnetic fields generated by changing electric fields are generally very weak and therefore difficult to detect.

To justify his hypothesis, Maxwell searched for other phenomena that might be explained by it. He turned his attention to the motion of rapidly oscillating (accelerating) charges, such as those in a conducting rod connected to an alternating voltage. Such charges are accelerated and, according to Maxwell's predictions, generate changing electric and magnetic fields. The changing fields cause electromagnetic disturbances that travel through space as waves, similar to the spreading water waves created by a pebble thrown into a pool. The waves sent out by the oscillating charges are fluctuating electric and magnetic fields, so they are called *electromagnetic waves*. From Faraday's law and from Maxwell's own generalization of Ampère's law, Maxwell calculated the speed of the waves to be equal to the speed of light, $c = 3 \times 10^8$ m/s. He concluded that visible light and other electromagnetic waves consist of fluctuating electric and magnetic fields traveling through empty space, with each varying field inducing the other! His was truly one of the greatest discoveries of science, on a par with Newton's discovery of the laws of motion. Like Newton's laws, it had a profound influence on later scientific developments.

## 21.9 Hertz's Confirmation of Maxwell's Predictions

**LEARNING OBJECTIVE**

1. Discuss Heinrich Hertz's experimental confirmation of Maxwell's predictions about electromagnetic waves.

In 1887, after Maxwell's death, Heinrich Hertz (1857–1894) was the first to generate and detect electromagnetic waves in a laboratory setting, using $LC$ circuits. In such a circuit a charged capacitor is connected to an inductor, as in Figure 21.16.

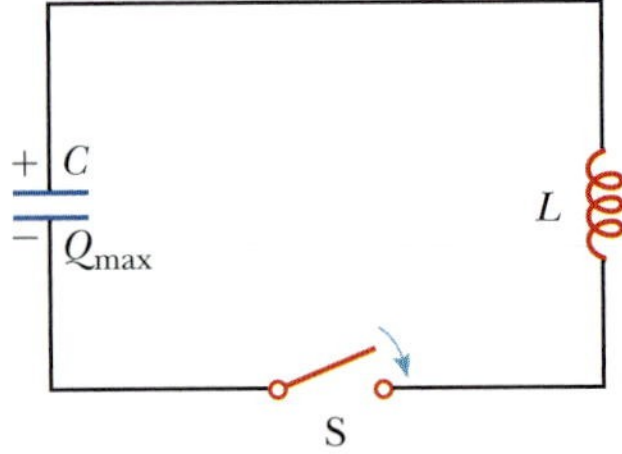


**Figure 21.16** A simple $LC$ circuit. The capacitor has an initial charge of $Q_{max}$, and the switch is closed at $t = 0$.

When the switch is closed, oscillations occur in the current in the circuit and in the charge on the capacitor. If the resistance of the circuit is neglected, no energy is dissipated and the oscillations continue.

In the following analysis, we neglect the resistance in the circuit. We assume the capacitor has an initial charge of $Q_{max}$ and the switch is closed at $t = 0$. When the capacitor is fully charged, the total energy in the circuit is stored in the electric field of the capacitor and is equal to $Q_{max}^2/2C$. At this time, the current is zero, so no energy is stored in the inductor. As the capacitor begins to discharge, the energy stored in its electric field decreases. At the same time, the current increases and energy equal to $LI^2/2$ is now stored in the magnetic field of the inductor. Thus, energy is transferred from the electric field of the capacitor to the magnetic field of the inductor. When the capacitor is fully discharged, it stores no energy. At this time, the current reaches its maximum value and all the energy is stored in the inductor. The process then repeats in the reverse direction. The energy continues to transfer between the inductor and the capacitor, corresponding to oscillations in the current and charge.

As we saw in Section 21.6, the frequency of oscillation of an $LC$ circuit is called the *resonance frequency* of the circuit and is given by

$$f_0 = \frac{1}{2\pi\sqrt{LC}}$$

The circuit Hertz used in his investigations of electromagnetic waves is similar to that just discussed and is shown schematically in Figure 21.17. An induction coil (a large coil of wire) is connected to two metal spheres with a narrow gap between them to form a capacitor. Oscillations are initiated in the circuit by short voltage pulses sent via the coil to the spheres, charging one positive, the other negative. Because $L$ and $C$ are quite small in this circuit, the frequency of oscillation is quite high, $f \approx 100$ MHz. This circuit is called a transmitter because it produces electromagnetic waves.

Several meters from the transmitter circuit, Hertz placed a second circuit, the receiver, which consisted of a single loop of wire connected to two spheres. It had its own effective inductance, capacitance, and natural frequency of oscillation. Hertz found that energy was being sent from the transmitter to the receiver when the resonance frequency of the receiver was adjusted to match that of the transmitter. The energy transfer was detected when the voltage across the spheres in the receiver circuit became high enough to produce ionization in the air, which caused sparks to appear in the air gap separating the spheres. Hertz's experiment is analogous to the mechanical phenomenon in which a tuning fork picks up the vibrations from another, identical tuning fork.

Hertz hypothesized that the energy transferred from the transmitter to the receiver is carried in the form of waves, now recognized as electromagnetic waves. In a series of experiments, he also showed that the radiation generated by the transmitter exhibits wave properties: interference, diffraction, reflection, refraction, and polarization. As you will see shortly, all these properties are exhibited by light. It became evident that Hertz's electromagnetic waves had the same known properties of light waves and differed only in frequency and wavelength. Hertz effectively confirmed Maxwell's theory by showing that Maxwell's mysterious electromagnetic waves existed and had all the properties of light waves.

Perhaps the most convincing experiment Hertz performed was the measurement of the speed of waves from the transmitter, accomplished as follows: waves of known frequency from the transmitter were reflected from a metal sheet so that an interference pattern was set up, much like the standing-wave pattern on a stretched string. As we learned in our discussion of standing waves, the distance between nodes is $\lambda/2$, so Hertz was able to determine the wavelength $\lambda$. Using the relationship $v = \lambda f$, he found that $v$ was close to $3 \times 10^8$ m/s, the known speed of visible light. Hertz's experiments thus provided the first evidence in support of Maxwell's theory.

The transmitter consists of two spherical electrodes connected to an induction coil, which provides short voltage surges to the spheres, setting up oscillations in the discharge.

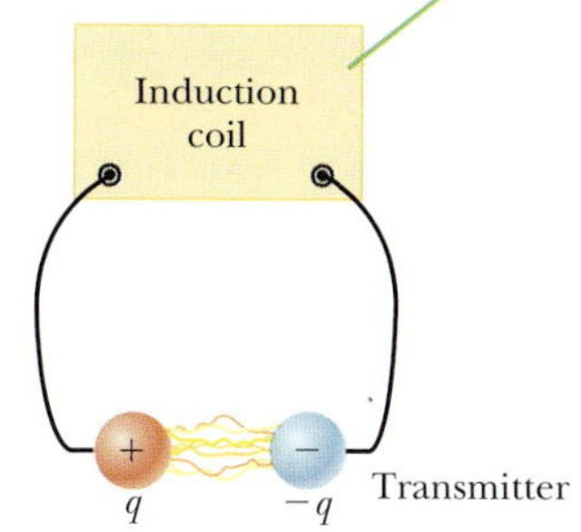


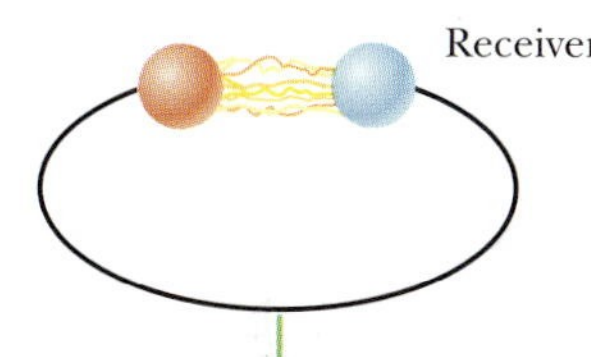


The receiver is a nearby loop of wire containing a second spark gap.

**Figure 21.17** A schematic diagram of Hertz's apparatus for generating and detecting electromagnetic waves.

**Heinrich Rudolf Hertz**
**German Physicist (1857–1894)**
Hertz made his most important discovery of radio waves in 1887. After finding that the speed of a radio wave was the same as that of light, Hertz showed that radio waves, like light waves, could be reflected, refracted, and diffracted. Hertz died of blood poisoning at the age of 36. During his short life, he made many contributions to science. The hertz, equal to one complete vibration or cycle per second, is named after him.

# 21.10 Production of Electromagnetic Waves by an Antenna

**LEARNING OBJECTIVES**

1. Discuss the radiation of energy by the accelerated charges in an AC circuit.
2. Discuss the transverse nature of electromagnetic waves.

In the previous section we found that the energy stored in an $LC$ circuit is continually transferred between the electric field of the capacitor and the magnetic field of the inductor. This energy transfer, however, continues for prolonged periods of time only when the changes occur slowly. If the current alternates rapidly, the circuit loses some of its energy in the form of electromagnetic waves. In fact, electromagnetic waves are radiated by *any* circuit carrying an alternating current. The fundamental mechanism responsible for this radiation is the acceleration of a charged particle. **Whenever a charged particle accelerates, it radiates energy.**

**APPLICATION**
Radio-Wave Transmission

An alternating voltage applied to the wires of an antenna forces electric charges in the antenna to oscillate. This common technique for accelerating charged particles is the source of the radio waves emitted by the broadcast antenna of a radio station.

Figure 21.18 illustrates the production of an electromagnetic wave by oscillating electric charges in an antenna. Two metal rods are connected to an AC source, which causes charges to oscillate between the rods. The output voltage of the generator is sinusoidal. At $t = 0$, the upper rod is given a maximum positive charge and the bottom rod an equal negative charge, as in Figure 21.18a. The electric field near the antenna at this instant is also shown in the figure. As the charges oscillate, the rods become less charged, the field near the rods decreases in strength, and the downward-directed maximum electric field produced at $t = 0$ moves away from the rod. When the charges are neutralized, as in Figure 21.18b, the electric field has dropped to zero, after an interval equal to one-quarter of the period of oscillation. Continuing in this fashion, the upper rod soon obtains a maximum negative charge and the lower rod becomes positive, as in Figure 21.18c, resulting in an electric field directed upward. This occurs after an interval equal to one-half the period of oscillation. The oscillations continue as indicated in Figure 21.18d. Note that the electric field near the antenna oscillates in phase with the charge distribution: the field points down when the upper rod is positive and up when the upper rod is negative. Further, the magnitude of the field at any instant depends on the amount of charge on the rods at that instant.

**Tip 21.1 Accelerated Charges Produce Electromagnetic Waves**

Stationary charges produce only electric fields, whereas charges in uniform motion (i.e., constant velocity) produce electric and magnetic fields, but no electromagnetic waves. In contrast, accelerated charges produce electromagnetic waves as well as electric and magnetic fields. An accelerating charge also radiates energy.

As the charges continue to oscillate (and accelerate) between the rods, the electric field set up by the charges moves away from the antenna in all directions at the speed of light. Figure 21.18 shows the electric field pattern on one side of the antenna at certain times during the oscillation cycle. As you can see, one cycle of charge oscillation produces one full wavelength in the electric field pattern.

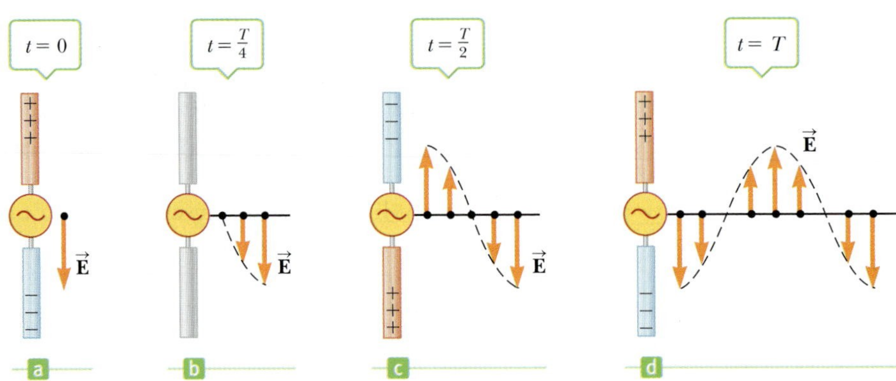


**Figure 21.18** An electric field set up by oscillating charges in an antenna. The field moves away from the antenna at the speed of light.

Because the oscillating charges create a current in the rods, a magnetic field is also generated when the current in the rods is upward, as shown in Figure 21.19. The magnetic field lines circle the antenna (recall right-hand rule number 2) and are perpendicular to the electric field at all points. As the current changes with time, the magnetic field lines spread out from the antenna. At great distances from the antenna, the strengths of the electric and magnetic fields become very weak. At these distances, however, it is necessary to take into account the facts that (1) a changing magnetic field produces an electric field and (2) a changing electric field produces a magnetic field, as predicted by Maxwell. These induced electric and magnetic fields are in phase: at any point, the two fields reach their maximum values at the same instant. This synchrony is illustrated at one instant of time in Figure 21.20. Note that (1) the $\vec{\mathbf{E}}$ and $\vec{\mathbf{B}}$ fields are perpendicular to each other and (2) both fields are perpendicular to the direction of motion of the wave. This second property is characteristic of transverse waves. Hence, we see that **an electromagnetic wave is a transverse wave**.

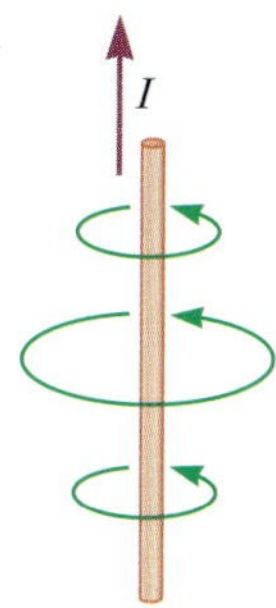
I

**Figure 21.19** Magnetic field lines around an antenna carrying a changing current.

## 21.11 Properties of Electromagnetic Waves

### LEARNING OBJECTIVES

1. Discuss the concept of a plane wave.
2. Relate the speed of light to the permeability and permittivity of the propagation medium.
3. Relate the speed of light to the ratio of the electric and magnetic field magnitudes in an electromagnetic wave.
4. Discuss the fundamental properties of electromagnetic waves.
5. Evaluate the average intensity of an electromagnetic wave and the momentum delivered by the wave to perfectly absorbing or reflecting surfaces.
6. Calculate thermal and mechanical effects due to the interaction of electromagnetic waves and matter.

We have seen that Maxwell's detailed analysis predicted the existence and properties of electromagnetic waves. In this section we summarize what we know about electromagnetic waves thus far and consider some additional properties. In our discussion here and in future sections, we will often make reference to a type of wave called a **plane wave**. A plane electromagnetic wave is a wave traveling from a very distant source. Figure 21.20 pictures such a wave at a given instant of time. In this case the oscillations of the electric and magnetic fields take place in planes perpendicular

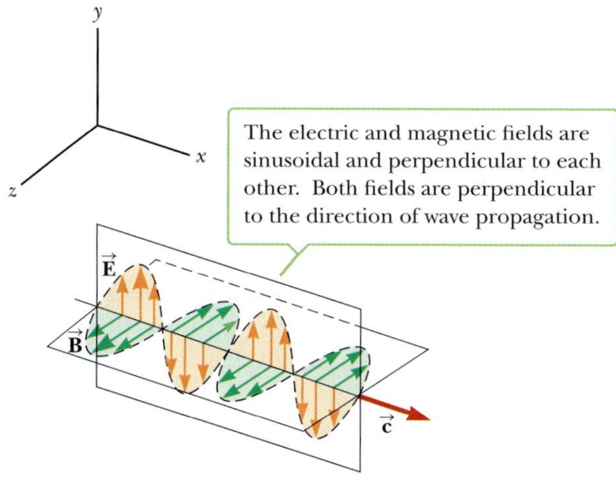
y
x
z
The electric and magnetic fields are sinusoidal and perpendicular to each other. Both fields are perpendicular to the direction of wave propagation.
E
B
c

**Figure 21.20** An electromagnetic wave sent out by oscillating charges in an antenna, represented at one instant of time and far from the antenna, moving in the positive $x$-direction with speed $c$.

to the $x$-axis and are therefore perpendicular to the direction of travel of the wave. Because of the latter property, electromagnetic waves are transverse waves. In the figure the electric field $\vec{\mathbf{E}}$ is in the $y$-direction and the magnetic field $\vec{\mathbf{B}}$ is in the $z$-direction. Light propagates in a direction perpendicular to these two fields. That direction is determined by yet another right-hand rule: (1) point the fingers of your right hand in the direction of $\vec{\mathbf{E}}$, (2) curl them in the direction of $\vec{\mathbf{B}}$, and (3) the right thumb then points in the direction of propagation of the wave.

Electromagnetic waves travel with the speed of light. In fact, it can be shown that the speed of an electromagnetic wave is related to the permeability and permittivity of the medium through which it travels. Maxwell found this relationship for free space to be

Speed of light ▶

$$c = \frac{1}{\sqrt{\mu_0 \epsilon_0}} \qquad [21.24]$$

where $c$ is the speed of light, $\mu_0 = 4\pi \times 10^{-7}\ \mathrm{N{\cdot}s^2/C^2}$ is the permeability constant of vacuum, and $\epsilon_0 = 8.854\ 19 \times 10^{-12}\ \mathrm{C^2/N{\cdot}m^2}$ is the permittivity of free space. Substituting these values into Equation 21.24, we find that

$$c = 2.997\ 92 \times 10^8\ \mathrm{m/s} \qquad [21.25]$$

**Tip 21.2 *E* Stronger than *B*?**

The relationship $E = Bc$ makes it appear that the electric fields associated with light are much larger than the magnetic fields. That is not the case: The units are different, so the quantities can't be directly compared. The two fields contribute equally to the energy of a light wave.

Because electromagnetic waves travel at the same speed as light in vacuum, scientists concluded (correctly) that **light is an electromagnetic wave**.

Maxwell also proved the following relationship for electromagnetic waves:

$$\frac{E}{B} = c \qquad [21.26]$$

which states that the ratio of the magnitude of the electric field to the magnitude of the magnetic field equals the speed of light.

Electromagnetic waves carry energy as they travel through space, and this energy can be transferred to objects placed in their paths. The average rate at which energy passes through an area perpendicular to the direction of travel of a wave, or the average power per unit area, is called the **intensity *I*** of the wave and is given by

$$I = \frac{E_{\max} B_{\max}}{2\mu_0} \qquad [21.27]$$

where $E_{\max}$ and $B_{\max}$ are the *maximum* values of $E$ and $B$. The quantity $I$ is analogous to the intensity of sound waves introduced in Chapter 14. From Equation 21.26, we see that $E_{\max} = cB_{\max} = B_{\max}/\sqrt{\mu_0 \epsilon_0}$. Equation 21.27 can therefore also be expressed as

$$I = \frac{E_{\max}^2}{2\mu_0 c} = \frac{c}{2\mu_0} B_{\max}^2 \qquad [21.28]$$

Note that in these expressions we use the *average* power per unit area. A detailed analysis would show that the energy carried by an electromagnetic wave is shared equally by the electric and magnetic fields.

Light is an electromagnetic wave and transports energy and momentum ▶

Electromagnetic waves have an average intensity given by Equation 21.28. When the waves strike an area $A$ of an object's surface for a given time $\Delta t$, energy $U = IA\Delta t$ is transferred to the surface. Momentum is transferred, as well. Hence, pressure is exerted on a surface when an electromagnetic wave impinges on it. In what follows, we assume the electromagnetic wave transports a total energy $U$ to a surface in a time $\Delta t$. If the surface absorbs all the incident energy $U$ in this time, Maxwell showed that the total momentum $\vec{\mathbf{p}}$ delivered to this surface has a magnitude

$$p = \frac{U}{c} \qquad \text{(complete absorption)} \qquad [21.29]$$

If the surface is a perfect reflector, then the momentum transferred in a time $\Delta t$ for normal incidence is twice that given by Equation 21.29. This is analogous to a molecule of gas bouncing off the wall of a container in a perfectly elastic collision. If the molecule is initially traveling in the positive $x$-direction at velocity $v$ and after the collision is traveling in the negative $x$-direction at velocity $-v$, its change in momentum is given by $\Delta p = mv - (-mv) = 2mv$. Light bouncing off a perfect reflector is a similar process, so for complete reflection,

$$p = \frac{2U}{c} \quad \text{(complete reflection)} \qquad [21.30]$$

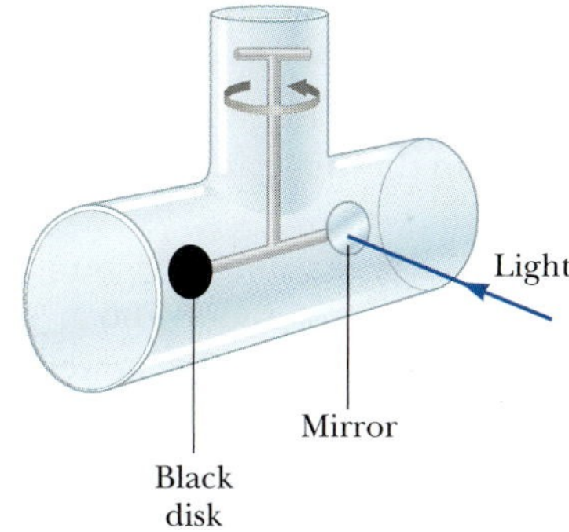


**Figure 21.21** An apparatus for measuring the radiation pressure of light. In practice, the system is contained in a high vacuum.

Although radiation pressures are very small (about $5 \times 10^{-6}$ N/m$^2$ for direct sunlight), they have been measured with a device such as the one shown in Figure 21.21. Light is allowed to strike a mirror and a black disk that are connected to each other by a horizontal bar suspended from a fine fiber. Light striking the black disk is completely absorbed, so *all* the momentum of the light is transferred to the disk. Light striking the mirror head-on is totally reflected; hence, the momentum transfer to the mirror is twice that transmitted to the disk. As a result, the horizontal bar supporting the disks twists counterclockwise as seen from above. The bar comes to equilibrium at some angle under the action of the torques caused by radiation pressure and the twisting of the fiber. The radiation pressure can be determined by measuring the angle at which equilibrium occurs. The apparatus must be placed in a high vacuum to eliminate the effects of air currents. It's interesting that similar experiments demonstrate that electromagnetic waves carry angular momentum, as well.

In summary, electromagnetic waves traveling through free space have the following properties:

◀ Some properties of electromagnetic waves

1. Electromagnetic waves travel at the speed of light.
2. Electromagnetic waves are transverse waves because the electric and magnetic fields are perpendicular to the direction of propagation of the wave and to each other.
3. The ratio of the electric field to the magnetic field in an electromagnetic wave equals the speed of light.
4. Electromagnetic waves carry both energy and momentum, which can be delivered to a surface.

## APPLYING PHYSICS 21.3 Solar System Dust

In the interplanetary space in the solar system, there is a large amount of dust. Although interplanetary dust can in theory have a variety of sizes—from molecular size upward—why are there very few dust particles smaller than about 0.2 $\mu$m in the solar system? *Hint:* The solar system originally contained dust particles of all sizes.

**EXPLANATION** Dust particles in the solar system are subject to two forces: the gravitational force toward the Sun and the force from radiation pressure, which is directed away from the Sun. The gravitational force is proportional to the cube of the radius of a spherical dust particle because it is proportional to the mass ($\rho V$) of the particle. The radiation pressure is proportional to the square of the radius because it depends on the cross-sectional area of the particle. For large particles, the gravitational force is larger than the force of radiation pressure, and the weak attraction to the Sun causes such particles to move slowly toward it. For small particles, less than about 0.2 $\mu$m, the larger force from radiation pressure sweeps them out of the solar system. ■

### Quick Quiz

**21.7** In an apparatus such as the one in Figure 21.21, suppose the black disk is replaced by one with half the radius. Which of the following are different after the disk is replaced? (a) radiation pressure on the disk (b) radiation force on the disk (c) radiation momentum delivered to the disk in a given time interval

## EXAMPLE 21.8 A Hot Tin Roof (Solar-Powered Homes)

**GOAL** Calculate some basic properties of light and relate them to thermal radiation.

**PROBLEM** Assume the Sun delivers an average power per unit area of about $1.00 \times 10^3$ W/m$^2$ to Earth's surface. **(a)** Calculate the total power incident on a flat tin roof 8.00 m by 20.0 m. Assume the radiation is incident *normal* (perpendicular) to the roof. **(b)** The tin roof reflects some light, and convection, conduction, and radiation transport the rest of the thermal energy away until some equilibrium temperature is established. If the roof is a perfect blackbody and rids itself of one-half of the incident radiation through thermal radiation, what's its equilibrium temperature? Assume the ambient temperature is 298 K.

(Example 21.8) A solar home.

**SOLUTION**

**(a)** Calculate the power delivered to the roof.

Multiply the intensity by the area to get the power:

$$P = IA = (1.00 \times 10^3 \text{ W/m}^2)(8.00 \text{ m} \times 20.0 \text{ m})$$

$$= \boxed{1.60 \times 10^5 \text{ W}}$$

**(b)** Find the equilibrium temperature of the roof.

Substitute into Stefan's law. Only one-half the incident power should be substituted, and twice the area of the roof (both the top and the underside of the roof count).

$$P = \sigma e A(T^4 - T_0^{\,4})$$

$$T^4 = T_0^{\,4} + \frac{P}{\sigma e A}$$

$$= (298 \text{ K})^4 + \frac{(0.500)(1.60 \times 10^5 \text{ W/m}^2)}{(5.67 \times 10^{-8} \text{ W/m}^2 \cdot \text{K}^4)(1)(3.20 \times 10^2 \text{ m}^2)}$$

$$T = 333 \text{ K} = \boxed{6.0 \times 10^1\,^\circ\text{C}}$$

**REMARKS** If the incident power could *all* be converted to electric power, it would be more than enough for the average home. Unfortunately, solar energy isn't easily harnessed, and the prospects for large-scale conversion are not as bright as they may appear from this simple calculation. For example, the conversion efficiency from solar to electrical energy is far less than 100%; 10–20% is typical for photovoltaic cells. Roof systems for using solar energy to raise the temperature of water with efficiencies of around 50% have been built. Other practical problems must be considered, however, such as overcast days, geographic location, and energy storage.

**QUESTION 21.8** Does the angle the roof makes with respect to the horizontal affect the amount of power absorbed by the roof? Explain.

**EXERCISE 21.8** A spherical satellite orbiting Earth is lighted on one side by the Sun, with intensity 1 340 W/m$^2$. If the radius of the satellite is 1.00 m, what power is incident upon it? *Note:* The satellite effectively intercepts radiation only over a cross section, an area equal to that of a disk, $\pi r^2$.

**ANSWERS** $4.21 \times 10^3$ W

## EXAMPLE 21.9 Clipper Ships of Space

**GOAL** Relate the intensity of light to its mechanical effect on matter.

**PROBLEM** Aluminized Mylar film is a highly reflective, lightweight material that could be used to make sails for spacecraft driven by the light of the Sun. Suppose a sail with area 1.00 km$^2$ is orbiting the Sun at a distance of $1.50 \times 10^{11}$ m. The sail has a mass of $5.00 \times 10^3$ kg and is tethered to a payload of mass $2.00 \times 10^4$ kg. **(a)** If the intensity of sunlight is $1.34 \times 10^3$ W/m$^2$ and the sail is oriented perpendicular to the incident light, what radial force is exerted on the sail? **(b)** About how long would it take to change the radial speed of the sail by 1.00 km/s? Assume the sail is perfectly reflecting. **(c)** Suppose the light were supplied by a large, powerful laser beam instead of the Sun. (Such systems have been proposed.) Calculate the peak electric and magnetic fields of the laser light.

**STRATEGY** Equation 21.30 gives the momentum imparted when light strikes an object and is totally reflected. The change in this momentum with time is a force.

For part (b), use Newton's second law to obtain the acceleration. The velocity kinematics equation then yields the necessary time to achieve the desired change in speed. Part (c) follows from Equation 21.27 and $E = Bc$.

**SOLUTION**

**(a)** Find the force exerted on the sail.

Write Equation 21.30 and substitute $U = P\Delta t = IA\,\Delta t$ for the energy delivered to the sail:

$$\Delta p = \frac{2U}{c} = \frac{2P\Delta t}{c} = \frac{2IA\Delta t}{c}$$

Divide both sides by $\Delta t$, obtaining the force $\Delta p/\Delta t$ exerted by the light on the sail:

$$F = \frac{\Delta p}{\Delta t} = \frac{2IA}{c} = \frac{2(1\,340\text{ W/m}^2)(1.00 \times 10^6\text{ m}^2)}{3.00 \times 10^8\text{ m/s}}$$

$$= \boxed{8.93\text{ N}}$$

**(b)** Find the time it takes to change the radial speed by 1.00 km/s.

Substitute the force into Newton's second law and solve for the acceleration of the sail:

$$a = \frac{F}{m} = \frac{8.93\text{ N}}{2.50 \times 10^4\text{ kg}} = 3.57 \times 10^{-4}\text{ m/s}^2$$

Apply the kinematics velocity equation:

$$v = at + v_0$$

Solve for $t$:

$$t = \frac{v - v_0}{a} = \frac{1.00 \times 10^3\text{ m/s}}{3.57 \times 10^{-4}\text{ m/s}^2} = \boxed{2.80 \times 10^6\text{ s}}$$

**(c)** Calculate the peak electric and magnetic fields if the light is supplied by a laser.

Solve Equation 21.28 for $E_{\text{max}}$:

$$I = \frac{E_{\text{max}}^2}{2\mu_0 c} \quad \rightarrow \quad E_{\text{max}} = \sqrt{2\mu_0 cI}$$

$$E_{\text{max}} = \sqrt{2(4\pi \times 10^{-7}\text{ N}\cdot\text{s}^2/\text{C}^2)(3.00 \times 10^8\text{ m/s})(1.34 \times 10^3\text{ W/m}^2)}$$

$$= \boxed{1.01 \times 10^3\text{ N/C}}$$

Obtain $B_{\text{max}}$ using $E_{\text{max}} = B_{\text{max}}c$:

$$B_{\text{max}} = \frac{E_{\text{max}}}{c} = \frac{1.01 \times 10^3\text{ N/C}}{3.00 \times 10^8\text{ m/s}} = \boxed{3.37 \times 10^{-6}\text{ T}}$$

**REMARKS** The answer to part (b) is a little over a month. While the acceleration is very low, there are no fuel costs, and within a few months the velocity can change sufficiently to allow the spacecraft to reach any planet in the solar system. Such spacecraft may be useful for certain purposes and are highly economical, but require a considerable amount of patience.

**QUESTION 21.9** By what factor will the force exerted by the Sun's light be changed when the spacecraft is twice as far from the Sun? (a) no change (b) $\frac{1}{2}$ (c) $\frac{1}{4}$ (d) $\frac{1}{8}$

**EXERCISE 21.9** A laser has a power of 22.0 W and a beam radius of 0.500 mm. (a) Find the intensity of the laser. (b) Suppose you were floating in space and pointed the laser beam away from you. What would your acceleration be? Assume your total mass, including equipment, is 72.0 kg and the force is directed through your center of mass. *Hint:* The change in momentum is the same as in the nonreflective case. (c) Calculate your acceleration if it were due to the gravity of a space station with mass $1.00 \times 10^6$ kg and center of mass 100.0 m away. (d) Calculate the peak electric and magnetic fields of the laser.

**ANSWERS** (a) $2.80 \times 10^7\text{ W/m}^2$ (b) $1.02 \times 10^{-9}\text{ m/s}^2$ (c) $6.67 \times 10^{-9}\text{ m/s}^2$ (d) $1.45 \times 10^5$ N/C, $4.84 \times 10^{-4}$ T.
**Remark** If you were planning to use your laser welding torch as a thruster to get you back to the station, don't bother, because the force of gravity is stronger. Better yet, get somebody to toss you a line.

## 21.12 The Spectrum of Electromagnetic Waves

**LEARNING OBJECTIVE**

1. Describe and discuss the electromagnetic spectrum.

All electromagnetic waves travel in a vacuum with the speed of light, $c$. These waves transport energy and momentum from some source to a receiver. In 1887 Hertz successfully generated and detected the radio-frequency electromagnetic

Wearing sunglasses lacking ultraviolet (UV) protection is worse for your eyes than wearing no sunglasses at all. Sunglasses without protection absorb some visible light, causing the pupils to dilate. This allows more UV light to enter the eye, increasing the damage to the lens of the eye over time. Without the sunglasses, the pupils constrict, reducing both visible and dangerous UV radiation. Be cool: wear sunglasses with UV protection.

waves predicted by Maxwell. Maxwell himself had recognized as electromagnetic waves both visible light and the infrared radiation discovered in 1800 by William Herschel. It is now known that other forms of electromagnetic waves exist that are distinguished by their frequencies and wavelengths.

Because all electromagnetic waves travel through free space with a speed $c$, their frequency $f$ and wavelength $\lambda$ are related by the important expression

$$c = f\lambda \qquad [21.31]$$

The various types of electromagnetic waves are presented in Figure 21.22. Notice the wide and overlapping range of frequencies and wavelengths. For instance, an AM radio wave with a frequency of 1.50 MHz (a typical value) has a wavelength of

$$\lambda = \frac{c}{f} = \frac{3.00 \times 10^8 \text{ m/s}}{1.50 \times 10^6 \text{ s}^{-1}} = 2.00 \times 10^2 \text{ m}$$

The following abbreviations are often used to designate short wavelengths and distances:

$$1 \text{ micrometer } (\mu\text{m}) = 10^{-6} \text{ m}$$

$$1 \text{ nanometer (nm)} = 10^{-9} \text{ m}$$

$$1 \text{ angstrom (Å)} = 10^{-10} \text{ m}$$

The wavelengths of visible light, for example, range from 0.4 $\mu$m to 0.7 $\mu$m, or 400 nm to 700 nm, or 4 000 Å to 7 000 Å.

**Quick Quiz**

**21.8** Which of the following statements are true about light waves? (a) The higher the frequency, the longer the wavelength. (b) The lower the frequency, the longer the wavelength. (c) Higher-frequency light travels faster than lower-frequency light. (d) The shorter the wavelength, the higher the frequency. (e) The lower the frequency, the shorter the wavelength.

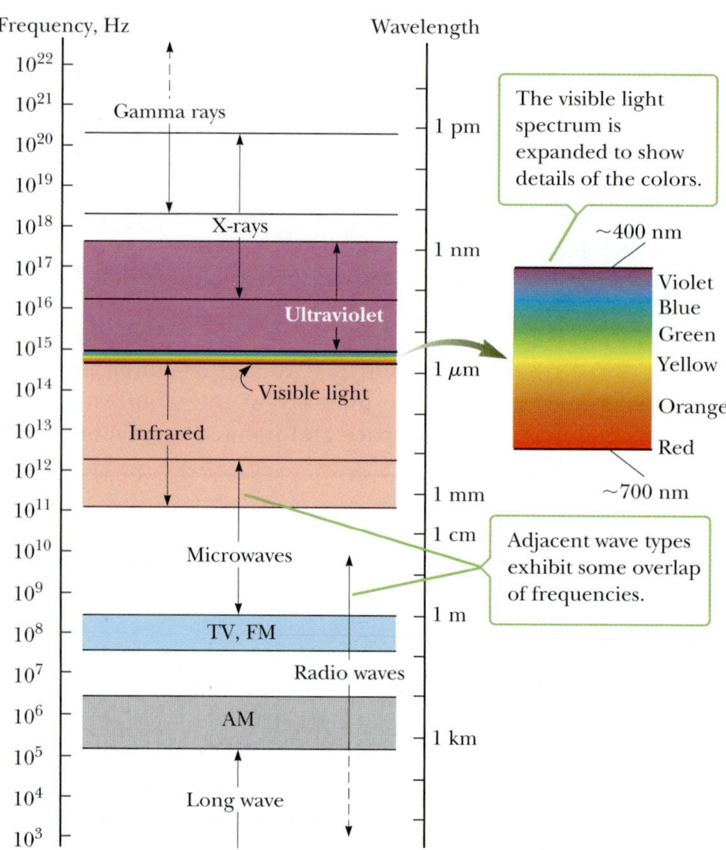


**Figure 21.22** The electromagnetic spectrum.

Brief descriptions of the wave types follow, in order of decreasing wavelength. There is no sharp division between one kind of electromagnetic wave and the next. All forms of electromagnetic radiation are produced by accelerating charges.

**Radio waves,** which were discussed in Section 21.10, are the result of charges accelerating through conducting wires. They are, of course, used in radio and television communication systems.

**Microwaves** (short-wavelength radio waves) have wavelengths ranging between about 1 mm and 30 cm and are generated by electronic devices. Their short wavelengths make them well suited for the radar systems used in aircraft navigation and for the study of atomic and molecular properties of matter. Microwave ovens are an interesting domestic application of these waves. It has been suggested that solar energy might be harnessed by beaming microwaves to Earth from a solar collector in space.

**Infrared waves** (sometimes incorrectly called "heat waves"), produced by hot objects and molecules, have wavelengths ranging from about 1 mm to the longest wavelength of visible light, $7 \times 10^{-7}$ m. They are readily absorbed by most materials. The infrared energy absorbed by a substance causes it to get warmer because the energy agitates the atoms of the object, increasing their vibrational or translational motion. The result is a rise in temperature. Infrared radiation has many practical and scientific applications, including physical therapy, infrared photography, and the study of the vibrations of atoms.

**Visible light,** the most familiar form of electromagnetic waves, may be defined as the part of the spectrum that is detected by the human eye. Light is produced by the rearrangement of electrons in atoms and molecules. The wavelengths of visible light are classified as colors ranging from violet ($\lambda \approx 4 \times 10^{-7}$ m) to red ($\lambda \approx 7 \times 10^{-7}$ m). The eye's sensitivity is a function of wavelength and is greatest at a wavelength of about $5.6 \times 10^{-7}$ m (yellow green).

**Ultraviolet (UV) light** covers wavelengths ranging from about $4 \times 10^{-7}$ m (400 nm) down to $6 \times 10^{-10}$ m (0.6 nm). The Sun is an important source of ultraviolet light (which is the main cause of suntans). Most of the ultraviolet light from the Sun is absorbed by atoms in the upper atmosphere, or stratosphere, which is fortunate, because UV light in large quantities has harmful effects on humans. One important constituent of the stratosphere is ozone ($O_3$), produced from reactions of oxygen with ultraviolet radiation. The resulting ozone shield causes lethal high-energy ultraviolet radiation to warm the stratosphere.

**X-rays** are electromagnetic waves with wavelengths from about $10^{-8}$ m (10 nm) down to $10^{-13}$ m ($10^{-4}$ nm). The most common source of x-rays is the acceleration of high-energy electrons bombarding a metal target. X-rays are used as a diagnostic tool in medicine and as a treatment for certain forms of cancer. Because x-rays easily penetrate and damage or destroy living tissues and organisms, care must be taken to avoid unnecessary exposure and overexposure.

**Gamma rays**—electromagnetic waves emitted by radioactive nuclei—have wavelengths ranging from about $10^{-10}$ m to less than $10^{-14}$ m. They are highly penetrating and cause serious damage when absorbed by living tissues. Accordingly, those working near such radiation must be protected by garments containing heavily absorbing materials, such as layers of lead.

When astronomers observe the same celestial object using detectors sensitive to different regions of the electromagnetic spectrum, striking variations in the object's features can be seen. Figure 21.23 (page 760) shows images of the Crab Nebula made in three different wavelength ranges. The Crab Nebula is the remnant of a supernova explosion that was seen on Earth in 1054 A.D. (Compare with Fig. 8.31.)

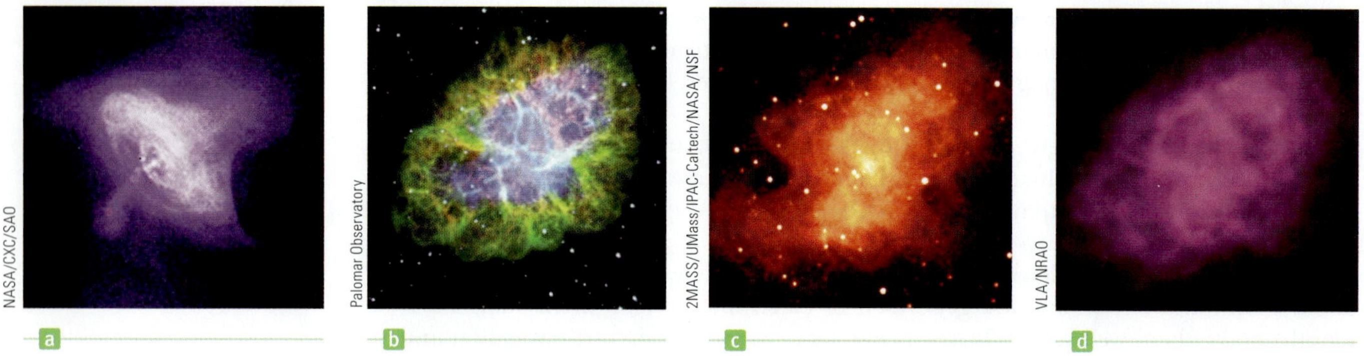


**Figure 21.23** Observations in different parts of the electromagnetic spectrum show different features of the Crab Nebula. (a) X-ray image. (b) Optical image. (c) Infrared. (d) Radio image.

## APPLYING PHYSICS 21.4 Light and Wound Treatment BIO

An important issue in human health is wound management. Chronic wounds affect five million to seven million people in the United States at an annual cost of more than twenty billion dollars. Low-level laser therapy has been shown to facilitate the healing and closure of wounds.

Infrared light increases the generation of adenosine triphosphate (ATP) in mitochondria and may stimulate the activation of genes and enzymes associated with cellular respiration. (ATP molecules provide energy for a variety of important cell functions.) Infrared light may also increase the concentration of reactive oxygen molecules, which could increase communication between the nucleus, cytosol, and the mitochondria. This mechanism may enhance and accelerate the healing process.

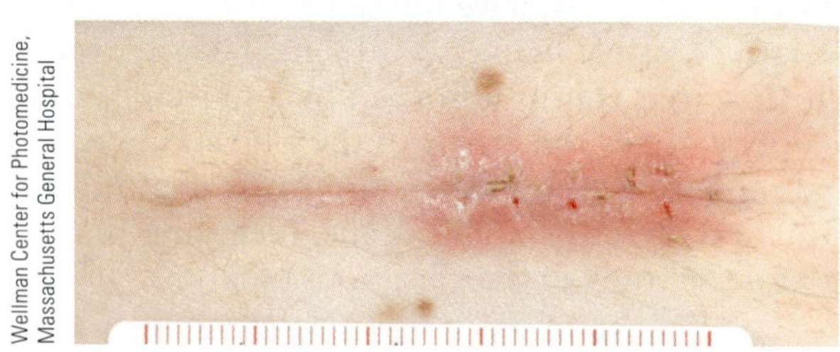


**Figure 21.24** After bringing the sides of this wound together with deep sutures, closure on the left side was achieved with light-activated technology, whereas on the right side closure was carried out with sutures. This photo, taken at the end of two weeks, shows that healing was enhanced with light activation.

Green laser light can also be used to stimulate the body's repair mechanisms via a different process. A pink dye, called "rose bengal," is applied to the tissue, which is then exposed to the laser light for a few minutes. When the dye absorbs the light, it causes cross-linkages between collagen molecules in the tissue. The cross-linked molecules promote the closing of the tissue while reducing or eliminating the formation of scar tissue. Figure 21.24 shows the contrast between tissue receiving the normal treatment and tissue irradiated by lasers. The technique is also being studied for application to damaged peripheral nerves, blood vessels, and other tissues, such as incisions made in the cornea during eye surgery. ■

## APPLYING PHYSICS 21.5 The Sun and the Evolution of the Eye BIO

The center of sensitivity of our eyes coincides with the center of the wavelength distribution of the Sun. Is this an amazing coincidence?

**EXPLANATION** This fact is not a coincidence; rather, it's the result of biological evolution. Humans have evolved with vision most sensitive to wavelengths that are strongest from the Sun. If aliens from another planet ever arrived at Earth, their eyes would have the center of sensitivity at wavelengths different from ours. If their sun were a red dwarf, for example, the alien's eyes would be most sensitive to red light. ■

# 21.13 The Doppler Effect for Electromagnetic Waves

### LEARNING OBJECTIVE

1. Describe the Doppler effect for electromagnetic waves and discuss its role in scientific discoveries.

As we saw in Section 14.6, sound waves exhibit the Doppler effect when the observer, the source, or both are moving relative to the medium of propagation.

Recall that in the Doppler effect, the observed frequency of the wave is larger or smaller than the frequency emitted by the source of the wave.

A Doppler effect also occurs for electromagnetic waves, but it differs from the Doppler effect for sound waves in two ways. First, in the Doppler effect for sound waves, motion relative to the medium is most important because sound waves require a medium in which to propagate. In contrast, the medium of propagation plays no role in the Doppler effect for electromagnetic waves because the waves require no medium in which to propagate. Second, the speed of sound that appears in the equation for the Doppler effect for sound depends on the reference frame in which it is measured. In contrast, as we'll see in Chapter 26, the speed of electromagnetic waves has the same value in all coordinate systems that are either at rest or moving at constant velocity with respect to one another.

The single equation that describes the Doppler effect for electromagnetic waves is given by the approximate expression

$$f_O \approx f_S\left(1 \pm \frac{u}{c}\right) \qquad \text{if } u \ll c \qquad [21.32]$$

where $f_O$ is the observed frequency, $f_S$ is the frequency emitted by the source, $u$ is the *relative* speed of the observer and source, and $c$ is the speed of light in a vacuum. Note that Equation 21.32 is valid only if $u$ is much smaller than $c$. Further, it can also be used for sound as long as the relative velocity of the source and observer is much less than the velocity of sound. The positive sign in the equation must be used when the source and observer are moving toward each other, whereas the negative sign must be used when they are moving away from each other. Thus, we anticipate an increase in the observed frequency if the source and observer are approaching each other and a decrease if the source and observer recede from each other.

Astronomers have made important discoveries using Doppler observations on light reaching Earth from distant galaxies. Such measurements have shown that the more distant a galaxy is from Earth, the more its light is shifted toward the red end of the spectrum. This *cosmological red shift* is evidence that the Universe is expanding. The stretching and expanding of space, like a rubber sheet being pulled in all directions, is consistent with Einstein's theory of general relativity. A given star or galaxy, however, can have a peculiar motion toward or away from Earth. For example, Doppler effect measurements made with the Hubble Space Telescope have shown that a galaxy labeled M87 is rotating, with one edge moving toward us and the other moving away. Its measured speed of rotation was used to identify a supermassive black hole located at its center.

## SUMMARY

### 21.1 Resistors in an AC Circuit

If an AC circuit consists of a generator and a resistor, the current in the circuit is in phase with the voltage, which means that the current and voltage reach their maximum values at the same time.

In discussions of voltages and currents in AC circuits, **rms values** of voltages are usually used. One reason is that AC ammeters and voltmeters are designed to read rms values. The rms values of currents and voltages ($I_{\text{rms}}$ and $\Delta V_{\text{rms}}$) are related to the maximum values of these quantities ($I_{\text{max}}$ and $\Delta V_{\text{max}}$) as follows:

$$I_{\text{rms}} = \frac{I_{\text{max}}}{\sqrt{2}} \quad \text{and} \quad \Delta V_{\text{rms}} = \frac{\Delta V_{\text{max}}}{\sqrt{2}} \qquad [21.2, 21.3]$$

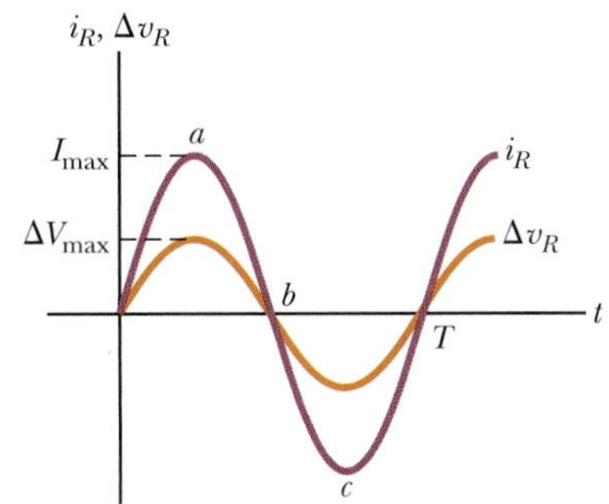


The voltage across a resistor and the current are in phase: they simultaneously reach their maximum values, their minimum values, and their zero values.

The rms voltage across a resistor is related to the rms current in the resistor by **Ohm's law**:

$$\Delta V_{R,\text{rms}} = I_{\text{rms}}R \qquad [21.4a]$$

## 21.2 Capacitors in an AC Circuit

If an AC circuit consists of a generator and a capacitor, the voltage lags behind the current by 90°. This means that the voltage reaches its maximum value one-quarter of a period after the current reaches its maximum value.

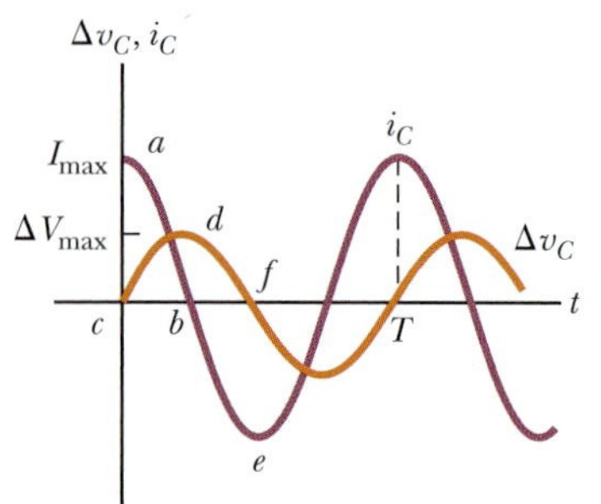


The voltage across a capacitor reaches its maximum value 90° after the current reaches its maximum value, so the voltage "lags" the current.

The impeding effect of a capacitor on current in an AC circuit is given by the **capacitive reactance** $X_C$, defined as

$$X_C \equiv \frac{1}{2\pi f C} \qquad [21.5]$$

where $f$ is the frequency of the AC voltage source.

The rms voltage across and the rms current in a capacitor are related by

$$\Delta V_{C,\text{rms}} = I_{\text{rms}} X_C \qquad [21.6]$$

## 21.3 Inductors in an AC Circuit

If an AC circuit consists of a generator and an inductor, the voltage leads the current by 90°. This means the voltage reaches its maximum value one-quarter of a period before the current reaches its maximum value.

The effective impedance of a coil in an AC circuit is measured by a quantity called the **inductive reactance** $X_L$, defined as

$$X_L \equiv 2\pi f L \qquad [21.8]$$

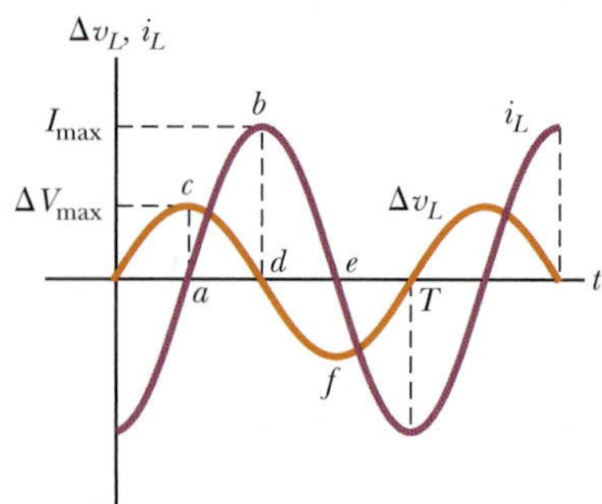


The voltage across an inductor reaches its maximum value 90° before the current reaches its maximum value, so the voltage "leads" the current.

The rms voltage across a coil is related to the rms current in the coil by

$$\Delta V_{L,\text{rms}} = I_{\text{rms}} X_L \qquad [21.9]$$

## 21.4 The *RLC* Series Circuit

In an *RLC* series AC circuit, the maximum applied voltage $\Delta V$ is related to the maximum voltages across the resistor ($\Delta V_R$), capacitor ($\Delta V_C$), and inductor ($\Delta V_L$) by

$$\Delta V_{\text{max}} = \sqrt{\Delta V_R{}^2 + (\Delta V_L - \Delta V_C)^2} \qquad [21.10]$$

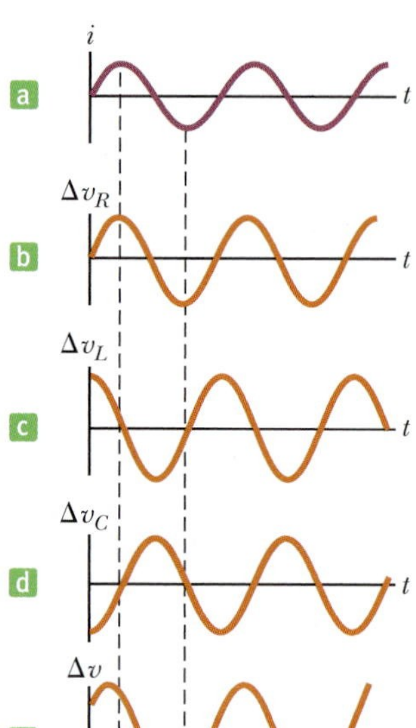


The time dependence of voltage differences of different circuit elements of an *RLC* series circuit are shown in these graphs. Notice that $\Delta v_R$ is in phase with the current, $\Delta v_L$ leads the current, and $\Delta v_C$ lags the current.

If an AC circuit contains a resistor, an inductor, and a capacitor connected in series, the limit they place on the current is given by the **impedance Z** of the circuit, defined as

$$Z \equiv \sqrt{R^2 + (X_L - X_C)^2} \qquad [21.13]$$

The relationship between the maximum voltage supplied to an *RLC* series AC circuit and the maximum current in the circuit, which is the same in every element, is

$$\Delta V_{\text{max}} = I_{\text{max}} Z \qquad [21.14]$$

In an *RLC* series AC circuit, the applied rms voltage and current are out of phase. The **phase angle** $\phi$ between the current and voltage is given by

$$\tan \phi = \frac{X_L - X_C}{R} \qquad [21.15]$$

## 21.5 Power in an AC Circuit

The **average power** delivered by the voltage source in an *RLC* series AC circuit is

$$P_{\text{av}} = I_{\text{rms}} \Delta V_{\text{rms}} \cos \phi \qquad [21.17]$$

where the constant $\cos \phi$ is called the **power factor**.

## 21.6 Resonance in a Series *RLC* Circuit

In general, the rms current in a series *RLC* circuit can be written

$$I_{\text{rms}} = \frac{\Delta V_{\text{rms}}}{Z} = \frac{\Delta V_{\text{rms}}}{\sqrt{R^2 + (X_L - X_C)^2}} \qquad [21.18]$$

The current has its *maximum* value when the impedance has its *minimum* value, corresponding to $X_L = X_C$ and $Z = R$. The frequency $f_0$ at which this happens is called the **resonance frequency** of the circuit, given by

$$f_0 = \frac{1}{2\pi\sqrt{LC}} \quad [21.19]$$

## 21.7 The Transformer

If the primary winding of a transformer has $N_1$ turns and the secondary winding consists of $N_2$ turns and then an input AC voltage $\Delta V_1$ is applied to the primary, the induced voltage in the secondary winding is given by

$$\Delta V_2 = \frac{N_2}{N_1}\Delta V_1 \quad [21.22]$$

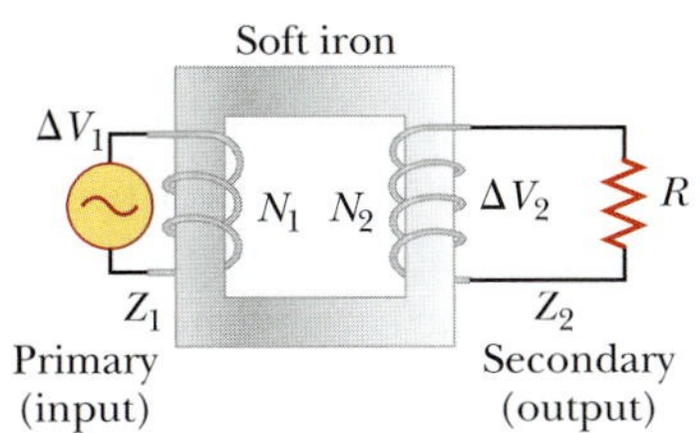


An AC voltage $\Delta V_1$ is applied to the primary coil, and the output voltage $\Delta V_2$ is observed across the load resistance $R$.

When $N_2$ is greater than $N_1$, $\Delta V_2$ exceeds $\Delta V_1$ and the transformer is referred to as a *step-up transformer*. When $N_2$ is less than $N_1$, making $\Delta V_2$ less than $\Delta V_1$, we have a *step-down transformer*. In an ideal transformer, the power output equals the power input.

$$I_1\,\Delta V_1 = I_2\,\Delta V_2 \quad [21.23]$$

## 21.8–21.13 Electromagnetic Waves and Their Properties

**Electromagnetic waves** were predicted by James Clerk Maxwell and experimentally confirmed by Heinrich Hertz. These waves are created by accelerating electric charges and have the following properties:

1. Electromagnetic waves are transverse waves because the electric and magnetic fields are perpendicular to the direction of propagation of the waves.
2. Electromagnetic waves travel at the speed of light.
3. The ratio of the electric field to the magnetic field at a given point in an electromagnetic wave equals the speed of light:

$$\frac{E}{B} = c \quad [21.26]$$

4. Electromagnetic waves carry energy as they travel through space. The average power per unit area is the intensity $I$, given by

$$I = \frac{E_{max}B_{max}}{2\mu_0} = \frac{E_{max}^2}{2\mu_0 c} = \frac{c}{2\mu_0}B_{max}^2 \quad [21.27, 21.28]$$

where $E_{max}$ and $B_{max}$ are the maximum values of the electric and magnetic fields.

5. Electromagnetic waves transport linear and angular momentum as well as energy. The momentum $p$ delivered in time $\Delta t$ at normal incidence to an object that completely absorbs light energy $U$ is given by

$$p = \frac{U}{c} \quad \text{(complete absorption)} \quad [21.29]$$

If the surface is a perfect reflector, the momentum delivered in time $\Delta t$ at normal incidence is twice that given by Equation 21.29:

$$p = \frac{2U}{c} \quad \text{(complete reflection)} \quad [21.30]$$

6. The speed $c$, frequency $f$, and wavelength $\lambda$ of an electromagnetic wave are related by

$$c = f\lambda \quad [21.31]$$

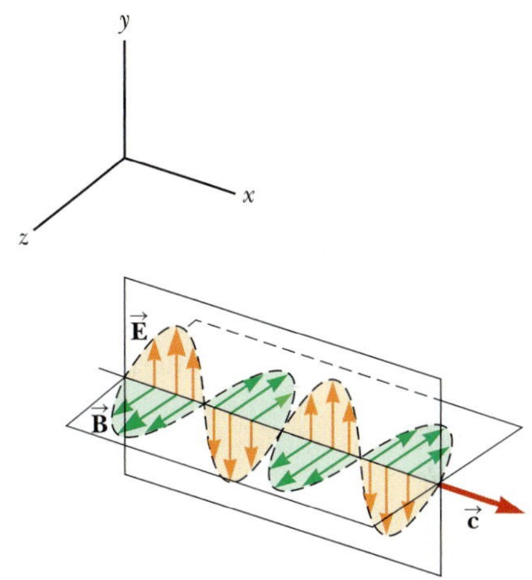


An electromagnetic wave sent out by oscillating charges in an antenna, represented at one instant of time and far from the antenna, moving in the positive $x$-direction with speed $c$.

The **electromagnetic spectrum** includes waves covering a broad range of frequencies and wavelengths. These waves have a variety of applications and characteristics, depending on their frequencies or wavelengths. The frequency of a given wave can be shifted by the relative velocity of observer and source, with the observed frequency $f_O$ given by

$$f_O \approx f_S\left(1 \pm \frac{u}{c}\right) \quad \text{if } u \ll c \quad [21.32]$$

where $f_S$ is the frequency of the source, $u$ is the *relative* speed of the observer and source, and $c$ is the speed of light in a vacuum. The positive sign is used when the source and observer approach each other, the negative sign when they recede from each other.

## WARM-UP EXERCISES

ENHANCED WebAssign The warm-up exercises in this chapter may be assigned online in Enhanced WebAssign.

1. **Math Review** The Cartesian components of a vector are (1.50 m, 2.50 m). Find (a) the magnitude and (b) the polar coordinate angle of the vector. (See Sections 1.7 and 1.8.)

2. **Physics Review** A potential difference of 12.0 V is measured across a $1.10 \times 10^3$ Ω resistor. Determine (a) the current through the resistor and (b) the power dissipated by the resistor. (See Sections 17.4 and 17.6.)

3. **Physics Review** Determine the frequency of a sound wave if it has a speed is 343 m/s and a wavelength of 1.50 m.

4. An rms potential difference of $1.20 \times 10^2$ V is measured across an AC circuit element where the rms current is 8.50 A. Determine the maximum (a) current through and (b) potential difference across the element. (See Section 21.1.)

5. An AC power source has an rms voltage of $1.20 \times 10^2$ V and operates at a frequency of 60.0 Hz. If a purely capacitive circuit is made from the power source and a 0.470 $\mu$F capacitor, determine (a) the capacitive reactance and (b) the rms current through the capacitor. (See Section 21.2.)

6. An AC power source has an rms voltage of $1.20 \times 10^2$ V and operates at a frequency of 60.0 Hz. If a purely inductive circuit is made from the power source and a 47.0 H inductor, determine (a) the inductive reactance and (b) the rms current through the inductor. (See Section 21.3.)

7. An AC power source has an rms voltage of $1.20 \times 10^2$ V and operates at a frequency of 60.0 Hz. If a series *RLC* circuit is made from the power source, a 0.850 $\mu$F capacitor, a 13.0 H inductor and a $1.50 \times 10^3$ Ω resistor, determine (a) the capacitive reactance, (b) the inductive reactance, (c) the circuit impedance, and (d) the maximum current in the circuit. (See Section 21.4.)

8. A series *RLC* circuit has an inductive reactance of $5.50 \times 10^2$ Ω, a capacitive reactance of $2.75 \times 10^2$ Ω, and an equivalent resistance of $1.70 \times 10^2$ Ω. Determine (a) the circuit impedance, (b) the phase angle between the current and the voltage, and (c) the power factor of the circuit. (See Sections 21.4 and 21.5.)

9. An analog receiver tunes in a radio station when the circuit's *RLC* resonance frequency matches the frequency transmitted by the station. For a receiver tuned to $6.90 \times 10^2$ kHz with an equivalent inductance of 10.0 $\mu$H, determine the equivalent capacitance. (See Section 21.6.)

10. An AC transformer powering a neon sign has potential differences of $1.20 \times 10^2$ V and $9.00 \times 10^3$ V across the primary coil and high-voltage secondary coils, respectively. If the current through the secondary coil is 30.0 mA, determine the current through the primary coil. (See Section 21.7.)

11. Suppose the Sun delivers an average power of $1.00 \times 10^3$ W/m$^2$ to Earth's surface. Determine the average pressure on a perfectly reflecting surface pointed directly at the Sun. (See Section 21.11.)

## CONCEPTUAL QUESTIONS

ENHANCED WebAssign The conceptual questions in this chapter may be assigned online in Enhanced WebAssign.

1. Despite the advent of digital television, some viewers still use "rabbit ears" atop their sets (Fig. CQ21.1) instead of purchasing cable television service or satellite dishes. Certain orientations of the receiving antenna on a television set give better reception than others. Furthermore, the best orientation varies from station to station. Explain.

**Figure CQ21.1**

2. (a) Does the phase angle in an *RLC* series circuit depend on frequency? (b) What is the phase angle for the circuit when the inductive reactance equals the capacitive reactance?

3. If the fundamental source of a sound wave is a vibrating object, what is the fundamental source of an electromagnetic wave?

4. Receiving radio antennas can be in the form of conducting lines or loops. What should the orientation of each of these antennas be relative to a broadcasting antenna that is vertical?

5. In radio transmission a radio wave serves as a carrier wave, and the sound signal is superimposed on the carrier wave. In amplitude modulation (AM) radio, the amplitude of the carrier wave varies according to the sound wave. The U.S. Navy sometimes uses flashing lights to send Morse code between neighboring ships, a process that has similarities to radio broadcasting.

(a) Is this process AM or FM? (b) What is the carrier frequency? (c) What is the signal frequency? (d) What is the broadcasting antenna? (e) What is the receiving antenna?

6. When light (or other electromagnetic radiation) travels across a given region, (a) what is it that oscillates? (b) What is it that is transported?

7. In space sailing, which is a proposed alternative for transport to the planets, a spacecraft carries a very large sail. Sunlight striking the sail exerts a force, accelerating the spacecraft. Should the sail be absorptive or reflective to be most effective?

8. What does a radio wave do to the charges in the receiving antenna to provide a signal for your car radio?

9. Does a wire connected to a battery emit an electromagnetic wave?

10. Suppose a creature from another planet had eyes that were sensitive to infrared radiation. Describe what it would see if it looked around the room that you are now in. That is, what would be bright and what would be dim?

11. Why should an infrared photograph of a person look different from a photograph taken using visible light?

12. If a high-frequency current is passed through a solenoid containing a metallic core, the core becomes warm due to induction. Explain why the temperature of the material rises in this situation.

13. What is the advantage of transmitting power at high voltages?

14. Why is the sum of the maximum voltages across each of the elements in a series *RLC* circuit usually greater than the maximum applied voltage? Doesn't this violate Kirchhoff's loop rule?

15. If the resistance in an *RLC* circuit remains the same, but the capacitance and inductance are each doubled, how will the resonance frequency change?

16. An inductor and a resistor are connected in series across an AC generator, as shown in Figure CQ21.16. Immediately after the switch is closed, which of the following statements is true? (a) The current is $\Delta V/R$. (b) The voltage across the inductor is zero. (c) The current in the circuit is zero. (d) The voltage across the resistor is $\Delta V$. (e) The voltage across the inductor is half its maximum value.

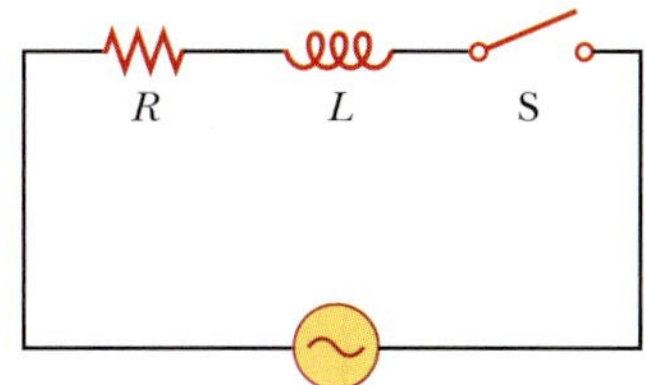


**Figure CQ21.16**

17. A capacitor and a resistor are connected in series across an AC generator, as shown in Figure CQ21.17. After the switch is closed, which of the following statements is true? (a) The voltage across the capacitor lags the current by 90°. (b) The voltage across the resistor is out of phase with the current. (c) The voltage across the capacitor leads the current by 90°. (d) The current decreases as the frequency of the generator is increased, but its peak voltage remains the same. (e) None of these

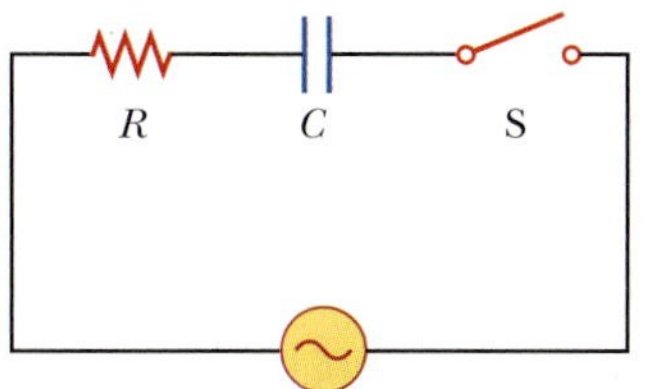


**Figure CQ21.17**

18. What is the impedance of a series RLC circuit at resonance? (a) $X_L$ (b) $X_C$ (c) $R$ (d) $X_L - X_C$ (e) 0

19. Which of the following statements are true regarding electromagnetic waves traveling through a vacuum? More than one statement may be correct. (a) All waves have the same wavelength. (b) All waves have the same frequency. (c) All waves travel at $3.00 \times 10^8$ m/s. (d) The electric and magnetic fields associated with the waves are perpendicular to each other and to the direction of wave propagation. (e) The speed of the waves depends on their frequency.

## PROBLEMS

ENHANCED WebAssign The problems in this chapter may be assigned online in Enhanced WebAssign.

**1.** denotes straightforward problem; **2.** denotes intermediate problem; **3.** denotes challenging problem

**1.** (boxed) denotes full solution available in *Student Solutions Manual/ Study Guide*

**1.** (shaded) denotes problems most often assigned in Enhanced WebAssign

BIO denotes biomedical problems

GP denotes guided problems

M denotes Master It tutorial available in Enhanced WebAssign

Q|C denotes asking for quantitative and conceptual reasoning

S denotes symbolic reasoning problem

W denotes Watch It video solution available in Enhanced WebAssign

### 21.1 Resistors in an AC Circuit

1. (a) What is the resistance of a lightbulb that uses an average power of 75.0 W when connected to a 60.0-Hz power source having a maximum voltage of 170 V? (b) What is the resistance of a 100-W lightbulb?

2. **Q|C** A certain lightbulb is rated at 60.0 W when operating at an rms voltage of 120 V. (a) What is the peak voltage applied across the bulb? (b) What is the resistance of the bulb? (c) Does a 100-W bulb have greater or less resistance than a 60.0-W bulb? Explain.

3. The current in the circuit shown in Figure P21.3 equals 60.0% of the peak current at $t$ = 7.00 ms. What is the lowest source frequency that gives this current?

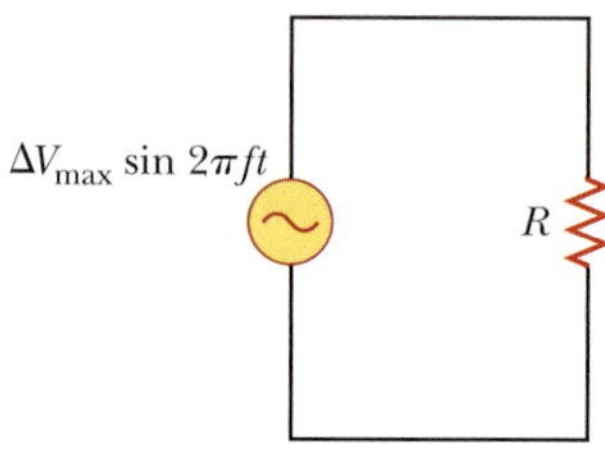


**Figure P21.3**

4. Figure P21.4 shows three lamps connected to a 120-V AC (rms) household supply voltage. Lamps 1 and 2 have 150-W bulbs; lamp 3 has a 100-W bulb. For each bulb, find (a) the rms current and (b) the resistance.

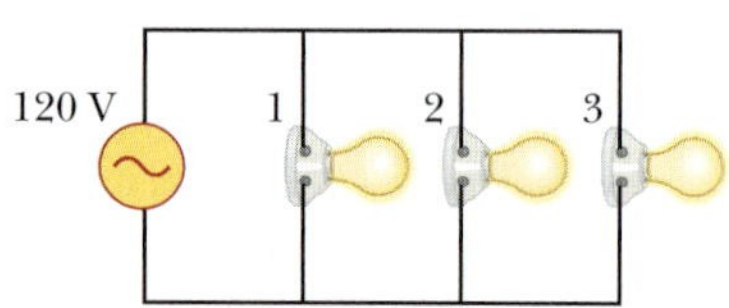


**Figure P21.4**

5. An audio amplifier, represented by the AC source and the resistor $R$ in Figure P21.5, delivers alternating voltages at audio frequencies to the speaker. If the source puts out an alternating voltage of 15.0 V (rms), the resistance $R$ is 8.20 Ω, and the speaker is equivalent to a resistance of 10.4 Ω, what is the time-averaged power delivered to the speaker?

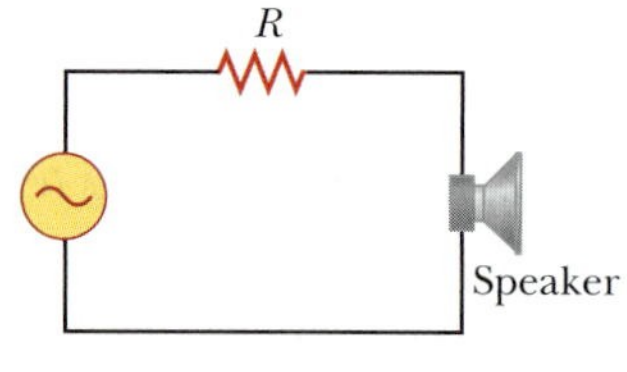


**Figure P21.5**

6. **GP** The output voltage of an AC generator is given by $\Delta v$ = (170 V) sin $(60\pi t)$. The generator is connected across a 20.0-Ω resistor. By inspection, what are the (a) maximum voltage and (b) frequency? Find the (c) rms voltage across the resistor, (d) rms current in the resistor, (e) maximum current in the resistor, (f) power delivered to the resistor, and (g) current when $t$ = 0.005 0 s. (h) Should the argument of the sine function be in degrees or radians?

## 21.2 Capacitors in an AC Circuit

7. (a) For what frequencies does a 22.0-μF capacitor have a reactance below 175 Ω? (b) What is the reactance of a 44.0-μF capacitor over this same frequency range?

8. What is the maximum current delivered to a circuit containing a 2.20-μF capacitor when it is connected across (a) a North American outlet having $\Delta V_{rms}$ = 120 V and $f$ = 60.0 Hz and (b) a European outlet having $\Delta V_{rms}$ = 240 V and $f$ = 50.0 Hz?

9. **W** When a 4.0-μF capacitor is connected to a generator whose rms output is 30 V, the current in the circuit is observed to be 0.30 A. What is the frequency of the source?

10. **Q|C** An AC generator with an output rms voltage of 36.0 V at a frequency of 60.0 Hz is connected across a 12.0-μF capacitor. Find the (a) capacitive reactance, (b) rms current, and (c) maximum current in the circuit. (d) Does the capacitor have its maximum charge when the current takes its maximum value? Explain.

11. **M** What maximum current is delivered by an AC source with $\Delta V_{max}$ = 48.0 V and $f$ = 90.0 Hz when connected across a 3.70-μF capacitor?

12. A generator delivers an AC voltage of the form $\Delta v$ = (98.0 V) sin $(80\pi t)$ to a capacitor. The maximum current in the circuit is 0.500 A. Find the (a) rms voltage of the generator, (b) frequency of the generator, (c) rms current, (d) reactance, and (e) value of the capacitance.

## 21.3 Inductors in an AC Circuit

13. An inductor has a 54.0-Ω reactance when connected to a 60.0-Hz source. The inductor is removed and then connected to a 50.0-Hz source that produces a 100-V rms voltage. What is the maximum current in the inductor?

14. An AC generator has an output rms voltage of 78.0 V at a frequency of 80.0 Hz. If the generator is connected across a 25.0-mH inductor, find the (a) inductive reactance, (b) rms current, and (c) maximum current in the circuit.

15. **M** In a purely inductive AC circuit as shown in Figure P21.15, $\Delta V_{max}$ = 100 V. (a) The maximum current is 7.50 A at 50.0 Hz. Calculate the inductance $L$. (b) At what angular frequency $\omega$ is the maximum current 2.50 A?

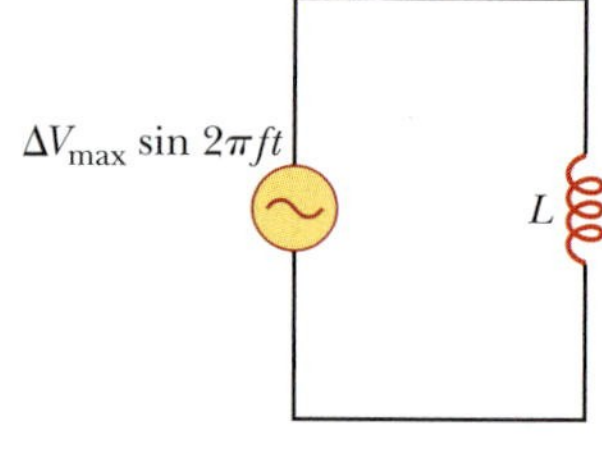


**Figure P21.15**

16. **GP** The output voltage of an AC generator is given by $\Delta v$ = $(1.20 \times 10^2$ V) sin $(30\pi t)$. The generator is connected across a 0.500-H inductor. Find the (a) frequency of the generator, (b) rms voltage across the inductor, (c) inductive reactance, (d) rms current in the inductor, (e) maximum current in the inductor, and (f) average power delivered to the inductor. (g) Find an expression for the instantaneous current. (h) At what time after $t$ = 0 does the instantaneous current first reach 1.00 A? (Use the inverse sine function.)

17. Determine the maximum magnetic flux through an inductor connected to a standard outlet ($\Delta V_{rms}$ = 120 V, $f$ = 60.0 Hz).

## 21.4 The *RLC* Series Circuit

18. A sinusoidal voltage $\Delta v = (80.0 \text{ V}) \sin(150t)$ is applied to a series *RLC* circuit with $L = 80.0$ mH, $C = 125.0$ $\mu$F, and $R = 40.0$ $\Omega$. (a) What is the impedance of the circuit? (b) What is the maximum current in the circuit?

19. A 40.0-$\mu$F capacitor is connected to a 50.0-$\Omega$ resistor and a generator whose rms output is 30.0 V at 60.0 Hz. Find (a) the rms current in the circuit, (b) the rms voltage drop across the resistor, (c) the rms voltage drop across the capacitor, and (d) the phase angle for the circuit.

20. M An inductor ($L$ = 400 mH), a capacitor ($C$ = 4.43 $\mu$F), and a resistor ($R$ = 500 $\Omega$) are connected in series. A 50.0-Hz AC generator connected in series to these elements produces a maximum current of 250 mA in the circuit. (a) Calculate the required maximum voltage $\Delta V_{max}$. (b) Determine the phase angle by which the current leads or lags the applied voltage.

21. A resistor ($R = 9.00 \times 10^2$ $\Omega$), a capacitor ($C$ = 0.250 $\mu$F), and an inductor ($L$ = 2.50 H) are connected in series across a $2.40 \times 10^2$-Hz AC source for which $\Delta V_{max} = 1.40 \times 10^2$ V. Calculate (a) the impedance of the circuit, (b) the maximum current delivered by the source, and (c) the phase angle between the current and voltage. (d) Is the current leading or lagging the voltage?

22. A 50.0-$\Omega$ resistor, a 0.100-H inductor, and a 10.0-$\mu$F capacitor are connected in series to a 60.0-Hz source. The rms current in the circuit is 2.75 A. Find the rms voltages across (a) the resistor, (b) the inductor, (c) the capacitor, and (d) the *RLC* combination. (e) Sketch the phasor diagram for this circuit.

23. An *RLC* circuit consists of a 150-$\Omega$ resistor, a 21.0-$\mu$F capacitor, and a 460-mH inductor connected in series with a 120-V, 60.0-Hz power supply. (a) What is the phase angle between the current and the applied voltage? (b) Which reaches its maximum earlier, the current or the voltage?

24. An AC source operating at 60 Hz with a maximum voltage of 170 V is connected in series with a resistor ($R$ = 1.2 k$\Omega$) and an inductor ($L$ = 2.8 H). (a) What is the maximum value of the current in the circuit? (b) What are the maximum values of the potential difference across the resistor and the inductor? (c) When the current is at a maximum, what are the magnitudes of the potential differences across the resistor, the inductor, and the AC source? (d) When the current is zero, what are the magnitudes of the potential difference across the resistor, the inductor, and the AC source?

25. A person is working near the secondary of a transformer, as shown in Figure P21.25. The primary voltage is 120 V (rms) at 60.0 Hz. The capacitance $C_s$, which is the stray capacitance between the hand and the secondary winding, is 20.0 pF. Assuming the person has a body resistance to ground of $R_b$ = 50.0 k$\Omega$, determine the rms voltage across the body. *Hint:* Redraw the circuit with the secondary of the transformer as a simple AC source.

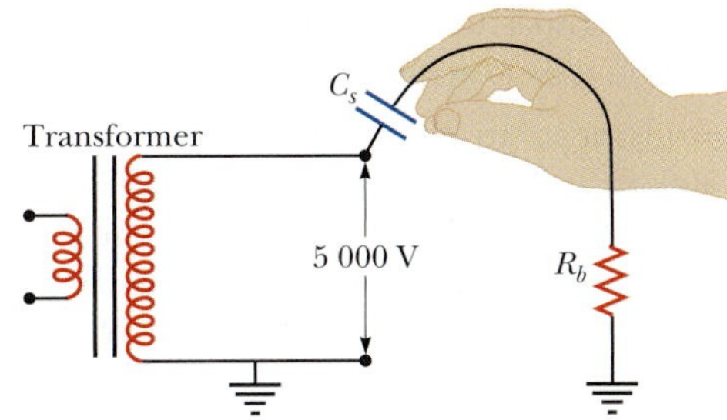


**Figure P21.25**

26. Q|C A 60.0-$\Omega$ resistor is connected in series with a 30.0-$\mu$F capacitor and a generator having a maximum voltage of $1.20 \times 10^2$ V and operating at 60.0 Hz. Find the (a) capacitive reactance of the circuit, (b) impedance of the circuit, and (c) maximum current in the circuit. (d) Does the voltage lead or lag the current? (e) How will putting an inductor in series with the existing capacitor and resistor affect the current? Explain.

27. GP A series AC circuit contains a resistor, an inductor of 150 mH, a capacitor of 5.00 $\mu$F, and a generator with $\Delta V_{max}$ = 240 V operating at 50.0 Hz. The maximum current in the circuit is 100 mA. Calculate (a) the inductive reactance, (b) the capacitive reactance, (c) the impedance, (d) the resistance in the circuit, and (e) the phase angle between the current and the generator voltage.

28. At what frequency does the inductive reactance of a 57.0-$\mu$H inductor equal the capacitive reactance of a 57.0-$\mu$F capacitor?

29. W An AC source with a maximum voltage of 150 V and $f$ = 50.0 Hz is connected between points $a$ and $d$ in Figure P21.29. Calculate the rms voltages between points (a) $a$ and $b$, (b) $b$ and $c$, (c) $c$ and $d$, and (d) $b$ and $d$.

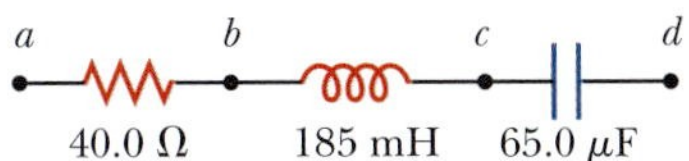


**Figure P21.29**

## 21.5 Power in an AC Circuit

30. An AC source operating at 60 Hz with a maximum voltage of 170 V is connected in series with a resistor ($R$ = 1.2 k$\Omega$) and a capacitor ($C$ = 2.5 $\mu$F). (a) What is the maximum value of the current in the circuit? (b) What are the maximum values of the potential difference across the resistor and the capacitor? (c) When the current is zero, what are the magnitudes of the potential difference across the resistor, the capacitor, and the AC source? How much charge is on the capacitor at this instant? (d) When the current is at a maximum, what are the magnitudes of the potential differences across the

resistor, the capacitor, and the AC source? How much charge is on the capacitor at this instant?

**31.** A multimeter in an *RL* circuit records an rms current of 0.500 A and a 60.0-Hz rms generator voltage of 104 V. A wattmeter shows that the average power delivered to the resistor is 10.0 W. Determine (a) the impedance in the circuit, (b) the resistance *R*, and (c) the inductance *L*.

**32.** An AC voltage of the form $\Delta v = (90.0 \text{ V}) \sin(350t)$ is applied to a series *RLC* circuit. If $R = 50.0\ \Omega$, $C = 25.0\ \mu\text{F}$, and $L = 0.200$ H, find the (a) impedance of the circuit, (b) rms current in the circuit, and (c) average power delivered to the circuit.

**33.** W An AC voltage of the form $\Delta v = 100 \sin(1\,000t)$, where $\Delta v$ is in volts and *t* is in seconds, is applied to a series *RLC* circuit. Assume the resistance is 400 Ω, the capacitance is 5.00 μF, and the inductance is 0.500 H. Find the average power delivered to the circuit.

**34.** A series *RLC* circuit has a resistance of 22.0 Ω and an impedance of 80.0 Ω. If the rms voltage applied to the circuit is 160 V, what average power is delivered to the circuit?

**35.** An inductor and a resistor are connected in series. When connected to a 60-Hz, 90-V (rms) source, the voltage drop across the resistor is found to be 50 V (rms) and the power delivered to the circuit is 14 W. Find (a) the value of the resistance and (b) the value of the inductance.

**36.** Q|C Consider a series *RLC* circuit with $R = 25\ \Omega$, $L = 6.0$ mH, and $C = 25\ \mu\text{F}$. The circuit is connected to a 10-V (rms), 600-Hz AC source. (a) Is the sum of the voltage drops across *R*, *L*, and *C* equal to 10 V (rms)? (b) Which is greatest, the power delivered to the resistor, to the capacitor, or to the inductor? (c) Find the average power delivered to the circuit.

## 21.6 Resonance in a Series *RLC* Circuit

**37.** M An *RLC* circuit is used in a radio to tune into an FM station broadcasting at $f = 99.7$ MHz. The resistance in the circuit is $R = 12.0\ \Omega$, and the inductance is $L = 1.40\ \mu\text{H}$. What capacitance should be used?

**38.** The resonant frequency of a certain series *RLC* circuit is 2.84 kHz, and the value of its capacitance is 6.50 μF. What is the value of the resonant frequency when the capacitance of the circuit is 9.80 μF?

**39.** The AM band extends from approximately 500 kHz to 1 600 kHz. If a 2.0-μH inductor is used in a tuning circuit for a radio, what are the extremes that a capacitor must reach to cover the complete band of frequencies?

**40.** Consider a series *RLC* circuit with $R = 15\ \Omega$, $L = 200$ mH, $C = 75\ \mu\text{F}$, and a maximum voltage of 150 V. (a) What is the impedance of the circuit at resonance? (b) What is the resonance frequency of the circuit? (c) When will the current be greatest: at resonance, at 10% below the resonant frequency, or at 10% above the resonant frequency? (d) What is the rms current in the circuit at a frequency of 60 Hz?

**41.** Two electrical oscillators are used in a heterodyne metal detector to detect buried metal objects (see Fig. P21.41). The detector uses two identical electrical oscillators in the form of *LC* circuits having resonant frequencies of 725 kHz. When the signals from the two oscillating circuits are combined, the beat frequency is zero because each has the same resonant frequency. However, when the coil of one circuit encounters a buried metal object, the inductance of this circuit increases by 1.000%, while that of the second is unchanged. Determine the beat frequency that would be detected in this situation.

Andy Ryan/Stone/Getty Images

**Figure P21.41**

**42.** A series circuit contains a 3.00-H inductor, a 3.00-μF capacitor, and a 30.0-Ω resistor connected to a 120-V (rms) source of variable frequency. Find the power delivered to the circuit when the frequency of the source is (a) the resonance frequency, (b) one-half the resonance frequency, (c) one-fourth the resonance frequency, (d) two times the resonance frequency, and (e) four times the resonance frequency. From your calculations, can you draw a conclusion about the frequency at which the maximum power is delivered to the circuit?

## 21.7 The Transformer

**43.** The primary coil of a transformer has $N_1 = 250$ turns, and its secondary coil has $N_2 = 1\,500$ turns. If the input voltage across the primary coil is $\Delta v = (170 \text{ V}) \sin \omega t$, what rms voltage is developed across the secondary coil?

**44.** A step-down transformer is used for recharging the batteries of portable devices. The turns ratio $N_2/N_1$ for a particular transformer used in a CD player is 1:13. When used with 120-V (rms) household service, the transformer draws an rms current of 250 mA. Find the (a) rms output voltage of the transformer and (b) power delivered to the CD player.

**45.** An AC power generator produces 50 A (rms) at 3 600 V. The voltage is stepped up to 100 000 V by an ideal transformer, and the energy is transmitted through a long-distance power line that has a resistance of 100 Ω. What percentage of the power delivered by the generator is dissipated as heat in the power line?

**46.** W A transformer is to be used to provide power for a computer disk drive that needs 6.0 V (rms) instead of the 120 V (rms) from the wall outlet. The number of turns in the primary is 400, and it delivers 500 mA (the secondary current) at an output voltage of 6.0 V (rms). (a) Should the transformer have more turns in the secondary compared with the primary, or fewer turns? (b) Find the current in the primary. (c) Find the number of turns in the secondary.

**47.** A transformer on a pole near a factory steps the voltage down from 3 600 V (rms) to 120 V (rms). The transformer is to deliver 1 000 kW to the factory at 90% efficiency. Find (a) the power delivered to the primary, (b) the current in the primary, and (c) the current in the secondary.

**48.** A transmission line that has a resistance per unit length of $4.50 \times 10^{-4}\ \Omega/\text{m}$ is to be used to transmit 5.00 MW over 400 miles ($6.44 \times 10^5$ m). The output voltage of the generator is 4.50 kV (rms). (a) What is the line loss if a transformer is used to step up the voltage to 500 kV (rms)? (b) What fraction of the input power is lost to the line under these circumstances? (c) What difficulties would be encountered on attempting to transmit the 5.00 MW at the generator voltage of 4.50 kV (rms)?

## 21.10 Production of Electromagnetic Waves by an Antenna

## 21.11 Properties of Electromagnetic Waves

**49.** The U.S. Navy has long proposed the construction of extremely low frequency (ELF waves) communications systems; such waves could penetrate the oceans to reach distant submarines. Calculate the length of a quarter-wavelength antenna for a transmitter generating ELF waves of frequency 75 Hz. How practical is this antenna?

**50.** (a) The distance to Polaris, the North Star, is approximately $6.44 \times 10^{18}$ m. If Polaris were to burn out today, how many years would it take to see it disappear? (b) How long does it take sunlight to reach Earth? (c) How long does it take a microwave signal to travel from Earth to the Moon and back? (The distance from Earth to the Moon is $3.84 \times 10^5$ km.)

**51.** QIC The Earth reflects approximately 38.0% of the incident sunlight from its clouds and surface. (a) Given that the intensity of solar radiation at the top of the atmosphere is 1 370 W/m$^2$, find the radiation pressure on the Earth, in pascals, at the location where the Sun is straight overhead. (b) State how this quantity compares with normal atmospheric pressure at the Earth's surface, which is 101 kPa.

**52.** Experimenters at the National Institute of Standards and Technology have made precise measurements of the speed of light using the fact that, in vacuum, the speed of electromagnetic waves is $c = 1/\sqrt{\mu_0 \epsilon_0}$, where the constants $\mu_0 = 4\pi \times 10^{-7}\ \text{N} \cdot \text{s}^2/\text{C}^2$ and $\epsilon_0 = 8.854 \times 10^{-12}\ \text{C}^2/\text{N} \cdot \text{m}^2$. What value (to four significant figures) does this formula give for the speed of light in vacuum?

**53.** BIO Oxygenated hemoglobin absorbs weakly in the red (hence its red color) and strongly in the near infrared, whereas deoxygenated hemoglobin has the opposite absorption. This fact is used in a "pulse oximeter" to measure oxygen saturation in arterial blood. The device clips onto the end of a person's finger and has two light-emitting diodes—a red (660 nm) and an infrared (940 nm)—and a photocell that detects the amount of light transmitted through the finger at each wavelength. (a) Determine the frequency of each of these light sources. (b) If 67% of the energy of the red source is absorbed in the blood, by what factor does the amplitude of the electromagnetic wave change? *Hint:* The intensity of the wave is equal to the average power per unit area as given by Equation 21.28.

**54.** BIO **Operation of the pulse oximeter (see previous problem).** The transmission of light energy as it passes through a solution of light-absorbing molecules is described by the Beer–Lambert law

$$I = I_0 10^{-\epsilon CL} \quad \text{or} \quad \log_{10}\left(\frac{I}{I_0}\right) = -\epsilon CL$$

which gives the decrease in intensity $I$ in terms of the distance $L$ the light has traveled through a fluid with a concentration $C$ of the light-absorbing molecule. The quantity $\epsilon$ is called the extinction coefficient, and its value depends on the frequency of the light. (It has units of m$^2$/mol.) Assume the extinction coefficient for 660-nm light passing through a solution of oxygenated hemoglobin is identical to the coefficient for 940-nm light passing through deoxygenated hemoglobin. Also assume 940-nm light has zero absorption ($\epsilon = 0$) in oxygenated hemoglobin and 660-nm light has zero absorption in deoxygenated hemoglobin. If 33% of the energy of the red source and 76% of the infrared energy is transmitted through the blood, what is the fraction of hemoglobin that is oxygenated?

**55.** The Sun delivers an average power of 1 370 W/m$^2$ to the top of Earth's atmosphere. Find the magnitudes of $\vec{\mathbf{E}}_{\text{max}}$ and $\vec{\mathbf{B}}_{\text{max}}$ for the electromagnetic waves at the top of the atmosphere.

**56.** QIC A laser beam is used to levitate a metal disk against the force of Earth's gravity. (a) Derive an equation giving the required intensity of light, $I$, in terms of the mass $m$ of the disk, the gravitational acceleration $g$, the speed of light $c$, and the cross-sectional area of the disk $A$. Assume the disk is perfectly reflecting and the beam is directed perpendicular to the disk. (b) If the disk has mass 5.00 g and radius 4.00 cm, find the necessary light intensity. (c) Give two reasons why using light pressure as propulsion near Earth's surface is impractical.

**57.** A microwave oven is powered by an electron tube called a magnetron that generates electromagnetic waves of frequency 2.45 GHz. The microwaves enter the oven and are reflected by the walls. The standing-wave pattern produced in the oven can cook food unevenly, with hot spots in the food at antinodes and cool spots at nodes, so a turntable is often used to rotate the food and distribute the energy. If a microwave oven is used with a cooking dish in a fixed position, the antinodes can appear as burn marks on foods such as carrot strips or cheese. The separation distance between the burns is measured to be 6.00 cm. Calculate the speed of the microwaves from these data.

58. Consider a bright star in our night sky. Assume its distance from the Earth is 20.0 light-years (ly) and its power output is $4.00 \times 10^{28}$ W, about 100 times that of the Sun. (a) Find the intensity of the starlight at the Earth. (b) Find the power of the starlight the Earth intercepts. One light-year is the distance traveled by light through a vacuum in one year.

### 21.12 The Spectrum of Electromagnetic Waves

**59.** What are the wavelengths of electromagnetic waves in free space that have frequencies of (a) $5.00 \times 10^{19}$ Hz and (b) $4.00 \times 10^{9}$ Hz?

**60.** BIO A diathermy machine, used in physiotherapy, generates electromagnetic radiation that gives the effect of "deep heat" when absorbed in tissue. One assigned frequency for diathermy is 27.33 MHz. What is the wavelength of this radiation?

**61.** What are the wavelength ranges in (a) the AM radio band (540–1 600 kHz) and (b) the FM radio band (88–108 MHz)?

62. An important news announcement is transmitted by radio waves to people who are 100 km away, sitting next to their radios, and by sound waves to people sitting across the newsroom, 3.0 m from the newscaster. Who receives the news first? Explain. Take the speed of sound in air to be 343 m/s.

**63.** Infrared spectra are used by chemists to help identify an unknown substance. Atoms in a molecule that are bound together by a particular bond vibrate at a predictable frequency, and light at that frequency is absorbed strongly by the atom. In the case of the C═O double bond, for example, the oxygen atom is bound to the carbon by a bond that has an effective spring constant of 2 800 N/m. If we assume the carbon atom remains stationary (it is attached to other atoms in the molecule), determine the resonant frequency of this bond and the wavelength of light that matches that frequency. Verify that this wavelength lies in the infrared region of the spectrum. (The mass of an oxygen atom is $2.66 \times 10^{-26}$ kg.)

### 21.13 The Doppler Effect for Electromagnetic Waves

64. A spaceship is approaching a space station at a speed of $1.8 \times 10^{5}$ m/s. The space station has a beacon that emits green light with a frequency of $6.0 \times 10^{14}$ Hz. (a) What is the frequency of the beacon observed on the spaceship? (b) What is the change in frequency? (Carry five digits in these calculations.)

**65.** While driving at a constant speed of 80 km/h, you are passed by a car traveling at 120 km/h. If the frequency of light emitted by the taillights of the car that passes you is $4.3 \times 10^{14}$ Hz, what frequency will you observe? What is the change in frequency?

66. A speeder tries to explain to the police that the yellow warning lights she was approaching on the side of the road looked green to her because of the Doppler shift. How fast would she have been traveling if yellow light of wavelength 580 nm had been shifted to green with a wavelength of 560 nm? *Note:* For speeds less than $0.03c$, Equation 21.32 will lead to a value for the observed frequency accurate to approximately two significant digits.

### Additional Problems

**67.** A 25.0-mW laser beam of diameter 2.00 mm is reflected at normal incidence by a perfectly reflecting mirror. Calculate the radiation pressure on the mirror.

68. The intensity of solar radiation at the top of Earth's atmosphere is 1 370 W/m$^3$. Assuming 60% of the incoming solar energy reaches Earth's surface and assuming you absorb 50% of the incident energy, make an order-of-magnitude estimate of the amount of solar energy you absorb in a 60-minute sunbath.

**69.** A 200-Ω resistor is connected in series with a 5.0-μF capacitor and a 60-Hz, 120-V rms line. If electrical energy costs \$0.080/kWh, how much does it cost to leave this circuit connected for 24 h?

70. S In an *RLC* series circuit that includes a source of alternating current operating at fixed frequency and voltage, the resistance $R$ is equal to the inductive reactance. If the plate separation of the parallel-plate capacitor is reduced to one-half its original value, the current in the circuit doubles. Find the initial capacitive reactance in terms of $R$.

71. As a way of determining the inductance of a coil used in a research project, a student first connects the coil to a 12.0-V battery and measures a current of 0.630 A. The student then connects the coil to a 24.0-V (rms), 60.0-Hz generator and measures an rms current of 0.570 A. What is the inductance?

72. (a) What capacitance will resonate with a one-turn loop of inductance 400 pH to give a radar wave of wavelength 3.0 cm? (b) If the capacitor has square parallel plates separated by 1.0 mm of air, what should the edge length of the plates be? (c) What is the common reactance of the loop and capacitor at resonance?

73. M A dish antenna with a diameter of 20.0 m receives (at normal incidence) a radio signal from a distant source, as shown in Figure P21.73. The radio signal is a continuous sinusoidal wave with amplitude $E_{max} = 0.20$ μV/m. Assume the antenna absorbs all the radiation that falls

on the dish. (a) What is the amplitude of the magnetic field in this wave? (b) What is the intensity of the radiation received by the antenna? (c) What is the power received by the antenna?

**Figure P21.73**

74. A particular inductor has appreciable resistance. When the inductor is connected to a 12-V battery, the current in the inductor is 3.0 A. When it is connected to an AC source with an rms output of 12 V and a frequency of 60 Hz, the current drops to 2.0 A. What are (a) the impedance at 60 Hz and (b) the inductance of the inductor?

75. One possible means of achieving space flight is to place a perfectly reflecting aluminized sheet into Earth's orbit and to use the light from the Sun to push this solar sail. Suppose such a sail, of area $6.00 \times 10^4$ m$^2$ and mass 6 000 kg, is placed in orbit facing the Sun. (a) What force is exerted on the sail? (b) What is the sail's acceleration? (c) How long does it take this sail to reach the Moon, $3.84 \times 10^8$ m away? Ignore all gravitational effects and assume a solar intensity of 1 340 W/m$^2$. *Hint:* The radiation pressure by a reflected wave is given by 2 (average power per unit area)/$c$.

76. BIO Q|C The U.S. Food and Drug Administration limits the radiation leakage of microwave ovens to no more than 5.0 mW/cm$^2$ at a distance of 2.0 in. A typical cell phone, which also transmits microwaves, has a peak output power of about 2.0 W. (a) Approximating the cell phone as a point source, calculate the radiation intensity of a cell phone at a distance of 2.0 in. How does the answer compare with the maximum allowable microwave oven leakage? (b) The distance from your ear to your brain is about 2 in. What would the radiation intensity in your brain be if you used a Bluetooth headset, keeping the phone in your pocket, 1.0 m away from your brain? Most headsets are so-called Class 2 devices with a maximum output power of 2.5 mW.

Light is bent (refracted) as it passes through water, with different wavelengths bending by different amounts, a phenomenon called dispersion. Together with reflection, these physical phenomena lead to the creation of a rainbow when light passes through small, suspended droplets of water.

Dave Walker/Shutterstock.com

# 22 Reflection and Refraction of Light



Light has a dual nature. In some experiments it acts like a particle, while in others it acts like a wave. In this and the next two chapters, we concentrate on the aspects of light that are best understood through the wave model. First we discuss the reflection of light at the boundary between two media and the refraction (bending) of light as it travels from one medium into another. We use these ideas to study the refraction of light as it passes through lenses and the reflection of light from mirrored surfaces. In Chapter 25, we describe how lenses and mirrors can be used to view objects with telescopes and microscopes and how lenses are used in photography. The ability to manipulate light has greatly enhanced our capacity to investigate and understand the nature of the Universe.

## 22.1 The Nature of Light

**LEARNING OBJECTIVES**

1. Discuss the dual nature of light as both a wave and as particles called photons.
2. State the equation yielding the energy of a photon.

Until the beginning of the 19th century, light was modeled as a stream of particles emitted by a source that stimulated the sense of sight on entering the eye. The chief architect of the particle theory of light was Newton. With this theory, he provided simple explanations of some known experimental facts concerning the nature of light, namely, the laws of reflection and refraction.

Most scientists accepted Newton's particle theory of light. During Newton's lifetime, however, another theory was proposed. In 1678 Dutch physicist and astronomer Christian Huygens (1629–1695) showed that a wave theory of light could also explain the laws of reflection and refraction.

The wave theory didn't receive immediate acceptance, for several reasons. First, all the waves known at the time (sound, water, and so on) traveled through some sort of medium, but light from the Sun could travel to Earth through empty space. Further, it was argued that if light were some form of wave, it would bend around obstacles; hence, we should be able to see around corners. It is now known that light does indeed bend around the edges of objects. This phenomenon, known as *diffraction*, is difficult to observe because light waves have such short wavelengths. Even though experimental evidence for the diffraction of light was discovered by Francesco Grimaldi (1618–1663) around 1660, for more than a century most scientists rejected the wave theory and adhered to Newton's particle theory, probably due to Newton's great reputation as a scientist.

Photo Researchers

**Christian Huygens**
**(1629–1695), Dutch Physicist and Astronomer**
Huygens is best known for his contributions to the fields of optics and dynamics. To Huygens, light was a vibratory motion in the ether, spreading out and producing the sensation of light when impinging on the eye. On the basis of this theory, he deduced the laws of reflection and refraction and explained the phenomenon of double refraction.

The first clear demonstration of the wave nature of light was provided in 1801 by Thomas Young (1773–1829), who showed that under appropriate conditions, light exhibits interference behavior. Light waves emitted by a single source and traveling along two different paths can arrive at some point and combine and cancel each other by destructive interference. Such behavior couldn't be explained at that time by a particle model because scientists couldn't imagine how two or more particles could come together and cancel one another.

The most important development in the theory of light was the work of Maxwell, who predicted in 1865 that light was a form of high-frequency electromagnetic wave (Chapter 21). His theory also predicted that these waves should have a speed of $3 \times 10^8$ m/s, in agreement with the measured value.

Although the classical theory of electricity and magnetism explained most known properties of light, some subsequent experiments couldn't be explained by the assumption that light was a wave. The most striking experiment was the *photoelectric effect* (which we examine more closely in Chapter 27), discovered by Hertz. Hertz found that clean metal surfaces emit charges when exposed to ultraviolet light.

In 1905, Einstein published a paper that formulated the theory of light quanta ("particles") and explained the photoelectric effect. He reached the conclusion that light was composed of corpuscles, or discontinuous quanta of energy. These corpuscles or quanta are now called *photons* to emphasize their particle-like nature. According to Einstein's theory, the energy of a photon is proportional to the frequency of the electromagnetic wave associated with it, or

$$E = hf \qquad [22.1]$$

◀ Energy of a photon

where $h = 6.63 \times 10^{-34}$ J · s is *Planck's constant*. This theory retains some features of both the wave and particle theories of light. As we discuss later, the photoelectric effect is the result of energy transfer from a single photon to an electron in the metal. This means the electron interacts with one photon of light as if the electron had been struck by a particle. Yet the photon has wave-like characteristics, as implied by the fact that a frequency is used in its definition.

In view of these developments, light must be regarded as having a *dual nature*: **In some experiments light acts as a wave and in others it acts as a particle.** Classical electromagnetic wave theory provides adequate explanations of light propagation and of the effects of interference, whereas the photoelectric effect and other experiments involving the interaction of light with matter are best explained by assuming light is a particle.

So in the final analysis, is light a wave or a particle? The answer is neither and both: light has a number of physical properties, some associated with waves and others with particles.

# 22.2 Reflection and Refraction

**LEARNING OBJECTIVES**

1. State and apply the ray approximation.
2. Apply the law of reflection of light.
3. Discuss the physical meaning of the refraction of light.

When light traveling in one medium encounters a boundary leading into a second medium, the processes of reflection and refraction can occur. In **reflection** part of the light encountering the second medium bounces off that medium. In **refraction** the light passing into the second medium bends through an angle with respect to the normal to the boundary. Often, both processes occur at the same time, with part of the light being reflected and part refracted. To study reflection and refraction we need a way of thinking about beams of light, and this is given by the ray approximation.

The rays are straight lines perpendicular to the wave fronts and pointing in the direction of the wave motion.

Rays

Wave fronts

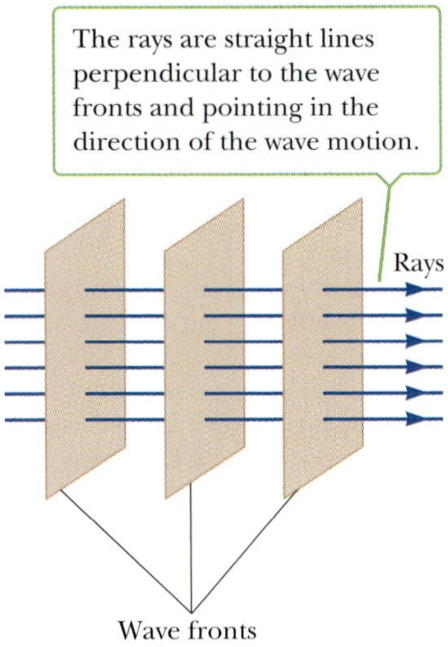

**Figure 22.1** A plane wave traveling to the right.

## The Ray Approximation in Geometric Optics

An important property of light that can be understood based on common experience is the following: **light travels in a straight-line path in a homogeneous medium, until it encounters a boundary between two different materials.** When light strikes a boundary, it is reflected from that boundary, passes into the material on the other side of the boundary, or partially does both.

The preceding observation leads us to use what is called the **ray approximation** to represent beams of light. As shown in Figure 22.1, a ray of light is an imaginary line drawn along the direction of travel of the light beam. For example, a beam of sunlight passing through a darkened room traces out the path of a light ray. We also make use of the concept of wave fronts of light. A **wave front** is a surface passing through the points of a wave that have the same phase and amplitude. For instance, the wave fronts in Figure 22.1 could be surfaces passing through the crests of waves. The rays, corresponding to the direction of wave motion, are straight lines perpendicular to the wave fronts. When light rays travel in parallel paths, the wave fronts are planes perpendicular to the rays.

## Reflection of Light

When a light ray traveling in a transparent medium encounters a boundary leading into a second medium, part of the incident ray is reflected back into the first medium. Figure 22.2a shows several rays of a beam of light incident on a smooth, mirror-like reflecting surface. The reflected rays are parallel to one another, as indicated in the figure. The reflection of light from such a smooth surface is called **specular reflection**. On the other hand, if the reflecting surface is rough, as in Figure 22.2b, the surface reflects the rays in a variety of directions. Reflection from any rough surface is known as **diffuse reflection**. A surface behaves as a smooth surface as long as its variations are small compared with the wavelength of the incident light. Figures 22.2c and 22.2d are photographs of specular and diffuse reflection of laser light, respectively.

**APPLICATION**
Seeing the Road on a Rainy Night

As an example, consider the two types of reflection from a road surface that someone might observe while driving at night. When the road is dry, light from oncoming vehicles is scattered off the road in different directions (diffuse reflection) and the road is clearly visible. On a rainy night when the road is wet, the road's irregularities are filled with water. Because the wet surface is smooth, the light undergoes specular reflection. This means that the light is reflected straight ahead, and the driver of a car sees only what is directly in front of him. Light from the side never reaches the driver's eye. In this book we concern ourselves only with specular reflection, and we use the term *reflection* to mean specular reflection.

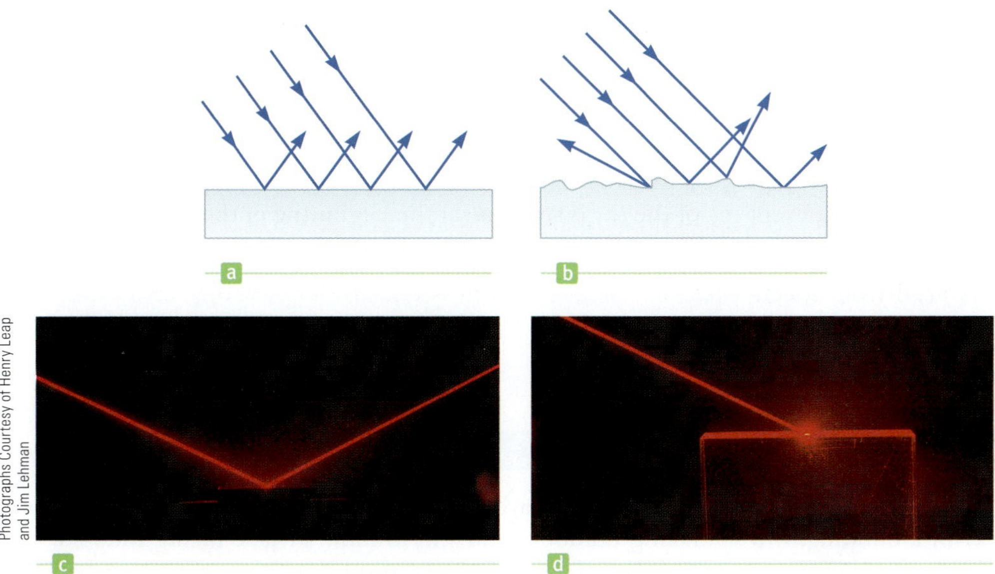

a
b
c
d
Photographs Courtesy of Henry Leap and Jim Lehman

**Figure 22.2** A schematic representation of (a) specular reflection, where the reflected rays are all parallel to one another, and (b) diffuse reflection, where the reflected rays travel in random directions. (c, d) Photographs of specular and diffuse reflection, made with laser light.

**Quick Quiz**

**22.1** Which part of Figure 22.3, (a) or (b), better shows specular reflection of light from the roadway?

© Charles D. Winters/Cengage Learning
a
b

**Figure 22.3** (Quick Quiz 22.1)

Consider a light ray traveling in air and incident at some angle on a flat, smooth surface, as in Figure 22.4. The incident and reflected rays make angles $\theta_1$ and $\theta_1'$, respectively, **with a line perpendicular to the surface** at the point where the incident ray strikes the surface. We call this line the *normal* to the surface. Experiments show that **the angle of reflection equals the angle of incidence**:

$$\theta_1' = \theta_1 \quad [22.2]$$

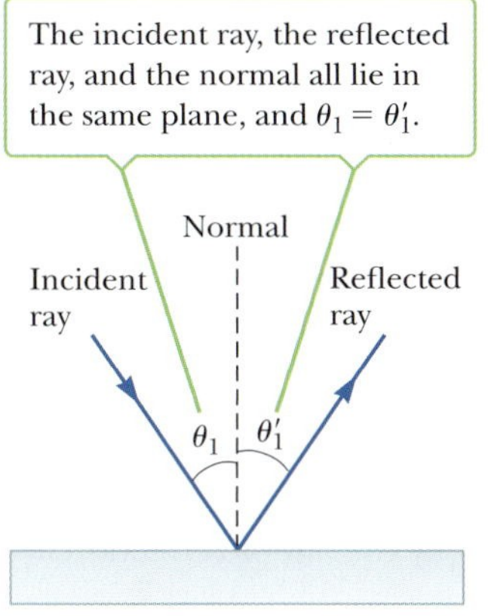

The incident ray, the reflected ray, and the normal all lie in the same plane, and $\theta_1 = \theta_1'$.
Normal
Incident ray
Reflected ray
$\theta_1$
$\theta_1'$

**Figure 22.4** The wave under reflection model.

You may have noticed a common occurrence in photographs of individuals: their eyes appear to be glowing red. "Red-eye" occurs when a photographic flash is used and the flash unit is close to the camera lens. Light from the flash unit enters the eye and is reflected back along its original path from the retina. This type of reflection back along the original direction is called *retroreflection*. If the flash unit and lens are close together, retroreflected light can enter the lens. Most of the light reflected from the retina is red due to the blood vessels at the back of the eye, giving the red-eye effect in the photograph.

**BIO APPLICATION**
Red Eyes in Flash Photographs

## APPLYING PHYSICS 22.1 The Colors of Water Ripples at Sunset

An observer on the west-facing beach of a large lake is watching the beginning of a sunset. The water is very smooth except for some areas with small ripples. The observer notices that some areas of the water are blue and some are pink. Why does the water appear to be different colors in different areas?

**EXPLANATION** The different colors arise from specular and diffuse reflection. The smooth areas of the water will specularly reflect the light from the west, which is the pink light from the sunset. The areas with small ripples will reflect the light diffusely, so light from all parts of the sky will be reflected into the observer's eyes. Because most of the sky is still blue at the beginning of the sunset, these areas will appear to be blue. ■

## APPLYING PHYSICS 22.2 Double Images

When standing outside in the Sun close to a single-pane window looking to the darker interior of a building, why can you often see two images of yourself, one superposed on the other?

**EXPLANATION** Reflection occurs whenever there is an interface between two different media. For the glass in the window, there are two such surfaces, the window surface facing outdoors and the window surface facing indoors. Each of these interfaces results in an image. You will notice that one image is slightly smaller than the other, because the reflecting surface is farther away. ■

## EXAMPLE 22.1 The Double-Reflecting Light Ray

**GOAL** Calculate a resultant angle from two reflections.

**PROBLEM** Two mirrors make an angle of 120° with each other, as in Figure 22.5. A ray is incident on mirror $M_1$ at an angle of 65° to the normal. Find the angle the ray makes with the normal to $M_2$ after it is reflected from both mirrors.

**STRATEGY** Apply the law of reflection twice. Given the incident ray at angle $\theta_{inc}$, find the final resultant angle, $\beta_{ref}$.

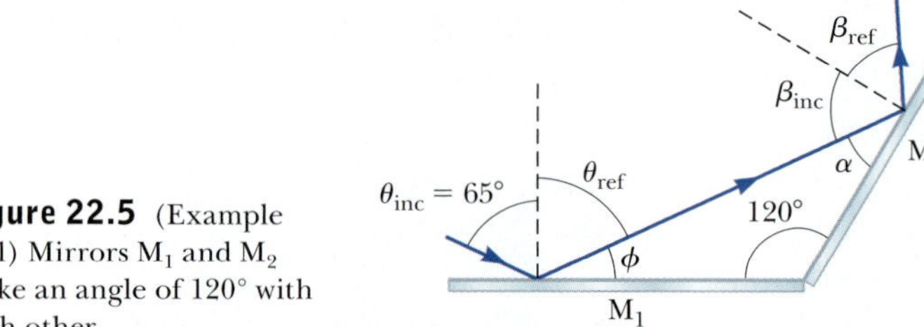


**Figure 22.5** (Example 22.1) Mirrors $M_1$ and $M_2$ make an angle of 120° with each other.

**SOLUTION**

Apply the law of reflection to $M_1$ to find the angle of reflection, $\theta_{ref}$:

$$\theta_{ref} = \theta_{inc} = 65°$$

Find the angle $\phi$ that is the complement of the angle $\theta_{ref}$:

$$\phi = 90° - \theta_{ref} = 90° - 65° = 25°$$

Find the unknown angle $\alpha$ in the triangle of $M_1$, $M_2$, and the ray traveling from $M_1$ to $M_2$, using the fact that the three angles sum to 180°:

$$180° = 25° + 120° + \alpha \quad \rightarrow \quad \alpha = 35°$$

The angle $\alpha$ is complementary to the angle of incidence, $\beta_{inc}$, for $M_2$:

$$\alpha + \beta_{inc} = 90° \quad \rightarrow \quad \beta_{inc} = 90° - 35° = 55°$$

Apply the law of reflection a second time, obtaining $\beta_{ref}$:

$$\beta_{ref} = \beta_{inc} = 55°$$

**REMARKS** Notice the heavy reliance on elementary geometry and trigonometry in these reflection problems.

**QUESTION 22.1** In general, what is the relationship between the incident angle $\theta_{inc}$ and the final reflected angle $\beta_{ref}$ when the angle between the mirrors is 90.0°? (a) $\theta_{inc} + \beta_{ref} = 90.0°$ (b) $\theta_{inc} - \beta_{ref} = 90.0°$ (c) $\theta_{inc} + \beta_{ref} = 180°$

**EXERCISE 22.1** Repeat the problem if the angle of incidence is 55° and the second mirror makes an angle of 100° with the first mirror.

**ANSWER** 45°

## Refraction of Light

When a ray of light traveling through a transparent medium encounters a boundary leading into another transparent medium, as in Figure 22.6a, part of the ray is reflected and part enters the second medium. The ray that enters the second medium is bent at the boundary and is said to be *refracted.* The incident ray, the reflected ray, the refracted ray, and the normal at the point of incidence all lie in the same plane. The **angle of refraction,** $\theta_2$, in Figure 22.6a depends on the properties of the two media and on the angle of incidence, through the relationship

$$\frac{\sin \theta_2}{\sin \theta_1} = \frac{v_2}{v_1} = \text{constant} \qquad [22.3]$$

where $v_1$ is the speed of light in medium 1 and $v_2$ is the speed of light in medium 2. Note that the angle of refraction is also measured with respect to the normal. In Section 22.7 we derive the laws of reflection and refraction using Huygens' principle.

Experiment shows that **the path of a light ray through a refracting surface is reversible.** For example, the ray in Figure 22.6a travels from point $A$ to point $B$. If the ray originated at $B$, it would follow the same path to reach point $A$, but the reflected ray would be in the glass.

**Quick Quiz**

**22.2** If beam 1 is the incoming beam in Figure 22.6b, which of the other four beams are due to reflection? Which are due to refraction?

When light moves from a material in which its speed is high to a material in which its speed is lower, the angle of refraction $\theta_2$ is less than the angle of incidence. The refracted ray therefore bends toward the normal, as shown in Figure 22.7a (page 778). If the ray moves from a material in which it travels slowly to a material in which it travels more rapidly, $\theta_2$ is greater than $\theta_1$, so the ray bends away from the normal, as shown in Figure 22.7b.

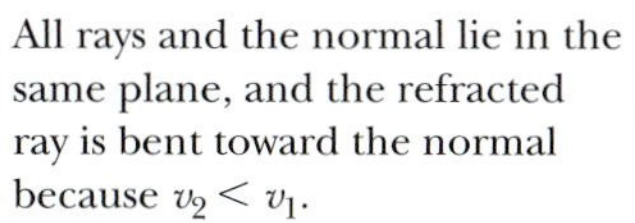

All rays and the normal lie in the same plane, and the refracted ray is bent toward the normal because $v_2 < v_1$.


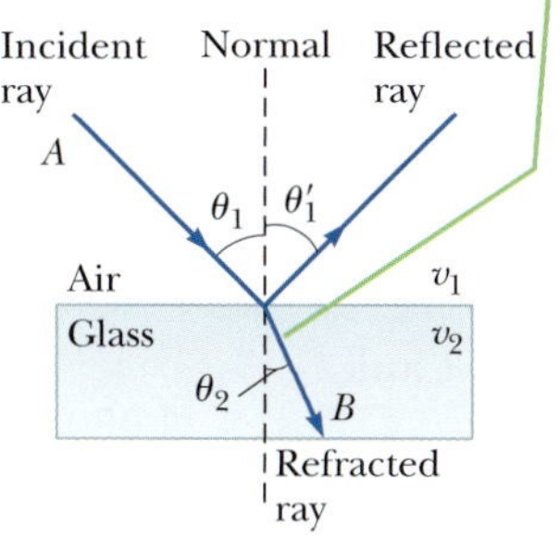


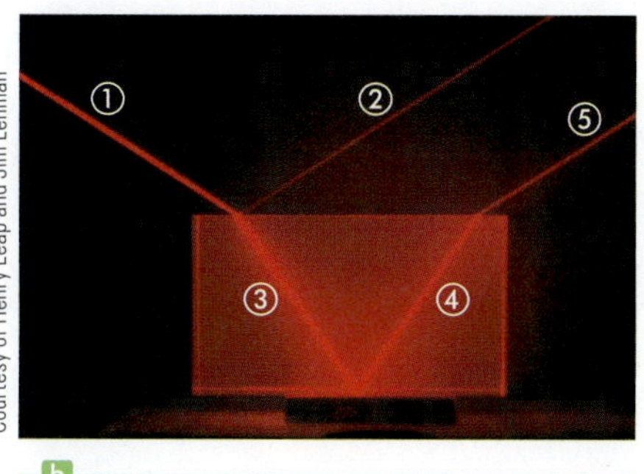


**Figure 22.6** (a) The wave under refraction model. (b) Light incident on the Lucite block refracts both when it enters the block and when it leaves the block.

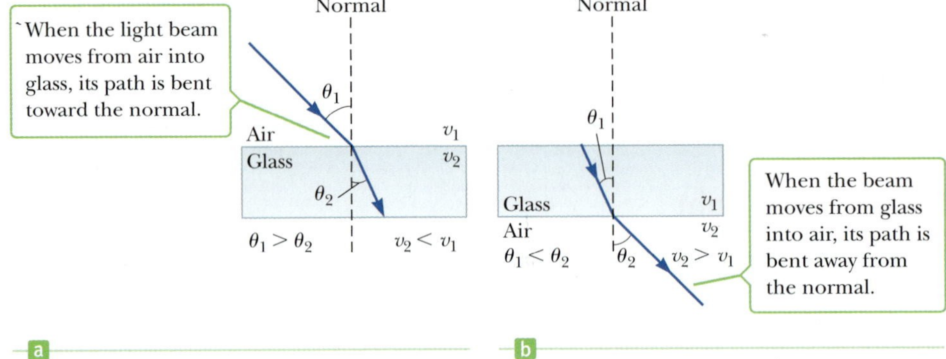

Normal
When the light beam moves from air into glass, its path is bent toward the normal.
$\theta_1$
Air
$v_1$
Glass
$v_2$
$\theta_2$
$\theta_1 > \theta_2$
$v_2 < v_1$
a
Normal
$\theta_1$
Glass
$v_1$
Air
$v_2$
$\theta_1 < \theta_2$
$\theta_2$
$v_2 > v_1$
When the beam moves from glass into air, its path is bent away from the normal.
b

**Figure 22.7** The refraction of light as it (a) moves from air into glass and (b) moves from glass into air.

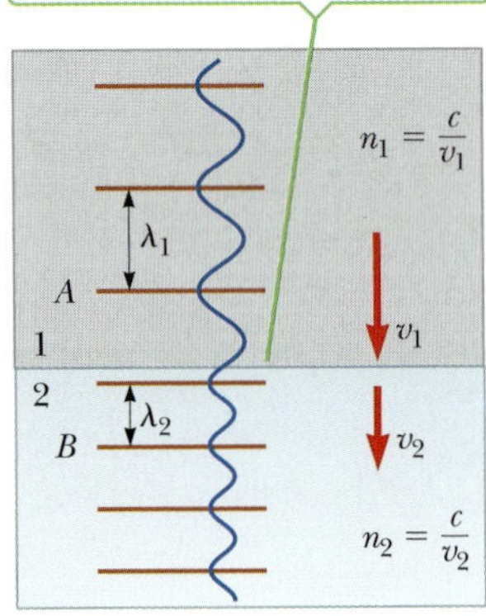

As a wave moves from medium 1 to medium 2, its wavelength changes but its frequency remains constant.
$n_1 = \frac{c}{v_1}$
$\lambda_1$
A
1
$v_1$
2
$\lambda_2$
B
$v_2$
$n_2 = \frac{c}{v_2}$

**Figure 22.8** A wave travels from medium 1 to medium 2, in which it moves with lower speed.

# 22.3 The Law of Refraction

**LEARNING OBJECTIVES**

1. Define the index of refraction.
2. Derive the wavelength of light in different media.
3. Apply Snell's Law.

When light passes from one transparent medium to another, it's refracted because the speed of light is different in the two media.[1] The **index of refraction,** $n$, of a medium is defined as the ratio $c/v$;

$$n \equiv \frac{\text{speed of light in vacuum}}{\text{speed of light in a medium}} = \frac{c}{v} \qquad [22.4]$$

From this definition, we see that the index of refraction is a dimensionless number that is greater than or equal to 1 because $v$ is always less than $c$. Further, $n$ is equal to one for vacuum. Table 22.1 lists the indices of refraction for various substances.

**As light travels from one medium to another, its frequency doesn't change.** To see why, consider Figure 22.8. Wave fronts pass an observer at point $A$ in medium 1 with a certain frequency and are incident on the boundary between

**Table 22.1** Indices of Refraction for Various Substances, Measured with Light of Vacuum Wavelength $\lambda_0 = 589$ mn

| Substance | Index of Refraction | Substance | Index of Refraction |
|---|---|---|---|
| **Solids at 20°C** | | **Liquids at 20°C** | |
| Diamond (C) | 2.419 | Benzene | 1.501 |
| Fluorite ($CaF_2$) | 1.434 | Carbon disulfide | 1.628 |
| Fused quartz ($SiO_2$) | 1.458 | Carbon tetrachloride | 1.461 |
| Glass, crown | 1.52 | Ethyl alcohol | 1.361 |
| Glass, flint | 1.66 | Glycerine | 1.473 |
| Ice ($H_2O$) (at 0°C) | 1.309 | Water | 1.333 |
| Polystyrene | 1.49 | | |
| Sodium chloride (NaCl) | 1.544 | **Gases at 0°C, 1 atm** | |
| Zircon | 1.923 | Air | 1.000 293 |
| | | Carbon dioxide | 1.000 45 |

[1]The speed of light varies between media because the time lags caused by the absorption and reemission of light as it travels from atom to atom depend on the particular electronic structure of the atoms constituting each material.

medium 1 and medium 2. The frequency at which the wave fronts pass an observer at point $B$ in medium 2 must equal the frequency at which they arrive at point $A$. If not, the wave fronts would either pile up at the boundary or be destroyed or created at the boundary. Because neither of these events occurs, the frequency must remain the same as a light ray passes from one medium into another.

Therefore, because the relation $v = f\lambda$ must be valid in both media and because $f_1 = f_2 = f$, we see that

$$v_1 = f\lambda_1 \text{ and } v_2 = f\lambda_2$$

Because $v_1 \neq v_2$, it follows that $\lambda_1 \neq \lambda_2$. A relationship between the index of refraction and the wavelength can be obtained by dividing these two equations and making use of the definition of the index of refraction given by Equation 22.4:

$$\frac{\lambda_1}{\lambda_2} = \frac{v_1}{v_2} = \frac{c/n_1}{c/n_2} = \frac{n_2}{n_1} \quad [22.5]$$

which gives

$$\lambda_1 n_1 = \lambda_2 n_2 \quad [22.6]$$

Let medium 1 be the vacuum so that $n_1 = 1$. It follows from Equation 22.6 that the index of refraction of any medium can be expressed as the ratio

$$n = \frac{\lambda_0}{\lambda_n} \quad [22.7]$$

where $\lambda_0$ is the wavelength of light in vacuum and $\lambda_n$ is the wavelength in the medium having index of refraction $n$. Figure 22.9 is a schematic representation of this reduction in wavelength when light passes from a vacuum into a transparent medium.

We are now in a position to express Equation 22.3 in an alternate form. If we substitute Equation 22.5 into Equation 22.3, we get

$$n_1 \sin\theta_1 = n_2 \sin\theta_2 \quad [22.8]$$

◄ Snell's law of refraction

The experimental discovery of this relationship is usually credited to Willebrørd Snell (1591–1626) and is therefore known as **Snell's law of refraction**.

**Tip 22.1 An Inverse Relationship**

The index of refraction is *inversely* proportional to the wave speed. Therefore, as the wave speed $v$ decreases, the index of refraction, $n$, *increases.*

**Tip 22.2 The Frequency Remains the Same**

The *frequency* of a wave does *not* change as the wave passes from one medium to another. Both the wave speed and the wavelength *do* change, but the frequency remains the same.

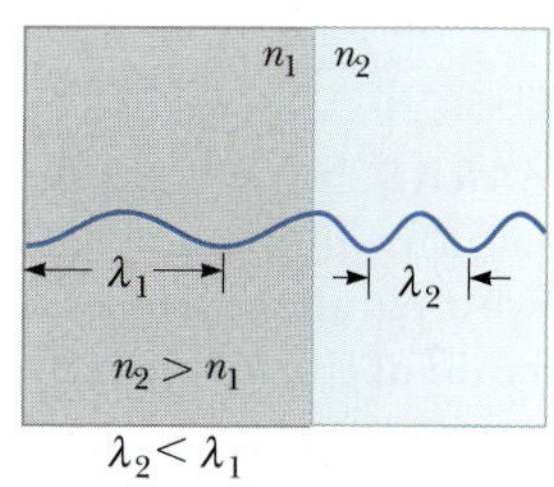


**Figure 22.9** A schematic diagram of the *reduction* in wavelength when light travels from a medium with a low index of refraction to one with a higher index of refraction.

## Quick Quiz

**22.3** A material has an index of refraction that increases continuously from top to bottom. Of the three paths shown in Figure 22.10, which path will a light ray follow as it passes through the material?

**22.4** As light travels from a vacuum ($n = 1$) to a medium such as glass ($n > 1$), which of the following properties remains the same, the (a) wavelength, (b) wave speed, or (c) frequency?

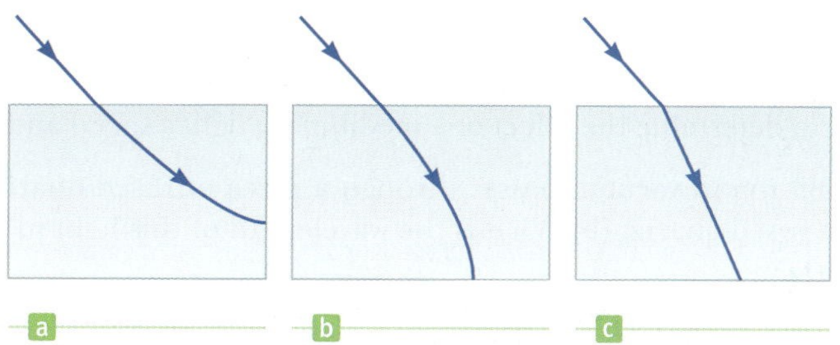


**Figure 22.10** (Quick Quiz 22.3)

## EXAMPLE 22.2 Angle of Refraction for Glass

**GOAL** Apply Snell's law.

**PROBLEM** A light ray of wavelength 589 nm (produced by a sodium lamp) traveling through air is incident on a smooth, flat slab of crown glass at an angle $\theta_1$ of 30.0° to the normal, as sketched in Figure 22.11. **(a)** Find the angle of refraction, $\theta_2$. **(b)** At what angle $\theta_3$ does the ray leave the glass as it re-enters the air? **(c)** How does the answer for $\theta_3$ change if the ray enters water below the slab instead of the air?

**STRATEGY** Substitute quantities into Snell's law and solve for the unknown angles of refraction.

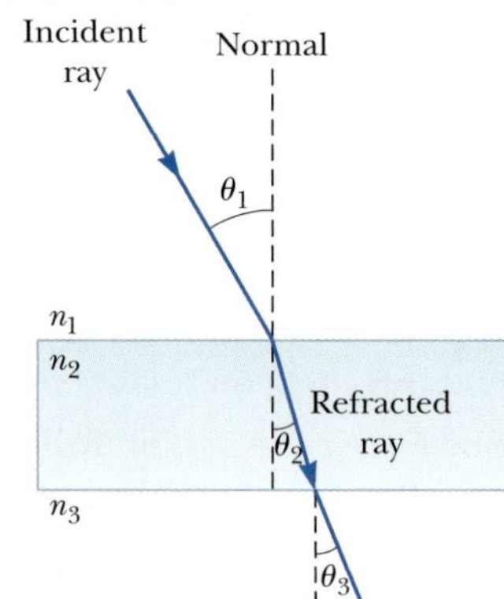


**Figure 22.11** (Example 22.2) Refraction of light by glass.

**SOLUTION**

**(a)** Find the angle of refraction, $\theta_2$.

Solve Snell's law (Eq. 22.8) for sin $\theta_2$:

$$\textbf{(1)}\quad \sin\theta_2 = \frac{n_1}{n_2}\sin\theta_1$$

From Table 22.1, find $n_1 = 1.00$ for air and $n_2 = 1.52$ for crown glass. Substitute these values into Equation (1) and take the inverse sine of both sides:

$$\sin\theta_2 = \left(\frac{1.00}{1.52}\right)(\sin 30.0°) = 0.329$$

$$\theta_2 = \sin^{-1}(0.329) = \boxed{19.2°}$$

**(b)** At what angle $\theta_3$ does the ray leave the glass as it re-enters the air?

Write Equation (1), replacing $\theta_3$ with $\theta_2$ and $\theta_1$ with $\theta_2$:

$$\textbf{(2)}\quad \sin\theta_3 = \frac{n_2}{n_3}\sin\theta_2 = \frac{1.52}{1.00}\sin(19.2°) = 0.500$$

Take the inverse sine of both sides to find $\theta_3$:

$$\theta_3 = \sin^{-1}(0.500) = \boxed{30.0°}$$

**(c)** How does the answer for $\theta_3$ change if the ray enters water below the slab instead of air?

Write Equation (2) and substitute a different value for $n_3$:

$$\sin\theta_3 = \frac{n_2}{n_3}\sin\theta_2 = \frac{1.52}{1.333}\sin(19.2°) = 0.375$$

$$\theta_3 = \sin^{-1}(0.375) = \boxed{22.0°}$$

**REMARKS** Notice that the light ray bends toward the normal when it enters a material of a higher index of refraction, and away from the normal when entering a material with a lower index of refraction. In passing through a slab of material with parallel surfaces, for example, from air to glass and back to air, the final direction of the ray is parallel to the direction of the incident ray. The only effect in that case is a lateral displacement of the light ray.

**QUESTION 22.2** If the glass is replaced by a transparent material with smaller index of refraction, will the refraction angle $\theta_2$ be (a) smaller, (b) larger, or (c) unchanged?

**EXERCISE 22.2** Suppose a light ray in air ($n = 1.00$) enters a cube of material ($n = 2.50$) at a 45.0° angle with respect to the normal and then exits the bottom of the cube into water ($n = 1.333$). At what angle to the normal does the ray leave the slab?

**ANSWER** 32.0°

## EXAMPLE 22.3 Light in Fused Quartz

**GOAL** Use the index of refraction to determine the effect of a medium on light's speed and wavelength.

**PROBLEM** Light of wavelength 589 nm in vacuum passes through a piece of fused quartz of index of refraction $n = 1.458$. **(a)** Find the speed of light in fused quartz. **(b)** What is the wavelength of this light in fused quartz? **(c)** What is the frequency of the light in fused quartz?

**STRATEGY** Substitute values into Equations 22.4 and 22.7.

**SOLUTION**

**(a)** Find the speed of light in fused quartz.

Obtain the speed from Equation 22.4:

$$v = \frac{c}{n} = \frac{3.00 \times 10^8 \text{ m/s}}{1.458} = \boxed{2.06 \times 10^8 \text{ m/s}}$$

**(b)** What is the wavelength of this light in fused quartz?

Use Equation 22.7 to calculate the wavelength:

$$\lambda_n = \frac{\lambda_0}{n} = \frac{589 \text{ nm}}{1.458} = \boxed{404 \text{ nm}}$$

**(c)** What is the frequency of the light in fused quartz?

The frequency in quartz is the same as in vacuum. Solve $c = f\lambda$ for the frequency:

$$f = \frac{c}{\lambda} = \frac{3.00 \times 10^8 \text{ m/s}}{589 \times 10^{-9} \text{ m}} = \boxed{5.09 \times 10^{14} \text{ Hz}}$$

**REMARKS** It's interesting to note that the speed of light in vacuum, $3.00 \times 10^8$ m/s, is an upper limit for the speed of material objects. In our treatment of relativity in Chapter 26, we will find that this upper limit is consistent with experimental observations. However, it's possible for a particle moving in a medium to have a speed that exceeds the speed of light in that medium. For example, it's theoretically possible for a particle to travel through fused quartz at a speed greater than $2.06 \times 10^8$ m/s, but it must still have a speed less than $3.00 \times 10^8$ m/s.

**QUESTION 22.3** True or False: If light with wavelength $\lambda$ in glass passes into water with index $n_w$, the new wavelength of the light is $\lambda/n_w$.

**EXERCISE 22.3** Light with wavelength 589 nm passes through crystalline sodium chloride. In this medium, find (a) the speed of light, (b) the wavelength, and (c) the frequency of the light.

**ANSWER** (a) $1.94 \times 10^8$ m/s (b) 381 nm (c) $5.09 \times 10^{14}$ Hz

## EXAMPLE 22.4 Refraction of Laser Light in a Digital Videodisc (DVD)

**GOAL** Apply Snell's law together with geometric constraints.

**PROBLEM** A DVD is a video recording consisting of a spiral track about 1.0 $\mu$m wide with digital information. (See Fig. 22.12a.) The digital information consists of a series of pits that are "read" by a laser beam sharply focused on a track in the information layer. If the width $a$ of the beam at the information layer must equal 1.0 $\mu$m to distinguish individual tracks and the width $w$ of the beam as it enters the plastic is 0.700 0 mm, find the angle $\theta_1$ at which the conical beam should enter the plastic. (See Fig. 22.12b.) Assume the plastic has a thickness $t$ = 1.20 mm and an index of refraction $n$ = 1.55. Note that this system is relatively immune to small dust particles degrading the video quality because particles would have to be as large as 0.700 mm to obscure the beam at the point where it enters the plastic.

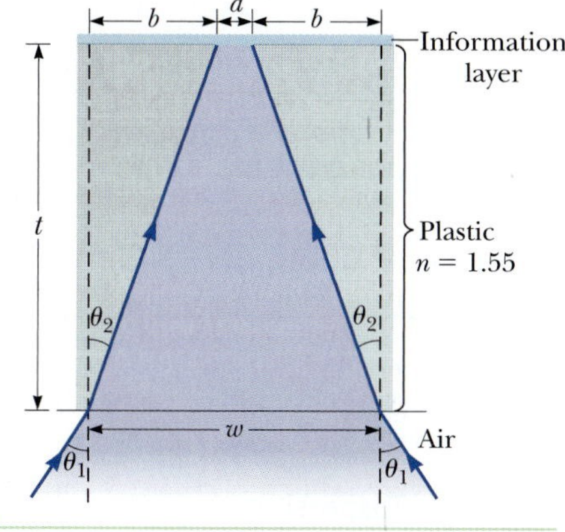


**Figure 22.12** (Example 22.4) A micrograph of a DVD surface showing tracks and pits along each track. (b) Cross section of a cone-shaped laser beam used to read a DVD.

**STRATEGY** Use right-triangle trigonometry to determine the angle $\theta_2$ and then apply Snell's law to obtain the angle $\theta_1$.

*(Continued)*

**SOLUTION**

From the top and bottom of Figure 22.12b, obtain an equation relating $w$, $b$, and $a$:

$$w = 2b + a$$

Solve this equation for $b$ and substitute given values:

$$b = \frac{w - a}{2} = \frac{700.0 \times 10^{-6}\,\text{m} - 1.0 \times 10^{-6}\,\text{m}}{2} = 349.5\ \mu\text{m}$$

Now use the tangent function to find $\theta_2$:

$$\tan\theta_2 = \frac{b}{t} = \frac{349.5\ \mu\text{m}}{1.20 \times 10^3\ \mu\text{m}} \quad \rightarrow \quad \theta_2 = 16.2°$$

Finally, use Snell's law to find $\theta_1$:

$$n_1 \sin\theta_1 = n_2 \sin\theta_2$$

$$\sin\theta_1 = \frac{n_2 \sin\theta_2}{n_1} = \frac{1.55 \sin 16.2°}{1.00} = 0.432$$

$$\theta_1 = \sin^{-1}(0.432) = 25.6°$$

**REMARKS** Despite its apparent complexity, the problem isn't that different from Example 22.2.

**QUESTION 22.4** Suppose the plastic were replaced by a material with a higher index of refraction. How would the width of the beam at the information layer be affected? (a) It would remain the same. (b) It would decrease. (c) It would increase.

**EXERCISE 22.4** Suppose you wish to redesign the system to decrease the initial width of the beam from 0.700 0 mm to 0.600 0 mm but leave the incident angle $\theta_1$ and all other parameters the same as before, except the index of refraction for the plastic material ($n_2$) and the angle $\theta_2$. What index of refraction should the plastic have?

**ANSWER** 1.79

## 22.4 Dispersion and Prisms

**LEARNING OBJECTIVES**

1. Define dispersion.
2. Describe how prisms work.

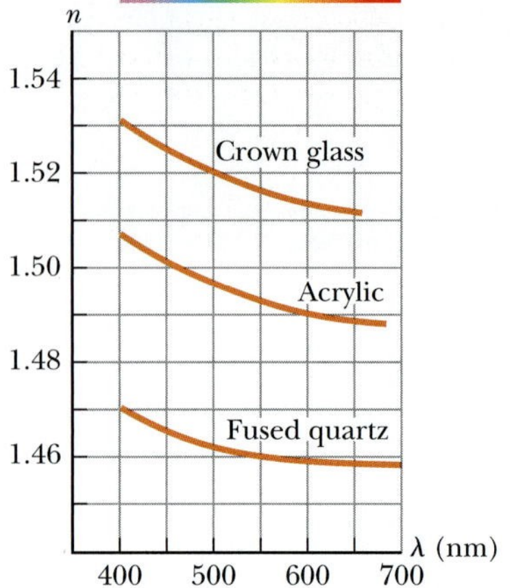


**Figure 22.13** Variations of index of refraction in the visible spectrum with respect to vacuum wavelength for three materials.

In Table 22.1 we presented values for the index of refraction of various materials. If we make careful measurements, however, we find that the index of refraction in anything but vacuum depends on the wavelength of light. The dependence of the index of refraction on wavelength is called **dispersion**. Figure 22.13 is a graphical representation of this variation in the index of refraction with wavelength. Because $n$ is a function of wavelength, Snell's law indicates that **the angle of refraction made when light enters a material depends on the wavelength of the light**. As seen in the figure, the index of refraction for a material usually decreases with increasing wavelength. This means that violet light ($\lambda \cong 400$ nm) refracts more than red light ($\lambda \cong 650$ nm) when passing from air into a material.

To understand the effects of dispersion on light, consider what happens when light strikes a prism, as in Figure 22.14a. A ray of light of a single wavelength that is incident on the prism from the left emerges bent away from its original direction of travel by an angle $\delta$, called the **angle of deviation**. Now suppose a beam of white light (a combination of all visible wavelengths) is incident on a prism. Because of dispersion, the different colors refract through different angles of deviation, as illustrated in Figure 22.14b. The rays that emerge from the second face of the prism spread out in a series of colors known as a visible **spectrum**, as shown in Figure 22.15. These colors, in order of decreasing wavelength, are red, orange, yellow, green, blue, and violet. Violet light deviates the most, red light the least, and the remaining colors in the visible spectrum fall between these extremes.

**Tip 22.3 Dispersion**

Light of shorter wavelength, such as violet light, refracts more than light of longer wavelengths, such as red light.

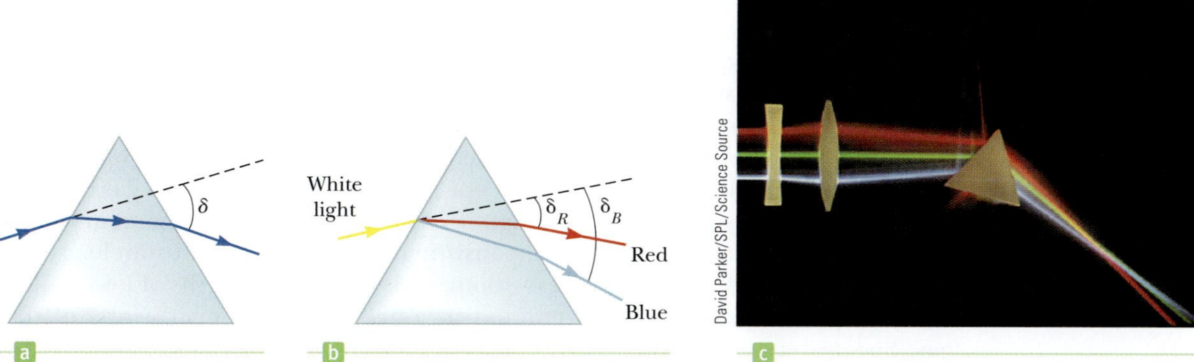


**Figure 22.14** (a) A prism refracts a light ray and deviates the light through the angle δ. (b) When light is incident on a prism, the blue light is bent more than the red light. (c) Light of different colors passes through a prism and two lenses. Note that as the light passes through the prism, different wavelengths are refracted at different angles.

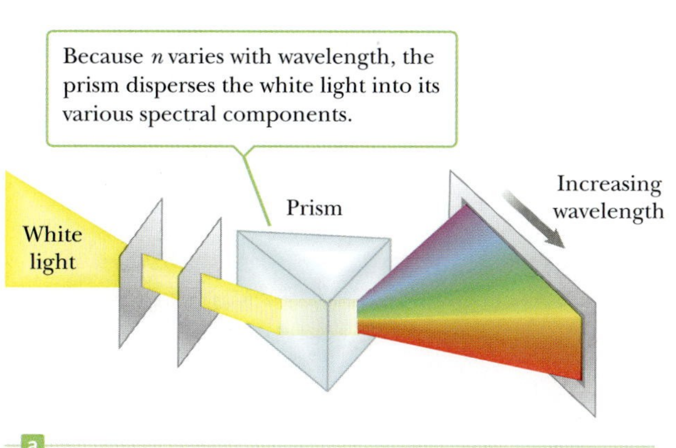


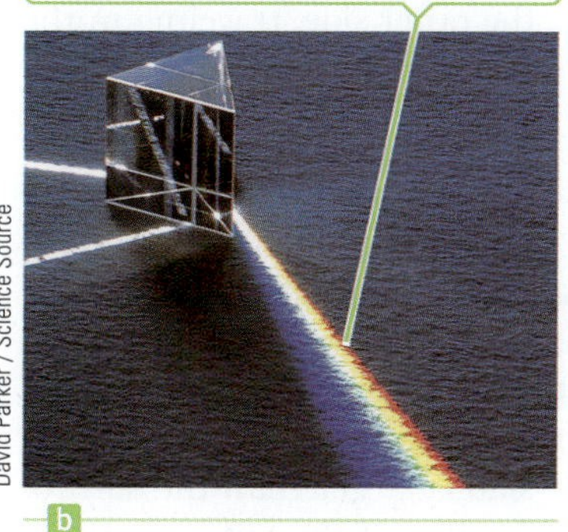

Different colors of light that pass through a prism are refracted at different angles because the index of refraction of the glass depends on wavelength. Violet light bends the most, red light the least.
David Parker / Science Source
b


**Figure 22.15** (a) Dispersion of white light by a prism. (b) White light enters a glass prism at the upper left.

Prisms are often used in an instrument known as a **prism spectrometer**, the essential elements of which are shown in Figure 22.16a. This instrument is commonly used to study the wavelengths emitted by a light source, such as a sodium vapor lamp. Light from the source is sent through a narrow, adjustable slit and lens to produce a parallel, or collimated, beam. The light then passes through the prism and is dispersed into a spectrum. The refracted light is observed through a telescope. The experimenter sees different colored images of the slit through the

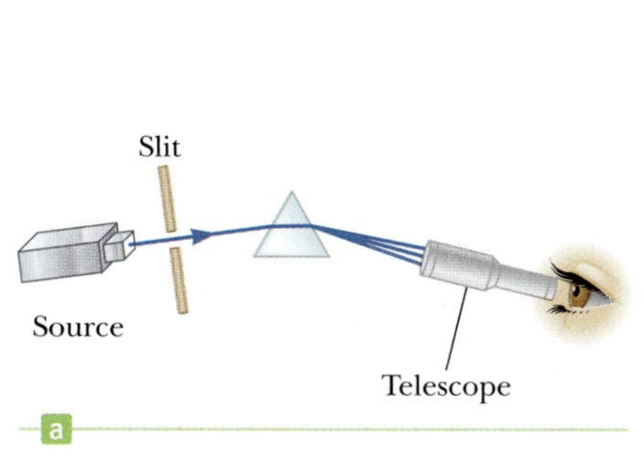


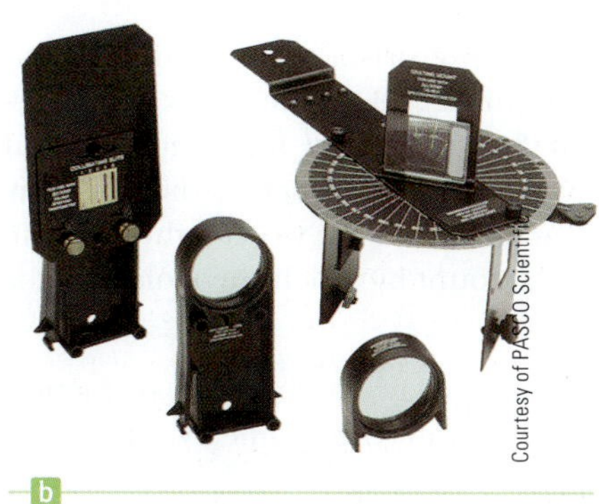


**Figure 22.16** (a) A diagram of a prism spectroscope. The colors in the spectrum are viewed through a telescope. (b) A prism spectrometer with interchangeable components. A spectrometer is a spectroscope with a scale or detector that measures wavelengths.

eyepiece of the telescope. The telescope can be moved or the prism can be rotated to view the various wavelengths, which have different angles of deviation. Figure 22.16b shows one type of prism spectrometer used in undergraduate laboratories.

**APPLICATION**
Identifying Gases with a Spectrometer

All hot, low-pressure gases emit their own characteristic spectra, so one use of a prism spectrometer is to identify gases. For example, sodium emits only two wavelengths in the visible spectrum: two closely spaced yellow lines. (The bright line-like images of the slit seen in a spectroscope are called *spectral lines.*) A gas emitting these, and only these, colors can be identified as sodium. Likewise, mercury vapor has its own characteristic spectrum, consisting of four prominent wavelengths—orange, green, blue, and violet lines—along with some wavelengths of lower intensity. The particular wavelengths emitted by a gas serve as "fingerprints" of that gas. Spectral analysis, which is the measurement of the wavelengths emitted or absorbed by a substance, is a powerful general tool in many scientific areas. As examples, chemists and biologists use infrared spectroscopy to identify molecules, astronomers use visible-light spectroscopy to identify elements on distant stars, and geologists use spectral analysis to identify minerals.

## APPLYING PHYSICS 22.3 | Dispersion

When a beam of light enters a glass prism, which has nonparallel sides, the rainbow of color exiting the prism is a testimonial to the dispersion occurring in the glass. Suppose a beam of light enters a slab of material with parallel sides. When the beam exits the other side, traveling in the same direction as the original beam, is there any evidence of dispersion?

**EXPLANATION** Due to dispersion, light at the violet end of the spectrum exhibits a larger angle of refraction on entering the glass than light at the red end. All colors of light return to their original direction of propagation as they refract back out into the air. As a result, the outgoing beam is white. The net shift in the position of the violet light along the edge of the slab is larger than the shift of the red light, however, so one edge of the outgoing beam has a bluish tinge to it (it appears blue rather than violet because the eye is not very sensitive to violet light), whereas the other edge has a reddish tinge. This effect is indicated in Figure 22.17. The colored edges of the outgoing beam of white light are evidence of dispersion. ■

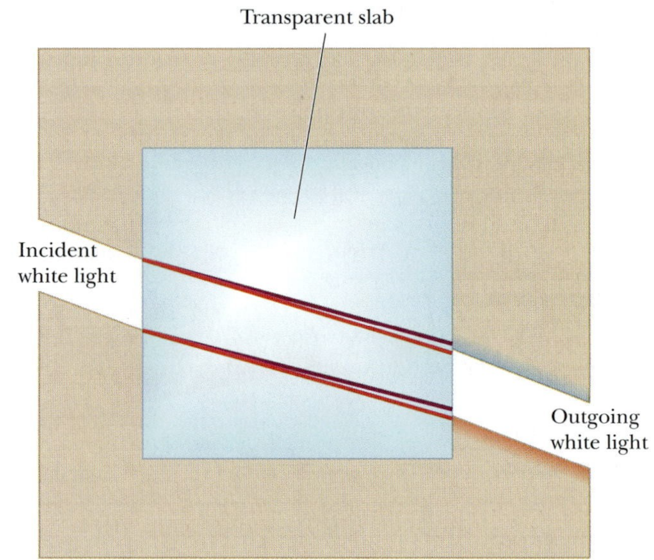


**Figure 22.17** (Applying Physics 22.3)

## EXAMPLE 22.5 | Light Through a Prism

**GOAL** Calculate the consequences of dispersion.

**PROBLEM** A beam of light is incident on a prism of a certain glass at an angle of $\theta_1 = 30.0°$, as shown in Figure 22.18. If the index of refraction of the glass for violet light is 1.80, find **(a)** $\theta_2$, the angle of refraction at the air–glass interface, **(b)** $\phi_2$, the angle of incidence at the glass–air interface, and **(c)** $\phi_1$, the angle of refraction when the violet light exits the prism. **(d)** What is the value of $\Delta y$, the amount by which the violet light is displaced vertically?

**STRATEGY** This problem requires Snell's law to find the refraction angles and some elementary geometry and trigonometry based on Figure 22.18.

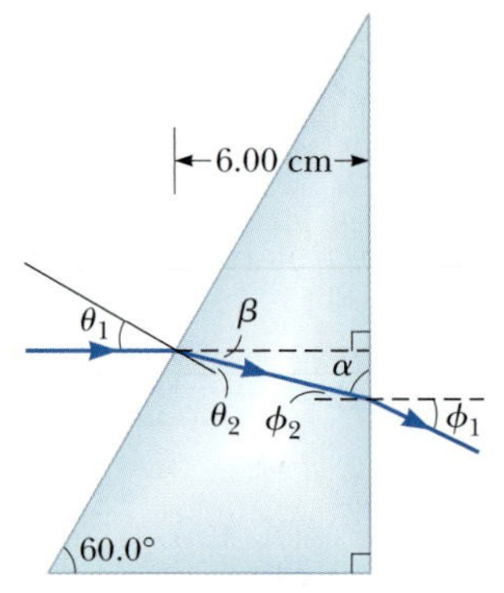


**Figure 22.18** (Example 22.5)

**SOLUTION**

**(a)** Find $\theta_2$, the angle of refraction at the air–glass interface.

Use Snell's law to find the first angle of refraction:

$$n_1 \sin \theta_1 = \sin \theta_2 \rightarrow (1.00) \sin 30.0 = (1.80) \sin \theta_2$$

$$\theta_2 = \sin^{-1}\left(\frac{0.500}{1.80}\right) = \boxed{16.1°}$$

**(b)** Find $\phi_2$, the angle of incidence at the glass–air interface.

Compute the angle $\beta$:

$$\beta = 30.0° - \theta_2 = 30.0° - 16.1° = 13.9°$$

Compute the angle $\alpha$ using the fact that the sum of the interior angles of a triangle equals 180°:

$$180° = 13.9° + 90° + \alpha \rightarrow \alpha = 76.1°$$

The incident angle $\phi_2$ at the glass–air interface is complementary to $\alpha$:

$$\phi_2 = 90° - \alpha = 90° - 76.1° = \boxed{13.9°}$$

**(c)** Find $\phi_1$, the angle of refraction when the violet light exits the prism.

Apply Snell's law:

$$\phi_1 = \left(\frac{1}{n_1}\right) \sin^{-1}(n_2 \sin \phi_2)$$

$$= \left(\frac{1}{1.00}\right) \sin^{-1}[(1.80) \sin 13.9°] = \boxed{25.6°}$$

**(d)** What is the value of $\Delta y$, the amount by which the violet light is displaced vertically?

Use the tangent function to find the vertical displacement:

$$\tan \beta = \frac{\Delta y}{\Delta x} \rightarrow \Delta y = \Delta x \tan \beta$$

$$\Delta y = (6.00 \text{ cm}) \tan (13.9°) = \boxed{1.48 \text{ cm}}$$

**REMARKS** The same calculation for red light is left as an exercise. The violet light is bent more and displaced farther down the face of the prism. Notice that a theorem in geometry about parallel lines and the angles created by a transverse line give $\phi_2 = \beta$ immediately, which would have saved some calculation. In general, however, this tactic might not be available.

**QUESTION 22.5** On passing through the prism, will yellow light bend through a larger angle or smaller angle than the violet light? (a) Yellow light bends through a larger angle. (b) Yellow light bends through a smaller angle. (c) The angles are the same.

**EXERCISE 22.5** Repeat parts (a) through (d) of the example for red light passing through the prism, given that the index of refraction for red light is 1.72.

**ANSWERS** (a) 16.9° (b) 13.1° (c) 22.9° (d) 1.40 cm

## 22.5 The Rainbow

**LEARNING OBJECTIVE**

1. Describe how the dispersion of light forms rainbows.

The dispersion of light into a spectrum is demonstrated most vividly in nature through the formation of a rainbow, often seen by an observer positioned between the Sun and a rain shower. To understand how a rainbow is formed, consider Figure 22.19 (page 786). A ray of light passing overhead strikes a drop of water in the atmosphere and is refracted and reflected as follows: It is first refracted at the front surface of the drop, with the violet light deviating the most and the red light

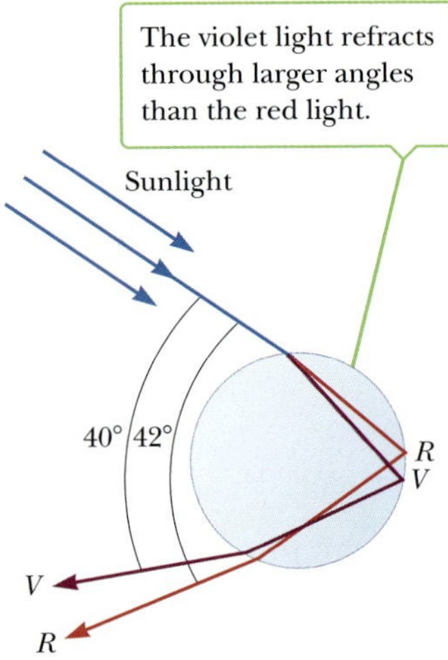


**Figure 22.19** Refraction of sunlight by a spherical raindrop.

the least. At the back surface of the drop, the light is reflected and returns to the front surface, where it again undergoes refraction as it moves from water into air. The rays leave the drop so that the angle between the incident white light and the returning violet ray is 40° and the angle between the white light and the returning red ray is 42°. This small angular difference between the returning rays causes us to see the bow as explained in the next paragraph.

Now consider an observer viewing a rainbow, as in Figure 22.20a. If a raindrop high in the sky is being observed, the red light returning from the drop can reach the observer because it is deviated the most, but the violet light passes over the observer because it is deviated the least. Hence, the observer sees this drop as being red. Similarly, a drop lower in the sky would direct violet light toward the observer and appear to be violet. (The red light from this drop would strike the ground and not be seen.) The remaining colors of the spectrum would reach the observer from raindrops lying between these two extreme positions. Figure 22.20b shows a beautiful rainbow and a secondary rainbow with its colors reversed.

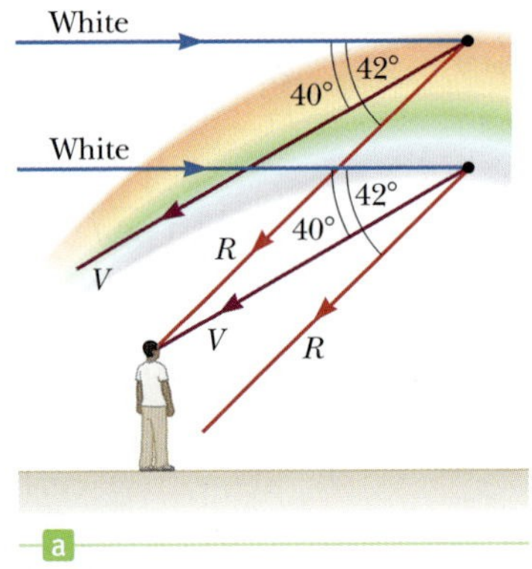

**Figure 22.20** The formation of a rainbow seen by an observer standing with the Sun behind his back. (b) This photograph of a rainbow shows a distinct secondary rainbow with the colors reversed.

## 22.6 Huygens' Principle

**LEARNING OBJECTIVE**

1. State Huygens' principle, and apply it to the understanding of reflection and refraction.

The laws of reflection and refraction can be deduced using a geometric method proposed by Huygens in 1678. Huygens assumed light is a form of wave motion rather than a stream of particles. He had no knowledge of the nature of light or of its electromagnetic character. Nevertheless, his simplified wave model is adequate for understanding many practical aspects of the propagation of light.

Huygens' principle is a geometric construction for determining at some instant the position of a new wave front from knowledge of the wave front that preceded it. (A wave front is a surface passing through those points of a wave which have the same phase and amplitude. For instance, a wave front could be a surface passing through the crests of waves.) In Huygens' construction, **all points on a given wave front are taken as point sources for the production of spherical secondary waves, called wavelets, that propagate in the forward direction with speeds characteristic of waves in that medium. After some time has elapsed, the new position of the wave front is the surface tangent to the wavelets.**

Huygens' principle ▶

Figure 22.21 illustrates two simple examples of Huygens' construction. First, consider a plane wave moving through free space, as in Figure 22.21a. At $t = 0$, the wave front is indicated by the plane labeled $AA'$. In Huygens' construction, each point on this wave front is considered a point source. For clarity, only a few points

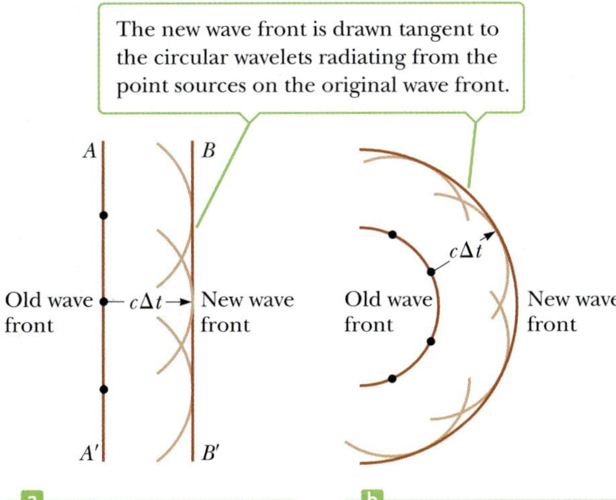


**Figure 22.21** Huygens' constructions for (a) a plane wave propagating to the right and (b) a spherical wave.

on $AA'$ are shown. With these points as sources for the wavelets, we draw circles of radius $c\,\Delta t$, where $c$ is the speed of light in vacuum and $\Delta t$ is the period of propagation from one wave front to the next. The surface drawn tangent to these wavelets is the plane $BB'$, which is parallel to $AA'$. In a similar manner, Figure 22.21b shows Huygens' construction for an outgoing spherical wave.

## Huygens' Principle Applied to Reflection and Refraction

The laws of reflection and refraction were stated earlier in the chapter without proof. We now derive these laws using Huygens' principle. Figure 22.22a illustrates the law of reflection. The line $AA'$ represents a wave front of the incident light. As ray 3 travels from $A'$ to $C$, ray 1 reflects from $A$ and produces a spherical wavelet of radius $AD$. (Recall that the radius of a Huygens wavelet is $v\,\Delta t$.) Because the two wavelets having radii $A'C$ and $AD$ are in the same medium, they have the same speed $v$, so $AD = A'C$. Meanwhile, the spherical wavelet centered at $B$ has spread only half as far as the one centered at $A$ because ray 2 strikes the surface later than ray 1.

From Huygens' principle, we find that the reflected wave front is $CD$, a line tangent to all the outgoing spherical wavelets. The remainder of our analysis depends on geometry, as summarized in Figure 22.22b. Note that the right triangles $ADC$ and $AA'C$ are congruent because they have the same hypotenuse, $AC$, and because $AD = A'C$. From the figure, we have

$$\sin\theta_1 = \frac{A'C}{AC} \qquad \text{and} \qquad \sin\theta_1' = \frac{AD}{AC}$$

The right-hand sides are equal, so $\sin\theta_1 = \sin\theta_1'$, and it follows that $\theta_1 = \theta_1'$, which is the law of reflection.

Huygens' principle and Figure 22.23a (page 788) can be used to derive Snell's law of refraction. In the time interval $\Delta t$, ray 1 moves from $A$ to $B$ and ray 2 moves from $A'$ to $C$. The radius of the outgoing spherical wavelet centered at $A$ is equal

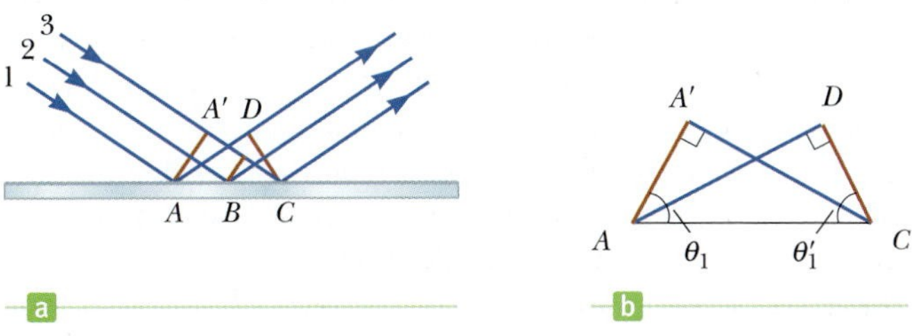


**Figure 22.22** (a) Huygens' construction for proving the law of reflection. (b) Triangle $ADC$ is congruent to triangle $AA'C$.

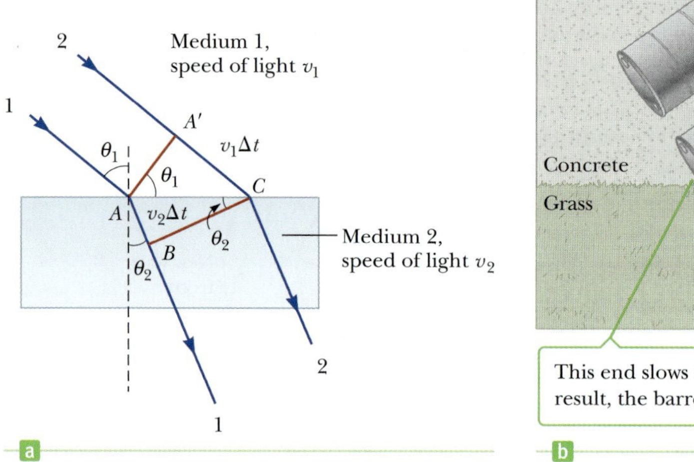


**Figure 22.23** (a) Huygens' construction for proving the law of refraction. (b) Overhead view of a barrel rolling from concrete onto grass.

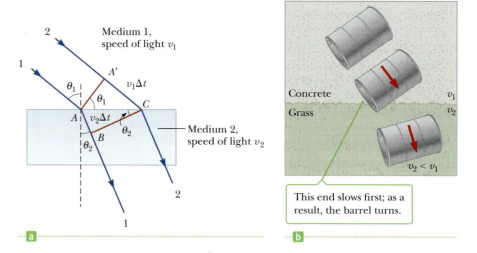

to $v_2\,\Delta t$. The distance $A'C$ is equal to $v_1\,\Delta t$. Geometric considerations show that angle $A'AC$ equals $\theta_1$ and angle $ACB$ equals $\theta_2$. From triangles $AA'C$ and $ACB$, we find that

$$\sin\theta_1 = \frac{v_1\Delta t}{AC} \quad \text{and} \quad \sin\theta_2 = \frac{v_2\Delta t}{AC}$$

If we divide the first equation by the second, we get

$$\frac{\sin\theta_1}{\sin\theta_2} = \frac{v_1}{v_2}$$

From Equation 22.4, though, we know that $v_1 = c/n_1$ and $v_2 = c/n_2$. Therefore,

$$\frac{\sin\theta_1}{\sin\theta_2} = \frac{c/n_1}{c/n_2} = \frac{n_2}{n_1}$$

and it follows that

$$n_1\sin\theta_1 = n_2\sin\theta_2$$

which is the law of refraction.

A mechanical analog of refraction is shown in Figure 22.23b. When the left end of the rolling barrel reaches the grass, it slows down, while the right end remains on the concrete and moves at its original speed. This difference in speeds causes the barrel to pivot, changing the direction of its motion.

## 22.7 Total Internal Reflection

**LEARNING OBJECTIVE**

1. Define total internal reflection and apply it in basic physical contexts.

Courtesy of Henry Leap and Jim Lehman

This photograph shows nonparallel light rays entering a glass prism. The bottom two rays undergo total internal reflection at the longest side of the prism. The top three rays are refracted at the longest side as they leave the prism.

An interesting effect called *total internal reflection* can occur when light encounters the boundary between a medium with a *higher* index of refraction and one with a *lower* index of refraction. Consider a light beam traveling in medium 1 and meeting the boundary between medium 1 and medium 2, where $n_1$ is greater than $n_2$ (Fig. 22.24). Possible directions of the beam are indicated by rays 1 through 5. Note that the refracted rays are bent away from the normal because $n_1$ is greater than $n_2$. At some particular angle of incidence $\theta_c$, called the **critical angle**, the refracted light ray moves parallel to the boundary so that $\theta_2 = 90°$ (Fig. 22.24b). *For angles of incidence greater than* $\theta_c$, the beam is entirely reflected at the boundary, as is ray 5 in Figure 22.24a. This ray is reflected as though it had struck a perfectly reflecting surface. It and all rays like it obey the law of reflection: the angle of incidence equals the angle of reflection.

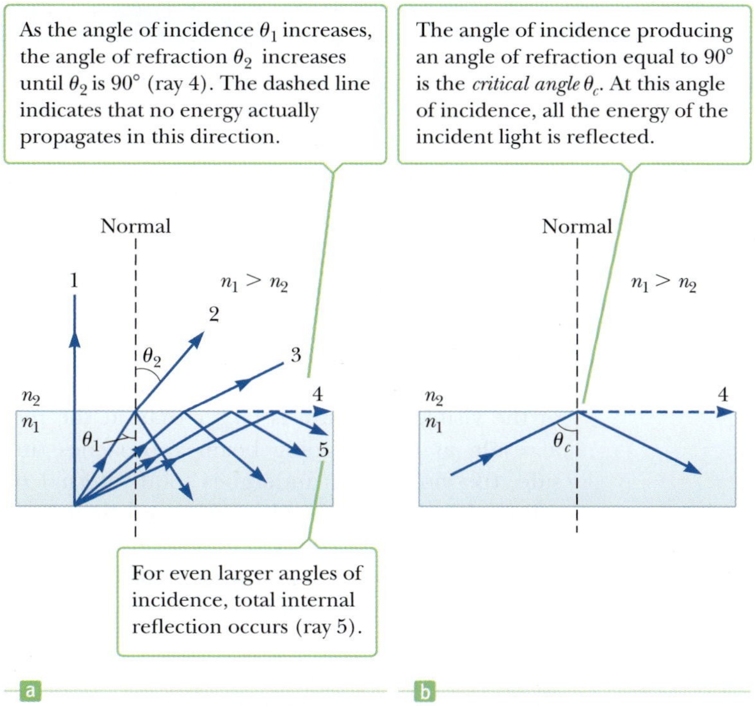

As the angle of incidence $\theta_1$ increases, the angle of refraction $\theta_2$ increases until $\theta_2$ is 90° (ray 4). The dashed line indicates that no energy actually propagates in this direction.
The angle of incidence producing an angle of refraction equal to 90° is the *critical angle* $\theta_c$. At this angle of incidence, all the energy of the incident light is reflected.
Normal
$n_1 > n_2$
1
2
3
4
5
$\theta_2$
$\theta_1$
$n_2$
$n_1$
For even larger angles of incidence, total internal reflection occurs (ray 5).
$\theta_c$
a
b

**Figure 22.24** (a) Rays from a medium with index of refraction $n_1$ travel to a medium with index of refraction $n_2$, where $n_1 > n_2$. (b) Ray 4 is singled out.

We can use Snell's law to find the critical angle. When $\theta_1 = \theta_c$ and $\theta_2 = 90°$, Snell's law (Eq. 22.8) gives

$$n_1 \sin \theta_c = n_2 \sin 90° = n_2$$

$$\sin \theta_c = \frac{n_2}{n_1} \quad \text{for } n_1 > n_2 \qquad [22.9]$$

Equation 22.9 can be used only when $n_1$ is greater than $n_2$ because **total internal reflection occurs only when light is incident on the boundary of a medium having a lower index of refraction than the medium in which it's traveling.** If $n_1$ were less than $n_2$, Equation 22.9 would give $\sin \theta_c > 1$, which is an absurd result because the sine of an angle can never be greater than 1.

When medium 2 is air, the critical angle is small for substances with large indices of refraction, such as diamond, where $n = 2.42$ and $\theta_c = 24.0°$. By comparison, for crown glass, $n = 1.52$ and $\theta_c = 41.0°$. This property, combined with proper faceting, causes a diamond to sparkle brilliantly.

**APPLICATION**
Submarine Periscopes

A prism and the phenomenon of total internal reflection can alter the direction of travel of a light beam. Figure 22.25 illustrates two such possibilities. In one case the light beam is deflected by 90° (Fig. 22.25a), and in the second case the path of the beam is reversed (Fig. 22.25b). A common application of total internal reflection is a submarine periscope. In this device two prisms are arranged as in Figure 22.25c so that an incident beam of light follows the path shown and the user can "see around corners."

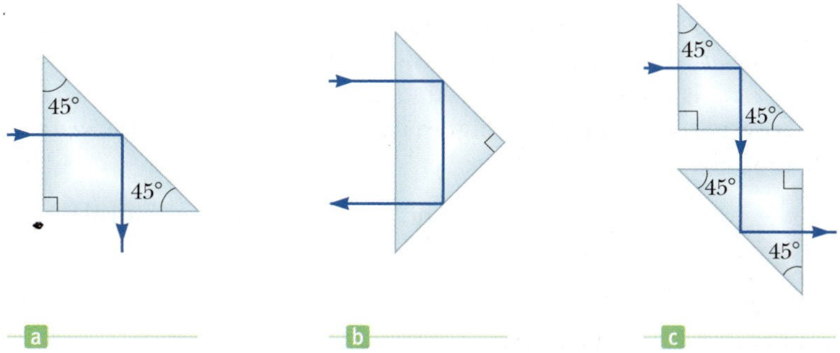

**Figure 22.25** Internal reflection in a prism. (a) The ray is deviated by 90°. (b) The direction of the ray is reversed. (c) Two prisms used as a periscope.

## APPLYING PHYSICS 22.4 | Total Internal Reflection and Dispersion

A beam of white light is incident on the curved edge of a semicircular piece of glass, as shown in Figure 22.26. The light enters the curved surface along the normal, so it shows no refraction. It encounters the straight side of the glass at the center of curvature of the curved side and refracts into the air. The incoming beam is moved clockwise (so that the angle $\theta$ increases) such that the beam always enters along the normal to the curved side and encounters the straight side at the center of curvature of the curved side. Why does the refracted beam become redder as it approaches a direction parallel to the straight side?

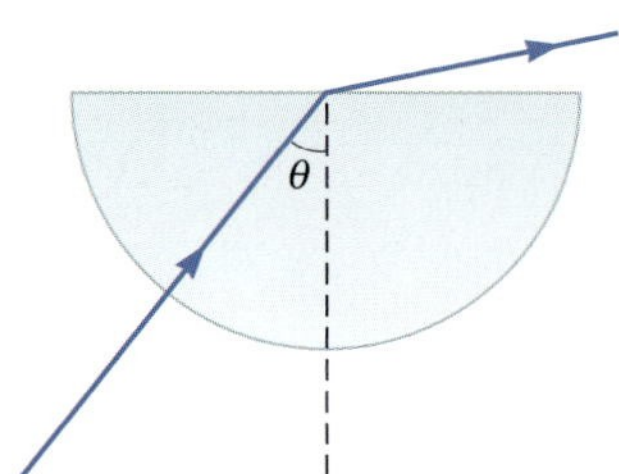


**Figure 22.26** (Applying Physics 22.4)

**EXPLANATION** When the outgoing beam approaches the direction parallel to the straight side, the incident angle is approaching the critical angle for total internal reflection. Dispersion occurs as the light passes out of the glass. The index of refraction for light at the violet end of the visible spectrum is larger than at the red end. As a result, as the outgoing beam approaches the straight side, the violet light undergoes total internal reflection, followed by the other colors. The red light is the last to undergo total internal reflection, so just before the outgoing light disappears, it's composed of light from the red end of the visible spectrum. ■

## EXAMPLE 22.6 | A View from the Fish's Eye

**GOAL** Apply the concept of total internal reflection.

**PROBLEM** **(a)** Find the critical angle for a water–air boundary. **(b)** Use the result of part (a) to predict what a fish will see (Fig. 22.27) if it looks up toward the water surface at angles of 40.0°, 48.6°, and 60.0°.

**STRATEGY** After finding the critical angle by substitution, use the fact that the path of a light ray is reversible: at a given angle, wherever a light beam can go is also where a light beam can come from, along the same path.

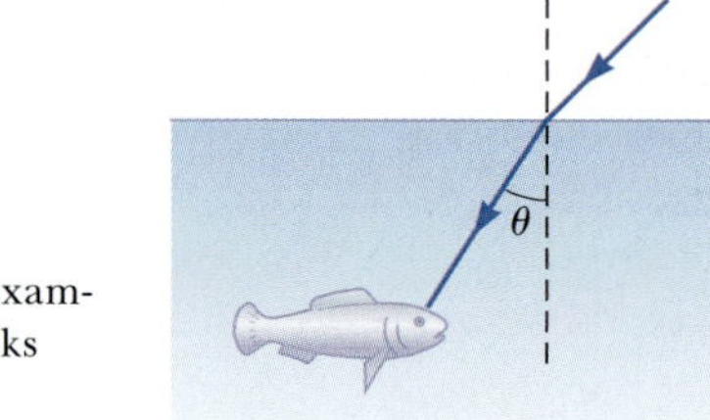


**Figure 22.27** (Example 22.6) A fish looks upward toward the water's surface.

**SOLUTION**

**(a)** Find the critical angle for a water–air boundary.

Substitute into Equation 22.9 to find the critical angle:

$$\sin\theta_c = \frac{n_2}{n_1} = \frac{1.00}{1.333} = 0.750$$

$$\theta_c = \sin^{-1}(0.750) = \boxed{48.6^\circ}$$

**(b)** Predict what a fish will see if it looks up toward the water surface at angles of 40.0°, 48.6°, and 60.0°.

At an angle of 40.0°, a beam of light from underwater will be refracted at the surface and enter the air above. Because the path of a light ray is reversible (Snell's law works both going and coming), light from above can follow the same path and be perceived by the fish. At an angle of 48.6°, the critical angle for water, light from underwater is bent so that it travels along the surface. So light following the same path in reverse can reach the fish only by skimming along the water surface before being refracted toward the fish's eye. At angles greater than the critical angle of 48.6°, a beam of light shot toward the surface will be completely reflected down toward the bottom of the pool. Reversing the path, the fish sees a reflection of some object on the bottom.

**QUESTION 22.6** If the water is replaced by a transparent fluid with a higher index of refraction, is the critical angle of the fluid–air boundary (a) larger, (b) smaller, or (c) the same as for water?

**EXERCISE 22.6** Suppose a layer of oil with $n = 1.50$ coats the surface of the water. What is the critical angle for total internal reflection for light traveling in the oil layer and encountering the oil–water boundary?

**ANSWER** 62.7°

## Fiber Optics

Another interesting application of total internal reflection is the use of solid glass or transparent plastic rods to "pipe" light from one place to another. As indicated in Figure 22.28, light is confined to traveling within the rods, even around gentle curves, as a result of successive internal reflections. Such a light pipe can be quite flexible if thin fibers are used rather than thick rods. If a bundle of parallel fibers is used to construct an optical transmission line, images can be transferred from one point to another.

**Figure 22.28** Light travels in a curved transparent rod by multiple internal reflections.

Very little light intensity is lost in these fibers as a result of reflections on the sides. Any loss of intensity is due essentially to reflections from the two ends and absorption by the fiber material. Fiber-optic devices are particularly useful for viewing images produced at inaccessible locations. Physicians often use fiber-optic cables to aid in the diagnosis and correction of certain medical problems without the intrusion of major surgery. For example, a fiber-optic cable can be threaded through the esophagus and into the stomach to look for ulcers. In this application the cable consists of two fiber-optic lines: one to transmit a beam of light into the stomach for illumination and the other to allow the light to be transmitted out of the stomach. The resulting image can, in some cases, be viewed directly by the physician, but more often is displayed on a television monitor or saved in digital form. In a similar way, fiber-optic cables can be used to examine the colon or to help physicians perform surgery without the need for large incisions.

**BIO APPLICATION**
Fiber Optics in Medical Diagnosis and Surgery

The field of fiber optics has revolutionized the entire communications industry. Billions of kilometers of optical fiber have been installed in the United States to carry high-speed Internet traffic, radio and television signals, and telephone calls. The fibers can carry much higher volumes of telephone calls and other forms of communication than electrical wires because of the higher frequency of the infrared light used to carry the information on optical fibers. Optical fibers are also preferable to copper wires because they are insulators and don't pick up stray electric and magnetic fields or electronic "noise."

**APPLICATION**
Fiber Optics in Telecommunications

Dennis O'Clair/Tony Stone Images/Getty Images

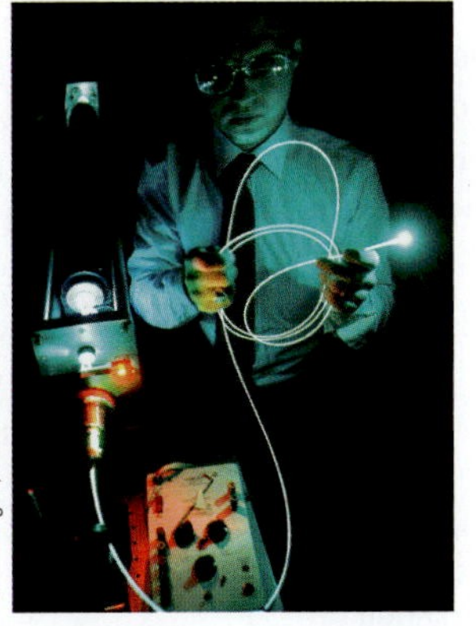
Hank Morgan/Science Source

*(Left)* Strands of glass optical fibers are used to carry voice, video, and data signals in telecommunication networks. *(Right)* A bundle of optical fibers is illuminated by a laser.

### APPLYING PHYSICS 22.5 | Design of an Optical Fiber

An optical fiber consists of a transparent core surrounded by cladding, which is a material with a lower index of refraction than the core (Fig. 22.29). A cone of angles, called the acceptance cone, is at the entrance to the fiber. Incoming light at angles within this cone will be transmitted through the fiber, whereas light entering the core from angles outside the cone will not be transmitted. The figure shows a light ray entering the fiber just within the acceptance cone and undergoing total internal reflection at the interface between the core and the cladding. If it is

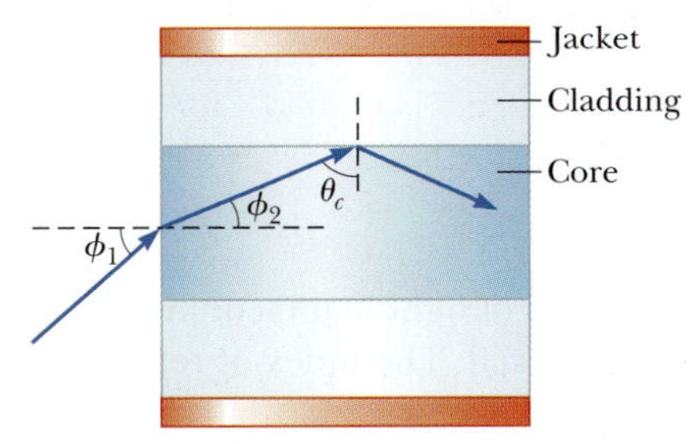

**Figure 22.29** (Applying Physics 22.5)

*(Continued)*

technologically difficult to produce light so that it enters the fiber from a small range of angles, how could you adjust the indices of refraction of the core and cladding to increase the size of the acceptance cone? Would you design the indices to be farther apart or closer together?

**EXPLANATION** The acceptance cone would become larger if the critical angle ($\theta_c$ in the figure) could be made smaller. This requires making the index of refraction of the cladding material smaller so that the indices of refraction of the core and cladding material are farther apart. ■

## SUMMARY

### 22.1 The Nature of Light

Light has a dual nature. In some experiments it acts like a wave, in others like a particle, called a photon by Einstein. The energy of a photon is proportional to its frequency,

$$E = hf \quad [22.1]$$

where $h = 6.63 \times 10^{-34}$ J · s is *Planck's constant.*

### 22.2 Reflection and Refraction

In the reflection of light off a flat, smooth surface, the angle of incidence, $\theta_1$, with respect to a line perpendicular to the surface is equal to the angle of reflection, $\theta_1'$:

$$\theta_1' = \theta_1 \quad [22.2]$$

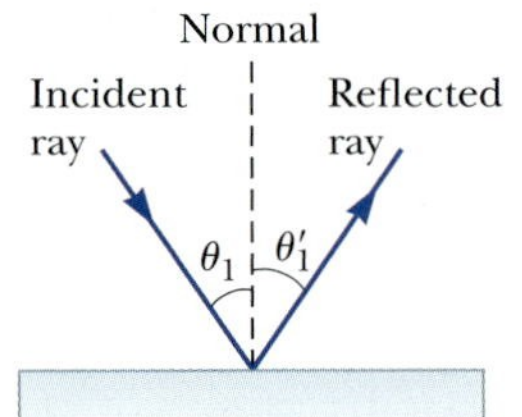


The wave under reflection model.

Light that passes into a transparent medium is bent at the boundary and is said to be *refracted*. The angle of refraction is the angle the ray makes with respect to a line perpendicular to the surface after it has entered the new medium.

### 22.3 The Law of Refraction

The **index of refraction** of a material, $n$, is defined as

$$n = \frac{c}{v} \quad [22.4]$$

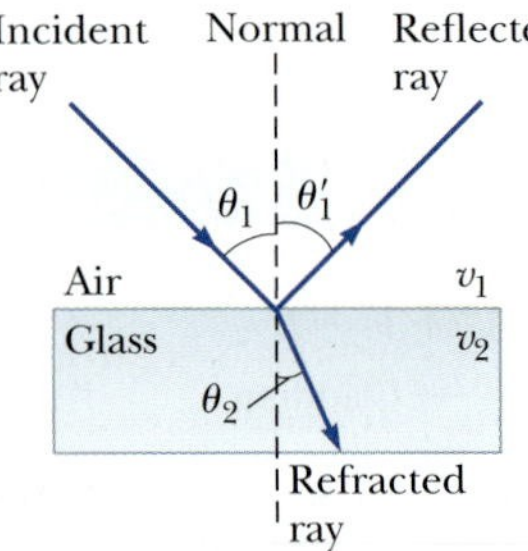


The wave under refraction model.

where $c$ is the speed of light in a vacuum and $v$ is the speed of light in the material. The index of refraction of a material is also

$$n = \frac{\lambda_0}{\lambda_n} \quad [22.7]$$

where $\lambda_0$ is the wavelength of the light in vacuum and $\lambda_n$ is its wavelength in the material.

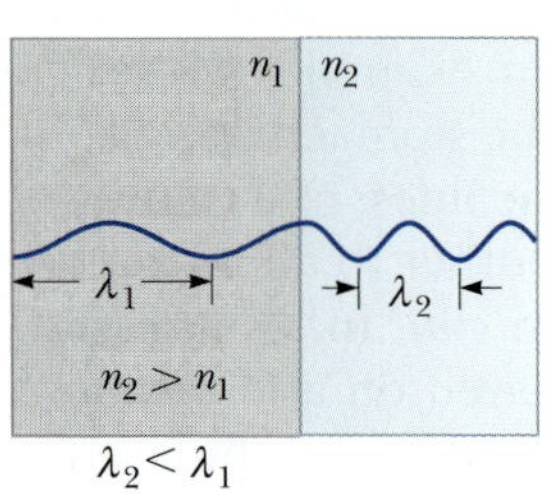


When entering a material with higher index of refraction, the wavelength of light is reduced.

The **law of refraction**, or **Snell's law**, states that

$$n_1 \sin \theta_1 = n_2 \sin \theta_2 \quad [22.8]$$

where $n_1$ and $n_2$ are the indices of refraction in the two media. The incident ray, the reflected ray, the refracted ray, and the normal to the surface all lie in the same plane.

### 22.4 Dispersion and Prisms

### 22.5 The Rainbow

The index of refraction of a material depends on the wavelength of the incident light, an effect called *dispersion*. Light at the violet end of the spectrum exhibits a larger angle of refraction on entering glass than light at the red end. Rainbows are a consequence of dispersion.

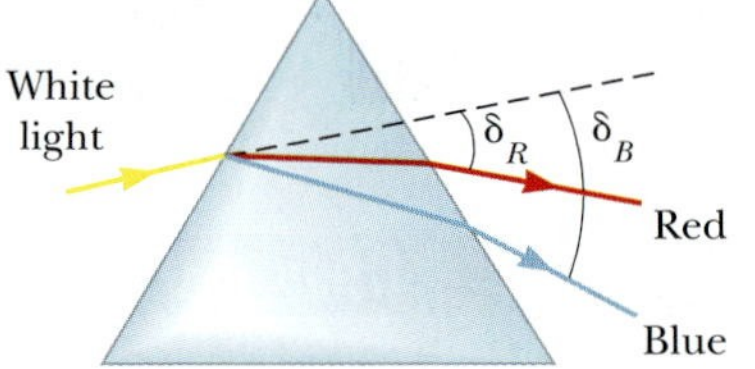


Due to dispersion, blue light is bent more than red light.

### 22.6 Huygens' Principle

**Huygens' principle** states that all points on a wave front are point sources for the production of spherical secondary waves called wavelets. These wavelets propagate forward at a speed characteristic of waves in a particular medium. After some time has elapsed, the new position of the wave front is the surface tangent to the wavelets. This principle can be used to deduce the laws of reflection and refraction.

## 22.7 Total Internal Reflection

When light propagating in a medium with index of refraction $n_1$ is incident on the boundary of a region with index of refraction $n_2$, and $n_1 > n_2$, then total internal reflection can occur if the angle of incidence equals or exceeds a **critical angle** $\theta_c$ given by

$$\sin \theta_c = \frac{n_2}{n_1} \qquad (n_1 > n_2) \qquad [22.9]$$

Total internal reflection is used in the optical fibers that carry data at high speed around the world.

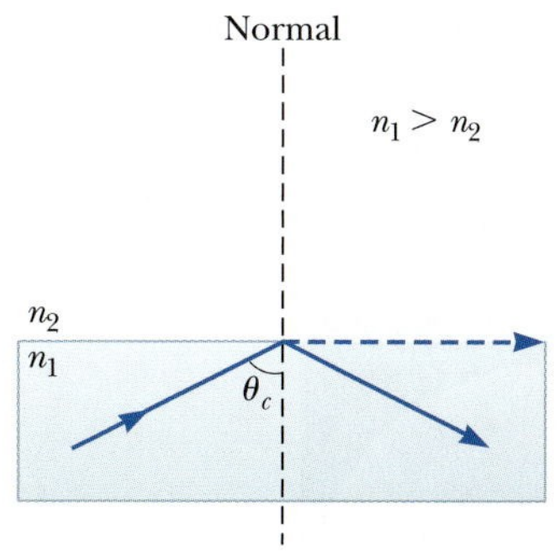


A ray undergoes total internal reflection at angles equal to or greater than the critical angle.

## WARM-UP EXERCISES

ENHANCED WebAssign The warm-up exercises in this chapter may be assigned online in Enhanced WebAssign.

1. How many 800-nm photons does it take to have the same total energy as four 200-nm photons? (See Section 22.1.)
2. Two mirrors are set up at an angle of 90.0°. If light is incident on the first mirror at an angle of 30.0° from the normal, at what angle is it incident on the second mirror? (See Section 22.2.)
3. A ray of light in air is incident at an angle of 30.0° on a glass slide with index of refraction 1.73. (a) At what angle is the ray refracted? (b) If the wavelength of the light in vacuum is 564 nm, find its wavelength in the glass. (See Section 22.3.)
4. A source emits monochromatic light of wavelength 495 nm in air. When the light passes through a liquid, its wavelength reduces to 434 nm. (a) What is the liquid's index of refraction? (b) Find the speed of light in the liquid. (See Section 22.3.)
5. Carbon disulfide ($n$ = 1.63) is poured into a container made of crown glass ($n$ = 1.52). What is the critical angle for total internal reflection of a light ray in the liquid when it is incident on the liquid-to-glass surface? (See Section 22.7.)

## CONCEPTUAL QUESTIONS

ENHANCED WebAssign The conceptual questions in this chapter may be assigned online in Enhanced WebAssign.

1. Why does the arc of a rainbow appear with red on top and violet on the bottom?
2. A ray of light is moving from a material having a high index of refraction into a material with a lower index of refraction. (a) Is the ray bent toward the normal or away from it? (b) If the wavelength is 600 nm in the material with the high index of refraction, is it greater, smaller, or the same in the material with the lower index of refraction? (c) How does the frequency change as the light moves between the two materials? Does it increase, decrease, or remain the same?
3. A light ray travels through three parallel slabs having different indices of refraction as in Figure CQ22.3. The rays shown are only the refracted rays. Rank the materials according to the size of their indices of refraction, from largest to smallest.

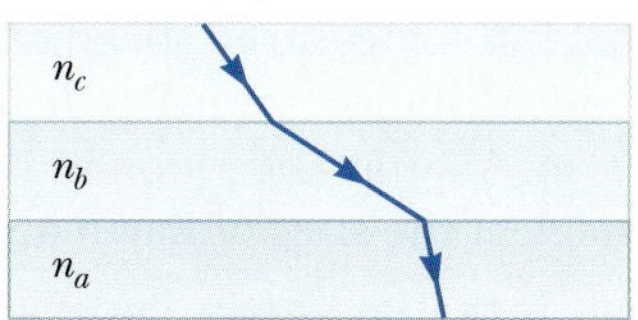


**Figure CQ22.3**

4. Under what conditions is a mirage formed? On a hot day, what are we seeing when we observe a mirage of a water puddle on the road?
5. Explain why a diamond loses most of its sparkle when submerged in carbon disulfide.
6. A type of mirage called a *pingo* is often observed in Alaska. Pingos occur when the light from a small hill passes to an observer by a path that takes the light over a body of water warmer than the air. What is seen is the hill and an inverted image directly below it. Explain how these mirages are formed.
7. In dispersive materials, the angle of refraction for a light ray depends on the wavelength of the light. Does the angle of reflection from the surface of the material depend on the wavelength? Why or why not?
8. The level of water in a clear, colorless glass can easily be observed with the naked eye. The level of liquid helium in a clear glass vessel is extremely difficult to see with the naked eye. Explain. *Hint:* The index of refraction of liquid helium is close to that of air.
9. Suppose you are told that only two colors of light ($X$ and $Y$) are sent through a glass prism and that $X$ is bent more than $Y$. Which color travels more slowly in the prism?

**10.** Is it possible to have total internal reflection for light incident from air on water? Explain.

**11.** Figure CQ22.11 shows a pencil partially immersed in a cup of water. Why does the pencil appear to be bent?

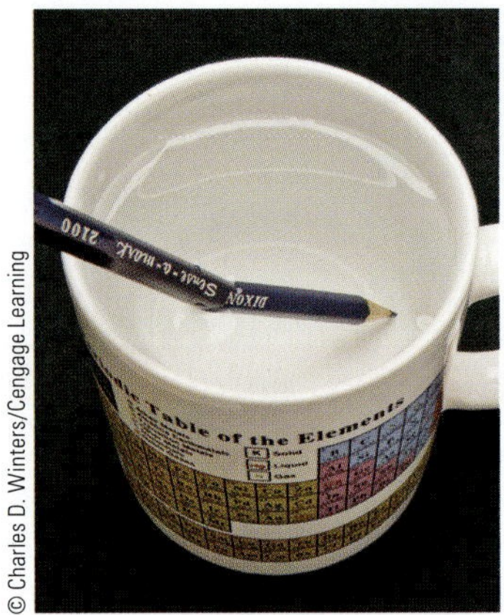


**Figure CQ22.11**

**12.** Try this simple experiment on your own. Take two opaque cups, place a coin at the bottom of each cup near the edge, and fill one cup with water. Next, view the cups at some angle from the side so that the coin in water is just visible as shown on the left in Figure CQ22.12. Notice that the coin in air is not visible as shown on the right in Figure CQ22.12. Explain this observation.

**Figure CQ22.12**

**13.** Why do astronomers looking at distant galaxies talk about looking backward in time?

**14.** Light can travel from air into water. Some possible paths for the light ray in the water are shown in Figure CQ22.14. Which path will the light most likely follow? (a) *A* (b) *B* (c) *C* (d) *D* (e) *E*

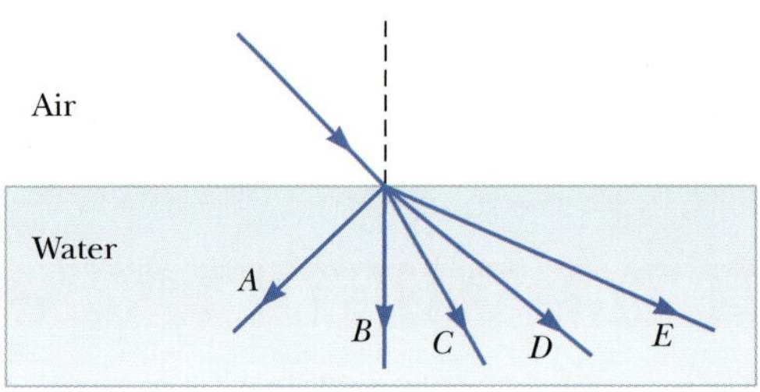


**Figure CQ22.14**

**15.** A light ray containing both blue and red wavelengths is incident at an angle on a slab of glass. Which of the sketches in Figure CQ22.15 represents the most likely outcome? (a) *A* (b) *B* (c) *C* (d) *D* (e) none of these

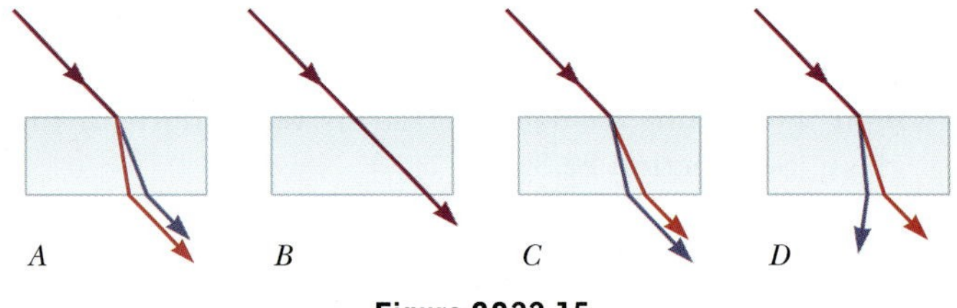


**Figure CQ22.15**

## PROBLEMS

ENHANCED WebAssign The problems in this chapter may be assigned online in Enhanced WebAssign.

**1.** denotes straightforward problem; **2.** denotes intermediate problem;
**3.** denotes challenging problem
**1.** denotes full solution available in *Student Solutions Manual/ Study Guide*
**1.** denotes problems most often assigned in Enhanced WebAssign

BIO denotes biomedical problems
GP denotes guided problems
M denotes Master It tutorial available in Enhanced WebAssign
Q|C denotes asking for quantitative and conceptual reasoning
S denotes symbolic reasoning problem
W denotes Watch It video solution available in Enhanced WebAssign

### 22.1 The Nature of Light

**1.** During the Apollo XI Moon landing, a retroreflecting panel was erected on the Moon's surface. The speed of light can be found by measuring the time it takes a laser beam to travel from Earth, reflect from the panel, and return to Earth. If this interval is found to be 2.51 s, what is the measured speed of light? Take the center-to-center distance from Earth to the Moon to be $3.84 \times 10^8$ m. Assume the Moon is directly overhead and do not neglect the sizes of Earth and the Moon.

**2.** Q|C (a) What is the energy in joules of an x-ray photon with wavelength $1.00 \times 10^{-10}$ m? (b) Convert the energy to electron volts. (c) If more penetrating x-rays are desired, should the wavelength be increased or decreased? (d) Should the frequency be increased or decreased?

**3.** M Find the energy of (a) a photon having a frequency of $5.00 \times 10^{17}$ Hz and (b) a photon having a wavelength of $3.00 \times 10^{2}$ nm. Express your answers in units of electron volts, noting that 1 eV $= 1.60 \times 10^{-19}$ J.

**4.** QIC (a) Calculate the wavelength of light in vacuum that has a frequency of $5.45 \times 10^{14}$ Hz. (b) What is its wavelength in benzene? (c) Calculate the energy of one photon of such light in vacuum. Express the answer in electron volts. (d) Does the energy of the photon change when it enters the benzene? Explain.

**5.** Find the speed of light in (a) water, (b) crown glass, and (c) diamond.

**6.** S (a) Find a symbolic expression for the wavelength $\lambda$ of a photon in terms of its energy $E$, Planck's constant $h$, and the speed of light $c$. (b) What does the equation say about the wavelengths of higher-energy photons?

**7.** A ray of light travels from air into another medium, making an angle of $\theta_1 = 45.0°$ with the normal as in Figure P22.7. Find the angle of refraction $\theta_2$ if the second medium is (a) fused quartz, (b) carbon disulfide, and (c) water.

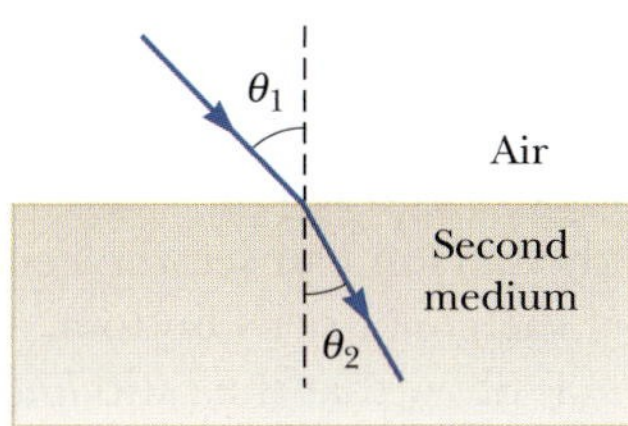


**Figure P22.7**

## 22.2 Reflection and Refraction

## 22.3 The Law of Refraction

**8.** W The two mirrors in Figure P22.8 meet at a right angle. The beam of light in the vertical plane $P$ strikes mirror 1 as shown. (a) Determine the distance the reflected light beam travels before striking mirror 2. (b) In what direction does the light beam travel after being reflected from mirror 2?

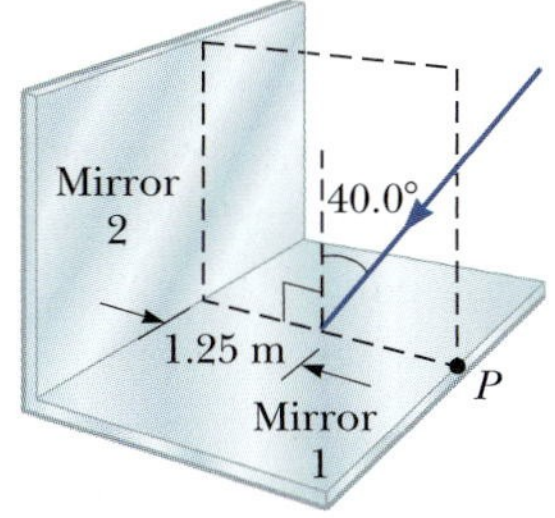


**Figure P22.8**

**9.** An underwater scuba diver sees the Sun at an apparent angle of 45.0° from the vertical. What is the actual direction of the Sun?

**10.** Two plane mirrors are at right angles to each other as shown by the side view in Figure P22.10. A light ray is incident on mirror 1 at an angle $\theta$ with the vertical. Using the law of reflection and geometry, show that after the ray is reflected off of both mirrors, the outgoing reflected ray is parallel to the incident ray.

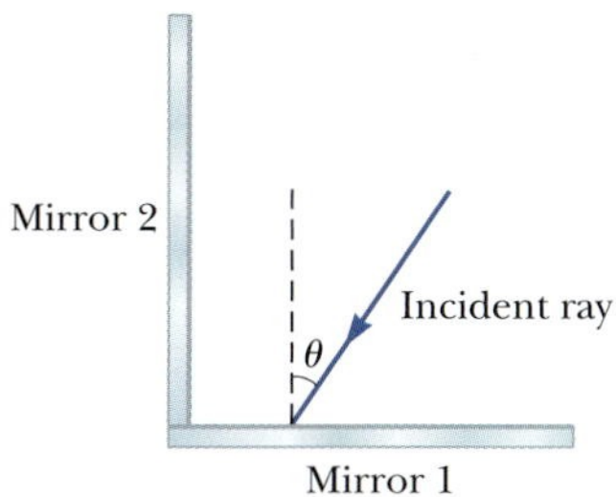


**Figure P22.10**

**11.** A laser beam is incident at an angle of 30.0° to the vertical onto a solution of corn syrup in water. If the beam is refracted to 19.24° to the vertical, (a) what is the index of refraction of the syrup solution? Suppose the light is red, with wavelength 632.8 nm in a vacuum. Find its (b) wavelength, (c) frequency, and (d) speed in the solution.

**12.** Light containing wavelengths of 400 nm, 500 nm, and 650 nm is incident from air on a block of crown glass at an angle of 25.0°. (a) Are all colors refracted alike, or is one color bent more than the others? (b) Calculate the angle of refraction in each case to verify your answer.

**13.** A ray of light is incident on the surface of a block of clear ice at an angle of 40.0° with the normal. Part of the light is reflected, and part is refracted. Find the angle between the reflected and refracted light.

**14.** Two plane mirrors are at an angle of $\theta_1 = 50.0°$ with each other as in the side view shown in Figure P22.14. If a horizontal ray is incident on mirror 1, at what angle $\theta_2$ does the outgoing reflected ray make with the surface of mirror 2?

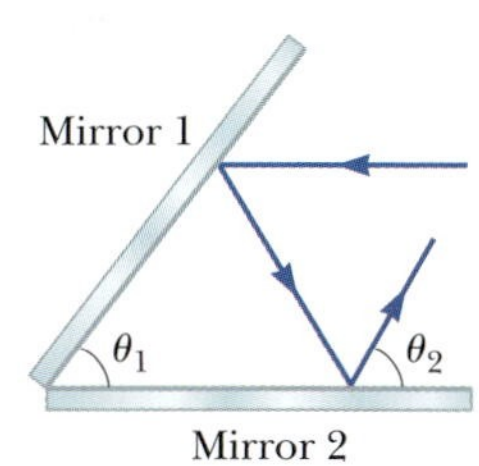


**Figure P22.14**

**15.** The light emitted by a helium–neon laser has a wavelength of 632.8 nm in air. As the light travels from air into zircon, find its (a) speed, (b) wavelength, and (c) frequency, all in the zircon.

**16.** Figure P22.16 shows a light ray traveling in a slab of crown glass surrounded by air. The ray is incident on the right surface at an angle of 55° with the normal and then reflects from points $A$, $B$, and $C$. (a) At which of these points does part of the ray enter the air? (b) If the glass slab is surrounded by carbon disulfide, at which point does part of the ray enter the carbon disulfide?

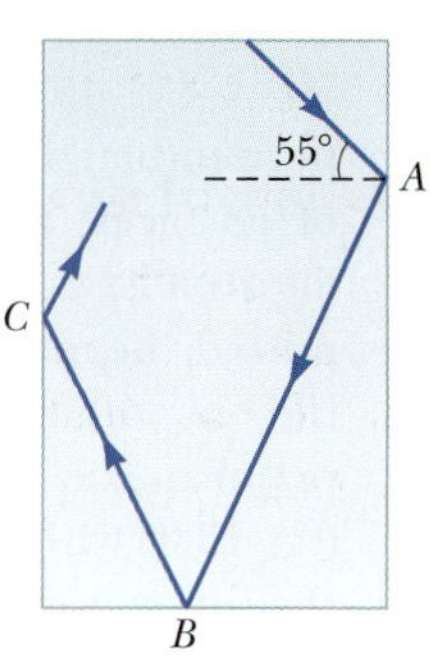


**Figure P22.16**

17. How many times will the incident beam shown in Figure P22.17 be reflected by each of the parallel mirrors?

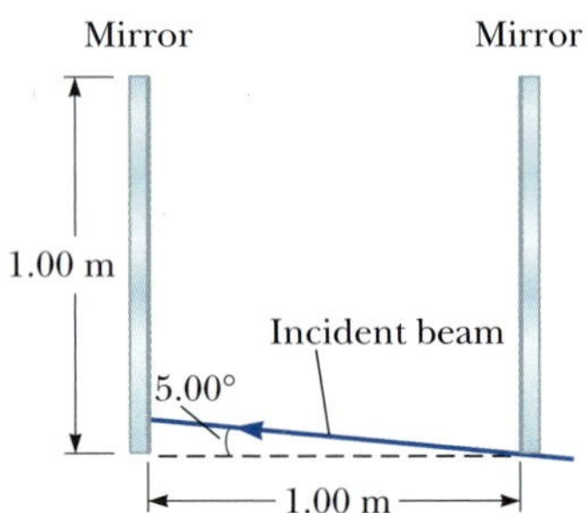


Figure P22.17

18. Q|C A ray of light strikes a flat, 2.00-cm-thick block of glass ($n$ = 1.50) at an angle of 30.0° with respect to the normal (Fig. P22.18). (a) Find the angle of refraction at the top surface. (b) Find the angle of incidence at the bottom surface and the refracted angle. (c) Find the lateral distance $d$ by which the light beam is shifted. (d) Calculate the speed of light in the glass and (e) the time required for the light to pass through the glass block. (f) Is the travel time through the block affected by the angle of incidence? Explain.

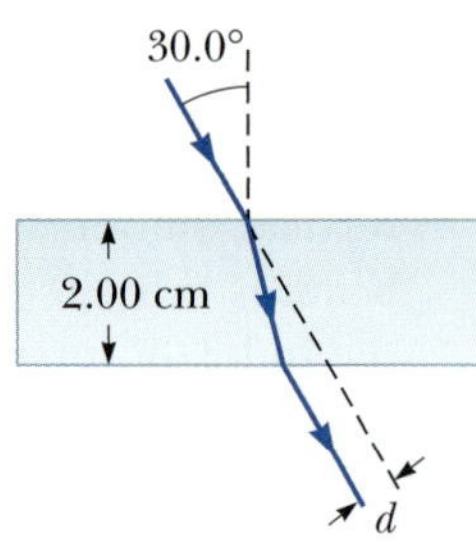


Figure P22.18

19. M The light beam shown in Figure P22.19 makes an angle of 20.0° with the normal line $NN'$ in the linseed oil. Determine the angles $\theta$ and $\theta'$. (The refractive index for linseed oil is 1.48.)

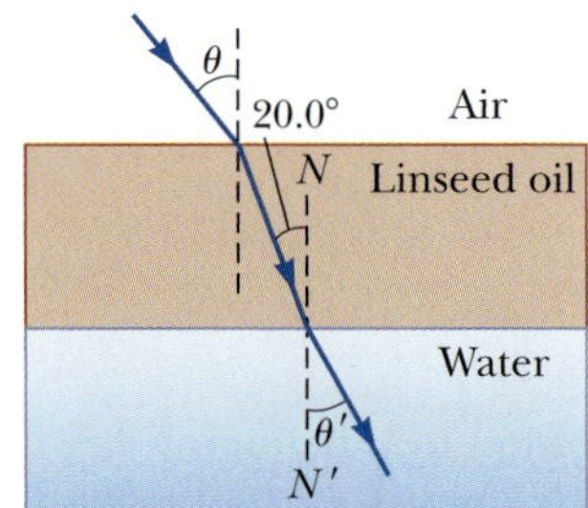


Figure P22.19

20. A laser beam is incident on a 45°–45°–90° prism perpendicular to one of its faces, as shown in Figure P22.20. The transmitted beam that exits the hypotenuse of the prism makes an angle of $\theta$ = 15.0° with the direction of the incident beam. Find the index of refraction of the prism.

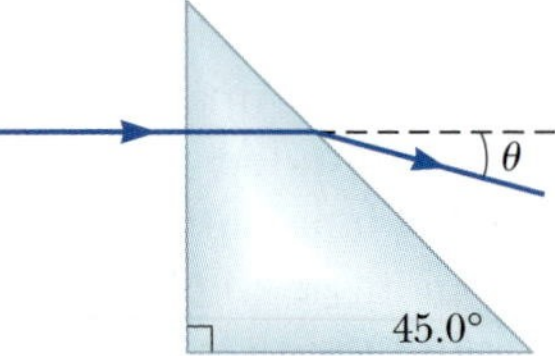


Figure P22.20

21. A block of crown glass is immersed in water as in Figure P22.21. A light ray is incident on the top face at an angle of $\theta_1$ = 42.0° with the normal and exits the block at point $P$. (a) Find the vertical distance $y$ from the top of the block to $P$. (b) Find the angle of refraction $\theta_2$ of the light ray leaving the block at $P$.

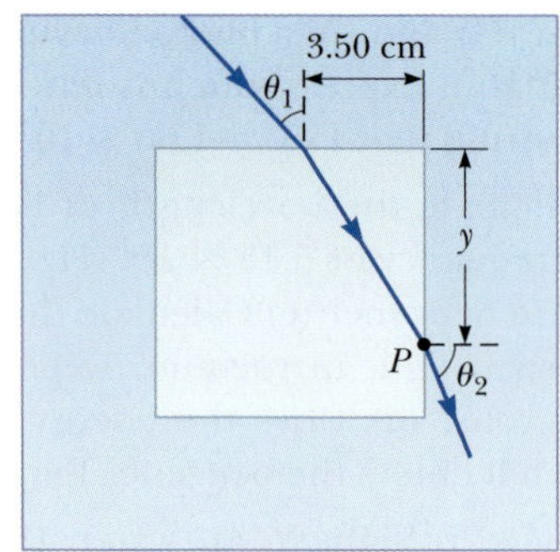


Figure P22.21

22. W BIO A narrow beam of ultrasonic waves reflects off the liver tumor in Figure P22.22. If the speed of the wave is 10.0% less in the liver than in the surrounding medium, determine the depth of the tumor.

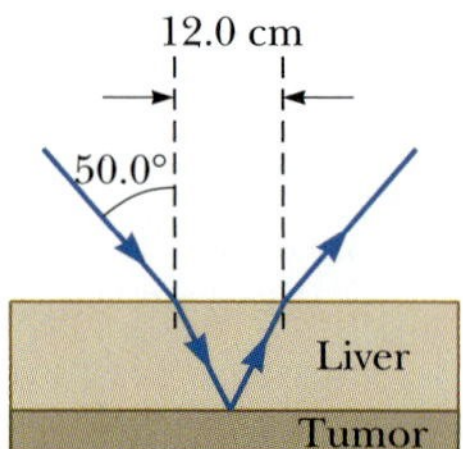


Figure P22.22

23. S A person looking into an empty container is able to see the far edge of the container's bottom, as shown in Figure P22.23a. The height of the container is $h$, and its width is $d$. When the container is completely filled with a fluid of index of refraction $n$ and viewed from the same angle, the person can see the center of a coin at the middle of the container's bottom, as shown in Figure P22.23b. (a) Show that the ratio $h/d$ is given by

$$\frac{h}{d} = \sqrt{\frac{n^2 - 1}{4 - n^2}}$$

(b) Assuming the container has a width of 8.00 cm and is filled with water, use the expression above to find the height of the container.

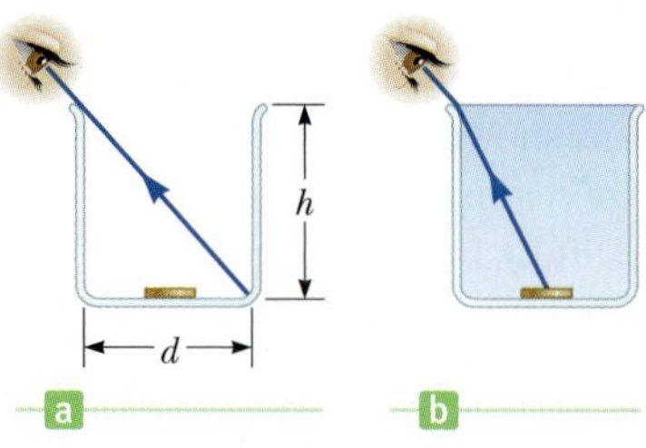


Figure P22.23

24. GP A submarine is $3.00 \times 10^2$ m horizontally from shore and $1.00 \times 10^2$ m beneath the surface of the water. A laser beam is sent from the submarine so that the beam strikes the surface of the water $2.10 \times 10^2$ m from the shore. A building stands on the shore, and

the laser beam hits a target at the top of the building. The goal is to find the height of the target above sea level. (a) Draw a diagram of the situation, identifying the two triangles that are important to finding the solution. (b) Find the angle of incidence of the beam striking the water–air interface. (c) Find the angle of refraction. (d) What angle does the refracted beam make with respect to the horizontal? (e) Find the height of the target above sea level.

25. S A beam of light both reflects and refracts at the surface between air and glass, as shown in Figure P22.25. If the index of refraction of the glass is $n_g$, find the angle of incidence, $\theta_1$, in the air that would result in the reflected ray and the refracted ray being perpendicular to each other. *Hint:* Remember the identity $\sin(90° - \theta) = \cos\theta$.

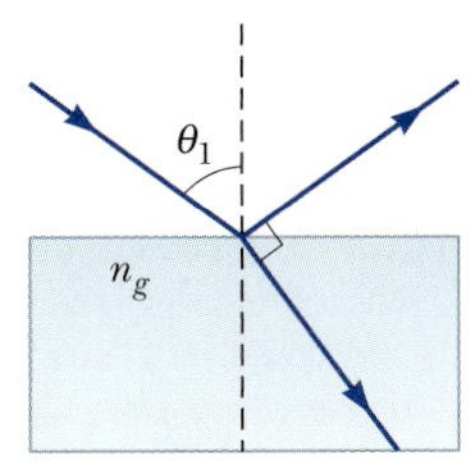

Figure P22.25

26. Figure P22.26 shows a light ray incident on a series of slabs having different refractive indices, where $n_1 < n_2 < n_3 < n_4$. Notice that the path of the ray steadily bends toward the normal. If the variation in $n$ were continuous, the path would form a smooth curve. Use this idea and a ray diagram to explain why you can see the Sun at sunset after it has fallen below the horizon.

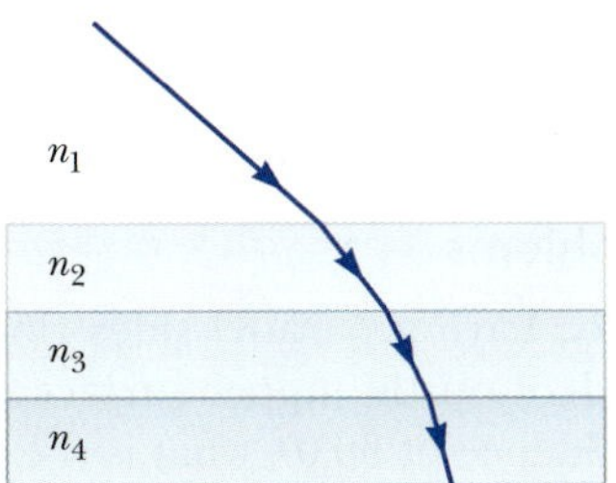

Figure P22.26

27. M An opaque cylindrical tank with an open top has a diameter of 3.00 m and is completely filled with water. When the afternoon Sun reaches an angle of 28.0° above the horizon, sunlight ceases to illuminate the bottom of the tank. How deep is the tank?

## 22.4 Dispersion and Prisms

28. W A certain kind of glass has an index of refraction of 1.650 for blue light of wavelength 430 nm and an index of 1.615 for red light of wavelength 680 nm. If a beam containing these two colors is incident at an angle of 30.00° on a piece of this glass, what is the angle between the two beams inside the glass?

29. The index of refraction for red light in water is 1.331 and that for blue light is 1.340. If a ray of white light enters the water at an angle of incidence of 83.00°, what are the underwater angles of refraction for the (a) blue and (b) red components of the light?

30. The index of refraction for crown glass is 1.512 at a wavelength of 660 nm (red), whereas its index of refraction is 1.530 at a wavelength of 410 nm (violet). If both wavelengths are incident on a slab of crown glass at the same angle of incidence, 60.0°, what is the angle of refraction for each wavelength?

31. A light beam containing red and violet wavelengths is incident on a slab of quartz at an angle of incidence of 50.00°. The index of refraction of quartz is 1.455 at 660 nm (red light), and its index of refraction is 1.468 at 410 nm (violet light). Find the dispersion of the slab, which is defined as the difference in the angles of refraction for the two wavelengths.

32. The index of refraction for violet light in silica flint glass is 1.66 and that for red light is 1.62. What is the angular dispersion of visible light passing through an equilateral prism of apex angle 60.0° if the angle of incidence is 50.0°? (See Fig. P22.32.)

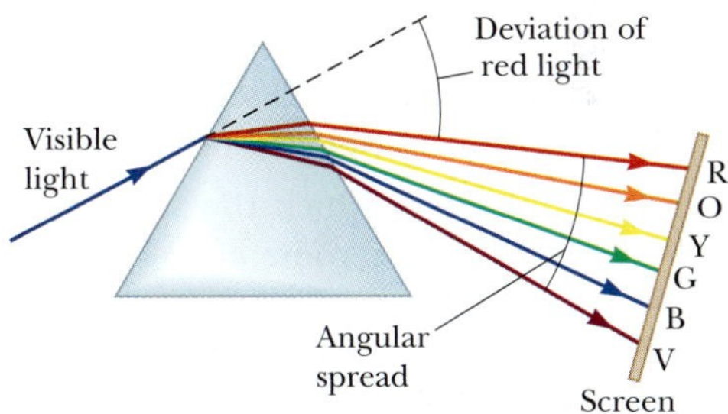

Figure P22.32

33. M A ray of light strikes the midpoint of one face of an equiangular (60°–60°–60°) glass prism ($n = 1.5$) at an angle of incidence of 30°. (a) Trace the path of the light ray through the glass and find the angles of incidence and refraction at each surface. (b) If a small fraction of light is also reflected at each surface, what are the angles of reflection at the surfaces?

## 22.7 Total Internal Reflection

34. For light of wavelength 589 nm, calculate the critical angles for the following substances when surrounded by air: (a) fused quartz, (b) polystyrene, and (c) sodium chloride.

35. Repeat Problem 34, but this time assume the quartz, polystyrene, and sodium chloride are surrounded by water.

36. M A beam of light is incident from air on the surface of a liquid. If the angle of incidence is 30.0° and the angle of refraction is 22.0°, find the critical angle for the liquid when surrounded by air.

37. A plastic light pipe has an index of refraction of 1.53. For total internal reflection, what is the minimum angle of incidence if the pipe is in (a) air and (b) water?

**38.** Determine the maximum angle $\theta$ for which the light rays incident on the end of the light pipe in Figure P22.38 are subject to total internal reflection along the walls of the pipe. Assume the light pipe has an index of refraction of 1.36 and the outside medium is air.

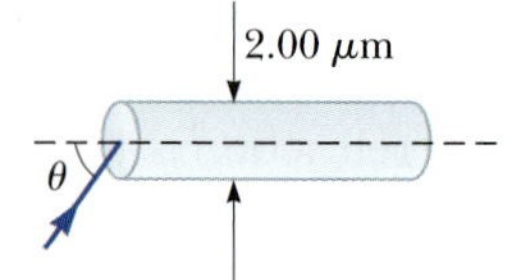

**Figure P22.38**

**39.** A light ray is incident normally to the long face (the hypotenuse) of a 45°–45°–90° prism surrounded by air, as shown in Figure 22.26b. Calculate the minimum index of refraction of the prism for which the ray will totally internally reflect at each of the two sides making the right angle.

**40.** Q|C A beam of laser light with wavelength 612 nm is directed through a slab of glass having index of refraction 1.78. (a) For what minimum incident angle would a ray of light undergo total internal reflection? (b) If a layer of water is placed over the glass, what is the minimum angle of incidence on the glass–water interface that will result in total internal reflection at the water–air interface? (c) Does the thickness of the water layer or glass affect the result? (d) Does the index of refraction of the intervening layer affect the result?

**41.** A room contains air in which the speed of sound is 343 m/s. The walls of the room are made of concrete, in which the speed of sound is 1 850 m/s. (a) Find the critical angle for total internal reflection of sound at the concrete–air boundary. (b) In which medium must the sound be traveling in order to undergo total internal reflection? (c) "A bare concrete wall is a highly efficient mirror for sound." Give evidence for or against this statement.

**42.** GP Q|C Consider a light ray traveling between air and a diamond cut in the shape shown in Figure P22.42. (a) Find the critical angle for total internal reflection for light in the diamond incident on the interface between the diamond and the outside air. (b) Consider the light ray incident normally on the top surface of the diamond as shown in Figure P22.42. Show that the light traveling toward point $P$ in the diamond is totally reflected. (c) If the diamond is immersed in water, find the critical angle at the diamond–water interface. (d) When the diamond is immersed in water, does the light ray entering the top surface in Figure P22.42 undergo total internal reflection at $P$? Explain. (e) If the light ray entering the diamond remains vertical as shown in Figure P22.42, which way should the diamond in the water be rotated about an axis perpendicular to the page through $O$ so that light will exit the diamond at $P$? (f) At what angle of rotation in part (e) will light first exit the diamond at point $P$?

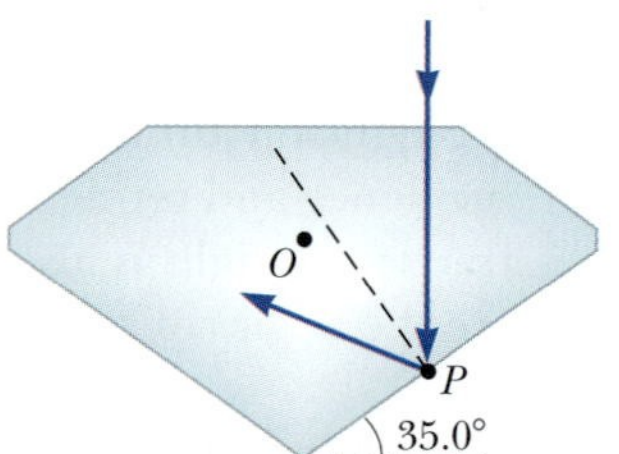

**Figure P22.42**

**43.** The light beam in Figure P22.43 strikes surface 2 at the critical angle. Determine the angle of incidence, $\theta_1$.

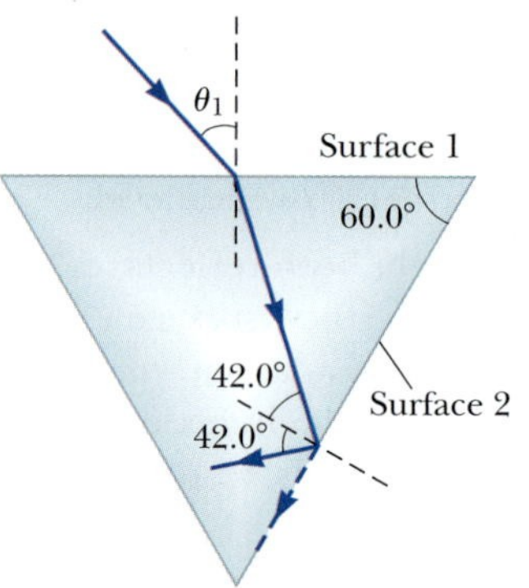

**Figure P22.43**

**44.** W A jewel thief hides a diamond by placing it on the bottom of a public swimming pool. He places a circular raft on the surface of the water directly above and centered over the diamond, as shown in Figure P22.44. If the surface of the water is calm and the pool is 2.00 m deep, find the minimum diameter of the raft that would prevent the diamond from being seen.

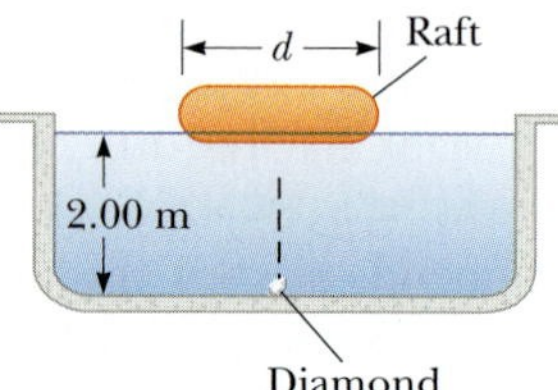

**Figure P22.44**

## Additional Problems

**45.** A layer of ice having parallel sides floats on water. If light is incident on the upper surface of the ice at an angle of incidence of 30.0°, what is the angle of refraction in the water?

**46.** Q|C A ray of light is incident at an angle 30.0° on a plane slab of flint glass surrounded by water. (a) Find the refraction angle. (b) Suppose the index of refraction of the surrounding medium can be adjusted, but the incident angle of the light remains the same. As the index of refraction of the medium approaches that of the glass, what happens to the refraction angle? (c) What happens to the refraction angle when the medium's index of refraction exceeds that of the glass?

**47.** When a man stands near the edge of an empty drainage ditch of depth 2.80 m, he can barely see the boundary between the opposite wall and bottom of the ditch as in Figure P22.47a. The distance from his eyes to the ground is 1.85 m. (a) What is the horizontal distance $d$ from the man to the edge of the drainage ditch? (b) After the drainage ditch is filled with water as in Figure P22.47b, what is the maximum distance $x$ the man can stand from the edge and still see the same boundary?

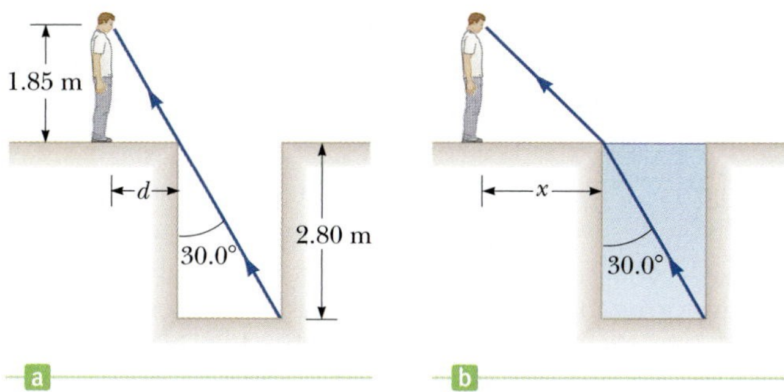

Figure P22.47

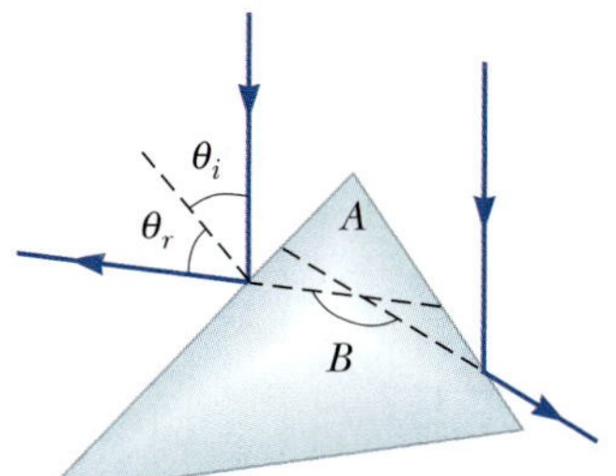

Figure P22.51

48. A light ray of wavelength 589 nm is incident at an angle $\theta$ on the top surface of a block of polystyrene surrounded by air, as shown in Figure P22.48. (a) Find the maximum value of $\theta$ for which the refracted ray will undergo total internal reflection at the left vertical face of the block. (b) Repeat the calculation for the case in which the polystyrene block is immersed in water. (c) What happens if the block is immersed in carbon disulfide?

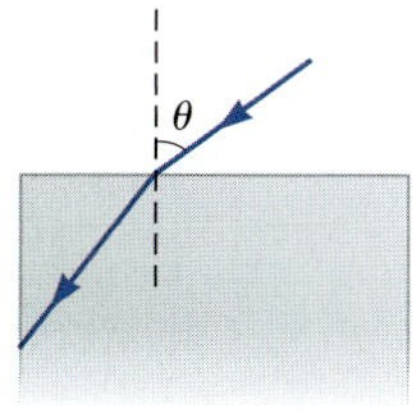

Figure P22.48

49. As shown in Figure P22.49, a light ray is incident normal to one face of a 30°–60°–90° block of flint glass (a prism) that is immersed in water. (a) Determine the exit angle $\theta_3$ of the ray. (b) A substance is dissolved in the water to increase the index of refraction $n_2$. At what value of $n_2$ does total internal reflection cease at point $P$?

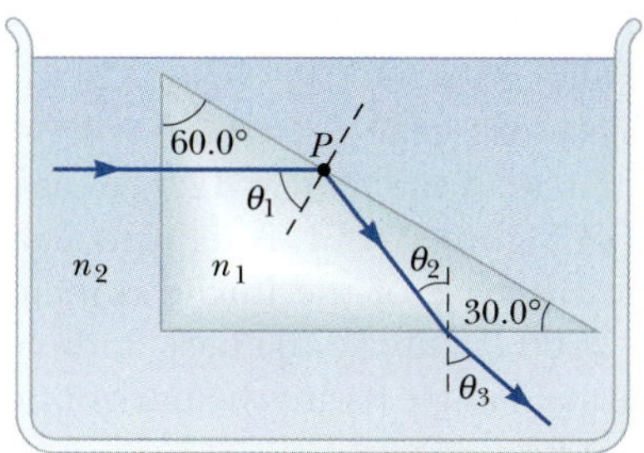

Figure P22.49

50. A narrow beam of light is incident from air onto a glass surface with index of refraction 1.56. Find the angle of incidence for which the corresponding angle of refraction is one-half the angle of incidence. *Hint:* You might want to use the trigonometric identity $\sin 2\theta = 2 \sin \theta \cos \theta$.

51. One technique for measuring the angle of a prism is shown in Figure P22.51. A parallel beam of light is directed onto the apex of the prism so that the beam reflects from opposite faces of the prism. Show that the angular separation of the two reflected beams is given by $B = 2A$.

52. An optical fiber with index of refraction $n$ and diameter $d$ is surrounded by air. Light is sent into the fiber along its axis, as shown in Figure P22.52. (a) Find the smallest outside radius $R$ permitted for a bend in the fiber if no light is to escape. (b) Does the result for part (a) predict reasonable behavior as $d$ approaches zero? As $n$ increases? As $n$ approaches unity? (c) Evaluate $R$, assuming the diameter of the fiber is 100 $\mu$m and its index of refraction is 1.40.

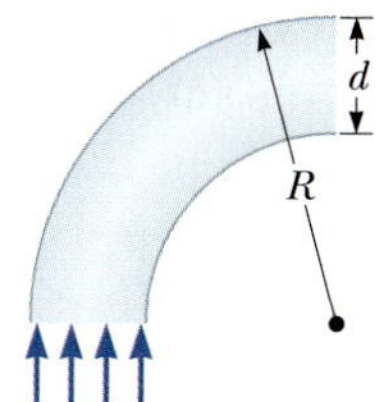

Figure P22.52

53. A piece of wire is bent through an angle $\theta$. The bent wire is partially submerged in benzene (index of refraction = 1.50) so that, to a person looking along the dry part, the wire appears to be straight and makes an angle of 30.0° with the horizontal. Determine the value of $\theta$.

54. A light ray traveling in air is incident on one face of a right-angle prism with index of refraction $n = 1.50$, as shown in Figure P22.54, and the ray follows the path shown in the figure. Assuming $\theta = 60.0°$ and the base of the prism is mirrored, determine the angle $\phi$ made by the outgoing ray with the normal to the right face of the prism.

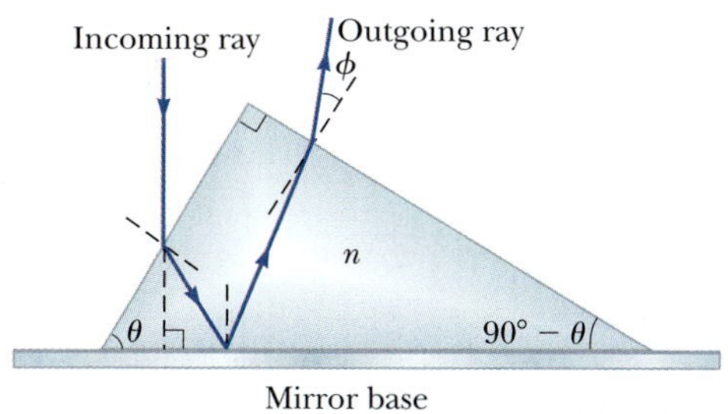

Figure P22.54

55. A transparent cylinder of radius $R = 2.00$ m has a mirrored surface on its right half, as shown in Figure P22.55 (page 800). A light ray traveling in air is incident on the left side of the cylinder. The incident light ray and the exiting light ray are parallel, and $d = 2.00$ m. Determine the index of refraction of the material.

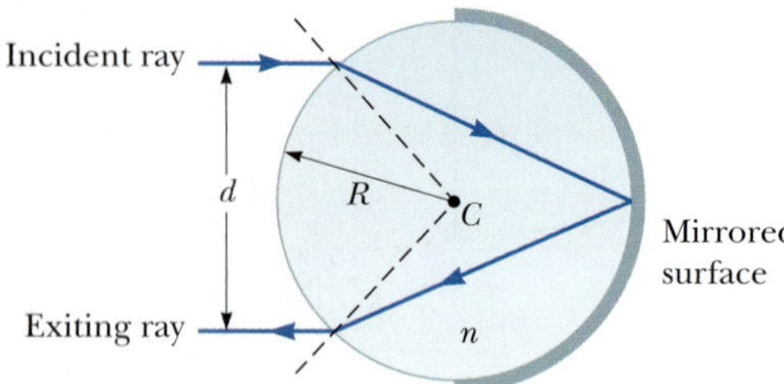


**Figure P22.55**

56. A laser beam strikes one end of a slab of material, as in Figure P22.56. The index of refraction of the slab is 1.48. Determine the number of internal reflections of the beam before it emerges from the opposite end of the slab.

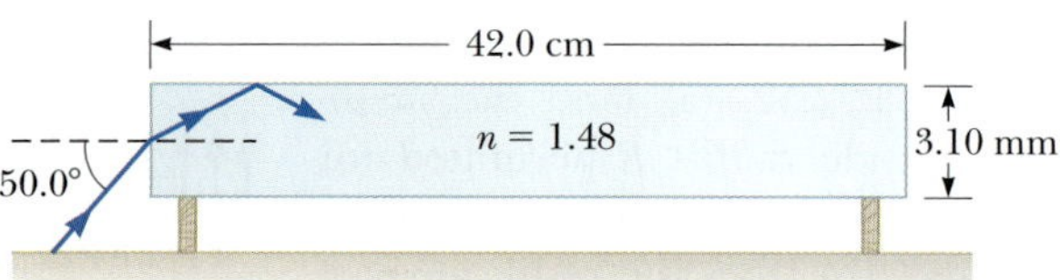


**Figure P22.56**

57. A light ray enters a rectangular block of plastic at an angle $\theta_1 = 45.0°$ and emerges at an angle $\theta_2 = 76.0°$, as shown in Figure P22.57. (a) Determine the index of refraction of the plastic. (b) If the light ray enters the plastic at a point $L = 50.0$ cm from the bottom edge, how long does it take the light ray to travel through the plastic?

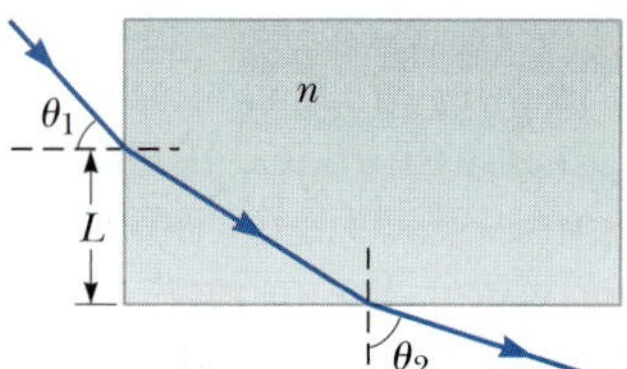


**Figure P22.57**

58. Students allow a narrow beam of laser light to strike a water surface. They arrange to measure the angle of refraction for selected angles of incidence and record the data shown in the following table:

| Angle of Incidence (degrees) | Angle of Refraction (degrees) |
|---|---|
| 10.0 | 7.5 |
| 20.0 | 15.1 |
| 30.0 | 22.3 |
| 40.0 | 28.7 |
| 50.0 | 35.2 |
| 60.0 | 40.3 |
| 70.0 | 45.3 |
| 80.0 | 47.7 |

Use the data to verify Snell's law of refraction by plotting the sine of the angle of incidence versus the sine of the angle of refraction. From the resulting plot, deduce the index of refraction of water.

59. Figure P22.59 shows the path of a beam of light through several layers with different indices of refraction. (a) If $\theta_1 = 30.0°$, what is the angle $\theta_2$ of the emerging beam? (b) What must the incident angle $\theta_1$ be to have total internal reflection at the surface between the medium with $n = 1.20$ and the medium with $n = 1.00$?

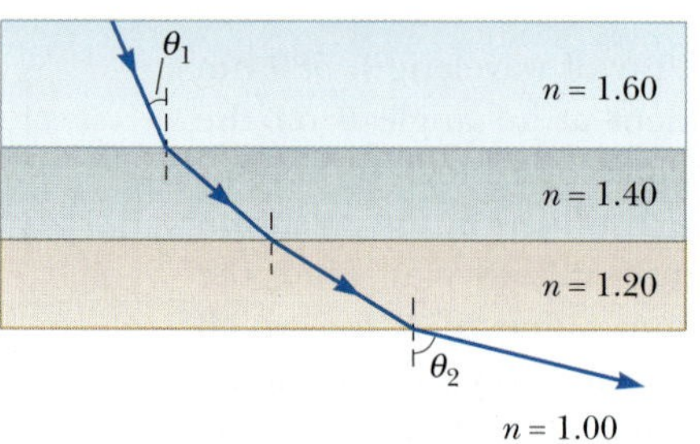


**Figure P22.59**

60. Three sheets of plastic have unknown indices of refraction. Sheet 1 is placed on top of sheet 2, and a laser beam is directed onto the sheets from above so that it strikes the interface at an angle of 26.5° with the normal. The refracted beam in sheet 2 makes an angle of 31.7° with the normal. The experiment is repeated with sheet 3 on top of sheet 2, and with the same angle of incidence, the refracted beam makes an angle of 36.7° with the normal. If the experiment is repeated again with sheet 1 on top of sheet 3, what is the expected angle of refraction in sheet 3? Assume the same angle of incidence.

61. A thick piece of Lucite ($n = 1.50$) is has the shape of a quarter circle of radius $R = 12.0$ cm as shown in the side view of Figure P22.61. A light ray traveling in air parallel to the base of the Lucite is incident at a distance $h = 6.00$ cm above the base and emerges out of the Lucite at an angle $\theta$ with the horizontal. Determine the value of $\theta$.

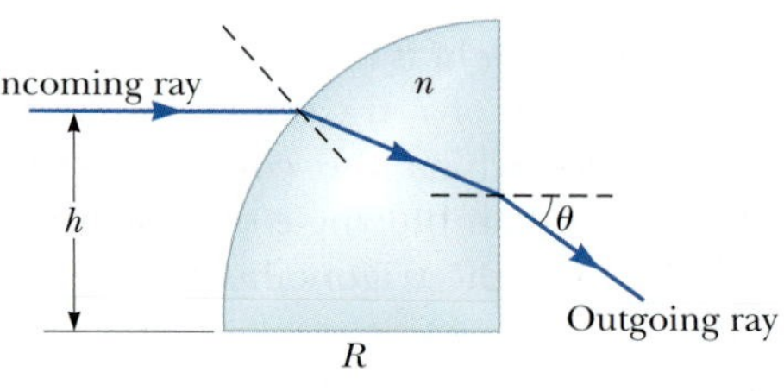


**Figure P22.61**

Gail Mooney/Masterfile Corporation

Funhouse mirrors distort images because the curved surfaces essentially change the angle of incidence of incoming rays, change that differs depending on the mirror's shape in a given location. In every case, however, the angle of reflection equals the angle of incidence.

# Mirrors and Lenses 23

The development of the technology of mirrors and lenses led to a revolution in the progress of science. These devices, relatively simple to construct from cheap materials, led to microscopes and telescopes, extending human sight and opening up new pathways to knowledge, from microbes to distant planets.

This chapter covers the formation of images when plane and spherical light waves fall on plane and spherical surfaces. Images can be formed by reflection from mirrors or by refraction through lenses. In our study of mirrors and lenses, we continue to assume light travels in straight lines (the ray approximation), ignoring diffraction.



## 23.1 Flat Mirrors

**LEARNING OBJECTIVE**

1. Discuss and apply the properties of flat mirrors.

We begin by examining the flat mirror. Consider a point source of light placed at $O$ in Figure 23.1, a distance $p$ in front of a flat mirror. The distance $p$ is called the **object distance**. Light rays leave the source and are reflected from the mirror. After reflection, the rays diverge (spread apart), but they appear to the viewer to come from a point $I$ behind the mirror. Point $I$ is called the **image** of the object at $O$. Regardless of the system under study, **images are formed at the point where rays of light actually intersect or where they appear to originate**. Because the rays in the figure appear to originate at $I$, which is a distance $q$ behind the mirror, that is the location of the image. The distance $q$ is called the **image distance**.

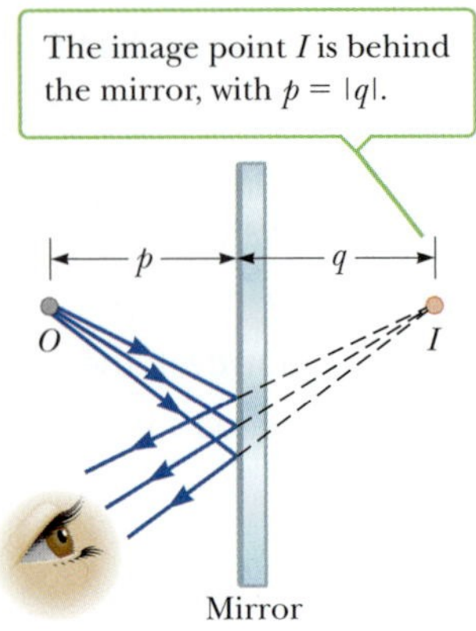


**Figure 23.1** An image formed by reflection from a flat mirror. The image at point $I$ is virtual. In Section 23.3, it will be shown that $q$ must be taken as negative for virtual images: the object distance $p$, therefore, equals the absolute value of the image distance $q$.

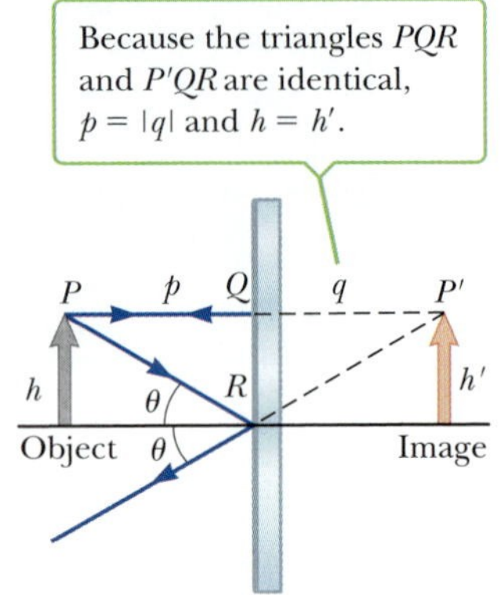


**Figure 23.2** A geometric construction to locate the image of an object placed in front of a flat mirror.

Images are classified as real or virtual. In the formation of a *real image*, light actually passes through the image point. For a *virtual image*, light doesn't pass through the image point, but appears to come (diverge) from there. The image formed by the flat mirror in Figure 23.1 is a virtual image. In fact, the images seen in flat mirrors are always virtual (for real objects). Real images can be displayed on a screen (as at a movie), but virtual images cannot.

We examine some of the properties of the images formed by flat mirrors by using the simple geometric techniques. To find out where an image is formed, it's necessary to follow at least two rays of light as they reflect from the mirror as in Figure 23.2. One of those rays starts at $P$, follows the horizontal path $PQ$ to the mirror, and reflects back on itself. The second ray follows the oblique path $PR$ and reflects as shown. An observer to the left of the mirror would trace the two reflected rays back to the point from which they appear to have originated: point $P'$. A continuation of this process for points other than $P$ on the object would result in a virtual image (drawn as a yellow arrow) to the right of the mirror. Because triangles $PQR$ and $P'QR$ are identical, $PQ = P'Q$. Hence, we conclude that **the image formed by an object placed in front of a flat mirror is as far behind the mirror as the object is in front of the mirror**. Geometry also shows that the object height $h$ equals the image height $h'$. The **lateral magnification** $M$ is defined as

$$M \equiv \frac{\text{image height}}{\text{object height}} = \frac{h'}{h} \quad [23.1]$$

Equation 23.1 is a general definition of the lateral magnification of any type of mirror. For a flat mirror, $M = 1$ because $h' = h$.

In summary, the image formed by a flat mirror has the following properties:

1. The image is as far behind the mirror as the object is in front.
2. The image is unmagnified, virtual, and upright. (By *upright*, we mean that if the object arrow points upward, as in Figure 23.2, so does the image arrow. The opposite of an upright image is an inverted image.)

Finally, note that a flat mirror produces an image having an *apparent* left–right reversal. You can see this reversal standing in front of a mirror and raising your right hand. Your image in the mirror raises the left hand. Likewise, your hair appears to be parted on the opposite side, and a mole on your right cheek appears to be on your image's left cheek.

**Quick Quiz**

**23.1** In the overhead view of Figure 23.3, the image of the stone seen by observer 1 is at $C$. Where does observer 2 see the image: at $A$, at $B$, at $C$, at $D$, at $E$, or not at all?

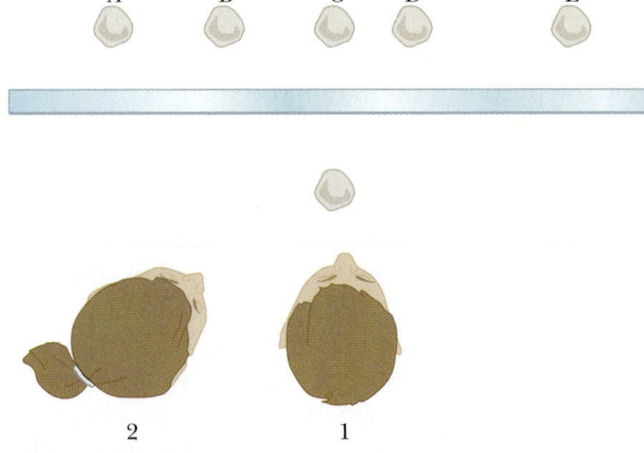


**Figure 23.3** (Quick Quiz 23.1)

**Tip 23.1 Magnification ≠ Enlargement**

Note that the word *magnification*, as used in optics, doesn't always mean *enlargement* because the image could be smaller than the object.

## EXAMPLE 23.1 "Mirror, Mirror, on the Wall"

**GOAL** Apply the properties of a flat mirror.

**PROBLEM** A man 1.80 m tall stands in front of a mirror and sees his full height, no more and no less. If his eyes are 0.14 m from the top of his head, what is the minimum height of the mirror?

**STRATEGY** Figure 23.4 shows two rays of light, one from the man's feet and the other from the top of his head, reflecting off the mirror and entering his eye. The ray from his feet just strikes the bottom of the mirror, so if the mirror were longer, it would be too long, and if shorter, the ray would not be reflected. The angle of incidence and the angle of reflection are equal, labeled $\theta$. This means the two triangles, $ABD$ and $DBC$, are identical because they are right triangles with a common side ($DB$) and two identical angles $\theta$. Use this key fact and the small isosceles triangle $FEC$ to solve the problem.

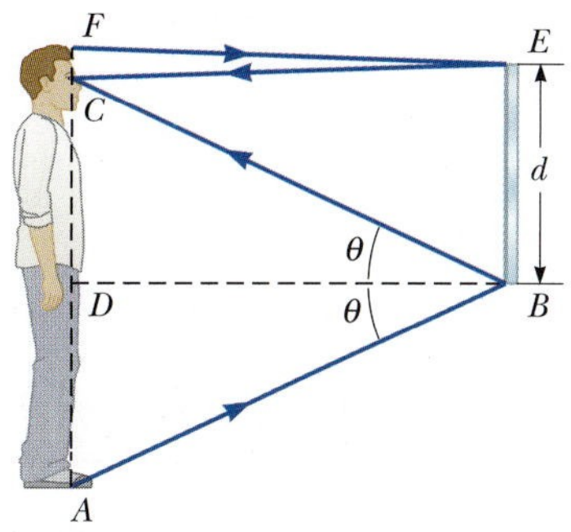


**Figure 23.4** (Example 23.1)

**SOLUTION**

We need to find $BE$, which equals $d$. Relate this length to lengths on the man's body:

$$\textbf{(1)}\quad BE = DC + \tfrac{1}{2}CF$$

We need the lengths $DC$ and $CF$. Set the sum of sides opposite the identical angles $\theta$ equal to $AC$:

$$\textbf{(2)}\quad AD + DC = AC = (1.80 - 0.14) = 1.66 \text{ m}$$

$AD = DC$, so substitute into Equation (2) and solve for $DC$:

$$AD + DC = 2DC = 1.66 \text{ m} \quad \rightarrow \quad DC = 0.83 \text{ m}$$

$CF$ is given as 0.14 m. Substitute this value and $DC$ into Equation (1):

$$BE = d = DC + \tfrac{1}{2}CF = 0.83 \text{ m} + \tfrac{1}{2}(0.14 \text{ m}) = \boxed{0.90 \text{ m}}$$

**REMARKS** The mirror must be exactly equal to half the height of the man for him to see only his full height and nothing more or less. Notice that the answer doesn't depend on his distance from the mirror.

**QUESTION 23.1** Would a taller man be able to see his full height in the same mirror?

**EXERCISE 23.1** How large should the mirror be if he wants to see only the upper third of his full height?

**ANSWER** 0.30 m

---

Most rearview mirrors in cars have a day setting and a night setting. The night setting greatly diminishes the intensity of the image so that lights from trailing cars will not blind the driver. To understand how such a mirror works, consider Figure 23.5. The mirror is a wedge of glass with a reflecting metallic coating on the back side. When the mirror is in the day setting, as in Figure 23.5a, light from an object behind the car strikes the mirror at point 1. Most of the light enters the

**APPLICATION**

Day and Night Settings for Rearview Mirrors

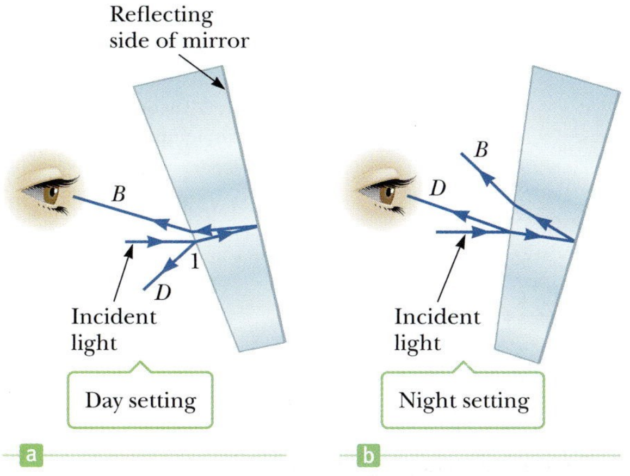


**Figure 23.5** Cross-sectional views of a rearview mirror. (a) With the day setting, the silvered back surface of the mirror reflects a bright ray $B$ into the driver's eyes. (b) With the night setting, the glass of the unsilvered front surface of the mirror reflects a dim ray $D$ into the driver's eyes.

wedge, is refracted, and reflects from the back of the mirror to return to the front surface, where it is refracted again as it reenters the air as ray $B$ (for *bright*). In addition, a small portion of the light is reflected at the front surface, as indicated by ray $D$ (for *dim*). This dim reflected light is responsible for the image observed when the mirror is in the night setting, as in Figure 23.5b. Now the wedge is rotated so that the path followed by the bright light (ray $B$) doesn't lead to the eye. Instead, the dim light reflected from the front surface travels to the eye, and the brightness of trailing headlights doesn't become a hazard.

## APPLYING PHYSICS 23.1 Illusionist's Trick

The professor in the box shown in Figure 23.6 appears to be balancing himself on a few fingers with both of his feet elevated from the floor. He can maintain this position for a long time, and appears to defy gravity. How do you suppose this illusion was created?

**EXPLANATION** This trick is an example of an optical illusion, used by magicians, that makes use of a mirror. The box the professor is standing in is a cubical open frame that contains a flat, vertical mirror through a diagonal plane. The professor straddles the mirror so that one leg is in front of the mirror and the other leg is behind it, out of view. When he raises his front leg, that leg's reflection rises also, making it appear both his feet are off the ground, creating the illusion that he's floating in the air. In fact, he supports himself with the leg behind the mirror, which remains in contact with the ground. ■

Courtesy of Henry Leap and Jim Lehman

**Figure 23.6** (Applying Physics 23.1)

# 23.2 Images Formed by Concave Mirrors

**LEARNING OBJECTIVES**

1. Use ray diagrams to understand relationships between the radius, focal length, object position, and image position for concave mirrors.
2. Define the focal length.
3. State the mirror equation.

A **spherical mirror**, as its name implies, has the shape of a segment of a sphere. Figure 23.7 shows a spherical mirror with a silvered inner, concave surface; this type of mirror is called a **concave mirror**. The mirror has radius of curvature $R$,

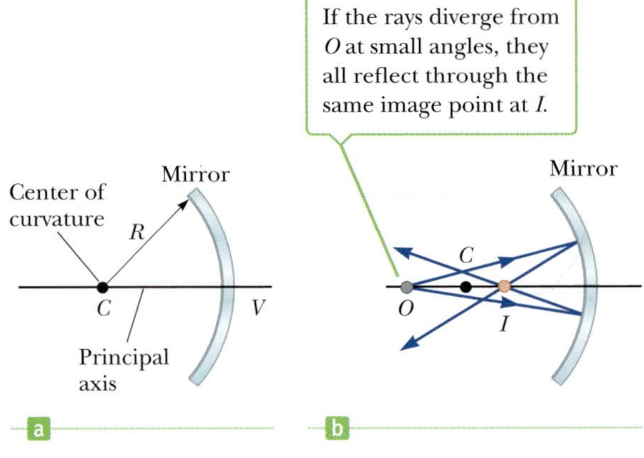


**Figure 23.7** (a) A concave mirror of radius $R$. The center of curvature, $C$, is located on the principal axis. (b) A point object placed at $O$ in front of a concave spherical mirror of radius $R$, where $O$ is any point on the principal axis farther than $R$ from the surface of the mirror, forms a real image at $I$.

and its center of curvature is at point $C$. Point $V$ is the center of the spherical segment, and a line drawn from $C$ to $V$ is called the **principal axis** of the mirror.

Now consider a point source of light placed at point $O$ in Figure 23.7b, on the principal axis and outside point $C$. Several diverging rays originating at $O$ are shown. After reflecting from the mirror, these rays converge to meet at $I$, called the **image point**. The rays then continue and diverge from $I$ as if there were an object there. As a result, a real image is formed. **Whenever reflected light actually passes through a point, the image formed there is real.**

We often assume all rays that diverge from the object make small angles with the principal axis. All such rays reflect through the image point, as in Figure 23.7b. Rays that make a large angle with the principal axis, as in Figure 23.8, converge to other points on the principal axis, producing a blurred image. This effect, called **spherical aberration**, is present to some extent with any spherical mirror and will be discussed in Section 23.7.

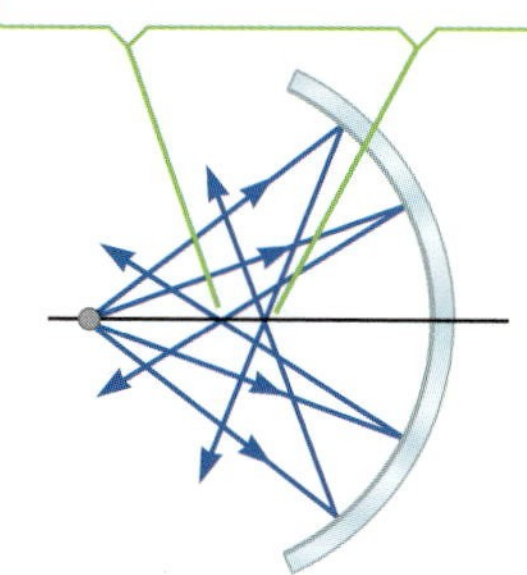
The reflected rays intersect at different points on the principal axis.

**Figure 23.8** A spherical concave mirror exhibits *spherical aberration* when light rays make large angles with the principal axis.

We can use the geometry shown in Figure 23.9 to calculate the image distance $q$ from the object distance $p$ and radius of curvature $R$. By convention, these distances are measured from point $V$. The figure shows two rays of light leaving the tip of the object. One ray passes through the center of curvature, $C$, of the mirror, hitting the mirror head-on (perpendicular to the mirror surface) and reflecting back on itself. The second ray strikes the mirror at point $V$ and reflects as shown, obeying the law of reflection. The image of the tip of the arrow is at the point where the two rays intersect. From the largest triangle in Figure 23.9, we see that $\tan\theta = h/p$; the light-blue triangle gives $\tan\theta = -h'/q$. The negative sign has been introduced to satisfy our convention that $h'$ is negative when the image is inverted with respect to the object, as it is here. From Equation 23.1 and these results, we find that the magnification of the mirror is

$$M = \frac{h'}{h} = -\frac{q}{p} \qquad [23.2]$$

From two other triangles in the figure, we get

$$\tan\alpha = \frac{h}{p - R} \quad \text{and} \quad \tan\alpha = -\frac{h'}{R - q}$$

from which we find that

$$\frac{h'}{h} = -\frac{R - q}{p - R} \qquad [23.3]$$

If we compare Equation 23.2 with Equation 23.3, we see that

$$\frac{R - q}{p - R} = \frac{q}{p}$$

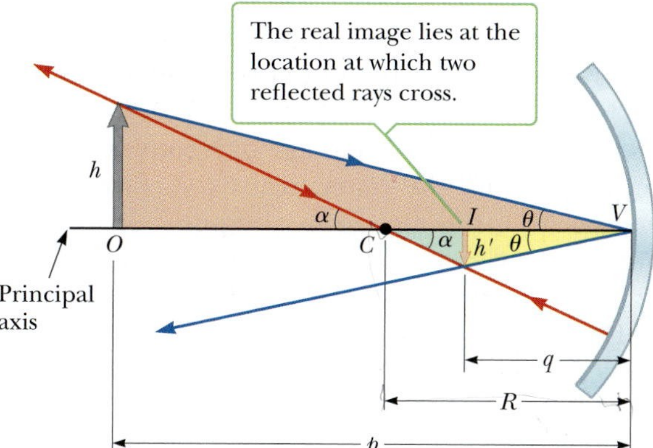
The real image lies at the location at which two reflected rays cross.
h
α
I
θ
V
O
C
α
h'
θ
Principal axis
q
R
p

**Figure 23.9** The image formed by a spherical concave mirror, where the object at $O$ lies outside the center of curvature, $C$.

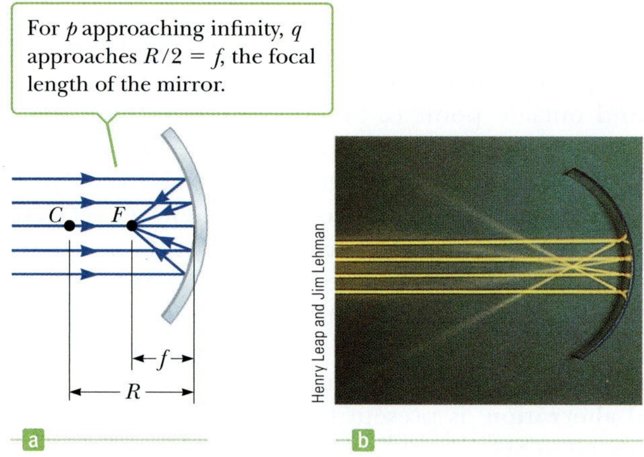
For $p$ approaching infinity, $q$ approaches $R/2 = f$, the focal length of the mirror.
C
F
f
R
a
Henry Leap and Jim Lehman
b

**Figure 23.10** (a) Light rays from a distant object ($p = \infty$) reflect from a concave mirror through the focal point $F$. (b) A photograph of the reflection of parallel rays from a concave mirror.

Simple algebra reduces this equation to

Mirror equation ▶

$$\frac{1}{p} + \frac{1}{q} = \frac{2}{R} \qquad [23.4]$$

This expression is called the **mirror equation.**

If the object is very far from the mirror—if the object distance $p$ is great enough compared with $R$ that $p$ can be said to approach infinity—then $1/p \approx 0$, and we see from Equation 23.4 that $q \approx R/2$. In other words, when the object is very far from the mirror, **the image point is halfway between the center of curvature and the center of the mirror**, as in Figure 23.10a. The incoming rays are essentially parallel in that figure because the source is assumed to be very far from the mirror. In this special case we call the image point the **focal point** $F$ and the image distance the **focal length** $f$, where

Focal length ▶

$$f = \frac{R}{2} \qquad [23.5]$$

The mirror equation can therefore be expressed in terms of the focal length:

$$\frac{1}{p} + \frac{1}{q} = \frac{1}{f} \qquad [23.6]$$

Note that rays from objects at infinity are always focused at the focal point.

**Tip 23.2 Focal Point ≠ Focus Point**

The focal point is *not* the point at which light rays focus to form an image. The focal point of a mirror is determined *solely* by its curvature; it doesn't depend on the location of any object.

## 23.3 Convex Mirrors and Sign Conventions

**LEARNING OBJECTIVES**

1. Understand the optical properties of convex mirrors.
2. Master the sign conventions for concave and convex mirrors.
3. Analyze concave and convex mirrors, finding image positions, magnifications, and stating whether images are upright or inverted, real or virtual.

Figure 23.11 shows the formation of an image by a **convex mirror**, which is silvered so that light is reflected from the outer, convex surface. It is sometimes called a **diverging mirror** because the rays from any point on the object diverge after reflection, as though they were coming from some point behind the mirror. The image in Figure 23.11 is virtual rather than real because it lies behind the mirror at the point the reflected rays appear to originate. In general, the image formed by a convex mirror is upright, virtual, and smaller than the object.

We won't derive any equations for convex spherical mirrors. If we did, we would find that the equations developed for concave mirrors can be used with convex

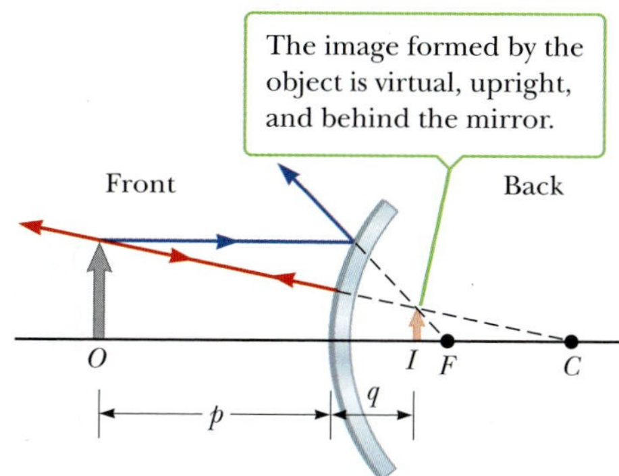


**Figure 23.11** Formation of an image by a spherical, convex mirror.

mirrors if particular sign conventions are used. We call the region in which light rays move the *front side* of the mirror, and the other side, where virtual images are formed, the *back side*. For example, in Figures 23.9 and 23.11, the side to the left of the mirror is the front side and the side to the right is the back side. Figure 23.12 is helpful for understanding the rules for object and image distances, and Table 23.1 summarizes the sign conventions for all the necessary quantities. Notice that when the quantities $p$, $q$, and $f$ (and $R$) are located where the light is—in front of the mirror—they are positive, whereas when they are located behind the mirror (where the light isn't), they are negative.

**Tip 23.3 Positive Is Where the Light Is**

The quantities $p$, $q$, and $f$ are all positive when they are located where the light is—in front of the mirror as indicated in Figure 23.12.

## Ray Diagrams for Mirrors

We can conveniently determine the positions and sizes of images formed by mirrors by constructing *ray diagrams* similar to the ones we have been using. This kind of graphical construction tells us the overall nature of the image and can be used to check parameters calculated from the mirror and magnification equations. Making a ray diagram requires knowing the position of the object and the location of the center of curvature. To locate the image, three rays are constructed (rather than only the two we have been constructing so far), as shown by the examples in Figure 23.13 (page 808). All three rays start from the same object point; for these examples, the tip of the arrow was chosen. For the concave mirrors in Figures 23.13a and 23.13b, the rays are drawn as follows:

1. Ray 1 is drawn parallel to the principal axis and is reflected back through the focal point $F$.
2. Ray 2 is drawn through the focal point and is reflected parallel to the principal axis.
3. Ray 3 is drawn through the center of curvature, $C$, and is reflected back on itself.

Note that rays actually go in all directions from the object; we choose to follow those moving in a direction that simplifies our drawing.

The intersection of any *two* of these rays at a point locates the image. The third ray serves as a check of our construction. The image point obtained in

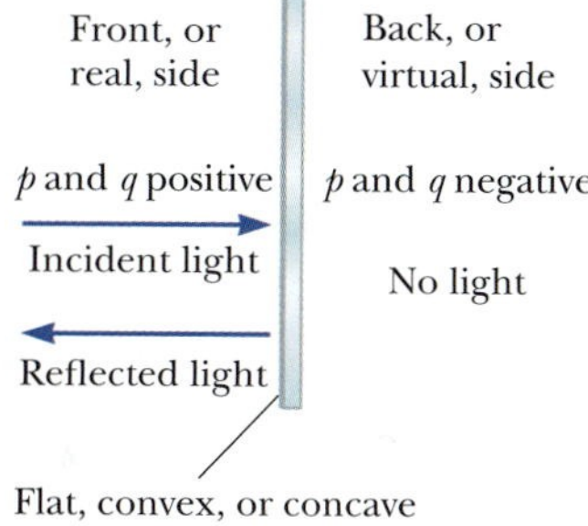


**Figure 23.12** A diagram describing the signs of $p$ and $q$ for convex and concave mirrors.

**Table 23.1** Sign Conventions for Mirrors

| Quantity | Symbol | In Front | In Back | Upright Image | Inverted Image |
|---|---|---|---|---|---|
| Object location | $p$ | + | − | | |
| Image location | $q$ | + | − | | |
| Focal length | $f$ | + | − | | |
| Image height | $h'$ | | | + | − |
| Magnification | $M$ | | | + | − |

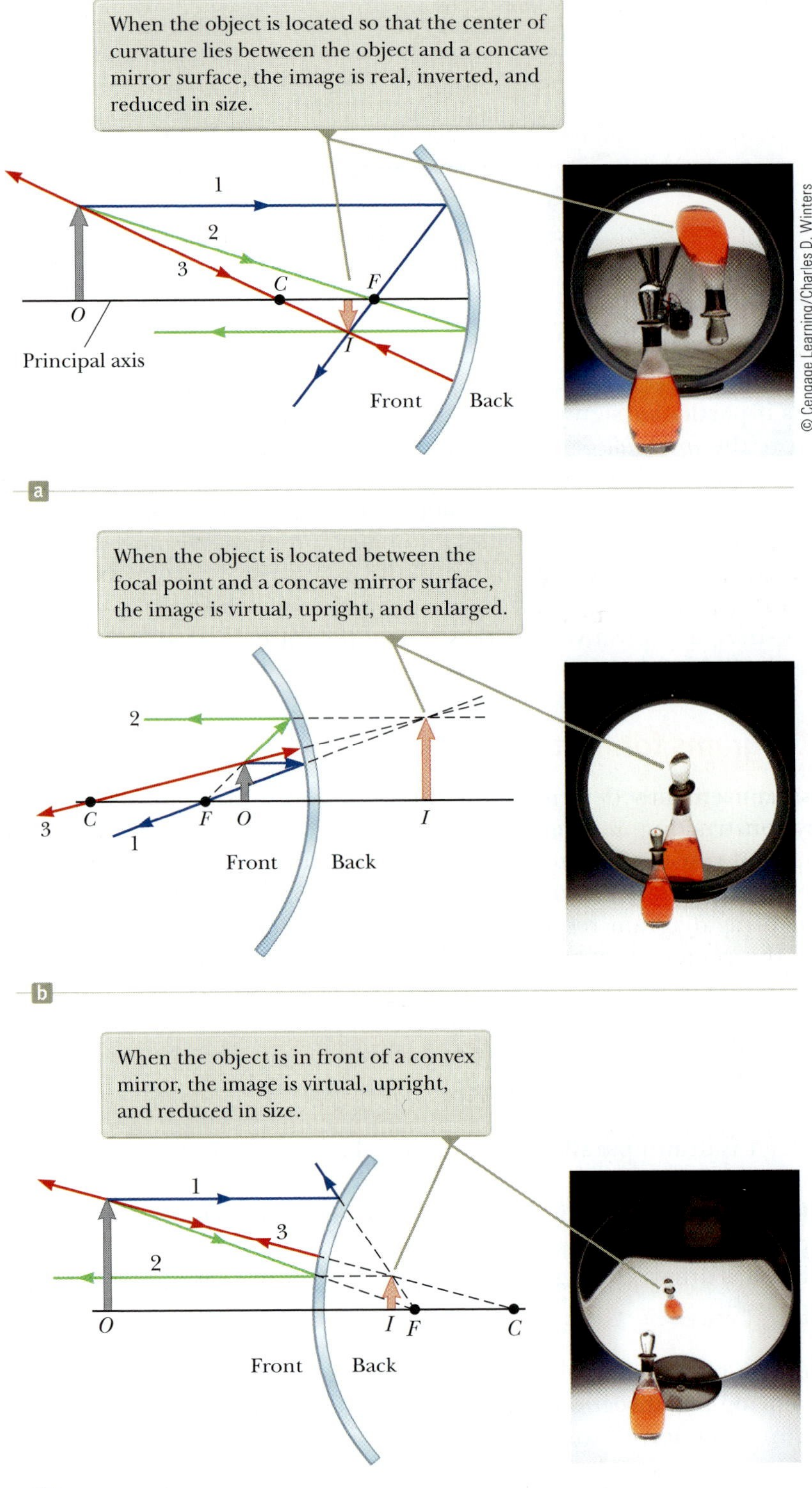

When the object is located so that the center of curvature lies between the object and a concave mirror surface, the image is real, inverted, and reduced in size.
1
2
3
C
F
O
I
Principal axis
Front
Back
© Cengage Learning/Charles D. Winters
a
When the object is located between the focal point and a concave mirror surface, the image is virtual, upright, and enlarged.
2
3
C
F
O
1
I
Front
Back
b
When the object is in front of a convex mirror, the image is virtual, upright, and reduced in size.
1
3
2
O
I
F
C
Front
Back
c

**Figure 23.13** Ray diagrams for spherical mirrors and corresponding photographs of the images of bottles.

this fashion must always agree with the value of $q$ calculated from the mirror formula.

In the case of a concave mirror, note what happens as the object is moved closer to the mirror. The real, inverted image in Figure 23.13a moves to the left as the object approaches the focal point. When the object is at the focal point, the image is infinitely far to the left. When the object lies between the focal point and the mirror surface, as in Figure 23.13b, however, the image is virtual and upright.

With the convex mirror shown in Figure 23.13c, the image of a real object is always virtual and upright. As the object distance increases, the virtual image shrinks and approaches the focal point as $p$ approaches infinity. You should construct a ray diagram to verify these statements.

The image-forming characteristics of curved mirrors obviously determine their uses. For example, suppose you want to design a mirror that will help people shave or apply cosmetics. For this, you need a concave mirror that puts the user inside the focal point, such as the mirror in Figure 23.13b. With that mirror, the image is upright and greatly enlarged. In contrast, suppose the primary purpose of a mirror is to observe a large field of view. In that case you need a convex mirror such as the one in Figure 23.13c. The diminished size of the image means that a fairly large field of view is seen in the mirror. Mirrors like this one are often placed in stores to help employees watch for shoplifters. A second use of such a mirror is as a side-view mirror on a car (Fig. 23.14). This kind of mirror is usually placed on the passenger side of the car and carries the warning "Objects are closer than they appear." Without such warning, a driver might think she is looking into a flat mirror, which doesn't alter the size of the image. She could be fooled into believing that a truck is far away because it looks small, when it's actually a large semi very close behind her, but diminished in size because of the image formation characteristics of the convex mirror.

**Figure 23.14** A convex side-view mirror on a vehicle produces an upright image that is smaller than the object. The smaller image means that the object is closer than its apparent distance as observed in the mirror.

## APPLYING PHYSICS 23.2 Concave Versus Convex

A virtual image can be anywhere behind a concave mirror. Why is there a maximum distance at which the image can exist behind a *convex* mirror?

**EXPLANATION** Consider the concave mirror first and imagine two different light rays leaving a tiny object and striking the mirror. If the object is at the focal point, the light rays reflecting from the mirror will be parallel to the mirror axis. They can be interpreted as forming a virtual image infinitely far away behind the mirror. As the object is brought closer to the mirror, the reflected rays will diverge through larger and larger angles, resulting in their extensions converging closer and closer to the back of the mirror. When the object is brought right up to the mirror, the image is right behind the mirror. When the object is much closer to the mirror than the focal length, the mirrors acts like a flat mirror and the image is just as far behind the mirror as the object is in front of it. The image can therefore be anywhere from infinitely far away to right at the surface of the mirror. For the convex mirror, an object at infinity produces a virtual image at the focal point. As the object is brought closer, the reflected rays diverge more sharply and the image moves closer to the mirror. As a result, the virtual image is restricted to the region between the mirror and the focal point. ■

## APPLYING PHYSICS 23.3 Reversible Waves

Large trucks often have a sign on the back saying, "If you can't see my mirror, I can't see you." Explain this sign.

**EXPLANATION** The trucking companies are making use of the principle of the reversibility of light rays. For an image of you to be formed in the driver's mirror, there must be a pathway for rays of light to reach the mirror, allowing the driver to see your image. If you can't see the mirror, this pathway doesn't exist. ■

## EXAMPLE 23.2 Images Formed by a Concave Mirror

**GOAL** Calculate properties of a concave mirror.

**PROBLEM** Assume a certain concave, spherical mirror has a focal length of 10.0 cm. **(a)** Locate the image and find the magnification for an object distance of 25.0 cm. Determine whether the image is real or virtual, inverted or upright, and larger or smaller. Do the same for object distances of **(b)** 10.0 cm and **(c)** 5.00 cm.

**STRATEGY** For each part, substitute into the mirror and magnification equations. Part (b) involves a limiting process because the answers are infinite. Notice that when the magnification $M$ is positive, the image is upright, and when $M$ is negative, the image is inverted. Similarly, when $q$ is positive the image is real, and when $q$ is negative the image is virtual.

*(Continued)*

**SOLUTION**

**(a)** Find the image position for an object distance of 25.0 cm. Calculate the magnification and describe the image.

Use the mirror equation to find the image distance:

$$\frac{1}{p} + \frac{1}{q} = \frac{1}{f}$$

Substitute and solve for $q$. According to Table 23.1, $p$ and $f$ are positive.

$$\frac{1}{25.0\text{ cm}} + \frac{1}{q} = \frac{1}{10.0\text{ cm}}$$

$$q = \boxed{16.7\text{ cm}}$$

Because $q$ is positive, the image is in front of the mirror and is real. The magnification is given by substituting into Equation 23.2:

$$M = -\frac{q}{p} = -\frac{16.7\text{ cm}}{25.0\text{ cm}} = \boxed{-0.668}$$

The image is smaller than the object because $|M| < 1$, and it is inverted because $M$ is negative. (See Fig. 23.13a.)

**(b)** Locate the image when the object distance is 10.0 cm. Calculate the magnification and describe the image.

The object is at the focal point. Substitute $p = 10.0$ cm and $f = 10.0$ cm into the mirror equation:

$$\frac{1}{10.0\text{ cm}} + \frac{1}{q} = \frac{1}{10.0\text{ cm}}$$

$$\frac{1}{q} = 0 \quad \rightarrow \quad \boxed{q = \infty}$$

Because $M = -q/p$, the magnification is also infinite.

**(c)** Locate the image when the object distance is 5.00 cm. Calculate the magnification and describe the image.

Once again, substitute into the mirror equation:

$$\frac{1}{5.00\text{ cm}} + \frac{1}{q} = \frac{1}{10.0\text{ cm}}$$

$$\frac{1}{q} = \frac{1}{10.0\text{ cm}} - \frac{1}{5.00\text{ cm}} = -\frac{1}{10.0\text{ cm}}$$

$$q = \boxed{-10.0\text{ cm}}$$

The image is virtual (behind the mirror) because $q$ is negative. Use Equation 23.2 to calculate the magnification:

$$M = -\frac{q}{p} = -\left(\frac{-10.0\text{ cm}}{5.00\text{ cm}}\right) = \boxed{2.00}$$

The image is larger (magnified by a factor of 2) because $|M| > 1$, and upright because $M$ is positive. (See Fig. 23.13b.)

**REMARKS** Note the characteristics of an image formed by a concave, spherical mirror. When the object is outside the focal point, the image is inverted and real; at the focal point, the image is formed at infinity; inside the focal point, the image is upright and virtual.

**QUESTION 23.2** What location does the image approach as the object gets arbitrarily far away from the mirror? (a) infinity (b) the focal point (c) the radius of curvature of the mirror (d) the mirror itself

**EXERCISE 23.2** If the object distance is 20.0 cm, find the image distance and the magnification of the mirror.

**ANSWER** $q = 20.0$ cm, $M = -1.00$

## EXAMPLE 23.3 Images Formed by a Convex Mirror

**GOAL** Calculate properties of a convex mirror.

**PROBLEM** An object 3.00 cm high is placed 20.0 cm from a convex mirror with a focal length of magnitude 8.00 cm. Find **(a)** the position of the image, **(b)** the magnification of the mirror, and **(c)** the height of the image.

**STRATEGY** This problem again requires only substitution into the mirror and magnification equations. Multiplying the object height by the magnification gives the image height.

**SOLUTION**

**(a)** Find the position of the image.

Because the mirror is convex, its focal length is negative. Substitute into the mirror equation:

$$\frac{1}{p} + \frac{1}{q} = \frac{1}{f}$$

$$\frac{1}{20.0\text{ cm}} + \frac{1}{q} = \frac{1}{-8.00\text{ cm}}$$

Solve for $q$:

$$q = -5.71\text{ cm}$$

**(b)** Find the magnification of the mirror.

Substitute into Equation 23.2:

$$M = -\frac{q}{p} = -\left(\frac{-5.71\text{ cm}}{20.0\text{ cm}}\right) = 0.286$$

**(c)** Find the height of the image.

Multiply the object height by the magnification:

$$h' = hM = (3.00\text{ cm})(0.286) = 0.858\text{ cm}$$

**REMARKS** The negative value of $q$ indicates the image is virtual, or behind the mirror, as in Figure 23.13c. The image is upright because $M$ is positive.

**QUESTION 23.3** True or False: A convex mirror can produce only virtual images.

**EXERCISE 23.3** Suppose the object is moved so that it is 4.00 cm from the same mirror. Repeat parts (a) through (c).

**ANSWERS** (a) −2.67 cm (b) 0.668 (c) 2.00 cm; the image is upright and virtual.

## EXAMPLE 23.4 The Face in the Mirror

**GOAL** Find a focal length from a magnification and an object distance.

**PROBLEM** When a woman stands with her face 40.0 cm from a cosmetic mirror, the upright image is twice as tall as her face. What is the focal length of the mirror?

**STRATEGY** To find $f$ in this example, we must first find $q$, the image distance. Because the problem states that the image is upright, the magnification must be positive (in this case, $M = +2$), and because $M = -q/p$, we can determine $q$.

**SOLUTION**

Obtain $q$ from the magnification equation:

$$M = -\frac{q}{p} = 2$$

$$q = -2p = -2(40.0\text{ cm}) = -80.0\text{ cm}$$

Because $q$ is negative, the image is on the opposite side of the mirror and hence is virtual. Substitute $q$ and $p$ into the mirror equation and solve for $f$:

$$\frac{1}{40.0\text{ cm}} - \frac{1}{80.0\text{ cm}} = \frac{1}{f}$$

$$f = 80.0\text{ cm}$$

**REMARKS** The positive sign for the focal length tells us that the mirror is concave, a fact we already knew because the mirror magnified the object. (A convex mirror would have produced a smaller image.)

*(Continued)*

**QUESTION 23.4** If she moves the mirror closer to her face, what happens to the image? (a) It becomes inverted and smaller. (b) It remains upright and becomes smaller. (c) It becomes inverted and larger. (d) It remains upright and becomes larger.

**EXERCISE 23.4** Suppose a fun-house spherical mirror makes you appear to be one-third your normal height. If you are 1.20 m away from the mirror, find its focal length. Is the mirror concave or convex?

**ANSWERS** −0.600 m, convex

# 23.4 Images Formed by Refraction

**LEARNING OBJECTIVES**

1. State the generalized equations relating object distance, image distance, and radius of curvature for images formed by transparent spherical surfaces.
2. Master the sign conventions for refracting surfaces.
3. Analyze the images formed by refracting surfaces.

In this section we describe how images are formed by refraction at a spherical surface. Consider two transparent media with indices of refraction $n_1$ and $n_2$, where the boundary between the two media is a spherical surface of radius $R$ (Fig. 23.15). We assume the medium to the right has a higher index of refraction than the one to the left: $n_2 > n_1$. That would be the case for light entering a curved piece of glass from air or for light entering the water in a fishbowl from air. The rays originating at the object location $O$ are refracted at the spherical surface and then converge to the image point $I$. We can begin with Snell's law of refraction and use simple geometric techniques to show that the object distance, image distance, and radius of curvature are related by the equation

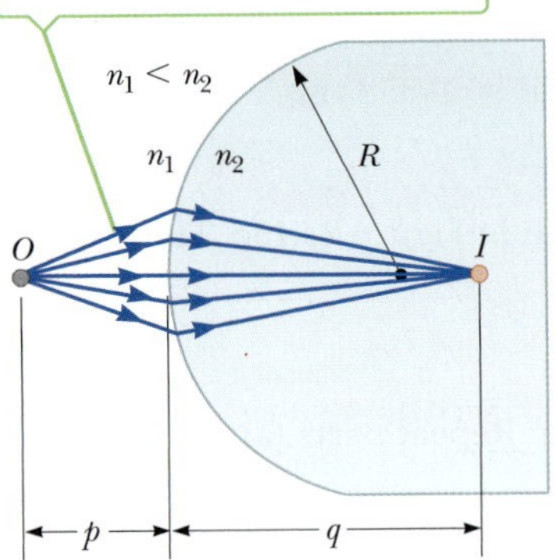
Rays making small angles with the principal axis diverge from a point object at O and pass through the image point I.
$n_1 < n_2$
$n_1$
$n_2$
R
O
I
p
q

**Figure 23.15** An image formed by refraction at a spherical surface.

$$\frac{n_1}{p} + \frac{n_2}{q} = \frac{n_2 - n_1}{R} \qquad [23.7]$$

Further, the magnification of a refracting surface is

$$M = \frac{h'}{h} = -\frac{n_1 q}{n_2 p} \qquad [23.8]$$

As with mirrors, certain sign conventions hold, depending on circumstances. First note that real images are formed by refraction on the side of the surface *opposite* the side from which the light comes, in contrast to mirrors, where real images are formed on the *same* side of the reflecting surface. This makes sense because light reflects off mirrors, so any real images must form on the same side the light comes from. With a transparent medium, the rays pass through and naturally form real images on the opposite side. We define the side of the surface where light rays originate as the front side. The other side is called the back side. Because of the difference in location of real images, the refraction sign conventions for $q$ and $R$ are the opposite of those for reflection. For example, $p$, $q$, and $R$ are all positive in Figure 23.15. The sign conventions for spherical refracting surfaces are summarized in Table 23.2.

**Table 23.2** Sign Conventions for Refracting Surfaces

| Quantity | Symbol | In Front | In Back | Upright Image | Inverted Image |
|---|---|---|---|---|---|
| Object location | $p$ | + | − | | |
| Image location | $q$ | − | + | | |
| Radius | $R$ | − | + | | |
| Image height | $h'$ | | | + | − |

## APPLYING PHYSICS 23.4 Underwater Vision BIO

Why does a person with normal vision see a blurry image if the eyes are opened underwater with no goggles or diving mask in use?

**EXPLANATION** The eye presents a spherical refraction surface. The eye normally functions so that light entering from the air is refracted to form an image in the retina located at the back of the eyeball. The difference in the index of refraction between water and the eye is smaller than the difference in the index of refraction between air and the eye. Consequently, light entering the eye from the water doesn't undergo as much refraction as does light entering from the air, and the image is formed behind the retina. A diving mask or swimming goggles have no optical action of their own; they are simply flat pieces of glass or plastic in a rubber mount. They do, however, provide a region of air adjacent to the eyes so that the correct refraction relationship is established and images will be in focus. ■

## Flat Refracting Surfaces

If the refracting surface is flat, then $R$ approaches infinity and Equation 23.7 reduces to

$$\frac{n_1}{p} = -\frac{n_2}{q}$$

$$q = -\frac{n_2}{n_1}p \qquad [23.9]$$

From Equation 23.9, we see that the sign of $q$ is opposite that of $p$. Consequently, **the image formed by a flat refracting surface is on the same side of the surface as the object.** This statement is illustrated in Figure 23.16 for the situation in which $n_1$ is greater than $n_2$, where a virtual image is formed between the object and the surface. Note that the refracted ray bends *away* from the normal in this case because $n_1 > n_2$.

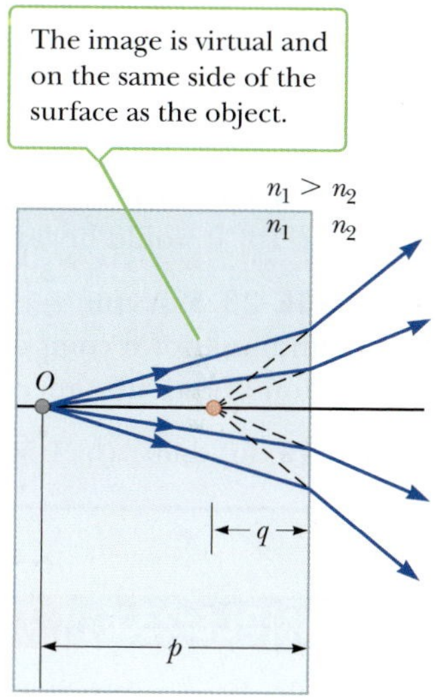

**Figure 23.16** The image formed by a flat refracting surface.

### Quick Quiz

**23.2** A person spearfishing from a boat sees a fish located 3 m from the boat at an apparent depth of 1 m. To spear the fish, should the person aim (a) at, (b) above, or (c) below the image of the fish?

**23.3** True or False: (a) The image of an object placed in front of a concave mirror is always upright. (b) The height of the image of an object placed in front of a concave mirror must be smaller than or equal to the height of the object. (c) The image of an object placed in front of a convex mirror is always upright and smaller than the object.

## EXAMPLE 23.5 Gaze into the Crystal Ball

**GOAL** Calculate the properties of an image created by a spherical lens.

**PROBLEM** A coin 2.00 cm in diameter is embedded in a solid glass ball of radius 30.0 cm (Fig. 23.17). The index of refraction of the ball is 1.50, and the coin is 20.0 cm from the surface. Find the position of the image of the coin and the height of the coin's image.

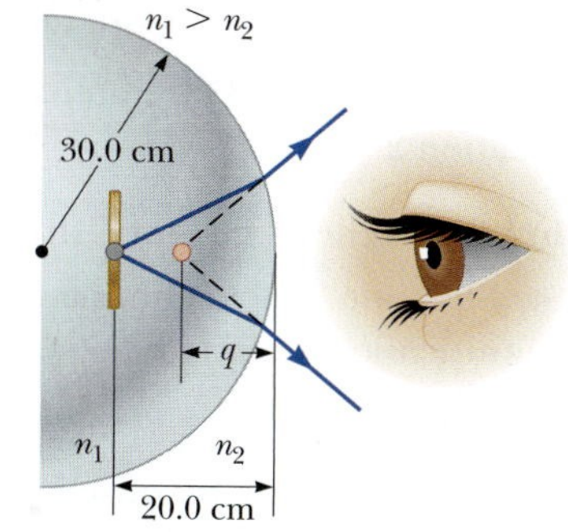


**Figure 23.17** (Example 23.5) A coin embedded in a glass ball forms a virtual image between the coin and the surface of the glass.

**STRATEGY** Because the rays are moving from a medium of high index of refraction (the glass ball) to a medium of lower index of refraction (air), the rays originating at the coin are refracted away from the normal at the surface and diverge outward. The image is formed in the glass and is virtual. Substitute into Equations 23.7 and 23.8 for the image position and magnification, respectively.

*(Continued)*

**SOLUTION**

Apply Equation 23.7 and take $n_1 = 1.50$, $n_2 = 1.00$, $p = 20.0$ cm, and $R = -30.0$ cm:

$$\frac{n_1}{p} + \frac{n_2}{q} = \frac{n_2 - n_1}{R}$$

$$\frac{1.50}{20.0\text{ cm}} + \frac{1.00}{q} = \frac{1.00 - 1.50}{-30.0\text{ cm}}$$

Solve for $q$:

$$q = -17.1\text{ cm}$$

To find the image height, use Equation 23.8 for the magnification:

$$M = -\frac{n_1 q}{n_2 p} = -\frac{1.50(-17.1\text{ cm})}{1.00(20.0\text{ cm})} = \frac{h'}{h}$$

$$h' = 1.28h = (1.28)(2.00\text{ cm}) = 2.56\text{ cm}$$

**REMARKS** The negative sign on $q$ indicates that the image is in the same medium as the object (the side of incident light), in agreement with our ray diagram, and therefore must be virtual. The positive value for $M$ means that the image is upright.

**QUESTION 23.5** How would the final answer be affected if the ball and observer were immersed in water? (a) It would be smaller. (b) It would be larger. (c) There would be no change.

**EXERCISE 23.5** A coin is embedded 20.0 cm from the surface of a similar ball of transparent substance having radius 30.0 cm and unknown composition. If the coin's image is virtual and located 15.0 cm from the surface, find the (a) index of refraction of the substance and (b) magnification.

**ANSWERS** (a) 2.00 (b) 1.50

## EXAMPLE 23.6 The One That Got Away

**GOAL** Calculate the properties of an image created by a flat refractive surface.

**PROBLEM** A small fish is swimming at a depth $d$ below the surface of a pond (Fig. 23.18). **(a)** What is the *apparent depth* of the fish as viewed from directly overhead? **(b)** If the fish is 12 cm long, how long is its image?

**STRATEGY** In this example the refracting surface is flat, so $R$ is infinite. Hence, we can use Equation 23.9 to determine the location of the image, which is the apparent location of the fish.

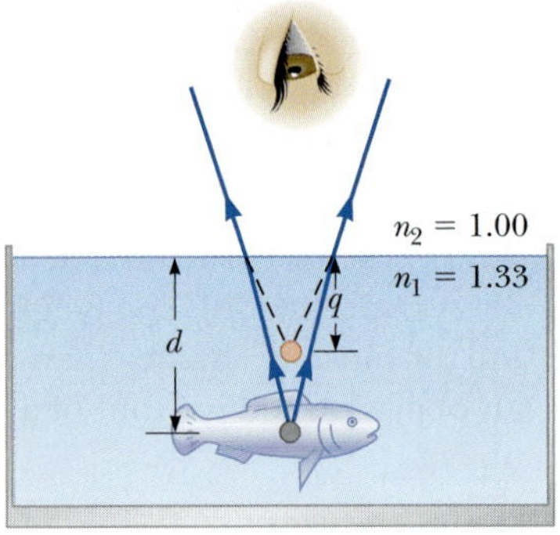

**Figure 23.18** (Example 23.6) The apparent depth $q$ of the fish is less than the true depth $d$.

**SOLUTION**

**(a)** Find the apparent depth of the fish.

Substitute $n_1 = 1.33$ for water and $p = d$ into Equation 23.9:

$$q = -\frac{n_2}{n_1}p = -\frac{1}{1.33}d = -0.752d$$

**(b)** What is the size of the fish's image?

Use Equation 23.9 to eliminate $q$ from Equation 23.8, the magnification equation:

$$M = \frac{h'}{h} = -\frac{n_1 q}{n_2 p} = -\frac{n_1\left(-\frac{n_2}{n_1}p\right)}{n_2 p} = 1$$

$$h' = h = 12\text{ cm}$$

**REMARKS** Again, because $q$ is negative, the image is virtual, as indicated in Figure 23.18. The apparent depth is approximately three-fourths the actual depth. For instance, if $d = 4.0$ m, then $q = -3.0$ m.

**QUESTION 23.6** Suppose a similar experiment is carried out with an object immersed in oil ($n = 1.5$) the same distance below the surface. How does the apparent depth of the object compare with its apparent depth when immersed in water? (a) The apparent depth is unchanged. (b) The apparent depth is larger. (c) The apparent depth is smaller.

**EXERCISE 23.6** A spear fisherman estimates that a trout is 1.5 m below the water's surface. What is the actual depth of the fish?

**ANSWER** 2.0 m

## 23.5 Atmospheric Refraction

**LEARNING OBJECTIVE**

1. Discuss images formed by atmospheric refraction, such as mirages.

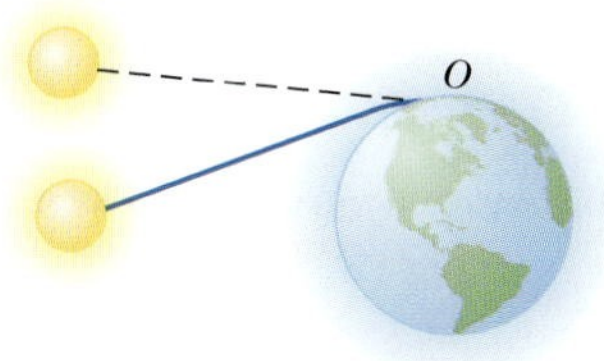


**Figure 23.19** Because light is refracted by Earth's atmosphere, an observer at $O$ sees the Sun even though it has fallen below the horizon.

Images formed by refraction in our atmosphere lead to some interesting phenomena. One such phenomenon that occurs daily is the visibility of the Sun at dusk even though it has passed below the horizon. Figure 23.19 shows why it occurs. Rays of light from the Sun strike Earth's atmosphere (represented by the shaded area around the planet) and are bent as they pass into a medium that has an index of refraction different from that of the almost empty space in which they have been traveling. The bending in this situation differs somewhat from the bending we have considered previously in that it is gradual and continuous as the light moves through the atmosphere toward an observer at point $O$. This is because the light moves through layers of air that have a continuously changing index of refraction. When the rays reach the observer, the eye follows them back along the direction from which they appear to have come (indicated by the dashed path in the figure). The end result is that the Sun appears to be above the horizon even after it has fallen below it.

The **mirage** is another phenomenon of nature produced by refraction in the atmosphere. A mirage can be observed when the ground is so hot that the air directly above it is warmer than the air at higher elevations. The desert is a region in which such circumstances prevail, but mirages are also seen on heated roadways during the summer. The layers of air at different heights above Earth have different densities and different refractive indices. The effect these differences can have is pictured in Figure 23.20a. The observer sees the sky and a cactus in two different ways. One group of light rays reaches the observer by the straight-line path $A$, and the eye traces these rays back to see the cactus in the normal fashion. In addition, a second group of rays travels along the curved path $B$. These rays are directed toward the ground and are then bent as a result of refraction. As a consequence,

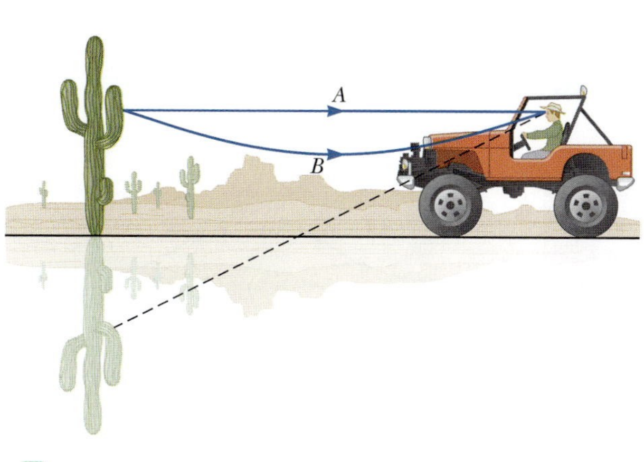

**Figure 23.20** (a) A mirage is produced by the bending of light rays in the atmosphere when there are large temperature differences between the ground and the air. (b) Notice the reflection of the cars in this photograph of a mirage. The road looks like it's flooded with water, but it is actually dry.

the observer also sees an inverted image of the cactus and the background of the sky as he traces the rays back to the point at which they appear to have originated. Because both an upright image and an inverted image are seen when the image of a cactus or other object is observed in a reflecting pool of water, the observer unconsciously calls on this past experience and concludes that the sky is reflected by a pool of water in front of the cactus.

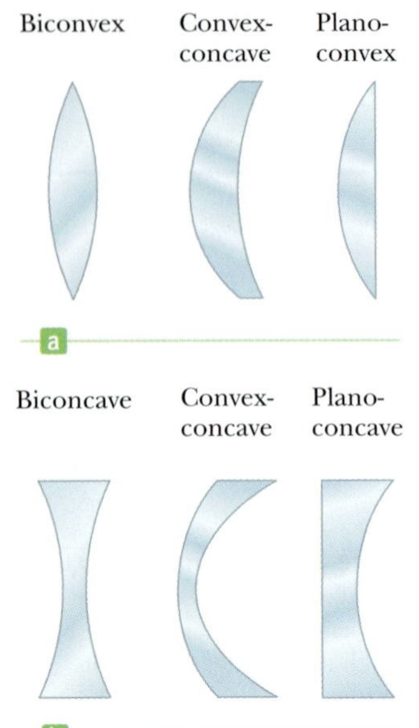


**Figure 23.21** Various lens shapes. (a) Converging lenses have positive focal lengths and are thickest at the middle. (b) Diverging lenses have negative focal lengths and are thickest at the edges.

## 23.6 Thin Lenses

**LEARNING OBJECTIVES**

1. Discuss converging and diverging thin lenses and their properties.
2. Master the sign conventions for thin lenses.
3. Apply the lens-maker's equation to calculating the physical properties of thin lenses.
4. Analyze systems involving more than one thin lens or having a lens and a mirror.

A typical **thin lens** consists of a piece of glass or plastic, ground so that each of its two refracting surfaces is a segment of either a sphere or a plane. Lenses are commonly used to form images by refraction in optical instruments, such as cameras, telescopes, and microscopes. The equation that relates object and image distances for a lens is virtually identical to the mirror equation derived earlier, and the method used to derive it is also similar.

Figure 23.21 shows some representative shapes of lenses. Notice that we have placed these lenses in two groups. Those in Figure 23.21a are thicker at the center than at the rim, and those in Figure 23.21b are thinner at the center than at the rim. The lenses in the first group are examples of **converging lenses**, and those in the second group are **diverging lenses**. The reason for these names will become apparent shortly.

As we did for mirrors, it is convenient to define a point called the **focal point** for a lens. For example, in Figure 23.22a, a group of rays parallel to the axis passes

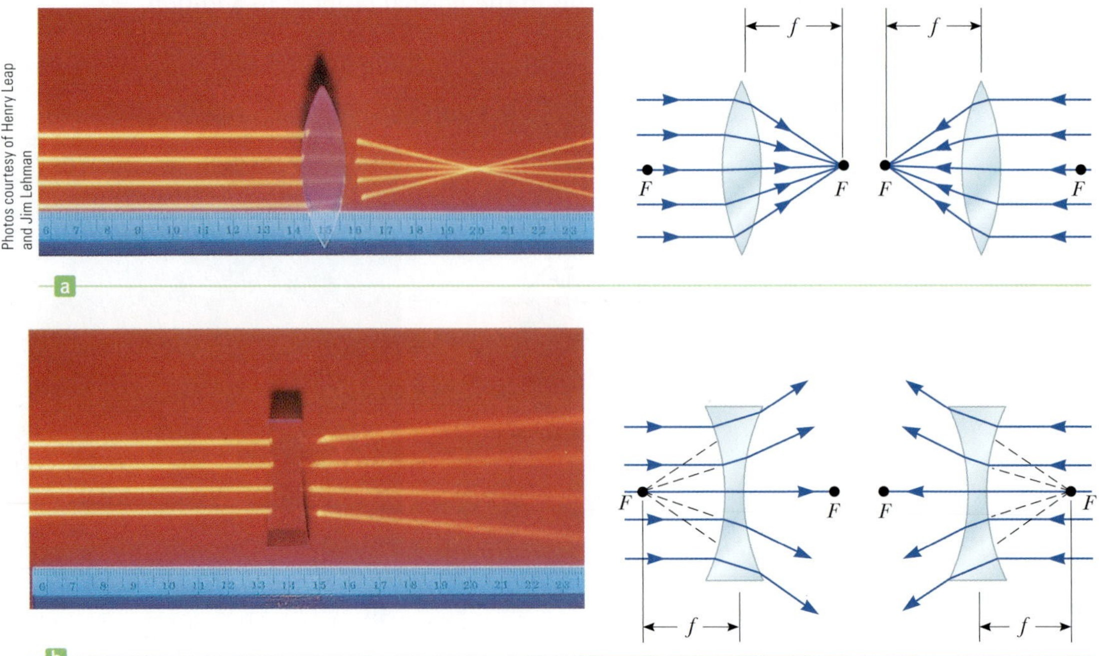


**Figure 23.22** (*Left*) Photographs of the effects of converging and diverging lenses on parallel rays. (*Right*) The focal points of the (a) biconvex lens and (b) biconcave lens.

through the focal point $F$ after being converged by the lens. The distance from the focal point to the lens is called the **focal length $f$. The focal length is the image distance that corresponds to an infinite object distance.** Recall that we are considering the lens to be very thin. As a result, it makes no difference whether we take the focal length to be the distance from the focal point to the surface of the lens or the distance from the focal point to the center of the lens because the difference between these two lengths is negligible. A thin lens has *two* focal points, as illustrated in Figure 23.22, one on each side of the lens. One focal point corresponds to parallel rays traveling from the left and the other corresponds to parallel rays traveling from the right.

Rays parallel to the axis diverge after passing through a lens of biconcave shape, shown in Figure 23.22b. In this case the focal point is defined to be the point where the diverged rays appear to originate, labeled $F$ in the figure. Figures 23.22a and 23.22b indicate why the names *converging* and *diverging* are applied to these lenses.

Now consider a ray of light passing through the center of a lens. Such a ray is labeled ray 1 in Figure 23.23. For a thin lens, a ray passing through the center is undeflected. Ray 2 in the same figure is parallel to the principal axis of the lens (the horizontal axis passing through $O$), and as a result it passes through the focal point $F$ after refraction. Rays 1 and 2 intersect at the point that is the tip of the image arrow.

We first note that the tangent of the angle $\alpha$ can be found by using the blue and gold shaded triangles in Figure 23.23:

$$\tan \alpha = \frac{h}{p} \quad \text{or} \quad \tan \alpha = -\frac{h'}{q}$$

From this result, we find that

$$M = \frac{h'}{h} = -\frac{q}{p} \qquad [23.10]$$

The equation for magnification by a lens is the same as the equation for magnification by a mirror. We also note from Figure 23.23 that

$$\tan \theta = \frac{PQ}{f} \quad \text{or} \quad \tan \theta = -\frac{h'}{q-f}$$

The height $PQ$ used in the first of these equations, however, is the same as $h$, the height of the object. Therefore,

$$\frac{h}{f} = -\frac{h'}{q-f}$$

$$\frac{h'}{h} = -\frac{q-f}{f}$$

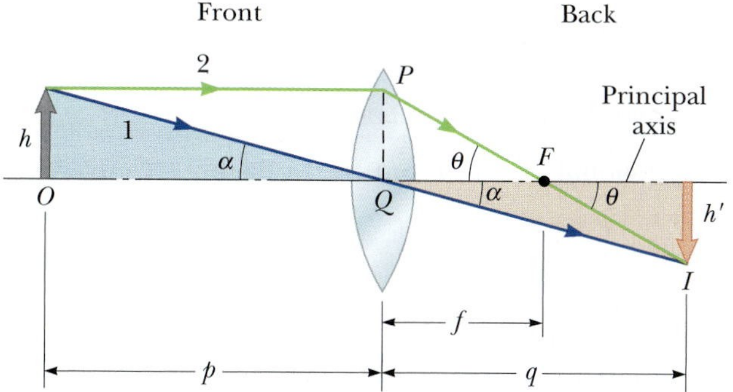


**Figure 23.23** A geometric construction for developing the thin-lens equation.

Using the latter equation in combination with Equation 23.10 gives

$$\frac{q}{p} = \frac{q - f}{f}$$

which reduces to

Thin-lens equation ▶

$$\frac{1}{p} + \frac{1}{q} = \frac{1}{f} \qquad [23.11]$$

This equation, called the **thin-lens equation**, can be used with both converging and diverging lenses if we adhere to a set of sign conventions. Figure 23.24 is useful for obtaining the signs of $p$ and $q$, and Table 23.3 gives the complete sign conventions for lenses. Note that **a converging lens has a positive focal length** under this convention and **a diverging lens has a negative focal length**. Hence, the names *positive* and *negative* are often given to these lenses.

The focal length for a lens in air is related to the curvatures of its front and back surfaces and to the index of refraction $n$ of the lens material by

Lens-maker's equation ▶

$$\frac{1}{f} = (n - 1)\left(\frac{1}{R_1} - \frac{1}{R_2}\right) \qquad [23.12]$$

Front, or virtual, side

Back, or real, side

$p$ positive
$q$ negative

$p$ negative
$q$ positive

Incident light

Refracted light

Converging or diverging lens

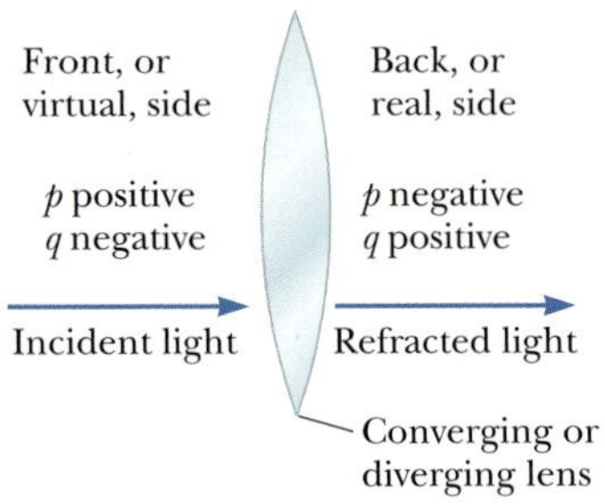

**Figure 23.24** A diagram for obtaining the signs of $p$ and $q$ for a thin lens or a refracting surface.

where $R_1$ is the radius of curvature of the front surface of the lens and $R_2$ is the radius of curvature of the back surface. (As with mirrors, we arbitrarily call the side from which the light approaches the *front* of the lens.) Table 23.3 gives the sign conventions for $R_1$ and $R_2$. Equation 23.12, called the **lens-maker's equation**, enables us to calculate the focal length from the known properties of the lens.

## Ray Diagrams for Thin Lenses

Ray diagrams are essential for understanding the overall image formation by a thin lens or a system of lenses. They should also help clarify the sign conventions already discussed. Figure 23.25 illustrates this method for three single-lens situations. To locate the image formed by a converging lens (Figs. 23.25a and 23.25b), the following three rays are drawn from the top of the object:

1. The first ray is drawn parallel to the principal axis. After being refracted by the lens, this ray passes through (or appears to come from) one of the focal points.
2. The second ray is drawn through the center of the lens. This ray continues in a straight line.
3. The third ray is drawn through the other focal point and emerges from the lens parallel to the principal axis.

A similar construction is used to locate the image formed by a diverging lens, as shown in Figure 23.25c. The point of intersection of *any two* of the rays in these diagrams can be used to locate the image. The third ray serves as a check on construction.

For the converging lens in Figure 23.25a, where the object is *outside* the front focal point ($p > f$), the ray diagram shows that the image is real and inverted.

**Tip 23.4 Positive Is Again Where the Light Is**

For lenses, $p$ and $q$ are positive where the light is, where the object or image is real. For real objects, the light originates with the object in front of the lens, so $p$ is positive there as indicated in Figure 23.24. If the image forms in back of the lens, $q$ is positive there, as well.

**Tip 23.5 We *Choose* Only a Few Rays**

Although our ray diagrams in Figure 23.25 only show three rays leaving an object, an infinite number of rays can be drawn between the object and its image.

**Table 23.3** Sign Conventions for Thin Lenses

| Quantity | Symbol | In Front | In Back | Convergent | Divergent |
|---|---|---|---|---|---|
| Object location | $p$ | + | − | | |
| Image location | $q$ | − | + | | |
| Lens radii | $R_1, R_2$ | − | + | | |
| Focal length | $f$ | | | + | − |

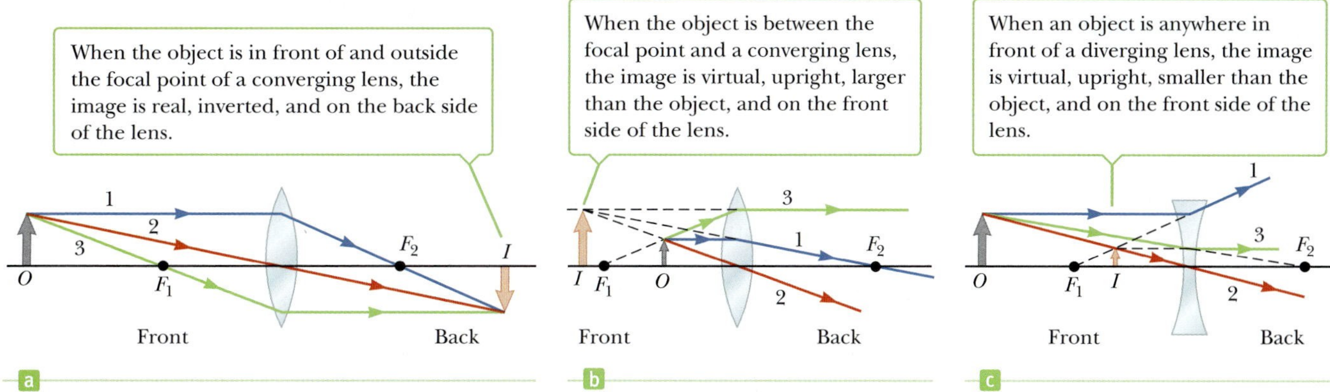


**Figure 23.25** Ray diagrams for locating the image formed by a thin lens.

When the real object is *inside* the front focal point ($p < f$), as in Figure 23.25b, the image is virtual and upright. For the diverging lens of Figure 23.25c, the image is virtual and upright.

## Quick Quiz

**23.4** A clear plastic sandwich bag filled with water can act as a crude converging lens in air. If the bag is filled with air and placed under water, is the effective lens (a) converging or (b) diverging?

**23.5** In Figure 23.25a the blue object arrow is replaced by one that is much taller than the lens. How many rays from the object will strike the lens?

**23.6** An object is placed to the left of a converging lens. Which of the following statements are true, and which are false? (a) The image is always to the right of the lens. (b) The image can be upright or inverted. (c) The image is always smaller or the same size as the object.

Your success in working lens or mirror problems will be determined largely by whether you make sign errors when substituting into the lens or mirror equations. The only way to ensure you don't make sign errors is to become adept at using the sign conventions. The best way to do so is to work a multitude of problems on your own and construct confirming ray diagrams. Watching an instructor or reading the example problems is no substitute for practice.

## APPLYING PHYSICS 23.5 Vision and Diving Masks BIO

Diving masks often have a lens built into the glass faceplate for divers who don't have perfect vision. This lens allows the individual to dive without the necessity of glasses because the faceplate performs the necessary refraction to produce clear vision. Normal glasses have lenses that are curved on both the front and rear surfaces. The lenses in a diving-mask faceplate often have curved surfaces only on the inside of the glass. Why is this design desirable?

**SOLUTION** The main reason for curving only the inner surface of the lens in the diving-mask faceplate is to enable the diver to see clearly while underwater and in the air. If there were curved surfaces on both the front and the back of the diving lens, there would be two refractions. The lens could be designed so that these two refractions would give clear vision while the diver is in air. When the diver went underwater, however, the refraction between the water and the glass at the first interface would differ because the index of refraction of water is different from that of air. Consequently, the diver's vision wouldn't be clear underwater. ■

## EXAMPLE 23.7 Images Formed by a Converging Lens

**GOAL** Calculate geometric quantities associated with a converging lens.

**PROBLEM** A converging lens of focal length 10.0 cm forms images of an object situated at various distances. **(a)** If the object is placed 30.0 cm from the lens, locate the image, state whether it's real or virtual, and find its magnification. **(b)** Repeat the problem when the object is at 10.0 cm and **(c)** again when the object is 5.00 cm from the lens.

**STRATEGY** All three problems require only substitution into the thin-lens equation and the associated magnification equation, Equations 23.10 and 23.11, respectively. The conventions of Table 23.3 must be followed.

**Figure 23.26** (Example 23.7)

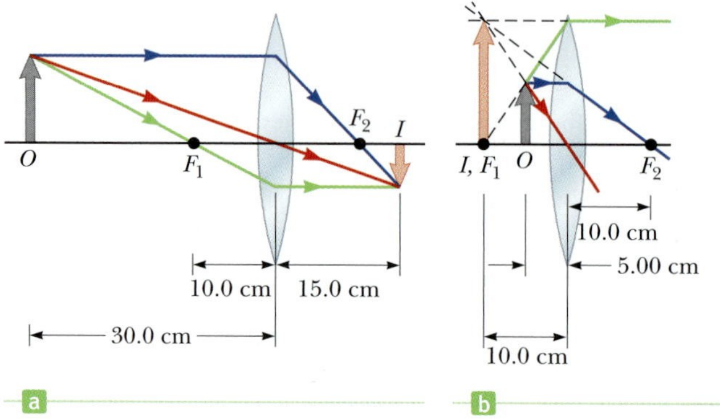


**SOLUTION**

**(a)** Find the image distance and describe the image when the object is placed at 30.0 cm.

The ray diagram is shown in Figure 23.26a. Substitute values into the thin-lens equation to locate the image:

$$\frac{1}{p} + \frac{1}{q} = \frac{1}{f}$$

$$\frac{1}{30.0 \text{ cm}} + \frac{1}{q} = \frac{1}{10.0 \text{ cm}}$$

Solve for $q$, the image distance. It's positive, so the image is real and on the far side of the lens:

$$q = +15.0 \text{ cm}$$

The magnification of the lens is obtained from Equation 23.10. $M$ is negative and less than 1 in absolute value, so the image is inverted and smaller than the object:

$$M = -\frac{q}{p} = -\frac{15.0 \text{ cm}}{30.0 \text{ cm}} = -0.500$$

**(b)** Repeat the problem, when the object is placed at 10.0 cm.

Locate the image by substituting into the thin-lens equation:

$$\frac{1}{10.0 \text{ cm}} + \frac{1}{q} = \frac{1}{10.0 \text{ cm}} \quad \rightarrow \quad \frac{1}{q} = 0$$

This equation is satisfied only in the limit as $q$ becomes infinite. Similarly, $M$ becomes infinite, as well.

$$q \rightarrow \infty$$

**(c)** Repeat the problem when the object is placed 5.00 cm from the lens.

See the ray diagram shown in Figure 23.26b. Substitute into the thin-lens equation to locate the image:

$$\frac{1}{5.00 \text{ cm}} + \frac{1}{q} = \frac{1}{10.0 \text{ cm}}$$

Solve for $q$, which is negative, meaning the image is on the same side as the object and is virtual:

$$q = -10.0 \text{ cm}$$

Substitute the values of $p$ and $q$ into the magnification equation. $M$ is positive and larger than 1, so the image is upright and double the object size:

$$M = -\frac{q}{p} = -\left(\frac{-10.0 \text{ cm}}{5.00 \text{ cm}}\right) = +2.00$$

**REMARKS** The ability of a lens to magnify objects led to the inventions of reading glasses, microscopes, and telescopes.

**QUESTION 23.7** If the lens is used to form an image of the Sun on a screen, how far from the lens should the screen be located?

**EXERCISE 23.7** Suppose the image of an object is upright and magnified 1.75 times when the object is placed 15.0 cm from a lens. Find (a) the location of the image and (b) the focal length of the lens.

**ANSWERS** (a) −26.3 cm (virtual, on the same side as the object) (b) 34.9 cm

## EXAMPLE 23.8 The Case of a Diverging Lens

**GOAL** Calculate geometric quantities associated with a diverging lens.

**PROBLEM** Repeat the problem of Example 23.7 for a *diverging* lens having a focal length of magnitude 10.0 cm.

**STRATEGY** Once again, substitution into the thin-lens equation and the associated magnification equation, together with the conventions in Table 23.3, solve the various parts. The only difference is the negative focal length.

**Figure 23.27** (Example 23.8)

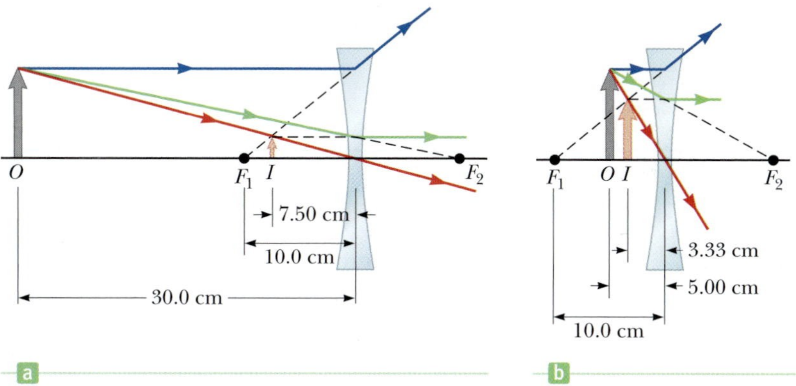


**SOLUTION**

**(a)** Locate the image and its magnification if the object is at 30.0 cm.

The ray diagram is given in Figure 23.27a. Apply the thin-lens equation with $p = 30.0$ cm to locate the image:

$$\frac{1}{p} + \frac{1}{q} = \frac{1}{f}$$

$$\frac{1}{30.0 \text{ cm}} + \frac{1}{q} = -\frac{1}{10.0 \text{ cm}}$$

Solve for $q$, which is negative and hence virtual:

$$q = \boxed{-7.50 \text{ cm}}$$

Substitute into Equation 23.10 to get the magnification. Because $M$ is positive and has absolute value less than 1, the image is upright and smaller than the object:

$$M = -\frac{q}{p} = -\left(\frac{-7.50 \text{ cm}}{30.0 \text{ cm}}\right) = \boxed{+0.250}$$

**(b)** Locate the image and find its magnification if the object is 10.0 cm from the lens.

Apply the thin-lens equation, taking $p = 10.0$ cm:

$$\frac{1}{10.0 \text{ cm}} + \frac{1}{q} = -\frac{1}{10.0 \text{ cm}}$$

Solve for $q$ (once again, the result is negative, so the image is virtual):

$$q = \boxed{-5.00 \text{ cm}}$$

Calculate the magnification. Because $M$ is positive and has absolute value less than 1, the image is upright and smaller than the object:

$$M = -\frac{q}{p} = -\left(\frac{-5.00 \text{ cm}}{10.0 \text{ cm}}\right) = \boxed{+0.500}$$

*(Continued)*

**(c)** Locate the image and find its magnification when the object is at 5.00 cm.

The ray diagram is given in Figure 23.27b. Substitute $p = 5.00$ cm into the thin-lens equation to locate the image:

$$\frac{1}{5.00\text{ cm}} + \frac{1}{q} = -\frac{1}{10.0\text{ cm}}$$

Solve for $q$. The answer is negative, so once again the image is virtual:

$$q = -3.33\text{ cm}$$

Calculate the magnification. Because $M$ is positive and less than 1, the image is upright and smaller than the object:

$$M = -\left(\frac{-3.33\text{ cm}}{5.00\text{ cm}}\right) = +0.666$$

**REMARKS** Notice that in every case the image is virtual, hence on the same side of the lens as the object. Further, the image is smaller than the object. For a diverging lens and a real object, this is *always* the case, as can be proven mathematically.

**QUESTION 23.8** Can a diverging lens be used as a magnifying glass? Explain.

**EXERCISE 23.8** Repeat the calculation, finding the position of the image and the magnification if the object is 20.0 cm from the lens.

**ANSWERS** $q = -6.67$ cm, $M = 0.334$

## Combinations of Thin Lenses

Many useful optical devices require two lenses. Handling problems involving two lenses is not much different from dealing with a single-lens problem twice. First, the image produced by the first lens is calculated as though the second lens were not present. The light then approaches the second lens *as if* it had come from the image formed by the first lens. Hence, **the image formed by the first lens is treated as the object for the second lens.** The image formed by the second lens is the final image of the system. If the image formed by the first lens lies on the back side of the second lens, the image is treated as a virtual object for the second lens, so $p$ is negative. The same procedure can be extended to a system of three or more lenses. The overall magnification of a system of thin lenses is the *product* of the magnifications of the separate lenses. It's also possible to combine thin lenses and mirrors as shown in Example 23.10.

### EXAMPLE 23.9 Two Lenses in a Row

**GOAL** Calculate geometric quantities for a sequential pair of lenses.

**PROBLEM** Two converging lenses are placed 20.0 cm apart, as shown in Figure 23.28a, with an object 30.0 cm in front of lens 1 on the left. **(a)** If lens 1 has a focal length of 10.0 cm, locate the image formed by this lens and determine its magnification. **(b)** If lens 2 on the right has a focal length of 20.0 cm, locate the final image formed and find the total magnification of the system.

**STRATEGY** We apply the thin-lens equation to each lens. The image formed by lens 1 is treated as the object for lens 2. Also, we use the fact that the total magnification of the system is the product of the magnifications produced by the separate lenses.

**SOLUTION**

**(a)** Locate the image and determine the magnification of lens 1.

See the ray diagram, Figure 23.28b. Apply the thin-lens equation to lens 1:

$$\frac{1}{30.0\text{ cm}} + \frac{1}{q} = \frac{1}{10.0\text{ cm}}$$

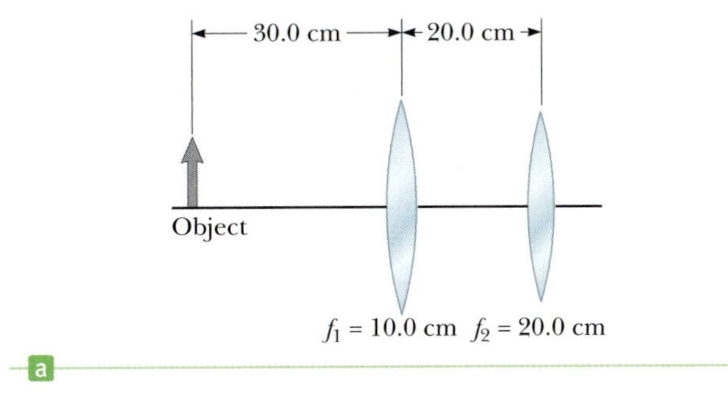


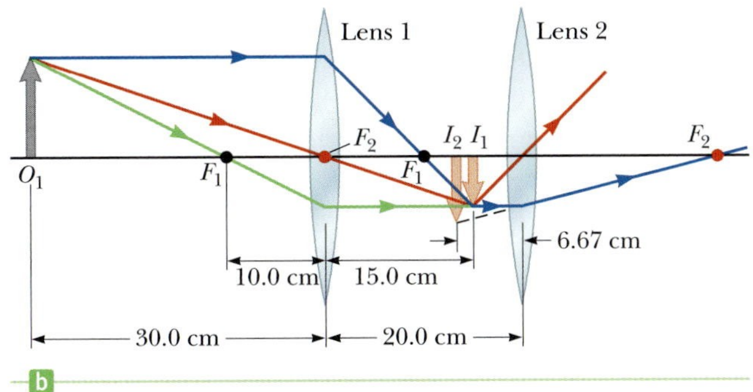


**Figure 23.28** (Example 23.9)

Solve for $q$, which is positive and hence to the right of the first lens:

$$q = +15.0 \text{ cm}$$

Compute the magnification of lens 1:

$$M_1 = -\frac{q}{p} = -\frac{15.0 \text{ cm}}{30.0 \text{ cm}} = -0.500$$

**(b)** Locate the final image and find the total magnification.

The image formed by lens 1 becomes the object for lens 2. Compute the object distance for lens 2:

$$p = 20.0 \text{ cm} - 15.0 \text{ cm} = 5.00 \text{ cm}$$

Once again apply the thin-lens equation to lens 2 to locate the final image:

$$\frac{1}{5.00 \text{ cm}} + \frac{1}{q} = \frac{1}{20.0 \text{ cm}}$$

$$q = -6.67 \text{ cm}$$

Calculate the magnification of lens 2:

$$M_2 = -\frac{q}{p} = -\frac{(-6.67 \text{ cm})}{5.00 \text{ cm}} = +1.33$$

Multiply the two magnifications to get the overall magnification of the system:

$$M = M_1M_2 = (-0.500)(1.33) = -0.665$$

**REMARKS** The negative sign for $M$ indicates that the final image is inverted and smaller than the object because the absolute value of $M$ is less than 1. Because $q$ is negative, the final image is virtual.

**QUESTION 23.9** If lens 2 is moved so it is 40 cm away from lens 1, would the final image be upright or inverted?

**EXERCISE 23.9** If the two lenses in Figure 23.28 are separated by 10.0 cm, locate the final image and find the magnification of the system. *Hint:* The object for the second lens is virtual!

**ANSWERS** 4.00 cm behind the second lens, $M = -0.400$

## EXAMPLE 23.10 Thin Lens and a Concave Mirror

**GOAL** Solve a problem involving both a lens and a mirror.

**PROBLEM** An object is placed 20.0 cm to the right of a concave mirror with focal length 12.0 cm and 30.0 cm to the left of a converging lens with focal length 10.0 cm, as in Figure 23.29. Locate **(a)** the image formed by the lens alone, **(b)** the image created by the mirror alone, and **(c)** the image created by both the mirror and lens. **(d)** The mirror is moved so that it is 6.00 cm away from the object. Locate the image formed by the mirror and lens.

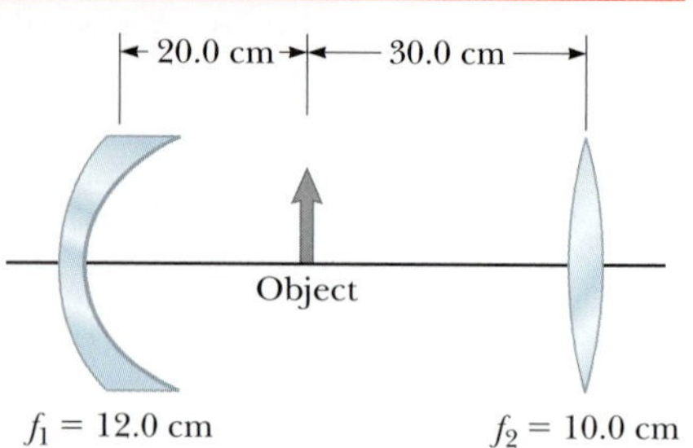


**Figure 23.29** (Example 23.10)

**STRATEGY** Part (a) is a simple application of the thin-lens equation, Equation 23.11. Part (b) can be calculated from Equation 23.6. Using the image formed by the mirror as the object for the lens, find the image location asked for in part (c), for light that first reflects off the mirror before passing through the lens.

**SOLUTION**

**(a)** Locate the image formed by the lens alone.

Apply Equation 23.11:

$$\frac{1}{p_2} + \frac{1}{q_2} = \frac{1}{f_2}$$

Substitute values and solve for the image position, $q_2$:

$$\frac{1}{q_2} = \frac{1}{f_2} - \frac{1}{p_2} = \frac{1}{10.0 \text{ cm}} - \frac{1}{30.0 \text{ cm}} = \frac{1}{15.0 \text{ cm}}$$

$$q_2 = \boxed{15.0 \text{ cm}}$$

**(b)** Locate the image created by the mirror alone.

Apply Equation 23.6:

$$\frac{1}{p_1} + \frac{1}{q_1} = \frac{1}{f_1}$$

Substitute values for $f_1$ and $p_1$, which are both positive, and solve for $q_1$:

$$\frac{1}{q_1} = \frac{1}{f_1} - \frac{1}{p_1} = \frac{1}{12.0 \text{ cm}} - \frac{1}{20.0 \text{ cm}} = \frac{1}{30.0 \text{ cm}}$$

$$q_1 = \boxed{30.0 \text{ cm}}$$

**(c)** Locate the image created by both the mirror and lens.

Apply Equation 23.11 to the image found in part (b), which becomes a real object for the lens, noticing that the image formed by the mirror is 20.0 cm from the lens:

$$\frac{1}{q_{2f}} = \frac{1}{f_2} - \frac{1}{p_{2f}} = \frac{1}{10.0 \text{ cm}} - \frac{1}{20.0 \text{ cm}} = \frac{1}{20.0 \text{ cm}}$$

$$q_{2f} = \boxed{20.0 \text{ cm}}$$

**(d)** The mirror is moved so that it is 6.00 cm to the left of the object. Locate the image formed by the mirror and lens.

The object is now much closer to the mirror. Find the new location of the image created by the mirror:

$$\frac{1}{q_1} = \frac{1}{f_1} - \frac{1}{p_1} = \frac{1}{12.0 \text{ cm}} - \frac{1}{6.0 \text{ cm}} = -\frac{1}{12.0 \text{ cm}}$$

$$q_1 = -12.0 \text{ cm}$$

The image created by the mirror is virtual and therefore behind the mirror. However, it acts like a real object for the lens. Apply Equation 23.11 with $p_2 = 30.0 \text{ cm} + 18.0 \text{ cm} = 48.0 \text{ cm}$:

$$\frac{1}{q_2} = \frac{1}{f_2} - \frac{1}{p_2} = \frac{1}{10.0 \text{ cm}} - \frac{1}{48.0 \text{ cm}} = \frac{19}{2.40 \times 10^2 \text{ cm}}$$

$$q_2 = \boxed{12.6 \text{ cm}}$$

**REMARKS** There are two final images created to the right of the lens, as expected. As the mirror is moved closer to the object, the final image due to both the mirror and lens moves closer to the lens. The image of part (a) is inverted; however, the image of part (c) goes through two inversions, hence is upright. The virtual mirror image of part (d) is upright, so its lens image is inverted.

**QUESTION 23.10** Is it possible to have a virtual object for a mirror? Explain, giving an example.

**EXERCISE 23.10** The same mirror and lens are repositioned so that the mirror is 24.0 cm to the left of the lens and the object is 20.0 cm to the right of the lens. Locate the image of (a) the lens alone, (b) the first image formed by the mirror, and (c) the final, second image formed by the lens.

**ANSWERS** (a) 20.0 cm to the left of the lens (b) 6.00 cm behind the mirror (c) 15.0 cm to the right of the lens

# 23.7 Lens and Mirror Aberrations

**LEARNING OBJECTIVE**

1. Discuss defects in lenses and mirrors and their consequent aberrations.

One of the basic problems of systems containing mirrors and lenses is the imperfect quality of the images, which is largely the result of defects in shape and form. The simple theory of mirrors and lenses assumes rays make small angles with the principal axis and all rays reaching the lens or mirror from a point source are focused at a single point, producing a sharp image. This is not always true in the real world. Where the approximations used in this theory do not hold, imperfect images are formed.

If one wishes to analyze image formation precisely, it is necessary to trace each ray, using Snell's law, at each refracting surface. This procedure shows that there is no single point image; instead, the image is blurred. The departures of real (imperfect) images from the ideal predicted by the simple theory are called **aberrations.** Two common types of aberrations are spherical aberration and chromatic aberration.

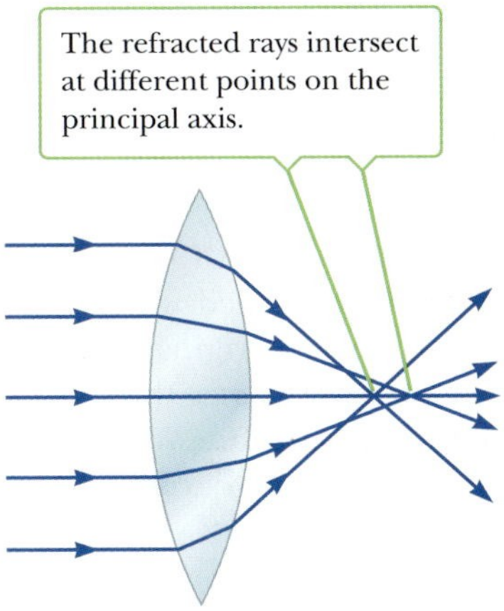

The refracted rays intersect at different points on the principal axis.

**Figure 23.30** Spherical aberration used by a converging lens. Does a diverging lens use spherical aberration?

## Spherical Aberration

Spherical aberration results from the fact that the focal points of light rays passing far from the principal axis of a spherical lens (or mirror) are different from the focal points of rays with the same wavelength passing near the axis. Figure 23.30 illustrates spherical aberration for parallel rays passing through a converging lens. Rays near the middle of the lens are imaged farther from the lens than rays at the edges. Hence, there is no single focal length for a spherical lens.

Most cameras are equipped with an adjustable aperture to control the light intensity and, when possible, reduce spherical aberration. (An aperture is an opening that controls the amount of light transmitted through the lens.) As the aperture size is reduced, sharper images are produced because only the central portion of the lens is exposed to the incident light when the aperture is very small. At the same time, however, progressively less light is imaged. To compensate for this loss, a longer exposure time is used. An example of the results obtained with small apertures is the sharp image produced by a pinhole camera, with an aperture size of approximately 0.1 mm.

In the case of mirrors used for very distant objects, one can eliminate, or at least minimize, spherical aberration by employing a parabolic rather than spherical surface. Parabolic surfaces are not used in many applications, however, because they are very expensive to make with high-quality optics. Parallel light rays incident on such a surface focus at a common point. Parabolic reflecting surfaces are used in many astronomical telescopes to enhance the image quality. They are also used in flashlights, in which a nearly parallel light beam is produced from a small lamp placed at the focus of the reflecting surface.

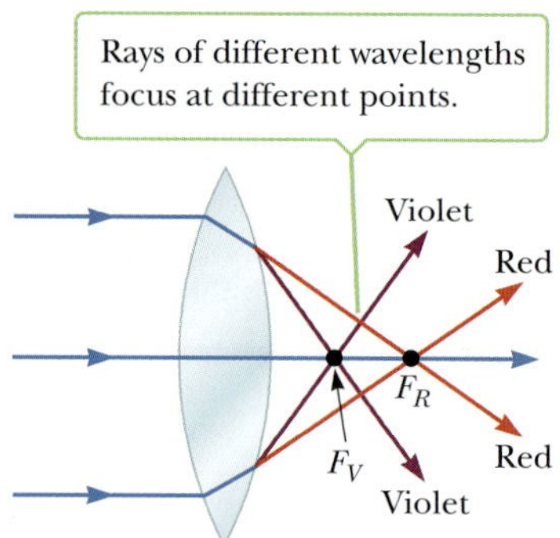


**Figure 23.31** Chromatic aberration produced by a converging lens.

## Chromatic Aberration

Different wavelengths of light refracted by a lens focus at different points, which gives rise to chromatic aberration. In Chapter 22 we described how the index of refraction of a material varies with wavelength. When white light passes through a lens, for example, violet light rays are refracted more than red light rays (see Fig. 23.31), so the focal length for red light is greater than for violet light. Other wavelengths (not shown in the figure) would have intermediate focal points. Chromatic aberration for a diverging lens is opposite that for a converging lens. Chromatic aberration can be greatly reduced by a combination of converging and diverging lenses. Chromatic aberration isn't a problem with mirrors, because all wavelengths of light are reflected at the same angle.

# SUMMARY

## 23.1 Flat Mirrors

Images are formed where rays of light intersect or where they appear to originate. A **real image** is formed when light intersects, or passes through, an image point. In a **virtual image** the light doesn't pass through the image point, but appears to diverge from it.

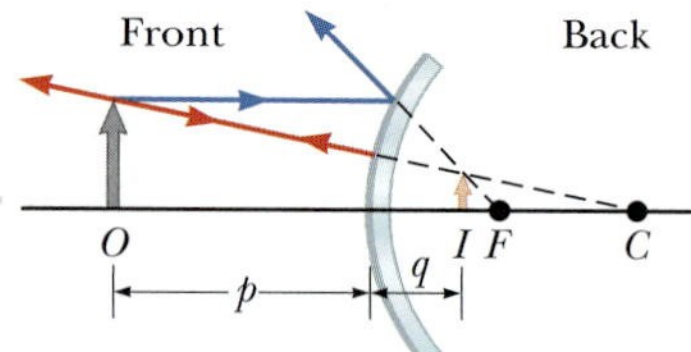


The image of a convex mirror is virtual, upright, and behind the mirror.

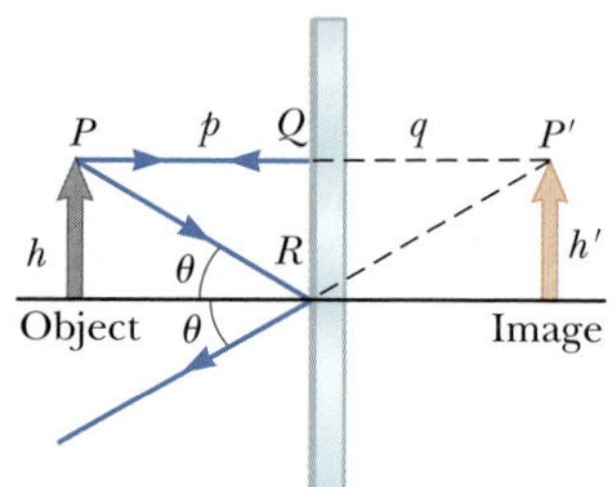


A geometric construction to locate the image of an object placed in front of a flat mirror. Because the triangles $PQR$ and $P'QR$ are identical, $p = |q|$ and $h = h'$.

The image formed by a flat mirror has the following properties: **1.** The image is as far behind the mirror as the object is in front of it. **2.** The image is unmagnified, virtual, and upright.

## 23.2 Images Formed by Concave Mirrors

## 23.3 Convex Mirrors and Sign Conventions

The **magnification** $M$ of a spherical mirror is defined as the ratio of the **image height** $h'$ to the **object height** $h$, which is the negative of the ratio of the image distance $q$ to the object distance $p$:

$$M = \frac{h'}{h} = -\frac{q}{p} \qquad [23.2]$$

The **object distance** and **image distance** for a spherical mirror of radius $R$ are related by the **mirror equation**:

$$\frac{1}{p} + \frac{1}{q} = \frac{1}{f} \qquad [23.6]$$

where $f = R/2$ is the **focal length** of the mirror.

Equations 23.2 and 23.6 hold for both concave and convex mirrors, subject to the sign conventions given in Table 23.1.

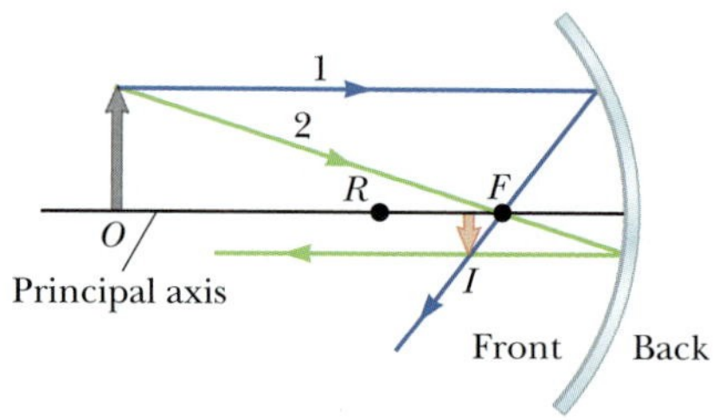


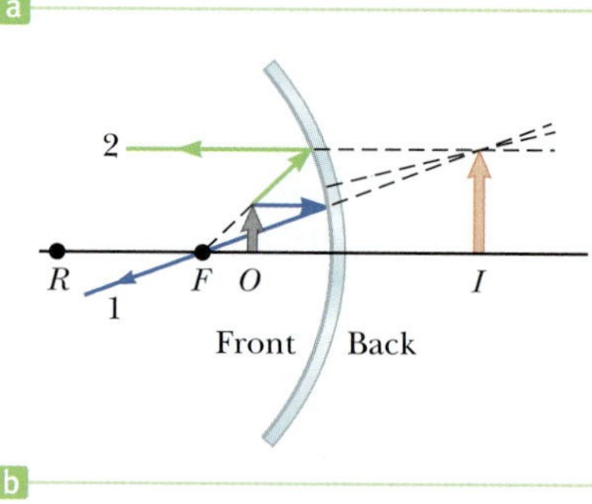


(a) The image of a concave mirror is real and inverted when the object is outside the focal point, i.e. $p > f$. The image is larger than the object when $f < p < R$, and smaller than the object when $p > R$. (b) The image of a concave mirror is virtual, upright, and larger than the object when $p < f$.

## 23.4 Images Formed by Refraction

An image can be formed by refraction at a spherical surface of radius $R$. The object and image distances for refraction from such a surface are related by

$$\frac{n_1}{p} + \frac{n_2}{q} = \frac{n_2 - n_1}{R} \qquad [23.7]$$

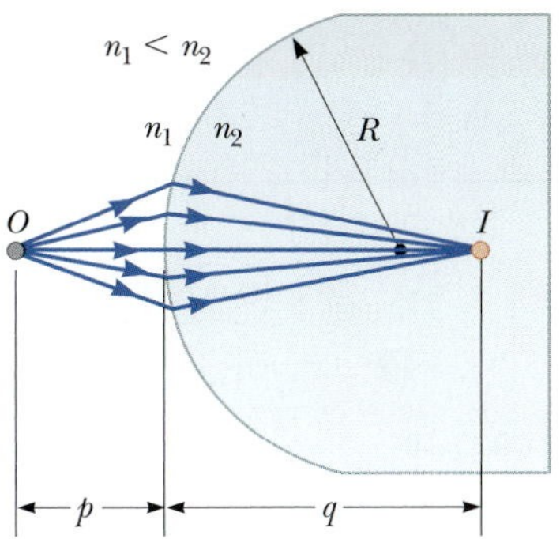


An image formed by refraction at a spherical surface. Rays making small angles with the principal axis diverge from a point object at *O* and pass through the image point *I*.

The **magnification of a refracting surface** is

$$M = \frac{h'}{h} = -\frac{n_1 q}{n_2 p} \qquad [23.8]$$

where the object is located in the medium with index of refraction $n_1$ and the image is formed in the medium with index of refraction $n_2$. Equations 23.7 and 23.8 are subject to the sign conventions of Table 23.2.

## 23.6 Thin Lenses

The **magnification of a thin lens** is

$$M = \frac{h'}{h} = -\frac{q}{p} \qquad [23.10]$$

The object and image distances of a thin lens are related by the **thin-lens equation**:

$$\frac{1}{p} + \frac{1}{q} = \frac{1}{f} \qquad [23.11]$$

Equations 23.10 and 23.11 are subject to the sign conventions of Table 23.3.

## 23.7 Lens and Mirror Aberrations

**Aberrations** are responsible for the formation of imperfect images by lenses and mirrors. **Spherical aberration** results from the focal points of light rays far from the principal axis of a spherical lens or mirror being different from those of rays passing through the center. **Chromatic aberration** arises because light rays of different wavelengths focus at different points when refracted by a lens.

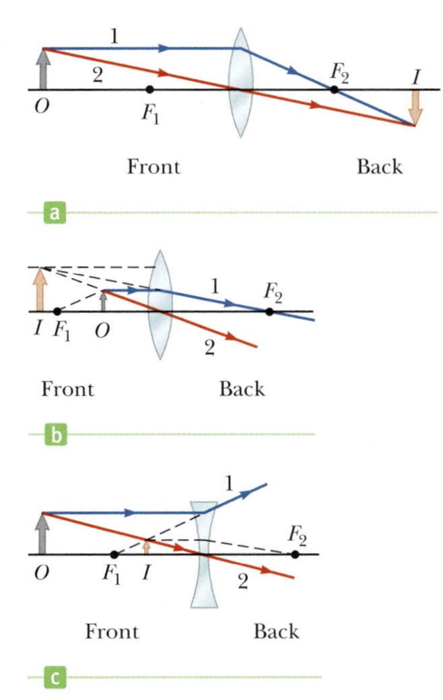


Ray diagrams for locating the image of an object. (a) The object is outside the focal point of a converging lens. (b) The object is inside the focal point of a converging lens. (c) The object is outside the focal point of a diverging lens.

## WARM-UP EXERCISES

ENHANCED WebAssign The warm-up exercises in this chapter may be assigned online in Enhanced WebAssign.

**1.** A woman stands 1.50 m in front of a full-length flat mirror. Find (a) the position of her image, with proper sign, (b) the magnification of her image, and (c) whether it is real or virtual. (See Section 23.1.)

**2.** An object is 20.0 cm away from a concave mirror with focal length 15.0 cm. Find (a) the image distance, *q*, and (b) its magnification. State whether the image is (c) real or virtual, (d) upright or inverted. (See Section 23.3.)

**3.** An object is placed 16.0 cm away from a convex mirror with a focal length of magnitude 6.00 cm. (a) Is the sign of the focal length negative or positive? Find (b) the image distance *q*, and (c) the magnification. State whether the image is (d) real or virtual, (e) upright or inverted. (See Section 23.3.)

**4.** A coin is at the bottom of a pool of water 1.20 m deep. Find (a) the object distance, (b) the image distance, and (c) the magnification of the image. (d) Is the image real or virtual? (See Section 23.4.)

**5.** A gold coin is embedded in solid ball of clear plastic of radius 25.0 cm, 10.0 cm from the surface. A person looks directly at the coin along a line going through the coin and to the center of the ball, as in Example 23.5. If the plastic has index of refraction 1.70, find (a) the image distance *q*, and (b) the coin's magnification. (See Section 23.4.)

**6.** A thin, convergent lens has a focal length of 8.00 cm. If the object distance is 24.0 cm, find (a) the image distance, and (b) the magnification. State whether the image is (c) real or virtual, (d) upright or inverted. (See Section 23.6.)

**7.** A real object is 10.0 cm to the left of a thin, diverging lens having a focal length of magnitude 16.0 cm. (a) Is the sign of the focal length negative or positive? Find (b) the image distance, and (c) the magnification. State whether the image is (d) real or virtual, (e) upright or inverted. (See Section 23.6.)

## CONCEPTUAL QUESTIONS

WebAssign The conceptual questions in this chapter may be assigned online in Enhanced WebAssign.

1. Tape a picture of yourself on a bathroom mirror. Stand several centimeters away from the mirror. Can you focus your eyes on *both* the picture taped to the mirror *and* your image in the mirror *at the same time*? So where is the image of yourself?

2. Why does a clear stream always appear to be shallower than it actually is?

3. A flat mirror creates a virtual image of your face. Suppose the flat mirror is combined with another optical element. Can the mirror form a real image in such a combination?

4. Explain why a mirror cannot give rise to chromatic aberration.

5. A common mirage is formed when the air gets gradually cooler as the height above the ground increases. What might happen if the air grows gradually warmer as the height increases? This often happens over bodies of water or snow-covered ground; the effect is called *looming*.

6. A virtual image is often described as an image through which light rays don't actually travel, as they do for a real image. Can a virtual image be photographed?

7. Suppose you want to use a converging lens to project the image of two trees onto a screen. One tree is a distance $x$ from the lens; the other is at $2x$, as in Figure CQ23.7. You adjust the screen so that the near tree is in focus. If you now want the far tree to be in focus, do you move the screen toward or away from the lens?

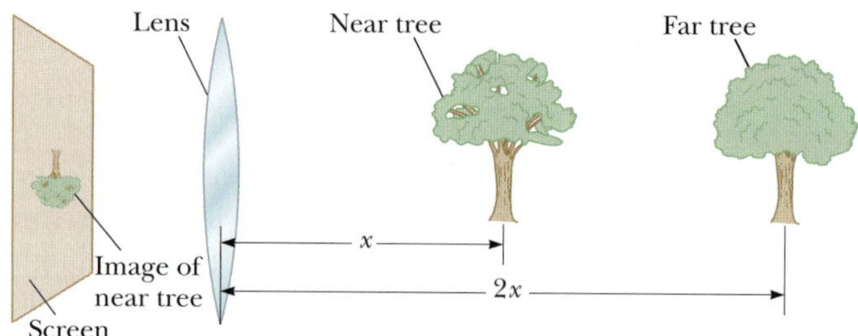


**Figure CQ23.7**

8. Lenses used in eyeglasses, whether converging or diverging, are always designed such that the middle of the lens curves away from the eye. Why?

9. In a Jules Verne novel, a piece of ice is shaped into a magnifying lens to focus sunlight to start a fire. Is that possible?

10. If a cylinder of solid glass or clear plastic is placed above the words LEAD OXIDE and viewed from the side, as shown in Figure CQ23.10, the word LEAD appears inverted, but the word OXIDE does not. Explain.

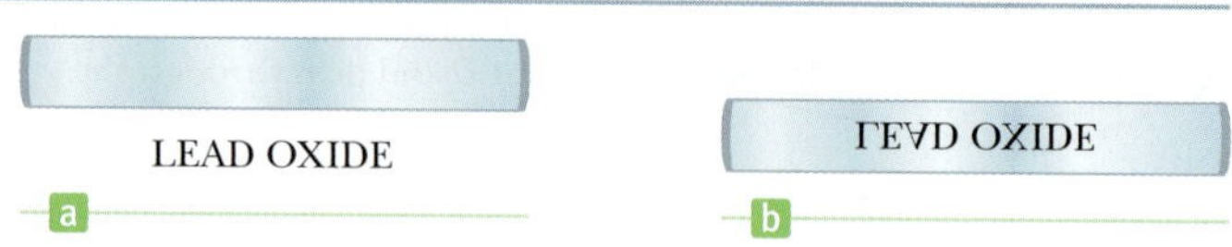


**Figure CQ23.10**

11. Can a converging lens be made to diverge light if placed in a liquid? How about a converging mirror?

12. Light from an object passes through a lens and forms a visible image on a screen. If the screen is removed, would you be able to see the image (a) if you remained in your present position and (b) if you could look at the lens along its axis, beyond the original position of the screen?

13. Why does the focal length of a mirror not depend on the mirror material when the focal length of a lens does depend on the lens material?

14. An inverted image of an object is viewed on a screen from the side facing a converging lens. An opaque card is then introduced covering only the upper half of the lens. What happens to the image on the screen? (a) Half the image would disappear. (b) The entire image would appear and remain unchanged. (c) Half the image would disappear and be dimmer. (d) The entire image would appear, but would be dimmer.

15. Why do some emergency vehicles have the symbol ƎƆИA⅃UBMA written on the front?

16. A person spear fishing from a boat sees a stationary fish a few meters away in a direction about 30° below the horizontal. To spear the fish, and assuming the spear does not change direction when it enters the water, should the person (a) aim above where he sees the fish, (b) aim below the fish, or (c) aim precisely at the fish?

17. An object, represented by a gray arrow, is placed in front of a plane mirror. Which of the diagrams in Figure CQ23.17 best describes the image, represented by the pink arrow?

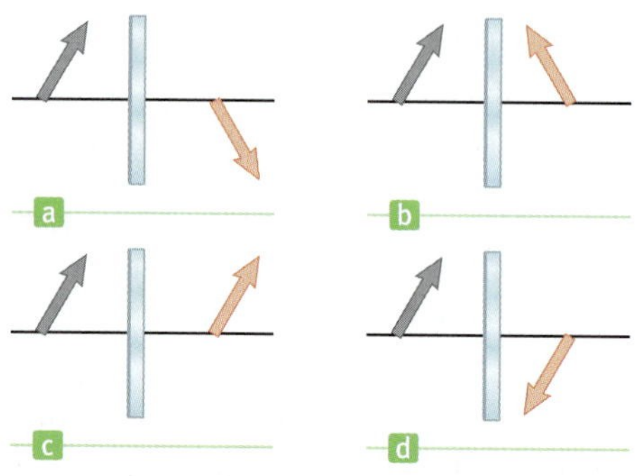


**Figure CQ23.17**

## PROBLEMS

ENHANCED WebAssign The problems in this chapter may be assigned online in Enhanced WebAssign.

**1.** denotes straightforward problem; **2.** denotes intermediate problem; **3.** denotes challenging problem

**1.** denotes full solution available in *Student Solutions Manual/ Study Guide*

**1.** denotes problems most often assigned in Enhanced WebAssign

BIO denotes biomedical problems

GP denotes guided problems

M denotes Master It tutorial available in Enhanced WebAssign

QIC denotes asking for quantitative and conceptual reasoning

S denotes symbolic reasoning problem

W denotes Watch It video solution available in Enhanced WebAssign

### 23.1 Flat Mirrors

**1.** (a) Does your bathroom mirror show you older or younger than your actual age? (b) Compute an order-of-magnitude estimate for the age difference, based on data you specify.

**2.** QIC Two plane mirrors stand facing each other, 3.00 m apart, and a woman stands between them. The woman faces one of the mirrors from a distance of 1.00 m and holds her left arm out to the side of her body with the palm of her left hand facing the closer mirror. (a) What is the apparent position of the closest image of her left hand, measured perpendicularly from the surface of the mirror in front of her? (b) Does it show the palm of her hand or the back of her hand? (c) What is the position of the next image? (d) Does it show the palm of her hand or the back of her hand? (e) What is the position of the third closest image? (f) Does it show the palm of her hand or the back of her hand? (g) Which of the images are real and which are virtual?

**3.** A person walks into a room that has, on opposite walls, two plane mirrors producing multiple images. Find the distances from the person to the first three images seen in the left-hand mirror when the person is 5.00 ft from the mirror on the left wall and 10.0 ft from the mirror on the right wall.

**4.** In a church choir loft, two parallel walls are 5.30 m apart. The singers stand against the north wall. The organist faces the south wall, sitting 0.800 m away from it. So that she can see the choir, a flat mirror 0.600 m wide is mounted on the south wall, straight in front of the organist. What width of the north wall can she see? *Hint:* Draw a top-view diagram to justify your answer.

**5.** A periscope (Fig. P23.5) is useful for viewing objects that cannot be seen directly. It can be used in submarines and when watching golf matches or parades from behind a crowd of people. Suppose the object is a distance $p_1$ from the upper mirror and the centers of the two flat mirrors are separated by a distance $h$. (a) What is the distance of the final image from the lower mirror? (b) Is the final image real or virtual? (c) Is it upright or inverted? (d) What is its magnification? (e) Does it appear to be left–right reversed?

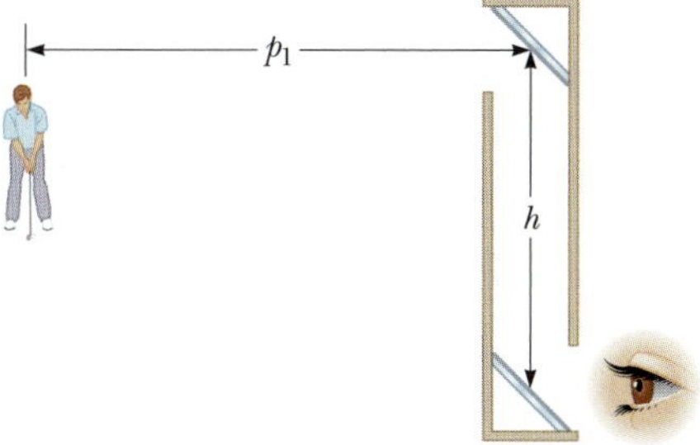


**Figure P23.5**

### 23.2 Images Formed by Concave Mirrors

### 23.3 Convex Mirrors and Sign Conventions

In the following problems, algebraic signs are not given. We leave it to you to determine the correct sign to use with each quantity, based on an analysis of the problem and the sign conventions in Table 23.1.

**6.** A dentist uses a mirror to examine a tooth that is 1.00 cm in front of the mirror. The image of the tooth is formed 10.0 cm behind the mirror. Determine (a) the mirror's radius of curvature and (b) the magnification of the image.

**7.** A convex spherical mirror, whose focal length has a magnitude of 15.0 cm, is to form an image 10.0 cm behind the mirror. (a) Where should the object be placed? (b) What is the magnification of the mirror?

**8.** BIO To fit a contact lens to a patient's eye, a *keratometer* can be used to measure the curvature of the cornea—the front surface of the eye. This instrument places an illuminated object of known size at a known distance $p$ from the cornea, which then reflects some light from the object, forming an image of it. The magnification $M$ of the image is measured by using a small viewing telescope that allows a comparison of the image formed by the cornea with a second calibrated image projected into the field of view by a prism arrangement. Determine the radius of curvature of the cornea when $p = 30.0$ cm and $M = 0.013\,0$.

**9.** A virtual image is formed 20.0 cm from a concave mirror having a radius of curvature of 40.0 cm. (a) Find the position of the object. (b) What is the magnification of the mirror?

10. Q|C While looking at her image in a cosmetic mirror, Dina notes that her face is highly magnified when she is close to the mirror, but as she backs away from the mirror, her image first becomes blurry, then disappears when she is about 30 cm from the mirror, and then inverts when she is beyond 30 cm. Based on these observations, what can she conclude about the properties of the mirror?

11. W A 2.00-cm-high object is placed 3.00 cm in front of a concave mirror. If the image is 5.00 cm high and virtual, what is the focal length of the mirror?

12. A dedicated sports car enthusiast polishes the inside and outside surfaces of a hubcap that is a section of a sphere. When he looks into one side of the hubcap, he sees an image of his face 30.0 cm in back of it. He then turns the hubcap over, keeping it the same distance from his face. He now sees an image of his face 10.0 cm in back of the hubcap. (a) How far is his face from the hubcap? (b) What is the magnitude of the radius of curvature of the hubcap?

13. A concave makeup mirror is designed so that a person 25 cm in front of it sees an upright image magnified by a factor of two. What is the radius of curvature of the mirror?

14. A concave mirror has a focal length of 30.0 cm. (a) What is its radius of curvature? Locate and describe the properties of the image when the object distance is (b) 100 cm and (c) 10.0 cm.

15. M A man standing 1.52 m in front of a shaving mirror produces an inverted image 18.0 cm in front of it. How close to the mirror should he stand if he wants to form an upright image of his chin that is twice the chin's actual size?

16. A convex mirror has a focal length of magnitude 8.0 cm. (a) If the image is virtual, what is the object location for which the magnitude of the image distance is one third the magnitude of the object distance? (b) Find the magnification of the image and state whether it is upright or inverted.

17. M At an intersection of hospital hallways, a convex spherical mirror is mounted high on a wall to help people avoid collisions. The magnitude of the mirror's radius of curvature is 0.550 m. (a) Locate the image of a patient located 10.0 m from the mirror. (b) Indicate whether the image is upright or inverted. (c) Determine the magnification of the image.

18. Q|C A concave mirror has a radius of curvature of 24.0 cm. (a) Determine the object position for which the resulting image is upright and larger than the object by a factor of 3.00. (b) Draw a ray diagram to determine the position of the image. (c) Is the image real or virtual?

19. Q|C A spherical mirror is to be used to form an image, five times as tall as an object, on a screen positioned 5.0 m from the mirror. (a) Describe the type of mirror required. (b) Where should the mirror be positioned relative to the object?

20. Q|C A ball is dropped from rest 3.00 m directly above the vertex of a concave mirror having a radius of 1.00 m and lying in a horizontal plane. (a) Describe the motion of the ball's image in the mirror. (b) At what time do the ball and its image coincide?

## 23.4 Images Formed by Refraction

21. A cubical block of ice 50.0 cm on an edge is placed on a level floor over a speck of dust. Locate the image of the speck, when viewed from directly above, if the index of refraction of ice is 1.309.

22. A goldfish is swimming inside a spherical bowl of water having an index of refraction $n = 1.333$. Suppose the goldfish is $p = 10.0$ cm from the wall of a bowl of *radius* $|R| = 15.0$ cm, as in Figure P23.22. Neglecting the refraction of light caused by the wall of the bowl, determine the apparent distance of the goldfish from the wall according to an observer outside the bowl.

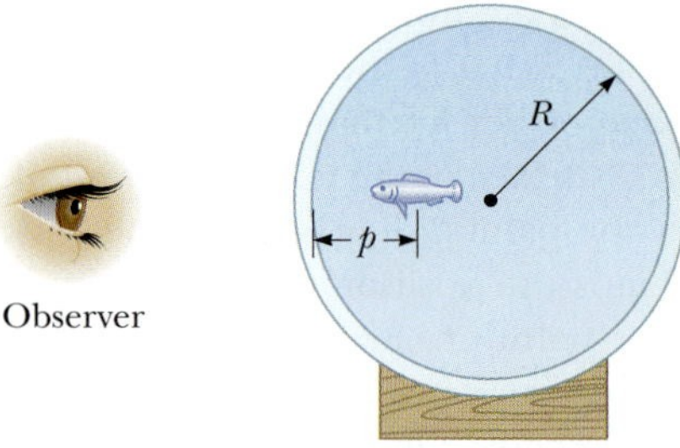


**Figure P23.22**

23. A paperweight is made of a solid glass hemisphere with index of refraction 1.50. The radius of the circular cross section is 4.0 cm. The hemisphere is placed on its flat surface, with the center directly over a 2.5-mm-long line drawn on a sheet of paper. What length of line is seen by someone looking vertically down on the hemisphere?

24. W The top of a swimming pool is at ground level. If the pool is 2.00 m deep, how far below ground level does the bottom of the pool appear to be located when (a) the pool is completely filled with water? (b) When it is filled halfway with water?

25. A transparent sphere of unknown composition is observed to form an image of the Sun on its surface opposite the Sun. What is the refractive index of the sphere material?

26. A flint glass plate ($n = 1.66$) rests on the bottom of an aquarium tank. The plate is 8.00 cm thick (vertical dimension) and covered with water ($n = 1.33$) to a depth of 12.0 cm. Calculate the apparent thickness of the plate as viewed from above the water. (Assume nearly normal incidence of light rays.)

27. Q|C A jellyfish is floating in a water-filled aquarium 1.00 m behind a flat pane of glass 6.00 cm thick and having an index of refraction of 1.50. (a) Where is the image of the jellyfish located? (b) Repeat the problem when the glass is so thin that its thickness can be neglected. (c) How does the thickness of the glass affect the answer to part (a)?

28. S Figure P23.28 shows a curved surface separating a material with index of refraction $n_1$ from a material with index $n_2$. The surface forms an image $I$ of object $O$. The ray shown in red passes through the surface along a radial line. Its angles of incidence and refraction are both zero, so its direction does not change at the surface. For the ray shown in blue, the direction changes according to $n_1 \sin \theta_1 = n_2 \sin \theta_2$. For paraxial rays, we assume $\theta_1$ and $\theta_2$ are small, so we may write $n_1 \tan \theta_1 = n_2 \tan \theta_2$. The magnification is defined as $M = h'/h$. Prove that the magnification is given by $M = -n_1q/n_2p$.

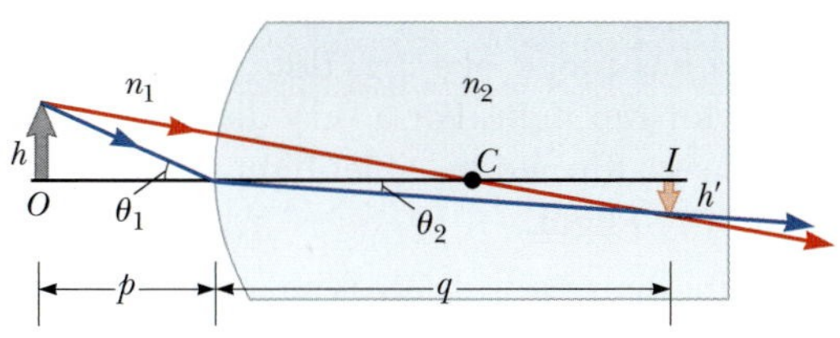


Figure P23.28

## 23.6 Thin Lenses

29. BIO A contact lens is made of plastic with an index of refraction of 1.50. The lens has an outer radius of curvature of +2.00 cm and an inner radius of curvature of +2.50 cm. What is the focal length of the lens?

30. An object is placed 50.0 cm from a screen. (a) Where should a converging lens of focal length 10.0 cm be placed to form an image on the screen? (b) Find the magnification of the lens.

31. A converging lens has a focal length of 10.0 cm. Locate the images for object distances of (a) 20.0 cm, (b) 10.0 cm, and (c) 5.00 cm, if they exist. For each case, state whether the image is real or virtual, upright or inverted, and find the magnification.

32. An object is placed 20.0 cm from a concave spherical mirror having a focal length of magnitude 40.0 cm. (a) Use graph paper to construct an accurate ray diagram for this situation. (b) From your ray diagram, determine the location of the image. (c) What is the magnification of the image? (d) Check your answers to parts (b) and (c) using the mirror equation.

33. A diverging lens has a focal length of magnitude 20.0 cm. (a) Locate the images for object distances of (i) 40.0 cm, (ii) 20.0 cm, and (iii) 10.0 cm. For each case, state whether the image is (b) real or virtual and (c) upright or inverted. (d) For each case, find the magnification.

34. Q|C A diverging lens has a focal length of 20.0 cm. Use graph paper to construct accurate ray diagrams for object distances of (a) 40.0 cm and (b) 10.0 cm. In each case determine the location of the image from the diagram and the image magnification, and state whether the image is upright or inverted. (c) Estimate the magnitude of uncertainty in locating the points in the graph. Are your answers and the uncertainty consistent with the algebraic answers found in Problem 33?

35. A transparent photographic slide is placed in front of a converging lens with a focal length of 2.44 cm. An image of the slide is formed 12.9 cm from the slide. How far is the lens from the slide if the image is (a) real? (b) Virtual?

36. W The nickel's image in Figure P23.36 has twice the diameter of the nickel when the lens is 2.84 cm from the nickel. Determine the focal length of the lens.

Figure P23.36

37. The projection lens in a certain slide projector is a single thin lens. A slide 24.0 mm high is to be projected so that its image fills a screen 1.80 m high. The slide-to-screen distance is 3.00 m. (a) Determine the focal length of the projection lens. (b) How far from the slide should the lens of the projector be placed to form the image on the screen?

38. M An object is located 20.0 cm to the left of a diverging lens having a focal length $f = -32.0$ cm. Determine (a) the location and (b) the magnification of the image. (c) Construct a ray diagram for this arrangement.

39. A converging lens is placed 30.0 cm to the right of a diverging lens of focal length 10.0 cm. A beam of parallel light enters the diverging lens from the left, and the beam is again parallel when it emerges from the converging lens. Calculate the focal length of the converging lens.

40. Q|C S (a) Use the thin-lens equation to derive an expression for $q$ in terms of $f$ and $p$. (b) Prove that for a real object and a diverging lens, the image must always be virtual. *Hint:* Set $f = -|f|$ and show that $q$ must be less than zero under the given conditions. (c) For a real object and converging lens, what inequality involving $p$ and $f$ must hold if the image is to be real?

41. W Two converging lenses, each of focal length 15.0 cm, are placed 40.0 cm apart, and an object is placed 30.0 cm in front of the first lens. Where is the final image formed, and what is the magnification of the system?

42. Object $O_1$ is 15.0 cm to the left of a converging lens with a 10.0-cm focal length. A second lens is positioned 10.0 cm to the right of the first lens and is observed to form a virtual image at the position of the original object $O_1$. (a) What is the focal length of the second lens? (b) What is the overall magnification of this system? (c) What is the nature (i.e., real or virtual, upright or inverted) of the final image?

43. M A 1.00-cm-high object is placed 4.00 cm to the left of a converging lens of focal length 8.00 cm. A diverging lens of focal length −16.00 cm is 6.00 cm to the right of the converging lens. Find the position and height of the final image. Is the image inverted or upright? Real or virtual?

44. Two converging lenses having focal lengths of $f_1$ = 10.0 cm and $f_2$ = 20.0 cm are placed $d$ = 50.0 cm apart, as shown in Figure P23.44. The final image is to be located between the lenses, at the position $x$ = 31.0 cm indicated. (a) How far to the left of the first lens should the object be positioned? (b) What is the overall magnification of the system? (c) Is the final image upright or inverted?

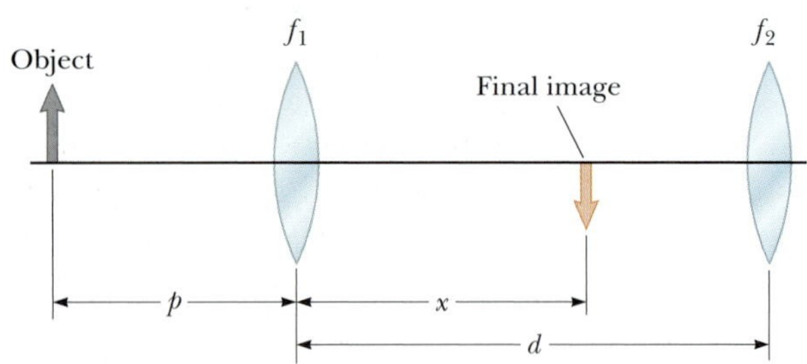


**Figure P23.44**

45. Lens $L_1$ in Figure P23.45 has a focal length of 15.0 cm and is located a fixed distance in front of the film plane of a camera. Lens $L_2$ has a focal length of 13.0 cm, and its distance $d$ from the film plane can be varied from 5.00 cm to 10.0 cm. Determine the range of distances for which objects can be focused on the film.

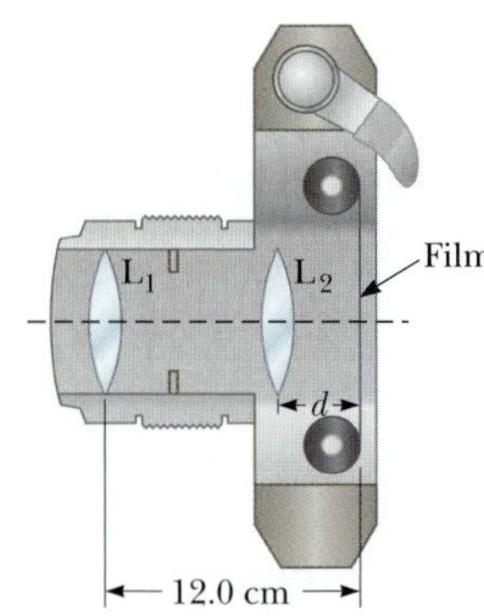


**Figure P23.45**

46. GP An object is placed 15.0 cm from a first converging lens of focal length 10.0 cm. A second converging lens with focal length 5.00 cm is placed 10.0 cm to the right of the first converging lens. (a) Find the position $q_1$ of the image formed by the first converging lens. (b) How far from the second lens is the image of the first lens? (c) What is the value of $p_2$, the object position for the second lens? (d) Find the position $q_2$ of the image formed by the second lens. (e) Calculate the magnification of the first lens. (f) Calculate the magnification of the second lens. (g) What is the total magnification for the system? (h) Is the final image real or virtual? Is it upright or inverted (compared to the original object for the lens system)?

## Additional Problems

47. An object placed 10.0 cm from a concave spherical mirror produces a real image 8.00 cm from the mirror. If the object is moved to a new position 20.0 cm from the mirror, what is the position of the image? Is the final image real or virtual?

48. S A real object's distance from a converging lens is five times the focal length. (a) Determine the location of the image $q$ in terms of the focal length $f$. (b) Find the magnification of the image. (c) Is the image real or virtual? Is it upright or inverted? Is the image on the same side of the lens as the object or on the opposite side?

49. The magnitudes of the radii of curvature are 32.5 cm and 42.5 cm for the two faces of a biconcave lens. The glass has index of refraction 1.53 for violet light and 1.51 for red light. For a very distant object, locate (a) the image formed by violet light and (b) the image formed by red light.

50. A diverging lens ($n$ = 1.50) is shaped like that in Figure 23.25c. The radius of the first surface is 15.0 cm, and that of the second surface is 10.0 cm. (a) Find the focal length of the lens. Determine the positions of the images for object distances of (b) infinity, (c) $3|f|$, (d) $|f|$, and (e) $|f|/2$.

51. The lens and the mirror in Figure P23.51 are separated by 1.00 m and have focal lengths of +80.0 cm and −50.0 cm, respectively. If an object is placed 1.00 m to the left of the lens, where will the final image be located? State whether the image is upright or inverted, and determine the overall magnification.

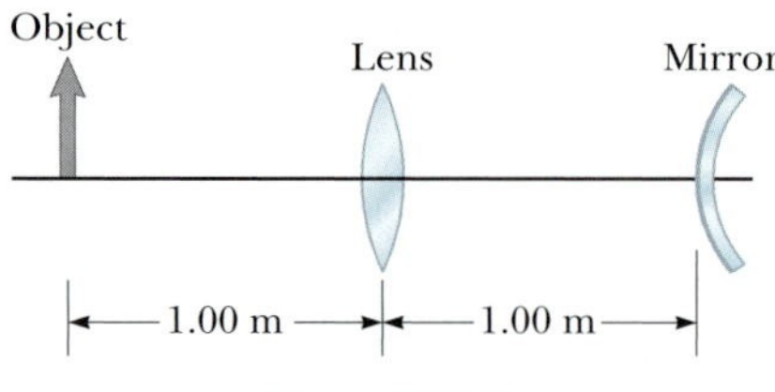


**Figure P23.51**

52. The object in Figure P23.52 is midway between the lens and the mirror, which are separated by a distance $d$ = 25.0 cm. The magnitude of the mirror's radius of curvature is 20.0 cm, and the lens has a focal length of −16.7 cm. (a) Considering only the light that leaves the object and travels first toward the mirror, locate the final image formed by this system. (b) Is the image

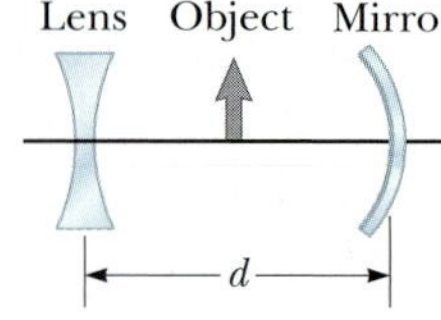


**Figure P23.52**

real or virtual? (c) Is it upright or inverted? (d) What is the overall magnification of the image?

53. A parallel beam of light enters a glass hemisphere perpendicular to the flat face, as shown in Figure P23.53. The radius of the hemisphere is $R = 6.00$ cm, and the index of refraction is $n = 1.56$. Determine the point at which the beam is focused. (Assume paraxial rays; i.e., assume all rays are located close to the principal axis.)

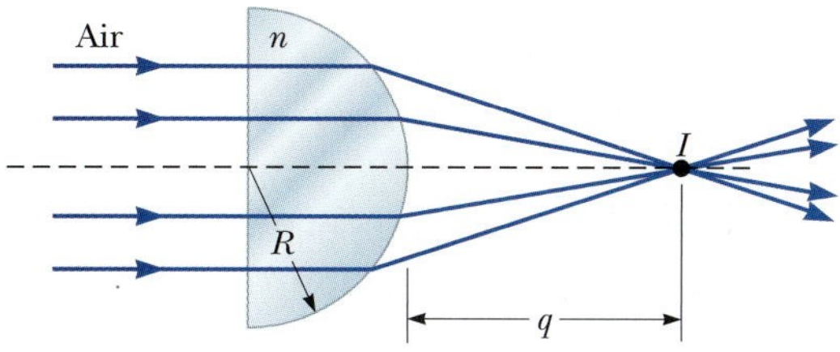


Figure P23.53

54. Two rays traveling parallel to the principal axis strike a large plano-convex lens having a refractive index of 1.60 (Fig. P23.54). If the convex face is spherical, a ray near the edge does not pass through the focal point (spherical aberration occurs). Assume this face has a radius of curvature of $R = 20.0$ cm and the two rays are at distances $h_1 = 0.500$ cm and $h_2 = 12.0$ cm from the principal axis. Find the difference $\Delta x$ in the positions where each crosses the principal axis.

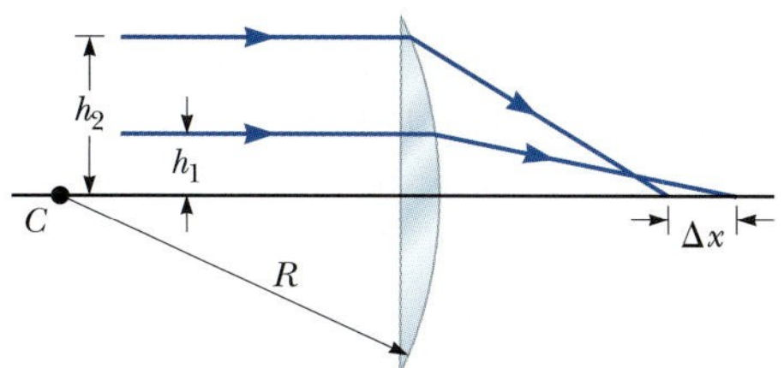


Figure P23.54

55. To work this problem, use the fact that the image formed by the first surface becomes the object for the second surface. Figure P23.55 shows a piece of glass with index of refraction $n = 1.50$ surrounded by air. The ends are hemispheres with radii $R_1 = 2.00$ cm and $R_2 = 4.00$ cm, and the centers of the hemispherical ends are separated by a distance of $d = 8.00$ cm. A point object is in air, a distance $p = 1.00$ cm from the left end of the glass. (a) Locate the image of the object due to refraction at the two spherical surfaces. (b) Is the image real or virtual?

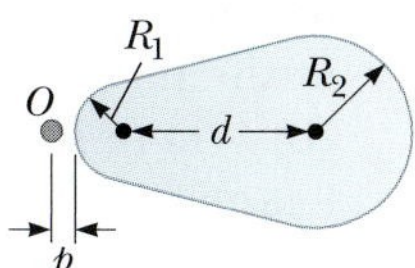


Figure P23.55

56. **S** Consider two thin lenses, one of focal length $f_1$ and the other of focal length $f_2$, placed in contact with each other, as shown in Figure P23.56. Apply the thin-lens equation to each of these lenses and combine the results to show that this combination of lenses behaves like a thin lens having a focal length $f$ given by $1/f = 1/f_1 + 1/f_2$. Assume the thicknesses of the lenses can be ignored in comparison to the other distances involved.

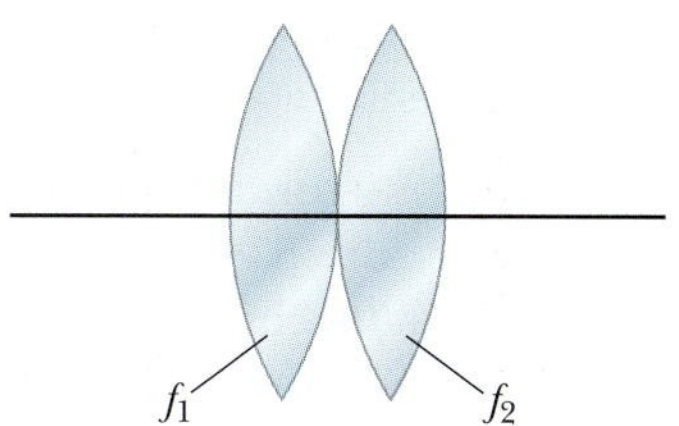


Figure P23.56

57. An object 2.00 cm high is placed 40.0 cm to the left of a converging lens having a focal length of 30.0 cm. A diverging lens having a focal length of −20.0 cm is placed 110 cm to the right of the converging lens. (a) Determine the final position and magnification of the final image. (b) Is the image upright or inverted? (c) Repeat parts (a) and (b) for the case in which the second lens is a converging lens having a focal length of +20.0 cm.

58. A "floating strawberry" illusion can be produced by two parabolic mirrors, each with a focal length of 7.5 cm, facing each other so that their centers are 7.5 cm apart (Fig. P23.58). If a strawberry is placed on the bottom mirror, an image of the strawberry forms at the small opening at the center of the top mirror. Show that the final image forms at that location and describe its characteristics. *Note:* A flashlight beam shone on these *images* has a very startling effect: Even at a glancing angle, the incoming light beam is seemingly reflected off the *images* of the strawberry! Do you understand why?

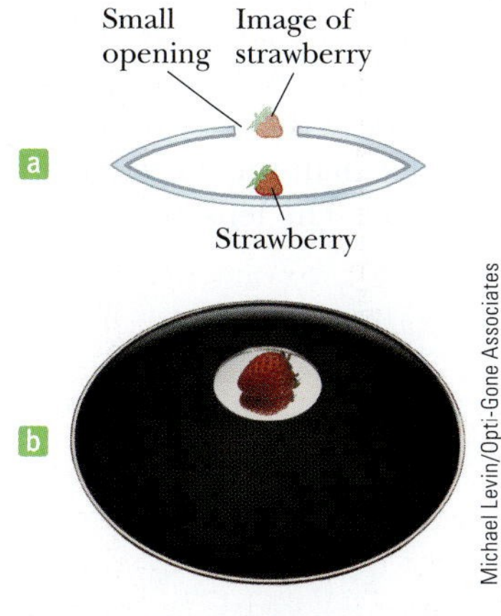


Figure P23.58

59. Figure P23.59 shows a converging lens with radii $R_1 = 9.00$ cm and $R_2 = -11.00$ cm, in front of a concave spherical mirror of radius $R = 8.00$ cm. The focal points ($F_1$ and $F_2$) for the thin lens and the center of

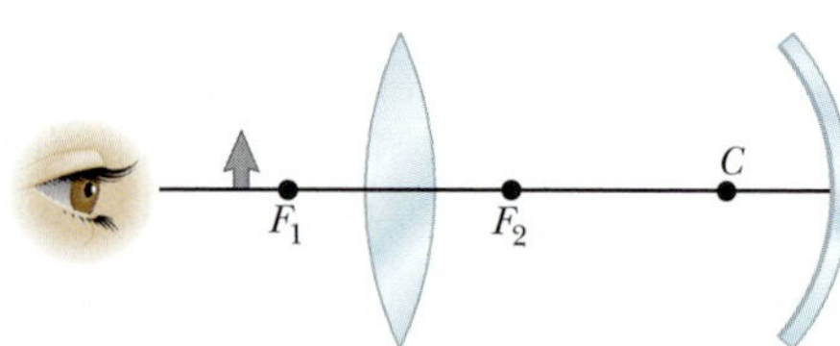


Figure P23.59

curvature ($C$) of the mirror are also shown. (a) If the focal points $F_1$ and $F_2$ are 5.00 cm from the vertex of the thin lens, what is the index of refraction of the lens? (b) If the lens and mirror are 20.0 cm apart and an object is placed 8.00 cm to the left of the lens, what is the position of the final image and its magnification as seen by the eye in the figure? (c) Is the final image inverted or upright? Explain.

60. **S** Find the object distances (in terms of $f$) for a thin converging lens of focal length $f$ if (a) the image is real and the image distance is four times the focal length and (b) the image is virtual and the absolute value of the image distance is three times the focal length. (c) Calculate the magnification of the lens for cases (a) and (b).

61. The lens-maker's equation for a lens with index $n_1$ immersed in a medium with index $n_2$ takes the form

$$\frac{1}{f} = \left(\frac{n_1}{n_2} - 1\right)\left(\frac{1}{R_1} - \frac{1}{R_2}\right)$$

A thin diverging glass (index = 1.50) lens with $R_1 = -3.00$ m and $R_2 = -6.00$ m is surrounded by air. An arrow is placed 10.0 m to the left of the lens. (a) Determine the position of the image. Repeat part (a) with the arrow and lens immersed in (b) water (index = 1.33) (c) a medium with an index of refraction of 2.00. (d) How can a lens that is diverging in air be changed into a converging lens?

62. An observer to the right of the mirror–lens combination shown in Figure P23.62 sees two real images that are the same size and in the same location. One image is upright, and the other is inverted. Both images are 1.50 times larger than the object. The lens has a focal length of 10.0 cm. The lens and mirror are separated by 40.0 cm. Determine the focal length of the mirror. (Don't assume the figure is drawn to scale.)

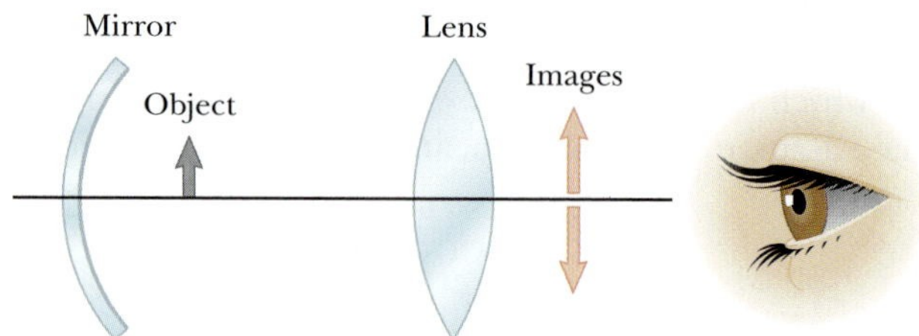


Figure P23.62

63. The lens-maker's equation applies to a lens immersed in a liquid if $n$ in the equation is replaced by $n_1/n_2$. Here $n_1$ refers to the refractive index of the lens material and $n_2$ is that of the medium surrounding the lens. (a) A certain lens has focal length of 79.0 cm in air and a refractive index of 1.55. Find its focal length in water. (b) A certain mirror has focal length of 79.0 cm in air. Find its focal length in water.

64. A certain Christmas tree ornament is a silver sphere having a diameter of 8.50 cm. (a) If the size of an image created by reflection in the ornament is three-fourth's the reflected object's actual size, determine the object's location. (b) Use a principal-ray diagram to determine whether the image is upright or inverted.

65. **M** A glass sphere ($n = 1.50$) with a radius of 15.0 cm has a tiny air bubble 5.00 cm above its center. The sphere is viewed looking down along the extended radius containing the bubble. What is the apparent depth of the bubble below the surface of the sphere?

66. An object 10.0 cm tall is placed at the zero mark of a meterstick. A spherical mirror located at some point on the meterstick creates an image of the object that is upright, 4.00 cm tall, and located at the 42.0-cm mark of the meterstick. (a) Is the mirror convex or concave? (b) Where is the mirror? (c) What is the mirror's focal length?

Dec Hogan/Shutterstock.com

The colors in many of a hummingbird's feathers are not due to pigment. The *iridescence* that makes the brilliant colors that often appear on the bird's throat and belly is due to an interference effect caused by structures in the feathers. The colors vary with the viewing angle.

# Wave Optics 24



Colors swirl on a soap bubble as it drifts through the air on a summer day, and vivid rainbows reflect from the filth of oil films in the puddles of a dirty city street. Beachgoers, covered with thin layers of oil, wear their coated sunglasses that absorb half the incoming light. In laboratories scientists determine the precise composition of materials by analyzing the light they give off when hot, and in observatories around the world, telescopes gather light from distant galaxies, filtering out individual wavelengths in bands and thereby determining the speed of expansion of the Universe.

Understanding how these rainbows are made and how certain scientific instruments can determine wavelengths is the domain of *wave optics*. Light can be viewed as either a particle or a wave. Geometric optics, the subject of the previous chapter, depends on the particle nature of light. Wave optics depends on the wave nature of light. The three primary topics we examine in this chapter are interference, diffraction, and polarization. These phenomena can't be adequately explained with ray optics, but can be understood if light is viewed as a wave.

## 24.1 Conditions for Interference

**LEARNING OBJECTIVES**

1. **Discuss the two conditions that facilitate observations of interference in light waves.**
2. **Define coherent and incoherent light.**

In our discussion of interference of mechanical waves in Chapter 13, we found that two waves could add together either constructively or destructively. In constructive interference the amplitude of the resultant wave is greater than that of either of the individual waves, whereas in destructive interference, the resultant amplitude is less than that of either individual wave. Light waves also interfere with one another. Fundamentally, all interference associated with light waves arises when the electromagnetic fields that constitute the individual waves combine.

Interference effects in light waves aren't easy to observe because of the short wavelengths involved (about $4 \times 10^{-7}$ m to about $7 \times 10^{-7}$ m). The following two conditions, however, facilitate the observation of interference between two sources of light:

Conditions facilitating the observation of interference ▶

1. The sources are **coherent**, which means that the waves they emit must maintain a constant phase with respect to one another.
2. The waves have identical wavelengths.

Two sources (producing two traveling waves) are needed to create interference. To produce a stable interference pattern, the individual waves must maintain a constant phase with one another. When this situation prevails, the sources are said to be coherent. The sound waves emitted by two side-by-side loudspeakers driven by a single amplifier can produce interference because the two speakers respond to the amplifier in the same way at the same time: they are in phase.

If two light sources are placed side by side, however, no interference effects are observed because the light waves from one source are emitted independently of the waves from the other source; hence, the emissions from the two sources don't maintain a constant phase relationship with each other during the time of observation. An ordinary light source undergoes random changes about once every $10^{-8}$ s. Therefore, the conditions for constructive interference, destructive interference, and intermediate states have durations on the order of $10^{-8}$ s. The result is that no interference effects are observed because the eye can't follow such short-term changes. Ordinary light sources are said to be **incoherent**.

An older method for producing two coherent light sources is to pass light from a single wavelength (monochromatic) source through a narrow slit and then allow the light to fall on a screen containing two other narrow slits. The first slit is needed to create a single wave front that illuminates both slits coherently. The light emerging from the two slits is coherent because a single source produces the original light beam and the slits serve only to separate the original beam into two parts. Any random change in the light emitted by the source will occur in the two separate beams at the same time, and interference effects can be observed.

Currently it's much more common to use a laser as a coherent source to demonstrate interference. A laser produces an intense, coherent, monochromatic beam over a width of several millimeters. The laser may therefore be used to illuminate multiple slits directly, and interference effects can be easily observed in a fully lighted room. The principles of operation of a laser are explained in Chapter 28.

## 24.2 Young's Double-Slit Experiment

**LEARNING OBJECTIVES**

1. Describe Young's double-slit experiment. Explain how interference creates the observed fringes.
2. Apply the conditions for constructive and destructive interferences to a Young's experiment.

Thomas Young first demonstrated interference in light waves from two sources in 1801. Figure 24.1a (page 837) is a schematic diagram of the apparatus used in this experiment. (Young used pinholes rather than slits in his original experiments.) Light is incident on a screen containing a narrow slit $S_0$. The light waves emerging from this slit arrive at a second screen that contains two narrow, parallel slits $S_1$ and $S_2$. These slits serve as a pair of coherent light sources because waves emerging from them originate from the same wave front and therefore are always in

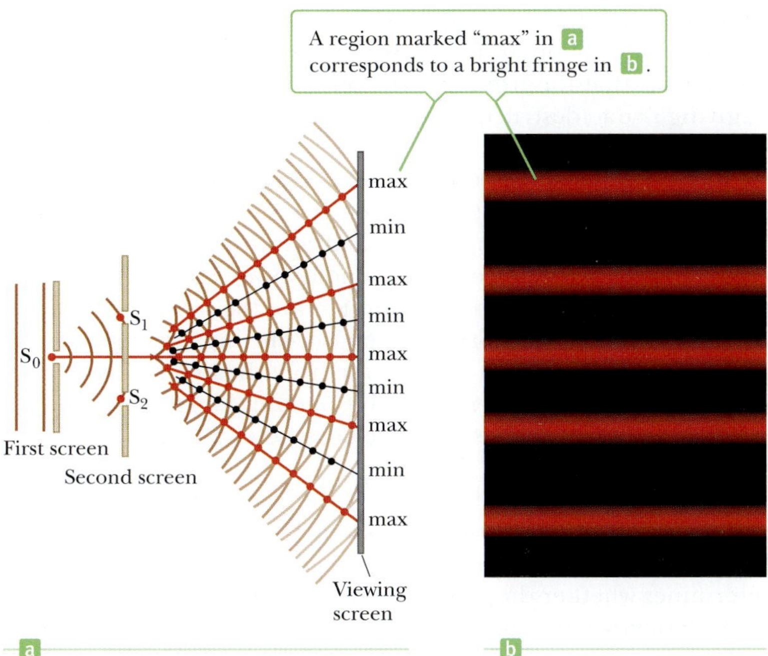


**Figure 24.1** (a) A diagram of Young's double-slit experiment. The narrow slits act as sources of waves. Slits $S_1$ and $S_2$ behave as coherent sources that produce an interference pattern on screen C. (The drawing is not to scale.) (b) The fringe pattern formed on screen C could look like this.

phase. The light from the two slits produces a visible pattern on screen C consisting of a series of bright and dark parallel bands called **fringes** (Fig. 24.1b). When the light from slits $S_1$ and $S_2$ arrives at a point on the screen so that constructive interference occurs at that location, a bright fringe appears. When the light from the two slits combines destructively at any location on the screen, a dark fringe results. Figure 24.2 is a photograph of an interference pattern produced by two coherent vibrating sources in a water tank.

Figure 24.3 is a schematic diagram of some of the ways in which the two waves can combine at screen C of Figure 24.1. In Figure 24.3a two waves, which leave the two slits in phase, strike the screen at the central point $P$. Because these waves travel equal distances, they arrive in phase at $P$, and as a result, constructive interference occurs there and a bright fringe is observed. In Figure 24.3b the two light waves again start in phase, but the upper wave has to travel one wavelength farther to reach point $Q$ on the screen. Because the upper wave falls behind the lower one by exactly one wavelength, the two waves still arrive in phase at $Q$, so a second

**Figure 24.2** An interference pattern involving water waves is produced by two vibrating sources at the water's surface. The pattern is analogous to that observed in Young's double-slit experiment. Note the regions of constructive and destructive interference.

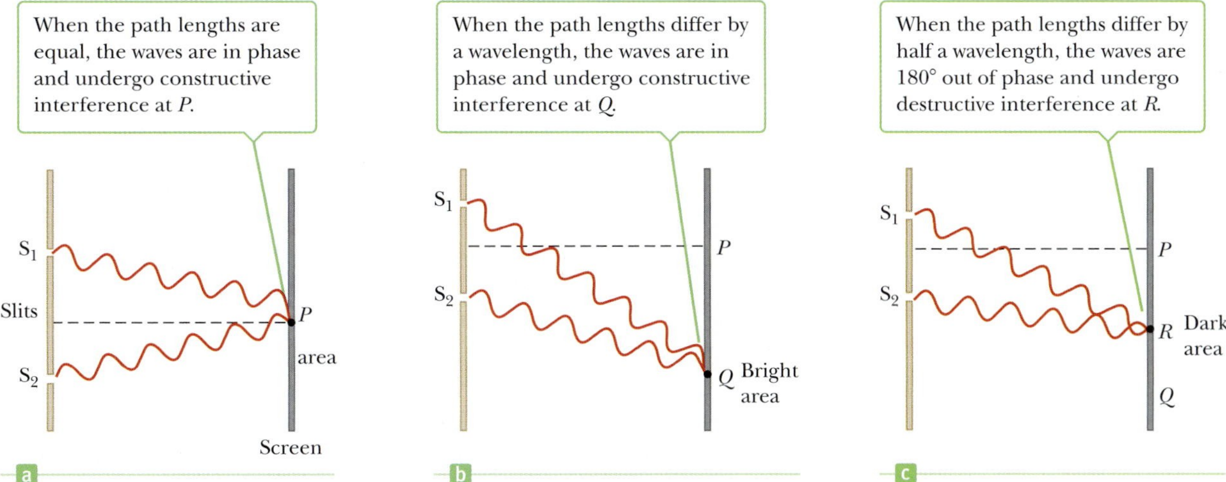

When the path lengths are equal, the waves are in phase and undergo constructive interference at P.
$S_1$
Slits
$S_2$
P
area
Screen
a
When the path lengths differ by a wavelength, the waves are in phase and undergo constructive interference at Q.
$S_1$
$S_2$
P
Q
Bright area
b
When the path lengths differ by half a wavelength, the waves are 180° out of phase and undergo destructive interference at R.
$S_1$
$S_2$
P
R
Dark area
Q
c


**Figure 24.3** Waves leave the slits and combine at various points on the viewing screen. (All figures not to scale.)

bright fringe appears at that location. Now consider point $R$, midway between $P$ and $Q$, in Figure 24.3c. At $R$, the upper wave has fallen half a wavelength behind the lower wave. This means that the trough of the bottom wave overlaps the crest of the upper wave, giving rise to destructive interference. As a result, a dark fringe can be observed at $R$.

We can describe Young's experiment quantitatively with the help of Figure 24.4. Consider point $P$ on the viewing screen; the screen is positioned a perpendicular distance $L$ from the screen containing slits $S_1$ and $S_2$, which are separated by distance $d$, and $r_1$ and $r_2$ are the distances the secondary waves travel from slit to screen. We assume the waves emerging from $S_1$ and $S_2$ have the same constant frequency, have the same amplitude, and start out in phase. The light intensity on the screen at $P$ is the result of light from both slits. A wave from the lower slit, however, travels farther than a wave from the upper slit by the amount $d \sin \theta$. This distance is called the **path difference** $\delta$ (lowercase Greek delta), where

Path difference ▶

$$\delta = r_2 - r_1 = d \sin \theta \qquad [24.1]$$

Equation 24.1 assumes the two waves travel in parallel lines, which is approximately true because $L$ is much greater than $d$. As noted earlier, the value of this path difference determines whether the two waves are in phase when they arrive at $P$. If the path difference is either zero or some integral multiple of the wavelength, the two waves are in phase at $P$ and constructive interference results. Therefore, the condition for bright fringes, or **constructive interference**, at $P$ is

Condition for constructive interference (two slits) ▶

$$\delta = d \sin \theta_{\text{bright}} = m\lambda \qquad m = 0, \pm 1, \pm 2, \ldots \qquad [24.2]$$

The number $m$ is called the **order number**. The central bright fringe that appears at $\theta_{\text{bright}} = 0$ ($m = 0$) is called the *zeroth-order maximum*. The first maximum on either side, where $m = \pm 1$, is called the *first-order maximum*, and so forth.

When $\delta$ is an odd multiple of $\lambda/2$, the two waves arriving at $P$ are 180° out of phase and give rise to destructive interference. Therefore, the condition for dark fringes, or **destructive interference**, at $P$ is

Condition for destructive interference (two slits) ▶

$$\delta = d \sin \theta_{\text{dark}} = (m + \tfrac{1}{2})\lambda \qquad m = 0, \pm 1, \pm 2, \ldots \qquad [24.3]$$

If $m = 0$ in this equation, the path difference is $\delta = \lambda/2$, which is the condition for the location of the first dark fringe on either side of the central (bright) maximum. Likewise, if $m = 1$, the path difference is $\delta = 3\lambda/2$, which is the condition for the second dark fringe on each side, and so forth.

It's useful to obtain expressions for the positions of the bright and dark fringes measured vertically from $O$ to $P$. In addition to our assumption that $L >> d$, we assume $d >> \lambda$. These assumptions can be valid because, in practice,

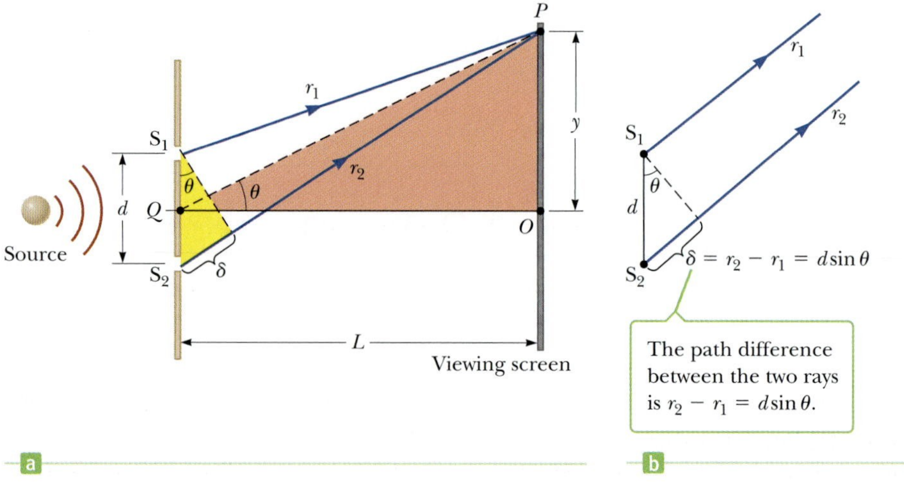


**Figure 24.4** A geometric construction that describes Young's double-slit experiment. (This figure is not drawn to scale.)

$L$ is often on the order of 1 m, $d$ is a fraction of a millimeter, and $\lambda$ is a fraction of a micrometer for visible light. Under these conditions $\theta$ is small, so we can use the approximation $\sin\theta \cong \tan\theta$. Then, from triangle $OPQ$ in Figure 24.4, we see that

$$y = L\tan\theta \approx L\sin\theta \qquad [24.4]$$

Solving Equation 24.2 for $\sin\theta$ and substituting the result into Equation 24.4, we find that the positions of the *bright fringes*, measured from $O$, are

$$y_{\text{bright}} = \frac{\lambda L}{d}m \qquad m = 0, \pm 1, \pm 2, \ldots \qquad [24.5]$$

Using Equations 24.3 and 24.4, we find that the *dark fringes* are located at

$$y_{\text{dark}} = \frac{\lambda L}{d}\left(m + \tfrac{1}{2}\right) \qquad m = 0, \pm 1, \pm 2, \ldots \qquad [24.6]$$

As we will show in Example 24.1, Young's double-slit experiment provides a method for measuring the wavelength of light. In fact, Young used this technique to do just that. In addition, his experiment gave the wave model of light a great deal of credibility. It was inconceivable that particles of light coming through the slits could cancel each other in a way that would explain the dark fringes.

**Tip 24.1 Small-Angle Approximation: Size Matters!**

The small-angle approximation $\sin\theta \cong \tan\theta$ is true to three-digit precision only for angles less than about 4°.

Reflection, interference, and diffraction can be seen in this aerial photograph of waves in the sea.

## APPLYING PHYSICS 24.1 A Smoky Young's Experiment

Consider a double-slit experiment in which a laser beam is passed through a pair of very closely spaced slits and a clear interference pattern is displayed on a distant screen. Now suppose you place smoke particles between the double slit and the screen. With the presence of the smoke particles, will you see the effects of interference in the space between the slits and the screen, or will you see only the effects on the screen?

**EXPLANATION** You will see the interference pattern both on the screen and in the area filled with smoke between the slits and the screen. There will be bright lines directed toward the bright areas on the screen and dark lines directed toward the dark areas on the screen. This is because Equations 24.5 and 24.6 depend on the distance to the screen, $L$, which can take any value. ■

## APPLYING PHYSICS 24.2 Television Signal Interference

Suppose you are watching television by means of an antenna rather than a cable system. If an airplane flies near your location, you may notice wavering ghost images in the television picture. What might cause this phenomenon?

**EXPLANATION** Your television antenna receives two signals: the direct signal from the transmitting antenna and a signal reflected from the surface of the airplane. As the airplane changes position, there are some times when these two signals are in phase and other times when they are out of phase. As a result, the intensity of the combined signal received at your antenna will vary. The wavering of the ghost images of the picture is evidence of this variation. ■

### Quick Quiz

**24.1** In a two-slit interference pattern projected on a screen, are the fringes equally spaced on the screen (a) everywhere, (b) only for large angles, or (c) only for small angles?

**24.2** If the distance between the slits is doubled in Young's experiment, what happens to the width of the central maximum? (a) The width is doubled. (b) The width is unchanged. (c) The width is halved.

**24.3** A Young's double-slit experiment is performed with three different colors of light: red, green, and blue. Rank the colors by the distance between adjacent bright fringes, from smallest to largest. (a) red, green, blue (b) green, blue, red (c) blue, green, red

## ■ EXAMPLE 24.1 Measuring the Wavelength of a Light Source

**GOAL** Show how Young's experiment can be used to measure the wavelength of coherent light.

**PROBLEM** A screen is separated from a double-slit source by 1.20 m. The distance between the two slits is 0.030 0 mm. The second-order bright fringe ($m = 2$) is measured to be 4.50 cm from the centerline. Determine **(a)** the wavelength of the light and **(b)** the distance between adjacent bright fringes.

**STRATEGY** Equation 24.5 relates the positions of the bright fringes to the other variables, including the wavelength of the light. Substitute into this equation and solve for $\lambda$. Taking the difference between $y_{m+1}$ and $y_m$ results in a general expression for the distance between bright fringes.

**SOLUTION**

**(a)** Determine the wavelength of the light.

Solve Equation 24.5 for the wavelength and substitute the values $m = 2$, $y_2 = 4.50 \times 10^{-2}$ m, $L = 1.20$ m, and $d = 3.00 \times 10^{-5}$ m:

$$\lambda = \frac{y_2 d}{mL} = \frac{(4.50 \times 10^{-2}\,\text{m})(3.00 \times 10^{-5}\,\text{m})}{2(1.20\,\text{m})}$$

$$= 5.63 \times 10^{-7}\,\text{m} = \boxed{563\,\text{nm}}$$

**(b)** Determine the distance between adjacent bright fringes.

Use Equation 24.5 to find the distance between *any* adjacent bright fringes (here, those characterized by $m$ and $m + 1$):

$$\Delta y = y_{m+1} - y_m = \frac{\lambda L}{d}(m+1) - \frac{\lambda L}{d}m = \frac{\lambda L}{d}$$

$$= \frac{(5.63 \times 10^{-7}\,\text{m})(1.20\,\text{m})}{3.00 \times 10^{-5}\,\text{m}} = \boxed{2.25\,\text{cm}}$$

**REMARKS** This calculation depends on the angle $\theta$ being small because the small-angle approximation was implicitly used. The measurement of the position of the bright fringes yields the wavelength of light, which in turn is a signature of atomic processes, as is discussed in the chapters on modern physics. This kind of measurement therefore helped open the world of the atom.

**QUESTION 24.1** True or False: A larger slit creates a larger separation between interference fringes.

**EXERCISE 24.1** Suppose the same experiment is run with a different light source. If the first-order maximum is found at 1.85 cm from the centerline, what is the wavelength of the light?

**ANSWER** 463 nm

## 24.3 Change of Phase Due to Reflection

**LEARNING OBJECTIVE**

1. Describe and discuss the conditions for which a reflected electromagnetic wave undergoes a 180° change of phase.

Young's method of producing two coherent light sources involves illuminating a pair of slits with a single source. Another simple, yet ingenious, arrangement for producing an interference pattern with a single light source is known as *Lloyd's mirror*. A point source of light is placed at point $S$, close to a mirror, as illustrated in Figure 24.5. Light waves can reach the viewing point $P$ either by the direct path $SP$ or by the path involving reflection from the mirror. The reflected ray can be treated as a ray originating at the source $S'$ behind the mirror. Source $S'$, which is the image of $S$, can be considered a virtual source.

At points far from the source, an interference pattern due to waves from $S$ and $S'$ is observed, just as for two real coherent sources. The positions of the dark and bright fringes, however, are *reversed* relative to the pattern obtained from two real coherent sources (Young's experiment). This is because the coherent sources $S$ and $S'$ differ in phase by 180°, a phase change produced by reflection.

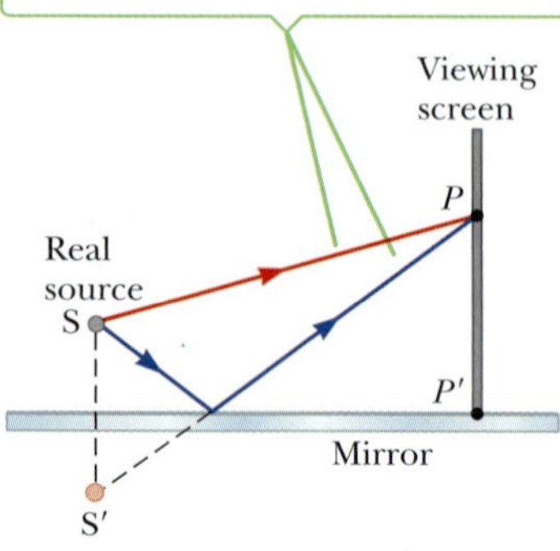

An interference pattern is produced on a screen at P as a result of the combination of the direct ray (red) and the reflected ray (blue). The reflected ray undergoes a phase change of 180°.
Viewing screen
P
Real source
S
P′
Mirror
S′

**Figure 24.5** Lloyd's mirror.

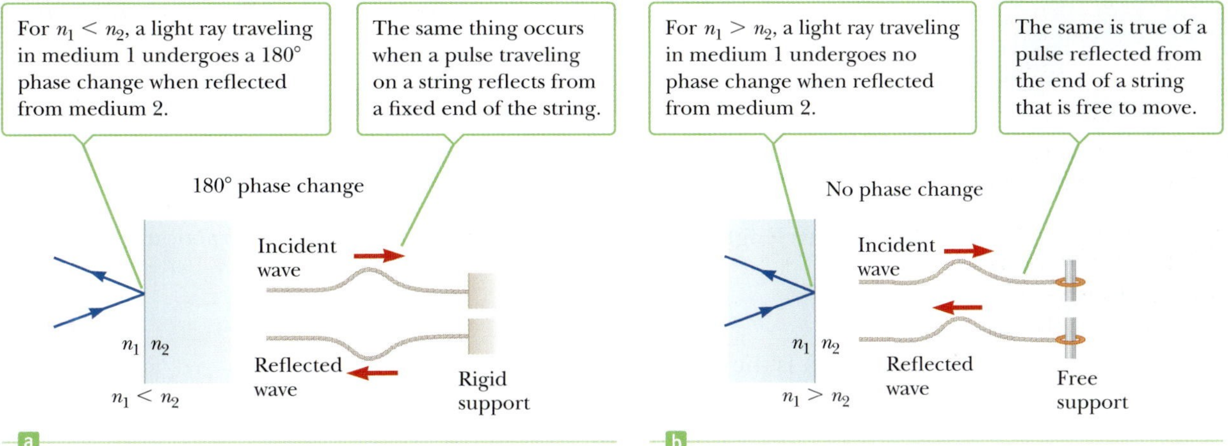
For $n_1 < n_2$, a light ray traveling in medium 1 undergoes a 180° phase change when reflected from medium 2.
The same thing occurs when a pulse traveling on a string reflects from a fixed end of the string.
For $n_1 > n_2$, a light ray traveling in medium 1 undergoes no phase change when reflected from medium 2.
The same is true of a pulse reflected from the end of a string that is free to move.
180° phase change
Incident wave
Reflected wave
Rigid support
$n_1$ $n_2$
$n_1 < n_2$
No phase change
Incident wave
Reflected wave
Free support
$n_1$ $n_2$
$n_1 > n_2$
a
b

**Figure 24.6** Comparisons of reflections of light waves and waves on strings.

To illustrate the point further, consider $P'$, the point where the mirror intersects the screen. This point is equidistant from $S$ and $S'$. If path difference alone were responsible for the phase difference, a bright fringe would be observed at $P'$ (because the path difference is zero for this point), corresponding to the central fringe of the two-slit interference pattern. Instead, we observe a *dark* fringe at $P$, from which we conclude that a 180° phase change must be produced by reflection from the mirror. In general, **an electromagnetic wave undergoes a phase change of 180° upon reflection from a medium that has an index of refraction higher than the one in which the wave was traveling.**

An analogy can be drawn between reflected light waves and the reflections of a transverse wave on a stretched string when the wave meets a boundary, as in Figure 24.6. The pulse on a string undergoes a phase change of 180° when it is reflected from the boundary of a denser string or from a rigid support and undergoes no phase change when it is reflected from the boundary of a less dense string or free support. Similarly, an electromagnetic wave undergoes a 180° phase change when reflected from the boundary of a medium with index of refraction higher than the one in which it has been traveling. There is no phase change when the wave is reflected from a boundary leading to a medium of lower index of refraction. The transmitted wave that crosses the boundary also undergoes no phase change.

## 24.4 Interference in Thin Films

### LEARNING OBJECTIVES

1. State the conditions for constructive and destructive interference in systems involving thin films.
2. Apply the thin film interference conditions to systems involving zero, one- or two-phase reversals.
3. Describe the phenomenon of Newton's rings.

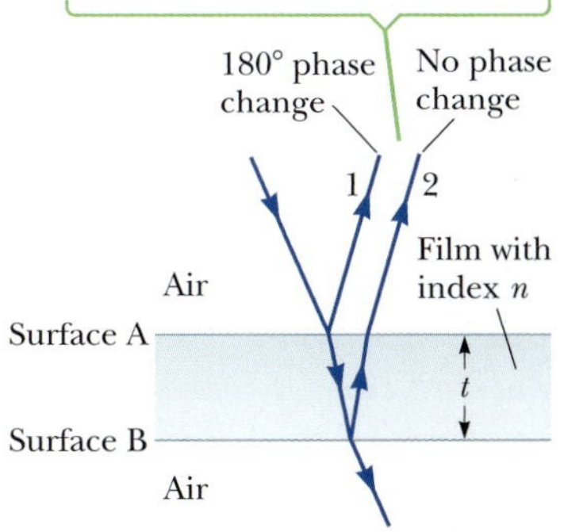
Interference in light reflected from a thin film is due to a combination of rays 1 and 2 reflected from the upper and lower surfaces of the film.
180° phase change
No phase change
1
2
Film with index $n$
Air
Surface A
$t$
Surface B
Air

**Figure 24.7** Light passes through a thin film.

Interference effects are commonly observed in thin films, such as the thin surface of a soap bubble or thin layers of oil on water. The varied colors observed when incoherent white light is incident on such films result from the interference of waves reflected from the two surfaces of the film.

Consider a film of uniform thickness $t$ and index of refraction $n$, as in Figure 24.7. Assume the light rays traveling in air are nearly normal to the two surfaces of the

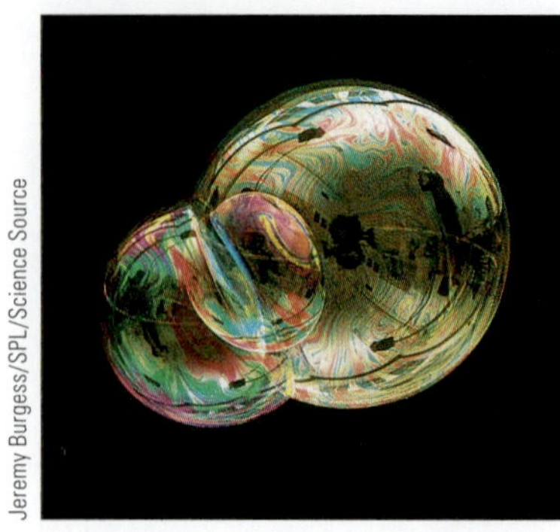

Jeremy Burgess/SPL/Science Source

The colors observed in soap bubbles are due to interference between light rays reflected from the front and back of the thin film of soap making up the bubble. The color depends on the thickness of the film, ranging from black where the film is at its thinnest to magenta where it is thickest.

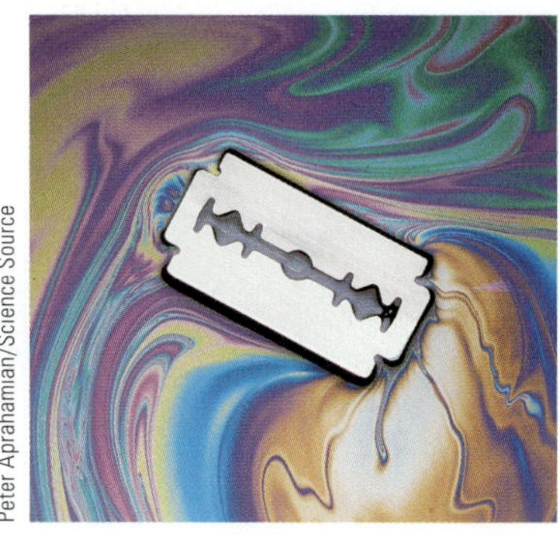

Peter Aprahamian/Science Source

A thin film of oil on water displays interference, evidenced by the pattern of colors when white light is incident on the film. Variations in the film's thickness produce the intersecting color pattern. The razor blade gives you an idea of the size of the colored bands.

film. To determine whether the reflected rays interfere constructively or destructively, we first note the following facts:

1. An electromagnetic wave traveling from a medium of index of refraction $n_1$ toward a medium of index of refraction $n_2$ undergoes a 180° phase change on reflection when $n_2 > n_1$. There is no phase change in the reflected wave if $n_2 < n_1$.
2. The wavelength of light $\lambda_n$ in a medium with index of refraction $n$ is

$$\lambda_n = \frac{\lambda}{n} \qquad [24.7]$$

where $\lambda$ is the wavelength of light in vacuum.

We apply these rules to the film of Figure 24.7. According to the first rule, ray 1, which is reflected from the upper surface A, undergoes a phase change of 180° with respect to the incident wave. Ray 2, which is reflected from the lower surface B, undergoes no phase change with respect to the incident wave. Therefore, ray 1 is 180° out of phase with respect to ray 2, which is equivalent to a path difference of $\lambda_n/2$. We must also consider, though, that ray 2 travels an extra distance of $2t$ before the waves recombine in the air above the surface. For example, if $2t = \lambda_n/2$, rays 1 and 2 recombine in phase and constructive interference results. In general, the condition for *constructive interference* in thin films is

$$2t = (m + \tfrac{1}{2})\lambda_n \qquad m = 0, 1, 2, \ldots \qquad [24.8]$$

This condition takes into account two factors: (1) the difference in path length for the two rays (the term $m\lambda_n$) and (2) the 180° phase change upon reflection (the term $\lambda_n/2$). Because $\lambda_n = \lambda/n$, we can write Equation 24.8 in the form

$$2nt = (m + \tfrac{1}{2})\lambda \qquad m = 0, 1, 2, \ldots \qquad [24.9]$$

If the extra distance $2t$ traveled by ray 2 is a multiple of $\lambda_n$, the two waves combine out of phase and the result is destructive interference. The general equation for *destructive interference* in thin films is

$$2nt = m\lambda \qquad m = 0, 1, 2, \ldots \qquad [24.10]$$

**Equations 24.9 and 24.10 for constructive and destructive interferences are valid when there is only one phase reversal.** This will occur when the media above and below the thin film both have indices of refraction greater than the film or when both have indices of refraction less than the film. Figure 24.7 is a case in point: the air ($n = 1$) that is both above and below the film has an index of refraction less than that of the film. As a result, there is a phase reversal on reflection off the top layer of the film but not the bottom, and Equations 24.9 and 24.10 apply. **If the film is placed between two different media, one of lower refractive index than the film and one of higher refractive index, Equations 24.9 and 24.10 are reversed: Equation 24.9 is used for destructive interference and Equation 24.10 for constructive interference.** In this case either there is a phase change of 180° for both ray 1 reflecting from surface A and ray 2 reflecting from surface B, as in Figure 24.9 of Example 24.3, or there is no phase change for either ray, which would be the case if the incident ray came from underneath the film. Hence, the net change in relative phase due to the reflections is *zero*.

## Quick Quiz

**24.4** Suppose Young's experiment is carried out in air, and then, in a second experiment, the apparatus is immersed in water. In what way does the distance between bright fringes change? (a) They move farther apart. (b) They move closer together. (c) There is no change.

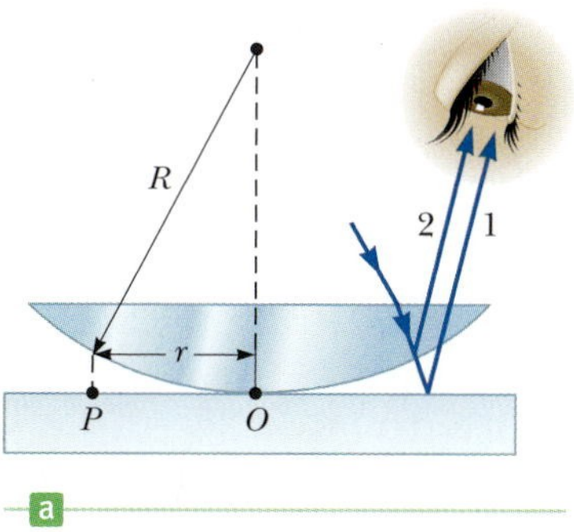


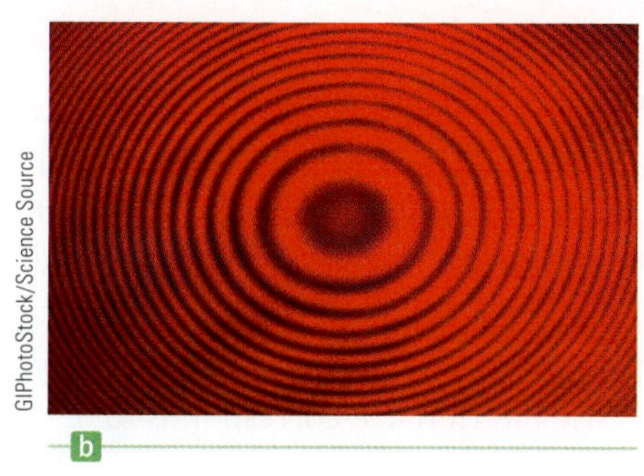


**Figure 24.8** (a) The combination of rays reflected from the glass plate and the curved surface of the lens gives rise to an interference pattern known as Newton's rings. (b) A photograph of Newton's rings.

## Newton's Rings

Another method for observing interference in light waves is to place a plano-convex lens on top of a flat glass surface, as in Figure 24.8a. With this arrangement, the air film between the glass surfaces varies in thickness from zero at the point of contact to some value $t$ at $P$. If the radius of curvature $R$ of the lens is much greater than the distance $r$ and the system is viewed from above using light of wavelength $\lambda$, a pattern of light and dark rings is observed (Fig. 24.8b). These circular fringes, discovered by Newton, are called **Newton's rings**. The interference is due to the combination of ray 1, reflected from the plate, with ray 2, reflected from the lower surface of the lens. Ray 1 undergoes a phase change of 180° on reflection because it is reflected from a boundary leading into a medium of higher refractive index, whereas ray 2 undergoes no phase change because it is reflected from a medium of lower refractive index. Hence, the conditions for constructive and destructive interference are given by Equations 24.9 and 24.10, respectively, with $n = 1$ because the "film" is air. The contact point at $O$ is dark, as seen in Figure 24.8b, because there is no path difference and the total phase change is due only to the 180° phase change upon reflection. Using the geometry shown in Figure 24.8a, we can obtain expressions for the radii of the bright and dark bands in terms of the radius of curvature $R$ and vacuum wavelength $\lambda$. For example, the dark rings have radii of $r \approx \sqrt{m\lambda R/n}$.

**Tip 24.2 The Two Tricks of Thin Films**

Be sure to include *both* effects—path length and phase change—when you analyze an interference pattern from a thin film.

One important use of Newton's rings is in the testing of optical lenses. A circular pattern like that in Figure 24.8b is achieved only when the lens is ground to a perfectly spherical curvature. Variations from such symmetry produce distorted patterns that also give an indication of how the lens must be reground and repolished to remove imperfections.

**APPLICATION**

Checking for Imperfections in Optical Lenses

## ■ PROBLEM-SOLVING STRATEGY

### Thin-Film Interference

*The following steps are recommended in addressing thin-film interference problems:*

1. Identify the thin film causing the interference, and the indices of refraction in the film and in the media on either side of it.
2. Determine the number of phase reversals: zero, one, or two.
3. Consult the following table, which contains Equations 24.9 and 24.10, and select the correct column for the problem in question:

| Equation ($m$ = 0, 1, . . .) | 1 Phase Reversal | 0 or 2 Phase Reversals |
|---|---|---|
| $2nt = (m + \frac{1}{2})\lambda$ **[24.9]** | Constructive | Destructive |
| $2nt = m\lambda$ **[24.10]** | Destructive | Constructive |

4. Substitute values in the appropriate equations, as selected in the previous step.

## EXAMPLE 24.2 Interference in a Soap Film

**GOAL** Study constructive interference effects in a thin film.

**PROBLEM** **(a)** Calculate the minimum thickness of a soap-bubble film ($n = 1.33$) that will result in constructive interference in the reflected light if the film is illuminated by light with wavelength 602 nm in free space. **(b)** Recalculate the minimum thickness for constructive interference when the soap-bubble film is on top of a glass slide with $n = 1.50$.

**STRATEGY** In part (a) there is only one inversion, so the condition for constructive interference is $2nt = (m + \frac{1}{2})\lambda$. The minimum film thickness for constructive interference corresponds to $m = 0$ in this equation. Part (b) involves two inversions, so $2nt = m\lambda$ is required.

**SOLUTION**

**(a)** Calculate the minimum thickness of the soap-bubble film that will result in constructive interference.

Solve $2nt = \lambda/2$ for the thickness $t$ and substitute:

$$t = \frac{\lambda}{4n} = \frac{602 \text{ nm}}{4(1.33)} = \boxed{113 \text{ nm}}$$

**(b)** Find the minimum soap-film thickness when the film is on top of a glass slide with $n = 1.50$.

Write the condition for constructive interference, when two inversions take place:

$$2nt = m\lambda$$

Solve for $t$ and substitute:

$$t = \frac{m\lambda}{2n} = \frac{1 \cdot (602 \text{ nm})}{2(1.33)} = \boxed{226 \text{ nm}}$$

**REMARKS** The different colors in a soap bubble result from the thickness of the soap layer varying from one place to another. The swirling is caused by the changing thickness of the layer with time.

**QUESTION 24.2** A soap film looks red in one area and violet in a nearby area. In which area is the soap film thicker?

**EXERCISE 24.2** What other film thicknesses in part (a) will produce constructive interference?

**ANSWERS** 339 nm, 566 nm, 792 nm, and so on

## EXAMPLE 24.3 Nonreflective Coatings for Solar Cells and Optical Lenses

**GOAL** Study destructive interference effects in a thin film when there are two inversions.

**PROBLEM** Semiconductors such as silicon are used to fabricate solar cells, devices that generate electric energy when exposed to sunlight. Solar cells are often coated with a transparent thin film, such as silicon monoxide (SiO; $n = 1.45$), to minimize reflective losses (Fig. 24.9). A silicon solar cell ($n = 3.50$) is coated with a thin film of silicon monoxide for this purpose. Assuming normal incidence, determine the minimum thickness of the film that will produce the least reflection at a wavelength of 552 nm.

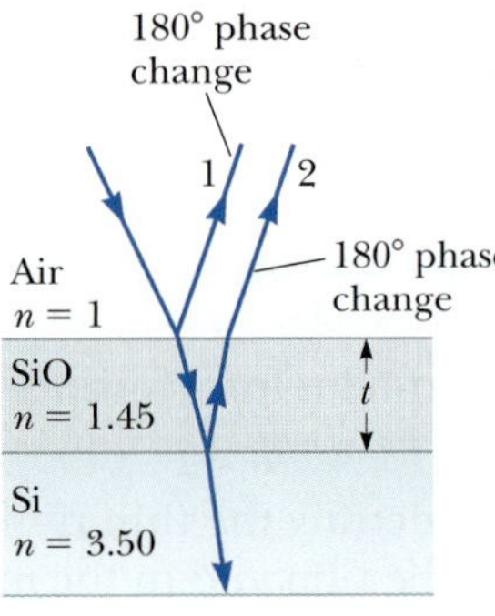


**Figure 24.9** (Example 24.3) Reflective losses from a silicon solar cell are minimized by coating it with a thin film of silicon monoxide (SiO).

**STRATEGY** Reflection is least when rays 1 and 2 in Figure 24.9 meet the condition for destructive interference. Note that *both* rays undergo 180° phase changes on reflection. The condition for a reflection *minimum* is therefore $2nt = \lambda/2$.

**SOLUTION**

Solve $2nt = \lambda/2$ for $t$, the required thickness:

$$t = \frac{\lambda}{4n} = \frac{552 \text{ nm}}{4(1.45)} = \boxed{95.2 \text{ nm}}$$

**REMARKS** Typically, such coatings reduce the reflective loss from 30% (with no coating) to 10% (with a coating), thereby increasing the cell's efficiency because more light is available to create charge carriers in the cell. In reality the coating is never perfectly nonreflecting because the required thickness is wavelength dependent and the incident light covers a wide range of wavelengths.

**QUESTION 24.3** To minimize reflection of a smaller wavelength, should the thickness of the coating be thicker or thinner?

**EXERCISE 24.3** Glass lenses used in cameras and other optical instruments are usually coated with one or more transparent thin films, such as magnesium fluoride ($MgF_2$), to reduce or eliminate unwanted reflection. Carl Zeiss developed this method; his first coating was $1.00 \times 10^2$ nm thick, on glass. Using $n = 1.38$ for $MgF_2$, what visible wavelength would be eliminated by destructive interference in the reflected light?

**ANSWER** 552 nm

## EXAMPLE 24.4 Interference in a Wedge-Shaped Film

**GOAL** Calculate interference effects when the film has variable thickness.

**PROBLEM** A pair of glass slides 10.0 cm long and with $n = 1.52$ are separated on one end by a hair, forming a triangular wedge of air, as illustrated in Figure 24.10. When coherent light from a helium–neon laser with wavelength 633 nm is incident on the film from above, 15.0 dark fringes per centimeter are observed. How thick is the hair?

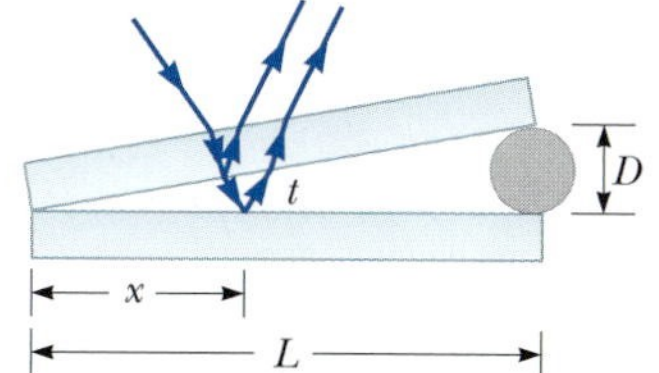


**Figure 24.10** (Example 24.4) Interference bands in reflected light can be observed by illuminating a wedge-shaped film with monochromatic light. The dark areas in the interference pattern correspond to positions of destructive interference.

**STRATEGY** The interference pattern is created by the thin film of air having variable thickness. The pattern is a series of alternating bright and dark parallel bands. A dark band corresponds to destructive interference, and there is one phase reversal, so $2nt = m\lambda$ should be used. We can also use the similar triangles in Figure 24.10 to obtain the relation $t/x = D/L$. We can find the thickness for any $m$, and if the position $x$ can also be found, this last equation gives the diameter of the hair, $D$.

**SOLUTION**

Solve the destructive-interference equation for the thickness of the film, $t$, with $n = 1$ for air:

$$t = \frac{m\lambda}{2}$$

If $d$ is the distance from one dark band to the next, then the $x$-coordinate of the $m$th band is a multiple of $d$:

$$x = md$$

By dimensional analysis, $d$ is just the inverse of the number of bands per centimeter.

$$d = \left(15.0 \frac{\text{bands}}{\text{cm}}\right)^{-1} = 6.67 \times 10^{-2} \frac{\text{cm}}{\text{band}}$$

Now use similar triangles and substitute all the information:

$$\frac{t}{x} = \frac{m\lambda/2}{md} = \frac{\lambda}{2d} = \frac{D}{L}$$

Solve for $D$ and substitute given values:

$$D = \frac{\lambda L}{2d} = \frac{(633 \times 10^{-9}\,\text{m})(0.100\,\text{m})}{2(6.67 \times 10^{-4}\,\text{m})} = \boxed{4.75 \times 10^{-5}\,\text{m}}$$

**REMARKS** Some may be concerned about interference caused by light bouncing off the top and bottom of, say, the upper glass slide. It's unlikely, however, that the thickness of the slide will be half an integer multiple of the wavelength of the helium–neon laser (for some very large value of $m$). In addition, in contrast to the air wedge, the thickness of the glass doesn't vary.

**QUESTION 24.4** If the air wedge is filled with water, how is the distance between dark bands affected? Explain.

**EXERCISE 24.4** The air wedge is replaced with water, with $n = 1.33$. Find the distance between dark bands when the helium–neon laser light hits the glass slides.

**ANSWER** $5.02 \times 10^{-4}$ m

## APPLYING PHYSICS 24.3 Perfect Mirrors

When light hits a metallic mirror, electrons in the metal move in response to the electromagnetic fields, absorbing some of the light's energy. For many applications, such as directing high-intensity laser light, that reduction in intensity is undesirable. A dielectric mirror, on the other hand, is made of glass or plastic and doesn't conduct electricity. To improve reflectance, thin layers of different dielectric materials are stacked on the glass surface. If the thicknesses and dielectric constants are chosen properly, light reflected off one layer combines constructively with light reflected from the layer underneath, increasing the mirror's reflectance. Nearly perfect mirrors can be constructed using several thin dielectric layers. The mirrors can be designed to reflect a particular wavelength or a range of wavelengths.

Dielectric mirrors, first developed at MIT in 1998, have an enormous number of applications. One of the most important is the OmniGuide fiber, a hollow tube the size of a spaghetti noodle that can guide light without any significant loss of intensity. (Ordinary optical fibers tend to heat up.) Such fibers have up to forty concentric layers of plastic and glass and are highly flexible. Using an OmniGuide fiber, an intense laser light can be safely guided into the human body during surgery to remove tumors and other diseased tissues without harming the healthy surrounding tissue. ■

# 24.5 Using Interference to Read CDs and DVDs

**LEARNING OBJECTIVES**

1. Discuss the application of interference principles to the reading of CDs and DVDs.
2. Apply interference principles to CDs and DVDs.

**APPLICATION**
The Physics of CDs and DVDs

Compact discs (CDs) and digital videodiscs (DVDs) provide high-density storage of text, graphics, and movies; and high-quality sound recordings. The data on these discs are stored digitally as a series of zeros and ones, and these zeros and ones are read by laser light reflected from the disc. Strong reflections (constructive interference) from the disc are chosen to represent zeros, and weak reflections (destructive interference) represent ones.

To see in more detail how thin-film interference plays a crucial role in reading CDs and DVDs, consider Figure 24.11. This figure shows a photomicrograph of several DVD tracks, which consist of a sequence of pits (when viewed from the top or label side of the disc) of varying length formed in a reflecting-metal information layer. A cross-sectional view of a CD as shown in Figure 24.12 reveals that the pits appear as bumps to the laser beam, which shines on the metallic layer through a clear plastic coating from below.

Andrew Syred / Science Source

**Figure 24.11** A photomicrograph of adjacent tracks on a digital video disc (DVD). The information encoded in these pits and smooth areas is read by a laser beam.

As the disk rotates, the laser beam reflects off the sequence of bumps and lower areas into a photodetector, which converts the fluctuating reflected light intensity into an electrical string of zeros and ones. To make the light fluctuations more pronounced and easier to detect, the pit depth $t$ is made equal to one-quarter of a

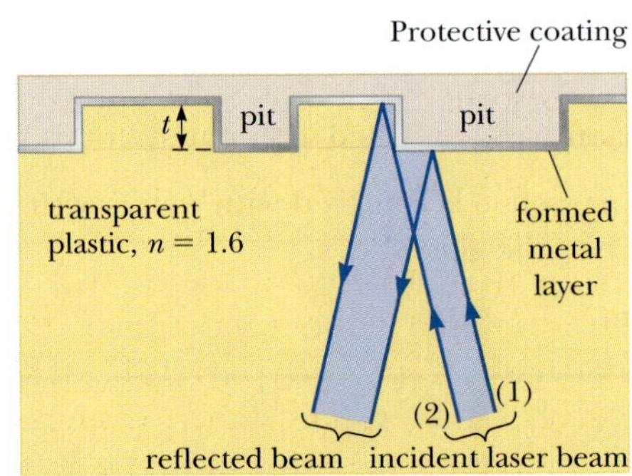


**Figure 24.12** Cross section of a CD showing metallic pits of depth $t$ and a laser beam detecting the edge of a pit.

wavelength of the laser light in the plastic. When the beam hits a rising or falling bump edge, part of the beam reflects from the top of the bump and part from the lower adjacent area, ensuring destructive interference and very low intensity when the reflected beams combine at the detector. Bump edges are read as ones, and flat bump tops and intervening flat plains are read as zeros.

In Example 24.5 the pit depth for a standard CD, using an infrared laser of wavelength 780 nm, is calculated. DVDs use shorter wavelength lasers of 635 nm, so the track separation, pit depth, and minimum pit length are all smaller. These differences allow a DVD to store about 30 times more information than a CD.

### EXAMPLE 24.5 Pit Depth in a CD

**GOAL** Apply interference principles to a CD.

**PROBLEM** Find the pit depth in a CD that has a plastic transparent layer with index of refraction of 1.60 and is designed for use in a CD player using a laser with a wavelength of $7.80 \times 10^2$ nm in air.

**STRATEGY** (See Fig. 24.12.) Rays 1 and 2 both reflect from the metal layer, which acts like a mirror, so there is no phase difference due to reflection between those rays. There is, however, the usual phase difference caused by the extra distance $2t$ traveled by ray 2. The wavelength is $\lambda/n$, where $n$ is the index of refraction in the substance.

**SOLUTION**

Use the appropriate condition for destructive interference in a thin film:

$$2t = \frac{\lambda}{2n}$$

Solve for the thickness $t$ and substitute:

$$t = \frac{\lambda}{4n} = \frac{7.80 \times 10^2 \text{ nm}}{(4)(1.60)} = 1.22 \times 10^2 \text{ nm}$$

**REMARKS** Different CD systems have different tolerances for scratches. Anything that changes the reflective properties of the disk can affect the readability of the disk.

**QUESTION 24.5** True or False: Given two plastics with different indices of refraction, the material with the larger index of refraction will have a larger pit depth.

**EXERCISE 24.5** Repeat the example for a laser with wavelength 635 nm.

**ANSWER** 99.2 nm

## 24.6 Diffraction

**LEARNING OBJECTIVES**

1. Describe the physical origins of diffraction.
2. Discuss the appearance and ordering of maxima and minima in a diffraction pattern.

Suppose a light beam is incident on two slits, as in Young's double-slit experiment. If the light truly traveled in straight-line paths after passing through the slits, as in Figure 24.13a (page 848), the waves wouldn't overlap and no interference pattern would be seen. Instead, Huygens' principle requires that the waves spread out from the slits, as shown in Figure 24.13b. In other words, the light bends from a straight-line path and enters the region that would otherwise be shadowed. This spreading out of light from its initial line of travel is called **diffraction**.

In general, diffraction occurs when waves pass through small openings, around obstacles, or by sharp edges. For example, when a single narrow slit is placed between a distant light source (or a laser beam) and a screen, the light produces

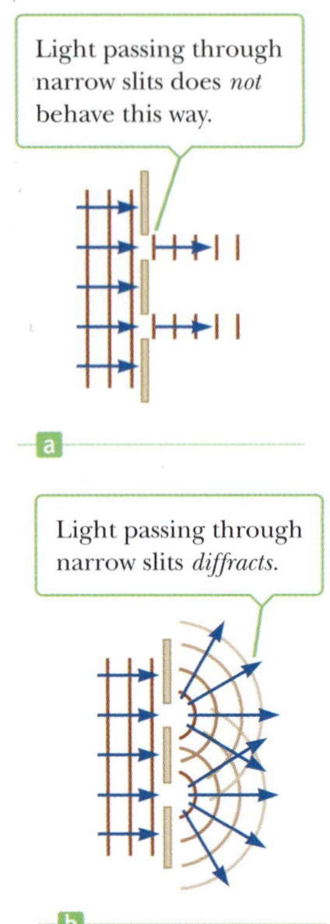
Light passing through narrow slits does *not* behave this way.
a
Light passing through narrow slits *diffracts.*
b

**Figure 24.13** (a) If light did not spread out after passing through the slits, no interference would occur. (b) The light from the two slits overlaps as it spreads out, filling the expected shadowed regions with light and producing interference fringes.

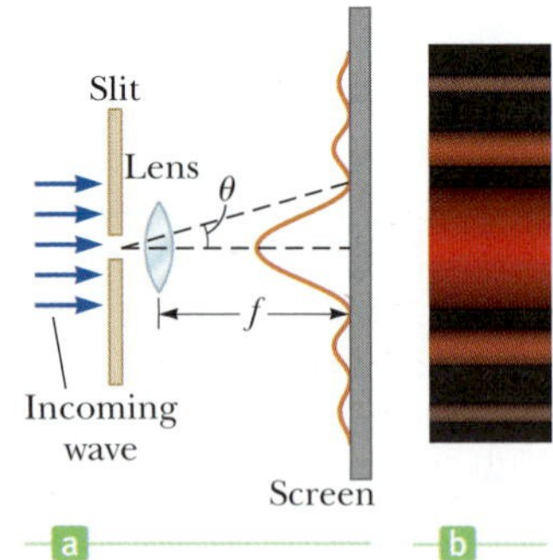
Slit
Lens
θ
f
Incoming wave
Screen
a
b

**Figure 24.16** (a) The Fraunhofer diffraction pattern of a single slit. The parallel rays are brought into focus on the screen with a converging lens. The pattern consists of a central bright region flanked by much weaker maxima. (This drawing is not to scale.) (b) A photograph of a single-slit Fraunhofer diffraction pattern.

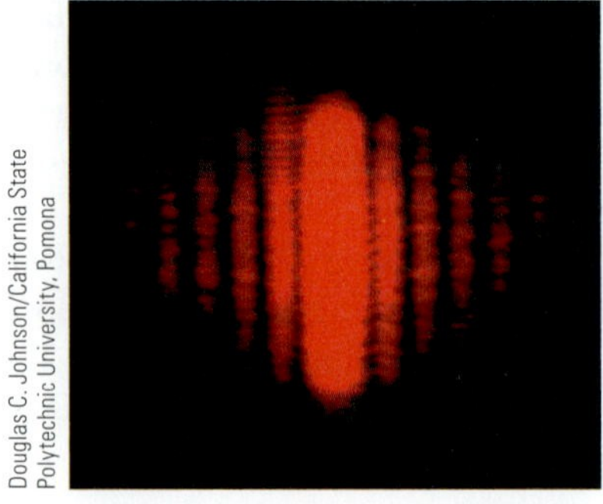

**Figure 24.14** The diffraction pattern that appears on a screen when light passes through a narrow vertical slit. The pattern consists of a broad central band and a series of less intense and narrower side bands.

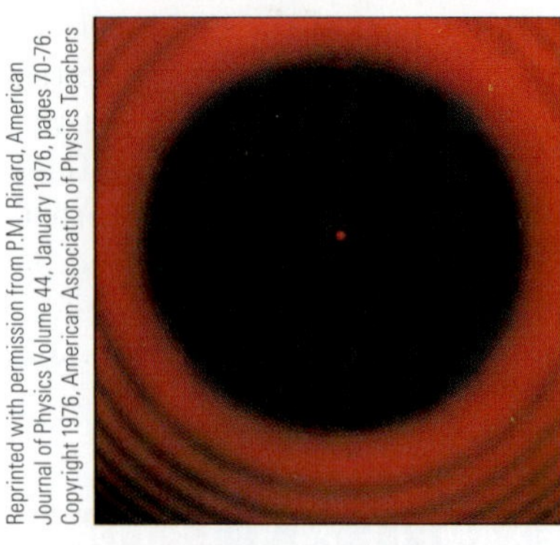

**Figure 24.15** The diffraction pattern of a penny placed midway between the screen and the source. Notice the bright spot at the center.

a diffraction pattern like that in Figure 24.14. The pattern consists of a broad, intense central band flanked by a series of narrower, less intense secondary bands (called **secondary maxima**) and a series of dark bands, or **minima**. This phenomenon can't be explained within the framework of geometric optics, which says that light rays traveling in straight lines should cast a sharp image of the slit on the screen.

Figure 24.15 shows the diffraction pattern and shadow of a penny. The pattern consists of the shadow, a bright spot at its center, and a series of bright and dark circular bands of light near the edge of the shadow. The bright spot at the center (called the *Fresnel bright spot*) is explained by Augustin Fresnel's wave theory of light, which predicts constructive interference at this point for certain locations of the penny. From the viewpoint of geometric optics, there shouldn't be any bright spot: the center of the pattern would be completely screened by the penny.

One type of diffraction, called **Fraunhofer diffraction,** occurs when the rays leave the diffracting object in parallel directions. Fraunhofer diffraction can be achieved experimentally either by placing the observing screen far from the slit or by using a converging lens to focus the parallel rays on a nearby screen, as in Figure 24.16a. A bright fringe is observed along the axis at $\theta = 0$, with alternating dark and bright fringes on each side of the central bright fringe. Figure 24.16b is a photograph of a single-slit Fraunhofer diffraction pattern.

## 24.7 Single-Slit Diffraction

### LEARNING OBJECTIVES

1. State the condition for destructive interference in a single slit experiment.
2. Apply the single-slit destructive interference condition to a single slit experiment.

Until now we have assumed slits have negligible width, acting as line sources of light. In this section we determine how their nonzero widths are the basis for understanding the nature of the Fraunhofer diffraction pattern produced by a single slit.

We can deduce some important features of this problem by examining waves coming from various portions of the slit, as shown in Figure 24.17. According to Huygens' principle, **each portion of the slit acts as a source of waves. Hence, light from one portion of the slit can interfere with light from another portion,** and the resultant intensity on the screen depends on the direction $\theta$.

To analyze the diffraction pattern, it's convenient to divide the slit into halves, as in Figure 24.17. All the waves that originate at the slit are in phase. Consider waves 1 and 3, which originate at the bottom and center of the slit, respectively. Wave 1 travels farther than wave 3 by an amount equal to the path difference $(a/2)\sin\theta$, where $a$ is the width of the slit. Similarly, the path difference between waves 3 and 5 is $(a/2)\sin\theta$. If this path difference is exactly half of a wavelength (corresponding to a phase difference of 180°), the two waves cancel each other, and destructive interference results. This is true, in fact, for any two waves that originate at points separated by half the slit width because the phase difference between two such points is 180°. Therefore, waves from the upper half of the slit interfere *destructively* with waves from the lower half of the slit when

$$\frac{a}{2}\sin\theta = \frac{\lambda}{2}$$

or when

$$\sin\theta = \frac{\lambda}{a}$$

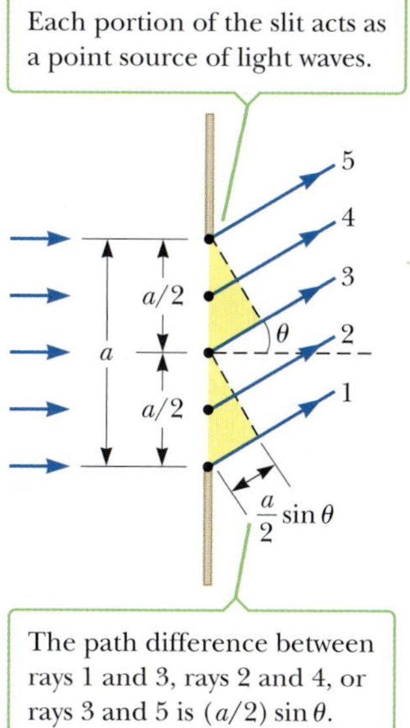


**Figure 24.17** Diffraction of light by a narrow slit of width $a$. (This drawing is not to scale, and the waves are assumed to converge at a distant point.)

If we divide the slit into four parts rather than two and use similar reasoning, we find that the screen is also dark when

$$\sin\theta = \frac{2\lambda}{a}$$

Continuing in this way, we can divide the slit into six parts and show that darkness occurs on the screen when

$$\sin\theta = \frac{3\lambda}{a}$$

Therefore, the general condition for **destructive interference** for a single slit of width $a$ is

$$\sin\theta_{\text{dark}} = m\frac{\lambda}{a} \qquad m = \pm1, \pm2, \pm3, \ldots \qquad [24.11]$$

Equation 24.11 gives the values of $\theta$ for which the diffraction pattern has zero intensity, where a dark fringe forms. The equation tells us nothing about the variation in intensity along the screen, however. The general features of the intensity distribution along the screen are shown in Figure 24.18. A broad central bright fringe is flanked by much weaker bright fringes alternating with dark fringes. The various dark fringes (points of zero intensity) occur at the values of $\theta$ that satisfy Equation 24.11. The points of constructive interference lie approximately halfway between the dark fringes. Note that the central bright fringe is twice as wide as the weaker maxima having $m > 1$.

**Tip 24.3 The Same, but Different**

Although Equations 24.2 and 24.11 have the same form, they have different meanings. Equation 24.2 describes the *bright* regions in a two-slit interference pattern, whereas Equation 24.11 describes the *dark* regions in a single-slit interference pattern.

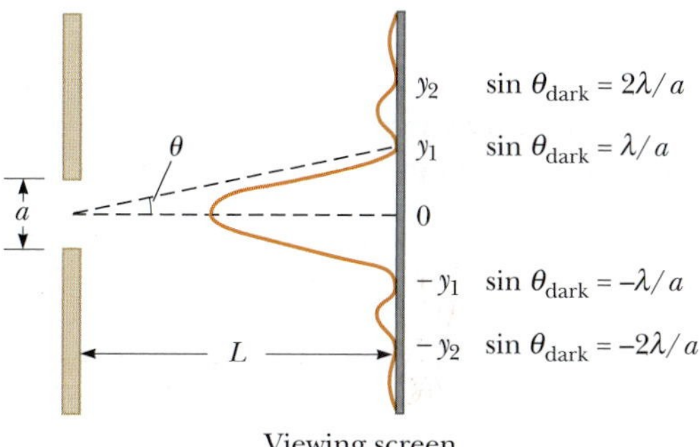


**Figure 24.18** Positions of the minima for the Fraunhofer diffraction pattern of a single slit of width $a$. (This drawing is not to scale.)

**Quick Quiz**

**24.5** In a single-slit diffraction experiment, as the width of the slit is made smaller, does the width of the central maximum of the diffraction pattern (a) becomes smaller, (b) become larger, or (c) remain the same?

## APPLYING PHYSICS 24.4 Diffraction of Sound Waves

If a classroom door is open even a small amount, you can hear sounds coming from the hallway, yet you can't see what is going on in the hallway. How can this difference be explained?

**EXPLANATION** The space between the slightly open door and the wall is acting as a single slit for waves. Sound waves have wavelengths larger than the width of the slit, so sound is effectively diffracted by the opening and the central maximum spreads throughout the room. Light wavelengths are much smaller than the slit width, so there is virtually no diffraction for the light. You must have a direct line of sight to detect the light waves. ■

## EXAMPLE 24.6 A Single-Slit Experiment

**GOAL** Find the positions of the dark fringes in single-slit diffraction.

**PROBLEM** Light of wavelength $5.80 \times 10^2$ nm is incident on a slit of width 0.300 mm. The observing screen is placed 2.00 m from the slit. Find the positions of the first dark fringes and the width of the central bright fringe.

**STRATEGY** This problem requires substitution into Equation 24.11 to find the sines of the angles of the first dark fringes. The positions can then be found with the tangent function because for small angles $\sin\theta \approx \tan\theta$. The extent of the central maximum is defined by these two dark fringes.

**SOLUTION**

The first dark fringes that flank the central bright fringe correspond to $m = \pm 1$ in Equation 24.11:

$$\sin\theta = \pm\frac{\lambda}{a} = \pm\frac{5.80 \times 10^{-7}\text{ m}}{0.300 \times 10^{-3}\text{ m}} = \pm 1.93 \times 10^{-3}$$

Use the triangle in Figure 24.18 to relate the position of the fringe to the tangent function:

$$\tan\theta = \frac{y_1}{L}$$

Because $\theta$ is very small, we can use the approximation $\sin\theta \approx \tan\theta$ and then solve for $y_1$:

$$\sin\theta \approx \tan\theta \approx \frac{y_1}{L}$$

$$y_1 \approx L\sin\theta = (2.00\text{ m})(\pm 1.93 \times 10^{-3}) = \boxed{\pm\, 3.86 \times 10^{-3}\text{ m}}$$

Compute the distance between the positive and negative first-order maxima, which is the width $w$ of the central maximum:

$$w = +3.86 \times 10^{-3}\text{ m} - (-3.86 \times 10^{-3}\text{ m}) = \boxed{7.72 \times 10^{-3}\text{ m}}$$

**REMARKS** Note that this value of $w$ is much greater than the width of the slit. As the width of the slit is *increased*, however, the diffraction pattern *narrows*, corresponding to smaller values of $\theta$. In fact, for large values of $a$, the maxima and minima are so closely spaced that the only observable pattern is a large central bright area resembling the geometric image of the slit. Because the width of the geometric image increases as the slit width increases, the narrowest image occurs when the geometric and diffraction widths are equal.

**QUESTION 24.6** Suppose the entire apparatus is immersed in water. If the same wavelength of light (in air) is incident on the slit immersed in water, is the resulting central maximum larger or smaller? Explain.

**EXERCISE 24.6** Determine the width of the first-order bright fringe in the example, when the apparatus is in air.

**ANSWER** 3.86 mm

# 24.8 The Diffraction Grating

**LEARNING OBJECTIVES**

1. Describe the effects of a diffraction grating on incident plane waves.
2. Discuss the order number of maxima in a diffraction pattern.
3. Apply the condition for maxima in the interference pattern of a diffraction grating to optical systems.

The diffraction grating, a useful device for analyzing light sources, consists of a large number of equally spaced parallel slits. A grating can be made by scratching parallel lines on a glass plate with a precision machining technique. The clear panes between scratches act like slits. A typical grating contains several thousand lines per centimeter. For example, a grating ruled with 5 000 lines/cm has a slit spacing $d$ equal to the reciprocal of that number; hence, $d = (1/5\ 000)$ cm $= 2 \times 10^{-4}$ cm.

Figure 24.19 is a schematic diagram of a section of a plane diffraction grating. A plane wave is incident from the left, normal to the plane of the grating. The intensity of the pattern on the screen is the result of the combined effects of interference and diffraction. Each slit causes diffraction, and the diffracted beams in turn interfere with one another to produce the pattern. Moreover, each slit acts as a source of waves, and all waves start in phase at the slits. For some arbitrary direction $\theta$ measured from the horizontal, however, the waves must travel *different* path lengths before reaching a particular point $P$ on the screen. In Figure 24.19, note that the path difference between waves from any two adjacent slits is $d \sin \theta$. If this path difference equals one wavelength or some integral multiple of a wavelength, waves from all slits will be in phase at $P$ and a bright line will be observed at that point. Therefore, the condition for **maxima** in the interference pattern at the angle $\theta$ is

$$d \sin \theta_{\text{bright}} = m\lambda \qquad m = 0, \pm 1, \pm 2, \ldots \qquad [24.12]$$

◀ Condition for maxima in the interference pattern of a diffraction grating

Light emerging from a slit at an angle other than that for a maximum interferes nearly completely destructively with light from some other slit on the grating. All such pairs will result in little or no transmission in that direction, as illustrated in Figure 24.20 (page 852).

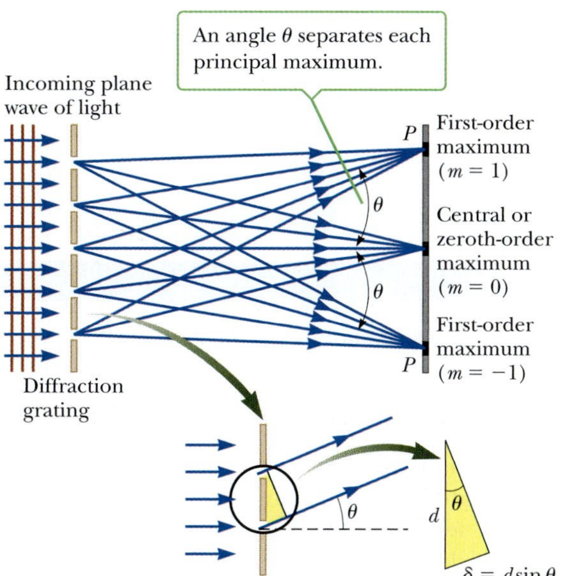


**Figure 24.19** A side view of a diffraction grating. The slit separation is $d$, and the path difference between adjacent slits is $d \sin \theta$.

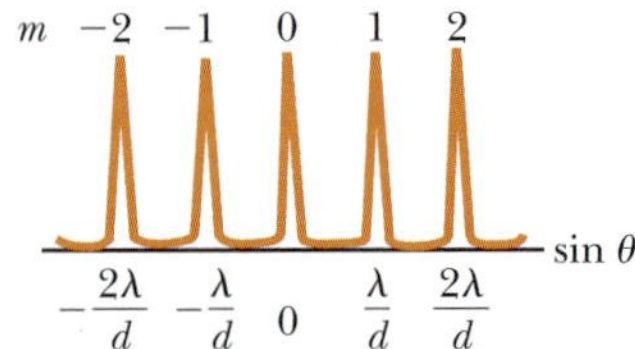


**Figure 24.20** Intensity versus sin $\theta$ for the diffraction grating. The zeroth-, first-, and second-order principal maxima are shown.

Equation 24.12 can be used to calculate the wavelength from the grating spacing and the angle of deviation, $\theta$. The integer $m$ is the **order number** of the diffraction pattern. If the incident radiation contains several wavelengths, each wavelength deviates through a specific angle, which can be found from Equation 24.12. All wavelengths are focused at $\theta = 0$, corresponding to $m = 0$. This point is called the *zeroth-order maximum*. The *first-order maximum*, corresponding to $m = 1$, is observed at an angle that satisfies the relationship $\sin \theta = \lambda/d$; the *second-order maximum*, corresponding to $m = 2$, is observed at a larger angle $\theta$, and so on. Figure 24.20 is a sketch of the intensity distribution for some of the orders produced by a diffraction grating. Note the sharpness of the principal maxima and the broad range of the dark areas, a pattern in direct contrast to the broad bright fringes characteristic of the two-slit interference pattern.

A simple arrangement that can be used to measure the angles in a diffraction pattern is shown in Figure 24.21. This setup is a form of a diffraction-grating spectrometer. The light to be analyzed passes through a slit and is formed into a parallel beam by a lens. The light then strikes the grating at a 90° angle. The diffracted light leaves the grating at angles that satisfy Equation 24.12. A telescope is used to view the image of the slit. The wavelength can be determined by measuring the angles at which the images of the slit appear for the various orders.

**Quick Quiz**

**24.6** If laser light is reflected from a phonograph record or a compact disc, a diffraction pattern appears. The pattern arises because both devices contain parallel tracks of information that act as a reflection diffraction grating. Which device, record or compact disc, results in diffraction maxima that are farther apart?

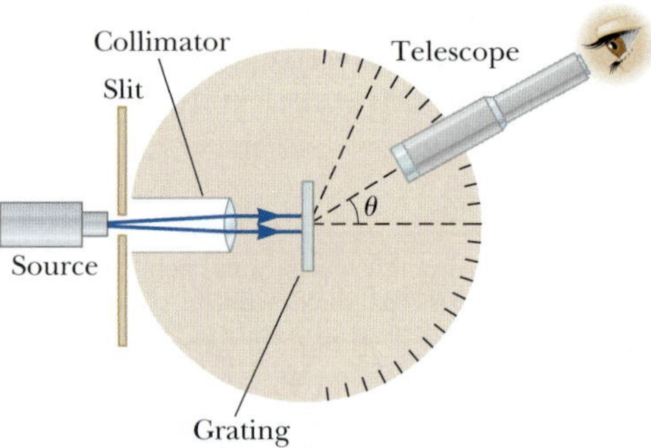


**Figure 24.21** A diagram of a diffraction grating spectrometer. The collimated beam incident on the grating is diffracted into the various orders at the angles $\theta$ that satisfy the equation $d \sin \theta = m\lambda$, where $m = 0, \pm 1, \pm 2, \ldots$

## APPLYING PHYSICS 24.5 Prism vs. Grating

When white light enters through an opening in an opaque box and exits through an opening on the other side of the box, a spectrum of colors appears on the wall. From this observation, how would you be able to determine whether the box contains a prism or a diffraction grating?

**EXPLANATION** The determination could be made by noticing the order of the colors in the spectrum relative to the direction of the original beam of white light. For a prism, in which the separation of light is a result of dispersion, the violet light will be refracted more than the red light. Hence, the order of the spectrum from a prism will be from red, closest to the original direction, to violet. For a diffraction grating, the angle of diffraction increases with wavelength, so the spectrum from the diffraction grating will have colors in the order from violet, closest to the original direction, to red. Further, the diffraction grating will produce *two* first-order spectra on either side of the grating, whereas the prism will produce only a single spectrum. ■

## APPLYING PHYSICS 24.6 Rainbows from a Compact Disc

White light reflected from the surface of a CD has a multicolored appearance, as shown in Figure 24.22. The observation depends on the orientation of the disc relative to the eye and the position of the light source. Explain how all this works.

Carlos E. Santa Maria/Shutterstock.com

**Figure 24.22** (Applying Physics 24.5) Compact discs act as diffraction gratings when observed under white light.

**EXPLANATION** The surface of a CD has a spiral-shaped track (with a spacing of approximately 1 $\mu$m) that acts as a reflection grating. The light scattered by these closely spaced parallel tracks interferes constructively in certain directions that depend on both the wavelength and the direction of the incident light. Any one section of the disc serves as a diffraction grating for white light, sending beams of constructive interference for different colors in different directions. The different colors you see when viewing one section of the disc change as the light source, the disc, or you move to change the angles of incidence or diffraction. ■

## Use of a Diffraction Grating in CD Tracking

**APPLICATION**
Tracking Information on a CD

If a CD player is to reproduce sound faithfully, the laser beam must follow the spiral track of information perfectly. Sometimes the laser beam can drift off track, however, and without a feedback procedure to let the player know that is happening, the fidelity of the music can be greatly reduced.

Figure 24.23 shows how a diffraction grating is used in a three-beam method to keep the beam on track. The central maximum of the diffraction pattern reads the information on the CD track, and the two first-order maxima steer the beam. The grating is designed so that the first-order maxima fall on the smooth surfaces on either side of the information track. Both of these reflected beams have their own detectors, and because both beams are reflected from smooth surfaces, they should have the same strong intensity when they are detected. If the central beam wanders off the track, however, one of the steering beams will begin to strike bumps on the information track and the amount of light reflected will decrease. This information is then used by electronic circuits to drive the main beam back to its desired location.

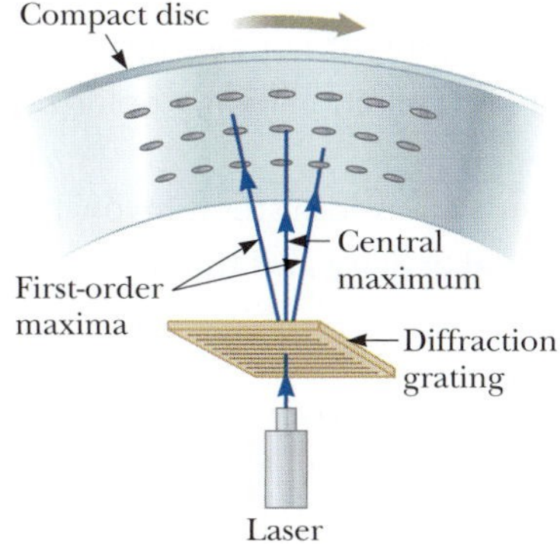


**Figure 24.23** The laser beam in a CD player is able to follow the spiral track by using three beams produced with a diffraction grating.

## EXAMPLE 24.7 A Diffraction Grating

**GOAL** Calculate different-order principal maxima for a diffraction grating.

**PROBLEM** Monochromatic light from a helium–neon laser ($\lambda = 632.8$ nm) is incident normally on a diffraction grating containing $6.00 \times 10^3$ lines/cm. Find the angles at which one would observe the first-order maximum, the second-order maximum, and so forth.

*(Continued)*

**STRATEGY** Find the slit separation by inverting the number of lines per centimeter, then substitute values into Equation 24.12.

**SOLUTION**

Invert the number of lines per centimeter to obtain the slit separation:

$$d = \frac{1}{6.00 \times 10^3\,\text{cm}^{-1}} = 1.67 \times 10^{-4}\,\text{cm} = 1.67 \times 10^3\,\text{nm}$$

Substitute $m = 1$ into Equation 24.12 to find the sine of the angle corresponding to the first-order maximum:

$$\sin\theta_1 = \frac{\lambda}{d} = \frac{632.8\,\text{nm}}{1.67 \times 10^3\,\text{nm}} = 0.379$$

Take the inverse sine of the preceding result to find $\theta_1$:

$$\theta_1 = \sin^{-1} 0.379 = \boxed{22.3°}$$

Repeat the calculation for $m = 2$:

$$\sin\theta_2 = \frac{2\lambda}{d} = \frac{2(632.8\,\text{nm})}{1.67 \times 10^3\,\text{nm}} = 0.758$$

$$\theta_2 = \boxed{49.3°}$$

Repeat the calculation for $m = 3$:

$$\sin\theta_3 = \frac{3\lambda}{d} = \frac{3(632.8\,\text{nm})}{1.67 \times 10^3\,\text{nm}} = 1.14$$

Because $\sin\theta$ can't exceed 1, there is no solution for $\theta_3$.

**REMARKS** The foregoing calculation shows that there can only be a finite number of principal maxima. In this case only zeroth-, first-, and second-order maxima would be observed.

**QUESTION 24.7** Does a diffraction grating with more lines have a smaller or larger separation between adjacent principal maxima?

**EXERCISE 24.7** Suppose light with wavelength $7.80 \times 10^2$ nm is used instead and the diffraction grating has $3.30 \times 10^3$ lines per centimeter. Find the angles of all the principal maxima.

**ANSWERS** 0°, 14.9°, 31.0°, 50.6°

## 24.9 Polarization of Light Waves

**LEARNING OBJECTIVES**

1. Discuss the concept of polarization of light waves.
2. Describe three processes for obtaining linearly polarized light from an unpolarized source.
3. Apply Malus's and Brewster's Laws to systems involving unpolarized light.
4. Discuss the concept of optical activity and the practical applications of liquid crystals.

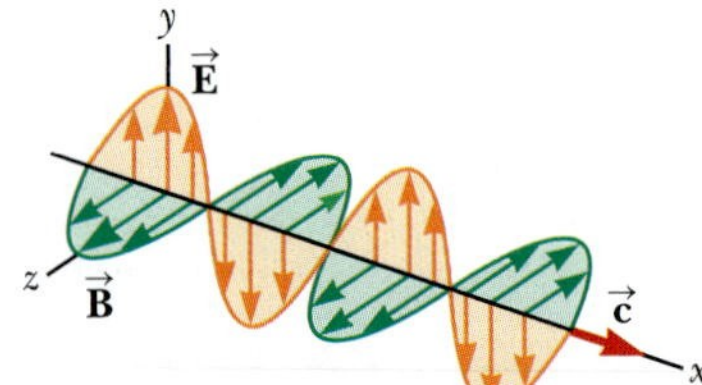


**Figure 24.24** A schematic diagram of a polarized electromagnetic wave propagating in the $x$-direction. The electric field vector $\vec{E}$ vibrates in the $xy$-plane, whereas the magnetic field vector $\vec{B}$ vibrates in the $xz$-plane.

In Chapter 21 we described the transverse nature of electromagnetic waves. Figure 24.24 shows that the electric and magnetic field vectors associated with an electromagnetic wave are at right angles to each other and also to the direction of wave propagation. The phenomenon of polarization, described in this section, is firm evidence of the transverse nature of electromagnetic waves.

An ordinary beam of light consists of a large number of electromagnetic waves emitted by the atoms or molecules of the light source. The vibrating charges associated with the atoms act as tiny antennas. Each atom produces a wave with its own orientation of $\vec{E}$, as in Figure 24.24, corresponding to the direction of atomic vibration. Because all directions of vibration are possible, however, the resultant electromagnetic wave is a superposition of waves produced by the individual atomic sources. The result is an **unpolarized** light wave, represented schematically in Figure 24.25a. The direction of wave propagation shown in the figure is perpendicular to the page. Note that *all* directions of the electric field vector are equally probable and lie in a plane (such as the plane of this page) perpendicular to the direction of propagation.

A wave is said to be **linearly polarized** if the resultant electric field $\vec{\mathbf{E}}$ vibrates in the same direction *at all times* at a particular point, as in Figure 24.25b. (Sometimes such a wave is described as *plane polarized* or simply *polarized*.) The wave in Figure 24.24 is an example of a wave that is linearly polarized in the $y$-direction. As the wave propagates in the $x$-direction, $\vec{\mathbf{E}}$ is always in the $y$-direction. The plane formed by $\vec{\mathbf{E}}$ and the direction of propagation is called the *plane of polarization* of the wave. In Figure 24.24 the plane of polarization is the $xy$-plane.

It's possible to obtain a linearly polarized beam from an unpolarized beam by removing all waves from the beam except those with electric field vectors that oscillate in a single plane. We now discuss three processes for doing this: (1) selective absorption, (2) reflection, and (3) scattering.

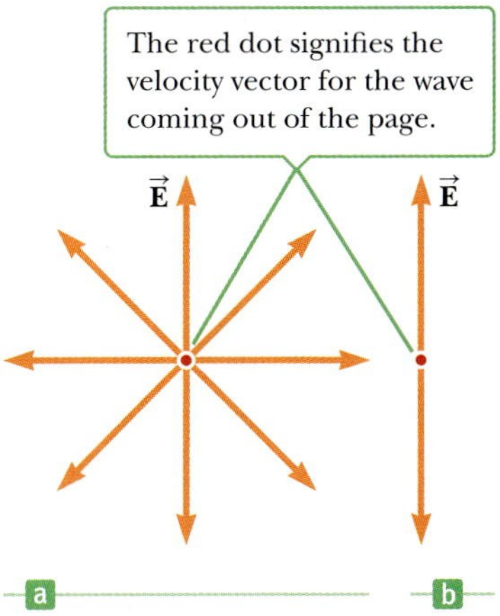

The red dot signifies the velocity vector for the wave coming out of the page.

**Figure 24.25** (a) An unpolarized light beam viewed along the direction of propagation. The transverse electric field vector can vibrate in any direction with equal probability. (b) A linearly polarized light beam with the electric field vector vibrating in the vertical direction.

## Polarization by Selective Absorption

The most common technique for polarizing light is to use a material that transmits waves having electric field vectors that vibrate in a plane parallel to a certain direction and absorbs those waves with electric field vectors vibrating in directions perpendicular to that direction.

In 1932 E. H. Land discovered a material, which he called **Polaroid**, that polarizes light through selective absorption by oriented molecules. This material is fabricated in thin sheets of long-chain hydrocarbons, which are stretched during manufacture so that the molecules align. After a sheet is dipped into a solution containing iodine, the molecules become good electrical conductors. Conduction takes place primarily along the hydrocarbon chains, however, because the valence electrons of the molecules can move easily only along those chains. (Recall that valence electrons are "free" electrons that can move easily through the conductor.) As a result, the molecules readily *absorb* light having an electric field vector parallel to their lengths and *transmit* light with an electric field vector perpendicular to their lengths. It's common to refer to the direction perpendicular to the molecular chains as the **transmission axis**. In an ideal polarizer all light with $\vec{\mathbf{E}}$ parallel to the transmission axis is transmitted and all light with $\vec{\mathbf{E}}$ perpendicular to the transmission axis is absorbed.

Polarizing material reduces the intensity of light passing through it. In Figure 24.26 an unpolarized light beam is incident on the first polarizing sheet, called the **polarizer**; the transmission axis is as indicated. The light that passes through this sheet is polarized vertically, and the transmitted electric field vector is $\vec{\mathbf{E}}_0$. A second polarizing sheet, called the **analyzer**, intercepts this beam with its transmission axis at an angle of $\theta$ to the axis of the polarizer. The component of $\vec{\mathbf{E}}_0$ that is perpendicular to the axis of the analyzer is completely absorbed. The component of $\vec{\mathbf{E}}_0$ that is parallel to the analyzer axis, $E_0 \cos\theta$, is allowed to pass through the

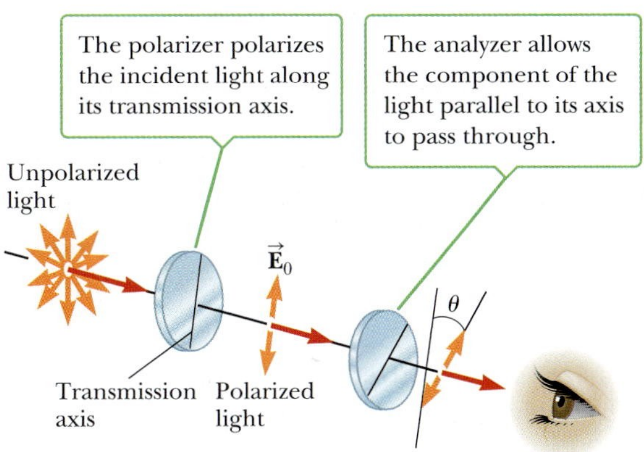
The polarizer polarizes the incident light along its transmission axis.
The analyzer allows the component of the light parallel to its axis to pass through.
Unpolarized light
$\vec{\mathbf{E}}_0$
$\theta$
Transmission axis
Polarized light

**Figure 24.26** Two polarizing sheets whose transmission axes make an angle $\theta$ with each other. Only a fraction of the polarized light incident on the analyzer is transmitted.

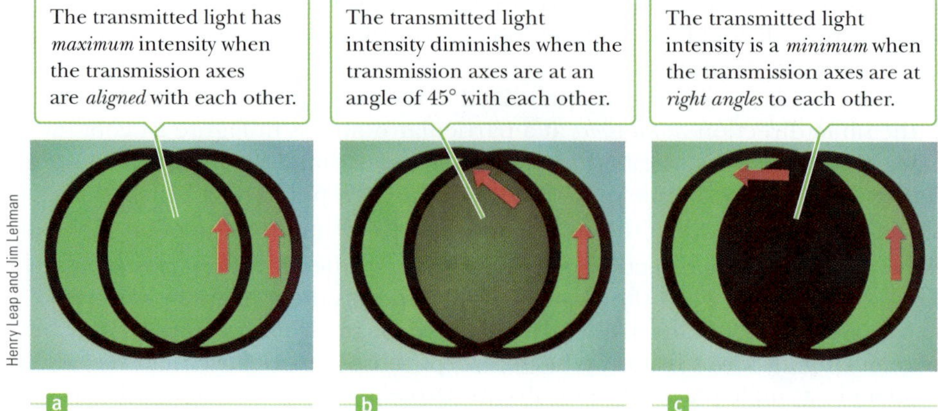

The transmitted light has *maximum* intensity when the transmission axes are *aligned* with each other.
The transmitted light intensity diminishes when the transmission axes are at an angle of 45° with each other.
The transmitted light intensity is a *minimum* when the transmission axes are at *right angles* to each other.
Henry Leap and Jim Lehman
a
b
c


**Figure 24.27** The intensity of light transmitted through two polarizers depends on the relative orientations of their transmission axes.

analyzer. Because the intensity of the transmitted beam varies as the *square* of its amplitude $E$, we conclude that the intensity of the (polarized) beam transmitted through the analyzer varies as

Malus's law ▶

$$I = I_0 \cos^2 \theta \tag{24.13}$$

where $I_0$ is the intensity of the polarized wave incident on the analyzer. This expression, known as **Malus's law**, applies to any two polarizing materials having transmission axes at an angle of $\theta$ to each other. Note from Equation 24.13 that the transmitted intensity is a maximum when the transmission axes are parallel ($\theta = 0$ or $180°$) and is a minimum (complete absorption by the analyzer) when the transmission axes are perpendicular to each other. This variation in transmitted intensity through a pair of polarizing sheets is illustrated in Figure 24.27.

When unpolarized light of intensity $I_0$ is sent through a single ideal polarizer, the transmitted linearly polarized light has intensity $I_0/2$. This fact follows from Malus's law because the average value of $\cos^2 \theta$ is one-half.

## APPLYING PHYSICS 24.7 Polarizing Microwaves

A polarizer for microwaves can be made as a grid of parallel metal wires about 1 cm apart. Is the electric field vector for microwaves transmitted through this polarizer parallel or perpendicular to the metal wires?

**EXPLANATION** Electric field vectors parallel to the metal wires cause electrons in the metal to oscillate parallel to the wires. Thus, the energy from the waves with these electric field vectors is transferred to the metal by accelerating the electrons and is eventually transformed to internal energy through the resistance of the metal. Waves with electric field vectors perpendicular to the metal wires are not able to accelerate electrons and pass through the wires. Consequently, the electric field polarization is perpendicular to the metal wires. ■

## EXAMPLE 24.8 Polarizer

**GOAL** Understand how polarizing materials affect light intensity.

**PROBLEM** Unpolarized light is incident upon three polarizers. The first polarizer has a vertical transmission axis, the second has a transmission axis rotated 30.0° with respect to the first, and the third has a transmission axis rotated 75.0° relative to the first. If the initial light intensity of the beam is $I_b$, calculate the light intensity after the beam passes through **(a)** the second polarizer and **(b)** the third polarizer.

**STRATEGY** After the beam passes through the first polarizer, it is polarized and its intensity is cut in half. Malus's law can then be applied to the second and third polarizers. The angle used in Malus's law must be relative to the immediately preceding transmission axis.

**SOLUTION**

**(a)** Calculate the intensity of the beam after it passes through the second polarizer.

The incident intensity is $I_b/2$. Apply Malus's law to the second polarizer:

$$I_2 = I_0 \cos^2 \theta = \frac{I_b}{2} \cos^2 (30.0°) = \frac{I_b}{2}\left(\frac{\sqrt{3}}{2}\right)^2 = \boxed{\frac{3}{8} I_b}$$

**(b)** Calculate the intensity of the beam after it passes through the third polarizer.

The incident intensity is now $3I_b/8$. Apply Malus's law to the third polarizer.

$$I_3 = I_2 \cos^2 \theta = \frac{3}{8} I_b \cos^2 (45.0°) = \frac{3}{8} I_b \left(\frac{\sqrt{2}}{2}\right)^2 = \boxed{\frac{3}{16} I_b}$$

**REMARKS** Notice that the angle used in part (b) was not 75.0°, but 75.0° − 30.0° = 45.0°. The angle is always with respect to the previous polarizer's transmission axis because the polarizing material physically determines what direction the transmitted electric fields can have.

**QUESTION 24.8** At what angle relative to the previous polarizer must an additional polarizer be placed so as to completely block the light?

**EXERCISE 24.8** The polarizers are rotated so that the second polarizer has a transmission axis of 40.0° with respect to the first polarizer, and the third polarizer has an angle of 90.0° with respect to the first. If $I_b$ is the intensity of the original unpolarized light, what is the intensity of the beam after it passes through (a) the second polarizer and (b) the third polarizer? (c) What is the final transmitted intensity if the second polarizer is removed?

**ANSWERS** (a) $0.293I_b$ (b) $0.121I_b$ (c) 0

## Polarization by Reflection

When an unpolarized light beam is reflected from a surface, the reflected light is completely polarized, partially polarized, or unpolarized, depending on the angle of incidence. If the angle of incidence is either 0° or 90° (a normal or grazing angle), the reflected beam is unpolarized. For angles of incidence between 0° and 90°, however, the reflected light is polarized to some extent. For one particular angle of incidence the reflected beam is completely polarized.

Suppose an unpolarized light beam is incident on a surface, as in Figure 24.28a. The beam can be described by two electric field components, one parallel to the surface (represented by dots) and the other perpendicular to the first component and to the direction of propagation (represented by orange arrows). It is found

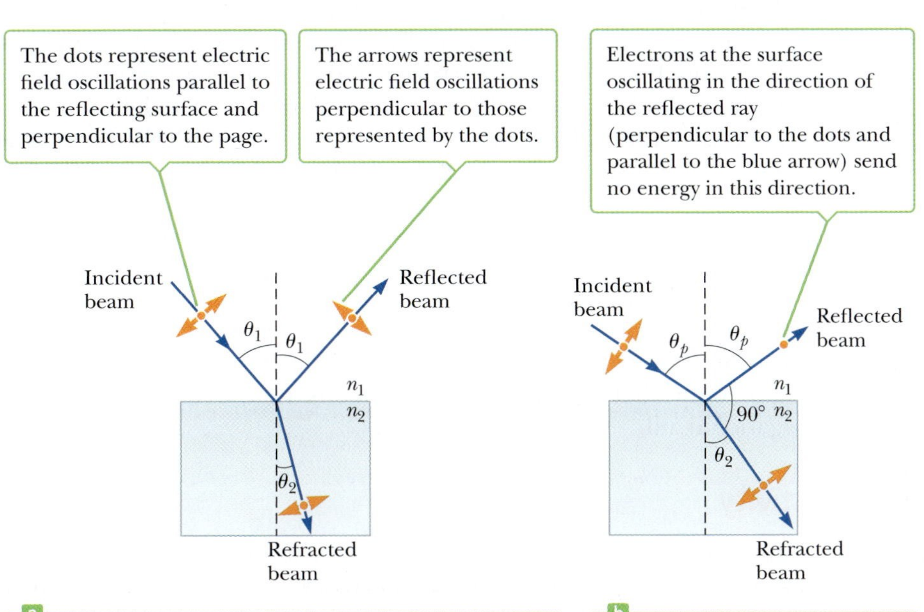

The dots represent electric field oscillations parallel to the reflecting surface and perpendicular to the page.
The arrows represent electric field oscillations perpendicular to those represented by the dots.
Electrons at the surface oscillating in the direction of the reflected ray (perpendicular to the dots and parallel to the blue arrow) send no energy in this direction.
Incident beam
Reflected beam
$\theta_1$
$\theta_1$
$n_1$
$n_2$
$\theta_2$
Refracted beam
a
Incident beam
Reflected beam
$\theta_p$
$\theta_p$
$n_1$
90°
$n_2$
$\theta_2$
Refracted beam
b

**Figure 24.28** (a) When unpolarized light is incident on a reflecting surface, the reflected and refracted beams are partially polarized. (b) The reflected beam is completely polarized when the angle of incidence equals the polarizing angle $\theta_p$, satisfying the equation $n = \tan \theta_p$.

that the parallel component reflects more strongly than the other components, and the result is a partially polarized beam. In addition, the refracted beam is also partially polarized.

Now suppose the angle of incidence, $\theta_1$, is varied until the angle between the reflected and refracted beams is 90° (Fig. 24.28b). At this particular angle of incidence, called the **polarizing angle** $\theta_p$, the reflected beam is completely polarized, with its electric field vector parallel to the surface, while the refracted beam is partially polarized.

An expression relating the polarizing angle to the index of refraction of the reflecting surface can be obtained by the use of Figure 24.28b. From this figure, we see that at the polarizing angle, $\theta_p + 90° + \theta_2 = 180°$, so $\theta_2 = 90° - \theta_p$. Using Snell's law and taking $n_1 = n_{air} = 1.00$ and $n_2 = n$ yields

$$n = \frac{\sin \theta_1}{\sin \theta_2} = \frac{\sin \theta_p}{\sin \theta_2}$$

Because $\sin \theta_2 = \sin (90° - \theta_p) = \cos \theta_p$, the expression for $n$ can be written

Brewster's law ►

$$n = \frac{\sin \theta_p}{\cos \theta_p} = \tan \theta_p \qquad [24.14]$$

Equation 24.14 is called **Brewster's law**, and the polarizing angle $\theta_p$ is sometimes called **Brewster's angle** after its discoverer, Sir David Brewster (1781–1868). For example, Brewster's angle for crown glass (where $n = 1.52$) has the value $\theta_p = \tan^{-1}(1.52) = 56.7°$. Because $n$ varies with wavelength for a given substance, Brewster's angle is also a function of wavelength.

**APPLICATION**
Polaroid Sunglasses

Polarization by reflection is a common phenomenon. Sunlight reflected from water, glass, or snow is partially polarized. If the surface is horizontal, the electric field vector of the reflected light has a strong horizontal component. Sunglasses made of polarizing material reduce the glare, which *is* the reflected light. The transmission axes of the lenses are oriented vertically to absorb the strong horizontal component of the reflected light. Because the reflected light is mostly polarized, most of the glare can be eliminated without removing most of the normal light.

## Polarization by Scattering

When light is incident on a system of particles, such as a gas, the electrons in the medium can absorb and reradiate part of the light. The absorption and reradiation of light by the medium, called **scattering**, is what causes sunlight reaching an observer on Earth from straight overhead to be polarized. You can observe this effect by looking directly up through a pair of sunglasses made of polarizing glass. Less light passes through at certain orientations of the lenses than at others.

Figure 24.29 illustrates how the sunlight becomes polarized. The left side of the figure shows an incident unpolarized beam of sunlight on the verge of striking an air molecule. When the beam strikes the air molecule, it sets the electrons of the molecule into vibration. These vibrating charges act like those in an antenna except that they vibrate in a complicated pattern. The horizontal part of the electric field vector in the incident wave causes the charges to vibrate horizontally, and the vertical part of the vector simultaneously causes them to vibrate vertically. A horizontally polarized wave is emitted by the electrons as a result of their horizontal motion, and a vertically polarized wave is emitted parallel to Earth as a result of their vertical motion.

Scientists have found that bees and homing pigeons use the polarization of sunlight as a navigational aid.

The scattered light traveling perpendicular to the incident light is plane-polarized because the vertical vibrations of the charges in the air molecule send no light in this direction.

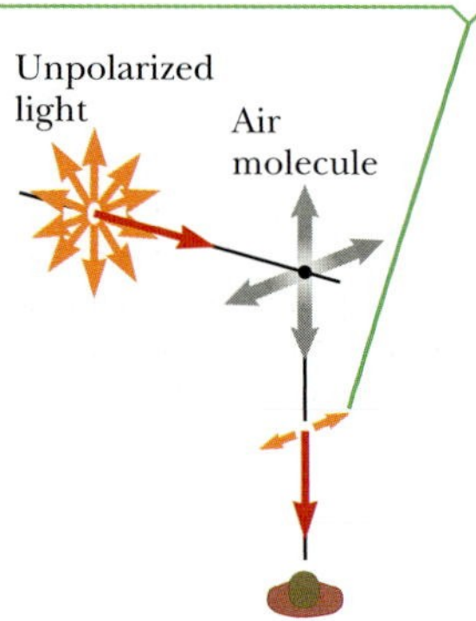


**Figure 24.29** The scattering of unpolarized sunlight by air molecules.

## Optical Activity

Many important practical applications of polarized light involve the use of certain materials that display the property of **optical activity**. A substance is said to be optically active if it rotates the plane of polarization of transmitted light. Suppose

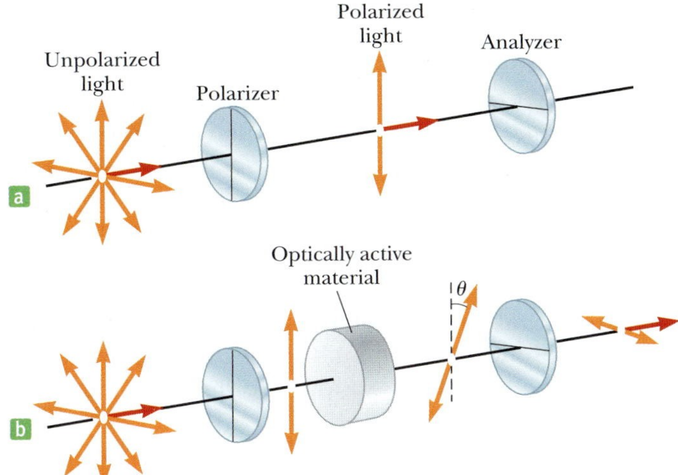


**Figure 24.30** (a) When crossed polarizers are used, none of the polarized light can pass through the analyzer. (b) An optically active material rotates the direction of polarization through the angle $\theta$, enabling some of the polarized light to pass through the analyzer.

unpolarized light is incident on a polarizer from the left, as in Figure 24.30a. The transmitted light is polarized vertically, as shown. If this light is then incident on an analyzer with its axis perpendicular to that of the polarizer, no light emerges from it. If an optically active material is placed between the polarizer and analyzer, as in Figure 24.30b, the material causes the direction of the polarized beam to rotate through the angle $\theta$. As a result, some light is able to pass through the analyzer. The angle through which the light is rotated by the material can be found by rotating the polarizer until the light is again extinguished. It is found that the angle of rotation depends on the length of the sample and, if the substance is in solution, on the concentration. One optically active material is a solution of common sugar, dextrose. A standard method for determining the concentration of a sugar solution is to measure the rotation produced by a fixed length of the solution.

**APPLICATION**

Finding the Concentrations of Solutions by Means of Their Optical Activity

Optical activity occurs in a material because of an asymmetry in the shape of its constituent molecules. For example, some proteins are optically active because of their spiral shapes. Other materials, such as glass and plastic, become optically active when placed under stress. If polarized light is passed through an unstressed piece of plastic and then through an analyzer with an axis perpendicular to that of the polarizer, none of the polarized light is transmitted. If the plastic is placed under stress, however, the regions of greatest stress produce the largest angles of rotation of polarized light, and a series of light and dark bands are observed in the transmitted light. Engineers often use this property in the design of structures ranging from bridges to small tools. A plastic model is built and analyzed under different load conditions to determine positions of potential weakness and failure under stress. If the design is poor, patterns of light and dark bands will indicate the points of greatest weakness, and the design can be corrected at an early stage. Figure 24.31 shows examples of stress patterns in plastic.

**Figure 24.31** A plastic model of an arch structure under load conditions observed between perpendicular polarizers. Such patterns are useful in the optimum design of architectural components.

**APPLICATION**
Liquid Crystal Displays (LCDs)

## Liquid Crystals

An effect similar to rotation of the plane of polarization is used to create the familiar displays on pocket calculators, wristwatches, notebook computers, and so forth. The properties of a unique substance called a liquid crystal make these displays (called LCDs, for *liquid crystal displays*) possible. As its name implies, a **liquid crystal** is a substance with properties intermediate between those of a crystalline solid and those of a liquid; that is, the molecules of the substance are more orderly than those in a liquid, but less orderly than those in a pure crystalline solid. The forces that hold the molecules together in such a state are just barely strong enough to enable the substance to maintain a definite shape, so it is reasonable to call it a solid. Small inputs of mechanical or electrical energy, however, can disrupt these weak bonds and make the substance flow, rotate, or twist.

To see how liquid crystals can be used to create a display, consider Figure 24.32a. The liquid crystal is placed between two glass plates in the pattern shown, and electrical contacts, indicated by the thin lines, are made. When a voltage is applied across any segment in the display, that segment turns dark. In this fashion any number between 0 and 9 can be formed by the pattern, depending on the voltages applied to the seven segments.

To see why a segment can be changed from dark to light by the application of a voltage, consider Figure 24.32b, which shows the basic construction of a portion of the display. The liquid crystal is placed between two glass substrates that are packaged between two pieces of Polaroid material with their transmission axes perpendicular. A reflecting surface is placed behind one of the pieces of Polaroid. First consider what happens when light falls on this package and no voltages are applied to the liquid crystal, as shown in Figure 24.32b. Incoming light is polarized by the polarizer on the left and then falls on the liquid crystal. As the light passes through the crystal, its plane of polarization is rotated by 90°, allowing it to pass through the polarizer on the right. It reflects from the reflecting surface and

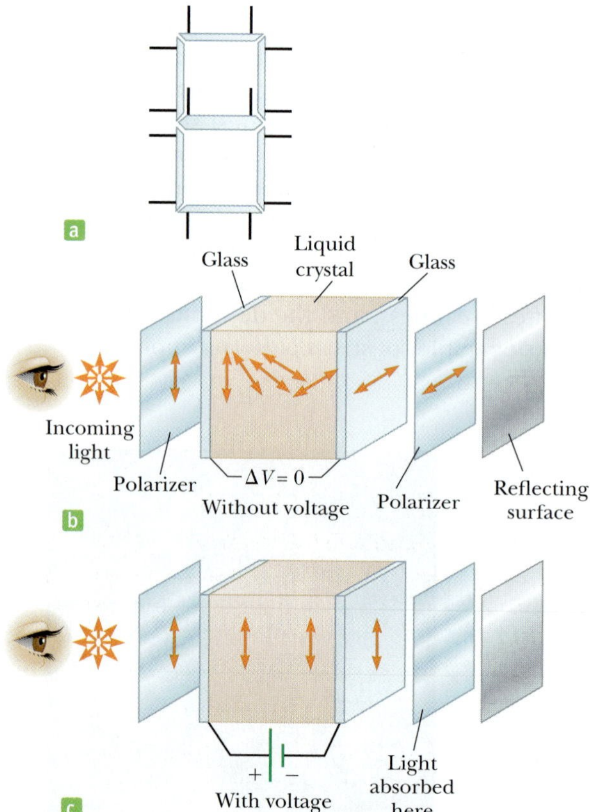


**Figure 24.32** (a) The light-segment pattern of a liquid crystal display. (b) Rotation of a polarized light beam by a liquid crystal when the applied voltage is zero. (c) Molecules of the liquid crystal align with the electric field when a voltage is applied.

retraces its path through the crystal. Thus, an observer to the left of the crystal sees the segment as being bright. When a voltage is applied as in Figure 24.32c, the molecules of the liquid crystal don't rotate the plane of polarization of the light. In this case the light is absorbed by the polarizer on the right, and none is reflected back to the observer to the left of the crystal. As a result, the observer sees this segment as black. Changing the applied voltage to the crystal in a precise pattern at precise times can make the pattern tick off the seconds on a watch, display a letter on a computer display, and so forth.

## SUMMARY

### 24.1 Conditions for Interference

**Interference** occurs when two or more light waves overlap at a given point. A sustained interference pattern is observed if (1) the sources are coherent (that is, they maintain a constant phase relationship with one another), (2) the sources have identical wavelengths, and (3) the superposition principle is applicable.

### 24.2 Young's Double-Slit Experiment

In **Young's double-slit experiment** two slits separated by distance $d$ are illuminated by a single-wavelength light source. An interference pattern consisting of bright and dark fringes is observed on a screen a distance $L$ from the slits. The condition for **bright fringes** (constructive interference) is

$$d \sin \theta_{\text{bright}} = m\lambda \qquad m = 0, \pm 1, \pm 2, \ldots \qquad [24.2]$$

The number $m$ is called the **order number** of the fringe.

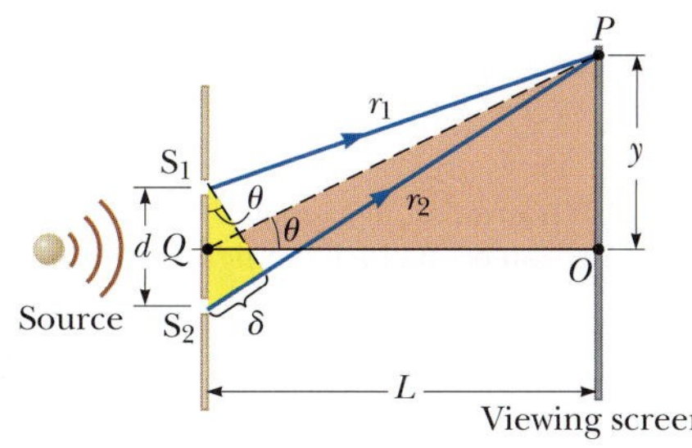


A geometric construction that describes Young's double-slit experiment. (This figure is not drawn to scale.)

The condition for **dark fringes** (destructive interference) is

$$d \sin \theta_{\text{dark}} = (m + \tfrac{1}{2})\lambda \qquad m = 0, \pm 1, \pm 2, \ldots \qquad [24.3]$$

The position $y_m$ of the bright fringes on the screen can be determined by using the relation $\sin \theta \approx \tan \theta = y_m/L$, which is true for small angles. This relation can be substituted into Equations 24.2 and 24.3, yielding the location of the bright fringes:

$$y_{\text{bright}} = \frac{\lambda L}{d} m \qquad m = 0, \pm 1, \pm 2, \ldots \qquad [24.5]$$

A similar expression can be derived for the dark fringes. This equation can be used either to locate the maxima or to determine the wavelength of light by measuring $y_m$.

### 24.3 Change of Phase Due to Reflection

### 24.4 Interference in Thin Films

An electromagnetic wave undergoes a phase change of 180° on reflection from a medium with an index of refraction higher than that of the medium in which the wave is traveling. There is no change when the wave, traveling in a medium with higher index of refraction, reflects from a medium with a lower index of refraction.

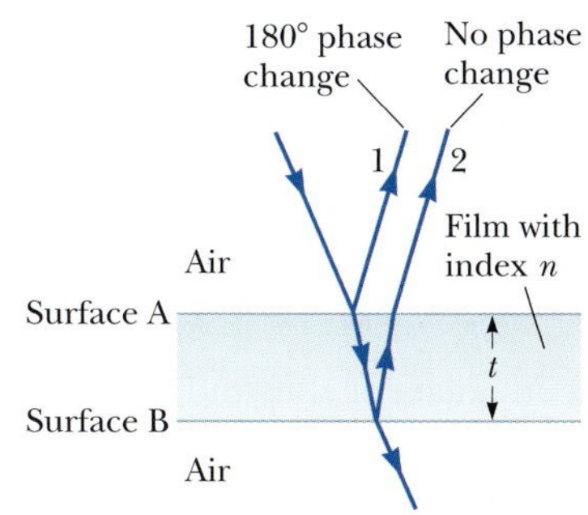


Interference in light reflected from a thin film is due to a combination of rays 1 and 2 reflected from the upper and lower surfaces of the film.

The wavelength $\lambda_n$ of light in a medium with index of refraction $n$ is

$$\lambda_n = \frac{\lambda}{n} \qquad [24.7]$$

where $\lambda$ is the wavelength of the light in free space. Light encountering a thin film of thickness $t$ will reflect off the top and bottom of the film, each ray undergoing a possible phase change as described above. The two rays recombine, and bright and dark fringes will be observed, with the conditions of interference given by the following table:

| **Equation ($m = 0, 1, \ldots$)** | | **1 Phase Reversal** | **0 or 2 Phase Reversals** |
|---|---|---|---|
| $2nt = (m + \frac{1}{2})\lambda$ | [24.9] | Constructive | Destructive |
| $2nt = m\lambda$ | [24.10] | Destructive | Constructive |

### 24.6 Diffraction

### 24.7 Single-Slit Diffraction

Diffraction occurs when waves pass through small openings, around obstacles, or by sharp edges. The **diffraction pattern** produced by a single slit on a distant screen consists of a central bright maximum flanked by less bright fringes alternating with dark regions. The angles $\theta$ at which the diffraction pattern has zero intensity (regions of destructive interference) are described by

$$\sin\theta_{\text{dark}} = m\frac{\lambda}{a} \qquad m = \pm1, \pm2, \pm3, \ldots \qquad [24.11]$$

where $\lambda$ is the wavelength of the light incident on the slit and $a$ is the width of the slit.

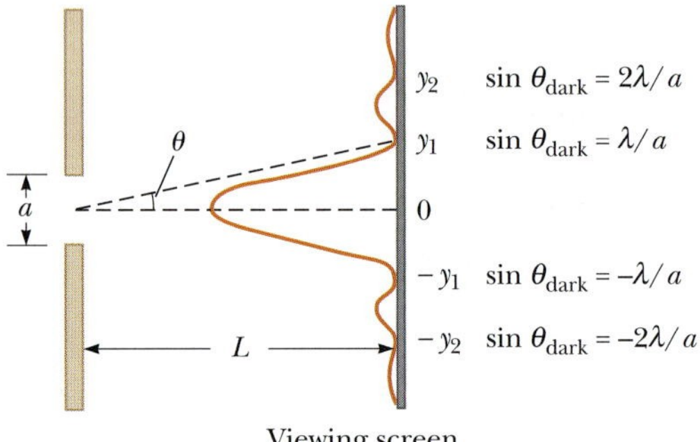


Positions of the minima for the Fraunhofer diffraction pattern of a single slit of width $a$. (This drawing is not to scale.)

## 24.8 The Diffraction Grating

A **diffraction grating** consists of many equally spaced, identical slits. The condition for **maximum intensity** in the interference pattern of a diffraction grating is

$$d\sin\theta_{\text{bright}} = m\lambda \qquad m = 0, \pm1, \pm2, \ldots \qquad [24.12]$$

where $d$ is the spacing between adjacent slits and $m$ is the order number of the diffraction pattern. A diffraction grating can be made by putting a large number of evenly spaced scratches on a glass slide. The number of such lines per centimeter is the inverse of the spacing $d$.

## 24.9 Polarization of Light Waves

Unpolarized light can be polarized by selective absorption, reflection, or scattering. A material can polarize light if it transmits waves having electric field vectors that vibrate in a plane parallel to a certain direction and absorbs waves with electric field vectors vibrating in directions perpendicular to that direction. When unpolarized light passes through a polarizing sheet, its intensity is reduced by half and the light becomes polarized. When this light passes through a second polarizing sheet with transmission axis at an angle of $\theta$ with respect to the transmission axis of the first sheet, the transmitted intensity is given by

$$I = I_0\cos^2\theta \qquad [24.13]$$

where $I_0$ is the intensity of the light after passing through the first polarizing sheet.

In general, light reflected from an amorphous material, such as glass, is partially polarized. Reflected light is completely polarized, with its electric field parallel to the surface, when the angle of incidence produces a 90° angle between the reflected and refracted beams. This angle of incidence, called the **polarizing angle** $\theta_p$, satisfies **Brewster's law,** given by

$$n = \tan\theta_p \qquad [24.14]$$

where $n$ is the index of refraction of the reflecting medium.

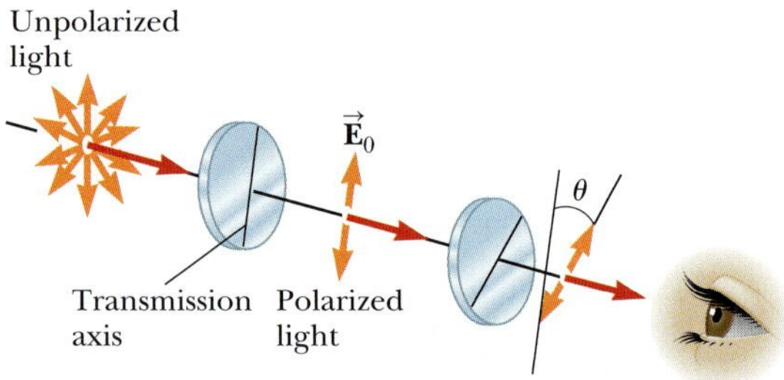


Two polarizing sheets, with transmission axes at angle $\theta$, transmit only a fraction of the incident light.

## WARM-UP EXERCISES

ENHANCED WebAssign The warm-up exercises in this chapter may be assigned online in Enhanced WebAssign.

1. **Physics Review** Two speakers, one at the origin and the other facing it at $x$ = 1.00 m, are driven by the same oscillator at a frequency of 655 Hz. If the speed of sound is 343 m/s, find (a) the wavelength of sound waves produced by the speakers and (b) the location of a point of destructive interference closest to 0.500 m on the interval [0, 0.500 m]. (See Section 14.7.)
2. **Physics Review** (a) Calculate the frequency of monochromatic light having a wavelength in air of $5.30 \times 10^2$ nm. Find (b) the wavelength and (c) the frequency of the same light when inside diamond. (See Sections 21.12 and 22.3.)
3. In a Young's double-slit experiment, 633-nm laser light illuminates two slits separated by a distance of $3.50 \times 10^{-5}$ m. (a) Find the angle at which the first order maximum occurs. (b) If the interference pattern is displayed on a screen 1.50 m away, determine the distance from the centerline to the first order maximum. Find also

(c) the angle at which the second-order ($m = 2$) minimum occurs and (d) the distance of that minimum from the centerline. (See Section 24.2.)

4. For a $4.76 \times 10^{14}$-Hz light wave, determine the wavelength (a) in space where the index of refraction is 1.00 and (b) in water where the index of refraction is 1.33. (See Section 24.4.)

5. Suppose the thin-film coating ($n = 1.17$) on an eyeglass lens ($n = 1.33$) is designed to eliminate reflection of 535-nm light. Determine (a) the number of phase inversions that occur, (b) the wavelength of light in the thin-film coating and (c) the minimum required coating thickness. Take the index of refraction for air to be 1.00. (See Section 24.4.)

6. Suppose 633-nm laser light illuminates a single slit of width $2.50 \times 10^{-4}$ m. Determine the angle from the laser beam's path to the third dark fringe. (See Section 24.7.)

7. Light from a 543-nm laser beam is normally incident on a diffraction grating with $4.50 \times 10^5$ lines/m. Determine (a) the spacing between slits in the grating and (b) the angle to the third-order maximum. (See Section 24.8.)

8. An unpolarized beam of light with intensity of 50.0 W/m$^2$ passes through a polarizer with a vertical transmission axis. Determine the intensity of the light beam after passing through the polarizer. (See Section 24.9.)

9. A beam of light has an intensity of 70.0 W/m$^2$ and is polarized along the horizontal axis. Determine the intensity of the light beam after passing through a polarizer with a transmission axis oriented 30.0° from the horizontal. (See Section 24.9.)

10. Determine Brewster's angle for light reflected from water with an index of refraction equal to 1.33. (See Section 24.9.)

## CONCEPTUAL QUESTIONS

ENHANCED WebAssign The conceptual questions in this chapter may be assigned online in Enhanced WebAssign.

1. Your automobile has two headlights. What sort of interference pattern do you expect to see from them? Why?

2. Holding your hand at arm's length, you can readily block direct sunlight from your eyes. Why can you not block sound from your ears this way?

3. Consider a dark fringe in an interference pattern at which almost no light energy is arriving. Light from both slits is arriving at this point, but the waves cancel. Where does the energy go?

4. If Young's double-slit experiment were performed under water, how would the observed interference pattern be affected?

5. In a laboratory accident, you spill two liquids onto water, neither of which mixes with the water. They both form thin films on the water surface. As the films spread and become very thin, you notice that one film becomes bright and the other black in reflected light. Why might that be?

6. If white light is used in Young's double-slit experiment rather than monochromatic light, how does the interference pattern change?

7. A lens with outer radius of curvature $R$ and index of refraction $n$ rests on a flat glass plate, and the combination is illuminated from white light from above. Is there a dark spot or a light spot at the center of the lens? What does it mean if the observed rings are noncircular?

8. Fingerprints left on a piece of glass such as a windowpane can show colored spectra like that from a diffraction grating. Why?

9. In everyday experience, why are radio waves polarized, whereas light is not?

10. Suppose reflected white light is used to observe a thin, transparent coating on glass as the coating material is gradually deposited by evaporation in a vacuum. Describe some color changes that might occur during the process of building up the thickness of the coating.

11. Would it be possible to place a nonreflective coating on an airplane to cancel radar waves of wavelength 3 cm?

12. Certain sunglasses use a polarizing material to reduce the intensity of light reflected from shiny surfaces, such as water or the hood of a car. What orientation of the transmission axis should the material have to be most effective?

13. Why is it so much easier to perform interference experiments with a laser than with an ordinary light source?

14. A soap film is held vertically in air and is viewed in reflected light as in Figure CQ24.14. Explain why the film appears to be dark at the top.

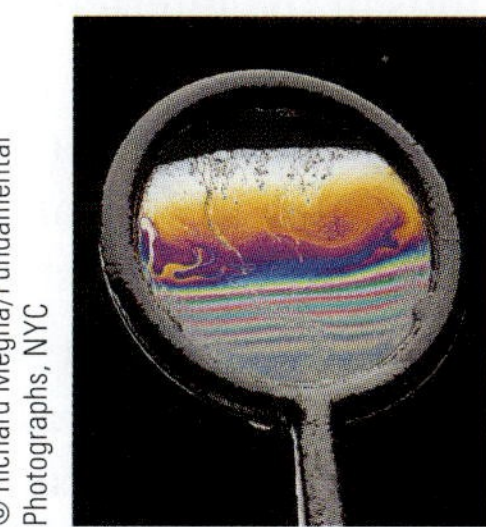



**Figure CQ24.14**

15. A plane monochromatic light wave is incident on a double-slit as illustrated in Figure 24.4. As the slit separation decreases, what happens to the separation between the interference fringes on the screen? (a) It decreases. (b) It increases. (c) It remains the same.

(d) It may increase or decrease, depending on the wavelength of the light. (e) More information is required.

**16.** A plane monochromatic light wave is incident on a double-slit as illustrated in Figure 24.4. If the viewing screen is moved away from the double slit, what happens to the separation between the interference fringes on the screen? (a) It increases. (b) It decreases. (c) It remains the same. (d) It may increase or decrease, depending on the wavelength of the light. (e) More information is required.

## PROBLEMS

ENHANCED WebAssign The problems in this chapter may be assigned online in Enhanced WebAssign.

**1.** denotes straightforward problem; **2.** denotes intermediate problem;
**3.** denotes challenging problem
**1.** denotes full solution available in *Student Solutions Manual/ Study Guide*
**1.** denotes problems most often assigned in Enhanced WebAssign

BIO denotes biomedical problems
GP denotes guided problems
M denotes Master It tutorial available in Enhanced WebAssign
QIC denotes asking for quantitative and conceptual reasoning
S denotes symbolic reasoning problem
W denotes Watch It video solution available in Enhanced WebAssign

### 24.2 Young's Double-Slit Experiment

**1.** W A laser beam is incident on two slits with a separation of 0.200 mm, and a screen is placed 5.00 m from the slits. If the bright interference fringes on the screen are separated by 1.58 cm, what is the wavelength of the laser light?

**2.** In a Young's double-slit experiment, a set of parallel slits with a separation of 0.100 mm is illuminated by light having a wavelength of 589 nm, and the interference pattern is observed on a screen 4.00 m from the slits. (a) What is the difference in path lengths from each of the slits to the location of a third-order bright fringe on the screen? (b) What is the difference in path lengths from the two slits to the location of the third dark fringe on the screen, away from the center of the pattern?

**3.** A pair of narrow, parallel slits separated by 0.250 mm is illuminated by the green component from a mercury vapor lamp ($\lambda = 546.1$ nm). The interference pattern is observed on a screen 1.20 m from the plane of the parallel slits. Calculate the distance (a) from the central maximum to the first bright region on either side of the central maximum and (b) between the first and second dark bands in the interference pattern.

**4.** Light of wavelength 620 nm falls on a double slit, and the first bright fringe of the interference pattern is seen at an angle of 15.0° from the central maximum. Find the separation between the slits.

**5.** In a location where the speed of sound is 354 m/s, a 2 000-Hz sound wave impinges on two slits 30.0 cm apart. (a) At what angle is the first maximum located? (b) If the sound wave is replaced by 3.00-cm microwaves, what slit separation gives the same angle for the first maximum? (c) If the slit separation is 1.00 $\mu$m, what frequency of light gives the same first maximum angle?

**6.** A double slit separated by 0.058 0 mm is placed 1.50 m from a screen. (a) If yellow light of wavelength 588 nm strikes the double slit, what is the separation between the zeroth-order and first-order maxima on the screen? (b) If blue light of wavelength 412 nm strikes the double slit, what is the separation between the second-order and fourth-order maxima?

**7.** M Two radio antennas separated by $d = 300$ m, as shown in Figure P24.7, simultaneously broadcast identical signals at the same wavelength. A car travels due north along a straight line at position $x = 1\,000$ m from the center point between the antennas, and its radio receives the signals. (a) If the car is at the position of the second maximum after that at point $O$ when it has traveled a distance of $y = 400$ m northward, what is the wavelength of the signals? (b) How much farther must the car travel from this position to encounter the next minimum in reception? *Hint:* Do not use the small-angle approximation in this problem.

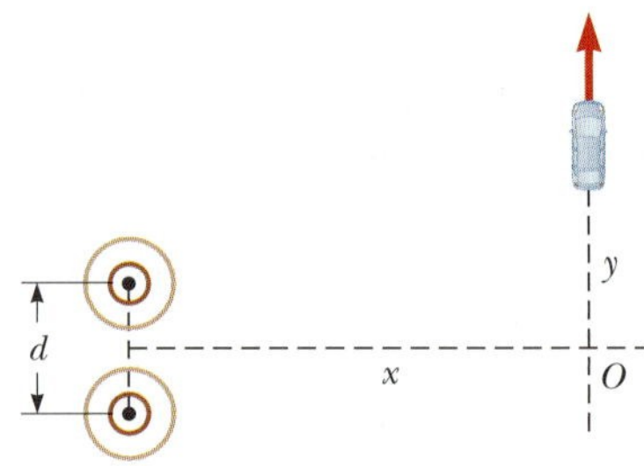


**Figure P24.7**

**8.** Light of wavelength $6.0 \times 10^2$ nm falls on a double slit, and the first bright fringe of the interference pattern is observed to make an angle of 12° with the horizontal. Find the separation between the slits.

**9.** Monochromatic light falls on a screen 1.75 m from two slits separated by 2.10 mm. The first- and second-order bright fringes are separated by 0.552 mm. What is the wavelength of the light?

**10.** A pair of slits, separated by 0.150 mm, is illuminated by light having a wavelength of $\lambda = 643$ nm. An interference pattern is observed on a screen 140 cm from the slits. Consider a point on the screen located at $y = 1.80$ cm from the central maximum of this pattern. (a) What is the path difference $\delta$ for the two slits at the location $y$? (b) Express this path difference in terms of the wavelength. (c) Will the interference correspond to a maximum, a minimum, or an intermediate condition?

**11.** A riverside warehouse has two open doors, as in Figure P24.11. Its interior is lined with a sound-absorbing material. A boat on the river sounds its horn. To person A, the sound is loud and clear. To person B, the sound is barely audible. The principal wavelength of the sound waves is 3.00 m. Assuming person B is at the position of the first minimum, determine the distance between the doors, center to center.

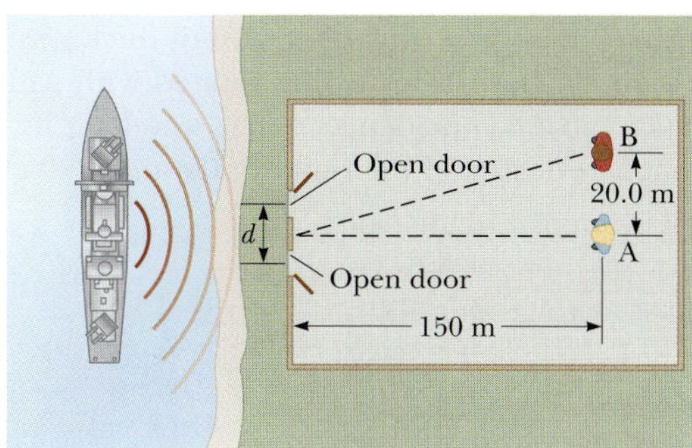


**Figure P24.11**

**12.** **Q|C** A student sets up a double-slit experiment using monochromatic light of wavelength $\lambda$. The distance between the slits is equal to $25\lambda$. (a) Find the angles at which the $m = 1$, 2, and 3 maxima occur on the viewing screen. (b) At what angles do the first three dark fringes occur? (c) Why are the answers so evenly spaced? Is the spacing even for all orders? Explain.

**13.** Radio waves from a star, of wavelength 250 m, reach a radio telescope by two separate paths, as shown in Figure P24.13. One is a direct path to the receiver, which is situated on the edge of a cliff by the ocean. The second is by reflection off the water. The first minimum of destructive interference occurs when the star is $\theta = 25.0°$ above the horizon. Find the height of the cliff. (Assume no phase change on reflection.)

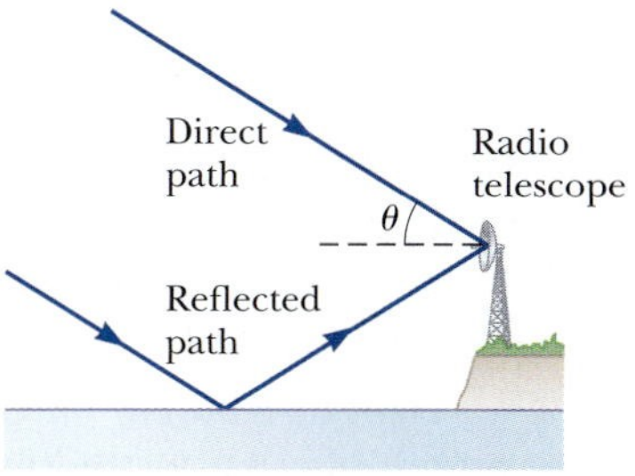


**Figure P24.13**

**14.** **GP** Monochromatic light of wavelength $\lambda$ is incident on a pair of slits separated by $2.40 \times 10^{-4}$ m, and forms an interference pattern on a screen placed 1.80 m away from the slits. The first-order bright fringe is 4.52 mm from the center of the central maximum. (a) Draw a picture, labeling the angle $\theta$ and the legs of the right triangle associated with the first-order bright fringe. (b) Compute the tangent of the angle $\theta$ associated with the first-order bright fringe. (c) Find the angle corresponding to the first-order bright fringe and compute the sine of that angle. Are the sine and tangent of the angle comparable in value? Does your answer always hold true? (d) Calculate the wavelength of the light. (e) Compute the angle of the fifth-order bright fringe. (f) Find its position on the screen.

**15.** Waves from a radio station have a wavelength of 300 m. They travel by two paths to a home receiver 20.0 km from the transmitter. One path is a direct path, and the second is by reflection from a mountain directly behind the home receiver. What is the minimum distance from the mountain to the receiver that produces destructive interference at the receiver? (Assume that no phase change occurs on reflection from the mountain.)

## 24.3 Change of Phase Due to Reflection

## 24.4 Interference in Thin Films

**16.** **Q|C** A soap bubble ($n = 1.33$) having a wall thickness of 120 nm is floating in air. (a) What is the wavelength of the visible light that is most strongly reflected? (b) Explain how a bubble of different thickness could also strongly reflect light of this same wavelength. (c) Find the two smallest film thicknesses larger than the one given that can produce strongly reflected light of this same wavelength.

**17.** A thin layer of liquid methylene iodide ($n = 1.756$) is sandwiched between two flat, parallel plates of glass ($n = 1.50$). What is the minimum thickness of the liquid layer if normally incident light with $\lambda = 6.00 \times 10^2$ nm in air is to be strongly reflected?

**18.** **S** A thin film of oil ($n = 1.25$) is located on smooth, wet pavement. When viewed from a direction perpendicular to the pavement, the film reflects most strongly red light at 640 nm and reflects no green light at 512 nm. (a) What is the minimum thickness of the oil film? (b) Let $m_1$ correspond to the order of the constructive interference and $m_2$ to the order of the destructive interference. Obtain a relationship between $m_1$ and $m_2$ that is consistent with the given data.

**19.** A thin film of glass ($n = 1.52$) of thickness 0.420 $\mu$m is viewed under white light at near normal incidence. What wavelength of *visible* light is most strongly reflected by the film when surrounded by air?

**20.** **W** A transparent oil with index of refraction 1.29 spills on the surface of water (index of refraction 1.33), producing a maximum of reflection with normally incident orange light (wavelength 600 nm in air). Assuming the maximum occurs in the first order, determine the thickness of the oil slick.

**21.** M A possible means for making an airplane invisible to radar is to coat the plane with an antireflective polymer. If radar waves have a wavelength of 3.00 cm and the index of refraction of the polymer is $n = 1.50$, how thick would you make the coating?

**22.** Q|C An oil film ($n = 1.45$) floating on water is illuminated by white light at normal incidence. The film is $2.80 \times 10^2$ nm thick. Find (a) the wavelength and color of the light in the visible spectrum most strongly reflected and (b) the wavelength and color of the light in the visible spectrum most strongly transmitted. Explain your reasoning.

**23.** Q|C Astronomers observe the chromosphere of the Sun with a filter that passes the red hydrogen spectral line of wavelength 656.3 nm, called the $H_\alpha$ line. The filter consists of a transparent dielectric of thickness $d$ held between two partially aluminized glass plates. The filter is kept at a constant temperature. (a) Find the minimum value of $d$ that will produce maximum transmission of perpendicular $H_\alpha$ light if the dielectric has an index of refraction of 1.378. (b) If the temperature of the filter increases above the normal value increasing its thickness, what happens to the transmitted wavelength? (c) The dielectric will also pass what near-visible wavelength? One of the glass plates is colored red to absorb this light.

**24.** Two rectangular optically flat plates ($n = 1.52$) are in contact along one end and are separated along the other end by a 2.00-$\mu$m-thick spacer (Fig. P24.24). The top plate is illuminated by monochromatic light of wavelength 546.1 nm. Calculate the number of dark parallel bands crossing the top plate (including the dark band at zero thickness along the edge of contact between the plates).

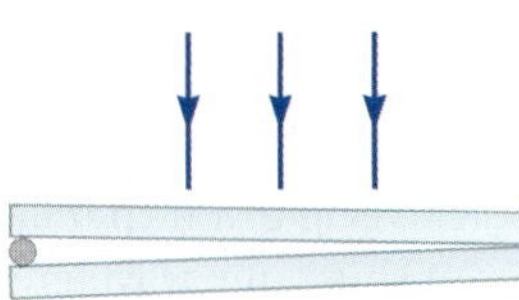

**Figure P24.24**
Problems 24 and 25.

**25.** An investigator finds a fiber at a crime scene that he wishes to use as evidence against a suspect. He gives the fiber to a technician to test the properties of the fiber. To measure the diameter of the fiber, the technician places it between two flat glass plates at their ends as in Figure P24.24. When the plates, of length 14.0 cm, are illuminated from above with light of wavelength 650 nm, she observes bright interference bands separated by 0.580 mm. What is the diameter of the fiber?

**26.** A plano-convex lens with radius of curvature $R = 3.0$ m is in contact with a flat plate of glass. A light source and the observer's eye are both close to the normal, as shown in Figure 24.8a. The radius of the 50th bright Newton's ring is found to be 9.8 mm. What is the wavelength of the light produced by the source?

**27.** A plano-convex lens rests with its curved side on a flat glass surface and is illuminated from above by light of wavelength 500 nm. (See Fig. 24.8.) A dark spot is observed at the center, surrounded by 19 concentric dark rings (with bright rings in between). How much thicker is the air wedge at the position of the 19th dark ring than at the center?

**28.** Nonreflective coatings on camera lenses reduce the loss of light at the surfaces of multilens systems and prevent internal reflections that might mar the image. Find the minimum thickness of a layer of magnesium fluoride ($n = 1.38$) on flint glass ($n = 1.66$) that will cause destructive interference of reflected light of wavelength 550 nm near the middle of the visible spectrum.

**29.** A thin film of glycerin ($n = 1.473$) of thickness 524 nm with air on both sides is illuminated with white light at near normal incidence. What wavelengths will be strongly reflected in the range 300 nm to 700 nm?

**30.** Q|C A lens made of glass ($n_g = 1.52$) is coated with a thin film of $MgF_2$ ($n_s = 1.38$) of thickness $t$. Visible light is incident normally on the coated lens as in Figure P24.30. (a) For what minimum value of $t$ will the reflected light of wavelength 540 nm (in air) be missing? (b) Are there other values of $t$ that will minimize the reflected light at this wavelength? Explain.

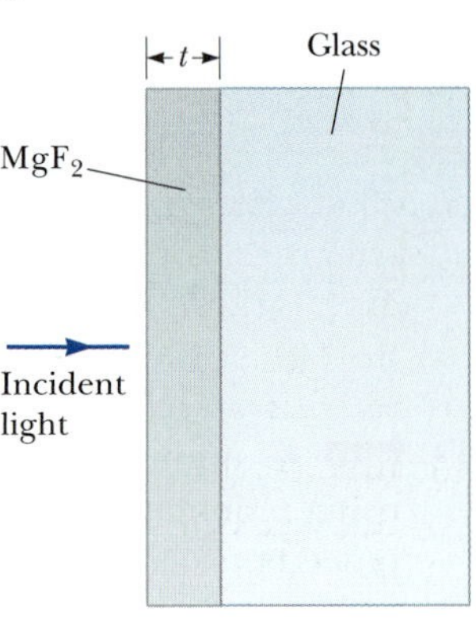


**Figure P24.30**

## 24.7 Single-Slit Diffraction

**31.** Light of wavelength $5.40 \times 10^2$ nm passes through a slit of width 0.200 mm. (a) Find the width of the central maximum on a screen located 1.50 m from the slit. (b) Determine the width of the first-order bright fringe.

**32.** W Light of wavelength 600 nm falls on a 0.40-mm-wide slit and forms a diffraction pattern on a screen 1.5 m away. (a) Find the position of the first dark band on each side of the central maximum. (b) Find the width of the central maximum.

**33.** M Light of wavelength 587.5 nm illuminates a slit of width 0.75 mm. (a) At what distance from the slit should a screen be placed if the first minimum in the diffraction pattern is to be 0.85 mm from the central maximum? (b) Calculate the width of the central maximum.

**34.** Microwaves of wavelength 5.00 cm enter a long, narrow window in a building that is otherwise essentially opaque to the incoming waves. If the window is 36.0 cm wide, what is the distance from the central maximum to the first-order minimum along a wall 6.50 m from the window?

35. A beam of monochromatic light is diffracted by a slit of width 0.600 mm. The diffraction pattern forms on a wall 1.30 m beyond the slit. The width of the central maximum is 2.00 mm. Calculate the wavelength of the light.

36. A screen is placed 50.0 cm from a single slit that is illuminated with light of wavelength 680 nm. If the distance between the first and third minima in the diffraction pattern is 3.00 mm, what is the width of the slit?

37. A slit of width 0.50 mm is illuminated with light of wavelength 500 nm, and a screen is placed 120 cm in front of the slit. Find the widths of the first and second maxima on each side of the central maximum.

38. The second-order dark fringe in a single-slit diffraction pattern is 1.40 mm from the center of the central maximum. Assuming the screen is 85.0 cm from a slit of width 0.800 mm and assuming monochromatic incident light, calculate the wavelength of the incident light.

## 24.8 The Diffraction Grating

39. Three discrete spectral lines occur at angles of 10.1°, 13.7°, and 14.8°, respectively, in the first-order spectrum of a diffraction-grating spectrometer. (a) If the grating has 3 660 slits/cm, what are the wavelengths of the light? (b) At what angles are these lines found in the second-order spectra?

40. Intense white light is incident on a diffraction grating that has 600 lines/mm. (a) What is the highest order in which the complete visible spectrum can be seen with this grating? (b) What is the angular separation between the violet edge (400 nm) and the red edge (700 nm) of the first-order spectrum produced by the grating?

41. M The hydrogen spectrum has a red line at 656 nm and a violet line at 434 nm. What angular separation between these two spectral lines is obtained with a diffraction grating that has 4 500 lines/cm?

42. Consider an array of parallel wires with uniform spacing of 1.30 cm between centers. In air at 20.0°C, ultrasound with a frequency of 37.2 kHz from a distant source is incident perpendicular to the array. (Take the speed of sound to be 343 m/s.) (a) Find the number of directions on the other side of the array in which there is a maximum of intensity. (b) Find the angle for each of these directions relative to the direction of the incident beam.

43. A helium–neon laser ($\lambda$ = 632.8 nm) is used to calibrate a diffraction grating. If the first-order maximum occurs at 20.5°, what is the spacing between adjacent grooves in the grating?

44. W White light is spread out into its spectral components by a diffraction grating. If the grating has 2 000 lines per centimeter, at what angle does red light of wavelength 640 nm appear in the first-order spectrum?

45. Light from an argon laser strikes a diffraction grating that has 5 310 grooves per centimeter. The central and first-order principal maxima are separated by 0.488 m on a wall 1.72 m from the grating. Determine the wavelength of the laser light.

46. S Take red light at 700 nm and violet at 400 nm as the ends of the visible spectrum and consider the continuous spectrum of white light formed by a diffraction grating with a spacing of *d* meters between adjacent lines. Show that the interval $\theta_{v2} \le \theta \le \theta_{r2}$ of the continuous spectrum in second order must overlap the interval $\theta_{v3} \le \theta \le \theta_{r3}$ of the third-order spectrum. *Note:* $\theta_{v2}$ is the angle of the violet light in second order, and $\theta_{r2}$ is the angle made by red light in second order.

47. Sunlight is incident on a diffraction grating that has 2 750 lines/cm. The second-order spectrum over the visible range (400–700 nm) is to be limited to 1.75 cm along a screen that is a distance *L* from the grating. What is the required value of *L*?

48. GP A diffraction grating has $4.200 \times 10^3$ rulings per centimeter. The screen is 2.000 m from the grating. In parts (a) through (e), round each result to four digits, using the rounded values for subsequent calculations. (a) Compute the value of *d*, the distance between adjacent rulings. Express the answer in meters. (b) Calculate the angle of the second-order maximum made by the 589.0-nm wavelength of sodium. (c) Find the position of this second-order maximum on the screen. (d) Repeat parts (b) and (c) for the second-order maximum of the 589.6-nm wavelength of sodium, finding the angle and position on the screen. (e) What is the distance between the two second-order maxima on the screen? (f) Now find the distance between the two second-order maxima without rounding, carrying all digits in calculator memory. How many significant digits are in agreement? What can you conclude about rounding and the use of intermediate answers in this case?

49. Light of wavelength 500 nm is incident normally on a diffraction grating. If the third-order maximum of the diffraction pattern is observed at 32.0°, (a) what is the number of rulings per centimeter for the grating? (b) Determine the total number of primary maxima that can be observed in this situation.

50. Light containing two different wavelengths passes through a diffraction grating with 1 200 slits/cm. On a screen 15.0 cm from the grating, the third-order maximum of the shorter wavelength falls midway between the central maximum and the first side maximum for the longer wavelength. If the neighboring maxima of the longer wavelength are 8.44 mm apart on the screen, what are the wavelengths in the light? *Hint:* Use the small-angle approximation.

## 24.9 Polarization of Light Waves

51. The angle of incidence of a light beam in air onto a reflecting surface is continuously variable. The reflected ray is found to be completely polarized when the angle of incidence is 48.0°. (a) What is the index of refraction of the reflecting material? (b) If some of the incident light (at an angle of 48.0°) passes into the material below the surface, what is the angle of refraction?

52. Unpolarized light passes through two Polaroid sheets. The transmission axis of the analyzer makes an angle of 35.0° with the axis of the polarizer. (a) What fraction of the original unpolarized light is transmitted through the analyzer? (b) What fraction of the original light is absorbed by the analyzer?

53. The index of refraction of a glass plate is 1.52. What is the Brewster's angle when the plate is (a) in air and (b) in water? (See Problem 57.)

54. At what angle above the horizon is the Sun if light from it is completely polarized upon reflection from water?

55. A light beam is incident on a piece of fused quartz ($n = 1.458$) at the Brewster's angle. Find (a) the value of Brewster's angle and (b) the angle of refraction for the transmitted ray.

56. The critical angle for total internal reflection for sapphire surrounded by air is 34.4°. Calculate the Brewster's angle for sapphire if the light is incident from the air.

57. Equation 24.14 assumes the incident light is in air. If the light is incident from a medium of index $n_1$ onto a medium of index $n_2$, follow the procedure used to derive Equation 24.14 to show that $\tan \theta_p = n_2/n_1$.

58. Plane-polarized light is incident on a single polarizing disk, with the direction of $E_0$ parallel to the direction of the transmission axis. Through what angle should the disk be rotated so that the intensity in the transmitted beam is reduced by a factor of (a) 2.00, (b) 4.00, and (c) 6.00?

59. **M** Three polarizing plates whose planes are parallel are centered on a common axis. The directions of the transmission axes relative to the common vertical direction are shown in Figure P24.59. A linearly polarized beam of light with plane of polarization parallel to the vertical reference direction is incident from the left onto the first disk with intensity $I_i = 10.0$ units (arbitrary). Calculate the transmitted intensity $I_f$ when $\theta_1 = 20.0°$, $\theta_2 = 40.0°$, and $\theta_3 = 60.0°$. *Hint:* Make repeated use of Malus's law.

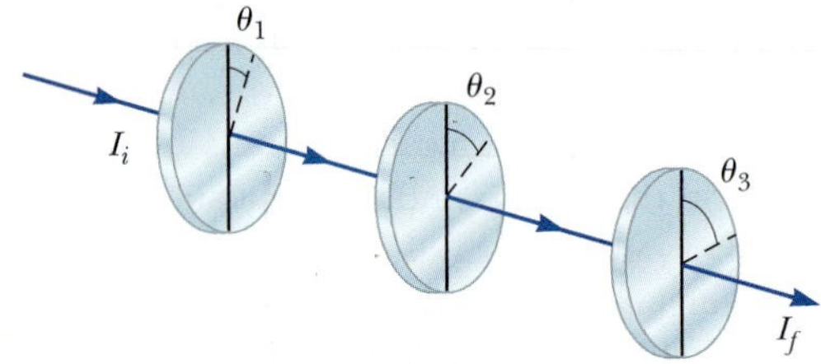


**Figure P24.59** Problems 59 and 70.

60. Light of intensity $I_0$ and polarized parallel to the transmission axis of a polarizer is incident on an analyzer. (a) If the transmission axis of the analyzer makes an angle of 45° with the axis of the polarizer, what is the intensity of the transmitted light? (b) What should the angle between the transmission axes be to make $I/I_0 = 1/3$?

61. **BIO** Light with a wavelength in vacuum of 546.1 nm falls perpendicularly on a biological specimen that is 1.000 $\mu$m thick. The light splits into two beams polarized at right angles, for which the indices of refraction are 1.320 and 1.333, respectively. (a) Calculate the wavelength of each component of the light while it is traversing the specimen. (b) Calculate the phase difference between the two beams when they emerge from the specimen.

## Additional Problems

62. Light from a helium–neon laser ($\lambda = 632.8$ nm) is incident on a single slit. What is the maximum width of the slit for which no diffraction minima are observed?

63. **Q|C** Laser light with a wavelength of 632.8 nm is directed through one slit or two slits and allowed to fall on a screen 2.60 m beyond. Figure P24.63 shows the pattern on the screen, with a centimeter ruler below it. Did the light pass through one slit or two slits? Explain how you can tell. If the answer is one slit, find its width. If the answer is two slits, find the distance between their centers.

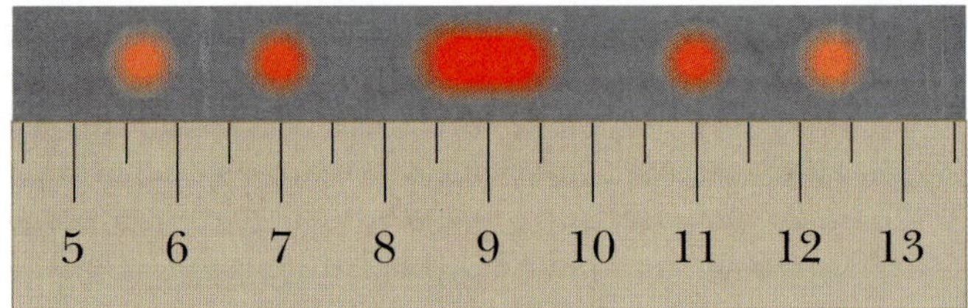


**Figure P24.63**

64. **S** In a Young's interference experiment, the two slits are separated by 0.150 mm and the incident light includes two wavelengths: $\lambda_1 = 540$ nm (green) and $\lambda_2 = 450$ nm (blue). The overlapping interference patterns are observed on a screen 1.40 m from the slits. (a) Find a relationship between the orders $m_1$ and $m_2$ that determines where a bright fringe of the green light coincides with a bright fringe of the blue light. (The order $m_1$ is associated with $\lambda_1$, and $m_2$ is associated with $\lambda_2$.) (b) Find the minimum values of $m_1$ and $m_2$ such that the overlapping of the bright fringes will occur and find the position of the overlap on the screen.

65. Light of wavelength 546 nm (the intense green line from a mercury source) produces a Young's interference pattern in which the second minimum from the central maximum is along a direction that makes an angle of 18.0 min of arc with the axis through the

central maximum. What is the distance between the parallel slits?

66. The two speakers are placed 35.0 cm apart. A single oscillator makes the speakers vibrate in phase at a frequency of 2.00 kHz. At what angles, measured from the perpendicular bisector of the line joining the speakers, would a distant observer hear maximum sound intensity? Minimum sound intensity? (Take the speed of sound to be 340 m/s.)

67. Interference effects are produced at point $P$ on a screen as a result of direct rays from a 500-nm source and reflected rays off a mirror, as shown in Figure P24.67. If the source is $L = 100$ m to the left of the screen and $h = 1.00$ cm above the mirror, find the distance $y$ (in millimeters) to the first dark band above the mirror.

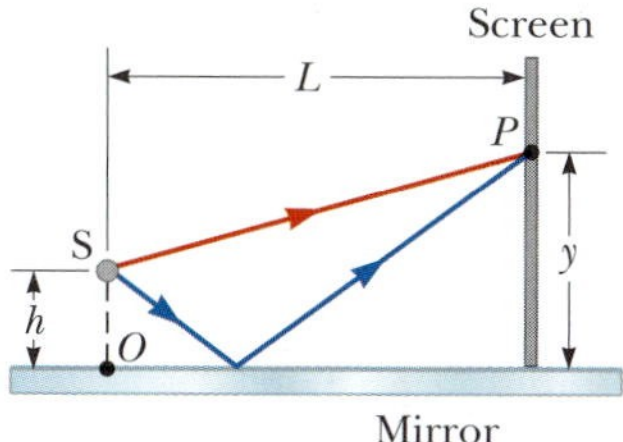


**Figure P24.67**

68. BIO Many cells are transparent and colorless. Structures of great interest in biology and medicine can be practically invisible to ordinary microscopy. An *interference microscope* reveals a difference in refractive index as a shift in interference fringes to indicate the size and shape of cell structures. The idea is exemplified in the following problem: An air wedge is formed between two glass plates in contact along one edge and slightly separated at the opposite edge. When the plates are illuminated with monochromatic light from above, the reflected light has 85 dark fringes. Calculate the number of dark fringes that appear if water ($n = 1.33$) replaces the air between the plates.

69. Figure P24.69 shows a radio-wave transmitter and a receiver, both $h = 50.0$ m above the ground and $d = 600$ m apart. The receiver can receive signals directly from the transmitter and indirectly from signals that bounce off the ground. If the ground is level between the transmitter and receiver and a $\lambda/2$ phase shift occurs upon reflection, determine the longest wavelengths that interfere (a) constructively and (b) destructively.

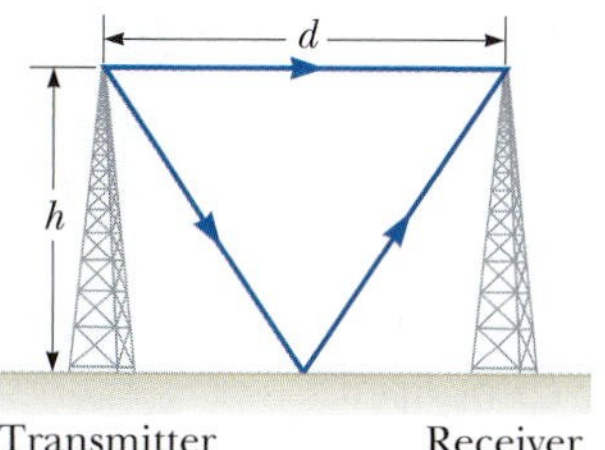


**Figure P24.69**

70. Three polarizers, centered on a common axis and with their planes parallel to one another, have transmission axes oriented at angles of $\theta_1$, $\theta_2$, and $\theta_3$ from the vertical, as shown in Figure P24.59. Light of intensity $I_i$, polarized with its plane of polarization oriented vertically, is incident from the left onto the first polarizer. What is the ratio $I_f/I_i$ of the final transmitted intensity to the incident intensity if (a) $\theta_1 = 45°$, $\theta_2 = 90°$, and $\theta_3 = 0°$? (b) $\theta_1 = 0°$, $\theta_2 = 45°$, and $\theta_3 = 90°$?

71. The transmitting antenna on a submarine is 5.00 m above the water when the ship surfaces. The captain wishes to transmit a message to a receiver on a 90.0-m-tall cliff at the ocean shore. If the signal is to be completely polarized by reflection off the ocean surface, how far must the ship be from the shore?

72. S A plano-convex lens (flat on one side, convex on the other) with index of refraction $n$ rests with its curved side (radius of curvature $R$) on a flat glass surface of the same index of refraction with a film of index $n_{\text{film}}$ between them. The lens is illuminated from above by light of wavelength $\lambda$. Show that the dark Newton rings that appear have radii of

$$r \approx \sqrt{m\lambda R/n_{\text{film}}}$$

where $m$ is an integer.

73. A diffraction pattern is produced on a screen 140 cm from a single slit, using monochromatic light of wavelength 500 nm. The distance from the center of the central maximum to the first-order maximum is 3.00 mm. Calculate the slit width. *Hint:* Assume that the first-order maximum is halfway between the first- and second-order minima.

74. A flat piece of glass is supported horizontally above the flat end of a 10.0-cm-long metal rod that has its lower end rigidly fixed. The thin film of air between the rod and the glass is observed to be bright when illuminated by light of wavelength 500 nm. As the temperature is slowly increased by 25.0°C, the film changes from bright to dark and back to bright 200 times. What is the coefficient of linear expansion of the metal?

The twin Keck telescopes atop the summit of Hawaii's Mauna Kea volcano are the world's largest optical and infrared telescopes. Each telescope has a primary mirror 10 m in diameter comprised of 36 hexagonal segments that function together as a single piece of reflective glass. Far from city lights, high on a dormant volcano in dry, unpolluted air, the telescopes are ideally placed to probe the mysteries of the Universe.

# 25 Optical Instruments



We use devices made from lenses, mirrors, and other optical components every time we put on a pair of eyeglasses or contact lenses, take a photograph, look at the sky through a telescope, and so on. In this chapter we examine how optical instruments work. For the most part, our analyses involve the laws of reflection and refraction and the procedures of geometric optics. To explain certain phenomena, however, we must use the wave nature of light.

## 25.1 The Camera

**LEARNING OBJECTIVES**

1. Describe the components and operation of a single-lens camera.
2. Define the *f*-number and discuss its effects on a camera's image.

The single-lens photographic camera is a simple optical instrument having the features shown in Figure 25.1. It consists of an opaque box, a converging lens that produces a real image, and a photographic film behind the lens to receive the image. Digital cameras differ in that the image is formed on a charge-coupled device (CCD) or a complementary metal-oxide semiconductor (CMOS) sensor instead of on film. Both the CCD and the CMOS image sensors convert the image into digital form, which can then be stored in the camera's memory.

Focusing a camera is accomplished by varying the distance between the lens and sensor, with an adjustable bellows in antique cameras and other mechanisms in contemporary models. For proper focusing, which leads to sharp images, the lens-to-sensor distance depends on the object distance as well as on the focal length of the lens. The shutter, located behind the lens, is a mechanical device that is opened for selected time intervals. With this arrangement, moving objects can be photographed by using short exposure times and dark scenes (with low light levels) by using long exposure times. If this adjustment were not available, it would be impossible to take stop-action photographs. A rapidly moving vehicle, for example, could move far enough while the shutter was open to produce a blurred image. Another major cause of blurred images is movement of the *camera* while the shutter is open. To prevent such movement, you should mount the camera on a tripod or use short exposure times. Typical shutter speeds (that is, exposure times) are 1/30 s, 1/60 s, 1/125 s, and 1/250 s. Stationary objects are often shot with a shutter speed of 1/60 s.

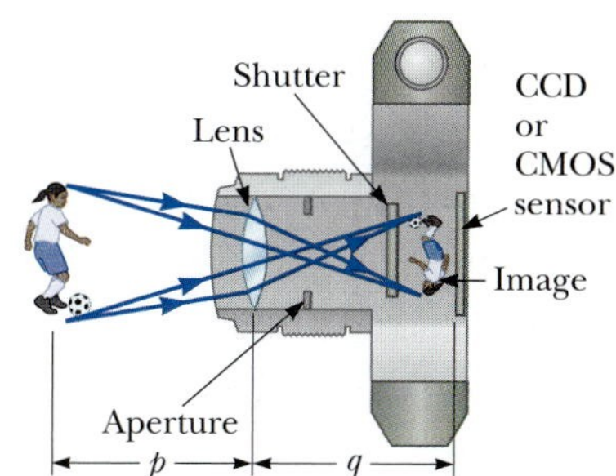


**Figure 25.1** Cross-sectional view of a simple digital camera. The CCD or CMOS image sensor is the light-sensitive component of the camera. In a nondigital camera the light from the lens falls onto photographic film. In reality $p >> q$.

Most cameras also have an aperture of adjustable diameter to further control the intensity of the light reaching the sensor. When an aperture of small diameter is used, only light from the central portion of the lens reaches the sensor, reducing spherical aberration.

The intensity $I$ of the light reaching the sensor is proportional to the area of the lens. Because this area in turn is proportional to the square of the lens diameter $D$, the intensity is also proportional to $D^2$. Light intensity is a measure of the rate at which energy is received by the sensor per unit area of the image. Because the area of the image is proportional to $q^2$ in Figure 25.1 and $q \approx f$ (when $p >> f$, so that $p$ can be approximated as infinite), we conclude that the intensity is also proportional to $1/f^2$. Therefore, $I \propto D^2/f^2$. The brightness of the image formed on the sensor depends on the light intensity, so we see that it ultimately depends on both the focal length $f$ and diameter $D$ of the lens. The ratio $f/D$ is called the ***f*-number** (or focal ratio) of a lens:

$$f\text{-number} \equiv \frac{f}{D} \qquad [25.1]$$

The *f*-number is often given as a description of the lens "speed." A lens with a low *f*-number is a "fast" lens. Extremely fast lenses, which have an *f*-number as low as approximately 1.2, are expensive because of the difficulty of keeping aberrations acceptably small with light rays passing through a large area of the lens. Camera lenses are often marked with a range of *f*-numbers, such as 1.4, 2, 2.8, 4, 5.6, 8, and 11. Any one of these settings can be selected by adjusting the aperture, which changes the value of $D$. Increasing the setting from one *f*-number to the next-higher value (for example, from 2.8 to 4) decreases the area of the aperture by a factor of 2. The lowest *f*-number setting on a camera corresponds to a wide-open aperture and the use of the maximum possible lens area.

Simple cameras usually have a fixed focal length and fixed aperture size, with an *f*-number of about 11. This high value for the *f*-number allows for a large **depth of field** and means that objects at a wide range of distances from the lens form reasonably sharp images on the sensor. In other words, the camera doesn't have to be focused. Most cameras with variable *f*-numbers adjust them automatically.

## 25.2 The Eye BIO

### LEARNING OBJECTIVES

1. Identify the essential parts of the eye and discuss the eye's ability to adjust to varying light conditions.
2. Define near point and far point and discuss the focusing process of accommodation.
3. Describe several focusing conditions of the eye and the characteristics of optical lenses used to correct them.
4. Define the power of a lens.
5. Calculate the required power of corrective optical lenses.

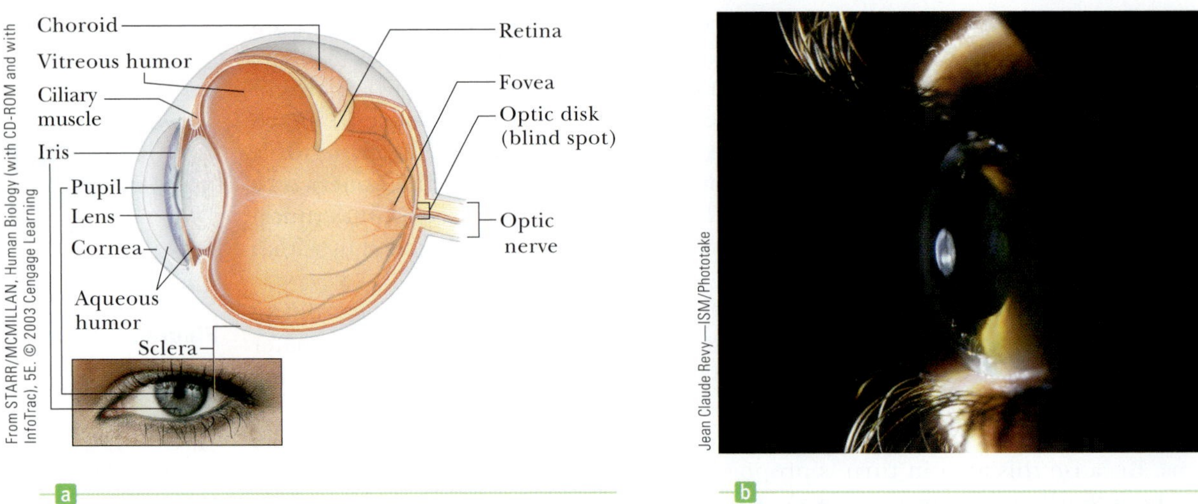


**Figure 25.2** (a) Essential parts of the eye. Can you correlate the essential parts of the eye with those of the simple camera in Figure 25.1? (b) Close-up photograph of the human cornea.

Like a camera, a normal eye focuses light and produces a sharp image. The mechanisms by which the eye controls the amount of light admitted and adjusts to produce correctly focused images, however, are far more complex, intricate, and effective than those in even the most sophisticated camera. In all respects the eye is a physiological wonder.

Figure 25.2a shows the essential parts of the eye. Light entering the eye passes through a transparent structure called the *cornea,* behind which are a clear liquid (the *aqueous humor*), a variable aperture (the *pupil,* which is an opening in the *iris*), and the *crystalline lens.* Most of the refraction occurs at the outer surface of the eye, where the cornea is covered with a film of tears. Relatively little refraction occurs in the crystalline lens because the aqueous humor in contact with the lens has an average index of refraction close to that of the lens. The iris, which is the colored portion of the eye, is a muscular diaphragm that controls pupil size. The iris regulates the amount of light entering the eye by dilating the pupil in low-light conditions and contracting the pupil under conditions of bright light. The *f*-number of the eye ranges from about 2.8 to 16.

The cornea–lens system focuses light onto the back surface of the eye—the *retina*—which consists of millions of sensitive receptors called *rods* and *cones.* When stimulated by light, these structures send impulses to the brain via the optic nerve, converting them into our conscious view of the world. The process by which the brain performs this conversion is not well understood and is the subject of much speculation and research. Unlike film in a camera, the rods and cones chemically adjust their sensitivity according to the prevailing light conditions. This adjustment, which takes about 15 minutes, is responsible for the experience of "getting used to the dark" in such places as movie theaters. Iris aperture control, which takes less than a second, helps protect the retina from overload in the adjustment process.

The eye focuses on an object by varying the shape of the pliable crystalline lens through an amazing process called **accommodation**. An important component in accommodation is the *ciliary muscle,* which is situated in a circle around the rim of the lens. Thin filaments, called *zonules,* run from this muscle to the edge of the lens. When the eye is focused on a distant object, the ciliary muscle is relaxed, tightening the zonules that attach the ciliary muscle to the edge of the lens. The force of the zonules causes the lens to flatten, increasing its focal length. For an object distance of infinity, the focal length of the eye is equal to the fixed distance between lens and retina, about 1.7 cm. The eye focuses on nearby objects by tensing the ciliary muscle, which relaxes the zonules. This action allows the lens to bulge a bit and its focal length decreases, resulting in the image being focused on the retina. All these lens adjustments take place so swiftly that we are not even aware of the change. In this respect even the finest electronic camera is a toy compared with the eye.

There is a limit to accommodation because objects that are very close to the eye produce blurred images. The **near point** is the closest distance for which the lens can accommodate to focus light on the retina. This distance usually increases with age and has an average value of 25 cm. Typically, at age 10 the near point of the eye is about 18 cm. This increases to about 25 cm at age 20, 50 cm at age 40, and 500 cm or greater at age 60. The **far point** of the eye represents the farthest distance for which the lens of the relaxed eye can focus light on the retina. A person with normal vision is able to see very distant objects, such as the Moon, and so has a far point at infinity.

## Conditions of the Eye

When the eye suffers a mismatch between the focusing power of the lens–cornea system and the length of the eye, so that light rays reach the retina before they converge to form an image, as in Figure 25.3a, the condition is known as **farsightedness** (or *hyperopia*). A farsighted person can usually see faraway objects clearly but not nearby objects. Although the near point of a normal eye is approximately 25 cm, the near point of a farsighted person is much farther than that. The eye of a farsighted person tries to focus by accommodation, by shortening its focal length. Accommodation works for distant objects, but because the focal length of the farsighted eye is longer than normal, the light from nearby objects can't be brought to a sharp focus before it reaches the retina, causing a blurred image. The condition can be corrected by placing a converging lens in front of the eye, as in Figure 25.3b. The lens refracts the incoming rays more toward the principal axis before entering the eye, allowing them to converge and focus on the retina.

**Nearsightedness** (or *myopia*) is another mismatch condition in which a person is able to focus on nearby objects, but not faraway objects. In the case of *axial myopia*, nearsightedness is caused by the lens being too far from the retina. It is also possible to have *refractive myopia*, in which the lens–cornea system is too powerful for the normal length of the eye. The far point of the nearsighted eye is not at infinity and may be less than 1 meter. The maximum focal length of the nearsighted eye is insufficient to produce a sharp image on the retina, and rays from a distant object converge to a focus in front of the retina. They then continue past that point, diverging before they finally reach the retina and produce a blurred image (Fig. 25.4a, page 874).

**BIO APPLICATION**

Using Optical Lenses to Correct for Defects

Nearsightedness can be corrected with a diverging lens, as shown in Figure 25.4b. The lens refracts the rays away from the principal axis before they enter the eye, allowing them to focus on the retina.

Beginning with middle age, most people lose some of their accommodation ability as the ciliary muscle weakens and the lens hardens. Unlike farsightedness, which is a

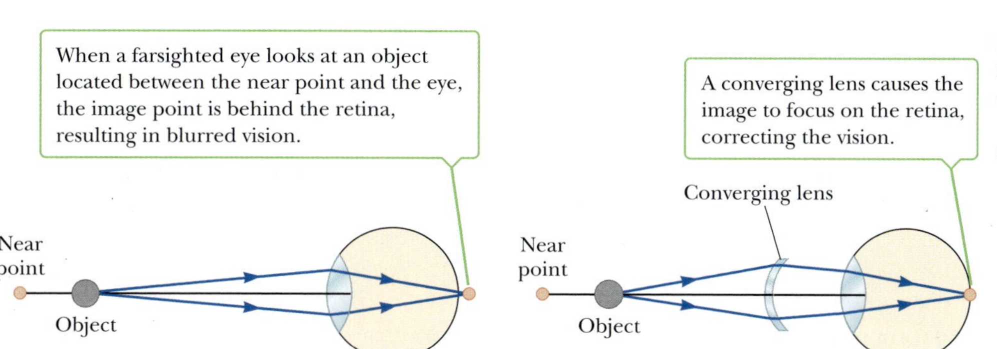
When a farsighted eye looks at an object located between the near point and the eye, the image point is behind the retina, resulting in blurred vision.
Near point
Object
a
A converging lens causes the image to focus on the retina, correcting the vision.
Converging lens
Near point
Object
b

**Figure 25.3** (a) An uncorrected farsighted eye. (b) A farsighted eye corrected with a converging lens. (The object is assumed to be very small in these figures.)

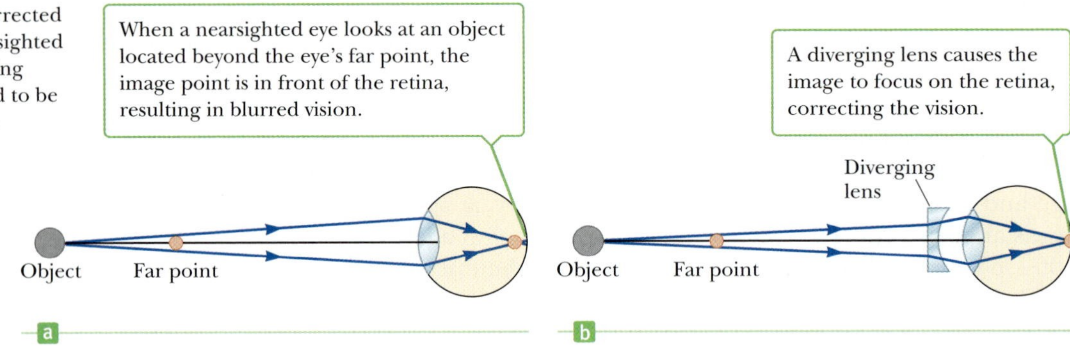
When a nearsighted eye looks at an object located beyond the eye's far point, the image point is in front of the retina, resulting in blurred vision.
A diverging lens causes the image to focus on the retina, correcting the vision.
Diverging lens
Object
Far point
Object
Far point
a
b

**Figure 25.4** (a) An uncorrected nearsighted eye. (b) A nearsighted eye corrected with a diverging lens. (The object is assumed to be very small in these figures.)

mismatch of focusing power and eye length, **presbyopia** (literally, "old-age vision") is due to a reduction in accommodation ability. This means the cornea and lens aren't able to bring nearby objects into focus on the retina. The symptoms are the same as with farsightedness, and the condition can be corrected with converging lenses.

In the eye defect known as **astigmatism**, light from a point source produces a line image on the retina. This condition arises when either the cornea or the lens (or both) is not perfectly symmetric. Astigmatism can be corrected with lenses having different curvatures in two mutually perpendicular directions.

Optometrists and ophthalmologists usually prescribe lenses measured in **diopters**:

> The **power** $P$ of a lens in diopters equals the inverse of the focal length in meters: $P = 1/f$.

For example, a converging lens with a focal length of +20 cm has a power of +5.0 diopters, and a diverging lens with a focal length of −40 cm has a power of −2.5 diopters. (Although the symbol $P$ is the same as for mechanical power, there is no relationship between the two concepts.)

The position of the lens relative to the eye causes differences in power, but they usually amount to less than one-quarter diopter, which isn't noticeable to most patients. As a result, practicing optometrists deal in increments of one-quarter diopter. Neglecting the eye–lens distance is equivalent to doing the calculation for a contact lens, which rests directly on the eye.

## EXAMPLE 25.1 Prescribing a Corrective Lens for a Farsighted Patient BIO

**GOAL** Apply geometric optics to correct farsightedness.

**PROBLEM** The near point of a patient's eye is 50.0 cm. **(a)** What focal length must a corrective lens have to enable the eye to clearly see an object 25.0 cm away? Neglect the eye–lens distance. **(b)** What is the power of this lens? **(c)** Repeat the problem, taking into account that, for typical eyeglasses, the corrective lens is 2.00 cm in front of the eye.

**STRATEGY** This problem requires substitution into the thin-lens equation (Eq. 23.11) and then using the definition of lens power in terms of diopters. The object is at 25.0 cm, but the lens must form an image at the patient's near point, 50.0 cm, the closest point at which the patient's eye can see clearly. In part (c) 2.00 cm must be subtracted from both the object distance and the image distance to account for the position of the lens.

**SOLUTION**

**(a)** Find the focal length of the corrective lens, neglecting its distance from the eye.

Apply the thin-lens equation:

$$\frac{1}{p} + \frac{1}{q} = \frac{1}{f}$$

Substitute $p = 25.0$ cm and $q = -50.0$ cm (the latter is negative because the image must be virtual) on the same side of the lens as the object:

$$\frac{1}{25.0 \text{ cm}} + \frac{1}{-50.0 \text{ cm}} = \frac{1}{f}$$

Solve for $f$. The focal length is positive, corresponding to a converging lens.

$$f = \boxed{50.0\text{ cm}}$$

**(b)** What is the power of this lens?

The power is the reciprocal of the focal length in meters:

$$P = \frac{1}{f} = \frac{1}{0.500\text{ m}} = \boxed{+\,2.00\text{ diopters}}$$

**(c)** Repeat the problem, noting that the corrective lens is actually 2.00 cm in front of the eye.

Substitute the corrected values of $p$ and $q$ into the thin-lens equation:

$$\frac{1}{p} + \frac{1}{q} = \frac{1}{23.0\text{ cm}} + \frac{1}{(-48.0\text{ cm})} = \frac{1}{f}$$

$$f = 44.2\text{ cm}$$

Compute the power:

$$P = \frac{1}{f} = \frac{1}{0.442\text{ m}} = \boxed{+\,2.26\text{ diopters}}$$

**REMARKS** Notice that the calculation in part (c), which doesn't neglect the eye–lens distance, results in a difference of 0.26 diopter.

**QUESTION 25.1** True or False: The larger the distance to a near point, the larger the power of the required corrective lens.

**EXERCISE 25.1** Suppose a lens is placed in a device that determines its power as 2.75 diopters. Find (a) the focal length of the lens and (b) the minimum distance at which a patient will be able to focus on an object if the patient's near point is 60.0 cm. Neglect the eye–lens distance.

**ANSWERS** (a) 36.4 cm (b) 22.7 cm

## EXAMPLE 25.2 A Corrective Lens for Nearsightedness

**GOAL** Apply geometric optics to correct nearsightedness.

**PROBLEM** A particular nearsighted patient can't see objects clearly when they are beyond 25 cm (the far point of the eye). **(a)** What focal length should the prescribed contact lens have to correct this problem? **(b)** Find the power of the lens, in diopters. Neglect the distance between the eye and the corrective lens.

**STRATEGY** The purpose of the lens in this instance is to take objects at infinity and create an image of them at the patient's far point. Apply the thin-lens equation.

**SOLUTION**

**(a)** Find the focal length of the corrective lens.

Apply the thin-lens equation for an object at infinity and image at 25.0 cm:

$$\frac{1}{p} + \frac{1}{q} = \frac{1}{\infty} + \frac{1}{(-25.0\text{ cm})} = \frac{1}{f}$$

$$f = \boxed{-25.0\text{ cm}}$$

**(b)** Find the power of the lens in diopters.

$$P = \frac{1}{f} = \frac{1}{-0.250\text{ m}} = \boxed{-\,4.00\text{ diopters}}$$

**REMARKS** The focal length is negative, consistent with a diverging lens. Notice that the power is also negative and has the same numeric value as the sum on the left side of the thin-lens equation.

**QUESTION 25.2** True or False: The shorter the distance to a patient's far point, the more negative the power of the required corrective lens.

**EXERCISE 25.2** (a) What power lens would you prescribe for a patient with a far point of 35.0 cm? Neglect the eye–lens distance. (b) Repeat, assuming an eye-corrective lens distance of 2.00 cm.

**ANSWERS** (a) −2.86 diopters (b) −3.03 diopters

**APPLYING PHYSICS 25.1 Vision of the Invisible Man**

A classic science fiction story, *The Invisible Man* by H. G. Wells, tells of a man who becomes invisible by changing the index of refraction of his body to that of air. Students who know how the eye works have criticized this story; they claim that the invisible man would be unable to see. On the basis of your knowledge of the eye, would he be able to see?

**EXPLANATION** He wouldn't be able to see. For the eye to see an object, incoming light must be refracted at the cornea and lens to form an image on the retina. If the cornea and lens have the same index of refraction as air, refraction can't occur and an image wouldn't be formed. ■

**Quick Quiz**

**25.1** Two campers wish to start a fire during the day. One camper is nearsighted and one is farsighted. Whose glasses should be used to focus the Sun's rays onto some paper to start the fire? (a) either camper's (b) the nearsighted camper's (c) the farsighted camper's

## 25.3 The Simple Magnifier

**LEARNING OBJECTIVES**

1. Define the angular magnification of a simple magnifier.
2. Apply geometric optics to systems involving a simple magnifier.

The size of the image formed on the retina depends on the angle $\theta$ subtended at the eye.

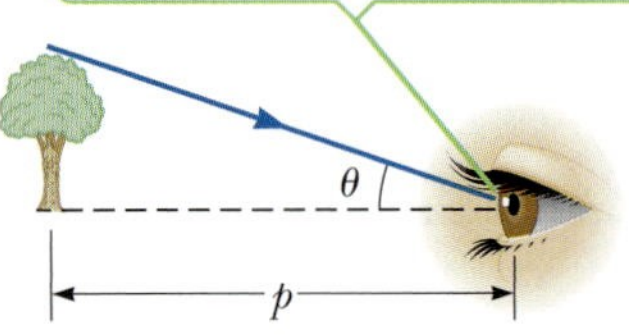


**Figure 25.5** An observer looks at an object at distance $p$.

The **simple magnifier** is one of the most basic of all optical instruments because it consists only of a single converging lens. As the name implies, this device is used to increase the apparent size of an object. Suppose an object is viewed at some distance $p$ from the eye, as in Figure 25.5. Clearly, the size of the image formed at the retina depends on the angle $\theta$ subtended by the object at the eye. As the object moves closer to the eye, $\theta$ increases and a larger image is observed. A normal eye, however, can't focus on an object closer than about 25 cm, the near point (Fig. 25.6a). (Try it!) Therefore, $\theta$ is a maximum at the near point.

To further increase the apparent angular size of an object, a converging lens can be placed in front of the eye with the object positioned at point $O$, just inside the focal point of the lens, as in Figure 25.6b. At this location, the lens forms a virtual, upright, and enlarged image, as shown. The lens allows the object to be viewed closer to the eye than is otherwise possible. We define the **angular magnification** $m$ as the ratio of the angle subtended by a small object when the lens is in use (angle $\theta$ in Fig. 25.6b) to the angle subtended by the object placed at the near point with no lens in use (angle $\theta_0$ in Fig. 25.6a):

Angular magnification with the object at the near point ▶

$$m \equiv \frac{\theta}{\theta_0} \qquad [25.2]$$

For the case in which the lens is held close to the eye, the angular magnification is a maximum when the image formed by the lens is at the near point of

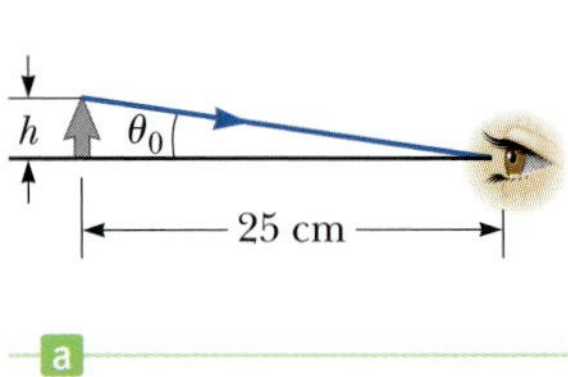


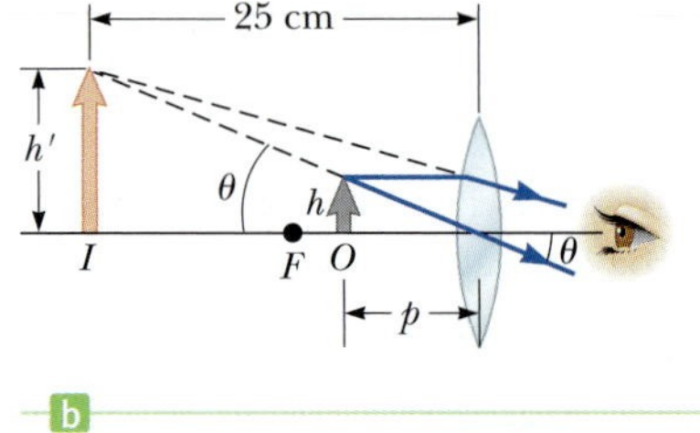


**Figure 25.6** (a) An object placed at the near point ($p = 25$ cm) subtends an angle of $\theta_0 \approx h/25$ at the eye. (b) An object placed near the focal point of a converging lens produces a magnified image, which subtends an angle of $\theta \approx h'/25$ at the eye. Note that in this situation $q = -25$ cm.

the eye, which corresponds to $q = -25$ cm (see Fig. 25.6b). The object distance corresponding to this image distance can be calculated from the thin-lens equation:

$$\frac{1}{p} + \frac{1}{-25 \text{ cm}} = \frac{1}{f} \qquad [25.3]$$

$$p = \frac{25f}{25 + f}$$

Here, $f$ is the focal length of the magnifier in centimeters. From Figures 25.6a and 25.6b, the small-angle approximation gives

$$\tan \theta_0 \approx \theta_0 \approx \frac{h}{25} \quad \text{and} \quad \tan \theta \approx \theta \approx \frac{h}{p} \qquad [25.4]$$

Equation 25.2 therefore becomes

$$m_{\text{max}} = \frac{\theta}{\theta_0} = \frac{h/p}{h/25} = \frac{25}{p} = \frac{25}{25f/(25 + f)}$$

so that

$$m_{\text{max}} = 1 + \frac{25 \text{ cm}}{f} \qquad [25.5]$$

The maximum angular magnification given by Equation 25.5 is the ratio of the angular size seen with the lens to the angular size seen without the lens, with the object at the near point of the eye. Although the normal eye can focus on an image formed anywhere between the near point and infinity, it's most relaxed when the image is at infinity (Sec. 25.2). For the image formed by the magnifying lens to appear at infinity, the object must be placed at the focal point of the lens so that $p = f$. In this case Equation 25.4 becomes

$$\theta_0 \approx \frac{h}{25} \quad \text{and} \quad \theta \approx \frac{h}{f}$$

and the angular magnification is

$$m = \frac{\theta}{\theta_0} = \frac{25 \text{ cm}}{f} \qquad [25.6]$$

With a single lens, it's possible to achieve angular magnifications up to about 4 without serious aberrations. Magnifications up to about 20 can be achieved by using one or two additional lenses to correct for aberrations.

## EXAMPLE 25.3 Magnification of a Lens

**GOAL** Compute magnifications of a lens when the image is at the near point and when it's at infinity.

**PROBLEM** **(a)** What is the maximum angular magnification of a lens with a focal length of 10.0 cm? **(b)** What is the angular magnification of this lens when the eye is relaxed? Assume an eye–lens distance of zero.

**STRATEGY** The maximum angular magnification occurs when the image formed by the lens is at the near point of the eye. Under these circumstances, Equation 25.5 gives us the maximum angular magnification. In part (b) the eye is relaxed only if the image is at infinity, so Equation 25.6 applies.

**SOLUTION**

**(a)** Find the maximum angular magnification of the lens.

Substitute into Equation 25.5:

$$m_{\text{max}} = 1 + \frac{25 \text{ cm}}{f} = 1 + \frac{25 \text{ cm}}{10.0 \text{ cm}} = 3.5$$

*(Continued)*

**(b)** Find the magnification of the lens when the eye is relaxed.

When the eye is relaxed, the image is at infinity, so substitute into Equation 25.6:

$$m = \frac{25 \text{ cm}}{f} = \frac{25 \text{ cm}}{10.0 \text{ cm}} = \boxed{2.5}$$

**QUESTION 25.3** For greater magnification, should a lens with a larger or smaller focal length be selected?

**EXERCISE 25.3** What focal length would be necessary if the lens were to have a maximum angular magnification of 4.0?

**ANSWER** 8.3 cm

## 25.4 The Compound Microscope

**LEARNING OBJECTIVES**

1. Discuss the optical components, characteristics, and limitations of a compound microscope.
2. Evaluate the magnification of a compound microscope.

A simple magnifier provides only limited assistance with inspection of the minute details of an object. Greater magnification can be achieved by combining two lenses in a device called a compound microscope, a schematic diagram of which is shown in Figure 25.7a. The instrument consists of two lenses: an objective with a very short focal length $f_o$ (where $f_o < 1$ cm) and an ocular lens, or eyepiece, with a focal length $f_e$ of a few centimeters. The two lenses are separated by a distance $L$ that is much greater than either $f_o$ or $f_e$.

The basic approach used to analyze the image formation properties of a microscope is that of two lenses in a row: the image formed by the first becomes the object for the second. The object $O$ placed just outside the focal length of the objective forms a real, inverted image at $I_1$ that is at or just inside the focal point of the eyepiece. This image is much enlarged. (For clarity, the enlargement of $I_1$ is not shown in Fig. 25.7a.) The eyepiece, which serves as a simple magnifier, uses the

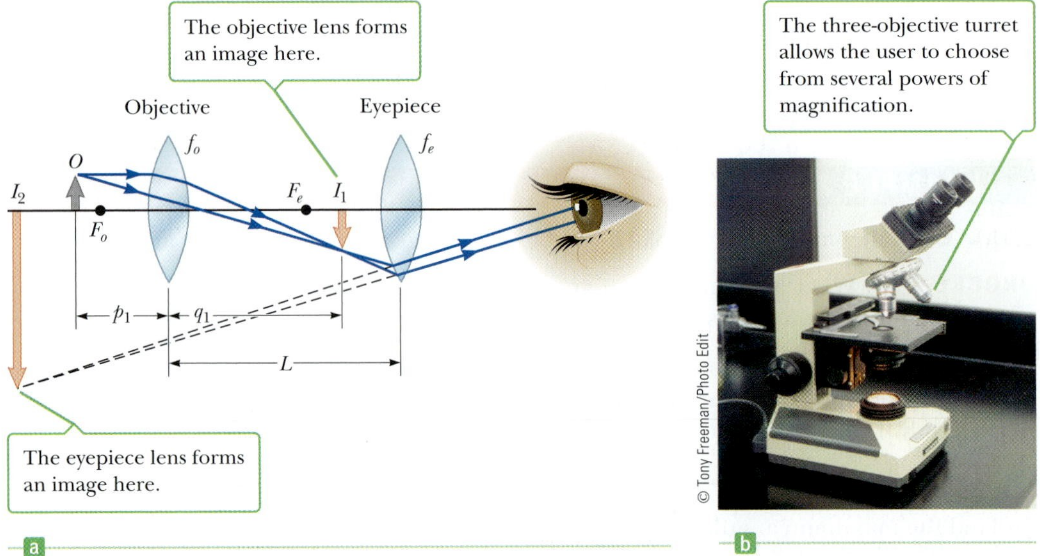
The objective lens forms an image here.
Objective
Eyepiece
$f_o$
$f_e$
$O$
$I_2$
$F_o$
$F_e$
$I_1$
$p_1$
$q_1$
$L$
The eyepiece lens forms an image here.
The three-objective turret allows the user to choose from several powers of magnification.
© Tony Freeman/Photo Edit
a
b

**Figure 25.7** (a) A diagram of a compound microscope, which consists of an objective and an eyepiece, or ocular lens. (b) A compound microscope. Combinations of eyepieces with different focal lengths and different objectives can produce a wide range of magnifications.

image at $I_1$ as its object and produces an image at $I_2$. The image seen by the eye at $I_2$ is virtual, inverted, and very much enlarged.

The lateral magnification $M_1$ of the first image is $-q_1/p_1$. Note that $q_1$ is approximately equal to $L$ because the object is placed close to the focal point of the objective lens, which ensures that the image formed will be far from the objective lens. Further, because the object is very close to the focal point of the objective lens, $p_1 \approx f_o$. Therefore, the lateral magnification of the objective is

$$M_1 = -\frac{q_1}{p_1} \approx -\frac{L}{f_o}$$

From Equation 25.6, the angular magnification of the eyepiece for an object (corresponding to the image at $I_1$) placed at the focal point is found to be

$$m_e = \frac{25 \text{ cm}}{f_e}$$

The overall magnification of the compound microscope is defined as the product of the lateral and angular magnifications:

$$m = M_1 m_e = -\frac{L}{f_o}\left(\frac{25 \text{ cm}}{f_e}\right) \quad [25.7]$$

◀ Magnification of a microscope

The negative sign indicates that the image is inverted with respect to the object.

The microscope has extended our vision into the previously unknown realm of incredibly small objects, and the capabilities of this instrument have increased steadily with improved techniques in precision grinding of lenses. A natural question is whether there is any limit to how powerful a microscope could be. For example, could a microscope be made powerful enough to allow us to see an atom? The answer to this question is no, as long as visible light is used to illuminate the object. To be seen, the object under a microscope must be at least as large as a wavelength of light. An atom is many times smaller than the wavelength of visible light, so its mysteries must be probed via other techniques.

The wavelength dependence of the "seeing" ability of a wave can be illustrated by water waves set up in a bathtub in the following way. Imagine that you vibrate your hand in the water until waves with a wavelength of about 6 in. are moving along the surface. If you fix a small object, such as a toothpick, in the path of the waves, you will find that the waves are not appreciably disturbed by the toothpick, but continue along their path. Now suppose you fix a larger object, such as a toy sailboat, in the path of the waves. In this case the waves are considerably disturbed by the object. The toothpick was much smaller than the wavelength of the waves, and as a result the waves didn't "see" it. The toy sailboat, however, is about the same size as the wavelength of the waves and hence creates a disturbance. Light waves behave in this same general way. The ability of an optical microscope to view an object depends on the size of the object relative to the wavelength of the light used to observe it. Hence, it will never be possible to observe atoms or molecules with such a microscope because their dimensions are so small ($\approx$0.1 nm) relative to the wavelength of the light ($\approx$500 nm).

## EXAMPLE 25.4 Microscope Magnifications

**GOAL** Understand the critical factors involved in determining the magnifying power of a microscope.

**PROBLEM** A certain microscope has two interchangeable objectives. One has a focal length of 2.0 cm, and the other has a focal length of 0.20 cm. Also available are two eyepieces of focal lengths 2.5 cm and 5.0 cm. If the length of the microscope is 18 cm, compute the magnifications for the following combinations: the 2.0-cm objective and 5.0-cm eyepiece, the 2.0-cm objective and 2.5-cm eyepiece, and the 0.20-cm objective and 5.0-cm eyepiece.

**STRATEGY** The solution consists of substituting into Equation 25.7 for three different combinations of lenses.

*(Continued)*

**SOLUTION**

Apply Equation 25.7 and combine the 2.0-cm objective with the 5.0-cm eyepiece:

$$m = -\frac{L}{f_o}\left(\frac{25 \text{ cm}}{f_e}\right) = -\frac{18 \text{ cm}}{2.0 \text{ cm}}\left(\frac{25 \text{ cm}}{5.0 \text{ cm}}\right) = \boxed{-45}$$

Combine the 2.0-cm objective with the 2.5-cm eyepiece:

$$m = -\frac{18 \text{ cm}}{2.0 \text{ cm}}\left(\frac{25 \text{ cm}}{2.5 \text{ cm}}\right) = \boxed{-9.0 \times 10^1}$$

Combine the 0.20-cm objective with the 5.0-cm eyepiece:

$$m = -\frac{18 \text{ cm}}{0.20 \text{ cm}}\left(\frac{25 \text{ cm}}{5.0 \text{ cm}}\right) = \boxed{-450}$$

**REMARKS** Much higher magnifications can be achieved, but the resolution starts to fall, resulting in fuzzy images that don't convey any details. (See Section 25.6 for further discussion of this point.)

**QUESTION 25.4** True or False: A shorter focal length for either the eyepiece or objective lens will result in greater magnification.

**EXERCISE 25.4** Combine the 0.20-cm objective with the 2.5-cm eyepiece and find the magnification.

**ANSWER** $-9.0 \times 10^2$

## 25.5 The Telescope

**LEARNING OBJECTIVES**

1. Describe the optical components of reflecting and refracting telescopes and discuss the advantages and disadvantages of each.
2. Evaluate the angular magnification of a telescope.

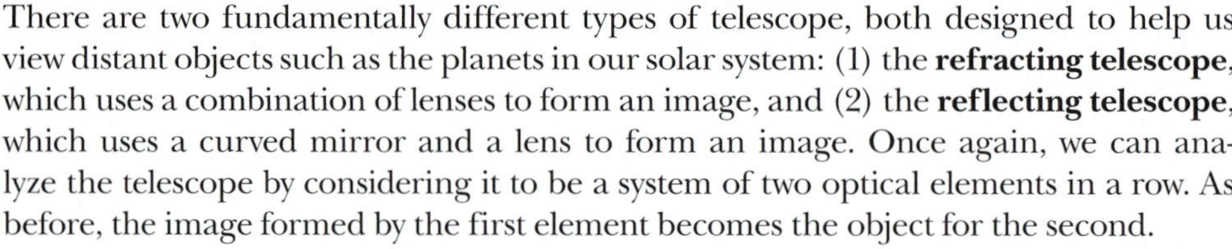

There are two fundamentally different types of telescope, both designed to help us view distant objects such as the planets in our solar system: (1) the **refracting telescope**, which uses a combination of lenses to form an image, and (2) the **reflecting telescope**, which uses a curved mirror and a lens to form an image. Once again, we can analyze the telescope by considering it to be a system of two optical elements in a row. As before, the image formed by the first element becomes the object for the second.

NASA

The Hubble Space Telescope enables us to see both further into space and further back in time than ever before.

In the refracting telescope two lenses are arranged so that the objective forms a real, inverted image of the distant object very near the focal point of the eyepiece (Fig. 25.8a). Further, the image at $I_1$ is formed at the focal point of the objective because the object is essentially at infinity. Hence, the two lenses are separated by the distance $f_o + f_e$, which corresponds to the length of the telescope's tube. Finally, at $I_2$, the eyepiece forms an enlarged image of the image at $I_1$.

The angular magnification of the telescope is given by $\theta/\theta_o$, where $\theta_o$ is the angle subtended by the object at the objective and $\theta$ is the angle subtended by the final image. From the triangles in Figure 25.8a, and for small angles, we have

$$\theta_o \approx \frac{h'}{f_e} \quad \text{and} \quad \theta_o \approx \frac{h'}{f_o}$$

Therefore, the angular magnification of the telescope can be expressed as

Angular magnification of a telescope ▶

$$m = \frac{\theta}{\theta_o} = \frac{h'/f_e}{h'/f_o} = \frac{f_o}{f_e} \qquad [25.8]$$

This equation says that the angular magnification of a telescope equals the ratio of the objective focal length to the eyepiece focal length. Here again, the angular magnification is the ratio of the angular size seen with the telescope to the angular size seen with the unaided eye.

In some applications—for instance, the observation of relatively nearby objects such as the Sun, the Moon, or planets—angular magnification is important. Stars, however, are so far away that they always appear as small points of light regardless

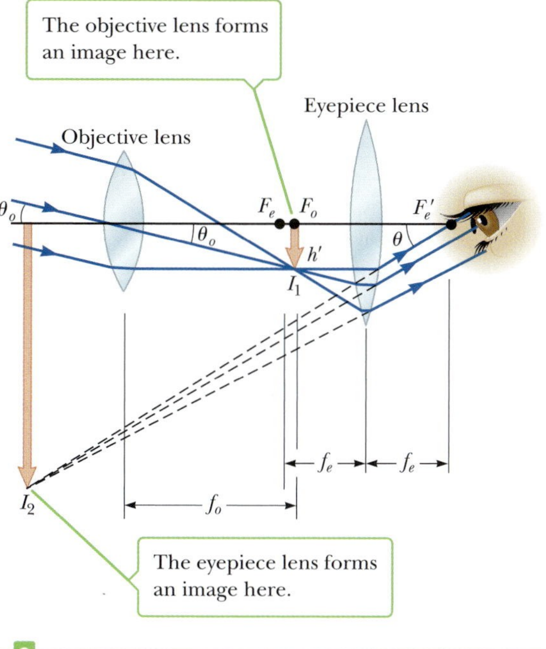

© Tony Freeman/Photo Edit

**Figure 25.8** (a) A diagram of a refracting telescope, with the object at infinity. (b) A refracting telescope.

of how much angular magnification is used. The large research telescopes used to study very distant objects must have great diameters to gather as much light as possible. It's difficult and expensive to manufacture such large lenses for refracting telescopes. In addition, the heaviness of large lenses leads to sagging, which is another source of aberration.

These problems can be partially overcome by replacing the objective lens with a reflecting, concave mirror, usually having a parabolic shape so as to avoid spherical aberration. Figure 25.9 shows the design of a typical reflecting telescope. Incoming light rays pass down the barrel of the telescope and are reflected by a parabolic mirror at the base. These rays converge toward point $A$ in the figure, where an image would be formed on a photographic plate or another detector. Before this image is formed, however, a small, flat mirror at $M$ reflects the light toward an opening in the side of the tube that passes into an eyepiece. This design is said to have a *Newtonian focus*, after its developer. Note that in the reflecting telescope the light never passes through glass (except in the small eyepiece). As a result, problems associated with chromatic aberration are virtually eliminated.

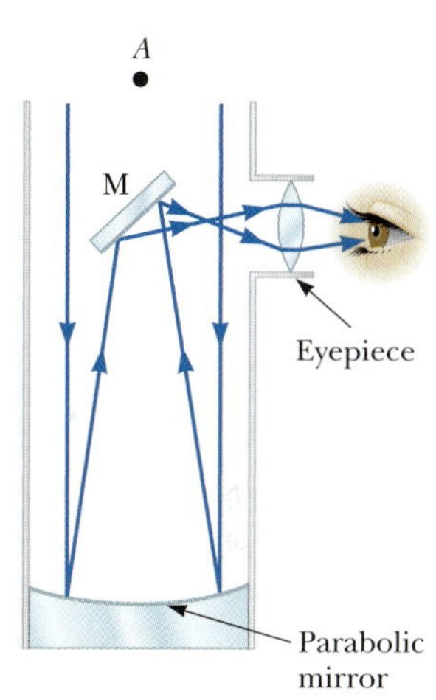


**Figure 25.9** A reflecting telescope with a Newtonian focus.

The largest optical telescopes in the world are the two 10-m-diameter Keck reflectors on Mauna Kea in Hawaii. The largest single-mirrored reflecting telescope in the United States is the 5-m-diameter instrument on Mount Palomar in California. (See Fig. 25.10.) In contrast, the largest refracting telescope in the world, at the Yerkes Observatory in Williams Bay, Wisconsin, has a diameter of only 1 m.

Courtesy of Palomar Observatory/California Institute of Technology

**Figure 25.10** The Hale telescope at Mount Palomar Observatory. Just before taking the elevator up to the prime-focus cage, a first-time observer is always told, "Good viewing! And, if you should fall, try to miss the mirror."

## ■ EXAMPLE 25.5 Hubble Power

**GOAL** Understand magnification in telescopes.

**PROBLEM** The Hubble Space Telescope is 13.2 m long, but has a secondary mirror that increases its effective focal length to 57.8 m. (See Fig. 25.11.) The telescope doesn't have an eyepiece because various instruments, not a human eye, record the collected light. It can, however, produce images several thousand times larger than they would appear with the unaided human eye. What focal-length eyepiece used with the Hubble mirror system would produce a magnification of $8.00 \times 10^3$?

**STRATEGY** Equation 25.8 for telescope magnification can be solved for the eyepiece focal length. The equation for finding the angular magnification of a reflector is the same as that for a refractor.

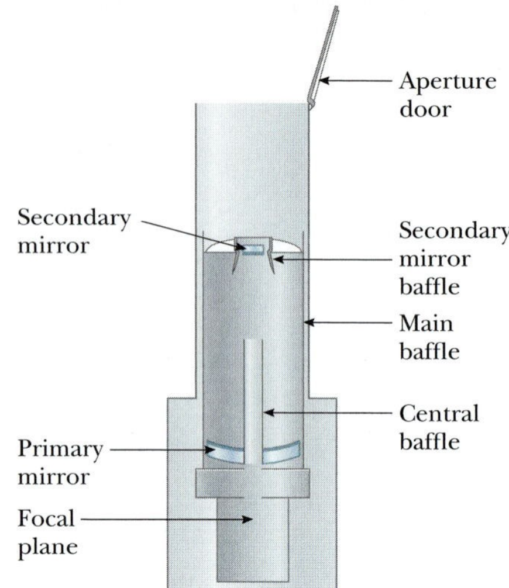


**Figure 25.11** A schematic of the Hubble Space Telescope.

**SOLUTION**

Solve for $f_e$ in Equation 25.8 and substitute values:

$$m = \frac{f_o}{f_e} \quad \rightarrow \quad f_e = \frac{f_o}{m} = \frac{57.8 \text{ m}}{8.00 \times 10^3} = 7.23 \times 10^{-3} \text{ m}$$

**REMARKS** The light-gathering power of a telescope and the length of the baseline over which light is gathered are in fact more important than a telescope's magnification, because these two factors contribute to the resolution of the image. A high-resolution image can always be magnified so its details can be examined. A low resolution image, however, is often fuzzy when magnified. (See Section 25.6.)

**QUESTION 25.5** Can greater magnification of a telescope be achieved by increasing the focal length of the mirror? What effect will increasing the focal length of the eyepiece have on the magnification?

**EXERCISE 25.5** The Hale telescope on Mount Palomar has a focal length of 16.8 m. Find the magnification of the telescope in conjunction with an eyepiece having a focal length of 5.00 mm.

**ANSWER** $3.36 \times 10^3$

# 25.6 Resolution of Single-Slit and Circular Apertures

### LEARNING OBJECTIVES

1. State and discuss the physical origins of Rayleigh's criterion.
2. Determine the limiting angle for slit and circular apertures.
3. Evaluate the resolution limitations of optical instruments and the resolving power of a diffraction grating.

The ability of an optical system such as the eye, a microscope, or a telescope to distinguish between closely spaced objects is limited because of the wave nature of light. To understand this difficulty, consider Figure 25.12, which shows two light sources far from a narrow slit of width $a$. The sources can be taken as two point

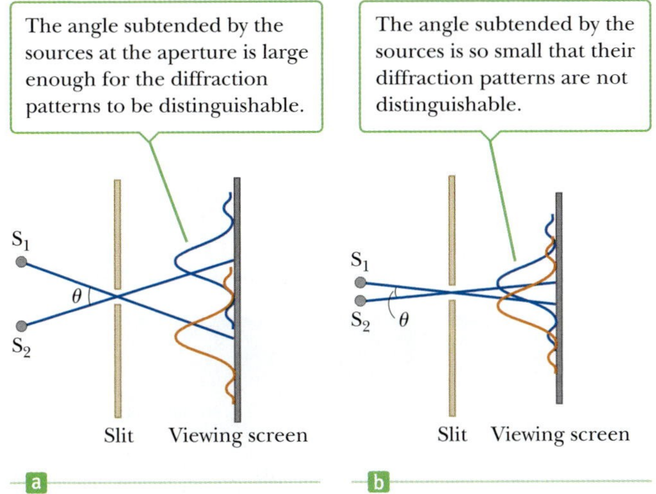
The angle subtended by the sources at the aperture is large enough for the diffraction patterns to be distinguishable.
The angle subtended by the sources is so small that their diffraction patterns are not distinguishable.
S1
S2
θ
Slit
Viewing screen
a
b

**Figure 25.12** Two point sources far from a narrow slit each produce a diffraction pattern. (a) The sources are separated by a large angle. (b) The sources are separated by a small angle. (Notice that the angles are greatly exaggerated. The drawing is not to scale.)

sources $S_1$ and $S_2$ that are *not* coherent. For example, they could be two distant stars. If no diffraction occurred, two distinct bright spots (or images) would be observed on the screen at the right in the figure. Because of diffraction, however, each source is imaged as a bright central region flanked by weaker bright and dark rings. What is observed on the screen is the sum of two diffraction patterns, one from $S_1$ and the other from $S_2$.

If the two sources are separated so that their central maxima don't overlap, as in Figure 25.12a, their images can be distinguished and are said to be *resolved.* If the sources are close together, however, as in Figure 25.12b, the two central maxima may overlap and the images are *not resolved.* To decide whether two images are resolved, the following condition is often applied to their diffraction patterns:

When the central maximum of one image falls on the first minimum of another image, the images are said to be just resolved. This limiting condition of resolution is known as **Rayleigh's criterion**.

◀ Rayleigh's criterion

Figure 25.13 (page 884) shows diffraction patterns in three situations. In Fig. 25.13a, the sources are sufficiently separated, so the images are resolved. As the sources are brought closer together, as in Figure 25.13b, the central maximum of one image is centered on the first minimum of the other, so by Rayleigh's criterion, the images are just resolved. Finally, when the sources are very close to each other, their images are not resolved (Fig. 25.13c).

From Rayleigh's criterion, we can determine the minimum angular separation $\theta_{\min}$ subtended by the source at the slit so that the images will be just resolved. In Chapter 24 we found that the first minimum in a single-slit diffraction pattern occurs at the angle that satisfies the relationship

$$\sin\theta = \frac{\lambda}{a}$$

where $a$ is the width of the slit. According to Rayleigh's criterion, this expression gives the smallest angular separation for which the two images can be resolved. Because $\lambda << a$ in most situations, $\sin\theta$ is small and we can use the approximation $\sin\theta \approx \theta$. Therefore, the limiting angle of resolution for a slit of width $a$ is

$$\theta_{\min} \approx \frac{\lambda}{a} \qquad [25.9]$$

◀ Limiting angle for a slit

where $\theta_{\min}$ is in radians. Hence, the angle subtended by the two sources at the slit must be *greater* than $\lambda/a$ if the images are to be resolved.

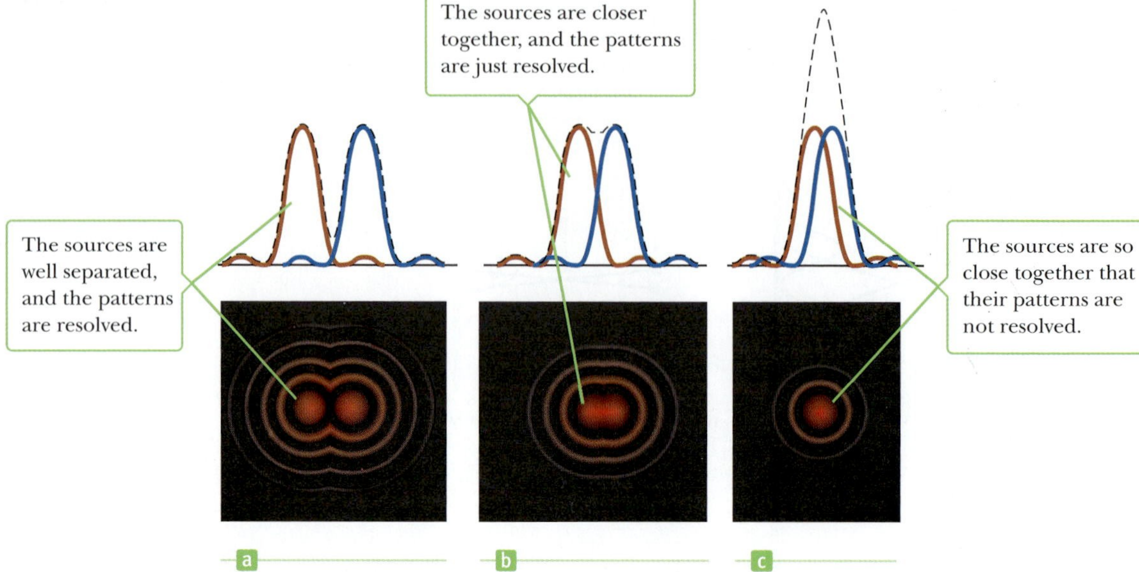

The sources are closer together, and the patterns are just resolved.
The sources are well separated, and the patterns are resolved.
The sources are so close together that their patterns are not resolved.
a
b
c


**Figure 25.13** The diffraction patterns of two point sources (solid curves) and the resultant pattern (dashed curve) for three angular separations of the sources.

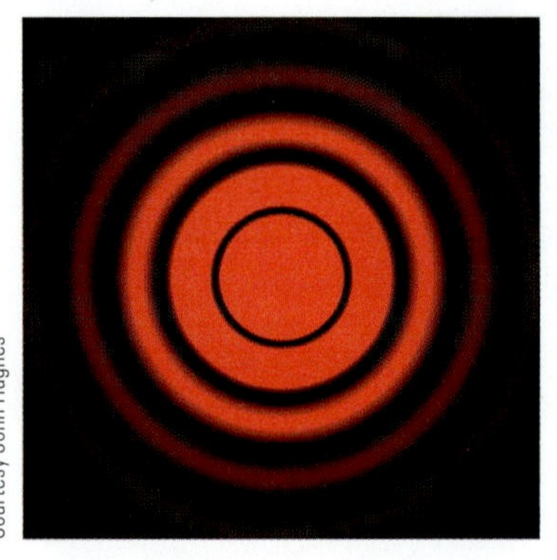


**Figure 25.14** The diffraction pattern of a circular aperture consists of a central bright disk surrounded by concentric bright and dark rings.

Many optical systems use circular apertures rather than slits. The diffraction pattern of a circular aperture (Fig. 25.14) consists of a central circular bright region surrounded by progressively fainter rings. Analysis shows that the limiting angle of resolution of the circular aperture is

$$\theta_{\text{min}} = 1.22 \frac{\lambda}{D} \qquad [25.10]$$

where $D$ is the diameter of the aperture. Note that Equation 25.10 is similar to Equation 25.9 except for the factor 1.22, which arises from a complex mathematical analysis of diffraction from a circular aperture.

**Quick Quiz**

**25.2** Suppose you are observing a binary star with a telescope and are having difficulty resolving the two stars. Which color filter will better help resolve the stars? (a) blue (b) red (c) neither because colored filters have no effect on resolution

## APPLYING PHYSICS 25.2 Cat's Eyes BIO

Cats' eyes have vertical pupils in dim light. Which would cats be most successful at resolving at night, headlights on a distant car or vertically separated running lights on a distant boat's mast having the same separation as the car's headlights?

**EXPLANATION** The effective slit width in the vertical direction of the cat's eye is larger than that in the horizontal direction. Thus, it has more resolving power for lights separated in the vertical direction and would be more effective at resolving the mast lights on the boat. ■

## EXAMPLE 25.6 Resolution of a Microscope

**GOAL** Study limitations on the resolution of a microscope.

**PROBLEM** Sodium light of wavelength 589 nm is used to view an object under a microscope. The aperture of the objective has a diameter of 0.90 cm. **(a)** Find the limiting angle of resolution for this microscope. **(b)** Using visible light of any wavelength you desire, find the best limit of resolution for this microscope. **(c)** Water of index of refraction 1.33 now fills the space between the object and the objective. What effect would this water have on the resolving power of the microscope, using 589-nm light?

**STRATEGY** Parts (a) and (b) require substitution into Equation 25.10. Because the wavelength appears in the numerator, violet light, with the shortest visible wavelength, gives the maximum resolution. In part (c) the only difference is that the wavelength changes to $\lambda/n$, where $n$ is the index of refraction of water.

**SOLUTION**

**(a)** Find the limiting angle of resolution for this microscope.

Substitute into Equation 25.10 to obtain the limiting angle of resolution:

$$\theta_{\text{min}} = 1.22\frac{\lambda}{D} = 1.22\left(\frac{589 \times 10^{-9}\text{ m}}{0.90 \times 10^{-2}\text{ m}}\right)$$

$$= 8.0 \times 10^{-5}\text{ rad}$$

**(b)** Calculate the microscope's best limit of resolution.

To obtain the best resolution, substitute the shortest visible wavelength available, which is violet light, of wavelength $4.0 \times 10^2$ nm:

$$\theta_{\text{min}} = 1.22\frac{\lambda}{D} = 1.22\left(\frac{4.0 \times 10^{-7}\text{ m}}{0.90 \times 10^{-2}\text{ m}}\right)$$

$$= 5.4 \times 10^{-5}\text{ rad}$$

**(c)** What effect does water between the object and the objective lens have on the resolution, with 589-nm light?

Calculate the wavelength of the sodium light in the water:

$$\lambda_w = \frac{\lambda_a}{n} = \frac{589\text{ nm}}{1.33} = 443\text{ nm}$$

Substitute this wavelength into Equation 25.10 to get the resolution:

$$\theta_{\text{min}} = 1.22\left(\frac{443 \times 10^{-9}\text{ m}}{0.90 \times 10^{-2}\text{ m}}\right) = 6.0 \times 10^{-5}\text{ rad}$$

**REMARKS** In each case any two points on the object subtending an angle of less than the limiting angle $\theta_{\text{min}}$ at the objective cannot be distinguished in the image. Consequently, it may be possible to see a cell but then be unable to clearly see smaller structures within the cell. Obtaining an increase in resolution is the motivation behind placing a drop of oil on the slide for certain objective lenses.

**QUESTION 25.6** Does having two eyes instead of one improve the human ability to resolve distant objects? In general, would more widely spaced eyes increase visual resolving power? Explain.

**EXERCISE 25.6** Suppose oil with $n = 1.50$ fills the space between the object and the objective for this microscope. Calculate the limiting angle $\theta_{\text{min}}$ for sodium light of wavelength 589 nm in air.

**ANSWER** $5.3 \times 10^{-5}$ rad

## EXAMPLE 25.7 Resolving Craters on the Moon

**GOAL** Calculate the resolution of a telescope.

**PROBLEM** The Hubble Space Telescope has an aperture of diameter 2.40 m. **(a)** What is its limiting angle of resolution at a wavelength of $6.00 \times 10^2$ nm? **(b)** What's the smallest crater it could resolve on the Moon? (The Moon's distance from Earth is $3.84 \times 10^8$ m.)

**STRATEGY** After substituting into Equation 25.10 to find the limiting angle, use $s = r\theta$ to compute the minimum size of crater that can be resolved.

**SOLUTION**

**(a)** What is the limiting angle of resolution at a wavelength of $6.00 \times 10^2$ nm?

Substitute $D = 2.40$ m and $\lambda = 6.00 \times 10^{-7}$ m into Equation 25.10:

$$\theta_{\text{min}} = 1.22\frac{\lambda}{D} = 1.22\left(\frac{6.00 \times 10^{-7}\text{ m}}{2.40\text{ m}}\right)$$

$$= 3.05 \times 10^{-7}\text{ rad}$$

*(Continued)*

**(b)** What's the smallest lunar crater the Hubble Space Telescope can resolve?

The two opposite sides of the crater must subtend the minimum angle. Use the arc length formula:

$$s = r\theta = (3.84 \times 10^8 \text{ m})(3.05 \times 10^{-7} \text{ rad}) = \boxed{117 \text{ m}}$$

**REMARKS** The distance is so great and the angle so small that using the arc length of a circle is justified because the circular arc is very nearly a straight line. The Hubble Space Telescope has produced several gigabytes of data every day since it first began operation.

**QUESTION 25.7** Is the resolution of a telescope better at the red end of the visible spectrum or the violet end?

**EXERCISE 25.7** The Hale telescope on Mount Palomar has a diameter of 5.08 m (200 in.). (a) Find the limiting angle of resolution for a wavelength of $6.00 \times 10^2$ nm. (b) Calculate the smallest crater diameter the telescope can resolve on the Moon. (c) The answers appear better than what the Hubble can achieve. Why are the answers misleading?

**ANSWERS** (a) $1.44 \times 10^{-7}$ rad (b) 55.3 m (c) Although the numbers are better than Hubble's, the Hale telescope must contend with the effects of atmospheric turbulence, so the smaller space-based telescope actually obtains far better results.

---

It's interesting to compare the resolution of the Hale telescope with that of a large radio telescope, such as the system at Arecibo, Puerto Rico, which has a diameter of 1 000 ft (305 m). This telescope detects radio waves at a wavelength of 0.75 m. The corresponding minimum angle of resolution can be calculated as $3.0 \times 10^{-3}$ rad (10 min 19 s of arc), which is more than 10 000 times larger than the calculated minimum angle for the Hale telescope.

With such relatively poor resolution, why is Arecibo considered a valuable astronomical instrument? Unlike its optical counterparts, Arecibo can see through clouds of dust. The center of our Milky Way galaxy is obscured by such dust clouds, which absorb and scatter visible light. Radio waves easily penetrate the clouds, so radio telescopes allow direct observations of the galactic core.

## Resolving Power of the Diffraction Grating

The diffraction grating studied in Chapter 24 is most useful for making accurate wavelength measurements. Like the prism, it can be used to disperse a spectrum into its components. Of the two devices, the grating is better suited to distinguishing between two closely spaced wavelengths. We say that the grating spectrometer has a higher *resolution* than the prism spectrometer. If $\lambda_1$ and $\lambda_2$ are two nearly equal wavelengths between which the spectrometer can just barely distinguish, the **resolving power** of the grating is defined as

$$R \equiv \frac{\lambda}{\lambda_2 - \lambda_1} = \frac{\lambda}{\Delta\lambda} \qquad [25.11]$$

where $\lambda \approx \lambda_1 \approx \lambda_2$ and $\Delta\lambda = \lambda_2 - \lambda_1$. From this equation, it's clear that a grating with a high resolving power can distinguish small differences in wavelength. Further, if $N$ lines of the grating are illuminated, it can be shown that the resolving power in the $m$th-order diffraction is given by

Resolving power of a grating ▶

$$R = Nm \qquad [25.12]$$

So, the resolving power $R$ increases with the order number $m$ and is large for a grating with a great number of illuminated slits. Note that for $m = 0$, $R = 0$, which signifies that *all wavelengths are indistinguishable* for the zeroth-order maximum. (All wavelengths fall at the same point on the screen.) Consider, however, the second-order diffraction pattern of a grating that has 5 000 rulings illuminated by the light source. The resolving power of such a grating in second order is $R = 5\,000 \times 2 = 10\,000$. Therefore, the *minimum* wavelength separation between two spectral lines that can be just resolved, assuming a mean wavelength of 600 nm, is calculated from Equation 25.12 to be $\Delta\lambda = \lambda/R = 6 \times 10^{-2}$ nm. For the third-order principal maximum, $R = 15\,000$ and $\Delta\lambda = 4 \times 10^{-2}$ nm, and so on.

### EXAMPLE 25.8 Light from Sodium Atoms

**GOAL** Find the necessary resolving power to distinguish spectral lines.

**PROBLEM** Two bright lines in the spectrum of sodium have wavelengths of 589.00 nm and 589.59 nm, respectively. **(a)** What must the resolving power of a grating be so as to distinguish these wavelengths? **(b)** To resolve these lines in the second-order spectrum, how many lines of the grating must be illuminated?

**STRATEGY** This problem requires little more than substituting into Equations 25.11 and 25.12.

**SOLUTION**

**(a)** What must the resolving power of a grating be in order to distinguish the given wavelengths?

Substitute into Equation 25.11 to find $R$:

$$R = \frac{\lambda}{\Delta\lambda} = \frac{589.00 \text{ nm}}{589.59 \text{ nm} - 589.00 \text{ nm}} = \frac{589 \text{ nm}}{0.59 \text{ nm}}$$

$$= 1.0 \times 10^3$$

**(b)** To resolve these lines in the second-order spectrum, how many lines of the grating must be illuminated?

Solve Equation 25.12 for $N$ and substitute:

$$N = \frac{R}{m} = \frac{1.0 \times 10^3}{2} = 5.0 \times 10^2 \text{ lines}$$

**REMARKS** The ability to resolve spectral lines is particularly important in experimental atomic physics.

**QUESTION 25.8** True or False: If two diffraction gratings differ only by the number of lines, the grating with the larger number of lines can yield a greater resolving power.

**EXERCISE 25.8** Due to a phenomenon called electron spin, when the lines of a spectrum are examined at high resolution, each line is actually found to be two closely spaced lines called a doublet. An example is the doublet in the hydrogen spectrum having wavelengths of 656.272 nm and 656.285 nm. (a) What must be the resolving power of a grating so as to distinguish these wavelengths? (b) How many lines of the grating must be illuminated to resolve these lines in the third-order spectrum?

**ANSWERS** (a) $5.0 \times 10^4$ (b) $1.7 \times 10^4$ lines

## 25.7 The Michelson Interferometer

**LEARNING OBJECTIVE**

1. Describe the optical components and operating principles of the Michelson interferometer.

The Michelson interferometer is an optical instrument having great scientific importance. Invented by American physicist A. A. Michelson (1852–1931), it is an ingenious device that splits a light beam into two parts and then recombines them to form an interference pattern. The interferometer is used to make accurate length measurements.

Figure 25.15 (page 888) is a schematic diagram of an interferometer. A beam of light provided by a monochromatic source is split into two rays by a partially silvered mirror M inclined at an angle of 45° relative to the incident light beam. One ray is reflected vertically upward to mirror $M_1$, and the other ray is transmitted horizontally through mirror M to mirror $M_2$. Hence, the two rays travel separate paths, $L_1$ and $L_2$. After reflecting from mirrors $M_1$ and $M_2$, the two rays eventually recombine to produce an interference pattern, which can be viewed through a telescope. The glass plate P, equal in thickness to mirror M, is placed in the path of the horizontal ray to ensure that the two rays travel the same distance through glass.

The interference pattern for the two rays is determined by the difference in their path lengths. When the two rays are viewed as shown, the image of $M_2$ is at

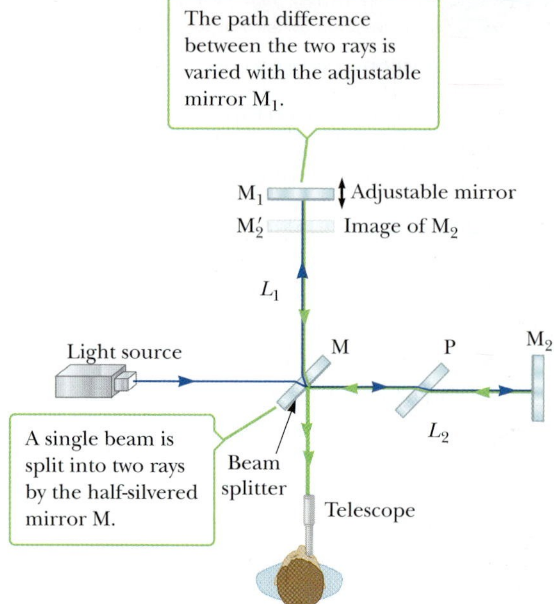


**Figure 25.15** A diagram of the Michelson interferometer.

$M_2'$, parallel to $M_1$. Hence, the space between $M_2'$ and $M_1$ forms the equivalent of a parallel air film. The effective thickness of the air film is varied by using a finely threaded screw to move mirror $M_1$ in the direction indicated by the arrows in Figure 25.15. If one of the mirrors is tipped slightly with respect to the other, the thin film between the two is wedge shaped and an interference pattern consisting of parallel fringes is set up, as described in Example 24.4. Now suppose we focus on one of the dark lines with the crosshairs of a telescope. As mirror $M_1$ is moved to lengthen the path $L_1$, the thickness of the wedge increases. When the thickness increases by $\lambda/4$, the destructive interference that initially produced the dark fringe has changed to constructive interference, and we now observe a bright fringe at the location of the crosshairs. The term *fringe shift* is used to describe the change in a fringe from dark to light or from light to dark. Successive light and dark fringes are formed each time $M_1$ is moved a distance of $\lambda/4$. The wavelength of light can be measured by counting the number of fringe shifts for a measured displacement of $M_1$. Conversely, if the wavelength is accurately known (as with a laser beam), the mirror displacement can be determined to within a fraction of the wavelength. Because the interferometer can measure displacements precisely, it is often used to make highly accurate measurements of the dimensions of mechanical components.

If the mirrors are perfectly aligned rather than tipped with respect to each other, the path difference differs slightly for different angles of view. This arrangement results in an interference pattern that resembles Newton's rings. The pattern can be used in a fashion similar to that for tipped mirrors. An observer pays attention to the center spot in the interference pattern. For example, suppose the spot is initially dark, indicating that destructive interference is occurring. If $M_1$ is now moved a distance of $\lambda/4$, this central spot changes to a light region, corresponding to a fringe shift.

# SUMMARY

## 25.1 The Camera

The light-concentrating power of a lens of focal length $f$ and diameter $D$ is determined by the **$f$-number**, defined as

$$f\text{-number} \equiv \frac{f}{D} \qquad [25.1]$$

The smaller the $f$-number of a lens, the brighter the image formed.

## 25.2 The Eye

**Hyperopia** (farsightedness) is a defect of the eye that occurs either when the eyeball is too short or when the ciliary muscle cannot change the shape of the lens enough to form a properly focused image. **Myopia** (nearsightedness) occurs either when the eye is longer than normal or when the maximum focal length of the lens is insufficient to produce a clearly focused image on the retina.

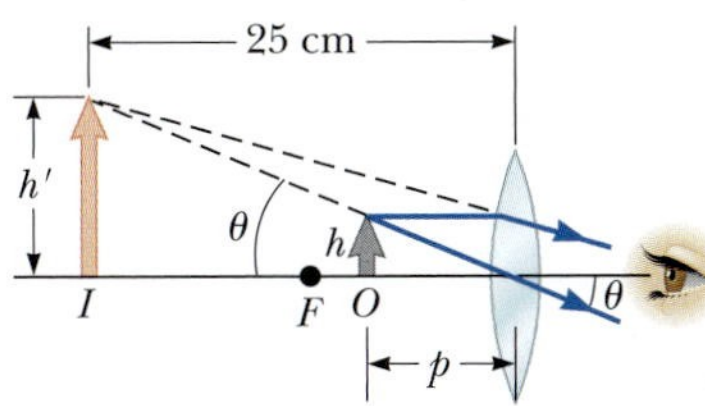


An object placed near the focal point of a converging lens produces a magnified image, which subtends an angle of $\theta \approx h'/25$ at the eye. Note that in this situation $q = -25$ cm.

The **power** of a lens in **diopters** is the inverse of the focal length in meters.

## 25.3 The Simple Magnifier

The **angular magnification of a lens** is defined as

$$m \equiv \frac{\theta}{\theta_0} \quad [25.2]$$

where $\theta$ is the angle subtended by an object at the eye with a lens in use and $\theta_0$ is the angle subtended by the object when it is placed at the near point of the eye and no lens is used. The **maximum angular magnification of a lens** is

$$m_{\text{max}} = 1 + \frac{25 \text{ cm}}{f} \quad [25.5]$$

When the eye is relaxed, the angular magnification is

$$m = \frac{25 \text{ cm}}{f} \quad [25.6]$$

## 25.4 The Compound Microscope

The overall **magnification of a compound microscope** of length $L$ is the product of the magnification produced by the objective, of focal length $f_o$, and the magnification produced by the eyepiece, of focal length $f_e$:

$$m = -\frac{L}{f_o}\left(\frac{25 \text{ cm}}{f_e}\right) \quad [25.7]$$

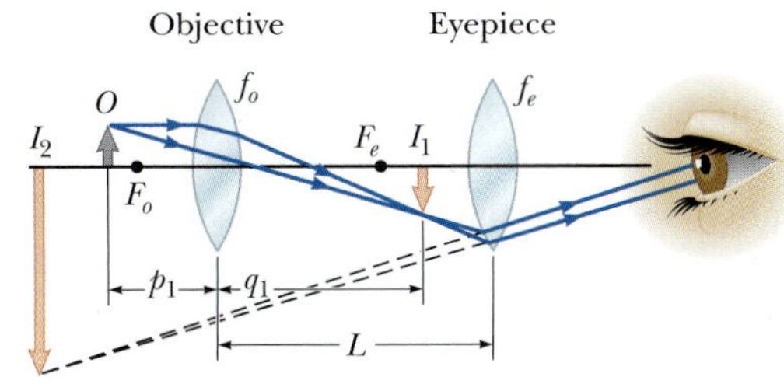


A ray diagram for a compound microscope, which consists of an objective and an eyepiece, or ocular lens.

## 25.5 The Telescope

The **angular magnification of a telescope** is

$$m = \frac{f_o}{f_e} \quad [25.8]$$

where $f_o$ is the focal length of the objective and $f_e$ is the focal length of the eyepiece.

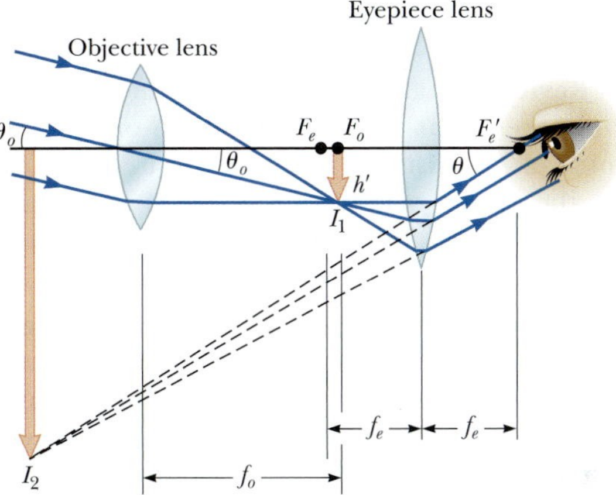


A ray diagram for a refracting telescope, with the object at infinity.

## 25.6 Resolution of Single-Slit and Circular Apertures

Two images are said to be **just resolved** when the central maximum of the diffraction pattern for one image falls on the first minimum of the other image. This limiting condition of resolution is known as **Rayleigh's criterion**. The limiting angle of resolution for a **slit** of width $a$ is

$$\theta_{\text{min}} \approx \frac{\lambda}{a} \quad [25.9]$$

The limiting angle of resolution of a **circular aperture** is

$$\theta_{\text{min}} = 1.22\frac{\lambda}{D} \quad [25.10]$$

where $D$ is the diameter of the aperture.

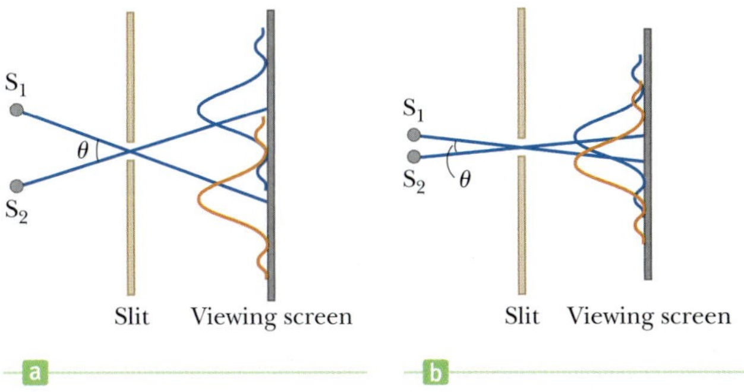


Two point sources far from a narrow slit each produce a diffraction pattern. (a) The sources are separated by a large angle. (b) The sources are separated by a small angle.

If $\lambda_1$ and $\lambda_2$ are two nearly equal wavelengths between which a grating spectrometer can just barely distinguish, the **resolving power** $R$ of the grating is defined as

$$R \equiv \frac{\lambda}{\lambda_2 - \lambda_1} = \frac{\lambda}{\Delta\lambda} \qquad [25.11]$$

where $\lambda \approx \lambda_1 \approx \lambda_2$ and $\Delta\lambda = \lambda_2 - \lambda_1$. The **resolving power** of a diffraction grating in the $m$th order is

$$R = Nm \qquad [25.12]$$

where $N$ is the number of illuminated rulings on the grating.

## WARM-UP EXERCISES

ENHANCED WebAssign The warm-up exercises in this chapter may be assigned online in Enhanced WebAssign.

1. **Physics Review** A real object is 12.0 cm in front of a thin, convergent lens with a focal length of 10.0 cm. Determine (a) the distance from the lens to the image and (b) the image magnification. (c) Is the image upright or inverted? (d) Is the image real or virtual? (See Section 23.6.)
2. **Physics Review** A thin, diverging lens has a focal length of magnitude 15.0 cm. An object is placed 10.0 cm in front of the lens. Find (a) the image distance and (b) the image magnification. (c) Is the image upright or inverted? (d) Real or virtual? (See Section 23.6.)
3. A large telephoto camera lens has a focal length of 0.825 m and an *f*-number of 4. Determine the lens diameter. (See Section 25.1.)
4. A prescription lens has a focal length of 25.0 cm. What is the power of the lens in diopters? (See Section 25.2.)
5. A man can see no farther than 42.0 cm without corrective eyeglasses. (a) Is the man nearsighted or farsighted? Find (b) the focal length of the appropriate corrective lens and (c) the power of the lens in diopters. (See Section 25.2.)
6. A woman can see clearly only when objects are 48.0 cm or farther away from her. (a) Is she nearsighted or farsighted? Find (b) the focal length of the appropriate corrective lens and (c) the power of the lens in diopters. (See Section 25.2.)
7. A single-lens magnifier has a maximum angular magnification of 9.33. Determine (a) the lens's focal length (in cm) and (b) the magnification when used with a relaxed eye. (See Section 25.3.)
8. A compound microscope has objective and eyepiece lenses of focal lengths 0.80 cm and 4.0 cm, respectively. If the microscope length is 15 cm, what is the magnification of the microscope? (See Section 25.4.)
9. Determine the angular magnification of a small, handheld telescope if its objective lens has a focal length of 25.0 cm and its eyepiece has a focal length of 2.20 cm. (See Section 25.5.)
10. Determine the approximate limiting angle of resolution (in rad) for 645-nm light incident on a single slit having a width of $1.12 \times 10^{-3}$ m. (See Section 25.6.)
11. Two stars have an angular separation of $7.50 \times 10^{-8}$ rad when viewed in the night sky from Earth. Determine the minimum diameter of a telescope's circular aperture if the two stars are to be angularly resolved at a wavelength of 532 nm. (See Section 25.6.)
12. What minimum resolving power must a diffraction grating have to distinguish between two emissions near a wavelength of 632 nm and separated by 2.00 nm? (See Section 25.6.)

## CONCEPTUAL QUESTIONS

ENHANCED WebAssign The conceptual questions in this chapter may be assigned online in Enhanced WebAssign.

1. A lens is used to examine an object across a room. Is the lens probably being used as a simple magnifier? Explain in terms of focal length, the image, and magnification.
2. A laser beam is incident at a shallow angle on a horizontal machinist's ruler that has a finely calibrated scale. The engraved rulings on the scale give rise to a diffraction pattern on a vertical screen. Discuss how you can use this technique to obtain a measure of the wavelength of the laser light.
3. BIO The optic nerve and the brain invert the image formed on the retina. Why don't we see everything upside down?
4. Suppose you are observing the interference pattern formed by a Michelson interferometer in a laboratory and a joking colleague holds a lit match in the light path of one arm of the interferometer. Will this match have an effect on the interference pattern?
5. If you want to examine the fine detail of an object with a magnifying glass with a power of +20.0 diopters, where should the object be placed so as to observe a magnified image of the object?
6. BIO Compare and contrast the eye and a camera. What parts of the camera correspond to the iris, the retina, and the cornea of the eye?

7. If you want to use a converging lens to set fire to a piece of paper, why should the light source be farther from the lens than its focal point?
8. Large telescopes are usually reflecting rather than refracting. List some reasons for this choice.
9. Explain why it is theoretically impossible to see an object as small as an atom regardless of the quality of the light microscope being used.
10. Which is most important in the use of a camera photoflash unit, the intensity of the light (the energy per unit area per unit time) or the product of the intensity and the time of the flash, assuming the time is less than the shutter speed?
11. BIO A patient has a near point of 1.25 m. Is she nearsighted or farsighted? Should the corrective lens be converging or diverging?
12. A lens with a certain power is used as a simple magnifier. If the power of the lens is doubled, does the angular magnification increase or decrease?
13. A laser produces a beam a few millimeters wide, with uniform intensity across its width. A hair is stretched vertically across the front of the laser to cross the beam. (a) How is the diffraction pattern it produces on a distant screen related to that of a vertical slit equal in width to the hair? (b) How could you determine the width of the hair from measurements of its diffraction pattern?
14. BIO During LASIK eye surgery (laser-assisted *in situ* keratomileusis), the shape of the cornea is modified by vaporizing some of its material. If the surgery is performed to correct for nearsightedness, how does the cornea need to be reshaped?
15. If you increase the aperture diameter of a camera by a factor of 3, how is the intensity of the light striking the film affected? (a) It increases by a factor of 3. (b) It decreases by a factor of 3. (c) It increases by a factor of 9. (d) It decreases by a factor of 9. (e) Increasing the aperture doesn't affect the intensity.

## PROBLEMS

ENHANCED WebAssign The problems in this chapter may be assigned online in Enhanced WebAssign.

1. denotes straightforward problem; 2. denotes intermediate problem;
3. denotes challenging problem
[1.] denotes full solution available in *Student Solutions Manual/ Study Guide*
1. denotes problems most often assigned in Enhanced WebAssign

BIO denotes biomedical problems
GP denotes guided problems
M denotes Master It tutorial available in Enhanced WebAssign
Q|C denotes asking for quantitative and conceptual reasoning
S denotes symbolic reasoning problem
W denotes Watch It video solution available in Enhanced WebAssign

### 25.1 The Camera

1. A lens has a focal length of 28 cm and a diameter of 4.0 cm. What is the $f$-number of the lens?
2. A certain camera has $f$-numbers that range from 1.2 to 22. If the focal length of the lens is 55 mm, what is the range of aperture diameters for the camera?
3. A photographic image of a building is 0.092 0 m high. The image was made with a lens with a focal length of 52.0 mm. If the lens was 100 m from the building when the photograph was made, determine the height of the building.
4. The image area of a typical 35 mm slide is 23.5 mm by 35.0 mm. If a camera's lens has a focal length of 55.0 mm and forms an image of the constellation Orion, which is 20° across, will the full image fit on a 35-mm slide?
5. A camera is being used with a correct exposure at $f/4$ and a shutter speed of $\frac{1}{15}$ s. In addition to the $f$-numbers listed in Section 25.1, this camera has $f$-numbers $f/1$, $f/1.4$, and $f/2$. To photograph a rapidly moving subject, the shutter speed is changed to $\frac{1}{125}$ s. Find the new $f$-number setting needed on this camera to maintain satisfactory exposure.
6. (a) Use conceptual arguments to show that the intensity of light (energy per unit area per unit time) reaching the film in a camera is proportional to the square of the reciprocal of the $f$-number as

$$I \propto \frac{1}{(f/D)^2}$$

(b) The correct exposure time for a camera set to $f/1.8$ is (1/500) s. Calculate the correct exposure time if the $f$-number is changed to $f/4$ under the same lighting conditions. *Note:* "$f/4$," on a camera, means "an $f$-number of 4."
7. A certain type of film requires an exposure time of 0.010 s with an $f/11$ lens setting. Another type of film requires twice the light energy to produce the same level of exposure. What $f$-number does the second type of film need with the 0.010-s exposure time?
8. A certain camera lens has a focal length of 175 mm. Its position can be adjusted to produce images when the lens is between 180 mm and 210 mm from the plane of the film. Over what range of object distances is the lens useful?

## 25.2 The Eye

**9.** BIO The near point of a person's eye is 60.0 cm. To see objects clearly at a distance of 25.0 cm, what should be the (a) focal length and (b) power of the appropriate corrective lens? (Neglect the distance from the lens to the eye.)

**10.** BIO GP A patient can't see objects closer than 40.0 cm and wishes to clearly see objects that are 20.0 cm from his eye. (a) Is the patient nearsighted or farsighted? (b) If the eye–lens distance is 2.00 cm, what is the minimum object distance $p$ from the lens? (c) What image position with respect to the lens will allow the patient to see the object? (d) Is the image real or virtual? Is the image distance $q$ positive or negative? (e) Calculate the required focal length. (f) Find the power of the lens in diopters. (g) If a contact lens is to be prescribed instead, find $p$, $q$, and $f$, and the power of the lens.

**11.** BIO M The accommodation limits for Nearsighted Nick's eyes are 18.0 cm and 80.0 cm. When he wears his glasses, he is able to see faraway objects clearly. At what minimum distance is he able to see objects clearly?

**12.** BIO W A certain child's near point is 10.0 cm; her far point (with eyes relaxed) is 125 cm. Each eye lens is 2.00 cm from the retina. (a) Between what limits, measured in diopters, does the power of this lens–cornea combination vary? (b) Calculate the power of the eyeglass lens the child should use for relaxed distance vision. Is the lens converging or diverging?

**13.** BIO An individual is nearsighted; his near point is 13.0 cm and his far point is 50.0 cm. (a) What lens power is needed to correct his nearsightedness? (b) When the lenses are in use, what is this person's near point?

**14.** BIO Q|C A patient has a near point of 45.0 cm and far point of 85.0 cm. (a) Can a single lens correct the patient's vision? Explain the patient's options. (b) Calculate the power lens needed to correct the near point so that the patient can see objects 25.0 cm away. Neglect the eye–lens distance. (c) Calculate the power lens needed to correct the patient's far point, again neglecting the eye–lens distance.

**15.** BIO An artificial lens is implanted in a person's eye to replace a diseased lens. The distance between the artificial lens and the retina is 2.80 cm. In the absence of the lens, an image of a distant object (formed by refraction at the cornea) falls 5.33 cm behind the implanted lens. The lens is designed to put the image of the distant object on the retina. What is the power of the implanted lens? *Hint:* Consider the image formed by the cornea to be a virtual object.

**16.** BIO A person is to be fitted with bifocals. She can see clearly when the object is between 30 cm and 1.5 m from the eye. (a) The upper portions of the bifocals (Fig. P25.16) should be designed to enable her to see distant objects clearly. What power should they have? (b) The lower portions of the bifocals should enable her to see objects located 25 cm in front of the eye. What power should they have?

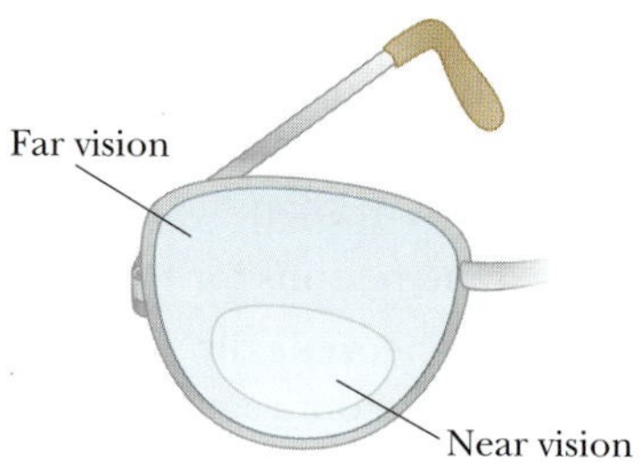

**Figure P25.16**

**17.** BIO A nearsighted woman can't see objects clearly beyond 40.0 cm (her far point). If she has no astigmatism and contact lenses are prescribed, what power and type of lens are required to correct her vision?

**18.** BIO Q|C A person sees clearly wearing eyeglasses that have a power of −4.00 diopters when the lenses are 2.00 cm in front of the eyes. (a) What is the focal length of the lens? (b) Is the person nearsighted or farsighted? (c) If the person wants to switch to contact lenses placed directly on the eyes, what lens power should be prescribed?

## 25.3 The Simple Magnifier

**19.** A stamp collector uses a lens with 7.5-cm focal length as a simple magnifier. The virtual image is produced at the normal near point (25 cm). (a) How far from the lens should the stamp be placed? (b) What is the expected angular magnification?

**20.** A lens that has a focal length of 5.00 cm is used as a magnifying glass. (a) To obtain maximum magnification and an image that can be seen clearly by a normal eye, where should the object be placed? (b) What is the angular magnification?

**21.** W A biology student uses a simple magnifier to examine the structural features of an insect's wing. The wing is held 3.50 cm in front of the lens, and the image is formed 25.0 cm from the eye. (a) What is the focal length of the lens? (b) What angular magnification is achieved?

**22.** A jeweler's lens of focal length 5.0 cm is used as a magnifier. With the lens held near the eye, determine (a) the angular magnification when the object is at the focal point of the lens and (b) the angular magnification when the image formed by the lens is at the near point of the eye (25 cm). (c) What is the object distance giving the maximum magnification?

**23.** A leaf of length $h$ is positioned 71.0 cm in front of a converging lens with a focal length of 39.0 cm. An observer views the image of the leaf from a position 1.26 m behind the lens, as shown in Figure P25.23. (a) What is the magnitude of the lateral magnification (the ratio of the image size to the object size) produced

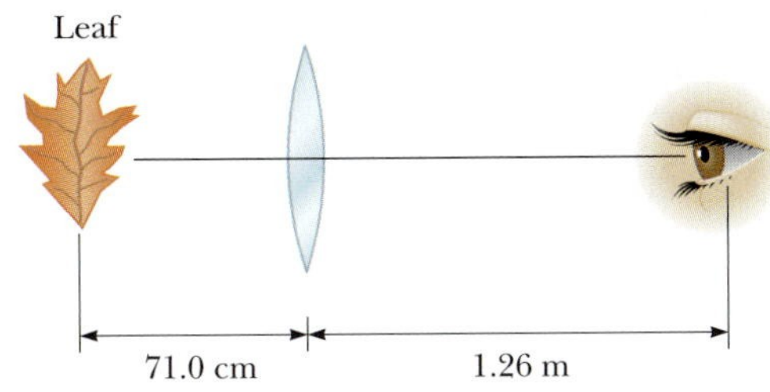


**Figure P25.23**

by the lens? (b) What angular magnification is achieved by viewing the image of the leaf rather than viewing the leaf directly?

**24.** (a) What is the maximum angular magnification of an eyeglass lens having a focal length of 18.0 cm when used as a simple magnifier? (b) What is the magnification of this lens when the eye is relaxed?

## 25.4 The Compound Microscope

## 25.5 The Telescope

**25.** The desired overall magnification of a compound microscope is 140×. The objective alone produces a lateral magnification of 12×. Determine the required focal length of the eyepiece.

**26.** The distance between the eyepiece and the objective lens in a certain compound microscope is 20.0 cm. The focal length of the objective is 0.500 cm, and that of the eyepiece is 1.70 cm. Find the overall magnification of the microscope.

**27.** Find the magnification of a telescope that uses a 2.75-diopter objective lens and a 35.0-diopter eyepiece.

**28.** BIO A microscope has an objective lens with a focal length of 16.22 mm and an eyepiece with a focal length of 9.50 mm. With the length of the barrel set at 29.0 cm, the diameter of a red blood cell's image subtends an angle of 1.43 mrad with the eye. If the final image distance is 29.0 cm from the eyepiece, what is the actual diameter of the red blood cell? *Hint:* To solve this question, go back to basics and use the thin-lens equation.

**29.** A certain telescope has an objective mirror with an aperture diameter of 200 mm and a focal length of 2 000 mm. It captures the image of a nebula on photographic film at its prime focus with an exposure time of 1.50 min. To produce the same light energy per unit area on the film, what is the required exposure time to photograph the same nebula with a smaller telescope that has an objective with a 60.0-mm diameter and a 900-mm focal length?

**30.** S (a) Find an equation for the length $L$ of a refracting telescope in terms of the focal length of the objective $f_o$ and the magnification $m$. (b) A knob adjusts the eyepiece forward and backward. Suppose the telescope is in focus with an eyepiece giving a magnification of 50.0. By what distance must the eyepiece be adjusted when the eyepiece is replaced, with a resulting magnification of $1.00 \times 10^2$? Must the eyepiece be adjusted backward or forward? Assume the objective lens has a focal length of 2.00 m.

**31.** Suppose an astronomical telescope is being designed to have an angular magnification of 34.0. If the focal length of the objective lens being used is 86.0 cm, find (a) the required focal length of the eyepiece and (b) the distance between the two lenses for a relaxed eye. *Hint:* For a relaxed eye, the image formed by the objective lens is at the focal point of the eyepiece.

**32.** A certain telescope has an objective of focal length 1 500 cm. If the Moon is used as an object, a 1.0-cm-long image formed by the objective corresponds to what distance, in miles, on the Moon? Assume $3.8 \times 10^8$ m for the Earth–Moon distance.

**33.** S Astronomers often take photographs with the objective lens or mirror of a telescope alone, without an eyepiece. (a) Show that the image size $h'$ for a telescope used in this manner is given by $h' = fh/(f - p)$, where $h$ is the object size, $f$ is the objective focal length, and $p$ is the object distance. (b) Simplify the expression in part (a) if the object distance is much greater than the objective focal length. (c) The "wingspan" of the International Space Station is 108.6 m, the overall width of its solar panel configuration. When it is orbiting at an altitude of 407 km, find the width of the image formed by a telescope objective of focal length 4.00 m.

**34.** BIO W An elderly sailor is shipwrecked on a desert island, but manages to save his eyeglasses. The lens for one eye has a power of +1.20 diopters, and the other lens has a power of +9.00 diopters. (a) What is the magnifying power of the telescope he can construct with these lenses? (b) How far apart are the lenses when the telescope is adjusted for minimum eyestrain?

**35.** M A person decides to use an old pair of eyeglasses to make some optical instruments. He knows that the near point in his left eye is 50.0 cm and the near point in his right eye is 100 cm. (a) What is the maximum angular magnification he can produce in a telescope? (b) If he places the lenses 10.0 cm apart, what is the maximum overall magnification he can produce in a microscope? *Hint:* Go back to basics and use the thin-lens equation to solve part (b).

**36.** Q|C Galileo devised a simple terrestrial telescope that produces an upright image. It consists of a converging objective lens and a diverging eyepiece at opposite ends of the telescope tube. For distant objects, the tube length is the objective focal length less the absolute value of the eyepiece focal length. (a) Does the user of the telescope see a real or virtual image? (b) Where is the final image? (c) If a telescope is to be constructed with a tube of length 10.0 cm and a magnification of 3.00, what are the focal lengths of the objective and eyepiece?

## 25.6 Resolution of Single-Slit and Circular Apertures

**37.** A converging lens with a diameter of 30.0 cm forms an image of a satellite passing overhead. The satellite has two green lights (wavelength 500 nm) spaced 1.00 m apart. If the lights can just be resolved according to the Rayleigh criterion, what is the altitude of the satellite?

**38.** While flying at an altitude of 9.50 km, you look out the window at various objects on the ground. If your ability to distinguish two objects is limited only by diffraction, find the smallest separation between two objects on the ground that are distinguishable. Assume your pupil has a diameter of 4.0 mm and take $\lambda = 575$ nm.

**39.** M To increase the resolving power of a microscope, the object and the objective are immersed in oil ($n = 1.5$). If the limiting angle of resolution without the oil is 0.60 $\mu$rad, what is the limiting angle of resolution with the oil? *Hint:* The oil changes the wavelength of the light.

**40.** BIO (a) Calculate the limiting angle of resolution for the eye, assuming a pupil diameter of 2.00 mm, a wavelength of 500 nm *in air*, and an index of refraction for the eye of 1.33. (b) What is the maximum distance from the eye at which two points separated by 1.00 cm could be resolved?

**41.** A vehicle with headlights separated by 2.00 m approaches an observer holding an infrared detector sensitive to radiation of wavelength 885 nm. What aperture diameter is required in the detector if the two headlights are to be resolved at a distance of 10.0 km?

**42.** Two stars located 23 light-years from Earth are barely resolved using a reflecting telescope having a mirror of diameter 68 cm. Assuming $\lambda = 575$ nm and assuming that the resolution is limited only by diffraction, find the separation between the stars.

**43.** Suppose a 5.00-m-diameter telescope were constructed on the Moon, where the absence of atmospheric distortion would permit excellent viewing. If observations were made using 500-nm light, what minimum separation between two objects could just be resolved on Mars at closest approach (when Mars is $8.0 \times 10^7$ km from the Moon)?

**44.** W A spy satellite circles Earth at an altitude of 200 km and carries out surveillance with a special high-resolution telescopic camera having a lens diameter of 35 cm. If the angular resolution of this camera is limited by diffraction, estimate the separation of two small objects on Earth's surface that are just resolved in yellow-green light ($\lambda = 550$ nm).

**45.** A 15.0-cm-long grating has 6 000 slits per centimeter. Can two lines of wavelengths 600.000 nm and 600.003 nm be separated with this grating? Explain.

**46.** The $H_\alpha$ line in hydrogen has a wavelength of 656.20 nm. This line differs in wavelength from the corresponding spectral line in deuterium (the heavy stable isotope of hydrogen) by 0.18 nm. (a) Determine the minimum number of lines a grating must have to resolve these two wavelengths in the first order. (b) Repeat part (a) for the second order.

## 25.7 The Michelson Interferometer

**47.** Light of wavelength 550 nm is used to calibrate a Michelson interferometer. With the use of a micrometer screw, the platform on which one mirror is mounted is moved 0.180 mm. How many fringe shifts are counted?

**48.** Q|C Monochromatic light is beamed into a Michelson interferometer. The movable mirror is displaced 0.382 mm, causing the central spot in the interferometer pattern to change from bright to dark and back to bright $N = 1\,700$ times. (a) Determine the wavelength of the light. What color is it? (b) If monochromatic red light is used instead and the mirror is moved the same distance, would $N$ be larger or smaller? Explain.

**49.** BIO An interferometer is used to measure the length of a bacterium. The wavelength of the light used is 650 nm. As one arm of the interferometer is moved from one end of the cell to the other, 310 fringe shifts are counted. How long is the bacterium?

**50.** Mirror $M_1$ in Figure 25.15 is displaced a distance $\Delta L$. During this displacement, 250 fringe shifts are counted. The light being used has a wavelength of 632.8 nm. Calculate the displacement $\Delta L$.

**51.** A thin sheet of transparent material has an index of refraction of 1.40 and is 15.0 $\mu$m thick. When it is inserted in the light path along one arm of an interferometer, how many fringe shifts occur in the pattern? Assume the wavelength (in a vacuum) of the light used is 600 nm. *Hint:* The wavelength will change within the material.

**52.** M The Michelson interferometer can be used to measure the index of refraction of a gas by placing an evacuated transparent tube in the light path along one arm of the device. Fringe shifts occur as the gas is slowly added to the tube. Assume 600-nm light is used, the tube is 5.00 cm long, and 160 fringe shifts occur as the pressure of the gas in the tube increases to atmospheric pressure. What is the index of refraction of the gas? *Hint:* The fringe shifts occur because the wavelength of the light changes inside the gas-filled tube.

## Additional Problems

**53.** The Yerkes refracting telescope has a 1.00-m-diameter objective lens of focal length 20.0 m. Assume it is used with an eyepiece of focal length 2.50 cm. (a) Determine the magnification of the planet Mars as seen through the telescope. (b) Are the observed Martian polar caps right side up or upside down?

**54.** Estimate the minimum angle subtended at the eye of a hawk flying at an altitude of 50 m necessary to recognize a mouse on the ground.

55. An American standard analog television picture (non-HDTV), also known as NTSC, is composed of approximately 485 visible horizontal lines of varying light intensity. Assume your ability to resolve the lines is limited only by the Rayleigh criterion, the pupils of your eyes are 5.00 mm in diameter, and the average wavelength of the light coming from the screen is 550 nm. Calculate the ratio of the minimum viewing distance to the vertical dimension of the picture such that you will not be able to resolve the lines.

56. BIO A person with a nearsighted eye has near and far points of 16 cm and 25 cm, respectively. (a) Assuming a lens is placed 2.0 cm from the eye, what power must the lens have to correct this condition? (b) Suppose contact lenses placed directly on the cornea are used to correct the person's eyesight. What is the power of the lens required in this case, and what is the new near point? *Hint:* The contact lens and the eyeglass lens require slightly different powers because they are at different distances from the eye.

57. BIO The near point of an eye is 75.0 cm. (a) What should be the power of a corrective lens prescribed to enable the eye to see an object clearly at 25.0 cm? (b) If, using the corrective lens, the person can see an object clearly at 26.0 cm but not at 25.0 cm, by how many diopters did the lens grinder miss the prescription?

58. BIO If a typical eyeball is 2.00 cm long and has a pupil opening that can range from about 2.00 mm to 6.00 mm, what are (a) the focal length of the eye when it is focused on objects 1.00 m away, (b) the smallest *f*-number of the eye when it is focused on objects 1.00 m away, and (c) the largest *f*-number of the eye when it is focused on objects 1.00 m away?

59. BIO M A cataract-impaired lens in an eye may be surgically removed and replaced by a manufactured lens. The focal length required for the new lens is determined by the lens-to-retina distance, which is measured by a sonarlike device, and by the requirement that the implant provide for correct distance vision. (a) If the distance from lens to retina is 22.4 mm, calculate the power of the implanted lens in diopters. (b) Since there is no accommodation and the implant allows for correct distance vision, a corrective lens for close work or reading must be used. Assume a reading distance of 33.0 cm, and calculate the power of the lens in the reading glasses.

60. BIO If the aqueous humor of the eye has an index of refraction of 1.34 and the distance from the vertex of the cornea to the retina is 2.00 cm, what is the radius of curvature of the cornea for which distant objects will be focused on the retina? (For simplicity, assume all refraction occurs in the aqueous humor.)

61. A Boy Scout starts a fire by using a lens from his eyeglasses to focus sunlight on kindling 5.0 cm from the lens. The Boy Scout has a near point of 15 cm. When the lens is used as a simple magnifier, (a) what is the maximum magnification that can be achieved and (b) what is the magnification when the eye is relaxed? *Caution:* The equations derived in the text for a simple magnifier assume a "normal" eye.

62. A laboratory (astronomical) telescope is used to view a scale that is 300 cm from the objective, which has a focal length of 20.0 cm; the eyepiece has a focal length of 2.00 cm. Calculate the angular magnification when the telescope is adjusted for minimum eyestrain. *Note:* The object is not at infinity, so the simple expression $m = f_o/f_e$ is not sufficiently accurate for this problem. Also, assume small angles, so that $\tan\theta \approx \theta$.

Albert Einstein revolutionized modern physics. He explained the random motion of pollen grains, which proved the existence of atoms, and the photoelectric effect, which showed that light was a particle as well as a wave. With Satyendra Nath Bose he predicted a new form of matter, the Bose-Einstein condensate, which was discovered in the laboratory forty years after his death. His theory of special relativity made clear the foundations of space and time, and established a key relationship between mass and energy. His theory of gravitation, general relativity, led to a deeper understanding of planetary motion, the structure and evolution of stars, and the expanding universe. The equation in this photo, loosely translated, says that the average curvature of spacetime is zero in empty space.

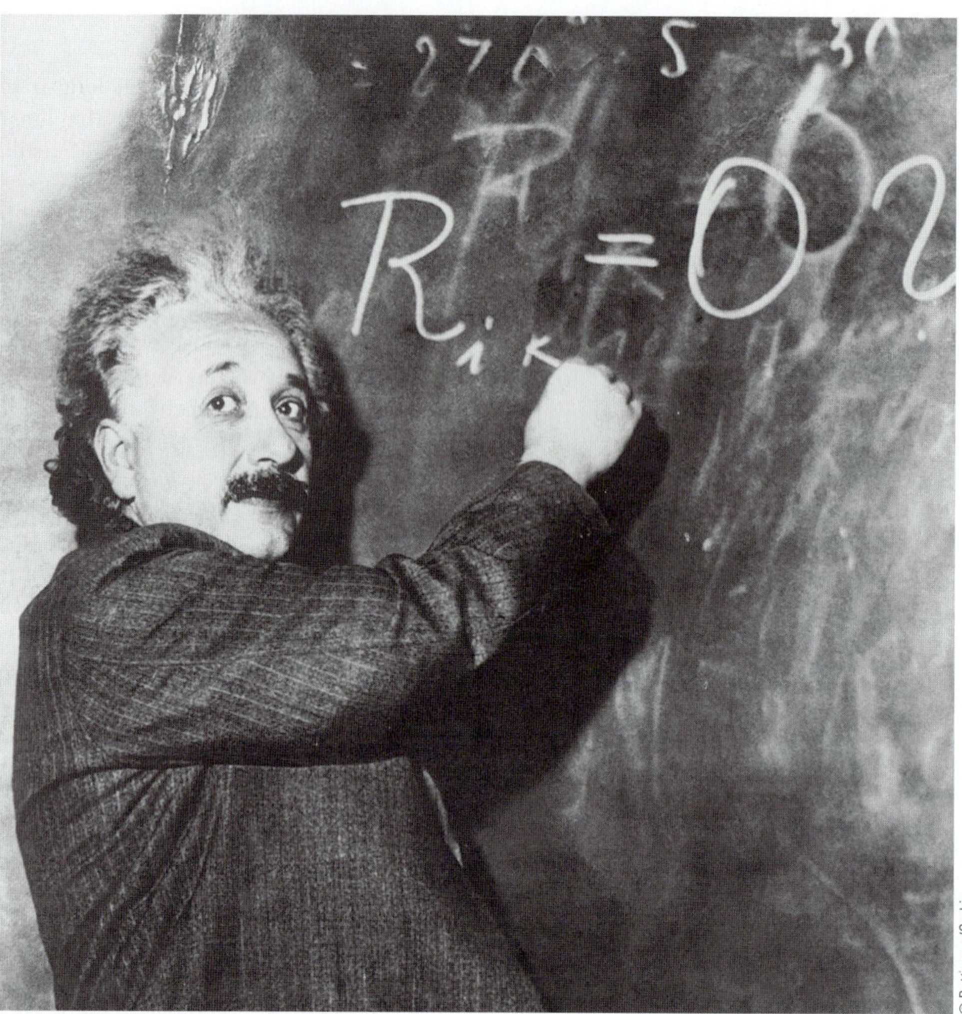



# 26 Relativity



Most of our everyday experiences and observations have to do with objects that move at speeds much less than the speed of light. Newtonian mechanics was formulated to describe the motion of such objects, and its formalism is quite successful in describing a wide range of phenomena that occur at low speeds. It fails, however, when applied to particles having speeds approaching that of light.

Experimentally, for example, it's possible to accelerate an electron to a speed of $0.99c$ (where $c$ is the speed of light) by using a potential difference of several million volts. According to Newtonian mechanics, if the potential difference is increased by a factor of 4, the electron's kinetic energy is four times greater and its speed should double to $1.98c$. Experiments, however, show that the speed of the electron—as well as the speed of any other particle that has mass—always remains *less* than the speed of light, regardless of the size of the accelerating voltage.

The existence of a universal speed limit has far-reaching consequences. It means that the usual concepts of force, momentum, and energy no longer apply for rapidly moving objects. A less obvious consequence is that observers moving at different speeds will measure different time intervals and displacements between the same two events. Relating the measurements made by different observers is the subject of relativity.

## 26.1 Galilean Relativity

**LEARNING OBJECTIVE**

1. State and discuss the principle of Galilean relativity.

To describe a physical event, it's necessary to choose a *frame of reference.* When you perform an experiment in a laboratory, for example, you select a coordinate system, or frame of reference, that is at rest with respect to the laboratory. Suppose, instead, you choose to do an experiment in the back of a truck moving at a constant velocity $\vec{\mathbf{v}}$. You can then select a moving frame of reference that's at rest with respect to the truck. If you found Newton's first law to be valid in that frame, would an observer at rest with respect to the Earth agree with you?

According to the principle of Galilean relativity, **the laws of mechanics must be the same in all inertial frames of reference**. Inertial frames of reference are those in which Newton's laws are valid. In these frames, objects move in straight lines at constant speed unless acted on by a nonzero net force, thus the name "inertial frame" because objects observed from these frames obey Newton's first law, the law of inertia. For the situation described in the previous paragraph, the laboratory coordinate system and the coordinate system of the moving car are both inertial frames of reference. Consequently, if the laws of mechanics are found to be true in the laboratory, then the person in the car must also observe the same laws.

Consider a truck in motion, moving with a constant velocity, as in Figure 26.1a. If a passenger in the truck throws a ball straight up in the air, the passenger observes that the ball moves in a vertical path. The motion of the ball is precisely the same as it would be if the ball were thrown while at rest on Earth. The law of gravity and the equations of motion under constant acceleration are obeyed whether the truck is at rest or in uniform motion.

Now consider the same experiment when viewed by another observer at rest on Earth. This stationary observer views the path of the ball in the truck to be a parabola, as in Figure 26.1b. Further, according to this observer, the ball has a velocity to the right equal to the velocity of the truck. Although the two observers disagree on the shape of the ball's path, both agree that the motion of the ball obeys the law of gravity and Newton's laws of motion, and they even agree on how long the ball is in the air. We draw the following important conclusion: **There is no preferred frame of reference for describing the laws of mechanics.**

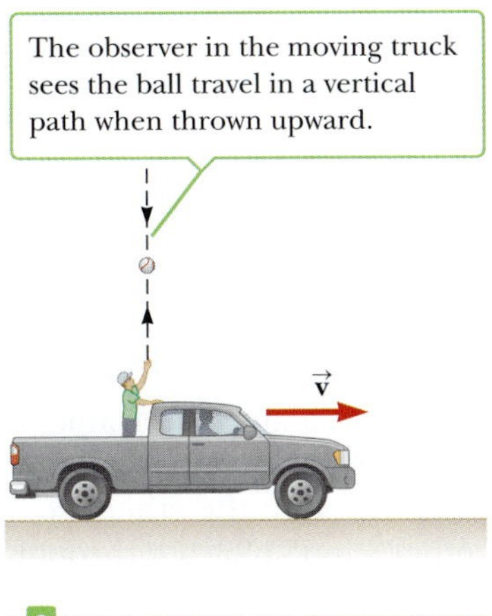
The observer in the moving truck sees the ball travel in a vertical path when thrown upward.
$\vec{\mathbf{v}}$
a

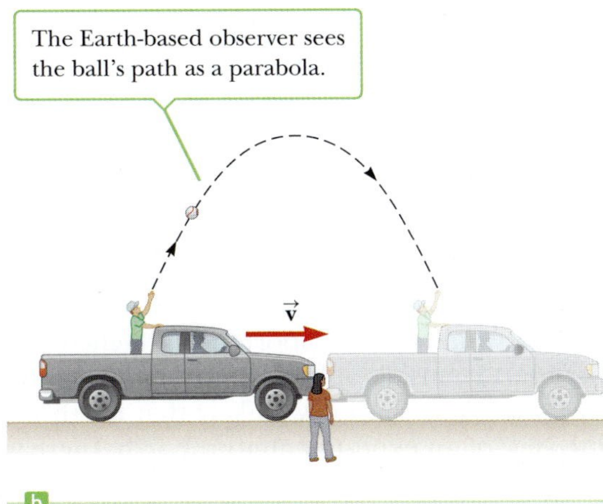
The Earth-based observer sees the ball's path as a parabola.
$\vec{\mathbf{v}}$
b

**Figure 26.1** Two observers watch the path of a thrown ball and obtain different results.

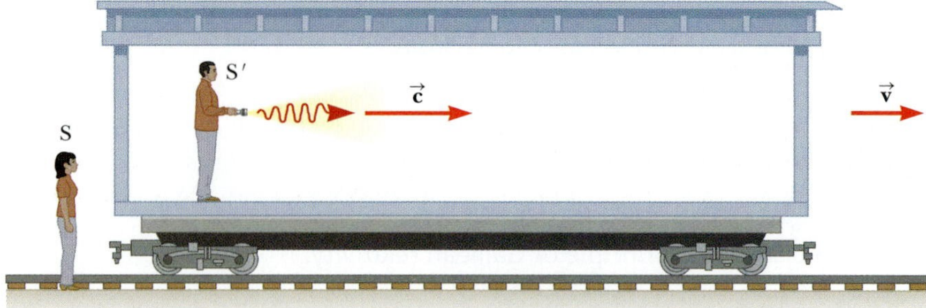
S′
$\vec{c}$
$\vec{v}$
S

**Figure 26.2** A pulse of light is sent out by a person in a moving boxcar. According to Galilean relativity, the speed of the pulse should be $\vec{c} + \vec{v}$ relative to a stationary observer.

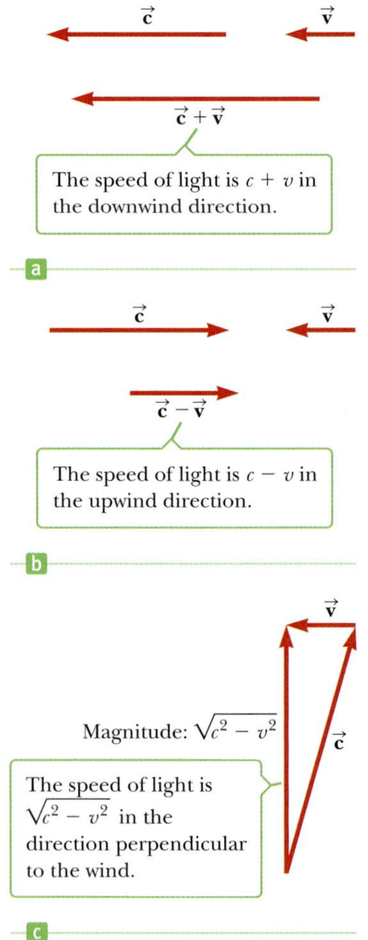
$\vec{c}$
$\vec{v}$
$\vec{c} + \vec{v}$
The speed of light is $c + v$ in the downwind direction.
a
$\vec{c}$
$\vec{v}$
$\vec{c} - \vec{v}$
The speed of light is $c - v$ in the upwind direction.
b
$\vec{v}$
Magnitude: $\sqrt{c^2 - v^2}$
$\vec{c}$
The speed of light is $\sqrt{c^2 - v^2}$ in the direction perpendicular to the wind.
c

**Figure 26.3** If the speed of the ether wind relative to Earth is $v$ and $c$ is the speed of light relative to the ether, the speed of light relative to Earth is (a) $c + v$ in the down-wind direction, (b) $c - v$ in the upwind direction, and (c) $\sqrt{c^2 - v^2}$ in the direction perpendicular to the wind. The Michelson–Morley experiment, however, disproved the ether wind hypothesis, leading to Einstein's postulate that the speed of light in vacuum has the same value regardless of the motion of an inertial observer.

## 26.2 The Speed of Light

**LEARNING OBJECTIVES**

1. Discuss the speed of light in the context of Galilean relativity and their predictions.
2. Explain the discredited concept of the luminiferous ether.
3. Describe the Michelson-Morley experiment and its consequences for the speed of light as measured by different observers, and what the experiment implies about the ether.

It's natural to ask whether the concept of Galilean relativity in mechanics also applies to experiments in electricity, magnetism, optics, and other areas. Experiments indicate that the answer is no. Further, if we assume the laws of electricity and magnetism are the same in all inertial frames, a paradox concerning the speed of light immediately arises. According to electromagnetic theory, the speed of light always has the fixed value of $2.997\ 924\ 58 \times 10^8$ m/s in free space. According to Galilean relativity, however, the speed of the pulse relative to the stationary observer S outside the boxcar in Figure 26.2 should be $c + v$. Hence, Galilean relativity is inconsistent with Maxwell's well-tested theory of electromagnetism.

Electromagnetic theory predicts that light waves must propagate through free space with a speed equal to the speed of light. The theory doesn't require the presence of a medium for wave propagation, however. This is in contrast to other types of waves, such as water or sound waves, that do require a medium to support the disturbances. In the 19th century, physicists thought that electromagnetic waves also required a medium to propagate. They proposed that such a medium existed and gave it the name **luminiferous ether**. The ether was assumed to be present everywhere, even in empty space, and light waves were viewed as ether oscillations. Further, the ether would have to be a massless but rigid medium with no effect on the motion of planets or other objects. These concepts are indeed strange. In addition, it was found that the troublesome laws of electricity and magnetism would take on their simplest forms in a special frame of reference at *rest* with respect to the ether. This frame was called the *absolute frame*. The laws of electricity and magnetism would be valid in this absolute frame, but they would have to be modified in any reference frame moving with respect to the absolute frame.

As a result of the importance attached to the ether and the absolute frame, it became of considerable interest in physics to prove by experiment that they existed. Because it was considered likely that Earth was in motion through the ether, from the view of an experimenter on Earth, there was an "ether wind" blowing through the laboratory. A direct method for detecting the ether wind would use an apparatus fixed to Earth to measure the wind's influence on the speed of light. If $v$ is the speed of the ether relative to Earth, then the speed of light should have its maximum value, $c + v$, when propagating downwind, as shown in Figure 26.3a. Likewise, the speed of light should have its minimum value, $c - v$, when propagating upwind, as in Figure 26.3b, and an intermediate value, $(c^2 - v^2)^{1/2}$, in the direction perpendicular to the ether wind, as in Figure 26.3c. If the Sun were assumed to be at rest in the ether, then the velocity of the ether wind would be equal to the orbital velocity of Earth around the Sun, which has a magnitude of approximately $3 \times 10^4$ m/s. Because $c = 3 \times 10^8$ m/s, it should be possible to detect a change in speed of about 1 part in $10^4$ for measurements in the upwind or downwind direction.

**Albert Einstein**
**German–American Physicist (1879–1955)**

One of the greatest physicists of all time, Einstein was born in Ulm, Germany. In 1905 at the age of 26 he published four scientific papers that revolutionized physics. Two of these papers introduced the special theory of relativity, considered by many to be his most important work. In 1916, in an exciting race with mathematician David Hilbert, Einstein published his theory of gravity, called the general theory of relativity. The most dramatic prediction of this theory is the degree to which light is deflected by a gravitational field. Measurements made by astronomers on bright stars in the vicinity of the eclipsed Sun in 1919 confirmed Einstein's prediction, and as a result Einstein became a world celebrity. Einstein was deeply disturbed by the development of quantum mechanics in the 1920s despite his own role as a scientific revolutionary. In particular, he could never accept the probabilistic view of events in nature that is a central feature of quantum theory. The last few decades of his life were devoted to an unsuccessful search for a unified theory that would combine gravitation and electromagnetism.

## The Michelson–Morley Experiment

The most famous experiment designed to detect these small changes in the speed of light was first performed in 1881 by Albert A. Michelson (1852–1931) and later repeated under various conditions by Michelson and Edward W. Morley (1838–1923). The experiment was designed to determine the velocity of Earth relative to the hypothetical ether. The experimental tool used was the Michelson interferometer, shown in Figure 26.4. Arm 2 is aligned along the direction of Earth's motion through space. Earth's moving through the ether at speed $v$ is equivalent to the ether flowing past Earth in the opposite direction with speed $v$. This ether wind blowing in the direction opposite the direction of Earth's motion should cause the speed of light measured in the Earth frame to be $c - v$ as the light approaches mirror $M_2$ and $c + v$ after reflection, where $c$ is the speed of light in the ether frame.

The two beams reflected from $M_1$ and $M_2$ recombine, and an interference pattern consisting of alternating dark and bright fringes is formed. The interference pattern was observed while the interferometer was rotated through an angle of 90°. This rotation supposedly would change the speed of the ether wind along the direction of arm 1. The effect of such rotation should have been to cause the fringe pattern to shift slightly but measurably, but measurements failed to show any change in the interference pattern! Even though the Michelson–Morley experiment was repeated at different times of the year when the ether wind was expected to change direction, the results were always the same: **no fringe shift of the magnitude required was ever observed**.

The negative results of the Michelson–Morley experiment not only contradicted the ether hypothesis, but also showed that it was impossible to measure the absolute velocity of Earth with respect to the ether frame. As we will see in the next section, however, Einstein suggested a postulate in the special theory of relativity that places quite a different interpretation on these negative results. In later years, when more was known about the nature of light, the idea of an ether that permeates all space was discarded. **Light is now understood to be an electromagnetic wave that requires no medium for its propagation.**

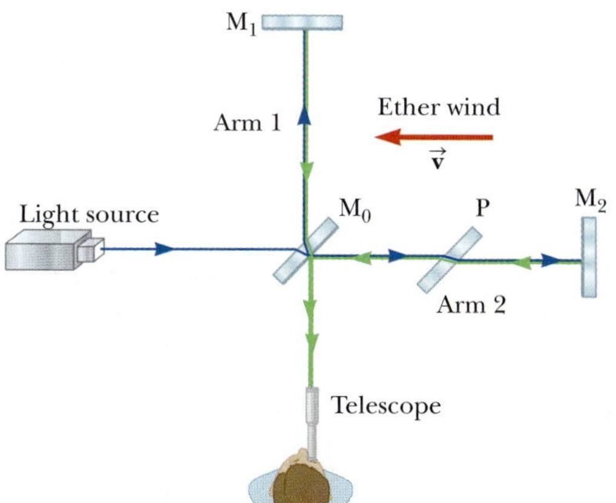


**Figure 26.4** According to the ether wind theory, the speed of light should be $c - v$ as the beam approaches mirror $M_2$ and $c + v$ after reflection.

# 26.3 Einstein's Principle of Relativity

**LEARNING OBJECTIVE**

1. State the two postulates of special relativity.

In 1905 Albert Einstein proposed a theory that explained the result of the Michelson–Morley experiment and completely altered our notions of space and time. He based his special theory of relativity on two postulates:

Postulates of relativity ▶

1. **The principle of relativity:** All the laws of physics are the same in all inertial frames.
2. **The constancy of the speed of light:** The speed of light in a vacuum has the same value, $c = 2.997\ 924\ 58 \times 10^8$ m/s, in all inertial reference frames, regardless of the velocity of the observer or the velocity of the source emitting the light.

The first postulate asserts that *all* the laws of physics are the same in all reference frames moving with constant velocity relative to each other. This postulate is a sweeping generalization of the principle of Galilean relativity, which refers only to the laws of mechanics. From an experimental point of view, Einstein's principle of relativity means that *any* kind of experiment—mechanical, thermal, optical, or electrical—performed in a laboratory at rest must give the same result when performed in a laboratory moving at a constant speed past the first one. Hence, no preferred inertial reference frame exists, and it is impossible to detect absolute motion.

Although postulate 2 was a brilliant theoretical insight on Einstein's part in 1905, it has since been confirmed experimentally in many ways. Perhaps the most direct demonstration involves measuring the speed of photons emitted by particles traveling at 99.99% of the speed of light. The measured photon speed in this case agrees to five significant figures with the speed of light in empty space.

The null result of the Michelson–Morley experiment can be readily understood within the framework of Einstein's theory. According to his principle of relativity, the premises of the Michelson–Morley experiment were incorrect. In the process of trying to explain the expected results, we stated that when light traveled against the ether wind its speed was $c - v$. If, however, the state of motion of the observer or of the source has no influence on the value found for the speed of light, the measured value must always be $c$. Likewise, the light makes the return trip after reflection from the mirror at a speed of $c$, not at a speed of $c + v$. Thus, the motion of Earth does not influence the fringe pattern observed in the Michelson–Morley experiment, and a null result should be expected.

If we accept Einstein's theory of relativity, we must conclude that uniform relative motion is unimportant when measuring the speed of light. At the same time, we have to adjust our commonsense notions of space and time and be prepared for some rather bizarre consequences.

**Quick Quiz**

**26.1** True or False: If you were traveling in a spaceship at a speed of $c/2$ relative to Earth and you fired a laser beam in the direction of the spaceship's motion, the light from the laser would travel at a speed of $3c/2$ relative to Earth.

# 26.4 Consequences of Special Relativity

**LEARNING OBJECTIVES**

1. Discuss the concept of simultaneity and why Einstein abandoned it.
2. Use the second postulate of relativity to derive the phenomenon of time dilation. Discuss its experimental proof.

3. Calculate the elapsed time as measured by observers in different states of relative motion.
4. Explain the twin paradox.
5. Derive the phenomenon of length contraction.
6. Calculate lengths measured by observers moving at different relative velocities.

Almost everyone who has dabbled even superficially in science is aware of some of the startling predictions that arise because of Einstein's approach to relative motion. As we examine some of the consequences of relativity in this section, we'll find that they conflict with some of our basic notions of space and time. We restrict our discussion to the concepts of length, time, and simultaneity, which are quite different in relativistic mechanics from what they are in Newtonian mechanics. For example, in relativistic mechanics the distance between two points and the time interval between two events depend on the frame of reference in which they are measured. **In relativistic mechanics there is no such thing as absolute length or absolute time.** Further, **events at different locations that are observed to occur simultaneously in one frame are not observed to be simultaneous in another frame moving uniformly past the first.**

◀ Length and time measurements depend on the frame of reference

## Simultaneity and the Relativity of Time

A basic premise of Newtonian mechanics is that a universal time scale exists that is the same for all observers. Newton and his followers simply took simultaneity for granted. In his special theory of relativity, Einstein abandoned that assumption.

Einstein devised the following thought experiment to illustrate this point. A boxcar moves with uniform velocity, and two lightning bolts strike its ends, as in Figure 26.5a, leaving marks on the boxcar and the ground. The marks on the boxcar are labeled $A'$ and $B'$, and those on the ground are labeled $A$ and $B$. An observer at $O'$ moving with the boxcar is midway between $A'$ and $B'$, and an observer on the ground at $O$ is midway between $A$ and $B$. The events recorded by the observers are the striking of the boxcar by the two lightning bolts.

The light signals recording the instant when the two bolts struck reach observer $O$ at the same time, as indicated in Figure 26.5b. This observer realizes that the signals have traveled at the same speed over equal distances and so rightly concludes that the events at $A$ and $B$ occurred simultaneously. Now consider the same events as viewed by observer $O'$. By the time the signals have reached observer $O$, observer $O'$ has moved as indicated in Figure 26.5b. Thus, the signal from $B'$ has already swept past $O'$, but the signal from $A'$ has not yet reached $O'$. In other words, $O'$ sees the signal from $B'$ before seeing the signal from $A'$. According to Einstein, *the two observers must find that light travels at the same speed.* Therefore, observer $O'$ concludes that the lightning struck the front of the boxcar before it struck the back.

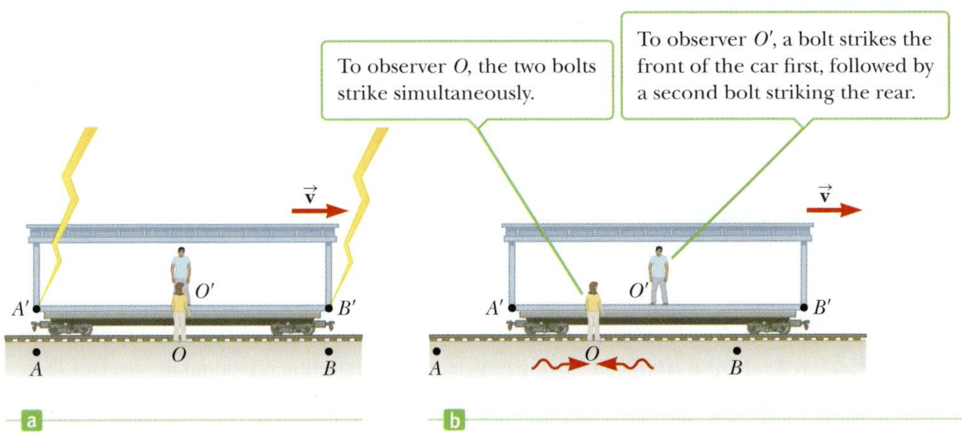

To observer O, the two bolts strike simultaneously.
To observer O′, a bolt strikes the front of the car first, followed by a second bolt striking the rear.
v
A′
O′
B′
A
O
B
a
b


**Figure 26.5** (a) Two lightning bolts strike the ends of a moving boxcar. (b) The leftward-traveling light signal has already passed $O'$, but the rightward-traveling signal has not yet reached $O'$.

This thought experiment clearly demonstrates that the two events that appear to be simultaneous to observer $O$ do not appear to be simultaneous to observer $O'$. In other words,

> Two events that are simultaneous in one reference frame are in general not simultaneous in a second frame moving relative to the first. Simultaneity depends on the state of motion of the observer and is therefore not an absolute concept.

**Tip 26.1 Who's Right?**

Which of two inertial observers is correct regarding the simultaneity of two events? Despite different conclusions, both are correct in their own inertial reference frames.

At this point, you might wonder which observer is right concerning the two events. The answer is that *both* are correct because the principle of relativity states that **there is no preferred inertial frame of reference**. Although the two observers reach different conclusions, both are correct in their own reference frames because the concept of simultaneity is not absolute. In fact, this is the central point of relativity: Any inertial frame of reference can be used to describe events and do physics.

## Time Dilation

We can illustrate that observers in different inertial frames may measure different time intervals between a pair of events by considering a vehicle moving to the right with a speed $v$ as in Figure 26.6a. A mirror is fixed to the ceiling of the vehicle, and an observer $O'$ at rest in this system holds a laser a distance $d$ below the mirror. At some instant, the laser emits a pulse of light directed toward the mirror (event 1), and at some later time after reflecting from the mirror, the pulse arrives back at the laser (event 2). Observer $O'$ carries a clock and uses it to measure the time interval $\Delta t_p$ between these two events, which she views as occurring at the same place. (The subscript $p$ stands for *proper*, as we'll see in a moment.) Because the light pulse has a speed $c$, the time it takes it to travel from point $A$ to the mirror and back to point $A$ is

$$\Delta t_p = \frac{\text{distance traveled}}{\text{speed}} = \frac{2d}{c} \qquad [26.1]$$

The time interval $\Delta t_p$ measured by $O'$ requires only a single clock located at the same place as the laser in this frame.

Now consider the same set of events as viewed by $O$ in a second frame, as shown in Figure 26.6b. According to this observer, the mirror and laser are moving to the right with a speed $v$, and, as a result, the sequence of events appears different. By the time the light from the laser reaches the mirror, the mirror has moved to the

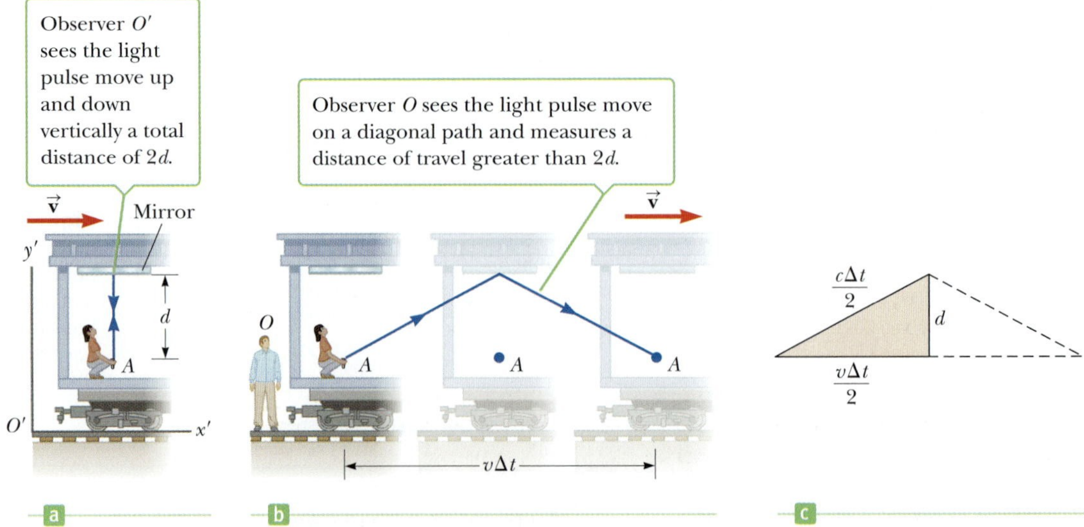
Observer $O'$ sees the light pulse move up and down vertically a total distance of $2d$.
Observer $O$ sees the light pulse move on a diagonal path and measures a distance of travel greater than $2d$.
$\vec{v}$
Mirror
$y'$
$d$
$A$
$O'$
$x'$
$O$
$A$
$A$
$A$
$\vec{v}$
$v\Delta t$
$\frac{c\Delta t}{2}$
$d$
$\frac{v\Delta t}{2}$
a
b
c

**Figure 26.6** (a) A mirror is fixed to a moving vehicle, and a light pulse is sent out by observer $O'$ at rest in the vehicle. (b) Relative to a stationary observer $O$ standing alongside the vehicle, the mirror and $O'$ move with a speed $v$. (c) The right triangle for calculating the relationship between $\Delta t$ and $\Delta t_p$.

right a distance $v\Delta t/2$, where $\Delta t$ is the time it takes the light pulse to travel from point $A$ to the mirror and back to point $A$ as measured by $O$. In other words, $O$ concludes that because of the motion of the vehicle, if the light is to hit the mirror, it must leave the laser at an angle with respect to the vertical direction. Comparing Figures 26.6a and 26.6b, we see that the light must travel farther in (b) than in (a). (Note that neither observer "knows" that he or she is moving. Each is at rest in his or her own inertial frame.)

According to the second postulate of the special theory of relativity, both observers must measure $c$ for the speed of light. Because the light travels farther in the frame of $O$, it follows that the time interval $\Delta t$ measured by $O$ is longer than the time interval $\Delta t_p$ measured by $O'$. To obtain a relationship between these two time intervals, it is convenient to examine the right triangle shown in Figure 26.6c. The Pythagorean theorem gives

$$\left(\frac{c\Delta t}{2}\right)^2 = \left(\frac{v\Delta t}{2}\right)^2 + d^2$$

Solving for $\Delta t$ yields

$$\Delta t = \frac{2d}{\sqrt{c^2 - v^2}} = \frac{2d}{c\sqrt{1 - v^2/c^2}}$$

Because $\Delta t_p = 2d/c$, we can express this result as

$$\Delta t = \frac{\Delta t_p}{\sqrt{1 - v^2/c^2}} = \gamma\,\Delta t_p \qquad [26.2]$$

◄ Time dilation

where

$$\gamma = \frac{1}{\sqrt{1 - v^2/c^2}} \qquad [26.3]$$

Because $\gamma$ is always greater than 1, Equation 26.2 says that **the time interval $\Delta t$ between two events measured by an observer moving with respect to a clock[1] is longer than the time interval $\Delta t_p$ between the same two events measured by an observer at rest with respect to the clock**. Consequently, $\Delta t > \Delta t_p$, and the proper time interval is expanded or dilated by the factor $\gamma$. Hence, this effect is known as **time dilation**.

For example, suppose the observer at rest with respect to the clock measures the time required for the light flash to leave the laser and return. We assume the measured time interval in this frame of reference, $\Delta t_p$, is 1 s. (This would require a very tall vehicle.) Now we find the time interval as measured by observer $O$ moving with respect to the same clock. If observer $O$ is traveling at half the speed of light ($v = 0.500c$), then $\gamma = 1.15$ and, according to Equation 26.2, $\Delta t = \gamma\,\Delta t_p = 1.15(1.00\text{ s}) = 1.15\text{ s}$. Therefore, when observer $O'$ claims that 1.00 s has passed, observer $O$ claims that 1.15 s has passed. Observer $O$ considers the clock of $O'$ to be reading too low a value for the elapsed time between the two events and says that the clock of $O'$ is "running slow." From this phenomenon, we may conclude the following:

> A clock moving past an observer at speed $v$ runs more slowly than an identical clock at rest with respect to the observer by a factor of $\gamma^{-1}$.

◄ A clock in motion runs more slowly than an identical stationary clock

The time interval $\Delta t_p$ in Equations 26.1 and 26.2 is called the **proper time**. In general, **proper time is the time interval between two events as measured by an observer who sees the events occur at the same position.**

Although you may have realized it by now, it's important to spell out that relativity is a scientific democracy: the view of $O'$ that $O$ is actually the one moving

[1]Actually, Figure 26.6 shows the clock moving and not the observer, but that is equivalent to observer $O$ moving to the left with velocity $\vec{v}$ with respect to the clock.

with speed $v$ to the left and that the clock of $O$ is running more slowly is just as valid as the view of $O$. The principle of relativity requires that the views of two observers in uniform relative motion be equally valid and capable of being checked experimentally.

**Tip 26.2 Proper Time Interval**

You must be able to correctly identify the observer who measures the proper time interval. The proper time interval between two events is the time interval measured by an observer for whom the two events take place at the same position.

We have seen that moving clocks run slow by a factor of $\gamma^{-1}$. This is true for ordinary mechanical clocks as well as for the light clock just described. In fact we can generalize these results by stating that all physical processes, including chemical and biological ones, slow down relative to a clock when those processes occur in a frame moving with respect to the clock. For example, the heartbeat of an astronaut moving through space would keep time with a clock inside the spaceship. Both the astronaut's clock and heartbeat would be slowed down relative to a clock back on Earth (although the astronaut would have no sensation of life slowing down in the spaceship).

Time dilation is a very real phenomenon that has been verified by various experiments involving the ticking of natural clocks. An interesting example of time dilation involves the observation of *muons,* unstable elementary particles that are very similar to electrons, having the same charge, but 207 times the mass. Muons can be produced by the collision of cosmic radiation with atoms high in the atmosphere. These particles have a lifetime of 2.2 $\mu$s when measured in a reference frame at rest with respect to them. If we take 2.2 $\mu$s as the average lifetime of a muon and assume that their speed is close to the speed of light, say $0.99c$, we find that these particles can travel only about 650 m before they decay (Fig. 26.7a). Hence, they could never reach Earth from the upper atmosphere where they are produced. Experiments, however, show that a large number of muons *do* reach Earth, and the phenomenon of time dilation explains how. Relative to an observer on Earth, the muons have a lifetime equal to $\gamma\tau_p$, where $\tau_p = 2.2\ \mu\text{s}$ is the lifetime in a frame of reference traveling with the muons. For example, for $v = 0.99c$, $\gamma \approx 7.1$, and $\gamma\tau_p \approx 16\ \mu\text{s}$. Hence, the average distance muons travel as measured by an observer on Earth is $\gamma v\tau_p \approx 4\ 800$ m, as indicated in Figure 26.7b. Consequently, muons can reach Earth's surface.

In 1976 experiments with muons were conducted at the laboratory of the European Council for Nuclear Research (CERN) in Geneva. Muons were injected into a large storage ring, reaching speeds of about $0.999\ 4c$. Electrons produced by the decaying muons were detected by counters around the ring, enabling scientists to measure the decay rate and hence the lifetime of the muons. The lifetime of the moving muons was measured to be about 30 times as long as that of stationary muons to within two parts in a thousand, in agreement with the prediction of relativity.

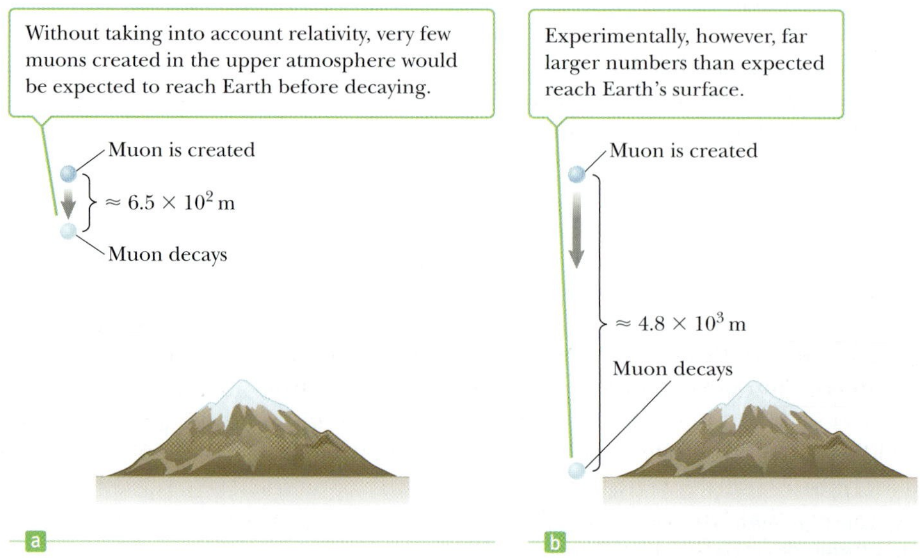
Without taking into account relativity, very few muons created in the upper atmosphere would be expected to reach Earth before decaying.
Muon is created
≈ 6.5 × 10² m
Muon decays
Experimentally, however, far larger numbers than expected reach Earth's surface.
Muon is created
≈ 4.8 × 10³ m
Muon decays
a
b

**Figure 26.7** (a) A muon created in the upper atmosphere and moving at $0.99c$ relative Earth would ordinarily travel only 650 m, on the average, before decaying after $2.2 \times 10^{-6}$ s. (b) Because of time dilation, an Earth observer measures a longer muon lifetime, so the muons travel an average of $4.8 \times 10^3$ m before decaying. Consequently, far more muons are observed reaching Earth's surface than expected. From the muons' point of view, by contrast, their lifetime is only $2.2 \times 10^{-6}$ s on the average, but the distance between them and the Earth is contracted, again making it possible for more of them to reach the surface before decaying.

**Quick Quiz**

**26.2** Suppose you're an astronaut being paid according to the time you spend traveling in space. You take a long voyage traveling at a speed near that of light. Upon your return to Earth, you're asked how you'd like to be paid: according to the time elapsed on a clock on Earth or according to your ship's clock. To maximize your paycheck, which should you choose? (a) The Earth clock (b) The ship's clock (c) Either clock because it doesn't make a difference

### EXAMPLE 26.1 Pendulum Periods

**GOAL** Apply the concept of time dilation.

**PROBLEM** The period of a pendulum is measured to be 3.00 s in the inertial frame of the pendulum at Earth's surface. What is the period as measured by an observer moving at a speed of $0.950c$ with respect to the pendulum?

**STRATEGY** Here, we're given the period of the clock as measured by an observer in the rest frame of the clock, so that's a proper time interval $\Delta t_p$. We want to know how much time passes as measured by an observer in a frame moving relative to the clock, which is $\Delta t$. Substitution into Equation 26.2 then solves the problem.

**SOLUTION**

Substitute the proper time and relative speed into Equation 26.2:

$$\Delta t = \frac{\Delta t_p}{\sqrt{1 - v^2/c^2}} = \frac{3.00 \text{ s}}{\sqrt{1 - \dfrac{(0.950c)^2}{c^2}}} = \boxed{9.61 \text{ s}}$$

**REMARKS** The moving observer considers the *pendulum* to be moving, and moving clocks are observed to run more slowly: while the pendulum oscillates once in 3 s for an observer in the rest frame of the clock, it takes nearly 10 s to oscillate once according the moving observer.

**QUESTION 26.1** Suppose a mass-spring system with the same period as the pendulum is placed in the observer's spaceship. When the spaceship is traveling at a speed of $0.950c$ relative to an observer on Earth, what is the period of the pendulum as measured by the Earth observer?

**EXERCISE 26.1** What is the period of the pendulum as measured by a third observer moving at $0.900c$?

**ANSWER** 6.88 s

Confusion arises in problems like Example 26.1 because movement is relative: from the point of view of someone in the pendulum's rest frame, the pendulum is standing still (except, of course, for the swinging motion), whereas to someone in a frame that is moving with respect to the pendulum, it's the pendulum that's doing the moving. To keep it straight, always focus on the observer making the measurement and ask yourself whether the clock being used to measure time is moving with respect to that observer. If the answer is no, then the observer is in the rest frame of the clock and measures the clock's proper time. If the answer is yes, then the time measured by the observer will be dilated, or larger than the clock's proper time. This confusion of perspectives led to the famous "twin paradox."

## The Twin Paradox

An intriguing consequence of time dilation is the so-called twin paradox (Fig. 26.8, page 906). Consider an experiment involving a set of twins named Speedo and Goslo. When they are 20 years old, Speedo, the more adventuresome of the two, sets out on an epic journey to Planet X, located 20 light-years from Earth. Further, his spaceship is capable of reaching a speed of $0.95c$ relative to the inertial frame of his twin brother back home. After reaching Planet X, Speedo becomes homesick and immediately returns to Earth at the same speed of $0.95c$. Upon his return, Speedo is shocked to discover that Goslo has aged $2D/v = 2(20 \text{ ly})/(0.95 \text{ ly/yr}) = 42$ years and is now 62 years old. Speedo, on the other hand, has aged only 13 years.

As Speedo (left) leaves his brother on Earth, both twins are the same age.
When Speedo returns from his journey, Goslo is much older than Speedo.
a
b


**Figure 26.8** The twin paradox. Speedo takes a journey to Planet X 20 light-years away and returns to the Earth.

Some wrongly consider *this* the paradox; that twins could age at different rates and end up after a period of time having very different ages. Although contrary to our common sense, that isn't the paradox at all. The paradox is that, from Speedo's point of view, *he* was at rest while Goslo (on Earth) sped away from *him* at $0.95c$ and returned later. So Goslo's clock was moving relative to Speedo and hence running slow compared with Speedo's clock. The conclusion: Speedo, not Goslo, should be the older of the twins!

To resolve this apparent paradox, consider a third observer moving at a constant speed of $0.5c$ relative to Goslo. To the third observer, Goslo never changes inertial frames: his speed relative to the third observer is always the same. The third observer notes, however, that Speedo accelerates during his journey, *changing reference frames in the process.* From the third observer's perspective, it's clear that there is something very different about the motion of Goslo when compared with that of Speedo. The roles played by Goslo and Speedo are not symmetric, so it isn't surprising that time flows differently for each. Further, because Speedo accelerates, he is in a noninertial frame of reference and is technically outside the bounds of special relativity (although there are methods for dealing with accelerated motion in relativity). Only Goslo, who is in a single inertial frame, can apply the simple time-dilation formula to Speedo's trip. Goslo finds that instead of aging 42 years, Speedo ages only $(1 - v^2/c^2)^{1/2}(42 \text{ years}) = 13$ years. Of these 13 years, Speedo spends 6.5 years traveling to Planet X and 6.5 years returning, for a total travel time of 13 years, in agreement with our earlier statement.

■ *Quick Quiz*

**26.3** True or False: People traveling near the speed of light relative to Earth would measure their lifespans and find them, on the average, longer than the average human lifespan as measured on Earth.

## Length Contraction

The measured distance between two points depends on the frame of reference of the observer. The **proper length** $L_p$ of an object is **the length of the object as measured by an observer at rest relative to the object**. The length of an object measured in a reference frame that is moving with respect to the object is always less than the proper length. This effect is known as **length contraction**.

To understand length contraction quantitatively, consider a spaceship traveling with a speed $v$ from one star to another as seen by two observers, one on Earth and

**Tip 26.3 The Proper Length**

You must be able to correctly identify the observer who measures the proper length. The proper length between two points in space is the length measured by an observer at rest with respect to the length. Very often, the proper time interval and the proper length are not measured by the same observer.

the other in the spaceship. The observer at rest on Earth (and also assumed to be at rest with respect to the two stars) measures the distance between the stars to be $L_p$. According to this observer, the time it takes the spaceship to complete the voyage is $\Delta t = L_p/v$. Because of time dilation, the space traveler, using his spaceship clock, measures a smaller time of travel: $\Delta t_p = \Delta t/\gamma$. The space traveler claims to be at rest and sees the destination star moving toward the spaceship with speed $v$. Because the space traveler reaches the star in time $\Delta t_p$, he concludes that the distance $L$ between the stars is shorter than $L_p$. The distance measured by the space traveler is

$$L = v\,\Delta t_p = v\frac{\Delta t}{\gamma}$$

Because $L_p = v\,\Delta t$, it follows that

$$L = \frac{L_p}{\gamma} = L_p\sqrt{1 - v^2/c^2} \qquad [26.4]$$

According to this result, illustrated in Figure 26.9, if an observer at rest with respect to an object measures its length to be $L_p$, an observer moving at a speed $v$ relative to the object will find it to be shorter than its proper length by the factor $\sqrt{1 - v^2/c^2}$. Note that **length contraction takes place only along the direction of motion**.

Time-dilation and length contraction effects have interesting applications for future space travel to distant stars. For the star to be reached in a fraction of a human lifetime, the trip must be taken at very high speeds. According to an Earth-bound observer, the time for a spacecraft to reach the destination star will be dilated compared with the time interval measured by travelers. As discussed in the treatment of the twin paradox, the travelers will be younger than their twins when they return to Earth. Therefore, by the time the travelers reach the star, they will have aged by some number of years, while their partners back on Earth will have aged a larger number of years, the exact ratio depending on the speed of the spacecraft. At a spacecraft speed of $0.94c$, this ratio is about 3:1.

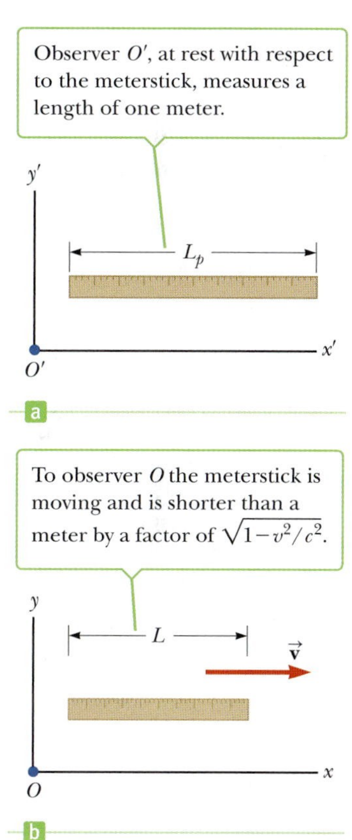
Observer $O'$, at rest with respect to the meterstick, measures a length of one meter.
$y'$
$L_p$
$x'$
$O'$
a
To observer $O$ the meterstick is moving and is shorter than a meter by a factor of $\sqrt{1-v^2/c^2}$.
$y$
$L$
$\vec{v}$
$x$
$O$
b

**Figure 26.9** The length of a meterstick is measured by two observers.

**Quick Quiz**

**26.4** You are packing for a trip to another star, and on your journey you will be traveling at a speed of $0.99c$. Can you sleep in a smaller cabin than usual, because you will be shorter when you lie down? Explain your answer.

**26.5** You observe a rocket moving away from you. **(i)** Compared with its length when it was at rest on the ground, will you measure its length to be (a) shorter, (b) longer, or (c) the same? Compared to the passage of time measured by the watch on your wrist, is the passage of time on the rocket's clock (d) faster, (e) slower, or (f) the same? **(ii)** Answer the same questions from part (i) if the rocket turns around and comes toward you.

## EXAMPLE 26.2 Speedy Plunge

**GOAL** Apply the concept of length contraction to a distance.

**PROBLEM** **(a)** An observer on Earth sees a spaceship at an altitude of 4 350 km moving downward toward Earth with a speed of $0.970c$. What is the distance from the spaceship to Earth as measured by the spaceship's captain? **(b)** After firing her engines, the captain measures her ship's altitude as 267 km, whereas the observer on Earth measures it to be 625 km. What is the speed of the spaceship at this instant?

*(Continued)*

**STRATEGY** To the captain, Earth is rushing toward her ship at $0.970c$; hence, the distance between her ship and Earth is contracted. Substitution into Equation 26.9 yields the answer. In part (b) use the same equation, substituting the distances and solving for the speed.

**SOLUTION**

**(a)** Find the distance from the ship to Earth as measured by the captain.

Substitute into Equation 26.4, getting the altitude as measured by the captain in the ship:

$$L = L_p\sqrt{1 - v^2/c^2} = (4\,350\text{ km})\sqrt{1 - (0.970c)^2/c^2}$$
$$= 1.06 \times 10^3\text{ km}$$

**(b)** What is the subsequent speed of the spaceship if the Earth observer measures the distance from the ship to Earth as 625 km and the captain measures it as 267 km?

Apply the length-contraction equation:

$$L = L_p\sqrt{1 - v^2/c^2}$$

Square both sides of this equation and solve for $v$:

$$L^2 = L_p^{\,2}(1 - v^2/c^2) \quad \rightarrow \quad 1 - v^2/c^2 = \left(\frac{L}{L_p}\right)^2$$
$$v = c\sqrt{1 - (L/L_p)^2} = c\sqrt{1 - (267\text{ km}/625\text{ km})^2}$$
$$v = 0.904c$$

**REMARKS** The proper length is always the length measured by an observer at rest with respect to that length.

**QUESTION 26.2** As a spaceship approaches an observer at nearly the speed of light, the captain directs a beam of yellow light at the observer. What would the observer report upon seeing the light? (a) Its wavelength would be shifted toward the red end of the spectrum. (b) Its wavelength would correspond to yellow light. (c) Its wavelength would be shifted toward the blue end of the spectrum.

**EXERCISE 26.2** Suppose the observer on the ship measures the distance from Earth as 50.0 km, whereas the observer on Earth measures the distance as 125 km. At what speed is the spacecraft approaching Earth?

**ANSWER** $0.917c$

Length contraction occurs only in the direction of the observer's motion. No contraction occurs perpendicular to that direction. For example, a spaceship at rest relative to an observer may have the shape of an equilateral triangle, but if it passes the observer at relativistic speed in a direction parallel to its base, the base will shorten while the height remains the same. Hence the observer will report that the craft has the form of an isosceles triangle. An observer traveling with the ship will still observe it to be equilateral.

## 26.5 Relativistic Momentum

**LEARNING OBJECTIVES**

1. Generalize the concept of momentum to relativity.
2. Through calculation, compare the classical and relativistic concepts of momentum.

Properly describing the motion of particles within the framework of special relativity requires generalizing Newton's laws of motion and the definitions of momentum and energy. These generalized definitions reduce to the classical (nonrelativistic) definitions when $v$ is much less than $c$.

First, recall that conservation of momentum states that when two objects collide, the total momentum of the system remains constant, assuming the objects are isolated, reacting only with each other. When analyzing such collisions from rapidly moving inertial frames, however, it is found that momentum is not conserved if the classical definition of momentum, $p = mv$, is used. To have momentum conservation in all inertial frames—even those moving at an appreciable fraction of $c$—the definition of momentum must be modified to read

$$p \equiv \frac{mv}{\sqrt{1 - v^2/c^2}} = \gamma mv \qquad [26.5]$$

◀ Relativistic momentum

where $v$ is the speed of the particle and $m$ is its mass as measured by an observer at rest with respect to the particle. Note that when $v$ is much less than $c$, the denominator of Equation 26.5 approaches 1, so that $p$ approaches $mv$. Therefore, the relativistic equation for momentum reduces to the classical expression when $v$ is small compared with $c$.

## ■ EXAMPLE 26.3 The Relativistic Momentum of an Electron

**GOAL** Contrast the classical and relativistic definitions of momentum.

**PROBLEM** An electron, which has a mass of $9.11 \times 10^{-31}$ kg, moves with a speed of $0.750c$. Find the classical (nonrelativistic) momentum and compare it with its relativistic counterpart $p_{\text{rel}}$.

**STRATEGY** Substitute into the classical definition to get the classical momentum, then multiply by the gamma factor to obtain the relativistic version.

**SOLUTION**

First, compute the classical (nonrelativistic) momentum with $v = 0.750c$:

$$p = mv = (9.11 \times 10^{-31}\text{ kg})(0.750 \times 3.00 \times 10^{8}\text{ m/s})$$
$$= 2.05 \times 10^{-22}\text{ kg}\cdot\text{m/s}$$

Multiply this result by $\gamma$ to obtain the relativistic momentum:

$$p_{\text{rel}} = \frac{mv}{\sqrt{1 - v^2/c^2}} = \frac{2.05 \times 10^{-22}\text{ kg}\cdot\text{m/s}}{\sqrt{1 - (0.750c/c)^2}}$$
$$= 3.10 \times 10^{-22}\text{ kg}\cdot\text{m/s}$$

**REMARKS** The (correct) relativistic result is 50% greater than the classical result. In subsequent calculations no notational distinction will be made between classical and relativistic momentum. For problems involving relative speeds of $0.2c$, the answer using the classical expression is about 2% below the correct answer.

**QUESTION 26.3** A particle with initial momentum $p_i$ doubles its speed. How does its final momentum $p_f$ compare with its initial momentum? (a) $p_f > 2p_i$ (b) $p_f = 2p_i$ (c) $p_f < 2p_i$

**EXERCISE 26.3** Repeat the calculation for a proton traveling at $0.600c$.

**ANSWERS** $p = 3.01 \times 10^{-19}\text{ kg}\cdot\text{m/s}$, $p_{\text{rel}} = 3.76 \times 10^{-19}\text{ kg}\cdot\text{m/s}$

## 26.6 Relative Velocity in Special Relativity

**LEARNING OBJECTIVES**

1. Contrast the expression for relative velocity in Galilean relativity to the corresponding expression in special relativity.
2. Calculate one-dimensional relative velocities in special relativity.

In Galilean relativity, if a man in a spaceship traveling at velocity $v$ shines a laser straight ahead, the light beam would be expected to travel at velocity $c + v$ relative to an observer on Earth. The null result of the Michelson–Morley experiment, however, indicates the laser light will still travel at speed $c$ relative to that same observer, although the light's frequency increases. (When light's frequency increases, the light is said to be "blue-shifted." When the frequency decreases, the light is said to be "red-shifted.") Evidently, some new formula must be derived to allow the comparison of velocities by observers moving at high relative speeds.

The procedure is very similar to that used in Chapter 3 for nonrelativistic relative velocity. Here, an observer in a spaceship will be labeled B and the Earth observer E, with B moving in the positive $x$-direction at speed $v_{BE}$ with respect to the Earth observer E. The goal is to find the relationship between their independent measurements of an object A. Given $v_{AE}$, the velocity of A according to observer E, what is $v_{AB}$, the velocity of A relative to observer B? According to Galilean relativity, the answer derived in Chapter 3 is

Relative velocity in Galilean relativity ▶

$$v_{AB} = v_{AE} - v_{BE} \qquad [26.6]$$

For velocities that are about 10% of the speed of light and greater, relativistic effects start becoming appreciable, so the relativistic expression for relative velocity should be used[2]:

Relative velocity in special relativity ▶

$$v_{AB} = \frac{v_{AE} - v_{BE}}{1 - \dfrac{v_{AE}v_{BE}}{c^2}} \qquad [26.7]$$

Notice that when either $v_{AE}$ or $v_{BE}$ is much smaller than $c$, this expression agrees with the Galilean (nonrelativistic) relationship, as it should. Equation 26.7 is useful for determining velocities measured by B in the moving frame of reference when the velocity measured by observer E in the rest frame is known. On the other hand, when the velocity in question is measured by observer B and the task is to find the velocity measured by observer E, Equation 26.7 must be algebraically solved for $v_{AE}$. The resulting expression is

Relativistic addition of velocities ▶

$$v_{AE} = \frac{v_{AB} + v_{BE}}{1 + \dfrac{v_{AB}v_{BE}}{c^2}} \qquad [26.8]$$

Ignoring the expression in the denominator, Equation 26.8 has the expected form: if observer B is moving with respect to observer E at velocity $v_{BE}$ and fires a projectile at velocity $v_{AB}$ relative to himself, then adding $v_{AB}$ and $v_{BE}$ should give the velocity of the projectile $v_{AE}$ as measured by the Earth observer. Special relativity contributes the denominator of Equation 26.8, an equation often called the relativistic addition of velocities.

Using Equation 26.8, the speed of an object projected forward from a moving vehicle as measured by an observer on Earth can be calculated. Suppose, for example, that observer B is moving at $v_{BE}$ with respect to the Earth observer and directs the beam of a laser in front of his rapidly moving spacecraft. Here $v_{AB}$ is the velocity of light relative to observer B on the spacecraft. The speed of the light $v_{AE}$ as measured by the Earth observer is, therefore,

$$v_{AE} = \frac{v_{AB} + v_{BE}}{1 + \dfrac{v_{AB}v_{BE}}{c^2}} = \frac{c + v_{BE}}{1 + \dfrac{cv_{BE}}{c^2}} = \frac{c\left(1 + \dfrac{v_{BE}}{c}\right)}{1 + \dfrac{v_{BE}}{c}} = c$$

This calculation shows that the derived velocity transformation is consistent with the experimental result proving the speed of light is the same for all observers.

[2] The derivation of Equation 26.7 requires the use of Lorentz transformations and will not be presented in this textbook.

## ■ EXAMPLE 26.4 Urgent Course Correction Required!

**GOAL** Apply the concept of relative velocity in relativity.

**PROBLEM** Suppose that Alice's spacecraft is traveling at $0.600c$ in the positive $x$-direction, as measured by a nearby Earth-based observer at rest, while Bob is traveling in his own vehicle directly toward Alice in the negative $x$-direction at velocity $-0.800c$ relative the same Earth observer. What's the velocity of Alice according to Bob?

**STRATEGY** Alice's spacecraft is the object of interest that both the Earth observer and Bob are tracking. We're given the Earth observer's velocity measurements and wish to find Bob's measurement. Use the relative velocity equation for relativity, Equation 26.7, with $v_{AB}$ corresponding to the measurement of Alice's velocity made by Bob and $v_{AE}$ is the measurement of Alice's velocity according to the Earth observer. Notice that the velocity of Bob's frame is in the negative $x$-direction, so $v_{BE} < 0$.

**SOLUTION**

Write Equation 26.7:

$$v_{AB} = \frac{v_{AE} - v_{BE}}{1 - \dfrac{v_{AE}v_{BE}}{c^2}}$$

Substitute values:

$$v_{AB} = \frac{0.600c - (-0.800c)}{1 - \dfrac{(0.600c)(-0.800c)}{c^2}} = \frac{1.400c}{1 - (-0.480)} = 0.946c$$

**REMARKS** Notice that care was taken to use the correct signs. Common sense might lead us to believe that Bob would measure Alice's velocity as $1.40c$, but as the calculation shows, Bob measures Alice's velocity as less than that of light.

**QUESTION 26.4** What is Bob's velocity according to Alice?

**EXERCISE 26.4** Suppose yet another observer, Ray, reports that Alice's velocity is only $0.400c$. What is Ray's velocity according to the Earth observer?

**ANSWER** $0.263c$

# 26.7 Relativistic Energy and the Equivalence of Mass and Energy

**LEARNING OBJECTIVES**

1. Define the rest energy, kinetic energy, and total energy in special relativity.
2. Discuss the equivalence of mass and energy.
3. From the definitions, derive an equation relating energy to momentum. Specialize it to the case of photons.
4. Apply relativistic energy and momentum to the understanding of nuclear reactions.

We have seen that the definition of momentum required generalization to make it compatible with the principle of relativity. Likewise, the definition of kinetic energy requires modification in relativistic mechanics. Einstein found that the correct expression for the **kinetic energy** of an object is

$$KE = \gamma mc^2 - mc^2 \qquad [26.9]$$

◄ Kinetic energy

The constant term $mc^2$ in Equation 26.9, which is independent of the speed of the object, is called the **rest energy** of the object, $E_R$:

$$E_R = mc^2 \qquad [26.10]$$

◄ Rest energy

The term $\gamma mc^2$ in Equation 26.9 depends on the object's speed and is the sum of the kinetic and rest energies. We define $\gamma mc^2$ to be the **total energy** $E$, so

$$\text{Total energy} = \text{kinetic energy} + \text{rest energy}$$

or, using Equation 26.9,

$$E = KE + mc^2 = \gamma mc^2 \qquad [26.11]$$

Because $\gamma = (1 - v^2/c^2)^{-1/2}$, we can also express the total energy $E$ as

Total energy ▶

$$E = \frac{mc^2}{\sqrt{1 - v^2/c^2}} \qquad [26.12]$$

This is Einstein's famous mass-energy equivalence equation.[3]

The relation $E = \gamma mc^2 = KE + mc^2$ shows the amazing result that **a stationary particle with zero kinetic energy has an energy proportional to its mass**. Further, a small mass corresponds to an enormous amount of energy because the proportionality constant between mass and energy is large: $c^2 = 9 \times 10^{16}\ \text{m}^2/\text{s}^2$. The equation $E_R = mc^2$, as Einstein first suggested, indirectly implies that the mass of a particle may be completely convertible to energy and that pure energy—for example, electromagnetic energy—may be converted to particles having mass. That is indeed the case, as has been shown in the laboratory many times in interactions involving matter and antimatter.

On a larger scale, nuclear power plants produce energy by the fission of uranium, which involves the conversion of a small amount of the mass of the uranium into energy. The Sun, too, converts mass into energy and continually loses mass in pouring out a tremendous amount of electromagnetic energy in all directions.

It's extremely interesting that although we have been talking about the interconversion of mass and energy for particles, the expression $E = mc^2$ is universal and applies to all objects, processes, and systems: a hot object has slightly more mass and is slightly more difficult to accelerate than an identical cold object because it has more thermal energy, and a stretched spring has more elastic potential energy and more mass than an identical unstretched spring. A key point, however, is that these changes in mass are often far too small to measure. Our best bet for measuring mass changes is in nuclear transformations, where a measurable fraction of the mass is converted into energy.

**Quick Quiz**

**26.6** True or False: Because the speed of a particle cannot exceed the speed of light, there is an upper limit to its momentum and kinetic energy.

## Energy and Relativistic Momentum

Often the momentum or energy of a particle rather than its speed is measured, so it's useful to find an expression relating the total energy $E$ to the relativistic momentum $p$. We can do so by using the expressions $E = \gamma mc^2$ and $p = \gamma mv$. By squaring these equations and subtracting, we can eliminate $v$. The result, after some algebra, is

$$E^2 = p^2c^2 + (mc^2)^2 \qquad [26.13]$$

When the particle is at rest, $p = 0$, so $E = E_R = mc^2$. In this special case the total energy equals the rest energy. For the case of particles that have zero mass, such as photons (massless, chargeless particles of light), we set $m = 0$ in Equation 26.10 and find that

$$E = pc \qquad [26.14]$$

This equation is an exact expression relating energy and momentum for photons, which always travel at the speed of light.

[3]Although this expression doesn't look exactly like the famous equation $E = mc^2$, it used to be common to write $m = \gamma m_0$ (Einstein himself wrote it that way), where $m$ is the effective mass of an object moving at speed $v$ and $m_0$ is the mass of that object as measured by an observer at rest with respect to the object. Then our $E = \gamma mc^2$ becomes the familiar $E = mc^2$. It is currently unfashionable to use $m = \gamma m_0$.

In dealing with subatomic particles, it's convenient to express their energy in electron volts (eV) because the particles are given energy when accelerated through an electrostatic potential difference. The conversion factor is

$$1 \text{ eV} = 1.60 \times 10^{-19} \text{ J}$$

For example, the mass of an electron is $9.11 \times 10^{-31}$ kg. Hence, the rest energy of the electron is

$$m_e c^2 = (9.11 \times 10^{-31} \text{ kg})(3.00 \times 10^8 \text{ m/s})^2 = 8.20 \times 10^{-14} \text{ J}$$

Converting to eV, we have

$$m_e c^2 = (8.20 \times 10^{-14} \text{ J})(1 \text{ eV}/1.60 \times 10^{-19} \text{ J}) = 0.511 \text{ MeV}$$

where 1 MeV $= 10^6$ eV. Because we frequently use the expression $E = \gamma mc^2$ in nuclear physics and because $m$ is usually in atomic mass units, u, it is useful to have the conversion factor 1 u $= 931.494 \text{ MeV}/c^2$. Using this factor makes it easy, for example, to find the rest energy in MeV of the nucleus of a uranium atom with a mass of 235.043 924 u:

$$E_R = mc^2 = (235.043\,924 \text{ u})(931.494 \text{ MeV/u} \cdot c^2)(c^2) = 2.189\,42 \times 10^5 \text{ MeV}$$

**Quick Quiz**

**26.7** A photon is reflected from a mirror. True or False: (a) Because a photon has zero mass, it does not exert a force on the mirror. (b) Although the photon has energy, it can't transfer any energy to the surface because it has zero mass. (c) The photon carries momentum, and when it reflects off the mirror, it undergoes a change in momentum and exerts a force on the mirror. (d) Although the photon carries momentum, its change in momentum is zero when it reflects from the mirror, so it can't exert a force on the mirror.

## EXAMPLE 26.5 A Speedy Electron

**GOAL** Compute a total energy and a relativistic kinetic energy.

**PROBLEM** An electron moves with a speed $v = 0.850c$. Find its total energy and kinetic energy in mega electron volts (MeV) and compare the latter to the classical kinetic energy ($10^6$ eV = 1 MeV).

**STRATEGY** Substitute into Equation 26.12 to get the total energy and subtract the rest mass energy to obtain the kinetic energy.

**SOLUTION**

Substitute values into Equation 26.12 to obtain the total energy:

$$E = \frac{m_e c^2}{\sqrt{1 - v^2/c^2}} = \frac{(9.11 \times 10^{-31} \text{ kg})(3.00 \times 10^8 \text{ m/s})^2}{\sqrt{1 - (0.850c/c)^2}}$$

$$= 1.56 \times 10^{-13} \text{ J} = (1.56 \times 10^{-13} \text{ J})\left(\frac{1.00 \text{ eV}}{1.60 \times 10^{-19} \text{ J}}\right)$$

$$= \boxed{0.975 \text{ MeV}}$$

The kinetic energy is obtained by subtracting the rest energy from the total energy:

$$KE = E - m_e c^2 = 0.975 \text{ MeV} - 0.511 \text{ MeV} = \boxed{0.464 \text{ MeV}}$$

Calculate the classical kinetic energy:

$$KE_{\text{classical}} = \tfrac{1}{2} m_e v^2$$

$$= \tfrac{1}{2}(9.11 \times 10^{-31} \text{ kg})(0.850 \times 3.00 \times 10^8 \text{ m/s})^2$$

$$= 2.96 \times 10^{-14} \text{ J} = 0.185 \text{ MeV}$$

*(Continued)*

**REMARKS** Notice the large discrepancy between the relativistic kinetic energy and the classical kinetic energy.

**QUESTION 26.5** According to an observer, the speed $v$ of a particle with kinetic energy $KE_i$ increases to $2v$. How does the final kinetic energy $KE_f$ compare with the initial kinetic energy? (a) $KE_f > 4KE_i$ (b) $KE_f = 4KE_i$ (c) $KE_f < 4KE_i$

**EXERCISE 26.5** Calculate the total energy and the kinetic energy in MeV of a proton traveling at $0.600c$. (The rest energy of a proton is approximately 938 MeV.)

**ANSWERS** $E = 1.17 \times 10^3$ MeV, $KE = 2.3 \times 10^2$ MeV

## EXAMPLE 26.6 The Conversion of Mass to Kinetic Energy in Uranium Fission

**GOAL** Understand the production of energy from nuclear sources.

**PROBLEM** The fission, or splitting, of uranium was discovered in 1938 by Lise Meitner, who successfully interpreted some curious experimental results found by Otto Hahn as due to fission. (Hahn received the Nobel Prize.) The fission of ${}^{235}_{92}\text{U}$ begins with the absorption of a slow-moving neutron that produces an unstable nucleus of ${}^{236}\text{U}$. The ${}^{236}\text{U}$ nucleus then quickly decays into two heavy fragments moving at high speed, as well as several neutrons. Most of the kinetic energy released in such a fission is carried off by the two large fragments. **(a)** For the typical fission process,

$$ {}^{1}_{0}\text{n} + {}^{235}_{92}\text{U} \rightarrow {}^{141}_{56}\text{Ba} + {}^{92}_{36}\text{Kr} + 3{}^{1}_{0}\text{n} $$

calculate the kinetic energy in MeV carried off by the fission fragments, neglecting the kinetic energy of the reactants. **(b)** What percentage of the initial energy is converted into kinetic energy? The atomic masses involved, given in atomic mass units, are

$$ {}^{1}_{0}\text{n} = 1.008\ 665\ \text{u} \qquad {}^{235}_{92}\text{U} = 235.043\ 923\ \text{u} $$
$$ {}^{141}_{56}\text{Ba} = 140.903\ 496\ \text{u} \qquad {}^{92}_{36}\text{Kr} = 91.907\ 936\ \text{u} $$

**STRATEGY** This problem is an application of the conservation of relativistic energy. Write the conservation law as a sum of kinetic energy and rest energy, and solve for the final kinetic energy.

**SOLUTION**

**(a)** Calculate the final kinetic energy for the given process.

Apply the conservation of relativistic energy equation, assuming $KE_{\text{initial}} = 0$:

$$ (KE + mc^2)_{\text{initial}} = (KE + mc^2)_{\text{final}} $$
$$ 0 + m_{\text{n}}c^2 + m_{\text{U}}c^2 = m_{\text{Ba}}c^2 + m_{\text{Kr}}c^2 + 3m_{\text{n}}c^2 + KE_{\text{final}} $$

Solve for $KE_{\text{final}}$ and substitute, converting to MeV in the last step:

$$ KE_{\text{final}} = [(m_{\text{n}} + m_{\text{U}}) - (m_{\text{Ba}} + m_{\text{Kr}} + 3m_{\text{n}})]c^2 $$
$$ \begin{aligned} KE_{\text{final}} &= (1.008\ 665\ \text{u} + 235.043\ 923\ \text{u})c^2 \\ &\quad - [140.903\ 496\ \text{u} + 91.907\ 936\ \text{u} + 3(1.008\ 665\ \text{u})]c^2 \\ &= (0.215\ 161\ \text{u})(931.494\ \text{MeV/u}\cdot c^2)(c^2) \\ &= \boxed{200.421\ \text{MeV}} \end{aligned} $$

**(b)** What percentage of the initial energy is converted into kinetic energy?

Compute the total energy, which is the initial energy:

$$ \begin{aligned} E_{\text{initial}} &= 0 + m_{\text{n}}c^2 + m_{\text{U}}c^2 \\ &= (1.008\ 665\ \text{u} + 235.043\ 923\ \text{u})c^2 \\ &= (236.052\ 59\ \text{u})(931.494\ \text{MeV/u}\cdot c^2)(c^2) \\ &= 2.198\ 82 \times 10^5\ \text{MeV} \end{aligned} $$

Divide the kinetic energy by the total energy and multiply by 100%:

$$ \frac{200.421\ \text{MeV}}{2.198\ 82 \times 10^5\ \text{MeV}} \times 100\% = \boxed{9.115 \times 10^{-2}\%} $$

**REMARKS** This calculation shows that nuclear reactions liberate only about one-tenth of 1% of the rest energy of the constituent particles. Some fusion reactions result in a percent yield several times as large.

**QUESTION 26.6** Why is so little of the mass converted to other forms of energy?

**EXERCISE 26.6** In a fusion reaction light elements combine to form a heavier element. Deuterium, which is also called heavy hydrogen, has an extra neutron in its nucleus. Two such particles can fuse into a heavier form of hydrogen, called tritium, plus an ordinary hydrogen atom. The reaction is

$$ {}^{2}_{1}\text{D} + {}^{2}_{1}\text{D} \rightarrow {}^{3}_{1}\text{T} + {}^{1}_{1}\text{H} $$

(a) Calculate the energy released in the form of kinetic energy, assuming for simplicity the initial kinetic energy is zero. (b) What percentage of the rest mass is converted to energy? The atomic masses involved are

$${}^{2}_{1}\text{D} = 2.014\ 102\ \text{u} \qquad {}^{3}_{1}\text{T} = 3.016\ 049\ \text{u} \qquad {}^{1}_{1}\text{H} = 1.007\ 825\ \text{u}$$

**ANSWERS** (a) 4.033 37 MeV (b) 0.107 5%

## 26.8 General Relativity

**LEARNING OBJECTIVES**

1. Differentiate between inertial mass and gravitational mass.
2. State the two postulates of general relativity.
3. Explain the principle of equivalence that led Einstein to develop the theory.
4. Discuss the relationship between spacetime curvature and energy in general relativity.
5. Give examples of the observational evidence for general relativity, and of some of its predictions, such as black holes.

Special relativity relates observations of inertial observers. Einstein sought a more general theory that would address accelerating systems. His search was motivated in part by the following curious fact: mass determines the inertia of an object and also the strength of the gravitational field. The mass involved in inertia is called inertial mass, $m_i$, whereas the mass responsible for the gravitational field is called the gravitational mass, $m_g$. These masses appear in Newton's law of gravitation and in the second law of motion:

$$\text{Gravitational property} \qquad F_g = G\frac{m_g m_g'}{r^2}$$

$$\text{Inertial property} \qquad F_i = m_i a$$

The value for the gravitational constant $G$ was chosen to make the magnitudes of $m_g$ and $m_i$ numerically equal. Regardless of how $G$ is chosen, however, the strict proportionality of $m_g$ and $m_i$ has been established experimentally to an extremely high degree: a few parts in $10^{12}$. It appears that gravitational mass and inertial mass may indeed be exactly equal: $m_i = m_g$.

In Einstein's view the remarkable coincidence that $m_g$ and $m_i$ were exactly equal was evidence for an intimate connection between the two concepts. He pointed out that no mechanical experiment (such as releasing a mass) could distinguish between the two situations illustrated in Figures 26.10a and 26.10b (page 916). In each case a mass released by the observer undergoes a downward acceleration of $g$ relative to the floor.

Einstein carried this idea further and proposed that *no* experiment, mechanical or otherwise, could distinguish between the two cases. This extension to include all phenomena (not just mechanical ones) has interesting consequences. For example, suppose a light pulse is sent horizontally across the box, as in Figure 26.10c. The trajectory of the light pulse bends downward as the box accelerates upward to meet it. Einstein proposed that a beam of light should also be bent downward by a gravitational field (Fig. 26.10d).

The two postulates of Einstein's **general relativity** are as follows:

1. All the laws of nature have the same form for observers in any frame of reference, accelerated or not.
2. In the vicinity of any given point, a gravitational field is equivalent to an accelerated frame of reference without a gravitational field. (This is the *principle of equivalence*.)

The second postulate implies that gravitational mass and inertial mass are completely equivalent, not just proportional. What were thought to be two different types of mass are actually identical.

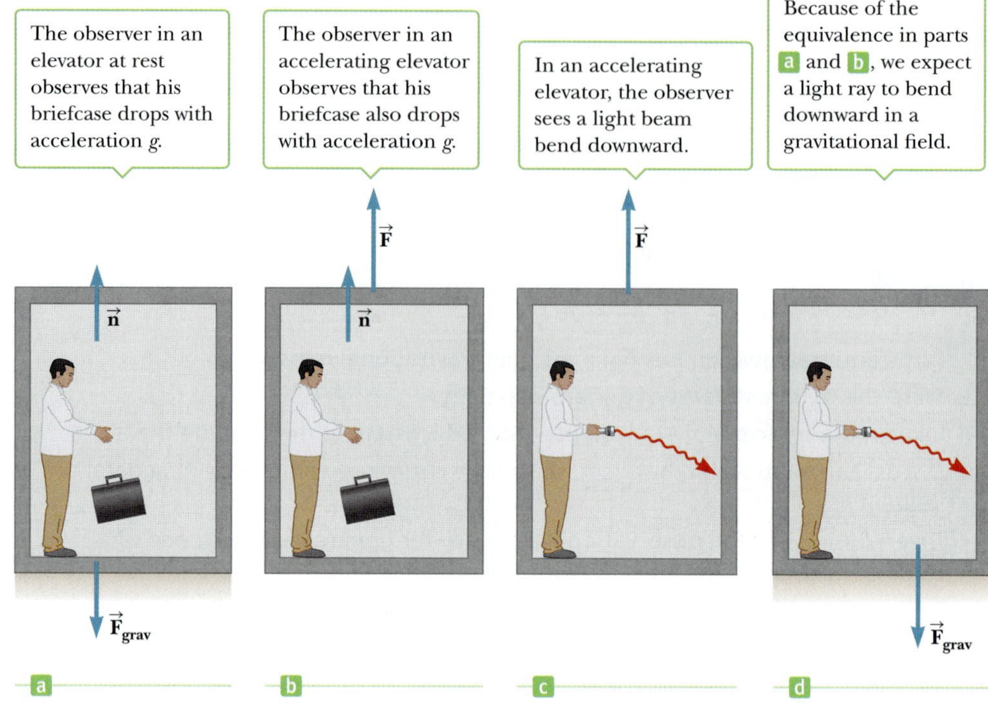

The observer in an elevator at rest observes that his briefcase drops with acceleration g.
The observer in an accelerating elevator observes that his briefcase also drops with acceleration g.
In an accelerating elevator, the observer sees a light beam bend downward.
Because of the equivalence in parts a and b, we expect a light ray to bend downward in a gravitational field.
n
F
n
F
F grav
F grav
a
b
c
d


**Figure 26.10** (a) The observer in the cubicle is at rest in a uniform gravitational field $\vec{\mathbf{g}}$. He experiences a normal force $\vec{\mathbf{n}}$. (b) Now the observer is in a region where gravity is negligible, but an external force $\vec{\mathbf{F}}$ acts on the frame of reference, producing an acceleration with magnitude $g$. Again, the man experiences a normal force $\vec{\mathbf{n}}$ that accelerates him along with the cubicle. According to Einstein, the frames of reference in (a) and (b) are equivalent in every way. No local experiment could distinguish between them. (c) The observer turns on his pocket flashlight. Because of the acceleration of the cubicle, the beam would appear to bend toward the floor, just as a tossed ball would. (d) Given the equivalence of the frames, the same phenomenon should be observed in the presence of a gravity field.

One interesting effect predicted by general relativity is that time scales are altered by gravity. A clock in the presence of gravity runs more slowly than one in which gravity is negligible. As a consequence, light emitted from atoms in a strong gravity field, such as the Sun's, is observed to have a lower frequency than the same light emitted by atoms in the laboratory. This gravitational shift has been detected in spectral lines emitted by atoms in massive stars. It has also been verified on Earth by comparing the frequencies of gamma rays emitted from nuclei separated vertically by about 20 m.

The second postulate suggests a gravitational field may be "transformed away" at any point if we choose an appropriate accelerated frame of reference: a freely falling one. Einstein developed an ingenious method of describing the acceleration necessary to make the gravitational field "disappear." He specified a certain quantity, the *curvature of spacetime*, that describes the gravitational effect at every point. In fact, the curvature of spacetime completely replaces Newton's gravitational theory. According to Einstein, there is no such thing as a gravitational force. Rather, the presence of a mass causes a curvature of spacetime in the vicinity of the mass. Planets going around the Sun follow the natural contours of the spacetime, much as marbles roll around inside a bowl. The fundamental equation of general relativity can be roughly stated as a proportion as follows:

$$\text{Average curvature of spacetime} \propto \text{energy density}$$

Einstein pursued a new theory of gravity in large part because of a discrepancy in the orbit of Mercury as calculated from Newton's second law. The closest approach of Mercury to the Sun, called the perihelion, changes position slowly over time. Newton's theory accounted for all but 43 arc seconds per century; Einstein's general relativity explained the discrepancy.

The most dramatic test of general relativity came shortly after the end of World War I. Einstein's theory predicts that a star would bend a light ray by a certain precise amount. Sir Arthur Eddington mounted an expedition to Africa and, during a solar eclipse, confirmed that starlight bent on passing the Sun in an amount matching the prediction of general relativity (Fig. 26.11). When this discovery was announced, Einstein became an international celebrity.

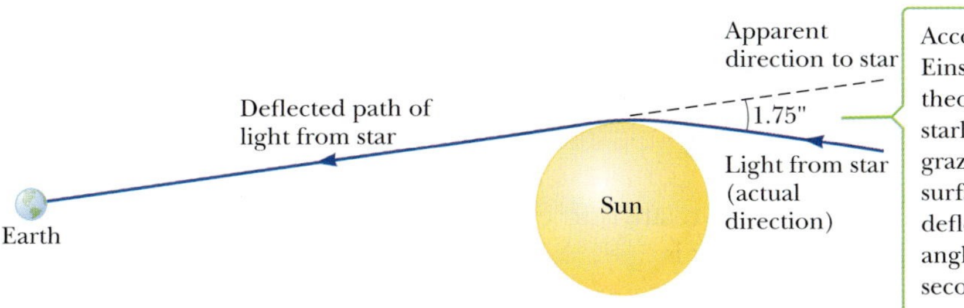
Apparent direction to star
Deflected path of light from star
1.75"
Light from star (actual direction)
Sun
Earth
According to Einstein's general theory of relativity, starlight just grazing the Sun's surface should be deflected by an angle of 1.75 arc seconds.

**Figure 26.11** Deflection of starlight passing near the Sun. Because of this effect, the Sun and other remote objects can act as a *gravitational lens.*

General relativity also predicts that a large star can exhaust its nuclear fuel and collapse to a very small volume, turning into a **black hole**. Here the curvature of spacetime is so extreme that all matter and light within a certain radius becomes trapped. This radius, called the *Schwarzschild radius* or *event horizon*, is about 3 km for a black hole with the mass of our Sun. At the black hole's center may lurk a *singularity*, a point of infinite density and curvature where spacetime comes to an end.

There is strong evidence for the existence of a black hole having a mass of millions of Suns at the center of our galaxy.

## APPLYING PHYSICS 26.1 Faster Clocks in a "Mile-High City"

Atomic clocks are extremely accurate; in fact, an error of 1 s in 3 million years is typical. This error can be described as about one part in $10^{14}$. On the other hand, the atomic clock in Boulder, Colorado, is often 15 ns faster than the atomic clock in Washington, D.C., after only one day. This error is about one part in $6 \times 10^{12}$, which is about 17 times larger than the typical error. If atomic clocks are so accurate, why does a clock in Boulder not remain synchronous with one in Washington, D.C.?

**EXPLANATION** According to the general theory of relativity, the passage of time depends on gravity: clocks run more slowly in strong gravitational fields. Washington, D.C., is at an elevation very close to sea level, whereas Boulder is about a mile higher in altitude, so the gravitational field at Boulder is weaker than at Washington, D.C. As a result, an atomic clock runs more rapidly in Boulder than in Washington, D.C. (This effect has been verified by experiment.) ■

## SUMMARY

### 26.3 Einstein's Principle of Relativity

The two basic postulates of the **special theory of relativity** are as follows:

1. The laws of physics are the same in all inertial frames of reference.
2. The speed of light is the same for all inertial observers, independently of their motion or of the motion of the source of light.

### 26.4 Consequences of Special Relativity

Some of the consequences of the special theory of relativity are as follows:

1. Clocks in motion relative to an observer slow down, a phenomenon known as **time dilation**. The relationship between time intervals in the moving and at-rest systems is

$$\Delta t = \gamma\, \Delta t_p \qquad [26.2]$$

where $\Delta t$ is the time interval measured in the system in relative motion with respect to the clock,

$$\gamma = \frac{1}{\sqrt{1 - v^2/c^2}} \qquad [26.3]$$

and $\Delta t_p$ is the proper time interval measured in the system moving with the clock.

2. The length of an object in motion is *contracted* in the direction of motion. The equation for **length contraction** is

$$L = L_p\sqrt{1 - v^2/c^2} \qquad [26.4]$$

where $L$ is the length measured by an observer in motion relative to the object and $L_p$ is the proper length measured by an observer for whom the object is at rest.

3. Events that are simultaneous for one observer are not simultaneous for another observer in motion relative to the first.

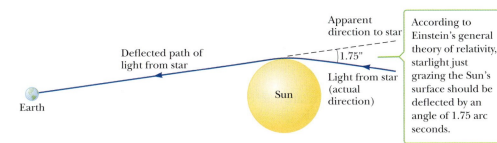

## 26.5 Relativistic Momentum

The relativistic expression for the **momentum** of a particle moving with velocity $v$ is

$$p \equiv \frac{mv}{\sqrt{1 - v^2/c^2}} = \gamma mv \qquad [26.5]$$

## 26.6 Relative Velocity in Special Relativity

When observers E and B are in relative motion, they will measure different velocities for some third object. Those measurements are related by

$$v_{AB} = \frac{v_{AE} - v_{BE}}{1 - \dfrac{v_{AE}v_{BE}}{c^2}} \qquad [26.7]$$

At low relative velocities compared to that of light, this expression agrees with the Galilean form, Equation 26.6. Equation 26.7 can be inverted to give the equation of relativistic velocity addition:

$$v_{AE} = \frac{v_{AB} + v_{BE}}{1 + \dfrac{v_{AB}v_{BE}}{c^2}} \qquad [26.8]$$

Equation 26.8 gives the intuitive answer agreeing with everyday experience when $v << c$ (i.e., $v_{AE} = v_{AB} + v_{BE}$).

## 26.7 Relativistic Energy and the Equivalence of Mass and Energy

The relativistic expression for the **kinetic energy** of an object is

$$KE = \gamma mc^2 - mc^2 \qquad [26.9]$$

where $mc^2$ is the **rest energy** of the object, $E_R$.

The **total energy** of a particle is

$$E = \frac{mc^2}{\sqrt{1 - v^2/c^2}} \qquad [26.12]$$

Einstein's famous mass-energy equivalence equation results when $v = 0$.

The relativistic momentum is related to the total energy through the equation

$$E^2 = p^2c^2 + (mc^2)^2 \qquad [26.13]$$

## WARM-UP EXERCISES

ENHANCED WebAssign The warm-up exercises in this chapter may be assigned online in Enhanced WebAssign.

1. **Physics Review** A jet aircraft is flying horizontally at 365 m/s in pursuit of a hijacked passenger plane at the same height, flying horizontally at 243 m/s. If both aircraft are traveling in the positive $x$-direction, find the velocity of the plane relative the jet. (See Section 3.5.)
2. **Physics Review** A white rhinoceros of mass 3 450 kg is running across the African savannah at 15.3 m/s. Calculate the magnitude of the rhino's (a) momentum and (b) kinetic energy. (See Sections 6.1 and 5.2.)
3. **Physics Review** A block of mass $m$ = 3.00 kg and second block of mass $M$ = 6.00 kg, both moving along the $x$-axis, collide elastically. If the initial total energy of the system is 123 J, and the final speed of mass $m$ is 4.00 m/s, find the final speed of mass $M$. (See Sections 5.5 and 6.3.)
4. A pendulum in a spaceship traveling through space at a speed of 0.910$c$ relative to Earth has a period 8.00 s as measured by an astronaut in the spaceship. (a) Calculate the $\gamma$-factor. (b) What period is as measured by an observer on Earth? (See Section 26.4.)
5. A spaceship is hurtling towards Earth at 0.680$c$. An observer on Earth notes the spaceship's position just as it crosses the orbit of the moon, $3.84 \times 10^5$ km. (a) Calculate the $\gamma$-factor. (b) What is the distance to Earth as measured by the captain of the spaceship? (See Section 6.4.)
6. A proton is traveling at 0.970$c$ relative the Earth in a linear accelerator. (a) Calculate the proton's nonrelativistic momentum. (b) Calculate the $\gamma$-factor of the particle. (c) What is the particle's momentum as measured by scientists operating the accelerator? (See Section 26.5.)
7. A spaceship traveling in the positive $x$-direction at 0.500$c$ relative a space station shoots a projectile out in front of it at a speed of 0.250$c$ relative the spaceship. Find the speed of the projectile according to an observer in the space station, using (a) Galilean relativity, and (b) special relativity. (See Section 26.6.)
8. Calculate the rest energy of 12.0 g of water. (See Section 26.7.)
9. (a) Convert $5.34 \times 10^{-14}$ J to mega electron volts (MeV). (b) Convert 3.76 MeV to joules. (See Section 26.7.)
10. A deuteron, consisting of a proton and neutron and having mass $3.34 \times 10^{-27}$ kg, is traveling at 0.990$c$ relative the Earth in a linear accelerator. Calculate the deuteron's (a) rest energy, (b) $\gamma$-factor, (c) total energy, and (d) kinetic energy. (See Section 26.7.)

## CONCEPTUAL QUESTIONS

ENHANCED WebAssign The conceptual questions in this chapter may be assigned online in Enhanced WebAssign.

1. A spacecraft with the shape of a sphere of diameter $D$ moves past an observer on Earth with a speed of $0.5c$. What shape does the observer measure for the spacecraft as it moves past?
2. What two speed measurements will two observers in relative motion always agree upon?
3. The speed of light in water is $2.30 \times 10^8$ m/s. Suppose an electron is moving through water at $2.50 \times 10^8$ m/s. Does this particle speed violate the principle of relativity?
4. With regard to reference frames, how does general relativity differ from special relativity?
5. Give a physical argument that shows it is impossible to accelerate an object of mass $m$ to the speed of light, even with a continuous force acting on it.
6. It is said that Einstein, in his teenage years, asked the question, "What would I see in a mirror if I carried it in my hands and ran at a speed near that of light?" How would you answer this question?
7. List some ways our day-to-day lives would change if the speed of light were only 50 m/s.
8. Two identically constructed clocks are synchronized. One is put into orbit around Earth, and the other remains on Earth. (a) Which clock runs more slowly? (b) When the moving clock returns to Earth, will the two clocks still be synchronized? Discuss from the standpoints of both special and general relativity.
9. Photons of light have zero mass. How is it possible that they have momentum?
10. Imagine an astronaut on a trip to Sirius, which lies 8 light-years from Earth. Upon arrival at Sirius, the astronaut finds that the trip lasted 6 years. If the trip was made at a constant speed of $0.8c$, how can the 8-light-year distance be reconciled with the 6-year duration?
11. Explain why, when defining the length of a rod, it is necessary to specify that the positions of the ends of the rod are to be measured simultaneously.
12. Suppose a photon, proton, and electron all have the same total energy $E$. Rank the magnitude of their momenta from smallest to greatest. (a) photon, electron, proton (b) proton, photon, electron (c) electron, photon, proton (d) electron, proton, photon (e) proton, electron, photon

## PROBLEMS

ENHANCED WebAssign The problems in this chapter may be assigned online in Enhanced WebAssign.

**1.** denotes straightforward problem; **2.** denotes intermediate problem;
**3.** denotes challenging problem
**[1.]** denotes full solution available in *Student Solutions Manual/ Study Guide*
**1.** denotes problems most often assigned in Enhanced WebAssign

BIO denotes biomedical problems
GP denotes guided problems
M denotes Master It tutorial available in Enhanced WebAssign
QIC denotes asking for quantitative and conceptual reasoning
S denotes symbolic reasoning problem
W denotes Watch It video solution available in Enhanced WebAssign

### 26.4 Consequences of Special Relativity

**[1.]** If astronauts could travel at $v = 0.950c$, we on Earth would say it takes $(4.20/0.950) = 4.42$ years to reach Alpha Centauri, 4.20 light-years away. The astronauts disagree. (a) How much time passes on the astronauts' clocks? (b) What is the distance to Alpha Centauri as measured by the astronauts?

**2.** QIC A meterstick moving at $0.900c$ relative to the Earth's surface approaches an observer at rest with respect to the Earth's surface. (a) What is the meterstick's length as measured by the observer? (b) Qualitatively, how would the answer to part (a) change if the observer started running toward the meterstick?

**3.** The length of a moving spaceship is 28.0 m according to an astronaut on the spaceship. If the spaceship is contracted by 15.0 cm according to an Earth observer, what is the speed of the spaceship?

**4.** BIO W An astronaut at rest on Earth has a heart rate of 70 beats/min. When the astronaut is traveling in a spaceship at $0.90c$, what will this rate be as measured by (a) an observer also in the ship and (b) an observer at rest on Earth?

**5.** M The average lifetime of a pi meson in its own frame of reference (i.e., the proper lifetime) is $2.6 \times 10^{-8}$ s. If the meson moves with a speed of $0.98c$, what is (a) its mean lifetime as measured by an observer on Earth, and (b) the average distance it travels before decaying, as measured by an observer on Earth? (c) What distance would it travel if time dilation did not occur?

**[6.]** BIO An astronaut is traveling in a space vehicle that has a speed of $0.500c$ relative to Earth. The astronaut measures his pulse rate at 75.0 per minute. Signals generated by the astronaut's pulse are radioed to Earth when the vehicle is moving perpendicular to a line that connects the vehicle with an Earth observer. (a) What pulse rate does the Earth observer measure? (b) What would be the pulse rate if the speed of the space vehicle were increased to $0.990c$?

7. GP A muon formed high in Earth's atmosphere travels toward Earth at a speed $v = 0.990c$ for a distance of 4.60 km as measured by an observer at rest with respect to Earth. It then decays into an electron, a neutrino, and an antineutrino. (a) How long does the muon survive according to an observer at rest on Earth? (b) Compute the gamma factor associated with the muon. (c) How much time passes according to an observer traveling with the muon? (d) What distance does the muon travel according to an observer traveling with the muon? (e) A third observer traveling toward the muon at $c/2$ measures the lifetime of the particle. According to this observer, is the muon's lifetime shorter or longer than the lifetime measured by the observer at rest with respect to Earth? Explain.

8. A deep-space probe moves away from Earth with a speed of $0.80c$. An antenna on the probe requires 3.0 s, in probe time, to rotate through 1.0 rev. How much time is required for 1.0 rev according to an observer on Earth?

9. The proper length of one spaceship is three times that of another. The two spaceships are traveling in the same direction and, while both are passing overhead, an Earth observer measures the two spaceships to have the same length. If the slower spaceship has a speed of $0.350c$ with respect to Earth, determine the speed of the faster spaceship.

10. A car traveling at 35.0 m/s takes 26.0 minutes to travel a certain distance according to the driver's clock in the car. How long does the trip take according to an observer at rest on Earth? *Hint:* The following approximation is helpful: $[1 - x]^{-\frac{1}{2}} \approx 1 + \frac{1}{2}x$ for $x << 1$.

11. W A supertrain of proper length 100 m travels at a speed of $0.95c$ as it passes through a tunnel having proper length 50 m. As seen by a trackside observer, is the train ever completely within the tunnel? If so, by how much?

12. A box is cubical with sides of proper lengths $L_1 = L_2 = L_3$, as shown in Figure P26.12, when viewed in its own rest frame. If this block moves parallel to one of its edges with a speed of $0.80c$ past an observer, (a) what shape does it appear to have to this observer? (b) What is the length of each side as measured by the observer?

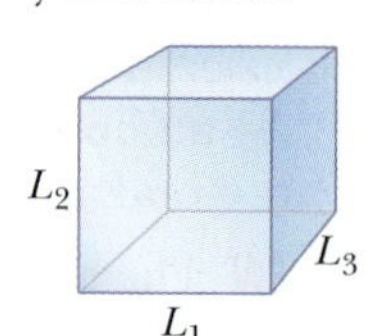

**Figure P26.12**

## 26.5 Relativistic Momentum

13. Q|C An electron has a momentum with magnitude three times the magnitude of its classical momentum. (a) Find the speed of the electron. (b) How would your result change if the particle were a proton?

14. Q|C Calculate the classical momentum of a proton traveling at $0.990c$, neglecting relativistic effects. (b) Repeat the calculation while including relativistic effects. (c) Does it make sense to neglect relativity at such speeds?

15. An unstable particle at rest breaks up into two fragments of *unequal mass*. The mass of the lighter fragment is equal to $2.50 \times 10^{-28}$ kg and that of the heavier fragment is $1.67 \times 10^{-27}$ kg. If the lighter fragment has a speed of $0.893c$ after the breakup, what is the speed of the heavier fragment?

## 26.6 Relative Velocity in Special Relativity

16. Spaceship $R$ is moving to the right at a speed of $0.70c$ with respect to Earth. A second spaceship, $L$, moves to the left at the same speed with respect to Earth. What is the speed of $L$ with respect to $R$?

17. An electron moves to the right with a speed of $0.90c$ relative to the laboratory frame. A proton moves to the left with a speed of $0.70c$ relative to the electron. Find the speed of the proton relative to the laboratory frame.

18. A spaceship travels at $0.750c$ relative to Earth. If the spaceship fires a small rocket in the forward direction, how fast (relative to the ship) must it be fired for it to travel at $0.950c$ relative to Earth?

19. W Spaceship $A$ moves away from Earth at a speed of $0.800c$ (Fig. P26.19). Spaceship $B$ pursues at a speed of $0.900c$ relative to Earth. Observers on Earth see $B$ overtaking $A$ at a relative speed of $0.100c$. With what speed is $B$ overtaking $A$ as seen by the crew of spaceship $B$?

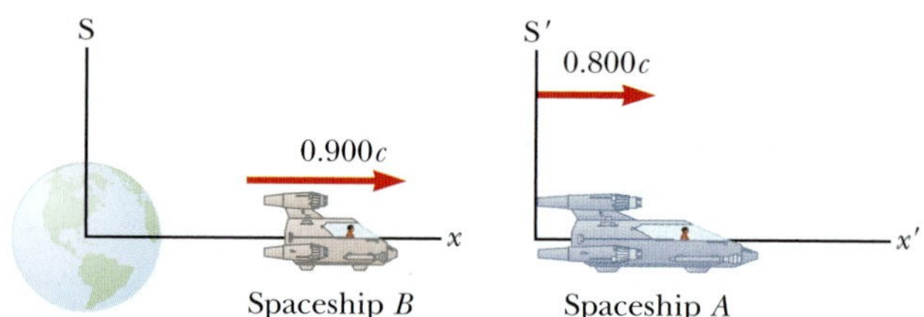

**Figure P26.19**

20. A pulsar is a stellar object that emits light in short bursts. Suppose a pulsar with a speed of $0.950c$ approaches Earth, and a rocket with a speed of $0.995c$ heads toward the pulsar. (Both speeds are measured in Earth's frame of reference.) If the pulsar emits 10.0 pulses per second in its own frame of reference, at what rate are the pulses emitted in the rocket's frame of reference?

21. M A rocket moves with a velocity of $0.92c$ to the right with respect to a stationary observer $A$. An observer $B$ moving relative to observer $A$ finds that the rocket is moving with a velocity of $0.95c$ to the left. What is the velocity of observer $B$ relative to observer $A$? (*Hint:* Consider observer $B$'s velocity in the frame of reference of the rocket.)

## 26.7 Relativistic Energy and the Equivalence of Mass and Energy

22. W A proton moves with a speed of $0.950c$. Calculate (a) its rest energy, (b) its total energy, and (c) its kinetic energy.

23. Protons in an accelerator at the Fermi National Laboratory near Chicago are accelerated to a total energy that is 400 times their rest energy. (a) What is the speed of these protons in terms of $c$? (b) What is their kinetic energy in MeV?

24. **S** A proton in a large accelerator has a kinetic energy of 175 GeV. (a) Compare this kinetic energy to the rest energy of the proton, and find an approximate expression for the proton's kinetic energy. (b) Find the speed of the proton.

25. What speed must a particle attain before its kinetic energy is double the value predicted by the nonrelativistic expression $KE = \frac{1}{2}mv^2$?

26. Determine the energy required to accelerate an electron from (a) $0.500c$ to $0.900c$ and (b) $0.900c$ to $0.990c$.

27. A spaceship of mass $2.40 \times 10^6$ kg is to be accelerated to a speed of $0.700c$. (a) What minimum amount of energy does this acceleration require from the spaceship's fuel, assuming perfect efficiency? (b) How much fuel would it take to provide this much energy if all the rest energy of the fuel could be transformed to kinetic energy of the spaceship?

28. An unstable particle with a mass equal to $3.34 \times 10^{-27}$ kg is initially at rest. The particle decays into two fragments that fly off with velocities of $0.987c$ and $-0.868c$, respectively. Find the masses of the fragments. *Hint:* Conserve both mass–energy and momentum.

29. **QIC** Starting with the definitions of relativistic energy and momentum, show that $E^2 = p^2c^2 + m^2c^4$ (Eq. 26.13).

30. **GP** Consider the reaction ${}^{235}_{92}\text{U} + {}^{1}_{0}\text{n} \rightarrow {}^{148}_{57}\text{La} + {}^{87}_{35}\text{Br} + {}^{1}_{0}\text{n}$. (a) Write the conservation of relativistic energy equation symbolically in terms of the rest energy and the kinetic energy, setting the initial total energy equal to the final total energy. (b) Using values from Appendix B, find the total mass of the initial particles. (c) Using the values given below, find the total mass of the particles after the reaction takes place. (d) Subtract the final particle mass from the initial particle mass. (e) Convert the answer to part (d) to MeV, obtaining the kinetic energy of the daughter particles. Neglect the kinetic energy of the reactants. *Note:* Lanthanum-148 has atomic mass 147.932 236 u; bromine-87 has atomic mass 86.920 711 19 u.

## Additional Problems

31. Consider electrons accelerated to a total energy of 20.0 GeV in the 3.00-km-long Stanford Linear Accelerator. (a) What is the $\gamma$ factor for the electrons? (b) How long does the accelerator appear to the electrons? Electron mass energy: 0.511 MeV.

32. An electron has a speed of $0.750c$. (a) Find the speed of a proton that has the same kinetic energy as the electron. (b) Find the speed of a proton that has the same momentum as the electron.

33. The rest energy of an electron is 0.511 MeV. The rest energy of a proton is 938 MeV. Assume both particles have kinetic energies of 2.00 MeV. Find the speed of (a) the electron and (b) the proton. (c) By how much does the speed of the electron exceed that of the proton? *Note:* Perform the calculations in MeV; don't convert the energies to joules. The answer is sensitive to rounding.

34. **QIC** A spring of force constant $k$ is compressed by a distance $x$ from its equilibrium length. (a) Does the mass of the spring change when the spring is compressed? Explain. (b) Find an expression for the change in mass of the spring in terms of $k$, $x$, and $c$. (c) What is the change in mass if the force constant is $2.0 \times 10^2$ N/m and $x = 15$ cm?

35. A star is 5.00 ly from the Earth. At what speed must a spacecraft travel on its journey to the star such that the Earth–star distance measured in the frame of the spacecraft is 2.00 ly?

36. An electron has a total energy equal to five times its rest energy. (a) What is its momentum? (b) Repeat for a proton.

37. **M** An astronaut wishes to visit the Andromeda galaxy, making a one-way trip that will take 30.0 years in the spaceship's frame of reference. Assume the galaxy is 2.00 million light-years away and his speed is constant. (a) How fast must he travel relative to Earth? (b) What will be the kinetic energy of his spacecraft, which has mass of $1.00 \times 10^6$ kg? (c) What is the cost of this energy if it is purchased at a typical consumer price for electric energy, 13.0 cents per kWh? The following approximation will prove useful:

$$\frac{1}{\sqrt{1+x}} \approx 1 - \frac{x}{2} \qquad \text{for } x << 1$$

38. An alarm clock is set to sound in 10 h. At $t = 0$, the clock is placed in a spaceship moving with a speed of $0.75c$ (relative to Earth). What distance, as determined by an Earth observer, does the spaceship travel before the alarm clock sounds?

39. Owen and Dina are at rest in frame S′, which is moving with a speed of $0.600c$ with respect to frame S. They play a game of catch while Ed, at rest in frame S, watches the action (Fig. P26.39). Owen throws the ball to Dina with a speed of $0.800c$ (according to Owen) and their separation (measured in S′) is equal to $1.80 \times 10^{12}$ m.

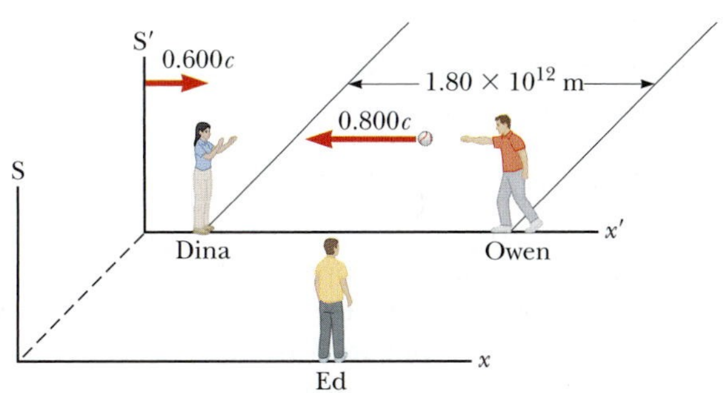


**Figure P26.39**

(a) According to Dina, how fast is the ball moving? (b) According to Dina, what time interval is required for the ball to reach her? According to Ed, (c) how far apart are Owen and Dina, and (d) how fast is the ball moving?

40. An observer in a coasting spacecraft moves toward a mirror at speed $v$ relative to the reference frame labeled by S in Figure P26.40. The mirror is stationary with respect to S. A light pulse emitted by the spacecraft travels toward the mirror and is reflected back to the spacecraft. The spacecraft is a distance $d$ from the mirror (as measured by observers in S) at the moment the light pulse leaves the spacecraft. What is the total travel time of the pulse as measured by observers in (a) the S frame and (b) the spacecraft?

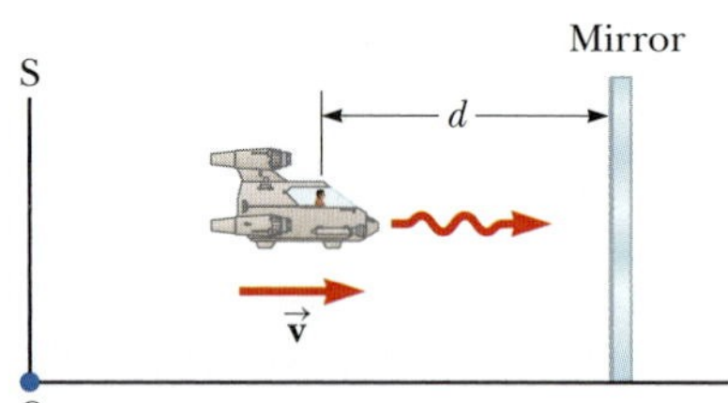


**Figure P26.40**

41. A spaceship of proper length 300 m takes 0.75 $\mu$s to pass an Earth observer. Determine the speed of this spaceship as measured by the Earth observer.

42. The cosmic rays of highest energy are protons that have kinetic energy on the order of $10^{13}$ MeV. (a) From the point of view of the proton, how many kilometers across is the galaxy? (b) How long would it take a proton of this energy to travel across the Milky Way galaxy, having a diameter $\sim 10^5$ light-years, as measured in the proton's frame?

43. **M** The nonrelativistic expression for the momentum of a particle, $p = mv$, can be used if $v << c$. For what speed does the use of this formula give an error in the momentum of (a) 1.00% and (b) 10.0%?

44. (a) Show that a potential difference of $1.02 \times 10^6$ V would be sufficient to give an electron a speed equal to twice the speed of light if Newtonian mechanics remained valid at high speeds. (b) What speed would an electron actually acquire in falling through a potential difference equal to $1.02 \times 10^6$ V?

45. The muon is an unstable particle that spontaneously decays into an electron and two neutrinos. In a reference frame in which the muons are stationary, if the number of muons at $t = 0$ is $N_0$, the number at time $t$ is given by $N = N_0 e^{-t/\tau}$, where $\tau$ is the mean lifetime, equal to 2.2 $\mu$s. Suppose the muons move at a speed of $0.95c$ and there are $5.0 \times 10^4$ muons at $t = 0$. (a) What is the observed lifetime of the muons? (b) How many muons remain after traveling a distance of 3.0 km?

46. Imagine that the entire Sun collapses to a sphere of radius $R_g$ such that the work required to remove a small mass $m$ from the surface would be equal to its rest energy $mc^2$. This radius is called the *gravitational radius* for the Sun. Find $R_g$. (It is believed that the ultimate fate of very massive stars is to collapse beyond their gravitational radii into black holes.)

47. The identical twins Speedo and Goslo join a migration from Earth to Planet X, which is 20.0 light-years away in a reference frame in which both planets are at rest. The twins, of the same age, depart at the same time on different spacecraft. Speedo's craft travels steadily at $0.950c$, Goslo's at $0.750c$. Calculate the age difference between the twins after Goslo's spacecraft lands on Planet X. Which twin is the older?

48. **M** An interstellar space probe is launched from Earth. After a brief period of acceleration, it moves with a constant velocity, 70.0% of the speed of light. Its nuclear-powered batteries supply the energy to keep its data transmitter active continuously. The batteries have a lifetime of 15.0 years as measured in a rest frame. (a) How long do the batteries on the space probe last as measured by mission control on Earth? (b) How far is the probe from Earth when its batteries fail as measured by mission control? (c) How far is the probe from Earth as measured by its built-in trip odometer when its batteries fail? (d) For what total time after launch are data received from the probe by mission control? Note that radio waves travel at the speed of light and fill the space between the probe and Earth at the time the battery fails.

49. An observer moving at a speed of $0.995c$ relative to a rod (Fig. P26.49) measures its length to be 2.00 m and sees its length to be oriented at 30.0° with respect to its direction of motion. (a) What is the proper length of the rod? (b) What is the orientation angle in a reference frame moving with the rod?

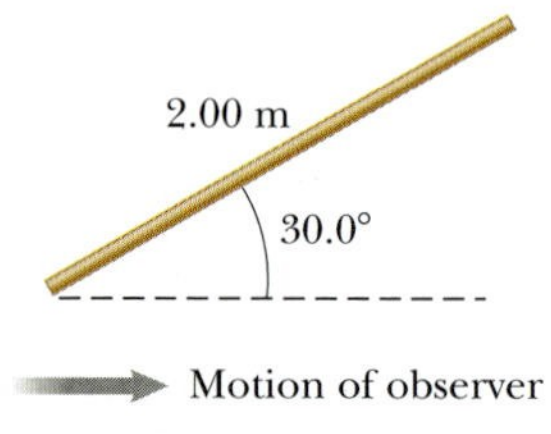


**Figure P26.49**

50. An alien spaceship traveling $0.600c$ toward Earth launches a landing craft with an advance guard of purchasing agents. The lander travels in the same direction with a velocity $0.800c$ relative to the spaceship. As observed on Earth, the spaceship is 0.200 light-years from Earth when the lander is launched. (a) With what velocity is the lander observed to be approaching by observers on Earth? (b) What is the distance to Earth at the time of lander launch, as observed by the aliens on the mother ship? (c) How long does it take the lander to reach Earth as observed by the aliens on the mother ship? (d) If the lander has a mass of $4.00 \times 10^5$ kg, what is its kinetic energy as observed in Earth's reference frame?

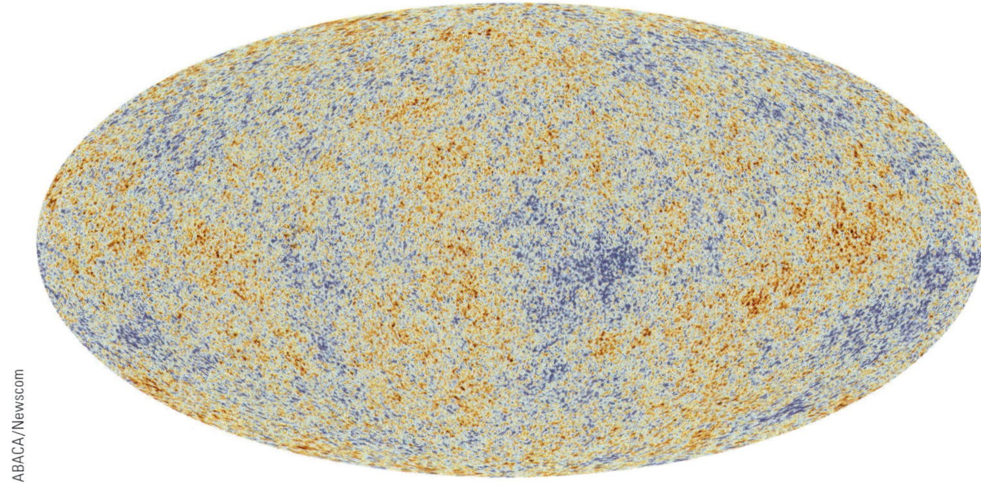

ABACA/Newscom

This is the distribution of the cosmic microwave background as observed by the Planck space-based telescope, leftover blackbody radiation created 380 000 years after the Big Bang. The temperature of this thermal radiation is about 2.7 K. The false color image represents differences from that temperature by about one part in one hundred thousand, giving clues to structure formation in the early universe that led to the stars, galaxies, and clusters of galaxies observed in the universe today.

# Quantum Physics 27



Although many problems were resolved by the theory of relativity in the early part of the 20th century, many others remained unsolved. Attempts to explain the behavior of matter on the atomic level with the laws of classical physics were consistently unsuccessful. Various phenomena, such as the electromagnetic radiation emitted by a heated object (blackbody radiation), the emission of electrons by illuminated metals (the photoelectric effect), and the emission of sharp spectral lines by gas atoms in an electric discharge tube couldn't be understood within the framework of classical physics. Between 1900 and 1930, however, a modern version of mechanics called *quantum mechanics* or *wave mechanics* was highly successful in explaining the behavior of atoms, molecules, and nuclei.

The earliest ideas of quantum theory were introduced by Planck, and most of the subsequent mathematical developments, interpretations, and improvements were made by a number of distinguished physicists, including Einstein, Bohr, Schrödinger, de Broglie, Heisenberg, Born, and Dirac. In this chapter we introduce the underlying ideas of quantum theory and the wave-particle nature of matter, and discuss some simple applications of quantum theory, including the photoelectric effect, the Compton effect, and x-rays.

## 27.1 Blackbody Radiation and Planck's Hypothesis

**LEARNING OBJECTIVES**

1. Define thermal radiation and discuss its physical origin.
2. Define blackbody and explain the reason for the definition.

3. Sketch a typical curve for the intensity of blackbody radiation as a function of wavelength.
4. State Wien's displacement law. Explain its origin from the graph of the intensity of blackbody radiation versus wavelength and how the graph changes with temperature.
5. State the key assumption in Planck's theory of blackbody radiation: the quantization of energy.

**Tip 27.1 Expect the Unexpected**

Our life experiences take place in the macroscopic world, where quantum effects are not evident. Quantum effects can be even more bizarre than relativistic effects. As Nobel Prize–winning physicist Richard Feynman once said, "Nobody understands quantum mechanics."

© Bettmann/CORBIS

**Max Planck**
**German Physicist (1858–1947)**
Planck introduced the concept of a "quantum of action" (Planck's constant *h*) in an attempt to explain the spectral distribution of blackbody radiation, which formed the foundations for quantum theory. In 1918 he was awarded the Nobel Prize in Physics for this discovery of the quantized nature of energy.

An object at any temperature emits electromagnetic radiation, called **thermal radiation**. Stefan's law, discussed in Section 11.5, describes the total power radiated. The spectrum of the radiation depends on the temperature and properties of the object. At low temperatures, the wavelengths of the thermal radiation are mainly in the infrared region and hence not observable by the eye. As the temperature of an object increases, the object eventually begins to glow red. At sufficiently high temperatures, it appears to be white, as in the glow of the hot tungsten filament of a lightbulb. A careful study of thermal radiation shows that it consists of a continuous distribution of wavelengths from the infrared, visible, and ultraviolet portions of the spectrum.

From a classical viewpoint, thermal radiation originates from accelerated charged particles near the surface of an object; such charges emit radiation, much as small antennas do. The thermally agitated charges can have a distribution of frequencies, which accounts for the continuous spectrum of radiation emitted by the object. By the end of the 19th century, it had become apparent that the classical theory of thermal radiation was inadequate. The basic problem was in understanding the observed distribution energy as a function of wavelength in the radiation emitted by a blackbody. By definition, a **blackbody** is an ideal system that absorbs *all* radiation incident on it. A good approximation of a blackbody is a small hole leading to the inside of a hollow object, as shown in Figure 27.1. The nature of the radiation emitted through the small hole leading to the cavity depends *only on the temperature* of the cavity walls and not at all on the material composition of the object, its shape, or other factors.

Experimental data for the distribution of energy in blackbody radiation at three temperatures are shown in Figure 27.2. The radiated energy varies with wavelength and temperature. As the temperature of the blackbody increases, the total amount of energy (area under the curve) it emits increases. Also, with increasing temperature, the peak of the distribution shifts to shorter wavelengths. This shift obeys the following relationship, called **Wien's displacement law**,

$$\lambda_{\max} T = 0.289\,8 \times 10^{-2}\ \text{m} \cdot \text{K} \qquad [27.1]$$

where $\lambda_{\max}$ is the wavelength at which the curve peaks and $T$ is the absolute temperature of the object emitting the radiation.

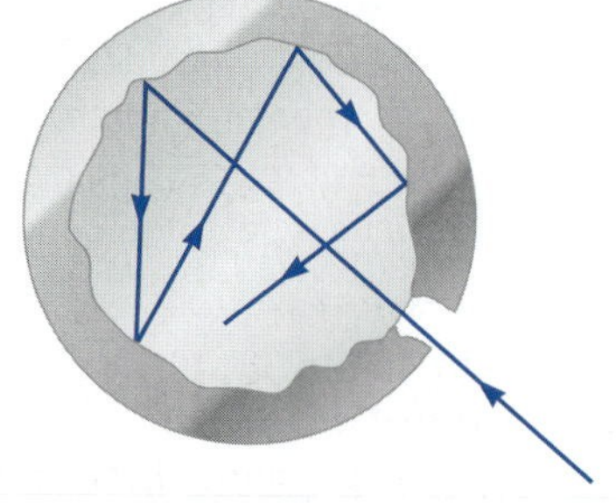

**Figure 27.1** An opening in the cavity of a body is a good approximation of a blackbody. As light enters the cavity through the small opening, part is reflected and part is absorbed on each reflection from the interior walls. After many reflections, essentially all the incident energy is absorbed.

## APPLYING PHYSICS 27.1 | Star Colors

If you look carefully at stars in the night sky, you can distinguish three main colors: red, white, and blue. What causes these particular colors?

**EXPLANATION** These colors result from the different surface temperatures of stars. A relatively cool star, with a surface temperature of 3 000 K, has a radiation curve similar to the middle curve in Figure 27.2. The peak in this curve is above the visible wavelengths, 0.4 $\mu$m to 0.7 $\mu$m, beyond the wavelength of red light, so significantly more radiation is emitted within the visible range at the red end than the blue end of the spectrum. Consequently, the star appears reddish in color, similar to the red glow from the burner of an electric range. ■

A hotter star has a radiation curve more like the upper curve in Figure 27.2. In this case the star emits significant radiation throughout the visible range, and the combination of all colors causes the star to look white. Such is the case with our own Sun, with a surface temperature of 5 800 K. For much hotter stars, the peak can be shifted so far below the visible range that significantly more blue radiation is emitted than red, so the star appears bluish in color. Stars cooler than the Sun tend to have orange or red colors. The surface temperature of a star can be obtained by first finding the wavelength corresponding to the maximum in the intensity versus wavelength curve, then substituting that wavelength into Wien's law. For example, if the wavelength were $2.30 \times 10^{-7}$ m, the surface temperature would be given by

$$T = \frac{0.289\,8 \times 10^{-2}\ \text{m}\cdot\text{K}}{\lambda_{\text{max}}} = \frac{0.289\,8 \times 10^{-2}\ \text{m}\cdot\text{K}}{2.30 \times 10^{-7}\ \text{m}} = 1.26 \times 10^{4}\ \text{K}$$

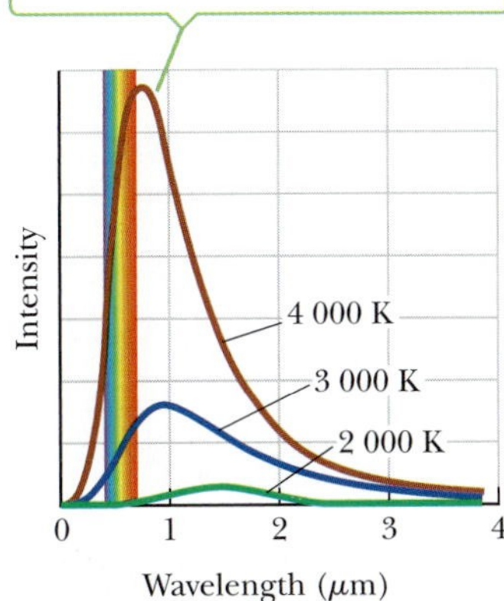


**Figure 27.2** Intensity of blackbody radiation versus wavelength at three temperatures. The visible range of wavelengths is between 0.4 $\mu$m and 0.7 $\mu$m. At approximately 6 000 K, the peak is in the center of the visible wavelengths, and the object appears white.

Attempts to use classical ideas to explain the shapes of the curves shown in Figure 27.2 failed. Figure 27.3 shows an experimental plot of the blackbody radiation spectrum (blue curve), together with the theoretical picture of what this curve should look like based on classical theories (rust-colored curve). At long wavelengths, classical theory is in good agreement with the experimental data. At short wavelengths, however, major disagreement exists between classical theory and experiment. As $\lambda$ approaches zero, classical theory erroneously predicts that the intensity should go to infinity, when the experimental data show it should approach zero.

In 1900 Planck developed a formula for blackbody radiation that was in complete agreement with experiments at all wavelengths, leading to a curve shown by the blue line in Figure 27.3. Planck hypothesized that blackbody radiation was produced by submicroscopic charged oscillators, which he called *resonators.* He assumed the walls of a glowing cavity were composed of billions of these resonators, although their exact nature was unknown. The resonators were allowed to have only certain discrete energies $E_n$, given by

$$E_n = nhf \qquad [27.2]$$

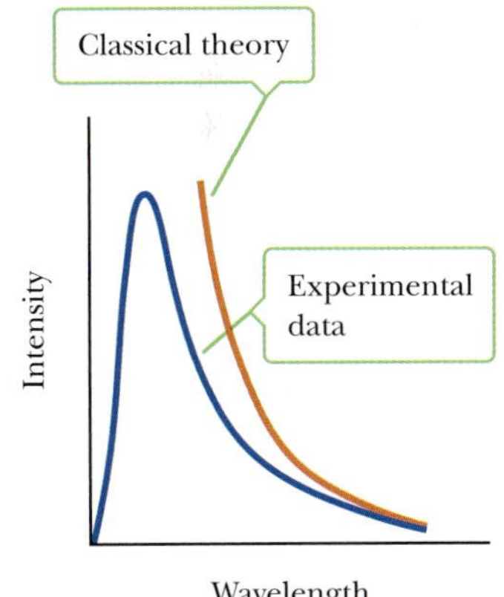


**Figure 27.3** Comparison of experimental data with the classical theory of blackbody radiation. Planck's theory matches the experimental data perfectly.

where $n$ is a positive integer called a **quantum number**, $f$ is the frequency of vibration of the resonator, and $h$ is a constant known as **Planck's constant**, which has the value

$$h = 6.626 \times 10^{-34}\ \text{J}\cdot\text{s} \qquad [27.3]$$

Because the energy of each resonator can have only discrete values given by Equation 27.2, we say the energy is *quantized.* Each discrete energy value represents a different *quantum state*, with each value of $n$ representing a specific quantum state. (When the resonator is in the $n = 1$ quantum state, its energy is $hf$; when it is in the $n = 2$ quantum state, its energy is $2hf$; and so on.)

The key point in Planck's theory is the assumption of quantized energy states. This radical departure from classical physics is the "quantum leap" that led to a totally new understanding of nature. It's shocking: it's like saying a pitched baseball can have only a fixed number of different speeds, and no speeds in between those fixed values. The fact that energy can assume only certain, discrete values instead of any one of a continuum of values is the single most important difference between quantum theory and the classical theories of Newton and Maxwell.

# 27.2 The Photoelectric Effect and the Particle Theory of Light

**LEARNING OBJECTIVES**

1. Describe the photoelectric effect.
2. Discuss non-classical features of the photoelectric effect and their explanations using the photon theory of light.
3. Apply the photoelectric effect equation to quantum systems.

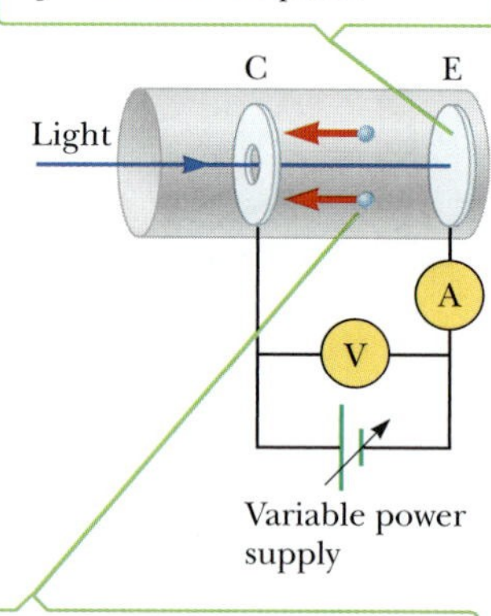
When light strikes plate E (the emitter), photoelectrons are ejected from the plate.
C
E
Light
A
V
Variable power supply
Electrons moving from plate E to plate C (the collector) create a current in the circuit, registered at the ammeter, A.

**Figure 27.4** A circuit diagram for studying the photoelectric effect.

In the latter part of the 19th century, experiments showed that light incident on certain metallic surfaces caused the emission of electrons from the surfaces. This phenomenon is known as the **photoelectric effect**, and the emitted electrons are called **photoelectrons**. The first discovery of this phenomenon was made by Hertz, who was also the first to produce the electromagnetic waves predicted by Maxwell.

Figure 27.4 is a schematic diagram of a photoelectric effect apparatus. An evacuated glass tube known as a photocell contains a metal plate E (the emitter) connected to the negative terminal of a variable power supply. Another metal plate, C (the collector), is maintained at a positive potential by the power supply. When the tube is kept in the dark, the ammeter reads zero, indicating that there is no current in the circuit. When plate E is illuminated by light having a wavelength shorter than some particular wavelength that depends on the material used to make plate E, however, a current is detected by the ammeter, indicating a flow of charges across the gap between E and C. This current arises from photoelectrons emitted from the negative plate E and collected at the positive plate C.

Figure 27.5 is a plot of the photoelectric current versus the potential difference $\Delta V$ between E and C for two light intensities. At large values of $\Delta V$, the current reaches a maximum value. In addition, the current increases as the incident light intensity increases, as you might expect. Finally, when $\Delta V$ is negative—that is, when the power supply in the circuit is reversed to make E positive and C negative—the current drops to a low value because most of the emitted photoelectrons are repelled by the now negative plate C. In this situation, only those electrons having a kinetic energy greater than the magnitude of $e\,\Delta V$ reach C, where $e$ is the charge on the electron.

When $\Delta V$ is equal to or more negative than $-\Delta V_s$, the **stopping potential**, no electrons reach C, and the current is zero. The stopping potential is *independent* of the radiation intensity. The maximum kinetic energy of the photoelectrons is related to the stopping potential through the relationship

$$KE_{\max} = e\Delta V_s \qquad [27.4]$$

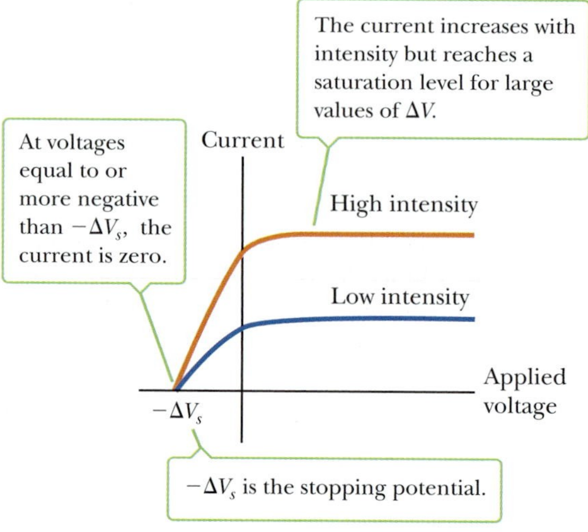
The current increases with intensity but reaches a saturation level for large values of ΔV.
At voltages equal to or more negative than −ΔV_s, the current is zero.
Current
High intensity
Low intensity
Applied voltage
−ΔV_s
−ΔV_s is the stopping potential.

**Figure 27.5** Photoelectric current versus applied potential difference for two light intensities.

Several features of the photoelectric effect can't be explained with classical physics or with the wave theory of light:

- No electrons are emitted if the incident light frequency falls below some **cut-off frequency** $f_c$, also called the threshold frequency, which is characteristic of the material being illuminated. This fact is inconsistent with the wave theory, which predicts that the photoelectric effect should occur at *any* frequency, provided the light intensity is sufficiently high.
- The maximum kinetic energy of the photoelectrons is independent of light intensity. According to wave theory, light of higher intensity should carry more energy into the metal per unit time and therefore eject photoelectrons having higher kinetic energies.
- The maximum kinetic energy of the photoelectrons increases with increasing light frequency. The wave theory predicts no relationship between photoelectron energy and incident light frequency.
- Electrons are emitted from the surface almost instantaneously (less than $10^{-9}$ s after the surface is illuminated), even at low light intensities. Classically, we expect the photoelectrons to require some time to absorb the incident radiation before they acquire enough kinetic energy to escape from the metal.

A successful explanation of the photoelectric effect was given by Einstein in 1905, the same year he published his special theory of relativity. As part of a general paper on electromagnetic radiation, for which he received the Nobel Prize in Physics in 1921, Einstein extended Planck's concept of quantization to electromagnetic waves. He suggested that a tiny packet of light energy or **photon** would be emitted when a quantized oscillator made a jump from an energy state $E_n = nhf$ to the next lower state $E_{n-1} = (n - 1)hf$. Conservation of energy would require the decrease in oscillator energy, $hf$, to be equal to the photon's energy $E$, so that

$$E = hf \qquad [27.5]$$

◄ Energy of a photon

where $h$ is Planck's constant and $f$ is the frequency of the light, which is equal to the frequency of Planck's oscillator.

The key point here is that the light energy lost by the emitter, $hf$, stays sharply localized in a tiny packet or particle called a photon. In Einstein's model a photon is so localized that it can give *all* its energy $hf$ to a single electron in the metal. According to Einstein, the maximum kinetic energy for these liberated photoelectrons is

$$KE_{\text{max}} = hf - \phi \qquad [27.6]$$

◄ Photoelectric effect equation

where $\phi$ is called the **work function** of the metal. The work function, which represents the minimum energy with which an electron is bound in the metal, is on the order of a few electron volts. Table 27.1 lists work functions for various metals.

With the photon theory of light, we can explain the previously mentioned features of the photoelectric effect that cannot be understood using concepts of classical physics:

- Photoelectrons are created by absorption of a single photon, so the energy of that photon must be greater than or equal to the work function, else no photoelectrons will be produced. This explains the cutoff frequency.
- From Equation 27.6, $KE_{\text{max}}$ depends only on the frequency of the light and the value of the work function. Light intensity is immaterial because absorption of a single photon is responsible for the electron's change in kinetic energy.
- Equation 27.6 is linear in the frequency, so $KE_{\text{max}}$ increases with increasing frequency.

**Table 27.1** Work Functions of Selected Metals

| Metal | $\phi$ (eV) |
|---|---|
| Ag | 4.73 |
| Al | 4.08 |
| Cu | 4.70 |
| Fe | 4.50 |
| Na | 2.46 |
| Pb | 4.14 |
| Pt | 6.35 |
| Zn | 4.31 |

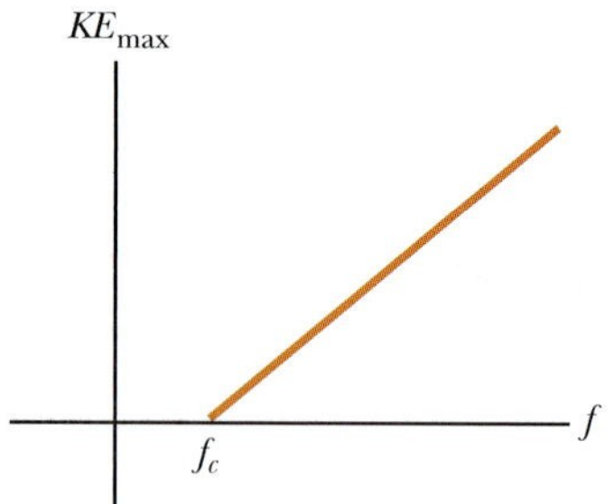


**Figure 27.6** A sketch of $KE_{max}$ versus the frequency of incident light for photoelectrons in a typical photoelectric effect experiment. Photons with frequency less than $f_c$ don't have sufficient energy to eject an electron from the metal.

- Electrons are emitted almost instantaneously, regardless of intensity, because the light energy is concentrated in packets rather than spread out in waves. If the frequency is high enough, no time is needed for the electron to gradually acquire sufficient energy to escape the metal.

Experimentally, a linear relationship is observed between $f$ and $KE_{max}$, as sketched in Figure 27.6. The intercept on the horizontal axis, corresponding to $KE_{max} = 0$, gives the cutoff frequency below which no photoelectrons are emitted, regardless of light intensity. The cutoff *wavelength* $\lambda_c$ can be derived from Equation 27.6:

$$KE_{max} = hf_c - \phi = 0 \quad \rightarrow \quad h\frac{c}{\lambda_c} - \phi = 0$$

$$\lambda_c = \frac{hc}{\phi} \qquad [27.7]$$

where $c$ is the speed of light. Wavelengths *greater* than $\lambda_c$ incident on a material with work function $\phi$ don't result in the emission of photoelectrons.

## EXAMPLE 27.1 Photoelectrons from Sodium

**GOAL** Understand the quantization of light and its role in the photoelectric effect.

**PROBLEM** A sodium surface is illuminated with light of wavelength 0.300 $\mu$m. The work function for sodium is 2.46 eV. Calculate **(a)** the energy of each photon in electron volts, **(b)** the maximum kinetic energy of the ejected photoelectrons, and **(c)** the cutoff wavelength for sodium.

**STRATEGY** Parts (a), (b), and (c) require substitution of values into Equations 27.5, 27.6, and 27.7, respectively.

**SOLUTION**

**(a)** Calculate the energy of each photon.

Obtain the frequency from the wavelength:

$$c = f\lambda \quad \rightarrow \quad f = \frac{c}{\lambda} = \frac{3.00 \times 10^8 \text{ m/s}}{0.300 \times 10^{-6} \text{ m}}$$

$$f = 1.00 \times 10^{15} \text{ Hz}$$

Use Equation 27.5 to calculate the photon's energy:

$$E = hf = (6.63 \times 10^{-34} \text{ J} \cdot \text{s})(1.00 \times 10^{15} \text{ Hz})$$

$$= 6.63 \times 10^{-19} \text{ J}$$

$$= (6.63 \times 10^{-19} \text{ J})\left(\frac{1.00 \text{ eV}}{1.60 \times 10^{-19} \text{ J}}\right) = \boxed{4.14 \text{ eV}}$$

**(b)** Find the maximum kinetic energy of the photoelectrons.

Substitute into Equation 27.6:

$$KE_{max} = hf - \phi = 4.14 \text{ eV} - 2.46 \text{ eV} = \boxed{1.68 \text{ eV}}$$

**(c)** Compute the cutoff wavelength.

Convert $\phi$ from electron volts to joules:

$$\phi = 2.46 \text{ eV} = (2.46 \text{ eV})(1.60 \times 10^{-19} \text{ J/eV})$$

$$= 3.94 \times 10^{-19} \text{ J}$$

Find the cutoff wavelength using Equation 27.7:

$$\lambda_c = \frac{hc}{\phi} = \frac{(6.63 \times 10^{-34} \text{ J} \cdot \text{s})(3.00 \times 10^8 \text{ m/s})}{3.94 \times 10^{-19} \text{ J}}$$

$$= 5.05 \times 10^{-7} \text{ m} = \boxed{505 \text{ nm}}$$

**REMARKS** The cutoff wavelength is in the yellow-green region of the visible spectrum.

**QUESTION 27.1** True or False: Suppose in a given photoelectric experiment the frequency of light is larger than the cutoff frequency. The magnitude of the stopping potential times the electron charge, then, is larger than the energy of the incident photons.

**EXERCISE 27.1** (a) What minimum-frequency light will eject photoelectrons from a copper surface? (b) If this frequency is tripled, find the maximum kinetic energy (in eV) of the resulting photoelectrons.

**ANSWERS** (a) $1.13 \times 10^{15}$ Hz (b) 9.40 eV

## Photocells

The photoelectric effect has many interesting applications using a device called the *photocell.* The photocell shown in Figure 27.4 produces a current in the circuit when light of sufficiently high frequency falls on the cell, but it doesn't allow a current in the dark. This device is used in streetlights: a photoelectric control unit in the base of the light activates a switch that turns off the streetlight when ambient light strikes it. Many garage-door systems and elevators use a light beam and a photocell as a safety feature in their design. When the light beam strikes the photocell, the electric current generated is sufficiently large to maintain a closed circuit. When an object or a person blocks the light beam, the current is interrupted, which signals the door to open.

**APPLICATION**
Photocells

# 27.3 X-Rays

**LEARNING OBJECTIVES**

1. Discuss the production of x-rays by electron bombardment of a metal. State their typical wavelengths.
2. Discuss the intensity of a typical x-ray spectrum and its dependence on wavelength. Explain the discrete and continuous portions of the spectrum.
3. Define threshold voltage and bremsstrahlung.
4. Derive the shortest wavelength that can be produced by electron bombardment of a given metal.

X-rays were discovered in 1895 by Wilhelm Röntgen and much later identified as electromagnetic waves, following a suggestion by Max von Laue in 1912. X-rays have higher frequencies than ultraviolet radiation and can penetrate most materials with relative ease. Typical x-ray wavelengths are about 0.1 nm, which is on the order of the atomic spacing in a solid. As a result, they can be diffracted by the regular atomic spacings in a crystal lattice, which act as a diffraction grating. The x-ray diffraction pattern of a well-ordered protein crystal is shown in Figure 27.7.

X-rays are produced when high-speed electrons are suddenly slowed down, such as when a metal target is struck by electrons that have been accelerated through a potential difference of several thousand volts. Figure 27.8a shows a schematic

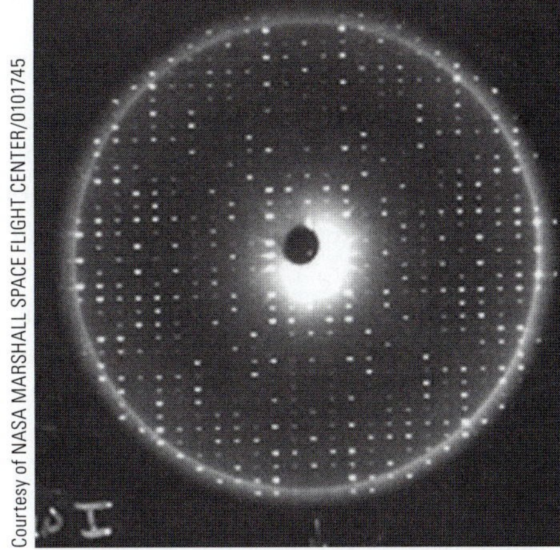

Courtesy of NASA MARSHALL SPACE FLIGHT CENTER/0101745

**Figure 27.7** An x-ray diffraction pattern of a well-ordered protein crystal can be analyzed mathematically with powerful computers, leading to a determination of the structure of the protein. Once the structure is known, molecules can be designed that fit the active site of the protein. Such molecules can be used in developing therapeutic drugs that deactivate a given protein without affecting other biological systems.

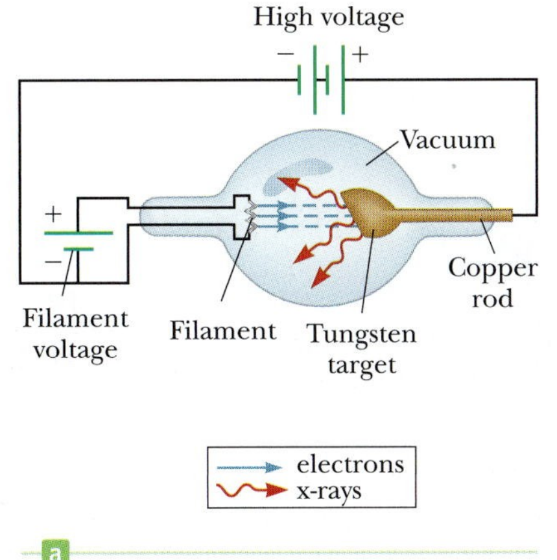


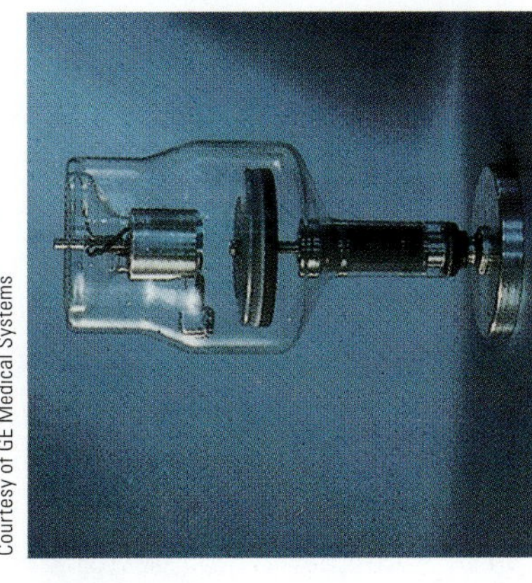


Courtesy of GE Medical Systems

**Figure 27.8** (a) Diagram of an x-ray tube. (b) Photograph of an x-ray tube.

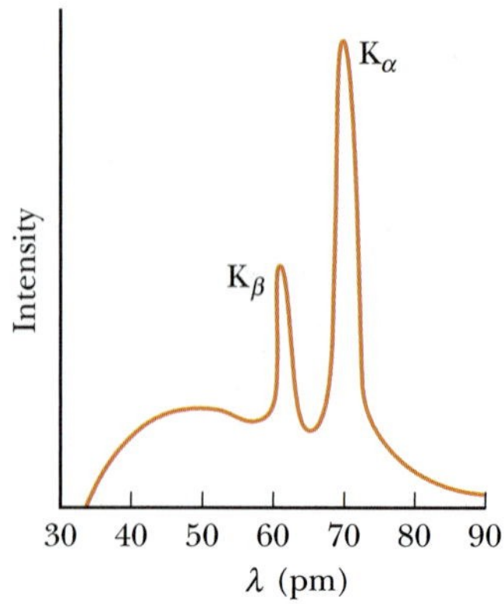


**Figure 27.9** The x-ray spectrum of a metal target consists of a broad continuous spectrum plus a number of sharp lines, which are due to *characteristic x-rays*. The data shown were obtained when 35-keV electrons bombarded a molybdenum target. Note that 1 pm $= 10^{-12}$ m $= 10^{-3}$ nm.

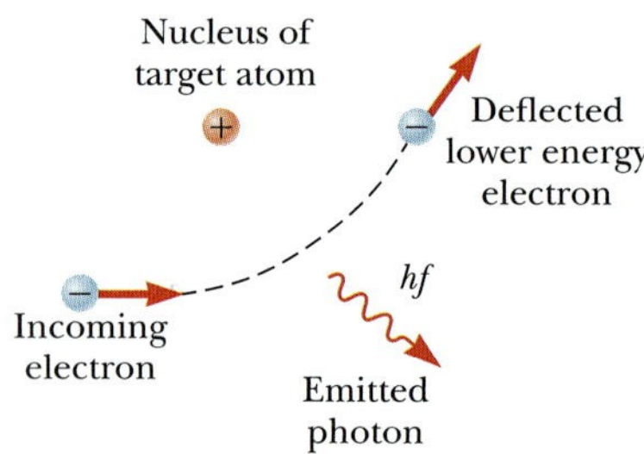


**Figure 27.10** An electron passing near a charged target atom experiences an acceleration, and a photon is emitted in the process.

diagram of an x-ray tube. A current in the filament causes electrons to be emitted, and these freed electrons are accelerated toward a dense metal target, such as tungsten, which is held at a higher potential than the filament.

Figure 27.9 represents a plot of x-ray intensity versus wavelength for the spectrum of radiation emitted by an x-ray tube. Note that the spectrum has two distinct components. One component is a continuous broad spectrum that depends on the voltage applied to the tube. Superimposed on this component is a series of sharp, intense lines that depend on the nature of the target material. To observe these sharp lines, which represent radiation emitted by the target atoms as their electrons undergo rearrangements, the accelerating voltage must exceed a certain value, called the **threshold voltage**. We discuss threshold voltage further in Chapter 28. The continuous radiation is sometimes called **bremsstrahlung**, a German word meaning "braking radiation," because electrons emit radiation when they undergo an acceleration inside the target.

Figure 27.10 illustrates how x-rays are produced when an electron passes near a charged target nucleus. As the electron passes close to a positively charged nucleus in the target material, it is deflected from its path because of its electrical attraction to the nucleus; hence, it undergoes an acceleration. An analysis from classical physics shows that any charged particle will emit electromagnetic radiation when it is accelerated. (An example of this phenomenon is the production of electromagnetic waves by accelerated charges in a radio antenna, as described in Chapter 21.) According to quantum theory, this radiation must appear in the form of photons. Because the radiated photon shown in Figure 27.10 carries energy, the electron must lose kinetic energy because of its encounter with the target nucleus. An extreme example consists of the electron losing all of its energy in a single collision. In this case the initial energy of the electron ($e\Delta V$) is transformed completely into the energy of the photon ($hf_{\max}$). In equation form,

$$e\,\Delta V = hf_{\max} = \frac{hc}{\lambda_{\min}} \qquad \textbf{[27.8]}$$

where $e\Delta V$ is the energy of the electron after it has been accelerated through a potential difference of $\Delta V$ volts and $e$ is the charge on the electron. This equation says that the shortest wavelength radiation that can be produced is

$$\lambda_{\min} = \frac{hc}{e\,\Delta V} \qquad \textbf{[27.9]}$$

Not all the radiation produced has this particular wavelength because many of the electrons aren't stopped in a single collision. The result is the production of the continuous spectrum of wavelengths.

**APPLICATION**
Using X-Rays to Study the Work of Master Painters

Interesting insights into the process of painting and revising a masterpiece are being revealed by x-rays. Long-wavelength x-rays are absorbed in varying degrees by some paints, such as those having lead, cadmium, chromium, or cobalt as a base. The x-ray interactions with the paints give contrast because the different elements in the paints have different electron densities. Also, thicker layers will absorb more than thin layers. To examine a painting by an old master, a film is placed behind it while it is x-rayed from the front. Ghost outlines of earlier paintings and earlier forms of the final masterpiece are sometimes revealed when the film is developed.

## 27.4 Diffraction of X-Rays by Crystals

**LEARNING OBJECTIVES**

1. Discuss x-ray diffraction and state Bragg's law.
2. Apply Bragg's law to crystalline systems.

In Chapter 24 we described how a diffraction grating could be used to measure the wavelength of light. In principle the wavelength of *any* electromagnetic wave can be measured if a grating having a suitable line spacing can be found. The spacing between lines must be approximately equal to the wavelength of the radiation to

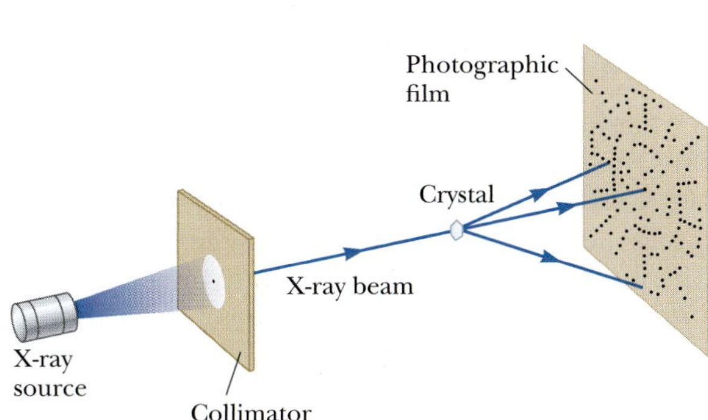


**Figure 27.11** Schematic diagram of the technique used to observe the diffraction of x-rays by a single crystal. The array of spots formed on the film by the diffracted beams is called a Laue pattern. (See Fig. 27.7.)

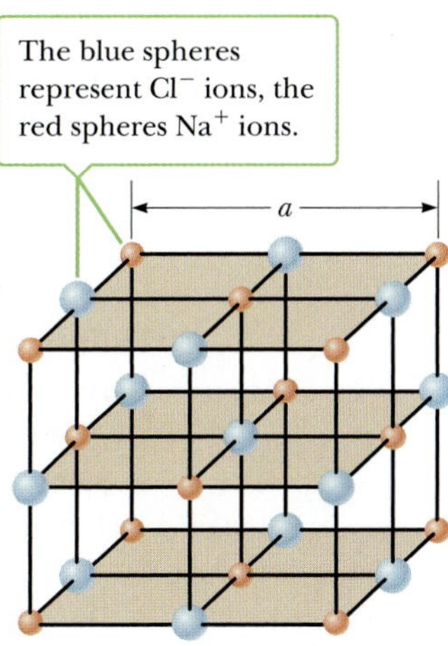

The blue spheres represent Cl⁻ ions, the red spheres Na⁺ ions.
a


**Figure 27.12** A model of the cubic crystalline structure of sodium chloride. The length of the cube edge is $a = 0.563$ nm.

be measured. X-rays are electromagnetic waves with wavelengths on the order of 0.1 nm. It would be impossible to construct a grating with such a small spacing. As noted in the previous section, however, the regular array of atoms in a crystal could act as a three-dimensional grating for observing the diffraction of x-rays.

One experimental arrangement for observing x-ray diffraction is shown in Figure 27.11. A narrow beam of x-rays with a continuous wavelength range is incident on a crystal such as sodium chloride. The diffracted radiation is very intense in certain directions, corresponding to constructive interference from waves reflected from layers of atoms in the crystal. The diffracted radiation is detected by a photographic film and forms an array of spots known as a *Laue pattern.* The crystal structure is determined by analyzing the positions and intensities of the various spots in the pattern.

The arrangement of atoms in a crystal of NaCl is shown in Figure 27.12. The smaller red spheres represent $Na^+$ ions, and the larger blue spheres represent $Cl^-$ ions. The spacing between successive $Na^+$ (or $Cl^-$) ions in this cubic structure, denoted by the symbol $a$ in Figure 27.12, is approximately 0.563 nm.

A careful examination of the NaCl structure shows that the ions lie in various planes. The shaded areas in Figure 27.12 represent one example, in which the atoms lie in equally spaced planes. Now suppose an x-ray beam is incident at grazing angle $\theta$ on one of the planes, as in Figure 27.13. The beam can be reflected from both the upper and lower plane of atoms. The geometric construction in Figure 27.13, however, shows that the beam reflected from the lower surface travels farther than the beam reflected from the upper surface by a distance of $2d \sin \theta$. The two portions of the reflected beam will combine to produce constructive interference when this path difference equals some integral multiple of the wavelength $\lambda$. The condition for constructive interference is given by

$$2d \sin \theta = m\lambda \qquad m = 1, 2, 3, \ldots \qquad [27.10]$$

◀ Bragg's law

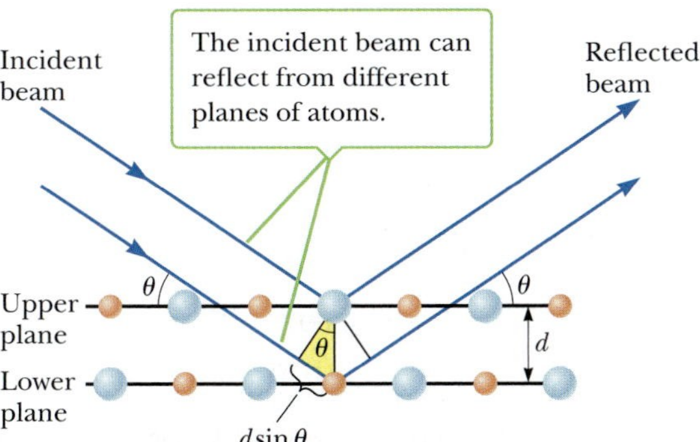


**Figure 27.13** A two-dimensional depiction of the reflection of an x-ray beam from two parallel crystalline planes separated by a distance $d$. The beam reflected from the lower plane travels farther than the one reflected from the upper plane by an amount equal to $2d \sin \theta$.

This condition is known as **Bragg's law**, after W. L. Bragg (1890–1971), who first derived the relationship. If the wavelength and diffraction angle are measured, Equation 27.10 can be used to calculate the spacing between atomic planes.

The technique of x-ray diffraction has been used to determine the atomic arrangement of complex organic molecules such as proteins. Proteins are large molecules containing thousands of atoms that help regulate chemical processes in cells. Some proteins are amazing catalysts, speeding up the slow room-temperature reactions in cells by 17 orders of magnitude. To understand this incredible biochemical reactivity, it is important to determine the structure of these intricate molecules.

The main technique used to determine the molecular structure of proteins, DNA, and RNA is x-ray diffraction using x-rays of wavelength of about 1 Å (1 Å = 0.1 nm = $1 \times 10^{-10}$ m). This technique allows the experimenter to "see" individual atoms that are separated by about this distance in molecules. Because the biochemical x-ray diffraction sample is prepared in crystal form, the *geometry* (position of the bright spots in space) of the diffraction pattern is determined by the regular three-dimensional crystal lattice arrangement of molecules in the sample. The *intensities* of the bright diffraction spots are determined by the atoms and their electronic distributions in the fundamental building block of the crystal: the unit cell. Using complicated computational techniques, investigators can essentially deduce the molecular structure by matching the observed intensities of diffracted beams with a series of assumed atomic positions that determine the atomic structure and electron density of the molecule.

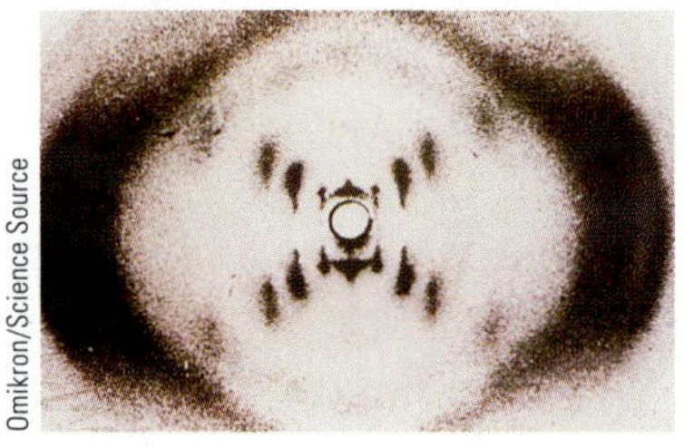
Omikron/Science Source

**Figure 27.14** An x-ray diffraction photograph of DNA taken by Rosalind Franklin. The cross pattern of spots was a clue that DNA has a helical structure.

**Figure 27.15** The double-helix structure of DNA.

Crystallizing a large molecule such as DNA and obtaining its x-ray diffraction pattern is highly challenging. In the case of DNA, obtaining sufficiently pure crystals is especially difficult because there exist two crystalline forms, A and B, which arise in mixed form during preparation. These two forms result in diffraction patterns that can't be easily deciphered. In 1951 Rosalind Franklin, a researcher at King's College in London, developed an ingenious method of separating the two forms and managed to obtain excellent x-ray diffraction images of pure crystalline DNA in B-form. With these images she determined that the helical shape of DNA consisted of two interwoven strands, with the sugar-phosphate backbone on the outside of the molecule, refuting prior models that had the backbone on the inside. One of her images is shown in Figure 27.14.

James Watson and Francis Crick used Franklin's work to uncover further details of the molecule and its function in heredity, particularly in regards to the internal structure. Attached to each sugar-phosphate unit of each strand is one of four base molecules: adenine, cytosine, guanine, or thymine. The bases are arranged sequentially along the strand, with patterns in the sequence of bases acting as codes for proteins that carry out various functions for a given organism. The bases in one strand bind to those in the other strand, forming a double helix. A model of the double helix is shown in Figure 27.15. In 1962 Watson, Crick, and a colleague of Franklin's, Maurice Wilkins, received the Nobel Prize for Physiology and Medicine for their work on understanding the structure and function of DNA. Franklin would have shared the prize, but died in 1958 of cancer at the age of thirty-eight. (The prize is not awarded posthumously.)

## EXAMPLE 27.2 X-Ray Diffraction from Calcite

**GOAL** Understand Bragg's law and apply it to a crystal.

**PROBLEM** If the spacing between certain planes in a crystal of calcite ($CaCO_3$) is 0.314 nm, find the grazing angles at which first- and third-order interference will occur for x-rays of wavelength 0.070 0 nm.

**STRATEGY** Solve Bragg's law for sin $\theta$ and substitute, using the inverse-sine function to obtain the angle.

**SOLUTION**

Find the grazing angle corresponding to $m = 1$, for first-order interference:

$$\sin\theta = \frac{m\lambda}{2d} = \frac{(0.070\,0\text{ nm})}{2(0.314\text{ nm})} = 0.111$$

$$\theta = \sin^{-1}(0.111) = 6.37^\circ$$

Repeat the calculation for third-order interference ($m = 3$):

$$\sin\theta = \frac{m\lambda}{2d} = \frac{3(0.070\,0\text{ nm})}{2(0.314\text{ nm})} = 0.334$$

$$\theta = \sin^{-1}(0.334) = 19.5°$$

**REMARKS** Notice there is little difference between this kind of problem and a Young's slit experiment.

**QUESTION 27.2** True or False: A smaller grazing angle implies a smaller distance between planes in the crystal lattice.

**EXERCISE 27.2** X-rays of wavelength 0.060 0 nm are scattered from a crystal with a grazing angle of 11.7°. Assume $m = 1$ for this process. Calculate the spacing between the crystal planes.

**ANSWER** 0.148 nm

# 27.5 The Compton Effect

**LEARNING OBJECTIVES**

1. Describe the Compton effect in terms of conservation of energy and momentum.
2. Apply the Compton effect to systems of photons and electrons.

Further justification for the photon nature of light came from an experiment conducted by Arthur H. Compton in 1923. In his experiment Compton directed an x-ray beam of wavelength $\lambda_0$ toward a block of graphite. He found that the scattered x-rays had a slightly longer wavelength $\lambda$ than the incident x-rays and hence the energies of the scattered rays were lower. The amount of energy reduction depended on the angle at which the x-rays were scattered. The change in wavelength $\Delta\lambda$ between a scattered x-ray and an incident x-ray is called the **Compton shift.**

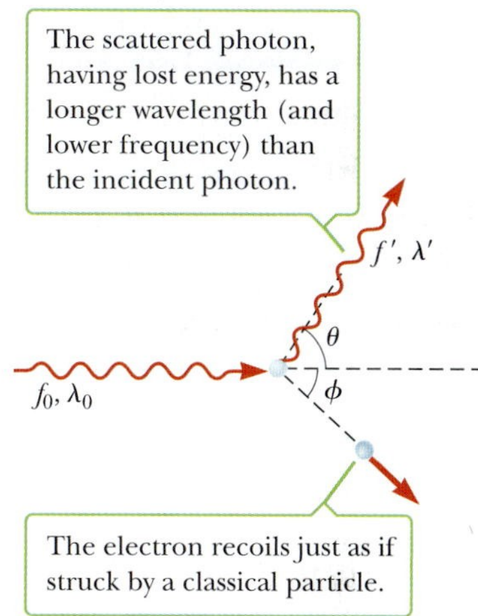
The scattered photon, having lost energy, has a longer wavelength (and lower frequency) than the incident photon.
$f', \lambda'$
$\theta$
$f_0, \lambda_0$
$\phi$
The electron recoils just as if struck by a classical particle.

**Figure 27.16** Diagram representing Compton scattering of a photon by an electron.

To explain this effect, Compton assumed that if a photon behaves like a particle, its collision with other particles is similar to a collision between two billiard balls. Hence, the x-ray photon carries both measurable *energy* and *momentum*, and these two quantities must be conserved in a collision. If the incident photon collides with an electron initially at rest, as in Figure 27.16, the photon transfers some of its energy and momentum to the electron. As a consequence, the energy and frequency of the scattered photon are lowered and its wavelength increases. Applying relativistic energy and momentum conservation to the collision described in Figure 27.16, the shift in wavelength of the scattered photon is given by

$$\Delta\lambda = \lambda - \lambda_0 = \frac{h}{m_e c}(1 - \cos\theta) \qquad [27.11]$$

◀ The Compton shift formula

**Arthur Holly Compton**
**American Physicist (1892–1962)**
Compton attended Wooster College and Princeton University. He became director of the laboratory at the University of Chicago, where experimental work concerned with sustained chain reactions was conducted. This work was of central importance to the construction of the first atomic bomb. Because of his discovery of the Compton effect and his work with cosmic rays, he shared the 1927 Nobel Prize in Physics with Charles Wilson.

where $m_e$ is the mass of the electron and $\theta$ is the angle between the directions of the scattered and incident photons. The quantity $h/m_e c$ is called the **Compton wavelength** and has a value of 0.002 43 nm. The Compton wavelength is very small relative to the wavelengths of visible light, so the shift in wavelength would be difficult to detect if visible light were used. Further, note that the Compton shift depends on the scattering angle $\theta$ and not on the wavelength. Experimental results for x-rays scattered from various targets obey Equation 27.11 and strongly support the photon concept.

## Quick Quiz

**27.1** True or False: When a photon scatters off an electron, the photon loses energy.

**27.2** An x-ray photon is scattered by an electron. Does the frequency of the scattered photon relative to that of the incident photon (a) increase, (b) decrease, or (c) remain the same?

**27.3** A photon of energy $E_0$ strikes a free electron, with the scattered photon of energy $E$ moving in the direction opposite that of the incident photon. In this Compton effect interaction, what is the resulting kinetic energy of the electron? (a) $E_0$ (b) $E$ (c) $E_0 - E$ (d) $E_0 + E$

## EXAMPLE 27.3 Scattering X-Rays

**GOAL** Understand Compton scattering and its effect on the photon's energy.

**PROBLEM** X-rays of wavelength $\lambda_i = 0.200\,000$ nm are scattered from a block of material. The scattered x-rays are observed at an angle of 45.0° to the incident beam. **(a)** Calculate the wavelength of the x-rays scattered at this angle. **(b)** Compute the fractional change in the energy of a photon in the collision.

**STRATEGY** To find the wavelength of the scattered x-ray photons, substitute into Equation 27.11 to obtain the wavelength shift, then add the result to the initial wavelength, $\lambda_i$. In part (b), calculating the fractional change in energy involves calculating the energy of the x-ray photon before and after, using $E = hf = hc/\lambda$. Taking the difference and dividing by the initial energy yields the desired fractional change in energy. Here, however, a symbolic expression is derived that relates energy terms and wavelengths.

**SOLUTION**

**(a)** Calculate the wavelength of the x-rays.

Substitute into Equation 27.11 to obtain the wavelength shift:

$$\Delta\lambda = \frac{h}{m_e c}(1 - \cos\theta)$$

$$= \frac{6.63 \times 10^{-34}\ \mathrm{J\cdot s}}{(9.11 \times 10^{-31}\ \mathrm{kg})(3.00 \times 10^{8}\ \mathrm{m/s})}(1 - \cos 45.0°)$$

$$= 7.11 \times 10^{-13}\ \mathrm{m} = 0.000\,711\ \mathrm{nm}$$

Add this shift to the original wavelength to obtain the wavelength of the scattered photon:

$$\lambda_f = \Delta\lambda + \lambda_i = 0.200\,711\ \mathrm{nm}$$

**(b)** Find the fraction of energy lost by the photon in the collision.

Rewrite the energy $E$ in terms of wavelength, using $c = f\lambda$:

$$E = hf = h\frac{c}{\lambda}$$

Compute $\Delta E/E$ using this expression:

$$\frac{\Delta E}{E} = \frac{E_f - E_i}{E_i} = \frac{hc/\lambda_f - hc/\lambda_i}{hc/\lambda_i}$$

Cancel $hc$ and rearrange terms:

$$\frac{\Delta E}{E} = \frac{1/\lambda_f - 1/\lambda_i}{1/\lambda_i} = \frac{\lambda_i}{\lambda_f} - 1 = \frac{\lambda_i - \lambda_f}{\lambda_f} = -\frac{\Delta\lambda}{\lambda_f}$$

Substitute values from part (a):

$$\frac{\Delta E}{E} = -\frac{0.000\,711\ \mathrm{nm}}{0.200\,711\ \mathrm{nm}} = -3.54 \times 10^{-3}$$

**REMARKS** It is also possible to find this answer by substituting into the energy expression at an earlier stage, but the algebraic derivation is more elegant and instructive because it shows how changes in energy are related to changes in wavelength.

**QUESTION 27.3** The incident photon loses energy. Where does it go?

**EXERCISE 27.3** Repeat the example for a photon with wavelength $3.00 \times 10^{-2}$ nm that scatters at an angle of 60.0°.

**ANSWERS** (a) $3.12 \times 10^{-2}$ nm (b) $\Delta E/E = -3.88 \times 10^{-2}$

# 27.6 The Dual Nature of Light and Matter

**LEARNING OBJECTIVES**

1. State de Broglie's hypothesis and discuss the dual nature of light and matter.
2. Apply de Broglie's hypothesis to quantum and classical objects.

Phenomena such as the photoelectric effect and the Compton effect offer evidence that when light (or other forms of electromagnetic radiation) and matter interact, the light behaves as if it were composed of particles having energy $hf$ and momentum $h/\lambda$. In other contexts, however, light acts like a wave, exhibiting interference and diffraction effects. This apparent duality can be partly explained by considering the energies of photons in different contexts. For example, photons with frequencies in the radio wavelengths carry very little energy, and it may take $10^{10}$ such photons to create a signal in an antenna. These photons therefore act together like a wave to create the effect. Gamma rays, on the other hand, are so energetic that a single gamma ray photon can be detected.

In his doctoral dissertation in 1924, Louis de Broglie postulated that **because photons have wave and particle characteristics, perhaps all forms of matter have both properties**. This highly revolutionary idea had no experimental confirmation at that time. According to de Broglie, electrons, just like light, have a dual particle–wave nature.

◀ De Broglie's hypothesis

In Chapter 26 we found that the relationship between energy and momentum for a photon, which has a rest energy of zero, is $p = E/c$. We also know from Equation 27.5 that the energy of a photon is

$$E = hf = \frac{hc}{\lambda} \qquad [27.12]$$

Consequently, the momentum of a photon can be expressed as

$$p = \frac{E}{c} = \frac{hc}{c\lambda} = \frac{h}{\lambda} \qquad [27.13]$$

◀ Momentum of a photon

From this equation, we see that the photon wavelength can be specified by its momentum, or $\lambda = h/p$. De Broglie suggested that *all* material particles with momentum $p$ should have a characteristic wavelength $\lambda = h/p$. Because the momentum of a particle of mass $m$ and speed $v$ is $mv = p$, the **de Broglie wavelength** of a particle is

$$\lambda = \frac{h}{p} = \frac{h}{mv} \qquad [27.14]$$

◀ De Broglie wavelength

Further, de Broglie postulated that the frequencies of matter waves (waves associated with particles having nonzero rest energy) obey the Einstein relationship for photons, $E = hf$, so that

$$f = \frac{E}{h} \qquad [27.15]$$

◀ Frequency of matter waves

The dual nature of matter is quite apparent in Equations 27.14 and 27.15 because each contains both particle concepts ($mv$ and $E$) and wave concepts ($\lambda$ and $f$). The fact that these relationships had been established experimentally for photons made the de Broglie hypothesis that much easier to accept. The Davisson–Germer experiment in 1927 confirmed de Broglie's hypothesis by showing that electrons

SPL/Getty Images

**Louis de Broglie**
**French Physicist (1892–1987)**
De Broglie attended the Sorbonne in Paris, where he changed his major from history to theoretical physics. He was awarded the Nobel Prize in Physics in 1929 for his discovery of the wave nature of electrons.

scattering off crystals form a diffraction pattern. The regularly spaced planes of atoms in crystalline regions of a nickel target act as a diffraction grating for electron matter waves.

**Quick Quiz**

**27.4** True or False: As the momentum of a particle of mass $m$ increases, its wavelength increases.

**27.5** A nonrelativistic electron and a nonrelativistic proton are moving and have the same de Broglie wavelength. Which of the following are also the same for the two particles? (a) speed (b) kinetic energy (c) momentum (d) frequency

## EXAMPLE 27.4 The Electron Versus the Baseball

**GOAL** Apply the de Broglie hypothesis to a quantum and a classical object.

**PROBLEM** **(a)** Compare the de Broglie wavelength for an electron ($m_e = 9.11 \times 10^{-31}$ kg) moving at a speed equal to $1.00 \times 10^7$ m/s with that of a baseball of mass 0.145 kg pitched at 45.0 m/s. **(b)** Compare these wavelengths with that of an electron traveling at $0.999c$.

**STRATEGY** This problem is a matter of substitution into Equation 27.14 for the de Broglie wavelength. In part (b) the relativistic momentum must be used.

**SOLUTION**

**(a)** Compare the de Broglie wavelengths of the electron and the baseball.

Substitute data for the electron into Equation 27.14:

$$\lambda_e = \frac{h}{m_e v} = \frac{6.63 \times 10^{-34}\ \text{J}\cdot\text{s}}{(9.11 \times 10^{-31}\ \text{kg})(1.00 \times 10^{7}\ \text{m/s})}$$

$$= 7.28 \times 10^{-11}\ \text{m}$$

Repeat the calculation with the baseball data:

$$\lambda_b = \frac{h}{m_b v} = \frac{6.63 \times 10^{-34}\ \text{J}\cdot\text{s}}{(0.145\ \text{kg})(45.0\ \text{m/s})} = 1.02 \times 10^{-34}\ \text{m}$$

**(b)** Find the wavelength for an electron traveling at $0.999c$.

Replace the momentum in Equation 27.14 with the relativistic momentum:

$$\lambda_e = \frac{h}{m_e v/\sqrt{1 - v^2/c^2}} = \frac{h\sqrt{1 - v^2/c^2}}{m_e v}$$

Substitute:

$$\lambda_e = \frac{(6.63 \times 10^{-34}\ \text{J}\cdot\text{s})\sqrt{1 - (0.999c)^2/c^2}}{(9.11 \times 10^{-31}\ \text{kg})(0.999 \cdot 3.00 \times 10^{8}\ \text{m/s})}$$

$$= 1.09 \times 10^{-13}\ \text{m}$$

**REMARKS** The electron wavelength corresponds to that of x-rays in the electromagnetic spectrum. The baseball, by contrast, has a wavelength much smaller than any aperture through which the baseball could possibly pass, so we couldn't observe any of its diffraction effects. It is generally true that the wave properties of large-scale objects can't be observed. Notice that even at extreme relativistic speeds, the electron wavelength is still far larger than the baseball's.

**QUESTION 27.4** How does doubling the speed of a particle affect its wavelength? Is your answer always true? Explain.

**EXERCISE 27.4** Find the de Broglie wavelength of a proton ($m_p = 1.67 \times 10^{-27}$ kg) moving at $1.00 \times 10^7$ m/s.

**ANSWER** $3.97 \times 10^{-14}$ m

## Application: The Electron Microscope

BIO **APPLICATION**
Electron Microscopes

A practical device that relies on the wave characteristics of electrons is the **electron microscope**. A *transmission* electron microscope, used for viewing flat, thin samples, is shown in Figure 27.17. In many respects it is similar to an optical microscope, but the electron microscope has a much greater resolving power because it can accelerate electrons to very high kinetic energies, giving them very short wavelengths. No microscope can resolve details that are significantly smaller than the wavelength of the radiation used to illuminate the object. Typically, the wavelengths of electrons in an electron microscope are smaller than the visible wavelengths by a factor of about $10^{-5}$.

The electron beam in an electron microscope is controlled by electrostatic or magnetic deflection, which acts on the electrons to focus the beam to an image. Due to limitations in the electromagnetic lenses used, however, the improvement in resolution over light microscopes is only about a factor of 1 000, two orders

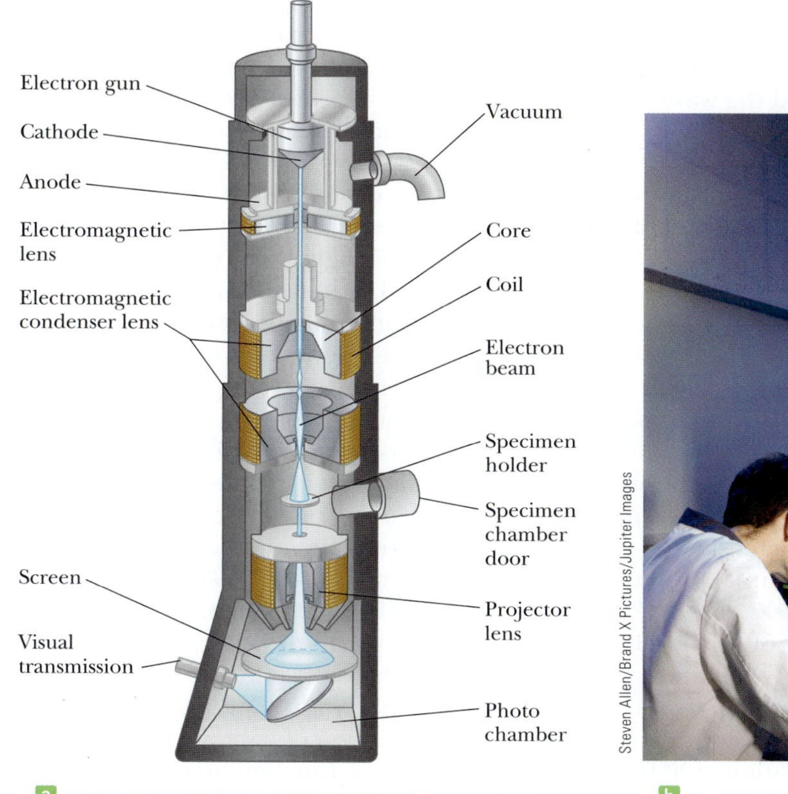


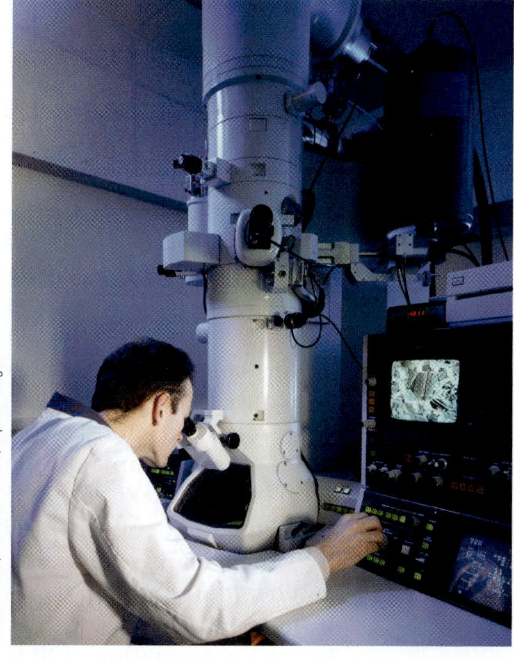


**Figure 27.17** (a) Diagram of a transmission electron microscope for viewing a thin, sectioned sample. The "lenses" that control the electron beam are magnetic deflection coils. (b) An electron microscope.

of magnitude smaller than that implied by the electron wavelength. Rather than examining the image through an eyepiece as in an optical microscope, the viewer looks at an image formed on a fluorescent screen. (The viewing screen must be fluorescent because otherwise the image produced wouldn't be visible.)

**APPLYING PHYSICS 27.2 | X-Ray Microscopes?**

Electron microscopes (Fig. 27.17) take advantage of the wave nature of particles. Electrons are accelerated to high speeds, giving them a short de Broglie wavelength. Imagine an electron microscope using electrons with a de Broglie wavelength of 0.2 nm. Why don't we design a microscope using 0.2-nm *photons* to do the same thing?

**EXPLANATION** Because electrons are charged particles, they interact electrically with the sample in the microscope and scatter according to the shape and density of various portions of the sample, providing a means of viewing the sample. Photons of wavelength 0.2 nm are uncharged and in the x-ray region of the spectrum. They tend to simply pass through the thin sample without interacting. ■

## 27.7 The Wave Function

**LEARNING OBJECTIVE**

1. Discuss the Schrödinger wave equation and the wave function.

De Broglie's revolutionary idea that particles should have a wave nature soon moved out of the realm of skepticism to the point where it was viewed as a necessary concept in understanding the subatomic world. In 1926 Austrian–German physicist Erwin Schrödinger proposed a wave equation that described how matter waves change in space and time. The Schrödinger wave equation represents a key element in the theory of quantum mechanics. It's as important in quantum mechanics as Newton's laws in classical mechanics. Schrödinger's equation has been successfully applied to the hydrogen atom and to many other microscopic systems.

**Erwin Schrödinger**
**Austrian Theoretical Physicist (1887–1961)**
Schrödinger is best known as the creator of wave mechanics, a less cumbersome theory than the equivalent matrix mechanics developed by Werner Heisenberg. In 1933 Schrödinger left Germany and eventually settled at the Dublin Institute of Advanced Study, where he spent 17 happy, creative years working on problems in general relativity, cosmology, and the application of quantum physics to biology. In 1956 he returned home to Austria and his beloved Tyrolean mountains, where he died in 1961.

Solving Schrödinger's equation (beyond the level of this course) determines a quantity $\Psi$ called the **wave function**. Each particle is represented by a wave function $\Psi$ that depends on both position and time. Once $\Psi$ is found, $\Psi^2$ gives us information on the **probability** (per unit volume) of finding the particle in any given region. To understand this, we return to Young's experiment involving coherent light passing through a double slit.

First, recall from Chapter 21 that the intensity of a light beam is proportional to the square of the electric field strength $E$ associated with the beam: $I \propto E^2$. According to the wave model of light, there are certain points on the viewing screen where the net electric field is zero as a result of destructive interference of waves from the two slits. Because $E$ is zero at these points, the intensity is also zero and the screen is dark there. Likewise, at points on the screen where constructive interference occurs, $E$ is large, as is the intensity; hence, these locations are bright.

Now consider the same experiment when light is viewed as having a particle nature. The number of photons reaching a point on the screen per second increases as the intensity (brightness) increases. Consequently, the number of photons that strike a unit area on the screen each second is proportional to the square of the electric field, or $N \propto E^2$. From a probabilistic point of view, a photon has a high probability of striking the screen at a point where the intensity (and $E^2$) is high and a low probability of striking the screen where the intensity is low.

When describing particles rather than photons, $\Psi$ rather than $E$ plays the role of the amplitude. Using an analogy with the description of light, we make the following interpretation of $\Psi$ for particles: If $\Psi$ is a wave function used to describe a single particle, the value of $\Psi^2$ at some location at a given time is proportional to the probability per unit volume of finding the particle at that location at that time. Adding all the values of $\Psi^2$ in a given region gives the probability of finding the particle in that region.

# 27.8 The Uncertainty Principle

**LEARNING OBJECTIVES**

1. State two uncertainty principles and discuss their physical origins.
2. Apply uncertainty principles to quantum systems.

If you were to measure the position and speed of a particle at any instant, you would always be faced with experimental uncertainties in your measurements. According to classical mechanics, no fundamental barrier to an ultimate refinement of the apparatus or experimental procedures exists. In other words, it's possible, in principle, to make such measurements with arbitrarily small uncertainty. Quantum theory predicts, however, that such a barrier does exist. In 1927 Werner Heisenberg (1901–1976) introduced this notion, which is now known as the **uncertainty principle**:

> If a measurement of the position of a particle is made with precision $\Delta x$ and a simultaneous measurement of linear momentum is made with precision $\Delta p_x$, the product of the two uncertainties can never be smaller than $h/4\pi$:
>
> $$\Delta x\,\Delta p_x \geq \frac{h}{4\pi} \qquad [27.16]$$

In other words, **it is physically impossible to measure simultaneously the exact position and exact linear momentum of a particle**. If $\Delta x$ is very small, then $\Delta p_x$ is large, and vice versa.

To understand the physical origin of the uncertainty principle, consider the following thought experiment introduced by Heisenberg. Suppose you wish to measure the position and linear momentum of an electron as accurately as possible. You might be able to do so by viewing the electron with a powerful light microscope. For you to see the electron and determine its location, at least one photon of light must bounce off the electron, as shown in Figure 27.18a, and pass through the microscope into your eye, as shown in Figure 27.18b. When it strikes the electron, however, the photon transfers some unknown amount of its momentum to the electron. Thus, in the process of locating the electron very accurately (that is, by making $\Delta x$ very small), the light that enables you to succeed in your measurement changes the electron's momentum to some undeterminable extent (making $\Delta p_x$ very large).

The incoming photon has momentum $h/\lambda$. As a result of the collision, the photon transfers part or all of its momentum along the $x$-axis to the electron. Therefore, the *uncertainty* in the electron's momentum after the collision is as great as the momentum of the incoming photon: $\Delta p_x = h/\lambda$. Further, because the photon also has wave properties, we expect to be able to determine the electron's position

**Werner Heisenberg**
**German Theoretical Physicist (1901–1976)**
Heisenberg obtained his Ph.D. in 1923 at the University of Munich, where he studied under Arnold Sommerfeld. While physicists such as de Broglie and Schrödinger tried to develop physical models of the atom, Heisenberg developed an abstract mathematical model called *matrix mechanics* to explain the wavelengths of spectral lines. Heisenberg made many other significant contributions to physics, including the prediction of two forms of molecular hydrogen, theoretical models of the nucleus of an atom, and his famous uncertainty principle, for which he received the Nobel Prize in Physics in 1932.

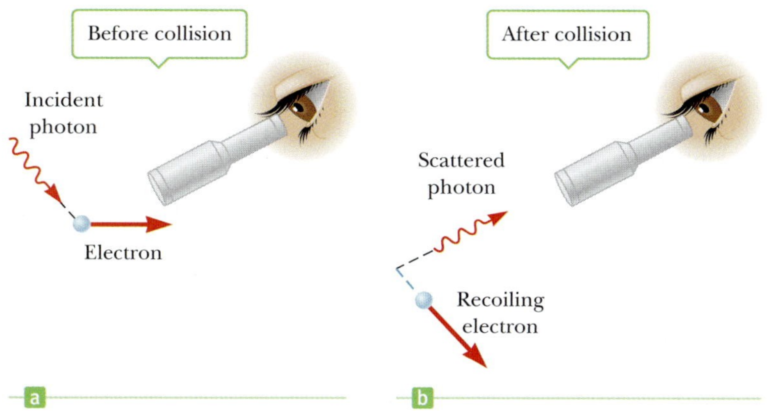


**Figure 27.18** A thought experiment for viewing an electron with a powerful microscope. (a) The electron is viewed before colliding with the photon. (b) The electron recoils (is disturbed) as the result of the collision with the photon.

to within one wavelength of the light being used to view it, so $\Delta x = \lambda$. Multiplying these two uncertainties gives

$$\Delta x\,\Delta p_x = \lambda\left(\frac{h}{\lambda}\right) = h$$

The value $h$ represents the minimum in the product of the uncertainties. Because the uncertainty can always be greater than this minimum, we have

$$\Delta x\,\Delta p_x \geq h$$

Apart from the numerical factor $1/4\pi$ introduced by Heisenberg's more precise analysis, this inequality agrees with Equation 27.16.

Another form of the uncertainty relationship sets a limit on the accuracy with which the energy $E$ of a system can be measured in a finite time interval $\Delta t$:

$$\Delta E\,\Delta t \geq \frac{h}{4\pi} \qquad [27.17]$$

It can be inferred from this relationship that the energy of a particle cannot be measured with complete precision in a very short interval of time. Thus, when an electron is viewed as a particle, the uncertainty principle tells us that (a) its position and velocity cannot both be known precisely at the same time and (b) its energy can be uncertain for a period given by $\Delta t = h/(4\pi\,\Delta E)$.

## EXAMPLE 27.5 Locating an Electron

**GOAL** Apply Heisenberg's position–momentum uncertainty principle.

**PROBLEM** The speed of an electron is measured to be $5.00 \times 10^3$ m/s to an accuracy of 0.003 00%. Find the minimum uncertainty in determining the position of this electron.

**STRATEGY** After computing the momentum and its uncertainty, substitute into Heisenberg's uncertainty principle, Equation 27.16.

**SOLUTION**

Calculate the momentum of the electron:

$$p_x = m_e v = (9.11 \times 10^{-31}\text{ kg})(5.00 \times 10^3\text{ m/s})$$
$$= 4.56 \times 10^{-27}\text{ kg}\cdot\text{m/s}$$

The uncertainty in $p_x$ is 0.003 00% of this value:

$$\Delta p_x = 0.000\,030\,0 p_x = (0.000\,030\,0)(4.56 \times 10^{-27}\text{ kg}\cdot\text{m/s})$$
$$= 1.37 \times 10^{-31}\text{ kg}\cdot\text{m/s}$$

Now calculate the uncertainty in position using this value of $\Delta p_x$ and Equation 27.17:

$$\Delta x\,\Delta p_x \geq \frac{h}{4\pi} \quad\rightarrow\quad \Delta x \geq \frac{h}{4\pi\,\Delta p_x}$$

$$\Delta x \geq \frac{6.626 \times 10^{-34}\text{ J}\cdot\text{s}}{4\pi(1.37 \times 10^{-31}\text{ kg}\cdot\text{m/s})} = 0.385 \times 10^{-3}\text{ m}$$

$$= 0.385\text{ mm}$$

**REMARKS** Notice that this isn't an exact calculation: the uncertainty in position can take any value as long as it's greater than or equal to the value given by the uncertainty principle.

**QUESTION 27.5** True or False: The uncertainty in the position of a proton in the helium nucleus is, on average, less than the uncertainty of a proton in a uranium atom.

**EXERCISE 27.5** Suppose an electron is found somewhere in an atom of diameter $1.25 \times 10^{-10}$ m. Estimate the uncertainty in the electron's momentum (in one dimension).

**ANSWER** $\Delta p \geq 4.22 \times 10^{-25}\text{ kg}\cdot\text{m/s}$

# SUMMARY

## 27.1 Blackbody Radiation and Planck's Hypothesis

The characteristics of **blackbody radiation** can't be explained with classical concepts. The peak of a blackbody radiation curve is given by **Wien's displacement law**,

$$\lambda_{max} T = 0.289\,8 \times 10^{-2}\ \mathrm{m \cdot K} \qquad [27.1]$$

where $\lambda_{max}$ is the wavelength at which the curve peaks and $T$ is the absolute temperature of the object emitting the radiation.

Planck first introduced the quantum concept when he assumed the subatomic oscillators responsible for blackbody radiation could have only discrete amounts of energy given by

$$E_n = nhf \qquad [27.2]$$

where $n$ is a positive integer called a **quantum number** and $f$ is the frequency of vibration of the resonator.

## 27.2 The Photoelectric Effect and the Particle Theory of Light

The **photoelectric effect** is a process whereby electrons are ejected from a metal surface when light is incident on that surface. Einstein provided a successful explanation of this effect by extending Planck's quantum hypothesis to electromagnetic waves. In this model, light is viewed as a stream of particles called photons, each with energy $E = hf$, where $f$ is the light frequency and $h$ is **Planck's constant.** The maximum kinetic energy of the ejected photoelectrons is

$$KE_{max} = hf - \phi \qquad [27.6]$$

where $\phi$ is the **work function** of the metal.

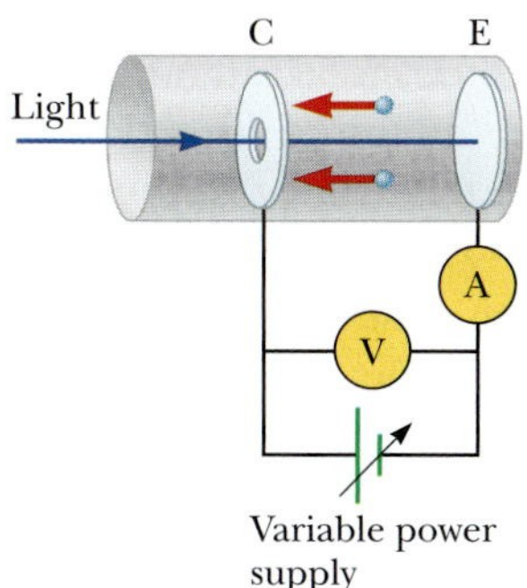


A circuit diagram for studying the photoelectric effect.

## 27.3 X-Rays

## 27.4 Diffraction of X-Rays by Crystals

**X-rays** are produced when high-speed electrons are suddenly decelerated. When electrons have been accelerated through a voltage $V$, the shortest-wavelength radiation that can be produced is

$$\lambda_{min} = \frac{hc}{e\,\Delta V} \qquad [27.9]$$

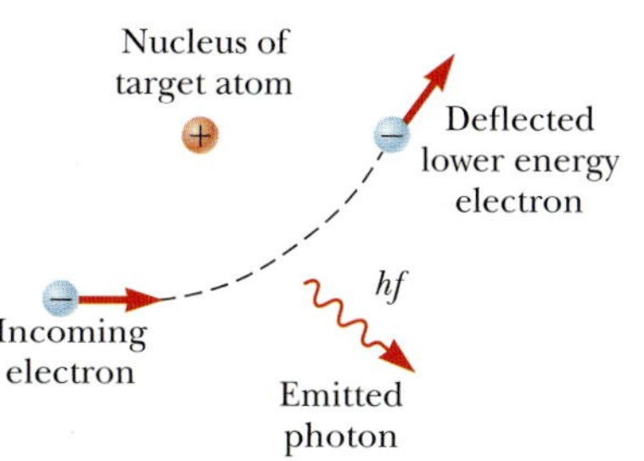


An electron passing near a charged target atom experiences an acceleration, and a photon is emitted in the process.

The regular array of atoms in a crystal can act as a diffraction grating for x-rays and for electrons. The condition for constructive interference of the diffracted rays is given by **Bragg's law**:

$$2d \sin\theta = m\lambda \qquad m = 1, 2, 3, \ldots \qquad [27.10]$$

Bragg's law bears a similarity to the equation for the diffraction pattern of a double slit.

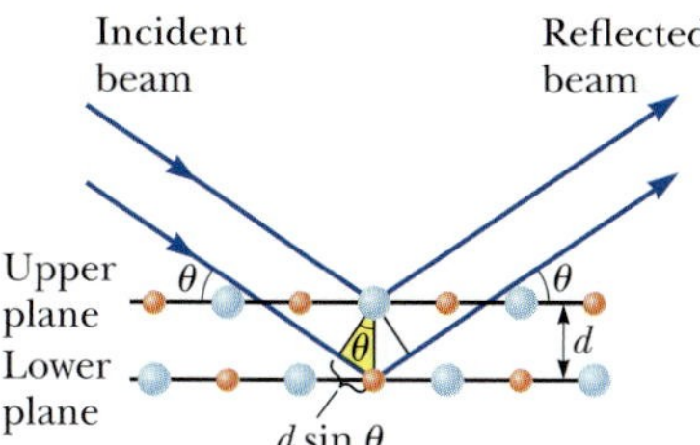


A two-dimensional depiction of the reflection of an x-ray beam from two parallel crystalline planes separated by a distance $d$. The beam reflected from the lower plane travels farther than the one reflected from the upper plane by an amount equal to $2d \sin\theta$.

## 27.5 The Compton Effect

X-rays from an incident beam are scattered at various angles by electrons in a target such as carbon. In such a scattering event, a shift in wavelength is observed for the scattered x-rays. This phenomenon is known as the **Compton shift**. Conservation of momentum and energy applied to a photon–electron collision yields the following expression for the shift in wavelength of the scattered x-rays:

$$\Delta\lambda = \lambda - \lambda_0 = \frac{h}{m_e c}(1 - \cos\theta) \qquad [27.11]$$

Here, $m_e$ is the mass of the electron, $c$ is the speed of light, and $\theta$ is the scattering angle.

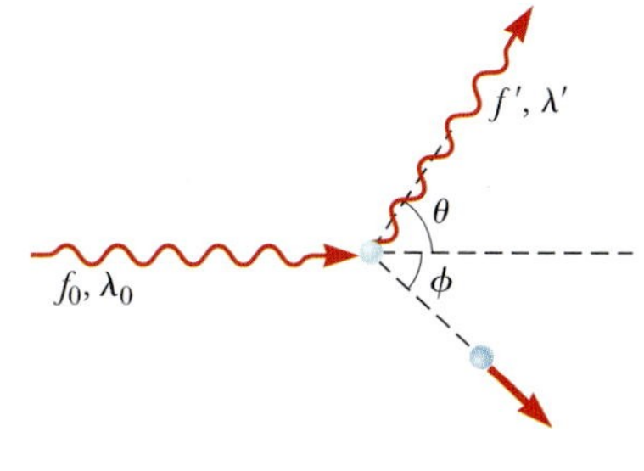


Diagram representing Compton scattering of a photon by an electron.

## 27.6 The Dual Nature of Light and Matter

Light exhibits both a particle and a wave nature. De Broglie proposed that *all* matter has both a particle and a wave nature. The **de Broglie wavelength** of any particle of mass $m$ and speed $v$ is

$$\lambda = \frac{h}{p} = \frac{h}{mv} \qquad [27.14]$$

De Broglie also proposed that the frequencies of the waves associated with particles obey the Einstein relationship $E = hf$.

## 27.7 The Wave Function

In the theory of **quantum mechanics,** each particle is described by a quantity $\Psi$ called the **wave function**. The probability per unit volume of finding the particle at a particular point at some instant is proportional to $\Psi^2$. Quantum mechanics has been highly successful in describing the behavior of atomic and molecular systems.

## 27.8 The Uncertainty Principle

According to Heisenberg's **uncertainty principle**, it is impossible to measure simultaneously the exact position and exact momentum of a particle. If $\Delta x$ is the uncertainty in the measured position and $\Delta p_x$ the uncertainty in the momentum, the product $\Delta x\,\Delta p_x$ is given by

$$\Delta x\,\Delta p_x \geq \frac{h}{4\pi} \qquad [27.16]$$

Also,

$$\Delta E\,\Delta t \geq \frac{h}{4\pi} \qquad [27.17]$$

where $\Delta E$ is the uncertainty in the energy of the particle and $\Delta t$ is the uncertainty in the time it takes to measure the energy.

## WARM-UP EXERCISES

ENHANCED WebAssign The warm-up exercises in this chapter may be assigned online in Enhanced WebAssign.

1. **Physics Review**. Determine the frequency of light at (a) $3.90 \times 10^2$ nm and (b) $7.00 \times 10^2$ nm. (See Section 21.12.)
2. **Physics Review** A diffraction grating has $6.00 \times 10^2$ grooves per mm. Determine (a) the spacing between the grooves in meters and (b) the angle of the second-order maximum if 589-nm light from a sodium lamp is normally incident on the grating. (See Section 21.8.)
3. **Physics Review** Calculate the speed of a proton, initially at rest, after it accelerates through a potential difference of $2.00 \times 10^4$ V. (See Section 16.1.)
4. Suppose the visible surface of a star has a temperature of 5 780 K. Determine the wavelength in nm at which the star's blackbody radiation has a peak intensity. (See Section 27.1.)
5. In a photoelectric experiment, a metal is irradiated with light of energy 3.56 eV. If a stopping potential of 1.10 V is required, what is the work function of the metal?
6. A sodium surface is illuminated with 305-nm light. Determine the energy of each photon in (a) joules and (b) electron volts. (c) Determine the maximum kinetic energy in electron volts of the ejected photoelectrons. (See Section 27.2.)
7. Electrons are accelerated through a potential difference of $1.00 \times 10^4$ V. If these electrons collide with a metal surface, determine the shortest wavelength in nm of the produced x-rays. (See Section 27.3.)
8. Table salt (NaCl) is a crystal having planes of atoms regularly separated by about 0.282 nm. Determine the grazing angle for which first-order constructive interference will occur for x-rays of wavelength $2.50 \times 10^{-2}$ nm. (See Section 27.4.)
9. A photon scatters off an electron at an angle of $1.80 \times 10^{2\circ}$ with respect to its initial direction of motion. What is the change in nm of the photon's wavelength? (See Section 27.5.)
10. An electron is accelerated from rest through a potential difference of $1.20 \times 10^2$ V. Determine the electron's (a) speed (ignore relativistic effects) and (b) de Broglie wavelength. (See Section 27.6.)
11. The position of an electron is determined to within 2.50 nm. What is the minimum uncertainty in the electron's speed? (See Section 27.8.)

## CONCEPTUAL QUESTIONS

WebAssign The conceptual questions in this chapter may be assigned online in Enhanced WebAssign.

1. If you observe objects inside a very hot kiln, why is it difficult to discern the shapes of the objects?
2. Why is an electron microscope more suitable than an optical microscope for "seeing" objects of atomic size?
3. Are blackbodies black?
4. Why is it impossible to simultaneously measure the position and velocity of a particle with infinite accuracy?
5. All objects radiate energy. Why, then, are we not able to see all the objects in a dark room?
6. Is light a wave or a particle? Support your answer by citing specific experimental evidence.
7. In the photoelectric effect, explain why the stopping potential depends on the frequency of the light but not on the intensity.
8. Which has more energy, a photon of ultraviolet radiation or a photon of yellow light?
9. Why does the existence of a cutoff frequency in the photoelectric effect favor a particle theory of light rather than a wave theory?
10. What effect, if any, would you expect the temperature of a material to have on the ease with which electrons can be ejected from it via the photoelectric effect?
11. The cutoff frequency of a material is $f_0$. Are electrons emitted from the material when (a) light of frequency greater than $f_0$ is incident on the material? Or (b) Less than $f_0$?
12. The brightest star in the constellation Lyra is the bluish star Vega, whereas the brightest star in Boötes is the reddish star Arcturus. How do you account for the difference in color of the two stars?
13. If the photoelectric effect is observed in one metal, can you conclude that the effect will also be observed in another metal under the same conditions? Explain.
14. The atoms in a crystal lie in planes separated by a few tenths of a nanometer. Can a crystal be used to produce a diffraction pattern with visible light as it does for x-rays? Explain your answer with reference to Bragg's law.
15. Is an electron a wave or a particle? Support your answer by citing some experimental results.
16. If matter has a wave nature, why is this wave-like characteristic not observable in our daily experiences?

## PROBLEMS

WebAssign The problems in this chapter may be assigned online in Enhanced WebAssign.

**1.** denotes straightforward problem; **2.** denotes intermediate problem;
**3.** denotes challenging problem
**1.** denotes full solution available in *Student Solutions Manual/ Study Guide*
**1.** denotes problems most often assigned in Enhanced WebAssign

BIO denotes biomedical problems
GP denotes guided problems
M denotes Master It tutorial available in Enhanced WebAssign
Q|C denotes asking for quantitative and conceptual reasoning
S denotes symbolic reasoning problem
W denotes Watch It video solution available in Enhanced WebAssign

### 27.1 Blackbody Radiation and Planck's Hypothesis

1. W (a) What is the surface temperature of Betelgeuse, a red giant star in the constellation of Orion, which radiates with a peak wavelength of about 970 nm? (b) Rigel, a bluish-white star in Orion, radiates with a peak wavelength of 145 nm. Find the temperature of Rigel's surface.
2. (a) Lightning produces a maximum air temperature on the order of $10^4$ K, whereas (b) a nuclear explosion produces a temperature on the order of $10^7$ K. Use Wien's displacement law to find the order of magnitude of the wavelength of the thermally produced photons radiated with greatest intensity by each of these sources. Name the part of the electromagnetic spectrum where you would expect each to radiate most strongly.
3. BIO The temperature of a student's skin is 33.0°C. At what wavelength does the radiation emitted from the skin reach its peak?
4. The radius of our Sun is $6.96 \times 10^8$ m, and its total power output is $3.85 \times 10^{26}$ W. (a) Assuming the Sun's surface emits as a black body, calculate its surface temperature. (b) Using the result of part (a), find $\lambda_{max}$ for the Sun.
5. Calculate the energy in electron volts of a photon having a wavelength (a) in the microwave range, 5.00 cm, (b) in the visible light range, 500 nm, and (c) in the x-ray range, 5.00 nm.
6. Suppose a star with radius $8.50 \times 10^8$ m has a peak wavelength of 685 nm in the spectrum of its emitted radiation. (a) Find the energy of a photon with this wavelength. (b) What is the surface temperature of the star? (c) At what rate is energy emitted from the star in the form of radiation? Assume the star is a blackbody ($e = 1$). (d) Using the answer to part (a), estimate the rate at which photons leave the surface of the star.

**7.** M Calculate the energy, in electron volts, of a photon whose frequency is (a) 620 THz, (b) 3.10 GHz, and (c) 46.0 MHz.

**8.** BIO W The threshold of dark-adapted (scotopic) vision is $4.0 \times 10^{-11}$ W/m$^2$ at a central wavelength of 500 nm. If light with this intensity and wavelength enters the eye when the pupil is open to its maximum diameter of 8.5 mm, how many photons per second enter the eye?

## 27.2 The Photoelectric Effect and the Particle Theory of Light

**9.** When light of wavelength 350 nm falls on a potassium surface, electrons having a maximum kinetic energy of 1.31 eV are emitted. Find (a) the work function of potassium, (b) the cutoff wavelength, and (c) the frequency corresponding to the cutoff wavelength.

**10.** The work function for zinc is 4.31 eV. (a) Find the cutoff wavelength for zinc. (b) What is the lowest frequency of light incident on zinc that releases photoelectrons from its surface? (c) If photons of energy 5.50 eV are incident on zinc, what is the maximum kinetic energy of the ejected photoelectrons?

**11.** GP The work function for platinum is 6.35 eV. (a) Convert the value of the work function from electron volts to joules. (b) Find the cutoff frequency for platinum. (c) What maximum wavelength of light incident on platinum releases photoelectrons from the platinum's surface? (d) If light of energy 8.50 eV is incident on zinc, what is the maximum kinetic energy of the ejected photoelectrons? Give the answer in electron volts. (e) For photons of energy 8.50 eV, what stopping potential would be required to arrest the current of photoelectrons?

**12.** Q|C Lithium, beryllium, and mercury have work functions of 2.30 eV, 3.90 eV, and 4.50 eV, respectively. Light with a wavelength of $4.00 \times 10^2$ nm is incident on each of these metals. (a) Which of these metals emit photoelectrons in response to the light? Why? (b) Find the maximum kinetic energy for the photoelectrons in each case.

**13.** When light of wavelength 254 nm falls on cesium, the required stopping potential is 3.00 V. If light of wavelength 436 nm is used, the stopping potential is 0.900 V. Use this information to plot a graph like that shown in Figure 27.6, and from the graph determine the cutoff frequency for cesium and its work function.

**14.** W Two light sources are used in a photoelectric experiment to determine the work function for a particular metal surface. When green light from a mercury lamp ($\lambda$ = 546.1 nm) is used, a stopping potential of 0.376 V reduces the photocurrent to zero. (a) Based on this measurement, what is the work function for this metal? (b) What stopping potential would be observed when using the yellow light from a helium discharge tube ($\lambda$ = 587.5 nm)?

## 27.3 X-Rays

**15.** The extremes of the x-ray portion of the electromagnetic spectrum range from approximately $1.0 \times 10^{-8}$ m to $1.0 \times 10^{-13}$ m. Find the minimum accelerating voltages required to produce wavelengths at these two extremes.

**16.** Q|C Calculate the minimum-wavelength x-ray that can be produced when a target is struck by an electron that has been accelerated through a potential difference of (a) 15.0 kV and (b) $1.00 \times 10^2$ kV. (c) What happens to the minimum wavelength as the potential difference increases?

**17.** What minimum accelerating voltage is required to produce an x-ray with a wavelength of 70.0 pm?

## 27.4 Diffraction of X-Rays by Crystals

**18.** When x-rays of wavelength of 0.129 nm are incident on the surface of a crystal having a structure similar to that of NaCl, a first-order maximum is observed at 8.15°. Calculate the interplanar spacing of the crystal based on this information.

**19.** M Potassium iodide has an interplanar spacing of $d$ = 0.296 nm. A monochromatic x-ray beam shows a first-order diffraction maximum when the grazing angle is 7.6°. Calculate the x-ray wavelength.

**20.** Q|C The first-order diffraction maximum is observed at 12.6° for a crystal having an interplanar spacing of 0.240 nm. How many other orders can be observed in the diffraction pattern, and at what angles do they appear? Why is there an upper limit to the number of observed orders?

**21.** X-rays of wavelength 0.140 nm are reflected from a certain crystal, and the first-order maximum occurs at an angle of 14.4°. What value does this give for the interplanar spacing of the crystal?

## 27.5 The Compton Effect

**22.** X-rays are scattered from a target at an angle of 55.0° with the direction of the incident beam. Find the wavelength shift of the scattered x-rays.

**23.** M A 0.001 60-nm photon scatters from a free electron. For what (photon) scattering angle does the recoiling electron have kinetic energy equal to the energy of the scattered photon?

**24.** S A photon having wavelength $\lambda$ scatters off a free electron at $A$ (Fig. P27.24), producing a second photon having wavelength $\lambda'$. This photon then scatters off another free electron at $B$, producing a third photon having wavelength $\lambda''$ and moving in a direction directly opposite the original photon as shown in the figure. Determine the value of $\Delta\lambda = \lambda'' - \lambda$.

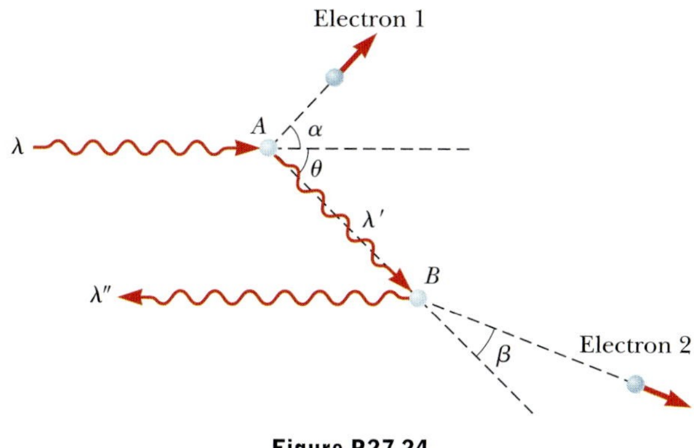


**Figure P27.24**

25. A 0.110-nm photon collides with a stationary electron. After the collision, the electron moves forward and the photon recoils backwards. Find (a) the momentum and (b) the kinetic energy of the electron.

26. In a Compton scattering experiment, an x-ray photon scatters through an angle of 17.4° from a free electron that is initially at rest. The electron recoils with a speed of 2 180 km/s. Calculate (a) the wavelength of the incident photon and (b) the angle through which the electron scatters.

## 27.6 The Dual Nature of Light and Matter

27. (a) If the wavelength of an electron is $5.00 \times 10^{-7}$ m, how fast is it moving? (b) If the electron has a speed equal to $1.00 \times 10^{7}$ m/s, what is its wavelength?

28. W Calculate the de Broglie wavelength of a proton moving at (a) $2.00 \times 10^{4}$ m/s and (b) $2.00 \times 10^{7}$ m/s.

29. De Broglie postulated that the relationship $\lambda = h/p$ is valid for relativistic particles. What is the de Broglie wavelength for a (relativistic) electron having a kinetic energy of 3.00 MeV?

30. (a) An electron has a kinetic energy of 3.00 eV. Find its wavelength. (b) A photon has energy 3.00 eV. Find its wavelength.

31. The resolving power of a microscope is proportional to the wavelength used. A resolution of $1.0 \times 10^{-11}$ m (0.010 nm) would be required in order to "see" an atom. (a) If electrons were used (electron microscope), what minimum kinetic energy would be required of the electrons? (b) If photons were used, what minimum photon energy would be needed to obtain $1.0 \times 10^{-11}$ m resolution?

32. QIC A nonrelativistic particle of mass $m$ and charge $q$ is accelerated from rest through a potential difference $\Delta V$. (a) Use conservation of energy to find a symbolic expression for the momentum of the particle in terms of $m$, $q$, and $\Delta V$. (b) Write a symbolic expression for the de Broglie wavelength using the result of part (a). (c) If an electron and proton go through the same potential difference but in opposite directions, which particle will have the shorter wavelength?

## 27.7 The Wave Function

## 27.8 The Uncertainty Principle

33. In the ground state of hydrogen, the uncertainty in the position of the electron is roughly 0.10 nm. If the speed of the electron is approximately the same as the uncertainty in its speed, about how fast is it moving?

34. An electron is located on a pinpoint having a diameter of 2.5 $\mu$m. What is the minimum uncertainty in the speed of the electron?

35. M An electron and a 0.020 0-kg bullet each have a velocity of magnitude 500 m/s, accurate to within 0.010 0%. Within what lower limit could we determine the position of each object along the direction of the velocity?

36. Suppose Fuzzy, a quantum mechanical duck, lives in a world in which $h = 2\pi$ J · s. Fuzzy has a mass of 2.00 kg and is initially known to be within a pond 1.00 m wide. (a) What is the minimum uncertainty in the duck's speed? (b) Assuming this uncertainty in speed to prevail for 5.00 s, determine the uncertainty in Fuzzy's position after this time.

37. The average lifetime of a muon is about 2 $\mu$s. Estimate the minimum uncertainty in the energy of a muon.

38. QIC (a) Show that the kinetic energy of a nonrelativistic particle can be written in terms of its momentum as $KE = p^2/2m$. (b) Use the results of part (a) to find the minimum kinetic energy of a proton confined within a nucleus having a diameter of $1.0 \times 10^{-15}$ m.

## Additional Problems

39. A microwave photon in the x-band region has a wavelength of 3.00 cm. Find (a) the momentum, (b) the frequency, and (c) the energy of the photon in electron volts.

40. Find the speed of an electron having a de Broglie wavelength equal to its Compton wavelength. *Hint:* This electron is relativistic.

41. A 2.0-kg object falls from a height of 5.0 m to the ground. If all the gravitational potential energy of this mass could be converted to visible light of wavelength $5.0 \times 10^{-7}$ m, how many photons would be produced?

42. An x-ray tube is operated at 50 000 V. (a) Find the minimum wavelength of the radiation emitted by this tube. (b) If the radiation is directed at a crystal, the first-order maximum in the reflected radiation occurs when the grazing angle is 2.5°. What is the spacing between reflecting planes in the crystal?

43. QIC Figure P27.43 (page 946) shows the spectrum of light emitted by a firefly. (a) Determine the temperature of a blackbody that would emit radiation peaked at the same frequency. (b) Based on your result, explain whether firefly radiation is blackbody radiation.

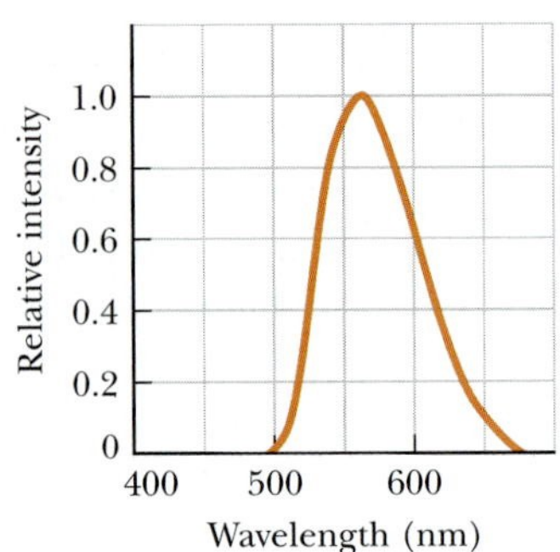


**Figure P27.43**

44. Johnny Jumper's favorite trick is to step out of his 16th-story window and fall 50.0 m into a pool. A news reporter takes a picture of 75.0-kg Johnny just before he makes a splash, using an exposure time of 5.00 ms. Find (a) Johnny's de Broglie wavelength at this moment, (b) the uncertainty of his kinetic energy measurement during such a period of time, and (c) the percent error caused by such an uncertainty.

45. M Photons of wavelength 450 nm are incident on a metal. The most energetic electrons ejected from the metal are bent into a circular arc of radius 20.0 cm by a magnetic field with a magnitude of $2.00 \times 10^{-5}$ T. What is the work function of the metal?

46. An electron initially at rest recoils after a head-on collision with a 6.20-keV photon. Determine the kinetic energy acquired by the electron.

47. A light source of wavelength $\lambda$ illuminates a metal and ejects photoelectrons with a maximum kinetic energy of 1.00 eV. A second light source of wavelength $\lambda/2$ ejects photoelectrons with a maximum kinetic energy of 4.00 eV. What is the work function of the metal?

48. Red light of wavelength 670 nm produces photoelectrons from a certain photoemissive material. Green light of wavelength 520 nm produces photoelectrons from the same material with 1.50 times the maximum kinetic energy. What is the material's work function?

49. How fast must an electron be moving if all its kinetic energy is lost to a single x-ray photon (a) at the high end of the x-ray electromagnetic spectrum with a wavelength of $1.00 \times 10^{-8}$ m and (b) at the low end of the x-ray electromagnetic spectrum with a wavelength of $1.00 \times 10^{-13}$ m?

50. Q|C From the scattering of sunlight, J. J. Thomson calculated the classical radius of the electron as having the value $2.82 \times 10^{-15}$ m. Sunlight with an intensity of 500 W/m$^2$ falls on a disk with this radius. Assume light is a classical wave and the light striking the disk is completely absorbed. (a) Calculate the time interval required to accumulate 1.00 eV of energy. (b) Explain how your result for part (a) compares with the observation that photoelectrons are emitted promptly (within $10^{-9}$ s).

©lilying1982/Shutterstock.com

Burning gunpowder transfers energy to the atoms of color-producing chemicals, exciting their electrons to higher energy states. In returning to the ground state, the electrons emit light of specific colors, resulting in spectacular fireworks displays. Strontium produces red, and sodium produces yellow/orange.

# Atomic Physics 28

A hot gas emits light of certain characteristic wavelengths that can be used to identify it, much as a fingerprint can identify a person. For a given atom, these characteristic emitted wavelengths can be understood using physical quantities called quantum numbers. The simplest atom is hydrogen, and understanding it can lead to understanding the structure of other atoms and their combinations. The fact that no two electrons in an atom can have the same set of quantum numbers—the Pauli exclusion principle—is extremely important in understanding the properties of complex atoms and the arrangement of elements in the periodic table. Knowledge of atomic structure can be used to describe the mechanisms involved in the production of x-rays and the operation of a laser, among many other applications.



## 28.1 Early Models of the Atom

**LEARNING OBJECTIVES**

1. Describe Thomson's model of the atom and Rutherford's experiment, which disproved it.
2. Discuss Rutherford's planetary model of the atom. State two weaknesses of that model.

The model of the atom in the days of Newton was a tiny, hard, indestructible sphere. Although this model was a good basis for the kinetic theory of gases, new models had to be devised when later experiments revealed the electronic nature of atoms. J. J. Thomson (1856–1940) suggested a model of the atom as a volume of positive charge with electrons embedded throughout the volume, much like the seeds in a watermelon (Fig. 28.1, page 948).

In 1911 Ernest Rutherford (1871–1937) and his students Hans Geiger and Ernest Marsden performed a critical experiment showing that Thomson's model couldn't be correct. In this experiment a beam of positively charged **alpha particles** was projected against a thin metal foil, as in Figure 28.2a (page 948). Most of the alpha

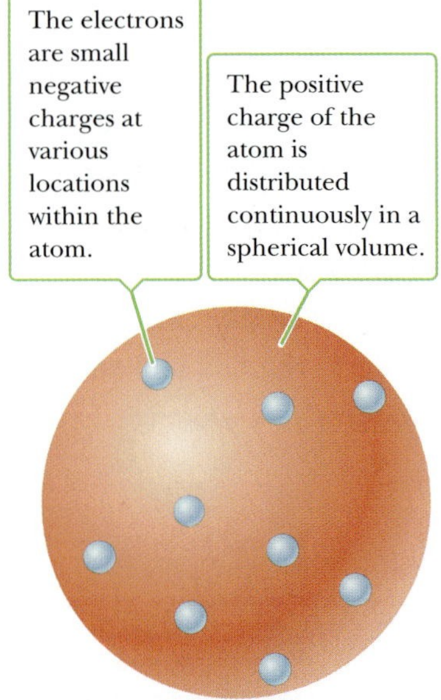
The electrons are small negative charges at various locations within the atom.
The positive charge of the atom is distributed continuously in a spherical volume.

**Figure 28.1** Thomson's model of the atom.

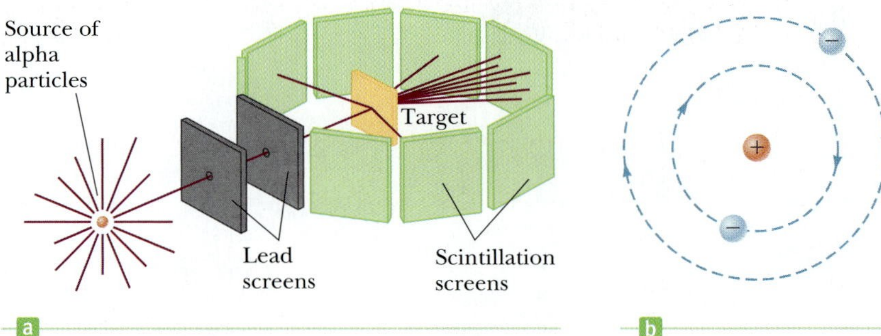
Source of alpha particles
Target
Lead screens
Scintillation screens
a
b

**Figure 28.2** (a) Geiger and Marsden's technique for observing the scattering of alpha particles from a thin foil target. The source is a naturally occurring radioactive substance, such as radium. (b) Rutherford's planetary model of the atom.

particles passed through the foil as if it were empty space, but a few particles were scattered through large angles, some even traveling backward.

Such large deflections weren't expected. In Thomson's model a positively charged alpha particle would never come close enough to a large positive charge to cause any large-angle deflections. Rutherford explained these results by assuming the positive charge in an atom was concentrated in a region called the **nucleus** that was small relative to the size of the atom. Any electrons belonging to the atom were visualized as orbiting the nucleus, much as planets orbit the Sun, as shown in Figure 28.2b. The alpha particles used in Rutherford's experiments were later identified as the nuclei of helium atoms.

There were two basic difficulties with Rutherford's planetary model. First, an atom emits certain discrete characteristic frequencies of electromagnetic radiation and no others; the Rutherford model was unable to explain this phenomenon. Second, the electrons in Rutherford's model undergo a centripetal acceleration. According to Maxwell's theory of electromagnetism, centripetally accelerated charges revolving with frequency $f$ should radiate electromagnetic waves of the same frequency. As the electron radiates energy, the radius of its orbit steadily decreases and its frequency of revolution increases. This process leads to an ever-increasing frequency of emitted radiation and a rapid collapse of the atom as the electron spirals into the nucleus.

Rutherford's model of the atom gave way to that of Niels Bohr, which explained the characteristic radiation emitted from atoms. Bohr's theory, in turn, was supplanted by quantum mechanics. Both the latter theories are based on studies of atomic spectra: the special pattern in the wavelengths of emitted light that is unique for every different element.

**Sir Joseph John Thomson**
**English Physicist (1856–1940)**
Thomson, usually considered the discoverer of the electron, opened up the field of subatomic particle physics with his extensive work on the deflection of cathode rays (electrons) in an electric field. He received the 1906 Nobel Prize in Physics for his discovery of the electron.

## 28.2 Atomic Spectra

**LEARNING OBJECTIVES**

1. Discuss the physical origins of an element's emission and absorption spectra.
2. Discuss the Rydberg equation and the emission spectrum of hydrogen.

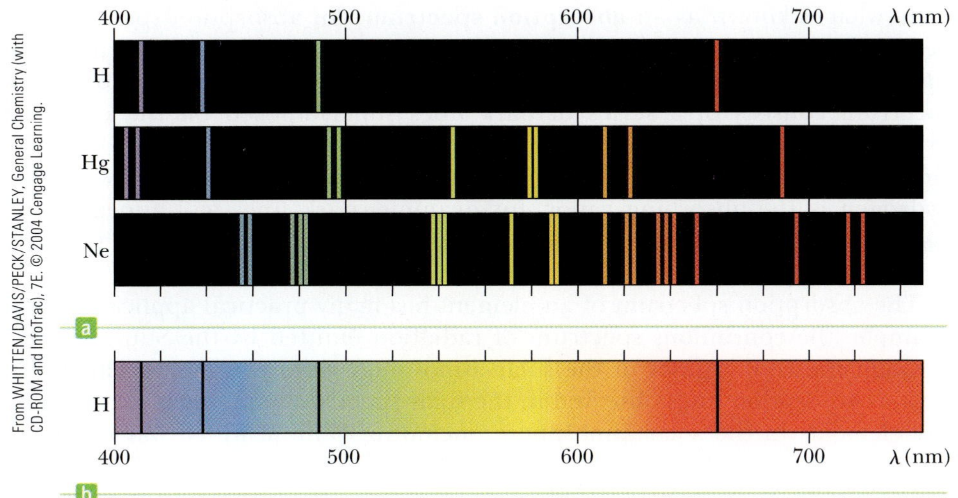




**Figure 28.3** Visible spectra. (a) Line spectra produced by emission in the visible range for the elements hydrogen, mercury, and neon. (b) The absorption spectrum for hydrogen. The dark absorption lines occur at the same wavelengths as the emission lines for hydrogen shown in (a).

Suppose an evacuated glass tube is filled with hydrogen (or some other gas) at low pressure. If a voltage applied between metal electrodes in the tube is great enough to produce an electric current in the gas, the tube emits light having a color that depends on the gas inside. (That's how a neon sign works.) When the emitted light is analyzed with a spectrometer, discrete bright lines are observed, each having a different wavelength, or color. Such a series of spectral lines is called an **emission spectrum**. The wavelengths contained in such a spectrum are characteristic of the element emitting the light. Because no two elements emit the same line spectrum, this phenomenon represents a reliable technique for identifying elements in a gaseous substance. Several emission spectra are shown in Figure 28.3a.

The emission spectrum of hydrogen shown in Figure 28.4 includes four prominent lines that occur at wavelengths of 656.3 nm, 486.1 nm, 434.1 nm, and 410.2 nm. In 1885 Johann Balmer (1825–1898) found that the wavelengths of these and less prominent lines can be described by the simple empirical equation

$$\frac{1}{\lambda} = R_H\left(\frac{1}{2^2} - \frac{1}{n^2}\right) \quad [28.1]$$

◀ Balmer series

where $n$ may have integral values of 3, 4, 5, . . . , and $R_H$ is a constant, called the **Rydberg constant**. If the wavelength is in meters, then $R_H$ has the value

$$R_H = 1.097\,373\,2 \times 10^7 \text{ m}^{-1} \quad [28.2]$$

◀ Rydberg constant

The first line in the Balmer series, at 656.3 nm, corresponds to $n = 3$ in Equation 28.1, the line at 486.1 nm corresponds to $n = 4$, and so on. In addition to the Balmer series of spectral lines, the Lyman series was subsequently discovered in the far ultraviolet, with the radiated wavelengths described by a similar equation, with $2^2$ in Equation 28.1 replaced by $1^2$ and the integer $n$ greater than 1. The Paschen series corresponded to longer wavelengths than the Balmer series, with the $2^2$ in Equation 28.1 replaced by $3^2$ and $n > 3$. These models, together with many other observations, can be combined to yield the Rydberg equation,

$$\frac{1}{\lambda} = R_H\left(\frac{1}{m^2} - \frac{1}{n^2}\right) \quad [28.3]$$

◀ Rydberg equation

where $m$ and $n$ are positive integers and $n > m$.

In addition to emitting light at specific wavelengths, an element can absorb light at specific wavelengths. The spectral lines corresponding to this process

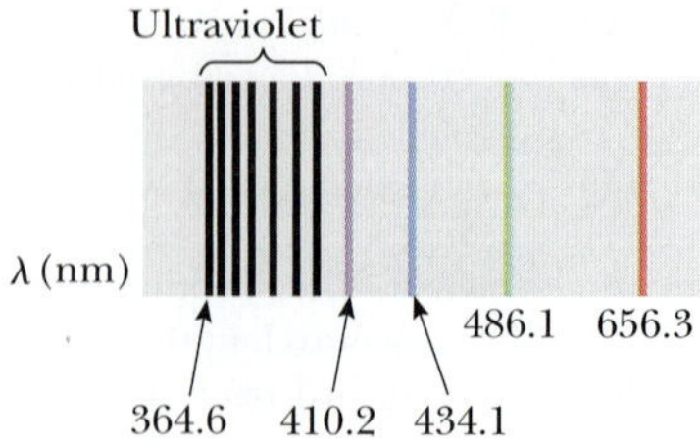


**Figure 28.4** The Balmer series of spectral lines for atomic hydrogen, with several lines marked with the wavelength in nanometers. The line labeled 364.6 is the shortest-wavelength line and is in the ultraviolet region of the electromagnetic spectrum. The other labeled lines are in the visible region.

form what is known as an **absorption spectrum**. An absorption spectrum can be obtained by passing a continuous radiation spectrum (one containing all wavelengths) through a vapor of the element being analyzed. The absorption spectrum consists of a series of dark lines superimposed on the otherwise bright, continuous spectrum. Each line in the absorption spectrum of a given element coincides with a line in the emission spectrum of the element. If hydrogen is the absorbing vapor, for example, dark lines will appear at the visible wavelengths 656.3 nm, 486.1 nm, 434.1 nm, and 410.2 nm, as shown in Figures 28.3b and 28.4.

The absorption spectrum of an element has many practical applications. For example, the continuous spectrum of radiation emitted by the Sun must pass through the cooler gases of the solar atmosphere before reaching Earth. The various absorption lines observed in the solar spectrum have been used to identify elements in the solar atmosphere, including helium, which was previously unknown.

## APPLYING PHYSICS 28.1 Thermal or Spectral

On observing a yellow candle flame, your laboratory partner claims that the light from the flame originates from excited sodium atoms in the flame. You disagree, stating that because the candle flame is hot, the radiation must be thermal in origin. Before the disagreement becomes more intense, how could you determine who is correct?

**EXPLANATION** A simple determination could be made by observing the light from the candle flame through a spectrometer, which is a slit and diffraction grating combination discussed in Chapter 25. If the spectrum of the light is continuous, it's probably thermal in origin. If the spectrum shows discrete lines, it's atomic in origin. The results of the experiment show that the light is indeed thermal, originating from random molecular motion in the candle flame. ■

## APPLYING PHYSICS 28.2 Auroras

At extreme northern latitudes, the aurora borealis provides a beautiful and colorful display in the night sky. A similar display, called the aurora australis, occurs near the southern polar region. What is the origin of the various colors seen in the auroras?

**EXPLANATION** The aurora results from high-speed particles interacting with Earth's magnetic field and entering the atmosphere. When these particles collide with molecules in the atmosphere, they excite the molecules just as does the voltage in the spectrum tubes discussed earlier in this section. In response the molecules emit colors of light according to the characteristic spectra of their atomic constituents. For our atmosphere, the primary constituents are nitrogen and oxygen, which provide the red, blue, and green colors of the aurora. ■

# 28.3 The Bohr Model

### LEARNING OBJECTIVES

1. State the basic assumptions of Bohr's model of hydrogen and derive the energy levels from them.
2. Define ground state and ionization energy.
3. Relate the energy levels obtained from the Bohr model to the Rydberg equation.
4. Apply the Bohr model to hydrogen and hydrogen-like atoms.
5. State Bohr's correspondence principle.

At the beginning of the 20th century, it wasn't understood why atoms of a given element emitted and absorbed only certain wavelengths. In 1913 Bohr provided an explanation of the spectra of hydrogen that includes some features of the currently accepted theory. His model of the hydrogen atom included the following basic assumptions:

1. The electron moves in circular orbits about the proton under the influence of the Coulomb force of attraction, as in Figure 28.5. The Coulomb force produces the electron's centripetal acceleration.
2. Only certain electron orbits are stable and allowed. In these orbits no energy in the form of electromagnetic radiation is emitted, so the total energy of the atom remains constant.
3. Radiation is emitted by the hydrogen atom when the electron "jumps" from a more energetic initial state to a less energetic state. The "jump" can't be visualized or treated classically. The frequency $f$ of the radiation emitted in the jump is related to the change in the atom's energy, given by

$$E_i - E_f = hf \qquad [28.4]$$

where $E_i$ is the energy of the initial state, $E_f$ is the energy of the final state, $h$ is Planck's constant, and $E_i > E_f$. The frequency of the radiation is *independent of the frequency of the electron's orbital motion.*
4. The circumference of an electron's orbit must contain an integral number of de Broglie wavelengths,

$$2\pi r = n\lambda \qquad n = 1, 2, 3, \ldots$$

(See Fig. 28.6.) Because the de Broglie wavelength of an electron is given by $\lambda = h/m_e v$, we can write the preceding equation as

$$m_e v r = n\hbar \qquad n = 1, 2, 3, \ldots \qquad [28.5]$$

where $\hbar = h/2\pi$.

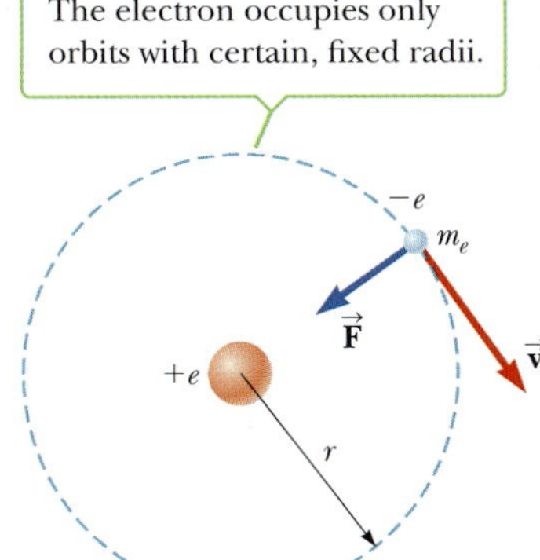


**Figure 28.5** Diagram representing Bohr's model of the hydrogen atom.

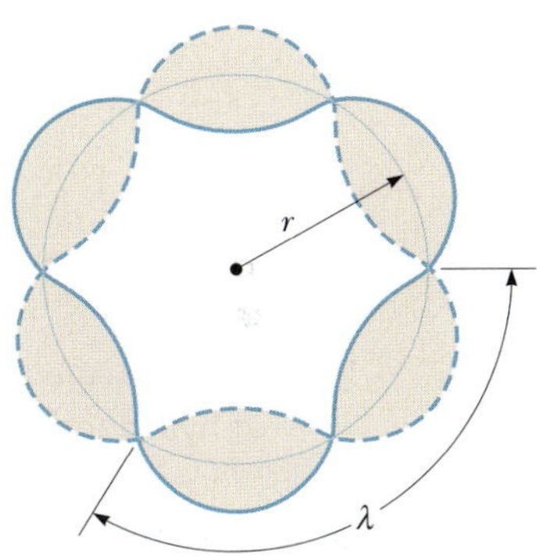


**Figure 28.6** Standing-wave pattern for an electron wave in a stable orbit of hydrogen. There are three full wavelengths in this orbit.

With these four assumptions, we can calculate the allowed energies and emission wavelengths of the hydrogen atom using the model pictured in Figure 28.5, in which the electron travels in a circular orbit of radius $r$ with an orbital speed $v$. The electrical potential energy of the atom is

$$PE = k_e \frac{q_1 q_2}{r} = k_e \frac{(-e)(e)}{r} = -k_e \frac{e^2}{r}$$

where $k_e$ is the Coulomb constant. Assuming the nucleus is at rest, the total energy $E$ of the atom is the sum of the kinetic and potential energy:

$$E = KE + PE = \tfrac{1}{2} m_e v^2 - k_e \frac{e^2}{r} \qquad [28.6]$$

By Newton's second law, the electric force of attraction on the electron, $k_e e^2/r^2$, must equal $m_e a_r$, where $a_r = v^2/r$ is the centripetal acceleration of the electron, so

$$m_e \frac{v^2}{r} = k_e \frac{e^2}{r^2} \qquad [28.7]$$

Multiply both sides of this equation by $r/2$ to get an expression for the kinetic energy:

$$\tfrac{1}{2} m_e v^2 = \frac{k_e e^2}{2r} \qquad [28.8]$$

Combining this result with Equation 28.6 gives an expression for the energy of the atom,

$$E = -\frac{k_e e^2}{2r} \qquad [28.9]$$

where the negative value of the energy indicates that the electron is bound to the proton.

**Niels Bohr**
**Danish Physicist (1885–1962)**
Bohr was an active participant in the early development of quantum mechanics and provided much of its philosophical framework. During the 1920s and 1930s he headed the Institute for Advanced Studies in Copenhagen, where many of the world's best physicists came to exchange ideas. Bohr was awarded the 1922 Nobel Prize in Physics for his investigation of the structure of atoms and of the radiation emanating from them.

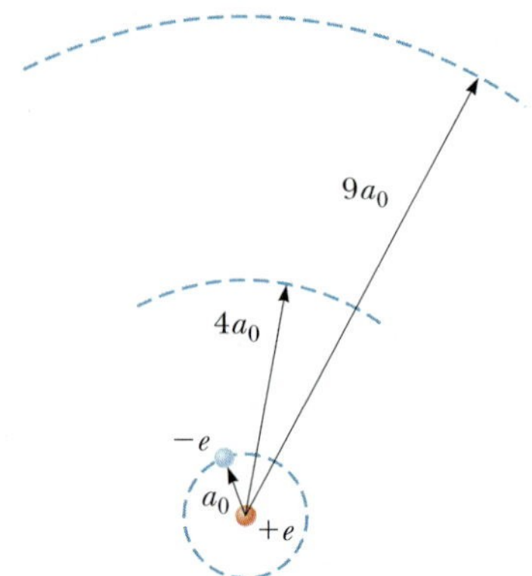


**Figure 28.7** The first three circular orbits predicted by the Bohr model of the hydrogen atom. The electron is shown in the ground state orbit.

An expression for $r$ can be obtained by solving Equations 28.5 and 28.7 for $v^2$ and equating the results:

$$v^2 = \frac{n^2\hbar}{m_e^2 r^2} = \frac{k_e e^2}{m_e r}$$

$$r_n = \frac{n^2\hbar^2}{m_e k_e e^2} \qquad n = 1, 2, 3, \ldots \qquad [28.10]$$

This equation is based on the assumption that the **electron can exist only in certain allowed orbits determined by the integer *n*.**

The orbit with the smallest radius, called the **Bohr radius**, $a_0$, corresponds to $n = 1$ and has the value

$$a_0 = \frac{\hbar^2}{mk_e e^2} = 0.052\,9 \text{ nm} \qquad [28.11]$$

A general expression for the radius of any orbit in the hydrogen atom is obtained by substituting Equation 28.11 into Equation 28.10:

$$r_n = n^2 a_0 = n^2(0.052\,9 \text{ nm}) \qquad [28.12]$$

The first three Bohr orbits for hydrogen are shown in Figure 28.7.

Equation 28.10 can then be substituted into Equation 28.9 to give the following expression for the energies of the quantum states:

The energy levels of hydrogen ▶

$$E_n = -\frac{m_e k_e^2 e^4}{2\hbar^2}\left(\frac{1}{n^2}\right) \qquad n = 1, 2, 3, \ldots \qquad [28.13]$$

If we substitute numerical values into Equation 28.13, we obtain

$$E_n = -\frac{13.6}{n^2} \text{ eV} \qquad [28.14]$$

The lowest-energy state, or **ground state**, corresponds to $n = 1$ and has an energy $E_1 = -m_e k_e^2 e^4/2\hbar^2 = -13.6$ eV. The next state, corresponding to $n = 2$, has an energy $E_2 = E_1/4 = -3.40$ eV, and so on. An energy level diagram showing the energies of these stationary states and the corresponding quantum numbers is given in Figure 28.8. The uppermost level shown, corresponding to $E = 0$ and $n \to \infty$, represents the state for which the electron is completely removed from the atom. In this state the electron's *KE* and *PE* are both zero, which means that the electron is at rest infinitely far away from the proton. The minimum energy required to ionize the atom—that is, to completely remove the electron—is called the **ionization energy**. The ionization energy for hydrogen is 13.6 eV.

Equations 28.4 and 28.13 and the third Bohr postulate show that if the electron jumps from one orbit with quantum number $n_i$ to a second orbit with quantum number $n_f$, it emits a photon of frequency $f$ given by

$$f = \frac{E_i - E_f}{h} = \frac{m_e k_e^2 e^4}{4\pi\hbar^3}\left(\frac{1}{n_f^2} - \frac{1}{n_i^2}\right) \qquad [28.15]$$

where $n_f < n_i$.

To convert this equation into one analogous to the Rydberg equation, substitute $f = c/\lambda$ and divide both sides by $c$, obtaining

$$\frac{1}{\lambda} = R_{\mathrm{H}}\left(\frac{1}{n_f^2} - \frac{1}{n_i^2}\right) \qquad [28.16]$$

where

$$R_{\mathrm{H}} = \frac{m_e k_e^2 e^4}{4\pi c\hbar^3} \qquad [28.17]$$

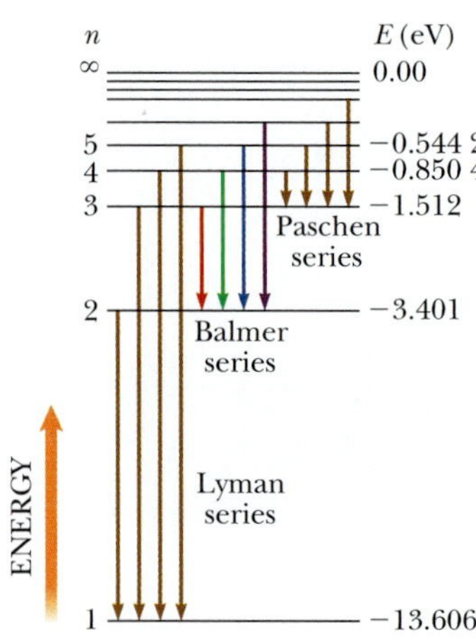


**Figure 28.8** An energy level diagram for hydrogen. Quantum numbers are given on the left, and energies (in electron volts) are given on the right. Vertical arrows represent the four lowest-energy transitions for each of the spectral series shown. The colored arrows for the Balmer series indicate that this series results in visible light.

Substituting the known values of $m_e$, $k_e$, $e$, $c$, and $\hbar$ verifies that this theoretical value for the Rydberg constant is in excellent agreement with the experimentally derived value in Equations 12.1 through 12.3. When Bohr demonstrated this agreement, it was recognized as a major accomplishment of his theory.

We can use Equation 28.16 to evaluate the wavelengths for the various series in the hydrogen spectrum. For example, in the Balmer series, $n_f = 2$ and $n_i = 3, 4, 5, \ldots$ (Eq. 28.1). The energy level diagram for hydrogen shown in Figure 28.8 indicates the origin of the spectral lines. The transitions between levels are represented by vertical arrows. Note that whenever a transition occurs between a state designated by $n_i$ to one designated by $n_f$ (where $n_i > n_f$), a photon with a frequency $(E_i - E_f)/h$ is emitted. This process can be interpreted as follows: the lines in the visible part of the hydrogen spectrum arise when the electron jumps from the third, fourth, or even higher orbit to the second orbit. The Bohr theory successfully predicts the wavelengths of all the observed spectral lines of hydrogen.

**Tip 28.1 Energy Depends on *n* Only for Hydrogen**

Because all other quantities in Equation 28.13 are constant, the energy levels of a hydrogen atom depend only on the quantum number *n*. For more complicated atoms, the energy levels depend on other quantum numbers as well.

## EXAMPLE 28.1 The Balmer Series for Hydrogen

**GOAL** Calculate the wavelength, frequency, and energy of a photon emitted during an electron transition in an atom.

**PROBLEM** The Balmer series for the hydrogen atom corresponds to electronic transitions that terminate in the state with quantum number $n = 2$, as shown in Figure 28.9. **(a)** Find the longest-wavelength photon emitted in the Balmer series and determine its frequency and energy. **(b)** Find the shortest-wavelength photon emitted in the same series.

**STRATEGY** This problem is a matter of substituting values into Equation 28.16. The frequency can then be obtained from $c = f\lambda$ and the energy from $E = hf$. The longest-wavelength photon corresponds to the one that is emitted when the electron jumps from the $n_i = 3$ state to the $n_f = 2$ state. The shortest-wavelength photon corresponds to the one that is emitted when the electron jumps from $n_i = \infty$ to the $n_f = 2$ state.

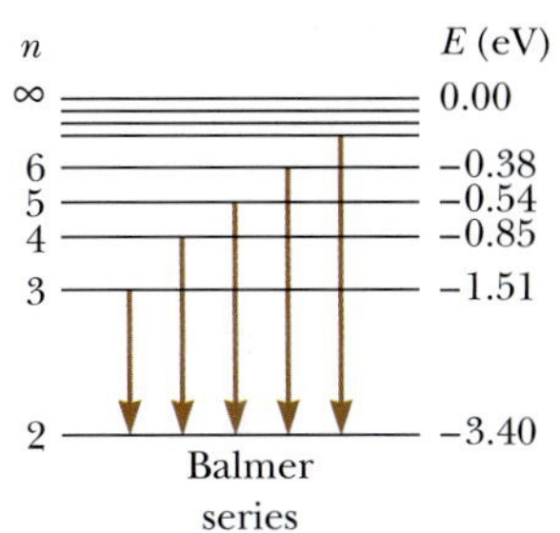


**Figure 28.9** (Example 28.1) Transitions responsible for the Balmer series for the hydrogen atom. All transitions terminate at the $n = 2$ level.

**SOLUTION**

**(a)** Find the longest-wavelength photon emitted in the Balmer series and determine its frequency and energy.

Substitute into Equation 28.16, with $n_i = 3$ and $n_f = 2$:

$$\frac{1}{\lambda} = R_H\left(\frac{1}{n_f^2} - \frac{1}{n_i^2}\right) = R_H\left(\frac{1}{2^2} - \frac{1}{3^2}\right) = \frac{5R_H}{36}$$

Take the reciprocal and substitute, finding the wavelength:

$$\lambda = \frac{36}{5R_H} = \frac{36}{5(1.097 \times 10^7\ \text{m}^{-1})} = 6.563 \times 10^{-7}\ \text{m}$$

$$= \boxed{656.3\ \text{nm}}$$

Now use $c = f\lambda$ to obtain the frequency:

$$f = \frac{c}{\lambda} = \frac{2.998 \times 10^8\ \text{m/s}}{6.563 \times 10^{-7}\ \text{m}} = \boxed{4.568 \times 10^{14}\ \text{Hz}}$$

Calculate the photon's energy by substituting into Equation 27.5:

$$E = hf = (6.626 \times 10^{-34}\ \text{J} \cdot \text{s})(4.568 \times 10^{14}\ \text{Hz}) = 3.027 \times 10^{-19}\ \text{J}$$

$$= 3.027 \times 10^{-19}\ \text{J}\left(\frac{1\ \text{eV}}{1.602 \times 10^{-19}\ \text{J}}\right) = \boxed{1.892\ \text{eV}}$$

**(b)** Find the shortest-wavelength photon emitted in the Balmer series.

Substitute into Equation 28.16, with $1/n_i \rightarrow 0$ as $n_i \rightarrow \infty$ and $n_f = 2$:

$$\frac{1}{\lambda} = R_H\left(\frac{1}{n_f^2} - \frac{1}{n_i^2}\right) = R_H\left(\frac{1}{2^2} - 0\right) = \frac{R_H}{4}$$

Take the reciprocal and substitute, finding the wavelength:

$$\lambda = \frac{4}{R_H} = \frac{4}{(1.097 \times 10^7\ \text{m}^{-1})} = 3.646 \times 10^{-7}\ \text{m}$$

$$= \boxed{364.6\ \text{nm}}$$

*(Continued)*

**REMARKS** The first wavelength is in the red region of the visible spectrum. We could also obtain the energy of the photon by using Equation 28.4 in the form $hf = E_3 - E_2$, where $E_2$ and $E_3$ are the energy levels of the hydrogen atom, calculated from Equation 28.14. Note that this photon is the lowest-energy photon in the Balmer series because it involves the smallest energy change. The second photon, the most energetic, is in the ultraviolet region.

**QUESTION 28.1** What is the upper-limit energy of a photon that can be emitted from hydrogen due to the transition of an electron between energy levels? Explain.

**EXERCISE 28.1** (a) Calculate the energy of the shortest-wavelength photon emitted in the Balmer series for hydrogen. (b) Calculate the wavelength of the photon emitted when an electron transits from $n = 4$ to $n = 2$.

**ANSWERS** (a) 3.40 eV (b) 486 nm

## Bohr's Correspondence Principle

In our study of relativity in Chapter 26, we found that Newtonian mechanics can't be used to describe phenomena that occur at speeds approaching the speed of light. Newtonian mechanics is a special case of relativistic mechanics and applies only when $v$ is much smaller than $c$. Similarly, **quantum mechanics is in agreement with classical physics when the energy differences between quantized levels are very small**. This principle, first set forth by Bohr, is called the **correspondence principle**.

## Hydrogen-like Atoms

The analysis used in the Bohr theory is also successful when applied to *hydrogen-like* atoms. An atom is said to be hydrogen-like when it contains only one electron. Examples are singly ionized helium, doubly ionized lithium, and triply ionized beryllium. The results of the Bohr theory for hydrogen can be extended to hydrogen-like atoms by substituting $Ze^2$ for $e^2$ in the hydrogen equations, where $Z$ is the atomic number of the element. For example, Equations 28.13 and 28.16 through 28.17 become

$$E_n = -\frac{m_e k_e^2 Z^2 e^4}{2\hbar^2}\left(\frac{1}{n^2}\right) \qquad n = 1, 2, 3, \ldots \qquad [28.18]$$

and

$$\frac{1}{\lambda} = \frac{m_e k_e^2 Z^2 e^4}{4\pi c\hbar^3}\left(\frac{1}{n_f^2} - \frac{1}{n_i^2}\right) \qquad [28.19]$$

Although many attempts were made to extend the Bohr theory to more complex, multi-electron atoms, the results were unsuccessful. Even today, only approximate methods are available for treating multi-electron atoms.

### Quick Quiz

**28.1** Consider a hydrogen atom and a singly ionized helium atom. Which atom has the lower ground state energy? (a) Hydrogen (b) Helium (c) The ground state energy is the same for both.

## EXAMPLE 28.2 Singly Ionized Helium

**GOAL** Apply the modified Bohr theory to a hydrogen-like atom.

**PROBLEM** Singly ionized helium, $He^+$, a hydrogen-like system, has one electron in the orbit corresponding to $n = 1$ when the atom is in its ground state. Find **(a)** the energy of the system in the ground state in electron volts and **(b)** the radius of the ground-state orbit.

**STRATEGY** Part (a) requires substitution into the modified Bohr model, Equation 28.18. In part (b) modify Equation 28.10 for the radius of the Bohr orbits by replacing $e^2$ by $Ze^2$, where $Z$ is the number of protons in the nucleus.

**SOLUTION**

**(a)** Find the energy of the system in the ground state.

Write Equation 28.18 for the energies of a hydrogen-like system:

$$E_n = -\frac{m_e k_e^2 Z^2 e^4}{2\hbar^2}\left(\frac{1}{n^2}\right)$$

Substitute the constants and convert to electron volts:

$$E_n = -\frac{Z^2(13.6 \text{ eV})}{n^2}$$

Substitute $Z = 2$ (the atomic number of helium) and $n = 1$ to obtain the ground state energy:

$$E_1 = -4(13.6 \text{ eV}) = -54.4 \text{ eV}$$

**(b)** Find the radius of the ground state.

Generalize Equation 28.10 to a hydrogen-like atom by substituting $Ze^2$ for $e^2$:

$$r_n = \frac{n^2\hbar^2}{m_e k_e Z e^2} = \frac{n^2}{Z}(a_0) = \frac{n^2}{Z}(0.052\ 9 \text{ nm})$$

For our case, $n = 1$ and $Z = 2$:

$$r_1 = 0.026\ 5 \text{ nm}$$

**REMARKS** Notice that for higher Z, the energy of a hydrogen-like atom is lower (more negative), which means that the electron is more tightly bound than in hydrogen. The result is a smaller atom, as seen in part (b).

**QUESTION 28.2** When an electron undergoes a transition from a higher to lower state in singly ionized helium, how will the energy of the emitted photon compare with the analogous transition in hydrogen? Explain.

**EXERCISE 28.2** Repeat the problem for the first excited state of doubly ionized lithium ($Z = 3$, $n = 2$).

**ANSWERS** (a) $E_2 = -30.6$ eV (b) $r_2 = 0.070\ 5$ nm

Bohr's theory was extended in an ad hoc manner so as to include further details of atomic spectra. All these modifications were replaced with the theory of quantum mechanics, developed independently by Werner Heisenberg and Erwin Schrödinger.

## 28.4 Quantum Mechanics and the Hydrogen Atom

**LEARNING OBJECTIVES**

1. Identify the four quantum numbers and state their permitted values.
2. Discuss the concepts of electron spin and electron clouds.
3. Apply the permitted ranges of quantum numbers to the hydrogen atom.

One of the first great achievements of quantum mechanics was the solution of the wave equation for the hydrogen atom. Although the details of the solution are beyond the level of this book, the solution and its implications for atomic structure can be described.

According to quantum mechanics, the energies of the allowed states are in exact agreement with the values obtained by the Bohr theory (Eq. 28.13) when the allowed energies depend only on the principal quantum number $n$.

In addition to the principal quantum number, two other quantum numbers emerged from the solution of the Schrödinger wave equation: the **orbital quantum number,** $\ell$, and the **orbital magnetic quantum number,** $m_\ell$.

The effect of the magnetic quantum number $m_\ell$ can be observed in spectra when magnetic fields are present, which results in a splitting of individual spectral lines into several lines. This splitting is called the *Zeeman effect.* Figure 28.10 (page 956) shows

A

B

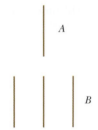

**Figure 28.10** A single line (*A*) can split into three separate lines (*B*) in a magnetic field.

a single spectral line being split into three closely spaced lines. This indicates that the energy of an electron is slightly modified when the atom is immersed in a magnetic field.

The allowed ranges of the values of these quantum numbers are as follows:

- The value of $n$ can range from 1 to $\infty$ in integer steps.
- The value of $\ell$ can range from 0 to $n - 1$ in integer steps.
- The value of $m_\ell$ can range from $-\ell$ to $\ell$ in integer steps.

From these rules, it can be seen that for a given value of $n$, there are $n$ possible values of $\ell$, whereas for a given value of $\ell$, there are $2\ell + 1$ possible values of $m_\ell$. For example, if $n = 1$, there is only 1 value of $\ell$, which is $\ell = 0$. Because $2\ell + 1 = 2 \cdot 0 + 1 = 1$, there is only one value of $m_\ell$, which is $m_\ell = 0$. If $n = 2$, the value of $\ell$ may be 0 or 1; if $\ell = 0$, then $m_\ell = 0$; but if $\ell = 1$, then $m_\ell$ may be 1, 0, or $-1$. Table 28.1 summarizes the rules for determining the allowed values of $\ell$ and $m_\ell$ for a given value of $n$.

For historical reasons, **all states with the same principal quantum number $n$ are said to form a shell.** Shells are identified by the letters K, L, M, . . . , which designate the states for which $n = 1, 2, 3$, and so forth. **The states with given values of $n$ and $\ell$ are said to form a subshell.** The letters *s, p, d, f, g*, . . . are used to designate the states for which $\ell = 0, 1, 2, 3, 4, \ldots$. These notations are summarized in Table 28.2.

States that violate the rules given in Table 28.1 can't exist. One state that cannot exist, for example, is the 2*d* state, which would have $n = 2$ and $\ell = 2$. This state is not allowed because the highest allowed value of $\ell$ is $n - 1$, or 1 in this case. So for $n = 2$, 2*s* and 2*p* are allowed states, but 2*d*, 2*f*, . . . are not. For $n = 3$, the allowed states are 3*s*, 3*p*, and 3*d*.

In general, for a given value of $n$, there are $n^2$ states with distinct pairs of values of $\ell$ and $m_\ell$.

**Quick Quiz**

**28.2** When the principal quantum number is $n = 5$, how many different values of (a) $\ell$ and (b) $m_\ell$ are possible? (c) How many states have distinct pairs of values of $\ell$ and $m_\ell$?

## Spin

In high-resolution spectrometers, close examination of one of the prominent lines of sodium vapor shows that it is, in fact, two very closely spaced lines. The wavelengths of these lines occur in the yellow region of the spectrum at 589.0 nm and 589.6 nm. This kind of splitting is referred to as **fine structure**. In 1925, when this doublet was first noticed, atomic theory couldn't explain it, so Samuel Goudsmit

**Table 28.1** Three Quantum Numbers for the Hydrogen Atom

| Quantum Number | Name | Allowed Values | Number of Allowed States |
|---|---|---|---|
| $n$ | Principal quantum number | $1, 2, 3, \ldots$ | Any number |
| $\ell$ | Orbital quantum number | $0, 1, 2, \ldots, n - 1$ | $n$ |
| $m_\ell$ | Orbital magnetic quantum number | $-\ell, -\ell + 1, \ldots, 0, \ldots, \ell - 1, \ell$ | $2\ell + 1$ |

**Table 28.2** Shell and Subshell Notation

| $n$ | Shell Symbol | $\ell$ | Subshell Symbol |
|---|---|---|---|
| 1 | K | 0 | *s* |
| 2 | L | 1 | *p* |
| 3 | M | 2 | *d* |
| 4 | N | 3 | *f* |
| 5 | O | 4 | *g* |
| 6 | P | 5 | *h* |
| . . . | | . . . | |

and George Uhlenbeck, following a suggestion by Austrian physicist Wolfgang Pauli, proposed the introduction of a fourth quantum number to describe atomic energy levels, $m_s$, called the **spin magnetic quantum number**. Spin isn't found in the solutions of Schrödinger's equations; rather, it naturally arises in the Dirac equation, derived in 1927 by Paul Dirac. This equation is important in relativistic quantum theory.

In describing the spin quantum number, it's convenient (but technically incorrect) to think of the electron as spinning on its axis as it orbits the nucleus, just as Earth spins on its axis as it orbits the Sun. Unlike the spin of a world, however, there are only two ways in which the electron can spin as it orbits the nucleus, as shown in Figure 28.11. If the direction of spin is as shown in Figure 28.11a, the electron is said to have "spin up." If the direction of spin is reversed as in Figure 28.11b, the electron is said to have "spin down." The energy of the electron is slightly different for the two spin directions, and this energy difference accounts for the sodium doublet. The quantum numbers associated with electron spin are $m_s = \frac{1}{2}$ for the spin-up state and $m_s = -\frac{1}{2}$ for the spin-down state. As we see in Example 28.3, this new quantum number doubles the number of allowed states specified by the quantum numbers $n$, $\ell$, and $m_\ell$.

For each electron, there are two spin states. A subshell corresponding to a given factor of $\ell$ can contain no more than $2(2\ell + 1)$ electrons. This number is used because electrons in a subshell must have unique pairs of the quantum numbers $(m_\ell, m_s)$. There are $2\ell + 1$ different magnetic quantum numbers $m_\ell$ and two different spin quantum numbers $m_s$, making $2(2\ell + 1)$ unique pairs $(m_\ell, m_s)$. For example, the $p$ subshell ($\ell = 1$) is filled when it contains $2(2 \cdot 1 + 1) = 6$ electrons. This fact can be extended to include all four quantum numbers, as will be important to us later when we discuss the *Pauli exclusion principle.*

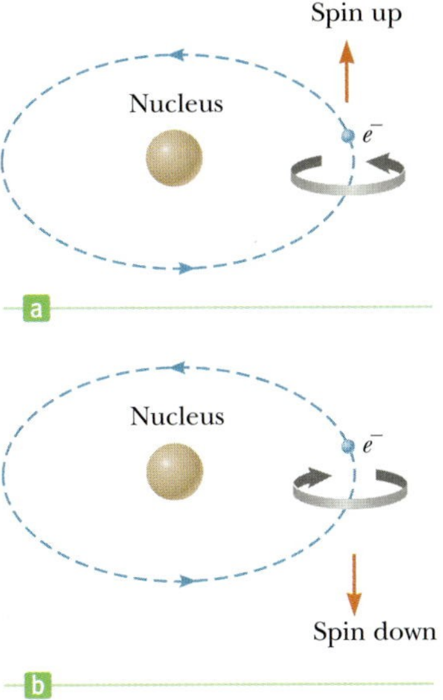


**Figure 28.11** As an electron moves in its orbit about the nucleus, its spin can be either (a) up or (b) down.

**Tip 28.2 The Electron Isn't Actually Spinning**

The electron is *not* physically spinning. Electron spin is a purely quantum effect that gives the electron an angular momentum *as if* it were physically spinning.

## EXAMPLE 28.3 The $n = 2$ Level of Hydrogen

**GOAL** Count and tabulate distinct quantum states and determine their energy based on atomic energy level.

**PROBLEM** **(a)** Determine the number of states with a unique set of values for $\ell$, $m_\ell$, and $m_s$ in the hydrogen atom for $n = 2$. **(b)** Tabulate the distinct possible quantum states, including spin. **(c)** Calculate the energies of these states in the absence of a magnetic field, disregarding small differences caused by spin.

**STRATEGY** This problem is a matter of counting, following the quantum rules for $n$, $\ell$, $m_\ell$, and $m_s$. "Unique" means that no other quantum state has the same set of numbers. The energies—disregarding spin or Zeeman splitting in magnetic fields—are all the same because all states have the same principal quantum number, $n = 2$.

**SOLUTION**

**(a)** Determine the number of states with a unique set of values for $\ell$ and $m_\ell$ in the hydrogen atom for $n = 2$.

Determine the different possible values of $\ell$ for $n = 2$:

$0 \le \ell \le n - 1$, so for $n = 2$, $0 \le \ell \le 1$ and $\ell = 0$ or 1

Find the different possible values of $m_\ell$ for $\ell = 0$:

$-\ell \le m_\ell \le \ell$, so $-0 \le m_\ell \le 0$ implies that $m_\ell = 0$

List the distinct pairs of $(\ell, m_\ell)$ for $\ell = 0$:

There is only one: $(\ell, m_\ell) = (0, 0)$.

Find the different possible values of $m_\ell$ for $\ell = 1$:

$-\ell \le m_\ell \le \ell$, so $-1 \le m_\ell \le 1$ implies that $m_\ell = -1$, 0, or 1

List the distinct pairs of $(\ell, m_\ell)$ for $\ell = 1$:

There are three: $(\ell, m_\ell) = (1, -1)$, $(1, 0)$, and $(1, 1)$.

*(Continued)*

Sum the results for $\ell = 0$ and $\ell = 1$ and multiply by 2 to account for the two possible spins of each state:

$$\text{Number of states} = 2(1 + 3) = \boxed{8}$$

**(b)** Tabulate the different possible sets of quantum numbers.

Use the results of part (a) and recall that the spin quantum number is always $+\frac{1}{2}$ or $-\frac{1}{2}$.

| $n$ | $\ell$ | $m_\ell$ | $m_s$ |
|---|---|---|---|
| 2 | 1 | −1 | $-\frac{1}{2}$ |
| 2 | 1 | −1 | $\frac{1}{2}$ |
| 2 | 1 | 0 | $-\frac{1}{2}$ |
| 2 | 1 | 0 | $\frac{1}{2}$ |
| 2 | 1 | 1 | $-\frac{1}{2}$ |
| 2 | 1 | 1 | $\frac{1}{2}$ |
| 2 | 0 | 0 | $-\frac{1}{2}$ |
| 2 | 0 | 0 | $\frac{1}{2}$ |

**(c)** Calculate the energies of these states.

The common energy of all the states, disregarding Zeeman splitting and spin, can be found with Equation 28.14:

$$E_n = -\frac{13.6\text{ eV}}{n^2} \rightarrow E_2 = -\frac{13.6\text{ eV}}{2^2} = \boxed{-3.40\text{ eV}}$$

**REMARKS** Although these states normally have the same energy, the application of a magnetic field causes them to take slightly different energies centered around the energy corresponding to $n = 2$. In addition, the slight difference in energy due to spin state was neglected.

**QUESTION 28.3** Which of the four quantum numbers are never negative?

**EXERCISE 28.3** (a) Determine the number of states with a unique pair of values for $\ell$, $m_\ell$, and $m_s$ in the $n = 3$ level of hydrogen. (b) Determine the energies of those states, disregarding any splitting effects.

**ANSWERS** (a) 18 (b) $E_3 = -1.51$ eV

## Electron Clouds

The solution of the wave equation, as discussed in Section 27.7, yields a wave function $\Psi$ that depends on the quantum numbers $n$, $\ell$, and $m_\ell$. Recall that if $p$ is a point and $V_p$ a very small volume containing that point, then $\Psi^2 V_p$ is approximately the probability of finding the electron inside the volume $V_p$. Figure 28.12 gives the probability per unit length of finding the electron at various distances from the nucleus in the 1*s* state of hydrogen ($n = 1$, $\ell = 0$, and $m_\ell = 0$). Note that the curve peaks at a value of $r = 0.052\,9$ nm, the Bohr radius for the first ($n = 1$) electron orbit in hydrogen. This peak means that there is a maximum probability of finding the electron in a small interval of a given, fixed length centered at that distance from the nucleus. As the curve indicates, however, there is also a probability of finding the electron in such a small interval centered at any other distance from the nucleus. In quantum mechanics the electron is not confined to a particular orbital distance from the nucleus, as assumed in the Bohr model. The electron may be found at various distances from the nucleus, but finding it in a small interval centered on the Bohr radius has the greatest probability. Quantum mechanics also predicts that the wave function for the hydrogen atom in the ground state is spherically symmetric; hence, the electron can be found in a spherical region surrounding the nucleus. This is in contrast to the Bohr theory, which confines the position of the electron to points in a plane. The quantum mechanical result is often interpreted by viewing the electron as a cloud surrounding the nucleus. An attempt at picturing this cloud-like behavior is shown in Figure 28.13. The densest regions of the cloud represent those locations where the electron is most likely to be found.

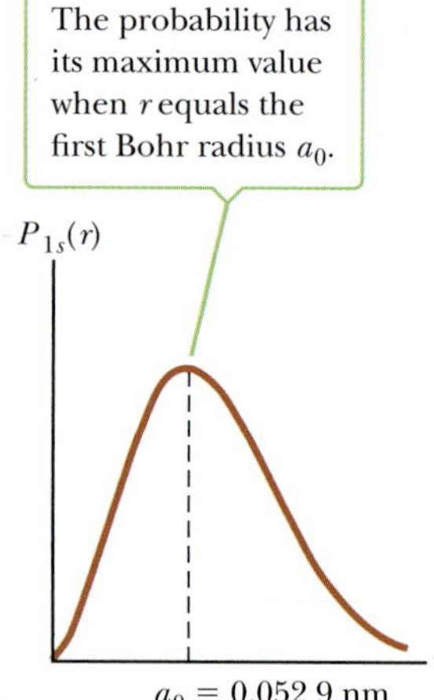
The probability has its maximum value when $r$ equals the first Bohr radius $a_0$.
$P_{1s}(r)$
$r$
$a_0 = 0.052\,9$ nm

**Figure 28.12** The probability per unit length of finding the electron versus distance from the nucleus for the hydrogen atom in the 1*s* (ground) state.

If a similar analysis is carried out for the $n = 2$, $\ell = 0$ state of hydrogen, a peak of the probability curve is found at $4a_0$, whereas for the $n = 3$, $\ell = 0$ state, the curve peaks at $9a_0$. In general, quantum mechanics predicts a most probable electron distance to the nucleus that is in agreement with the location predicted by the Bohr theory.

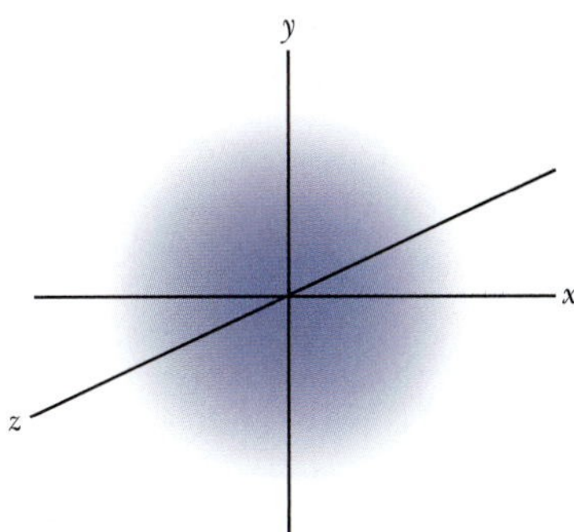


**Figure 28.13** The spherical electron cloud for the hydrogen atom in its 1*s* state.

# 28.5 The Exclusion Principle and the Periodic Table

**LEARNING OBJECTIVE**

1. State the Pauli exclusion principle and describe its importance in understanding the periodic table.

The state of an electron in a hydrogen atom is specified by four quantum numbers: $n$, $\ell$, $m_\ell$, and $m_s$. As it turns out, the state of any electron in any other atom can also be specified by this same set of quantum numbers.

How many electrons in an atom can have a particular set of quantum numbers? This important question was answered by Pauli in 1925 in a powerful statement known as the **Pauli exclusion principle**:

◀ The Pauli exclusion principle

> No two electrons in an atom can ever have the same set of values for the set of quantum numbers $n$, $\ell$, $m_\ell$, and $m_s$.

The Pauli exclusion principle explains the electronic structure of complex atoms as a succession of filled levels with different quantum numbers increasing in energy, where the outermost electrons are primarily responsible for the chemical properties of the element. If this principle weren't valid, every electron would end up in the lowest energy state of the atom and the chemical behavior of the elements would be grossly different. Nature as we know it would not exist, and *we* would not exist to wonder about it!

**Tip 28.3 The Exclusion Principle Is More General**

The exclusion principle stated here is a limited form of the more general exclusion principle, which states that no two *fermions* (particles with spin $\frac{1}{2}, \frac{3}{2} \ldots$) can be in the same quantum state.

As a general rule, the order that electrons fill an atom's subshell is as follows. Once one subshell is filled, the next electron goes into the vacant subshell that is lowest in energy. If the atom were not in the lowest energy state available to it, it would radiate energy until it reached that state. A subshell is filled when it contains $2(2\ell + 1)$ electrons. This rule is based on the analysis of quantum numbers to be described later. Following the rule, shells and subshells can contain numbers of electrons according to the pattern given in Table 28.3.

The exclusion principle can be illustrated by examining the electronic arrangement in a few of the lighter atoms. *Hydrogen* has only one electron, which, in its ground state, can be described by either of two sets of quantum numbers: 1, 0, 0, $\frac{1}{2}$ or 1, 0, 0, $-\frac{1}{2}$. The electronic configuration of this atom is often designated as $1s^1$.

Keystone/Getty Images

**Wolfgang Pauli**
**Austrian Theoretical Physicist (1900–1958)**
The extremely talented Pauli first gained public recognition at the age of 21 with a masterful review article on relativity. In 1945 he received the Nobel Prize in Physics for his discovery of the exclusion principle. Among his other major contributions were the explanation of the connection between particle spin and statistics, the theory of relativistic quantum electrodynamics, the neutrino hypothesis, and the hypothesis of nuclear spin.

**Table 28.3** Number of Electrons in Filled Subshells and Shells

| Shell | Subshell | Number of Electrons in Filled Subshell | Number of Electrons in Filled Shell |
|---|---|---|---|
| K ($n = 1$) | $s(\ell = 0)$ | 2 | 2 |
| L ($n = 2$) | $s(\ell = 0)$ | 2 | 8 |
| | $p(\ell = 1)$ | 6 | |
| | $s(\ell = 0)$ | 2 | |
| M ($n = 3$) | $p(\ell = 1)$ | 6 | 18 |
| | $d(\ell = 2)$ | 10 | |
| | $s(\ell = 0)$ | 2 | |
| N ($n = 4$) | $p(\ell = 1)$ | 6 | 32 |
| | $d(\ell = 2)$ | 10 | |
| | $f(\ell = 3)$ | 14 | |

The notation 1$s$ refers to a state for which $n = 1$ and $\ell = 0$, and the superscript indicates that one electron is present in this level.

Neutral *helium* has two electrons. In the ground state, the quantum numbers for these two electrons are 1, 0, 0, $\frac{1}{2}$ and 1, 0, 0, $-\frac{1}{2}$. No other possible combinations of quantum numbers exist for this level, and we say that the K shell is filled. The helium electronic configuration is designated as $1s^2$.

Neutral *lithium* has three electrons. In the ground state, two of them are in the 1$s$ subshell and the third is in the 2$s$ subshell because the latter subshell is lower in energy than the 2$p$ subshell. Hence, the electronic configuration for lithium is $1s^22s^1$.

A list of electronic ground-state configurations for a number of atoms is provided in Table 28.4. In 1871 Dmitry Mendeleyev (1834–1907), a Russian chemist, arranged the elements known at that time into a table according to their atomic masses and chemical similarities. The first table Mendeleyev proposed contained many blank spaces, and he boldly stated that the gaps were there only because those elements had not yet been discovered. By noting the column in which these missing elements should be located, he was able to make rough predictions about their chemical properties. Within 20 years of this announcement, those elements were indeed discovered.

The elements in our current version of the periodic table are still arranged so that all those in a vertical column have similar chemical properties. For example, consider the elements in the last column: He (helium), Ne (neon), Ar (argon), Kr (krypton), Xe (xenon), and Rn (radon). The outstanding characteristic of these elements is that they don't normally take part in chemical reactions—joining with other atoms to form molecules—and are therefore classified as inert. They are called the *noble gases.* We can partially understand their behavior by looking at the electronic configurations shown in Table 28.4. The element helium has the electronic configuration $1s^2$. In other words, one shell is filled. The electrons in this filled shell are considerably separated in energy from the next available level, the 2$s$ level.

**Table 28.4** Electronic Configurations of Some Elements

| Z | Symbol | Ground-State Configuration | Ionization Energy (eV) | Z | Symbol | Ground-State Configuration | Ionization Energy (eV) |
|---|---|---|---|---|---|---|---|
| 1 | H | $1s^1$ | 13.595 | 19 | K | [Ar] $4s^1$ | 4.339 |
| 2 | He | $1s^2$ | 24.581 | 20 | Ca | $4s^2$ | 6.111 |
| | | | | 21 | Sc | $3d4s^2$ | 6.54 |
| 3 | Li | [He] $2s^1$ | 5.390 | 22 | Ti | $3d^24s^2$ | 6.83 |
| 4 | Be | $2s^2$ | 9.320 | 23 | V | $3d^34s^2$ | 6.74 |
| 5 | B | $2s^22p^1$ | 8.296 | 24 | Cr | $3d^54s^1$ | 6.76 |
| 6 | C | $2s^22p^2$ | 11.256 | 25 | Mn | $3d^54s^2$ | 7.432 |
| 7 | N | $2s^22p^3$ | 14.545 | 26 | Fe | $3d^64s^2$ | 7.87 |
| 8 | O | $2s^22p^4$ | 13.614 | 27 | Co | $3d^74s^2$ | 7.86 |
| 9 | F | $2s^22p^5$ | 17.418 | 28 | Ni | $3d^84s^2$ | 7.633 |
| 10 | Ne | $2s^22p^6$ | 21.559 | 29 | Cu | $3d^{10}4s^1$ | 7.724 |
| | | | | 30 | Zn | $3d^{10}4s^2$ | 9.391 |
| 11 | Na | [Ne] $3s^1$ | 5.138 | 31 | Ga | $3d^{10}4s^24p^1$ | 6.00 |
| 12 | Mg | $3s^2$ | 7.644 | 32 | Ge | $3d^{10}4s^24p^2$ | 7.88 |
| 13 | Al | $3s^23p^1$ | 5.984 | 33 | As | $3d^{10}4s^24p^3$ | 9.81 |
| 14 | Si | $3s^23p^2$ | 8.149 | 34 | Se | $3d^{10}4s^24p^4$ | 9.75 |
| 15 | P | $3s^23p^3$ | 10.484 | 35 | Br | $3d^{10}4s^24p^5$ | 11.84 |
| 16 | S | $3s^23p^4$ | 10.357 | 36 | Kr | $3d^{10}4s^24p^6$ | 13.996 |
| 17 | Cl | $3s^23p^5$ | 13.01 | | | | |
| 18 | Ar | $3s^23p^6$ | 15.755 | | | | |

*Note:* The bracket notation is used as a shorthand method to avoid repetition in indicating inner-shell electrons. Thus, [He] represents $1s^2$, [Ne] represents $1s^22s^22p^6$, [Ar] represents $1s^22s^22p^63s^23p^6$, and so on.

The electronic configuration for neon is $1s^22s^22p^6$. Again, the outer shell is filled and there is a large difference in energy between the $2p$ level and the $3s$ level. Argon has the configuration $1s^22s^22p^63s^23p^6$. Here, the $3p$ subshell is filled and there is a wide gap in energy between the $3p$ subshell and the $3d$ subshell. Through all the noble gases, the pattern remains the same: a noble gas is formed when either a shell or a subshell is filled, and there is a large gap in energy before the next possible level is encountered.

The elements in the first column of the periodic table are called the *alkali metals* and are highly active chemically. Referring to Table 28.4, we can understand why these elements interact so strongly with other elements. These alkali metals all have a single outer electron in an $s$ subshell. This electron is shielded from the nucleus by all the electrons in the inner shells. Consequently, it's only loosely bound to the atom and can readily be accepted by other atoms that bind it more tightly to form molecules.

The elements in the seventh column of the periodic table are called the *halogens* and are also highly active chemically. All these elements are lacking one electron in a subshell, so they readily accept electrons from other atoms to form molecules.

**Quick Quiz**

**28.3** Krypton (atomic number 36) has how many electrons in its next-to-outer shell ($n = 3$)? (a) 2 (b) 4 (c) 8 (d) 18

## 28.6 Characteristic X-Rays

**LEARNING OBJECTIVES**

1. Describe the physical origins of characteristic x-rays.
2. Determine the energies and wavelengths of characteristic x-rays.

X-rays are emitted when a metal target is bombarded with high-energy electrons. The x-ray spectrum typically consists of a broad continuous band and a series of intense sharp lines that are dependent on the type of metal used for the target, as shown in Figure 28.14. These discrete lines, called **characteristic x-rays**, were discovered in 1908, but their origin remained unexplained until the details of atomic structure were developed.

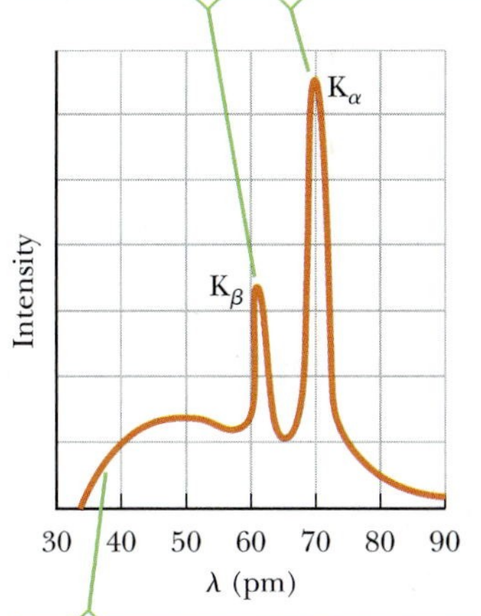

The peaks represent *characteristic x-rays*. Their appearance depends on the target material.
$K_\alpha$
$K_\beta$
Intensity
30 40 50 60 70 80 90
λ (pm)
The continuous curve represents *bremsstrahlung*. The shortest wavelength depends on the accelerating voltage.

**Figure 28.14** The x-ray spectrum of a metal target. The data shown were obtained when 35-keV electrons bombarded a molybdenum target. Note that 1 pm = $10^{-12}$ m = 0.001 nm.

The first step in the production of characteristic x-rays occurs when a bombarding electron collides with an electron in an inner shell of a target atom with sufficient energy to remove the electron from the atom. The vacancy created in the shell is filled when an electron in a higher level drops down into the lower-energy level containing the vacancy. The time it takes for that to happen is very short, less than $10^{-9}$ s. The transition is accompanied by the emission of a photon with energy equaling the difference in energy between the two levels. Typically, the energy of such transitions is greater than 1 000 eV, and the emitted x-ray photons have wavelengths in the range of 0.01 nm to 1 nm.

We assume the incoming electron has dislodged an atomic electron from the innermost shell, the K shell. If the vacancy is filled by an electron dropping from the next-higher shell, the L shell, the photon emitted in the process is referred to as the $K_\alpha$ line on the curve of Figure 28.14. If the vacancy is filled by an electron dropping from the M shell, the line produced is called the $K_\beta$ line.

Other characteristic x-ray lines are formed when electrons drop from upper levels to vacancies other than those in the K shell. For example, L lines are produced when vacancies in the L shell are filled by electrons dropping from higher shells. An $L_\alpha$ line is produced as an electron drops from the M shell to the L shell, and an $L_\beta$ line is produced by a transition from the N shell to the L shell.

We can estimate the energy of the emitted x-rays as follows. Consider two electrons in the K shell of an atom whose atomic number is $Z$. Each electron partially

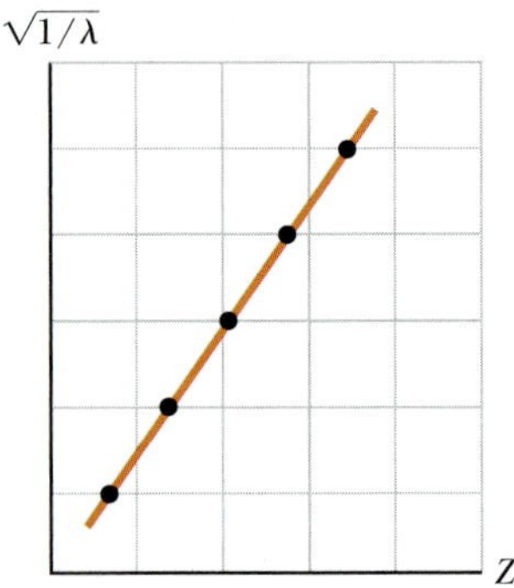


**Figure 28.15** A Moseley plot of $\sqrt{1/\lambda}$ versus $Z$, where $\lambda$ is the wavelength of the $K_\alpha$ x-ray line of the element of atomic number $Z$.

shields the other from the charge of the nucleus, $Ze$, so each is subject to an effective nuclear charge $Z_{\text{eff}} = (Z - 1)e$. We can now use a modified form of Equation 28.18 to estimate the energy of either electron in the K shell (with $n = 1$). We have

$$E_K = -m_e Z_{\text{eff}}^2 \frac{k_e^2 e^4}{2\hbar^2} = -Z_{\text{eff}}^2 E_0$$

where $E_0$ is the ground-state energy. Substituting $Z_{\text{eff}} = Z - 1$ gives

$$E_K = -(Z - 1)^2(13.6 \text{ eV}) \qquad [28.20]$$

As Example 28.4 shows, we can estimate the energy of an electron in an L or an M shell in a similar fashion. Taking the energy difference between these two levels, we can then calculate the energy and wavelength of the emitted photon.

In 1914 Henry G. J. Moseley plotted the $Z$ values for a number of elements against $\sqrt{1/\lambda}$, where $\lambda$ is the wavelength of the $K_\alpha$ line for each element. He found that such a plot produced a straight line, as in Figure 28.15, which is consistent with our rough calculations of the energy levels based on Equation 28.20. From his plot, Moseley was able to determine the $Z$ values of other elements, providing a periodic chart in excellent agreement with the known chemical properties of the elements.

## EXAMPLE 28.4 Characteristic X-Rays

**GOAL** Calculate the energy and wavelength of characteristic x-rays.

**PROBLEM** Estimate the energy and wavelength of the characteristic x-ray emitted from a tungsten target when an electron drops from an M shell ($n = 3$ state) to a vacancy in the K shell ($n = 1$ state).

**STRATEGY** Develop two estimates, one for the electron in the K shell ($n = 1$) and one for the electron in the M shell ($n = 3$). For the K-shell estimate, we can use Equation 28.20. For the M-shell estimate, we need a new equation. There is 1 electron in the K shell (because one is missing) and there are 8 in the L shell, making 9 electrons shielding the nuclear charge. Therefore $Z_{\text{eff}} = 74 - 9$ and $E_M = -Z_{\text{eff}}^2 E_3$, where $E_3$ is the energy of the $n = 3$ level in hydrogen. The difference $E_M - E_K$ is the energy of the photon.

**SOLUTION**

Use Equation 28.20 to estimate the energy of an electron in the K shell of tungsten, atomic number $Z = 74$:

$$E_K = -(74 - 1)^2(13.6 \text{ eV}) = -72\,500 \text{ eV}$$

Estimate the energy of an electron in the M shell in the same way:

$$E_M = -Z_{\text{eff}}^2 E_3 = -(Z - 9)^2 \frac{E_0}{3^2} = -(74 - 9)^2 \frac{(13.6 \text{ eV})}{9}$$

$$= -6\,380 \text{ eV}$$

Calculate the difference in energy between the M and K shells:

$$E_M - E_K = -6\,380 \text{ eV} - (-72\,500 \text{ eV}) = \boxed{66\,100 \text{ eV}}$$

Find the wavelength of the emitted x-ray:

$$\Delta E = hf = h\frac{c}{\lambda} \quad \rightarrow \quad \lambda = \frac{hc}{\Delta E}$$

$$\lambda = \frac{(6.63 \times 10^{-34} \text{ J} \cdot \text{s})(3.00 \times 10^8 \text{ m/s})}{(6.61 \times 10^4 \text{ eV})(1.60 \times 10^{-19} \text{ J/eV})}$$

$$= 1.88 \times 10^{-11} \text{ m} = \boxed{0.018\,8 \text{ nm}}$$

**REMARKS** These estimates depend on the amount of shielding of the nuclear charge, which can be difficult to determine.

**QUESTION 28.4** Could a transition from the L shell to the K shell ever result in a more energetic photon than a transition from the M to the K shell? Discuss.

**EXERCISE 28.4** Repeat the problem for a $2p$ electron transiting from the L shell to the K shell. (For technical reasons, the L shell electron must have $\ell = 1$, so a single $1s$ electron and two $2s$ electrons shield the nucleus.)

**ANSWERS** (a) $5.54 \times 10^4$ eV (b) 0.022 4 nm

# 28.7 Atomic Transitions and Lasers

**LEARNING OBJECTIVES**

1. Define the stimulated absorption of photons by atoms.
2. Define what is meant by excited states of atomic electrons.
3. Define the spontaneous emission of photons, and contrast it with the stimulated emission of photons.
4. Define population inversion and describe how a laser works on the atomic level.

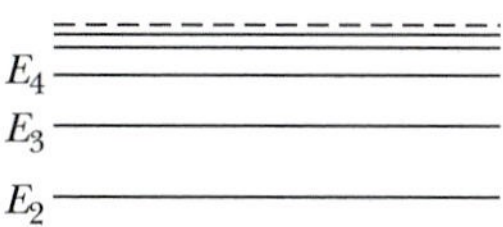

$E_4$
$E_3$
$E_2$
$E_1$

**Figure 28.16** Energy level diagram of an atom with various allowed states. The lowest-energy state, $E_1$, is the ground state. All others are excited states.

We have seen that an atom will emit radiation only at certain frequencies that correspond to the energy separation between the various allowed states. Consider an atom with many allowed energy states, labeled $E_1$, $E_2$, $E_3$, . . . , as in Figure 28.16. When light is incident on the atom, only those photons with energy $hf$ matching the energy separation $\Delta E$ between two levels can be absorbed. A schematic diagram representing this **stimulated absorption process** is shown in Figure 28.17. At ordinary temperatures, most of the atoms in a sample are in the ground state. If a vessel containing many atoms of a gas is illuminated with a light beam containing all possible photon frequencies (that is, a continuous spectrum), only those photons of energies $E_2 - E_1$, $E_3 - E_1$, $E_4 - E_1$, and so on can be absorbed. As a result of this absorption, some atoms are raised to various allowed higher-energy levels, called **excited states**.

Once an atom is in an excited state, there is a constant probability that it will jump back to a lower level by emitting a photon, as shown in Figure 28.18. This process is known as **spontaneous emission**. Typically, an atom will remain in an excited state for only about $10^{-8}$ s.

A third process that is important in lasers, **stimulated emission**, was predicted by Einstein in 1917. Suppose an atom is in the excited state $E_2$, as in Figure 28.19 (page 964), and a photon with energy $hf = E_2 - E_1$ is incident on it. The incoming photon increases the probability that the excited atom will return to the ground state and thereby emit a second photon having the same energy $hf$. Note that two identical photons result from stimulated emission: the incident photon and the emitted photon. *The emitted photon is exactly in phase with the incident photon.* These photons can stimulate other atoms to emit photons in a chain of similar processes.

The intense, coherent (in-phase) light in a laser (*l*ight *a*mplification by *s*timulated *e*mission of *r*adiation) is a result of stimulated emission. In a laser, voltages can be used to put more electrons in excited states than in the ground state. This process is called **population inversion**. The excited state of the system must be a *metastable state,* which means that its lifetime must be relatively long. When that

Philippe Plailly/SPL/Science Source

Scientist checking the performance of an experimental laser-cutting device mounted on a robot arm. The laser is being used to cut through a metal plate.

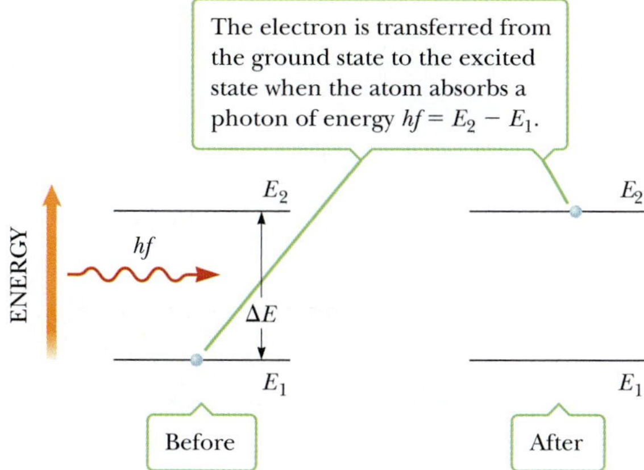

The electron is transferred from the ground state to the excited state when the atom absorbs a photon of energy $hf = E_2 - E_1$.
ENERGY
$hf$
$E_2$
$\Delta E$
$E_1$
$E_2$
$E_1$
Before
After

**Figure 28.17** Diagram representing the process of *stimulated absorption* of a photon by an atom.

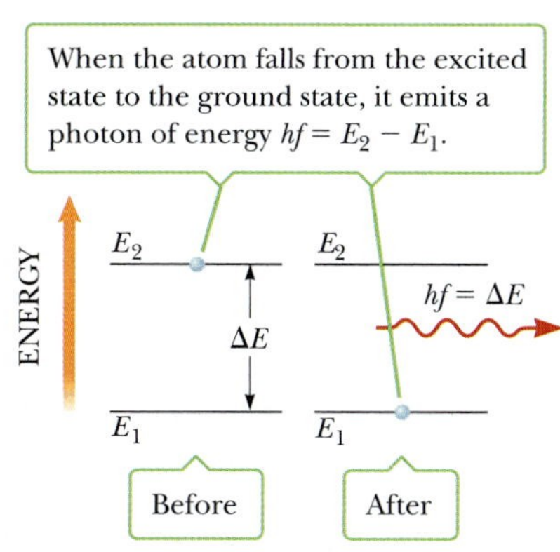

When the atom falls from the excited state to the ground state, it emits a photon of energy $hf = E_2 - E_1$.
ENERGY
$E_2$
$\Delta E$
$E_1$
$E_2$
$hf = \Delta E$
$E_1$
Before
After

**Figure 28.18** Diagram representing the process of *spontaneous emission* of a photon by an atom that is initially in the excited state $E_2$.

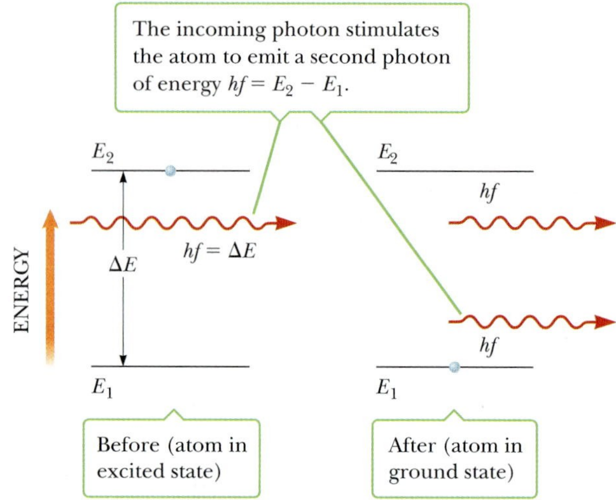

The incoming photon stimulates the atom to emit a second photon of energy $hf = E_2 - E_1$.
$E_2$
$hf = \Delta E$
$\Delta E$
ENERGY
$E_1$
$hf$
$hf$
$E_2$
$E_1$
Before (atom in excited state)
After (atom in ground state)

**Figure 28.19** Diagram representing the process of *stimulated emission* of a photon by an incoming photon of energy $hf$. Initially, the atom is in the excited state.

**Figure 28.20** Energy level diagram for the neon atom in a helium–neon laser.

is the case, stimulated emission will occur before spontaneous emission. Finally, the photons produced must be retained in the system for a while so that they can stimulate the production of still more photons. This step can be done with mirrors, one of which is partly transparent.

Figure 28.20 is an energy level diagram for the neon atom in a helium–neon gas laser. The mixture of helium and neon is confined to a glass tube sealed at the ends by mirrors. A high voltage applied to the tube causes electrons to sweep through it, colliding with the atoms of the gas and raising them into excited states. Neon atoms are excited to state $E_3^*$ through this process and also as a result of collisions with excited helium atoms. When a neon atom makes a transition to state $E_2$, it stimulates emission by neighboring excited atoms. The result is the production of coherent light at a wavelength of 632.8 nm. Figure 28.21 summarizes the steps in the production of a laser beam.

**APPLICATION**
Laser Technology

Lasers that cover wavelengths in the infrared, visible, and ultraviolet regions of the spectrum are now available. Applications include the surgical "welding" of detached retinas, "lasik" surgery, precision surveying and length measurement, a potential source for inducing nuclear fusion reactions, precision cutting of metals and other materials, and telephone communication along optical fibers.

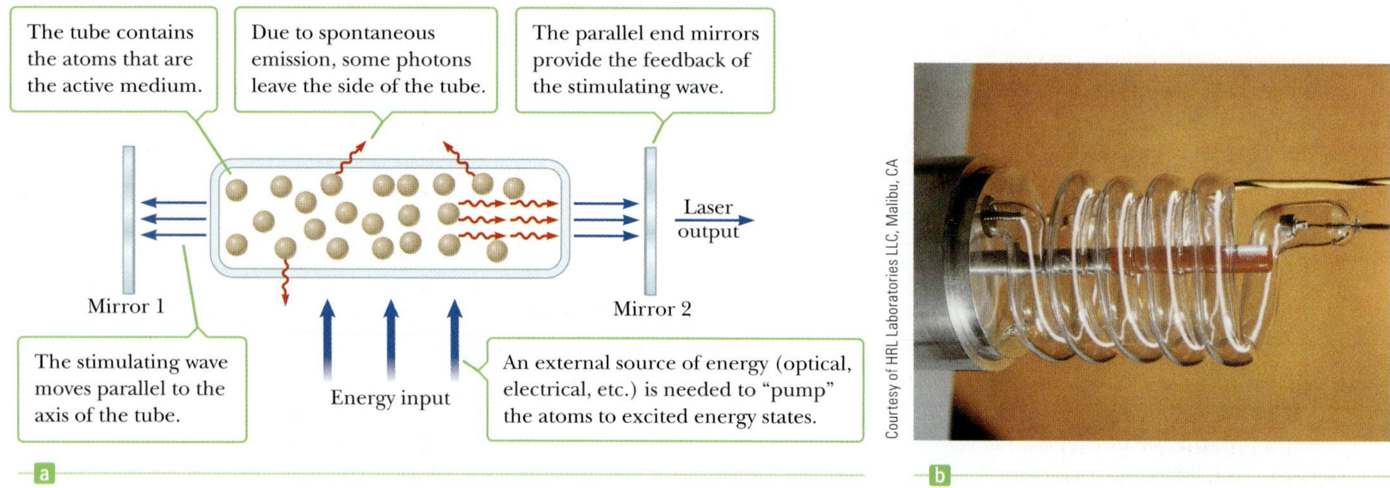

The tube contains the atoms that are the active medium.
Due to spontaneous emission, some photons leave the side of the tube.
The parallel end mirrors provide the feedback of the stimulating wave.
Laser output
Mirror 1
Mirror 2
The stimulating wave moves parallel to the axis of the tube.
Energy input
An external source of energy (optical, electrical, etc.) is needed to "pump" the atoms to excited energy states.
Courtesy of HRL Laboratories LLC, Malibu, CA
a
b

**Figure 28.21** (a) Steps in the production of a laser beam. (b) Photograph of the first ruby laser, showing the flash lamp surrounding the ruby rod.

# SUMMARY

## 28.3 The Bohr Model

The **Bohr model** of the atom is successful in describing the spectra of atomic hydrogen and hydrogen-like ions. One basic assumption of the model is that the electron can exist only in certain orbits such that its angular momentum $mvr$ is an integral multiple of $\hbar$, where $\hbar$ is Planck's constant divided by $2\pi$. Assuming circular orbits and a Coulomb force of attraction between electron and proton, the energies of the quantum states for hydrogen are

$$E_n = -\frac{m_e k_e^2 e^4}{2\hbar^2}\left(\frac{1}{n^2}\right) \qquad n = 1, 2, 3, \ldots \qquad [28.13]$$

where $k_e$ is the Coulomb constant, $e$ is the charge on the electron, and $n$ is an integer called a **quantum number**.

If the electron in the hydrogen atom jumps from an orbit having quantum number $n_i$ to an orbit having quantum number $n_f$, it emits a photon of frequency $f$, given by

$$f = \frac{E_i - E_f}{h} = \frac{m_e k_e^2 e^4}{4\pi\hbar^3}\left(\frac{1}{n_f^2} - \frac{1}{n_i^2}\right) \qquad [28.15]$$

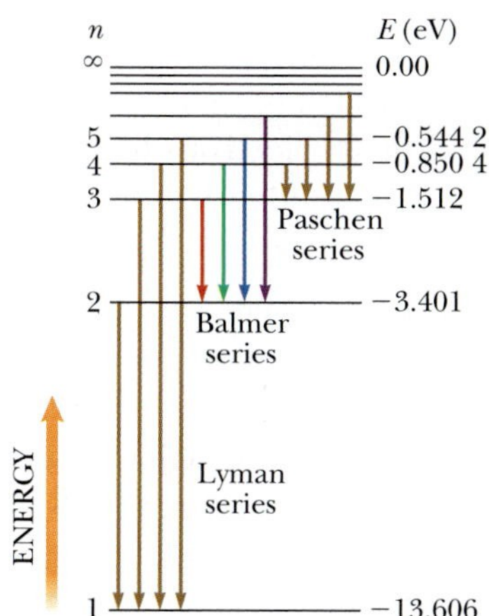


Energy level diagram for hydrogen. Vertical arrows represent the lowest-energy transitions for each of the spectral series shown.

Bohr's **correspondence principle** states that quantum mechanics is in agreement with classical physics when the quantum numbers for a system are very large.

The Bohr theory can be generalized to hydrogen-like atoms, such as singly ionized helium or doubly ionized lithium. This modification consists of replacing $e^2$ by $Ze^2$ wherever it occurs.

## 28.4 Quantum Mechanics and the Hydrogen Atom

One of the many successes of quantum mechanics is that the quantum numbers $n$, $\ell$, and $m_\ell$ associated with atomic structure arise directly from the mathematics of the theory. The quantum number $n$ is called the **principal quantum number**, $\ell$ is the **orbital quantum number**, and $m_\ell$ is the **orbital magnetic quantum number.** These quantum numbers can take only certain values: $1 \le n < \infty$ in integer steps, $0 \le \ell \le n - 1$, and $-\ell \le m_\ell \le \ell$. In addition, a fourth quantum number, called the **spin magnetic quantum number** $m_s$, is needed to explain a fine doubling of lines in atomic spectra, with $m_s = \pm\frac{1}{2}$.

## 28.5 The Exclusion Principle and the Periodic Table

An understanding of the periodic table of the elements became possible when Pauli formulated the **exclusion principle**, which states that no two electrons in the same atom can have the same values for the set of quantum numbers $n$, $\ell$, $m_\ell$, and $m_s$. A particular set of these quantum numbers is called a quantum state. The exclusion principle explains how different energy levels in atoms are populated. Once one subshell is filled, the next electron goes into the vacant subshell that is lowest in energy. Atoms with similar configurations in their outermost shell have similar chemical properties and are found in the same column of the periodic table.

## 28.6 Characteristic X-Rays

**Characteristic x-rays** are produced when a bombarding electron collides with an electron in an inner shell of an atom with sufficient energy to remove the electron from the atom. The vacancy is filled when an electron from a higher level drops down into the level containing the vacancy, emitting a photon in the x-ray part of the spectrum in the process.

## 28.7 Atomic Transitions and Lasers

When an atom is irradiated by light of all different wavelengths, it will only absorb wavelengths equal to the difference in energy of two of its energy levels. This phenomenon, called **stimulated absorption**, places an atom's electrons into **excited states**. Atoms in an excited state have a probability of returning to a lower level of excitation by **spontaneous emission**. The wavelengths that can be emitted are the same as the wavelengths that can be absorbed. If an atom is in an excited state and a photon with energy $hf = E_2 - E_1$ is incident on it, the probability of emission of a second photon of this energy is greatly enhanced. The emitted photon is exactly in phase with the incident photon. This process is called **stimulated emission**. The emitted and original photon can then stimulate more emission, creating an amplifying effect.

**Lasers** are monochromatic, coherent light sources that work on the principle of **stimulated emission** of radiation from a system of atoms.

## WARM-UP EXERCISES

ENHANCED WebAssign The warm-up exercises in this chapter may be assigned online in Enhanced WebAssign.

1. **Physics Review** Determine the (a) energy in joules and (b) wavelength in nm of a photon with energy of 4.05 eV. (See Section 27.2.)
2. Determine the energies in eV of the (a) third and (b) fourth energy levels of the hydrogen atom. (See Section 28.3.)
3. An electron in the $n = 5$ energy level of hydrogen undergoes a transition to the $n = 3$ energy level. Determine (a) the energy in eV, (b) the energy in joules, and (c) the frequency of the emitted photon. (See Section 28.3.)
4. The so-called Lyman-$\alpha$ photon is the lowest energy photon in the Lyman series of hydrogen and results from an electron transitioning from the $n = 2$ to the $n = 1$ energy level. Determine (a) the energy in eV, (b) in joules, and (c) the wavelength in nm of the Lyman-$\alpha$ line. (See Section 28.3.)
5. Determine the orbital radius in nm of an electron in hydrogen's (a) $n = 2$, and (c) $n = 4$ energy levels. (See Section 28.3.)
6. Singly-ionized helium ($He^+$) is a hydrogen-like atom. Determine the energy in eV required to raise a $He^+$ electron from the $n = 1$ to the $n = 2$ energy level. (See Section 28.3.)
7. Hydrogen's single electron can occupy any of the atom's distinct quantum states. Determine the number of distinct quantum states in the (a) $n = 1$, (b) $n = 2$, and (c) $n = 3$ energy levels. (See Section 28.4.)
8. For an electron in a $3d$ state, determine (a) the principle quantum number and (b) the orbital quantum number. (c) How many different magnetic quantum numbers are possible for electrons in that state? (d) What total number of electrons could occupy that state? (See Section 28.4.)
9. (a) Identify the number of electrons in the ground-state outer shell of atomic oxygen (atomic number 8). (b) How many electrons are in the ground-state outer shell of aluminum? (See Section 28.5 and Table 28.4.)
10. Estimate the energy in eV of a K shell electron in lead ($Z = 82$) (See Section 28.7.)

## CONCEPTUAL QUESTIONS

ENHANCED WebAssign The conceptual questions in this chapter may be assigned online in Enhanced WebAssign.

1. In the hydrogen atom, the quantum number $n$ can increase without limit. Because of this fact, does the frequency of possible spectral lines from hydrogen also increase without limit?
2. Does the light emitted by a neon sign constitute a continuous spectrum or only a few colors? Defend your answer.
3. In an x-ray tube, if the energy with which the electrons strike the metal target is increased, the wavelengths of the characteristic x-rays do not change. Why not?
4. An energy of about 21 eV is required to excite an electron in a helium atom from the $1s$ state to the $2s$ state. The same transition for the $He^+$ ion requires approximately twice as much energy. Explain.
5. Is it possible for a spectrum from an x-ray tube to show the continuous spectrum of x-rays without the presence of the characteristic x-rays?
6. Suppose the electron in the hydrogen atom obeyed classical mechanics rather than quantum mechanics. Why should such a hypothetical atom emit a continuous spectrum rather than the observed line spectrum?
7. When a hologram is produced, why must the system (including light source, object, beam splitter, and so on) be held motionless within a quarter of the light's wavelength?
8. Why are three quantum numbers needed to describe the state of a one-electron atom (ignoring spin)?
9. Describe how the structure of atoms would differ if the Pauli exclusion principle were not valid. What consequences would follow, both at the atomic level and in the world at large?
10. Can the electron in the ground state of hydrogen absorb a photon of energy less than 13.6 eV? Can it absorb a photon of energy greater than 13.6 eV? Explain.
11. Why do lithium, potassium, and sodium exhibit similar chemical properties?
12. List some ways in which quantum mechanics altered our view of the atom pictured by the Bohr theory.
13. It is easy to understand how two electrons (one with spin up, one with spin down) can fill the $1s$ shell for a helium atom. How is it possible that eight more electrons can fit into the $2s$, $2p$ level to complete the $1s2s^22p^6$ shell for a neon atom?
14. The ionization energies for Li, Na, K, Rb, and Cs are 5.390, 5.138, 4.339, 4.176, and 3.893 eV, respectively. Explain why these values are to be expected in terms of the atomic structures.
15. Why is stimulated emission so important in the operation of a laser?

## PROBLEMS

**WebAssign** The problems in this chapter may be assigned online in Enhanced WebAssign.

**1.** denotes straightforward problem; **2.** denotes intermediate problem;
**3.** denotes challenging problem
**1.** denotes full solution available in *Student Solutions Manual/ Study Guide*
**1.** denotes problems most often assigned in Enhanced WebAssign

BIO denotes biomedical problems
GP denotes guided problems
M denotes Master It tutorial available in Enhanced WebAssign
Q|C denotes asking for quantitative and conceptual reasoning
S denotes symbolic reasoning problem
W denotes Watch It video solution available in Enhanced WebAssign

### 28.1 Early Models of the Atom

### 28.2 Atomic Spectra

**1.** Q|C The wavelengths of the Lyman series for hydrogen are given by

$$\frac{1}{\lambda} = R_{\mathrm{H}}\left(1 - \frac{1}{n^2}\right) \qquad n = 2, 3, 4, \ldots$$

(a) Calculate the wavelengths of the first three lines in this series. (b) Identify the region of the electromagnetic spectrum in which these lines appear.

**2.** Q|C The wavelengths of the Paschen series for hydrogen are given by

$$\frac{1}{\lambda} = R_{\mathrm{H}}\left(\frac{1}{3^2} - \frac{1}{n^2}\right) \qquad n = 4, 5, 6, \ldots$$

(a) Calculate the wavelengths of the first three lines in this series. (b) Identify the region of the electromagnetic spectrum in which these lines appear.

**3.** The "size" of the *atom* in Rutherford's model is about $1.0 \times 10^{-10}$ m. (a) Determine the attractive electrostatic force between an electron and a proton separated by this distance. (b) Determine (in eV) the electrostatic potential energy of the atom.

**4.** An isolated atom of a certain element emits light of wavelength 520 nm when the atom falls from its fifth excited state into its second excited state. The atom emits a photon of wavelength 410 nm when it drops from its sixth excited state into its second excited state. Find the wavelength of the light radiated when the atom makes a transition from its sixth to its fifth excited state.

**5.** Q|C The "size" of the atom in Rutherford's model is about $1.0 \times 10^{-10}$ m. (a) Determine the speed of an electron moving about the proton using the attractive electrostatic force between an electron and a proton separated by this distance. (b) Does this speed suggest that Einsteinian relativity must be considered in studying the atom? (c) Compute the de Broglie wavelength of the electron as it moves about the proton. (d) Does this wavelength suggest that wave effects, such as diffraction and interference, must be considered in studying the atom?

**6.** W In a Rutherford scattering experiment, an $\alpha$-particle (charge $= +2e$) heads directly toward a gold nucleus (charge $= +79e$). The $\alpha$-particle had a kinetic energy of 5.0 MeV when very far ($r \to \infty$) from the nucleus. Assuming the gold nucleus to be fixed in space, determine the distance of closest approach. *Hint:* Use conservation of energy with $PE = k_e q_1 q_2 / r$.

### 28.3 The Bohr Model

**7.** M A hydrogen atom is in its first excited state ($n = 2$). Using the Bohr theory of the atom, calculate (a) the radius of the orbit, (b) the linear momentum of the electron, (c) the angular momentum of the electron, (d) the kinetic energy, (e) the potential energy, and (f) the total energy.

**8.** W For a hydrogen atom in its ground state, use the Bohr model to compute (a) the orbital speed of the electron, (b) the kinetic energy of the electron, and (c) the electrical potential energy of the atom.

**9.** S Show that the speed of the electron in the $n$th Bohr orbit in hydrogen is given by

$$v_n = \frac{k_e e^2}{n\hbar}$$

**10.** A photon is emitted when a hydrogen atom undergoes a transition from the $n = 5$ state to the $n = 3$ state. Calculate (a) the wavelength, (b) the frequency, and (c) the energy (in electron volts) of the emitted photon.

**11.** A hydrogen atom emits a photon of wavelength 656 nm. From what energy orbit to what lower-energy orbit did the electron jump?

**12.** Following are four possible transitions for a hydrogen atom

I. $n_i = 2;\ n_f = 5$ II. $n_i = 5;\ n_f = 3$
III. $n_i = 7;\ n_f = 4$ IV. $n_i = 4;\ n_f = 7$

(a) Which transition will emit the shortest-wavelength photon? (b) For which transition will the atom gain the most energy? (c) For which transition(s) does the atom lose energy?

**13.** M What is the energy of a photon that, when absorbed by a hydrogen atom, could cause an electronic transition from (a) the $n = 2$ state to the $n = 5$ state and (b) the $n = 4$ state to the $n = 6$ state?

**14.** W A hydrogen atom initially in its ground state ($n = 1$) absorbs a photon and ends up in the state for which

$n = 3$. (a) What is the energy of the absorbed photon? (b) If the atom eventually returns to the ground state, what photon energies could the atom emit?

15. The Balmer series for the hydrogen atom corresponds to electronic transitions that terminate in the state with quantum number $n = 2$ as shown in Figure P28.15. Consider the photon of longest wavelength corresponding to a transition shown in the figure. Determine (a) its energy and (b) its wavelength. Consider the spectral line of shortest wavelength corresponding to a transition shown in the figure. Find (c) its photon energy and (d) its wavelength. (e) What is the shortest possible wavelength in the Balmer series?

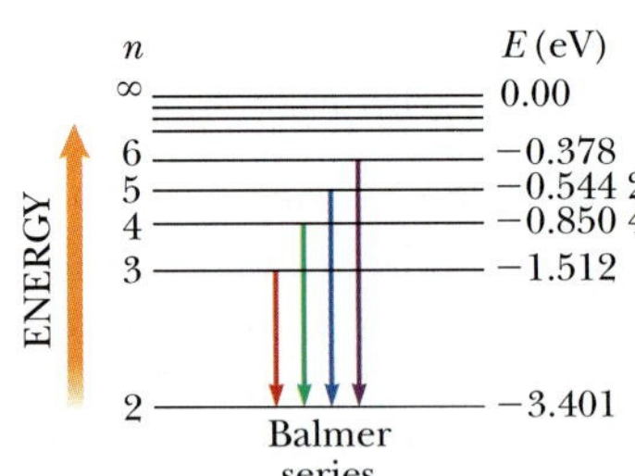


**Figure P28.15**

16. S A particle of charge $q$ and mass $m$, moving with a constant speed $v$, perpendicular to a constant magnetic field $B$, follows a circular path. If in this case the angular momentum about the center of this circle is quantized so that $mvr = 2n\hbar$, show that the allowed radii for the particle are

$$r_n = \sqrt{\frac{2n\hbar}{qB}}$$

where $n = 1, 2, 3, \ldots$.

17. M (a) If an electron makes a transition from the $n = 4$ Bohr orbit to the $n = 2$ orbit, determine the wavelength of the photon created in the process. (b) Assuming that the atom was initially at rest, determine the recoil speed of the hydrogen atom when this photon is emitted.

18. Consider a large number of hydrogen atoms, with electrons all initially in the $n = 4$ state. (a) How many different wavelengths would be observed in the emission spectrum of these atoms? (b) What is the longest wavelength that could be observed? (c) To which series does the wavelength found in (b) belong?

19. A photon with energy 2.28 eV is absorbed by a hydrogen atom. Find (a) the minimum $n$ for a hydrogen atom that can be ionized by such a photon and (b) the speed of the electron released from the state in part (a) when it is far from the nucleus.

20. (a) Calculate the angular momentum of the Moon due to its orbital motion about Earth. In your calculation use $3.84 \times 10^8$ m as the average Earth–Moon distance and $2.36 \times 10^6$ s as the period of the Moon in its orbit. (b) If the angular momentum of the Moon obeys Bohr's quantization rule ($L = n\hbar$), determine the value of the quantum number $n$. (c) By what fraction would the Earth–Moon radius have to be increased to increase the quantum number by 1?

21. An electron is in the second excited orbit of hydrogen, corresponding to $n = 3$. Find (a) the radius of the orbit and (b) the wavelength of the electron in this orbit.

22. S (a) Write an expression relating the kinetic energy $KE$ of the electron and the potential energy $PE$ in the Bohr model of the hydrogen atom. (b) Suppose a hydrogen atom absorbs a photon of energy $E$, resulting in the transfer of the electron to a higher-energy level. Express the resulting change in the potential energy of the system in terms of $E$. (c) What is the change in the electron's kinetic energy during this process?

23. The orbital radii of a hydrogen-like atom is given by the equation

$$r_n = \frac{n^2\hbar^2}{Zm_ek_ee^2}$$

What is the radius of the first Bohr orbit in (a) $He^+$, (b) $Li^{2+}$, and (c) $Be^{3+}$?

24. GP Consider a Bohr model of doubly ionized lithium. (a) Write an expression similar to Equation 28.14 for the energy levels of the sole remaining electron. (b) Find the energy corresponding to $n = 4$. (c) Find the energy corresponding to $n = 2$. (d) Calculate the energy of the photon emitted when the electron transits from the fourth energy level to the second energy level. Express the answer both in electron volts and in joules. (e) Find the frequency and wavelength of the emitted photon. (f) In what part of the spectrum is the emitted light?

25. A general expression for the energy levels of one-electron atoms and ions is

$$E_n = -\frac{\mu k_e^2 q_1^2 q_2^2}{2\hbar^2 n^2}$$

Here $\mu$ is the reduced mass of the atom, given by $\mu = m_1m_2/(m_1 + m_2)$, where $m_1$ is the mass of the electron and $m_2$ is the mass of the nucleus; $k_e$ is the Coulomb constant; and $q_1$ and $q_2$ are the charges of the electron and the nucleus, respectively. The wavelength for the $n = 3$ to $n = 2$ transition of the hydrogen atom is 656.3 nm (visible red light). What are the wavelengths for this same transition in (a) positronium, which consists of an electron and a positron, and (b) singly ionized helium? *Note:* A positron is a positively charged electron.

26. S Using the concept of standing waves, de Broglie was able to derive Bohr's stationary orbit postulate. He assumed a confined electron could exist only in states where its de Broglie waves form standing-wave patterns, as in Figure 28.6. Consider a particle confined in a box of length $L$ to be equivalent to a string of length $L$ and

fixed at both ends. Apply de Broglie's concept to show that (a) the linear momentum of this particle is quantized with $p = mv = nh/2L$ and (b) the allowed states correspond to particle energies of $E_n = n^2E_0$, where $E_0 = h^2/(8mL^2)$.

## 28.4 Quantum Mechanics and the Hydrogen Atom

**27.** List the possible sets of quantum numbers for electrons in the 3*d* subshell.

**28.** When the principal quantum number is $n = 4$, how many different values of (a) $\ell$ and (b) $m_\ell$ are possible?

**29.** The $\rho$-meson has a charge of $-e$, a spin quantum number of 1, and a mass 1 507 times that of the electron. If the electrons in atoms were replaced by $\rho$-mesons, list the possible sets of quantum numbers for $\rho$-mesons in the 3*d* subshell.

## 28.5 The Exclusion Principle and the Periodic Table

**30.** W (a) Write out the electronic configuration of the ground state for nitrogen ($Z = 7$). (b) Write out the values for the possible set of quantum numbers $n$, $\ell$, $m_\ell$, and $m_s$ for the electrons in nitrogen.

**31.** A certain element has its outermost electron in a 3*p* subshell. It has valence +3 because it has three more electrons than a certain noble gas. What element is it?

**32.** Two electrons in the same atom have $n = 3$ and $\ell = 1$. (a) List the quantum numbers for the possible states of the atom. (b) How many states would be possible if the exclusion principle did not apply to the atom?

**33.** Zirconium ($Z = 40$) has two electrons in an incomplete *d* subshell. (a) What are the values of $n$ and $\ell$ for each electron? (b) What are all possible values of $m_\ell$ and $m_s$? (c) What is the electron configuration in the ground state of zirconium?

## 28.6 Characteristic X-Rays

**34.** A tungsten target is struck by electrons that have been accelerated from rest through a 40.0-kV potential difference. Find the shortest wavelength of the radiation emitted.

**35.** A bismuth target is struck by electrons, and x-rays are emitted. Estimate (a) the M- to L-shell transitional energy for bismuth and (b) the wavelength of the x-ray emitted when an electron falls from the M shell to the L shell.

**36.** When an electron drops from the M shell ($n = 3$) to a vacancy in the K shell ($n = 1$), the measured wavelength of the emitted x-ray is found to be 0.101 nm. Identify the element.

**37.** M The K series of the discrete spectrum of tungsten contains wavelengths of 0.018 5 nm, 0.020 9 nm, and 0.021 5 nm. The K-shell ionization energy is 69.5 keV. Determine the ionization energies of the L, M, and N shells.

## Additional Problems

**38.** In a hydrogen atom, what is the principal quantum number of the electron orbit with a radius closest to 1.0 $\mu$m?

**39.** (a) How much energy is required to cause an electron in hydrogen to move from the $n = 1$ state to the $n = 2$ state? (b) If the electrons gain this energy by collision between hydrogen atoms in a high-temperature gas, find the minimum temperature of the heated hydrogen gas. The thermal energy of the heated atoms is given by $3k_BT/2$, where $k_B$ is the Boltzmann constant.

**40.** A pulsed ruby laser emits light at 694.3 nm. For a 14.0-ps pulse containing 3.00 J of energy, find (a) the physical length of the pulse as it travels through space and (b) the number of photons in it. (c) If the beam has a circular cross section 0.600 cm in diameter, what is the number of photons per cubic millimeter?

**41.** An electron in chromium moves from the $n = 2$ state to the $n = 1$ state without emitting a photon. Instead, the excess energy is transferred to an outer electron (one in the $n = 4$ state), which is then ejected by the atom. In this Auger (pronounced "ohjay") process, the ejected electron is referred to as an Auger electron. (a) Find the change in energy associated with the transition from $n = 2$ into the vacant $n = 1$ state using Bohr theory. Assume only one electron in the K shell is shielding part of the nuclear charge. (b) Find the energy needed to ionize an $n = 4$ electron, assuming 22 electrons shield the nucleus. (c) Find the kinetic energy of the ejected (Auger) electron. (All answers should be in electron volts.)

**42.** QIC S As the Earth moves around the Sun, its orbits are quantized. (a) Follow the steps of Bohr's analysis of the hydrogen atom to show that the allowed radii of the Earth's orbit are given by

$$r_n = \frac{n^2\hbar^2}{GM_S M_E^2}$$

where $n$ is an integer quantum number, $M_S$ is the mass of the Sun, and $M_E$ is the mass of the Earth. (b) Calculate the numerical value of $n$ for the Sun–Earth system. (c) Find the distance between the orbit for quantum number $n$ and the next orbit out from the Sun corresponding to the quantum number $n + 1$. (d) Discuss the significance of your results from parts (b) and (c).

**43.** BIO M A laser used in eye surgery emits a 3.00-mJ pulse in 1.00 ns, focused to a spot 30.0 $\mu$m in diameter on the retina. (a) Find (in SI units) the power per unit area at the retina. (This quantity is called the *irradiance*.) (b) What energy is delivered per pulse to an area of molecular size (say, a circular area 0.600 nm in diameter)?

**44.** An electron has a de Broglie wavelength equal to the diameter of a hydrogen atom in its ground state. (a) What is the kinetic energy of the electron? (b) How does this energy compare with the magnitude of the ground-state energy of the hydrogen atom?

45. S Use Bohr's model of the hydrogen atom to show that when the atom makes a transition from the state $n$ to the state $n - 1$, the frequency of the emitted light is given by

$$f = \frac{2\pi^2 m k_e^2 e^4}{h^3}\left[\frac{2n - 1}{(n - 1)^2 n^2}\right]$$

46. Suppose the ionization energy of an atom is 4.100 eV. In this same atom, we observe emission lines that have wavelengths of 310.0 nm, 400.0 nm, and 1 378 nm. Use this information to construct the energy level diagram with the least number of levels. Assume that the higher energy levels are closer together.

Peter Ginter/Science Faction/Getty Images

Technicians prepare the vacuum chamber of the ASDEX Upgrade Fusion Reactor (Axially Symmetric Divertor Experiment), where plasma is heated to over sixty million degrees, leading to the fusion of deuterium and tritium and the release of energy. Such experimental tokamaks may lead to working fusion reactors.

# Nuclear Physics 29

In this chapter we discuss the properties and structure of the atomic nucleus. We start by describing the basic properties of nuclei and follow with a discussion of the phenomenon of radioactivity. Finally, we explore nuclear reactions and the various processes by which nuclei decay.



## 29.1 Some Properties of Nuclei

**LEARNING OBJECTIVES**

1. Define a nuclei's atomic, neutron, and mass numbers and describe the notation used to represent them.
2. Define the unified mass unit and the term isotope.
3. Discuss the size of nuclei and the concept of nuclear stability.

All nuclei are composed of two types of particles: protons and neutrons. The only exception is the ordinary hydrogen nucleus, which is a single proton. In describing some of the properties of nuclei, such as their charge, mass, and radius, we make use of the following quantities:

- the **atomic number** $Z$, which equals the number of protons in the nucleus
- the **neutron number** $N$, which equals the number of neutrons in the nucleus
- the **mass number** $A$, which equals the number of nucleons in the nucleus (*nucleon* is a generic term used to refer to either a proton or a neutron)

The symbol we use to represent nuclei is ${}^{A}_{Z}\text{X}$, where X represents the chemical symbol for the element. For example, ${}^{27}_{13}\text{Al}$ has the mass number 27 and the atomic number 13; therefore, it contains 13 protons and 14 neutrons. When no confusion

**Ernest Rutherford**
**New Zealand Physicist (1871–1937)**
Rutherford was awarded the 1908 Nobel Prize in Chemistry for studying radioactivity and for discovering that atoms can be broken apart by alpha rays. "On consideration, I realized that this scattering backward must be the result of a single collision, and when I made calculations I saw that it was impossible to get anything of that order of magnitude unless you took a system in which the greater part of the mass of the atom was concentrated in a minute nucleus. It was then that I had the idea of an atom with a minute massive center carrying a charge."

**Table 29.1** Masses of the Proton, Neutron, and Electron in Various Units

| | Mass | | |
|---|---|---|---|
| **Particle** | **kg** | **u** | **MeV/$c^2$** |
| Proton | $1.672\,6 \times 10^{-27}$ | 1.007 276 | 938.28 |
| Neutron | $1.675\,0 \times 10^{-27}$ | 1.008 665 | 939.57 |
| Electron | $9.109 \times 10^{-31}$ | $5.486 \times 10^{-4}$ | 0.511 |

is likely to arise, we often omit the subscript $Z$ because the chemical symbol can always be used to determine $Z$.

The nuclei of all atoms of a particular element must contain the same number of protons, but they may contain different numbers of neutrons. Nuclei that are related in this way are called **isotopes. The isotopes of an element have the same *Z* value, but different *N* and *A* values.** The natural abundances of isotopes can differ substantially. For example, ${}^{11}_{6}C$, ${}^{12}_{6}C$, ${}^{13}_{6}C$, and ${}^{14}_{6}C$ are four isotopes of carbon. The natural abundance of the ${}^{12}_{6}C$ isotope is about 98.9%, whereas that of the ${}^{13}_{6}C$ isotope is only about 1.1%. Some isotopes don't occur naturally, but can be produced in the laboratory through nuclear reactions. Even the simplest element, hydrogen, has isotopes: ${}^{1}_{1}H$, hydrogen; ${}^{2}_{1}H$, deuterium; and ${}^{3}_{1}H$, tritium.

## Charge and Mass

The proton carries a single positive charge $+e = 1.602\,177\,33 \times 10^{-19}$ C, the electron carries a single negative charge $-e$, and the neutron is electrically neutral. Because the neutron has no charge, it's difficult to detect. The proton is about 1 836 times as massive as the electron, and the masses of the proton and the neutron are almost equal (Table 29.1).

Definition of the unified mass unit, u ▶

For atomic masses, it is convenient to define the **unified mass unit, u**, in such a way that the mass of one atom of the isotope ${}^{12}C$ is exactly 12 u, where 1 u = $1.660\,559 \times 10^{-27}$ kg. The proton and neutron each have a mass of about 1 u, and the electron has a mass that is only a small fraction of an atomic mass unit.

Because the rest energy of a particle is given by $E_R = mc^2$, it is often convenient to express the particle's mass in terms of its energy equivalent. For one atomic mass unit, we have an energy equivalent of

$$E_R = mc^2 = (1.660\,559 \times 10^{-27}\text{ kg})(2.997\,92 \times 10^8\text{ m/s})^2$$
$$= 1.492\,431 \times 10^{-10}\text{ J} = 931.494\text{ MeV}$$

In calculations nuclear physicists often express *mass* in terms of the unit MeV/$c^2$, where

$$1\text{ u} = 931.494\text{ MeV}/c^2$$

## The Size of Nuclei

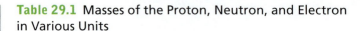

The size and structure of nuclei were first investigated in the scattering experiments of Rutherford, discussed in Section 28.1. Using the principle of conservation of energy, Rutherford found an expression for how close an alpha particle moving directly toward the nucleus can come to the nucleus before being turned around by Coulomb repulsion.

In such a head-on collision, the kinetic energy of the incoming alpha particle must be converted completely to electrical potential energy when the particle stops at the point of closest approach and turns around (Fig. 29.1). If we equate the initial kinetic energy of the alpha particle to the maximum electrical potential energy of the system (alpha particle plus target nucleus), we have

$$\tfrac{1}{2}mv^2 = k_e\frac{q_1q_2}{r} = k_e\frac{(2e)(Ze)}{d}$$

2e $\vec{v}$ $v = 0$ Ze d

An alpha particle approaches a nucleus no closer than distance $d$ because of the repulsive electric force between them.

**Figure 29.1** An alpha particle on a head-on collision course with a nucleus of charge $Ze$.

where $d$ is the distance of closest approach. Solving for $d$, we get

$$d = \frac{4k_e Ze^2}{mv^2}$$

From this expression, Rutherford found that alpha particles approached to within $3.2 \times 10^{-14}$ m of a nucleus when the foil was made of gold, implying that the radius of the gold nucleus must be less than this value. For silver atoms, the distance of closest approach was $2 \times 10^{-14}$ m. From these results, Rutherford concluded that the positive charge in an atom is concentrated in a small sphere, which he called the nucleus, with radius no greater than about $10^{-14}$ m. Because such small lengths are common in nuclear physics, a convenient unit of length is the *femtometer* (fm), sometimes called the **fermi** and defined as

$$1 \text{ fm} \equiv 10^{-15}$$

**Figure 29.2** A nucleus can be visualized as a cluster of tightly packed spheres, each of which is a nucleon.

Since the time of Rutherford's scattering experiments, a multitude of other experiments have shown that most nuclei are approximately spherical and have an average radius given by

$$r = r_0 A^{1/3} \qquad [29.1]$$

where $r_0$ is a constant equal to $1.2 \times 10^{-15}$ m and $A$ is the total number of nucleons. Because the volume of a sphere is proportional to the cube of its radius, it follows from Equation 29.1 that the volume of a nucleus (assumed to be spherical) is directly proportional to $A$, the total number of nucleons. This relationship then suggests **all nuclei have nearly the same density**. Nucleons combine to form a nucleus *as though* they were tightly packed spheres (Fig. 29.2).

**Tip 29.1 Mass Number Is Not the Atomic Mass**

Don't confuse the mass number $A$ with the atomic mass. Mass number is an integer that specifies an isotope and has no units; it's simply equal to the number of nucleons. Atomic mass is an average of the masses of the isotopes of a given element and has units of u.

## Nuclear Stability

Given that the nucleus consists of a closely packed collection of protons and neutrons, you might be surprised that it can even exist. The very large repulsive electrostatic forces between protons should cause the nucleus to fly apart. Nuclei, however, are stable because of the presence of another, short-range (about 2-fm) force: the **nuclear force**, an attractive force that acts between all nuclear particles. The protons attract each other via the nuclear force, and at the same time they repel each other through the Coulomb force. The attractive nuclear force also acts between pairs of neutrons and between neutrons and protons.

The nuclear attractive force is stronger than the Coulomb repulsive force with-in the nucleus (at short ranges). If it were not, stable nuclei would not exist. Moreover, the strong nuclear force is nearly independent of charge. In other words, the nuclear forces associated with proton–proton, proton–neutron, and neutron–neutron interactions are approximately the same, apart from the additional repulsive Coulomb force for the proton–proton interaction.

There are about 260 stable nuclei; hundreds of others have been observed, but are unstable. A plot of $N$ versus $Z$ for a number of stable nuclei is given in Figure 29.3 (page 974). Note that light nuclei are most stable if they contain equal numbers of protons and neutrons so that $N = Z$, but heavy nuclei are more stable if $N > Z$. This difference can be partially understood by recognizing that as the number of protons increases, the strength of the Coulomb force increases, which tends to break the nucleus apart. As a result, more neutrons are needed to keep the nucleus stable because neutrons are affected only by the attractive nuclear forces. In effect, the additional neutrons "dilute" the nuclear charge density. Eventually, when $Z = 83$, the repulsive forces between protons cannot be compensated for by the addition of neutrons. Elements that contain more than 83 protons don't have stable nuclei, but, rather, decay or disintegrate into other particles in various amounts of time. The masses and some other properties of selected isotopes are provided in Appendix B.

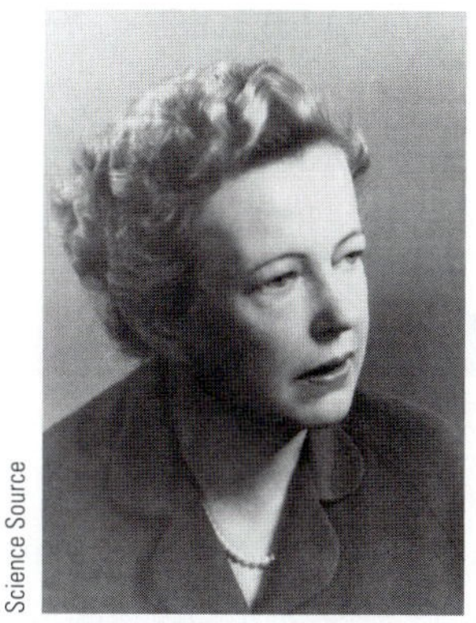

Science Source

**Maria Goeppert-Mayer**
**German Physicist (1906–1972)**
Goeppert-Mayer is best known for her development of the shell model of the nucleus, published in 1950. A similar model was simultaneously developed by Hans Jensen, a German scientist. Maria Goeppert-Mayer and Hans Jensen were awarded the Nobel Prize in Physics in 1963 for their extraordinary work in understanding the structure of the nucleus.

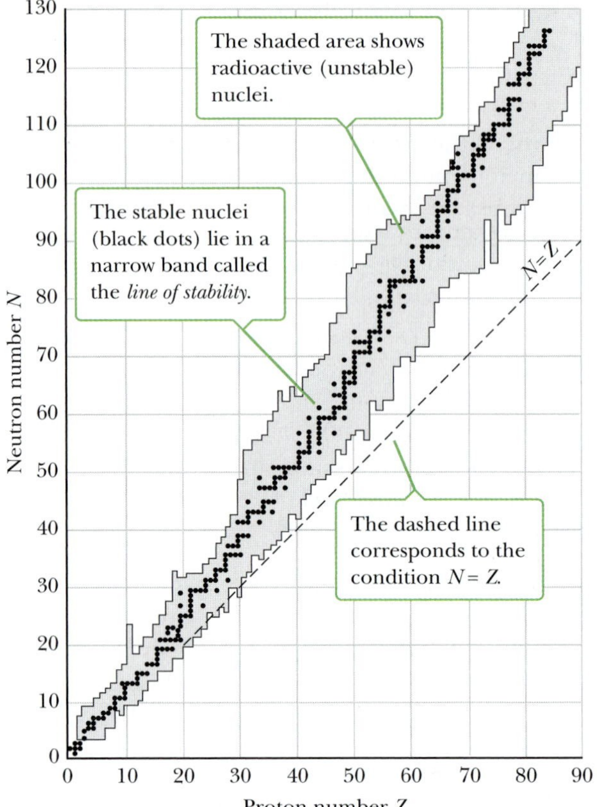


**Figure 29.3** A plot of the neutron number $N$ versus the proton number $Z$ for the stable nuclei (black dots).

## 29.2 Binding Energy

**LEARNING OBJECTIVES**

1. Discuss the concept of binding energy and its importance in nuclear reactions.
2. Apply the concept of binding energy to nuclei.

The total mass of a nucleus is always less than the sum of the masses of its nucleons. Also, because mass is another manifestation of energy, **the total energy of the bound system (the nucleus) is less than the combined energy of the separated nucleons**. This difference in energy is called the **binding energy** of the nucleus and can be thought of as the energy that must be added to a nucleus to break it apart into its separated neutrons and protons.

### EXAMPLE 29.1 The Binding Energy of the Deuteron

**GOAL** Calculate the binding energy of a nucleus.

**PROBLEM** The nucleus of the deuterium atom, called the deuteron, consists of a proton and a neutron. Calculate the deuteron's binding energy in MeV, given that its atomic mass, *the mass of a deuterium nucleus plus an electron*, is 2.014 102 u.

**STRATEGY** Calculate the sum of the masses of the individual particles and subtract the mass of the combined particle. The masses of the neutral atoms can be used instead of the nuclei because the electron masses cancel. Use the values from Appendix B. The mass of an atom given in Appendix B includes the mass of $Z$ electrons, where $Z$ is the atom's atomic number.

**SOLUTION**

To find the binding energy, first sum the masses of the hydrogen atom and neutron and subtract the mass of the deuteron:

$$\Delta m = (m_p + m_n) - m_d$$
$$= (1.007\ 825\ \text{u} + 1.008\ 665\ \text{u}) - 2.014\ 102\ \text{u}$$
$$= 0.002\ 388\ \text{u}$$

Using this mass difference, find the binding energy in MeV:

$$E_b = (0.002\ 388\ \text{u})\frac{931.5\ \text{MeV}}{1\ \text{u}} = \boxed{2.224\ \text{MeV}}$$

**REMARKS** This result tells us that to separate a deuteron into a proton and a neutron, it's necessary to add 2.224 MeV of energy to the deuteron to overcome the attractive nuclear force between the proton and the neutron. One way to supply the deuteron with this energy is to bombard it with energetic particles.

If the binding energy of a nucleus were zero, the nucleus would separate into its constituent protons and neutrons without the addition of any energy; that is, it would spontaneously break apart.

**QUESTION 29.1** Tritium and helium-3 have the same number of nucleons, but tritium has one proton and two neutrons whereas helium-3 has two protons and one neutron. Without doing a calculation, which nucleus has a greater binding energy? Explain.

**EXERCISE 29.1** Calculate the binding energy of ${}_2^3\text{He}$.

**ANSWER** 7.718 MeV

It's interesting to examine a plot of binding energy per nucleon, $E_b/A$, as a function of mass number for various stable nuclei (Fig. 29.4). Except for the lighter nuclei, the average binding energy per nucleon is about 8 MeV. Note that the curve peaks in the vicinity of $A = 60$, which means that nuclei with mass numbers greater or less than 60 are not as strongly bound as those near the middle of the periodic table. As we'll see later, this fact allows energy to be released in fission and fusion reactions. The curve is slowly varying for $A > 40$, which suggests the nuclear force saturates. In other words, a particular nucleon can interact with only a limited number of other nucleons, which can be viewed as the "nearest neighbors" in the close-packed structure illustrated in Figure 29.2.

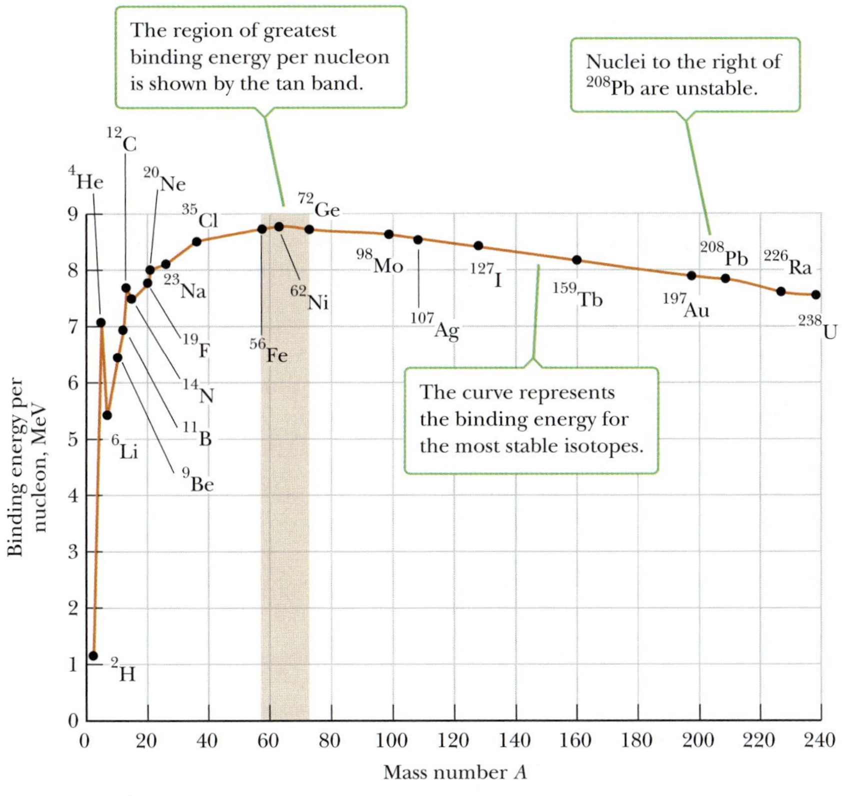


**Figure 29.4** Binding energy per nucleon versus the mass number $A$ for nuclei that are along the line of stability shown in Figure 29.3. Some representative nuclei appear as black dots with labels.

## APPLYING PHYSICS 29.1 Binding Nucleons and Electrons

Figure 29.4 shows a graph of the amount of energy required to remove a nucleon from the nucleus. The figure indicates that an approximately constant amount of energy is necessary to remove a nucleon above $A = 40$, whereas we saw in Chapter 28 that widely varying amounts of energy are required to remove an electron from the atom. What accounts for this difference?

**EXPLANATION** In the case of Figure 29.4, the approximately constant value of the nuclear binding energy is a

*(Continued)*

result of the short-range nature of the nuclear force. A given nucleon interacts only with its few nearest neighbors rather than with all the nucleons in the nucleus. Consequently, no matter how many nucleons are present in the nucleus, removing any nucleon involves separating it only from its nearest neighbors. The energy to do so is therefore approximately independent of how many nucleons are present. For the clearest comparison with the electron, think of averaging the energies required to remove all the electrons from an atom, from the outermost valence electron to the innermost K-shell electron. This average increases with increasing atomic number. The electrical force binding the electrons to the nucleus in an atom is a long-range force. An electron in an atom interacts with all the protons in the nucleus. When the nuclear charge increases, there is a stronger attraction between the nucleus and the electrons. Therefore, as the nuclear charge increases, more energy is necessary to remove an average electron. ■

Time Life Pictures/Mansell/Time Life Pictures/Getty Images

**Marie Curie**
**Polish Scientist (1867–1934)**
In 1903 Marie Curie shared the Nobel Prize in Physics with her husband, Pierre, and with Antoine Henri Becquerel for their studies of radioactive substances. In 1911 she was awarded a second Nobel Prize, this time in chemistry, for the discovery of radium and polonium. Marie Curie died of leukemia caused by years of exposure to radioactive substances. "I persist in believing that the ideas that then guided us are the only ones which can lead to the true social progress. We cannot hope to build a better world without improving the individual. Toward this end, each of us must work toward his own highest development, accepting at the same time his share of responsibility in the general life of humanity."

## 29.3 Radioactivity

**LEARNING OBJECTIVES**

1. Describe radioactivity and identify the three types of radiation.
2. Define the decay rate, the decay constant, and the half-life of a radioactive sample.
3. Apply decay concepts to radioactive substances.

In 1896 Becquerel accidentally discovered that uranium salt crystals emit an invisible radiation that can darken a photographic plate even if the plate is covered to exclude light. After several such observations under controlled conditions, he concluded that the radiation emitted by the crystals was of a new type, one requiring no external stimulation. This spontaneous emission of radiation was soon called **radioactivity**. Subsequent experiments by other scientists showed that other substances were also radioactive.

The most significant investigations of this type were conducted by Marie and Pierre Curie. After several years of careful and laborious chemical separation processes on tons of pitchblende, a radioactive ore, the Curies reported the discovery of two previously unknown elements, both of which were radioactive. These elements were named polonium and radium. Subsequent experiments, including Rutherford's famous work on alpha-particle scattering, suggested that radioactivity was the result of the decay, or disintegration, of unstable nuclei.

Three types of radiation can be emitted by a radioactive substance: alpha ($\alpha$) particles, in which the emitted particles are ${}^4_2$He nuclei; beta ($\beta$) particles, in which the emitted particles are either electrons or positrons; and gamma ($\gamma$) rays, in which the emitted "rays" are high-energy photons. A **positron** is a particle similar to the electron in all respects except that it has a charge of $+e$. (The positron is said to be the **antiparticle** of the electron.) The symbol $e^-$ is used to designate an electron, and $e^+$ designates a positron.

It's possible to distinguish these three forms of radiation by using the scheme described in Figure 29.5. The radiation from a radioactive sample is directed into a region with a magnetic field, and the beam splits into three components, two bending in opposite directions and the third not changing direction. From this

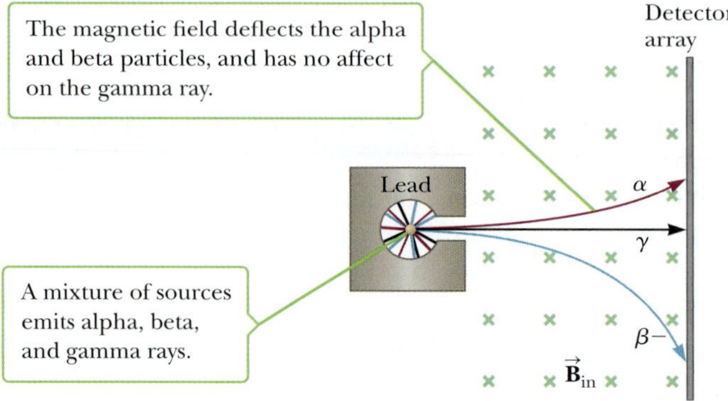


**Figure 29.5** The radiation from radioactive sources can be separated into three components by using a magnetic field to deflect the charged particles. The detector array at the right records the events.

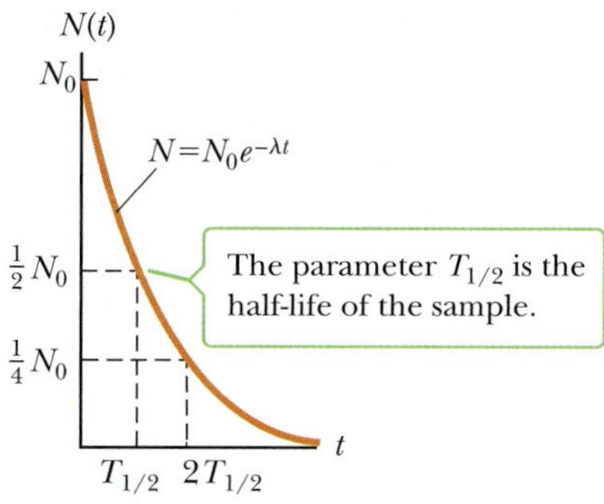


**Figure 29.6** Plot of the exponential decay law for radioactive nuclei. The vertical axis represents the number of radioactive nuclei present at any time $t$, and the horizontal axis is time.

simple observation, it can be concluded that the radiation of the undeflected beam (the gamma ray) carries no charge, the component deflected upward contains positively charged particles (alpha particles), and the component deflected downward contains negatively charged particles ($e^-$). If the beam includes a positron ($e^+$), it is deflected upward.

The three types of radiation have quite different penetrating powers. Alpha particles barely penetrate a sheet of paper, beta particles can penetrate a few millimeters of aluminum, and gamma rays can penetrate several centimeters of lead.

## The Decay Constant and Half-Life

Observation has shown that if a radioactive sample contains $N$ radioactive nuclei at some instant, the number of nuclei, $\Delta N$, that decay in a small time interval $\Delta t$ is proportional to $N$; mathematically,

$$\frac{\Delta N}{\Delta t} \propto N$$

or

$$\Delta N = -\lambda N \Delta t \qquad [29.2]$$

where $\lambda$ is a constant called the **decay constant**. The negative sign signifies that $N$ decreases with time; that is, $\Delta N$ is negative. The value of $\lambda$ for any isotope determines the rate at which that isotope will decay. **The decay rate, or activity *R*, of a sample is defined as the number of decays per second.** From Equation 29.2, we see that the decay rate is

$$R = \left|\frac{\Delta N}{\Delta t}\right| = \lambda N \qquad [29.3]$$

◀ Decay rate

Isotopes with a large $\lambda$ value decay rapidly; those with small $\lambda$ decay slowly.

A general decay curve for a radioactive sample is shown in Figure 29.6. It can be shown from Equation 29.2 (using calculus) that the number of nuclei present varies with time according to the equation

$$N = N_0 e^{-\lambda t} \qquad [29.4a]$$

where $N$ is the number of radioactive nuclei present at time $t$, $N_0$ is the number present at time $t = 0$, and $e = 2.718\ldots$ is Euler's constant. Processes that obey Equation 29.4a are sometimes said to undergo **exponential decay**.[1]

Another parameter that is useful for characterizing radioactive decay is the **half-life $T_{1/2}$. The half-life of a radioactive substance is the time it takes for half of a given number of radioactive nuclei to decay.** Using the concept of half-life, it can be shown that Equation 29.4a can also be written as

$$N = N_0\left(\frac{1}{2}\right)^n \qquad [29.4b]$$

[1]Other examples of exponential decay were discussed in Chapter 18 in connection with $RC$ circuits and in Chapter 20 in connection with $RL$ circuits.

where $n$ is the number of half-lives. The number $n$ can take any nonnegative value and need not be an integer. From the definition, it follows that $n$ is related to time $t$ and the half-life $T_{1/2}$ by

$$n = \frac{t}{T_{1/2}} \qquad [29.4c]$$

Setting $N = N_0/2$ and $t = T_{1/2}$ in Equation 29.4a gives

$$\frac{N_0}{2} = N_0 e^{-\lambda T_{1/2}}$$

Writing this expression in the form $e^{\lambda T_{1/2}} = 2$ and taking the natural logarithm of both sides, we get

$$T_{1/2} = \frac{\ln 2}{\lambda} = \frac{0.693}{\lambda} \qquad [29.5]$$

Equation 29.5 is a convenient expression relating the half-life to the decay constant. Note that after an elapsed time of one half-life, $N_0/2$ radioactive nuclei remain (by definition); after two half-lives, half of those will have decayed and $N_0/4$ radioactive nuclei will be left; after three half-lives, $N_0/8$ will be left; and so on.

The unit of activity $R$ is the **curie** (Ci), defined as

$$1 \text{ Ci} \equiv 3.7 \times 10^{10} \text{ decays/s} \qquad [29.6]$$

This unit was selected as the original activity unit because it is the approximate activity of 1 g of radium. The SI unit of activity is the **becquerel** (Bq):

$$1 \text{ Bq} = 1 \text{ decay/s} \qquad [29.7]$$

Therefore, $1 \text{ Ci} = 3.7 \times 10^{10}$ Bq. The most commonly used units of activity are the millicurie ($10^{-3}$ Ci) and the microcurie ($10^{-6}$ Ci).

**Tip 29.2 Two Half-Lives Don't Make a Whole Life**

A half-life is the time it takes for half of a given number of nuclei to decay. During a second half-life, half the remaining nuclei decay, so in two half-lives, three-quarters of the original material has decayed, not all of it.

**Quick Quiz**

**29.1** True or False: A radioactive atom always decays after two half-lives have elapsed.

**29.2** What fraction of a radioactive sample has decayed after three half-lives have elapsed? (a) 1/8 (b) 3/4 (c) 7/8 (d) none of these

**29.3** Suppose the decay constant of radioactive substance A is twice the decay constant of radioactive substance B. If substance B has a half-life of 4 h, what's the half-life of substance A? (a) 8 h (b) 4 h (c) 2 h

## EXAMPLE 29.2 The Activity of Radium

**GOAL** Calculate the activity of a radioactive substance at different times.

**PROBLEM** The half-life of the radioactive nucleus $^{226}_{88}\text{Ra}$ is $1.6 \times 10^3$ yr. If a sample initially contains $3.00 \times 10^{16}$ such nuclei, determine **(a)** the initial activity in curies, **(b)** the number of radium nuclei remaining after $4.8 \times 10^3$ yr, and **(c)** the activity at this later time.

**STRATEGY** For parts (a) and (c), find the decay constant and multiply it by the number of nuclei. Part (b) requires multiplying the initial number of nuclei by one-half for every elapsed half-life. (Essentially, this is an application of Eq. 29.4b.)

**SOLUTION**

**(a)** Determine the initial activity in curies.

Convert the half-life to seconds:

$$T_{1/2} = (1.6 \times 10^3 \text{ yr})(3.156 \times 10^7 \text{ s/yr}) = 5.0 \times 10^{10} \text{ s}$$

Substitute this value into Equation 29.5 to get the decay constant:

$$\lambda = \frac{0.693}{T_{1/2}} = \frac{0.693}{5.0 \times 10^{10} \text{ s}} = 1.4 \times 10^{-11} \text{ s}^{-1}$$

Calculate the activity of the sample at $t = 0$, using $R_0 = \lambda N_0$, where $R_0$ is the decay rate at $t = 0$ and $N_0$ is the number of radioactive nuclei present at $t = 0$:

$$R_0 = \lambda N_0 = (1.4 \times 10^{-11}\ \text{s}^{-1})(3.0 \times 10^{16}\ \text{nuclei}) = 4.2 \times 10^{5}\ \text{decays/s}$$

Convert to curies to obtain the activity at $t = 0$, using $1\ \text{Ci} = 3.7 \times 10^{10}$ decays/s:

$$R_0 = (4.2 \times 10^{5}\ \text{decays/s})\left(\frac{1\ \text{Ci}}{3.7 \times 10^{10}\ \text{decays/s}}\right) = 1.1 \times 10^{-5}\ \text{Ci} = \boxed{11\ \mu\text{Ci}}$$

**(b)** How many radium nuclei remain after $4.8 \times 10^3$ yr?

Calculate the number of half-lives, $n$:

$$n = \frac{4.8 \times 10^{3}\ \text{yr}}{1.6 \times 10^{3}\ \text{yr/half-life}} = 3.0\ \text{half-lives}$$

Multiply the initial number of nuclei by the number of factors of one-half:

$$\textbf{(1)}\quad N = N_0\left(\frac{1}{2}\right)^{n}$$

Substitute $N_0 = 3.0 \times 10^{16}$ and $n = 3.0$:

$$N = (3.0 \times 10^{16}\ \text{nuclei})\left(\frac{1}{2}\right)^{3.0} = \boxed{3.8 \times 10^{15}\ \text{nuclei}}$$

**(c)** Calculate the activity after $4.8 \times 10^3$ yr.

Multiply the number of remaining nuclei by the decay constant to find the activity $R$:

$$R = \lambda N = (1.4 \times 10^{-11}\ \text{s}^{-1})(3.8 \times 10^{15}\ \text{nuclei}) = 5.3 \times 10^{4}\ \text{decays/s} = \boxed{1.4\ \mu\text{Ci}}$$

**REMARKS** The activity is reduced by half every half-life, which is naturally the case because activity is proportional to the number of remaining nuclei. The precise number of nuclei at any time is never truly exact because particles decay according to a probability. The larger the sample, however, the more accurate the predictions from Equation 29.4.

**QUESTION 29.2** How would doubling the initial mass of radioactive material affect the initial activity? How would it affect the half-life?

**EXERCISE 29.2** Find (a) the number of remaining radium nuclei after $3.2 \times 10^3$ yr and (b) the activity at this time.

**ANSWERS** (a) $7.5 \times 10^{15}$ nuclei (b) $2.8\ \mu$Ci

## 29.4 The Decay Processes

**LEARNING OBJECTIVES**

1. Discuss, compare and contrast the three radioactive decay processes.
2. Apply decay concepts to radioactive nuclei.
3. Describe several practical uses of radioactivity.

As stated in the previous section, radioactive nuclei decay spontaneously via alpha, beta, and gamma decay. As we'll see in this section, these processes are very different from one another.

### Alpha Decay

If a nucleus emits an alpha particle ($^4_2$He), it loses two protons and two neutrons. Therefore, the neutron number $N$ of a single nucleus decreases by 2, $Z$ decreases by 2, and $A$ decreases by 4. The decay can be written symbolically as

$$^{A}_{Z}\text{X} \rightarrow {}^{A-4}_{Z-2}\text{Y} + {}^{4}_{2}\text{He} \qquad [29.8]$$

where X is called the **parent nucleus** and Y is known as the **daughter nucleus**. As examples, $^{238}U$ and $^{226}Ra$ are both alpha emitters and decay according to the schemes

$$^{238}_{92}\text{U} \rightarrow {}^{234}_{90}\text{Th} + {}^{4}_{2}\text{He} \qquad [29.9]$$

and

$$^{226}_{88}\text{Ra} \rightarrow {}^{222}_{86}\text{Rn} + {}^{4}_{2}\text{He} \qquad [29.10]$$

The half-life for $^{238}U$ decay is $4.47 \times 10^9$ years, and the half-life for $^{226}Ra$ decay is $1.60 \times 10^3$ years. In both cases, note that the mass number $A$ of the daughter nucleus is four less than that of the parent nucleus, and the atomic number $Z$ is reduced by two. The differences are accounted for in the emitted alpha particle (the $^4He$ nucleus).

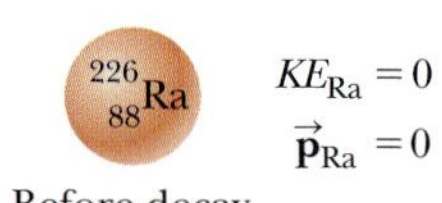

**Figure 29.7** The alpha decay of radium-226. The radium nucleus is initially at rest. After the decay, the radon nucleus has kinetic energy $KE_{Rn}$ and momentum $\vec{\mathbf{p}}_{Rn}$, and the alpha particle has kinetic energy $KE_\alpha$ and momentum $\vec{\mathbf{p}}_\alpha$.

The decay of $^{226}Ra$ is shown in Figure 29.7. When one element changes into another, as happens in alpha decay, the process is called **spontaneous decay** or transmutation. As a general rule, (1) the sum of the mass numbers $A$ must be the same on both sides of the equation, and (2) the sum of the atomic numbers $Z$ must be the same on both sides of the equation.

For alpha emission to occur, the mass of the parent must be greater than the combined mass of the daughter and the alpha particle. In the decay process, this excess mass is converted into energy of other forms and appears in the form of kinetic energy in the daughter nucleus and the alpha particle. Most of the kinetic energy is carried away by the alpha particle because it is much less massive than the daughter nucleus. This can be understood by first noting that a particle's kinetic energy and momentum $p$ are related as follows:

$$KE = \frac{p^2}{2m}$$

Because momentum is conserved, the two particles emitted in the decay of a nucleus at rest must have equal, but oppositely directed, momenta. As a result, the lighter particle, with the smaller mass in the denominator, has more kinetic energy than the more massive particle.

## APPLYING PHYSICS 29.2 | Energy and Half-Life

In comparing alpha decay energies from a number of radioactive nuclides, why is it found that the half-life of the decay goes down as the energy of the decay goes up?

**EXPLANATION** It should seem reasonable that the higher the energy of the alpha particle, the more likely it is to escape the confines of the nucleus. The higher probability of escape translates to a faster rate of decay, which appears as a shorter half-life. ■

## EXAMPLE 29.3 | Decaying Radium

**GOAL** Calculate the energy released during an alpha decay.

**PROBLEM** We showed that the $^{226}_{88}Ra$ nucleus undergoes alpha decay to $^{222}_{86}Rn$ (Eq. 29.10). Calculate the amount of energy liberated in this decay. Take the mass of $^{226}_{88}Ra$ to be 226.025 402 u, that of $^{222}_{86}Rn$ to be 222.017 571 u, and that of $^{4}_{2}He$ to be 4.002 602 u, as found in Appendix B.

**STRATEGY** The solution is a matter of subtracting the neutral masses of the daughter particles from the original mass of the radon atom.

**SOLUTION**

Compute the sum of the mass of the daughter particle, $m_d$, and the mass of the alpha particle, $m_\alpha$:

$$m_d + m_\alpha = 222.017\,571 \text{ u} + 4.002\,602 \text{ u} = 226.020\,173 \text{ u}$$

| | |
|---|---|
| Compute the loss of mass, $\Delta m$, during the decay by subtracting the previous result from $M_p$, the mass of the original particle: | $\Delta m = M_p - (m_d + m_\alpha) = 226.025\ 402\ \text{u} - 226.020\ 173\ \text{u}$ $= 0.005\ 229\ \text{u}$ |
| Change the loss of mass $\Delta m$ to its equivalent energy in MeV: | $E = (0.005\ 229\ \text{u})(931.494\ \text{MeV/u}) = \boxed{4.871\ \text{MeV}}$ |

**REMARKS** The potential barrier is typically higher than this value of the energy, but quantum tunneling permits the event to occur anyway.

**QUESTION 29.3** Convert the final answer to joules and estimate the energy produced by an Avogadro's number of such decays.

**EXERCISE 29.3** Calculate the energy released when $^{8}_{4}\text{Be}$ splits into two alpha particles. Beryllium-8 has an atomic mass of 8.005 305 u.

**ANSWER** 0.094 1 MeV

## Beta Decay

When a radioactive nucleus undergoes beta decay, the daughter nucleus has the same number of nucleons as the parent nucleus, but the atomic number is changed by 1:

$$^{A}_{Z}\text{X} \rightarrow {}^{A}_{Z+1}\text{Y} + \text{e}^- \qquad [29.11]$$

$$^{A}_{Z}\text{X} \rightarrow {}^{A}_{Z-1}\text{Y} + \text{e}^+ \qquad [29.12]$$

Again, note that the nucleon number and total charge are both conserved in these decays. As we will see shortly, however, these processes are not described completely by these expressions. A typical beta decay event is

$$^{14}_{6}\text{C} \rightarrow {}^{14}_{7}\text{N} + \text{e}^- \qquad [29.13]$$

The emission of electrons from a *nucleus* is surprising because, in all our previous discussions, we stated that the nucleus is composed of protons and neutrons only. This apparent discrepancy can be explained by noting that the emitted electron is created in the nucleus by a process in which a neutron is transformed into a proton. This process can be represented by

$$^{1}_{0}\text{n} \rightarrow {}^{1}_{1}\text{p} + \text{e}^- \qquad [29.14]$$

Consider the energy of the system of Equation 29.13 before and after decay. As with alpha decay, energy must be conserved in beta decay. The next example illustrates how to calculate the amount of energy released in the beta decay of $^{14}_{6}\text{C}$.

### EXAMPLE 29.4 The Beta Decay of Carbon-14

**GOAL** Calculate the energy released in a beta decay.

**PROBLEM** Find the energy liberated in the beta decay of $^{14}_{6}\text{C}$ to $^{14}_{7}\text{N}$, as represented by Equation 29.13. That equation refers to nuclei, whereas Appendix B gives the masses of neutral atoms. Adding six electrons to both sides of Equation 29.13 yields

$$^{14}_{6}\text{C atom} \rightarrow {}^{14}_{7}\text{N atom}$$

**STRATEGY** As in preceding problems, finding the released energy involves computing the difference in mass between the resultant particle(s) and the initial particle(s) and converting to MeV.

**SOLUTION**

| | |
|---|---|
| Obtain the masses of $^{14}_{6}\text{C}$ and $^{14}_{7}\text{N}$ from Appendix B and compute the difference between them: | $\Delta m = m_\text{C} - m_\text{N} = 14.003\ 242\ \text{u} - 14.003\ 074\ \text{u} = 0.000\ 168\ \text{u}$ |
| Find the liberated energy in MeV: | $E = (0.000\ 168\ \text{u})(931.494\ \text{MeV/u}) = \boxed{0.156\ \text{MeV}}$ |

**REMARKS** The calculated energy is generally more than the energy observed in this process. The discrepancy led to a crisis in physics because it appeared that energy wasn't conserved. As discussed below, this crisis was resolved by the discovery that another particle was also produced in the reaction.

*(Continued)*

**QUESTION 29.4** Is the binding energy per nucleon of the nitrogen-14 nucleus greater or less than the binding energy per nucleon for the carbon-14 nucleus? Justify your answer.

**EXERCISE 29.4** Calculate the maximum energy liberated in the beta decay of radioactive potassium to calcium: ${}^{40}_{19}\text{K} \rightarrow {}^{40}_{20}\text{Ca} + \text{e}^-$.

**ANSWER** 1.312 MeV

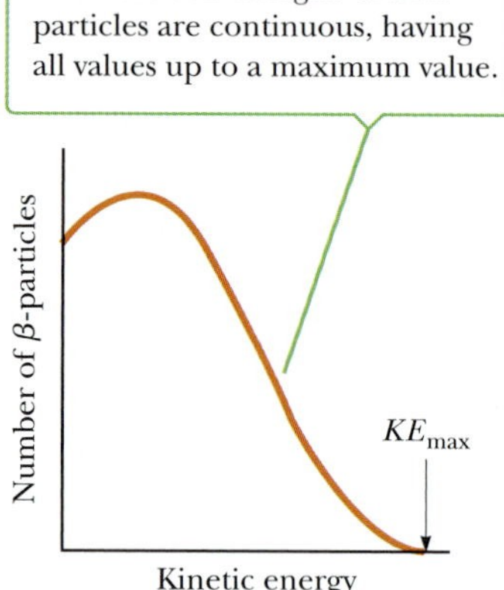

The observed energies of beta particles are continuous, having all values up to a maximum value.
Number of β-particles
KE max
Kinetic energy
a

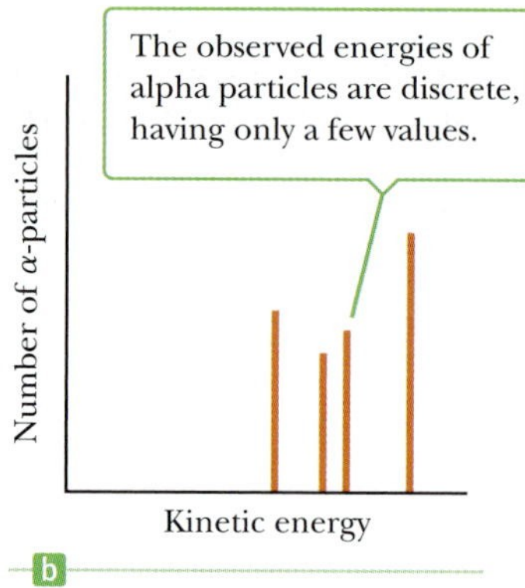

The observed energies of alpha particles are discrete, having only a few values.
Number of α-particles
Kinetic energy
b

**Figure 29.8** (a) Distribution of beta-particle energies in a typical beta decay. (b) Distribution of alpha-particle energies in a typical alpha decay.

From Example 29.4, we see that the energy released in the beta decay of $^{14}$C is approximately 0.16 MeV. As with alpha decay, we expect the electron to carry away virtually all this energy as kinetic energy because, apparently, it is the lightest particle produced in the decay. As Figure 29.8 shows, however, only a small number of electrons have this maximum kinetic energy, represented as $KE_{\text{max}}$ on the graph; most of the electrons emitted have kinetic energies lower than that predicted value. If the daughter nucleus and the electron aren't carrying away this liberated energy, where has the energy gone? As an additional complication, further analysis of beta decay shows that the principles of conservation of both angular momentum and linear momentum appear to have been violated!

In 1930 Pauli proposed that a third particle must be present to carry away the "missing" energy and to conserve momentum. Later, Enrico Fermi developed a complete theory of beta decay and named this particle the **neutrino** ("little neutral one") because it had to be electrically neutral and have little or no mass. Although it eluded detection for many years, the neutrino ($\nu$) was finally detected experimentally in 1956. The neutrino has the following properties:

- zero electric charge
- a mass much smaller than that of the electron, but probably not zero (Recent experiments suggest that the neutrino definitely has mass, but the value is uncertain, perhaps less than 1 eV/$c^2$.)
- a spin of $\frac{1}{2}$
- very weak interaction with matter, making it difficult to detect

With the introduction of the neutrino, we can now represent the beta decay process of Equation 29.13 in its correct form:

$$ {}^{14}_{6}\text{C} \rightarrow {}^{14}_{7}\text{N} + \text{e}^- + \overline{\nu} \qquad [29.15] $$

The bar in the symbol $\overline{\nu}$ indicates an **antineutrino**. To explain what an antineutrino is, we first consider the following decay:

$$ {}^{12}_{7}\text{N} \rightarrow {}^{12}_{6}\text{C} + \text{e}^+ + \nu \qquad [29.16] $$

Here, we see that when $^{12}$N decays into $^{12}$C, a particle is produced that is identical to the electron except that it has a positive charge of $+e$. This particle is called a **positron**. Because it is like the electron in all respects except charge, the positron is said to be the **antiparticle** of the electron. We discuss antiparticles further in Chapter 30; for now, it suffices to say that, **in beta decay, an electron and an antineutrino are emitted or a positron and a neutrino are emitted**.

Unlike beta decay, which results in a daughter particle with a variety of possible kinetic energies, alpha decays come in discrete amounts, as seen in Figure 29.8b. This is because the two daughter particles have momenta with equal magnitude and opposite direction and are each composed of a fixed number of nucleons.

**Tip 29.3 Mass Number of the Electron**

Another notation that is sometimes used for an electron is ${}^{\;0}_{-1}\text{e}$. This notation does not imply that the electron has zero rest energy. The mass of the electron is much smaller than that of the lightest nucleon, so we can approximate it as zero when we study nuclear decays and reactions.

## Gamma Decay

Very often a nucleus that undergoes radioactive decay is left in an excited energy state. The nucleus can then undergo a second decay to a lower energy state—perhaps even to the ground state—by emitting one or more high-energy photons. The process is similar to the emission of light by an atom. An atom emits radiation to release some extra energy when an electron "jumps" from a state of high energy to a state of lower energy. Likewise, the nucleus uses essentially the same method to release any extra energy it may have following a decay or some other nuclear event. In nuclear de-excitation, the "jumps" that release energy are made by protons or neutrons in the nucleus as they move from a higher energy level to a lower level. The photons emitted in the process are called **gamma rays**, which have very high energy relative to the energy of visible light.

A nucleus may reach an excited state as the result of a violent collision with another particle. It's more common, however, for a nucleus to be in an excited state as a result of alpha or beta decay. The following sequence of events typifies the gamma decay processes:

$$^{12}_{5}\text{B} \rightarrow {}^{12}_{6}\text{C}^* + e^- + \bar{\nu} \qquad [29.17]$$

$$^{12}_{6}\text{C}^* \rightarrow {}^{12}_{6}\text{C} + \gamma \qquad [29.18]$$

Equation 29.17 represents a beta decay in which $^{12}$B decays to $^{12}$C*, where the asterisk indicates that the carbon nucleus is left in an excited state following the decay. The excited carbon nucleus then decays to the ground state by emitting a gamma ray, as indicated by Equation 29.18. Note that gamma emission doesn't result in any change in either $Z$ or $A$.

**Enrico Fermi**
**Italian Physicist (1901–1954)**
Fermi was awarded the Nobel Prize in Physics in 1938 for producing the transuranic elements by neutron irradiation and for his discovery of nuclear reactions brought about by slow neutrons. He made many other outstanding contributions to physics, including his theory of beta decay, the free-electron theory of metals, and the development of the world's first fission reactor in 1942. Fermi was truly a gifted theoretical and experimental physicist. He was also well known for his ability to present physics in a clear and exciting manner. "Whatever Nature has in store for mankind, unpleasant as it may be, men must accept, for ignorance is never better than knowledge."

## Practical Uses of Radioactivity

**Carbon Dating** The beta decay of $^{14}$C given by Equation 29.15 is commonly used to date organic samples. Cosmic rays (high-energy particles from outer space) in the upper atmosphere cause nuclear reactions that create $^{14}$C from $^{14}$N. In fact, the ratio of $^{14}$C to $^{12}$C (by numbers of nuclei) in the carbon dioxide molecules of our atmosphere has a constant value of about $1.3 \times 10^{-12}$, as determined by measuring carbon ratios in tree rings. All living organisms have the same ratio of $^{14}$C to $^{12}$C because they continuously exchange carbon dioxide with their surroundings. When an organism dies, however, it no longer absorbs $^{14}$C from the atmosphere, so the ratio of $^{14}$C to $^{12}$C decreases as the result of the beta decay of $^{14}$C. It's therefore possible to determine the age of a material by measuring its activity per unit mass as a result of the decay of $^{14}$C. Through carbon dating, samples of wood, charcoal, bone, and shell have been identified as having lived from 1 000 to 25 000 years ago. This knowledge has helped researchers reconstruct the history of living organisms—including humans—during that time span.

**Smoke Detectors** Smoke detectors are frequently used in homes and industry for fire protection. Most of the common ones are the ionization-type that use radioactive materials. (See Fig. 29.9.) A smoke detector consists of an ionization chamber, a sensitive current detector, and an alarm. A weak radioactive source ionizes the air in the chamber of the detector, which creates charged particles. A voltage is maintained between the plates inside the chamber, setting up a small but detectable current in the external circuit. As long as the current is maintained, the alarm is deactivated. If smoke drifts into the chamber, though, the ions become attached to the smoke particles. These heavier particles do not drift as readily as do the lighter ions and cause a decrease in the detector current. The external circuit senses this decrease in current and sets off the alarm.

**APPLICATION**
Smoke Detectors

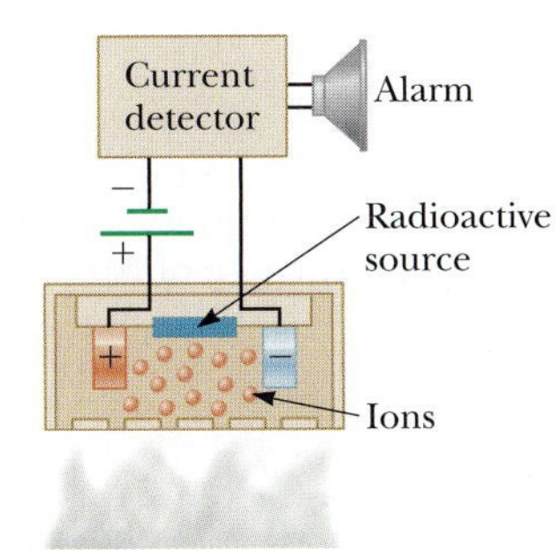


**Figure 29.9** An ionization-type smoke detector. Smoke entering the chamber reduces the detected current, causing the alarm to sound.

**Radon Detection** Radioactivity can also affect our daily lives in harmful ways. Soon after the discovery of radium by the Curies, it was found that air in contact with radium compounds becomes radioactive. It was then shown that this radioactivity came from the radium itself, and the product was therefore called "radium emanation." Rutherford and Frederick Soddy succeeded in condensing this "emanation," confirming that it was a real substance: the inert, gaseous element now called **radon** (Rn). Later, it was discovered that the air in uranium mines is radioactive because of the presence of radon gas. The mines must therefore be well ventilated to help protect the miners. Finally, the fear of radon pollution has moved from uranium mines into our own homes. Because certain types of rock, soil, brick, and concrete contain small quantities of radium, some of the resulting radon gas finds its way into our homes and other buildings. The most serious problems arise from leakage of radon from the ground into the structure. One practical remedy is to exhaust the air through a pipe just above the underlying soil or gravel directly to the outdoors by means of a small fan or blower.

**APPLICATION**
Radon Pollution

To use radioactive dating techniques, we need to recast some of the equations already introduced. We start by multiplying both sides of Equation 29.4 by $\lambda$:

$$\lambda N = \lambda N_0 e^{-\lambda t}$$

From Equation 29.3, we have $\lambda N = R$ and $\lambda N_0 = R_0$. Substitute these expressions into the above equation and divide through by $R_0$:

$$\frac{R}{R_0} = e^{-\lambda t}$$

where $R$ is the present activity and $R_0$ was the activity when the object in question was part of a living organism. We can solve for time by taking the natural logarithm of both sides of the foregoing equation:

$$\ln\left(\frac{R}{R_0}\right) = \ln\left(e^{-\lambda t}\right) = -\lambda t$$

$$t = -\frac{\ln\left(\frac{R}{R_0}\right)}{\lambda} \qquad [29.19]$$

## ■ EXAMPLE 29.5 Should We Report This Skeleton to Homicide?

**GOAL** Apply the technique of carbon-14 dating.

**PROBLEM** A 50.0-g sample of carbon taken from the pelvic bone of a skeleton is found to have a carbon-14 decay rate of 200.0 decays/min. It is known that carbon from a living organism has a decay rate of 15.0 decays/(min · g) and that $^{14}C$ has a half-life of 5 730 yr = $3.01 \times 10^9$ min. Find the age of the skeleton.

**STRATEGY** Calculate the original activity and the decay constant and then substitute those numbers and the current activity into Equation 29.19.

**SOLUTION**

Calculate the original activity $R_0$ from the decay rate and the mass of the sample:

$$R_0 = \left(15.0\ \frac{\text{decays}}{\text{min}\cdot\text{g}}\right)(50.0\text{ g}) = 7.50 \times 10^2\ \frac{\text{decays}}{\text{min}}$$

Find the decay constant from Equation 29.5:

$$\lambda = \frac{0.693}{T_{1/2}} = \frac{0.693}{3.01 \times 10^9\text{ min}} = 2.30 \times 10^{-10}\text{ min}^{-1}$$

$R$ is given, so now we substitute all values into Equation 29.19 to find the age of the skeleton:

$$t = -\frac{\ln\left(\frac{R}{R_0}\right)}{\lambda} = -\frac{\ln\left(\frac{200.0\text{ decays/min}}{7.50 \times 10^2\text{ decays/min}}\right)}{2.30 \times 10^{-10}\text{ min}^{-1}}$$

$$= \frac{1.32}{2.30 \times 10^{-10}\text{ min}^{-1}}$$

$$= 5.74 \times 10^9\text{ min} = \boxed{1.09 \times 10^4\text{ yr}}$$

**REMARKS** For much longer periods, other radioactive substances with longer half-lives must be used to develop estimates.

**QUESTION 29.5** Do the results of carbon dating depend on the mass of the original sample?

**EXERCISE 29.5** A sample of carbon of mass 7.60 g taken from an animal jawbone has an activity of 4.00 decays/min. How old is the jawbone?

**ANSWER** $2.77 \times 10^4$ yr

**Table 29.2** The Four Radioactive Series

| Series | Starting Isotope | Half-life (years) | Stable End Product |
|---|---|---|---|
| Uranium | ${}^{238}_{92}\text{U}$ | $4.47 \times 10^9$ | ${}^{206}_{82}\text{Pb}$ |
| Actinium | ${}^{235}_{92}\text{U}$ | $7.04 \times 10^8$ | ${}^{207}_{82}\text{Pb}$ |
| Thorium | ${}^{232}_{90}\text{Th}$ | $1.41 \times 10^{10}$ | ${}^{208}_{82}\text{Pb}$ |
| Neptunium | ${}^{237}_{93}\text{Np}$ | $2.14 \times 10^6$ | ${}^{209}_{82}\text{Pb}$ |

## 29.5 Natural Radioactivity

**LEARNING OBJECTIVE**

1. Discuss natural radioactivity and decay series.

Radioactive nuclei are generally classified into two groups: (1) unstable nuclei found in nature, which give rise to what is called **natural radioactivity**, and (2) nuclei produced in the laboratory through nuclear reactions, which exhibit **artificial radioactivity**.

Three series of naturally occurring radioactive nuclei exist (Table 29.2). Each starts with a specific long-lived radioactive isotope with half-life exceeding that of any of its descendants. The fourth series in Table 29.2 begins with $^{237}$Np, a transuranic element (an element having an atomic number greater than that of uranium) not found in nature. This element has a half-life of "only" $2.14 \times 10^6$ yr.

The two uranium series are somewhat more complex than the $^{232}$Th series (Fig. 29.10). Also, there are several other naturally occurring radioactive isotopes, such as $^{14}$C and $^{40}$K, that are not part of either decay series.

Natural radioactivity constantly supplies our environment with radioactive elements that would otherwise have disappeared long ago. For example, because the solar system is about $5 \times 10^9$ yr old, the supply of $^{226}$Ra (with a half-life of only 1 600 yr) would have been depleted by radioactive decay long ago were it not for the decay series that starts with $^{238}$U, with a half-life of $4.47 \times 10^9$ yr.

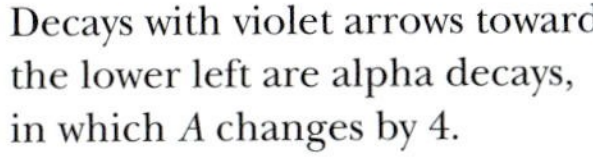

Decays with violet arrows toward the lower left are alpha decays, in which A changes by 4.


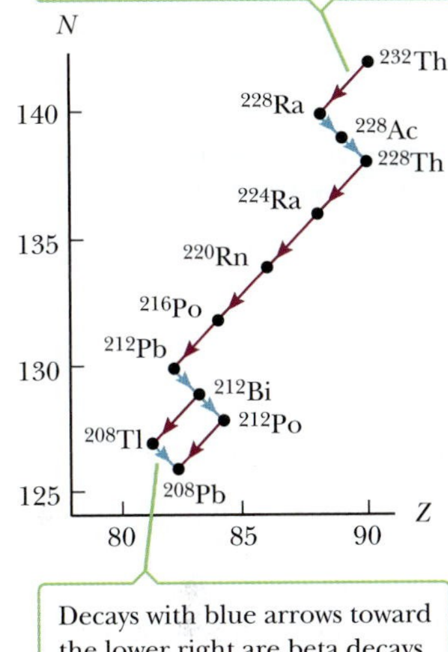


**Figure 29.10** Decay series beginning with $^{232}$Th.

## 29.6 Nuclear Reactions

**LEARNING OBJECTIVES**

1. Discuss the concept of nuclear reactions and the associated *Q* values.
2. Describe *Q* values for endothermic and exothermic reactions and define the threshold energy.

It is possible to change the structure of nuclei by bombarding them with energetic particles. Such changes are called **nuclear reactions**. Rutherford was the first to observe nuclear reactions, using naturally occurring radioactive sources for the bombarding particles. He found that protons were released when alpha particles were allowed to collide with nitrogen atoms. The process can be represented symbolically as

$$ {}^{4}_{2}\text{He} + {}^{14}_{7}\text{N} \rightarrow \text{X} + {}^{1}_{1}\text{H} \qquad [29.20] $$

This equation says that an alpha particle (${}^{4}_{2}$He) strikes a nitrogen nucleus and produces an unknown product nucleus (X) and a proton (${}^{1}_{1}$H). Balancing atomic numbers and mass numbers, as we did for radioactive decay, enables us to conclude that the unknown is characterized as ${}^{17}_{8}\text{X}$. Because the element with atomic number 8 is oxygen, we see that the reaction is

$$ {}^{4}_{2}\text{He} + {}^{14}_{7}\text{N} \rightarrow {}^{17}_{8}\text{O} + {}^{1}_{1}\text{H} \quad [29.21] $$

This nuclear reaction starts with two stable isotopes, helium and nitrogen, and produces two different stable isotopes, hydrogen and oxygen.

Since the time of Rutherford, thousands of nuclear reactions have been observed, particularly following the development of charged-particle accelerators in the 1930s. With today's advanced technology in particle accelerators and particle detectors, it is possible to achieve particle energies of at least 1 000 GeV = 1 TeV. These high-energy particles are used to create new particles whose properties are helping solve the mysteries of the nucleus (and indeed, of the Universe itself).

**Quick Quiz**

**29.4** Which of the following are possible reactions?

(a) ${}^{1}_{0}\text{n} + {}^{235}_{92}\text{U} \rightarrow {}^{140}_{54}\text{Xe} + {}^{94}_{38}\text{Sr} + 2({}^{1}_{0}\text{n})$

(b) ${}^{1}_{0}\text{n} + {}^{235}_{92}\text{U} \rightarrow {}^{132}_{50}\text{Sn} + {}^{101}_{42}\text{Mo} + 3({}^{1}_{0}\text{n})$

(c) ${}^{1}_{0}\text{n} + {}^{239}_{94}\text{Pu} \rightarrow {}^{127}_{53}\text{I} + {}^{93}_{41}\text{Nb} + 3({}^{1}_{0}\text{n})$

## EXAMPLE 29.6 The Discovery of the Neutron

**GOAL** Balance a nuclear reaction to determine an unknown decay product.

**PROBLEM** A nuclear reaction of significant note occurred in 1932 when Robert Chadwick, in England, bombarded a beryllium target with alpha particles. Analysis of the experiment indicated that the following reaction occurred:

$$ {}^{4}_{2}\text{He} + {}^{9}_{4}\text{Be} \rightarrow {}^{12}_{6}\text{C} + {}^{A}_{Z}\text{X} $$

What is ${}^{A}_{Z}\text{X}$ in this reaction?

**STRATEGY** Balancing mass numbers and atomic numbers yields the answer.

**SOLUTION**

Write an equation relating the atomic masses on either side:

$$ 4 + 9 = 12 + A \rightarrow A = 1 $$

Write an equation relating the atomic numbers:

$$ 2 + 4 = 6 + Z \rightarrow Z = 0 $$

Identify the particle:

$$ {}^{A}_{Z}\text{X} = {}^{1}_{0}\text{n (a neutron)} $$

**REMARKS** This was the first experiment to provide positive proof of the existence of neutrons.

**QUESTION 29.6** Where in nature is the reaction between helium and beryllium commonly found?

**EXERCISE 29.6** Identify the unknown particle in the reaction

$$ {}^{4}_{2}\text{He} + {}^{14}_{7}\text{N} \rightarrow {}^{17}_{8}\text{O} + {}^{A}_{Z}\text{X} $$

**ANSWER** ${}^{A}_{Z}\text{X} = {}^{1}_{1}\text{H}$ (a neutral hydrogen atom)

## Q Values

We have just examined some nuclear reactions for which mass numbers and atomic numbers must be balanced in the equations. We will now consider the energy involved in these reactions because energy is another important quantity that must be conserved.

We illustrate this procedure by analyzing the nuclear reaction

$$ {}^{2}_{1}\text{H} + {}^{14}_{7}\text{N} \rightarrow {}^{12}_{6}\text{C} + {}^{4}_{2}\text{He} \quad [29.22] $$

The total mass on the left side of the equation is the sum of the mass of $^2_1H$ (2.014 102 u) and the mass of $^{14}_7N$ (14.003 074 u), which equals 16.017 176 u. Similarly, the mass on the right side of the equation is the sum of the mass of $^{12}_6C$ (12.000 000 u) plus the mass of $^4_2He$ (4.002 602 u), for a total of 16.002 602 u. Thus, the total mass before the reaction is greater than the total mass after the reaction. The mass difference in the reaction is 16.017 176 u − 16.002 602 u = 0.014 574 u. This "lost" mass is converted to the kinetic energy of the nuclei present after the reaction. In energy units, 0.014 574 u is equivalent to 13.576 MeV of kinetic energy carried away by the carbon and helium nuclei.

The energy required to balance the equation is called the *Q* value of the reaction. In Equation 29.22, the *Q* value is 13.576 MeV. Nuclear reactions in which there is a release of energy—that is, positive *Q* values—are said to be **exothermic reactions.**

The energy balance sheet isn't complete, however, because we must also consider the kinetic energy of the incident particle before the collision. As an example, assume the deuteron in Equation 29.22 has a kinetic energy of 5 MeV. Adding this value to our *Q* value, we find that the carbon and helium nuclei have a total kinetic energy of 18.576 MeV following the reaction.

Now consider the reaction

$$^4_2He + ^{14}_7N \rightarrow ^{17}_8O + ^1_1H \qquad [29.23]$$

Before the reaction, the total mass is the sum of the masses of the alpha particle and the nitrogen nucleus: 4.002 602 u + 14.003 074 u = 18.005 676 u. After the reaction, the total mass is the sum of the masses of the oxygen nucleus and the proton: 16.999 133 u + 1.007 825 u = 18.006 958 u. In this case the total mass after the reaction is *greater* than the total mass before the reaction. The mass deficit is 0.001 282 u, equivalent to an energy deficit of 1.194 MeV. This deficit is expressed by the negative *Q* value of the reaction, −1.194 MeV. Reactions with negative *Q* values are called **endothermic reactions.** Such reactions won't take place unless the incoming particle has at least enough kinetic energy to overcome the energy deficit.

At first it might appear that the reaction in Equation 29.23 can take place if the incoming alpha particle has a kinetic energy of 1.194 MeV. In practice, however, the alpha particle must have more energy than that. If it has an energy of only 1.194 MeV, energy is conserved; careful analysis, though, shows that momentum isn't, which can be understood by recognizing that the incoming alpha particle has some momentum before the reaction. If its kinetic energy is only 1.194 MeV, however, the products (oxygen and a proton) would be created with zero kinetic energy and thus zero momentum. It can be shown that to conserve both energy and momentum, the incoming particle must have a minimum kinetic energy given by

$$KE_{\text{min}} = \left(1 + \frac{m}{M}\right)|Q| \qquad [29.24]$$

where $m$ is the mass of the incident particle, $M$ is the mass of the target, and the absolute value of the *Q* value is used. For the reaction given by Equation 29.23, we find that

$$KE_{\text{min}} = \left(1 + \frac{4.002\ 602}{14.003\ 074}\right)|-1.194\ \text{MeV}| = 1.535\ \text{MeV}$$

This minimum value of the kinetic energy of the incoming particle is called the **threshold energy.** The nuclear reaction shown in Equation 29.23 won't occur if the incoming alpha particle has a kinetic energy of less than 1.535 MeV, but can occur if its kinetic energy is equal to or greater than 1.535 MeV.

**Quick Quiz**

**29.5** If the *Q* value of an endothermic reaction is −2.17 MeV, the minimum kinetic energy needed in the reactant nuclei for the reaction to occur must be (a) equal to 2.17 MeV, (b) greater than 2.17 MeV, (c) less than 2.17 MeV, or (d) exactly half of 2.17 MeV.

# 29.7 Medical Applications of Radiation BIO

**LEARNING OBJECTIVE**

1. Describe the medical consequences of radiation exposure.
2. Describe some medical applications of radiation.

## Radiation Damage in Matter

Radiation absorbed by matter can cause severe damage. The degree and kind of damage depend on several factors, including the type and energy of the radiation and the properties of the absorbing material. Radiation damage in biological organisms is due primarily to ionization effects in cells. The normal function of a cell may be disrupted when highly reactive ions or radicals are formed as the result of ionizing radiation. For example, hydrogen and hydroxyl radicals produced from water molecules can induce chemical reactions that may break bonds in proteins and other vital molecules. Large acute doses of radiation are especially dangerous because damage to a great number of molecules in a cell may cause the cell to die. Also, cells that do survive the radiation may become defective, which can lead to cancer.

In biological systems it is common to separate radiation damage into two categories: somatic damage and genetic damage. **Somatic damage** is radiation damage to any cells except the reproductive cells. Such damage can lead to cancer at high radiation levels or seriously alter the characteristics of specific organisms. **Genetic damage** affects only reproductive cells. Damage to the genes in reproductive cells can lead to defective offspring. Clearly, we must be concerned about the effect of diagnostic treatments, such as x-rays and other forms of exposure to radiation.

Several units are used to quantify radiation exposure and dose. The **roentgen** (R) is defined as **the amount of ionizing radiation that will produce $2.08 \times 10^9$ ion pairs in 1 cm$^3$ of air under standard conditions**. Equivalently, the roentgen is **the amount of radiation that deposits $8.76 \times 10^{-3}$ J of energy into 1 kg of air.**

For most applications, the roentgen has been replaced by the **rad** (an acronym for *r*adiation *a*bsorbed *d*ose), defined as follows: **One rad is the amount of radiation that deposits $10^{-2}$ J of energy into 1 kg of absorbing material.**

Although the rad is a perfectly good physical unit, it's not the best unit for measuring the degree of biological damage produced by radiation because the degree of damage depends not only on the dose, but also on the *type* of radiation. For example, a given dose of alpha particles causes about ten times more biological damage than an equal dose of x-rays. The **RBE** (*r*elative *b*iological *e*ffectiveness) factor is defined as **the number of rads of x-radiation or gamma radiation that produces the same biological damage as 1 rad of the radiation being used**. The RBE factors for different types of radiation are given in Table 29.3. Note that the values are only approximate because they vary with particle energy and the form of damage.

Finally, the **rem** (*r*oentgen *e*quivalent in *m*an) is defined as the product of the dose in rads and the RBE factor:

$$\text{Dose in rem} = \text{dose in rads} \times \text{RBE}$$

According to this definition, 1 rem of any two kinds of radiation will produce the same amount of biological damage. From Table 29.3, we see that a dose of 1 rad of fast neutrons represents an effective dose of 10 rem and that 1 rad of x-radiation is equivalent to a dose of 1 rem.

Low-level radiation from natural sources, such as cosmic rays and radioactive rocks and soil, delivers a dose of about 0.13 rem/year per person. The upper limit of radiation dose recommended by the U.S. government (apart from background radiation and exposure related to medical procedures) is 0.5 rem/year. Many occupations involve higher levels of radiation exposure, and for individuals in these occupations, an upper limit of 5 rem/year has been set for whole-body exposure. Higher upper limits

**Table 29.3** RBE Factors for Several Types of Radiation

| Radiation | RBE Factor |
|---|---|
| X-rays and gamma rays | 1.0 |
| Beta particles | 1.0–1.7 |
| Alpha particles | 10–20 |
| Slow neutrons | 4–5 |
| Fast neutrons and protons | 10 |
| Heavy ions | 20 |

are permissible for certain parts of the body, such as the hands and forearms. An acute whole-body dose of 400 to 500 rem results in a mortality rate of about 50%. The most dangerous form of exposure is ingestion or inhalation of radioactive isotopes, especially those elements the body retains and concentrates, such as $^{90}$Sr. In some cases a dose of 1 000 rem can result from ingesting 1 mCi of radioactive material.

BIO APPLICATION
Irradiation of Food and Medical Equipment

Sterilizing objects by exposing them to radiation has been going on for years, but in recent years the methods used have become safer and more economical. Most bacteria, worms, and insects are easily destroyed by exposure to gamma radiation from radioactive cobalt. There is no intake of radioactive nuclei by an organism in such sterilizing processes as there is in the use of radioactive tracers. The process is highly effective in destroying *Trichinella* worms in pork, salmonella bacteria in chickens, insect eggs in wheat, and surface bacteria on fruits and vegetables that can lead to rapid spoilage. Recently, the procedure has been expanded to include the sterilization of medical equipment while in its protective covering. Surgical gloves, sponges, sutures, and so forth are irradiated while packaged. Also, bone, cartilage, and skin used for grafting are often irradiated to reduce the chance of infection.

## Tracing

BIO APPLICATION
Radioactive Tracers in Medicine

Radioactive particles can be used to trace chemicals participating in various reactions. One of the most valuable uses of radioactive tracers is in medicine. For example, $^{131}$I is an artificially produced isotope of iodine. (The natural, nonradioactive isotope is $^{127}$I.) Iodine, a necessary nutrient for our bodies, is obtained largely through the intake of seafood and iodized salt. The thyroid gland plays a major role in the distribution of iodine throughout the body. To evaluate the performance of the thyroid, the patient drinks a small amount of radioactive sodium iodide. Two hours later, the amount of iodine in the thyroid gland is determined by measuring the radiation intensity in the neck area.

A medical application of the use of radioactive tracers occurring in emergency situations is that of locating a hemorrhage inside the body. Often the location of the site cannot easily be determined, but radioactive chromium can identify the location with a high degree of precision. Chromium is taken up by red blood cells and carried uniformly throughout the body. The blood, however, will be dumped at a hemorrhage site, and the radioactivity of that region will increase markedly.

Tracing techniques are as wide ranging as human ingenuity can devise. Current applications range from checking the absorption of fluorine by teeth to checking contamination of food-processing equipment by cleansers to monitoring deterioration inside an automobile engine. In the last case a radioactive material is used in the manufacture of the pistons, and the oil is checked for radioactivity to determine the amount of wear on the pistons.

## Magnetic Resonance Imaging (MRI)

The heart of magnetic resonance imaging (MRI) is that when a nucleus having a magnetic moment is placed in an external magnetic field, its moment precesses about the magnetic field with a frequency proportional to the field. For example, a proton, with a spin of $\frac{1}{2}$, can occupy one of two energy states when placed in an

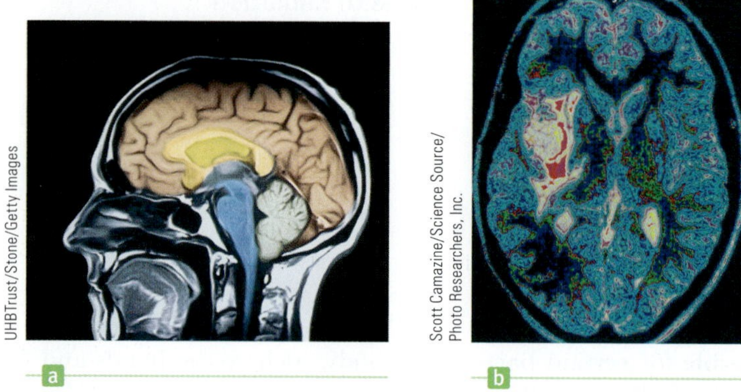


**Figure 29.11** Computer-enhanced MRI images of (a) a normal human brain and (b) a human brain with a glioma tumor.

external magnetic field. The lower-energy state corresponds to the case in which the spin is aligned with the field, whereas the higher-energy state corresponds to the case in which the spin is opposite the field. Transitions between these two states can be observed with a technique known as **nuclear magnetic resonance**. A DC magnetic field is applied to align the magnetic moments, and a second, weak oscillating magnetic field is applied perpendicular to the DC field. When the frequency of the oscillating field is adjusted to match the precessional frequency of the magnetic moments, the nuclei will "flip" between the two spin states. These transitions result in a net absorption of energy by the spin system, which can be detected electronically.

BIO APPLICATION
Magnetic Resonance Imaging (MRI)

In MRI, image reconstruction is obtained using spatially varying magnetic fields and a procedure for encoding each point in the sample being imaged. Two MRI images taken on a human head are shown in Figure 29.11. In practice, a computer-controlled pulse-sequencing technique is used to produce signals that are captured by a suitable processing device. The signals are then subjected to appropriate mathematical manipulations to provide data for the final image. The main advantage of MRI over other imaging techniques in medical diagnostics is that it causes minimal damage to cellular structures. Photons associated with the radio frequency signals used in MRI have energies of only about $10^{-7}$ eV. Because molecular bond strengths are much larger (on the order of 1 eV), the rf photons cause little cellular damage. In comparison, x-rays or $\gamma$-rays have energies ranging from $10^4$ to $10^6$ eV and can cause considerable cellular damage.

## SUMMARY

### 29.1 Some Properties of Nuclei

### 29.2 Binding Energy

Nuclei are represented symbolically as ${}^{A}_{Z}\text{X}$, where X represents the chemical symbol for the element. The quantity $A$ is the **mass number**, which equals the total number of nucleons (neutrons plus protons) in the nucleus. The quantity $Z$ is the **atomic number**, which equals the number of protons in the nucleus. Nuclei that contain the same number of protons but different numbers of neutrons are called **isotopes**. In other words, isotopes have the same $Z$ values but different $A$ values.

Most nuclei are approximately spherical, with an average radius given by

$$r = r_0 A^{1/3} \qquad [29.1]$$

where $r_0$ is a constant equal to $1.2 \times 10^{-15}$ m and $A$ is the mass number.

The total mass of a nucleus is always less than the sum of the masses of its individual nucleons. This mass difference $\Delta m$, multiplied by $c^2$, gives the **binding energy** of the nucleus.

### 29.3 Radioactivity

The spontaneous emission of radiation by certain nuclei is called **radioactivity**. There are three processes by which a radioactive substance can decay: alpha ($\alpha$) decay, in which the emitted particles are ${}^{4}_{2}\text{He}$ nuclei; beta ($\beta$) decay, in which the emitted particles are electrons or positrons; and gamma ($\gamma$) decay, in which the emitted particles are high-energy photons.

The **decay rate**, or **activity**, $R$, of a sample is given by

$$R = \left|\frac{\Delta N}{\Delta t}\right| = \lambda N \qquad [29.3]$$

where $N$ is the number of radioactive nuclei at some instant and $\lambda$ is a constant for a given substance called the **decay constant**.

Nuclei in a radioactive substance decay in such a way that the number of nuclei present varies with time according to the expression

$$N = N_0 e^{-\lambda t} \quad [29.4a]$$

where $N$ is the number of radioactive nuclei present at time $t$, $N_0$ is the number at time $t = 0$, and $e = 2.718\ldots$ is the base of the natural logarithms.

The **half-life** $T_{1/2}$ of a radioactive substance is the time required for half of a given number of radioactive nuclei to decay. The half-life is related to the decay constant by

$$T_{1/2} = \frac{0.693}{\lambda} \quad [29.5]$$

## 29.4 The Decay Processes

If a nucleus decays by alpha emission, it loses two protons and two neutrons. A typical alpha decay is

$$^{238}_{92}\text{U} \rightarrow {}^{234}_{90}\text{Th} + {}^{4}_{2}\text{He} \quad [29.9]$$

Note that in this decay, as in all radioactive decay processes, the sum of the $Z$ values on the left equals the sum of the $Z$ values on the right; the same is true for the $A$ values.

A typical beta decay is

$$^{14}_{6}\text{C} \rightarrow {}^{14}_{7}\text{N} + e^{-} + \bar{\nu} \quad [29.15]$$

When a nucleus undergoes beta decay, an **antineutrino** is emitted along with an electron, or a **neutrino** along with a positron. A neutrino has zero electric charge and a small mass (which may be zero) and interacts weakly with matter.

Nuclei are often in an excited state following radioactive decay, and they release their extra energy by emitting a high-energy photon called a **gamma ray** ($\gamma$). A typical gamma-ray emission is

$$^{12}_{6}\text{C}^* \rightarrow {}^{12}_{6}\text{C} + \gamma \quad [29.18]$$

where the asterisk indicates that the carbon nucleus was in an excited state before gamma emission.

## 29.6 Nuclear Reactions

**Nuclear reactions** can occur when a bombarding particle strikes another nucleus. A typical nuclear reaction is

$$^{4}_{2}\text{He} + {}^{14}_{7}\text{N} \rightarrow {}^{17}_{8}\text{O} + {}^{1}_{1}\text{H} \quad [29.21]$$

In this reaction, an alpha particle strikes a nitrogen nucleus, producing an oxygen nucleus and a proton. As in radioactive decay, atomic numbers and mass numbers balance on the two sides of the arrow.

Nuclear reactions in which energy is released are said to be **exothermic reactions** and are characterized by positive $Q$ values. Reactions with negative $Q$ values, called **endothermic reactions**, cannot occur unless the incoming particle has at least enough kinetic energy to overcome the energy deficit. To conserve both energy and momentum, the incoming particle must have a minimum kinetic energy, called the **threshold energy**, given by

$$KE_{\min} = \left(1 + \frac{m}{M}\right)|Q| \quad [29.24]$$

where $m$ is the mass of the incident particle and $M$ is the mass of the target atom.

## WARM-UP EXERCISES

ENHANCED WebAssign The warm-up exercises in this chapter may be assigned online in Enhanced WebAssign.

1. **Physics Review** Suppose a virus has a mass of $7.30 \times 10^{-18}$ kg. Calculate its rest energy in (a) joules and (b) MeV. (See Section 26.7.)
2. Determine the number of (a) electrons, (b) protons, and (c) neutrons in iron ($^{56}_{26}$Fe). (See Section 29.1.)
3. Estimate the radius of the nucleus of a carbon atom. (See Section 29.1.)
4. The atomic mass of oxygen is 15.999 u. Convert this mass to kilograms. (See Section 29.1.)
5. Convert the atomic mass of an oxygen atom to units of MeV/$c^2$. (See Section 29.1.)
6. Tritium ($^{3}_{1}$H), an isotope of hydrogen, has a nucleus composed of one proton and two neutrons and has an atomic mass of 3.016 049 u. Determine (a) the sum of the atomic masses of the three constituent particles, (b) the mass difference between the tritium nucleus and its constituent parts in atomic mass units, and (c) the binding energy in MeV. (Note: use 1 u = 931.5 MeV/$c^2$ in part (c).) (See Section 29.2.)
7. Uranium-235 is commonly used in nuclear power plants and has a half-life of about 704 million years ($2.22 \times 10^{16}$ s). Determine the decay constant for uranium-235. (See Section 29.3.)
8. The half-life of radium-224 is about 3.6 days ($3.11 \times 10^5$ s). (a) Determine the decay constant for radium-224. (b) What percentage of a sample remains undecayed after two weeks ($1.21 \times 10^6$ s)? (See Section 29.3.)
9. In the decay $^{234}_{90}\text{Th} \rightarrow {}^{A}_{Z}\text{Ra} + {}^{4}_{2}\text{He}$, identify (a) the mass number (by balancing mass numbers) and (b) the atomic number (by balancing atomic numbers) of the Ra nucleus. (See Section 29.4.)
10. What is the $Q$ value in MeV for the reaction $^9\text{Be} + \alpha \rightarrow {}^{12}\text{C} + \text{n}$ if the masses of each particle are $m_{^9\text{Be}}$ = 9.012 182 u, $m_\alpha$ = 4.002 603 u, $m_{^{12}\text{C}}$ = 12.000 000 u, and $m_\text{n}$ = 1.008 665 u? (Note: 1 u = 931.5 MeV/$c^2$.) (See Section 29.6.)

## CONCEPTUAL QUESTIONS

ENHANCED WebAssign The conceptual questions in this chapter may be assigned online in Enhanced WebAssign.

1. A student claims that a heavy form of hydrogen decays by alpha emission. How do you respond?
2. If a heavy nucleus that is initially at rest undergoes alpha decay, which has more kinetic energy after the decay, the alpha particle or the daughter nucleus?
3. Why do isotopes of a given element have different physical properties, such as mass, but the same chemical properties?
4. Why do nearly all the naturally occurring isotopes lie above the $N = Z$ line in Figure 29.3?
5. Consider two heavy nuclei X and Y having similar mass numbers. If X has the higher binding energy, which nucleus tends to be more unstable? Explain your answer.
6. Explain the main differences between alpha, beta, and gamma rays.
7. In beta decay, the energy of the electron or positron emitted from the nucleus lies somewhere in a relatively large range of possibilities. In alpha decay however, the alpha particle energy can only have discrete values. Why is there this difference?
8. What fraction of a radioactive sample has decayed after two half-lives have elapsed?
9. Pick any beta-decay process and show that the neutrino must have zero charge.
10. If film is kept in a box, alpha particles from a radioactive source outside the box cannot expose the film, but beta particles can. Explain.
11. In positron decay a proton in the nucleus becomes a neutron, and the positive charge is carried away by the positron. A neutron, though, has a larger rest energy than a proton. How is that possible?
12. An alpha particle has twice the charge of a beta particle. Why does the former deflect less than the latter when passing between electrically charged plates, assuming they both have the same speed?
13. Can carbon-14 dating be used to measure the age of a rock? Explain.

## PROBLEMS

ENHANCED WebAssign The problems in this chapter may be assigned online in Enhanced WebAssign.

1. denotes straightforward problem; 2. denotes intermediate problem;
3. denotes challenging problem
[1.] denotes full solution available in *Student Solutions Manual/ Study Guide*
1. denotes problems most often assigned in Enhanced WebAssign

BIO denotes biomedical problems
GP denotes guided problems
M denotes Master It tutorial available in Enhanced WebAssign
Q|C denotes asking for quantitative and conceptual reasoning
S denotes symbolic reasoning problem
W denotes Watch It video solution available in Enhanced WebAssign

*A list of atomic masses is given in Appendix B.*

### 29.1 Some Properties of Nuclei

1. Find the nuclear radii of the following nuclides: (a) ${}^{2}_{1}\text{H}$ (b) ${}^{60}_{27}\text{Co}$ (c) ${}^{197}_{79}\text{Au}$ (d) ${}^{239}_{94}\text{Pu}$.
2. Find the radius of a nucleus of (a) ${}^{4}_{2}\text{He}$ and (b) ${}^{238}_{92}\text{U}$.
3. Using $2.3 \times 10^{17}$ kg/m$^3$ as the density of nuclear matter, find the radius of a sphere of such matter that would have a mass equal to that of Earth. Earth has a mass equal to $5.98 \times 10^{24}$ kg and average radius of $6.37 \times 10^6$ m.
4. Consider the ${}^{65}_{29}\text{Cu}$ nucleus. Find approximate values for its (a) radius, (b) volume, and (c) density.
5. M An alpha particle ($Z = 2$, mass $= 6.64 \times 10^{-27}$ kg) approaches to within $1.00 \times 10^{-14}$ m of a carbon nucleus ($Z = 6$). What are (a) the maximum Coulomb force on the alpha particle, (b) the acceleration of the alpha particle at this time, and (c) the potential energy of the alpha particle at the same time?
6. Q|C Singly ionized carbon atoms are accelerated through 1 000 V and passed into a mass spectrometer to determine the isotopes present. (See Chapter 19.) The magnetic field strength in the spectrometer is 0.200 T. (a) Determine the orbital radii for the ${}^{12}\text{C}$ and the ${}^{13}\text{C}$ isotopes as they pass through the field. (b) Show that the ratio of the radii may be written in the form

$$\frac{r_1}{r_2} = \sqrt{\frac{m_1}{m_2}}$$

and verify that your radii in part (a) satisfy this formula.
7. W (a) Find the speed an alpha particle requires to come within $3.2 \times 10^{-14}$ m of a gold nucleus. (b) Find the energy of the alpha particle in MeV.
8. At the end of its life, a star with a mass of two times the Sun's mass is expected to collapse, combining its protons and electrons to form a neutron star. Such a star could be thought of as a gigantic atomic nucleus.

If a star of mass $2 \times 1.99 \times 10^{30}$ kg collapsed into neutrons ($m_n = 1.67 \times 10^{-27}$ kg), what would its radius be? Assume $r = r_0 A^{1/3}$.

## 29.2 Binding Energy

**9.** Find the average binding energy per nucleon of (a) $^{24}_{12}$Mg and (b) $^{85}_{37}$Rb.

**10.** Calculate the binding energy per nucleon for (a) $^2$H, (b) $^4$He, (c) $^{56}$Fe, and (d) $^{238}$U.

**11.** A pair of nuclei for which $Z_1 = N_2$ and $Z_2 = N_1$ are called *mirror isobars.* (The atomic and neutron numbers are interchangeable.) Binding-energy measurements on such pairs can be used to obtain evidence of the charge independence of nuclear forces. Charge independence means that the proton–proton, proton–neutron, and neutron–neutron forces are approximately equal. Calculate the difference in binding energy for the two mirror nuclei $^{15}_{8}$O and $^{15}_{7}$N.

**12.** W The peak of the stability curve occurs at $^{56}$Fe, which is why iron is prominent in the spectrum of the Sun and stars. Show that $^{56}$Fe has a higher binding energy per nucleon than its neighbors $^{55}$Mn and $^{59}$Co. Compare your results with Figure 29.4.

**13.** Q|C Two nuclei having the same mass number are known as *isobars.* (a) Calculate the difference in binding energy per nucleon for the isobars $^{23}_{11}$Na and $^{23}_{12}$Mg. (b) How do you account for this difference? (The mass of $^{23}_{12}$Mg = 22.994 127 u.)

**14.** Calculate the binding energy of the last neutron in the $^{43}_{20}$Ca nucleus. *Hint:* You should compare the mass of $^{43}_{20}$Ca with the mass of $^{42}_{20}$Ca plus the mass of a neutron. The mass of $^{42}_{20}$Ca = 41.958 622 u, whereas the mass of $^{43}_{20}$Ca = 42.958 770 u.

## 29.3 Radioactivity

**15.** Radon gas has a half-life of 3.83 days. If 3.00 g of radon gas is present at time $t = 0$, what mass of radon will remain after 1.50 days have passed?

**16.** BIO A drug tagged with $^{99}_{43}$Tc (half-life = 6.05 h) is prepared for a patient. If the original activity of the sample was $1.1 \times 10^4$ Bq, what is its activity after it has been on the shelf for 2.0 h?

**17.** GP The half-life of $^{131}$I is 8.04 days. (a) Convert the half-life to seconds. (b) Calculate the decay constant for this isotope. (c) Convert 0.500 $\mu$Ci to the SI unit the becquerel. (d) Find the number of $^{131}$I nuclei necessary to produce a sample with an activity of 0.500 $\mu$Ci. (e) Suppose the activity of a certain $^{131}$I sample is 6.40 mCi at a given time. Find the number of half-lives the sample goes through in 40.2 d and the activity at the end of that period.

**18.** Q|C Tritium has a half-life of 12.33 years. What fraction of the nuclei in a tritium sample will remain (a) after 5.00 yr? (b) After 10.0 yr? (c) After 123.3 yr? (d) According to Equation 29.4a, an infinite amount of time is required for the entire sample to decay. Discuss whether that is realistic.

**19.** After 2.00 days, the activity of a sample of an unknown type radioactive material has decreased to 84.2% of the initial activity. What is the half-life of this material?

**20.** BIO W After a plant or animal dies, its $^{14}$C content decreases with a half-life of 5 730 yr. If an archaeologist finds an ancient fire pit containing partially consumed firewood and the $^{14}$C content of the wood is only 12.5% that of an equal carbon sample from a present-day tree, what is the age of the ancient site?

**21.** M A freshly prepared sample of a certain radioactive isotope has an activity of 10.0 mCi. After 4.00 h, the activity is 8.00 mCi. (a) Find the decay constant and half-life of the isotope. (b) How many atoms of the isotope were contained in the freshly prepared sample? (c) What is the sample's activity 30 h after it is prepared?

**22.** A building has become accidentally contaminated with radioactivity. The longest-lived material in the building is strontium-90. (The atomic mass of $^{90}_{38}$Sr is 89.907 7 u.) If the building initially contained 5.0 kg of this substance and the safe level is less than 10.0 counts/min, how long will the building be unsafe?

## 29.4 The Decay Processes

**23.** Identify the missing nuclides in the following decays:
(a) $^{212}_{83}\text{Bi} \rightarrow ? + ^{4}_{2}\text{He}$
(b) $^{95}_{36}\text{Kr} \rightarrow ? + e^- + \bar{\nu}$
(c) $? \rightarrow ^{4}_{2}\text{He} + ^{140}_{58}\text{Ce}$

**24.** Complete the following radioactive decay formulas:
(a) $^{12}_{5}\text{B} \rightarrow ? + e^- + \bar{\nu}$
(b) $^{234}_{90}\text{Th} \rightarrow ^{230}_{88}\text{Ra} + ?$
(c) $? \rightarrow ^{14}_{7}\text{N} + e^- + \bar{\nu}$

**25.** The mass of $^{56}$Fe is 55.934 9 u, and the mass of $^{56}$Co is 55.939 9 u. Which isotope decays into the other and by what process?

**26.** Find the energy released in the alpha decay of $^{238}_{92}$U. The following mass value will be useful: $^{234}_{90}$Th has a mass of 234.043 583 u.

**27.** Determine which of the following suggested decays can occur spontaneously:
(a) $^{40}_{20}\text{Ca} \rightarrow e^+ + ^{40}_{19}\text{K}$ (b) $^{144}_{60}\text{Nd} \rightarrow ^{4}_{2}\text{He} + ^{140}_{58}\text{Ce}$

**28.** W $^{66}_{28}$Ni (mass = 65.929 1 u) undergoes beta decay to $^{66}_{29}$Cu (mass = 65.928 9 u). (a) Write the complete decay formula for this process. (b) Find the maximum kinetic energy of the emerging electrons.

**29.** A $^3$H (tritium) nucleus beta decays into $^3$He by creating an electron and an antineutrino according to the reaction

$$^{3}_{1}\text{H} \rightarrow ^{3}_{2}\text{He} + e^- + \bar{\nu}$$

Use Appendix B to determine the total energy released in this reaction.

30. A piece of charcoal used for cooking is found at the remains of an ancient campsite. A 1.00-kg sample of carbon from the wood has an activity equal to $5.00 \times 10^2$ decays per minute. Find the age of the charcoal. *Hint:* Living material has an activity equal to 15.0 decays/minute per gram of carbon present.

31. M A wooden artifact is found in an ancient tomb. Its carbon-14 ($^{14}_{6}C$) activity is measured to be 60.0% of that in a fresh sample of wood from the same region. Assuming the same amount of $^{14}C$ was initially present in the artifact as is now contained in the fresh sample, determine the age of the artifact.

## 29.6 Nuclear Reactions

32. A beam of 6.61-MeV protons is incident on a target of $^{27}_{13}Al$. Those protons that collide with the target produce the reaction

$$p + {}^{27}_{13}Al \rightarrow {}^{27}_{14}Si + n$$

($^{27}_{14}Si$ has a mass of 26.986 721 u.) Neglecting any recoil of the product nucleus, determine the kinetic energy of the emerging neutrons.

33. Identify the unknown particles X and X′ in the following nuclear reactions:
(a) $X + {}^{4}_{2}He \rightarrow {}^{24}_{12}Mg + {}^{1}_{0}n$
(b) ${}^{235}_{92}U + {}^{1}_{0}n \rightarrow {}^{90}_{38}Sr + X + 2{}^{1}_{0}n$
(c) $2{}^{1}_{1}H \rightarrow 2{}^{2}_{1}H + X + X'$

34. GP One method of producing neutrons for experimental use is to bombard $^{7}_{3}Li$ with protons. The neutrons are emitted according to the reaction

$${}^{1}_{1}H + {}^{7}_{3}Li \rightarrow {}^{7}_{4}Be + {}^{1}_{0}n$$

(a) Calculate the mass in atomic mass units of the particles on the left side of the equation. (b) Calculate the mass (in atomic mass units) of the particles on the right side of the equation. (c) Subtract the answer for part (b) from that for part (a) and convert the result to mega electron volts, obtaining the $Q$ value for this reaction. (d) Assuming lithium is initially at rest, the proton is moving at velocity $v$, and the resulting beryllium and neutron are both moving at velocity $V$ after the collision, write an expression describing conservation of momentum for this reaction in terms of the masses $m_p$, $m_{Be}$, $m_n$, and the velocities. (e) Write an expression relating the kinetic energies of particles before and after together with $Q$. (f) What minimum kinetic energy must the incident proton have if this reaction is to occur?

35. (a) Suppose $^{10}_{5}B$ is struck by an alpha particle, releasing a proton and a product nucleus in the reaction. What is the product nucleus? (b) An alpha particle and a product nucleus are produced when $^{13}_{6}C$ is struck by a proton. What is the product nucleus?

36. QIC Consider two reactions:

$$(1)\ n + {}^{2}_{1}H \rightarrow {}^{3}_{1}H$$

$$(2)\ {}^{1}_{1}H + {}^{2}_{1}H \rightarrow {}^{3}_{2}He$$

(a) Compute the $Q$ values for these reactions. Identify whether each reaction is exothermic or endothermic. (b) Which reaction results in more released energy? Why? (c) Assuming the difference is primarily due to the work done by the electric force, calculate the distance between the two protons in helium-3.

37. Natural gold has only one isotope, $^{197}_{79}Au$. If gold is bombarded with slow neutrons, $e^-$ particles are emitted. (a) Write the appropriate reaction equation. (b) Calculate the maximum energy of the emitted beta particles. The mass of $^{198}_{80}Hg$ is 197.966 75 u.

38. Complete the following nuclear reactions:
(a) $? + {}^{14}_{7}N \rightarrow {}^{1}_{1}H + {}^{17}_{8}O$ (b) ${}^{7}_{3}Li + {}^{1}_{1}H \rightarrow {}^{4}_{2}He + ?$

39. (a) Determine the product of the reaction ${}^{7}_{3}Li + {}^{4}_{2}He \rightarrow ? + n$ (b) What is the $Q$ value of the reaction?

## 29.7 Medical Applications of Radiation

40. BIO In terms of biological damage, how many rad of heavy ions are equivalent to 100 rad of x-rays?

41. BIO M A person whose mass is 75.0 kg is exposed to a whole-body dose of 25.0 rad. How many joules of energy are deposited in the person's body?

42. BIO A 200-rad dose of radiation is administered to a patient in an effort to combat a cancerous growth. Assuming all the energy deposited is absorbed by the growth, (a) calculate the amount of energy delivered per unit mass. (b) Assuming the growth has a mass of 0.25 kg and a specific heat equal to that of water, calculate its temperature rise.

43. A "clever" technician decides to heat some water for his coffee with an x-ray machine. If the machine produces 10 rad/s, how long will it take to raise the temperature of a cup of water by 50°C? Ignore heat losses during this time.

44. BIO An x-ray technician works 5 days per week, 50 weeks per year. Assume the technician takes an average of eight x-rays per day and receives a dose of 5.0 rem/yr as a result. (a) Estimate the dose in rem per x-ray taken. (b) How does this result compare with the amount of low-level background radiation the technician is exposed to?

45. BIO A patient swallows a radiopharmaceutical tagged with phosphorus-32 ($^{32}_{15}P$), a $\beta^-$ emitter with a half-life of 14.3 days. The average kinetic energy of the emitted electrons is 700 keV. If the initial activity of the sample is 1.31 MBq, determine (a) the number of electrons emitted in a 10-day period, (b) the total energy deposited in the body during the 10 days, and (c) the absorbed dose if the electrons are completely absorbed in 100 g of tissue.

46. BIO A particular radioactive source produces 100 mrad of 2-MeV gamma rays per hour at a distance of 1.0 m. (a) How long could a person stand at this distance before accumulating an intolerable dose of 1 rem? (b) Assuming the gamma radiation is emitted

uniformly in all directions, at what distance would a person receive a dose of 10 mrad/h from this source?

## Additional Problems

**47.** A radioactive sample contains 3.50 $\mu$g of pure $^{11}$C, which has a half-life of 20.4 min. (a) How many moles of $^{11}$C are present initially? (b) Determine the number of nuclei present initially. What is the activity of the sample (c) initially and (d) after 8.00 h?

**48.** Find the threshold energy that the incident neutron must have to produce the reaction: $^{1}_{0}\text{n} + ^{4}_{2}\text{He} \rightarrow ^{2}_{1}\text{H} + ^{3}_{1}\text{H}$.

**49.** M A 200.0-mCi sample of a radioactive isotope is purchased by a medical supply house. If the sample has a half-life of 14.0 days, how long will it keep before its activity is reduced to 20.0 mCi?

**50.** The $^{14}$C isotope undergoes beta decay according to the process given by Equation 29.15. Find the $Q$ value for this process.

**51.** Q|C In a piece of rock from the Moon, the $^{87}$Rb content is assayed to be $1.82 \times 10^{10}$ atoms per gram of material and the $^{87}$Sr content is found to be $1.07 \times 10^{9}$ atoms per gram. (The relevant decay is $^{87}\text{Rb} \rightarrow ^{87}\text{Sr} + e^{-}$. The half-life of the decay is $4.8 \times 10^{10}$ yr.) (a) Determine the age of the rock. (b) Could the material in the rock actually be much older? (c) What assumption is implicit in using the radioactive-dating method?

**52.** Many radioisotopes have important industrial, medical, and research applications. One of these is $^{60}$Co, which has a half-life of 5.2 years and decays by the emission of a beta particle (energy 0.31 MeV) and two gamma photons (energies 1.17 MeV and 1.33 MeV). A scientist wishes to prepare a $^{60}$Co sealed source that will have an activity of at least 10 Ci after 30 months of use. What is the minimum initial mass of $^{60}$Co required?

**53.** BIO A medical laboratory stock solution is prepared with an initial activity due to $^{24}$Na of 2.5 mCi/mL, and 10.0 mL of the stock solution is diluted at $t_0 = 0$ to a working solution whose total volume is 250 mL. After 48 h, a 5.0-mL sample of the working solution is monitored with a counter. What is the measured activity? *Note:* 1 mL = 1 milliliter.

**54.** After the sudden release of radioactivity from the Chernobyl nuclear reactor accident in 1986, the radioactivity of milk in Poland rose to 2 000 Bq/L due to iodine-131, with a half-life of 8.04 days. Radioactive iodine is particularly hazardous because the thyroid gland concentrates iodine. The Chernobyl accident caused a measurable increase in thyroid cancers among children in Belarus. (a) For comparison, find the activity of milk due to potassium. Assume 1 liter of milk contains 2.00 g of potassium, of which 0.011 7% is the isotope $^{40}$K, which has a half-life of $1.28 \times 10^{9}$ yr. (b) After what length of time would the activity due to iodine fall below that due to potassium?

**55.** The theory of nuclear astrophysics is that all the heavy elements like uranium are formed in the interior of massive stars. These stars eventually explode, releasing the elements into space. If we assume that at the time of explosion there were equal amounts of $^{235}$U and $^{238}$U, how long ago were the elements that formed our Earth released, given that the present $^{235}\text{U}/^{238}\text{U}$ ratio is 0.007? (The half-lives of $^{235}$U and $^{238}$U are $0.70 \times 10^{9}$ yr and $4.47 \times 10^{9}$ yr, respectively.)

**56.** A by-product of some fission reactors is the isotope $^{239}_{94}$Pu, which is an alpha emitter with a half-life of 24 000 years:

$$^{239}_{94}\text{Pu} \rightarrow ^{235}_{92}\text{U} + ^{4}_{2}\text{He}$$

Consider a sample of 1.0 kg of pure $^{239}_{94}$Pu at $t = 0$. Calculate (a) the number of $^{239}_{94}$Pu nuclei present at $t = 0$ and (b) the initial activity of the sample. (c) How long does the sample have to be stored if a "safe" activity level is 0.10 Bq?

The Compact Muon Solenoid (or CMS), a detector at the Large Hadron Collider (LHC) located near Geneva, Switzerland at CERN, is designed to search for the Higgs boson, particles associated with dark matter, and extra dimensions beyond the usual four. The CMS discovered a Higgs-like boson in July of 2012. The worker in the photo is performing winter maintenance.

George Grassie/ZUMAPRESS/Newscom

# 30 Nuclear Energy and Elementary Particles



In this concluding chapter we discuss the two means by which energy can be derived from nuclear reactions: fission, in which a nucleus of large mass number splits into two smaller nuclei, and fusion, in which two light nuclei fuse to form a heavier nucleus. In either case, there is a release of large amounts of energy that can be used destructively through bombs or constructively through the production of electric power. We end our study of physics by examining the known subatomic particles and the fundamental interactions that govern their behavior. We also discuss the current theory of elementary particles, which states that all matter in nature is constructed from only two families of particles: quarks and leptons. Finally, we describe how such models help us understand the evolution of the Universe.

## 30.1 Nuclear Fission

**LEARNING OBJECTIVES**

1. Describe nuclear fission and the sequence of events in a typical fission process.
2. Define a self-sustained nuclear reaction, the reproduction constant K, and the moderator, and explain their roles in reactor design.
3. Apply fission concepts to unstable nuclei.

**Nuclear fission** occurs when a heavy nucleus, such as $^{235}$U, splits, or fissions, into two smaller nuclei. In such a reaction **the total mass of the products is less than the original mass of the heavy nucleus**.

The fission of $^{235}$U by slow (low-energy) neutrons can be represented by the sequence of events

$$^{1}_{0}\text{n} + {}^{235}_{92}\text{U} \rightarrow {}^{236}_{92}\text{U}^* \rightarrow \text{X} + \text{Y} + \text{neutrons} \qquad [30.1]$$

where $^{236}U^*$ is an intermediate state that lasts only for about $10^{-12}$ s before splitting into nuclei X and Y, called **fission fragments**. Many combinations of X and Y satisfy the requirements of conservation of energy and charge. In the fission of uranium, about 90 different daughter nuclei can be formed. The process also results in the production of several (typically two or three) neutrons per fission event. On the average, 2.47 neutrons are released per event.

A typical reaction of this type is

$$^{1}_{0}n + ^{235}_{92}U \rightarrow ^{141}_{56}Ba + ^{92}_{36}Kr + 3^{1}_{0}n \qquad [30.2]$$

The fission fragments, barium and krypton, and the released neutrons have a great deal of kinetic energy following the fission event. Notice that the sum of the mass numbers, or number of nucleons, on the left $(1 + 235 = 236)$ is the same as the total number of nucleons on the right $(141 + 92 + 3 = 236)$. The total number of protons (92) is also the same on both sides. The energy $Q$ released through the disintegration in Equation 30.2 can be easily calculated using the data in Appendix B. The details of this calculation can be found in Chapter 26 (Example 26.5), with an answer of $Q = 200.422$ MeV.

The breakup of the uranium nucleus can be compared to what happens to a drop of water when excess energy is added to it. All the atoms in the drop have energy, but not enough to break up the drop. If enough energy is added to set the drop vibrating, however, it will undergo elongation and compression until the amplitude of vibration becomes large enough to cause the drop to break apart. In the uranium nucleus a similar process occurs (Fig. 30.1). The sequence of events is as follows:

◀ Sequence of events in a nuclear fission process

1. The $^{235}U$ nucleus captures a thermal (slow-moving) neutron.
2. The capture results in the formation of $^{236}U^*$, and the excess energy of this nucleus causes it to undergo violent oscillations.
3. The $^{236}U^*$ nucleus becomes highly elongated, and the force of repulsion between protons in the two halves of the dumbbell-shaped nucleus tends to increase the distortion.
4. The nucleus splits into two fragments, emitting several neutrons in the process.

Typically, the amount of energy released by the fission of a single heavy radioactive atom is about one hundred million times the energy released in the combustion of one molecule of the octane used in gasoline engines.

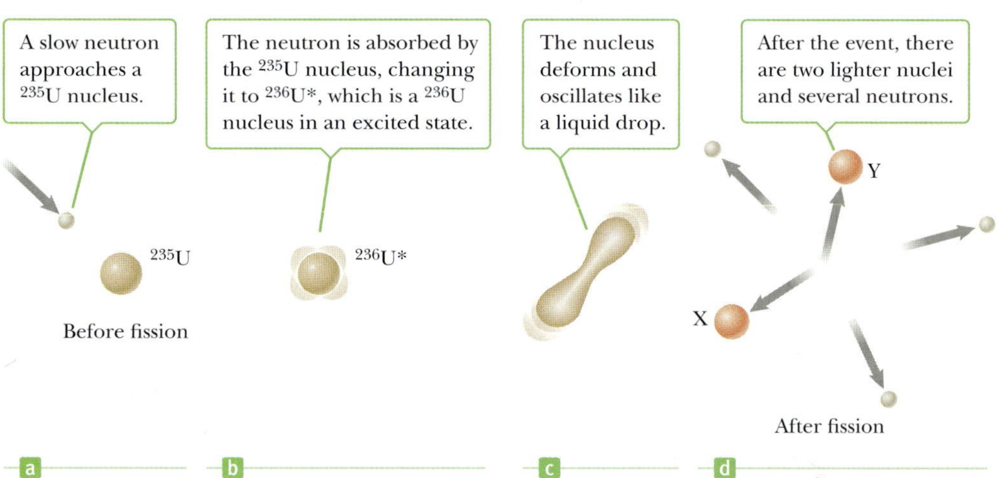

A slow neutron approaches a $^{235}U$ nucleus.
$^{235}U$
Before fission
The neutron is absorbed by the $^{235}U$ nucleus, changing it to $^{236}U^*$, which is a $^{236}U$ nucleus in an excited state.
$^{236}U^*$
The nucleus deforms and oscillates like a liquid drop.
After the event, there are two lighter nuclei and several neutrons.
Y
X
After fission
a
b
c
d


**Figure 30.1** A nuclear fission event as described by the liquid-drop model of the nucleus.

### APPLYING PHYSICS 30.1 | Unstable Products

If a heavy nucleus were to fission into only two product nuclei, they would be very unstable. Why?

**EXPLANATION** According to Figure 29.3, the ratio of the number of neutrons to the number of protons increases with $Z$. As a result, when a heavy nucleus splits in a fission reaction to two lighter nuclei, the lighter nuclei tend to have too many neutrons. The result is instability because the nuclei return to the curve in Figure 29.3 by decay processes that reduce the number of neutrons. ■

### EXAMPLE 30.1 | A Fission-Powered World

**GOAL** Relate raw material to energy output.

**PROBLEM** **(a)** Calculate the total energy released if 1.00 kg of $^{235}$U undergoes fission, taking the disintegration energy per event to be $Q = 208$ MeV. **(b)** How many kilograms of $^{235}$U would be needed to satisfy the world's annual energy consumption (about $4 \times 10^{20}$ J)?

**STRATEGY** In part (a), use the concept of a mole and Avogadro's number to obtain the total number of nuclei. Multiplying by the energy per reaction then gives the total energy released. Part (b) requires some light algebra.

**SOLUTION**

**(a)** Calculate the total energy released from 1.00 kg of $^{235}$U.

Find the total number of nuclei in 1.00 kg of uranium:

$$N = \left(\frac{6.02 \times 10^{23} \text{ nuclei/mol}}{235 \text{ g/mol}}\right)(1.00 \times 10^3 \text{ g})$$
$$= 2.56 \times 10^{24} \text{ nuclei}$$

Multiply $N$ by the energy yield per nucleus, obtaining the total disintegration energy:

$$E = NQ = (2.56 \times 10^{24} \text{ nuclei})\left(208 \frac{\text{MeV}}{\text{nucleus}}\right)$$
$$= 5.32 \times 10^{26} \text{ MeV}$$

**(b)** How many kilograms would provide for the world's annual energy needs?

Set the energy per kilogram, $E_{kg}$, times the number of kilograms, $N_{kg}$, equal to the total annual energy consumption. Solve for $N_{kg}$:

$$E_{kg}N_{kg} = E_{tot}$$
$$N_{kg} = \frac{E_{tot}}{E_{kg}} = \frac{4 \times 10^{20} \text{ J}}{(5.32 \times 10^{32} \text{ eV/kg})(1.60 \times 10^{-19} \text{ J/eV})}$$
$$= 5 \times 10^6 \text{ kg}$$

**REMARKS** The calculation implicitly assumes perfect conversion to usable power, which is never the case in real systems. There are sufficient easily recoverable reserves of uranium to supply the entire world's energy needs for about seven years at current levels of usage. Breeder reactor technology can greatly extend those reserves.

**QUESTION 30.1** Estimate the average mass of $^{235}$U needed to provide power for one family for one year.

**EXERCISE 30.1** How long can 1 kg of uranium-235 keep a 100-watt lightbulb burning if all its released energy is converted to electrical energy?

**ANSWER** ~ 30 000 yr

## Nuclear Reactors

The neutrons emitted when $^{235}$U undergoes fission can in turn trigger other nuclei to undergo fission, with the possibility of a chain reaction (Fig. 30.2). Calculations show that if the chain reaction isn't controlled, it will proceed too rapidly and possibly result in the sudden release of an enormous amount of energy (an explosion), even from only 1 g of $^{235}$U. If the energy in 1 kg of $^{235}$U were released, it would equal that released by the detonation of about 20 000 tons of

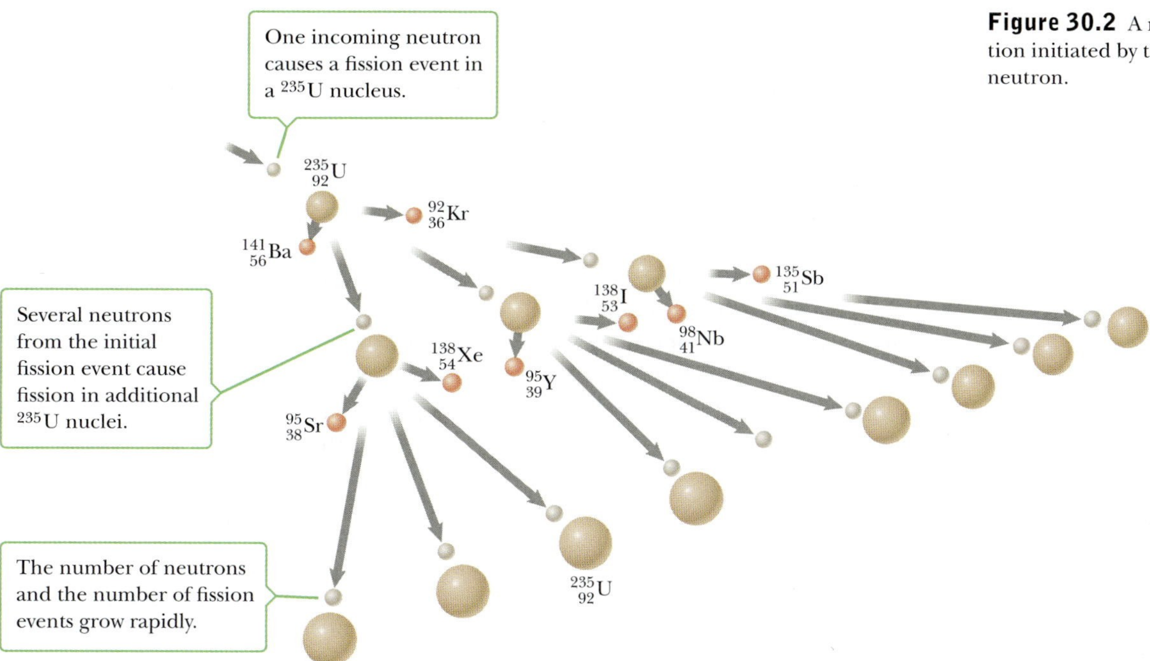
One incoming neutron causes a fission event in a $^{235}U$ nucleus.
$^{235}_{92}U$
$^{92}_{36}Kr$
$^{141}_{56}Ba$
$^{138}_{53}I$
$^{135}_{51}Sb$
Several neutrons from the initial fission event cause fission in additional $^{235}U$ nuclei.
$^{138}_{54}Xe$
$^{98}_{41}Nb$
$^{95}_{39}Y$
$^{95}_{38}Sr$
The number of neutrons and the number of fission events grow rapidly.
$^{235}_{92}U$

**Figure 30.2** A nuclear chain reaction initiated by the capture of a neutron.

TNT! An uncontrolled fission reaction, of course, is the principle behind the first nuclear bomb.

A nuclear reactor is a system designed to maintain what is called a **self-sustained chain reaction**, first achieved in 1942 by a team led by Enrico Fermi. Most reactors in operation today also use uranium as fuel. Natural uranium contains only about 0.7% of the $^{235}U$ isotope, with the remaining 99.3% being the $^{238}U$ isotope. This fact is important to the operation of a reactor because $^{238}U$ almost never undergoes fission. Instead, it tends to absorb neutrons, producing neptunium and plutonium. For this reason, reactor fuels must be artificially enriched so that they contain several percent of the $^{235}U$ isotope.

On average, about 2.5 neutrons are emitted in each fission event of $^{235}U$. To achieve a self-sustained chain reaction, one of these neutrons must be captured by another $^{235}U$ nucleus and cause it to undergo fission. A useful parameter for describing the level of reactor operation is the **reproduction constant *K*, defined as the average number of neutrons from each fission event that will cause another event**.

A self-sustained chain reaction is achieved when $K = 1$. Under this condition, the reactor is said to be **critical**. When $K$ is less than 1, the reactor is subcritical and the reaction dies out. When $K$ is greater than 1, the reactor is said to be supercritical and a runaway reaction occurs. In a nuclear reactor used to furnish power to a utility company, it is necessary to maintain a $K$ value close to 1.

The basic design of a nuclear reactor is shown in Figure 30.3. The fuel elements consist of enriched uranium. The size of the reactor is important in reducing neutron leakage: a large reactor has a smaller surface-to-volume ratio and smaller leakage than a smaller reactor.

It's also important to regulate the neutron energies because slow neutrons are far more likely to cause fissions than fast neutrons in $^{235}U$. Further, $^{238}U$ doesn't absorb slow neutrons. For the chain reaction to continue, the neutrons must, therefore, be slowed down. This slowing is accomplished by surrounding the fuel with a substance called a **moderator**, such as graphite (carbon) or heavy water ($D_2O$). Most modern reactors use heavy water. Collisions in the moderator slow the neutrons and enhance the fissioning of $^{235}U$.

The power output of a fission reactor is controlled by the control rods depicted in Figure 30.3. These rods are made of materials like cadmium that readily absorb neutrons.

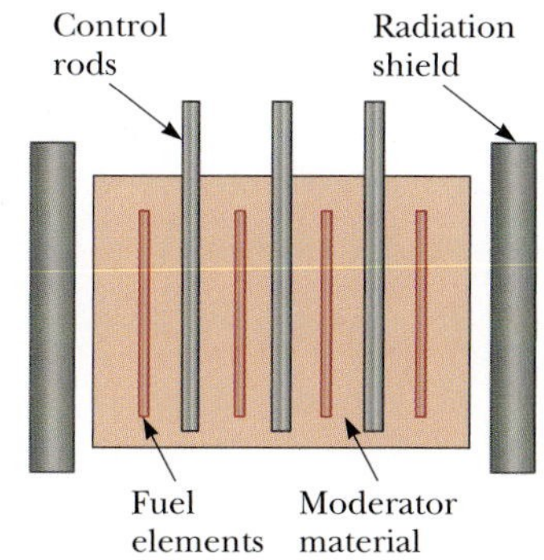
Control rods
Radiation shield
Fuel elements
Moderator material

**Figure 30.3** Cross section of a reactor core showing the control rods, fuel elements containing enriched fuel, and moderating material, all surrounded by a radiation shield.

**APPLICATION**
Nuclear Reactor Design

Fissions in a nuclear reactor heat molten sodium (or water, depending on the system), which is pumped through a heat exchanger. There, the thermal energy is transferred to water in a secondary system. The water is converted to steam, which drives a turbine-generator to create electric power.

Fission reactors are extremely safe. According to the *Oak Ridge National Laboratory Review*, "The health risk of living within 8 km (5 miles) of a nuclear reactor for 50 years is no greater than the risk of smoking 1.4 cigarettes, drinking 0.5 L of wine, traveling 240 km by car, flying 9 600 km by jet, or having one chest x-ray in a hospital. Each activity, by itself, is estimated to increase a person's chance of dying in any given year by one in a million."

The safety issues associated with nuclear power reactors are complex and often emotional and overblown. All sources of energy have associated risks. Coal, for example, exposes workers to health hazards (including radioactive radon) and produces atmospheric pollution (including greenhouse gases and highly radioactive ash). Solar panels introduce toxic substances into the environment such as cadmium and silicon tetrachloride, and in addition occupy significant land areas relative their power production capabilities. In each case the risks must be weighed against the benefits and the availability and cost of the energy source.

Given concerns about greenhouse gasses and global climate change, studying alternative forms of energy production is extremely important. Among green energy alternatives, nuclear fission is the most robust, providing reliable power with little down time. Despite advances in technology and reductions in cost, solar and wind power remain very expensive and intermittent.

The known sources of uranium ore are sufficient to supply all humanity's power requirements for several years at the current rate of usage if burned in conventional reactors. (See Problem 8.) Breeder reactors, on the other hand, convert nuclear waste into nuclear fuel and can also change thorium to an isotope of uranium suitable for a reactor. Using that technology, current reserves would likely be sufficient for well over a thousand years, even without the discovery and exploitation of new sources of uranium or thorium. The largest untapped source is the 4.5 billion metric tons of uranium dissolved in Earth's oceans. Several methods of extracting this uranium economically are currently under study. The dissolved uranium is steadily replenished by Earth's rivers and could potentially provide reactor fuel almost indefinitely. (See Problem 10.)

## 30.2 Nuclear Fusion

**LEARNING OBJECTIVES**

1. Describe the process of nuclear fusion and the proton-proton cycle in stars.
2. Discuss the potential application of fusion to power reactors and state Lawson's criterion.
3. Apply fusion concepts to nuclei.

**When two light nuclei combine to form a heavier nucleus, the process is called nuclear fusion.** Because the mass of the final nucleus is less than the sum of the masses of the original nuclei, there is a loss of mass, accompanied by a release of energy. Although fusion power plants have not yet been developed, a worldwide effort is under way to harness the energy from fusion reactions in the laboratory.

### Fusion in the Sun

All stars generate their energy through fusion processes. About 90% of stars, including the Sun, fuse hydrogen, whereas some older stars fuse helium or other heavier elements. The energy produced by fusion increases the pressure inside the star and prevents its collapse due to gravity.

Two conditions must be met before fusion reactions in the star can sustain its energy needs. First, the temperature must be high enough (about $10^7$ K for hydrogen) to allow the kinetic energy of the positively charged hydrogen nuclei to overcome their mutual Coulomb repulsion as they collide. Second, the density of nuclei must be high enough to ensure a high rate of collision.

It's interesting to note that a quantum effect is key in making sunshine. Temperatures inside stars like the Sun are not high enough to allow colliding protons to overcome the Coulomb repulsion. In a certain percentage of collisions, however, the nuclei pass through the barrier anyway, an example of *quantum tunneling*.

The **proton–proton cycle** is a series of three nuclear reactions that are believed to be the stages in the liberation of energy in the Sun and other stars rich in hydrogen. An overall view of the proton–proton cycle is that four protons combine to form an alpha particle and two positrons, with the release of 25 MeV of energy in the process.

The specific steps in the proton–proton cycle are

$$^1_1\text{H} + ^1_1\text{H} \rightarrow ^2_1\text{D} + e^+ + \nu$$

and

$$^1_1\text{H} + ^2_1\text{D} \rightarrow ^3_2\text{He} + \gamma \qquad [30.3]$$

where D stands for deuterium, the isotope of hydrogen having one proton and one neutron in the nucleus. (It can also be written as $^2_1\text{H}$.) The second reaction is followed by either hydrogen-helium fusion or helium-helium fusion:

$$^1_1\text{H} + ^3_2\text{He} \rightarrow ^4_2\text{He} + e^+ + \nu$$

or

$$^3_2\text{He} + ^3_2\text{He} \rightarrow ^4_2\text{He} + 2(^1_1\text{H})$$

The energy liberated is carried primarily by gamma rays, positrons, and neutrinos, as can be seen from the reactions. The gamma rays are soon absorbed by the dense gas, raising its temperature. The positrons combine with electrons to produce gamma rays, which in turn are also absorbed by the gas within a few centimeters. The neutrinos, however, almost never interact with matter; hence, they escape from the star, carrying about 2% of the energy generated with them. These energy-liberating fusion reactions are called **thermonuclear fusion reactions**. The hydrogen (fusion) bomb, first exploded in 1952, is an example of an uncontrolled thermonuclear fusion reaction.

## Fusion Reactors

**APPLICATION**
Fusion Reactors

A great deal of effort is under way to develop a sustained and controllable fusion power reactor. Controlled fusion is often called the ultimate energy source because of the availability of water, its fuel source. For example, if deuterium, the isotope of hydrogen consisting of a proton and a neutron, were used as the fuel, 0.06 g of it could be extracted from 1 gal of water at a cost of about four cents. Hence, the fuel costs of even an inefficient reactor would be almost insignificant. An additional advantage of fusion reactors is that comparatively few radioactive by-products are formed. As noted in Equation 30.3, the end product of the fusion of hydrogen nuclei is safe, nonradioactive helium. Unfortunately, a thermonuclear reactor that can deliver a net power output over a reasonable time interval is not yet a reality, and many problems must be solved before a successful device is constructed.

The fusion reactions that appear most promising in the construction of a fusion power reactor involve deuterium (D) and tritium (T), which are isotopes of hydrogen. These reactions are

$$^2_1\text{D} + ^2_1\text{D} \rightarrow ^3_2\text{He} + ^1_0\text{n} \qquad Q = 3.27\text{ MeV}$$

$$^2_1\text{D} + ^2_1\text{D} \rightarrow ^3_1\text{T} + ^1_1\text{H} \qquad Q = 4.03\text{ MeV} \qquad [30.4]$$

and

$$^2_1\text{D} + ^3_1\text{T} \rightarrow ^4_2\text{He} + ^1_0\text{n} \qquad Q = 17.59 \text{ MeV}$$

where the $Q$ values refer to the amount of energy released per reaction. As noted earlier, deuterium is available in almost unlimited quantities from our lakes and oceans and is very inexpensive to extract. Tritium, however, is radioactive ($T_{1/2}$ = 12.3 yr) and undergoes beta decay to $^3$He. For this reason, tritium doesn't occur naturally to any great extent and must be artificially produced.

The fundamental challenge of nuclear fusion power is to give the nuclei enough kinetic energy to overcome the repulsive Coulomb force between them at close proximity. This step can be accomplished by heating the fuel to extremely high temperatures (about $10^8$ K, far greater than the interior temperature of the Sun). Such high temperatures are not easy to obtain in a laboratory or a power plant. At these high temperatures, the atoms are ionized and the system then consists of a collection of electrons and nuclei, commonly referred to as a *plasma*.

In addition to the high temperature requirements, two other critical factors determine whether or not a thermonuclear reactor will function: the **plasma ion density** $n$ and the **plasma confinement time** $\tau$, the time the interacting ions are maintained at a temperature equal to or greater than that required for the reaction to proceed. The density and confinement time must both be large enough to ensure that more fusion energy will be released than is required to heat the plasma.

Lawson's criterion ►

**Lawson's criterion** states that a net power output in a fusion reactor is possible under the following conditions:

$$n\tau \geq 10^{14} \text{ s/cm}^3 \qquad \text{Deuterium–tritium interaction} \qquad [30.5]$$

$$n\tau \geq 10^{16} \text{ s/cm}^3 \qquad \text{Deuterium–deuterium interaction}$$

The problem of plasma confinement time has yet to be solved. How can a plasma be confined at a temperature of $10^8$ K for times on the order of 1 s? Most fusion experiments use magnetic field confinement to contain a plasma. One device, called a **tokamak**, has a doughnut-shaped geometry (a toroid), as shown in Figure 30.4. This device uses a combination of two magnetic fields to confine the plasma inside the doughnut. A strong magnetic field is produced by the current in the windings, and a weaker magnetic field is produced by the current in the toroid. The resulting magnetic field lines are helical, as shown in the figure. In this configuration the field lines spiral around the plasma and prevent it from touching the walls of the vacuum chamber.

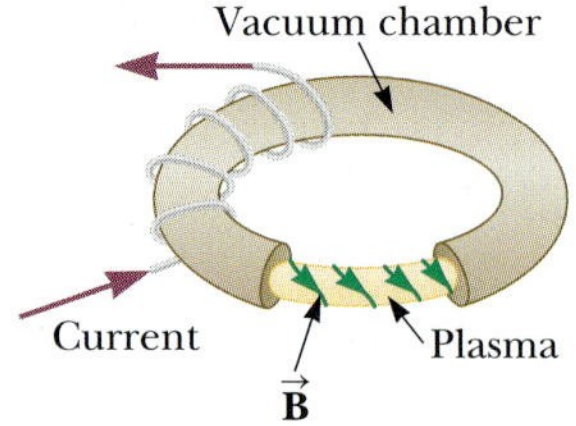


**Figure 30.4** Diagram of a tokamak used in the magnetic confinement scheme. The plasma is trapped within the spiraling magnetic field lines as shown.

In inertial confinement fusion, the fuel, typically deuterium and tritium, is put in the form of a small pellet. Directly or indirectly, powerful lasers deliver energy rapidly to the pellet, exploding off the outer layers and imploding the rest of the pellet, heating and compressing it. Shock waves form and meet at the center, greatly increasing the density and pressure and causing fusion reactions. The released energy then causes further fusion reactions. Fusion can also take place in a device the size of a TV set and in fact was invented by Philo Farnsworth, one of the pioneers of electronic television. In this method, called inertial electrostatic confinement, positively charged particles are rapidly attracted toward a negatively charged grid. Some of the positive particles then collide and fuse.

## EXAMPLE 30.2 Astrofuel on the Moon

**GOAL** Calculate the energy released in a fusion reaction.

**PROBLEM** Find the energy released in the reaction of helium-3 with deuterium:

$$^3_2\text{He} + ^2_1\text{D} \rightarrow ^4_2\text{He} + ^1_1\text{H}$$

**STRATEGY** The energy released is the difference between the mass energy of the reactants and the products.

**SOLUTION**

Add the masses on the left-hand side and subtract the masses on the right, obtaining $\Delta m$ in atomic mass units:

$$\begin{aligned}\Delta m &= m_{\text{He-3}} + m_{\text{D}} - m_{\text{He-4}} - m_{\text{H}} \\ &= 3.016\,029\text{ u} + 2.014\,102\text{ u} - 4.002\,603\text{ u} - 1.007\,825\text{ u} \\ &= 0.019\,703\text{ u}\end{aligned}$$

Convert the mass difference to an equivalent amount of energy in MeV:

$$E = (0.019\,703\text{ u})\left(\frac{931.5\text{ MeV}}{1\text{ u}}\right) = \boxed{18.35\text{ MeV}}$$

**REMARKS** The result is a large amount of energy per reaction. Helium-3 is rare on Earth but plentiful on the Moon, where it has become trapped in the fine dust of the lunar soil. Helium-3 has the advantage of producing more protons than neutrons (some neutrons are still produced by side reactions, such as D–D), but has the disadvantage of a higher ignition temperature. If fusion power plants using helium-3 became a reality, studies indicate that it would be economically advantageous to mine helium-3 robotically and return it to Earth. The energy return per dollar would be far greater than for mining coal or drilling for oil!

**QUESTION 30.2** How much energy, in joules, could be obtained from an Avogadro's number of helium-3–deuterium fusion reactions?

**EXERCISE 30.2** Find the energy yield in the fusion of two helium-3 nuclei:

$$^{3}_{2}\text{He} + ^{3}_{2}\text{He} \rightarrow\ ^{4}_{2}\text{He} + 2(^{1}_{1}\text{H})$$

**ANSWER** 12.88 MeV

# 30.3 Elementary Particles and the Fundamental Forces

**LEARNING OBJECTIVES**

1. Identify the three types of fundamental particles.
2. Identify and describe the four fundamental forces.

© Shelly Gazin/CORBIS

**Richard Feynman**
**American Physicist (1918–1988)**
Feynman, together with Julian S. Schwinger and Shinichiro Tomonaga, won the 1965 Nobel Prize in Physics for fundamental work in the principles of quantum electrodynamics. His many important contributions to physics include work on the first atomic bomb in the Manhattan project, the invention of simple diagrams to represent particle interactions graphically, the theory of the weak interaction of subatomic particles, a reformulation of quantum mechanics, and the theory of superfluid helium. Later he served on the commission investigating the *Challenger* tragedy, demonstrating the problem with the space shuttle's O-rings by dipping a scale-model O-ring in his glass of ice water and then shattering it with a hammer. He also contributed to physics education through the magnificent three-volume text *The Feynman Lectures on Physics.*

Besides the constituents of atoms—protons, electrons, and neutrons—numerous other particles can be found in high-energy experiments or observed in nature, subsequent to collisions involving cosmic rays. Unlike the highly stable protons and electrons, these particles decay rapidly, with half-lives ranging from $10^{-23}$ s to $10^{-6}$ s. There is very strong indirect evidence that most of these particles, including neutrons and protons, are combinations of more elementary particles called quarks. Quarks, leptons (the electron is an example), and the particles that convey forces (the photon is an example) are now thought to be the truly fundamental particles. The key to understanding the properties of elementary particles is the description of the forces of nature in which they participate.

All particles in nature are subject to four fundamental forces: the strong, electromagnetic, weak, and gravitational forces. The **strong force** is responsible for the tight binding of quarks to form neutrons and protons and for the nuclear force, a sort of residual strong force, binding neutrons and protons into nuclei. This force represents the "glue" that holds the nucleons together and is the strongest of all the fundamental forces. It is a very short-range force and is negligible for separations greater than about $10^{-15}$ m (the approximate size of the nucleus). The **electromagnetic force**, which is about $10^{-2}$ times the strength of the strong force, is responsible for the binding of atoms and molecules. It's a long-range force that decreases in strength as the inverse square of the separation between interacting particles. The **weak force** is a short-range nuclear force that is exhibited in the instability of certain nuclei. It's involved in the mechanism of beta decay, and its strength is only about $10^{-6}$ times that of the strong force. Finally, the **gravitational force** is a long-range force with a strength only about $10^{-43}$ times that of the strong force. Although this familiar interaction is the force that holds the planets, stars, and galaxies together, its effect on elementary particles

**Table 30.1** Particle Interactions

| Interaction (Force) | Relative Strength[a] | Range of Force | Mediating Field Particle |
|---|---|---|---|
| Strong | 1 | Short ($\approx$ 1 fm) | Gluon |
| Electromagnetic | $10^{-2}$ | Long ($\propto 1/r^2$) | Photon |
| Weak | $10^{-6}$ | Short ($\approx 10^{-3}$ fm) | $W^{\pm}$ and Z bosons |
| Gravitational | $10^{-43}$ | Long ($\propto 1/r^2$) | Graviton |

[a]For two quarks separated by $3 \times 10^{-17}$ m.

is negligible. The gravitational force is by far the weakest of all the fundamental forces.

Modern physics often describes the forces between particles in terms of the actions of field particles or quanta. In the case of the familiar electromagnetic interaction, the field particles are photons. In the language of modern physics, the electromagnetic force is *mediated* (carried) by photons, which are the quanta of the electromagnetic field. The strong force is mediated by field particles called *gluons*, the weak force is mediated by particles called the W and Z *bosons*, and the gravitational force is thought to be mediated by quanta of the gravitational field called *gravitons*. Forces between two particles are conveyed by an exchange of field quanta. This process is analogous to the covalent bond between two atoms created by an exchange or sharing of electrons. The electromagnetic interaction, for example, involves an exchange of photons.

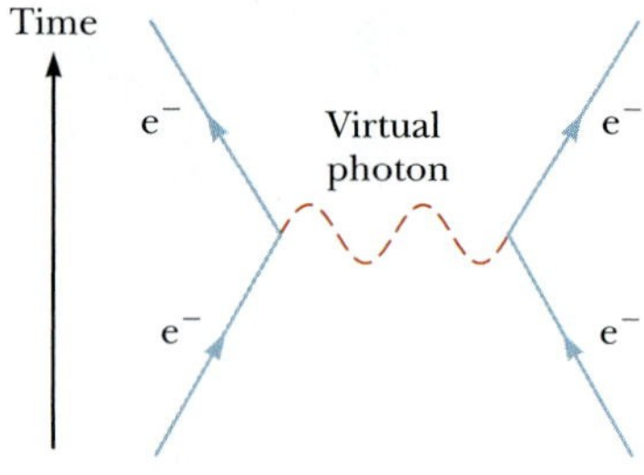


**Figure 30.5** Feynman diagram representing a photon mediating the electromagnetic force between two electrons.

The force between two particles can be understood in general with a simple illustration called a *Feynman diagram*, developed by Richard P. Feynman (1918–1988). Figure 30.5 is a Feynman diagram for the electromagnetic interaction between two electrons. In this simple case, a photon is the field particle that mediates the electromagnetic force between the electrons. The photon transfers energy and momentum from one electron to the other in the interaction. Such a photon, called a *virtual photon*, can never be detected directly because it is absorbed by the second electron very shortly after being emitted by the first electron. The existence of a virtual photon might be expected to violate the law of conservation of energy, but doesn't because of the time–energy uncertainty principle. Recall that the uncertainty principle says that the energy is uncertain or not conserved by an amount $\Delta E$ for a time $\Delta t$ such that $\Delta E\ \Delta t \approx \hbar$. If the exchange of the virtual photon happens quickly enough, the brief discrepancy in energy conservation is less than the minimum uncertainty in energy and the exchange is physically an acceptable process.

All the field quanta have been detected except for the graviton, which may never be found directly because of the weakness of the gravitational field. These interactions, their ranges, and their relative strengths are summarized in Table 30.1.

## 30.4 Positrons and Other Antiparticles

**LEARNING OBJECTIVES**

1. Discuss the existence and properties of antiparticles.
2. Discuss the medical technique of positron-emission tomography (PET).

In the 1920s theoretical physicist Paul Dirac (1902–1984) developed a version of quantum mechanics that incorporated special relativity. Dirac's theory accounted for the electron's spin and its magnetic moment, but had an apparent flaw in that it predicted negative energy states. The theory was rescued by positing the existence of an anti-electron having the same mass as an electron but the opposite charge, called a *positron*. The general and profound implication of Dirac's theory is that **for every particle, there is an antiparticle with the same mass as the particle, but the opposite charge**. An antiparticle is usually designated by a bar over the symbol for the particle. For example, $\bar{\text{p}}$ denotes the antiproton and $\bar{\nu}$ the antineutrino. In this book the notation $e^+$ is preferred for the positron. Practically every known elementary particle has a distinct antiparticle. Among the exceptions are the photon and the neutral pion ($\pi^0$), which are their own antiparticles.

In 1932 the positron was discovered by Carl Anderson in a cloud chamber experiment. To discriminate between positive and negative charges, he placed the cloud chamber in a magnetic field, causing moving charges to follow curved paths. He noticed that some of the electron-like tracks deflected in a direction corresponding to a positively charged particle: positrons.

**Tip 30.1 Antiparticles**

An antiparticle is not identified solely on the basis of opposite charge. Even neutral particles have antiparticles.

When a particle meets its antiparticle, both particles are annihilated, resulting in high-energy photons. The process of electron–positron annihilation is used in the medical diagnostic technique of positron-emission tomography (PET). The patient is injected with a glucose solution containing a radioactive substance that decays by positron emission. Examples of such substances are oxygen-15, nitrogen-13, carbon-11, and fluorine-18. The radioactive material is carried to the brain. When a decay occurs, the emitted positron annihilates with an electron in the brain tissue, resulting in two gamma ray photons. With the assistance of a computer, an image can be created of the sites in the brain where the glucose accumulates.

**BIO APPLICATION**

Positron-Emission Tomography (PET) Scanning

The images from a PET scan can point to a wide variety of disorders in the brain, including Alzheimer's disease. In addition, because glucose metabolizes more rapidly in active areas of the brain than in other parts of the body, a PET scan can indicate which areas of the brain are involved in various processes such as language, music, and vision.

# 30.5 Classification of Particles

**LEARNING OBJECTIVES**

1. Identify and classify elementary particles.
2. Compare and contrast mesons and baryons.
3. Describe basic lepton properties.

All particles other than those that transmit forces can be classified into two broad categories, hadrons and leptons, according to their interactions. The hadrons are composites of quarks, whereas the leptons are thought to be truly elementary, although there have been suggestions that they might also have internal structure.

## Hadrons

Particles that interact through the strong force are called *hadrons*. The two classes of hadrons, known as *mesons* and *baryons*, are distinguished by their masses and spins.

All mesons are known to decay finally into electrons, positrons, neutrinos, and photons. A good example of a meson is the pion ($\pi$), the lightest of the known mesons, with a mass of about 140 MeV/$c^2$ and a spin of 0. As seen in Table 30.2 (page 1006), the pion comes in three varieties, corresponding to three charge states: $\pi^+$, $\pi^-$, and $\pi^0$. Pions are highly unstable particles. For example, the $\pi^-$, which has a lifetime of about $2.6 \times 10^{-8}$ s, decays into a muon and an antineutrino. The $\mu^-$ muon, essentially a heavy electron with a lifetime of 2.2 $\mu$s, then decays into an electron, a neutrino, and an antineutrino. The sequence of decays is

$$\begin{aligned} \pi^- &\rightarrow \mu^- + \bar{\nu} \\ \mu^- &\rightarrow e^- + \nu + \bar{\nu} \end{aligned} \qquad [30.6]$$

© INTERFOTO/Alamy

**Paul Adrien Maurice Dirac**
**British physicist (1902–1984)**
Dirac was instrumental in the understanding of antimatter and in the unification of quantum mechanics and relativity. He made numerous contributions to the development of quantum physics and cosmology. In 1933 he won the Nobel Prize in Physics.

**Table 30.2** Some Particles and Their Properties

| Category | Particle Name | Symbol | Anti-particle | Mass ($MeV/c^2$) | $B$ | $L_e$ | $L_\mu$ | $L_\tau$ | $S$ | Lifetime(s) | Principal Decay Modes[a] |
|---|---|---|---|---|---|---|---|---|---|---|---|
| **Leptons** | Electron | $e^-$ | $e^+$ | 0.511 | 0 | +1 | 0 | 0 | 0 | Stable | |
| | Electron–neutrino | $\nu_e$ | $\bar{\nu}_e$ | $< 7 eV/c^2$ | 0 | +1 | 0 | 0 | 0 | Stable | |
| | Muon | $\mu^-$ | $\mu^+$ | 105.7 | 0 | 0 | +1 | 0 | 0 | $2.20 \times 10^{-6}$ | $e^-\bar{\nu}_e\nu_\mu$ |
| | Muon–neutrino | $\nu_\mu$ | $\bar{\nu}_\mu$ | $< 0.3$ | 0 | 0 | +1 | 0 | 0 | Stable | |
| | Tau | $\tau^-$ | $\tau^+$ | 1 784 | 0 | 0 | 0 | +1 | 0 | $< 4 \times 10^{-13}$ | $\mu^-\bar{\nu}_\mu\nu_\tau$, $e^-\bar{\nu}_e\nu_\tau$ |
| | Tau–neutrino | $\nu_\tau$ | $\bar{\nu}_\tau$ | $< 30$ | 0 | 0 | 0 | +1 | 0 | Stable | |
| **Hadrons** | | | | | | | | | | | |
| **Mesons** | Pion | $\pi^+$ | $\pi^-$ | 139.6 | 0 | 0 | 0 | 0 | 0 | $2.60 \times 10^{-8}$ | $\mu^+\nu_\mu$ |
| | | $\pi^0$ | Self | 135.0 | 0 | 0 | 0 | 0 | 0 | $0.83 \times 10^{-16}$ | $2\gamma$ |
| | Kaon | $K^+$ | $K^-$ | 493.7 | 0 | 0 | 0 | 0 | +1 | $1.24 \times 10^{-8}$ | $\mu^+\nu_\mu$, $\pi^+\pi^0$ |
| | | $K^0_S$ | $K^0_S$ | 497.7 | 0 | 0 | 0 | 0 | +1 | $0.89 \times 10^{-10}$ | $\pi^+\pi^-$, $2\pi^0$ |
| | | $K^0_L$ | $K^0_L$ | 497.7 | 0 | 0 | 0 | 0 | +1 | $5.2 \times 10^{-8}$ | $\pi^\pm e^\mp\bar{\nu}_e$, $3\pi^0$ |
| | | | | | | | | | | $\pi^\pm\mu^\mp\bar{\nu}_\mu$ | |
| | Eta | $\eta$ | Self | 548.8 | 0 | 0 | 0 | 0 | 0 | $< 10^{-18}$ | $2\gamma$, $3\pi$ |
| | | $\eta'$ | Self | 958 | 0 | 0 | 0 | 0 | 0 | $2.2 \times 10^{-21}$ | $\eta\pi^+\pi^-$ |
| **Baryons** | Proton | p | $\bar{p}$ | 938.3 | +1 | 0 | 0 | 0 | 0 | Stable | |
| | Neutron | n | $\bar{n}$ | 939.6 | +1 | 0 | 0 | 0 | 0 | 920 | $pe^-\bar{\nu}_e$ |
| | Lambda | $\Lambda^0$ | $\bar{\Lambda}^0$ | 1 115.6 | +1 | 0 | 0 | 0 | −1 | $2.6 \times 10^{-10}$ | $p\pi^-$, $n\pi 0$ |
| | Sigma | $\Sigma^+$ | $\bar{\Sigma}^-$ | 1 189.4 | +1 | 0 | 0 | 0 | −1 | $0.80 \times 10^{-10}$ | $p\pi^0$, $n\pi^+$ |
| | | $\Sigma^0$ | $\bar{\Sigma}^0$ | 1 192.5 | +1 | 0 | 0 | 0 | −1 | $6 \times 10^{-20}$ | $\Lambda^0\gamma$ |
| | | $\Sigma^-$ | $\bar{\Sigma}^+$ | 1 197.3 | +1 | 0 | 0 | 0 | −1 | $1.5 \times 10^{-10}$ | $n\pi^-$ |
| | Xi | $\Xi^0$ | $\bar{\Xi}^0$ | 1 315 | +1 | 0 | 0 | 0 | −2 | $2.9 \times 10^{-10}$ | $\Lambda^0\pi^0$ |
| | | $\Xi^-$ | $\Xi^+$ | 1 321 | +1 | 0 | 0 | 0 | −2 | $1.64 \times 10^{-10}$ | $\Lambda^0\pi^-$ |
| | Omega | $\Omega^-$ | $\Omega^+$ | 1 672 | +1 | 0 | 0 | 0 | −3 | $0.82 \times 10^{-10}$ | $\Xi^0\pi^-$, $\Lambda^0K^-$ |

[a]Notations in this column, such as $p\pi^-$, $n\pi^0$, mean two possible decay modes. In this case the two possible decays are $\Lambda^0 \rightarrow p + \pi^-$ and $\Lambda^0 \rightarrow n + \pi^0$.

Baryons have masses equal to or greater than the proton mass (the name *baryon* means "heavy" in Greek), and their spin is always a noninteger value ($\frac{1}{2}$ or $\frac{3}{2}$). Protons and neutrons are baryons, as are many other particles. With the exception of the proton, all baryons decay in such a way that the end products include a proton. For example, the baryon called the $\Xi$ hyperon first decays to a $\Lambda^0$ in about $10^{-10}$ s. The $\Lambda^0$ then decays to a proton and a $\pi^-$ in about $3 \times 10^{-10}$ s.

Today it is believed that hadrons are composed of quarks. Some of the important properties of hadrons are listed in Table 30.2.

## Leptons

Leptons (from the Greek *leptos,* meaning "small" or "light") are a group of particles that participate in the weak interaction. All leptons have a spin of $\frac{1}{2}$. Included in this group are electrons, muons, and neutrinos, which are all less massive than the lightest hadron. A muon is identical to an electron except that its mass is 207 times the electron mass. Although hadrons have size and structure, leptons appear to be truly elementary, with no structure down to the limit of resolution of experiment (about $10^{-19}$ m).

Unlike hadrons, the number of known leptons is small. Currently, scientists believe that there are only six leptons (each having an antiparticle): the electron, the muon, the tau, and a neutrino associated with each:

$$\begin{pmatrix} e^- \\ \nu_e \end{pmatrix} \quad \begin{pmatrix} \mu^- \\ \nu_\mu \end{pmatrix} \quad \begin{pmatrix} \tau^- \\ \nu_\tau \end{pmatrix}$$

The tau lepton, discovered in 1975, has a mass about twice that of the proton.

Although neutrinos have masses of about zero, there is strong indirect evidence that the electron neutrino has a nonzero mass of about 3 $\text{eV}/c^2$, or 1/180 000 of the electron mass. A firm knowledge of the neutrino's mass could have great significance in cosmological models and in our understanding of the future of the Universe.

# 30.6 Conservation Laws

**LEARNING OBJECTIVES**

1. State and describe the three empirical conservation laws: conservation of baryon number, conservation of lepton number, and conservation of strangeness.
2. Apply conservation laws to reactions involving elementary particles.

A number of conservation laws are important in the study of elementary particles. Although those described here have no theoretical foundation, they are supported by abundant empirical evidence.

## Baryon Number

The law of conservation of baryon number means that whenever a baryon is created in a reaction or decay, an antibaryon is also created. This information can be quantified by assigning a baryon number: $B = 1$ for all baryons, $B = -1$ for all antibaryons, and $B = 0$ for all other particles. Thus, the **law of conservation of baryon number** states that whenever a nuclear reaction or decay occurs, the sum of the baryon numbers before the process equals the sum of the baryon numbers after the process.

◀ Conservation of baryon number

Note that if the baryon number is absolutely conserved, the proton must be absolutely stable: if it were not for the law of conservation of baryon number, the proton could decay into a positron and a neutral pion. Such a decay, however, has never been observed. At present, we can only say that the proton has a half-life of at least $10^{31}$ years. (The estimated age of the Universe is about $10^{10}$ years.) In one version of a so-called grand unified theory, physicists predicted that the proton is actually unstable. According to this theory, the baryon number (sometimes called the *baryonic charge*) is not absolutely conserved, whereas electric charge is always conserved.

### EXAMPLE 30.3 Checking Baryon Numbers

**GOAL** Use conservation of baryon number to determine whether a given reaction can occur.

**PROBLEM** Determine whether the following reaction can occur based on the law of conservation of baryon number:

$$\text{p} + \text{n} \rightarrow \text{p} + \text{p} + \text{n} + \bar{\text{p}}$$

**STRATEGY** Count the baryons on both sides of the reaction, recalling that $B = 1$ for baryons and $B = -1$ for antibaryons.

**SOLUTION**

Count the baryons on the left: The neutron and proton are both baryons; hence,

$1 + 1 = 2.$

Count the baryons on the right: There are three baryons and one antibaryon, so

$1 + 1 + 1 + (-1) = 2;$ baryons conserved

**REMARKS** Baryon number is conserved in this reaction, so it can occur provided that the incoming proton has sufficient energy.

*(Continued)*

**QUESTION 30.3** True or False: A proton can't decay into a positron plus a neutrino.

**EXERCISE 30.3** Can the following reaction occur, based on the law of conservation of baryon number?

$$\mathrm{p} + \mathrm{n} \rightarrow \mathrm{p} + \mathrm{p} + \overline{\mathrm{p}}$$

**ANSWER** No. (Compute the baryon numbers on both sides and show that they're not equal.)

---

## Lepton Number

Conservation of lepton number

There are three conservation laws involving lepton numbers, one for each variety of lepton. The **law of conservation of electron-lepton number** states that the sum of the electron-lepton numbers before a reaction or decay must equal the sum of the electron-lepton numbers after the reaction or decay. The electron and the electron neutrino are assigned a positive electron-lepton number $L_e = 1$, the antileptons $e^+$ and $\overline{\nu}_e$ are assigned the electron-lepton number $L_e = -1$, and all other particles have $L_e = 0$. For example, consider neutron decay:

Neutron decay

$$\mathrm{n} \rightarrow \mathrm{p}^+ + \mathrm{e}^- + \overline{\nu}_e$$

Before the decay, the electron-lepton number is $L_e = 0$; after the decay, it is $0 + 1 + (-1) = 0$, so the electron-lepton number is conserved. It's important to recognize that baryon number must also be conserved. This can easily be seen by noting that before the decay $B = 1$, whereas after the decay $B = 1 + 0 + 0 = 1$.

Similarly, when a decay involves muons, the muon-lepton number $L_\mu$ is conserved. The $\mu^-$ and the $\nu_\mu$ are assigned $L_\mu = +1$, the antimuons $\mu^+$ and $\overline{\nu}_\mu$ are assigned $L_\mu = -1$, and all other particles have $L_\mu = 0$. Finally, the tau-lepton number $L_\tau$ is conserved, and similar assignments can be made for the $\tau$ lepton and its neutrino.

## EXAMPLE 30.4 Checking Lepton Numbers

**GOAL** Use conservation of lepton number to determine whether a given process is possible.

**PROBLEM** Determine which of the following decay schemes can occur on the basis of conservation of lepton number:

$$\textbf{(1)}\quad \mu^- \rightarrow \mathrm{e}^- + \overline{\nu}_e + \nu_\mu$$

$$\textbf{(2)}\quad \pi^+ \rightarrow \mu^+ + \nu_\mu + \nu_e$$

**STRATEGY** Count the leptons on each side and see if the numbers are equal.

**SOLUTION**

Because decay 1 involves both a muon and an electron, $L_\mu$ and $L_e$ must both be conserved. Before the decay, $L_\mu = +1$ and $L_e = 0$. After the decay, $L_\mu = 0 + 0 + 1 = +1$ and $L_e = +1 - 1 + 0 = 0$. Both lepton numbers are conserved, and on this basis, the decay mode is possible.

Before decay 2 occurs, $L_\mu = 0$ and $L_e = 0$. After the decay, $L_\mu = -1 + 1 + 0 = 0$, but $L_e = +1$. This decay isn't possible because the electron-lepton number is not conserved.

**QUESTION 30.4** Can a neutron decay into a positron and an electron? Explain.

**EXERCISE 30.4** Determine whether the decay $\tau^- \rightarrow \mu^- + \overline{\nu}_\mu$ can occur.

**ANSWER** No. (Compute lepton numbers on both sides and show that they're not equal in this case.)

---

### Quick Quiz

**30.1** Which of the following reactions cannot occur?
(a) $\mathrm{p} + \mathrm{p} \rightarrow \mathrm{p} + \mathrm{p} + \overline{\mathrm{p}}$ (b) $\mathrm{n} \rightarrow \mathrm{p} + \mathrm{e}^- + \overline{\nu}_e$
(c) $\mu^+ \rightarrow \mathrm{e}^+ + \nu_e + \overline{\nu}_\mu$ (d) $\pi^- \rightarrow \mu^- + \overline{\nu}_\mu$

**30.2** Which of the following reactions cannot occur?
(a) $\mathrm{p} + \overline{\mathrm{p}} \rightarrow 2\gamma$ (b) $\gamma + \mathrm{p} \rightarrow \mathrm{n} + \pi^0$
(c) $\pi^0 + \mathrm{n} \rightarrow \mathrm{K}^+ + \Sigma^-$ (d) $\pi^+ + \mathrm{p} \rightarrow \mathrm{K}^+ + \Sigma^+$

## Conservation of Strangeness

The K, Λ, and Σ particles exhibit unusual properties in their production and decay and hence are called *strange particles.*

One unusual property of strange particles is that they are always produced in pairs. For example, when a pion collides with a proton, two neutral strange particles are produced with high probability (Fig. 30.6) following the reaction:

$$\pi^- + p^+ \rightarrow K^0 + \Lambda^0$$

On the other hand, the reaction $\pi^- + p^+ \rightarrow K^0 + n$ has never occurred, even though it violates no known conservation laws and the energy of the pion is sufficient to initiate the reaction.

The second peculiar feature of strange particles is that although they are produced by the strong interaction at a high rate, they don't decay into particles that interact via the strong force at a very high rate. Instead, they decay very slowly, which is characteristic of the weak interaction. Their half-lives are in the range of $10^{-10}$ s to $10^{-8}$ s; most other particles that interact via the strong force have much shorter lifetimes on the order of $10^{-23}$ s.

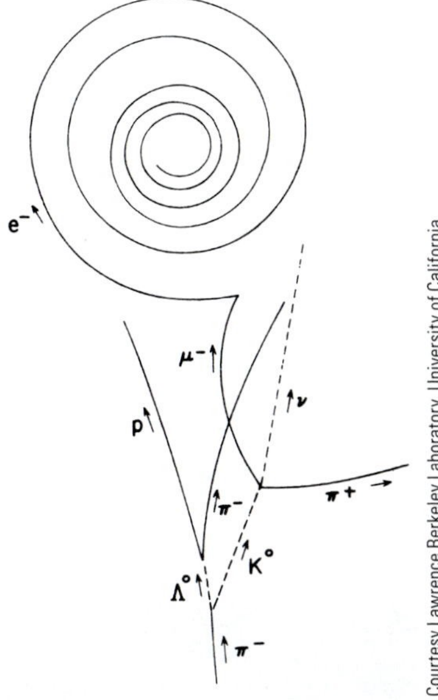



**Figure 30.6** This drawing represents tracks of many events obtained by analyzing a bubble-chamber photograph. The strange particles $\Lambda^0$ and $K^0$ are formed (at the bottom) as the $\pi^-$ interacts with a proton according to the interaction $\pi^- + p \rightarrow \Lambda^0 + K^0$. (Note that the neutral particles leave no tracks, as indicated by the dashed lines.) The $\Lambda^0$ and $K^0$ then decay according to the interactions $\Lambda^0 \rightarrow \pi + p$ and $K^0 \rightarrow \pi + \mu^- + \nu_\mu$.

To explain these unusual properties of strange particles, a law called *conservation of strangeness* was introduced, together with a new quantum number $S$ called **strangeness**. The strangeness numbers for some particles are given in Table 30.2. The production of strange particles in pairs is explained by assigning $S = +1$ to one of the particles and $S = -1$ to the other. All nonstrange particles are assigned strangeness $S = 0$. The **law of conservation of strangeness** states that whenever a nuclear reaction or decay occurs, the sum of the strangeness numbers before the process must equal the sum of the strangeness numbers after the process.

The slow decay of strange particles can be explained by assuming the strong and electromagnetic interactions obey the law of conservation of strangeness, whereas the weak interaction does not. Because the decay reaction involves the loss of one strange particle, it violates strangeness conservation and hence proceeds slowly via the weak interaction.

In checking reactions for proper strangeness conservation, the same procedure as with baryon number conservation and lepton number conservation is followed. Using Table 30.2, count the strangeness on each side. If the two results are equal, the reaction conserves strangeness.

### ■ APPLYING PHYSICS 30.2 Breaking Conservation Laws

A student claims to have observed a decay of an electron into two neutrinos traveling in opposite directions. What conservation laws would be violated by this decay?

**EXPLANATION** Several conservation laws would be violated. Conservation of electric charge would be violated because the negative charge of the electron has disappeared. Conservation of electron-lepton number would also be violated because there is one lepton before the decay and two afterward. If both neutrinos were electron neutrinos, electron-lepton number conservation would be violated in the final state. If one of the product neutrinos were other than an electron neutrino, however, another lepton conservation law would be violated because there were no other leptons in the initial state. Other conservation laws would be obeyed by this decay. Energy can be conserved; the rest energy of the electron appears as the kinetic energy (and possibly some small rest energy) of the neutrinos. The opposite directions of the two neutrinos' velocities allow for the conservation of momentum. Conservation of baryon number and conservation of other lepton numbers would also be upheld in this decay. ■

# 30.7 The Eightfold Way

**LEARNING OBJECTIVE**

1. Describe the eightfold way and its similarities with the periodic table.

Quantities such as spin, baryon number, lepton number, and strangeness are labels we associate with particles. Many classification schemes that group particles into families based on such labels have been proposed. First, consider the first eight baryons listed in Table 30.2, all having a spin of $\frac{1}{2}$. The family consists of the proton, the neutron, and six other particles. If we plot their strangeness versus their charge using a sloping coordinate system, as in Figure 30.7a, a fascinating pattern emerges: six of the baryons form a hexagon, and the remaining two are at the hexagon's center. (Particles with spin quantum number $\frac{1}{2}$ or $\frac{3}{2}$ are called *fermions*.)

Now consider the family of mesons listed in Table 30.2 with spins of zero. (Particles with spin quantum number 0 or 1 are called *bosons*.) If we count both particles and antiparticles, there are nine such mesons. Figure 30.7b is a plot of strangeness versus charge for this family. Again, a fascinating hexagonal pattern emerges. In this case the particles on the perimeter of the hexagon lie opposite their antiparticles, and the remaining three (which form their own antiparticles) are at its center. These and related symmetric patterns, called the **eightfold way**, were proposed independently in 1961 by Murray Gell-Mann and Yuval Ne'eman.

© Kevin Fleming/CORBIS

**Murray Gell-Mann**
**American Physicist (b. 1929)**
Gell-Mann was awarded the 1969 Nobel Prize in Physics for his theoretical studies dealing with subatomic particles.

The groups of baryons and mesons can be displayed in many other symmetric patterns within the framework of the eightfold way. For example, the family of spin-$\frac{3}{2}$ baryons contains ten particles arranged in a pattern like the tenpins in a bowling alley. After the pattern was proposed, one of the particles was missing; it had yet to be discovered. Gell-Mann predicted that the missing particle, which he called the *omega minus* ($\Omega^-$), should have a spin of $\frac{3}{2}$, a charge of $-1$, a strangeness of $-3$, and a mass of about 1 680 MeV/$c^2$. Shortly thereafter, in 1964, scientists at the Brookhaven National Laboratory found the missing particle through careful analyses of bubble-chamber photographs and confirmed all its predicted properties.

The patterns of the eightfold way in the field of particle physics have much in common with the periodic table. Whenever a vacancy (a missing particle or element) occurs in the organized patterns, experimentalists have a guide for their investigations.

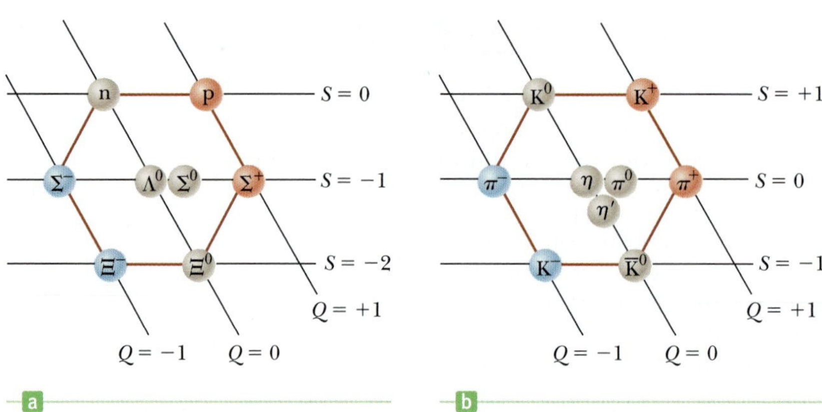


**Figure 30.7** (a) The hexagonal eightfold-way pattern for the eight spin-$\frac{1}{2}$ baryons. This strangeness versus charge plot uses a horizontal axis for the strangeness values $S$, but a sloping axis for the charge number $Q$. (b) The eightfold-way pattern for the nine spin-zero mesons.

# 30.8 Quarks and Color

**LEARNING OBJECTIVES**

1. Describe the quark model.
2. Discuss the color charge of quarks, the action of gluons, and the residual strong force.

Although leptons appear to be truly elementary particles without measurable size or structure, hadrons are more complex. There is strong evidence, including the scattering of electrons off nuclei, that hadrons are composed of more elementary particles called quarks.

## The Quark Model

According to the quark model, all hadrons are composite systems of two or three of six fundamental constituents called **quarks**, which rhymes with "sharks" (although some rhyme it with "forks"). These six quarks are given the arbitrary names *up, down, strange, charmed, bottom,* and *top,* designated by the letters u, d, s, c, b, and t.

Quarks have fractional electric charges, along with other properties, as shown in Table 30.3. Associated with each quark is an antiquark of opposite charge, baryon number, and strangeness. Mesons consist of a quark and an antiquark, whereas baryons consist of three quarks.

Table 30.4 lists the quark compositions of several mesons and baryons. Note that just two of the quarks, u and d, are contained in all hadrons encountered in ordinary matter (protons and neutrons). The third quark, s, is needed only to construct strange particles with a strangeness of either +1 or −1. Figure 30.8 is a pictorial representation of the quark compositions of several particles.

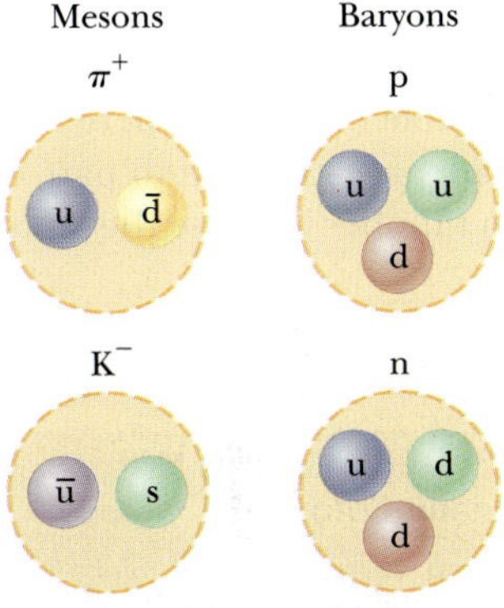


**Figure 30.8** Quark compositions of two mesons and two baryons. Note that the mesons on the left contain two quarks and that the baryons on the right contain three quarks.

**Table 30.3** Properties of Quarks and Antiquarks

| Quarks | | | | | | | | |
|---|---|---|---|---|---|---|---|---|
| **Name** | **Symbol** | **Spin** | **Charge** | **Baryon Number** | **Strange-ness** | **Charm** | **Bottom-ness** | **Top-ness** |
| Up | u | $\frac{1}{2}$ | $+\frac{2}{3}e$ | $\frac{1}{3}$ | 0 | 0 | 0 | 0 |
| Down | d | $\frac{1}{2}$ | $-\frac{1}{3}e$ | $\frac{1}{3}$ | 0 | 0 | 0 | 0 |
| Strange | s | $\frac{1}{2}$ | $-\frac{1}{3}e$ | $\frac{1}{3}$ | −1 | 0 | 0 | 0 |
| Charmed | c | $\frac{1}{2}$ | $+\frac{2}{3}e$ | $\frac{1}{3}$ | 0 | +1 | 0 | 0 |
| Bottom | b | $\frac{1}{2}$ | $-\frac{1}{3}e$ | $\frac{1}{3}$ | 0 | 0 | +1 | 0 |
| Top | t | $\frac{1}{2}$ | $+\frac{2}{3}e$ | $\frac{1}{3}$ | 0 | 0 | 0 | +1 |

| Antiquarks | | | | | | | | |
|---|---|---|---|---|---|---|---|---|
| **Name** | **Symbol** | **Spin** | **Charge** | **Baryon Number** | **Strange-ness** | **Charm** | **Bottom-ness** | **Top-ness** |
| Anti-up | $\bar{u}$ | $\frac{1}{2}$ | $-\frac{2}{3}e$ | $-\frac{1}{3}$ | 0 | 0 | 0 | 0 |
| Anti-down | $\bar{d}$ | $\frac{1}{2}$ | $+\frac{1}{3}e$ | $-\frac{1}{3}$ | 0 | 0 | 0 | 0 |
| Anti-strange | $\bar{s}$ | $\frac{1}{2}$ | $+\frac{1}{3}e$ | $-\frac{1}{3}$ | +1 | 0 | 0 | 0 |
| Anti-charmed | $\bar{c}$ | $\frac{1}{2}$ | $-\frac{2}{3}e$ | $-\frac{1}{3}$ | 0 | −1 | 0 | 0 |
| Anti-bottom | $\bar{b}$ | $\frac{1}{2}$ | $+\frac{1}{3}e$ | $-\frac{1}{3}$ | 0 | 0 | −1 | 0 |
| Anti-top | $\bar{t}$ | $\frac{1}{2}$ | $-\frac{2}{3}e$ | $-\frac{1}{3}$ | 0 | 0 | 0 | −1 |

**Table 30.4** Quark Composition of Several Hadrons

| Particle | Quark Composition |
|---|---|
| **Mesons** | |
| $\pi^+$ | $\bar{d}u$ |
| $\pi^-$ | $\bar{u}d$ |
| $K^+$ | $\bar{s}u$ |
| $K^-$ | $\bar{u}s$ |
| $K^0$ | $\bar{s}d$ |
| **Baryons** | |
| p | uud |
| n | udd |
| $\Lambda^0$ | uds |
| $\Sigma^+$ | uus |
| $\Sigma^0$ | uds |
| $\Sigma^-$ | dds |
| $\Xi^0$ | uss |
| $\Xi^-$ | dss |
| $\Omega^-$ | sss |

The charmed, bottom, and top quarks are more massive than the other quarks and occur in higher-energy interactions. Each has its own quantum number, called charm, bottomness, and topness, respectively. An example of a hadron formed from these quarks is the J/Ψ particle, also called charmonium, which is composed of a charmed quark and an anticharmed quark, $c\bar{c}$.

## APPLYING PHYSICS 30.3 Conservation of Meson Number

We have seen a law of conservation of lepton number and a law of conservation of baryon number. Why isn't there a law of conservation of meson number?

**EXPLANATION** We can answer this question from the point of view of creating particle–antiparticle pairs from available energy. If energy is converted to the rest energy of a lepton–antilepton pair, there is no net change in lepton number because the lepton has a lepton number of +1 and the antilepton −1. Energy could also be transformed into the rest energy of a baryon–antibaryon pair. The baryon has baryon number +1, the antibaryon −1, and there is no net change in baryon number.

Now suppose energy is transformed into the rest energy of a quark–antiquark pair. By definition in quark theory, a quark–antiquark pair is a meson. In this reaction, therefore, the number of mesons increases from zero to one, so meson number is not conserved. ■

**Tip 30.2 Color Is Not Really Color**

When we use the word *color* to describe a quark, it has nothing to do with visual sensation from light. It is simply a convenient name for a property analogous to electric charge.

## Color

Quarks have another property called **color** or **color charge**. This property isn't color in the visual sense; rather, it's just a label for something analogous to electric charge. Quarks are said to come in three colors: red, green, and blue. Antiquarks have the properties antired, antigreen, and antiblue.

Color was defined because some quark combinations appeared to violate the Pauli exclusion principle. An example is the omega-minus particle ($\Omega^-$), which consists of three strange quarks, sss, that are all spin up, giving a spin of $\frac{3}{2}$. Each strange quark is assumed to have a different color and hence is in a distinct quantum state, satisfying the exclusion principle.

In general, quark combinations must be "colorless." A meson consists of a quark of one color and an antiquark of the corresponding anticolor. Baryons must consist of one red, one green, and one blue quark, or their anticolors.

The theory of how quarks interact with one another by means of color charge is called **quantum chromodynamics**, or QCD, to parallel quantum electrodynamics (the theory of interactions among electric charges). The strong force between quarks is often called the **color force**. The force is carried by massless particles called **gluons** (which are analogous to photons for the electromagnetic force). According to QCD, there are eight gluons, all with color charge, and their antigluons. When a quark emits or absorbs a gluon, its color changes. For example, a blue quark that emits a gluon may become a red quark, and a red quark that absorbs this gluon becomes a blue quark. The color force between quarks is analogous to the electric force between charges: like colors repel and opposite colors attract. Therefore, two red quarks repel each other, but a red quark will be attracted to an antired quark. The attraction between quarks of opposite color to form a meson ($q\bar{q}$) is indicated in Figure 30.9a.

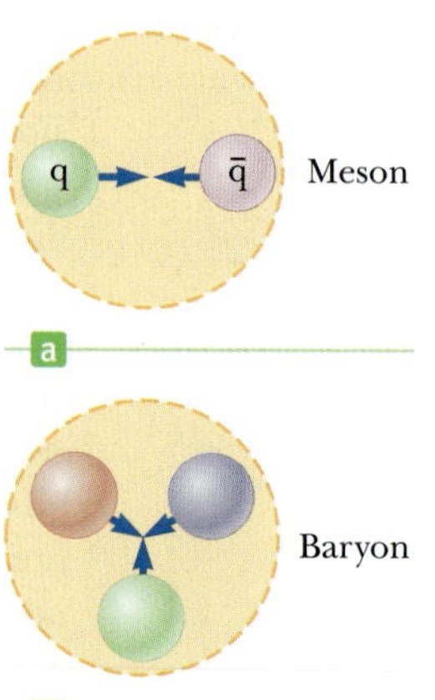


**Figure 30.9** (a) A green quark is attracted to an antigreen quark to form a meson with quark structure ($q\bar{q}$). (b) Three different-colored quarks attract one another to form a baryon.

Different-colored quarks also attract one another, but with less intensity than opposite colors of quark and antiquark. For example, a cluster of red, blue, and green quarks all attract one another to form baryons, as indicated in Figure 30.9b. Every baryon contains three quarks of three different colors.

Although the color force between two color-neutral hadrons (such as a proton and a neutron) is negligible at large separations, the strong color force between their constituent quarks does not exactly cancel at small separations of about 1 fm. **This residual strong force is in fact the nuclear force that binds protons and neutrons to form nuclei.** It is similar to the residual electromagnetic force that binds neutral atoms into molecules.

# 30.9 Electroweak Theory and the Standard Model

**LEARNING OBJECTIVES**

1. Discuss the importance of the weak interaction and its unusual properties.
2. Discuss the electroweak theory, the Standard Model and grand unification theories.

Recall that the weak interaction is an extremely short range force having an interaction distance of approximately $10^{-18}$ m. Such a short-range interaction implies that the quantized particles that carry the weak field (the spin 1 $W^+$, $W^-$, and $Z^0$ bosons) are extremely massive, as is indeed the case. These amazing bosons can be thought of as structureless, point-like particles as massive as krypton atoms! The weak interaction is responsible for the decay of the c, s, b, and t quarks into lighter, more stable u and d quarks, as well as the decay of the massive $\mu$ and $\tau$ leptons into (lighter) electrons. **The weak interaction is very important because it governs the stability of the basic particles of matter.**

A mysterious feature of the weak interaction is its lack of symmetry, especially when compared with the high degree of symmetry shown by the strong, electromagnetic, and gravitational interactions. For example, the weak interaction, unlike the strong interaction, is not symmetric under mirror reflection or charge exchange. (*Mirror reflection* means that all the quantities in a given particle reaction are exchanged as in a mirror reflection: left for right, an inward motion toward the mirror for an outward motion, and so forth. *Charge exchange* means that all the electric charges in a particle reaction are converted to their opposites: all positives to negatives and vice versa.) Not symmetric means that the reaction with all quantities changed occurs less frequently than the direct reaction. For example, the decay of the $K^0$, which is governed by the weak interaction, is not symmetric under charge exchange because the reaction $K^0 \rightarrow \pi^- + e^+ + \nu_e$ occurs much more frequently than the reaction $K^0 \rightarrow \pi^+ + e^- + \overline{\nu}_e$.

CERN

An engineer tests the electronics associated with a superconducting magnet in the Large Hadron Collider at the European Laboratory for Particle Physics, operated by CERN. (The acronym stands for Conseil Européen pour la Recherche Nucléaire, and has been retained although "Conseil," or "Council," has been replaced by "Organisation".)

The **electroweak theory** unifies the electromagnetic and weak interactions. This theory postulates that the weak and electromagnetic interactions have the same strength at very high particle energies and are different manifestations of a single unifying electroweak interaction. The photon and the three massive bosons ($W^{\pm}$ and $Z^0$) play key roles in the electroweak theory. The theory makes many concrete predictions, such as the prediction of the masses of the W and Z particles at about 82 GeV/$c^2$ and 93 GeV/$c^2$, respectively. These predictions have been experimentally verified.

The combination of the electroweak theory and QCD for the strong interaction forms what is referred to in high-energy physics as the **Standard Model**. Although the details of the Standard Model are complex, its essential ingredients can be summarized with the help of Figure 30.10. The strong force, mediated by gluons,

Matter and Energy

| FORCES | GAUGE BOSONS |
|---|---|
| Strong | Gluon |
| Electromagnetic | Photon |
| Weak | W and Z bosons |
| Gravity | Graviton |

FUNDAMENTAL PARTICLES

| Quarks | | | Charge: |
|---|---|---|---|
| u | c | t | $+\frac{2}{3}e$ |
| d | s | b | $-\frac{1}{3}e$ |

| Leptons | | | |
|---|---|---|---|
| e | $\mu$ | $\tau$ | $-e$ |
| $\nu_e$ | $\nu_\mu$ | $\nu_\tau$ | 0 |

Mass →

**Figure 30.10** The Standard Model of particle physics.

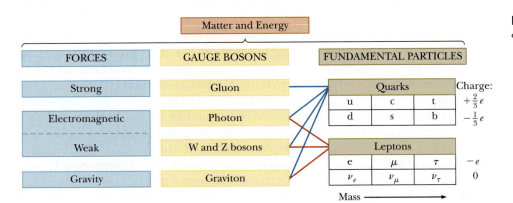

holds quarks together to form composite particles such as protons, neutrons, and mesons. Leptons participate only in the electromagnetic and weak interactions. The electromagnetic force is mediated by photons, and the weak force is mediated by W and Z bosons. Note that all fundamental forces are mediated by bosons (particles with spin 1) having properties given, to a large extent, by symmetries involved in the theories.

The Standard Model, however, doesn't answer all questions. A major question is why the photon has no mass although the W and Z bosons do. Because of this mass difference, the electromagnetic and weak forces are very different at low energies but become similar in nature at very high energies, where the rest energies of the W and Z bosons are insignificant fractions of their total energies. This behavior during the transition from high to low energies, called **symmetry breaking**, doesn't answer the question of the origin of particle masses. To resolve that problem, a hypothetical particle called the **Higgs boson**, which provides a mechanism for breaking the electroweak symmetry and bestowing different particle masses on different particles, has been proposed. The Standard Model, including the Higgs mechanism, provides a logically consistent explanation of the massive nature of the W and Z bosons. Although elusive for many years, scientists at CERN in July of 2012 reported observing a Higgs-like particle. In March of 2013 they announced that the particle they had observed, after further study, was in fact a certain kind of Higgs boson. Much further work, however, remains to be done before these particles can be fully understood and characterized.

Following the success of the electroweak theory, scientists attempted to combine it with QCD in a **grand unification theory** (GUT). In this model the electroweak force was merged with the strong color force to form a grand unified force. One version of the theory considers leptons and quarks as members of the same family that are able to change into each other by exchanging an appropriate particle. Many GUT theories predict that protons are unstable and will decay with a lifetime of about $10^{31}$ years, a period far greater than the age of the Universe. As yet, proton decays have not been observed.

AP Photo

**George Gamow**
**Russian Physicist (1904–1968)**
Gamow and two of his students, Ralph Alpher and Robert Herman, were the first to take the first half hour of the Universe seriously. In a mostly overlooked paper published in 1948, they made truly remarkable cosmological predictions. They correctly calculated the abundances of hydrogen and helium after the first half hour (75% H and 25% He) and predicted that radiation from the Big Bang should still be present and have an apparent temperature of about 5 K.

## 30.10 The Cosmic Connection

**LEARNING OBJECTIVES**

1. Discuss the Big Bang theory and the evolution of the four fundamental forces.
2. Discuss observations of the cosmic background radiation and its support for the Big Bang theory.

According to the Big Bang theory, the Universe erupted from an infinitely dense singularity about 15 billion to 20 billion years ago. The first few minutes after the Big Bang saw such extremes of energy that it is believed that all four interactions of physics were unified and all matter was contained in an undifferentiated "quark soup."

The evolution of the four fundamental forces from the Big Bang to the present is shown in Figure 30.11. During the first $10^{-43}$ s (the ultrahot epoch, with $T < 10^{32}$ K), the strong, electroweak, and gravitational forces were joined to form a completely unified force. In the first $10^{-35}$ s following the Big Bang (the hot epoch, with $T < 10^{29}$ K), gravity broke free of this unification and the strong and electroweak forces remained as one, described by a grand unification theory. During this period, particle energies were so great ($> 10^{16}$ GeV) that very massive particles as well as quarks, leptons, and their antiparticles existed. Then, after $10^{-35}$ s, the Universe rapidly expanded and cooled (the warm epoch, with $T < 10^{29}$ to $10^{15}$ K), the strong and electroweak forces parted company, and the grand unification scheme was broken. As the Universe continued to cool, the

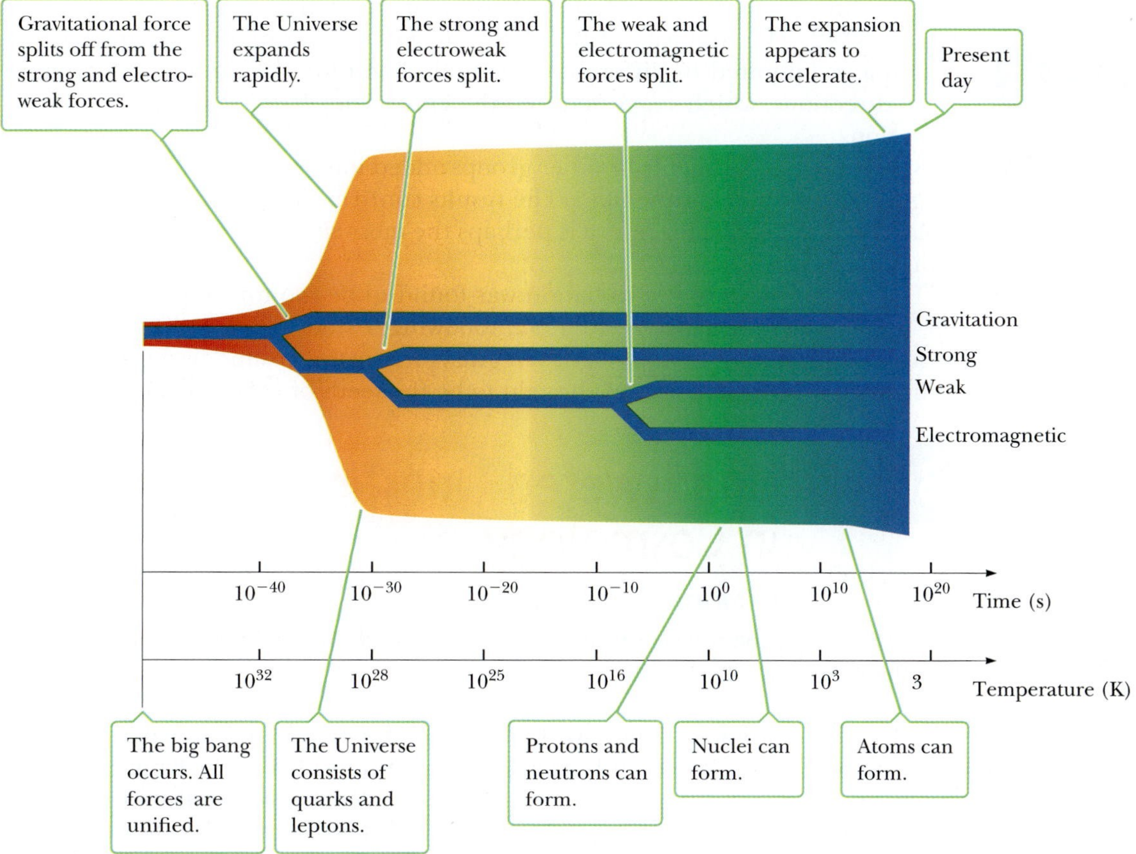


**Figure 30.11** A brief history of the Universe from the Big Bang to the present. The four forces became distinguishable during the first microsecond. Then, all the quarks combined to form particles that interact via the strong force. The leptons remained separate, however, and exist as individually observable particles to this day.

electroweak force split into the weak force and the electromagnetic force about $10^{-10}$ s after the Big Bang.

After a few minutes, protons condensed out of the hot soup. For half an hour, the Universe underwent thermonuclear detonation, exploding like a hydrogen bomb and producing most of the helium nuclei now present. The Universe continued to expand, and its temperature dropped. Until about 700 000 years after the Big Bang, the Universe was dominated by radiation. Energetic radiation prevented matter from forming single hydrogen atoms because collisions would instantly ionize any atoms that might form. Photons underwent continuous Compton scattering from the vast number of free electrons, resulting in a Universe that was opaque to radiation. By the time the Universe was about 700 000 years old, it had expanded and cooled to about 3 000 K. Protons could now bind to electrons to form neutral hydrogen atoms, and the Universe suddenly became transparent to photons. Radiation no longer dominated the Universe, and clumps of neutral matter grew steadily: first atoms, followed by molecules, gas clouds, stars, and finally galaxies.

## Observation of Radiation from the Primordial Fireball

In 1965 Arno A. Penzias (b. 1933) and Robert W. Wilson (b. 1936) of Bell Laboratories made an amazing discovery while testing a sensitive microwave receiver. A pesky signal producing a faint background hiss was interfering with their satellite communications experiments. Despite all their efforts, the signal remained. Ultimately, it became clear that they were observing microwave background radiation (at a wavelength of 7.35 cm) representing the leftover "glow" from the Big Bang.

**Figure 30.12** Robert W. Wilson (*left*) and Arno A. Penzias (*right*), with Bell Telephone Laboratories' horn-reflector antenna.

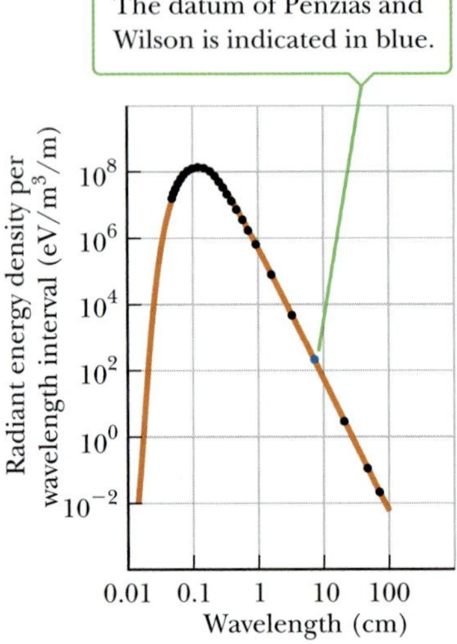


**Figure 30.13** Theoretical blackbody (rust-colored curve) and measured radiation spectra (black points) of the Big Bang. Most of the data were collected from the Cosmic Background Explorer (COBE) satellite.

The microwave horn that served as their receiving antenna is shown in Figure 30.12. The intensity of the detected signal remained unchanged as the antenna was pointed in different directions. The radiation had equal strength in all directions, which suggested that the entire Universe was the source of this radiation.

Subsequent experiments by other groups added intensity data at different wavelengths, as shown in Figure 30.13. The results confirm that the radiation is that of a blackbody at 2.9 K. This figure is perhaps the most clear-cut evidence for the Big Bang theory.

The cosmic background radiation was found to be too uniform to have led to the development of galaxies. In 1992, following a study using the Cosmic Background Explorer, or COBE, slight irregularities in the cosmic background were found. These irregularities are thought to be the seeds of galaxy formation.

# 30.11 Unanswered Questions in Cosmology

**LEARNING OBJECTIVE**

1. Discuss unanswered questions relating to dark matter and dark energy and their relevance to the fate of the universe.

In the past decade new data have raised questions that many consider to be the most important in science today. At issue is the composition of the Universe, which is closely tied to its ultimate fate. One of these questions concerns the rate at which stars orbit the galaxy, explained by a postulated material called **dark matter**. Although evidence for its existence was noted by Fritz Zwicky in 1933, only relatively recently has it become a dominant field of inquiry. The other question involves the accelerating expansion of the Universe discovered in 1998, attributed to an equally mysterious material called **dark energy**.

## Dark Matter

When the velocities of stars in our galaxy are measured, it is found they are traveling too fast to remain bound by gravity to the Milky Way, if the mass of the galaxy is due to that found in luminous stars. Figure 30.14a shows the velocity versus radial distance curve for bodies circling the Sun. As the distance from the Sun increases, the velocity of planetary bodies decreases, a consequence of the inverse square law of gravitation. Figure 30.14b, on the other hand, shows the velocity curve of stars in the Milky Way galaxy. The curve increases and flattens out but doesn't decline, meaning the stars are traveling much faster than expected if primarily under the influence of gravitation from visible stars. Traveling at higher than the expected galactic escape speed, the stars should leave the galaxy, yet remain in their orbits. Similar observations have been made of stars in other galaxies.

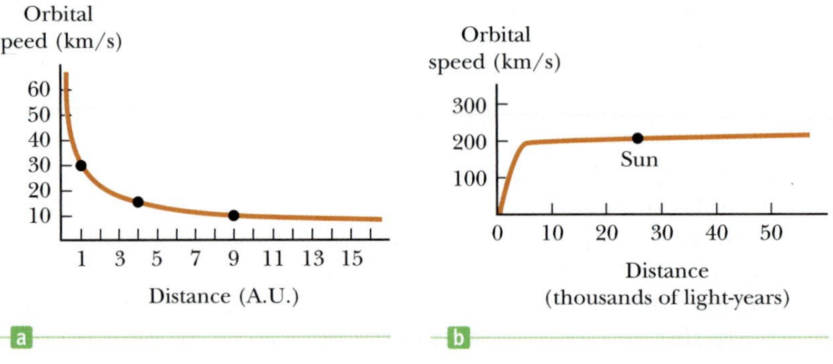


**Figure 30.14** (a) Velocity versus radial distance curve for bodies circling the Sun. (b) Velocity curve of stars in the Milky Way galaxy.

Two general theories have been advanced to account for the behavior of too-rapidly moving stars: either there is a new form of dark matter that has not been directly observed, or the law of gravitation must become stronger than an inverse square at long range. From the velocity profile of stars, 90% of the matter in the galaxy would consist of the hypothetical dark matter. Among the candidates for dark matter are neutrinos, which due to "neutrino oscillation," the spontaneous changing from one type of neutrino into another, are now thought to have mass. All stars emit enormous numbers of neutrinos every second, so if neutrinos had even a small mass, they could account for the dark matter. Another hypothetical candidate is a WIMP, a weakly interacting massive particle left over from the Big Bang. Because other galaxies have rotation curves similar to the Milky Way's, it may well be that dark matter predominates over ordinary matter in the Universe at large.

The leading alternate explanation for the galactic rotation curves is that Newton's law of gravitation doesn't hold over large distances. That theory, called MOND (Modified Newtonian Dynamics), has received a great deal of attention but thus far has not worked well enough to gain widespread acceptance. Some researchers have also tried to account for the rotation curves of galaxies by using Einstein's theory of gravity, general relativity. Finally, it is entirely possible that the correct theory may require both new kinds of matter and a modification of gravity theory.

## Dark Energy and the Accelerating Universe

By 1998 two groups of astronomers, one led by Brian Schmidt and Adam Riess and the other by Saul Perlmutter, had made highly accurate new measurements of the distances to other galaxies using Type 1a supernovae. Those observations showed that the Universe is both expanding and accelerating! The accelerated expansion can't be caused by normal matter nor by dark matter because they exert an attractive gravitational force. Instead, it is thought that a new kind of matter, called **dark energy**, exerts a repulsive force that causes the Universe to expand more rapidly than is predicted by Einstein's theory of general relativity. Figure 30.15 shows the theorized proportions of matter, dark matter, and dark energy. Normal atoms, the kind we're made of, comprise only about 4% of the Universe, whereas approximately 23% is dark matter and 73% is dark energy.

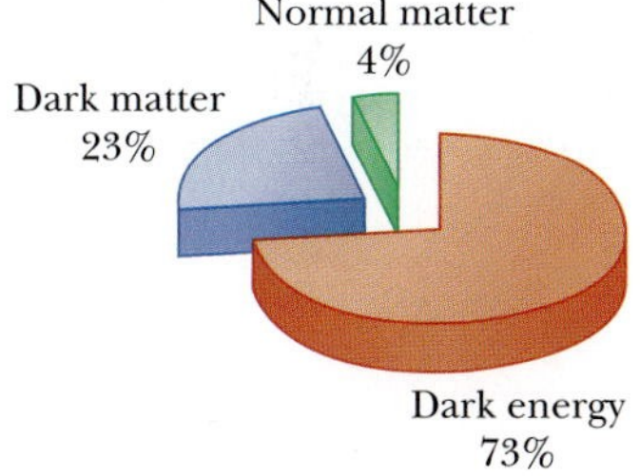


**Figure 30.15** The theorized composition of the Universe. Normal matter, as found on Earth and in the Sun, comprises only about 4% of the material in the Universe. The unknown material causing increased gravitational attraction on the galactic scale is called dark matter, whereas the similarly unknown material causing the accelerated expansion of the Universe is called dark energy.

Einstein introduced a cosmological constant into his theory of general relativity in order to explain why the Universe appeared not to change with time. The cosmological constant provided a repulsive force sufficient to prevent the matter of the Universe from collapsing under the attractive influence of gravity. When Edwin Hubble's observations of the red shift of galaxies led to the notion of a dynamically expanding universe, Einstein called the cosmological constant "the biggest blunder" of his life. That same cosmological constant can now produce a good model of the accelerating universe, turning his blunder into something of a triumph. The cosmological constant doesn't completely solve the mystery, however, because the origin of the cosmological constant hasn't been explained. As in the case of the galactic rotation curves, it isn't known whether the accelerating universe is a consequence of a new form of matter or energy, or an indication that the standard theories of cosmology derived from general relativity are in need of modification.

## The Evolution and Fate of the Universe

Unsolved questions remain about the origin and early evolution of the Universe. Although the Big Bang model explains why galaxies seem to be flying away from us, several observational problems have emerged that can't be fully explained by the Big Bang hypothesis alone.

First of all, the Universe, as measured by the temperature of the microwave background, is altogether too uniform. It's as if the entire Universe were in equilibrium.

For a system to be in equilibrium, its constituents must be able to exchange energy, arriving after a certain passage of time to a uniform temperature. How could this equilibrium be achieved, however, when different parts of the Universe are so far apart from each other that they could not possibly exchange energy? That mystery is called the horizon problem.

Second, the measurements of the cosmic microwave background strongly suggest that the Universe has a flat geometry. Figure 30.16 shows the standard three fates of the Universe, derived with Einstein's theory of general relativity by graphing the expansion factor, $R$, versus cosmic time. The expansion factor may be thought of as giving a measure of the size of the Universe, like a cosmic radius. A flat universe is expected to expand forever, although in the limit as time goes to infinity the expansion rate gradually slows to zero. A flat universe, however, is a state of unstable equilibrium, like a pencil standing on its point. With a small deviation one way or the other, either the Universe would collapse again as on the lower curve in Figure 30.16, or expand forever as in the upper curve in Figure 30.16. To be flat now, the Universe had to be flat also at the beginning of the Universe to extremely high accuracy. It is extremely improbable that the Universe was so finely tuned early in its evolution. That fine-tuning is called the flatness problem.

A third problem arises when particle theories are combined with cosmology. Studies of the standard model of particle physics in the early Universe show that large numbers of magnetic monopoles should have been created in the early Universe, so many that hundreds of thousands of them would pass through our bodies every second. Magnetic monopoles are tiny magnets consisting of an isolated north or south pole, and despite calculations predicting them to be common, they have never been observed. That is called the monopole problem.

In 1981 Alan Guth, now at MIT, proposed the inflationary model of the Universe to resolve these three problems with a single mechanism. In this model, an as yet unidentified field called the **inflaton field** caused the Universe to enter into a very rapid exponential inflation, expanding $10^{32}$ times in size in a tiny fraction of a second.

That accelerated expansion in the very early Universe would solve the monopole problem by making them so dilute that very few would exist in the observable Universe. Further, because the Universe was much smaller just prior to inflation, it would be in thermal equilibrium and, consequently, after the expansion would remain similar in all places and in all directions, solving the horizon problem. Finally, the rapid inflation would cause the curvature of spacetime to appear to flatten out, just as the Earth appears flat to those on its surface because only a very small portion of the entire Earth is visible in a given locality. That solves the flatness problem.

After the brief inflationary epoch the Universe could continue to expand normally. Whereas there is no definitive evidence that the inflationary Universe is

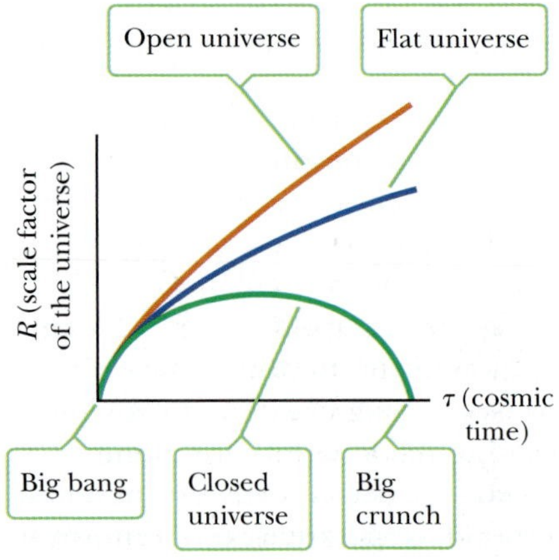


**Figure 30.16** The three fates of the Universe, according to Einstein's theory of general relativity. With a sufficient quantity of attractive matter, the universe would initially expand but eventually collapse back into a "big crunch." A flat universe would expand forever, with the expansion slowing to zero in the limit as cosmic time $\tau$ goes to infinity. A hyperbolic (or open) universe would accelerate forever. A dark energy universe, or equivalently, one due to a positive cosmological constant, would be similar to the hyperbolic universe but curve upward.

correct, it is currently the most accepted working hypothesis for how the early Universe evolved. Some researchers have attempted to combine inflation and dark energy into a single theory called "quintessence." To date, however, no single theory explaining the origins of either early inflation or later universal acceleration has found general support among cosmologists.

## 30.12 Problems and Perspectives

**LEARNING OBJECTIVE**

1. Consider and discuss the progress of science on the smallest and largest length scales.

While particle physicists have been exploring the realm of the very small, cosmologists have been exploring cosmic history back to the first microsecond of the Big Bang. Observation of the events that occur when two particles collide in an accelerator is essential in reconstructing the early moments in cosmic history. Perhaps the key to understanding the early Universe is first to understand the world of elementary particles.

Our understanding of physics at short and long distances is far from complete. Particle physicists are faced with many unanswered questions. Why is there so little antimatter in the Universe? Do neutrinos have a small mass, and if so, how much do they contribute to the "dark matter" holding the Universe together gravitationally? How can we understand the latest astronomical measurements, which show that the expansion of the Universe is accelerating and that there may be a kind of "antigravity force," or dark energy, acting between widely separated galaxies? Is it possible to unify the strong and electroweak theories in a logical and consistent manner? Why do quarks and leptons form three similar but distinct families? Are muons the same as electrons (apart from their different masses), or do they have subtle differences that have not been detected? Why are some particles charged and others neutral? Why do quarks carry a fractional charge? What determines the masses of the fundamental particles? The questions go on and on. Because of the rapid advances and new discoveries in the related fields of particle physics and cosmology, by the time you read this book some of these questions may have been resolved and others may have emerged.

An important question that remains is whether leptons and quarks have a substructure. Many physicists believe that the fundamental quantities are not infinitesimal points, but extremely tiny vibrating strings. Despite more than three decades of string theory research by thousands of physicists, however, a final Theory of Everything hasn't been found. Whether there is a limit to knowledge is an open question.

## SUMMARY

### 30.1 Nuclear Fission

In **nuclear fission** the total mass of the products is always less than the original mass of the reactants. Nuclear fission occurs when a heavy nucleus splits, or fissions, into two smaller nuclei. The lost mass is transformed into energy, electromagnetic radiation, and the kinetic energy of daughter particles.

A **nuclear reactor** is a system designed to maintain a self-sustaining chain reaction. Nuclear reactors using controlled fission events are currently being used to generate electric power. A useful parameter for describing the level of reactor operation is the reproduction constant $K$, which is the average number of neutrons from each fission event that will cause another event. A self-sustaining reaction is achieved when $K = 1$.

### 30.2 Nuclear Fusion

In nuclear fusion two light nuclei combine to form a heavier nucleus. This type of nuclear reaction occurs in the Sun, assisted by a quantum tunneling process that helps particles get through the Coulomb barrier.

Controlled fusion events offer the hope of plentiful supplies of energy in the future. The nuclear fusion reactor is considered by many scientists to be the ultimate energy source because its fuel is water. **Lawson's criterion** states that a fusion reactor will provide a net output power if the product of the plasma ion density $n$ and the plasma confinement time $\tau$ satisfies the following relationships:

$n\tau \geq 10^{14}$ s/cm$^3$ Deuterium–tritium interaction

$n\tau \geq 10^{16}$ s/cm$^3$ Deuterium–deuterium interaction [30.5]

## 30.3 Elementary Particles and the Fundamental Forces

There are four fundamental forces of nature: the **strong** (hadronic), **electromagnetic**, **weak**, and **gravitational** forces. The strong force is the force between nucleons that keeps the nucleus together. The weak force is responsible for beta decay. The electromagnetic and weak forces are now considered to be manifestations of a single force called the **electroweak** force.

Every fundamental interaction is said to be mediated by the exchange of field particles. The electromagnetic interaction is mediated by the photon, the weak interaction by the $W^{\pm}$ and $Z^0$ bosons, the gravitational interaction by gravitons, and the strong interaction by gluons.

## 30.4 Positrons and Other Antiparticles

An antiparticle and a particle have the same mass, but opposite charge, and may also have other properties with opposite values, such as lepton number and baryon number. It is possible to produce particle–antiparticle pairs in nuclear reactions if the available energy is greater than $2mc^2$, where $m$ is the mass of the particle (or antiparticle).

## 30.5 Classification of Particles

Particles other than photons are classified as hadrons or leptons. **Hadrons** interact primarily through the strong force. They have size and structure and hence are not elementary particles. There are two types of hadrons: *baryons* and *mesons*. Mesons have a baryon number of zero and have either zero or integer spin. Baryons, which generally are the most massive particles, have nonzero baryon numbers and spins of $\frac{1}{2}$ or $\frac{3}{2}$. The neutron and proton are examples of baryons.

**Leptons** have no known structure, down to the limits of current resolution (about $10^{-19}$ m). Leptons interact only through the weak and electromagnetic forces. There are six leptons: the electron, $e^-$; the muon, $\mu^-$; the tau, $\tau^-$; and their associated neutrinos, $\nu_e$, $\nu_\mu$, and $\nu_\tau$.

## 30.6 Conservation Laws

In all reactions and decays, quantities such as energy, linear momentum, angular momentum, electric charge, baryon number, and lepton number are strictly conserved. Certain particles have properties called **strangeness** and **charm**. These unusual properties are conserved only in those reactions and decays that occur via the strong force.

## 30.8 Quarks and Color

Recent theories postulate that all hadrons are composed of smaller units known as **quarks,** which have fractional electric charges and baryon numbers of $\frac{1}{3}$ and come in six "flavors": up, down, strange, charmed, top, and bottom. Each baryon contains three quarks, and each meson contains one quark and one antiquark.

According to the theory of **quantum chromodynamics**, quarks have a property called **color**, and the strong force between quarks is referred to as the **color force**. The color force increases as the distance between particles increases, so quarks are confined and are never observed in isolation. When two bound quarks are widely separated, a new quark–antiquark pair forms between them, and the single particle breaks into two new particles, each composed of a quark–antiquark pair.

## 30.10 The Cosmic Connection

Observation of background microwave radiation by Penzias and Wilson strongly confirmed that the Universe started with a Big Bang about 15 billion years ago and has been expanding ever since. The background radiation is equivalent to that of a blackbody at a temperature of about 3 K.

The cosmic microwave background has very small irregularities, corresponding to temperature variations of 0.000 3 K. Without these irregularities acting as nucleation sites, particles would never have clumped together to form galaxies and stars.

## WARM-UP EXERCISES

ENHANCED WebAssign The warm-up exercises in this chapter may be assigned online in Enhanced WebAssign.

1. **Physics Review** The first atomic bomb test released an amount of energy equivalent to approximately 17 kilotons of TNT. Determine (a) the energy in joules released by the explosion and (b) the mass converted into energy during this event. *Note:* One ton of TNT has an energy equivalent of $4.18 \times 10^9$ J. (See Section 26.7.)

2. Natural uranium ore contains about 0.720% of the fissile uranium-235 isotope. Suppose a sample of uranium ore contains $2.50 \times 10^{28}$ uranium nuclei. Determine the number of uranium-235 nuclei in the sample. (See Section 30.1.)

3. A typical uranium-235 fission event releases 208 MeV of energy. Determine (a) the energy released per event in joules and (b) the change in mass during the event. (See Section 30.1.)

4. The proton–proton cycle responsible for the Sun's $3.84 \times 10^{26}$ W power output yields about 26.7 MeV of energy for every four protons that are fused into a helium nucleus. Determine (a) the energy in joules

released during each proton–proton cycle fusion reaction, (b) the number of proton–proton cycles occurring per second in the Sun, and (c) the reduction in the Sun's mass each second due to this energy release. (See Section 30.2.)

**5.** Suppose a deuterium–deuterium fusion reactor is designed to have a plasma confinement time of 1.50 s. Determine the minimum ion density required to obtain a net power output from the reactor. (See Section 30.2.)

**6.** The annihilation of an electron and a positron, each with negligible kinetic energy, results in the production of two photons with the same energy. Determine (a) the energy of each photon in MeV and (b) the wavelength of each photon. (See Section 30.4.)

**7.** Determine the baryon number of the reaction $p + \bar{p} \rightarrow 2\gamma$. (See Sections 30.5 and 30.6.)

**8.** Determine (a) the baryon number and (b) the electron-lepton number of the reaction $\Omega^- \rightarrow \Xi^0 + e^- + \bar{\nu}_e$. (See Sections 30.5 and 30.6.)

**9.** Determine the muon-lepton number in the reaction $\mu^- \rightarrow e^- + \bar{\nu}_e + \nu_\mu$. (See Sections 30.5 and 30.6.)

**10.** Determine the value of strangeness in the reaction $\pi^- + p \rightarrow \Lambda^0 + K^0$. (See Sections 30.5 and 30.6.)

## CONCEPTUAL QUESTIONS

ENHANCED WebAssign The conceptual questions in this chapter may be assigned online in Enhanced WebAssign.

**1.** If high-energy electrons with de Broglie wavelengths smaller than the size of the nucleus are scattered from nuclei, the behavior of the electrons is consistent with scattering from very massive structures much smaller in size than the nucleus, namely, quarks. How is this behavior similar to a classic experiment that detected small structures in an atom?

**2.** What factors make a fusion reaction difficult to achieve?

**3.** Doubly charged baryons are known to exist. Why are there no doubly charged mesons?

**4.** Why would a fusion reactor produce less radioactive waste than a fission reactor?

**5.** Why didn't atoms exist until hundreds of thousands of years after the Big Bang?

**6.** Particles known as resonances have very short half-lives, on the order of $10^{-23}$ s. Would you guess that they are hadrons or leptons? Explain.

**7.** Describe the quark model of hadrons, including the properties of quarks.

**8.** In the theory of quantum chromodynamics, quarks come in three colors. How would you justify the statement, "All baryons and mesons are colorless"?

**9.** Describe the properties of baryons and mesons and the important differences between them.

**10.** Identify the particle decays in Table 30.2 that occur by the electromagnetic interaction. Justify your answer.

**11.** Kaons all decay into final states that contain no protons or neutrons. What is the baryon number of kaons?

**12.** Why is a neutron stable inside the nucleus? (In free space the neutron decays in 900 s.)

## PROBLEMS

ENHANCED WebAssign The problems in this chapter may be assigned online in Enhanced WebAssign.

**1.** denotes straightforward problem; **2.** denotes intermediate problem;
**3.** denotes challenging problem
**1.** denotes full solution available in *Student Solutions Manual/Study Guide*
**1.** denotes problems most often assigned in Enhanced WebAssign

BIO denotes biomedical problems
GP denotes guided problems
M denotes Master It tutorial available in Enhanced WebAssign
Q|C denotes asking for quantitative and conceptual reasoning
S denotes symbolic reasoning problem
W denotes Watch It video solution available in Enhanced WebAssign

### 30.1 Nuclear Fission

**1.** M If the average energy released in a fission event is 208 MeV, find the total number of fission events required to operate a 100-W lightbulb for 1.0 h.

**2.** W Find the energy released in the fission reaction

$$n + {}^{235}_{92}U \rightarrow {}^{98}_{40}Zr + {}^{135}_{52}Te + 3n$$

The atomic masses of the fission products are 97.912 0 u for ${}^{98}_{40}Zr$ and 134.908 7 u for ${}^{135}_{52}Te$.

**3.** Find the energy released in the fission reaction

$${}^{1}_{0}n + {}^{235}_{92}U \rightarrow {}^{88}_{38}Sr + {}^{136}_{54}Xe + 12{}^{1}_{0}n$$

**4.** According to one estimate, the first atomic bomb released an energy equivalent to 20 kilotons of TNT. If

1 ton of TNT releases about $4.0 \times 10^9$ J, how much uranium was lost through fission in this bomb? (Assume 208 MeV released per fission.)

**5.** M Assume ordinary soil contains natural uranium in amounts of 1 part per million by mass. (a) How much uranium is in the top 1.00 m of soil on a 1-acre (43 560-ft$^2$) plot of ground, assuming the specific gravity of soil is 4.00? (b) How much of the isotope $^{235}$U, appropriate for nuclear reactor fuel, is in this soil? *Hint:* See Appendix B for the percent abundance of $^{235}_{92}$U.

**6.** W A typical nuclear fission power plant produces about 1.00 GW of electrical power. Assume the plant has an overall efficiency of 40.0% and each fission produces 200 MeV of thermal energy. Calculate the mass of $^{235}$U consumed each day.

**7.** In order to minimize neutron leakage from a reactor, the ratio of the surface area to the volume must be as small as possible. Assume that a sphere of radius $a$ and a cube both have the same volume. Find the surface-to-volume ratio for (a) the sphere and (b) the cube. (c) Which of these reactor shapes would have the minimum leakage?

**8.** GP According to one estimate, there are $4.4 \times 10^6$ metric tons of world uranium reserves extractable at \$130/kg or less. About 0.7% of naturally occurring uranium is the fissionable isotope $^{235}$U. (a) Calculate the mass of $^{235}$U in this reserve in grams. (b) Find the number of moles of $^{235}$U and convert to a number of atoms. (c) Assuming 208 MeV is obtained from each reaction and all this energy is captured, calculate the total energy that can be extracted from the reserve in joules. (d) Assuming world power consumption to be constant at $1.5 \times 10^{13}$ J/s, how many years could the uranium reserves provide for all the world's energy needs using conventional reactors that don't generate nuclear fuel? (e) What conclusions can be drawn?

**9.** M An all-electric home uses approximately 2 000 kWh of electric energy per month. How much uranium-235 would be required to provide this house with its energy needs for one year? Assume 100% conversion efficiency and 208 MeV released per fission.

**10.** Q|C Seawater contains 3 mg of uranium per cubic meter. (a) Given that the average ocean depth is about 4 km and water covers two-thirds of Earth's surface, estimate the amount of uranium dissolved in the ocean. (b) Estimate how long this uranium could supply the world's energy needs at the current usage of $1.5 \times 10^{13}$ J/s. (c) Where does the dissolved uranium come from? Is it a renewable energy source? Can uranium from the ocean satisfy our energy requirements? Discuss. *Note:* Breeder reactors increase the efficiency of nuclear fuel use by approximately two orders of magnitude.

## 30.2 Nuclear Fusion

**11.** When a star has exhausted its hydrogen fuel, it may fuse other nuclear fuels. At temperatures above $1.0 \times 10^8$ K, helium fusion can occur. Write the equations for the following processes. (a) Two alpha particles fuse to produce a nucleus $A$ and a gamma ray. What is nucleus $A$? (b) Nucleus $A$ absorbs an alpha particle to produce a nucleus $B$ and a gamma ray. What is nucleus $B$? (c) Find the total energy released in the reactions given in parts (a) and (b). *Note:* The mass of $^8_4$Be = 8.005 305 u.

**12.** W Find the energy released in the fusion reaction

$$^1_1\text{H} + ^2_1\text{H} \rightarrow ^3_2\text{He} + \gamma$$

**13.** Find the energy released in the fusion reaction

$$^2_1\text{H} + ^2_1\text{H} \rightarrow ^3_1\text{H} + ^1_1\text{H}$$

**14.** GP Another series of nuclear reactions that can produce energy in the interior of stars is the cycle described below. This cycle is most efficient when the central temperature in a star is above $1.6 \times 10^7$ K. Because the temperature at the center of the Sun is only $1.5 \times 10^7$ K, the following cycle produces less than 10% of the Sun's energy. (a) A high-energy proton is absorbed by $^{12}$C. Another nucleus, $A$, is produced in the reaction, along with a gamma ray. Identify nucleus $A$. (b) Nucleus $A$ decays through positron emission to form nucleus $B$. Identify nucleus $B$. (c) Nucleus $B$ absorbs a proton to produce nucleus $C$ and a gamma ray. Identify nucleus $C$. (d) Nucleus $C$ absorbs a proton to produce nucleus $D$ and a gamma ray. Identify nucleus $D$. (e) Nucleus $D$ decays through positron emission to produce nucleus $E$. Identify nucleus $E$. (f) Nucleus $E$ absorbs a proton to produce nucleus $F$ plus an alpha particle. What is nucleus $F$? *Note:* If nucleus $F$ is not $^{12}$C—that is, the nucleus you started with—you have made an error and should review the sequence of events.

**15.** Assume a deuteron and a triton are at rest when they fuse according to the reaction

$$^2_1\text{H} + ^3_1\text{H} \rightarrow ^4_2\text{He} + ^1_0\text{n} + 17.6\text{ MeV}$$

Neglecting relativistic corrections, determine the kinetic energy acquired by the neutron.

**16.** Q|C A reaction that has been considered as a source of energy is the absorption of a proton by a boron-11 nucleus to produce three alpha particles:

$$^1_1\text{H} + ^{11}_5\text{B} \rightarrow 3(^4_2\text{He})$$

This reaction is an attractive possibility because boron is easily obtained from Earth's crust. A disadvantage is that the protons and boron nuclei must have large kinetic energies for the reaction to take place. This requirement contrasts to the initiation of uranium fission by slow neutrons. (a) How much energy is released in each reaction? (b) Why must the reactant particles have high kinetic energies?

## 30.4 Positrons and Other Antiparticles

**17.** A photon produces a proton–antiproton pair according to the reaction $\gamma \rightarrow p + \overline{p}$. What is the minimum possible frequency of the photon? What is its wavelength?

18. M A photon with an energy of 2.09 GeV creates a proton–antiproton pair in which the proton has a kinetic energy of 95.0 MeV. What is the kinetic energy of the antiproton?

19. A neutral pion at rest decays into two photons according to

$$\pi^0 \rightarrow \gamma + \gamma$$

Find the energy, momentum, and frequency of each photon.

### 30.6 Conservation Laws

20. For the following two reactions, the first may occur but the second cannot. Explain.

$$K^0 \rightarrow \pi^+ + \pi^- \text{ (can occur)}$$

$$\Lambda^0 \rightarrow \pi^+ + \pi^- \text{ (cannot occur)}$$

21. Each of the following reactions is forbidden. Determine a conservation law that is violated for each reaction.

(a) $p + \bar{p} \rightarrow \mu^+ + e^-$ (d) $p + p \rightarrow p + p + n$
(b) $\pi^- + p \rightarrow p + \pi^+$ (e) $\gamma + p \rightarrow n + \pi^0$
(c) $p + p \rightarrow p + \pi^+$

22. Determine which of the reactions below can occur. For those that cannot occur, determine the conservation law (or laws) that each violates.

(a) $p \rightarrow \pi^+ + \pi^0$ (d) $n \rightarrow p + e^- + \bar{\nu}_e$
(b) $p + p \rightarrow p + p + \pi^0$ (e) $\pi^+ \rightarrow \mu^+ + n$
(c) $\pi^+ \rightarrow \mu^+ + \nu_\mu$

23. Which of the following processes are allowed by the strong interaction, the electromagnetic interaction, the weak interaction, or no interaction at all?

(a) $\pi^- + p \rightarrow 2\eta^0$ (d) $\Omega^- \rightarrow \Xi^- + \pi^0$
(b) $K^- + n \rightarrow \Lambda^0 + \pi^-$ (e) $\eta^0 \rightarrow 2\gamma$
(c) $K^- \rightarrow \pi^- + \pi^0$

24. Q|C (a) Show that baryon number and charge are conserved in the following reactions of a pion with a proton:

$$(1)\ \pi^+ + p \rightarrow K^+ + \Sigma^+$$

$$(2)\ \pi^+ + p \rightarrow \pi^+ + \Sigma^+$$

(b) The first reaction is observed, but the second never occurs. Explain these observations. (c) Could the second reaction happen if it created a third particle? If so, which particles in Table 30.2 might make it possible? Would the reaction require less energy or more energy than the reaction of Equation (1)? Why?

25. Determine whether or not strangeness is conserved in the following decays and reactions.

(a) $\Lambda^0 \rightarrow p + \pi-$ (d) $\pi^- + p \rightarrow \pi^- + \Sigma^+$
(b) $\pi^- + p \rightarrow \Lambda^0 + K^0$ (e) $\Xi^- \rightarrow \Lambda^0 + \pi^-$
(c) $\bar{p} + p \rightarrow \bar{\Lambda}^0 + \Lambda^0$ (f) $\Xi^0 \rightarrow p + \pi^-$

### 30.8 Quarks and Color

26. The quark composition of the proton is uud, whereas that of the neutron is udd. Show that the charge, baryon number, and strangeness of these particles equal the sums of these numbers for their quark constituents.

27. M Find the number of electrons, and of each species of quark, in 1 L of water.

28. The quark compositions of the $K^0$ and $\Lambda^0$ particles are $d\bar{s}$ and uds, respectively. Show that the charge, baryon number, and strangeness of these particles equal the sums of these numbers for their quark constituents.

29. Identify the particles corresponding to the quark states (a) suu, (b) $\bar{u}d$, (c) $\bar{s}d$, and (d) ssd.

30. What is the electrical charge of the baryons with the quark compositions (a) $\overline{uud}$ and (b) $\overline{udd}$? What are these baryons called?

### Additional Problems

31. A $\Sigma^0$ particle traveling through matter strikes a proton. A $\Sigma^+$, a gamma ray, as well as a third particle, emerge. Use the quark model of each to determine the identity of the third particle.

32. Name at least one conservation law that prevents each of the following reactions from occurring.

(a) $\pi^- + p \rightarrow \Sigma^+ + \pi^0$
(b) $\mu^- \rightarrow \pi^- + \nu_e$
(c) $p \rightarrow \pi^+ + \pi^+ + \pi^-$

33. Find the energy released in the fusion reaction

$${}^1_1H + {}^3_2He \rightarrow {}^4_2He + e^+ + \nu$$

34. Occasionally, high-energy muons collide with electrons and produce two neutrinos according to the reaction $\mu^+ + e^- \rightarrow 2\nu$. What kind of neutrinos are they?

35. Fill in the missing particle. Assume that (a) occurs via the strong interaction while (b) and (c) involve the weak interaction.

(a) $K^+ + p \rightarrow ? + p$
(b) $\Omega^- \rightarrow ? + \pi^-$
(c) $K^+ \rightarrow ? + \mu^+ + \nu_\mu$

36. W Two protons approach each other with 70.4 MeV of kinetic energy and engage in a reaction in which a proton and a positive pion emerge at rest. What third particle, obviously uncharged and therefore difficult to detect, must have been created?

37. A 2.0-MeV neutron is emitted in a fission reactor. If it loses one-half its kinetic energy in each collision with a moderator atom, how many collisions must it undergo to reach an energy associated with a gas at a room temperature of 20.0°C?

38. Q|C The fusion reaction ${}^2_1D + {}^2_1D \rightarrow {}^3_2He + {}^1_0n$ releases 3.27 MeV of energy. If a fusion reactor operates strictly on the basis of this reaction, (a) how much energy could it produce by completely reacting 1 kg of deuterium? (b) At eight cents a kilowatt-hour, how much would the produced energy be worth? (c) Heavy water ($D_2O$) costs about $300 per kilogram. Neglecting the cost of separating the deuterium from the oxygen via electrolysis, how much does 1 kg of deuterium cost, if

derived from $D_2O$? (d) Would it be cost-effective to use deuterium as a source of energy? Discuss, assuming the cost of energy production is nine-tenths the value of energy produced.

39. (a) Show that about $1.0 \times 10^{10}$ J would be released by the fusion of the deuterons in 1.0 gal of water. Note that 1 of every 6 500 hydrogen atoms is a deuteron. (b) The average energy consumption rate of a person living in the United States is about $1.0 \times 10^4$ J/s (an average power of 10 kW). At this rate, how long would the energy needs of one person be supplied by the fusion of the deuterons in 1.0 gal of water? Assume the energy released per deuteron is 1.64 MeV.

40. The oceans have a volume of 317 million cubic miles and contain $1.32 \times 10^{21}$ kg of water. Of all the hydrogen nuclei in this water, 0.015 6% are deuterium. (a) If all of these deuterium nuclei were fused to helium via the first reaction in Equation 30.4, determine the total amount of energy that could be released. (b) Present world electric power consumption is about $7.00 \times 10^{12}$ W. If consumption were 100 times greater, how many years would the energy supply calculated in (a) last?

41. A $\pi$-meson at rest decays according to

$$\pi^- \rightarrow \mu^- + \bar{\nu}_\mu$$

What is the energy carried off by the neutrino? Assume the neutrino has no mass and moves off with the speed of light. Take $m_\pi c^2 = 139.6$ MeV and $m_\mu c^2 = 105.7$ MeV. *Note:* Use relativity; see Equation 26.13.

42. The reaction $\pi^- + p \rightarrow K^0 + \Lambda^0$ occurs with high probability, whereas the reaction $\pi^- + p \rightarrow K^0 + n$ never occurs. Analyze these reactions at the quark level. Show that the first reaction conserves the total number of each type of quark and the second reaction does not.

43. The sun radiates energy at the rate of $3.85 \times 10^{26}$ W. Suppose the net reaction

$$4p + 2e^- \rightarrow \alpha + 2\nu_e + 6\gamma$$

accounts for all the energy released. Calculate the number of protons fused per second. *Note:* Recall that an alpha particle is a helium-4 nucleus.

44. A $K^0$ particle at rest decays into a $\pi^+$ and a $\pi^-$. The mass of the $K^0$ is 497.7 MeV/$c^2$ and the mass of each pion is 139.6 MeV/$c^2$. What will be the speed of each of the pions?

# A.1 Mathematical Notation

Many mathematical symbols are used throughout this book. These symbols are described here, with examples illustrating their use.

## Equals Sign: =

The symbol = denotes the mathematical equality of two quantities. In physics, it also makes a statement about the relationship of different physical concepts. An example is the equation $E = mc^2$. This famous equation says that a given mass $m$, *measured in kilograms*, is equivalent to a certain amount of energy, $E$, *measured in joules*. The speed of light squared, $c^2$, can be considered a constant of proportionality, neccessary because the units chosen for given quantities are rather arbitrary, based on historical circumstances.

## Proportionality: $\propto$

The symbol $\propto$ denotes a proportionality. This symbol might be used when focusing on relationships rather than an exact mathematical equality. For example, we could write $E \propto m$, which says "the energy $E$ associated with an object is proportional to the mass $m$ of the object." Another example is found in kinetic energy, which is the energy associated with an object's motion, defined by $KE = \frac{1}{2}mv^2$, where $m$ is again the mass and $v$ is the speed. Both $m$ and $v$ are variables in this expression. Hence, the kinetic energy $KE$ is proportional to $m$, $KE \propto m$, and at the same time $KE$ is proportional to the speed squared, $KE \propto v^2$. Another term used here is "directly proportional." The density $\rho$ of an object is related to its mass and volume by $\rho = m/V$. Consequently, the density is said to be directly proportional to mass and inversely proportional to volume.

## Inequalities

The symbol $<$ means "is less than," and $>$ means "is greater than." For example, $\rho_{\text{Fe}} > \rho_{\text{Al}}$ means that the density of iron, $\rho_{\text{Fe}}$, is greater than the density of aluminum, $\rho_{\text{Al}}$. If there is a line underneath the symbol, there is the possibility of equality: $\leq$ means "less than or equal to," whereas $\geq$ means "greater than or equal to." Any particle's speed $v$, for example, is less than or equal to the speed of light, $c$: $v \leq c$.

Sometimes the size of a given quantity greatly differs from the size of another quantity. Simple inequality doesn't convey vast differences. For such cases, the symbol $<<$ means "is much less than" and $>>$ means "is much greater than." The mass of the Sun, $M_{\text{Sun}}$, is much greater than the mass of the Earth, $M_E$: $M_{\text{Sun}} >> M_E$. The mass of an electron, $m_e$, is much less than the mass of a proton, $m_p$: $m_e << m_p$.

## Approximately Equal: $\approx$

The symbol $\approx$ indicates that two quantities are approximately equal to each other. The mass of a proton, $m_p$, is approximately the same as the mass of a neutron, $m_n$. This relationship can be written $m_p \approx m_n$.

## Equivalence: $\equiv$

The symbol $\equiv$ means "is defined as," which is a different statement than a simple =. It means that the quantity on the left—usually a single quantity—is another way of expressing the quantity or quantities on the right. The classical momentum of an object, $p$, is defined to be the mass of the object $m$ times its velocity $v$, hence $p \equiv mv$. Because this equivalence is by definition, there is no possibility of $p$ being equal to something else. Contrast this case with that of the expression for the velocity $v$ of an object under constant acceleration, which is $v = at + v_0$. This equation would never be written with an equivalence sign because $v$ in this context is not a defined quantity; rather it is an equality that holds true only under the assumption of constant acceleration. The expression for the classical momentum, however, is always true by definition, so it would be appropriate to write $p \equiv mv$ the first time the concept is introduced. After the introduction of a term, an ordinary equals sign generally suffices.

## Differences: Δ

The Greek letter Δ (capital delta) is the symbol used to indicate the difference in a measured physical quantity, usually at two different times. The best example is a displacement along the $x$-axis, indicated by $\Delta x$ (read as "delta $x$"). Note that $\Delta x$ doesn't mean "the product of Δ and $x$." Suppose a person out for a morning stroll starts measuring her distance away from home when she is 10 m from her doorway. She then continues along a straight-line path and stops strolling 50 m from the door. Her change in position during the walk is $\Delta x = 50 \text{ m} - 10 \text{ m} = 40 \text{ m}$. In symbolic form, such displacements can be written

$$\Delta x = x_f - x_i$$

In this equation, $x_f$ is the final position and $x_i$ is the initial position. There are numerous other examples of differences in physics, such as the difference (or change) in momentum, $\Delta p = p_f - p_i$; the change in kinetic energy, $\Delta K = K_f - K_i$; and the change in temperature, $\Delta T = T_f - T_i$.

## Summation: Σ

In physics there are often contexts in which it's necessary to add several quantities. A useful abbreviation for representing such a sum is the Greek letter Σ (capital sigma). Suppose we wish to add a set of five numbers represented by $x_1$, $x_2$, $x_3$, $x_4$, and $x_5$. In the abbreviated notation, we would write the sum as

$$x_1 + x_2 + x_3 + x_4 + x_5 = \sum_{i=1}^{5} x_i$$

where the subscript $i$ on $x$ represents any one of the numbers in the set. For example, if there are five masses in a system, $m_1$, $m_2$, $m_3$, $m_4$, and $m_5$, the total mass of the system $M = m_1 + m_2 + m_3 + m_4 + m_5$ could be expressed as

$$M = \sum_{i=1}^{5} m_i$$

The $x$-coordinate of the center of mass of the five masses, meanwhile, could be written

$$x_{CM} = \frac{\sum_{i=1}^{5} m_i x_i}{M}$$

with similar expressions for the $y$- and $z$-coordinates of the center of mass.

## Absolute Value: | |

The magnitude of a quantity $x$, written $|x|$, is simply the absolute value of that quantity. The sign of $|x|$ is always positive, regardless of the sign of $x$. For example, if $x = -5$, then $|x| = 5$; if $x = 8$, then $|x| = 8$. In physics this sign is useful whenever the magnitude of a quantity is more important than any direction that might be implied by a sign.

## A.2 Scientific Notation

Many quantities in science have very large or very small values. The speed of light is about 300 000 000 m/s, and the ink required to make the dot over an *i* in this textbook has a mass of about 0.000 000 001 kg. It's very cumbersome to read, write, and keep track of such numbers because the decimal places have to be counted and because a number with one significant digit may require a large number of zeros. Scientific notation is a way of representing these numbers without having to write out so many zeros, which in general are only used to establish the magnitude of the number, not its accuracy. The key is to use powers of 10. The nonnegative powers of 10 are

$$
\begin{aligned}
10^0 &= 1 \\
10^1 &= 10 \\
10^2 &= 10 \times 10 = 100 \\
10^3 &= 10 \times 10 \times 10 = 1\,000 \\
10^4 &= 10 \times 10 \times 10 \times 10 = 10\,000 \\
10^5 &= 10 \times 10 \times 10 \times 10 \times 10 = 100\,000
\end{aligned}
$$

and so on. The number of decimal places following the first digit in the number and to the left of the decimal point corresponds to the power to which 10 is raised, called the **exponent** of 10. The speed of light, 300 000 000 m/s, can then be expressed as $3 \times 10^8$ m/s. Notice there are eight decimal places to the right of the leading digit, 3, and to the left of where the decimal point would be placed. Some representative numbers smaller than 1 are

$$
\begin{aligned}
10^{-1} &= \frac{1}{10} = 0.1 \\
10^{-2} &= \frac{1}{10 \times 10} = 0.01 \\
10^{-3} &= \frac{1}{10 \times 10 \times 10} = 0.001 \\
10^{-4} &= \frac{1}{10 \times 10 \times 10 \times 10} = 0.000\,1 \\
10^{-5} &= \frac{1}{10 \times 10 \times 10 \times 10 \times 10} = 0.000\,01
\end{aligned}
$$

In these cases, the number of decimal places to the right of the decimal point up to and including only the first nonzero digit equals the value of the (negative) exponent.

Numbers expressed as some power of 10 multiplied by another number between 1 and 10 are said to be in **scientific notation**. For example, Coulomb's constant, which is associated with electric forces, is given by 8 987 551 789 $\mathrm{N \cdot m^2/C^2}$ and is written in scientific notation as $8.987\,551\,789 \times 10^9\ \mathrm{N \cdot m^2/C^2}$. Newton's constant of gravitation is given by 0.000 000 000 066 731 $\mathrm{N \cdot m^2/kg^2}$, written in scientific notation as $6.673\,1 \times 10^{-11}\ \mathrm{N \cdot m^2/kg^2}$.

When numbers expressed in scientific notation are being multiplied, the following general rule is very useful:

$$10^n \times 10^m = 10^{n+m} \qquad \text{[A.1]}$$

where $n$ and $m$ can be any numbers (not necessarily integers). For example, $10^2 \times 10^5 = 10^7$. The rule also applies if one of the exponents is negative: $10^3 \times 10^{-8} = 10^{-5}$.

When dividing numbers expressed in scientific notation, note that

$$\frac{10^n}{10^m} = 10^n \times 10^{-m} = 10^{n-m} \qquad \textbf{[A.2]}$$

**Exercises**

With help from the above rules, verify the answers to the following:

1. $86\,400 = 8.64 \times 10^4$
2. $9\,816\,762.5 = 9.816\,762\,5 \times 10^6$
3. $0.000\,000\,039\,8 = 3.98 \times 10^{-8}$
4. $(4 \times 10^8)(9 \times 10^9) = 3.6 \times 10^{18}$
5. $(3 \times 10^7)(6 \times 10^{-12}) = 1.8 \times 10^{-4}$
6. $\dfrac{75 \times 10^{-11}}{5 \times 10^{-3}} = 1.5 \times 10^{-7}$
7. $\dfrac{(3 \times 10^6)(8 \times 10^{-2})}{(2 \times 10^{17})(6 \times 10^5)} = 2 \times 10^{-18}$

# A.3 Algebra

## A. Some Basic Rules

When algebraic operations are performed, the laws of arithmetic apply. Symbols such as $x$, $y$, and $z$ are usually used to represent quantities that are not specified, what are called the **unknowns**.

First, consider the equation

$$8x = 32$$

If we wish to solve for $x$, we can divide (or multiply) each side of the equation by the same factor without destroying the equality. In this case, if we divide both sides by 8, we have

$$\frac{8x}{8} = \frac{32}{8}$$

$$x = 4$$

Next consider the equation

$$x + 2 = 8$$

In this type of expression, we can add or subtract the same quantity from each side. If we subtract 2 from each side, we obtain

$$x + 2 - 2 = 8 - 2$$

$$x = 6$$

In general, if $x + a = b$, then $x = b - a$.

Now consider the equation

$$\frac{x}{5} = 9$$

If we multiply each side by 5, we are left with $x$ on the left by itself and 45 on the right:

$$\left(\frac{x}{5}\right)(5) = 9 \times 5$$

$$x = 45$$

In all cases, **whatever operation is performed on the left side of the equality must also be performed on the right side**.

The following rules for multiplying, dividing, adding, and subtracting fractions should be recalled, where $a$, $b$, and $c$ are three numbers:

| | Rule | Example |
|---|---|---|
| **Multiplying** | $\left(\frac{a}{b}\right)\left(\frac{c}{d}\right) = \frac{ac}{bd}$ | $\left(\frac{2}{3}\right)\left(\frac{4}{5}\right) = \frac{8}{15}$ |
| **Dividing** | $\frac{(a/b)}{(c/d)} = \frac{ad}{bc}$ | $\frac{2/3}{4/5} = \frac{(2)(5)}{(4)(3)} = \frac{10}{12} = \frac{5}{6}$ |
| **Adding** | $\frac{a}{b} \pm \frac{c}{d} = \frac{ad \pm bc}{bd}$ | $\frac{2}{3} - \frac{4}{5} = \frac{(2)(5) - (4)(3)}{(3)(5)} = -\frac{2}{15}$ |

Very often in physics we are called upon to manipulate symbolic expressions algebraically, a process most students find unfamiliar. It's very important, however, because substituting numbers into an equation too early can often obscure meaning. The following two examples illustrate how these kinds of algebraic manipulations are carried out.

## EXAMPLE

A ball is dropped from the top of a building 50.0 m tall. How long does it take the ball to fall to a height of 25.0 m?

**SOLUTION** First, write the general ballistics equation for this situation:

$$x = \tfrac{1}{2}at^2 + v_0t + x_0$$

Here, $a = -9.80 \text{ m/s}^2$ is the acceleration of gravity that causes the ball to fall, $v_0 = 0$ is the initial velocity, and $x_0 = 50.0$ m is the initial position. Substitute only the initial velocity, $v_0 = 0$, obtaining the following equation:

$$x = \tfrac{1}{2}at^2 + x_0$$

This equation must be solved for $t$. Subtract $x_0$ from both sides:

$$x - x_0 = \tfrac{1}{2}at^2 + x_0 - x_0 = \tfrac{1}{2}at^2$$

Multiply both sides by $2/a$:

$$\left(\frac{2}{a}\right)(x - x_0) = \left(\frac{2}{a}\right)\tfrac{1}{2}at^2 = t^2$$

It's customary to have the desired value on the left, so switch the equation around and take the square root of both sides:

$$t = \pm\sqrt{\left(\frac{2}{a}\right)(x - x_0)}$$

Only the positive root makes sense. Values could now be substituted to obtain a final answer.

## EXAMPLE

A block of mass $m$ slides over a frictionless surface in the positive $x$-direction. It encounters a patch of roughness having coefficient of kinetic friction $\mu_k$. If the rough patch has length $\Delta x$, find the speed of the block after leaving the patch.

**SOLUTION** Using the work-energy theorem, we have

$$\tfrac{1}{2}mv^2 - \tfrac{1}{2}mv_0^{\,2} = -\mu_k mg\,\Delta x$$

Add $\frac{1}{2}mv_0^2$ to both sides:

$$\tfrac{1}{2}mv^2 = \tfrac{1}{2}mv_0^{\,2} - \mu_k mg\,\Delta x$$

*(Continued)*

Multiply both sides by $2/m$:

$$v^2 = v_0^2 - 2\mu_k g\,\Delta x$$

Finally, take the square root of both sides. Because the block is sliding in the positive $x$-direction, the positive square root is selected.

$$v = \sqrt{v_0^2 - 2\mu_k g\,\Delta x}$$

### Exercises

In Exercises 1–4, solve for $x$:

| | Answers |
|---|---|
| **1.** $a = \dfrac{1}{1+x}$ | $x = \dfrac{1-a}{a}$ |
| **2.** $3x - 5 = 13$ | $x = 6$ |
| **3.** $ax - 5 = bx + 2$ | $x = \dfrac{7}{a-b}$ |
| **4.** $\dfrac{5}{2x+6} = \dfrac{3}{4x+8}$ | $x = -\dfrac{11}{7}$ |

**5.** Solve the following equation for $v_1$:

$$P_1 + \tfrac{1}{2}\rho v_1^2 = P_2 + \tfrac{1}{2}\rho v_2^2$$

Answer: $v_1 = \pm\sqrt{\dfrac{2}{\rho}(P_2 - P_1) + v_2^2}$

## B. Powers

When powers of a given quantity $x$ are multiplied, the following rule applies:

$$x^n x^m = x^{n+m} \qquad \text{[A.3]}$$

For example, $x^2x^4 = x^{2+4} = x^6$.

When dividing the powers of a given quantity, the rule is

$$\frac{x^n}{x^m} = x^{n-m} \qquad \text{[A.4]}$$

For example, $x^8/x^2 = x^{8-2} = x^6$.

A power that is a fraction, such as $\frac{1}{3}$, corresponds to a root as follows:

$$x^{1/n} = \sqrt[n]{x} \qquad \text{[A.5]}$$

For example, $4^{1/3} = \sqrt[3]{4} = 1.587\,4$. (A scientific calculator is useful for such calculations.)

Finally, any quantity $x^n$ raised to the $m$th power is

$$(x^n)^m = x^{nm} \qquad \text{[A.6]}$$

Table A.1 summarizes the rules of exponents.

**Table A.1** Rules of Exponents

| |
|---|
| $x^0 = 1$ |
| $x^1 = x$ |
| $x^n x^m = x^{n+m}$ |
| $x^n/x^m = x^{n-m}$ |
| $x^{1/n} = \sqrt[n]{x}$ |
| $(x^n)^m = x^{nm}$ |

### Exercises

Verify the following:

**1.** $3^2 \times 3^3 = 243$
**2.** $x^5x^{-8} = x^{-3}$
**3.** $x^{10}/x^{-5} = x^{15}$
**4.** $5^{1/3} = 1.709\,975$ (Use your calculator.)
**5.** $60^{1/4} = 2.783\,158$ (Use your calculator.)
**6.** $(x^4)^3 = x^{12}$

## C. Factoring

The following are some useful formulas for factoring an equation:

$$ax + ay + az = a(x + y + z) \qquad \text{common factor}$$

$$a^2 + 2ab + b^2 = (a + b)^2 \qquad \text{perfect square}$$

$$a^2 - b^2 = (a + b)(a - b) \qquad \text{difference of squares}$$

## D. Quadratic Equations

The general form of a quadratic equation is

$$ax^2 + bx + c = 0 \qquad \text{[A.7]}$$

where $x$ is the unknown quantity and $a$, $b$, and $c$ are numerical factors referred to as coefficients of the equation. This equation has two roots, given by

$$x = \frac{-b \pm \sqrt{b^2 - 4ac}}{2a} \qquad \text{[A.8]}$$

If $b^2 - 4ac > 0$, the roots are real.

### EXAMPLE

The equation $x^2 + 5x + 4 = 0$ has the following roots corresponding to the two signs of the square-root term:

$$x = \frac{-5 \pm \sqrt{5^2 - (4)(1)(4)}}{2(1)} = \frac{-5 \pm \sqrt{9}}{2} = \frac{-5 \pm 3}{2}$$

$$x_1 = \frac{-5 + 3}{2} = \boxed{-1} \qquad x_2 = \frac{-5 - 3}{2} = \boxed{-4}$$

where $x_1$ refers to the root corresponding to the positive sign and $x_2$ refers to the root corresponding to the negative sign.

### EXAMPLE

A ball is projected upwards at 16.0 m/s. Use the quadratic formula to determine the time necessary for it to reach a height of 8.00 m above the point of release.

**SOLUTION** From the discussion of ballistics in Chapter 2, we can write

$$\textbf{(1)} \quad x = \tfrac{1}{2}at^2 + v_0 t + x_0$$

The acceleration is due to gravity, given by $a = -9.80\ \text{m/s}^2$; the initial velocity is $v_0 = 16.0$ m/s; and the initial position is the point of release, taken to be $x_0 = 0$. Substitute these values into Equation (1) and set $x = 8.00$ m, arriving at

$$x = -4.90t^2 + 16.00t = 8.00$$

where units have been suppressed for mathematical clarity. Rearrange this expression into the standard form of Equation A.7:

$$-4.90t^2 + 16.00t - 8.00 = 0$$

The equation is quadratic in the time, $t$. We have $a = -4.9$, $b = 16$, and $c = -8.00$. Substitute these values into Equation A.8:

$$t = \frac{-16.0 \pm \sqrt{16^2 - 4(-4.90)(-8.00)}}{2(-4.90)} = \frac{-16.0 \pm \sqrt{99.2}}{-9.80}$$

$$= 1.63 \mp \frac{\sqrt{99.2}}{9.80} = \boxed{0.614\ \text{s}, 2.65\ \text{s}}$$

Both solutions are valid in this case, one corresponding to reaching the point of interest on the way up and the other to reaching it on the way back down.

### Exercises

Solve the following quadratic equations:

| | | **Answers** |
|---|---|---|
| **1.** $x^2 + 2x - 3 = 0$ | $x_1 = 1$ | $x_2 = -3$ |
| **2.** $2x^2 - 5x + 2 = 0$ | $x_1 = 2$ | $x_2 = \frac{1}{2}$ |
| **3.** $2x^2 - 4x - 9 = 0$ | $x_1 = 1 + \sqrt{22}/2$ | $x_2 = 1 - \sqrt{22}/2$ |

**4.** Repeat the ballistics example for a height of 10.0 m above the point of release.
Answer: $t_1 = 0.842$ s $\quad t_2 = 2.42$ s

## E. Linear Equations

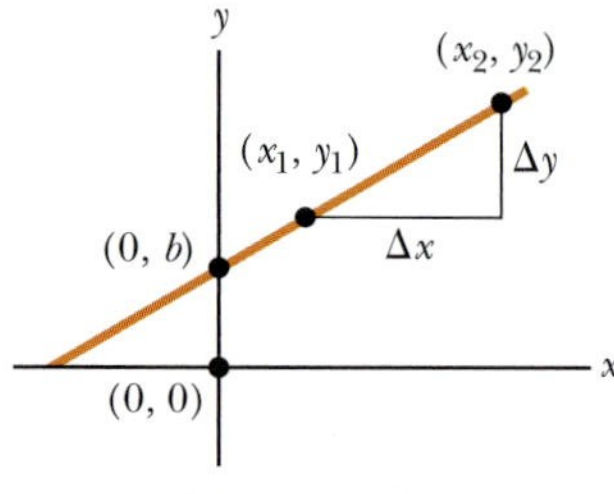


**Figure A.1**

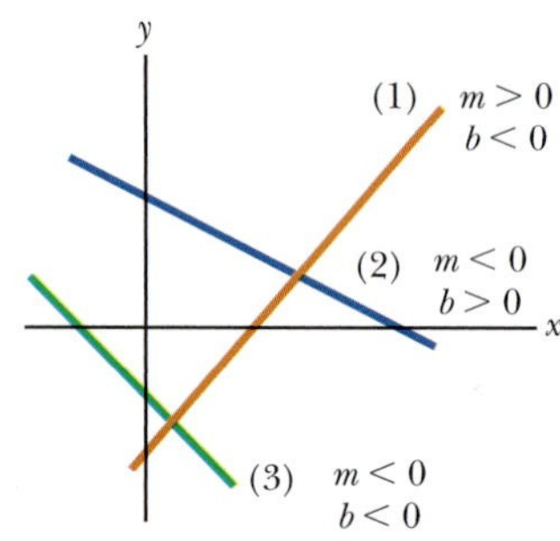


**Figure A.2**

A linear equation has the general form

$$y = mx + b \qquad \textbf{[A.9]}$$

where $m$ and $b$ are constants. This kind of equation is called linear because the graph of $y$ versus $x$ is a straight line, as shown in Figure A.1. The constant $b$, called the $y$-intercept, represents the value of $y$ at which the straight line intersects the $y$-axis. The constant $m$ is equal to the slope of the straight line. If any two points on the straight line are specified by the coordinates $(x_1, y_1)$ and $(x_2, y_2)$, as in Figure A.1, the slope of the straight line can be expressed as

$$\text{Slope} = \frac{y_2 - y_1}{x_2 - x_1} = \frac{\Delta y}{\Delta x} \qquad \textbf{[A.10]}$$

Note that $m$ and $b$ can have either positive or negative values. If $m > 0$, the straight line has a positive slope, as in Figure A.1. If $m < 0$, the straight line has a negative slope. In Figure A.1, both $m$ and $b$ are positive. Three other possible situations are shown in Figure A.2.

**EXAMPLE**

Suppose the electrical resistance of a metal wire is 5.00 Ω at a temperature of 20.0°C and 6.14 Ω at 80.0°C. Assuming the resistance changes linearly, what is the resistance of the wire at 60.0°C?

**SOLUTION** Find the equation of the line describing the resistance $R$ and then substitute the new temperature into it. Two points on the graph of resistance versus temperature, (20.0°C, 5.00 Ω) and (80.0°C, 6.14 Ω), allow computation of the slope:

$$\textbf{(1)} \quad m = \frac{\Delta R}{\Delta T} = \frac{6.14\ \Omega - 5.00\ \Omega}{80.0\text{°C} - 20.0\text{°C}} = 1.90 \times 10^{-2}\ \Omega/\text{°C}$$

Now use the point-slope formulation of a line, with this slope and (20.0°C, 5.00 Ω):

$$\textbf{(2)} \quad R - R_0 = m(T - T_0)$$

$$\textbf{(3)} \quad R - 5.00\ \Omega = (1.90 \times 10^{-2}\ \Omega/\text{°C})(T - 20.0\text{°C})$$

Finally, substitute $T = 60.0°$ into Equation (3) and solve for $R$, getting $R = 5.76\ \Omega$.

---

### Exercises

**1.** Draw graphs of the following straight lines:
**(a)** $y = 5x + 3$ **(b)** $y = -2x + 4$ **(c)** $y = -3x - 6$

**2.** Find the slopes of the straight lines described in Exercise 1.
**Answers: (a)** 5 **(b)** −2 **(c)** −3

**3.** Find the slopes of the straight lines that pass through the following sets of points: **(a)** (0, −4) and (4, 2) **(b)** (0, 0) and (2, −5) **(c)** (−5, 2) and (4, −2)
**Answers: (a)** 3/2 **(b)** −5/2 **(c)** −4/9

4. Suppose an experiment measures the following displacements (in meters) from equilibrium of a vertical spring due to attaching weights (in Newtons): (0.025 0 m, 22.0 N), (0.075 0 m, 66.0 N). Find the spring constant, which is the slope of the line in the graph of weight versus displacement.
**Answer:** 880 N/m

## F. Solving Simultaneous Linear Equations

Consider the equation $3x + 5y = 15$, which has two unknowns, $x$ and $y$. Such an equation doesn't have a unique solution. For example, note that ($x = 0$, $y = 3$), ($x = 5$, $y = 0$), and ($x = 2$, $y = 9/5$) are all solutions to this equation.

If a problem has two unknowns, a unique solution is possible only if we have two equations. In general, if a problem has $n$ unknowns, its solution requires $n$ equations. To solve two simultaneous equations involving two unknowns, $x$ and $y$, we solve one of the equations for $x$ in terms of $y$ and substitute this expression into the other equation.

**EXAMPLE**

Solve the following two simultaneous equations:

$$\textbf{(1)}\quad 5x + y = -8 \qquad \textbf{(2)}\quad 2x - 2y = 4$$

**SOLUTION** From Equation (2), we find that $x = y + 2$. Substitution of this value into Equation (1) gives

$$5(y + 2) + y = -8$$
$$6y = -18$$
$$y = -3$$
$$x = y + 2 = -1$$

**ALTERNATE SOLUTION** Multiply each term in Equation (1) by the factor 2 and add the result to Equation (2):

$$10x + 2y = -16$$
$$2x - 2y = 4$$
$$12x = -12$$
$$x = -1$$
$$y = x - 2 = -3$$

Two linear equations containing two unknowns can also be solved by a graphical method. If the straight lines corresponding to the two equations are plotted in a conventional coordinate system, the intersection of the two lines represents the solution. For example, consider the two equations

$$x - y = 2$$
$$x - 2y = -1$$

These equations are plotted in Figure A.3. The intersection of the two lines has the coordinates $x = 5$, $y = 3$, which represents the solution to the equations. You should check this solution by the analytical technique discussed above.

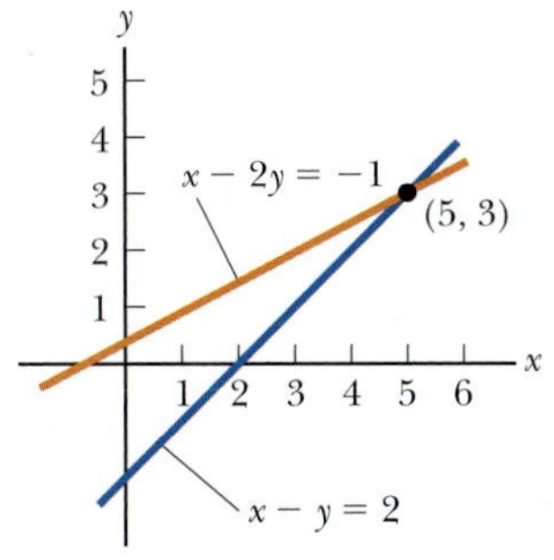

Figure A.3

**EXAMPLE**

A block of mass $m = 2.00$ kg travels in the positive $x$-direction at $v_i = 5.00$ m/s, while a second block, of mass $M = 4.00$ kg and leading the first block, travels in the positive $x$-direction at 2.00 m/s. The surface is frictionless. What are the blocks' final velocities if the collision is perfectly elastic?

*(Continued)*

**SOLUTION** As can be seen in Chapter 6, a perfectly elastic collision involves equations for the momentum and energy. With algebra, the energy equation, which is quadratic in $v$, can be recast as a linear equation. The two equations are given by

$$\textbf{(1)} \quad mv_i + MV_i = mv_f + MV_f$$

$$\textbf{(2)} \quad v_i - V_i = -(v_f - V_f)$$

Substitute the known quantities $v_i = 5.00$ m/s and $V_i = 2.00$ m/s into Equations (1) and (2):

$$\textbf{(3)} \quad 18 = 2v_f + 4V_f$$

$$\textbf{(4)} \quad 3 = -v_f + V_f$$

Multiply Equation (4) by 2 and add to Equation (3):

$$\begin{aligned} 18 &= 2v_f + 4V_f \\ 6 &= -2v_f + 2V_f \\ \hline 24 &= 6V_f \quad \rightarrow \quad V_f = \boxed{4.00 \text{ m/s}} \end{aligned}$$

Substituting the solution for $V_f$ back into Equation (4) yields $v_f = \boxed{1.00 \text{ m/s}}$.

---

### Exercises

Solve the following pairs of simultaneous equations involving two unknowns:

| | **Answers** |
|---|---|
| **1.** $x + y = 8$ <br> $x - y = 2$ | $x = 5, y = 3$ |
| **2.** $98 - T = 10a$ <br> $T - 49 = 5a$ | $T = 65.3, a = 3.27$ |
| **3.** $6x + 2y = 6$ <br> $8x - 4y = 28$ | $x = 2, y = -3$ |

## G. Logarithms and Exponentials

Suppose a quantity $x$ is expressed as a power of some quantity $a$:

$$x = a^y \qquad \textbf{[A.11]}$$

The number $a$ is called the **base** number. The **logarithm** of $x$ with respect to the base $a$ is equal to the exponent to which the base must be raised so as to satisfy the expression $x = a^y$:

$$y = \log_a x \qquad \textbf{[A.12]}$$

Conversely, the **antilogarithm** of $y$ is the number $x$:

$$x = \text{antilog}_a y \qquad \textbf{[A.13]}$$

The antilog expression is in fact identical to the exponential expression in Equation A.11, which is preferable for practical purposes.

In practice, the two bases most often used are base 10, called the *common* logarithm base, and base $e = 2.718\ldots$, called the *natural* logarithm base. When common logarithms are used,

$$y = \log_{10} x \qquad (\text{or } x = 10^y) \qquad \textbf{[A.14]}$$

When natural logarithms are used,

$$y = \ln x \qquad (\text{or } x = e^y) \qquad \textbf{[A.15]}$$

For example, $\log_{10} 52 = 1.716$, so $\text{antilog}_{10} 1.716 = 10^{1.716} = 52$. Likewise, $\ln_e 52 = 3.951$, so $\text{antiln}_e 3.951 = e^{3.951} = 52$.

In general, note that you can convert between base 10 and base $e$ with the equality

$$\ln x = (2.302\ 585)\log_{10} x \qquad \textbf{[A.16]}$$

Finally, some useful properties of logarithms are

$$\log(ab) = \log a + \log b \qquad \ln e = 1$$

$$\log(a/b) = \log a - \log b \qquad \ln e^a = a$$

$$\log(a^n) = n \log a \qquad \ln\left(\frac{1}{a}\right) = -\ln a$$

Logarithms in college physics are used most notably in the definition of decibel level. Sound intensity varies across several orders of magnitude, making it awkward to compare different intensities. Decibel level converts these intensities to a more manageable logarithmic scale.

**■ EXAMPLE (Logs)**

Suppose a jet testing its engines produces a sound intensity of $I = 0.750$ W at a given location in an airplane hangar. What decibel level corresponds to this sound intensity?

**SOLUTION** Decibel level $\beta$ is defined by

$$\beta = 10 \log\left(\frac{I}{I_0}\right)$$

where $I_0 = 1 \times 10^{-12}$ W/m² is the standard reference intensity. Substitute the given information:

$$\beta = 10 \log\left(\frac{0.750 \text{ W/m}^2}{10^{-12} \text{ W/m}^2}\right) = 119 \text{ dB}$$

**■ EXAMPLE (Antilogs)**

A collection of four identical machines creates a decibel level of $\beta = 87.0$ dB in a machine shop. What sound intensity would be created by only one such machine?

**SOLUTION** We use the equation of decibel level to find the total sound intensity of the four machines, and then we divide by 4. From Equation (1):

$$87.0 \text{ dB} = 10 \log\left(\frac{I}{10^{-12} \text{ W/m}^2}\right)$$

Divide both sides by 10 and take the antilog of both sides, which means, equivalently, to exponentiate:

$$10^{8.7} = 10^{\log(I/10^{-12})} = \frac{I}{10^{-12}}$$

$$I = 10^{-12} \cdot 10^{8.7} = 10^{-3.3} = 5.01 \times 10^{-4} \text{ W/m}^2$$

There are four machines, so this result must be divided by 4 to get the intensity of one machine:

$$I = 1.25 \times 10^{-4} \text{ W/m}^2$$

**■ EXAMPLE (Exponentials)**

The half-life of tritium is 12.33 years. (Tritium is the heaviest isotope of hydrogen, with a nucleus consisting of a proton and two neutrons.) If a sample contains 3.0 g of tritium initially, how much remains after 20.0 years?

**SOLUTION** The equation giving the number of nuclei of a radioactive substance as a function of time is

$$N = N_0\left(\frac{1}{2}\right)^n$$

*(Continued)*

where $N$ is the number of nuclei remaining, $N_0$ is the initial number of nuclei, and $n$ is the number of half-lives. Note that this equation is an exponential expression with a base of $\frac{1}{2}$. The number of half-lives is given by $n = t/T_{1/2} = 20.0 \text{ yr}/12.33 \text{ yr} = 1.62$. The fractional amount of tritium that remains after 20.0 yr is therefore

$$\frac{N}{N_0} = \left(\frac{1}{2}\right)^{1.62} = 0.325$$

Hence, of the original 3.00 g of tritium, $0.325 \times 3.00 \text{ g} = 0.975 \text{ g}$ remains.

## A.4 Geometry

Table A.2 gives the areas and volumes for several geometric shapes used throughout this text. These areas and volumes are important in numerous physics applications. A good example is the concept of pressure $P$, which is the force per unit area. As an equation, it is written $P = F/A$. Areas must also be calculated in problems involving the volume rate of fluid flow through a pipe using the equation of continuity, the tensile stress exerted on a cable by a weight, the rate of thermal energy transfer through a barrier, and the density of current through a wire. There are numerous other applications. Volumes are important in computing the buoyant force exerted by water on a submerged object, in calculating densities, and in determining the bulk stress of fluid or gas on an object, which affects its volume. Again, there are numerous other applications.

**Table A.2** Useful Information for Geometry

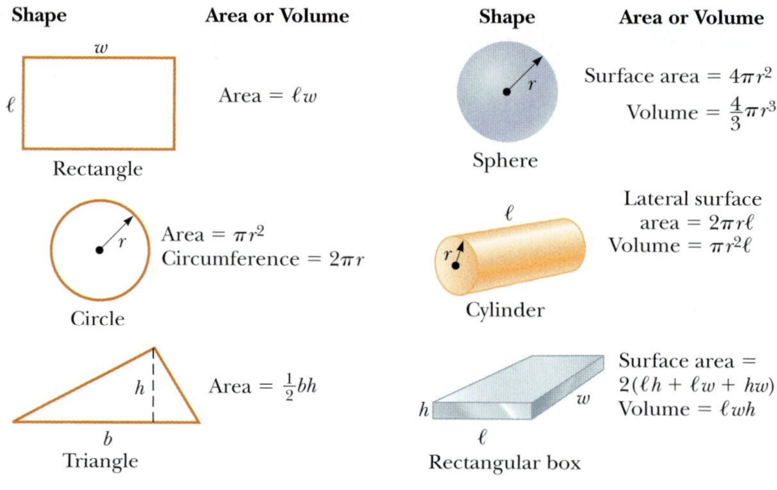

Shape
Area or Volume
Shape
Area or Volume
w
ℓ
Area = ℓw
Rectangle
r
Surface area = 4πr²
Volume = 4/3πr³
Sphere
r
Area = πr²
Circumference = 2πr
Circle
ℓ
r
Lateral surface area = 2πrℓ
Volume = πr²ℓ
Cylinder
h
b
Area = 1/2bh
Triangle
h
w
ℓ
Surface area = 2(ℓh + ℓw + hw)
Volume = ℓwh
Rectangular box


## A.5 Trigonometry

Some of the most basic facts concerning trigonometry are presented in Chapter 1, and we encourage you to study the material presented there if you are having trouble with this branch of mathematics. The most important trigonometric concepts include the Pythagorean theorem:

$$\Delta s^2 = \Delta x^2 + \Delta y^2 \qquad \text{[A.17]}$$

This equation states that the square distance along the hypotenuse of a right triangle equals the sum of the squares of the legs. It can also be used to find distances between points in Cartesian coordinates and the length of a vector, where $\Delta x$ is replaced by the $x$-component of the vector and $\Delta y$ is replaced by the $y$-component of the vector. If the vector $\vec{\mathbf{A}}$ has components $A_x$ and $A_y$, the magnitude $A$ of the vector satisfies

$$A^2 = A_x^2 + A_y^2 \qquad \text{[A.18]}$$

which has a form completely analogous to the form of the Pythagorean theorem. Also highly useful are the cosine and sine functions because they relate the length of a vector to its $x$- and $y$-components:

$$A_x = A\cos\theta \qquad \text{[A.19]}$$
$$A_y = A\sin\theta \qquad \text{[A.20]}$$

The direction $\theta$ of a vector in a plane can be determined by use of the tangent function:

$$\tan\theta = \frac{A_y}{A_x} \qquad \text{[A.21]}$$

A relative of the Pythagorean theorem is also frequently useful:

$$\sin^2\theta + \cos^2\theta = 1 \qquad \text{[A.22]}$$

Details on the above concepts can be found in the extensive discussions in Chapters 1 and 3. The following are some other useful trigonometric identities:

$$\sin\theta = \cos(90° - \theta)$$

$$\cos\theta = \sin(90° - \theta)$$

$$\sin 2\theta = 2\sin\theta\cos\theta$$

$$\cos 2\theta = \cos^2\theta - \sin^2\theta$$

$$\sin(\theta \pm \phi) = \sin\theta\cos\phi \pm \cos\theta\sin\phi$$

$$\cos(\theta \pm \phi) = \cos\theta\cos\phi \pm \sin\theta\sin\phi$$

The following relationships apply to any triangle, as shown in Figure A.4:

$$\alpha + \beta + \gamma = 180°$$

$$a^2 = b^2 + c^2 - 2bc\cos\alpha$$

$$b^2 = a^2 + c^2 - 2ac\cos\beta \qquad \text{law of cosines}$$

$$c^2 = a^2 + b^2 - 2ab\cos\gamma$$

$$\frac{a}{\sin\alpha} = \frac{b}{\sin\beta} = \frac{c}{\sin\gamma} \qquad \text{law of sines}$$

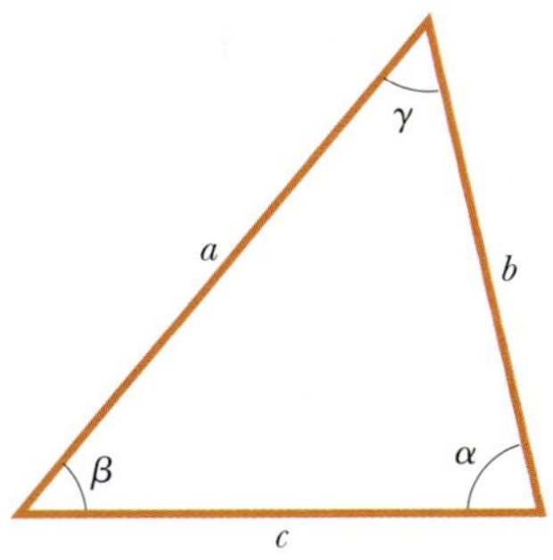

Figure A.4

# Appendix B An Abbreviated Table of Isotopes

| Atomic Number $Z$ | Element | Symbol | Chemical Atomic Mass (u) | Mass Number (* Indicates Radioactive) $A$ | Atomic Mass (u) | Percent Abundance | Half-Life (If Radioactive) $T_{1/2}$ |
|---|---|---|---|---|---|---|---|
| 0 | (Neutron) | n | | 1* | 1.008 665 | | 10.4 min |
| 1 | Hydrogen | H | 1.007 94 | 1 | 1.007 825 | 99.988 5 | |
| | Deuterium | D | | 2 | 2.014 102 | 0.011 5 | |
| | Tritium | T | | 3* | 3.016 049 | | 12.33 yr |
| 2 | Helium | He | 4.002 602 | 3 | 3.016 029 | 0.000 137 | |
| | | | | 4 | 4.002 603 | 99.999 863 | |
| 3 | Lithium | Li | 6.941 | 6 | 6.015 122 | 7.5 | |
| | | | | 7 | 7.016 004 | 92.5 | |
| 4 | Beryllium | Be | 9.012 182 | 7* | 7.016 929 | | 53.3 days |
| | | | | 9 | 9.012 182 | 100 | |
| 5 | Boron | B | 10.811 | 10 | 10.012 937 | 19.9 | |
| | | | | 11 | 11.009 306 | 80.1 | |
| 6 | Carbon | C | 12.010 7 | 10* | 10.016 853 | | 19.3 s |
| | | | | 11* | 11.011 434 | | 20.4 min |
| | | | | 12 | 12.000 000 | 98.93 | |
| | | | | 13 | 13.003 355 | 1.07 | |
| | | | | 14* | 14.003 242 | | 5 730 yr |
| 7 | Nitrogen | N | 14.006 7 | 13* | 13.005 739 | | 9.96 min |
| | | | | 14 | 14.003 074 | 99.632 | |
| | | | | 15 | 15.000 109 | 0.368 | |
| 8 | Oxygen | O | 15.999 4 | 15* | 15.003 065 | | 122 s |
| | | | | 16 | 15.994 915 | 99.757 | |
| | | | | 18 | 17.999 160 | 0.205 | |
| 9 | Fluorine | F | 18.998 403 2 | 19 | 18.998 403 | 100 | |
| 10 | Neon | Ne | 20.179 7 | 20 | 19.992 440 | 90.48 | |
| | | | | 22 | 21.991 385 | 9.25 | |
| 11 | Sodium | Na | 22.989 77 | 22* | 21.994 437 | | 2.61 yr |
| | | | | 23 | 22.989 770 | 100 | |
| | | | | 24* | 23.990 963 | | 14.96 h |
| 12 | Magnesium | Mg | 24.305 0 | 24 | 23.985 042 | 78.99 | |
| | | | | 25 | 24.985 837 | 10.00 | |
| | | | | 26 | 25.982 593 | 11.01 | |
| 13 | Aluminum | Al | 26.981 538 | 27 | 26.981 539 | 100 | |
| 14 | Silicon | Si | 28.085 5 | 28 | 27.976 926 | 92.229 7 | |
| 15 | Phosphorus | P | 30.973 761 | 31 | 30.973 762 | 100 | |
| | | | | 32* | 31.973 907 | | 14.26 days |
| 16 | Sulfur | S | 32.066 | 32 | 31.972 071 | 94.93 | |
| | | | | 35* | 34.969 032 | | 87.5 days |
| 17 | Chlorine | Cl | 35.452 7 | 35 | 34.968 853 | 75.78 | |
| | | | | 37 | 36.965 903 | 24.22 | |
| 18 | Argon | Ar | 39.948 | 40 | 39.962 383 | 99.600 3 | |
| 19 | Potassium | K | 39.098 3 | 39 | 38.963 707 | 93.258 1 | |
| | | | | 40* | 39.963 999 | 0.011 7 | $1.28 \times 10^9$ yr |
| 20 | Calcium | Ca | 40.078 | 40 | 39.962 591 | 96.941 | |
| 21 | Scandium | Sc | 44.955 910 | 45 | 44.955 910 | 100 | |
| 22 | Titanium | Ti | 47.867 | 48 | 47.947 947 | 73.72 | |

| Atomic Number $Z$ | Element | Symbol | Chemical Atomic Mass (u) | Mass Number (* Indicates Radioactive) $A$ | Atomic Mass (u) | Percent Abundance | Half-Life (If Radioactive) $T_{1/2}$ |
|---|---|---|---|---|---|---|---|
| 23 | Vanadium | V | 50.941 5 | 51 | 50.943 964 | 99.750 | |
| 24 | Chromium | Cr | 51.996 1 | 52 | 51.940 512 | 83.789 | |
| 25 | Manganese | Mn | 54.938 049 | 55 | 54.938 050 | 100 | |
| 26 | Iron | Fe | 55.845 | 56 | 55.934 942 | 91.754 | |
| 27 | Cobalt | Co | 58.933 200 | 59 | 58.933 200 | 100 | |
| | | | | 60* | 59.933 822 | | 5.27 yr |
| 28 | Nickel | Ni | 58.693 4 | 58 | 57.935 348 | 68.076 9 | |
| | | | | 60 | 59.930 790 | 26.223 1 | |
| 29 | Copper | Cu | 63.546 | 63 | 62.929 601 | 69.17 | |
| | | | | 65 | 64.927 794 | 30.83 | |
| 30 | Zinc | Zn | 65.39 | 64 | 63.929 147 | 48.63 | |
| | | | | 66 | 65.926 037 | 27.90 | |
| | | | | 68 | 67.924 848 | 18.75 | |
| 31 | Gallium | Ga | 69.723 | 69 | 68.925 581 | 60.108 | |
| | | | | 71 | 70.924 705 | 39.892 | |
| 32 | Germanium | Ge | 72.61 | 70 | 69.924 250 | 20.84 | |
| | | | | 72 | 71.922 076 | 27.54 | |
| | | | | 74 | 73.921 178 | 36.28 | |
| 33 | Arsenic | As | 74.921 60 | 75 | 74.921 596 | 100 | |
| 34 | Selenium | Se | 78.96 | 78 | 77.917 310 | 23.77 | |
| | | | | 80 | 79.916 522 | 49.61 | |
| 35 | Bromine | Br | 79.904 | 79 | 78.918 338 | 50.69 | |
| | | | | 81 | 80.916 291 | 49.31 | |
| 36 | Krypton | Kr | 83.80 | 82 | 81.913 485 | 11.58 | |
| | | | | 83 | 82.914 136 | 11.49 | |
| | | | | 84 | 83.911 507 | 57.00 | |
| | | | | 86 | 85.910 610 | 17.30 | |
| 37 | Rubidium | Rb | 85.467 8 | 85 | 84.911 789 | 72.17 | |
| | | | | 87* | 86.909 184 | 27.83 | $4.75 \times 10^{10}$ yr |
| 38 | Strontium | Sr | 87.62 | 86 | 85.909 262 | 9.86 | |
| | | | | 88 | 87.905 614 | 82.58 | |
| | | | | 90* | 89.907 738 | | 29.1 yr |
| 39 | Yttrium | Y | 88.905 85 | 89 | 88.905 848 | 100 | |
| 40 | Zirconium | Zr | 91.224 | 90 | 89.904 704 | 51.45 | |
| | | | | 91 | 90.905 645 | 11.22 | |
| | | | | 92 | 91.905 040 | 17.15 | |
| | | | | 94 | 93.906 316 | 17.38 | |
| 41 | Niobium | Nb | 92.906 38 | 93 | 92.906 378 | 100 | |
| 42 | Molybdenum | Mo | 95.94 | 92 | 91.906 810 | 14.84 | |
| | | | | 95 | 94.905 842 | 15.92 | |
| | | | | 96 | 95.904 679 | 16.68 | |
| | | | | 98 | 97.905 408 | 24.13 | |

*(Continued)*

| Atomic Number $Z$ | Element | Symbol | Chemical Atomic Mass (u) | Mass Number (* Indicates Radioactive) $A$ | Atomic Mass (u) | Percent Abundance | Half-Life (If Radioactive) $T_{1/2}$ |
|---|---|---|---|---|---|---|---|
| 43 | Technetium | Tc | | 98* | 97.907 216 | | $4.2 \times 10^6$ yr |
| | | | | 99* | 98.906 255 | | $2.1 \times 10^5$ yr |
| 44 | Ruthenium | Ru | 101.07 | 99 | 98.905 939 | 12.76 | |
| | | | | 100 | 99.904 220 | 12.60 | |
| | | | | 101 | 100.905 582 | 17.06 | |
| | | | | 102 | 101.904 350 | 31.55 | |
| | | | | 104 | 103.905 430 | 18.62 | |
| 45 | Rhodium | Rh | 102.905 50 | 103 | 102.905 504 | 100 | |
| 46 | Palladium | Pd | 106.42 | 104 | 103.904 035 | 11.14 | |
| | | | | 105 | 104.905 084 | 22.33 | |
| | | | | 106 | 105.903 483 | 27.33 | |
| | | | | 108 | 107.903 894 | 26.46 | |
| | | | | 110 | 109.905 152 | 11.72 | |
| 47 | Silver | Ag | 107.868 2 | 107 | 106.905 093 | 51.839 | |
| | | | | 109 | 108.904 756 | 48.161 | |
| 48 | Cadmium | Cd | 112.411 | 110 | 109.903 006 | 12.49 | |
| | | | | 111 | 110.904 182 | 12.80 | |
| | | | | 112 | 111.902 757 | 24.13 | |
| | | | | 113* | 112.904 401 | 12.22 | $9.3 \times 10^{15}$ yr |
| | | | | 114 | 113.903 358 | 28.73 | |
| 49 | Indium | In | 114.818 | 115* | 114.903 878 | 95.71 | $4.4 \times 10^{14}$ yr |
| 50 | Tin | Sn | 118.710 | 116 | 115.901 744 | 14.54 | |
| | | | | 118 | 117.901 606 | 24.22 | |
| | | | | 120 | 119.902 197 | 32.58 | |
| 51 | Antimony | Sb | 121.760 | 121 | 120.903 818 | 57.21 | |
| | | | | 123 | 122.904 216 | 42.79 | |
| 52 | Tellurium | Te | 127.60 | 126 | 125.903 306 | 18.84 | |
| | | | | 128* | 127.904 461 | 31.74 | $> 8 \times 10^{24}$ yr |
| | | | | 130* | 129.906 223 | 34.08 | $\leq 1.25 \times 10^{21}$ yr |
| 53 | Iodine | I | 126.904 47 | 127 | 126.904 468 | 100 | |
| | | | | 129* | 128.904 988 | | $1.6 \times 10^7$ yr |
| 54 | Xenon | Xe | 131.29 | 129 | 128.904 780 | 26.44 | |
| | | | | 131 | 130.905 082 | 21.18 | |
| | | | | 132 | 131.904 145 | 26.89 | |
| | | | | 134 | 133.905 394 | 10.44 | |
| | | | | 136* | 135.907 220 | 8.87 | $\geq 2.36 \times 10^{21}$ yr |
| 55 | Cesium | Cs | 132.905 45 | 133 | 132.905 447 | 100 | |
| 56 | Barium | Ba | 137.327 | 137 | 136.905 821 | 11.232 | |
| | | | | 138 | 137.905 241 | 71.698 | |
| 57 | Lanthanum | La | 138.905 5 | 139 | 138.906 349 | 99.910 | |
| 58 | Cerium | Ce | 140.116 | 140 | 139.905 434 | 88.450 | |
| | | | | 142* | 141.909 240 | 11.114 | $> 5 \times 10^{16}$ yr |
| 59 | Praseodymium | Pr | 140.907 65 | 141 | 140.907 648 | 100 | |
| 60 | Neodymium | Nd | 144.24 | 142 | 141.907 719 | 27.2 | |
| | | | | 144* | 143.910 083 | 23.8 | $2.3 \times 10^{15}$ yr |
| | | | | 146 | 145.913 112 | 17.2 | |

| Atomic Number $Z$ | Element | Symbol | Chemical Atomic Mass (u) | Mass Number (* Indicates Radioactive) $A$ | Atomic Mass (u) | Percent Abundance | Half-Life (If Radioactive) $T_{1/2}$ |
|---|---|---|---|---|---|---|---|
| 61 | Promethium | Pm | | 145* | 144.912 744 | | 17.7 yr |
| 62 | Samarium | Sm | 150.36 | 147* | 146.914 893 | 14.99 | $1.06 \times 10^{11}$ yr |
| | | | | 149* | 148.917 180 | 13.82 | $> 2 \times 10^{15}$ yr |
| | | | | 152 | 151.919 728 | 26.75 | |
| | | | | 154 | 153.922 205 | 22.75 | |
| 63 | Europium | Eu | 151.964 | 151 | 150.919 846 | 47.81 | |
| | | | | 153 | 152.921 226 | 52.19 | |
| 64 | Gadolinium | Gd | 157.25 | 156 | 155.922 120 | 20.47 | |
| | | | | 158 | 157.924 100 | 24.84 | |
| | | | | 160 | 159.927 051 | 21.86 | |
| 65 | Terbium | Tb | 158.925 34 | 159 | 158.925 343 | 100 | |
| 66 | Dysprosium | Dy | 162.50 | 162 | 161.926 796 | 25.51 | |
| | | | | 163 | 162.928 728 | 24.90 | |
| | | | | 164 | 163.929 171 | 28.18 | |
| 67 | Holmium | Ho | 164.930 32 | 165 | 164.930 320 | 100 | |
| 68 | Erbium | Er | 167.6 | 166 | 165.930 290 | 33.61 | |
| | | | | 167 | 166.932 045 | 22.93 | |
| | | | | 168 | 167.932 368 | 26.78 | |
| 69 | Thulium | Tm | 168.934 21 | 169 | 168.934 211 | 100 | |
| 70 | Ytterbium | Yb | 173.04 | 172 | 171.936 378 | 21.83 | |
| | | | | 173 | 172.938 207 | 16.13 | |
| | | | | 174 | 173.938 858 | 31.83 | |
| 71 | Lutecium | Lu | 174.967 | 175 | 174.940 768 | 97.41 | |
| 72 | Hafnium | Hf | 178.49 | 177 | 176.943 220 | 18.60 | |
| | | | | 178 | 177.943 698 | 27.28 | |
| | | | | 179 | 178.945 815 | 13.62 | |
| | | | | 180 | 179.946 549 | 35.08 | |
| 73 | Tantalum | Ta | 180.947 9 | 181 | 180.947 996 | 99.988 | |
| 74 | Tungsten (Wolfram) | W | 183.84 | 182 | 181.948 206 | 26.50 | |
| | | | | 183 | 182.950 224 | 14.31 | |
| | | | | 184* | 183.950 933 | 30.64 | $> 3 \times 10^{17}$ yr |
| | | | | 186 | 185.954 362 | 28.43 | |
| 75 | Rhenium | Re | 186.207 | 185 | 184.952 956 | 37.40 | |
| | | | | 187* | 186.955 751 | 62.60 | $4.4 \times 10^{10}$ yr |
| 76 | Osmium | Os | 190.23 | 188 | 187.955 836 | 13.24 | |
| | | | | 189 | 188.958 145 | 16.15 | |
| | | | | 190 | 189.958 445 | 26.26 | |
| | | | | 192 | 191.961 479 | 40.78 | |
| 77 | Iridium | Ir | 192.217 | 191 | 190.960 591 | 37.3 | |
| | | | | 193 | 192.962 924 | 62.7 | |
| 78 | Platinum | Pt | 195.078 | 194 | 193.962 664 | 32.967 | |
| | | | | 195 | 194.964 774 | 33.832 | |
| | | | | 196 | 195.964 935 | 25.242 | |
| 79 | Gold | Au | 196.966 55 | 197 | 196.966 552 | 100 | |
| 80 | Mercury | Hg | 200.59 | 199 | 198.968 262 | 16.87 | |
| | | | | 200 | 199.968 309 | 23.10 | |
| | | | | 201 | 200.970 285 | 13.18 | |
| | | | | 202 | 201.970 626 | 29.86 | |

*(Continued)*

| Atomic Number $Z$ | Element | Symbol | Chemical Atomic Mass (u) | Mass Number (* Indicates Radioactive) $A$ | Atomic Mass (u) | Percent Abundance | Half-Life (If Radioactive) $T_{1/2}$ |
|---|---|---|---|---|---|---|---|
| 81 | Thallium | Tl | 204.383 3 | 203 | 202.972 329 | 29.524 | |
| | | | | 205 | 204.974 412 | 70.476 | |
| | | (Th C″) | | 208* | 207.982 005 | | 3.053 min |
| | | (Ra C″) | | 210* | 209.990 066 | | 1.30 min |
| 82 | Lead | Pb | 207.2 | 204* | 203.973 029 | 1.4 | $\geq 1.4 \times 10^{17}$ yr |
| | | | | 206 | 205.974 449 | 24.1 | |
| | | | | 207 | 206.975 881 | 22.1 | |
| | | | | 208 | 207.976 636 | 52.4 | |
| | | (Ra D) | | 210* | 209.984 173 | | 22.3 yr |
| | | (Ac B) | | 211* | 210.988 732 | | 36.1 min |
| | | (Th B) | | 212* | 211.991 888 | | 10.64 h |
| | | (Ra B) | | 214* | 213.999 798 | | 26.8 min |
| 83 | Bismuth | Bi | 208.980 38 | 209 | 208.980 383 | 100 | |
| | | (Th C) | | 211* | 210.987 258 | | 2.14 min |
| 84 | Polonium | Po | | | | | |
| | | (Ra F) | | 210* | 209.982 857 | | 138.38 days |
| | | (Ra C′) | | 214* | 213.995 186 | | 164 μs |
| 85 | Astatine | At | | 218* | 218.008 682 | | 1.6 s |
| 86 | Radon | Rn | | 222* | 222.017 570 | | 3.823 days |
| 87 | Francium | Fr | | | | | |
| | | (Ac K) | | 223* | 223.019 731 | | 22 min |
| 88 | Radium | Ra | | 226* | 226.025 403 | | 1 600 yr |
| | | (Ms $Th_1$) | | 228* | 228.031 064 | | 5.75 yr |
| 89 | Actinium | Ac | | 227* | 227.027 747 | | 21.77 yr |
| 90 | Thorium | Th | 232.038 1 | | | | |
| | | (Rd Th) | | 228* | 228.028 731 | | 1.913 yr |
| | | (Th) | | 232* | 232.038 050 | 100 | $1.40 \times 10^{10}$ yr |
| 91 | Protactinium | Pa | 231.035 88 | 231* | 231.035 879 | | 32.760 yr |
| 92 | Uranium | U | 238.028 9 | 232* | 232.037 146 | | 69 yr |
| | | | | 233* | 233.039 628 | | $1.59 \times 10^{5}$ yr |
| | | (Ac U) | | 235* | 235.043 923 | 0.720 0 | $7.04 \times 10^{8}$ yr |
| | | | | 236* | 236.045 562 | | $2.34 \times 10^{7}$ yr |
| | | (UI) | | 238* | 238.050 783 | 99.274 5 | $4.47 \times 10^{9}$ yr |
| 93 | Neptunium | Np | | 237* | 237.048 167 | | $2.14 \times 10^{6}$ yr |
| 94 | Plutonium | Pu | | 239* | 239.052 156 | | $2.412 \times 10^{4}$ yr |
| | | | | 242* | 242.058 737 | | $3.73 \times 10^{6}$ yr |
| | | | | 244* | 244.064 198 | | $8.1 \times 10^{7}$ yr |

*Sources:* Chemical atomic masses are from T. B. Coplen, "Atomic Weights of the Elements 1999," a technical report to the International Union of Pure and Applied Chemistry, and published in *Pure and Applied Chemistry,* 73(4), 667–683, 2001. Atomic masses of the isotopes are from G. Audi and A. H. Wapstra, "The 1995 Update to the Atomic Mass Evaluation," *Nuclear Physics,* A595, vol. 4, 409–480, December 25, 1995. Percent abundance values are from K. J. R. Rosman and P. D. P. Taylor, "Isotopic Compositions of the Elements 1999," a technical report to the International Union of Pure and Applied Chemistry, and published in *Pure and Applied Chemistry,* 70(1), 217–236, 1998.

# Appendix C

# Some Useful Tables

**Table C.1** Mathematical Symbols Used in the Text and Their Meaning

| Symbol | Meaning |
|---|---|
| $=$ | is equal to |
| $\neq$ | is not equal to |
| $\equiv$ | is defined as |
| $\propto$ | is proportional to |
| $>$ | is greater than |
| $<$ | is less than |
| $>>$ | is much greater than |
| $<<$ | is much less than |
| $\approx$ | is approximately equal to |
| $\sim$ | is on the order of magnitude of |
| $\Delta x$ | change in $x$ or uncertainty in $x$ |
| $\Sigma x_i$ | sum of all quantities $x_i$ |
| $\|x\|$ | absolute value of $x$ (always a positive quantity) |

**Table C.2** Standard Symbols for Units

| Symbol | Unit | Symbol | Unit |
|---|---|---|---|
| A | ampere | kcal | kilocalorie |
| Å | angstrom | kg | kilogram |
| atm | atmosphere | km | kilometer |
| Bq | bequerel | kmol | kilomole |
| Btu | British thermal unit | L | liter |
| C | coulomb | lb | pound |
| °C | degree Celsius | ly | lightyear |
| cal | calorie | m | meter |
| cm | centimeter | min | minute |
| Ci | curie | mol | mole |
| d | day | N | newton |
| deg | degree (angle) | nm | nanometer |
| eV | electronvolt | Pa | pascal |
| °F | degree Fahrenheit | rad | radian |
| F | farad | rev | revolution |
| ft | foot | s | second |
| G | Gauss | T | tesla |
| g | gram | u | atomic mass unit |
| H | henry | V | volt |
| h | hour | W | watt |
| hp | horsepower | Wb | weber |
| Hz | hertz | yr | year |
| in. | inch | $\mu$m | micrometer |
| J | joule | $\Omega$ | ohm |
| K | kelvin | | |

**Table C.3** The Greek Alphabet

| | | | | | |
|---|---|---|---|---|---|
| Alpha | Α | α | Nu | Ν | ν |
| Beta | Β | β | Xi | Ξ | ξ |
| Gamma | Γ | γ | Omicron | Ο | ο |
| Delta | Δ | δ | Pi | Π | π |
| Epsilon | Ε | ϵ | Rho | Ρ | ρ |
| Zeta | Ζ | ζ | Sigma | Σ | σ |
| Eta | Η | η | Tau | Τ | τ |
| Theta | Θ | θ | Upsilon | Υ | υ |
| Iota | Ι | ι | Phi | Φ | ϕ |
| Kappa | Κ | κ | Chi | Χ | χ |
| Lambda | Λ | λ | Psi | Ψ | ψ |
| Mu | Μ | μ | Omega | Ω | ω |

**Table C.4** Physical Data Often Used[a]

| | |
|---|---|
| Average Earth–Moon distance | $3.84 \times 10^8$ m |
| Average Earth–Sun distance | $1.496 \times 10^{11}$ m |
| Equatorial radius of Earth | $6.38 \times 10^6$ m |
| Density of air (20°C and 1 atm) | 1.20 kg/m$^3$ |
| Density of water (20°C and 1 atm) | $1.00 \times 10^3$ kg/m$^3$ |
| Free-fall acceleration | 9.80 m/s$^2$ |
| Mass of Earth | $5.98 \times 10^{24}$ kg |
| Mass of Moon | $7.36 \times 10^{22}$ kg |
| Mass of Sun | $1.99 \times 10^{30}$ kg |
| Standard atmospheric pressure | $1.013 \times 10^5$ Pa |

[a] These are the values of the constants as used in the text.

**Table C.5** Some Fundamental Constants

| Quantity | Symbol | Value[a] |
|---|---|---|
| Atomic mass unit | u | $1.660\,538\,782\,(83) \times 10^{-27}$ kg<br>$931.494\,028\,(23)$ MeV/$c^2$ |
| Avogadro's number | $N_A$ | $6.022\,141\,79\,(30) \times 10^{23}$ particles/mol |
| Bohr magneton | $\mu_B = \dfrac{e\hbar}{2m_e}$ | $9.274\,009\,15\,(23) \times 10^{-24}$ J/T |
| Bohr radius | $a_0 = \dfrac{\hbar^2}{m_e e^2 k_e}$ | $5.291\,772\,085\,9\,(36) \times 10^{-11}$ m |
| Boltzmann's constant | $k_B = \dfrac{R}{N_A}$ | $1.380\,650\,4\,(24) \times 10^{-23}$ J/K |
| Compton wavelength | $\lambda_C = \dfrac{h}{m_e c}$ | $2.426\,310\,217\,5\,(33) \times 10^{-12}$ m |
| Coulomb constant | $k_e = \dfrac{1}{4\pi\epsilon_0}$ | $8.987\,551\,788\ldots \times 10^9$ N·m²/C² (exact) |
| Deuteron mass | $m_d$ | $3.343\,583\,20\,(17) \times 10^{-27}$ kg<br>$2.013\,553\,212\,724\,(78)$ u |
| Electron mass | $m_e$ | $9.109\,382\,15\,(45) \times 10^{-31}$ kg<br>$5.485\,799\,094\,3\,(23) \times 10^{-4}$ u<br>$0.510\,998\,910\,(13)$ MeV/$c^2$ |
| Electron volt | eV | $1.602\,176\,487\,(40) \times 10^{-19}$ J |
| Elementary charge | $e$ | $1.602\,176\,487\,(40) \times 10^{-19}$ C |
| Gas constant | $R$ | $8.314\,472\,(15)$ J/mol·K |
| Gravitational constant | $G$ | $6.674\,28\,(67) \times 10^{-11}$ N·m²/kg² |
| Neutron mass | $m_n$ | $1.674\,927\,211\,(84) \times 10^{-27}$ kg<br>$1.008\,664\,915\,97\,(43)$ u<br>$939.565\,346\,(23)$ MeV/$c^2$ |
| Nuclear magneton | $\mu_n = \dfrac{e\hbar}{2m_p}$ | $5.050\,783\,24\,(13) \times 10^{-27}$ J/T |
| Permeability of free space | $\mu_0$ | $4\pi \times 10^{-7}$ T·m/A (exact) |
| Permittivity of free space | $\epsilon_0 = \dfrac{1}{\mu_0 c^2}$ | $8.854\,187\,817\ldots \times 10^{-12}$ C²/N·m² (exact) |
| Planck's constant | $h$ | $6.626\,068\,96\,(33) \times 10^{-34}$ J·s |
| | $\hbar = \dfrac{h}{2\pi}$ | $1.054\,571\,628\,(53) \times 10^{-34}$ J·s |
| Proton mass | $m_p$ | $1.672\,621\,637\,(83) \times 10^{-27}$ kg<br>$1.007\,276\,466\,77\,(10)$ u<br>$938.272\,013\,(23)$ MeV/$c^2$ |
| Rydberg constant | $R_H$ | $1.097\,373\,156\,852\,7\,(73) \times 10^7$ m$^{-1}$ |
| Speed of light in vacuum | $c$ | $2.997\,924\,58 \times 10^8$ m/s (exact) |

*Note:* These constants are the values recommended in 2006 by CODATA, based on a least-squares adjustment of data from different measurements. For a more complete list, see P. J. Mohr, B. N. Taylor, and D. B. Newell, "CODATA Recommended Values of the Fundamental Physical Constants: 2006." *Rev. Mod. Phys.* **80**:2, 633–730, 2008.

[a]The numbers in parentheses represent the uncertainties of the last two digits.

# Appendix D

# *SI Units*

**Table D.1** SI Base Units

| Base Quantity | SI Base Unit Name | Symbol |
|---|---|---|
| Length | meter | m |
| Mass | kilogram | kg |
| Time | second | s |
| Electric current | ampere | A |
| Temperature | kelvin | K |
| Amount of substance | mole | mol |
| Luminous intensity | candela | cd |

**Table D.2** Derived SI Units

| Quantity | Name | Symbol | Expression in Terms of Base Units | Expression in Terms of Other SI Units |
|---|---|---|---|---|
| Plane angle | radian | rad | m/m | |
| Frequency | hertz | Hz | $s^{-1}$ | |
| Force | newton | N | $kg \cdot m/s^2$ | J/m |
| Pressure | pascal | Pa | $kg/m \cdot s^2$ | $N/m^2$ |
| Energy: work | joule | J | $kg \cdot m^2/s^2$ | $N \cdot m$ |
| Power | watt | W | $kg \cdot m^2/s^3$ | J/s |
| Electric charge | coulomb | C | $A \cdot s$ | |
| Electric potential (emf) | volt | V | $kg \cdot m^2/A \cdot s^3$ | W/A, J/C |
| Capacitance | farad | F | $A^2 \cdot s^4/kg \cdot m^2$ | C/V |
| Electric resistance | ohm | Ω | $kg \cdot m^2/A^2 \cdot s^3$ | V/A |
| Magnetic flux | weber | Wb | $kg \cdot m^2/A \cdot s^2$ | $V \cdot s$, $T \cdot m^2$ |
| Magnetic field intensity | tesla | T | $kg/A \cdot s^2$ | $Wb/m^2$ |
| Inductance | henry | H | $kg \cdot m^2/A^2 \cdot s^2$ | Wb/A |

# Answers to Quick Quizzes, Example Questions, Odd-Numbered Warm-Up Exercises, Conceptual Questions, and Problems

## CHAPTER 15

### Quick Quizzes

1. (b)
2. (b)
3. (c)
4. (a)
5. (c) and (d)
6. (a)
7. (c)
8. (b)
9. (d)
10. (b) and (d)

### Example Questions

1. $\frac{1}{4}$
2. (a)
3. Fourth quadrant
4. The suspended droplet would accelerate downward at twice the acceleration of gravity.
5. The angle $\phi$ would increase.
6. (c)
7. The charge on the inner surface would be negative.
8. Zero

### Warm-Up Exercises

1. (a) 8.32 N (b) 147°
3. (a) 7.50 N (b) −16.0 N (c) 17.7 N (d) −64.9°
5. (a) 2.97 N (b) 34.3°
7. (a) 2.85 N (b) 5.70 m/s$^2$
9. (a) 5.00 m (b) 53.1° (c) 5.75 N/C (d) 3.45 N/C (e) 4.60 N/C
11. (a) $-1.47 \times 10^{-14}$ N (b) $1.44 \times 10^{-14}$ N (c) −0.200 m/s$^2$
13. 198 N · m$^2$/C

### Conceptual Questions

1. Electrons have been removed from the glass object. Negative charge has been removed from the initially neutral rod, resulting in a net positive charge on the rod. The protons cannot be removed from the rod; protons are not mobile because they are within the nuclei of the atoms of the rod.
3. No. The charge on the metallic sphere resides on its outer surface, so the person is able to touch the surface without causing any charge transfer.
5. No. Life would be no different if electrons were + charged and protons were − charged. Opposite charges would still attract, and like charges would repel. The naming of + and − charge is merely a convention.
7. Move an object *A* with a net positive charge so it is near, but not touching, a neutral metallic object *B* that is insulated from the ground. The presence of *A* will polarize *B*, causing an excess negative charge to exist on the side nearest *A* and an excess positive charge of equal magnitude to exist on the side farthest from *A*. While *A* is still near *B*, touch *B* with your hand. Additional electrons will then flow from ground, through your body, and onto *B*. With *A* continuing to be near but not in contact with *B*, remove your hand from *B*, thus trapping the excess electrons on *B*. When *A* is now removed, *B* is left with excess electrons, or a net negative charge. By means of mutual repulsion, this negative charge will now spread uniformly over the entire surface of *B*.
9. She is not shocked. She becomes part of the dome of the Van de Graaff, and charges flow onto her body. They do not jump to her body via a spark, however, so she is not shocked.
11. The dry paper is initially neutral. The comb attracts the paper because its electric field causes the molecules of the paper to become polarized—the paper as a whole cannot be polarized because it is an insulator. Each molecule is polarized so that its unlike-charged side is closer to the charged comb than its like-charged side, so the molecule experiences a net attractive force toward the comb. Once the paper comes in contact with the comb, like charge can be transferred from the comb to the paper, and if enough of this charge is transferred, the like-charged paper is then repelled by the like-charged comb.
13. The electric shielding effect of conductors depends on the fact that there are two kinds of charge: positive and negative. As a result, charges can move within the conductor so that the combination of positive and negative charges establishes an electric field that exactly cancels the external field within the conductor and any cavities inside the conductor. There is only one type of gravitation charge, however, because there is no negative mass. As a result, gravitational shielding is not possible. A room cannot be gravitationally shielded because mass is always positive or zero, never negative.
15. You can only conclude that the net charge inside the Gaussian surface is positive.
17. (b)

### Problems

1. (a) $8.74 \times 10^{-8}$ N (b) repulsive
3. (a) $2.36 \times 10^{-5}$ C (b) The charges induce opposite charges in the bulkheads, but the induced charge in the bulkhead near ball *B* is greater because of ball *B*'s greater charge. The system therefore moves slowly toward the bulkhead closer to ball *B*.
5. (a) 36.8 N (b) $5.54 \times 10^{27}$ m/s$^2$
7. 49.3 N/m
9. (a) $2.2 \times 10^{-5}$ N (attraction) (b) $9.0 \times 10^{-7}$ N (repulsion)
11. $1.38 \times 10^{-5}$ N at 77.5° below the negative *x*-axis
13. 0.437 N at −85.3° from the +*x*-axis.
15. 7.2 nC
17. $2.07 \times 10^3$ N/C down
19. 3.15 N due north
21. (a) $\frac{mg}{|Q|}\sin\theta$ (b) $3.19 \times 10^3$ N/C down the incline
23. (a) $6.12 \times 10^{10}$ m/s$^2$ (b) 19.6 $\mu$s (c) 11.8 m (d) $1.20 \times 10^{-15}$ J
25. $E_x = (1 - \sqrt{2})k_e\frac{Q}{d^2}$, $E_y = \sqrt{2}k_e\frac{Q}{d^2}$
27. 1.8 m to the left of the −2.5-$\mu$C charge
29. zero
35. (a) 0 (b) 5 $\mu$C inside, −5 $\mu$C outside (c) 0 inside, −5 $\mu$C outside (d) 0 inside, −5 $\mu$C outside
37. $1.3 \times 10^{-3}$ C
39. (a) $2.54 \times 10^{-15}$ N (b) $1.59 \times 10^4$ N/C radially outward
41. (a) 858 N · m$^2$/C (b) 0 (c) 657 N · m$^2$/C

43. $-Q/\epsilon_0$ for $S_1$; 0 for $S_2$; $-2Q/\epsilon_0$ for $S_3$; 0 for $S_4$
45. (a) 0 (b) $k_e q/r^2$ outward
47. (a) −7.99 N/C (b) 0 (c) 1.44 N/C (d) 2.00 nC on the inner surface; 1.00 nC on the outer surface
49. 115 N
51. 24 N/C in the positive $x$-direction
53. (a) $E = 2k_e qb\,(a^2 + b^2)^{-3/2}$ in the positive $x$-direction (b) $E = k_e Qb(a^2 + b^2)^{-3/2}$ in the positive $x$-direction
55. $F_x = (0.354)k_e\dfrac{Q^2}{d^2}$; $F_y = (1.65)k_e\dfrac{Q^2}{d^2}$
57. $4.4 \times 10^5$ N/C
59. $1.14 \times 10^{-7}$ C on one sphere and $5.69 \times 10^{-8}$ C on the other
61. (a) 0 (b) $7.99 \times 10^7$ N/C (outward) (c) 0 (d) $7.34 \times 10^6$ N/C (outward)
63. (a) $1.00 \times 10^3$ N/C (b) $3.37 \times 10^{-8}$ s (c) accelerate at $1.76 \times 10^{14}$ m/s$^2$ in the direction opposite that of the electric field

## CHAPTER 16

### Quick Quizzes

1. (b)
2. (a)
3. (b)
4. (d)
5. (d)
6. (c)
7. (a)
8. (c)
9. (a) $C$ decreases. (b) $Q$ stays the same. (c) $E$ stays the same. (d) $\Delta V$ increases. (e) The energy stored increases.
10. (a) $C$ increases. (b) $Q$ increases. (c) $E$ stays the same. (d) $\Delta V$ remains the same. (e) The energy stored increases.
11. (a)

### Example Questions

1. True
2. True
3. False
4. (a)
5. Three
6. Each answer would be reduced by a factor of one-half.
7. One-quarter
8. The voltage drop is the smallest across the 24-$\mu$F capacitor and largest across the 3.0-$\mu$F capacitor.
9. The 3.0-$\mu$F capacitor
10. (c)
11. (b)
12. $C = \dfrac{\epsilon_0 A}{d}$

### Warm-Up Exercises

1. 1.96 Ω
3. (a) $9.00 \times 10^{-6}$ J (b) $-9.00 \times 10^{-6}$ J (c) −4.50 V
5. (a) $2.40 \times 10^{-9}$ V (b) $3.84 \times 10^{-28}$ J
7. (a) $2.70 \times 10^{-5}$ C (b) 24.0 V
9. (a) 5.00 $\mu$F (b) 1.20 $\mu$F
11. (a) 8.11 $\mu$F (b) $9.73 \times 10^{-5}$ C (c) 8.11 V (d) 3.89 V

### Conceptual Questions

1. (a) The proton moves in a straight line with constant acceleration in the direction of the electric field. (b) As its velocity increases, its kinetic energy increases, and the electric potential energy associated with the proton decreases.
3. The work done in pulling the capacitor plates farther apart is transferred into additional electric energy stored in the capacitor. The charge is constant, and the capacitance decreases, but the potential difference between the plates increases, which results in an increase in the stored electric energy.
5. If the power line makes electrical contact with the metal of the car, it will raise the potential of the car to 20 kV. It will also raise the potential of your body to 20 kV, because you are in contact with the car. In itself, this is not a problem. If you step out of the car, however, your body at 20 kV will make contact with the ground, which is at zero volts. As a result, a current will pass through your body, and you will likely be injured. Thus, it is best to stay in the car until help arrives.
7. If two points on a conducting object were at different potentials, then free charges in the object would move, and we would not have static conditions, in contradiction to the initial assumption. (Free positive charges would migrate from locations of higher to locations of lower potential. Free electrons would rapidly move from locations of lower to locations of higher potential.) All of the charges would continue to move until the potential became equal everywhere in the conductor.
9. (a) The capacitor often remains charged long after the voltage source is disconnected. This residual charge can be lethal. (b) The capacitor can be safely handled after discharging the plates by short-circuiting the device with a conductor, such as a screwdriver with an insulating handle.
11. $D > C > B > A$
13. Not all connections are simple combinations of series and parallel circuits. As an example of such a complex circuit, consider the network of five capacitors, $C_1$, $C_2$, $C_3$, $C_4$, and $C_5$ shown below

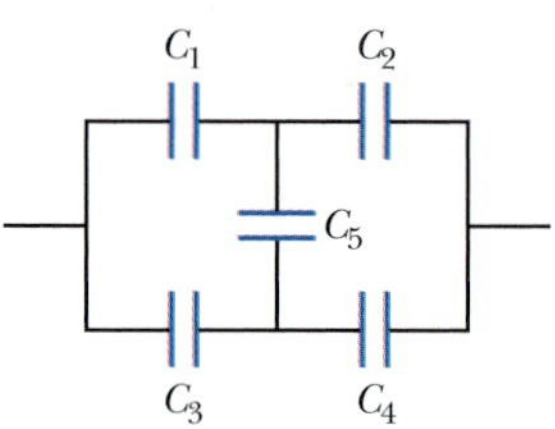


This combination cannot be reduced to a simple equivalent by the techniques of combining series and parallel capacitors.

### Problems

1. (a) $1.92 \times 10^{-18}$ J (b) $-1.92 \times 10^{-18}$ J (c) $2.05 \times 10^6$ m/s in the negative $x$-direction
3. $1.4 \times 10^{-20}$ J
5. $1.7 \times 10^6$ N/C
7. (a) $1.13 \times 10^5$ N/C (b) $1.80 \times 10^{-14}$ N (c) $4.37 \times 10^{-17}$ J
9. (a) The spring stretches by 4.36 cm. (b) Equilibrium: $x = 2.18$ cm; $A = 2.18$ cm (c) $\Delta V = -\dfrac{2kA^2}{Q}$
11. (a) $-5.75 \times 10^{-7}$ V (b) $-1.92 \times 10^{-7}$ V, $\Delta V = 3.84 \times 10^{-7}$ V (c) No. Unless fixed in place, the electron would move in the opposite direction, increasing its distance from points $A$ and $B$ and lowering the potential difference between them.
13. (a) $2.67 \times 10^6$ V (b) $2.13 \times 10^6$ V
15. (a) 103 V (b) $-3.85 \times 10^{-7}$ J; positive work must be done to separate the charges.
17. −11.0 kV
19. (a) $3.84 \times 10^{-14}$ J (b) $2.55 \times 10^{-13}$ J (c) $-2.17 \times 10^{-13}$ J (d) $8.08 \times 10^6$ m/s (e) $1.24 \times 10^7$ m/s
21. (a) 0.294 J (b) 271 m/s
23. $2.74 \times 10^{-14}$ m
25. (a) $1.1 \times 10^{-8}$ F (b) 27 C
27. (a) 1.36 pF (b) 16.3 pC (c) $8.00 \times 10^3$ V/m
29. (a) 11.1 kV/m toward the negative plate (b) 3.74 pF (c) 74.8 pC and −74.8 pC

**31.** (a) $5.90 \times 10^{-10}$ F (b) $3.54 \times 10^{-9}$ C (c) $2.00 \times 10^3$ N/C (d) $1.77 \times 10^{-8}$ C/m$^2$ (e) All the answers are reduced.
**33.** (a) 10.7 $\mu$C on each capacitor (b) 15.0 $\mu$C on the 2.50-$\mu$F capacitor and 37.5 $\mu$C on the 6.25-$\mu$F capacitor
**35.** (a) 2.67 $\mu$F (b) 24.0 $\mu$C on each 8.00-$\mu$F capacitor, 18.0 $\mu$C on the 6.00-$\mu$F capacitor, 6.00 $\mu$C on the 2.00-$\mu$F capacitor (c) 3.00 V across each capacitor
**37.** (a) 3.33 $\mu$F (b) 180 $\mu$C on the 3-$\mu$F and the 6-$\mu$F capacitors, 120 $\mu$C on the 2.00-$\mu$F and 4.00-$\mu$F capacitors (c) 60.0 V across the 3-$\mu$F and the 2-$\mu$F capacitors, 30.0 V across the 6-$\mu$F and the 4-$\mu$F capacitors
**39.** $Q_1 = 16.0\ \mu$C, $Q_5 = 80.0\ \mu$C, $Q_8 = 64.0\ \mu$C, $Q_4 = 32.0\ \mu$C
**41.** (a) $Q_{25} = 1.25$ mC, $Q_{40} = 2.00$ mC (b) $Q'_{25} = 288\ \mu$C, $Q'_{40} = 462\ \mu$C (c) $\Delta V = 11.5$ V
**43.** $Q'_1 = 3.33\ \mu$C, $Q'_2 = 6.67\ \mu$C
**45.** $3.24 \times 10^{-4}$ J
**47.** (a) 54.0 $\mu$J (b) 108 $\mu$J (c) 27.0 $\mu$J
**49.** (a) $\kappa = 3.4$. The material is probably nylon (see Table 16.1). (b) The voltage would lie somewhere between 25.0 V and 85.0 V.
**51.** (a) 8.1 nF (b) 2.4 kV
**53.** (a) volume $9.09 \times 10^{-16}$ m$^3$, area $4.54 \times 10^{-10}$ m$^2$ (b) $2.01 \times 10^{-13}$ F (c) $2.01 \times 10^{-14}$ C, $1.26 \times 10^5$ electronic charges
**59.** 0.188 m$^2$
**61.** 6.25 $\mu$F
**63.** 4.47 kV
**65.** 0.75 mC on $C_1$, 0.25 mC on $C_2$
**67.** Sphere A: 0.800 $\mu$C; sphere B: 1.20 $\mu$C

## CHAPTER 17

### Quick Quizzes

**1.** (d)
**2.** (b)
**3.** (c), (d)
**4.** (b)
**5.** (b)
**6.** (b)
**7.** (a)
**8.** (b)
**9.** (a)
**10.** (c)

### Example Questions

**1.** No. Such a current corresponds to the passage of one electron every 2 seconds. The average current, however, can have any value.
**2.** True
**3.** Higher
**4.** (b)
**5.** (a)
**6.** (c)

### Warm-Up Exercises

**1.** 10.4 W
**3.** (a) $3.39 \times 10^4$ m/s (b) $3.59 \times 10^4$ m/s
**5.** (a) $3.30 \times 10^{-6}$ m$^2$ (b) $2.24 \times 10^{-4}$ m/s
**7.** $3.9 \times 10^2$ m
**9.** (a) $1.20 \times 10^{-2}$ A (b) 0.144 W (c) 518 J

### Conceptual Questions

**1.** In the electrostatic case in which charges are stationary, the electric field inside a conductor must be zero. A nonzero field would produce a current (by interacting with the free electrons in the conductor), which would violate the condition of static equilibrium. In this chapter we deal with conductors that carry current, a nonelectrostatic situation. The current arises because of a potential difference applied between the ends of the conductor, which produces an internal electric field.
**3.** Because there are so many electrons in a conductor (approximately $10^{28}$ electrons/m$^3$), the average velocity of charges is very slow. When you connect a wire to a potential difference, you establish an electric field everywhere in the wire nearly instantaneously, to make electrons start drifting everywhere all at once.
**5.** We would conclude that the conductor is nonohmic.
**7.** A voltage is not something that "surges through" a completed circuit. A voltage is a potential difference that is applied across a device or a circuit. It would be more correct to say "1 ampere of electricity surged through the victim's body." Although this amount of current would have disastrous results on the human body, a value of 1 (ampere) doesn't sound as exciting for a newspaper article as 10 000 (volts). Another possibility is to write "10 000 volts of electricity were applied across the victim's body," which still doesn't sound quite as exciting.
**9.** The drift velocity might increase steadily as time goes on, because collisions between electrons and atoms in the wire would be essentially nonexistent and the conduction electrons would move with constant acceleration. The current would rise steadily without bound also, because $I$ is proportional to the drift velocity.
**11.** Once the switch is closed, the line voltage is applied across the bulb. As the voltage is applied across the cold filament when it is first turned on, the resistance of the filament is low, the current is high, and a relatively large amount of power is delivered to the bulb. As the filament warms, its resistance rises and the current decreases. As a result, the power delivered to the bulb decreases. The large current spike at the beginning of the bulb's operation is the reason that lightbulbs often fail just after they are turned on.

### Problems

**1.** (a) $3.00 \times 10^{20}$ electrons (b) the direction opposite to the current
**3.** 1.05 mA
**5.** 0.678 mm
**7.** 27 yr
**9.** (a) $55.85 \times 10^{-3}$ kg/mol (b) $1.41 \times 10^5$ mol/m$^3$ (c) $8.49 \times 10^{28}$ iron atoms/m$^3$ (d) $1.70 \times 10^{29}$ conduction electrons/m$^3$ (e) $2.21 \times 10^{-4}$ m/s
**11.** 32 V is 200 times larger than 0.16 V
**13.** (a) 13.0 Ω (b) 17.0 m
**15.** (a) 30 Ω (b) $4.7 \times 10^{-4}\ \Omega \cdot$ m
**17.** silver ($\rho = 1.59 \times 10^{-8}\ \Omega \cdot$ m)
**19.** 256 Ω
**23.** 2.0 A
**25.** 2.71 MΩ
**27.** 26 mA
**29.** (a) 3.0 A (b) 2.9 A
**31.** (a) 1.2 Ω (b) $8.0 \times 10^{-4}$ (a 0.080% increase)
**33.** (a) 8.33 A (b) 14.4 Ω
**35.** (a) 2.1 W (b) 3.42 W (c) The aluminum wire would not be as safe. If surrounded by thermal insulation, it would get much hotter than a copper wire.
**37.** 11.2 min
**39.** (a) $R_A = 576\ \Omega$; $R_B = 144\ \Omega$ (b) 4.80 s. (c) The charge is the same. It is at a location that is lower in potential. (d) 0.040 0 s (e) Energy enters the lightbulb by electric transmission and leaves by heat and electromagnetic radiation. (f) $1.98
**41.** 1.6 cm
**43.** 15.0 $\mu$W
**45.** (a) 116 V (b) 12.8 kW

47. $\frac{d_A}{d_B} = \sqrt{3}$
49. $2.16 \times 10^5$ C
51. (a) 1.50 Ω (b) 6.00 A
53. 1.1 km
55. $1.47 \times 10^{-6}\ \Omega \cdot$ m; differs by 2.0% from value in Table 17.1
57. (a) $3.06 (b) No. The circuit must be able to handle at least 26 A.
59. (a) 17.3 A (b) 22.4 MJ (c) $0.684
61. $3.77 \times 10^{28}/\text{m}^3$
63. 37 MΩ
65. 0.48 kg/s

## CHAPTER 18

### Quick Quizzes

1. True
2. Because of the battery's internal resistance, power is delivered to the battery material, raising its temperature.
3. (b)
4. (a)
5. (a)
6. (b)
7. *Parallel:* (a) unchanged (b) unchanged (c) increase (d) decrease
8. *Series:* (a) decrease (b) decrease (c) decrease (d) increase
9. (c)

### Example Questions

1. As the resistors get warmer, their resistance increases, reducing the current.
2. (b)
3. The 8.00-Ω resistor
4. The answers would be negative, but they would have the same magnitude as before.
5. Yes
6. (a)
7. (a)

### Warm-Up Exercises

1. (a) 9.00 A (b) 18.0 A
3. (a) 2.46 C (b) 0.183 s
5. (a) 2.50 μF (b) 0.400 μF
7. (a) 2.71 A (b) 8.12 V (c) 24.9 W
9. (a) 1.56 Ω (b) 37.0 Ω
11. (a) 1.94 Ω (b) 3.94 Ω (c) 3.05 A (d) 6.10 V (e) 5.90 V (f) 0.590 A
13. (a) $9.90 \times 10^{-3}$ s (b) $4.32 \times 10^{-4}$ C (c) $1.71 \times 10^{-4}$ C

### Conceptual Questions

1. No. When a battery serves as a source and supplies current to a circuit, the direction of the current is from the negative terminal of the battery to the positive one. However, when a source having a larger emf than the battery is used to charge the battery, the direction of the current is from the positive terminal of the battery to the negative one.
3. The total amount of energy delivered by the battery will be less than *W*. Recall that a battery can be considered an ideal, resistanceless battery in series with the internal resistance. When the battery is being charged, the energy delivered to it includes the energy necessary to charge the ideal battery, plus the energy that goes into raising the temperature of the battery due to $I^2r$ heating in the internal resistance. This latter energy is not available during discharge of the battery, when part of the reduced available energy again transforms into internal energy in the internal resistance, further reducing the available energy below *W*.
5. Connecting batteries in parallel does not increase the emf. A high-current device connected to two batteries in parallel can draw currents from both batteries. Thus, connecting the batteries in parallel increases the possible current output and, therefore, the possible power output.
7. The starter in the automobile draws a relatively large current from the battery. This large current causes a significant voltage drop across the internal resistance of the battery. As a result, the terminal voltage of the battery is reduced, and the headlights dim accordingly.
9. The bird is resting on a wire of fixed potential. In order to be electrocuted, a large potential difference is required between the bird's feet. The potential difference between the bird's feet is too small to harm the bird.
11. She will not be electrocuted if she holds onto only one high-voltage wire, because she is not completing a circuit. There is no potential difference across her body as long as she clings to only one wire. However, she should release the wire immediately once it breaks, because she will become part of a closed circuit when she reaches the ground or comes into contact with another object.
13. The junction rule is a statement of conservation of charge. It says that the amount of charge that enters a junction in some time interval must equal the charge that leaves the junction in that time interval. The loop rule is a statement of conservation of energy. It says that the increases and decreases in potential around a closed loop in a circuit must add to zero.

### Problems

1. 4.92 Ω
3. (a) Your circuit diagram will consist of two 0.800-Ω resistors in series with the 192-Ω resistance of the bulb. (b) 73.8 W
5. (a) 17.1 Ω (b) 1.99 A for 4.00 Ω and 9.00 Ω, 1.17 A for 7.00 Ω, 0.820 A for 10.0 Ω
7. 470 Ω and 220 Ω
9. (a) 5.68 V (b) 0.227 A
11. 55 Ω
13. 0.43 A
15. (a) Connect two 50-Ω resistors in parallel, and then connect this combination in series with a 20-Ω resistor. (b) Connect two 50-Ω resistors in parallel, connect two 20-V resistors in parallel, and then connect these two combinations in series with each other.
17. 0.714 A, 1.29 A, 12.6 V
19. 50.0 mA from *a* to *e*
21. (a) $I_1 = 1.00$ A upward in 200 Ω; $I_4 = 4.00$ A upward in 70.0 Ω; $I_2 = 3.00$ A upward in 80.0 Ω; $I_3 = 8.00$ A downward in 20.0 Ω; (b) 200 V
23. (a) 0.385 mA, 3.08 mA, 2.69 mA (b) 69.2 V, with *c* at the higher potential
25. (a) No. The only simplification is to note that the 2.0-Ω and 4.0-Ω resistors are in series and add to a resistance of 6.0 Ω. Likewise, the 5.0-Ω and 1.0-Ω resistors are in series and add to a resistance of 6.0 Ω. The circuit cannot be simplified any further. Kirchhoff's rules must be used to analyze the circuit. (b) $I_1 = 3.5$ A, $I_2 = 2.5$ A, $I_3 =$ 1.0 A
27. (a) No. The multiloop circuit cannot be simplified any further. Kirchhoff's rules must be used to analyze the circuit. (b) $I_{30} = 0.353$ A directed to the right, $I_5 = 0.118$ A directed to the right, $I_{20} = 0.471$ A directed to the left
29. $\Delta V_2 = 3.05$ V, $\Delta V_3 = 4.57$ V, $\Delta V_4 = 7.38$ V, $\Delta V_5 = 1.62$ V
31. (a) 1.88 s (b) $1.90 \times 10^{-4}$ C
33. (a) 5.00 s (b) 150 μC (c) 4.06 μA
35. (a) 0.432 s (b) 6.00 μF
37. 48 lightbulbs
39. (a) 6.25 A (b) 750 W

**41.** (a) $1.2 \times 10^{-9}$ C, $7.3 \times 10^{9}$ $K^+$ ions. Not large, only $1e/(290\ \text{Å})^2$. (b) $1.7 \times 10^{-9}$ C, $1.0 \times 10^{10}$ $Na^+$ ions (c) 0.83 mA (d) $7.5 \times 10^{-12}$ J
**43.** 11 nW
**45.** (a) 4.00 V (b) Point *a* is at the higher potential.
**47.** (a) 15 Ω (b) $I_1 = 1.0$ A, $I_2 = I_3 = 0.50$ A, $I_4 = 0.30$ A, and $I_5 = 0.20$ A (c) $(\Delta V)_{ac} = 6.0$ V, $(\Delta V)_{ce} = 1.2$ V, $(\Delta V)_{ed} = (\Delta V)_{fd} = 1.8$ V, $(\Delta V)_{cd} = 3.0$ V, $(\Delta V)_{db} = 6.0$ V (d) $P_{ac} = 6.0$ W, $P_{ce} = 0.60$ W, $P_{ed} = 0.54$ W, $P_{fd} = 0.36$ W, $P_{cd} = 1.5$ W, $P_{db} = 6.0$ W
**49.** 6.00 Ω, 3.00 Ω
**51.** (a) $R_{\text{eq}}^{\text{open}} = 3R$, $R_{\text{eq}}^{\text{closed}} = 2R$ (b) $P = \dfrac{\mathcal{E}^2}{3R}$, $P = \dfrac{\mathcal{E}^2}{2R}$
(c) Lamps A and B increase in brightness, lamp C goes out.
**53.** (a) 1.02 A down (b) 0.364 A down (c) 1.38 A up (d) 0 (e) 66.0 μC
**55.** (a) $R_x = R_2 - \frac{1}{4}R_1$ (b) $R_x = 2.8\ \Omega$ (inadequate grounding)
**59.** $P = \dfrac{(144\ \text{V}^2)R}{(R + 10.0\ \Omega)^2}$

$P_{\text{load}}$
3.6 W
10 Ω
$R_{\text{load}}$

**61.** (a) $6.0 \times 10^{2}$ W (b) 1.2 J (c) 37 m
**63.** 0.395 A; 1.50 V
**65.** (a) $2.41 \times 10^{-5}$ C (b) $1.61 \times 10^{-5}$ C (c) $1.61 \times 10^{-2}$ A

# CHAPTER 19

## Quick Quizzes

1. (b)
2. (c)
3. (c)
4. (a)
5. (a), (c)
6. (b)

## Example Questions

1. The force on the electron is opposite the direction of the force on the proton, and the acceleration of the electron is far greater than the acceleration on the proton due to the electron's lower mass.
2. A magnetic field always exerts a force perpendicular to the velocity of a charged particle, so it can change the particle's direction but not its speed.
3. Zero
4. As the angle approaches 90°, the magnitude of the force increases. After going beyond 90°, it decreases.
5. The magnitude of the momentum remains constant. The direction of momentum changes, unless the particle's velocity is parallel or antiparallel to the magnetic field.
6. 20 m
7. (a) to the left (b) to the right
8. Earth's magnetic field is extremely weak, so the current carried by the car would have to be correspondingly very large. Such a current would be impractical to generate, and if generated, it would heat and melt the wires carrying it.
9. The proton would accelerate downward due to gravity while circling.

## Warm-Up Exercises

**1.** (a) $1.19 \times 10^{3}$ N · m (b) $1.37 \times 10^{3}$ N · m
**3.** 0.200 T
**5.** $6.00 \times 10^{-3}$ N · m
**7.** $4.79 \times 10^{3}$ m/s
**9.** $6.03 \times 10^{-4}$ T

## Conceptual Questions

1. The set should be oriented such that the beam is moving either toward the east or toward the west.
3. The magnetic force on a moving charged particle is always perpendicular to the particle's direction of motion. There is no magnetic force on the charge when it moves parallel to the direction of the magnetic field. However, the force on a charged particle moving in an electric field is never zero and is always parallel to the direction of the field. Therefore, by projecting the charged particle in different directions, it is possible to determine the nature of the field.
5. In the figure, the magnetic field created by wire 1 at the position of wire 2 is into the paper. Hence, the magnetic force on wire 2 is in direction (current down) × (field into the paper) = (force to the right), away from wire 1. Now wire 2 creates a magnetic field into the page at the location of wire 1, so wire 1 feels force (current up) × (field into the paper) = (force to the left), away from wire 2.

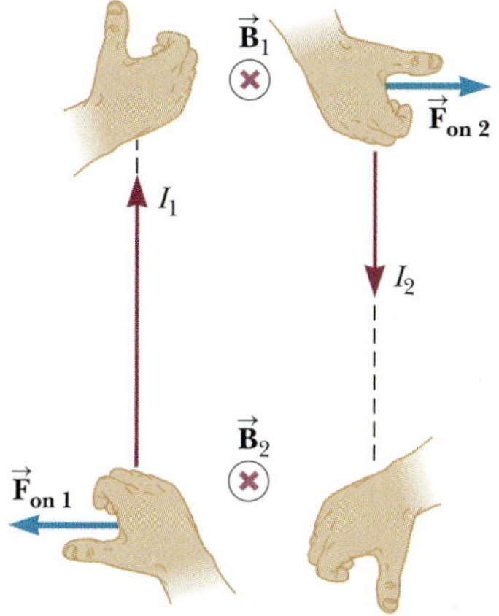


7. Such levitation could never occur. At the North Pole, where Earth's magnetic field is directed downward, toward the equivalent of a buried south pole, a coffin would be repelled if its south magnetic pole were directed downward. However, equilibrium would be only transitory, as any slight disturbance would upset the balance between the magnetic force and the gravitational force.
9. A compass does not detect currents in wires near light switches, for two reasons. The first is that, because the cable to the light switch contains two wires, one carrying current to the switch and the other carrying it away from the switch, the net magnetic field would be very small and would fall off rapidly with increasing distance. The second reason is that the current is alternating at 60 Hz. As a result, the magnetic field is oscillating at 60 Hz also. This frequency would be too fast for the compass to follow, so the effect on the compass reading would average to zero.
11. If you were moving along with the electrons, you would measure a zero current for the electrons, so they would not produce a magnetic field according to your observations. However, the fixed positive charges in the metal would now be moving backward relative to you, creating a current equivalent to the forward motion of the electrons when you were stationary. Thus, you would measure the same magnetic field as when you were stationary, but it would be due to the positive charges presumed to be moving from your point of view.
13. Each coil of the Slinky® will become a magnet, because a coil acts as a current loop. The sense of rotation of the current is the same in all coils, so each coil becomes a magnet with the

same orientation of poles. Thus, all of the coils attract, and the Slinky® will compress.

15. There is no net force on the wires, but there is a torque. To understand this distinction, imagine a fixed vertical wire and a free horizontal wire (see the figure below). The vertical wire carries an upward current and creates a magnetic field that circles the vertical wire itself. To the right, the magnetic field of the vertical wire points into the page, while on the left side it points out of the page, as indicated. Each segment of the horizontal wire (of length $\ell$) carries current that interacts with the magnetic field according to the equation $F = BI\ell \sin \theta$. Apply the right-hand rule on the right side: point the fingers of your right hand in the direction of the horizontal current and curl them into the page in the direction of the magnetic field. Your thumb points downward, the direction of the force on the right side of the wire. Repeating the process on the left side gives a force upward on the left side of the wire. The two forces are equal in magnitude and opposite in direction, so the net force is zero, but they create a net torque around the point where the wires cross.

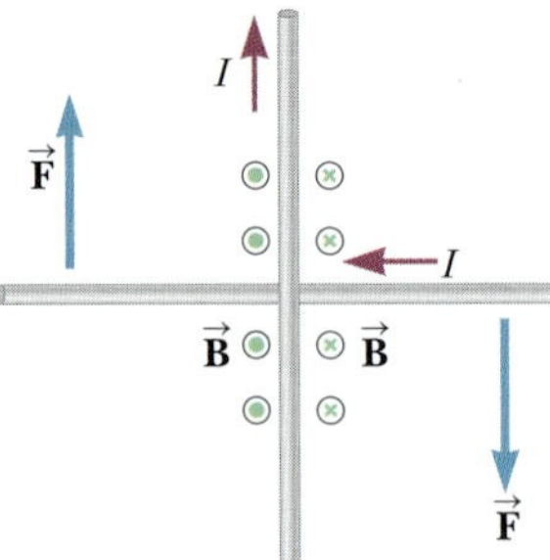


17. If they are projected in the same direction into the same magnetic field, the charges are of opposite sign.
19. (b)

## Problems

1. (a) west (b) zero deflection (c) up (d) down
3. (a) into the page (b) toward the right (c) toward the bottom of the page
5. (a) $1.44 \times 10^{-12}$ N (b) $8.62 \times 10^{14}$ m/s$^2$ (c) A force would be exerted on the electron that had the same magnitude as the force on a proton, but in the opposite direction because of its negative charge. (d) The magnitude of the acceleration of the electron would be much greater than that of the proton because the mass of the electron is much smaller. The electron's acceleration would also be in the opposite direction.
7. $2.83 \times 10^7$ m/s west
9. 48.9° or 131°
11. $F_g = 8.93 \times 10^{-30}$ N (downward), $F_e = 1.60 \times 10^{-17}$ N (upward), $F_m = 4.80 \times 10^{-17}$ N (downward)
13. $8.0 \times 10^{-3}$ T in the $+z$-direction
15. (a) into the page (b) toward the right (c) toward the bottom of the page
17. 7.50 N
19. 0.131 T (downward)
21. (a) The magnetic force and the force of gravity both act on the wire. When the magnetic force is upward and balances the downward force of gravity, the net force on the wire is zero, and the wire can move upward at constant velocity. (b) 0.20 T out of the page (c) If the field exceeds 0.20 T, the upward magnetic force exceeds the downward force of gravity, so the wire accelerates upward.
23. *ab*: 0, *bc*: 0.040 0 N in $-x$-direction, *cd*: 0.040 0 N in the $-z$-direction, *da*: 0.056 6 N parallel to the *xz*-plane and at 45° to both the $+x$- and the $+z$-directions
25. 118 $\mu$N · m
27. $9.05 \times 10^{-4}$ N · m, tending to make the left-hand side of the loop move toward you and the right-hand side move away.
29. (a) 0.56 A (b) 0.064 N · m
31. (a) 3.97° (b) $3.39 \times 10^{-3}$ N·m
33. (a) $8.51 \times 10^5$ m/s (b) $2.06 \times 10^{-5}$ m
37. 1.77 cm
39. $r = 3R/4$
41. (a) $2.08 \times 10^{-7}$ kg/C (b) $6.66 \times 10^{-26}$ kg (c) Calcium
43. 20.0 $\mu$T
45. $2.0 \times 10^{-10}$ A
47. 2.4 mm
49. 20.0 $\mu$T toward the bottom of page
51. 0.167 mT out of the page
53. (a) 4.00 m (b) 7.50 nT (c) 1.26 m (d) zero
55. (a) $3.50 \times 10^{-5}$ N/m (b) repulsive
57. 4.5 mm
59. 31.8 mA
61. (a) 920 turns (b) 12 cm
63. (a) 2.8 $\mu$T (b) 0.89 mA
65. (a) 5.24 $\mu$T (b) into the page (c) 7.20 cm
67. (a) 0.500 $\mu$T out of the page (b) 3.88 $\mu$T parallel to *xy*-plane and at 59.0° clockwise from $+x$-direction
69. (a) 1.33 m/s (b) The sign of the emf is independent of the charge.
71. 53 $\mu$T toward the bottom of the page, 20 $\mu$T toward the bottom of the page, and 0
73. (a) $-8.00 \times 10^{-21}$ kg·m/s (b) 8.90°
75. $1.41 \times 10^{-6}$ N

# CHAPTER 20

## Quick Quizzes

1. *b*, *c*, *a*
2. (c)
3. (b)
4. (a)
5. (b)
6. (b)

## Example Questions

1. False
2. $5.06 \times 10^{-3}$ V; the current would be in the opposite direction.
3. (a), (e)
4. A magnetic force directed to the right will be exerted on the bar.
5. Doubling the frequency doubles the maximum induced emf.
6. (b)
7. (a)
8. 72.6 A
9. (b)
10. False

## Warm-Up Exercises

1. (a) 0.233 Hz (b) 1.46 rad/s
3. $2.78 \times 10^{-2}$ s
5. (a) $6.40 \times 10^{-4}$ V (b) 0.533 A
7. (a) 135 V (b) 13.5 A
9. $1.60 \times 10^{-6}$ s

## Conceptual Questions

1. As the spacecraft moves through space, it is apparently moving from a region of one magnetic field strength to a region of a different magnetic field strength. The changing magnetic field through the coil induces an emf and a corresponding current in the coil.
3. According to Faraday's law, an emf is induced in a wire loop if the magnetic flux through the loop changes with time. In

this situation, an emf can be induced either by rotating the loop around an arbitrary axis or by changing the shape of the loop.

5. If the bar were moving to the left, the magnetic force on the negative charges in the bar would be upward, causing an accumulation of negative charges on the top and positive charges at the bottom. Hence, the electric field in the bar would be upward, as well.
7. If, for any reason, the magnetic field should change rapidly, a large emf could be induced in the bracelet. If the bracelet were not a continuous band, this emf would cause high-voltage arcs to occur at any gap in the band. If the bracelet were a continuous band, the induced emf would produce a large induced current and result in resistance heating of the bracelet.
11. As the aluminum plate moves into the field, eddy currents are induced in the metal by the changing magnetic field at the plate. The magnetic field of the electromagnet interacts with this current, producing a retarding force on the plate that slows it down. In a similar fashion, as the plate leaves the magnetic field, a current is induced, and once again there is an upward force to slow the plate.
13. Oscillating current in the solenoid produces an always-changing magnetic field. Vertical flux through the ring, alternately increasing and decreasing, produces current in it with a direction that is alternately clockwise and counter-clockwise. The current through the ring's resistance converts electrically transmitted energy into internal energy at the rate $I^2R$.

## Problems

1. $4.8 \times 10^{-3}$ T · m$^2$
3. (a) 0.177 T (b) 0
5. (a) $\Phi_{B,\text{net}} = 0$ (b) 0
7. (a) $3.1 \times 10^{-3}$ T · m$^2$ (b) $\Phi_{B,\text{net}} = 0$
9. (a) zero, since the flux through the loop A doesn't change (b) counterclockwise for loop B (c) clockwise for loop C
11. 94 mV
13. 33 mV
15. (a) from left to right (b) from right to left
17. (a) The current is zero, because the magnetic flux due to the current through the right side of the loop (into the page) is opposite the flux through the left side of the loop (out of the page), so net flux through the loop is always zero. (b) Clockwise, because the flux due to the long wire through the loop is out of the page, and the induced current in the loop must be clockwise to compensate for the increasing external flux.
19. (a) $3.77 \times 10^{-3}$ T (b) $9.42 \times 10^{-3}$ T (c) $7.07 \times 10^{-4}$ m$^2$ (d) $3.99 \times 10^{-6}$ Wb (e) $1.77 \times 10^{-5}$ V; the average induced emf is equal to the instantaneous in this case because the current increases steadily. (f) The induced emf is small, so the current in the 4-turn coil and its magnetic field will also be small.
21. 10.2 μV
23. 13.1 mV
25. 2.6 mV
27. (a) toward the east (b) $4.58 \times 10^{-4}$ V
29. 2.80 m/s
31. (a) 81.9 mV (b) 105 mA
33. $1.9 \times 10^{-11}$ V
35. (a) 60 V (b) 57 V (c) 0.13 s
37. (a) 18.1 μV (b) 0
39. 4.0 mH
41. (a) 2.0 mH (b) 38 A/s
43. (a) 5.33 A (b) 2.67 s (c) 8.00 s
45. (a) 2.4 V (b) 75 mH (c) 5.1 A (d) $\Delta V_R = -1.5$ V (e) $\Delta V_L = -0.9$ V
47. 1.92 Ω
49. (a) 0.208 mH (b) 0.936 mJ
51. (a) 18 J (b) 7.2 J
53. counterclockwise
55. (a) $F = N^2B^2w^2v/R$ to the left (b) 0 (c) $F = N^2B^2w^2v/R$ to the left
57. (a) 20.0 ms (b) 37.9 V (c) 1.52 mV (d) 51.8 mA
59. (a) 0.500 A (b) 2.00 W (c) 2.00 W
61. 115 kV
63. (a) 0.157 mV (end $B$ is positive) (b) 5.89 mV (end $A$ is positive)
65. (a) 9.00 A (b) 10.8 N (c) $b$ is at the higher potential (d) No
67. (b) A larger resistance would make the current smaller, so the loop must reach higher speed before the magnitude of the magnetic force will equal the gravitational force. (c) The magnetic force is proportional to the product of the field and the current, while the current itself is proportional to the field. If $B$ is cut in half, the speed must become four times larger to compensate and yield a magnetic force with magnitude equal to that of the gravitational force.

# CHAPTER 21

## Quick Quizzes

1. (c)
2. (b)
3. (b)
4. (a)
5. (a)
6. (b)
7. (b), (c)
8. (b), (d)

## Example Questions

1. False
2. False
3. True
4. True
5. The average power will be zero if the phase angle is 90° or −90° (i.e., when $R = 0$).
6. False
7. (d)
8. Yes. When the roof is perpendicular to the Sun's rays, the amount of energy intercepted by the roof is a maximum. At other angles the amount is smaller.
9. (c)

## Warm-Up Exercises

1. (a) 2.92 m (b) 59.0°
3. 229 Hz
5. (a) $5.64 \times 10^3$ Ω (b) $2.13 \times 10^{-2}$ A
7. (a) $3.12 \times 10^3$ Ω (b) $4.90 \times 10^3$ Ω (c) $2.33 \times 10^3$ Ω (d) $7.28 \times 10^{-2}$ A
9. $5.32 \times 10^{-9}$ F
11. $6.67 \times 10^{-6}$ N/m$^2$

## Conceptual Questions

1. Different stations have transmitting antennas at different locations. For best reception align your rabbit ears perpendicular to the straight-line path from your TV to the transmitting antenna. The transmitted signals are also polarized. The polarization direction of the wave can be changed by reflection from surfaces—including the atmosphere—and through Kerr rotation—a change in polarization axis when passing through an organic substance. In your home, the plane of polarization is determined by your surroundings, so antennas need to be adjusted to align with the polarization of the wave.

3. The fundamental source of an electromagnetic wave is an accelerating charge. For example, in a transmitting antenna of a radio station, charges are caused to move up and down at the frequency of the radio station. These moving charges set up electric and magnetic fields, the electromagnetic wave, in the space around the antenna.
5. (a) The flashing of the light according to Morse code is a drastic amplitude modulation (AM)—the amplitude is changing from a maximum to zero. In this sense, it is similar to the on-and-off binary code used in computers and compact disks. (b) The carrier frequency is that of the light, on the order of $10^{14}$ Hz. (c) The frequency of the signal depends on the skill of the signal operator, but it is on the order of a single hertz, as the light is flashed on and off. (d) The broadcasting antenna for this modulated signal is the filament of the lightbulb in the signal source. (e) The receiving antenna is the eye.
7. The sail should be as reflective as possible, so that the maximum momentum is transferred to the sail from the reflection of sunlight.
9. No. The wire will emit electromagnetic waves only if the current varies in time. The radiation is the result of accelerating charges, which can occur only when the current is not constant.
11. Photographs created using infrared light make warmer areas of the face brighter; visible light photographs capture mainly reflected ambient light, and are not affected by temperature. Consequently, bright areas in infrared light would not necessarily be bright in visible light. Pupils, for example, are very bright in the infrared but dark in visible light.
13. It is far more economical to transmit power at a high voltage than at a low voltage because the $I^2R$ loss on the transmission line is significantly lower at high voltage. Transmitting power at high voltage permits the use of step-down transformers to make "low" voltages and high currents available to the end user.
15. The resonance frequency is determined by the inductance and the capacitance in the circuit. If both $L$ and $C$ are doubled, the resonance frequency is reduced by a factor of two.
17. (a)
19. (c) and (d)

## Problems

1. (a) 193 Ω (b) 145 Ω
3. 14.6 Hz
5. 6.76 W
7. (a) $f > 41.3$ Hz (b) $X_C < 87.5$ Ω
9. $4.0 \times 10^2$ Hz
11. 100 mA
13. 3.14 A
15. (a) 0.042 4 H (b) 942 rad/s
17. 0.450 T · m$^2$
19. (a) 0.361 A (b) 18.1 V (c) 23.9 V (d) −53.0°
21. (a) 1.43 kΩ (b) 0.097 9 A (c) 51.1° (d) voltage leads current
23. (a) 17.4° (b) the voltage
25. 1.88 V
27. (a) 47.1 Ω (b) 637 Ω (c) 2.40 kΩ (d) 2.33 kΩ (e) −14.2°
29. (a) 103 V (b) 150 V (c) 127 V (d) 23.6 V
31. (a) 208 Ω (b) 40.0 Ω (c) 0.541 H
33. 8.00 W
35. (a) $1.8 \times 10^2$ Ω (b) 0.71 H
37. 1.82 pF
39. $C_{min} = 4.9$ nF, $C_{max} = 51$ nF
41. 3.60 kHz
43. 721 V
45. 0.18% is lost
47. (a) $1.1 \times 10^3$ kW (b) $3.1 \times 10^2$ A (c) $8.3 \times 10^3$ A
49. 1 000 km; there will always be better use for tax money.
51. (a) $6.30 \times 10^{-6}$ Pa (b) Atmospheric pressure is $1.60 \times 10^{10}$ times larger than the radiation pressure.
53. $f_{red} = 4.55 \times 10^{14}$ Hz, $f_{IR} = 3.19 \times 10^{14}$ Hz, $E_{max,f}/E_{max,i} = 0.57$
55. $B_{max} = 3.39 \times 10^{-6}$ T, $E_{max} = 1.02 \times 10^3$ V/m
57. $2.94 \times 10^8$ m/s
59. (a) 6.00 pm (b) 7.50 cm
61. (a) 188 m to 556 m (b) 2.78 m to 3.4 m
63. $5.2 \times 10^{13}$ Hz, 5.8 μm
65. $4.299\,999\,84 \times 10^{14}$ Hz; The frequency decreases by $1.6 \times 10^7$ Hz.
67. $5.31 \times 10^{-5}$ N/m$^2$
69. 1.7 cents
71. 99.6 mH
73. (a) $6.7 \times 10^{-16}$ T (b) $5.3 \times 10^{-17}$ W/m$^2$ (c) $1.7 \times 10^{-14}$ W
75. (a) 0.536 N (b) $8.93 \times 10^{-5}$ m/s$^2$ (c) 33.9 days

# CHAPTER 22

## Quick Quizzes

1. (a)
2. Beams 2 and 4 are reflected; beams 3 and 5 are refracted.
3. (b)
4. (c)

## Example Questions

1. (a)
2. (b)
3. False
4. (c)
5. (b)
6. (b)

## Warm-Up Exercises

1. 16
3. (a) 16.8° (b) 326 nm
5. 68.8°

## Conceptual Questions

1. The spectrum of the light sent back to you from a drop at the top of the rainbow arrives such that the red light (deviated by an angle of 42°) strikes the eye while the violet light (deviated by 40°) passes over your head. Thus, the top of the rainbow looks red. At the bottom of the rainbow, violet arrives at your eye and red light is deviated toward the ground. Thus, the bottom part of the rainbow appears violet.
3. $n_a > n_c > n_b$
5. On the one hand, a ball covered with mirrors sparkles by reflecting light from its surface. On the other hand, a faceted diamond lets in light at the top, reflects it by total internal reflection in the bottom half, and sends the light out through the top again. Because of its high index of refraction, the critical angle for diamond in air for total internal reflection, namely $\theta_c = \sin^{-1}(n_{air}/n_{diamond})$, is small. Thus, light rays enter through a large area and exit through a very small area with a much higher intensity. When a diamond is immersed in carbon disulfide, the critical angle is increased to $\theta_c = \sin^{-1}(n_{carbon\ disulfide}/n_{diamond})$. As a result, the light is emitted from the diamond over a larger area and appears less intense.
7. There is no dependence of the angle of reflection on wavelength, because the light does not enter deeply into the material during reflection—it reflects from the surface.
9. The color traveling slowest is bent the most. Thus, $X$ travels more slowly in the glass prism.
11. Light rays coming from parts of the pencil under water are bent away from the normal as they emerge into the air above.

The rays enter the eye (or camera) at angles closer to the horizontal, thus the parts of the pencil under water appear closer to the surface than they actually are, so the pencil appears bent.

**13.** Light travels through a vacuum at a speed of $3 \times 10^8$ m/s. Thus, an image we see from a distant star or galaxy must have been generated some time ago. For example, the star Altair is 16 light-years away; if we look at an image of Altair today, we know only what Altair looked like 16 years ago. This may not initially seem significant; however, astronomers who look at other galaxies can get an idea of what galaxies looked like when they were much younger. Thus, it does make sense to speak of "looking backward in time."

**15.** (c)

## Problems

**1.** $3.00 \times 10^8$ m/s
**3.** (a) $2.07 \times 10^3$ eV (b) 4.14 eV
**5.** (a) $2.25 \times 10^8$ m/s (b) $1.97 \times 10^8$ m/s (c) $1.24 \times 10^8$ m/s
**7.** (a) 29.0° (b) 25.7° (c) 32.0°
**9.** 19.5° above the horizontal
**11.** (a) 1.52 (b) 416 nm (c) $4.74 \times 10^{14}$ Hz (d) $1.97 \times 10^8$ m/s
**13.** 110.6°
**15.** (a) $1.56 \times 10^8$ m/s (b) 329.1 nm (c) $4.74 \times 10^{14}$ Hz
**17.** five times from the right-hand mirror and six times from the left
**19.** $\theta = 30.4°$, $\theta' = 22.3°$
**21.** (a) 4.84 cm (b) 67.5°
**23.** (b) 4.73 cm
**25.** $\theta = \tan^{-1}(n_g)$
**27.** 3.39 m
**29.** (a) $\theta_{\text{blue}} = 47.79°$ (b) $\theta_{\text{red}} = 48.22°$
**31.** $\theta_{\text{red}} - \theta_{\text{violet}} = 0.32°$
**33.** (a) $\theta_{1i} = 30°$, $\theta_{1r} = 19°$, $\theta_{2i} = 41°$, $\theta_{2r} = 77°$ (b) First surface: $\theta_{\text{reflection}} = 30°$; second surface: $\theta_{\text{reflection}} = 41°$
**35.** (a) 66.1° (b) 63.5° (c) 59.7°
**37.** (a) 40.8° (b) 60.6°
**39.** 1.414
**41.** (a) 10.7° (b) air (c) Sound falling on the wall from most directions is 100% reflected.
**43.** 27.5°
**45.** 22.0°
**47.** (a) 1.07 m (b) 1.65 m
**49.** (a) 38.5° (b) 1.44
**53.** 24.7°
**55.** 1.93
**57.** (a) 1.20 (b) 3.39 ns
**59.** (a) 53.1° (b) $\theta_1 \geq 38.7°$
**61.** 15.9°

# CHAPTER 23

## Quick Quizzes

**1.** At $C$
**2.** (c)
**3.** (a) False (b) False (c) True
**4.** (b)
**5.** An infinite number
**6.** (a) False (b) True (c) False

## Example Questions

**1.** No
**2.** (b)
**3.** True
**4.** (b)
**5.** (a)
**6.** (c)
**7.** The screen should be placed one focal length away from the lens.
**8.** No. The largest magnification would be 1, so a diverging lens would not make a good magnifying glass.
**9.** Upright
**10.** Yes, it's possible for a mirror to have a virtual object. Using the present system of mirror and lens, place the object to the right of the lens just outside the focal length. The image distance can be made as large as desired by moving the object toward the focal point, so at some point the lens's image will be located to the left of the mirror. That image behind the mirror forms a virtual object for the mirror.

## Warm-Up Exercises

**1.** (a) −1.50 m (b) 1.00 (c) virtual
**3.** (a) negative (b) −4.36 cm (c) 0.273 (d) virtual (e) upright
**5.** (a) −7.04 cm (b) 1.20
**7.** (a) negative (b) −6.15 cm (c) 0.615 (d) virtual (e) upright

## Conceptual Questions

**1.** You will not be able to focus your eyes on both the picture and your image at the same time. To focus on the picture, you must adjust your eyes so that an object several centimeters away (the picture) is in focus. Thus, you are focusing on the mirror surface. But your image in the mirror is as far behind the mirror as you are in front of it. Thus, you must focus your eyes beyond the mirror, twice as far away as the picture to bring the image into focus.

**3.** A single flat mirror forms a virtual image of an object due to two factors. First, the light rays from the object are necessarily diverging from the object, and second, the lack of curvature of the flat mirror cannot convert diverging rays to converging rays. If another optical element is first used to cause light rays to converge, then the flat mirror can be placed in the region in which the converging rays are present, and it will change the direction of the rays so that the real image is formed at a different location. For example, if a real image is formed by a convex lens, and the flat mirror is placed between the lens and the image position, the image formed by the mirror will be real.

**5.** An effect similar to a mirage is produced except the "mirage" is seen hovering in the air. Ghost lighthouses in the sky have been seen over bodies of water by this effect.

**7.** We consider the two trees to be two separate objects. The far tree is an object that is farther from the lens than the near tree. Thus, the image of the far tree will be closer to the lens than the image of the near tree. The screen must be moved closer to the lens to put the far tree in focus.

**9.** This is a possible scenario. When light crosses a boundary between air and ice, it will refract in the same manner as it does when crossing a boundary of the same shape between air and glass. Thus, a converging lens may be made from ice as well as glass. However, ice is such a strong absorber of infrared radiation that it is unlikely you will be able to start a fire with a small ice lens.

**11.** If a converging lens is placed in a liquid having an index of refraction larger than that of the lens material, the direction of refractions at the lens surfaces will be reversed, and the lens will diverge light. A mirror depends only on reflection that is independent of the surrounding material, so a converging mirror will be converging in any liquid.

**13.** The focal length for a mirror is determined by the law of reflection from the mirror surface. The law of reflection is independent of the material of which the mirror is made and of the surrounding medium. Thus, the focal length depends only on the radius of curvature and not on the material. The focal length of a lens depends on the indices of refraction of the lens material and surrounding medium. Thus, the focal length of a lens depends on the lens material.

15. Because when you look at the ƎƆИA⅃UᗺMA in your rear view mirror, the apparent left-right inversion clearly displays the name of the AMBULANCE behind you.
17. (b)

## Problems

1. (a) younger (b) on the order of $10^{-9}$ s
3. 10.0 ft, 30.0 ft, 40.0 ft
5. (a) $p_1 + h$, behind the lower mirror (b) virtual (c) upright (d) 1.00 (e) no
7. (a) The object should be placed 30.0 cm in front of the mirror. (b) 0.333
9. (a) The object is 10.0 cm in front of the mirror. (b) 2.00
11. 5.00 cm
13. 1.0 m
15. 8.05 cm
17. (a) 26.8 cm behind the mirror (b) upright (c) 0.026 8
19. (a) concave with focal length $f = 0.83$ m (b) Object must be 1.0 m in front of the mirror.
21. 38.2 cm below the upper surface of the ice
23. 3.8 mm
25. $n = 2.00$
27. (a) The image is 0.790 m from the outer surface of the glass pane, inside the water tank. (b) With a glass pane of negligible thickness, the image is 0.750 m inside the tank. (c) The thicker the glass, the greater the distance between the final image and the outer surface of the glass.
29. 20.0 cm
31. (a) 20.0 cm beyond the lens; real, inverted, $M = -1.00$ (b) No image is formed. Parallel rays leave the lens. (c) 10.0 cm in front of the lens; virtual, upright, $M = +2.00$
33. **(i)** (a) 13.3 cm in front of the lens (b) virtual (c) upright (d) +0.333 **(ii)** (a) 10.0 cm in front of the lens (b) virtual (c) upright (d) +0.500 **(iii)** (a) 6.67 cm in front of the lens (b) virtual (c) upright (d) +0.667
35. (a) either 9.63 cm or 3.27 cm (b) 2.10 cm
37. (a) 39.0 mm (b) 39.5 mm
39. 40.0 cm
41. 30.0 cm to the left of the second lens, $M = -3.00$
43. 7.47 cm in front of the second lens, 1.07 cm, virtual, upright
45. from 0.224 m to 18.2 m
47. real image, 5.71 cm in front of the mirror
49. (a) −34.7 cm (b) −36.1 cm
51. 160 cm to the left of the lens, inverted, $M = -0.800$
53. $q = 10.7$ cm
55. (a) 32.0 cm to the right of the second surface (b) real
57. (a) 20.0 cm to the right of the second lens, $M = -6.00$ (b) inverted (c) 6.67 cm to the right of the second lens, $M = -2.00$, inverted
59. (a) 1.99 (b) 10.0 cm to the left of the lens (c) inverted
61. (a) 5.45 m to the left of the lens (b) 8.24 m to the left of the lens (c) 17.1 m to the left of the lens (d) by surrounding the lens with a medium having a refractive index greater than that of the lens material
63. (a) 263 cm (b) 79.0 cm
65. 8.57 cm

# CHAPTER 24

## Quick Quizzes

1. (c)
2. (c)
3. (c)
4. (b)
5. (b)
6. The compact disc

## Example Questions

1. False
2. The soap film is thicker in the region that reflects red light.
3. The coating should be thinner.
4. The wavelength is smaller in water than in air, so the distance between dark bands is also smaller.
5. False
6. Because the wavelength becomes smaller in water, the angles to the first maxima become smaller, resulting in a smaller central maximum.
7. The separation between principal maxima will be larger.
8. The additional polarizer must make an angle of 90° with respect to the previous polarizer.

## Warm-Up Exercises

1. (a) 0.524 m (b) 0.369 m
3. (a) 1.04° (b) $2.71 \times 10^{-2}$ m (c) 2.59° (d) $6.79 \times 10^{-2}$ m
5. (a) two (b) $4.57 \times 10^{-7}$ m (c) $1.14 \times 10^{-7}$ m
7. (a) $2.22 \times 10^{-6}$ m (b) 47.1°
9. 52.5 W/m$^2$

## Conceptual Questions

1. You will *not* see an interference pattern from the automobile headlights, for two reasons. The first is that the headlights are not coherent sources and are therefore incapable of producing sustained interference. Also, the headlights are so far apart in comparison to the wavelengths emitted that, even if they were made into coherent sources, the interference maxima and minima would be too closely spaced to be observable.
3. The result of the double slit is to redistribute the energy arriving at the screen. Although there is no energy at the location of a dark fringe, there is four times as much energy at the location of a bright fringe as there would be with only a single narrow slit. The total amount of energy arriving at the screen is twice as much as with a single slit, as it must be according to the law of conservation of energy.
5. One of the materials has a higher index of refraction than water, and the other has a lower index. The material with the higher index will appear black as it approaches zero thickness. There will be a 180° phase change for the light reflected from the upper surface, but no such phase change for the light reflected from the lower surface, because the index of refraction for water on the other side is lower than that of the film. Thus, the two reflections will be out of phase and will interfere destructively. The material with index of refraction lower than water will have a phase change for the light reflected from both the upper and the lower surface, so that the reflections from the zero-thickness film will be back in phase and the film will appear bright.
7. Because the light reflecting at the lower surface of the film undergoes a 180° phase change, while light reflecting from the upper surface of the film does not undergo such a change, the central spot (where the film has near zero thickness) will be dark. If the observed rings are not circular, the curved surface of the lens does not have a true spherical shape.
9. For regional communication at the Earth's surface, radio waves are typically broadcast from currents oscillating in tall vertical towers. These waves have vertical planes of polarization. Light originates from the vibrations of atoms or electronic transitions within atoms, which represent oscillations in all possible directions. Thus, light generally is not polarized.
11. Yes. In order to do this, first measure the radar reflectivity of the metal of your airplane. Then choose a light, durable material that has approximately half the radar reflectivity of the metal in your plane. Measure its index of refraction,

and place onto the metal a coating equal in thickness to one-quarter of 3 cm, divided by that index. Sell the plane quickly, and then you can sell the supposed enemy new radars operating at 1.5 cm, which the coated metal will reflect with extra-high efficiency.

**13.** If you wish to perform an interference experiment, you need monochromatic coherent light. To obtain it, you must first pass light from an ordinary source through a prism or diffraction grating to disperse different colors into different directions. Using a single narrow slit, select a single color and make that light diffract to cover both slits for a Young's experiment. The procedure is much simpler with a laser because its output is already monochromatic and coherent.

**15.** (b)

## Problems

**1.** 632 nm
**3.** (a) 2.6 mm (b) 2.62 mm
**5.** (a) 36.2° (b) 5.08 cm (c) $5.08 \times 10^{14}$ Hz
**7.** (a) 55.7 m (b) 124 m
**9.** 662 nm
**11.** 11.3 m
**13.** 148 m
**15.** 75.0 m
**17.** 85.4 nm
**19.** 511 nm
**21.** 0.500 cm
**23.** (a) 238 nm (b) $\lambda$ will increase (c) 328 nm
**25.** 78.4 $\mu$m
**27.** 4.75 $\mu$m
**29.** 617 nm, 441 nm, 343 nm
**31.** (a) 8.10 mm (b) 4.05 mm
**33.** (a) 1.1 m (b) 1.7 mm
**35.** 462 nm
**37.** 1.20 mm, 1.20 mm
**39.** (a) 479 nm, 647 nm, 698 nm (b) 20.5°, 28.3°, 30.7°
**41.** 5.91° in first order; 13.2° in second order; and 26.5° in third order
**43.** 1.81 $\mu$m
**45.** 514 nm
**47.** 9.17 cm
**49.** (a) $3.53 \times 10^3$ grooves/cm (b) 11 maxima
**51.** (a) 1.11 (b) 42.0°
**53.** (a) 56.7° (b) 48.8°
**55.** (a) 55.6° (b) 34.4°
**59.** 6.89 units
**61.** (a) 413.7 nm, 409.7 nm (b) 8.6°
**63.** One slit 0.12 mm wide. The central maximum is twice as wide as the other maxima.
**65.** 0.156 mm
**67.** 2.50 mm
**69.** (a) 16.6 m (b) 8.28 m
**71.** 127 m
**73.** 0.350 mm

# CHAPTER 25

## Quick Quizzes

**1.** (c)
**2.** (a)

## Example Questions

**1.** True
**2.** True
**3.** A smaller focal length gives a greater magnification and should be selected.
**4.** True
**5.** Yes. Increasing the focal length of the mirror increases the magnification. Increasing the focal length of the eyepiece decreases the magnification.
**6.** More widely spaced eyes increase visual resolving power by effectively increasing the aperture size, $D$, in Equation 25.10. The limiting angle of resolution is thereby decreased, meaning finer details of distant objects can be resolved.
**7.** Resolution is better at the violet end of the visible spectrum.
**8.** True

## Warm-Up Exercises

**1.** (a) 60.0 cm (b) −5.00 (c) inverted (d) real
**3.** 0.206 m
**5.** (a) nearsighted (b) −42.0 cm (c) −2.38 diopters
**7.** (a) 3.00 cm (b) 8.33
**9.** 11.4
**11.** 8.65 m

## Conceptual Questions

**1.** The observer is *not* using the lens as a simple magnifier. For a lens to be used as a simple magnifier, the object distance must be less than the focal length of the lens. Also, a simple magnifier produces a virtual image at the normal near point of the eye, or at an image distance of about $q = -25$ cm. With a large object distance and a relatively short image distance, the magnitude of the magnification by the lens would be considerably less than one. Most likely, the lens in this example is part of a lens combination being used as a telescope.

**3.** The image formed on the retina by the lens and cornea is already inverted.

**5.** For a lens to operate as a simple magnifier, the object should be located just inside the focal point of the lens. If the power of the lens is +20.0 diopters, its focal length is

$f = (1.00 \text{ m})/P = (1.00 \text{ m})/{+20.0} = 0.050\,0 \text{ m} = 5.00 \text{ cm}$

The object should be placed slightly less than 5.00 cm in front of the lens.

**7.** You want a real image formed at the location of the paper. To form such an image, the object distance must be greater than the focal length of the lens.

**9.** In order for someone to see an object through a microscope, the wavelength of the light in the microscope must be smaller than the size of the object. An atom is much smaller than the wavelength of light in the visible spectrum, so an atom can never be seen with the use of visible light.

**11.** farsighted; converging

**13.** (a) The diffraction pattern of a hair is the same as the diffraction pattern produced by a single slit of the same width. (b) The central maximum is flanked by minima. Measure the width $2y$ of the central maximum between the minima bracketing it. Because the angle is small, you can use

$$\sin \theta_{\text{dark}} \approx \tan \theta_{\text{dark}}$$

$$m\lambda/a \approx y/L$$

to find the width $a$ of the hair.

**15.** (c)

## Problems

**1.** 7.0
**3.** 177 m
**5.** $f/1.4$
**7.** 8.0
**9.** (a) 42.9 cm (b)+2.33 diopters
**11.** 23.2 cm
**13.** (a) −2.00 diopters (b) 17.6 cm
**15.** +17.0 diopters
**17.** −2.50 diopters, a diverging lens
**19.** (a) 5.8 cm (b) $m = 4.3$

21. (a) 4.07 cm (b) $m = +7.14$
23. (a) $|M| = 1.22$ (b) $\theta/\theta_0 = 6.08$
25. 2.1 cm
27. 12.7
29. 3.38 min
31. (a) 2.53 cm (b) 88.5 cm
33. (b) $-fh/p$ (c) $-1.07$ mm
35. (a) 1.50 (b) 1.9
37. 492 km
39. 0.40 $\mu$rad
41. 5.40 mm
43. 9.8 km
45. No. A resolving power of $2.0 \times 10^5$ is needed, and that available is only $1.8 \times 10^5$.
47. $1.31 \times 10^3$
49. 50.4 $\mu$m
51. 40
53. (a) $8.00 \times 10^2$ (b) The image is inverted.
55. 15.4
57. (a) +2.67 diopters (b) 0.16 diopter too low
59. (a) +44.6 diopters (b) 3.03 diopters
61. (a) $m = 4.0$ (b) $m = 3.0$

## CHAPTER 26

### Quick Quizzes

1. False: the speed of light is $c$ for all observers.
2. (a)
3. False
4. No. From your perspective you're at rest with respect to the cabin, so you will measure yourself as having your normal length, and will require a normal-sized cabin.
5. **(i)** (a), (e) **(ii)** (a), (e)
6. False
7. (a) False (b) False (c) True (d) False

### Example Questions

1. 9.61 s
2. (c)
3. (a)
4. $-0.946c$
5. (a)
6. Very little of the mass is converted to other forms of energy in these reactions because the total number of neutrons and protons doesn't change. The energy liberated is only the energy associated with their interactions.

### Warm-Up Exercises

1. $-122$ m/s
3. 5.74 m/s
5. (a) 1.36 (b) $2.82 \times 10^5$ km
7. (a) $0.750c$ (b) $0.667c$
9. (a) 0.334 MeV (b) $6.02 \times 10^{-13}$ J

### Conceptual Questions

1. An ellipsoid. The dimension in the direction of motion would be measured to be less than $D$.
3. No. The principle of relativity implies that nothing can travel faster than the speed of light in a *vacuum*, which is equal to $3.00 \times 10^8$ m/s.
5. As the object approaches the speed of light, its kinetic energy grows without limit. It would take an infinite investment of work to accelerate the object to the speed of light.
7. For a wonderful fictional exploration of this question, get a "Mr. Tompkins" book by George Gamow. All of the relativity effects would be obvious in our lives. Time dilation and length contraction would both occur. Driving home in a hurry, you would push on the gas pedal not to increase your speed very much, but to make the blocks shorter. Big Doppler shifts in wave frequencies would make red lights look green as you approached and make car horns and radios useless. High-speed transportation would be very expensive, requiring huge fuel purchases, as well as dangerous, since a speeding car could knock down a building. When you got home, hungry for lunch, you would find that you had missed dinner; there would be a five-day delay in transit when you watch a live TV program originating in Australia. Finally, we would not be able to see the Milky Way, since the fireball of the Big Bang would surround us at the distance of Rigel or Deneb.
9. A photon transports energy. The relativistic equivalence of mass and energy is enough to give it momentum.
11. Suppose a railroad train is moving past you. One way to measure its length is this: you mark the tracks at the cowcatcher forming the front of the moving engine at 9:00:00 AM, while your assistant marks the tracks at the back of the caboose at the same time. Then you find the distance between the marks on the tracks with a tape measure. You and your assistant must make the marks simultaneously in your frame of reference, for otherwise the motion of the train would make its length different from the distance between marks.

### Problems

1. (a) 1.38 yr (b) 1.31 light-years
3. $0.103c$
5. (a) $1.3 \times 10^{-7}$ s (b) 38 m (c) 7.6 m
7. (a) $1.55 \times 10^{-5}$ s (b) 7.09 (c) $2.19 \times 10^{-6}$ s (d) 649 m (e) From the third observer's point of view, the muon is traveling faster, so according to the third observer, the muon's lifetime is longer than that measured by the observer at rest with respect to Earth.
9. $0.950c$
11. Yes, with 19 m to spare
13. (a) $\dfrac{2\sqrt{2}}{3}c = 0.943c = 2.83 \times 10^8$ m/s (b) The result would be the same.
15. $0.285c$
17. $0.54c$ to the right
19. $0.357c$
21. $0.998c$ toward the right
23. (a) $0.999\,997c$ (b) $3.74 \times 10^5$ MeV
25. $0.786c$
27. (a) $8.65 \times 10^{22}$ J (b) $9.61 \times 10^5$ kg
31. (a) $3.91 \times 10^4$ (b) 7.67 cm
33. (a) $0.979c$ (b) $0.065\,2c$ (c) $0.914c$
35. $0.917c$
37. (a) $v/c = 1 - 1.13 \times 10^{-10}$ (b) $5.99 \times 10^{27}$ J (c) $\$2.16 \times 10^{20}$
39. (a) $0.800c$ (b) $7.50 \times 10^3$ s (c) $1.44 \times 10^{12}$ m (d) $0.385c$
41. $0.80c$
43. (a) $v = 0.141c$ (b) $v = 0.436c$
45. (a) 7.0 $\mu$s (b) $1.1 \times 10^4$ muons
47. 5.5 yr; Goslo is older
49. (a) 17.4 m (b) 3.30° with respect to the direction of motion

## CHAPTER 27

### Quick Quizzes

1. True
2. (b)
3. (c)
4. False
5. (c)

### Example Questions

1. False
2. False
3. Some of the photon's energy is transferred to the electron.

4. Doubling the speed of a particle doubles its momentum, reducing the particle's wavelength by a factor of one-half. This answer is no longer true when the doubled speed is relativistic.
5. True

## Warm-Up Exercises

1. (a) $7.69 \times 10^{14}$ Hz (b) $4.29 \times 10^{14}$ Hz
3. $1.96 \times 10^{6}$ m/s
5. 2.46 eV
7. 0.124 nm
9. $4.85 \times 10^{-3}$ nm
11. $2.32 \times 10^{4}$ m/s

## Conceptual Questions

1. The shape of an object is normally determined by observing the light reflecting from its surface. In a kiln, the object will be very hot and will be glowing red. The emitted radiation is far stronger than the reflected radiation, and the thermal radiation emitted is only slightly dependent on the material from which the object is made. Unlike reflected light, the emitted light comes from all surfaces with equal intensity, so contrast is lost and the shape of the object is harder to discern.
3. The "blackness" of a blackbody refers to its ideal property of absorbing all radiation incident on it. If an observed room temperature object in everyday life absorbs all radiation, we describe it as (visibly) black. The black appearance, however, is due to the fact that our eyes are sensitive only to visible light. If we could detect infrared light with our eyes, we would see the object emitting radiation. If the temperature of the blackbody is raised, Wien's law tells us that the emitted radiation will move into the visible range of the spectrum. Thus, the blackbody could appear as red, white, or blue, depending on its temperature.
5. All objects do radiate energy, but at room temperature this energy is primarily in the infrared region of the electromagnetic spectrum, which our eyes cannot detect. (Pit vipers have sensory organs that *are* sensitive to infrared radiation; thus, they can seek out their warm-blooded prey in what we would consider absolute darkness.)
7. We can picture higher frequency light as a stream of photons of higher energy. In a collision, one photon can give all of its energy to a single electron. The kinetic energy of such an electron is measured by the stopping potential. The reverse voltage (stopping voltage) required to stop the current is proportional to the frequency of the incoming light. More intense light consists of more photons striking a unit area each second, but atoms are so small that one emitted electron never gets a "kick" from more than one photon. Increasing the intensity of the light will generally increase the size of the current, but it will not change the energy of the individual electrons that are ejected. Thus, the stopping potential remains constant.
9. Wave theory predicts that the photoelectric effect should occur at any frequency, provided that the light intensity is high enough. However, as seen in photoelectric experiments, the light must have sufficiently high frequency for the effect to occur.
11. (a) Electrons are emitted only if the photon frequency is greater than the cutoff frequency.
13. No. Suppose that the incident light frequency at which you first observed the photoelectric effect is above the cutoff frequency of the first metal, but less than the cutoff frequency of the second metal. In that case, the photoelectric effect would not be observed at all in the second metal.
15. An electron has both classical-wave and classical-particle characteristics. In single- and double-slit diffraction and interference experiments, electrons behave like classical waves. An electron has mass and charge. It carries kinetic energy and momentum in parcels of definite size, as classical particles do. At the same time it has a particular wavelength and frequency. Since an electron displays characteristics of both classical waves and classical particles, it is neither a classical wave nor a classical particle. It is customary to call it a *quantum particle*, but another invented term, such as "wavicle," could serve equally well.

## Problems

1. (a) $\approx 3\,000$ K (b) $\approx 20\,000$ K
3. 9.47 $\mu$m, which is in the infrared region of the spectrum
5. (a) $2.49 \times 10^{-5}$ eV (b) 2.49 eV (c) 249 eV
7. (a) 2.57 eV (b) $1.28 \times 10^{-5}$ eV (c) $1.91 \times 10^{-7}$ eV
9. (a) 2.24 eV (b) 555 nm (c) $5.41 \times 10^{14}$ Hz
11. (a) $1.02 \times 10^{-18}$ J (b) $1.53 \times 10^{15}$ Hz (c) 196 nm (d) 2.15 eV (e) 2.15 V
13. $4.8 \times 10^{14}$ Hz, 2.0 eV
15. $1.2 \times 10^{2}$ V and $1.2 \times 10^{7}$ V, respectively
17. 17.8 kV
19. 0.078 nm
21. 0.281 nm
23. 70.0°
25. (a) $1.18 \times 10^{-23}$ kg·m/s (b) 478 eV
27. (a) 1.46 km/s (b) $7.28 \times 10^{-11}$ m
29. $3.58 \times 10^{-13}$ m
31. (a) 15 keV (b) $1.2 \times 10^{2}$ keV
33. $\sim 10^{6}$ m/s
35. Within 1.16 mm for the electron, $5.28 \times 10^{-32}$ m for the bullet
37. $3 \times 10^{-29}$ J
39. (a) $2.21 \times 10^{-32}$ kg·m/s (b) $1.00 \times 10^{10}$ Hz (c) $4.14 \times 10^{-5}$ eV
41. $2.5 \times 10^{20}$ photons
43. (a) 5 200 K (b) Clearly, a firefly is not at this temperature, so this is not blackbody radiation.
45. 1.36 eV
47. 2.00 eV
49. (a) $0.022\,0c$ (b) $0.999\,2c$

# CHAPTER 28

## Quick Quizzes

1. (b)
2. (a) 5 (b) 9 (c) 25
3. (d)

## Example Questions

1. The energy associated with the quantum number $n$ increases with increasing quantum number $n$, going to zero in the limit of arbitrarily large $n$. A transition from a very high energy level to the ground state therefore results in the emission of photons approaching an energy of 13.6 eV, the same as the ionization energy.
2. The energy difference in singly-ionized helium will be four times that of the same transition in hydrogen. Energy levels in hydrogen-like atoms are proportional to $Z^2$, where $Z$ is the atomic number.
3. The quantum numbers $n$ and $\ell$ are never negative.
4. No. The M shell is at a higher energy; hence, transitions from the M shell to the K shell will always result in more energetic photons than any transition from the L shell to the K shell.

## Warm-Up Exercises

1. (a) $6.48 \times 10^{-19}$ J (b) 307 nm
3. (a) 0.967 eV (b) $1.55 \times 10^{-19}$ J (c) $2.33 \times 10^{14}$ Hz
5. (a) 0.212 nm (b) 0.846 nm
7. (a) 2 (b) 8 (c) 18
9. (a) 6 (b) 3

## Conceptual Questions

1. If the energy of the hydrogen atom were proportional to $n$ (or any power of $n$), then the energy would become infinite as $n$ grew to infinity. But the energy of the atom is inversely proportional to $n^2$. Thus, as $n$ grows to infinity, the energy of the atom approaches a value that is above the ground state by a finite amount, namely, the ionization energy 13.6 eV. As the electron falls from one bound state to another, its energy loss is always less than the ionization energy. The energy and frequency of any emitted photon are finite.
3. The characteristic x-rays originate from transitions within the atoms of the target, such as an electron from the L shell making a transition to a vacancy in the K shell. The vacancy is caused when an accelerated electron in the x-ray tube supplies energy to the K shell electron to eject it from the atom. If the energy of the bombarding electrons were to be increased, the K shell electron will be ejected from the atom with more remaining kinetic energy. But the energy difference between the K and L shell has not changed, so the emitted x-ray has exactly the same wavelength.
5. A continuous spectrum without characteristic x-rays is possible. At a low accelerating potential difference for the electron, the electron may not have enough energy to eject an electron from a target atom. As a result, there will be no characteristic x-rays. The change in speed of the electron as it enters the target will result in the continuous spectrum.
7. The hologram is an interference pattern between light scattered from the object and the reference beam. If anything moves by a distance comparable to the wavelength of the light (or more), the pattern will wash out. The effect is just like making the slits vibrate in Young's experiment, to make the interference fringes vibrate wildly so that a photograph of the screen displays only the average intensity everywhere.
9. If the Pauli exclusion principle were not valid, the elements and their chemical behavior would be grossly different because every electron would end up in the lowest energy level of the atom. All matter would therefore be nearly alike in its chemistry and composition, since the shell structures of each element would be identical. Most materials would have a much higher density, and the spectra of atoms and molecules would be very simple, resulting in the existence of less color in the world.
11. The three elements have similar electronic configurations, with filled inner shells plus a single electron in an $s$ orbital. Because atoms typically interact through their unfilled outer shells, and since the outer shells of these atoms are similar, the chemical interactions of the three atoms are also similar.
13. Each of the eight electrons must have at least one quantum number different from each of the others. They can differ (in $m_s$) by being spin-up or spin-down. They can differ (in $\ell$) in angular momentum and in the general shape of the wave function. Those electrons with $\ell = 1$ can differ (in $m_\ell$) in orientation of angular momentum.
15. Stimulated emission is the reason laser light is coherent and tends to travel in a well-defined parallel beam. When a photon passing by an excited atom stimulates that atom to emit a photon, the emitted photon is in phase with the original photon and travels in the same direction. As this process is repeated many times, an intense, parallel beam of coherent light is produced. Without stimulated emission, the excited atoms would return to the ground state by emitting photons at random times and in random directions. The resulting light would not have the useful properties of laser light.

## Problems

1. (a) 121.5 nm, 102.5 nm, 97.20 nm (b) Far ultraviolet
3. (a) $2.3 \times 10^{-8}$ N (b) −14 eV
5. (a) $1.6 \times 10^6$ m/s (b) No. $v/c = 5.3 \times 10^{-3} << 1$ (c) 0.45 nm (d) Yes. The wavelength is roughly the same size as the atom.
7. (a) 0.212 nm (b) $9.95 \times 10^{-25}$ kg·m/s (c) $2.11 \times 10^{-34}$ J·s (d) 3.40 eV (e) −6.80 eV (f) −3.40 eV
11. $E = -1.51$ eV ($n = 3$) to $E = -3.40$ eV ($n = 2$)
13. (a) 2.86 eV (b) 0.472 eV
15. (a) 1.89 eV (b) 658 nm (c) 3.02 eV (d) 412 nm (e) 366 nm
17. (a) 488 nm (b) 0.814 m/s
19. (a) 3 (b) 520 km/s
21. (a) 0.476 nm (b) 0.997 nm
23. (a) 0.026 5 nm (b) 0.017 6 nm (c) 0.013 2 nm
25. (a) 1.31 $\mu$m (b) 164 nm
27.

| $n$ | $\ell$ | $m_\ell$ | $m_s$ |
|---|---|---|---|
| 3 | 2 | 2 | $\frac{1}{2}$ |
| 3 | 2 | 2 | $-\frac{1}{2}$ |
| 3 | 2 | 1 | $\frac{1}{2}$ |
| 3 | 2 | 1 | $-\frac{1}{2}$ |
| 3 | 2 | 0 | $\frac{1}{2}$ |
| 3 | 2 | 0 | $-\frac{1}{2}$ |
| 3 | 2 | −1 | $\frac{1}{2}$ |
| 3 | 2 | −1 | $-\frac{1}{2}$ |
| 3 | 2 | −2 | $\frac{1}{2}$ |
| 3 | 2 | −2 | $-\frac{1}{2}$ |

29. Fifteen possible states, as summarized in the following table:

| $n$ | 3 | 3 | 3 | 3 | 3 | 3 | 3 | 3 | 3 | 3 | 3 | 3 | 3 | 3 | 3 |
|---|---|---|---|---|---|---|---|---|---|---|---|---|---|---|---|
| $\ell$ | 2 | 2 | 2 | 2 | 2 | 2 | 2 | 2 | 2 | 2 | 2 | 2 | 2 | 2 | 2 |
| $m_\ell$ | 2 | 2 | 2 | 1 | 1 | 1 | 0 | 0 | 0 | −1 | −1 | −1 | −2 | −2 | −2 |
| $m_s$ | 1 | 0 | −1 | 1 | 0 | −1 | 1 | 0 | −1 | 1 | 0 | −1 | 1 | 0 | −1 |

31. aluminum
33. (a) $n = 4$ and $\ell = 2$ (b) $m_\ell = (0, \pm1, \pm2)$, $m_s = \pm1/2$ (c) $1s^22s^22p^63s^23p^63d^{10}4s^24p^64d^25s^2$ = [Kr] $4d^25s^2$
35. (a) 14 keV (b) $8.9 \times 10^{-11}$ m
37. L shell: 11.7 keV; M shell: 10.0 keV; N shell: 2.30 keV
39. (a) 10.2 eV (b) $7.88 \times 10^4$ K
41. (a) 5.40 keV (b) 3.40 eV (c) 5.40 keV
43. (a) $4.24 \times 10^{15}$ W/m$^2$ (b) $1.20 \times 10^{-12}$ J

# CHAPTER 29

## Quick Quizzes

1. False
2. (c)
3. (c)
4. (a) and (b)
5. (b)

## Example Questions

1. Tritium has the greater binding energy. Unlike tritium, helium has two protons that exert a repulsive electrostatic force on each other. The helium-3 nucleus is therefore not as tightly bound as the tritium nucleus.
2. Doubling the initial mass of radioactive material doubles the initial activity. Doubling the mass has no effect on the half-life.
3. $7.794 \times 10^{-13}$ J, $4.69 \times 10^{11}$ J
4. The binding energy of carbon-14 should be greater than nitrogen-14 because nitrogen has more protons in its nucleus. The mutual repulsion of the protons means that the nitrogen nucleus would require less energy to break up than the carbon nucleus.
5. No
6. Reactions between helium and beryllium can be found in the Sun.

## Warm-Up Exercises

1. (a) 0.657 J (b) $4.11 \times 10^{12}$ MeV
3. $2.7 \times 10^{-15}$ m
5. $1.49 \times 10^4$ MeV/$c^2$
7. $3.12 \times 10^{-17}$ s$^{-1}$
9. (a) 230 (b) 88

## Conceptual Questions

1. An alpha particle contains two protons and two neutrons. Because a hydrogen nucleus contains only one proton, it cannot emit an alpha particle.
3. Isotopes of a given element correspond to nuclei with different numbers of neutrons. This will result in a variety of different physical properties for the nuclei, including the obvious one of mass. The chemical behavior, however, is governed by the element's electrons. All isotopes of a given element have the same number of electrons and, therefore, the same chemical behavior.
5. Nucleus Y will be more unstable. The nucleus with the higher binding energy requires more energy to be disassembled into its constituent parts and has less available energy to release in a decay.
7. In alpha decay, there are only two final particles: the alpha particle and the daughter nucleus. There are also two conservation principles: of energy and of momentum. As a result, the alpha particle must be ejected with a discrete energy to satisfy both conservation principles. However, beta decay is a three-particle decay: the beta particle, the neutrino (or antineutron), and the daughter nucleus. As a result, the energy and momentum can be shared in a variety of ways among the three particles while still satisfying the two conservation principles. This allows a continuous range of energies for the beta particle.
11. The larger rest energy of the neutron means that a free proton in space will not spontaneously decay into a neutron and a positron. When the proton is in the nucleus, however, the important question is that of the total rest energy of the nucleus. If it is energetically favorable for the nucleus to have one less proton and one more neutron, then the decay process will occur to achieve this lower energy.
13. Carbon dating cannot generally be used to estimate the age of a rock, because the rock was not alive to receive carbon, and hence radioactive carbon-14, from the environment. Only the ages of objects that were once alive can be estimated with carbon dating.

## Problems

1. (a) 1.5 fm (b) 4.7 fm (c) 7.0 fm (d) 7.4 fm
3. $1.8 \times 10^2$ m
5. (a) 27.6 N (b) $4.16 \times 10^{27}$ m/s$^2$ (c) 1.73 MeV
7. (a) $1.9 \times 10^7$ m/s (b) 7.1 MeV
9. (a) 8.26 MeV/nucleon (b) 8.70 MeV/nucleon
11. 3.54 MeV
13. (a) 0.210 MeV/nucleon greater for $^{23}_{11}$Na (b) attributable to less proton repulsion
15. 2.29 g
17. (a) $6.95 \times 10^5$ s (b) $9.97 \times 10^{-7}$ s$^{-1}$ (c) $1.9 \times 10^4$ decays/s (d) $1.9 \times 10^{10}$ nuclei (e) 5, 0.200 mCi
19. 8.06 days
21. (a) $5.58 \times 10^{-2}$ h$^{-1}$, 12.4 h (b) $2.39 \times 10^{13}$ nuclei (c) 1.9 mCi
23. (a) $^{208}_{81}$Tl (b) $^{95}_{37}$Rb (c) $^{144}_{60}$Nd
25. e$^+$ decay, $^{56}_{27}\text{Co} \rightarrow {}^{56}_{26}\text{Fe} + e^+ + \nu$
27. (a) cannot occur spontaneously (b) can occur spontaneously
29. 18.6 keV
31. $4.22 \times 10^3$ yr
33. (a) $^{21}_{10}$Ne (b) $^{144}_{54}$Xe (c) X = e$^+$, X′ = $\nu$
35. (a) $^{13}_{6}$C (b) $^{10}_{5}$B
37. (a) $^{197}_{79}\text{Au} + n \rightarrow {}^{198}_{80}\text{Hg} + e^- + \bar{\nu}$ (b) 7.89 MeV
39. (a) The product nucleus is $^{10}_{5}$B. (b) −2.79 MeV
41. 18.8 J
43. 24 d
45. (a) $8.97 \times 10^{11}$ electrons (b) 0.100 J (c) 100 rad
47. (a) $3.18 \times 10^{-7}$ mol (b) $1.91 \times 10^{17}$ nuclei (c) $1.08 \times 10^{14}$ Bq (d) $8.92 \times 10^6$ Bq
49. 46.5 d
51. (a) $4.0 \times 10^9$ yr (b) It could be no older. (c) The rock could be younger if some $^{87}$Sr were initially present.
53. 54 $\mu$Ci
55. 5.9 billion years

# CHAPTER 30

## Quick Quizzes

1. (a)
2. (b)

## Example Questions

1. 1 mg/yr
2. $1.77 \times 10^{12}$ J
3. True
4. No. That reaction violates conservation of baryon number.

## Warm-Up Exercises

1. (a) $7.11 \times 10^{13}$ J (b) $7.90 \times 10^{-4}$ kg
3. (a) $3.33 \times 10^{-11}$ J (b) $3.70 \times 10^{-28}$ kg
5. $6.67 \times 10^{15}$ cm$^{-3}$
7. 0
9. 1

## Conceptual Questions

1. The experiment described is a nice analogy to the Rutherford scattering experiment. In the Rutherford experiment, alpha particles were scattered from atoms and the scattering was consistent with a small structure in the atom containing the positive charge.
3. The largest quark charge is $2e/3$, so a combination of only two particles, a quark and an antiquark forming a meson, could not have an electric charge of $+2e$. Only particles containing three quarks, each with a charge of $2e/3$, can combine to produce a total charge of $2e$.
5. Until about 700 000 years after the Big Bang, the temperature of the Universe was high enough for any atoms that formed to be ionized by ambient radiation. Once the average radiation energy dropped below the hydrogen ionization energy of 13.6 eV, hydrogen atoms could form and remain as neutral atoms for relatively long period of time.
7. In the quark model, all hadrons are composed of smaller units called quarks. Quarks have a fractional electric charge and a baryon number of $\frac{1}{3}$. There are six flavors of quarks: up (u), down (d), strange (s), charmed (c), top (t), and bottom (b). All baryons contain three quarks, and all mesons contain one quark and one antiquark. Section 30.8 has a more detailed discussion of the quark model.
9. Baryons and mesons are hadrons, interacting primarily through the strong force. They are not elementary particles, being composed of either three quarks (baryons) or a quark and an antiquark (mesons). Baryons have a nonzero baryon number with a spin of either $\frac{1}{2}$ or $\frac{3}{2}$. Mesons have a baryon number of zero and a spin of either 0 or 1.
11. All stable particles other than protons and neutrons have baryon number zero. Since the baryon number must be conserved, and the final states of the kaon decay contain no protons or neutrons, the baryon number of all kaons must be *zero*.

## Problems

**1.** $1.1 \times 10^{16}$ fissions
**3.** 126 MeV
**5.** (a) 16.2 kg (b) 117 g
**7.** (a) $3/a$ (b) $3.72/a$ (c) The surface-to-volume ratio is lowest for the sphere; therefore, the sphere has the better shape for minimum leakage.
**9.** 1.01 g
**11.** (a) ${}^{8}_{4}\text{Be}$ (b) ${}^{12}_{6}\text{C}$ (c) 7.27 MeV
**13.** 4.03 MeV
**15.** 14.1 MeV
**17.** (a) $4.53 \times 10^{23}$ Hz (b) 0.662 fm
**19.** 67.5 MeV, 67.5 MeV/$c$, $1.63 \times 10^{22}$ Hz
**21.** (a) conservation of electron-lepton number and conservation of muon-lepton number (b) conservation of charge (c) conservation of baryon number (d) conservation of baryon number (e) conservation of charge
**23.** (a) Violates conservation of baryon number (not allowed). (b) This reaction may occur via the strong interaction. (c) This reaction may occur via the weak interaction only. (d) This reaction may occur by the weak interaction only. (e) This reaction may occur via the electromagnetic interaction.
**25.** (a) not conserved (b) conserved (c) conserved (d) not conserved (e) not conserved (f) not conserved
**27.** $3.34 \times 10^{26}$ electrons, $9.36 \times 10^{26}$ up quarks, $8.70 \times 10^{26}$ down quarks
**29.** (a) $\Sigma^+$ (b) $\pi^-$ (c) $K^0$ (d) $\Xi^-$
**31.** a neutron, udd
**33.** 18.8 MeV
**35.** (a) $K^+$ (b) $\Xi^0$ (c) $\pi^0$
**37.** 26
**39.** (b) 12 days
**41.** 29.8 MeV
**43.** $3.60 \times 10^{38}$ protons/s

# Index

Note: Page numbers followed by *f* and *t* refer to figures and tables respectively.

## F

## N



## O



## P

## Q

## PHYSICAL CONSTANTS

| Quantity | Symbol | Value | SI unit |
|---|---|---|---|
| Avogadro's number | $N_A$ | $6.02 \times 10^{23}$ | particles/mol |
| Bohr radius | $a_0$ | $5.29 \times 10^{-11}$ | m |
| Boltzmann's constant | $k_B$ | $1.38 \times 10^{-23}$ | J/K |
| Coulomb constant, $1/4\pi\epsilon_0$ | $k_e$ | $8.99 \times 10^9$ | $N \cdot m^2/C^2$ |
| Electron Compton wavelength | $h/m_e c$ | $2.43 \times 10^{-12}$ | m |
| Electron mass | $m_e$ | $9.11 \times 10^{-31}$ | kg |
| | | $5.49 \times 10^{-4}$ | u |
| | | $0.511\ MeV/c^2$ | |
| Elementary charge | $e$ | $1.60 \times 10^{-19}$ | C |
| Gravitational constant | $G$ | $6.67 \times 10^{-11}$ | $N \cdot m^2/kg^2$ |
| Mass of Earth | $M_E$ | $5.98 \times 10^{24}$ | kg |
| Mass of Moon | $M_M$ | $7.36 \times 10^{22}$ | kg |
| Molar volume of ideal gas at STP | $V$ | 22.4 | L/mol |
| | | $2.24 \times 10^{-2}$ | $m^3/mol$ |
| Neutron mass | $m_n$ | $1.674\ 93 \times 10^{-27}$ | kg |
| | | 1.008 665 | u |
| | | $939.565\ MeV/c^2$ | |
| Permeability of free space | $\mu_0$ | $1.26 \times 10^{-6}$ | $T \cdot m/A$ |
| | | ($4\pi \times 10^{-7}$ exactly) | |
| Permittivity of free space | $\epsilon_0$ | $8.85 \times 10^{-12}$ | $C^2/N \cdot m^2$ |
| Planck's constant | $h$ | $6.63 \times 10^{-34}$ | $J \cdot s$ |
| | $\hbar = h/2\pi$ | $1.05 \times 10^{-34}$ | $J \cdot s$ |
| Proton mass | $m_p$ | $1.672\ 62 \times 10^{-27}$ | kg |
| | | 1.007 276 | u |
| | | $938.272\ MeV/c^2$ | |
| Radius of Earth (at equator) | $R_E$ | $6.38 \times 10^6$ | m |
| Radius of Moon | $R_M$ | $1.74 \times 10^6$ | m |
| Rydberg constant | $R_H$ | $1.10 \times 10^7$ | $m^{-1}$ |
| Speed of light in vacuum | $c$ | $3.00 \times 10^8$ | m/s |
| Standard free-fall acceleration | $g$ | 9.80 | $m/s^2$ |
| Stefan-Boltzmann constant | $\sigma$ | $5.67 \times 10^{-8}$ | $W/m^2 \cdot K^4$ |
| Universal gas constant | $R$ | 8.31 | $J/mol \cdot K$ |

The values presented in this table are those used in computations in the text. Generally, the physical constants are known to much better precision.